WIRELESS COMMUNICATIONS

WIRELESS COMMUNICATIONS
ALGORITHMIC TECHNIQUES

Giorgio M. Vitetta
University of Modena and Reggio Emilia, Italy

Desmond P. Taylor
University of Canterbury, Christchurch, New Zealand

Giulio Colavolpe
University of Parma, Italy

Fabrizio Pancaldi
University of Modena and Reggio Emilia, Italy

Philippa A. Martin
University of Canterbury, Christchurch, New Zealand

WILEY

A John Wiley & Sons, Ltd., Publication

This edition first published 2013
©2013 John Wiley and Sons Ltd

Registered office
John Wiley & Sons Ltd, The Atrium, Southern Gate, Chichester, West Sussex, PO19 8SQ, United Kingdom

For details of our global editorial offices, for customer services and for information about how to apply for permission to reuse the copyright material in this book please see our website at www.wiley.com.

Library of Congress Cataloging-in-Publication Data

Wireless communications : algorithmic techniques / Giorgio Vitetta, Desmond
P. Taylor, Giulio Colavolpe, Fabrizio Pancaldi, Philippa A. Martin.
      pages cm
  Includes bibliographical references and index.
  ISBN 978-0-470-51239-5 (hardback)
  1. Wireless communication systems–Mathematics. 2. Algebra. I. Vitetta, Giorgio.
  TK5102.83.W57 2013
  621.38401'5181–dc23

                                                             2012047256

A catalogue record for this book is available from the British Library.

ISBN: 9780470512395

Set in 9/11pt Times by Laserwords Private Limited, Chennai, India.
Printed and bound in Singapore by Markono Print Media Pte Ltd

Contents

Preface

Digital radios have undergone an astonishing evolution in the last century. Born as a set of simple and power-hungry electrical and electromechanical devices for low data rate transmission of telegraph data in the Marconi age, they have transformed, thanks to substantial advances in electronic technology, into a set of small, reliable and sophisticated integrated devices supporting broadband multimedia communications. This, however, would not have been possible unless significant progress had been made in recent decades in the field of signal processing algorithms for baseband and passband signals. In fact, the core of any modern digital radio consists of a set of algorithms running over programmable electronic hardware. This book stems from the research and teaching activities of its co-authors in the field of algorithmic techniques for wireless communications. A huge body of technical literature has accumulated in the last four decades in this area, and an extensive coverage of all its important aspects in a single textbook is impossible. For this reason, we have selected a few important topics and, for ease of reading, organized them into two parts. Part I concerns digital modulation techniques, characterization and modeling of wireless channels, channel estimation, and channel equalization metrics and algorithms. Part II focuses on channel coding, coded modulation, and combined equalization and decoding. For each of these topics, we have tried to provide an advanced introduction, blending basic principles with advanced concepts and techniques which play an important role at the forefront of research in wireless communications. In addition, for each topic we have provided some historical notes, so that the reader can analyze it in the right perspective, understanding both its roots and its possible evolutionary paths.

From the outset our main goal has been to produce a textbook for beginning graduate and senior students, who are expected to have some basic knowledge in the fields of Fourier transform techniques, probability theory, random processes, sampling theory, linear filtering, vector spaces, matrix algebra and linear transformations. Some information about more advanced concepts in these fields is provided in the appendices of the book, which, for this reason, we believe to be self-contained.

This book can serve as a text in either one-semester or two-semester courses in digital communications and coding. A natural division is to cover Part I in the first semester and Part II in the second. An alternative one-semester course can cover a portion of the material of Part I (Chapters 1–4 and 6) and some basic material from Part II (Chapters 7–9).

The writing of this book has required a substantial commitment. We owe much to all those people who volunteered to read parts of it, correct mistakes and provide suggestions for enriching its technical content and improving its clarity of presentation. In particular, we are grateful to Francesco Montorsi, Fabio Gianaroli, Tommaso Foggi, Amina Piemontese, Nicolò Mazzali, Andrea Modenini and Alessandro Ugolini for their contributions. Our sincere thanks go also to the editorial staff of Wiley and, in particular, to Mark Hammond, Sarah Hinton, Jennifer Beal, and Susan Barclay, who have always supported us in the writing process.

We do hope that the uncountable hours devoted to this book will bear fruit in stimulating interest in the study of modern techniques for wireless communications.

List of Acronyms

A/D	analog-to-digital
ACE	approximate cycle extrinsic message degree
ACF	autocorrelation function
ADC	analog-to-digital converter
ADPS	angular delay power spectrum
AECM	alternating ECM
AG	algebraic geometry
AGN	additive Gaussian noise
AM-PM	amplitude modulation-phase modulation
AMMSE	approximated MMSE
AOA	angle of arrival
AOD	angle of departure
AP	access point
APP	a posteriori probabilities
AR	autoregressive
ARA	accumulate-repeat-accumulate
ARMA	autoregressive moving average
ASA	adaptive state allocation
ASIC	application specific integrated circuit
AWGN	additive white Gaussian noise
BCH	Bose–Chaudhuri–Hocquenghem
BCJR	Bahl–Cocke–Jelinek–Raviv
BCM	block-coded modulation
BCRB	Bayesian Cramér–Rao bound
BCRVB	Bayesian Cramér–Rao vector bound
BDFE	Bayesian decision feedback estimation
BEC	binary erasure channel
BEM	Bayesian EM
BEP	bit error probability
BER	bit error rate
BIBD	balanced incomplete block design
BICM-ID	BICM with iterative decoding
BICM	bit-interleaved coded modulation
BLAST	Bell Labs Layered Space-Time
BPA	belief propagation algorithm
BPSK	binary phase shift keying
BR	binary representation

BS	base station
BSC	binary symmetric channel
BTC	block turbo code
BW	Barnes–Wall
CA^m	convolutional accumulate-m
ccdf	complementary cumulative distribution function
CCI	co-channel interference
CCITT	Consultative Committee for International Telegraphy and Telephony
CCM	concatenated coded modulation
CCRB	constrained CRB
CDMA	code division multiple access
CDMA	code division multiple access
CE	convolutional encoder
CFO	carrier frequency offset
CIR	channel impulse response
CM	conditional maximization
CM	constant modulus
CNET	Centre National d'Etudes des Télécommunications
CP	cyclic prefix
CPE	continuous phase encoder
CPFSK	continuous-phase frequency shift keying
CPM	continuous phase modulation
CRB	Cramér–Rao bound
CRVB	Cramér–Rao vector bound
CSI	channel state information
D-BLAST	diagonal BLAST
DAB	Digital Audio Broadcasting
DC	direct current
DD	differential detector
DDFSD	delayed decision feedback sequence detection
DDFSE	delayed decision feedback sequence estimator
DE	density evolution
DFE	decision feedback equalizer
DFT	discrete Fourier transform
DLMS	delayed LMS
DMC	discrete memoryless channel
DMT	discrete multitone
DPC	differential parity check
DPSK	differentially encoded PSK
DPTCT	dual-pilot tone calibrated technique
DSF	directional scattering function
DVB-T	digital video broadcasting – terrestrial
ECM	expectation/conditional maximization
ECME	expectation/conditional maximization either
EDGE	enhanced data rate for GSM evolution
EGC	equal gain combining
EKF	extended Kalman filter
EM	expectation–maximization
EMVA	expectation maximization Viterbi algorithm
EXIT	extrinsic information transfer
FBA	forward–backward algorithm

FBF	feedback filter
FCC	face-centered cubic
FCT	Fourier continuous transform
FD-DFE	DFE in the FD
FD-LE	LE in the FD
FD-TDFE	decision feedback TE in the FD
FD-TLE	linear TE in the FD
FD	frequency domain
FDE	frequency-domain equalizer
FDM	frequency division multiplexing
FDTD	finite-difference time-domain
FER	frame error rate
FFAGC	feedforward automatic gain control
FFF	feedforward filter
FFSK	fast FSK
FFSR	feedforward signal regeneration
FFT	fast Fourier transform
FG	factor graph
FIM	Fisher information matrix
FIR	finite impulse response
FLA	fixed-lag algorithm
FM	frequency modulation
FS	Fourier series
FSK	frequency shift keying
FSSM	finite-state sequential machine
FTS	Fourier transform of a sequence
GBSBM	geometrically based single-bounce model
GBSM	geometrically based stochastic model
GCD	greatest common divisor
GFSK	Gaussian-filtered FSK
GMD	generalized minimum distance
GMSK	Gaussian-filtered MSK
GQ	Gaussian quadrature
GQR	Gaussian quadrature rule
GSM	Global System for Mobile
GTD	geometrical theory of diffraction
H-BLAST	horizontal BLAST
HCCC	hybrid concatenated convolutional code
HF	high frequency
IBI	interblock interference
ICC	International Conference on Communications
ICI	intercarrier interference
IDFT	inverse discrete Fourier transform
IF	intermediate frequency
IFCT	inverse Fourier continuous transform
IFFT	inverse fast Fourier transform
iid	independent and identically distributed
IR	impulse radio
IRA	irregular repeat-accumulate
IRWEF	input-redundancy weight enumerating function
ISI	intersymbol interference

KF	Kalman filter
KL	Karhunen–Loéve
LCM	least common multiple
LDPC	low-density parity check
LE	linear equalizer
LFSR	linear feedback shift register
LHS	left-hand side
LLR	log-likelihood ratio
LMMSE	linear MMSE
LMS	least mean square
LoR	logarithmic representation
LOS	line of sight
LR	likelihood ratio
LRP	least reliable position
LS	least squares
LSE	least squares estimator
LST	layered space-time
LT	Luby transform
LTVS	linear time-variant system
M-AM-PM	M-ary amplitude modulation-phase modulation
M-AM	M-ary amplitude modulation
M-ASK	M-ary amplitude shift keying
M-PSK	M-ary phase shift keying
M-QAM	M-ary quadrature amplitude modulation
MA	moving average
MAP	maximum a posteriori
MAPBD	MAP bit detection
MAPSD	MAP symbol detection
Max-Log-MAP	maximum logarithmic MAP
MC	multicarrier
MCM	multicarrier modulation
MCRB	modified CRB
MCRVB	modified Cramér–Rao vector bound
MDS	maximal distance separable
MF	matched filter
MFB	matched filter bound
MFED	matched filter and envelope detector
MFIM	modified FIM
MIMO	multiple-input multiple-output
MISO	multiple-input single-output
ML-BICM	multilevel BICM
ML	maximum likelihood
MLC	multilevel code
MLE	maximum likelihood estimator
MLSD	ML sequence detection
MLSE	maximum likelihood sequence estimation
MM	memoryless modulator
MMSE	minimum mean squared error
MP	message passing
MPA	max-product algorithm
MPCC	multiple parallel concatenated code

MRB	most reliable basis
MRC	maximal ratio combining
MRIP	most reliable independent position
MSA	min-sum algorithm
MSCC	multiple serially concatenated code
MSD	multistage decoder
MSDD	multiple-symbol differential detector
MSE	mean square error
MSK	minimum shift keying
MT	mobile terminal
MVU	minimum variance unbiased
NLMS	normalized LMS
NMSA	normalized MSA
NP	nondeterministic polynomial-time
NTT	Nippon Telegraph and Telephone Public Corporation
O-QAM	offset QAM
OFDM	orthogonal frequency division multiplexing
OSA	optimum soft output algorithm
OSD	order statistic decoding
OSTBC	orthogonal space-time block code
P/S	parallel-to-serial
PA	product accumulate
PAM	pulse amplitude modulation
PAN	personal area networks
PAPR	peak-to-average-power ratio
PAT	pilot-aided transmission
PBP	per-branch processing
PC	product code
PCCC	parallel concatenated convolutional code
PCM	pulse code modulation
PCS	personal communication system
pdf	probability density function
PDP	power delay profile
PEG	progressive edge growth
PEP	pairwise error probability
PIC	parallel interference cancelation
PLL	phase-locked loop
PoP	parity on parity
PR	polynomial representation
PSAM	pilot symbol assisted modulation
PSD	power spectral density
PSP	per-survivor processing
PSTN	public switched telephone network
QC	quasi-cyclic
QPSK	quaternary phase shift keying
QRD	QR decomposition
RA	repeat and accumulate
RBF	radial basis function
RC	raised cosine
REC	rectangular
RF	radio-frequency

RHS	right-hand side
RLS	recursive least squares
RM	Reed−Muller
rms	root mean square
RRNS	redundant residue number system
RS-BCJR	reduced-state BCJR
RS	Reed−Solomon
RSC	recursive systematic convolutional
RSSD	reduced-state sequence detection
S/P	serial-to-parallel
SA	search algorithm
SAGE	space-alternating generalized EM
SC	single-carrier
SCCC	serial concatenated convolutional code
SD	sphere decoding
SER	symbol error rate
SF	space-frequency
SFBC	SF block code
SFSK	sinusoidal frequency shift keying
SIC	serial interference cancelation
SiHo	soft-input hard-output
SIMO	single-input multiple-output
SISO	single-input single-output
SiSo	soft-input soft-output
SNR	signal-to-noise ratio
SOS	sum of sinusoids
SOVA	soft output VA
SPA	sum-product algorithm
SPC	single parity check
SRS	symbol rate sampling
SSA	soft output algorithm
SSB	single sideband modulation
ST	space-time
STBC	space-time block code
STC	space-time code
STF	space-time-frequency
STTC	space-time trellis code
SVD	singular value decomposition
TCM	trellis-coded modulation
TCT	tone calibration technique
TD	time domain
TDE	time-domain equalizer
TDL	tapped delay line
TDMA	time division multiple access
TE	turbo equalizer
TFM	tamed frequency modulation
THP	Tomlinson−Harashima precoding
TM II	second transmission mode
TOA	time of arrival
TPC	turbo product code
TSTC	threaded STC

TTIB	transparent tone-in-band
TVARCIR	time-variant angle-resolved channel impulse response
TVIR	time-variant impulse response
TVTF	time-variant transfer function
UBE	union bound estimate
UFI	uniform from the input
UMP-BPA	uniformly most powerful BPA
UP	uncorrelated path
US	uncorrelated scattering
UTD	uniform theory of diffraction
UWB	ultra wideband
V-BLAST	vertical BLAST
V-BLAST	vertical BLAST
VA	Viterbi algorithm
VCO	voltage controlled oscillator
VHF	very high frequency
VLMS	variable LMS
VR	Voronoi region
VSB	vestigial sideband modulation
VSS-LMS	variable step-size LMS
WF	whitening filter
WGN	white Gaussian noise
WLAN	wireless local area network
WMF	whitened matched filter
WSC	wide-sense cyclostationary
WSS	wide-sense stationary
WSTC	wrapped STC
ZF	zero forcing
ZP	zero padding

1

Introduction

The history of wireless communications stretches back many centuries. Many of the earliest systems were inherently *line of sight* (LOS) using such techniques as smoke signals, flashing lights and semaphore. For example, in Napoleonic times, the French had an elaborate, essentially countrywide, semaphore system, developed by Claude Chappe (1763–1805) and consisting of chains of relay stations [1, 2]. Possibly, the first non-LOS systems were the drum signaling techniques used by tribes in Africa.

Guglielmo Marconi (1874–1937) first demonstrated modern wireless technology, also known as radio, in 1895 [1, 3, 4]. The first such systems were in a sense digital since they used Morse code, which had been invented by Samuel Finley Breese Morse (1791–1872) for use in telegraphy. Speech communication, using analog modulation, followed only a few years later, and prior to the 1980s almost all wireless systems used analog transmission techniques. However, the widespread deployment of telephony based on *pulse code modulation* (PCM) and the development of digital satellite transmission and microwave relay systems fostered the development of digital transmission techniques. These systems are now being augmented and to a large extent supplanted for point-to-point communications by terrestrial digital wireless systems coupled with high speed backbone networks implemented using optical fiber. Satellite systems retain a very valuable niche in the area of wide area broadcasting to which they are well suited. They also retain an application in some data transfer systems, where delay is not of prime importance. Microwave relay systems are falling into disuse in many regions as they are replaced by fiber links.

The development of modern terrestrial wireless systems has been driven in large measure by the development of *cellular radio systems* [5, 6]. The cellular principle introduced the concept of *frequency reuse* over large spatial domains. This leads to a very efficient use of the available radio spectrum and allows for a very large number of simultaneous users of a given system. As a result the world is today moving to an untethered mobile wireless communications environment based on cellular-like system architectures.

AT&T deployed the first cellular system in Chicago in 1983 following several years of development. It used an analog transmission format and was completely saturated by 1984, the developers having grossly underestimated the public appetite for mobile phone services. Since then there has been an almost explosive growth of cellular radio, and this continues today. In the early 1990s the first digital cellular or second generation systems appeared. These provided increased capacity and performance using digital transmission formats coupled with improved digital signal techniques and hardware platforms. Today there are cellular systems based on both *time division multiple access* (TDMA) and *code division multiple access* (CDMA).

Wireless Communications: Algorithmic Techniques, First Edition.
Giorgio M. Vitetta, Desmond P. Taylor, Giulio Colavolpe, Fabrizio Pancaldi, Philippa A. Martin.
© 2013 John Wiley & Sons, Ltd. Published 2013 by John Wiley & Sons, Ltd.

The advent of digital cellular systems paved the way for mobile data services. There is now an increasing demand for these and, as a result, third generation cellular systems are being deployed. These provide for higher data rates and offer many new applications and services. In addition to cellular systems, there are numerous other wireless systems being developed and deployed. Moreover, there is now a convergence taking place to common transmission and networking environments for voice, data and multimedia communications. Consequently, there is an increasing demand for higher and higher data rates coupled with the requirement to make even more efficient use of the limited available radio spectrum.

Today there are numerous distinct wireless systems in use. These modern systems, while distinct, all use digital signaling formats and network architectures and there is a distinct trend toward convergence to a small number of these coupled with the ability to interwork between different systems and networks. Some of the systems that are currently deployed or being developed for deployment include the following:

1. Cellular telephone systems. While these ignited the wireless revolution, they are still undergoing development to improve their transmission rates and the range of applications to which they can cater.
2. Cordless telephones. These initially were developed to provide tetherless connections within the limited space of a single dwelling. However, with the development of CT-2 in North America followed by that of DECT in Europe [5, 6], their space has enlarged and there are signs of their convergence to the cellular telephone system.
3. *Wireless local area networks* (WLANs). These have seen a great deal of development in the past few years. Standardization of signaling formats to the IEEE 802.11b, 802.11a and 802.11g formats and their widespread use in unlicensed bands around 800 MHz, 2.4 GHz and 5 GHz has led to an almost explosive growth in mobile computing. This has fostered the development of networks of high data rate wireless *access points* (APs) interconnected by high speed backbone networks, thereby leading essentially to a cellular network architecture. In addition to the IEEE standards-based networks, there have been similar developments in Europe known as the Hiperlan I and II standards.
4. Broadband wireless access networks. These are in large measure based on the IEEE 802.16 standard [7] and are intended to provide high rate, wide area coverage similar to that of WLANs. These systems are just now beginning to be deployed, and it appears that they may subsume some of the functionality now provided by cellular networks.
5. Low-cost, low-power systems. Such systems, which include Bluetooth [8] and Zigbee [9], were initially intended to provide relatively low data rates with limited range and in small-scale networks. Bluetooth is primarily focused on so-called personal area networks (PANs) that support a very limited number of devices requiring limited data rates. Zigbee was developed primarily for use in sensor networks requiring low data rates with long-lived battery powered terminals.
6. *Ultra wideband* (UWB) systems. These are systems based at least initially on the concepts of *impulse radio* (IR) [10] and are characterized by percentage bandwidths in excess of 20% of the carrier frequency or by a bandwidth exceeding 500 MHz. Today there are two further basic system approaches, one based on spread spectrum and the other on multiband *orthogonal frequency division multiplexing* (OFDM). System deployment has only recently been licensed in North America and many of their applications are uncertain at this stage. However, it does appear that they may subsume many of the functions now provided by systems such as Bluetooth and Zigbee.

In addition to the system types mentioned above, there is today a trend toward *cognitive* or "smart" radios as first described by J. Mitola [11, 12]. Cognitive radio may be loosely thought of as overlay on a software-defined radio that causes a system to recognize its channel and interference environment and then to automatically adjust its parameters. There are many possible approaches to such systems and we will not make any attempt to categorize them here.

Finally, there are undoubtedly many wireless systems and applications that have not been mentioned here. Moreover, there are almost certainly others that have not yet been conceived. The world is moving rapidly to an untethered communications environment and there will be many new applications of both existing and new wireless systems appearing in the next few years.

This book is focused on the so-called *physical layer* of wireless communications systems. In particular, it is focused on techniques for mitigating the effects of the wireless channel including dispersion due to multipath propagation that causes *intersymbol interference* (ISI), adjacent and co-channel interference. It is also concerned with achieving high-rate, high-integrity communications in a power-efficient manner. The overall focus is the development and analysis of transmission techniques and algorithms for accomplishing this. The book considers both *single-input single-output* (SISO) systems and *multiple-input multiple-output* (MIMO) systems that utilize transmit and receiver diversity to achieve high-capacity signaling coupled with high-integrity transmission.

In the remainder of this introductory chapter, we will first provide an overview of both SISO and MIMO system architectures. We will then briefly describe the structure of the book and, finally, provide some suggestions for further reading.

1.1 Structure of a Digital Communication System

The overall focus here is on the structure of a digital communication system operating over a wireless channel. We will consider *conventional* systems, using a single antenna at the transmitter and, possibly, *diversity reception*, and MIMO systems. One of the most powerful techniques available to improve the performance and throughput of wireless transmission is that of *diversity*. In fact, diversity creates multiple copies of the transmitted signal at the receiver. In principle, these copies are uncorrelated, so that when one copy is deeply faded due to the wireless channel, the others are not. This allows for significant improvement in both the error performance and throughput of wireless transmission systems. The concept of diversity in receivers has been known for many years [13]; however, in recent years there has been much work in developing techniques to achieve diversity at the transmitter [14, 15] and to combine transmit and receive diversity through the use of *space-time coding* [16].

Systems that combine transmit and receive diversity are known as MIMO systems, which may in a sense be considered as the most general system architecture. Such systems include space-time coded systems [16] and the so-called *Bell Labs Layered Space-Time* (BLAST) [17] or *spatial multiplexing* architectures. The latter have been shown to allow for major increases in the available channel capacity [18] and a consequent increase in the efficiency of use of the available radio spectrum. Note that capacity provides a theoretical upper limit on the throughput that can be achieved in a given channel.

SISO systems that contain no diversity are clearly the simplest in structure. *Single-input multiple-output* (SIMO) systems encompass the classical architecture, providing diversity only at the receiver. *Multiple-input single-output* (MISO) systems provide only transmit diversity usually through the mechanism of space-time coding [16], which introduces both temporal and spatial correlation among multiple transmitted signal streams in such a manner that a single receiver can decode the multiple received signals and obtain the diversity effect introduced at the transmitter.

Generic system architectures are depicted for SIMO and MIMO systems in Figures 1.1 and 1.2, respectively. In the following chapters of this book a number of algorithmic techniques implemented in the various functional blocks forming the point-to-point wireless communication systems[1] illustrated in Figures 1.1 and 1.2 will be considered in detail. Here we confine ourselves to a more or less qualitative description of their various functions.

Let us consider the functions performed by the various system blocks, referring first to Figure 1.1, for simplicity. To begin, we consider the blocks over which a system designer does not usually

---

[1] Note that both systems are characterized by a single information source and a single destination; *multiuser* systems will not be investigated in the following.

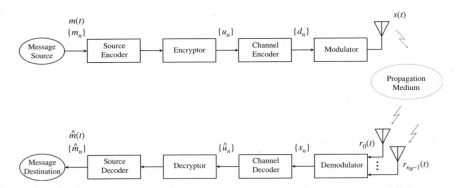

Figure 1.1 Block diagram of a conventional digital communication system with *diversity reception*.

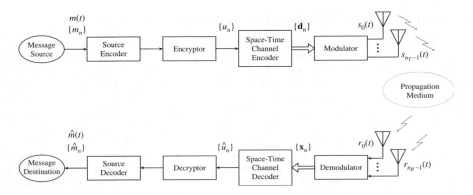

Figure 1.2 Block diagram of a *space-time* digital communication system.

have direct control, namely the *message source* and the *message destination*. The source generates a sequence of discrete[2] messages $\{m_n\}$ (where m_n denotes the nth message in the sequence). In the case where the source produces an analog signal, it is assumed that the source encoder accomplishes analog-to-digital conversion, producing a data stream or discrete message sequence. The message destination is relevant to the present discussion only because an appropriate *fidelity criterion* (i.e., a quality index), describing system performance, is usually defined for a given source–destination pair. Quality indexes commonly adopted to assess the performance of a digital communication system are the bit error probability and the symbol error probability.

A wireless communication system designer does not usually have complete control over the *communication channel*. With reference to Figures 1.1 and 1.2, this includes the *propagation medium* (i.e., the physical space through which the electromagnetic signal radiated by the transmit antenna travels), the final section of the transmitter (e.g., the transmit antenna and filtering/amplification stages preceding it), and the initial section of the receiver (e.g., the receive antenna and low noise amplifier and filter stages following it). In the present work, we will not focus on the details of the "channel" subsystem. Instead, we will limit ourselves to a mathematical description of its input–output behavior.

---

[2] This means that the message alphabet has a *finite cardinality*. Throughout the book, we will consider only discrete sources whose alphabet has this property.

As will be seen later, a wireless communication channel changes the shape of the transmitted signal, introducing linear (and, eventually, nonlinear) distortions and adding random noise.

The *distortions* due to a wireless channel can cause substantial changes in the temporal and spectral properties of transmitted signals. These often originate from the fact that electromagnetic waves do not propagate from the transmit to the receive antenna along a direct path, but are reflected and scattered by objects in the surrounding environment. As a result, receiver antennas collect *multiple copies* (echoes) of the same transmitted signal. These have usually traveled along distinct *paths*, with different prop-agation times, and generally arrive with different phases and amplitudes. As a result, in some spatial locations, these copies can interfere destructively, canceling each other, so that the useful component of the received signal *fades*. In other words, the presence of multiple paths generates the so-called *fading* phenomenon, representing one of the most significant impairments encountered in wireless system design. The oldest countermeasure to fading is known as *diversity reception*. This consists of equipping digital receivers with multiple antennas, which, when adequately spaced, collect *different* (i.e., distorted in different and, possibly, independent ways) replicas of the transmitted signal [19].

Any communication channel also adds *random noise*, which is generated by both external sources (e.g., cosmic and atmospheric signals, and interference) and by the electronic devices in the receiver. A brief discussion of its statistical properties will be provided later. At this point, we merely note that it usually has a Gaussian distribution and a white or constant power spectral density over the frequency bands of interest.

Let us summarize the functions of the other blocks of the *transmitter* (i.e., the source encoder), the encryptor, the channel encoder and the modulator:

- The *source encoder* processes the source message stream to remove its natural *redundancies*. This can result in appreciable reduction of the bit rate, sometimes achieved, however, at the price of an information loss. Despite this, the original message stream can be recovered by the source decoder at the receiver within some *specified fidelity*.
- The *encryptor*, if present, adds security coding to the data sequence generated by the source encoder. This result is achieved by a coding algorithm turning the unciphered data (usually called *plaintext*) into a new discrete sequence $\{u_n\}$ (called *ciphertext*). The encryption algorithm involves a parameter, called the *key*, knowledge of which at the receiver is essential to deciphering. One class of modern and well-known ciphering techniques, known as *public-key encryption*, relies on a double key mechanism, that is, on the use of a *public key* (potentially known to anyone) for enciphering and on a *private key* (known, in principle, only to the message destination) for deciphering [20].
- The *channel encoder* introduces an error-correction capability, so that most (possibly all) of the errors due to channel noise and distortion can be removed or corrected at the receiver. To achieve this target, the channel encoder introduces *memory* and *redundancy* into the coded sequence. The presence of redundancy is seen from the fact that, in a given time interval, the number of bits generated by the channel encoder is larger than the number of the information bits processed by it. Memory can be related to the fact that, generally speaking, each bit feeding the encoder influences multiple bits at its output. As discussed in Part II of this book, the receiver exploits both these properties to improve the reliability of its decisions.
- The *modulator* is fed by the symbol sequence $\{d_n\}$ (each symbol belongs to a multilevel alphabet) and generates an analog signal $s(t)$, which consists of the concatenation of waveforms belonging to some *finite alphabet of signals*. In practice, this device represents the interface between the stream of discrete data and the real communication medium. Therefore, it accomplishes multiple tasks (including frequency up-conversion) and power amplification and can incorporate transducers (e.g., multiple transmit antennas).

Let us now consider some of the subsystems in the *receiver*, namely the demodulator, the channel decoder, the decryptor and the source decoder. These units accomplish functions complementary to those of the corresponding blocks in the transmitter.

In general, the receiver has $n_R \geq 1$ antennas. The lth antenna (with $l = 0, 1, \ldots, n_R - 1$) feeds the *demodulator* with the noisy *radio-frequency* (RF) signal:

$$r_l(t) = z_l(t) + n_l(t), \tag{1.1}$$

where $z_l(t)$ and $n_l(t)$ represent the *useful signal component* (i.e., the response to $s(t)$ of the communication channel including the transmitter and the transmit/receive antennas in the absence of noise) and the *random noise* at the receive antenna terminals, respectively. The demodulator processes the waveforms $\{r_l(t)\}$ of (1.1), to extract a set of *synchronization parameters* (such as the phase and frequency of the carrier associated with $z_l(t)$, and timing information), and in many cases an *estimate* of the communication channel response. Then it uses this information to perform signal *detection* that generates a data sequence $\{x_n\}$. This contains either *hard* or *soft* information about the transmitted data. In the first case, if we focus on the data transmitted in the nth symbol interval, the demodulator generates a hard estimate or decision $\hat{d}_n$ on the value of the (coded) transmitted symbol d_n, whereas in the second case it produces information about the reliability (i.e., the *likelihood*) of each value that d_n can take.

The *channel decoder* exploits the information provided by the demodulator, to try to find the *most likely* data sequence $\{\hat{u}_n\}$ that has generated the coded sequence $\{d_n\}$. Note that the availability of soft information allows the decoder to improve the quality of its decisions with respect to the case of knowledge of hard information.

The task accomplished by the *decryptor* is the inverse to that of the encryptor. This task can be carried out successfully if both the ciphering algorithm and its key are known.

The *source decoder* processes an estimate of the binary data generated by the source encoder to generate a message in a proper format (the data sequence $\{\hat{m}_n\}$ or the analog signal $\hat{m}(t)$ in Figure 1.1) for the destination.

Finally, we note that the system of Figure 1.1 is characterized by a communication channel with a *single* input (corresponding to a single transmit antenna) and *multiple* outputs, to be processed by a receiver equipped with $n_R \geq 1$ antennas. For this reason, the communication system can be classified as SIMO. In particular, if $n_R = 1$, we have a SISO system.

The scheme illustrated in Figure 1.2 generalizes that of Figure 1.1, since it represents a system with $n_T > 1$ transmit antennas, resulting in a MIMO system. In such a system, the channel encoder, in response to the discrete data sequence $\{u_n\}$, generates a sequence of *vectors* $\{\mathbf{d}_n\}$, each consisting of n_T different elements. For any n, the kth element $d_n[k]$ (with $k = 0, 1, \ldots, n_T - 1$) of $\mathbf{d}_n$ is transmitted by the modulator as the RF signal $s_k(t)$ radiated by the kth antenna. Therefore, the redundancy and memory introduced by the encoder are spread over both *time* (as in the SIMO scenario described above) and *space* using *distinct transmit antennas* (*transmit diversity*). This is commonly referred to as *space-time* (ST) channel coding [16]. Generally speaking, each receive antenna observes a linear combination of all n_T transmitted signals. In fact, the noisy signal captured by the lth receive antenna can be expressed as:

$$r_l(t) = \sum_{k=0}^{n_T-1} z_{kl}(t) + n_l(t), \tag{1.2}$$

with $l = 0, 1, \ldots, n_R - 1$, where $z_{kl}(t)$ and $n_l(t)$ respectively represent the useful signal component (the channel response between the kth transmit and the lth receive antennas to $s_k(t)$ in the absence of noise) and the random noise collected by the antenna mentioned above. The demodulator processes the signals $\{r_l(t)\}$ of (1.2) and generates a sequence of n_T-dimensional vectors $\{\mathbf{x}_n\}$, whose elements contain, as in the previous case, hard or soft information about the sequence $\{\mathbf{d}_n\}$.

Recent studies have shown that the use of the spatial dimension in digital transmissions can substantially improve system robustness against channel fading and can allow an increase in the data rate transmitted within a given bandwidth. This explains the substantial research efforts on MIMO systems in the last decade [21, 22], to assess both their theoretical limits and to develop new digital transmission techniques for such systems. These studies have been followed by the development of *prototypes* of MIMO systems and, more recently, by the design of *application specific integrated circuits* (ASICs) for their low-cost implementation. This is illustrated by the so-called BLAST transmission technique, developed at Bell Labs by Gerard J. Foschini in 1996 [17]. In a BLAST system a data stream generated by a single source undergoes *spatial multiplexing*, that is, it is divided in n_T distinct substreams, each transmitted by a distinct antenna, using, however, the same time intervals and bandwidth as all the other antennas. At the receive side an array consisting of n_R antennas is used to collect the multiple linear combinations of the transmitted signals. Each receive antenna captures the superposition of all the n_T transmitted signals as in equation (1.2). Note that in a *rich scattering environment*, different antennas, having distinct spatial locations, receive different replicas of the same signal. This form of diversity allows the receiver to separate and detect, using sophisticated signal processing algorithms, the n_T transmitted signals, to reliably recover the overall transmitted data stream.

To assess the technical feasibility of the theoretical results derived by Foschini, in 1998 Bell Labs developed a BLAST prototype, having eight transmit antennas and 12 receive antennas. It clearly showed the possibility of achieving transmit data rates 10 times faster than those offered by traditional communication techniques in the same bandwidth [23]. On October 16, 2002, *Lucent Technologies* announced that Bell Labs had developed the prototypes of two chips for the use of the BLAST technology in mobile terminals and that the first lab tests had shown the possibility of transmitting at a rate of 19.2 Mbits/s, eight times faster than existing techniques under the same conditions.

Technically important results in the development of systems equipped with antenna arrays have also been obtained using the transmission technique known as MIMO-OFDM.[3] In this case spatial multiplexing is combined with *frequency division multiplexing* (FDM), so that spatial diversity is jointly exploited with spectral or frequency diversity. The last form of diversity arises due to the fact that, in a multipath channel, distinct spectral components of the transmitted signal undergo different phase/amplitude changes. Again in the development of MIMO-OFDM systems the derivation of many of theoretical results has been followed by the development of prototypes (e.g., see [24, 25, 26]) and, later, by the implementation of ASICs for modern wireless communications systems (e.g., in local area radio networks).

All this explains why today MIMO technology can be considered a mature technical solution for the design of digital communication systems.

In the following chapters of this book we will first focus on communication techniques employed in SISO and SIMO communication systems. We believe that a deep understanding of these techniques provides a solid foundation for the study of MIMO systems; this point will be stressed throughout the book, since various methodologies for the analysis and the design of MIMO systems will be presented as extensions of similar results derived for conventional systems, equipped with a single transmit antenna.

1.2 Plan of the Book

This book is divided into two parts. Part I concerns the wireless channel and the development of algorithms to process signals transmitted using uncoded transmission techniques. Part II deals with wireless systems that employ channel coding and develops algorithms to process signals that have been encoded prior to transmission. The use of coded transmission opens up the possibility of developing algorithms to jointly mitigate the distorting effect of the wireless channel and decode the information.

More specifically, in Part I, after describing the mathematical tools for both deterministic and stochastic descriptions of wireless channels in Chapter 2, an overview of the most important digital modulation techniques for radio communications is given in Chapter 3. In particular, we focus on both

---

[3] The OFDM technique is analyzed in detail in Chapter 3.

single carrier formats, such as passband *pulse amplitude modulation* and *continuous phase modulation*, and multicarrier formats, namely, *orthogonal frequency division multiplexing* signaling. We illustrate, for each class of signals, the structure of the modulated signals and their spectral properties. General rules for optimal signal detection are summarized in Chapter 4, to provide an overview of available techniques and of the analytical methods for estimating their performance. Detection over wireless channels may require estimation of channel properties, and, in particular, the *channel impulse response*. This is the subject of Chapter 5, which deals with both feedforward and iterative channel estimation techniques. Chapters 4 and 5 provide the necessary tools for the design of channel equalization algorithms, which are the subject of Chapter 6. There various algorithms are illustrated for the modulation formats described in Chapter 3. In particular, algorithm classification is done first on the basis of the modulation category (single carrier or multicarrier), and then on the basis of the available *channel state information* (CSI). As far as the last point is concerned, we consider three distinct possibilities: a receiver provided with perfect CSI knowledge; a receiver provided with statistical knowledge of CSI, but not performing explicit channel estimation; and a receiver performing joint estimation of data and CSI. Moreover, for single carrier modulations, equalization strategies operating in the time domain and in the frequency domain are considered.

In Part II we first discuss some essential results about the capacity of wireless channels (Chapter 7), showing the benefits of using multiple antennas at both transmitter and receiver. Then, in Chapter 8 an introduction to channel coding schemes and to coded modulations for wireless communication techniques is provided. Classical coding schemes, such as linear block codes and convolutional codes, are described in Chapter 9. For each class, we illustrate some well-known families of coding schemes and some important decoding techniques. In addition, some classical concatenated coding schemes are presented. Modern coding schemes, such as *turbo codes* and *low-density parity check codes*, are considered in Chapter 10. Again, coding and decoding algorithms are discussed, and some performance results are presented. The coding schemes and principles analyzed in Chapters 9 and 10 also provide the tools for understanding the signal space codes analyzed in Chapter 11. In particular, in that chapter we focus on *trellis coded modulation* (TCM), *bit-interleaved coded modulation* (BICM), and *modulation codes* based on *multilevel coding*, and finally on *space-time coding*, for both frequency-flat and frequency-selective fading channels. In a digital receiver equalization and decoding can be accomplished in a noniterative or in an *iterative fashion*, the latter possibility usually being in mobile scenarios. Some basic concepts from this modern research area are discussed in Chapter 12. Finally, appendices summarize various mathematical results (on Fourier transforms, linear systems, random variables and stochastic processes, etc.), that turn out to be extremely useful in both parts of the book.

1.3 Further Reading

A general introduction to digital communication techniques can be found in the textbooks [27–30]. Other introductory books, explicitly devoted to such techniques and to their applications in wireless communications, are [5, 6, 31–34]. A general introduction to the topic of channel coding theory is provided by the excellent book [35]. Channel coding schemes for wireless applications are investigated in [36, 37]. A study of various space-time processing and coding techniques can be found in the books [16, 38–40]. Books explicitly devoted to various algorithmic aspects of wireless communications are [41–43].

Part One

Modulation and Detection

2

Wireless Channels

2.1 Introduction

In wireless communication systems the channel introduces *random variations with time and/or frequency in both the amplitude and phase of the transmitted signal*. These phenomena are collectively known as *fading and dispersion* [19, 44–46]. The study of fading, dispersive channels is the main subject of this chapter.

Fading originates due to various causes. The most common is the presence of *multiple propagation paths*, that is, the existence of a number of paths along which an electromagnetic signal propagates from a transmitting antenna to the receiving one [19, 47, 48]. The presence of these multiple paths is normally due to *reflection*, *diffraction* and *scattering* caused by objects in the propagation medium and/or by its lack of homogeneity. Physical understanding of these phenomena requires study of the basic mechanisms governing the propagation of electromagnetic waves in the presence of obstacles with specific conductive or dielectric properties. This is outside the scope of this chapter: the reader can refer to [34, Chapter 4] and [6, Chapter 4] for an introduction to these topics. Here, when we consider the propagation medium, we will assume that the multipath propagation is due to the presence of a set of *scatterers*, each reflecting and/or dispersing the energy of an impinging electromagnetic signal.

When a communication system operates over a *time-dispersive* channel (a channel affected by multiple propagation paths), the distinct *echoes* of the transmitted signal captured by a receive antenna have different amplitudes and phases. These differences in general depend on *both frequency and time* and, if we neglect noise, the received signal consists of the linear combination of multiple echoes or replicas of the transmitted signal modified, in spectral content, according to the time variability of the medium.

Time variability manifests itself as *time selectivity* in the form of fluctuations in the intensity of the received signal. This variability is usually due to relative motion between receiver and transmitter and/or to environmental changes, producing changes in the characteristics of the various propagation paths. To mathematically describe channel behavior over a realistic time scale, channel models characterized by a set of fixed echo delays are commonly used and time variability is accounted for by assuming the echo amplitudes and phases are time-varying. Such models, however, do not provide a complete description of a channel, because of longer-term variations occurring in realistic propagation scenarios, which can entail significant changes in the *structure of the channel itself* (e.g., in the number of echoes and in their delays). These longer-term changes occur on time scales of minutes, tens of minutes, or hours, and are often due to meteorological factors or to the sun. In some cases, they may include daily, seasonal, or yearly phenomena or even the sunspot cycle.

Wireless Communications: Algorithmic Techniques, First Edition.
Giorgio M. Vitetta, Desmond P. Taylor, Giulio Colavolpe, Fabrizio Pancaldi, Philippa A. Martin.
© 2013 John Wiley & Sons, Ltd. Published 2013 by John Wiley & Sons, Ltd.

Fast changes in the intensity of the received signal (called *short-term fading*) are usually distinguished from those associated with slow changes (called *long-term fading*). This choice arises from an inaccurate, but useful, dichotomy regarding the time scale adopted in the observation of a communication channel. Both types of fading manifest themselves as a *time-continuous random process*, but they play different roles in wireless system design. The properties of short-term fading influence the choice of modulation and coding schemes, and of the receiver type, since they affect the structure of the received waveform and the presence of error correlation in data detection [49]. Long-term variations are also important, but they tend more to affect the *availability* of a wireless channel and, consequently, the *outage probability* of the system [45, 47, 48]. In fact, acting on the inner structure of a channel, they can cause the received signal to be significantly different from that for which a system design has been optimized. In fact, in the presence of appreciable variations, maintaining a minimum quality over the link, may require at the receiver a *signal-to-noise ratio* (SNR) larger than that achievable using the maximum available transmit power.

In this chapter we will focus only on short-term fading. We will assume that it is due exclusively to the presence of multiple, time-varying paths. We will not, however, forget that short-term fading models always have to be considered as conditioned on the "instantaneous" values of those parameters that are described by longer-term statistics. This needs to be kept in mind when we introduce the concept of *time stationarity* to describe, from a statistical perspective, channel variability with time. In fact, a channel affected by *fading that is statistically stationary over time* must be considered as a *local model*, that is, as a model that provides a short-term description, since its statistics can change appreciably over longer time intervals. Fortunately, it has been found that such locally stationary models are suitable representations of the actual behavior of fading channels commonly encountered in the study of wireless systems [19].

Let us now analyze the issues of *spatial variability* of a signal received over a fading channel. To grasp the essential aspects of this problem, let us consider Figure 2.1, which illustrates typical behavior (dotted curve) of the ratio (in decibels) between the received power P_R and the transmitted power P_T as a function of the transmitter–receiver separation, d, normalized to wavelength λ. This clearly illustrates the presence of *rapid fluctuations in the received signal power*. This is due to the presence of multiple echoes, associated with distinct propagation paths. These may interfere constructively, strengthening the received signal, or destructively, significantly attenuating it. Such effects, which are due to the mutual interference of multiple echoes, can change appreciably if a receiver moves only a fraction of a wavelength, since small variations in the path lengths may cause large changes in the phases of the associated echoes. This explains why such variations in the intensity of the received signal are usually called *small-scale propagation effects*. They in fact represent a manifestation of the *small-scale fading* affecting the communication channel. Note from the curve representing the ratio $(P_R/P_T)_{dB}$ versus d/λ, that another curve, showing the *average behavior*[1] of $(P_R/P_T)_{dB}$, can be extracted and, in the present case, is represented by the continuous line of Figure 2.1. In the literature this behavior is explained by introducing two phenomena characterizing wireless channels, known as *path loss* (or *propagation loss*) and *shadowing*. Both effects are classified as *large-scale propagation effects* and are considered as manifestations of so-called *large-scale fading* (or longer-term fading). However, the causes of these two phenomena are quite different. On the one hand, path loss is due to the spatial attenuation of the electromagnetic signal in the propagation medium and, analytically, is characterized by a monotonously decreasing dependence on d; this is represented by the dot-dashed curve shown in Figure 2.1. Shadowing, on the other hand, is caused by the presence of obstacles interposed between transmitter and receiver and the resulting attenuation, expressed in decibels, is represented by a slow random zero-mean fluctuation superimposed on the pass loss, as exemplified by the continuous curve of Figure 2.1. Another macroscopic difference between these phenomena

---

[1] In this case a *spatial average* is considered, since it is evaluated by processing multiple measurements extracted in the neighborhood of the point at which the average itself is evaluated. The way significant data are extracted in the estimation of the propagation loss from experimental measurements is analyzed, for instance, in [50 Sect. 2.2].

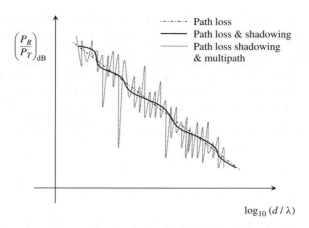

$\left(\dfrac{P_R}{P_T}\right)_{dB}$

------- Path loss
——— Path loss & shadowing
············ Path loss shadowing
              & multipath

$\log_{10}(d/\lambda)$

Figure 2.1 Typical behavior (dotted curve) of the ratio, expressed in decibels, between the received power P_R and the transmitted power P_T versus the transmitter–receiver distance d (normalized to the link wavelength λ) in a wireless communication system operating over a fading channel. The contributions due to path loss (dot-dashed curve) and to the joint effect of this loss and shadowing (continuous line) are also shown.

concerns the diversity in spatial scales over which appreciable variations appear. It is not difficult to show that significant changes in path loss occur due to a change of several wavelengths in d, whereas appreciable fluctuations in shadowing are perceived when the change in d is comparable with the size of the obstructing objects. In fact, typically, a significant change in path loss occurs when the variation of d is of the order of 100–1000 m, whereas one in shadowing requires a variation of the order of 10–100 m *outside buildings* (*outdoor* scenarios) and less *inside them* (*indoor* scenarios).

Note that the introduction of the small–large scale dichotomy in describing the spatial variability of fading is justified, analogously to what has been stated about time variability, by its technical usefulness. In fact, large-scale fading determines the *coverage area* of a wireless transmission and, hence the service availability in a given geographical region, whereas small-scale fading influences more the selection of signaling techniques and receiver design [19, 47, 49, 51]. This can be fully understood by observing that the number of multipath echoes, their time spread (due to different electrical path lengths) and their intensities significantly affect the structure of the received signal. Thus, in the following we will focus primarily on small-scale fading, providing only some brief hints about the analytical description of large-scale fading. Before doing that, however, it is worth pointing out that the mathematical description of these types of fading is substantially different. In fact, a description of the attenuation of received power with respect to that transmitted is commonly provided to describe the effects of large-scale fading. In contrast, in the analysis of small-scale fading, a channel is usually modeled as a *linear*, *time-varying* filter, whose behavior is fully described by proper functions, such as its impulse response and its frequency response, with particular statistical properties. In addition, we must not forget that the small-scale description is always "local", that is, conditioned with respect to the "instantaneous" locations of the transmitter and the receiver. Also, in most cases, appreciable variations in large-scale fading do not occur if the transmitter-receiver locations do not change substantially.

In general, it is not easy to derive a single mathematical model that allows accurate assessment of path loss in different propagation scenarios, because of the complexity of the propagation mechanisms. In a specific environment an accurate assessment of this parameter can be obtained by resorting to specific software packages implementing advanced mathematical methods or to a measurement campaign [52, 53]. These tools can be exploited when certain specifications must be precisely met in system design, for example in the selection of the locations of *base stations* (BSs) in a cellular mobile system operating within a geographic area over which coverage must be guaranteed [54]. A deeper study of these problems is outside the scope of this book: the interested reader can refer to Section 2.4.1, where some well-known methods for path loss estimation are listed and some references are provided. However, if the adequacy of specific design solutions must be assessed, simple models turn out to be extremely useful. In these cases, *if shadowing is neglected*, the *average[2] received power* $\bar{P}_R(d)$ in the *far field region*, for a given transmitted power P_T and at a transmitter–receiver distance d, can be assessed by evaluating the *average path loss* (or *propagation loss*) as:

$$L(d) \triangleq \frac{P_T}{\bar{P}_R(d)} \tag{2.1}$$

or, in decibels, as:

$$L(d)_{dB} \triangleq 10 \log_{10} L(d) = L(d_0)_{dB} + 10\, n \log_{10}\left(\frac{d}{d_0}\right), \tag{2.2}$$

where the parameters d_0 and n are the so-called *close-in reference distance* and *path loss exponent*, respectively. This does not include the random effects of shadowing, which can be accounted for by adding a random term X to the *right-hand side* (RHS) of (2.2), so that the total power loss in decibels is given by:

$$L_t(d)_{dB} = L(d)_{dB} + X. \tag{2.3}$$

It is commonly assumed that X is a Gaussian[3] random variable having zero mean and standard deviation σ_X (both in decibels); the value of the latter parameter reflects the intensity of the variations experienced in the average received power at a distance d, so that a smaller value of σ_X means that more accurate predictions of the overall path loss can be made. The assumption of Gaussianity for X implies that:

(a) at the receiver the power attenuation due to shadowing is *log-normally distributed* (*log-normal shadowing*), since from (2.1) and (2.3) it is easily seen that:

$$\bar{P}_R(d) = \frac{P_T}{L(d)} 10^{X/10}; \tag{2.4}$$

(b) the average power $\bar{P}_R(d)$ expressed in decibels has a normal distribution with mean given by (2.2).

We also note that the values of the parameters n, d_0 and σ_X in this path loss model depend on the scenario. In this book we consider channel models that can be applied to the analysis of systems operating in either *outdoor* or *indoor* scenarios [54].

Outdoor channel models are of great interest for cellular telephony systems. There the overall coverage area is divided into *cells* or *macrocells*, each having a radius of 1–10 km and served by a BS. The BS antennas are usually placed at a greater height than that of surrounding objects and radiate a power of 1–10 W [56]. In distinct macrocells, substantially different propagation environments can be encountered. This explains why, for instance, the *Global System for Mobile* (GSM) standard [57]

---

[2] As already mentioned, local averaging is accomplished spatially, to cancel the fast fluctuations due to small-scale fading.

[3] An electromagnetic signal usually undergoes multiple reflections/refractions before being received. Each such event introduces an attenuation, represented by a multiplicative coefficient. If all these coefficents are expressed in decibels, the overall attenuation is given by their sum, which, by the *central limit theorem* [55], can be modeled as a Gaussian random variable.

has proposed three different channel models, known as *rural area*, *hilly terrain* and *typical urban*, for system testing [58, pp. 17–19]. Note also that, in macrocellular environments, the LOS propagation path is often absent and this makes the prediction of path loss extremely difficult. In the recent past, the interest in new *personal communication systems* (PCSs) [59] has also fostered research on electromagnetic propagation in *urban microcells*, each consisting of a small area, with a radius of a few hundred meters. These exhibit channel properties that are substantially different than those of macrocells. This is due to the fact that BS antennas in microcells are placed below the roof line of surrounding buildings (typically at a height of 3–10 m) and BS powers are lower (0.1–1 W). For these reasons, the cell is shaped by the buildings themselves and electromagnetic waves propagate along shorter paths [56].

Indoor channel models are of interest, for instance, in the study of *cordless* telephony and WLANs operating in buildings devoted to different uses (offices, depots, stores, etc.) [60–63]. In the literature dealing with the radio coverage inside buildings, two distinct situations are considered. In the first, the transmitter is placed on the roof of a building different from that in which the receiver is operating, whereas in the second the transmitter and receiver are placed in the same building. In general, path loss prediction in indoor environments is not easier than in outdoor ones. For instance, to predict the intensity of the electromagnetic field inside an office building, several factors must be taken into account, including wall partitions (which may exhibit frequency-dependent behavior), the presence of multiple floors (if the transmitter and the receiver are on different floors), furniture, metallic pipes and ventilation ducts. In addition, the presence of multiple echoes, found in measurement campaigns in indoor environments, makes the fluctuations in the intensity of the received signal fast and, consequently, harder to predict in an indoor environment [64].

These considerations explain the large spread of parameter values adopted for the path loss model of (2.2) and (2.4). First, we note that the reference distance d_0 of (2.2) is associated with a circle in the far field of the radiating antenna and its value is small with respect with the usual link length of a system. Typical values of d_0 are 1 m, 100 m and 1 km for indoor, microcellular outdoor and macrocellular outdoor scenarios, respectively [56]. The use of the model of (2.2) also requires knowledge of both the path loss $L(d_0)_{\mathrm{dB}}$ at the reference distance and the exponent n. The former can be acquired through measurement or estimated assuming *free space* propagation[4] at distance d_0 [65], whereas the latter deserves more attention. In fact, different estimates of n have been measured in various propagation environments (e.g., see [52, 56, 60, 61, 65–73]). Typical ranges of n are summarized in Table 2.1 [5]. Note that these data have been extracted in different (indoor and outdoor) environments, using various antenna heights and at carrier frequencies of 0.9 GHz or 1.9 GHz.[5] It can be easily seen that, on the one hand, indoor environments are characterized by a large spread of the parameter n, due to the many factors influencing indoor propagation [72], and can on occasion be characterized by $n < 2$, because of a possible *waveguiding effect* [60, 68]. On the other hand, in outdoor environments n can take on values substantially larger than 2. Experimental data have also shown that, in a given environment, n tends to become larger with frequency (e.g., see [62] for indoor scenarios), and depends heavily on antenna heights (e.g., see [65, 75] for macrocellular and [73] for microcellular scenarios).

Finally, it is worth remembering that typical values of the shadowing parameter σ_X lie in the range 4–13 dB [5]. Moreover, in specific scenarios, n and σ_X can be easily extracted from experimental data via *linear regression* techniques (see [5, 56, 65, 71–73, 76] and [77, p. 1441]).

In this chapter, we will not consider further the mathematical characterization of path loss, and will concentrate, instead, on small-scale characterization. The chapter is organized as follows. Section 2.2 is devoted to the study of small-scale fading in SISO systems. This phenomenon is described first in *deterministic* terms, representing a wireless channel as a linear time-varying filter whose input–output behavior is described by proper *system functions*. This is followed by the *statistical* characterization via autocorrelation functions. In this framework, the properties of *wide-sense stationarity* and of

---

[4] A free space assumption means that $n = 2$ in (2.2) and $\sigma_X = 0$ in (2.4).

[5] The path loss depends significantly on the transmit frequency (e.g., see [74]).

Table 2.1 Typical values of the *path loss exponent n* in various environments

Environment	Range of n
Free space	2
Urban macrocells	3.7–6.5
Urban microcells	2.7–3.5
Office in a building (same floor)	1.6–3.5
Office in a building (different floors)	2–6
Warehouse	1.8–2.2
Factory	1.6–3.3
House	3

uncorrelated scattering, and some important statistics, such as the *power delay profile* and the *scattering function* of a channel, are introduced.

The study of various problems in analysis and design of wireless communication systems requires the availability of mathematical models describing short-term small-scale fading. Statistical modeling of SISO channels is investigated in Section 2.2.3, where the emphasis is on reduced-complexity models, that is, on those models in which randomness is described through a finite (and possibly small) set of random parameters. Many of the results concerning SISO channels are then extended to the case of MIMO channels in Section 2.3, where both matrix-based models and directional descriptions are provided.

Finally, some historical notes and suggestions for further reading are provided in Sections 2.4 and 2.5, respectively.

2.2 Mathematical Description of SISO Wireless Channels

As mentioned in the Introduction, small-scale fading in a wireless channel can be described in a mathematically rigorous fashion by modeling the channel as a *linear time-variant system* (LTVS). In fact, this allows the adoption of the so-called *system functions* for representing its input–output behavior, as discussed in Section 2.2.1. Their behavior, however, is unknown a priori, so that the system functions must be modeled as random processes with specific statistical characterization, as illustrated in Section 2.2.2.

2.2.1 Input–Output Characterization of a SISO Wireless Channel

2.2.1.1 General Case

The input–output behavior of any LTVS is fully described by its *time-variant impulse response* (TVIR), defined as the system response to an impulse or *Dirac delta function* delayed by τ sec. More specifically, the TVIR is defined as:

$$g(t, t - \tau) = \Upsilon_{RF}[\delta(t - \tau)], \tag{2.5}$$

where t represents *time*, τ represents *delay* (with respect to the reference instance $t = 0$) in the application of the impulse to the system, and the operator $\Upsilon_{RF}[\cdot]$ describes the transformation accomplished by the LTVS. Note that the *dependence of the system physical behavior on time* is explicitly indicated by the presence of t as the first variable in the argument of the TVIR. The dependence of the TVIR on the delay variable, τ, accounts for the *time dispersion* introduced by the channel, that is, for the generation of the multiple echoes of the transmitted signal.

In our analysis the channel is fed by a *real* RF signal $x_{RF}(t)$, having a central frequency f_c, and results in the *real* RF response:

$$y_{RF}(t) = \Upsilon_{RF}[x_{RF}(t)]. \tag{2.6}$$

It is not difficult to show that, given the TVIR of (2.5), $y_{RF}(t)$ in (2.6) can be expressed as [78]:

$$y_{RF}(t) = \int_{-\infty}^{+\infty} x_{RF}(t - \tau)\, g(t, \tau)\, d\tau. \tag{2.7}$$

This lends itself to a simple interpretation. In fact, it means that the input signal is delayed and multiplied by a differential scattering gain $g(t, \tau)\, d\tau$; this complex factor expresses the modulation due to the scatterers introducing a delay in the interval $(\tau, \tau + d\tau)$. For this reason, the system function $g(t, \tau)$ is also called *input delay-spread function* [78].

To simplify the study of system functions, it is useful to adopt an equivalent low-pass representation of the communication channel[6] [79]. We then let $x(t)$, $y(t)$ and $h(t, t - \tau)$ denote the *low-pass equivalent signals* (with respect to the reference frequency f_c) of $x_{RF}(t)$, $y_{RF}(t)$ and $g(t, t - \tau)$ respectively, that is, the complex signals such that:

$$x_{RF}(t) = \text{Re}\{x(t) \exp(j2\pi f_c t)\},$$

$$y_{RF}(t) = \text{Re}\{y(t) \exp(j2\pi f_c t)\},$$

and

$$g(t, t - \tau) = 2\text{Re}\{h(t, t - \tau) \exp(j2\pi f_c t)\}, \tag{2.8}$$

where $h(t, \tau)$ is the so-called *channel impulse response* (CIR) or *input delay spread function* (since it is the low-pass equivalent of $g(t, \tau)$ in (2.5)).[7] Then it can be shown that the RF input–output relationship (2.7) is equivalent to:

$$y(t) = \int_{-\infty}^{+\infty} x(t - \tau)\, h(t, \tau)\, d\tau = \int_{-\infty}^{+\infty} x(\tau)\, h(t, t - \tau)\, d\tau, \tag{2.9}$$

relating the low-pass signals $x(t)$, $y(t)$ and $h(t, \tau)$. Note that, as can be easily inferred from (2.9), $h(t, t - \tau)$ can also be defined as the response of the low-pass equivalent channel to the impulsive excitation $\delta(t - \tau)$, that is:

$$h(t, t - \tau) \triangleq \Upsilon_{BB}[\delta(t - \tau)], \tag{2.10}$$

where $\Upsilon_{BB}[\cdot]$ is the low-pass equivalent of the transformation accomplished by the channel, that is, a transformation such that $y(t) = \Upsilon_{BB}[x(t)]$.

To grasp the physical meaning of (2.9), it is useful to approximate the first integral as a sum. This can be done by discretizing the delay space in a uniform fashion and, in particular, generating the sequence $\tau_i \triangleq i\Delta\tau$, where $i \in \mathbb{Z}$ and $\Delta\tau$ is the discretization step. Then (2.9) can be approximated as:

$$y(t) \cong \sum_{i=-\infty}^{+\infty} x(t - \tau_i)\, h(t, \tau_i)\, \Delta\tau = \sum_{i=-\infty}^{+\infty} a_i(t)\, x(t - \tau_i), \tag{2.11}$$

---

[6] In the following pages we will always refer, unless explicitly stated otherwise, to the complex low-pass representation of signals and systems.

[7] The introduction of the factor 2 in (2.8) allows us to remove the factor of $1/2$ that would otherwise appear in the RHS of (2.9).

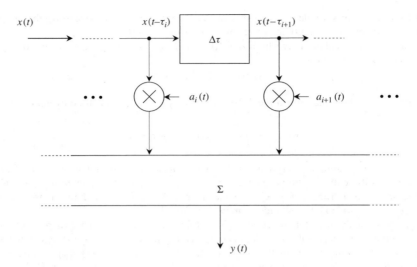

Figure 2.2 Representation of (2.11) (TDL model of a communication channel).

where:

$$a_i(t) \triangleq h(t, \tau_i)\, \Delta\tau. \tag{2.12}$$

The input–output relationship of (2.11) is summarized by the block diagram of Figure 2.2, representing the communication channel as a *tapped delay line* (TDL). In Section 2.2.3 we will show that, under certain assumptions, this model provides an *exact* description of a communication channel. In fact, $x(t)$ goes through a *delay line*, consisting of the serial concatenation of an infinite number of identical cells (each introducing a delay of $\Delta\tau$ sec). The ith delayed replica $x(t - \tau_i)$ of $x(t)$ feeds the ith *tap*, accomplishing a multiplication by the complex time-varying gain $a_i(t)$ (2.12). Then the tap outputs are summed, generating an approximation to $y(t)$ (2.9). In other words, the TDL model represents the channel output signal as the superposition of an infinite number of echoes, each having time-varying amplitude and phase.

The CIR is not the only system function that fully describes the input–output behavior of an LTVS. An equivalent description is provided by the so-called *time-variant transfer function* (TVTF) defined as [78]:

$$H(t, f) \triangleq \mathrm{FCT}_{\tau \to f}[h(t, \tau)] = \int_{-\infty}^{+\infty} h(t, \tau)\, \exp(-j2\pi f\tau)\, d\tau, \tag{2.13}$$

where $\mathrm{FCT}_{\tau \to f}[\cdot]$ denotes the *Fourier continuous transform* (FCT) evaluated with respect to the variable τ and leading to an explicit dependence on the *frequency* variable f. From (2.13) the inverse relationship:

$$h(t, \tau) = \mathrm{IFCT}_{f \to \tau}[H(t, f)] = \int_{-\infty}^{+\infty} H(t, f)\exp(j2\pi f\tau)\, df \tag{2.14}$$

is immediately inferred, where IFCT[·] denotes an *inverse Fourier continuous transform* (IFCT). Substituting this in (2.9) produces, after some manipulation:

$$y(t) = \int_{-\infty}^{+\infty} X(f)\, H(t, f)\, \exp(j2\pi ft)\, df, \tag{2.15}$$

which expresses the input–output relationship in a new form, involving $H(t, f)$, and the FCT $X(f)$ of $x(t)$.

Establishing a relationship between the FCT:

$$Y(f) \triangleq \mathrm{FCT}[y(t)] = \int_{-\infty}^{+\infty} y(t) \exp\left(-j2\pi ft\right) dt \tag{2.16}$$

of the output $y(t)$ of an LTVS and the FCT $X(f)$ of its input signal requires the introduction of a further system function. In fact, substituting (2.15) in the RHS of (2.16) yields, after some manipulation:

$$Y(f) = \int_{-\infty}^{+\infty} X(\alpha)\, \Gamma(f - \alpha, \alpha)\, d\alpha, \tag{2.17}$$

where

$$\Gamma(\nu, f) \triangleq \mathrm{FCT}_{t \to \nu}[H(t, f)] = \int_{-\infty}^{+\infty} H(t, f) \exp\left(-j2\pi\nu t\right) dt \tag{2.18}$$

is the so-called *output Doppler-spread function* [78]. Note that $\Gamma(\nu, f)$ depends on two spectral variables, namely the frequency f and the *Doppler shift*[8] ν, and that equation (2.17) expresses the output spectrum as a *convolution* between the input spectrum and $\Gamma(\nu, f)$ (2.18). A comparison of (2.17) with (2.9) illustrates a deep structural analogy between these two input–output relationships, referring, however, to different domains. In fact, the latter involves signals defined in the t and τ domains, whereas the former functions are defined in the ν and f domains. This analogy is usually stressed by stating that (2.17) is the *dual* of (2.9) and, in particular, that $\Gamma(\nu, f)$ is the *dual function* of the CIR $h(t, \tau)$ [80]. These considerations are important not only from a mathematical viewpoint, but also from a physical one, since they allow us to understand the real significance of $\Gamma(\nu, f)$. In fact, this function, by duality, could be introduced by evaluating the LTVS response, in the frequency domain, to an *impulsive input spectrum*. In particular, if $X(f) = \delta(f - f_0)$ is selected (corresponding to the time-domain choice $x(t) = \exp\left(j2\pi f_0 t\right)$), where f_0 is an arbitrary frequency, then (2.17) produces:

$$Y(f) = \int_{\alpha=-\infty}^{+\infty} \delta(\alpha - f_0)\, \Gamma(f - \alpha, \alpha)\, d\alpha = \Gamma(f - f_0, f_0). \tag{2.19}$$

Then, defining the variable $\upsilon \triangleq f - f_0$ (representing the *offset* of f with respect to the excitation frequency f_0), the last equality can be rewritten as:

$$Y(f_0 + \upsilon) = \Gamma(\upsilon, f_0), \tag{2.20}$$

which proves that $\Gamma(\nu, f)$ fully expresses the spectral content of the response of an LTVS to a complex exponential with frequency f_0. Thus, unlike what happens with *linear time-invariant systems*, the output signal contains spectral components characterized by $\upsilon \neq 0$, that is, by a frequency different from that (f_0) of the input signal, because of the Doppler effect affecting the communication channel.

Another important system function is the so-called *delay-Doppler-spread function* [78], which can be defined as the FCT, in the t variable, of $h(t, \tau)$, that is, as:

$$\gamma(\nu, \tau) \triangleq \mathrm{FCT}_{t \to \nu}[h(t, \tau)] = \int_{-\infty}^{+\infty} h(t, \tau)\, \exp\left(-j2\pi\nu t\right) dt, \tag{2.21}$$

for which the inverse relationship:

$$h(t, \tau) = \mathrm{IFCT}_{\nu \to t}[\gamma(\nu, \tau)] = \int_{-\infty}^{+\infty} \gamma(\nu, \tau)\, \exp\left(j2\pi\nu t\right) d\nu \tag{2.22}$$

---

[8] *Frequency shifts* in the spectral components of the transmitted signal are due to *time variations* of the channel, that is, to the so-called *Doppler effect*. These variations are parametrized in the variable t.

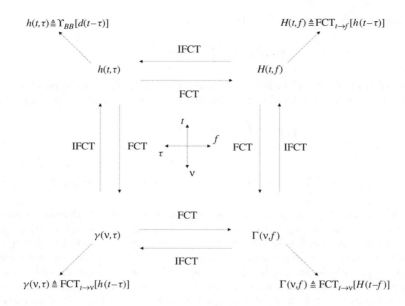

Figure 2.3 Relationships among the four system functions defined in Section 2.2.1.

holds. The input–output relationship involving $\gamma(v, \tau)$ can be easily derived by substituting (2.22) in (2.9). This leads to:

$$y(t) = \int_{\tau=-\infty}^{+\infty} \int_{v=-\infty}^{+\infty} \gamma(v, \tau) \, x(t - \tau) \, \exp(j2\pi vt) \, dv \, d\tau, \tag{2.23}$$

providing another interpretation of the generation mechanism of the channel output. In fact, it represents the response $y(t)$ as the superposition of an infinite number of replicas, each characterized by specific values of the delay τ and of the spectral shift v, and weighted by $\gamma(v, \tau)$ [78].

The relationships between different couples of the four system functions defined in this subsection are summarized in Figure 2.3. The specific structure of these functions for an L-ray multipath channel are illustrated in the following example:

Example 2.2.1 In the technical literature, a commonly adopted model for multipath fading channels assumes that the channel response is characterized by L distinct echoes of the transmitted signal, each described by a time-varying complex gain and a delay. The useful component of the baseband received signal is then given by:

$$y(t) = \sum_{i=0}^{L-1} a_i(t) \, x(t - \tau_i), \tag{2.24}$$

where $a_i(t)$ and τ_i represent the complex gain and the delay of the ith echo, respectively. It is usually assumed that $\tau_0 = 0$ and that the other delays $\{\tau_i, i = 1, 2, \dots, L - 1\}$ are in ascending order of value so that the maximum delay (i.e., τ_{L-1}) measures the extent of the *time dispersion* produced by the communication channel. Moreover, the delay parameters $\{\tau_i\}$ are usually modeled as being time-invariant (see Section 2.2.3) and their values are selected on the basis of the propagation scenario to which the channel model (2.24) refers. The tap gains $\{a_i(t), i = 0, 1, \dots, L - 1\}$, on the other hand, are modeled as *random processes* characterized by a given mean and autocorrelation functions, as shown in Section 2.2.2.

From (2.24) it is easy to see that the corresponding CIR is:

$$h(t, \tau) = \sum_{i=0}^{L-1} a_i(t) \, \delta(\tau - \tau_i), \qquad (2.25)$$

that is, it consists of the superposition of L distinct Dirac delta pulses in the delay variable τ, each weighted by a different time-varying gain. Applied to (2.25), the transformations summarized in Figure 2.3 yield the system functions:

$$H(t, f) = \sum_{i=0}^{L-1} a_i(t) \, \exp(-j2\pi f \tau_i), \qquad (2.26)$$

$$\Gamma(\nu, f) = \sum_{i=0}^{L-1} A_i(\nu) \, \exp(-j2\pi f \tau_i) \qquad (2.27)$$

and

$$\gamma(\nu, \tau) = \sum_{i=0}^{L-1} A_i(\nu) \, \delta(\tau - \tau_i), \qquad (2.28)$$

where $A_i(\nu) = \text{FCT}_{t \to \nu}[a_i(t)]$.

$\square$

The mathematical tools described in this chapter allow us to describe the behavior of any *linear* wireless channel. In many specific communication systems, however, the properties of the transmitted signal, *in terms of bandwidth and/or duration*, allow us to adopt a simpler representation of the channel, as discussed in Sections 2.2.1.2 and 2.2.1.3.

2.2.1.2   Frequency-Selective Channels

Let us assume that the transmitted signal $x(t)$ has a *finite duration* T_0 and that, within an interval of duration not exceeding T_0, the communication medium does not exhibit any significant change in its physical characteristics. In this case time variability can be neglected and we may represent the channel as a linear and *stationary* system with a CIR $h(\tau)$. For instance, under these assumptions, the L-ray model described in Example 2.2.1 may be characterized by the impulse response[9] (see (2.25))

$$h(\tau) = \sum_{i=0}^{L-1} a_i \, \delta(\tau - \tau_i), \qquad (2.29)$$

where the complex gains $\{a_i\}$ are *time-invariant*.

In this case only the effects of *time dispersion* can be perceived in the observation interval. Such dispersion accounts for *frequency selectivity*, that is, for the fact that the channel frequency response is not flat in the transmission bandwidth so that, generally speaking, distinct spectral components of the transmitted signal are affected differently. We also note that in a communication channel the effects of frequency selectivity are visible if the bandwidth B_x of $x(t)$ is not significantly smaller that the reciprocal of the maximum delay[10] τ_m of the echoes of the transmitted signal.[11] Otherwise, the channel is not only static, but also *frequency-flat*.

---

[9] A particular case of this model, corresponding to $L = 2$, was adopted by W. D. Rummler in 1979 [81] (see also [48, 82]).
[10] Note that $\tau_m = \tau_{L-1}$ for the communication channel described in Example 2.2.1.
[11] We will reformulate this statement referring to the so-called *coherence bandwidth* of a communication channel in Section 2.2.2.

2.2.1.3   Time-Selective Channels

Let us now consider the dual case to that illustrated in Section 2.2.1.2. We now assume that the channel variations are significant within the observation interval[12] and that the reciprocal of the bandwidth B_x of $x(t)$ is *significantly smaller than the time dispersive effect of the channel*, so that, if τ_m denotes the maximum delay as above, then:

$$B_x \ll 1/\tau_m. \tag{2.30}$$

Then, since $h(t, \tau) = 0$ for $\tau < 0$ and $\tau > \tau_m$, (2.9) can be rewritten as:

$$y(t) = \int_0^{\tau_m} x(t - \tau) \, h(t, \tau) \, d\tau. \tag{2.31}$$

From (2.30) it is immediately seen that, in any time interval of duration τ_m, $x(t)$ undergoes only small changes, so that $x(t - \tau) \cong x(t)$ for $0 \le \tau \le \tau_m$. Then (2.31) can be approximated as:

$$y(t) = x(t) \, a(t), \tag{2.32}$$

where

$$a(t) \triangleq \int_0^{\tau_m} h(t, \tau) \, d\tau. \tag{2.33}$$

Let us again consider the channel model of Example 2.2.1, to understand when its input–output relationship (2.24) becomes that of (2.32).

Example 2.2.2 If the delays $\{\tau_i, \ i = 1, 2, \ldots, L - 1\}$ in the model of (2.24) are in ascending order of value (and $\tau_0 = 0$), the maximum delay due to the channel τ_m is equal to τ_{L-1}. Therefore, if $B_x \ll 1/\tau_{L-1}$ (see (2.30)), the multiplicity of channel echoes collapses to a single echo characterized by a null delay and a complex gain:

$$a(t) \triangleq \sum_{i=0}^{L-1} a_i(t). \tag{2.34}$$

The general result expressed by (2.32) and (2.33) and the specific one given by (2.34) show that, if B_x is small enough, the effect of multiple echoes generated by the channels is perceived as that of a *single echo*, whose complex gain results from the superposition of all the echo gains. Moreover, it is not difficult to show that the CIR associated with the input–output relationship (2.32) is given by:

$$h(t, \tau) = \delta(\tau) \, a(t), \tag{2.35}$$

where $a(t)$ is expressed by (2.33).
□

Equation (2.32) describes a *time-selective* or, equivalently, a *frequency-dispersive* communication channel, since the time changes in the gain $a(t)$ entails both *time variations in the intensity of* $x(t)$ and its *spectral broadening*. In the technical literature it is often stated that this type of channel is *frequency-flat*, since all spectral components of the signal $x(t)$ are subject to the same distortion.

Note that the origin of spectral broadening can be easily understood by observing that, generally speaking, the scatterers of the propagation medium, even if they generate echoes with similar delays, are characterized by different speeds, resulting in distinct Doppler shifts. This means that a sinusoidal signal at frequency f_c transmitted over a time-selective channel is transformed in the superposition

---

[12] In the technical literature this condition can be formulated with reference to the *Doppler bandwidth* or to the *coherence time* of the communication channel, as illustrated in Section 2.2.2.

of a multiplicity of signals of the same type, characterized, however, by different frequencies, as illustrated in the following example.

Example 2.2.3 In a time-selective fading channel, scatterers can be grouped by putting together all those that produce similar Doppler shifts. Let us assume, for simplicity, that such shifts, occurring in a limited time interval (t_i, t_f), belong to the set $\{f_{D,i}, i = 0, 1, \ldots, L_D - 1\}$, consisting of L_D distinct and fixed frequencies (in other words, in the given interval the set of relative speeds between the mobile receiver and the scatterers undergoes negligible changes). Note that this scenario is the dual of that represented by equation (2.29) for a frequency-selective channel characterized by L *distinct echoes*. Then in the observation interval the fading distortion can be modeled as[13]:

$$a\,(t) \triangleq \sum_{i=0}^{L_D-1} a_i \, \exp\,(j2\pi f_{D,i}t), \tag{2.36}$$

where a_i is the complex gain associated with the set of scatterers producing a Doppler shift of $f_{D,i}$ Hertz. If the RF signal $x(t) = \cos(2\pi f_c t)$ is transmitted over the given channel, where f_c is the frequency with respect to which the complex envelope (2.36) has been derived, the RF channel response $y\,(t) = \mathrm{Re}\{\exp\,(j2\pi f_c t)\,a(t)\}$ is given by:

$$y\,(t) = \sum_{i=0}^{L_D-1} [b_i \, \cos(j2\pi(f_c + f_{D,i})t) - c_i \, \sin(j2\pi(f_c + f_{D,i})t)], \tag{2.37}$$

where $b_i \triangleq \mathrm{Re}\{a_i\}$ and $c_i \triangleq \mathrm{Im}\{a_i\}$. This shows that the received signal contains the frequencies $\{f_c + f_{D,i}, i = 0, 1, \ldots, L_D - 1\}$ and, consequently, illustrates the spreading of the signal spectrum beyond that which was transmitted.

□

Finally, we note that, in the most general case, a wireless channel is *selective* (*dispersive*) both *in frequency* (*time*) and *in time* (*frequency*). When this occurs, the channel is *doubly-selective*.

2.2.2   *Statistical Characterization of a SISO Wireless Channel*

In this subsection we consider the statistical characterization of wireless channels. In particular, after some general considerations, an analysis of some autocorrelation functions for frequency-selective channels and their duals (i.e., time-selective channels) is provided. This paves the way for a study of the most general case – that of doubly-selective channels.

2.2.2.1   **General Properties**

As mentioned in the Introduction, all the system functions described in the foregoing should be considered, for any wireless channel, as *random processes*. A complete characterization of a random process usually requires knowledge of the distribution functions of arbitrary order. Such knowledge, however, represents too ambitious an objective in any application. In practice, the statistical descriptions of wireless channels considered in the technical literature usually provide data about the *mean value function* and the *autocorrelation function* characterizing specific system functions. Autocorrelation functions, which represent specific second-order statistics, provide important information. In fact,

---

[13] We will reconsider this model in Section 2.2.3, where the problems of a proper selection of the Doppler shifts $\{f_{D,i}\}$ and that of the characterization of the random variables $\{a_i\}$ in the model (2.36) are discussed.

they allow us to determine the autocorrelation function of the channel response and, in particular, the spectral properties of the received signal, as illustrated in Section 3.9. Here, therefore, most attention is paid to the study of autocorrelation functions of the system functions introduced in Section 2.2.1. A mean value function can be evaluated for each system function describing a communication channel (impulse response, frequency response, etc.). In what follows we focus, however, for the sake of simplicity, on the mean function of a specific system function only – the TVIR $g(t, \tau)$ (2.5) – and do not consider the mean value functions for other possible system functions. We note that its mean:

$$g_d(t, \tau) \triangleq E\{g(t, \tau)\}, \tag{2.38}$$

describes the so-called *specular component* of a communication channel, that is, the *purely deterministic* component. This component can be fully characterized by resorting to all the other system functions illustrated in Section 3.9. A statistical description is required, on the other hand, primarily for the *purely random component* of the channel, which, in the TVIR case, is defined as:

$$g_r(t, \tau) \triangleq g(t, \tau) - g_d(t, \tau), \tag{2.39}$$

which clearly has *zero mean*. This decomposition of $g(t, \tau)$ is equivalent to representing a channel as the parallel combination of two subsystems (one deterministic, the other one purely random), both fed by the same signal $x(t)$. Since the analysis here aims to provide an understanding of the statistical characterization of wireless channels, it will be assumed in what follows, in the absence of explicit indications, that $g_d(t, \tau)$ of (2.38) is null, so that all the system functions consist of a *purely random component* only.

It is also important to note that, in principle, our study of specific channel statistics should refer to some *bandpass random processes*. However, for an arbitrary RF or bandpass random process $x_{RF}(t)$, it is straightforward to prove that its autocorrelation function $R_{x_{RF}}(t_1, t_2) \triangleq E\{x_{RF}(t_1)\, x_{RF}(t_2)\}$ can be expressed as [78]:

$$R_{x_{RF}}(t_1, t_2) = \frac{1}{2}\mathrm{Re}\{E\{x(t_1)\, x^*(t_2)\} \exp\left(j2\pi f_c(t_1 - t_2)\right)\}$$
$$+ \frac{1}{2}\mathrm{Re}\{E\{x(t_1)\, x(t_2)\} \exp\left(j2\pi f_c(t_1 + t_2)\right)\}, \tag{2.40}$$

where $x(t)$ is the complex envelope of $x_{RF}(t)$ with respect to the reference frequency f_c. The last result can be rewritten as:

$$R_{x_{RF}}(t_1, t_2) = \frac{1}{2}\mathrm{Re}\{R_x(t_1, t_2) \exp\left(j2\pi f_c(t_1 - t_2)\right)\}$$
$$+ \frac{1}{2}\mathrm{Re}\{\tilde{R}_x(t_1, t_2) \exp\left(j2\pi f_c(t_1 + t_2)\right)\}, \tag{2.41}$$

where the autocorrelation functions $R_x(t_1, t_2) \triangleq E\{x(t_1)\, x^*(t_2)\}$ and $\tilde{R}_x(t_1, t_2) \triangleq E\{x(t_1)x(t_2)\}$ refer to $x(t)$. In most applications we have that[14] $\tilde{R}_x(t_1, t_2) = 0$, so that only knowledge of $R_x(t_1, t_2)$ is required to determine $R_{x_{RF}}(t_1, t_2)$. In the following, we will always assume that this condition holds for all RF system functions, so that the evaluation of $R_x(t_1, t_2)$ only will be considered.

2.2.2.2 Frequency-Selective Channels

A time-dispersive channel is fully characterized by its *frequency response* $H(f) \triangleq \mathrm{FCT}[h(t)]$, which we assume to be a zero-mean random process (in the variable f) with correlation function:

$$R_H(f_1, f_2) \triangleq E\{H(f_1)\, H^*(f_2)\}. \tag{2.42}$$

---

[14] Note that this is a necessary condition for the *wide-sense stationarity* of $x_{RF}(t)$ (see (2.41)).

Let us now focus on the structure of this function for the communication channel characterized by the CIR (2.29).

Example 2.2.4 The frequency response associated with the impulse response (2.29) is given by:

$$H(f) = \sum_{i=0}^{L-1} a_i \exp(-j2\pi f \tau_i).$$
(2.43)

In the literature the delays $\{\tau_i, i = 0, 1, \ldots, L-1\}$ are *usually known*, whereas the complex gains $\{a_i, i = 0, 1, \ldots, L-1\}$ are *random*. Moreover, if $H(f)$ in (2.43) is *purely random*, these gains have zero mean, so that the autocorrelation function of $H(f)$ is given by (see (2.42)):

$$R_H(f_1, f_2) = \sum_{i=0}^{L-1}\sum_{k=0}^{L-1} R_{i,k} \exp[j2\pi(f_2\tau_k - f_1\tau_i)],$$
(2.44)

where $R_{i,k} \triangleq E\{a_i\, a_k^*\}$ is the *correlation* between a_i and a_k. This shows that $R_H(f_1, f_2)$ depends on f_1 and f_2 through their difference $(f_1 - f_2)$ if and only if $R_{i,k} = 0$ for $i \neq k$, that is, if and only if the random variables a_i and a_k are *uncorrelated*. In fact, if this occurs, (2.44) simplifies to:

$$R_H(f_1, f_2) = R_H(f_2 - f_1) = \sum_{i=0}^{L-1} \sigma_i^2 \exp[j2\pi\tau_i(f_2 - f_1)],$$
(2.45)

where $\sigma_i^2 \triangleq R_{i,i} = E\{|a_i|^2\}$, with $i = 0, 1, \ldots, L-1$. Note that:

(a) when (2.45) holds, the random process $H(f)$ is *wide-sense stationary* (WSS);
(b) for large L, if $H(f)$ is expressed, for a given f, as a linear combination of a large number of random variables, it can be modeled as a *Gaussian random process* by virtue of the *central limit theorem* [55];
(c) if the gains $\{a_i, i = 0, 1, \ldots, L-1\}$ are modeled as *jointly Gaussian* random variables, the property of Gaussianity holds independently of the specific value of L, and zero correlation among them is equivalent to *statistical independence*.
☐

Generally speaking, if in a frequency-selective channel the gains associated with echoes characterized by different delays are represented by uncorrelated random variables, the channel is said to exhibit *uncorrelated scattering* (US) or simply said to be US. The previous example shows that, if a purely random channel is US, then:

$$R_H(f_1, f_2) = R_H(f_2 - f_1).$$
(2.46)

It can also easily be proved that the autocorrelation function $R_h(\tau_1, \tau_2) \triangleq E\{h(\tau_1)\, h^*(\tau_2)\}$ of the impulse response $h(\tau)$ of a US channel is given by:

$$R_h(\tau_1, \tau_2) = \delta(\tau_1 - \tau_2)\, P_h(\tau_2),$$
(2.47)

where

$$P_h(\tau) \triangleq \text{FCT}[R_H(f)] = \int_{-\infty}^{+\infty} R_H(f)\, \exp(-j2\pi f \tau)\, df$$
(2.48)

is the so-called *power delay profile* (PDP)[15] of the communication channel. This can be used, as an alternative to (2.46), to define a US channel. The PDP function (2.48) provides important physical

---

[15] This is called the *delay power density spectrum* by P. A. Bello [78, p. 372].

information, since, being the FCT of an autocorrelation function in the variable f, it is in fact a *temporal power density*. In other words, this is a dual situation [80] with respect to that of time selectivity and, under this condition, the channel response exhibits the property of *wide-sense stationarity* in the frequency domain. Moreover, $P_h(\tau)$ represents the average distribution of power associated with the impulse response along the delay axis, so that it provides a clear indication of the intensity and the type of temporal dispersion introduced by a communication channel. For instance, the PDP associated with the autocorrelation (2.45) in Example 2.2.4 is given by:

$$P_h(\tau) = \sum_{i=0}^{L-1} \sigma_i^2 \, \delta(\tau - \tau_i) \tag{2.49}$$

and shows that the power associated with the CIR is *concentrated* around the echo delays $\{\tau_i, i = 0, 1, \ldots, L-1\}$.

Some parameters can be extracted by different PDPs to compare channels with different properties in terms of time dispersion. In particular, we mention the *mean excess delay*:

$$\tau_m \triangleq \int_{-\infty}^{+\infty} \tau \, p_h(\tau) \, d\tau, \tag{2.50}$$

where

$$p_h(\tau) \triangleq \frac{P_h(\tau)}{\int_{-\infty}^{+\infty} P_h(\tau) \, d\tau} \tag{2.51}$$

is a *normalized* version of $P_h(\tau)$, and the *root mean square* (rms) *delay spread*:

$$\tau_{ds} \triangleq \sqrt{\int_{-\infty}^{+\infty} \tau^2 \, p_h(\tau) \, d\tau - \tau_m^2}. \tag{2.52}$$

Note that τ_m and τ_{ds} represent a "mean value" and a "standard deviation" if $p_h(\tau)$ (2.51) is interpreted as the *probability density function* (pdf) of the echo delay; this explains why τ_{ds} measures the time dispersion of the communication channel. Expressions for τ_{ds} for some PDPs frequently used in the technical literature (e.g., see [83–87]) are illustrated in the following example.

Example 2.2.5 The expressions for $P_h(\tau)$ and the associated $R_H(f)$ for the *uniform* (U), *Gaussian* (G), *exponential* (E), *triangular* (T) and *truncated exponential* (TE) PDPs are summarized in the first two columns of Table 2.2. For each $P_h(\tau)$, (2.50) and (2.52) lead to the expression for τ_{ds} listed in the third column of the same table. For the TE case we have:

$$\tau_{ds} = \frac{\tau_0}{1 - e^{-\tau_M/\tau_0}} \sqrt{1 - \left[2 + \left(\frac{\tau_M}{\tau_0}\right)^2\right] e^{-\tau_M/\tau_0} + e^{-2\tau_M/\tau_0}}. \tag{2.53}$$

□

The parameter τ_{ds} depends strongly on the specific propagation environment: some typical values measured in indoor and outdoor scenarios can be found, for instance, in [6, p. 200].

Another important parameter characterizing a frequency-selective channel and related to τ_{ds} is the so-called *coherence bandwidth* B_c. This represents a statistical measure of the frequency interval over which the channel can be deemed approximately flat. That is, this parameter indicates the width of the spectral interval in which two sinusoids, having distinct frequencies, exhibit strong correlation in their amplitudes. If, however, the spectral separation between these sinusoidal components exceeds B_c, the

Table 2.2 Some relevant data for specific PDPs

PDP	$P_h(\tau)$	$R_H(f)$	τ_{ds}
U	$\dfrac{1}{\tau_0}\{u(\tau) - u(\tau - \tau_0)\}$	$\text{sinc}(f\,\tau_0)\,e^{-j\pi f\,\tau_0}$	$\dfrac{\tau_0}{\sqrt{12}}$
T	$\dfrac{1}{\tau_0}\left(1 - \dfrac{|\tau|}{\tau_0}\right)\text{rect}\left(\dfrac{\tau}{2\tau_0}\right)$	$\text{sinc}^2(f\,\tau_0)$	$\dfrac{\tau_0}{\sqrt{6}}$
E	$\dfrac{1}{\tau_0}\,e^{-\tau/\tau_0}u(\tau)$	$\dfrac{1}{1 + j2\pi f\,\tau_0}$	τ_0
G	$\dfrac{1}{\tau_0\sqrt{2\pi}}e^{-\tau^2/2\tau_0^2}$	$e^{-2(\pi f\,\tau_0)^2}$	τ_0
TE	$\dfrac{e^{-\tau/\tau_0}}{\tau_0(1 - e^{-\tau_M/\tau_0})}[u(\tau) - u(\tau - \tau_M)]$	$\dfrac{1 - e^{-\tau_M/\tau_0 - j2\pi f\tau_M}}{(1 - e^{-\tau_M/\tau_0})(1 + j2\pi f\,\tau_0)}$	eq. (2.53)

channel effects on the two signals can be substantially different. If B_c is defined as the width of the interval over which the normalized autocorrelation function of $H(f)$ does not take values smaller than 0.9, it can be expressed approximately as [6, p. 200]:

$$B_c \cong \frac{1}{50\tau_{ds}}. \tag{2.54}$$

Generally speaking, an exact relationship between B_c and τ_{ds} is not available, because the definition of B_c is not unique. However, B_c depends inversely on τ_{ds}, so that large fluctuations in the channel frequency response should be expected in the presence of a large time dispersion.

2.2.2.3 Time-Selective Channels

This case is the dual of the previous one and requires the statistical characterization of the multiplicative distortion $a(t)$ (see (2.31)). Because $a(t)$ is generated by the superposition of the complex gains associated with a number of scatterers, it is usually modeled as a *complex Gaussian process*, whose real and imaginary components, $a_R(t)$ and $a_I(t)$, are *independent and identically distributed* (iid). It is also usually assumed that $a(t)$ is WSS, so that its mean value $\eta_a(t) \triangleq E\{a(t)\}$ is constant and its autocorrelation function $R_a(t, \tau) \triangleq E\{a(t + \tau)\,a(t)^*\}$ depends only on τ. To illustrate the meaning of the WSS assumption, we consider the following example, with reference to the model of Example 2.2.3.

Example 2.2.6 Equation (2.36) represents $a(t)$, *in a limited observation interval*, as:

$$a(t) = \sum_{i=0}^{L_D-1} a_i \exp\left(j2\pi f_{D,i}t\right), \tag{2.55}$$

assuming L_D distinct Doppler shifts $\{f_{D,i}, i = 0, 1, \ldots, L_D - 1\}$ due to the channel. If, whatever the duration of the observation interval, the L_D complex gains $\{a_i, i = 0, 1, \ldots, L_D - 1\}$ are *jointly Gaussian* and have *zero mean*, it is easy to prove that $a(t)$ in (2.55) is a Gaussian random process with $\eta_a(t) = 0$. Moreover, it can be shown, by analogy with the dual scenario described in Example 2.2.4, that the autocorrelation function $R_a(t, \tau)$ depends on τ if and only if all the couples (a_i, a_k), with

$i \neq k$, are *uncorrelated*. In this case, it is found that:

$$R_a(t, \tau) = R_a(\tau) = \sum_{i=0}^{L_D-1} \sigma_i^2 \, \exp\,(j2\pi f_{D,i}\tau),$$

(2.56)

where $\sigma_i^2 \triangleq \mathrm{E}\{|a_i|^2\}$ $(i = 0, 1, \ldots, L_D - 1)$.

□

The last result lends itself to a simple generalization, since it can be proved (see, for instance, [88, pp. 69–70]) that $a(t)$ is WSS (i.e., the channel is WSS) if and only if in the propagation medium *scatterers producing distinct Doppler shifts are uncorrelated*. In the following we will always assume, in the absence of explicit indications, that $a(t)$ is a Gaussian and WSS random process, so that its full statistical characterization is provided by its mean $\eta_a(t) = A$ (usually $A = 0$) and its autocorrelation function $R_a(\tau)$, with $P_a \triangleq R_a(0)$.

Let us now focus on the *first-order characterization* of time-selective fading, through extracting the Gaussian random variable $a(\bar{t}) = a_R(\bar{t}) + ja_I(\bar{t})$ by sampling of $a(t)$ at $t = \bar{t}$. We first analyze the case with $A = 0$, and then that with $A \neq 0$. If $A = 0$, independently of the choice of $\bar{t}$, the independent Gaussian random variables, $a_R(\bar{t})$ and $a_I(\bar{t})$, are characterized by their joint pdf:

$$f_{a_R,a_I}(x, y) = \frac{1}{2\pi\sigma_a^2} \exp\left[-\frac{x^2 + y^2}{2\sigma_a^2}\right],$$

(2.57)

where $\sigma_a^2 = P_a/2$ is the variance of $a_R(\bar{t})$ (or, equivalently, of $a_I(\bar{t})$). Moreover, if $a(\bar{t})$ is represented in polar form with phase $\Theta \triangleq \angle a(\bar{t}) \in [0, 2\pi)$ and amplitude $R \triangleq |a(\bar{t})|$, it is easy to prove that:

(a) Θ and R are *statistically independent*;
(b) the pdf of Θ is uniform over $(0, 2\pi)$, whereas that of R is:

$$f_R\,(r) = u(r)\, \frac{r}{\sigma_a^2} \exp\left(-\frac{r^2}{2\sigma_a^2}\right),$$

(2.58)

that is, R has a *Rayleigh* distribution [55, p. 200]. For this reason, the case $A = 0$ is usually referred to as that of *Rayleigh fading*.

If we now assume that $A \neq 0$, so that $P_a = A^2 + 2\sigma_a^2$, it can be proved that the phase Θ of $a(\bar{t})$ is no longer uniform and statistically independent of the amplitude R (e.g., see [88, pp. 47–48]) and that R is characterized by the *Rice* pdf[16] [55]:

$$f_R(r) = u(r)\, \frac{r}{\sigma_a^2} \exp\left(-\frac{r^2 + A^2}{2\sigma_a^2}\right) I_0\left(\frac{A\,r}{\sigma_a^2}\right),$$

(2.59)

where $I_0(\cdot)$ is the *modified Bessel function* of the first kind and of zero order. In this case it is commonly stated that the channel is affected by *Rician fading*. Note that the condition $A \neq 0$ indicates the presence of a specular (i.e., LOS) component, often present in satellite and terrestrial (typically suburban and rural) scenarios, and that the relevance of the LOS component is described by the ratio of the power of the *deterministic* (LOS) *component* to the average power of the *random component* of $a(t)$, that is, by the parameter:

$$K \triangleq \frac{A^2}{2\sigma_a^2},$$

(2.60)

---

[16] Note that (2.59) becomes (2.58) if $A = 0$, since $I_0(0) = 1$.

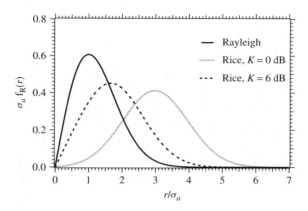

Figure 2.4 Representation of the Rice pdf (2.59) for $K_{dB} = -\infty$, 6 and 0 dB.

usually expressed in decibels ($K_{dB} \triangleq 10 \log_{10} K$). Note that Rayleigh fading corresponds to $K = 0$, whereas an additive white Gaussian noise channel ($\sigma_a^2 = 0$) is obtained if $K = +\infty$. The behavior of $f_R(r)$ (2.59) is shown in Figure 2.4, where the cases $K_{dB} = -\infty$ dB (Rayleigh fading), 6 dB and 0 dB are considered.

Rician and Rayleigh models for the fading amplitude are commonly adopted in the technical literature. However, various researchers have shown that, in some scenarios, better agreement between theory and experimental results is obtained if the so-called *Nakagami-m* pdf is employed for the statistical characterization of fading amplitude (see [88, p. 49] and [89–95]), that is, if (2.59) is replaced by:

$$f_R(r) = u(r)\frac{2}{\Gamma(m)}\left(\frac{m}{2\sigma_a^2}\right)^m r^{2m-1} \exp\left(-\frac{mr^2}{2\sigma_a^2}\right), \tag{2.61}$$

with $m \geq 1/2$, where m is the *order* of the pdf and $\Gamma(m)$ is the *gamma function* (giving $(m-1)!$ for m integer) [96]. It can be shown that:

(a) for $m = 1$, (2.61) reduces to (2.58) (corresponding to Rayleigh fading);
(b) when m increases, the Nakagami distribution changes its character from that of purely random fading to that of fading with an LOS component, so that it can represent a valid alternative to the Rice model (2.59);
(c) for large values of m, (2.61) can be approximated by a Gaussian pdf, similarly to what happens with the Rice pdf (2.60) for large K.

In various problems accurate knowledge of the *second-order* statistical characterization of $a(t)$ is needed. In this chapter we focus only on some aspects of this problem, since only the autocorrelation function (along with the corresponding power spectral density) of the fading distortion is taken into consideration. Further information about second-order statistics can be found in the book [88]. However, before tackling this problem from a mathematical viewpoint, it is useful to analyze the effects of the time-selective fading in the time and frequency domains. *Time-domain* effects can be easily inferred from (2.32), since this shows that fading produces random amplitude and phase variations in the useful component of the received signal, as seen for a specific scenario in the following example.

Example 2.2.7 Let us assume that time-selective fading is affecting a digital transmission with a symbol interval $T_s = 0.1$ ms and that the maximum Doppler frequency is $f_D = 100$ Hz (corresponding

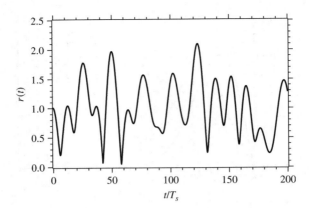

Figure 2.5 Time evolution of the fading amplitude $r(t)$ for a sample function of time-selective fading $a(t)$.

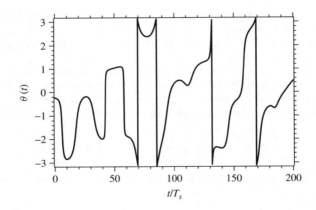

Figure 2.6 Time evolution of the fading phase $\theta(t)$ for a sample function of time-selective fading $a(t)$.

to a speed[17] $v = 30 \, m/s = 108 \, km/h$ of the mobile receiver if $f_c = 1$ GHz), so that $f_D T_s = 10^{-2}$. We also assume that the fading second-order statistics are those characterized by the so-called *Jakes model*, described in detail in Example 2.2.8. The amplitude $r(t)$ and the phase $\theta(t)$ (in radians) of a sample function of the random process[18] $a(t)$ are illustrated in Figures 2.5 and 2.6, respectively. These results show that the received signal undergoes substantial fluctuations in amplitude and phase, and that, in particular, during deep fades quick phase variations may occur.

☐

The effects of time-selective fading in the *frequency domain* can be understood by analyzing the channel response to the *deterministic* cosinusoidal signal $x_{RF}(t) = \cos[2\pi(f_c + F)t]$, whose complex envelope is $x(t) = \exp(j2\pi Ft)$ (here F denotes the signal frequency offset with respect to the

---

[17] The relationship between f_D and v is expressed by (2.64) in Example 2.2.8.
[18] A MATLAB-based implementation of the algorithm developed in [97] for Jakes' fading generation has been used.

reference frequency f_c). It is easy to show that the complex envelope $y(t)$ of the channel response (see (2.32)) is a random process with autocorrelation function $R_y(\tau) = R_a(\tau) \exp(j2\pi F\tau)$ and, consequently, with power spectral density:

$$S_y(f) \triangleq \text{FCT}[R_y(\tau)] = S_a(f) = |A|^2 \, \delta(f - F) + S_D(f - F), \tag{2.62}$$

where $S_D(f)$, the power spectral density of the random component of $a(t)$, is the so-called *Doppler power density spectrum*[19] or *Doppler spectrum*. This shows that, in the absence of an LOS component ($A = 0$), the transmitted (impulsive) spectrum $X(f) = \delta(f - F)$ leads to the received power spectral density $S_y(f) = S_D(f - F)$, having, generally speaking, a nonzero spectral width. In other words, time-selective fading results in spectral broadening, whose intensity depends on the mobile receiver speed, as illustrated in the following example.

Example 2.2.8 A model commonly adopted for time-selective fading is that derived by R. H. Clarke in 1968 [98], under the assumptions of *isotropic scattering* (i.e., the incoming electromagnetic energy is collected by the receive antenna uniformly in all directions) and of an *isotropic receive antenna* in a two-dimensional propagation scenario. The model is characterized by the Doppler spectrum:

$$S_D(\nu) = \frac{\sigma_D^2}{\pi B_D} \frac{1}{\sqrt{1 - (\nu/f_D)^2}} [u(f - f_D) - u(f + f_D)], \tag{2.63}$$

which is also known as *Jakes' spectrum*, since W. C. Jakes paid substantial attention to it in his well-known book [99]. Here $\sigma_D^2 = P_a$ and the *Doppler bandwidth* f_D, representing the maximum Doppler shift introduced by the communication channel, is given by:

$$f_D = \frac{v}{c} f_c, \tag{2.64}$$

where v is the mobile speed[20] and c is the speed of light. The autocorrelation function associated with $S_D(\nu)$ (2.63) is given by:

$$R_D(\tau) \triangleq \text{IFCT}[S_D(\nu)] = \sigma_D^2 \, J_0(2\pi f_D \, \tau), \tag{2.65}$$

where

$$J_0(x) = \frac{1}{2\pi} \int_{-\pi}^{\pi} \exp(jx \cos\theta) \, d\theta \tag{2.66}$$

is the *zero-order Bessel function* of the first kind. Figure 2.7 illustrates the *normalized Doppler spectrum*:

$$\bar{S}_D(\nu) \triangleq \frac{B_D}{\sigma_D^2} S_D(\nu) \tag{2.67}$$

as a function of the *normalized frequency* ν/f_D, and Figure 2.8 the *normalized autocorrelation*:

$$\bar{R}_D(\tau) \triangleq \frac{R_D(\tau)}{\sigma_D^2} \tag{2.68}$$

as a function of the *normalized time* τf_D. Note that the slow decrease of the autocorrelation function in Figure 2.8 is due to the singularities in the Doppler spectrum (2.63).

---

[19] The subscript D implies that this statistic characterizes, exactly like $R_a(\tau)$, the *Doppler* effect. For uniformity of notation we will adopt the notation $R_D(\tau)$ in place of $R_a(\tau)$ in the following.

[20] The model is usually derived under the assumptions of a static transmit antenna, a static scattering environment and a mobile receiver traveling at a speed v; see, for example, [88, pp. 61–65], where the model limits are also discussed.

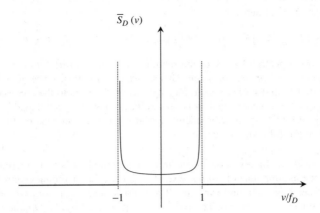

Figure 2.7   Normalized Doppler spectrum $\bar{S}_D(v)$ (2.67) (associated with the Doppler spectrum (2.63)) as a function of normalized frequency v/f_D.

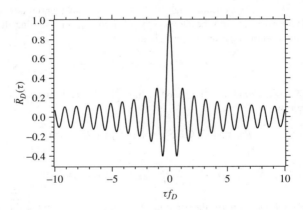

Figure 2.8   Normalized autocorrelation $\bar{R}_D(\tau)$ (2.65) (associated with the autocorrelation function (2.65)) as a function of normalized time $\tau\,f_D$.

The Jakes spectrum is commonly adopted in the analysis of wireless communications systems. From a physical viewpoint it is realistic when modeling fading for a *mobile terminal* (MT) held by a pedestrian. Its use for mobile communications is questionable if reception in a vehicle is considered. In fact, the latter scenario is characterized by moving scatterers (e.g., other vehicles) that can produce Doppler shifts even close to $\pm 2 f_D$, so that the Doppler spectrum can be substantially flatter than the derived one.

□

Finally, by analogy with what has been done with the frequency-selective case, some significant parameters can be extracted from $S_D(f)$; in particular, given the *mean Doppler shift*:

$$v_m \triangleq \frac{1}{P_a} \int_{-\infty}^{+\infty} v\,S_D(v)\,dv, \qquad (2.69)$$

the spectral width of $S_D(\nu)$ is measured by the *rms Doppler bandwidth*:

$$\nu_{rms} \triangleq \sqrt{\frac{1}{P_a} \int_{-\infty}^{+\infty} (\nu - \nu_m)^2 \, S_D(\nu) \, d\nu}. \tag{2.70}$$

To the latter two definitions the same considerations expressed for the *mean excess delay* (2.50) and the *rms delay* (2.52), respectively, apply. Note also that the parameter ν_{rms} provides an indication of the rate of change of $a(t)$ and that, if $S_D(\nu)$ is given by (2.63), it is found that $\nu_{rms} = B_D/\sqrt{2}$. In place of ν_{rms}, the so-called *coherence time* T_c, expressing the duration of the time interval over which $a(t)$ remains approximately constant, can be provided. Obviously, the larger the value of ν_{rms}, the shorter is T_c and, in particular, these two quantities can be related by the expression:

$$T_c = \frac{K_T}{\nu_{rms}}, \tag{2.71}$$

where K_T is a positive constant. It is worth pointing out, however, that the degree of the distortion introduced by time-selective fading is quite subjective, so that the definition of T_c is not unique and, consequently, different values for K_T can be adopted in distinct problems. For instance, the textbook [88] suggests $K_T = (2\pi)^{-1}$ (this leads to $T_c = (\pi B_D \sqrt{2})^{-1}$ with Jakes's fading), whereas [6] proposes the relationship:

$$T_c = \frac{9}{16\pi f_D}, \tag{2.72}$$

where f_D is the maximum Doppler shift. The last result assumes that the coherence time is defined as the maximum duration of the time interval over which the normalized autocorrelation $R_D(\tau)/R_D(0)$ does not drop below 0.5.

2.2.2.4   Doubly-Selective Channels

In considering time-selective and frequency-selective channels we have primarily been concerned with the channel multiplicative distortion $a(t)$ and the frequency response $H(f)$, respectively, and both system functions have been modeled as *Gaussian random processes*. It is not difficult to understand that, in the doubly-selective case, the property of Gaussianity is preserved for $H(t, f)$, so that a full statistical characterization of this system function is given by its mean value function:

$$H_d(t, f) \triangleq \mathrm{E}\{H(t, f)\} \tag{2.73}$$

and by its autocorrelation function:

$$R_H(t_1, t_2; f_1, f_2) \triangleq \mathrm{E}\{H(t_1, f_1) \, H^*(t_2, f_2)\}, \tag{2.74}$$

called the *time–frequency correlation function*. In the following we focus only on the case of Rayleigh fading, so that knowledge of the second statistic is sufficient for a complete statistical picture.

We now analyze the structure of $R_H(t_1, t_2; f_1, f_2)$ under the US and WSS hypotheses, and that of the autocorrelation functions of $h(t, \tau)$ and $\gamma(\nu, \tau)$ for the same cases. To begin, we note that acquiring an accurate knowledge of $R_H(t_1, t_2; f_1, f_2)$ is not easy, because it depends on four distinct variables, two in the time domain and two in the frequency domain. A substantial simplification in the mathematical structure of this function is introduced, however, if the US and/or WSS properties hold. In fact, if the channel is US (see (2.46)), then:

$$R_H(t_1, t_2; f_1, f_2) = R_H(t_1, t_2; f_1 - f_2); \tag{2.75}$$

whereas if the channel is WSS, then:

$$R_H(t_1, t_2; f_1, f_2) = R_H(t_1 - t_2; f_1, f_2). \tag{2.76}$$

If, as usually assumed in the following, both (2.75) and (2.76) hold (i.e., the channel is WSS-US), the dependence of $R_H(t_1, t_2; f_1, f_2)$ on its four variables becomes substantially simpler, since:

$$R_H(t_1, t_2; f_1, f_2) = R_H(t_1 - t_2; f_1 - f_2). \tag{2.77}$$

In other words, if the channel is WSS-US and Rayleigh, $H(t, f)$ is fully characterized from a statistical viewpoint by:

$$R_H(t; f) \triangleq \mathrm{E}\{H(t + t_0, f + f_0) \, H^*(t_0, f_0)\}, \tag{2.78}$$

where both the frequency f_0 and the instant t_0 are arbitrary quantities.

Let us investigate now the consequences of the WSS-US assumption for the system functions $h(t, \tau)$ and $\gamma(\nu, \tau)$. From (2.14) it is easy to see that the autocorrelation function of $h(t, \tau)$ can generally be expressed as:

$$\begin{aligned} R_h(t_1, t_2; \tau_1, \tau_2) &\triangleq \mathrm{E}\{h(t_1, \tau_1) \, h^*(t_2, \tau_2)\} \\ &= \int_{f_2=-\infty}^{+\infty} \int_{f_1=-\infty}^{+\infty} R_H(t_1, t_2; f_1, f_2) \exp\left(j2\pi(f_1\tau_1 - f_2\tau_2)\right) df_1 df_2 \,. \end{aligned} \tag{2.79}$$

If the channel is US, substituting the variable f_1 with $f \triangleq f_1 - f_2$ in the last integral of (2.79) leads, after some manipulation, to:

$$R_h(t_1, t_2; \tau_1, \tau_2) = P_h(t_1, t_2; \tau_1) \, \delta(\tau_1 - \tau_2), \tag{2.80}$$

where

$$P_h(t_1, t_2; \tau_1) \triangleq \int_{-\infty}^{+\infty} R_H(t_1, t_2; f) \, \exp\left(j2\pi f \tau_1\right) df \,. \tag{2.81}$$

If, instead, the channel is WSS (i.e., (2.76) holds), from (2.79) it is easily inferred that:

$$R_h(t_1, t_2; \tau_1, \tau_2) = R_h(t_1 - t_2; \tau_1, \tau_2). \tag{2.82}$$

Finally, if the channel is both WSS and US, we have that (see (2.80) and (2.82)):

$$R_h(t_1, t_2; \tau_1, \tau_2) = P_h(t_1 - t_2; \tau_1) \, \delta(\tau_1 - \tau_2), \tag{2.83}$$

where (see (2.81)):

$$P_h(t; \tau) \triangleq \int_{-\infty}^{+\infty} R_H(t; f) \, \exp\left(j2\pi f \tau\right) df \tag{2.84}$$

is the so-called *delay cross-power spectral density*. A detailed analysis of this function and a general integral expression for it can be found in [100]. From this density, the channel PDP $P_h(\tau)$ can be extracted as:

$$P_h(\tau) = P_h(0; \tau). \tag{2.85}$$

Note that the last equality is often given as a definition of the channel PDP (see, for instance, [78, p. 372, eq. (78)]).

Similarly, from (2.21) it can be easily seen that the autocorrelation function of $\gamma(v, \tau)$ is given by:

$$R_\gamma(v_1, v_2; \tau_1, \tau_2) \triangleq \mathrm{E}\{\gamma(v_1, \tau_1)\,\gamma^*(v_2, \tau_2)\}$$

$$= \int_{t_2=-\infty}^{+\infty} \int_{t_1=-\infty}^{+\infty} R_h(t_1, t_2; \tau_1, \tau_2) \exp\left(j2\pi\left(v_2 t_2 - v_1 t_1\right)\right) dt_1 dt_2. \qquad (2.86)$$

If the channel is both US and WSS, substituting (2.83) in (2.86) and substituting the variable t_1 with $t \triangleq t_1 - t_2$ yields:

$$R_\gamma(v_1, v_2; \tau_1, \tau_2) = \delta(\tau_1 - \tau_2)\,\delta(v_1 - v_2) \int_{-\infty}^{+\infty} P_h(t; \tau_1) \exp\left(-j2\pi v_1 t\right) dt. \qquad (2.87)$$

The last result shows that scatterers characterized by different delays (Doppler shifts) act in an uncorrelated fashion, since the channel is US (WSS). Finally, substituting (2.84) in (2.87) leads to:

$$R_\gamma(v_1, v_2; \tau_1, \tau_2) = \delta(\tau_1 - \tau_2)\,\delta(v_1 - v_2)\,S(v_1, \tau_1), \qquad (2.88)$$

where (see (2.84)):

$$S(v, \tau) \triangleq \int_{t=-\infty}^{+\infty} \int_{f=-\infty}^{+\infty} R_H(t; f) \exp\left[j2\pi(f\tau - vt)\right] df\, dt$$

$$= \underset{t \to v}{\mathrm{FCT}} \left[\underset{f \to \tau}{\mathrm{IFCT}} \left[R_H(t; f) \right] \right] = \underset{t \to v}{\mathrm{FCT}}[P_h(t; \tau)] \qquad (2.89)$$

is the so-called *delay-Doppler power density function* or *scattering function* [101]. This function plays a fundamental role in wireless communications and, therefore, deserves further comment. First, we note that it can be related to $R_H(t; f)$ as:

$$R_H(t; f) = \underset{v \to t}{\mathrm{IFCT}} \left[\underset{\tau \to f}{\mathrm{FCT}} \left[S(v, \tau) \right] \right] \qquad (2.90)$$

and that it can be interpreted as a measure of the power *scattered* by the communication channel for a given Doppler shift v and a given delay τ. Therefore, the scattering function can be determined theoretically or experimentally[21] for various scenarios by exploiting the fact that its structure is related to the distribution of the scatterers around a receiver, and, more precisely, that it represents the scatterer density versus the path delay and the Doppler shift (i.e., the path length and the azimuth angle measured with respect to the direction of motion, respectively, in a two-dimensional scenario). Moreover, from these considerations it is easily seen that the PDP $P_h(\tau)$ and the Doppler spectrum $S_D(v)$ can be derived from $S(v, \tau)$ as:

$$P_h(\tau) = \int_{-\infty}^{+\infty} S(v, \tau)\, dv \qquad (2.91)$$

and

$$S_D(v) = \int_{-\infty}^{+\infty} S(v, \tau)\, d\tau, \qquad (2.92)$$

respectively. It is also worth pointing out that, to simplify analysis of wireless communication systems, it is usually assumed that the scattering function can be factored as:

$$S(v, \tau) = P_h(\tau)\, S_D(v), \qquad (2.93)$$

---

[21] Measured scattering functions are available, for instance, in [102] for an urban environment, and in [103, 104] for a suburban environment.

so that (see (2.90), (2.62) and (2.48)):

$$R_H(t; f) = \underset{\nu \to t}{\text{IFCT}} [S_D(\nu)] \underset{\tau \to f}{\text{FCT}} [P_h(\tau)] = R_D(t) \, R_H(f), \tag{2.94}$$

where $R_D(t) \triangleq \text{IFCT}[S_D(\nu)]$ is the Doppler autocorrelation function. Moreover, substituting this result in (2.84) gives:

$$P_h(t; \tau) = R_D(t) \, P_h(\tau). \tag{2.95}$$

Note that adopting the *separability* assumption expressed by (2.93) is equivalent to assuming that the shape of the Doppler spectrum is independent of the scatterer delay τ. In other words, if we refer to the TDL model of Figure 2.2, equation (2.93) states that the time variability of all the channel taps is characterized by the same Doppler spectrum, that is, the time-varying gain (see (2.12)):

$$a_i(t) \triangleq h(t, \tau_i) \, \Delta\tau \tag{2.96}$$

of the ith echo has an average power proportional to $P_h(\tau_i)$ and an autocorrelation function proportional to $R_D(t)$.

We conclude our study of doubly-selective fading with the problem of channel classification. In this subsection it has been shown that the rms Doppler bandwidth ν_{rms} (2.93) and the *rms delay spread* τ_{ds} (2.52) quantify the spectral dispersion (time selectivity) and the time dispersion (frequency selectivity), respectively, in a wireless communication channel. In digital transmission, however, the absolute value of these parameters is not enough to assess the relevance of these dispersion/selection phenomena, since their effects inevitably also depend on the characteristics of the transmitted signal $x_{RF}(t)$ – or, more precisely, on its duration T_x and bandwidth B_x. For instance, as shown in Chapter 6, a digital receiver, when detecting a block of transmitted data, processes the received signal over an observation interval having a duration T_0 close to T_x. Therefore, it should be expected that the average error performance of a receiver designed under the assumption of a frequency-selective fading channel (i.e., *static* over the observation interval) is not appreciably affected by the presence of time variability if the product $\nu_{rms} T_x$ is small – substantially less than unity. This condition can also be rewritten as:

$$\nu_{rms} \, T_x < c_\nu, \tag{2.97}$$

where c_ν is a dimensionless positive constant whose value should be selected on the basis of the specific application. Based on the concept of duality, it is not difficult to understand that the average error performance of a digital receiver designed under the assumption of a *frequency-flat* (i.e., *time-selective*) fading channel is not substantially affected by the presence of time dispersion if:

$$B_x \, \tau_{ds} < c_\tau, \tag{2.98}$$

where c_τ is a dimensionless positive constant to which considerations similar to those expressed for c_ν apply. The inequalities (2.97) and (2.98) inevitably lead to the first quadrant of the (B_x, T_x) plane being partitioned into four regions (see Figure 2.9), in each of which the effects of frequency selectivity and time selectivity appear jointly, singularly or not at all. Note that the hyperbola delimiting, at least in part, three of the four regions derives from the fact that the bandwidth–duration product of an arbitrary signal is lower-bounded [79].

2.2.3   Reduced-Complexity Statistical Models for SISO Channels

2.2.3.1   Statistical Channel Modeling

As shown in Chapter 4, various difficulties encountered in the design of algorithms for data detection and the analysis of their performance can be often related to the structure of the useful signal available

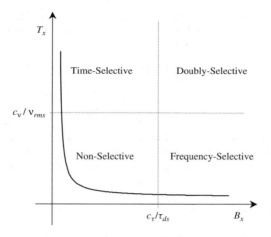

Figure 2.9 Schematic representation, in the bandwidth–duration plane (B_x, T_x) of the transmitted signal, illustrating the effects of fading on wireless communication.

at the output of the communication channel. In fact, in the presence of multipath fading, this structure can be substantially different from that of the transmitted signal. In many cases, these problems can be simplified by resorting to *reduced-dimensionality* (or simply *reduced*) *statistical channel models*. These models express randomness in time-continuous channel filtering as the effect of a *finite* number of stochastic parameters and, consequently, can have a substantially simpler structure than that of the actual channel. At the same time, however, they should have statistical properties similar to those of the actual channel.

Generally speaking, from an analytical viewpoint, the problem of developing a channel model which is statistically similar to a given WSS-US *Rayleigh wireless communication channel with factorizable scattering function* (see (2.93)) can be formulated as follows. If a given multipath fading channel with TVTF $H(t, f)$ is fed by a bandlimited signal $x(t)$ with bandwidth B and FCT $X(f)$, its output signal (see (2.15)):

$$y(t) = \int_{-B}^{B} X(f) \, H(t, f) \, \exp(j2\pi ft) \, df \tag{2.99}$$

is a zero-mean nonstationary Gaussian process having autocorrelation function:

$$R_y(t_1, t_2) \triangleq E\{y(t_1) \, y^*(t_2)\}$$

$$= \int_{f_1=-B}^{B} \int_{f_2=-B}^{B} X(f_1) \, X^*(f_2) \, R_H(f_1 - f_2, t_1 - t_2)$$

$$\cdot \exp[j2\pi(f_1 t_1 - f_2 t_2)] \, df_2 df_1$$

$$= R_D(t_1 - t_2) \int_{f_1=-B}^{B} \int_{f_2=-B}^{B} X(f_1) \, X^*(f_2) \, R_H(f_1 - f_2)$$

$$\cdot \exp[j2\pi(f_1 t_1 - f_2 t_2)] \, df_2 \, df_1. \tag{2.100}$$

Then developing a reduced-dimensionality model means devising a stochastic process $\tilde{H}(t, f)$ which depends on a *finite* (possibly small) number of random parameters and such that the random signal:

$$\tilde{y}(t) \triangleq \int_{-B}^{B} X(f) \, \tilde{H}(t, f) \, \exp(j2\pi ft) \, df \tag{2.101}$$

is *statistically equivalent*, within a given degree of accuracy, to $y(t)$ of (2.99) for any $x(t)$ in the set of bandlimited functions with bandwidth B. Statistical equivalence requires $\tilde{H}(t, f)$ to be a Gaussian process with zero mean and autocorrelation:

$$R_{\tilde{H}}(t_1, t_2; f_1, f_2) \triangleq E\{\tilde{H}(t_1, f_1)\, \tilde{H}^*(t_2, f_2)\}$$

$$= R_H(t_1 - t_2, f_1 - f_2)$$

$$= R_D(t_1 - t_2)R_H(f_1 - f_2) \qquad (2.102)$$

for $|f_1|$, $|f_2| \leq B$ (see equation (2.100)). Generally speaking, the second condition can be fulfilled by a reduced model only to a certain degree of accuracy.

In the following we focus on the problem of various channel representations that can be exploited to devise reduced channel models. In particular, we focus on:[22]

(a) TDL models with equally spaced taps;
(b) *Karhunen–Loève* (KL) models;
(c) models based on polynomial representations of system functions;
(d) models based on the approximation of channel autocorrelation functions via *Gaussian quadrature rules* (GQRs).

Before analyzing each class, it is useful to note that, generally speaking, channel models represent system functions as a superposition of multiple different components, so that they provide a *diversity representation* of the communication channel. In fact, each component in the system function results in a different portion of the useful received signal. Then a digital receiver which is able to separate each component from all the others can benefit from the *implicit diversity* provided by the multipath fading channel, that is, from its capacity for generating multiple (and differently distorted) replicas of the same transmitted signal. This viewpoint turns out to be useful in devising new detection algorithms, based on specific channel models and fully exploiting the channel diversity, thereby improving system performance appreciably.

2.2.3.2 Tapped Delay Line Model for Bandlimited Signals

This model belongs to the class of the so-called *sampling models*,[23] first devised by T. Kailath in [105, p. 20] and later extended by P. A. Bello in [78]. It is based on the assumption that the channel input signal $x(t)$ has a finite bandwidth B, so that the corresponding channel response $y(t)$ depends on its TVTF $H(f, t)$ for $f \in (-B, B)$ only (see (2.99)), that is, it depends on the function:

$$\hat{H}(t, f) \triangleq H(t, f)[u(f - B) - u(f + B)]. \qquad (2.103)$$

Let us define the function:

$$\hat{h}(t, \tau) \triangleq \text{IFCT}_{f \to \tau}[\hat{H}(t, f)] = \int_{-B}^{B} H(t, f) \exp{(j2\pi f \tau)}\, df \qquad (2.104)$$

and note that $\hat{h}(t, \tau)$, being rigorously bandlimited (if its dependence on τ is considered), can be represented via the *sampling theorem* (e.g., see [106, p. 20]) as:

$$\hat{h}(t, \tau) = \sum_{n=-\infty}^{+\infty} \hat{h}\left(t, \frac{n}{2B}\right) \text{sinc}\left(2B\left(\tau - \frac{n}{2B}\right)\right), \qquad (2.105)$$

---

[22] Other channel models are mentioned in Section 2.4, where a list of useful references on this topic is provided.
[23] The terminology is due to the fact that they result from the application of the *sampling theorem* [79].

so that (see (2.104)):

$$\hat{H}(t, f) \triangleq \text{FCT}_{\tau \to f}[\hat{h}(t, \tau)]$$

$$= \frac{1}{2B}[u(f - B) - u(f + B)] \sum_{n=-\infty}^{+\infty} \hat{h}\left(t, \frac{n}{2B}\right) \exp\left(-j2\pi n \frac{f}{2B}\right),$$ (2.106)

since FCT $[\text{sinc}(2Bt)] = (1/(2B)) [u(f - B) - u(f + B)]$. Then, substituting (2.106) in the RHS of (2.99) (in place of $H(t, f)$) yields, after some manipulation, the input–output relationship:

$$y(t) = \sum_{n=-\infty}^{+\infty} c_n(t) \, x \left(t - \frac{n}{2B}\right),$$ (2.107)

where

$$c_n(t) \triangleq \frac{1}{2B} \, \hat{h}\left(t, \frac{n}{2B}\right).$$ (2.108)

Equations (2.107) and (2.108) express the so-called TDL *model* [106]. This certainly admits the representation of Figure 2.2, provided that we set $\Delta \tau \triangleq 1/(2B)$ and $a_n(t) \triangleq c_n(t)$, with $c_n(t)$ given by (2.108). This result, however, unlike that expressed by (2.11), is *rigorous* and is suitable for representing *any linear wireless channel fed by a rigorously bandlimited signal*. It is also important to note that, generally speaking, the Gaussian random processes $\{c_n(t)\}$ are *statically correlated*. In fact, if we exploit Parseval's theorem [79], (2.108) can be rewritten (see also (2.104)) as:

$$c_n(t) \triangleq \frac{1}{2B} \int_{-B}^{B} H(t, f) \exp\left(j2\pi n \frac{f}{2B}\right) df$$

$$= \frac{1}{2B} \int_{-\infty}^{\infty} [(u(f - B) - u(f + B))H(t, f)] \exp\left(j2\pi f \frac{n}{2B}\right) df$$

$$= [\text{sinc}(2B\tau) \otimes h(t, \tau)]_{\tau = n/(2B)}$$

$$= \int_{-\infty}^{\infty} \text{sinc}(2B\alpha) \, h\left(t, \frac{n}{2B} - \alpha\right) d\alpha,$$ (2.109)

so that the cross-correlation $R_{n,m}^c(t, \tau)$ between $c_n(t)$ and $c_m(t)$, with $m \neq n$, can be expressed as:

$$R_{n,m}^c(t, \tau) \triangleq \text{E}\{c_n(t + \tau) \, c_m^*(t)\}$$

$$= \int_{\alpha=-\infty}^{\infty} \int_{\beta=-\infty}^{\infty} R_h\left(t + \tau, t; \frac{n}{2B} - \alpha, \frac{m}{2B} - \beta\right)$$

$$\cdot \text{sinc}(2B\alpha)\text{sinc}(2B\beta) \, d\beta \, d\alpha.$$ (2.110)

If the channel is WSS-US (see (2.83)), we then have that:

$$R_h\left(t + \tau, t; \frac{n}{2B} - \alpha, \frac{m}{2B} - \beta\right) = P_h\left(\tau; \frac{n}{2B} - \alpha\right) \delta\left(\frac{n - m}{2B} - \alpha + \beta\right),$$ (2.111)

where $P_h(t; \tau)$ is defined by (2.84), so that (2.110) can easily be put in the form:

$$R_{n,m}^c(\tau) = \left\{\left[\text{sinc}(2Bx) \, \text{sinc}\left(2B\left(x + \frac{m - n}{2B}\right)\right)\right] \otimes P_h(\tau; x)\right\}_{x = n/(2B)}.$$ (2.112)

If the scattering function is separable (see (2.95)), the latter result can be rewritten as:

$$R_{n,m}^c(\tau) = C_{m,n} \, R_D(\tau),$$ (2.113)

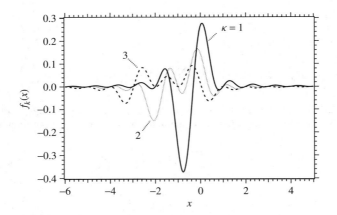

Figure 2.10   Representation of $f_k(x)$ (2.115) for $k = 1, 2$ and 3.

where

$$C_{m,n} \triangleq \left\{ \left[\operatorname{sinc}(2Bx) \operatorname{sinc}\left(2B\left(x + \frac{m-n}{2B}\right)\right) \right] \otimes P_h(x) \right\}_{x=n/(2B)}.$$

(2.114)

In Figure 2.10 a representation of the function:

$$f_k(x) \triangleq \operatorname{sinc}(y)\operatorname{sinc}(y+k)$$

(2.115)

appearing in (2.112) is given for $k \triangleq m - n = 1, 2$ and 3.

From this figure and (2.112) it can be inferred that $C_{m,n}$ in (2.114) (and, consequently, $R_{n,m}^c(\tau)$ in (2.112)) decreases as $|m - n|$ increases, that is, the channel taps become less correlated as their distance in the τ direction increases. We also note that setting $m = n$ in (2.112) produces the autocorrelation function of $c_n(t)$:

$$R_n^c(\tau) \triangleq \mathrm{E}\{c_n(t + \tau)\, c_n^*(t)\} = \{\operatorname{sinc}^2(2Bx) \otimes P_h(\tau; x)\}_{x=n/(2B)},$$

(2.116)

so that the corresponding Doppler power spectrum is given by:

$$S_n^c(\nu) \triangleq \mathrm{FCT}\{R_n^c(\tau)\} = \{\operatorname{sinc}^2(2Bx) \otimes S(\nu, \tau)\}_{x=n/(2B)},$$

(2.117)

where $S(\nu, \tau)$ is the channel scattering function. If $S(\nu, \tau)$ can be factored, we find (see (2.113)) that $R_n^c(\tau) = C_{n,n}\, R_D(\tau)$ and $S_n^c(\nu) = C_{n,n}\, S_D(f)$, so that, as expected, all the channel taps have the same Doppler spectrum (up to a constant).

Finally, it is important to point out that an exact representation of the output signal $y(t)$ according to (2.107) requires the use of a TDL model with an *infinite* number of taps. In the literature simplified channel models based on a TDL with a *finite number of statistically independent taps* are often used. These models can be derived from the rigorous model of (2.107) and (2.108) by restricting the range of the index n in (2.107) to $(0, 1, \ldots, L - 1)$ (where L is an integer whose value is proportional to the channel dispersion) and neglecting tap correlation. Such models, however, do not admit as rigorous a theoretical justification as that derived above. It is also important to point out that, in these cases, a reduced channel model can be devised by adopting a simplified representation of the channel tap gains $\{c_n(t), n = 0, 1, \ldots, L-1\}$ over a given time interval. This can be done, for instance, by exploiting the GQR models described in Section 2.2.3.4. *Independent* tap gains are also assumed in the so-called *bin model* [107, pp. 111–114], which also approximates a multipath channel having a continuous PDP as a TDL with fixed, but unequally spaced, delays.

2.2.3.3 Karhunen–Loève Model

The TVTF $H(t, f)$ of a *separable* doubly-selective channel (i.e., of a channel whose scattering function satisfies (2.93)) can also be expanded in a series of orthogonal functions $\{\phi_i(f), i = 1, 2, \ldots\}$ by means of the so-called KL representation [55] applied in the frequency domain [108]. The ith basis function $\phi_i(f)$ is the ith solution (eigenfunction) of the homogeneous Fredholm integral equation:

$$\int_{-B}^{B} R_H(f_1 - f_2)\, \phi_i(f_2)\, df_2 = \sigma_i^2\, \phi_i(f_1) \tag{2.118}$$

with eigenvalue σ_i^2. It can be shown [55] that the set $\{\phi_i(f), i = 0, 1, \ldots\}$ represents a *complete basis*[24] for $L^2(-B, B)$. Then $H(t, f)$ can be represented as:

$$H(t, f) = \sum_{i=0}^{+\infty} H_i(t)\phi_i(f), \tag{2.119}$$

where the time-varying coefficients:

$$H_i(t) = \int_{-B}^{B} H(f, t)\, \phi_i^*(f)\, df, \quad i = 0, 1, \ldots, \tag{2.120}$$

are complex WSS *independent* Gaussian processes having zero mean and autocorrelation $R_{H_i}(\tau) = \sigma_i^2\, R_D(\tau)$. Truncating the expansion of (2.119) to n_F terms yields the simplified-dimensionality model [109]:

$$\hat{H}_{n_F}(t, f) = \sum_{i=0}^{n_F-1} H_i(t)\, \phi_i(f), \tag{2.121}$$

which is zero-mean and Gaussian; however, it is not stationary in the variable f since, generally speaking, its autocorrelation function:

$$R_{\hat{H}_{n_F}}(t; f_1, f_2) \triangleq E\{\hat{H}_{n_F}(t + t_0, f_1)\hat{H}_{n_F}^*(t_0, f_2)\} = R_D(t) \sum_{i=0}^{n_F-1} \sigma_i^2\, \phi_i(f_1)\, \phi_i(f_2) \tag{2.122}$$

does not depend only on the frequency difference $(f_1 - f_2)$. The KL representation is an efficient representation of a stochastic process since only a small value n_F is usually required in order to produce an accurate representation of $H(f)$. Unfortunately, closed-form solutions of the integral equation (2.118) exist only in a few cases. This obstacle can be overcome by means of numerical integration techniques turning equation (2.118) into a classic eigenvalue problem solvable by means of standard computer routines [85, 109]. A significant drawback of the KL model is its physical interpretation, which is intuitively pleasant as a diversity representation of random channel [101], but is partially obscured by the irregular and unpredictable behavior of the complex functions $\{\phi_i(f)\}$. Finally, we note that the KL expansion, even if usually employed in the frequency domain (e.g., see [85, 108, 109]), can also be applied in the τ domain [86] or in the t domain [110].

---

[24] The completeness property of a basis is defined in Appendix D.

2.2.3.4   Power Series Models

In principle, any system function can be approximated by a *polynomial function*, in the time and/or frequency domains, thereby obtaining a reduced-complexity channel model. As originally proposed in the classic paper by P. A. Bello [78], polynomial approximations can be derived, for instance, by applying the well-known Taylor series representation. The statistical accuracy achievable depends on various factors, such as the order of the series approximation, the frequency/time size of the domain over which the function is represented and the rate of change in time/frequency of the communication channel. The use of this method in channel modeling can be easily exemplified by referring to the case of time-selective Rayleigh fading. In fact, in this case, a power series representation of the multiplicative distortion $a(t)$ over a limited time interval $(-T_0/2, T_0/2)$ is given by:

$$a(t) = \sum_{n=0}^{\infty} \hat{a}_n \left(\frac{t}{T_0} \right)^n, \tag{2.123}$$

where the nth coefficient $\hat{a}_n$ is given by:

$$\hat{a}_n = \frac{T_0^n}{n!} \frac{d^n \, a(t)}{dt^n} \bigg|_{t=0}. \tag{2.124}$$

It is then easy to show that the stochastic parameters $\{\hat{a}_n, n = 0, 1, \ldots\}$ are *correlated* jointly Gaussian random variables with zero mean and correlation[25] [111, 112]:

$$R_{n,m} \triangleq \mathrm{E}\{\hat{a}_n \, \hat{a}_m^*\} = \frac{(-1)^m \, T_0^{n+m}}{n! \, m!} \frac{d^{n+m} \, R_D(\tau)}{d\tau^{n+m}} \big|_{\tau=0}. \tag{2.125}$$

Then, in particular, if $R_D(\tau) = J_0(2\pi B_D \tau)$, we have that:

$$R_{n,m} = (-1)^{(n-m)/2} \frac{(n+m)!}{n! \, m!} \left[\left(\frac{n+m}{2} \right)! \right]^{-2} (\pi B_D T_0)^{n+m} \tag{2.126}$$

if $n + m$ is even and $R_{n,m} = 0$ otherwise, so that $R_{n,n} \triangleq \mathrm{E}\{|\hat{a}_n|^2\} = (2n)! \, (\pi B_D T_0)^{2n}$.

A reduced-complexity channel model can be derived from the representation (2.123) by truncating the power series to n_T terms. This produces an n_tth-order reduced-state model:

$$\hat{a}_{n_t}(t) = \sum_{n=0}^{n_t-1} \hat{a}_n \left(\frac{t}{T_0} \right)^n \tag{2.127}$$

of the process $a(t)$. This random process is nonstationary, has zero mean and autocorrelation function:

$$R_{\hat{a}_{n_t}}(t_1, t_2) = \sum_{n=0}^{n_t-1} \sum_{m=0}^{n_t-1} R_{n,m} \left(\frac{t_1}{T_0} \right)^n \left(\frac{t_2}{T_0} \right)^m \tag{2.128}$$

and can be used in any TDL model to represent the tap gains over a limited observation interval. Finally, we note that:

(a) specific cases of interest in the technical literature are those corresponding to $n_t = 0$ (known as *slow fading* [113]) and $n_t = 1$ (known as *linearly time-selective fading*),
(b) in a computer simulation of a time-selective channel based on the model (2.127), the channel parameters $\{\hat{a}_n, n = 0, 1, \ldots\}$ can be generated by a linear transformation of *independent* Gaussian random variables [55].

Further details about polynomial models and their applications can be found in [78, 111, 112, 114–116].

---

[25] Polynomial coefficients can be related to experimental data (e.g., see [48, pp. 36–37]).

2.2.3.5 Statistical Channel Modeling via Approximations of Channel Autocorrelation Functions

Reduced channel models can also be derived from numerically efficient approximations of channel autocorrelation functions; this idea has been exploited, for instance, in the so-called GQR models, proposed by E. Chiavaccini and G. M. Vitetta in [117] as a generalization of a channel representation first proposed in [118] for time-selective fading only.

GQR models can be easily understood by referring to a frequency-selective Rayleigh fading channel with $P_h(\tau)$ and autocorrelation function of its frequency response $H(f)$ (see (2.48)):

$$R_H(f) = \text{FCT}[P_h(\tau)] = \int_{-\infty}^{+\infty} P_h(\tau) \, \exp\left(j2\pi f\tau\right) \, d\tau. \tag{2.129}$$

If we assume that the minimum and maximum channel delay are τ_m and τ_M, respectively, and separate the real and imaginary parts of the integrand function, (2.129) can be rewritten as:

$$R_H(f) = \int_{\tau_m}^{\tau_M} P_h(\tau) \, \cos(2\pi f\tau)d\tau - j \int_{\tau_m}^{\tau_M} P_h(\tau) \, \sin(2\pi f\tau) \, d\tau. \tag{2.130}$$

Since $P_h(\tau)$ is a real and nonnegative function both integrals on the RHS of equation (2.130) can be evaluated numerically by means of GQRs [119, 120] for any $f \in (-2B, 2B)$, where B is the input signal bandwidth.[26] Applying the GQR technique to (2.130) results in:

$$R_H(f) \simeq \sum_{i=0}^{n_f-1} w_i \cos(2\pi f\tau_i) - j \sum_{i=0}^{n_f-1} w_i \sin(2\pi f\tau_i) = \sum_{i=0}^{n_f-1} w_i e^{-j2\pi f\tau_i} \triangleq R_{\tilde{H}_{n_f}}(f), \tag{2.131}$$

where $\{\tau_i, i = 0, 1, \dots, n_f - 1\}$ and $\{w_i, i = 1, 2, \dots, n_f\}$ are respectively the *nodes* and *weights* of the n_fth-order *Gaussian quadrature* (GQ) formula for the weight function $P_h(\tau)$. Expression (2.131) approximates $R_H(f)$ as the sum of n_f complex oscillations having positive amplitudes, null phases, and *characteristic "frequencies"* $\{\tau_i\}$ expressed by the GQR nodes. It is interesting to note that the approximate autocorrelation $R_{\tilde{H}_{n_f}}(f)$ in (2.131) is a *quasi-periodic* signal since the ratios $\{\tau_i/\tau_j, i < j\}$ are not necessarily rational numbers. It can be proved [120] that $R_{\tilde{H}_{n_f}}(f)$ in (2.131) converges locally to $R_H(f)$ as $n_f \to \infty$ under conditions that are satisfied for all the PDPs of usual interest in the technical literature, and, in particular, for the PDPs listed in Table 2.2.[27]

These considerations allow us to infer that the statistical behavior of $H(f)$ can be well approximated by a stationary Gaussian process $\tilde{H}_{n_f}(f)$ having zero mean and autocorrelation function $R_{\tilde{H}_{n_f}}(f)$ (2.131). Moreover, the autocorrelation structure suggests the WSS representation:

$$\tilde{H}_{n_f-1}(f) = \sum_{i=0}^{n_f-1} a_i \, e^{-j2\pi f\tau_i} \tag{2.132}$$

for $\tilde{H}_{n_f}(f)$, where $\{a_i, i = 0, 1, \dots, n_f - 1\}$ are *independent* Gaussian random variables having zero mean and variances $\{\text{E}\{|a_i|^2\} = w_i, i = 0, 1, \dots, n_f - 1\}$. Taking the IFCT of (2.132) with respect to the variable f produces the channel impulse response:

$$\tilde{h}_{n_f}(\tau) = \sum_{i=0}^{n_f-1} a_i \, \delta(\tau - \tau_i) \, . \tag{2.133}$$

---

[26] In equation (2.100) both f_1 and f_2 belong to the interval $(-B, B)$ ($x(t)$ is bandlimited to B Hertz) so that $(f_1 - f_2) \in (-2B, 2B)$.
[27] In all these cases the evaluation of nodes and weights of GQRs can be carried out via numerically efficient methods [119, 120].

The channel model (2.133) lends itself to an immediate physical interpretation: in the bandwidth of its input signal the channel behaves as a TDL with nonuniform spacing between successive taps. The tap complex gains $\{a_i\}$ are *random*, whereas the tap delays $\{\tau_i\}$ are *fixed* for a given PDP and *model order* n_f and depend only on the PDP shape. It is also worth noting that:

(a) the stochastic process $\tilde{H}_{n_f}(f)$ has been derived as a statistical approximation to $H(f)$ in the frequency range $(-B, B)$ only,
(b) in this frequency interval statistical equivalence between $H(f)$ and $\tilde{H}_{n_f}(f)$ is achieved in the limit as $n_f \to \infty$. Given n_f, the accuracy of the model (2.133) can be related to that of the GQ formula derived for the PDP $P_h(\tau)$ [117].

A dual approach can be followed to devise a reduced-complexity channel model for the multiplicative distortion $a(t)$ affecting a Rayleigh time-selective fading channel characterized by a Doppler spectrum $S_D(\nu)$ and the associated autocorrelation $R_D(\tau)$. This leads to the WSS representation [117, 118]:

$$a(t) \simeq \sum_{k=0}^{n_t-1} a_k \exp\left(j2\pi \nu_k t\right) \tag{2.134}$$

over the finite time interval $(-T_0/2, T_0/2)$, where $\{a_k, k = 0, 1, \ldots, n_t - 1\}$ are *independent* Gaussian random variables having zero mean and variances $\{E[|a_k|^2] = q_k, k = 0, 1, \ldots, n_t - 1\}$. In addition, $\{q_i\}$ and $\{\nu_i\}$ are respectively the weights and the nodes of the n_tth-order GQ formula for the numerical evaluation of:

$$R_D(\tau) = \int_{-\infty}^{+\infty} S_D(\nu) \exp\left(j2\pi \nu \tau\right) d\nu, \tag{2.135}$$

relating $R_D(\tau)$ to $S_D(\nu)$. For instance, if $R_D(\tau) = J_0(2\pi B\tau)$, then $q_i = 1/M$ and $\nu_i = B_D \cos(\pi(i + 0.5)/M)$ [117].

The results illustrated for frequency-selective and time-selective channels can be combined to derive a simple model for the TVTF $H(t, f)$ of a *separable* doubly-selective channel. This model represents the channel as a TDL characterized by n_f distinct delays and in which the delay of each tap consists of the superposition of n_t distinct complex oscillations. Further details can be found in [117].

2.3 Mathematical Description and Modeling of MIMO Wireless Channels

The description of SISO channels based on system functions and illustrated in Section 2.2.1 can be extended to MIMO scenarios characterized by *arrays* of arbitrary sizes. In fact, a MIMO channel can be thought of as a collection of SISO channels, each being associated with a distinct couple of transmit–receive antennas. An alternative to this approach to MIMO channel characterization is based on the concept of *directionality* [121, 122]. This leads to the definition of some new system functions depending on the *angle of arrival* (AOA) and *angle of departure* (AOD) on the receive and transmit sides, respectively. These functions, however, unlike those adopted in the first approach, describe the properties of the propagation medium *independently* of the specific configurations of the antenna arrays employed, but, given such configurations, allow the channel response to be determined for any input–output couple in a MIMO system. In this section both approaches to channel description are examined, and then the problem of channel statistical characterization is analyzed.

2.3.1 Input–Output Characterization of a MIMO Wireless Channel

We now focus on a MIMO communication system having n_R receive and n_T transmit antennas (or an $n_T \times n_R$ MIMO system for short), as illustrated in Figure 2.11. A full description of the input–output behavior of the overall MIMO channel is provided by the CIR set $\{h_{i,j}(t, \tau), i = 0, 1, \ldots, n_T - 1, j = 0, 1, \ldots, n_R - 1\}$, where $h_{i,j}(t, \tau)$ denotes the CIR between the ith transmit and jth receive antenna. In fact, the complex envelope of the useful signal $y_j(t)$ collected at the output of the jth receive antenna (with $j = 0, 1, \ldots, n_R - 1$) can be expressed as the superposition of the contributions from all the transmit antennas, that is, as (see (2.9)):

$$y_j(t) = \sum_{k=0}^{n_T-1} \int_{-\infty}^{+\infty} x_k(t - \tau)\, h_{k,j}(t, \tau)\, d\tau, \qquad (2.136)$$

where $x_k(t)$ is the complex envelope of the signal from the kth transmit antenna (with $k = 0, 1, \ldots, n_T - 1$). This can be expressed in a more compact vector form as:

$$\mathbf{y}(t) = \int_{-\infty}^{+\infty} \mathbf{H}(t, \tau)\, \mathbf{x}(t - \tau)\, d\tau, \qquad (2.137)$$

where $\mathbf{x}(t) \triangleq [x_0(t), x_1(t), \ldots, x_{n_T-1}(t)]^T$ and $\mathbf{y}(t) \triangleq [y_0(t), y_1(t), \ldots, y_{n_R-1}(t)]^T$ are the transmitted and the received signal vectors, respectively, and $\mathbf{H}(t, \tau)$ is an $n_R \times n_T$ matrix collecting all the elements of the CIR set and whose element on the jth row (with $j = 0, 1, \ldots, n_R - 1$) and on the ith column (with $i = 0, 1, \ldots, n_T - 1$) is $h_{i,j}(t, \tau)$. The vector function $\mathbf{H}(t, \tau)$ provides a full matrix description of a MIMO communication channel, similar to the CIR $h(t, \tau)$ in a SISO scenario.

TDL channel models, to which much attention has been already paid in the study of SISO links, can also be used in the study of MIMO systems, as illustrated in the following example.

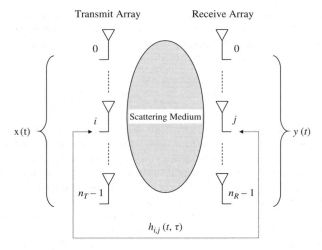

Figure 2.11 MIMO system characterized by n_T transmit and n_R receive antennas. The system function $h_{i,j}(t, \tau)$ denotes the CIR between the ith transmit and jth receive antenna.

Example 2.3.1 Let us assume that the scatterers in the propagation medium surrounding the antenna arrays can be grouped into L_h distinct *clusters* such that scatterers belonging to the same cluster are characterized by the same delay. Then the MIMO channel can be modeled as a vector TDL with L_h taps and the CIR matrix $\mathbf{H}(t, \tau)$ can be put in the form[28] (e.g., see [124]):

$$\mathbf{H}(t, \tau) \triangleq [h_{i,j}(t, \tau)] = \sum_{l=0}^{L_h-1} \mathbf{H}_l(t)\,\delta(\tau - \tau_l), \tag{2.138}$$

where $\mathbf{H}_l(t) = [h_{l,k}^r(t)]$ (with $l = 0, 1, \ldots, L_h - 1$) is an $n_R \times n_T$ matrix containing the complex gains corresponding to the lth matrix tap, characterized by the delay τ_l. In particular, $h_{k,l}^r(t)$ is the random complex gain associated with the lth echo or path between the kth transmit antenna ($k = 0, 1, \ldots, n_T - 1$) and the rth receive antenna ($r = 0, 1, \ldots, n_R - 1$). Substituting (2.138) in (2.137) produces the input–output relationship:

$$\mathbf{y}(t) = \sum_{l=0}^{L_h-1} \mathbf{H}_l(t)\,\mathbf{x}(t - \tau_l), \tag{2.139}$$

expressing the received signal vector as the superposition of the contributions coming from the L_h distinct clusters.

☐

As discussed below in the literature the case of frequency-flat fading is often considered for simplicity. Then (2.139) becomes:

$$\mathbf{y}(t) = \mathbf{H}(t)\,\mathbf{x}(t), \tag{2.140}$$

which, if we deem the channel static over the observation interval, simplifies to:

$$\mathbf{y}(t) = \mathbf{H}\,\mathbf{x}(t). \tag{2.141}$$

Note that, generally speaking, $\mathbf{H}$ can be represented in terms of its *singular value decomposition* (SVD) as $\mathbf{H} = \mathbf{U}\Lambda V$ [125], where $\mathbf{U}$ and $\mathbf{V}$ are $n_R \times n_R$ and $n_T \times n_T$ unitary matrices, respectively, and Λ is an $n_R \times n_T$ matrix having the singular values $\{\lambda_l, l = 0, 1, \ldots, \min(n_T, n_R) - 1\}$ of $\mathbf{H}$ on its main diagonal. This implies that $\mathbf{H}$ can be represented as [126]:

$$\mathbf{H} = \sum_{l=0}^{\min(n_T, n_R)-1} \lambda_l \mathbf{u}_l \mathbf{v}_l, \tag{2.142}$$

where $\mathbf{u}_l$ ($\mathbf{v}_l$) is the lth row (column) of $\mathbf{U}$ ($\mathbf{V}$). Note that each dyad on the RHS of (2.142) can be interpreted as an *independent mode of communication*, that is, a degree of freedom that the channel offers. The relevance of the mode is proportional to the magnitude of the associated eigenvalue. This representation of the channel can be very helpful in designing the correct processing to carry out at the transmitter for optimal signaling and at the receiver for mode separation.

An alternative approach to the characterization of the communication channel in a MIMO scenario is offered by a *directional description* of the propagation medium. Such a description represents an extension of that developed for SISO time-variant channels, as it incorporates into the system functions a dependence on the azimuth and the elevation angles of incidence (transmission) for the multipath components captured (sent) by receive (transmit) antennas [122, 127–130]. Such parameters can be gathered together in a vector Θ containing, in the most general case (i.e., in a *double directional* description of the channel), the azimuth angle φ_R and the elevation angle θ_R at the receive side, and

---

[28] Note that (2.138) provides a time-continuous model of a MIMO channel. A discrete-time TDL model can be found, for instance, in [123].

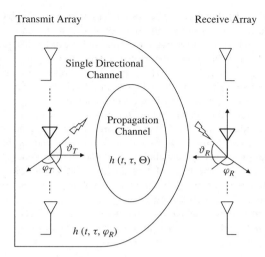

Figure 2.12   Directional description of a propagation scenario in which the transmit/receive arrays of a MIMO communication system operate.

the corresponding angular parameters (φ_T and θ_T, respectively) for the transmit side, as illustrated in Figure 2.12. For instance, following this approach, from the CIR $h(t, \tau)$ the so-called *time-variant angle-resolved channel impulse response* (TVARCIR) $h(t, \tau; \Theta)$ can be derived. This fully describes the directional behavior of a channel.

From a physical point of view, the above transformation of system functions has a well-defined meaning. In fact, as in the SISO case, the function $h(t, \tau; \Theta)$ still represents the response of a propagation scenario, at time epoch t, to a Dirac delta function applied τ seconds earlier. Its measurement assumes, however, that the impulsive signal is transmitted over the given channel only in the direction identified by the couple (φ_T, θ_T) and the corresponding response is extracted, at the receive side, collecting only the electromagnetic signal coming from the direction of arrival (φ_R, θ_R). In other words, the ideality of $h(t, \tau; \Theta)$ derives not only from being a response to an impulsive signal, but also from the fact that its definition involves the use of a couple of transmit/receive antennas both having a radiation pattern equal to a Dirac delta (i.e., an infinite gain).

From these considerations it is not difficult to infer that, as shown by Figure 2.12, double directional system functions characterize multipath propagation scenarios only, without referring in any way to the specific characteristics of the transmit/receive antennas. In particular, they do not take into account the antenna radiation patterns and their mathematical structure is not influenced by the presence of antenna arrays. For these reasons, on the one hand, a directional description of a communication channel could be used, in principle, in the study of SISO systems, even if, in this case, it would prove essentially trivial. On the other hand, the usefulness of the directional approach in the study of a MIMO system is because all the components of the TVIR matrix $\mathbf{H}(t, \tau) = [h_{i,j}(t, \tau)]$, defined in the previous subsection, can be univocally expressed through the single function $h(t, \tau; \Theta)$, provided that the array geometry and the antenna gains are known for both sides of the wireless link [122, 130]. Even if this result leads to a substantial simplification in the description of the input–output behavior of a MIMO channel, the use of directional functions opens a number of practical and theoretical problems. This can be clearly seen if we refer again to the system function $h(t, \tau; \Theta)$. We note that this function depends on six parameters in the double directional case, so that its measurement using proper channel sounding equipment [121], on the one hand, and its overall statistical characterization, on the other hand, are complicated problems.

The only way to simplify the directional description is to reduce the number of angular parameters. In the literature, for instance, directional models are often analyzed only in two-dimensional scenarios, so that the dependence on the transmit/receive elevation angles can be dropped [122, 131]. A further simplification is obtained if the directional description refers to only one end of the link (e.g., to the receive side), so that $h(t, \tau; \boldsymbol{\Theta})$ simplifies to $h(t, \tau; \varphi_R)$. In the following, for simplicity, we focus on only the latter case, since its results lend themselves to intuitively useful interpretations, and we drop the subscript R in the angular parameter to simplify the notation.

Before considering various directional system functions, it is important to point out that such functions can also be formulated in a *spatial*, or *aperture*, domain, instead of referring to the above angular domain [122]. The two domains, however, are closely related as the space and the angular parameters are dual quantities, exactly like the delay τ and the frequency f (or the time t and the Doppler frequency shift v). The meaning of this duality relationship can be fully understood by considering the physical phenomenon of interference involving the electromagnetic waves at the receive side of a radio link. In fact, when electromagnetic waves coming from different directions impinge on an antenna array, they combine constructively or destructively at distinct points of space, so that a spread in the angular domain[29] produces a spatial variation of the received field, that is, it generates *space-selective fading*. It is not surprising that:

(a) the relationship connecting the angular domain to the spatial one (and vice versa) is a Fourier transformation, involving the spatial coordinates on one side and some trigonometric functions of the angular parameters on the other [131],
(b) by analogy with the cases of frequency and time selectivity, an *rms angle spread* φ_{rms} and a *coherence distance* D_c, related by an inverse proportionality, can be defined to assess the relevance of space selectivity (further details can be found, for instance, in [38, p. 17]).

For SISO channels the four system functions shown in Figure 2.3 have been defined. The introduction of another couple of dual variables (one spatial x, the other angular φ) requires the use of $4 \cdot 2 = 8$ system functions for the description of single directional channels. As in the SISO case, the new system functions can be derived via Fourier transforms from the TVARCIR $h(t, \tau, \varphi)$, as sketched in Figure 2.13, where $F_x[\cdot]$ ($F_y^{-1}[\cdot]$) represents a shorthand notation for a FCT (IFCT) evaluated with respect to the variable x (y). Note that:

(a) all the system functions are *cyclically* related, as in the nondirectional case (see Figure 2.3)
(b) the spatial variable x, called the *aperture*, corresponds to the spatial dimension of the receive (or transmit) array and can take on any real value
(c) in the two-dimensional and single-directional scenario we consider in the following, the angular variable is defined as $\varphi = \sin \phi$, where ϕ is the arrival or departure azimuth[30] ($\phi \in [-\pi/2, \pi/2]$) and, consequently, in the evaluation of Fourier transforms, takes only values in the interval $[-1, 1]$.

Let us now summarize these relationships. The TVARCIR $h(t, \tau, \varphi)$ can be transformed with respect to each of the three variables on which it depends, so generating the *time-variant angle-resolved channel transfer function* $M(t, f, \varphi) \triangleq \text{TCF}_{\tau \to f}[h(t, \tau, \varphi)]$, the *time- and aperture-variant channel impulse response* $g(t, \tau, x) \triangleq \text{TCF}_{\varphi \to x}[h(t, \tau, \varphi)]$ and the *Doppler- and angle-resolved channel impulse response* $s(v, \tau, \varphi) \triangleq \text{TCF}_{t \to v}[h(t, \tau, \varphi)]$. From each of these functions, the other two functions can be generated by computing another Fourier transform with respect to one of the

---

[29] Here we refer to the AOAs of the multipath components, in this specific case. However, an angle spread at the transmitter can also be defined, since it refers to the spread in AODs of the multipath reaching the receiver.
[30] The azimuth variable ϕ covers only a half-plane, since in a linear array the interference due to an impinging electromagnetic wave does not change if the wave is replaced by its specular image with respect to the array axis [131].

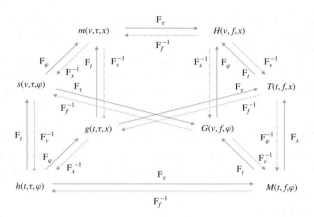

Figure 2.13 Representation of the relationships among the system functions adopted for the directional description of a propagation scenario.

remaining variables. For instance, from $M(t, f, \varphi)$ the *Doppler- and angle-resolved channel transfer function* $G(v, f, \varphi) \triangleq \text{TCF}_{t \to v}[M(t, f, \varphi)]$ and the *time- and aperture-variant transfer function* $T(t, f, x) \triangleq \text{TCF}_{\varphi \to x}[M(t, f, \varphi)]$ can be derived. Similarly, from $g(t, \tau, x)$ the *Doppler- and aperture-resolved channel impulse response* $m(v, \tau, x) \triangleq \text{TCF}_{t \to v}[g(t, \tau, x)]$ and, again, $T(t, f, x) = \text{TCF}_{\tau \to f}[g(t, \tau, x)]$ are obtained. Note also the following:

(a) From $s(v, \tau, \varphi)$ the functions $m(v, \tau, x)$ and $G(v, f, \varphi)$ can also be computed as $\text{TCF}_{\varphi \to x}$ $[s(v, \tau, \varphi)]$ and $\text{TCF}_{\tau \to f}[s(v, \tau, \varphi)]$, respectively.
(b) From each of the functions $m(v, \tau, x)$, $G(v, f, \varphi)$ and $T(t, f, x)$ the *Doppler- and aperture-resolved transfer function* can be derived as $H(v, f, x) \triangleq \text{TCF}_{\tau \to f}[m(v, \tau, x)]$ $= \text{TCF}_{\varphi \to x}[G(v, f, \varphi)] = \text{TCF}_{t \to v}[T(t, f, x)]$.
(c) Concatenating the previous definitions any system function different from $h(t, \tau, \varphi)$ can be computed as a multidimensional Fourier transform of $h(t, \tau, \varphi)$ itself. In this regard, it is useful to note that $T(t, f, x)$ can be expressed as $\text{TCF}_{\varphi \to x}[\text{TCF}_{\tau \to f}[h(t, \tau, \varphi)]]$. This shows that $T(t, f, x)$, representing an extension of $H(t, f)$ (2.13) to the aperture domain, can be defined as a two-dimensional Fourier transform of $h(t, \tau, \varphi)$ with respect to φ and τ, that is, more explicitly, as (e.g., see [131]):

$$T(t, f, x) = \int_{\tau=-\infty}^{+\infty} \int_{\varphi=-1}^{1} h(t, \tau, \varphi) \exp\left[-j2\pi\left(f\tau + \varphi\frac{x}{\lambda}\right)\right] d\tau \, d\varphi, \qquad (2.143)$$

where λ is the carrier wavelength.

Generally speaking, from the double directional TVARCIR $h(t, \tau; \boldsymbol{\Theta}) = h(t, \tau, \varphi_T, \tau_T, \varphi_R, \tau_R)$, the CIR matrix $\mathbf{H}(t, \tau) = [h_{i,j}(t, \tau)]$ of a MIMO channel can be derived if the transmit antenna gain[31] $g_T(\varphi_T, \tau_T)$ and the receive antenna gain $g_R(\varphi_R, \tau_R)$ are known. For instance, if we refer again to a two-dimensional, single-directional scenario and consider, for simplicity, a SISO link, the CIR $h(t, \tau)$ can be derived from $h(t, \tau, \varphi_R)$ and the receive antenna gain $g_R(\varphi_R)$ (if the directional description refers to the receive side) as (e.g., see [121, p. 52]):

$$h(t, \tau) = \int_{-\pi}^{\pi} h(t, \tau, \varphi_R) g_R(\varphi_R) d\varphi_R. \qquad (2.144)$$

---

[31] Here it is assumed, for simplicity, that all the antennas in the transmit (and in the receive) array are identical.

This result lends itself to a simple physical interpretation since it establishes that the CIR $h(t, \tau)$ can be seen as the superposition of multiple contributions, coming from different spatial directions and weighted by the receive antenna gain.

Finally, we note that, unfortunately, a double directional TVARCIR cannot be derived from $\mathbf{H}(t, \tau)$, since this matrix description encompasses the effects of transmit/receive antenna gains and does not provide enough information for a full directional description of a propagation scenario.

2.3.2 Statistical Characterization of a MIMO Wireless Channel

In this subsection we analyze the problem of the statistical characterization of the communication channel in a MIMO communication system. Both matrix and directional descriptions are considered.

2.3.2.1 Matrix Description

Generally speaking, the problem of the statistical characterization of the CIR matrix $\mathbf{H}(t, \tau)$ for an $n_T \times n_R$ MIMO system is very complicated. In the literature this problem is often investigated for a single-tap (i.e., time-selective) channel and, when considering a TDL model (2.138) with multiple taps, it is usually assumed that the channel taps $\{\mathbf{H}_l(t), l = 0, 1, \dots, L_h - 1\}$ are Gaussian[32] and statistically independent, so that the analysis of channel correlation is limited to the statistical relationships between complex gains belonging to the same tap. This problem can be tackled more easily if the channel is assumed static over the observation interval. As we will see, time variability can be accounted for later. Then (2.138) becomes:

$$\mathbf{H}(\tau) = \sum_{l=0}^{L_h-1} \mathbf{H}_l \, \delta(\tau - \tau_l), \tag{2.145}$$

so that the tap gain $\mathbf{H}_l$, for any l, is an $n_R \times n_T$ matrix of complex jointly Gaussian (correlated) random variables. To analyze the correlation properties of $\mathbf{H}_l$ we follow the approach of [133, p. 1220, column 2], by turning this matrix in an $n_R \cdot n_T$-dimensional vector $\mathbf{h}_l \triangleq \text{vec}\{\mathbf{H}_l\}$, generated from the ordered concatenation of the columns of $\mathbf{H}_l$. Then $\mathbf{h}_l$ is a Gaussian vector and a complete statistical characterization is provided by its average $\boldsymbol{\eta}_l \triangleq \text{E}\{\mathbf{h}_l\}$ and its covariance matrix:

$$\mathbf{C}_l \triangleq \text{E}\left\{(\mathbf{h}_l - \boldsymbol{\eta}_l)(\mathbf{h}_l - \boldsymbol{\eta}_l)^H\right\}, \tag{2.146}$$

where $(\cdot)^H$ indicates conjugate transpose. In the following, we assume that the channel is affected by Rayleigh fading (so that $\boldsymbol{\eta}_l = \mathbf{0}_{n_R n_T}$, where $\mathbf{0}_N$ is the N-dimensional null column vector) and that all the elements of $\mathbf{h}_l$ have the same variance σ_l^2. Then, given the *normalized correlation matrix*:

$$\boldsymbol{\Sigma}_l \triangleq \frac{1}{\sigma_l^2}\text{E}\left\{\mathbf{h}_l\mathbf{h}_l^H\right\} \tag{2.147}$$

and its *Cholesky factorization* [125]:

$$\boldsymbol{\Sigma}_l = \boldsymbol{\Sigma}_l^{1/2}(\boldsymbol{\Sigma}_l^{1/2}), \tag{2.148}$$

the gain vector $\mathbf{h}_l$ can be expressed as:

$$\mathbf{h}_l = \sigma_l \boldsymbol{\Sigma}_l^{1/2}\mathbf{a}_l, \tag{2.149}$$

---

[32] Gaussianity is a reasonable property if each group of scatterers consists of a large number of *independent* elements [132].

that is, as a linear transformation of a vector $\mathbf{a}_l$ consisting of $n_T n_R$ complex, independent and identically distributed Gaussian random variables, each having zero mean and unit variance. Note that (2.149) holds if all the elements of $\mathbf{h}_l$ have the same variance, that is, if the channel PDP is independent on the selected transmit/receive antenna couple. The case of power imbalance in the channel coefficient variances (i.e., in the *branch power ratio*) can be handled by applying the Cholesky factorization to the $n_T n_R \times n_T n_R$ matrix $\mathbf{\Gamma}_l$ resulting from the element-by-element product (i.e., Hadamard product) of $\mathbf{\Sigma}_l$ (2.156) and an $n_T n_R \times n_T n_R$ *power shaping matrix* $\mathbf{P}_l$, collecting the standard deviations of the complex gains $h_{k,l}^r$ (further details can be found in [133, p. 1220, column 2]). Then the resulting matrix $\mathbf{\Gamma}_l^{1/2}$ replaces the term $\sigma_l \mathbf{\Sigma}_l^{1/2}$ in (2.149).

The use of (2.149) requires the evaluation of $\mathbf{\Sigma}_l$ (2.147). This problem can be simplified if some additional assumptions are made about the channel statistical properties. To clarify this, let us define the $n_R \times n_R$ *normalized spatial correlation matrix of the receive antennas*:

$$\mathbf{R}_{R,l} \triangleq \frac{1}{\sigma_l^2} \mathrm{E}\left\{ \mathbf{h}_{l,k} \mathbf{h}_{l,k}^H \right\}, \tag{2.150}$$

where k is the transmit antenna index and $\mathbf{h}_{l,k}$ is the kth column of $\mathbf{H}_l$, and the $n_T \times n_T$ *normalized spatial correlation matrix of the transmit antennas*:

$$\mathbf{R}_{T,l} \triangleq \frac{1}{\sigma_l^2} \mathrm{E}\left\{ \mathbf{h}_l^r \left(\mathbf{h}_l^r\right)^H \right\}, \tag{2.151}$$

where r is the receive antenna index and $\mathbf{h}_l^r$ is the vector resulting from the transposition of the rth row of $\mathbf{H}_l$. Note that it has been implicitly assumed in (2.150) that $\mathbf{R}_{R,l}$ does not depend on k, and in (2.151) that $\mathbf{R}_{T,l}$ does not depend on r. The meaning of this assumption can be explained as follows. An $n_T \times n_R$ MIMO channel can be represented as the combination of n_T (n_R) SIMO (MISO) channels, where each of these SIMO (MISO) channels is associated with a distinct transmit (receive) antenna. Then stating that $\mathbf{R}_{R,l}$ ($\mathbf{R}_{T,l}$) is independent of the index k (r) is equivalent to stating the *statistical equivalence* of such SIMO (MISO) channels. Moreover, we note that the elements (r_1, r_2) of $\mathbf{R}_{R,l}$ and (k_1, k_2) of $\mathbf{R}_{T,l}$ are given by:

$$(\mathbf{R}_{R,l})_{r_1,r_2} = \frac{1}{\sigma_l^2} \mathrm{E}\left\{ h_{k,l}^{r_1} \left(h_{k,l}^{r_2}\right)^* \right\} \triangleq \rho_{R,l}(r_1, r_2) \tag{2.152}$$

and

$$(\mathbf{R}_{T,l})_{k_1,k_2} = \frac{1}{\sigma_l^2} \mathrm{E}\left\{ h_{k_1,l}^{r} \left(h_{k_2,l}^{r}\right)^* \right\} \triangleq \rho_{T,l}(k_1, k_2), \tag{2.153}$$

respectively, where $\rho_{R,l}(r_1, r_2)$ ($\rho_{T,l}(k_1, k_2)$) is the *normalized spatial correlation coefficient* between the receive (transmit) antennas r_1 (k_1) and r_2 (k_2). If the receive (transmit) antennas behave in an *uncorrelated* fashion, then $\mathbf{R}_{R,l} = \mathbf{I}_{n_R}$ ($\mathbf{R}_{T,l} = \mathbf{I}_{n_T}$), where $\mathbf{I}_N$ is the $N \times N$ identity matrix. It is also worth pointing out that, generally speaking, knowledge of the functions $\rho_{T,l}(k_1, k_2)$ and $\rho_{R,l}(r_1, r_2)$ does not provide enough information for the evaluation of $\mathbf{\Sigma}_l$ in (2.147), since this requires knowledge of the correlations between the channels (r_1, k_1) and (r_2, k_2) associated with disjoint couples of transmit–receive antennas (i.e., couples characterized by both $r_1 \neq r_2$ and $k_1 \neq k_2$). In other words, what is really needed for a full statistical characterization is the knowledge of the *normalized spatial correlation*:

$$\rho_l(r_1, r_2; k_1, k_2) \triangleq \frac{1}{\sigma_l^2} \mathrm{E}\left\{ h_{k_1,l}^{r_1} \left(h_{k_2,l}^{r_2}\right)^* \right\}, \tag{2.154}$$

where r_1 and r_2 (k_1 and k_2) are the indexes of the transmit (receive) antennas. In the literature the additional assumption that the transmit and receive arrays are "decoupled" is also often made, so that the equality (e.g., see [133]):

$$\rho_l(r_1, r_2; k_1, k_2) = \rho_{R,l}(r_1, r_2)\, \rho_{T,l}(k_1, k_2) \tag{2.155}$$

holds, that is, the correlation between the fading of the distinct transmit–receive pairs is the product of the corresponding receive correlation and transmit correlation. As shown below, this occurs if in the propagation scenario the AOAs at the receive side are *statistically independent* of the AODs at the transmit side[33] for each cluster of scatterers.

If (2.155) holds, $\boldsymbol{\Sigma}_l$ can be evaluated as:

$$\boldsymbol{\Sigma}_l = \mathbf{R}_{T,l} \otimes \mathbf{R}_{R,l}, \tag{2.156}$$

where $\otimes$ denotes the *Kronecker product* operator (defined in Appendix C). The latter result explains the name "Kronecker model"[34] usually given to a statistical MIMO channel model with this property [123].

Let us now apply these results to the case of a 2×2 communication channel.

Example 2.3.2 If $n_T = n_R = 2$, the normalized spatial correlation matrices $\mathbf{R}_{R,l}$ (2.150) and $\mathbf{R}_{T,l}$ (2.151) take the form (see [133, p. 1220]):

$$\mathbf{R}_{T,l} = \begin{bmatrix} 1 & \mu_l \\ \mu_l^* & 1 \end{bmatrix} \tag{2.157}$$

and

$$\mathbf{R}_{R,l} = \begin{bmatrix} 1 & \gamma_l \\ \gamma_l^* & 1 \end{bmatrix}, \tag{2.158}$$

respectively, where $\mu_l = \rho_{T,l}(1, 2) = \rho_{T,l}^*(2, 1)$, $\rho_l = \rho_{R,l}(1, 2) = \rho_{R,l}^*(2, 1)$. Then, substituting (2.157) and (2.158) in (2.156) yields the normalized correlation matrix:

$$\boldsymbol{\Sigma}_l = \begin{bmatrix} 1 & \gamma_l & \mu_l & \mu_l \gamma_l \\ \gamma_l^* & 1 & \mu_l \gamma_l^* & \mu_l \\ \mu_l^* & \mu_l^* \gamma_l & 1 & \gamma_l \\ \mu_l^* \gamma_l^* & \mu_l^* & \gamma_l^* & 1 \end{bmatrix}, \tag{2.159}$$

whose Cholesky factorization produces the lower triangular matrix:

$$\boldsymbol{\Sigma}_l^{1/2} = \begin{bmatrix} 1 & 0 & 0 & 0 \\ \gamma_l^* & a_l & 0 & 0 \\ \mu_l^* & 0 & b_l & 0 \\ \mu_l^* \gamma_l^* & \mu_l^* a_l & \gamma_l^* b_l & c_l \end{bmatrix}, \tag{2.160}$$

where $a_l \triangleq \sqrt{1 - |\gamma_l|^2}$, $b_l \triangleq \sqrt{1 - |\mu_l|^2}$ and $c_l \triangleq \sqrt{1 + |\mu_l|^2 |\gamma_l|^2 - |\mu_l|^2 - |\gamma_l|^2}$. Then the complex gain vector $\mathbf{h}_l = [h_{1,l}^1, h_{1,l}^2, h_{2,l}^1, h_{2,l}^2]^T$ can be generated via the linear transformation defined by (2.149), where $\boldsymbol{\Sigma}_l^{1/2}$ is expressed by (2.160) and $\mathbf{a}_l = [a_{1,l}, a_{2,l}, a_{3,l}, a_{3,l}]^T$ is a vector of complex Gaussian random variables, each having zero mean and unit variance.

□

---

[33] This is a reasonable hypothesis if the inter-antenna distances in each array are substantially shorter than the inter-array distance.

[34] In various papers (e.g., see [134, p. 1113], [135, p. 3598], [136, p. 92], [137, p. 821]) the generation of an $n_R \times n_T$ channel matrix $\mathbf{H}$ is described in a formally different fashion, that is, as

$$\mathbf{H} = \frac{1}{\text{tr}(\mathbf{R}_R)} \mathbf{R}_R^{1/2} \mathbf{G} \, \mathbf{R}_T^{1/2},$$

where $\mathbf{R}_R \triangleq \text{E}\{\mathbf{H} \, \mathbf{H}^H\}$ and $\mathbf{R}_T \triangleq \text{E}\{(\mathbf{H}^H \mathbf{H})^T\}$ are the receive and transmit marginal correlation matrices, $\text{tr}(\mathbf{X})$ denotes the trace of a matrix $\mathbf{X}$, and $\mathbf{G}$ is an $n_R \times n_T$ matrix consisting of iid complex Gaussian random variables, each having zero mean and unit variance.

If (2.155) holds, the evaluation of the coefficients $\rho_{R,l}(r_1, r_2)$ (2.152) and $\rho_{T,l}(k_1, k_2)$ (2.153) is needed for the computation of $\mathbf{\Sigma}_l$ via (2.156). Analytical expressions for these coefficients can be evaluated for the two-dimensional propagation scenario represented in Figure 2.14. In this case it is assumed that:

(a) the antennas of each array are uniformly spaced with inter-antenna spacings d_T (at the transmit side) and d_R (at the receive side) along a line (i.e., *uniform linear array* are used), and are *identical* and *omnidirectional*,

(b) the groups of scatterers are so far from the antenna arrays that the departure/arrival angles can be considered constant along the arrays [138–140].

If the parametric channel model proposed in [138, 141] is adopted, the matrix $\mathbf{H}_l$ of (2.145) can be expressed as:

$$\mathbf{H}_l = \beta_l \, \mathbf{a}_R(\theta_{R,l}) \, \mathbf{a}_T^H(\theta_{T,l}),\tag{2.161}$$

where $\theta_{R,l}$, $\theta_{T,l}$ and β_l are the AOA, the AOD and the *complex fading path gain* for the lth cluster, respectively. In addition, the n_R-dimensional vector $\mathbf{a}_R(\theta_{R,l})$, called the *receive array response vector* and expressing the array response to an electromagnetic wave impinging with an angle of arrival[35] $\theta_{R,l}$, and the n_T-dimensional vector $\mathbf{a}_T(\theta_{T,l})$, called the *transmit array response vector* and expressing the transmit array response corresponding to the departure angle $\theta_{T,l}$, are given by:

$$\mathbf{a}_R(\theta_{R,l}) = [1, \exp(j2\pi d_R \sin\theta_{R,l}/\lambda), \ldots, \exp(j2\pi d_R(n_R - 1)\sin\theta_{R,l}/\lambda)]^T\tag{2.162}$$

and

$$\mathbf{a}_T(\theta_{T,l}) = [1, \exp(j2\pi d_T \sin\theta_{T,l}/\lambda), \ldots, \exp(j2\pi d_T(n_T - 1)\sin\theta_{T,l}/\lambda)]^T,\tag{2.163}$$

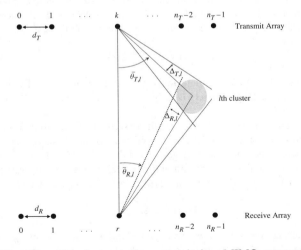

Figure 2.14   Two-dimensional propagation scenario in a MIMO communication system.

---

[35] The vector $\mathbf{a}_R(\theta_R)$ can be interpreted as a *spatial signature* (across the receive antenna array) induced by a planar wavefront arriving with an AOA θ_R and associated with a continuous wave at frequency f_c. Generally speaking, $\mathbf{a}_R(\theta_R)$ depends on the array geometry, the antenna gains and the AOA. The set of all array response vectors $\{\mathbf{a}_R(\theta_R), -\pi < \theta_R \le \pi\}$ is called *array manifold* and can be measured by moving a continuous wave source in azimuth around the array.

respectively. In addition, the parameter $\theta_{R,l}$ ($\theta_{T,l}$) is usually modeled as a random variable with mean $\bar{\theta}_{R,l}$ ($\bar{\theta}_{T,l}$) and standard deviation $2\Delta_{R,l}$ ($2\Delta_{T,l}$). Further, the *path gain* β_l is commonly modeled as a complex Gaussian random variable having zero mean, variance σ_l^2, and statistically independent real and imaginary components.

Substituting (2.162) and (2.163) in (2.161) yields:

$$h_{l,k}^r = \beta_l \exp\left(j2\pi(d_R r \sin\theta_{R,l} - d_T k \sin\theta_{T,l})/\lambda\right), \tag{2.164}$$

so that from (2.154) it easily seen that:

$$\rho_l(r_1, r_2; k_1, k_2) = \frac{1}{\sigma_l^2} \mathrm{E}\left\{|\beta_l|^2\right.$$

$$\left. \cdot \exp\left(j(2\pi/\lambda)[d_R(r_1 - r_2)\sin\theta_{R,l} + d_T(k_2 - k_1)\sin\theta_{T,l}]\right)\right\}. \tag{2.165}$$

If we assume now that the random variables β_l, $\theta_{R,l}$ e $\theta_{T,l}$ are *mutually independent*, (2.165) can be easily put in the form (2.155), with:

$$\rho_{R,l}(r_1, r_2) \triangleq \mathrm{E}\left\{\exp\left[j2\pi d_R(r_1 - r_2)\sin\theta_{R,l}/\lambda\right]\right\} \tag{2.166}$$

and

$$\rho_{T,l}(k_1, k_2) \triangleq \mathrm{E}\left\{\exp\left[j2\pi d_T(k_2 - k_1)\sin\theta_{T,l}/\lambda\right]\right\}. \tag{2.167}$$

The evaluation of $\rho_{T,l}(k_1, k_2)$ in (2.167) requires the pdf of $\theta_{T,l}$, and the evaluation of $\rho_{R,l}(r_1, r_2)$ in (2.166) requires the pdf of $\theta_{R,l}$. Analytical results[36] are available in the technical literature for a cosine raised to an even integer [143], uniform [144, 145], Gaussian [146], truncated Gaussian [147] and Laplacian [148] pdfs. Series representations of the spatial correlation coefficients in the cases of Laplacian and truncated Gaussian pdfs can be found in [149] and [150], respectively.

We note that the time variability of the channel can be easily accounted for in the model, if we assume that the elements of $\mathbf{H}_l(t)$ (with $l = 0, 1, \ldots, L_h - 1$) are all characterized by the same Doppler power spectrum $S_{D,l}(f)$. In fact, in this case, time variability can be easily incorporated in (2.149) by rewriting it as:

$$\mathbf{h}_l(t) = \sigma_l \mathbf{\Sigma}_l^{1/2} \mathbf{a}_l(t), \tag{2.168}$$

where the components of the $n_T n_R$-dimensional vector $\mathbf{a}_l(t)$ are mutually independent complex Gaussian random processes, each characterized by a zero mean and by a Doppler power spectrum $S_{D,l}(f)$.

Finally, it is important to point out that the results described above suffer from various limitations. For instance, the representation (2.149) of the tap vector $\mathbf{h}_l$ is unable to account for the well-known *pinhole* or *keyhole* phenomenon of channel degeneracy [135, 151–153]. In principle, this occurs when the scattering regions surrounding the transmit and the receive array are separated by a screen with a small hole in the middle, so that $\mathbf{H}_l$ can be modeled as the rank-one[37] matrix:

$$\mathbf{H}_l = \sigma_{cs}\, \mathbf{g}_{R,l}\mathbf{g}_{T,l}^T, \tag{2.169}$$

where σ_{cs} is the scattering cross-section of the keyhole, and $\mathbf{g}_{R,l}$ ($\mathbf{g}_{T,l}$) is an n_R (n_T) random column vector containing the channel coefficients from the keyhole (transmit array) to the receive array (keyhole). It is usually assumed that such vectors are statistically independent and each consist of iid zero-mean complex Gaussian random variables, so that a double-Rayleigh distribution is found for the amplitudes of the entries in $\mathbf{H}_l$ (e.g., see [154, p. 680]). From a physical viewpoint, the presence of

---

[36] Some experimental results can be found, for instance, in [142].
[37] In other words, the matrix $\mathbf{H}_l$ has only one nonnull singular value; see (2.142).

the keyhole effect can be related to scenarios where, even if rich scattering around the transmitter and receiver leads to low correlation of signals, other propagation effects, like waveguiding and diffraction, entail a rank reduction of the matrix transfer function. Recently, its existence has been confirmed by experimental results in [135]. A generalization of the Kronecker model, accounting for the keyhole effect, has been presented in [153].

Another relevant shortcoming of the Kronecker model is due to the fact that it neglects the statistical interdependence of both link ends, since it is derived under the assumption that correlation at the transmitter and at the receiver can be modeled independently, so that the transmitter does not affect the spatial properties of the received signal at all. However, this hypothesis is not reasonable in realistic indoor MIMO channels [155]. Recently, a novel approach to statistical modeling that combines the advantages of the "Kronecker model" with the so-called *virtual channel representation* of [140] has been proposed to overcome this problem [136].

Other comments concern the tap representation expressed by (2.161). In fact, this model is not flexible since every realization of $\mathbf{H}_l$ structured as (2.161) has rank 1. Generally speaking, channel rank is related to the *angular spread* of path clusters [147]. In particular, if this spread decreases, the correlation matrix $\mathbf{\Sigma}_l$ loses rank, and, consequently (see (2.149)), the rank of $\mathbf{H}_l$ decreases [156]. MIMO channels models in which the rank of the matrix $\mathbf{H}_l$ is controlled by the fading correlation at the antenna arrays have been proposed in [156] and [153].

Finally, we note that in this subsection our attention has been focused on the statistical characterization of the random matrix $\mathbf{H}_l$ through its covariance matrix. Other important statistical properties of this matrix concern the distribution of the singular values of its decomposition (2.142) and that of its squared norm $\|\mathbf{H}_l\|^2 = \mathrm{tr}(\mathbf{H}_l \mathbf{H}_l^H)$ (see (C.2)). The singular values of $\mathbf{H}_l$ can easily be related to the eigenvalues of $\mathbf{H}_l \mathbf{H}_l^H$, which have a Wishart-type distribution [157, 158]. Details of the moment generating, cumulative density and probability density functions of $\|\mathbf{H}_l\|_F^2$ can be found in [38, pp. 44–45] or [159].

2.3.2.2 Directional Models

The directional system functions defined in Section 2.3.1 are often modeled as Gaussian stochastic processes. Here we also assume that they have zero mean (Rayleigh fading), so that a complete statistical characterization only requires knowledge of their autocorrelation functions. In the following we consider some of these functions to formulate the properties of wide-sense stationarity and uncorrelated scattering (already defined for SISO channels in Section 2.2.2) in a directional scenario and to define new statistical properties referring to spatial or angular characteristics of the propagation scenario; further details of these properties can be found in [122, 130].

Wide-sense stationarity (WSS channel). If we refer to the Doppler- and angle-resolved channel transfer function $G(\nu, f, \varphi)$, this property entails that:

$$R_G(\nu_1, \nu_2; f_1, f_2; \varphi_1, \varphi_2) \triangleq \mathrm{E}\left[G(\nu_1, f_1, \varphi_1)G^*(\nu_2, f_2, \varphi_2)\right]$$
$$= \delta(\nu_2 - \nu_1)P_G(\nu_1; f_1, f_2; \varphi_1, \varphi_2), \tag{2.170}$$

since scatterers introducing different Doppler shifts are uncorrelated in a WSS channel.

Uncorrelated scattering (US channel). A US channel is characterized by the autocorrelation:

$$R_h(t_1, t_2; \tau_1, \tau_2; \varphi_1, \varphi_2) \triangleq \mathrm{E}\left[h(t_1, \tau_1, \varphi_1)\, h^*(t_2, \tau_2, \varphi_2)\right]$$
$$= \delta(\tau_2 - \tau_1)\, P_h(t_1, t_2; \tau_1; \varphi_1, \varphi_2) \tag{2.171}$$

of the TVARCIR; note that the last equality represents a simple extension of (2.80).

Wide-sense stationarity and *uncorrelated scattering* (WSS-US channel). If the latter two properties hold, the autocorrelation of the Doppler- and angle-resolved channel impulse response can be expressed as:

$$R_s(\nu_1, \nu_2; \tau_1, \tau_2; \varphi_1, \varphi_2) \triangleq E\left\{s(\nu_1, \tau_1, \varphi_1)\, s^*(\nu_2, \tau_2, \varphi_2)\right\}$$

$$= \delta(\tau_2 - \tau_1)\, \delta(\nu_2 - \nu_1)\, P_s(\nu_1; \tau_1; \varphi_1, \varphi_2), \qquad (2.172)$$

generalizing (2.88).

Uncorrelated paths (UP channel). This refers to the fact that signal components associated with distinct (departure or arrival) angles are uncorrelated; then the autocorrelation function of the TVARCIR takes the form:

$$R_h(t_1, t_2; \tau_1, \tau_2; \varphi_1, \varphi_2) = \delta(\varphi_2 - \varphi_1)\, P_h(t_1, t_2; \tau_1, \tau_2; \varphi_1). \qquad (2.173)$$

It can be shown that a UP channel is *stationary in space*,[38] that is, the autocorrelation function of the channel response across space depends on a couple of space locations though their difference vector only.

Wide-sense stationarity, uncorrelated scattering and *uncorrelated paths* (WSS-US-UP channel). In this case echoes characterized by distinct Doppler shifts or delays or angles are uncorrelated, so that (see (2.172)):

$$R_s(\nu_1, \nu_2; \tau_1, \tau_2; \varphi_1, \varphi_2) = \delta(\nu_2 - \nu_1)\, \delta(\tau_2 - \tau_1)\, \delta(\varphi_2 - \varphi_1)\, P_s(\nu_1, \tau_1, \varphi_1). \qquad (2.174)$$

Alternatively, we have:

$$R_h(t_1, t_2; \tau_1, \tau_2; \varphi_1, \varphi_2) = P_h(t_2 - t_1, \tau_1, \varphi_1)\, \delta(\tau_2 - \tau_1)\, \delta(\varphi_2 - \varphi_1). \qquad (2.175)$$

Note that the first- and second-order statistics of a WSS-US-UP channel are invariant with respect to translations in the time, frequency and aperture domains.

The functions $P_h(\Delta t, \tau, \varphi)$ and $P_s(\nu, \tau, \varphi)$ appearing in (2.175) and (2.174) are called the *delay-angle cross-power density* and the *directional scattering function* (DSF), respectively. They represent the extensions, to a directional scenario, of the *delay cross-power density* $P_h(\Delta t, \tau)$ (see (2.84)) and of the *scattering function* $S(\nu, \tau)$ (see (2.89)), respectively. Note that the DSF, unlike the SF, provides statistical information about the angular dispersion of the transmitted signal. Moreover, it can be shown that $P_s(\nu, \tau, \varphi)$ is related to the autocorrelation function $R_T(\Delta t, \Delta f, \Delta x) \triangleq E\{T^*(t, f, x)\, T(t + \Delta t, f + \Delta f, x + \Delta x)\}$ of the *time- and aperture-variant transfer function* $T(t, f, x)$ via a triple Fourier integral. In particular, in a two-dimensional, single-directional scenario, we have [131]:

$$R_T(\Delta t, \Delta f, \Delta x) = \int_{\nu = -B_D}^{B_D} \int_{\tau = \tau_m}^{\tau_M} \int_{\varphi = -1}^{1} P_s(\nu, \tau, \varphi) \exp\left[j2\pi \left(\nu\Delta t - \frac{\Delta x}{\lambda}\varphi - \tau\Delta f \right) \right] d\nu\, d\tau\, d\varphi. \qquad (2.176)$$

This extends (2.90), relating $R_H(t; f)$ to $S(\nu, \tau)$, to a directional scenario. Finally, we note that, if the phenomena of time dispersion and angular dispersion are *independent* of that of frequency dispersion (due to the relative motion between the transmitter and the receiver and/or the motion of the scatterers in the propagation medium), the DSF $P_s(\nu, \tau, \varphi)$ can be factored as:

$$P_s(\nu, \tau, \varphi) = S_D(\nu) P_h(\tau, \varphi), \qquad (2.177)$$

where $S_D(\nu)$ is the Doppler power spectrum and $P_h(\tau, \varphi)$ is the so-called *angular delay power spectrum* (ADPS). Generally speaking, the above three dispersion phenomena are correlated (e.g., see [127]) and therefore, the validity of (2.177) cannot usually be taken as a rigorous assumption, even

---

[38] In the literature this is also referred to as a *homogeneous* channel (e.g., see [38, p. 23]). Note that, in practice, a channel can be deemed homogeneous if spatial stationarity is observed over a multiple of several tens of its coherence distance D_c.

if turns out to be extremely useful in the development of reduced channel models [131]. Moreover, it is worth noting that $P_h(\tau, \varphi)$ cannot usually be factored, since time and angular dispersions are commonly correlated [129, 160, 161]. Despite this, various technical papers (e.g., see [129, 162]) propose factorizable models $P_h(\tau, \varphi)$ and provide a justification for the additional factorization in the Doppler domain.

2.3.3 Reduced-Complexity Statistical Modeling of MIMO Channels

In principle, any doubly selective SISO channel model described in Section 2.2.3 can be extended to a MIMO scenario if the channel statistics are independent of the specific transmit/receive antenna couple. For instance, if the channel PDP has this property, a TDL model (see Example 2.3.1) can be easily adopted for the MIMO CIR matrix $\mathbf{H}(t, \tau)$; in fact, under the above assumption, the characteristic delays are the same for any transmit/receive antenna couple and can be selected on the basis of experimental data acquired in specific communication scenarios, as in the so-called *Stanford University Interim* channel models for *fixed wireless access* systems [163], or by adopting a rigorous mathematical approach, such as the GQR-based technique of Section 2.2.3.5 [164] or the discretization techniques illustrated in [123, 138, 141]. This explains why the technical literature about MIMO channel modeling has mainly focused not on the problems of selecting a model structure, but on those of evaluating the correlation properties of model parameters for multipath and for single-tap (i.e., narrowband time-selective) fading channels, and of assessing the correlation impact on channel capacity (see Section 2.4.2 for further details).

Finally, we note that various directional channel models have been proposed; they have been developed on the basis of the specific properties of various radio environments, such as those illustrated in [165, 166], or by adopting a rigorous approach to the problem of approximating the autocorrelation properties of directional system functions, like the GQR-based models derived in [131].

Further information about the available technical literature on statistical channel models for MIMO channels can be found in Section 2.4.2.

2.4 Historical Notes

The study of wireless channels has a long history, dating back to the original development of long-range radio communications and the pioneering work of Guglielmo Marconi [3]. Knowledge in this area grew in the first half of the last century, but its rate of evolution increased after the Second World War. Indeed, during the 1940s and particularly the 1950s the propagation of radio signals over fading channels was intensively investigated and written about, even if not fully understood.[39] The expertise acquired in communications over the ionosphere using the *high frequency* (HF) band and troposphere using the *very high frequency* (VHF) band allowed the use of such channels for long-distance data transmission [13], specially for military applications, in the late 1950s and early 1960s. This in turn fostered research to optimize data transmission over multipath fading channels so as to achieve reliable communications. Preliminary results showed that the error performance achievable using traditional signaling techniques was not satisfactory for data exchange between computer systems. A solution to this problem was found in the mid-1960s, when the use of the first error-correcting schemes was suggested for these applications.

The study of multipath fading channels received a further stimulus from the advent of mobile phone systems in the 1970s. This motivated research to model both large-scale and small-scale fading for various propagation scenarios and progressively larger bandwidths. The long-standing interest of the scientific community in wireless channels explains the extent of the technical literature in this area. In

---

[39] Accounts of the state of knowledge during those years can be found in [167, 168], which are both included in [44].

this section we will focus our attention on two specific parts of the literature, namely that proposing *large-scale fading models*, and that dealing with the *statistical modeling* of small-scale fading. Both indoor and outdoor scenarios will be considered, but our interest will be mainly in the modeling of the multipath fading affecting *terrestrial mobile communications*. Ionospheric and tropospheric channels will not be considered.

2.4.1   Large-Scale Fading Models

The problem of estimating path loss in wireless channels has received considerable attention in the past two decades, since its solution allows the estimation of coverage areas in wireless cellular systems [52, 54], and provides data for locating BSs correctly and for accurate frequency planning [64]. Various *propagation models* for path loss prediction have been proposed in the literature; they can be classified as *empirical models* (also called *statistical models*) or *theoretical models* (often called *deterministic models*). *Hybrid models*, resulting from the combination of the two classes, are also available. Empirical models implicity incorporate various environmental factors and are characterized by good computational efficiency; however, their accuracy depends on the degree of similarity between the scenarios to which they are applied and those from which the experimental data for their derivation have been acquired. Deterministic models, which are based on the principles of physics, can be applied to any environment, but require substantial computational effort. In fact, good accuracy can be achieved only by processing a large quantity of environmental data using complicated algorithms. In the following, the available models for each class are listed for each of the following communication scenarios: macrocells, microcells and indoor environments.

2.4.1.1   Macrocellular Scenarios

In such scenarios an LOS component is normally absent and this makes path loss prediction extremely difficult. Historically, the first (and still famous) empirical model for path loss in macrocells was proposed by Y. Okumura, E. Ohmori and K. Fukuda in 1968 [169], and was based on signal strength measurements in radio links between BSs and MTs in Tokyo. The model, represented by a family of curves, can be used in the frequency interval 150–1500 MHz and for a link length in the interval 1–100 km. A few years later, in 1977, another contribution to empirical modeling was provided by K. Bullington, who generated a set of nomograms, which can be used to assess the received power in a point-to-point link, taking into account both the presence of the earth surface and its curvature [170]. In addition, Bullington provided approximate methods to assess the variations in the received power due to atmospheric conditions and to path obstructions introduced, for instance, by hills, buildings and trees. A further significant step was made in 1980 by M. Hata, who extracted from the curves provided in [169] an analytical expression for path loss [171] which can be applied to urban, suburban and open areas. Hata's expression can be used in the frequency interval 100–1500 MHz, and assumes a link length of 1–20 km, a BS height of 30–200 m and a mobile antenna height in the range 1–10 m.

These first results were followed by intense research activity and various measurement campaigns in the 1980s and in the 1990s, giving rise to various new models for specific scenarios, as discussed in detail in [54, 64, 107]. Here, due to space limitations, we mention only some relevant models, namely:

(a) the model developed by W. C. Y. Lee in 1982 whose parameters can be adjusted on the basis of additional measurement results to adapt it to specific local environments [50],
(b) the models proposed by M. Ibrahim and J. Parsons in 1983 to assess the path loss in urban scenarios in the frequency interval 150–450 MHz [172],
(c) the model derived in 1984 by F. Ikegami *et al.* to predict average field strength in urban streets [173],

(d) the model proposed by J. Walfisch and H. L. Bertoni in 1988 to describe propagation in urban environments (more specifically, from a downtown area with tall buildings) [174] and its more recent generalization proposed by H. K. Chung and H. L. Bertoni in 2002 [175],
(e) the model derived by V. Erceg *et al.* in 1999, using a set of experimental results acquired at 1.9 GHz by AT&T Wireless Services in 95 existing macrocells in several suburban areas of the USA [65] (the novelty of this model lies in the fact that the two most important parameters in the usual path loss formula (see (2.3)), the exponent n and the shadowing standard deviation σ_X, are both represented as Gaussian random variables, to model their changes from one macrocell to another),
(f) the COST 231–Walfisch–Ikegami model extensively used by the designers of public mobile land wireless networks in urban and suburban areas in which the building height is uniform [176],
(g) the models proposed by T. Kürner and A. Meier in 2002, which is useful in predicting radio coverage in outdoor and outdoor-to-indoor scenarios (in the last case, the coverage area is inside a building, but BSs are located outside it) at 1.8 GHz [177].

Finally, we note that other path loss models adopted by international institutions (e.g., the European Broadcasting Union and International Telecommunications Union) and by private companies (e.g., Ericsson) are also available.

2.4.1.2 Microcellular Scenarios

Various path loss models can be found in the literature for *microcells*, exhibiting a different behavior in signal propagation from that encountered in macrocells. In the following we mention some relevant deterministic and statistical models.

An important deterministic model was proposed by P. Harley in 1989 [76] to describe the received signal level in LOS conditions. According to this model, the path loss in decibels can be described by a broken line, consisting of two parts. The first part describes the propagation loss for points whose distance from a BS does not exceed a given *breakpoint* and its slope typically exhibits a path loss exponent close to 2. The second part is characterized by a substantially steeper slope. An alternative to an exact expression for the path loss is represented by the lower and upper bounds to this quantity provided in [178]. Some experimental results and a comparison with the above-mentioned models are illustrated in [66, 179]. A characterization similar to that of the LOS case can also be provided in non-LOS cases, for urban and suburban areas with perpendicular streets [180]. It is of interest to note that similar models can be derived, at least in the LOS case,[40] by resorting to *optical ray tracing theory*, under the assumption that reflected rays are dominant compared to diffracted ones. In this case the simplest model available is the so-called *two-ray model* [73, 77, 178]. Other more complicated models account for the presence of multiple reflected rays [180–182] and rays by *corner diffraction*. The latter phenomenon is considered in the models based on the *geometrical theory of diffraction* (GTD) [183], the *uniform theory of diffraction* (UTD) [184–188] and the *Fresnel–Kirchhoff theory of diffraction* [189, 190].

2.4.1.3 Indoor Scenarios

Both statistical and theoretical models are available for *indoor* scenarios. The bandwidth of interest in this case is 1.8–2 GHz, where most indoor communication systems operate. Statistical models are usually based on the approach illustrated in Section 2.1 (e.g., see [56, 68, 69]) and are expressed by

---

[40] The description of LOS scenarios is much more complicated. See, for instance, [76], which describes a prediction methodology based on experimental results acquired in a measurement campaign along various streets in Manhattan [181].

equation (2.2). Theoretical models result from various tools, such as *ray-tracing* and *finite-difference time-domain* (FDTD) techniques. Ray tracing methodologies evaluate the electromagnetic field at a point as the superposition of the contributions associated with all the *rays* coming from a given transmitter [191]. In the simplest models based on ray tracing only the free space path loss and reflection mechanisms are accounted for. More accurate models also take into consideration diffraction, diffuse scattering from walls and the transmission of electromagnetic waves through various materials. The numerical results generated by ray tracing techniques are approximate, since an exact evaluation of the intensity of the electromagnetic field can only be accomplished by solving Maxwell's equations. Despite this, accurate results are obtained when the observation point is many wavelengths away from the closest scatterers, and all the scatterers are large with respect to wavelength and are smooth. Even in this case, however, an accurate three-dimensional description of the area to be analyzed (and, consequently, a substantial quantity of data) and great computational effort are needed. Today, various software applications processing indoor or outdoor environmental data to generate signal intensity via ray tracing are available. They allow accurate planning of wireless systems operating not only in indoor environments [191–193], but also for outdoor scenarios [182, 187, 194–200].

In the FDTD method Maxwell's equations are solved by approximating them with a set of finite-difference equations, which are then solved on the nodes of a (regular or irregular) grid using iterative methods. Like ray tracing, the FDTD approach entails a substantial computational burden (proportional to the number of grid nodes and, consequently, to the size of the area being analyzed). The performance achievable is similar to that offered by ray tracing algorithms, even if these are preferred when the area to be analyzed is large.

Finally, we note that recently much attention has been paid to the assessment of the path loss in indoor scenarios for UWB communications. Details can be found in [201–203].

2.4.2   Small-Scale Fading Models

Research on statistical channel models for mobile communications focused primarily on SISO channels until the first half of the 1990s. Since then, the interest in space-time (ST) channel models accounting for the directional properties of wireless channels has increased due to the strong interest in the use of antenna arrays for enhancing capacity in digital communication systems. Today various classes of statistical fading models, explicitly designed for *theoretical analysis* or for *channel simulation*, exist. They can provide *discrete-time* or *continuous-time* representations of channel behavior. In addition, their derivation can be based on *standard mathematical techniques* (such as representation techniques for bandlimited deterministic signals or for stochastic processes with given second-order statistics) or can rely on *experimental data* acquired in specific propagation scenarios. Generally speaking, small-scale models provide *parametric representations* of continuous-time or discrete-time system functions of communication channels. From a historical perspective, the first class of statistical models available in the technical literature was aimed at representing the CIR or other channel system functions (e.g., the TVTF) as *a linear combination of some deterministic continuous-time basis functions*. In this class of models, combination coefficients are random processes or random variables having joint stochastic properties which ensure an approximate statistical equivalence of the model with the given real channel. Various models of this type were already available in the 1960s and were described in a unifying perspective[41] by P. A. Bello in his classic 1963 paper [78]. Since then, their use in communication system analysis and design has become widespread, and the number of available models has progressively increased, since new methods have been proposed for the derivation of function bases and for the representation of their random coefficients. In fact, bases of continuous-time functions can be derived in a number of ways, for example:

---

[41] The models described by Bello consist of two distinct classes, namely *sampling models* and *power series models*. Both classes, however, belong to the family of so-called *canonical models*, since they offer a simplified representation of linear time-varying channels in terms of *canonical* terms or building blocks.

(a) by resorting to *representation methods for bandlimited/time-limited system functions*, as illustrated
    by the equally-spaced TDL model proposed by T. Kailath in [105, p. 20], the *canonical represen-
    tation* due to A. M. Sayeed and B. Aazhang [204, 205], and the so-called *bin model* of [107, pp.
    111–114], [206],
(b) by exploiting the KL *expansion* [85, 86, 108, 110] or the GQR-based representation [117, 118],
    of a random process with given correlation properties (see Section 2.2.3).

These techniques were originally employed in the derivation of SISO communication models;
however, in principle, they can be easily applied to the *space-time characterization* of wireless channels
(see Section 2.3.3) for the analysis of MIMO systems. This idea is exemplified by the ST *canonical
signal representation* of [207, 208] and the directional GQR models of [131] derived from their
nondirectional counterparts [117, 204, 205], respectively.

Most of the above models provide a continuous-time representation of *doubly-selective fading*.
Other specific continuous-time models have been explicitly developed for the case of *frequency-flat
fading* since the 1960s [209]. The best-known SISO model is that originally derived by R. H. Clarke
[98] in 1968 (usually known as the Clarke–Jakes model) and partially based on previous work by
J. F. Ossanna, Jr. [210] and E. N. Gilbert [211] on the interference of reflected and scattered electromagnetic waves in wireless communications. Clarke's model was later generalized by T. Aulin in 1979
to the case where vertically polarized waves are not necessarily traveling horizontally in [212]. Note
also that one of the main assumptions in Clarke's classic model is *isotropic scattering* – a uniform distribution for the AOA of multipath components at the mobile station. In the presence of nonisotropic
scattering, however, the correlation function and the Doppler spectrum of the fading distortion are
strongly affected, as illustrated, for instance, in [213], where a von Mises angular distribution is
assumed for the impinging waves.

A totally different approach to the modeling of *frequency-flat fading* in SISO communications is
that based on *Markov models*. Further details can be found in [214–219].

It is also important to point out that, in the last decade, various models have also been proposed for
the continuous-time representation of *frequency-flat fading* in SIMO and MIMO communication links.
The simplest model is that representing the communication channel as a set of $n_T \times n_R$ statistically
independent Rayleigh SISO channels [18, 220]. However, more refined modeling takes into account
the *correlations* between distinct transmit/receive antenna couples. In the literature focusing on the
latter problem substantial attention has been paid to the evaluation of the correlation between channel
coefficients associated with distinct transmit–receive antenna couples (e.g., see [133, 136, 143, 144,
146–150, 153, 221]), or to the evaluation of the *spatial-temporal correlation function*, since this
enables the study of the basic impact of a random multipath fading channel on the performance of
space-time systems based on the use of antenna arrays (e.g., [222–225]). A different approach is due
to A. M. Sayeed [140], who has proposed a *virtual representation* relating the matrix describing the
real channel to that describing a *virtual channel*. A key property of this representation is that the
elements of the virtual channel matrix can be assumed to have *independent* entries without much loss
of accuracy.

MIMO channel models for frequency-flat fading have been largely exploited for evaluating the
capacity benefits deriving from the use of antenna arrays (e.g., see [133, 134, 137, 145, 153, 220, 224,
226–229]). Analytical results can also be compared with measured data to establish if the model is
realistic in specific propagation environments (e.g., [133, 155, 227, 230]). Note that a refined MIMO
model should also account for the so-called *pinhole* or *keyhole* phenomenon [135, 151–153].

Some of the above multipath fading models have also inspired other statistical models, which have
been developed in attempts to *fit measurement results in specific propagation scenarios*. This class of
models is illustrated by two well-known SISO channel models: that devised by G. L. Turin *et al.* in
1972 [231] for the description of urban multipath propagation, and the well-known Saleh–Valenzuela
model proposed in 1987 for indoor scenarios [68]. Both are based on an unequally spaced TDL
description of a wireless channel. However, Turin's model assumes that:

(a) the tap delays form a modified Poisson sequence;
(b) the tap phases are uniformly distributed over $(0, 2\pi]$ and are statistically independent; and
(c) the tap strengths are log-normally distributed and, generally speaking, correlated.

In contrast, Saleh and Valenzuela's model is based on the assumption that the multiple echoes arriving at the receive antenna appear in clusters. In addition, the clusters and the echoes within each cluster form Poisson arrival processes with different, but fixed, rates. It is also assumed that the echoes have uniformly distributed phases and independent Rayleigh amplitudes with variances decaying exponentially with cluster and echo delays. Modifications to Turin's model were proposed by H. Suzuki in 1977 to fit experimental data in urban environments [232].

All the above models provide efficient tools for a *theoretical analysis* of various effects of fading in wireless communication systems. They also provide the tools for the derivation of *discrete-time models*. These are useful for the derivation of signal processing algorithms employed in digital receivers (e.g., see [38, pp. 48–54] and the applications of the *basis expansion technique* proposed by G. B. Giannakis and C. Tepedelenlioğlu in [233]) and for the *efficient simulation of wireless channels*.[42] The first papers devoted to the latter application proposed different solutions for the simulation of SISO channels[43] [84, 236–238]; such models are based on various mathematical techniques that allow computationally efficient generation of random channel processes with given statistical properties. In particular, the model proposed by S. A. Fechtel in [236] applies a channel orthogonalization strategy to devise an efficient linear bandlimited channel simulator. This provides a good approximation to the physical channel dynamics and provides efficient representations for channels having quasi- or truly continuous delay profiles. The techniques proposed by P. Hoeher in [84] and by K.-W. Yip and T. S. Ng in [237] are based on the idea of approximating a WSS-US channel with given statistics via a *Monte Carlo* approach. For instance, in Hoeher's model[44] the multipath fading channel is represented as a TDL with N time-variant taps, each characterized by unit gain, but with *random delays, phases and Doppler shifts*. In this case Gaussianity of system functions is achieved asymptotically (i.e., for $N \to +\infty$). Finally, P. M. Crespo and J. Jiménez propose the use of a harmonic decomposition for approximating the time-variant behavior of CIR in [238].

Various simulation models have also been developed for *frequency-flat fading* channels. The approaches proposed can be classified into three families [240]. The first consists of simulating fading as a *sum of sinusoids* (SOS); see [88, pp. 165–173] and [99, 206, 241–243]. This idea was first proposed by W. C. Jakes[45] [99], who did not take into consideration, however, the problem of generating multiple uncorrelated fading waveforms. A solution to this problem was later proposed by P. Dent *et al.* in [241]. It is also interesting to note that a representation of flat fading as a superposition of fixed-frequency complex exponentials (having random amplitudes and phases) is also provided by GQR-based models [117, 118]. The second method is based on a combined use of filtering and *inverse discrete Fourier transform* (IDFT) processing and was originally illustrated by J. I. Smith in [244]. This algorithm was modified by D. C. Young and N. C. Beaulieu in [245] to reduce both its computational complexity and memory requirements. The third approach involves filtering discrete-time white Gaussian noise using recursive or nonrecursive digital filters, which leads to *moving average* (MA) [97], *autoregressive* (AR) [246–248], and *autoregressive moving average* (ARMA) [248, 249] models of flat fading. A hybrid model, representing fading as a filtered version

---

[42] Note that discrete-time (frequency-selective) TDL models have also been employed in the evaluation of channel capacity of MIMO systems (e.g., see [141, 156, 234]).

[43] In the past simulation models for SISO channel have also been formulated by referring to *continuous-time* representations of channel system functions; see, for instance, the simulation model for urban radio propagation proposed by H. Hashemi in 1979 [235] and based on the mathematical models developed by G. L. Turin and H. Suzuki.

[44] This model has recently been used for characterizing the aeronautical channel model by Haas [239].

[45] An in-depth analysis of the Jakes SOS simulator can be found in [243].

of two sinusoids in quadrature and whose amplitudes are AR processes, has recently been proposed in [240].

Efficient tools for the simulation of directional and MIMO wireless channels are available [123, 250–252]. In particular, a discrete-time MIMO channel model, based on the Knoecker correlation assumption and able to characterize *triply-selective* (i.e., selective in time, frequency and space) Rayleigh fading channels, has been derived in [123] (an analysis of its statistical properties can be found in [253]). A MIMO frequency-selective model based on a one-ring scattering[46] geometry has been derived in [251]. A flexibly configurable channel model for the simulation of mobile channels for communication systems exploiting channel directionality can be found in [250]. In [252] a stochastic TDL model for small-scale fading with a Rice-distributed envelope and temporal, spatial and spectral correlation is proposed and a channel simulator, achieving good accuracy at a reasonable computational complexity, is presented.

Another class of channel models, conceptually different from all the above models, consists of the so-called *geometrically based stochastic models* (GBSMs), that is, models whose random parameters are related to the evolving *geometry* of the propagation scenario[47] (i.e., the spatial distribution and mobility of scatterers). GBSMs rely on the availability of the *density function* of the spatial scatterers, which are usually placed in a two-dimensional region around a BS and/or an MT; such a region can extend to infinity or can have a specific convex shape, such as a circle or disk around the MT, or an ellipsis having the BS and the MT at its foci[48] (e.g., see [129, 223, 254–258]). Given the scatterer spatial density function, various spatial characteristics of the propagation scenario, such as the joint *time of arrival* (TOA) and AOA, the marginal AOA, and marginal TOA pdfs of multipath components, can be derived [259]. Therefore, this class of models is extremely useful in analyzing communication systems employing antenna arrays combined with spatial processing techniques to exploit spatial diversity and improve spectral efficiency [260]. Geometric models have been used to model single-antenna systems and systems with multiple antennas at one end of the radio link. Specific examples of these models can be found in [132, 143, 147, 148, 161, 222, 254, 258, 261–269]. Such models are based on different assumptions about the location of scatterers, their spatial distribution and apply to various propagation scenarios. Moreover, some of them, such as those proposed in [254, 258, 264], belong to the class of *geometrically based single-bounce models* (GBSBMs), since they rely on the assumption that received signal components interact with only a *single scatterer* in the channel.

GBSBMs for MIMO channels are also available in the technical literature. In particular, models assuming local scattering at one end of a wireless link can be found in [127, 156, 221, 223, 225, 251], while models based on scattering at both ends of a radio link are available in [127, 153, 165, 166, 224, 270]. It is also worth noting that the model proposed by A. F. Molisch in [127] can be classified as a "hybrid" directional model since it combines a geometric approach with physical arguments about relevant propagation effects.

GBSBMs can also be exploited to assess channel capacity when ST transmission techniques are employed. Study of the correlation properties and of their impact on the capacity of some of the above-mentioned GBSMs can be found in [134].

Finally, the following observations are of interest:

1. In the references mentioned above channel modeling often refers to the specific scenario of a fixed BS communicating with an MT. The modeling techniques proposed, however, can also be applied

---

[46] Geometrically based channel models are analyzed below.

[47] This approach to the description of channel behavior dates back to the early 1970s, and was pioneered by Jakes [99] and Lee [143].

[48] An elliptical region is suitable for microcell and picocell environments, where both the BS and the MT are surrounded by scatterers. On the other hand, an circular region is appropriate for macrocells, where the BS antenna is located higher than the MT antennas and only the MT antennas are assumed to be surrounded by scatterers.

or extended to the representation of *mobile-to-mobile* fading channels; further information on this problem can be found in [271] (where a MIMO scenario is considered) and the references therein.

2. Modeling of small-scale fading in UWB systems has received considerable attention in recent years, and recent results can be found in [201–203, 272].

3. The issue of *cross-polarization* in channel modeling and the use of the *cross-polarized antennas* are of relevant interest in MIMO systems; see [38, Chapters 2 and 3] and [273] (and references therein).

2.5   Further Reading

In this chapter a rich bibliography on the description and characterization of small- and large-scale fading has been provided. For this reason we limit ourselves here to pointing out some specialist books for further reading in this area. In particular, we mention [88] and [274], which offer an interesting overview of the characteristics of SISO channels. The most significant information-theoretic and communication aspects of SISO fading channels are analyzed in [275]. Some introductory material about ST propagation and MIMO channels can be found in [21, 38, 276], while [277] is completely devoted to ST wireless channels. In addition, there has been extensive work to develop models for both SISO and MIMO channels based on measured data. Much of this work is summarized in [22, 56, 278]. There continues to be interest in the measurement and modeling of wireless channels, but at present most work is focused on the adaptation of existing models to specific application scenarios.

3

Digital Modulation Techniques

3.1 Introduction

An information-carrying signal, before being transmitted in a communication system, usually undergoes processing, which modifies its characteristics, to make it suitable to send over a given channel. In particular, in digital wireless communications the message consists of a stream of binary data, which is intrinsically a *baseband discrete-time signal*, whereas the communication channel, as illustrated in Chapter 2, can always be modeled as a *passband time-continuous system*. Generally, the transformation of the discrete-time message into a time-continuous signal, having spectral properties suitable for transmission over a given communication channel, is accomplished through a process known as *modulation*.

This chapter considers digital modulation techniques that are of importance to wireless communications. In Section 3.2 a general mathematical model of a digital modulator is presented. The representation of the generated waveforms via an orthonormal basis and estimation of their spectral occupancy are discussed in Sections 3.3 and 3.4, respectively. Then a number of *single-carrier* (SC) modulation techniques are analyzed. For each of them the modulation process involves a single local oscillator, converting a baseband data signal into a radio frequency signal. In particular, in Sections 3.5 and 3.6 the mathematical structure of passband *pulse amplitude modulation* (PAM) signals and that of *continuous phase modulation* (CPM) signals, respectively, are considered. Then, in Section 3.7, a specific type of *multicarrier* (MC) signal, known as OFDM, is considered. In OFDM the information content of the message is distributed over a multiplicity of narrowband channels, all exploited at the same time (i.e., in a parallel fashion) and uniformly spaced in the frequency domain. In Section 3.8 we will also consider the design of multidimensional modulations based on sets of points, known as *lattices*. In Section 3.9 we will analyze the impact of a wireless channel on the spectral properties of digitally modulated signals. Finally, some suggestions for further reading are provided in Section 3.11.

3.2 General Structure of a Digital Modulator

A *digital modulator* is a device that converts a data sequence to an analog signal suitable for transmission over a physical channel. Here, it is assumed that digital modulators are fed by a WSS sequence $\{d_n\}$, consisting of M-ary data symbols, from some alphabet $A_d = \{0, 1, \ldots, M-1\}$. The data are generated by an information source at a given *signaling* or *symbol rate* $R_s = 1/T_s$, where T_s is the *signaling* or *symbol interval*. A modulator maps d_n, at the instant $t_n = nT_s$, to a *channel symbol* c_n, from an M-ary complex alphabet A_c. Then, the modulator associates $\{c_n\}$ with a baseband signal

Wireless Communications: Algorithmic Techniques, First Edition.
Giorgio M. Vitetta, Desmond P. Taylor, Giulio Colavolpe, Fabrizio Pancaldi, Philippa A. Martin.
© 2013 John Wiley & Sons, Ltd. Published 2013 by John Wiley & Sons, Ltd.

$s(t, \mathbf{c})$, which is then up-converted using a *carrier* at frequency f_c, always assumed to be substantially larger than R_s. This generates the RF signal:

$$s_{RF}(t, \mathbf{c}) = \text{Re}\{\, s(t, \mathbf{c}) \, \exp\, (j2\pi f_c t)\}, \tag{3.1}$$

where the vector $\mathbf{c} \triangleq [\dots, c_{-1}, c_0, c_1, \dots]$, which is generally infinite-dimensional, represents the sequence of channel symbols.

At first glance, expression (3.1), which always holds for SC signals, may look inapplicable to MC signals, since, in principle, the generation of such signals requires the availability of a bank of oscillators. However, as illustrated in Section 3.7, model (3.1) also holds for OFDM signals, since then the baseband section of the transmitter generates the multiple subcarriers via a *discrete Fourier transform* (DFT) algorithm.

The structure of the data signal $s(t, \mathbf{c})$ can easily be understood if we model its generation using a *finite-state sequential machine* (FSSM), whose inner state, for the modulation formats analyzed in what follows, is always defined by a finite set of discrete parameters. However, as shown in the following pages, it can be always represented by a single integer parameter; for this reason, the modulator state Δ_n at the beginning of the nth signaling interval can take one of N_s possible values, belonging to the alphabet $A_{st} = \{0, 1, \dots, N_s - 1\}$. At the instant $t_n = nT_s$ the modulator, on the basis of the *present* channel symbol c_n and the *present* state Δ_n, generates the signal $s(\Delta_n, c_n; t - nT_s)$, where $s(\Delta_n, c_n; t)$ belongs to a signal alphabet $A_s = \{s_k(t), k = 0, 1, \dots, N_w - 1\}$, consisting of $N_w \triangleq M \cdot N_s$ distinct functions, all having *finite energy*. Generally speaking, each of these functions has an arbitrary duration and takes on complex values.

The baseband signal $s(t, \mathbf{c})$ is given by the superposition of the waveforms $\{s(\Delta_n, c_n; t - nT_s)\}$, generated during consecutive symbol intervals as:

$$s(t, \mathbf{c}) = \sum_{n=-\infty}^{+\infty} s(\Delta_n, c_n; t - nT_s), \tag{3.2}$$

which expresses the *output equation* of the FSSM. Note that complete knowledge of the modulator behavior also requires knowledge of the *state equation*, that is, of the mathematical law:

$$\Delta_{n+1} = f(\Delta_n, c_n), \tag{3.3}$$

expressing the *next state* Δ_{n+1} as a function of c_n and Δ_n. A more immediate representation of the output and state equations is given by an oriented graph known as the *state diagram* of the FSSM. In this diagram each possible state is symbolically represented by a node and each possible transition from one state to another is indicated by an oriented branch connecting to the nodes associated with the pair of states. Moreover, each branch is labeled by the value of d_n producing the corresponding transition and by the output signal generated by the modulator for this event. Unfortunately, the state diagram does not readily lend itself to representing the FSSM *time evolution* resulting from the application of a data sequence $\{d_n\}$ of arbitrary duration. This result can be achieved, however, if we describe a FSSM behavior through its *trellis diagram*.[1] This diagram shows, at each instant $\{t_n = nT_s\}$, the set of possible values that the modulator present state can take. At the instant $t_n = nT_s$, M oriented branches emanate from each possible present state Δ_n. These represent the transitions to M (usually distinct) next states $\{\Delta_{n+1}\}$. As in the state diagram, a branch emanating from Δ_n, and associated with a specific value of d_n, is conventionally labeled by the couple $d_n/s(\Delta_n, c_n; t)$.

---

[1] Examples of state and trellis diagrams can be found in Section 3.6. Note also that for linear modulations the state diagram is trivial as the modulation has no memory and all states are reachable at any of the time instants $\{t_n = nT_s\}$.

Let us now assume that the modulator has a *single* state ($N_s = 1$) and that, consequently, it can select one of $N_w = M$ possible waveforms at the beginning of each symbol interval. In this case the signal $s(\Delta_n, c_n; t)$ depends on the present symbol c_n only, so that (3.2) simplifies to:

$$s(t, \mathbf{c}) = \sum_{n=-\infty}^{+\infty} s(c_n; t - nT_s). \tag{3.4}$$

In addition, if all the signals of the set A_s have time support within the interval $[0, T_s)$, the modulator is *memoryless* [32], since, for any n, the signal generated in the nth symbol interval, $[nT_s, (n + 1)T_s)$, depends only on the present symbol c_n.

It should also be noted that equations (3.1)–(3.4) describe the behavior of a digital modulator in *deterministic* terms. In practice, as already stated above, the data sequence $\{d_n\}$ is random, so that the state sequence $\{\Delta_n\}$ is a *discrete-time random process* belonging, if the data $\{d_n\}$ are statistically *independent*, to the class of so-called *Markov chains* [55, 279]; moreover, the signals $s_{RF}(t, \mathbf{c})$ (3.1) and $s(t, \mathbf{c})$ (3.2) (and (3.4)) are *time-continuous stochastic processes*.

When implementing a digital modulator whose behavior can be described as above, it is useful to note that the RF transmitted signal $s_{RF}(t, \mathbf{c})$ (3.1) can be rewritten as:

$$s_{RF}(t, \mathbf{c}) = s_I(t, \mathbf{c}) \cos(2\pi f_c t) - s_Q(t, \mathbf{c}) \sin(2\pi f_c t), \tag{3.5}$$

where its *in-phase component*:

$$s_I(t, \mathbf{c}) \triangleq \mathrm{Re}\,\{s(t, \mathbf{c})\} \tag{3.6}$$

and its *quadrature component*:

$$s_Q(t, \mathbf{c}) \triangleq \mathrm{Im}\,\{s(t, \mathbf{c})\} \tag{3.7}$$

are *baseband* signals (here $\mathrm{Re}\,\{x\}$ and $\mathrm{Im}\,\{x\}$ denote the real part and the imaginary part, respectively, of the complex quantity x). Equation (3.5) suggests that a practical way to generate $s_{RF}(t, \mathbf{c})$ is to up-convert $s_I(t, \mathbf{c})$ of (3.6) and $s_Q(t, \mathbf{c})$ of (3.7) separately and then to additively combine them. This is known as the *in-phase quadrature method of signal generation*.

In the following sections the structure of the complex envelope $s(t, \mathbf{c})$ (3.2) for various modulation formats will be described in detail. Before dealing with this, however, it is important to note that, in communication system design, the selection of a digital modulation is usually the result of a tradeoff among different requirements. The main factors influencing this are: (a) the need for good *spectral efficiency*, that is, minimal bandwidth occupancy for a given information transmission speed; (b) the requirement for good *energy efficiency*, that is, acceptably low energy consumption to achieve a given performance; and (c) reasonable *implementation complexity*, specially in mobile applications. Some additional factors are considered in specific applications. One of these, for example, is the need to use, because of the limited energy resources available for transmission, highly efficient (and, consequently, *nonlinear*) power amplifiers. In this case the fluctuations of the *envelope*:

$$s_{env}(t, \mathbf{c}) \triangleq |s(t, \mathbf{c})| = \sqrt{s_I^2(t, \mathbf{c}) + s_Q^2(t, \mathbf{c})} \tag{3.8}$$

of the modulated signal $s_{RF}(t, \mathbf{c})$ (3.1) (or, equivalently, (3.5)) become relevant. In fact, a *constant envelope* modulation (or, at least, a modulation with small fluctuations in its envelope) represents an important factor in the above selection process.

To assess unambiguously the needs from each modulation format in terms of complexity, energy expense and spectral occupancy, it is important to define some significant parameters, whose value must be estimated in each case.

As far as the *spectral occupancy* is concerned, the problem of assessing the bandwidth B of $s_{RF}(t, \mathbf{c})$ (3.1) is discussed in Section 3.4. Obviously, the bandwidth does not represent an absolute measure of the spectral efficiency of a system, since it must be compared with the transmission speed of the information bits, that is, with the *bit rate* R_b. Moreover, because of the pulsed nature of digital transmission, bandwidth is, strictly speaking, infinite. It is, then, necessary to establish suitable measures of required bandwidth in terms of the percentage of the symbol or bit energy within a given frequency interval.

To suitably define bandwidth, first note that the cardinality M of the alphabet A_d is usually an integer power of 2 (i.e., $M = 2^m$), so that the data $\{d_n\}$ input to the modulator must be put in one-to-one correspondence with the set of all the possible vectors consisting of m binary digits. Then, the *bit interval* T_b becomes[2]:

$$T_b \triangleq \frac{T_s}{m} = \frac{T_s}{\log M} \tag{3.9}$$

and, consequently, the *bit rate* R_b is given by:

$$R_b \triangleq \frac{1}{T_b}. \tag{3.10}$$

Since $R_s \triangleq 1/T_s$, from (3.9) and (3.10) it is easy to see that:

$$R_b = mR_s = R_s \log M. \tag{3.11}$$

In assessing the *energy efficiency* of a digital communication technique, a significant role is played by the parameter E_b, representing the *average energy consumption required for the transmission of a single information bit*. This parameter can be easily related to the average energy E_s spent by a digital modulator in a symbol interval to generate $s_{RF}(t, \mathbf{c})$. Since in each symbol interval $m = \log M$ bits are transmitted, we have that[3]:

$$E_b = \frac{E_s}{m} = \frac{E_s}{\log M}. \tag{3.12}$$

Using these quantities, we will establish appropriate bandwidth measures in Section 3.4.

3.3 Representation of Digital Modulated Waveforms on an Orthonormal Basis

The complexity of generating the complex envelope (3.2), or equivalently, the corresponding in-phase and quadrature components (see (3.6) and (3.7)), depends on the signal alphabet $A_s = \{s_k(t), k = 0, 1, \ldots, N_w - 1\}$, to which $s(\Delta_n, c_n; t)$ belongs for any n. To understand the structure of this alphabet and, possibly, to simplify the algorithm for its generation, it is useful to represent its elements in terms of an *orthonormal basis*. This basis, denoted by $B_s = \{\phi_l(t), l = 0, 1, \ldots, N - 1\}$ in what follows, can be generated by resorting to the so-called *Gram–Schmidt orthonormalization procedure* (see Section D.1.2) of the signal set A_s, that we assume to generate a subspace S of dimension N. Given B_s, the signals $s_k(t)$, with $k = 0, 1, \ldots, N_w - 1$, can be represented by the expansion (see (D.3)):

$$s_k(t) = \sum_{l=0}^{N-1} s_{k,l} \phi_l(t), \tag{3.13}$$

---

[2] In this and all subsequent chapters logarithms are to base 2 unless explicitly stated otherwise.
[3] As shown in the following pages, this equality holds for PAM and CPM signals, but does not apply directly to OFDM signals.

where, for the properties of orthonormal bases, the complex coefficient $s_{k,l}$ is given by the inner product of $s_k(t)$ and $\phi_l(t)$ (see (D.9)), that is:

$$s_{k,l} = (s_k, \phi_l) = \int_{t_i}^{t_f} s_k(t)\phi_l^*(t)\,dt, \tag{3.14}$$

with $k = 0, 1, \ldots, N_w - 1$, where the interval (t_i, t_f) is the time support of the signals belonging to B_s. Then $s_k(t)$ can be represented by its *image* $\mathbf{s}_k$, that is, the vector:

$$\mathbf{s}_k \triangleq [s_{k,0}, s_{k,1}, \ldots, s_{k,N-1}]^T, \tag{3.15}$$

which describes a point in the so-called *signal space*, in this case, a complex set of signals in the space $\mathbb{C}^N$. We note that the complex coefficients $s_{k,l}$, for $k = 0, 1, \ldots, N - 1$, can be replaced by the ordered pair $(s_{k,l}^R, s_{k,l}^I)$, where $s_{k,l}^R \triangleq \text{Re}\{s_{k,l}\}$ and $s_{k,l}^I \triangleq \text{Im}\{s_{k,l}\}$, and then $s_k(t)$ can be represented by a vector belonging to $\mathbb{R}^{2N}$. The set of points $\{\mathbf{s}_i, i = 0, 1, \ldots, N_w - 1\}$ is known as the *signal constellation* generated by the digital modulator.

An application of these concepts is given in the following example.

Example 3.3.1 Let us consider a *memoryless* digital modulator for which the kth waveform of the M-ary alphabet A_s is:

$$s_k(t) = \frac{1}{\sqrt{T_s}} \exp\left(j\frac{2\pi k}{M}\right) [u(t) - u(t - T_s)] \tag{3.16}$$

for $k = 0, 1, \ldots, M - 1$, where:

$$u(t) = \begin{cases} 1 & \text{for } t \geq 0 \\ 0 & \text{for } t < 0 \end{cases} \tag{3.17}$$

is the *unit step function*. The shape of the signals $\{s_k(t)\}$ (3.16) is illustrated in Figure 3.1 for the case of a quaternary alphabet ($M = 4$).

Note that, whatever the value of M, the signal $s_k(t)$ (3.16) can be represented as:

$$s_k(t) = s_k\phi(t), \tag{3.18}$$

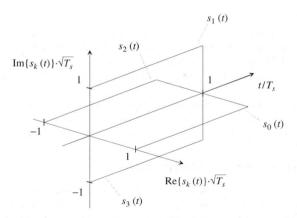

Figure 3.1 Graphical representation of the complex signals $\{s_k(t)\}$ (3.16) in the case of a quaternary alphabet ($M = 4$).

where

$$\phi(t) \triangleq \frac{1}{\sqrt{T_s}}[u(t) - u(t - T_s)] = \begin{cases} 1/\sqrt{T_s} & \text{for } 0 \leq t < T_s \\ 0 & \text{elsewhere} \end{cases} \tag{3.19}$$

is a unit energy signal and

$$s_k \triangleq \exp(j2\pi k/M). \tag{3.20}$$

Equation (3.18) shows that $\phi(t)$ represents a basis, consisting of a single element, for the alphabet A_s, and that the image of $s_k(t)$ (3.16), with $k = 0, 1, \dots, M - 1$, is the complex number s_k (3.20), or equivalently, the real two-dimensional vector $\mathbf{s}_k \triangleq [\cos(2\pi k/M), \sin(2\pi k/M)]$. Therefore, the signal constellation belongs to the space $\mathbb{C}$ and, as can be verified through its representation in the complex plane, consists of M points placed along a circle having unit radius and center at the origin of the reference plane.

The RF signal associated with the complex envelope $s_k(t)$ (3.16) is (see (3.1)):

$$s_{RF,k}(t) = \text{Re}\{s_k(t) \exp(j2\pi f_c t)\} = \phi(t) \cos\left(2\pi f_c t + \frac{2\pi k}{M}\right), \tag{3.21}$$

with $k = 0, 1, \dots, M - 1$, and can be generated by a *digital phase modulator*, that is, by a device introducing a *phase shift* that is a multiple of $2\pi/M$ radians on a cosinusoidal oscillation at frequency f_c. For this reason, the digital modulation characterized by the signals of (3.21) is known as[4] *M-ary phase shift keying* (*M*-PSK) [280]. We also note that, in this case, the modulator generates, starting at $t = nT_s$, for any n, one of the cosinusoidal signals $\{s_{RF,k}(t - nT_s), k = 0, 1, \dots, M - 1\}$ with no consideration of the signal generated in the previous symbol interval. Therefore, generally speaking, the modulated signal shows a *sharp phase change*, evidenced by the presence of a discontinuity of the first type in its shape at the beginning of each symbol interval.

It is easy to show that, if $f_c \gg R_s$, an orthonormal basis for the alphabet $A_{s,RF} \triangleq \{s_{RF,k}(t), k = 0, 1, \dots, M - 1\}$ is given by the pair of functions:

$$\phi_0(t) \triangleq \sqrt{2}\phi(t)\cos(2\pi f_c t), \tag{3.22}$$

$$\phi_1(t) \triangleq -\sqrt{2}\phi(t)\sin(2\pi f_c t), \tag{3.23}$$

and that the signal constellation associated with the signal alphabet $A_{s,RF}$ belongs to the two-dimensional space $\mathbb{R}^2$ and coincides with that associated with the alphabet A_s.

□

It is not difficult to understand that, if $N < N_w$, it is useful to generate the signals of A_s according to (3.13) and that, if this occurs, the smaller the number of functions in B_s (i.e., the dimension N of the subspace generated by the signals forming A_s), the more limited the modulator complexity. In fact, if N is small, generating $s(t, \mathbf{c})$ (3.2) requires the availability of only a small number of signals in B_s.

The representation of A_s by means of an orthonormal basis is a very useful way to analyze modulator complexity. It allows us to determine the minimum bandwidth occupancy for a modulated signal (see Section 3.4), and to analyze the detection process of the signal (see Section 4.4).

3.4 Bandwidth of Digital Modulations

The signals generated by the digital modulators described in this chapter are *wide-sense cyclostationary* (WSC) random processes with a period T_{cs} (see Appendix B), which is always a multiple of the symbol interval T_s. Therefore, given the modulated signal $s_{RF}(t, \mathbf{c})$ (3.1), characterized by a random

---

[4] Further details on this modulation format can be found in Section 3.5.2.

symbol vector $\mathbf{c}$ (whose elements belong to an M-ary constellation), a carrier frequency f_c and the complex envelope $s(t, \mathbf{c})$, its *average power spectral density* $S_{RF}(f)$ is given by:

$$S_{RF}(f) = \frac{1}{4}[S_s(f + f_c) + S_s(f - f_c)], \tag{3.24}$$

where $S_s(f)$ is the average *power spectral density* (PSD) of the complex envelope $s(t, \mathbf{c})$. The latter function is the FCT of the *average autocorrelation function* $R_s(\tau)$, that is:

$$S_s(f) = \int_{-\infty}^{+\infty} R_s(\tau) \exp(-j2\pi f \tau) \, d\tau, \tag{3.25}$$

with

$$R_s(\tau) \triangleq \frac{1}{T_{cs}} \int_0^{T_{cs}} R_s(t, \tau) \, dt \tag{3.26}$$

and

$$R_s(t, \tau) \triangleq E\{s(t + \tau, \mathbf{c}) \, s^*(t, \mathbf{c})\}. \tag{3.27}$$

Note that the only random quantities in (3.1) are the channel symbols $\{c_n\}$, so that the statistical average (denoted by the operator $E\{\cdot\}$) required in the computation of $R_s(t, \tau)$ according to (3.27) involves only these.

The average power P_{RF} of the RF signal $s_{RF}(t, \mathbf{c})$ (3.1) is given by:

$$P_{RF} = \int_{-\infty}^{+\infty} S_{RF}(f) \, df, \tag{3.28}$$

so that on substituting (3.24) into (3.28) it is found that:

$$P_{RF} = \frac{P_s}{2}, \tag{3.29}$$

where

$$P_s = \int_{-\infty}^{+\infty} S_s(f) \, df \tag{3.30}$$

is the average power of the complex envelope $s(t, \mathbf{c})$. The average energy E_s, spent by the transmitter in a symbol interval, is given by (see (3.29) and (3.30)):

$$E_s \triangleq P_{RF} \, T_s = \frac{P_s T_s}{2} = \frac{T_s}{2} \int_{-\infty}^{+\infty} S_s(f) \, df \tag{3.31}$$

and, consequently, the average energy expended per bit E_b, within a bit interval T_b, is:

$$E_b = \frac{E_s}{\log M} = \frac{T_s}{2 \log M} \int_{-\infty}^{+\infty} S_s(f) \, df, \tag{3.32}$$

if (3.9) holds, that is, the constellation of channel symbols is M-ary and $\log M$ information bits are transmitted in a symbol interval.[5]

Note that $S_s(f)$ (3.25) is usually expressed in Watt per Hertz, and has the dimensions of energy. Thus, when a representation of this function is needed, it is often more useful to consider the *normalized power spectral density*[6]:

$$S_{s,n}(f) \triangleq \frac{S_s(f)}{2E_s} = \frac{S_s(f)}{P_s T_s}, \tag{3.33}$$

which is *dimensionless*.

---

[5] As mentioned in the previous section, this assumption does not hold for an OFDM signal, because of the presence of a *cyclic prefix*. In fact, in this case, T_s is related to T_b not by (3.9), but by (3.276).

[6] Note that the quantity $2E_s$ in (3.33) represents the average energy consumption per symbol interval in generating $s(t, \mathbf{c})$.

In any practical application, the power spectrum of a digital signal is never strictly bandlimited, but the majority of the transmitted power is almost always contained in a limited frequency interval, centered on the carrier frequency f_c. This suggests the possibility of defining a *bandwidth* quantifying, *according to a specific rule*, the spectral occupancy of a signal whose RF spectral density $S_{RF}(f)$ is known. Various conventional definitions of bandwidth are available and the selection of a specific definition is dictated by system design considerations. Definitions that are commonly used are: (1) *null-to-null bandwidth*; (2) *fractional power-containment bandwidth*; (3) *bounded power-containment bandwidth*; (4) *equivalent noise bandwidth*.

The first definition requires measuring the width of the main spectral lobe of $S_{RF}(f)$; this can be simply evaluated when the first two nulls around the carrier frequency delimit the main lobe. Usually the majority of the transmitted power is contained within this interval.

The second definition refers to the frequency interval in which some fraction $(1 - \varepsilon)$ of the transmitted power is contained in the positive frequencies, where ε is a fixed and small positive quantity. In other words, the fractional power containment bandwidth $B_{1-\varepsilon}$ is implicitly defined by the equality:

$$2 \int_{f_c - B_{1-\varepsilon}/2}^{f_c + B_{1-\varepsilon}/2} S_{RF}(f) \, df \triangleq (1 - \varepsilon) \, P_{RF}, \tag{3.34}$$

where P_{RF} is given by (3.28). This definition is often adopted in the analysis of wireless communications systems, in which the overall available bandwidth is shared by multiple users transmitting in adjacent subbands. Then, if the bandwidth assigned to each user is $B_{1-\varepsilon}$, the fraction εP_{RF} of the power transmitted by a single user spills over adjacent channels, producing interference.

The bandwidth measured according to the third criterion (and henceforth denoted B_{BP}) represents the frequency interval outside of which the spectral density $S_{RF}(f)$ does not cross a reference threshold that is set X decibels below the value taken on by $S_{RF}(f)$ at the center of the bandwidth,[7] that is, at the frequency f_c. Typical values of the attenuation X are 35 dB and 50 dB, although larger values can be found in actual system specifications [281].

The bandwidth measured according to the last criterion, the equivalent noise bandwidth, provides an indication of the spread or width of a spectrum, but no information about its side lobes. It is defined as the parameter B_N satisfying the equality:

$$2B_N \, S_{RF}(f_c) \triangleq P_{RF}, \tag{3.35}$$

which states that, as shown in Figure 3.2, the overall area covered by two rectangular spectral shapes, each having a base and height of lengths B_N and $S_{RF}(f_c)$, respectively, centered at the frequencies $\pm f_c$, is equal to the area covered by the spectral density $S_{RF}(f)$ over the entire frequency axis, that is, the overall transmitted power P_{RF}. From (3.35) it can be seen that:

$$B_N \triangleq \frac{P_{RF}}{2 \, S_{RF}(f_c)}, \tag{3.36}$$

which can easily be reformulated with reference to the power spectrum of the complex envelope $s(t, \mathbf{c})$. In fact, if we note that $P_{RF} = P_s/2$ (see (3.29)) and that, if $f_c \gg R$, we have that (see (3.24)):

$$S_{RF}(f_c) = \frac{S_s(0)}{4}, \tag{3.37}$$

then equation (3.36) can be rewritten as:

$$B_N = \frac{P_s}{S_s(0)} \tag{3.38}$$

---

[7] This definition and the following one implicitly assume that $S_{RF}(f)$ takes on its maximum value at the carrier frequency f_c. If this is not the case, the reference frequency must be changed.

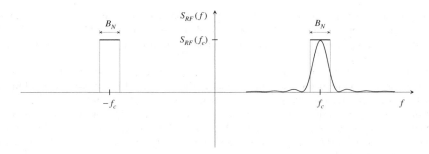

Figure 3.2 Geometrical interpretation of equality (3.35). The parameter B_N represents the equivalent noise bandwidth of a digital signal characterized by the PSD $S_{RF}(f)$, plotted on a linear scale.

or, taking (3.33) into consideration, as:

$$B_N = \frac{1}{T_s S_{s,n}(0)} = \frac{R_s}{S_{s,n}(0)}, \tag{3.39}$$

which lends itself to an immediate geometrical interpretation, similar to that illustrated in Figure 3.2, referring to $S_{RF}(f)$.

Other definitions and measures of the bandwidth of digital signals can be found in [281].

Given the spectral occupancy of a digital communication system, its *spectral efficiency* η_B can be evaluated. This parameter is defined as the *ratio between the transmitted bit rate and the RF bandwidth occupancy*:

$$\eta_B \triangleq \frac{R_b}{B}. \tag{3.40}$$

Although this ratio is, as a matter of fact, dimensionless, it is commonly expressed in bits per second per Hertz. In practice, the value of η_B for a specific modulation format depends on the definition adopted for B. In any comparison of digital modulations, spectral efficiencies should be always computed using the same definition of bandwidth. It is also important to note that the spectral efficiency achieved using various traditional transmission techniques in modern SISO and SIMO communication systems ranges from about 1–5 bit/sec/Hz to about 10–12 bit/sec/Hz [23]. The smaller values are typical, for instance, of cellular mobile systems, whereas the larger ones are usually found in microwave point-to-point wireless links. Substantially larger spectral efficiencies can of course be attained by resorting to MIMO communication techniques.[8]

The spectral occupancy of a digital modulation format can be related directly to the dimensionality of the set of functions generated by the corresponding modulator. This result is based on the theory developed by D. Slepian and H. O. Pollak that proved, in 1960, that a set of *real low-pass signals*, rigorously bandlimited to B Hertz can be generated using a basis of functions $\{\psi_i(t), i = 0, 1, \ldots\}$ which are *orthogonal* over both the time interval $(-\infty, +\infty)$ and $I_0 \triangleq (-T_0/2, T_0/2)$, where T_0 is some fixed duration [282]. In the basis, however, only the first N functions, with:

$$N \leq 2BT_0, \tag{3.41}$$

have their energy concentrated within the interval I_0, whereas the remaining functions take on significant values outside of this interval. This result can be exploited to derive an estimate of the *minimum* spectral efficiency of a given modulation format. In fact, let us focus on the transmission

---

[8] Spectral efficiencies of 20–40 bit/s/Hz have been attained in the prototypal BLAST system noted earlier.

of a binary data stream at a bit rate R_b, via a digital modulator that can generate, in each symbol interval, a signal belonging to the M-ary set of real functions $A_s = \{s_i(t), i = 0, 1, \ldots, M-1\}$, each having a bandwidth B. These functions generate a subspace S, also represented by the basis $B_s = \{\phi_i(t), i = 0, 1, \ldots, D-1\}$.

Intuitively, if we want the transmission accomplished by the digital modulator over P consecutive symbol intervals not to produce mutual interference at the receiver, it is necessary that all the signals generated by the time translations of B_s with a delay equal to iT_s seconds ($i = 0, 1, \ldots, P-1$) be *mutually orthogonal*. From a mathematical viewpoint this requires the generation of a set of PD orthogonal functions which are bandlimited to B Hertz and whose energy is primarily concentrated in an interval of duration PT_s seconds. Then substituting for the parameters N and T_0 with PD and PT_s, respectively, in (3.41) yields:

$$D \leq 2BT_s, \tag{3.42}$$

from which it can be inferred that:

$$B \geq \frac{D}{2T_s}. \tag{3.43}$$

If we now define the parameter:

$$D_b = \frac{D}{\log M}, \tag{3.44}$$

representing the number of dimensions employed for a transmitted bit, (3.43) can be rewritten as:

$$B \geq \frac{D_b R_b}{2}. \tag{3.45}$$

This leads to the estimate of the minimum bandwidth:

$$B_{\min} = \frac{D_b R_b}{2} \tag{3.46}$$

required by a digital signal employing D_b dimensions per bit and occurring at a rate R_b. This result, even if not exact, is significant, as it shows the importance of using modulation formats characterized by small values of D_b when limited bandwidth occupancy is required. Finally, we note that equation (3.45) can be used for both baseband and passband transmissions, but *real signals* must be always considered in the evaluation[9] of D_b.

3.5 Passband PAM

3.5.1 Signal Model

A digital passband PAM modulator is characterized by the following properties: (a) it has a single inner state, so that $s(\Delta_n, c_n; t - nT_s) = s(c_n; t - nT_s)$ (see (3.2)) and $N_w = M$; (b) all the signals belonging to its alphabet $A_s = \{s_k(t), k = 0, 1, \ldots, M-1\}$ are *proportional* to the same pulse shape $p(t)$; (c) the baseband signal $s(c_n; t - nT_s)$ depends *linearly* on the channel symbol c_n. Analytically this results in:

$$s(\Delta_n, c_n; t - nT_s) = K_c\, c_n\, p(t - nT_s), \tag{3.47}$$

where c_n belongs to some alphabet A_c, consisting of M distinct complex numbers and called the *constellation*; in addition, K_c is a real positive parameter depending on E_b, and on the alphabet of

---

[9] If complex signals are used to estimate the required dimensionality, then the number of estimated dimensions must be multiplied by 2.

symbols and its cardinality. In the following we will always assume that the signal $p(t)$, called the *modulator impulse response*, is *real* and has unit energy.

Now consider the structure of $s(t, \mathbf{c})$ (3.2). Substituting (3.47) into (3.2) gives:

$$s(t, \mathbf{c}) = K_c \sum_{n=-\infty}^{+\infty} c_n \, p(t - nT_s). \tag{3.48}$$

Then, if a_n and b_n respectively denote the real and imaginary parts of c_n, so that:

$$c_n = a_n + jb_n, \tag{3.49}$$

the in-phase $s_I(t, \mathbf{c})$ (3.6) and quadrature $s_Q(t, \mathbf{c})$ (3.7) components of the transmitted signal can respectively be expressed as:

$$s_I(t, \mathbf{c}) = K_c \sum_{n=-\infty}^{+\infty} a_n \, p(t - nT_s) \tag{3.50}$$

and

$$s_Q(t, \mathbf{c}) = K_c \sum_{n=-\infty}^{+\infty} b_n \, p(t - nT_s). \tag{3.51}$$

Moreover, if ρ_n and φ_n represent the amplitude and phase of c_n, that is:

$$c_n = \rho_n \exp(j\varphi_n), \tag{3.52}$$

the RF signal $s_{RF}(t, \mathbf{c})$ (3.1) can be expressed as (see (3.48)):

$$s_{RF}(t, \mathbf{c}) = K_c \sum_{n=-\infty}^{+\infty} p(t - nT_s) \, \rho_n \cos(2\pi f_c t + \varphi_n). \tag{3.53}$$

The latter result shows that in a passband PAM signal, for any n, the pulse $p(t - nT_s)$ is multiplied by a carrier, whose phase and amplitude are determined by the amplitude and phase, respectively, of the symbol c_n.

Finally, we note that, generally speaking, the envelope (see (3.8) and (3.50)–(3.51)):

$$s_{env}(t, \mathbf{c}) = K_c \sqrt{\left[\sum_{n=-\infty}^{+\infty} a_n \, p\left(t - nT_s\right) \right]^2 + \left[\sum_{n=-\infty}^{+\infty} b_n \, p\left(t - nT_s\right) \right]^2} \tag{3.54}$$

of the signal $s_{RF}(t, \mathbf{c})$ is not constant.

In practice, in order to limit the envelope fluctuations it can be very useful to insert a delay of $T_s/2$ seconds into the quadrature component $s_Q(t, \mathbf{c})$ (3.51), resulting in:

$$s_Q(t, \mathbf{c}) = K_c \sum_{n=-\infty}^{+\infty} b_n \, p(t - nT_s - T_s/2). \tag{3.55}$$

This choice ensures that the peaks and nulls associated with the pulses $\{p(t - nT_s)\}$ of $s_I(t, \mathbf{c})$ (3.50) do not occur at the same time instants as those of the pulses $\{p(t - T_s/2 - nT_s)\}$ of $s_Q(t, \mathbf{c})$ (3.55), with a consequent reduction in the fluctuations of $s_{env}(t, \mathbf{c})$ (3.8). A digital modulation characterized by the in-phase component (3.50) and by the quadrature component (3.55) is called *offset* PAM or *staggered* PAM.

3.5.2 *Constellation Selection*

Let us consider now some possible choices for the constellation A_c, under the assumption that its cardinality takes on *even* values. The most common choices belong to three distinct classes. The first class consists of all the constellations whose points are regularly placed along the circumference of a circle, usually having unit radius. In the *M*-ary case, the constellation is the set $A_c = \{s_l \triangleq \exp(j2\pi l/M), l = 0, 1, \ldots, M-1\}$ (see Figure 3.3(a)) and the resulting modulation is termed *M*-PSK, because each channel symbol produces only a phase variation of the pulse associated with it, leaving its amplitude unchanged.[10] Note that if $p(t)$ is a rectangular pulse lasting T_s seconds, that is:

$$p(t) = \frac{1}{\sqrt{T_s}}[u(t) - u(t - T_s)], \tag{3.56}$$

and $\rho_n = 1$ for any n, the RF transmitted signal becomes:

$$s_{RF}(t, \mathbf{c}) = \frac{K_c}{\sqrt{T_s}} \sum_{n=-\infty}^{+\infty} \cos(2\pi f_c t + \varphi_n)\,[u(t - nT_s) - u(t - (n+1)T_s)], \tag{3.57}$$

which represents the modulated signal as a series of cosinusoidal pulses, each having duration T_s seconds and amplitude $K_c/\sqrt{T_s}$. Therefore, in this case, the envelope $s_{env}(t, \mathbf{c})$ (3.54) is *rigorously constant*.

The second class consists of all the constellations in which the points are aligned, equally spaced and placed in symmetric couples with respect to the origin. In this case, if we assume that the distance between adjacent points is equal to 2, the constellation is $A_c = \{\pm 1, \pm 3, \ldots, \pm(M-1)\} = \{s_l = (M-1) - 2l, l = 0, 1, \ldots, M-1\}$ (see Figure 3.3(b)). The corresponding modulation is known as

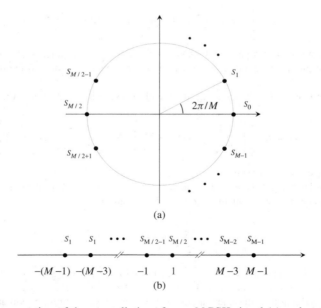

(a)

(b)

Figure 3.3   Representation of the constellations for an *M*-PSK signal (a) and an *M*-AM signal (b).

---

[10] In this class of signals *binary phase shift keying* (BPSK) and *quaternary phase shift keying* (QPSK) refer to the choices $M = 2$ and $M = 4$, respectively.

M-ary amplitude modulation (*M*-AM) or *M-ary amplitude shift keying* (*M*-ASK), since, as is seen from (3.53), each channel symbol produces an amplitude variation in the associated pulse, either leaving its phase unchanged or modifying it by π.

Finally, the third class consists of all those constellations in which the points are placed at the vertices of a lattice with square meshes and for each point there is another one placed symmetrically with respect to the origin. If the constellation consists of M points, generally speaking, the modulation is called *M-ary amplitude modulation-phase modulation* (*M*-AM-PM), since each channel symbol produces a variation in both amplitude and phase of the carrier associated with its pulse, as evidenced by (3.53). To this class of constellations belong two distinct subclasses, characterized by constellations of different shapes.

The first subclass is known as *M-ary quadrature amplitude modulation* (*M*-QAM), and is characterized by values of M equal to a power of 4. In this case, the symbols $\{a_n\}$ and $\{b_n\}$ belong to the same alphabet $\{\pm 1, \pm 3, \ldots, \pm(\sqrt{M} - 1)\}$, so that the constellation has a square shape, as illustrated in Figure 3.4(a) for $M = 4$, 16 and 64. It is not difficult to understand that an *M*-QAM modulation can be generated by superimposing two *independent* $\sqrt{M}$-ASK signals (one associated with the sequence $\{a_n\}$, the other with $\{b_n\}$) and characterized by quadrature oscillations. Note also that, unlike ASK and PSK, a QAM signal cannot be generated for any even value of M. This does not prevent, however, the construction of other rectangular constellations containing an even number of points which is not a power of 4. These constellations, which form the second subclass of interest, are called *cross-constellations*, for their shape,[11] as shown in Figure 3.4(b) for $M = 32$ and 128. With the help of this figure, the reader can easily verify, in any specific case, the validity of the *general rule* for $M \geq 32$, that an *M*-cross-constellation can be generated by placing, in a regular fashion, 4 groups of points, each consisting of $M/8$ elements, along the sides of an $M/2$-QAM constellation. The constellation structure inevitably influences the generation mechanism of the modulated signal. In fact, a cross signal, unlike QAM, cannot be represented as the combination of two independent ASK signals, since the choice of a_n is related to that of b_n, whatever the value of n. Therefore, despite the fact that both a_n and b_n can take on values belonging to the same alphabet $\{\pm 1, \pm 3, \ldots, \pm(\sqrt{M/2} + 1)\}$, not all possible pairs of these values can occur.

We note that the choices indicated above for the *M*-ary constellations of passband PAM signals are not optimal and exhaustive, since the constellation of PAM signals can be selected in different (and better) ways. A deeper discussion on the problem of the constellation selection in a multidimensional context is provided in Section 3.8.

When an *M*-ary constellation (with $M = 2^m$) is employed in a PAM transmission, a one-to-one correspondence (i.e., a *mapping*) rule associating each possible m-tuple of bits with a constellation point has to be established. The selected rule plays an important role, since it influences the average bit error performance, as will be discussed in Section 4.3. In most cases, a rule known as *Gray coding* is adopted; it can be employed[12] for any M and consists of assigning bits to the points of an *M*-ary constellation in such a way that nearest neighbors are labeled by blocks of m bits differing in a single bit only. This choice can be motivated by noting that a digital receiver, in case of a decision error, is likely to decide in favor of a constellation point which, among the possible choices, is the closest to the transmitted point in terms of Euclidean distance (see Section 4.3.3). When this occurs, a decision error on a channel symbol entails only a *single* bit error if Gray coding is adopted, so that the average bit error probability is actually lower than the average symbol error probability.

Another important issue related to the use of the constellations defined in this subsection is their property of *rotational invariance* – the fact that, because of their symmetry, they remain unchanged after a rotation of some specific angle values. For instance, an *M*-PSK constellation is invariant to

---

[11] Note that much of the published literature refers to both square and cross-constellations as QAM.
[12] Note that this can be done exactly only when $M = 2^q$, where q is an integer. Otherwise, *Gray coding* is only an approximation.

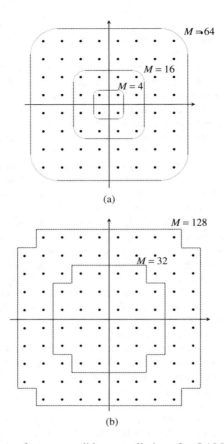

(a)

(b)

Figure 3.4   Representation of some possible constellations for QAM (a) and cross (b) signals.

rotations of any integer multiple of $2\pi/M$ radians, that is, it possesses an *M-fold rotational symmetry*. This geometric property of the constellation implies that a receiver can recognize the pattern of the transmitted signal points, but is unable to distinguish between the various symmetric phase orientations of the signal set, that is, it cannot autonomously solve the so-called *phase ambiguity* problem (for instance, an *M*-fold phase ambiguity is found in the *M*-PSK case). The following three different solutions are available to solve this problem:

(a) transmitting a constant *reference signal* of some kind along with the PAM signal,
(b) transmitting a specific *acquisition signal* and/or inserting a *synchronization sequence* (known to the receiver) in the data stream,
(c) employing the so-called *differential encoding* technique.

The latter solution offers the advantage that by properly encoding and decoding the signal points, proper bit detection can be accomplished regardless of rotational phase ambiguities. To show this, let us focus on the M-PSK case, for which the *encoding* algorithm can be formulated as[13]:

$$c_n = c_{n-1} \exp\left(j \frac{2\pi}{M} d_n \right), \tag{3.58}$$

which is equivalent to (see (3.52)):

$$\varphi_n - \varphi_{n-1} = \frac{2\pi}{M} d_n \tag{3.59}$$

since $\rho_n = \rho_{n-1} = 1$ for any n. Equation (3.59) shows that differential encoding maps the nth information symbol d_n into the phase change from the last transmitted symbol c_{n-1} to the following one c_n for any n. Then, at the receiver a decision $\hat{d}_n$ about d_n can be taken by computing the difference $(\hat{\varphi}_n - \hat{\varphi}_{n-1})$ between the phase estimates $\hat{\varphi}_n$ and $\hat{\varphi}_{n-1}$ of φ_n and φ_{n-1}, respectively, so that the effect of the above-mentioned phase ambiguity is canceled (in other words, the dependence of data decisions on the absolute phase of the carrier regenerated at the receiver is removed).

The differential encoding technique can be used with a variety of symmetric signal sets to resolve phase ambiguity, as shown in [283]. However, it has the following two drawbacks:

(a) The transmission of a block of $N + 1$ channel symbols $\{c_0, c_1, \ldots, c_N\}$ is required to send N information symbols $\{d_0, d_1, \ldots, d_{N-1}\}$, since the first symbol c_0 is required to establish a phase reference for the first datum d_0.
(b) A (low) performance penalty relative to uncoded performance has to be paid, since a single error in the sequence of phase estimates $\{\varphi_n\}$ typically entails two consecutive incorrect phase differences, that is, a couple of consecutive decision errors.

3.5.3 Data Block Transmission with Passband PAM Signals for Frequency-Domain Equalization

When passband PAM transmission is to be used with *frequency domain* (FD) equalization (see Section 6.2.2) in the receiver, the channel symbol stream, generated at a rate $f_s = 1/T_s$, is divided into nonoverlapping blocks, each of some length N. Let the vector $\mathbf{c}_N^{(l)} \triangleq [c_0^{(l)}, c_1^{(l)}, \ldots, c_{N-1}^{(l)}]^T$ denote the lth block. After serial-to-parallel conversion, a cyclic prefix $\mathbf{c}_p^{(l)} \triangleq [c_{N-N_{cp}}^{(l)}, c_{N-N_{cp}+1}^{(l)}, \ldots, c_{N-1}^{(l)}]^T$ of length N_{cp} is appended to $\mathbf{c}_N^{(l)}$ and this produces the cyclically extended block $\tilde{\mathbf{c}}_{N_T}^{(l)} \triangleq [(\mathbf{c}_p^{(l)})^T, (\mathbf{c}_N^{(l)})^T]^T = [c_{R_N[n]}^{(l)}, n = -N_{cp}, -N_{cp} + 1, \ldots, N - 1]^T$, where $R_N[\cdot]$ is the "modulo N operator" defined by:

$$R_N[n] = n - \left\lfloor \frac{n}{N} \right\rfloor N, \tag{3.60}$$

$\lfloor x \rfloor$ being the largest integer not exceeding the real variable x, and $N_T \triangleq N_{cp} + N$. The cyclically extended symbol sequence is input to the transmitter filter, which is characterized by an impulse response $p(t)$ having support $[0, L_p T_s]$. This produces the baseband transmitted signal:

$$s(t, \mathbf{c}) = \sum_{l=-\infty}^{+\infty} \sum_{n=-N_1}^{N-1} c_{R_N[n]}^{(l)} \, p(t - nT_s - lN_T T_s) \tag{3.61}$$

which is transmitted over the channel. Let us assume that: (a) the CIR lasts L_h symbol intervals; (b) the CIR is slowly time-varying and, in particular, undergoes negligible changes over each transmitted block

---

[13] In the technical literature, differentially encoded M-ary PSK is usually referred to as M-DPSK when noncoherent detection is employed.

(*quasi-static* channel), so that, during the transmission of $\tilde{\mathbf{c}}_{N_T}^{(l)}$, it can be denoted by the time-invariant impulse response $h^{(l)}(t)$; (c) the duration of the cyclic prefix is at least as long as the overall channel memory (i.e., $L_h + L_p \leq N_1$); (d) $p(t)$ is bandlimited to $B = 1/T_s$ Hertz. Under these conditions it can be proved [284] that the useful component of the baseband received signal $r(t)$ is given by:

$$z(t) = \frac{1}{\sqrt{NT_s}} \sum_{k=-N}^{N} P_k \, H_k^{(l)} \, C_k^{(l)} \exp\left(j \frac{2\pi k}{NT_s}(t - lN_T T_s) \right) \tag{3.62}$$

for $lN_T T_s \leq t < lN_T T_s + NT_s$, where $P_k \triangleq P(k/NT_s)/\sqrt{T_s}$, $H_k^{(l)} \triangleq H^{(l)}(k/NT_s)$, $P(f)$ and $H^{(l)}(f)$ are the FCTs of $p(t)$ and $h^{(l)}(t)$, respectively. In addition, $C_k^{(l)}$ is the kth component of the vector $\mathbf{C}_N^{(l)} \triangleq [C_0^{(l)}, C_1^{(l)}, \ldots, C_{N-1}^{(l)}]^T$, representing the frequency-domain symbols, resulting from the DFT of the data block $\mathbf{c}_N^{(l)}$, that is, $\mathbf{C}_N^{(l)} \triangleq \mathbf{Q}_N \, \mathbf{c}_N^{(l)}$, where $\mathbf{Q}_N = [q_{n,k}]$ is the N-point DFT matrix ($k, n = 0, 1, \ldots, N-1$) with $q_{n,k} = W_N^{kn}/\sqrt{N}$ and $W_N \triangleq \exp(-j2\pi/N)$. Thus, the useful portion of the received signal can be seen, for each data block, as the superposition of $2N + 1$ complex oscillations, the kth of which is characterized by the complex gain $P_k \, H_k^{(l)}$ and the frequency $f_k \triangleq k/(NT_s)$. Note that $C_k^{(l)} = C_{k+N}^{(l)}$ for any k, so that, in principle, a given frequency-domain symbol can be associated with two distinct oscillations.

3.5.4   Power Spectral Density of Linear Modulations

Let us now make the following assumptions on the complex envelope $s(t, \mathbf{c})$ (3.48) of a passband PAM signal:

1. The sequence $\{c_n\}$ is WSS with mean value $\eta_c \triangleq E\{c_n\}$ and autocorrelation function $R_c[k] \triangleq E\{c_{n+k} \, c_n^*\}$.
2. The channel symbols $\{c_n\}$ are *identically distributed*, that is, their probability function is independent of n.

It is not difficult to prove that the *mean value* $\eta_s(t)$ of $s(t, \mathbf{c})$ (3.48) is given by:

$$\eta_s(t) \triangleq E\{s(t, \mathbf{c})\} = \eta_c \, K_c \sum_{k=-\infty}^{+\infty} p(t - kT_s), \tag{3.63}$$

so that, in principle, $s(t, \mathbf{c})$ is not a WSS process unless $\eta_c = 0$ or:

$$\sum_{k=-\infty}^{+\infty} p(t - kT_s) \tag{3.64}$$

is a constant signal. However, it can be shown that, generally speaking, $s(t, \mathbf{c})$ is WSC with period $T_0 = T_s$ and that its average autocorrelation function and average PSD are given by:

$$R_s(\tau) = \frac{K_c^2}{T_s} \sum_{l=-\infty}^{+\infty} R_c[l] \int_{-\infty}^{+\infty} p(\alpha) \, p(\alpha + \tau - lT_s) \, d\alpha \tag{3.65}$$

and

$$S_s(f) = \frac{K_c^2}{T_s} \bar{S}_c(f) |P(f)|^2, \tag{3.66}$$

respectively, where

$$\bar{S}_c(f) \triangleq \sum_{l=-\infty}^{+\infty} R_c[l] \exp(-j2\pi l f T_s) \tag{3.67}$$

is the PSD of the symbol sequence $\{c_n\}$. These results show that the power spectrum of the transmitted signal depends on the spectral properties of the transmitter impulse response and on the correlation properties of the channel symbol sequence.

In addition, if we assume that the sequence $\{c_n\}$ consists of zero mean, iid random variables, its autocorrelation function becomes:

$$R_c[k] \triangleq \mathrm{E}\{c_{n+k}\, c_n^*\} = \sigma_c^2\, \delta[k], \tag{3.68}$$

where $\sigma_c^2 \triangleq \mathrm{E}\{|c_k|^2\}$ is the variance of channel symbols. Under these assumptions it is not difficult to show that the average power of $s(t, \mathbf{c})$ is:

$$P_s = \frac{\sigma_c^2 K_c^2}{T_s}, \tag{3.69}$$

and that the average energy per symbol interval is given by:

$$E_s = P_{RF} T_s = \frac{1}{2} P_s T_s = \frac{1}{2} \sigma_c^2 K_c^2, \tag{3.70}$$

so that:

$$K_c = \sqrt{\frac{2E_s}{\sigma_c^2}}. \tag{3.71}$$

If we assume the constellation points to be equally likely, we have $\sigma_c^2 = 1$, $\sigma_c^2 = (M^2 - 1)/3$ and $\sigma_c^2 = 2(M - 1)/3$ for M-PSK, M-ASK and M-QAM, respectively; then, it can be easily inferred from (3.71) that:

$$K_c = \sqrt{2E_s} \tag{3.72}$$

for M-PSK signaling:

$$K_c = \sqrt{\frac{6E_s}{M^2 - 1}} \tag{3.73}$$

for M-ASK signaling, and:

$$K_c = \sqrt{\frac{3E_s}{M - 1}} \tag{3.74}$$

for M-QAM signaling. As an example, let us apply these results to a specific modulation format.

Example 3.5.1 Consider a BPSK signal having:

$$p(t) \triangleq \frac{1}{\sqrt{T_s}} [u(t) - u(t - T_s)], \tag{3.75}$$

and whose symbols $\{c_n\}$ are zero mean, statistically independent, identically distributed and belong to the alphabet $\{\pm 1\}$. Then we have $K_c = \sqrt{2E_s}$ (see (3.72)), $\bar{S}_c(f) = 1$, and $P(f) = \sqrt{T_s}\,\mathrm{sinc}(fT_s)\exp(-j\pi fT_s)$, so that the PSD of the complex envelope $s(t, \mathbf{c})$ is (see (3.66)):

$$S_s(f) = 2E_s \,\mathrm{sinc}^2(fT_s) = 2E_s \frac{\sin^2(\pi fT_s)}{(\pi fT_s)^2}, \tag{3.76}$$

with the normalized version (see (3.33)):

$$S_{s,n}(f) = \mathrm{sinc}^2(fT_s) = \frac{\sin^2(\pi fT_s)}{(\pi fT_s)^2}. \tag{3.77}$$

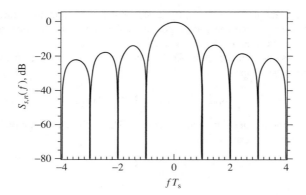

Figure 3.5 Representation of the normalized PSD $S_{s,n}(f)$ (3.77) on a log or decibel scale.

The function $S_{s,n}(f)$ is represented in Figure 3.5 on a log or decibel scale. This shows that the PSD contains a main lobe, occupying the frequency interval $(-1/T_s, 1/T_s)$ (so that the RF null-to-null bandwidth B_{NN} is equal to $2/T_s = 2R_s$) and an infinite number of side lobes decreasing at a rate of 6 dB per octave as $|f| \rightarrow +\infty$ due to the presence of $1/(fT_s)^2$ in (3.76). It can be shown that 90% of the energy or power in $s(t, \mathbf{c})$ is contained in the main lobe. In addition, it is not difficult to verify that the peaks of the first and the second side lobes are 13.5 and 17 dB, respectively, below the center of the main lobe.

The plot of Figure 3.5 can be employed to compute the *bounded power-containment bandwidth* B_{BP} of the given signal for a given threshold level X with respect to the central frequency. For instance, if $X = 15$ dB (see Figure 3.6) we have that $B_{BP} \cong 3.27\ R_s$. Similarly, if $X = 35$ dB and $X = 50$ dB are selected, it is found that $B_{BP} \cong 35.12\ R_s$ and $B_{BP} \cong 201.04\ R_s$, respectively [281].

$\square$

In the previous example, the PSD of the transmitted signal is not confined to a finite frequency interval because $p(t)$ has a limited duration, and consequently, unlimited bandwidth. If a rigorously

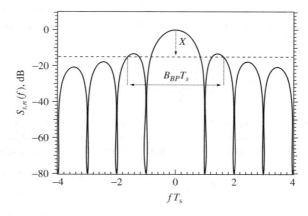

Figure 3.6 Representation of the normalized PSD $S_{s,n}(f)$ (3.77) and of the threshold level corresponding to an attenuation of $X = 15$ dB with respect to the central frequency, necessary to determine the bandwidth B_{BP} of the BPSK signal of Example 3.5.1.

bandlimited $S_s(f)$ is desired, the same property has to be shared by $p(t)$ or, more appropriately, $P(f)$. A possible choice, in the class of bandlimited signals, is represented by the signals having an FCT with the shape of a *raised cosine with roll-off factor* α. In this case, the pulse $p(t)$ is given by:

$$p(t) \triangleq \frac{1}{\sqrt{E_g}} g(t), \qquad (3.78)$$

where

$$g(t) \triangleq \operatorname{sinc}\left(\frac{t}{T_s}\right) \frac{\cos(\pi\alpha t/T_s)}{1-(2\alpha t/T_s)^2}, \qquad (3.79)$$

and

$$E_g = T_s\left(1-\frac{\alpha}{4}\right) \qquad (3.80)$$

is the energy of $g(t)$, and α, known as the *roll-off factor*, is a real parameter belonging to the interval $[0,1]$. The spectrum of $g(t)$ (3.79) is then given by:

$$G(f) \triangleq \operatorname{FCT}[g(t)]$$

$$= \begin{cases} T_s & \text{for } 0 \le |f| \le f_{1-\alpha} \\ \frac{T_s}{2}\left\{1+\cos\left[\frac{\pi T_s}{\alpha}\left(|f|-f_{1-\alpha}\right)\right]\right\} & \text{for } f_{1-\alpha} < |f| \le f_{1+\alpha} \\ 0 & \text{for } |f| > f_{1+\alpha} \end{cases} \qquad (3.81)$$

where

$$f_{1-\alpha} \triangleq \frac{1-\alpha}{2T_s} \qquad (3.82)$$

and

$$f_{1+\alpha} \triangleq \frac{1+\alpha}{2T_s}. \qquad (3.83)$$

If the pulse of (3.78) is selected, the *bandwidth* of the complex envelope $s(t,\mathbf{c})$ of the transmitted signal can be expressed as:

$$B = \varepsilon\, B_N, \qquad (3.84)$$

where

$$B_N \triangleq \frac{1}{2T_s} \qquad (3.85)$$

is the so-called *Nyquist bandwidth* and $\varepsilon \triangleq 1+\alpha$ is the so-called *excess bandwidth factor*. Note that the minimum bandwidth occupancy (B_N in baseband signaling, $2B_N$ at RF) is achieved for $\alpha=0$. In this case (3.79) reduces to:

$$g(t) = \operatorname{sinc}\left(\frac{t}{T_s}\right) \qquad (3.86)$$

and the spectrum (3.81) has a rectangular shape, since:

$$G(f) = T_s\,[\mathrm{u}(f-1/(2T_s))-\mathrm{u}(f+1/(2T_s))]. \qquad (3.87)$$

The signal $g(t)$ (3.79) and its corresponding FCT $G(f)$ (3.81) are illustrated in Figures 3.7 and 3.8, respectively, for $\alpha=0, 0.5$ and 1. Note that $g(t)$ takes on null values, because of the factor $\operatorname{sinc}(t/T_s)$ in (3.79), at all the instants which are multiples of T_s, excluding the origin, where, independently of α, $g(t)=1$. In other words, we have:

$$g(t)|_{t=kT_s} = \delta[k] \qquad (3.88)$$

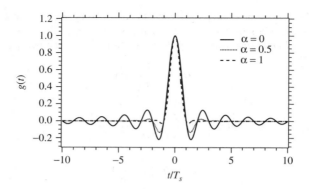

Figure 3.7   Representation of $g(t)$ (3.79) for $\alpha = 0$, 0.5 and 1.

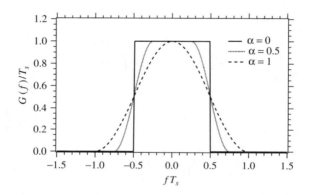

Figure 3.8   Representation of the pulse spectrum $G(f)$ (3.81) for $\alpha = 0$, 0.5 and 1.

for any integer k. The signal $g(t)$, being rigorously bandlimited for any α, always has *infinite duration*. Therefore, in practical applications, *causal* and *time-limited* approximations of this signal are used. These are generated by truncating the signal pulse, that is, limiting it to an interval which is symmetric with respect to the origin, and translating it to make it causal. When this is accomplished, it is important to remember that, as can easily be seen from Figure 3.7, the tails of $g(t)$ become more and more pronounced as the roll-off factor approaches 0. For this reason, if, independently of α, a given fraction of the overall energy of $p(t)$ (3.78) has to be captured, the truncation interval is inversely proportional to α.

Another important pulse shape $p(t)$, that is also strictly bandlimited and of unit energy, is given by:

$$p(t) = \frac{4\alpha}{\pi\sqrt{T_s}}\left[\cos\left(\frac{(1+\alpha)\pi t}{T_s}\right) + \frac{T_s}{4\alpha t}\sin\left(\frac{(1-\alpha)\pi t}{T_s}\right)\right]\frac{1}{1-(4\alpha t/T_s)^2}. \qquad (3.89)$$

The FCT of this pulse is the *square root of a raised cosine spectrum* with roll-off α, that is, it can be expressed as:

$$P(f) = \sqrt{G(f)}, \qquad (3.90)$$

where $G(f)$ is expressed by (3.81). The signal:

$$g(t) \triangleq \sqrt{T_s}\, p(t), \qquad (3.91)$$

with $p(t)$ given by (3.89), and the spectrum:

$$Q(f) \triangleq \frac{P(f)}{\sqrt{T_s}} \, , \tag{3.92}$$

with $P(f)$ given by (3.90), are illustrated in Figures 3.9 and 3.10, respectively, for $\alpha = 0$, 0.5 and 1. Finally, we note that $g(t)$ (3.91), even if its shape is similar to that expressed by (3.79) (see Figure 3.7), does not share at all the important properties of (3.88).

The results illustrated above for bandlimited signaling pulses can be exploited to develop some interesting considerations, as illustrated in the following examples.

Example 3.5.2 In M-ary passband signaling, the *minimum* bandwidth occupancy at RF[14] is $B = 1/T_s$. Therefore, the maximum achievable spectral efficiency is (see (3.40)):

$$\eta_B = \frac{\log M/T_s}{1/T_s} = \log M = m \text{ bit/sec/Hz,} \tag{3.93}$$

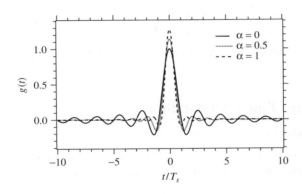

Figure 3.9 Representation of the pulse $g(t)$ (3.91) for $\alpha = 0$, 0.5 and 1.

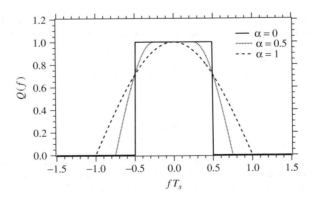

Figure 3.10 Representation of $Q(f)$ (3.92) for $\alpha = 0$, 0.5 and 1.

[14] This is achieved when the spectrum of the transmitter impulse response $p(t)$ is a raised cosine (or the root of a raised cosine) with roll-off factor $\alpha = 0$.

which increases with M. Unfortunately, this result is achieved at the price of a reduction in energy efficiency as M increases.

☐

Example 3.5.3 Consider 64-QAM signaling. This modulation format exploits a signal space with $D = 2$ dimensions to transmit one of $M = 64$ points in each symbol interval, so that (see (3.44)):

$$D_b = \frac{D}{\log M} = \frac{1}{3} \text{ dimensions/bit.} \tag{3.94}$$

Then (3.46) implies that the minimum spectral occupancy at RF is:

$$B_{\min} = \frac{N_b R_b}{2} = \frac{R_b}{6}. \tag{3.95}$$

To show that this result makes sense, let us select the signal $p(t)$ given by (3.78) as a modulator impulse response. This implies that the bandwidth B of the corresponding RF signal is given by (see (3.84)):

$$B = \frac{1 + \alpha}{2T_s} = \frac{R_b}{6}(1 + \alpha) \geq B_{\min}, \tag{3.96}$$

where α is the roll-off factor. In this case B takes on its minimum value ($B_{\min}$ in (3.95)), when $\alpha = 0$.

☐

3.6 Continuous Phase Modulation

3.6.1 Signal Model

A CPM signal can be generated, in principle, by keying a *voltage controlled oscillator* (VCO), having a free-running frequency of f_c Hertz, with the baseband ASK signal:

$$x(t, \mathbf{c}) = \sum_{k=-\infty}^{+\infty} c_k \, p(t - kT_s), \tag{3.97}$$

characterized by the channel symbol sequence $\{c_k\}$ (collected in the vector $\mathbf{c}$), the real pulse $p(t)$, called the *modulator frequency response*, and the symbol interval T_s. The VCO then generates the RF signal:

$$s_{RF}(t, \mathbf{c}) = \sqrt{\frac{2E_s}{T_s}} \cos(2\pi f_c t + \phi(t, \mathbf{c})), \tag{3.98}$$

where

$$\phi(t, \mathbf{c}) \triangleq 2\pi h \int_{-\infty}^{t} x(\tau, \mathbf{c}) \, d\tau, \tag{3.99}$$

E_s represents the average energy per symbol interval, f_c is the carrier frequency and h is a real positive parameter, characterizing the VCO behavior and known as the *modulation index*. The value of h is usually selected to be a rational fraction and, in this case, h can be expressed as:

$$h = \frac{2z}{p}, \tag{3.100}$$

where z and p are relatively prime integers.

From (3.98) and (3.99) it is easy to see that: (a) the complex envelope of $s_{RF}(t, \mathbf{c})$ is given by:

$$s(t, \mathbf{c}) = \sqrt{\frac{2E_s}{T_s}} \exp \ (j\phi(t, \mathbf{c}));\tag{3.101}$$

(b) the envelope $s_{env}(t, \mathbf{c})$ (see (3.8)) is strictly constant; and (c) the average transmit power at RF is $P_{RF} = E_s/T_s$, so that the average power P_s associated with $s(t, \mathbf{c})$ (3.101) is:

$$P_s = 2P_{RF} = \frac{2E_s}{T_s}.\tag{3.102}$$

Therefore, (3.98) and (3.101) can be rewritten, in more compact form, as:

$$s_{RF}(t, \mathbf{c}) = \sqrt{P_s} \cos(2\pi f_c t + \phi(t, \mathbf{c}))\tag{3.103}$$

and

$$s(t, \mathbf{c}) = \sqrt{P_s} \exp \ (j\phi(t, \mathbf{c})),\tag{3.104}$$

respectively.

Before analyzing the mathematical structure of a CPM signal in detail, we make some observations. First, we note that the transformation of the ASK signal $x(t, \mathbf{c})$, which can contain first-order discontinuities, in the phase $\phi(t, \mathbf{c})$, involves integration. This turns $x(t, \mathbf{c})$ into a (possibly piecewise) *continuous* signal or phase response $\phi(t, \mathbf{c})$ and explains the terminology adopted for the wide class of signals described by the relationships (3.98) and (3.99). It is also worth noting that, unlike PAM modulations, the dependence of $s_{RF}(t, \mathbf{c})$ (3.103), or equivalently of its complex envelope $s(t, \mathbf{c})$ (3.104), on the channel symbols $\{c_n\}$ is *nonlinear*.

In the following we make the following assumptions:

1. The channel symbols $\{c_k\}$ belong to the M-ary alphabet $\{\pm 1, \pm 3, \ldots, \pm(M-1)\}$ and are generated from the input information symbols $\{d_k\}$ according to the relationship:

$$c_k = 2d_k - (M-1).\tag{3.105}$$

2. The support of the frequency pulse $p(t)$ is the interval $[0, LT_s]$, where the real positive parameter L is the so-called *correlation length*.
3. The signal $p(t)$ is characterized by an overall area equal to $1/2$, that is:

$$\int_0^{LT_s} p(t) \ dt = \frac{1}{2}.\tag{3.106}$$

Let us rewrite the phase signal $\phi(t, \mathbf{c})$ in a different way, exploiting these assumptions. Substituting (3.97) into (3.99) produces:

$$\phi(t, \mathbf{c}) = 2\pi h \sum_{k=-\infty}^{+\infty} c_k \int_{-\infty}^{t-kT_s} p(\tau) \ d\tau.\tag{3.107}$$

We now define the waveform:

$$q(t) \triangleq \int_{-\infty}^{t} p(\tau) \ d\tau,\tag{3.108}$$

called the *modulator phase response*, and note that, for the above properties of $p(t)$, we have $q(t) = 0$ for $t < 0$ and $q(t) = 1/2$ for $t \geq LT_s$, as illustrated in the following example.

Example 3.6.1 Frequent choices for $p(t)$ are:

$$p_{L\text{-REC}}(t) \triangleq \frac{1}{2LT_s} \text{rect} \left(\frac{t - LT_s/2}{LT_s} \right) \tag{3.109}$$

and

$$p_{L\text{-RC}}(t) \triangleq \frac{1}{2LT_s} \left[1 - \cos \left(\frac{2\pi t}{LT_s} \right) \right] \text{rect} \left(\frac{t - LT_s/2}{LT_s} \right). \tag{3.110}$$

In the first case $p(t)$ is a *rectangular* (REC) pulse having a duration equal to L symbol intervals, and in the second it is a pulse with a *raised cosine* (RC) shape having the same duration. The corresponding phase responses $q_{L\text{-REC}}(t)$ and $q_{L\text{-RC}}(t)$ are then obtained as:

$$q_{L\text{-REC}}(t) \triangleq \int_{-\infty}^{t} p_{L\text{-REC}}(\tau) \, d\tau = \frac{t}{2LT_s} \tag{3.111}$$

and

$$q_{L\text{-RC}}(t) \triangleq \int_{-\infty}^{t} p_{L\text{-RC}}(\tau) \, d\tau = \frac{t}{2LT_s} - \frac{1}{4\pi} \sin \left(\frac{2\pi t}{LT_s} \right), \tag{3.112}$$

respectively, for $0 \le t < LT_s$. The pulses $p_{L\text{-REC}}(t)$ and $q_{L\text{-REC}}(t)$ are shown in Figure 3.11(a), while $p_{L\text{-RC}}(t)$ and $q_{L\text{-RC}}(t)$ are illustrated in Figure 3.11(b). It is interesting to note that: (a) in the L-REC case, $p_{L\text{-REC}}(t)$ (and, then, $x(\tau, \mathbf{c})$) is a discontinuous signal, whereas $q_{L\text{-REC}}(t)$ (and, consequently, $\phi(t, \mathbf{c})$) is piecewise continuous; (b) in both cases the phase response changes over an interval of duration LT_s seconds, at the end of which it reaches the value $1/2$.
□

If we now use definition (3.108), equation (3.107) can be rewritten as:

$$\phi(t, \mathbf{c}) = 2\pi h \sum_{k=-\infty}^{+\infty} c_k \, q(t - kT_s). \tag{3.113}$$

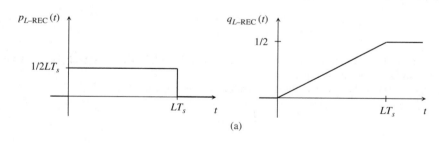

(a)

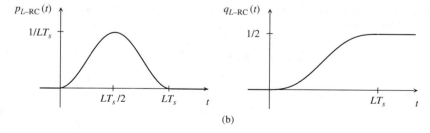

(b)

Figure 3.11 Representation of the frequency and phase responses for L-REC (a) and L-RC (b) CPM signals.

This shows that, at an arbitrary time instant t, the phase $\phi(t, \mathbf{c})$ is determined in general by the superposition of an infinite number of replicas of $q(t)$, each weighted by a channel symbol $\{c_k\}$. However, as shown by Example 3.6.1, the replicas $\{q(t - kT_s)\}$ do not have *finite duration*. In fact, the kth term $2\pi h c_k q(t - kT_s)$ starts to contribute to $\phi(t, \mathbf{c})$ at the instant $t = kT_s$ and never vanishes as time evolves, since it reaches the constant value:

$$2\pi h \cdot c_k \cdot \frac{1}{2} = \pi h c_k \qquad (3.114)$$

starting at the instant $t = (k + L)T_s$.

To acquire a deeper understanding of the structure of a CPM signal, let us now assume that transmission starts at $t = 0$ (so that $c_k = 0$ for $k < 0$) and focus on what happens in the nth symbol interval, that is, in the time interval $[nT_s, (n + 1)T_s)$, with $n \geq 0$. Then equation (3.113) simplifies to:

$$\phi(t, \mathbf{c}_n) = 2\pi h \sum_{k=0}^{n} c_k \, q(t - kT_s) \qquad (3.115)$$

with $\mathbf{c}_n \triangleq [c_0, c_1, \dots, c_n]^T$, since the future symbols $\{c_k, k = n + 1, n + 2, \dots\}$ do not provide any contribution. The RF signal $s_{RF}(t, \mathbf{c})$ (3.103) then becomes:

$$s_{RF}(t, \mathbf{c}_n) = \sqrt{P_s} \cos\left(2\pi f_c t + 2\pi h \sum_{k=0}^{n} c_k \, q\left(t - kT_s\right)\right). \qquad (3.116)$$

This shows that, given the carrier frequency f_c, the complete structure of the the CPM signal $s_{RF}(t, \mathbf{c})$ is defined by: (a) the modulation index h; (b) the cardinality M of the channel symbol alphabet; and (c) the phase response $q(t)$, or, equivalently, the frequency response $p(t)$ of the digital modulator.

In the following, the impact of each of these parameters on the CPM characteristics will be assessed. Our discussion of this problem starts from the analysis of the characteristics of $p(t)$ and, in particular, of its duration, that is, of its correlation length L. From previous considerations about $q(t)$ it can be easily seen that L represents the number of symbol intervals over which each symbol of the sequence $\{c_k\}$ causes a *time variation* of the phase $\phi(t, \mathbf{c}_n)$ of (3.115) or, equivalently, determines a variation in the *instantaneous frequency* of the VCO. In fact, as already discussed, for any k, in the interval $[kT_s, (k + L)T_s]$ the symbol c_k, induces a *phase variation* in $s_{RF}(t, \mathbf{c}_n)$ (3.116), with the *final value* of this variation being $\pi h c_k$. This value persists for $t > (k + L)T_s$ because of the properties of $q(t)$ and the structure of $\phi(t, \mathbf{c})$. For this reason, it is reasonable to partition the CPM signals into two large classes. The first class consists of the so-called *full-response modulations*, that is, all the signals having $L = 1$, whereas the second embraces the so-called *partial-response modulations*, all characterized by $L > 1$. Let us now analyze in detail how the structure of $\phi(t, \mathbf{c}_n)$ (3.115) simplifies in these two cases.

3.6.2 Full-Response CPM

3.6.2.1 Phase Structure in a Full-Response CPM

Let us consider the expression (3.115) (in the time interval $[nT_s, (n + 1)T_s)$) for $\phi(t, \mathbf{c}_n)$, under the assumption that $L = 1$. In this case, only the present symbol c_n produces a time-varying contribution to $\phi(t, \mathbf{c}_n)$, since, for $t \geq nT_s$, all the past symbols, $\{c_k, k = n - 1, n - 2, \dots\}$, yield a constant phase shift. Then we have:

$$\phi(t, \mathbf{c}_n) = \theta_n + 2\pi h c_n \, q(t - nT_s), \qquad (3.117)$$

for $nT_s \leq t < (n + 1)T_s$, where:

$$\theta_n \triangleq \pi h \sum_{k=0}^{n-1} c_k \qquad (3.118)$$

is the so-called *phase state* of the full-response modulator. The RF signal (3.116) can then be written as:

$$s_{RF}(t, \mathbf{c}_n) = \sqrt{P_s} \cos(2\pi f_c t + \theta_n + 2\pi h c_n \, q(t - nT_s)) \qquad (3.119)$$

in the same time interval. Equation (3.117) shows that, in the nth signaling interval, the signal phase depends on the pair $(\theta_n, \, c_n)$, i.e. on the phase state θ_n (summarizing the cumulative phase history), and on the present symbol c_n. We also note that (3.118) can be rewritten as:

$$\theta_n = \theta_{n-1} + \pi h c_{n-1} \qquad (3.120)$$

for $n \geq 1$, assuming, for simplicity, $\theta_0 = 0$. This recursive expression represents the *state equation* of the digital modulator, and, since phases differing by a multiple of 2π are *physically indistinguishable*, can be rewritten as[15]:

$$\theta_n = R_{2\pi}[\theta_{n-1} + \pi h c_{n-1}], \qquad (3.121)$$

where $R_{2\pi}[\cdot]$ is the "modulo 2π operator" defined by (see (3.60)):

$$R_{2\pi}[\theta] \triangleq \theta - \left\lfloor \frac{\theta}{2\pi} \right\rfloor 2\pi. \qquad (3.122)$$

We also note that, if h is a rational number (see (3.100)), θ_n (3.118) can be put in the form:

$$\theta_n = \left(\frac{2\pi}{p} \right) \cdot \left(z \sum_{k=0}^{n-1} c_k \right), \qquad (3.123)$$

in which the second factor takes on integer values only. Therefore, the phase state is always a multiple of $2\pi/p$, and, if the phase values are restricted to $[0, 2\pi)$, it can take on only the p distinct values of the set:

$$\Theta \triangleq \left\{ 0, \frac{2\pi}{p}, 2\frac{2\pi}{p}, \ldots, (p-1)\frac{2\pi}{p} \right\}. \qquad (3.124)$$

This shows that the CPM modulator can be described as an FSSM with $N_s = p$ states, its state being expressed by θ_n. Its state and output equations are given by (3.121) and (3.119), respectively. In addition, its time evolution can be represented by a p-state *trellis diagram*. Note that: (a) the trellis structure depends on M and h, but is independent of the shape of $p(t)$; (b) the state can also be represented by the integer parameter:

$$x_n \triangleq \theta_n \frac{p}{2\pi} \qquad (3.125)$$

taking on values in the set $\{0, 1, \ldots, p-1\}$, so that (3.121) can be represented as:

$$x_n = R_p[x_{n-1} + z c_{n-1}], \qquad (3.126)$$

involving integer quantities only. If the present state of the FSSM in the nth symbol interval is defined as:

$$\Delta_n \triangleq x_n, \qquad (3.127)$$

then the signal $s(\Delta_n, c_n; t - nT_s)$ of (3.2), for a full-response CPM signal (see (3.101), (3.117) and (3.125)) is given by:

$$s(\Delta_n, c_n; t - nT_s) = \sqrt{P_s} \exp\left[j \left(\frac{2\pi}{p} \Delta_n + 2\pi h c_n \, q\,(t - nT_s) \right) \right] \qquad (3.128)$$

for $nT_s \leq t < (n+1)T_s$.

---

[15] The mod 2π reduction is applicable to all the expressions defining the phase of a sinusoidal oscillation, even if not always indicated in an explicit fashion.

Let us now analyze the trellis structure for a specific subclass of CPM full-response signals.

Example 3.6.2 The state diagram and state trellis of any binary full-response CPM modulation having $h = 1/2$ are shown in Figure 3.12 (the signals labeling each state transition have been omitted for simplicity). Note that: (a) for $z = 1$ and $p = 4$, we have $N_s = 4$ distinct states, corresponding to the four phase states $\{0, \pi/2, \pi, 3\pi/2\}$; (b) in this case (3.123) becomes:

$$\theta_n = R_{2\pi} \left[\frac{\pi}{2} \cdot \sum_{k=0}^{n-1} c_k \right]. \tag{3.129}$$

Since the symbols $\{c_k\}$ are odd numbers, the sum in (3.129) takes on odd (even) values if n is odd (even). For this reason, if $\theta_0 = 0$, then $\theta_n \in \{0, \pi\}$ ($\{\pi/2, 3\pi/2\}$) for even (odd) n, as exemplified by Figure 3.12(b).
□

3.6.2.2 Some Specific Examples of Full-Response CPM

An important class of full-response signals consists of all 1-REC modulations, also known as *continuous-phase frequency shift keying* (CPFSK) modulations. We then have $q(t) = q_{1\text{-REC}}(t)$ (see (3.111)), so that (3.117) gives:

$$\phi(t, \mathbf{c}_n) = \theta_n + \pi h c_n \frac{t - n T_s}{T_s} \tag{3.130}$$

for $n T_s \leq t < (n + 1) T_s$. Then the instantaneous frequency $f_n(t)$ of the CPM signal in the nth signaling interval is given by:

$$f_n(t) \triangleq f_c + \frac{1}{2\pi} \frac{d}{dt} \phi(t, \mathbf{c}_n) = f_c + \frac{h}{2 T_s} c_n \tag{3.131}$$

and, consequently, is *constant* and *depends linearly* on c_n. For this reason, a 1-REC signal is a *frequency shift keying* (FSK) signal with spacing between adjacent tones of h/T_s.

The best-known form of the CPFSK class is the so-called *minimum shift keying* (MSK), characterized by $M = 2$ and $h = 1/2$ [285]. Thus, MSK is a binary continuous-phase FSK with $f_n \in \{f_c \pm 1/(4T_s)\}$, so that the tone spacing is equal to $1/(2T_s) = R_s/2$. It can be proved that this value is the *minimum* spacing ensuring, in a binary FSK, the use of a *coherent receiver for orthogonal signals*.

3.6.2.3 Phase Tree and Phase Cylinder

A useful tool for understanding the structure of the phase $\phi(t, \mathbf{c}_n)$ (3.117) of a full-response CPM signal is the so-called *phase tree*. This tree represents the possible trajectories of $\phi(t, \mathbf{c}_n)$ originating from a common node for $t = 0$ (usually, corresponding to $\theta_0 = 0$) for all possible values of $\mathbf{c}_n$. In this diagram, any trajectory, extending from $t = 0$ to $t = n T_s$, is a *continuous curve* and is uniquely determined by a sequence $\{c_k, k = 0, 1, \ldots, n - 1\}$ of symbols. This symbol sequence causes $\phi(t, \mathbf{c}_n)$ to evolve from the initial value $\theta_0 = 0$ to the final value $\theta_n = \pi h \sum_{k=0}^{n-1} c_k$ (see (3.118)). The phase trees for binary 1-REC and 1-RC signals are shown in Figure 3.13.

The phase tree cannot account for the identity of phase trajectories differing by multiples of 2π. Actually, such trajectories, although appearing distinct in a planar representation, correspond to the same signal $s_{RF}(t, \mathbf{c}_n)$ (3.119). For this reason, a more appropriate representation of phase trajectory is the so-called *phase cylinder*, which is generated by wrapping the plane of the phase tree around a cylinder in such a way that the straight lines corresponding to $\phi = 0$ and $\phi = 2\pi$ coincide. The

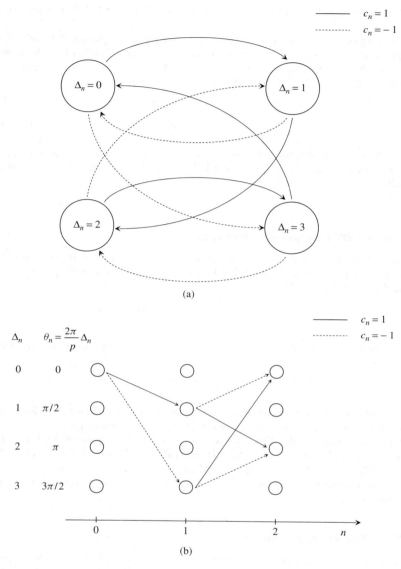

Figure 3.12   State diagram (a) and trellis diagram (b) for a full-response binary CPM with $h = 1/2$.

cylinder axis then indicates the direction of the evolution. From an analytical viewpoint, generating a phase cylinder means producing a three-dimensional representation of the curves defined by $z_x(t, \mathbf{c}_n) \triangleq s_I(t, \mathbf{c}_n)/\sqrt{P_s} = \cos[\phi(t, \mathbf{c}_n)]$ and $z_y(t, \mathbf{c}_n) \triangleq s_Q(t, \mathbf{c}_n)/\sqrt{P_s} = \sin[\phi(t, \mathbf{c}_n)]$ as time evolves, for any possible $\mathbf{c}_n$. The phase cylinder associated with the tree of Figure 3.13(a) for the case $h = 1/2$ (corresponding to MSK) is illustrated in Figure 3.14.

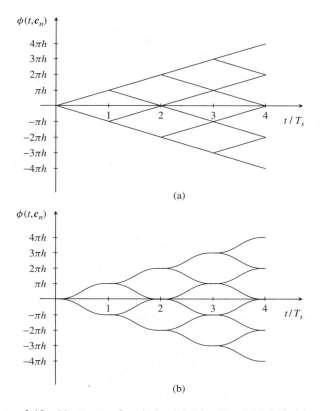

Figure 3.13   Phase trees for binary 1-REC (a) and 1-RC (b) signals.

3.6.3   Partial-Response CPM

3.6.3.1   Phase Structure in a Partial-Response CPM

Similarly to what we have done with full-response signals, we focus again on the expression for $\phi(t, \mathbf{c}_n)$ in (3.115), in the nth signaling interval, $nT_s \le t < (n+1)T_s$, under the assumption that $L > 1$. Within this interval, the time-varying part of $\phi(t, \mathbf{c}_n)$ depends not only on c_n, but also on the $L - 1$ previous symbols, that is, on the set $\{c_{n-(L-1)}, c_{n-(L-2)}, \ldots, c_{n-1}\}$, since each channel symbol causes a phase variation over L consecutive symbol intervals. Then (3.115) can be written in the form:

$$\phi(t, \mathbf{c}_n) = \theta_n + 2\pi h \sum_{k=n-(L-1)}^{n} c_k \, q(t - kT_s), \tag{3.132}$$

where

$$\theta_n \triangleq \pi h \sum_{k=0}^{n-L} c_k \tag{3.133}$$

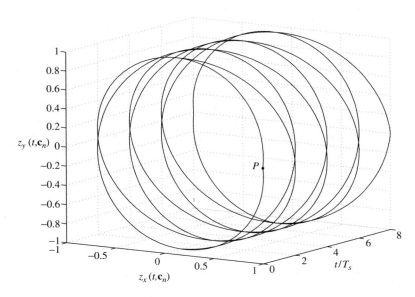

Figure 3.14 Representation of the MSK phase cylinder over eight consecutive symbol intervals. All the phase trajectories originate from the initial point P.

is the *phase state* of the partial-response modulator. Thus, in the nth symbol interval, the signal phase depends not only on (θ_n, c_n), as in the full-response case, but also on the vector $\mathbf{c}_{n-(L-1)}^{n-1} \triangleq [c_{n-(L-1)}, c_{n-(L-2)}, \ldots, c_{n-1}]^T$ of symbols, which defines the so-called *correlative state* of the CPM modulator. This state can be represented as a nonnegative integer number σ_n. In fact, the *bipolar* symbol c_n can be represented by the *unipolar symbol* d_n (see (3.105)), $d_n = (c_n + M - 1)/2$ belonging to the M-ary alphabet $\{0, 1, \ldots, M - 1\}$, so that, if we define:

$$\sigma_n \triangleq d_{n-(L-1)} + d_{n-(L-2)} \cdot M + \ldots + d_{n-1} \cdot M^{L-2}, \tag{3.134}$$

then $\sigma_n \in \{0, 1, \ldots, M^{L-1} - 1\}$.

It is of interest to note the following:

1. Equation (3.132) can also be rewritten as:

$$\phi(t, \mathbf{c}) = \theta_n + \gamma(t - nT_s, \sigma_n, c_n) \tag{3.135}$$

for $nT_s \leq t < (n + 1)T_s$, where

$$\gamma(t - nT_s, \sigma_n, c_n) \triangleq \sum_{k=-(L-1)}^{0} c_{n+k} \, q(t - kT_s - nT_s). \tag{3.136}$$

2. As in the full-response case, equation (3.133) can be put in a recursive form given by:

$$\theta_n = R_{2\pi}[\theta_{n-1} + \pi h c_{n-L}] \tag{3.137}$$

for $n \geq L$, or, equivalently, as:

$$x_n = R_p[x_{n-1} + z\, c_{n-L}], \tag{3.138}$$

where $x_n \triangleq \theta_n \, (p/2\pi)$ (see (3.125)).

3. If h is a rational fraction (see (3.100)), (3.133) can be rewritten as:

$$\theta_n = \frac{2\pi}{p} z \sum_{k=0}^{n-L} c_k,$$

showing that θ_n can take only the p distinct values of the set Θ (3.124). Therefore, the FSSM describing a partial-response modulator is characterized by a *state* defined by the ordered pair (θ_n, σ_n), which can take on, in each signaling interval, one of $N_s = p \cdot M^{L-1}$ possible values. The state equation of this FSSM consists of two parts, the first referring to the transition $\theta_n \to \theta_{n+1}$ and expressed by (3.137), the second given by:

$$\sigma_{n+1} = \frac{\sigma_n - d_{n-(L-1)}}{M} + d_n \cdot M^{L-2} \tag{3.139}$$

since the transformation of $\mathbf{c}_{n-(L-1)}^{n-1} = [c_{n-(L-1)}, c_{n-(L-2)}, \cdots, c_{n-1}]^T$ into $\mathbf{c}_{n-(L-2)}^{n} = [c_{n-(L-2)}, c_{n-(L-1)}, \cdots, c_n]^T$ requires the removal of the first element in the vector $\mathbf{c}_{n-(L-1)}^{n-1}$, a single step leftward shift and the insertion of c_n as last element.

4. If we refer to the baseband signal generated by the CPM modulator in the n-symbol interval (3.2), the output equation of the FSSM can be expressed as (see (3.101) and (3.135)):

$$s(\Delta_n, c_n; t - nT_s) = \sqrt{P_s} \exp\left[j\left(\frac{2\pi}{p} x_n + \gamma\left(t - nT_s, \sigma_n, c_n\right)\right)\right], \tag{3.140}$$

where the integer parameter:

$$\Delta_n \triangleq \sigma_n + x_n \cdot M^{L-1}, \tag{3.141}$$

whose values belong to the set $\{0, 1, \ldots, N_s - 1\}$, represents the overall state of the FSSM.

Let us now focus on a specific partial-response format.

Example 3.6.3 The trellis diagram for binary CPM signaling with $L = 3$ and $h = 1/2$ is illustrated in Figure 3.15. Note that, for $p = 4$, $M = 2$ and $L = 3$, the overall number of states is $N_s = 4 \cdot 2^2 = 16$; each state is defined by the triple $(\theta_n, c_{n-2}, c_{n-1})$ (with $\theta_n \in \{0, \pi/2, \pi, 3\pi/2\}$), admitting the integer representation Δ_n (3.141).

□

3.6.3.2 Gaussian FSK

An important class of partial-response CPM signals consists of the *Gaussian-filtered* FSK (GFSK) signals. These are characterized by the frequency response:

$$p_{GFSK}(t) \triangleq p_{CPFSK}(t) \otimes h_G(t), \tag{3.142}$$

where

$$p_{CPFSK}(t) = \frac{1}{2T_s}[u(t) - u(t - T_s)] \tag{3.143}$$

is the frequency response of CPFSK (corresponding to $p_{L\text{-REC}}(t)$ (3.109) with $L = 1$) and:

$$h_G(t) = B\sqrt{\frac{2\pi}{\ln 2}} \exp\left(-\frac{2\pi^2}{\ln 2} B^2 t^2\right) \tag{3.144}$$

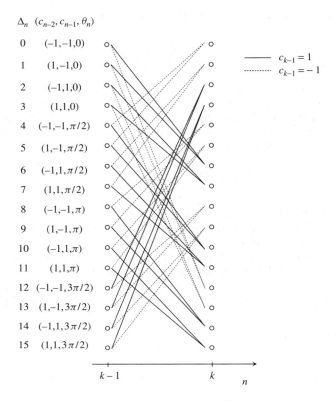

Δ_n $(c_{n-2}, c_{n-1}, \theta_n)$

0	$(-1,-1,0)$
1	$(1,-1,0)$
2	$(-1,1,0)$
3	$(1,1,0)$
4	$(-1,-1,\pi/2)$
5	$(1,-1,\pi/2)$
6	$(-1,1,\pi/2)$
7	$(1,1,\pi/2)$
8	$(-1,-1,\pi)$
9	$(1,-1,\pi)$
10	$(-1,1,\pi)$
11	$(1,1,\pi)$
12	$(-1,-1,3\pi/2)$
13	$(1,-1,3\pi/2)$
14	$(-1,1,3\pi/2)$
15	$(1,1,3\pi/2)$

$c_{k-1} = 1$
$c_{k-1} = -1$

$k-1$ k n

Figure 3.15 Trellis diagram of binary CPM signaling with $L = 3$ and $h = 1/2$.

is the impulse response of a *low-pass Gaussian filter* having the frequency response:

$$H_G(f) = \exp\left[-\frac{\ln 2}{2}\left(\frac{f}{B}\right)^2\right] \tag{3.145}$$

and 3-dB bandwidth B.

Substituting (3.143) and (3.144) into (3.142) leads to:

$$p_{GFSK}(t) = \frac{1}{2T_s}\left[Q\left(\frac{t-T_s}{\sigma}\right) - Q\left(\frac{t}{\sigma}\right)\right], \tag{3.146}$$

in terms of the Gaussian Q function $Q(x)$ (see Appendix F), with $\sigma \triangleq \sqrt{\ln 2}/(2\pi B)$. Figure 3.16 shows the shape of $p_{GFSK}(t)$ for four distinct values of the normalized bandwidth BT_s. Note that, on the one hand, if $BT_s \to +\infty$, $p_{GFSK}(t)$ approaches the rectangular pulse $p_{CPFSK}(t)$ (3.143). On the other hand, for finite values of BT_s, $p_{GFSK}(t)$ is a continuous function and has an *infinite duration*. In practice, for a given value of BT_s, it is possible to find a proper value of the parameter L such that the interval centered over $t = T_s/2$ and having duration LT_s contains most of the energy of $p_{GFSK}(t)$. In other words, $p_{GFSK}(t)$ is always *truncated*[16] in a symmetrical fashion with respect to its center, so that finding a *finite* value of the correlation length L affects only negligibly the characteristics of the transmitted signal.

---

[16] This truncation must be always followed by a renormalization, so that the truncated pulse satisfies (3.106).

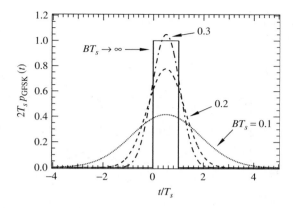

Figure 3.16 Representation of the impulse response $p_{GFSK}(t)$ (3.142) for $BT_s = 0.1, 0.2, 0.3$ and $+\infty$.

Table 3.1 Spectral occupancy of MSK and three distinct GMSK formats

$B_{1-\varepsilon}/R_b$	90%	99%	99.9%	99.99%
0.2 GMSK	0.52	0.79	0.99	1.22
0.25 GMSK	0.57	0.86	1.09	1.37
0.5 GMSK	0.69	1.04	1.33	2.08
MSK	0.78	1.20	2.76	6.00

Note that a reduction in the bandwidth B broadens $p_{GFSK}(t)$ and, consequently, reduces its rate of change. This results in slower fluctuations in the time evolution of $\phi(t, \mathbf{c}_n)$ (3.115) and, consequently, in a narrowing of the power spectrum of $s_{RF}(t, \mathbf{c}_n)$ (3.116). The penalty for this is, however, usually an increase in receiver complexity.

A specific case of GFSK signals is the so-called *Gaussian-filtered* MSK (GMSK)[17] [286], characterized, like MSK, by a binary alphabet ($M = 2$) and modulation index $h = 1/2$. GMSK signals exhibit good spectral efficiency. This property is shown by the numerical results listed in Table 3.1, summarizing some values of the RF bandwidth $B_{1-\varepsilon}$, normalized with respect to the bit rate R_b, for MSK and for three GMSK formats (each corresponding to a specific value of the normalized bandwidth BT_s for the Gaussian filter). Note that, as ε decreases, the GMSK bandwidth increases much more slowly than that of MSK. This is due to the fact that the rate of decrease of the spectral side lobes in GMSK is appreciably larger than for MSK. Further numerical results about the spectral occupancy of GMSK and MSK can be found in [286].

3.6.3.3   Phase Tree and Phase Cylinder

Even in the case of partial-response signals, the family of possible phase trajectories (see (3.115)) emerging from a common node at $t = 0$ in their planar representation leads, as in the full-response

---

[17] GSMK signaling has been adopted, for its constant envelope and spectral compactness, in the GSM standard [6, 57], defining the second generation of cellular phone systems. In the GSM system the normalized bandwidth BT_s of the Gaussian filter is set to 0.3 and the transmission speed to 270.833 kbit/sec.

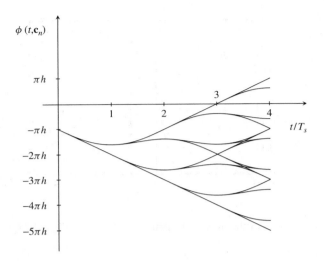

Figure 3.17   Phase tree for binary 3-RC CPM modulation.

case, to the so-called *phase tree*. This is shown in Figure 3.17, which depicts the phase tree[18] for binary 3-RC CPM. Note that, for a given correlation length L, two *distinct* phase trajectories splitting from a common state Δ_n at $t = nT_s$, because of a different choice for the value of c_n, cannot merge before $t = (n + L + 1)T_s$. In other words, merging requires that at least $L + 1$ symbol intervals have elapsed, since the time-varying contribution of c_n lasts L symbol intervals.

As in the full-response case, a more appropriate representation of the phase trajectories is provided by the so-called *phase cylinder*. This cylinder for a binary 3-RC CPM with $h = 1/2$ is shown in Figure 3.18.

3.6.4   Multi-h CPM

As noted earlier, MSK [285] is a special case of CPFSK with modulation index $h = 1/2$. When optimally detected, it has the same error rate performance as QPSK. In AWGN channels most of the binary CPM formats discussed above have essentially the same performance under optimal detection. Because of their inherent memory and trellis structure, the question then arises as to whether better performance is possible. It was noted that the performance of MSK requires detection over two bit intervals and that partial-response CPM formats require $L + 1$ intervals to achieve their best performance. Early work by W. Osborne and M. Luntz [287] showed that observing CPFSK over n bit intervals could lead to improved detection performance. They found that when using the optimal modulation index of $h = 0.715$ an improvement in error performance of almost 2 dB can be achieved by detecting each bit over an observation interval of 3–5 bit periods.

All of the CPMs have a trellis representation. The various full-response formats have trellises that have a forced merge every two intervals. On the other hand, the various partial-response formats are characterized by trellises that have forced merges only every $L + 1$ intervals, where L is the correlation length. However, even when full advantage is taken of the trellis structure in their decoding, only

---

[18] In the generation of each trajectory, the modulator has been initialized to the state $\Delta_0 = 0$, corresponding to the choice $\theta_0 = 0$, $c_{-1} = -1$ and $c_{-2} = -1$.

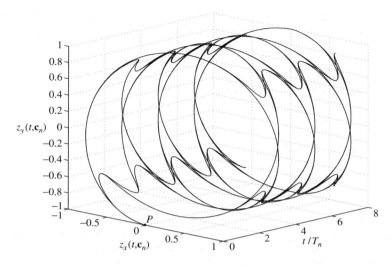

Figure 3.18   Phase cylinder for binary 3-RC CPM with $h = 1/2$. The representation is over eight consecutive symbol intervals. Note that all the phase trajectories originate from P.

very modest performance gains are possible. This is discussed in some detail in [288]. In 1975 [289], H. Miyakawa *et al.* described the cyclic use of a small set of modulation indexes $\{h_i\}_{i=1}^K$, where K is an integer, to provide increased distance in the phase tree of CPFSK and thereby improve error performance. Their scheme allowed for irrational values of the indexes, which required a tree description with an ever expanding set of nodes or states and was, therefore, not of great practical value. This problem was solved in [290], which proposed the use of a finite set of modulation indexes, $\{h_1, h_2, \ldots, h_K\}$, but restricted to being rational fractions of the form l_i/p and subject to the condition that:

$$\sum_{i=1}^{K} h_i = \frac{1}{p} \sum_{i=1}^{K} l_i \neq 1. \tag{3.147}$$

The resulting CPM formats are known as *multi-h signaling schemes*. Although not strictly necessary, almost all such schemes further restrict the indexes to $l_i/p < 1$. This ensures a continuous average power spectrum with no spectral lines [288] and a compact power spectrum having a main lobe of approximately the same width as MSK.

From the above discussion, we may now write a *multi-h* CPM *signal* in the form:

$$s_{RF}(t, \mathbf{c}_n) = \sqrt{\frac{2E_s}{T}} \cos\left(2\pi f_c t + 2\pi \sum_{k=0}^{n} h_{k'} c_k \, q\left(t - kT_s\right)\right) \tag{3.148}$$

for $nT_s \leq t \leq (n+1)T_s$, where the notation k' implies $k' = R_K[k]$ and thus a cyclic use of the modulation indexes in the defined set. We may then write the phase function for a multi-h signal in the nth signaling interval (i.e., for $nT_s \leq t \leq (n+1)T_s$) as:

$$\phi(t, \mathbf{c}) = 2\pi \sum_{k=0}^{n} h_{k'} c_k q(t - kT_s) = \frac{2\pi}{p} \sum_{k=0}^{n} l_{k'} c_k q(t - kT_s) \tag{3.149}$$

using the same notation as in previous subsections. In the special case of 1-REC signaling, the phase pulse $q(t)$ takes the linear form given by (3.111) and may be written in the nth interval as:

$$q_{\text{1-REC}}(t) = \frac{t - nT_s}{T_s}. \tag{3.150}$$

We may then write the phase of the modulated signal in the form:

$$\phi(t, \mathbf{c}) = \theta_n + \pi h_{n'} c_n \frac{t - nT_s}{T_s} = \theta_n + \pi \frac{l_{n'}}{p} c_n \frac{t - nT_s}{T_s}, \tag{3.151}$$

where the phase state θ_n is given by:

$$\theta_n = R_{2\pi} \left[\frac{\pi l_{n'}}{p} \sum_{k=0}^{n-1} c_k \right] \tag{3.152}$$

and $n' = R_K[n]$ represents the cyclic usage of the K modulation indexes.

It can be shown [288, 290] that the multi-h signaling format as defined above delays any forced merging in either the state or the phase trellis by up to $K + 1$ signaling intervals, when the set of modulation indexes satisfy condition (3.147). The multi-h signaling structure can in fact be regarded as a form of *coded modulation* and it can be shown [290] that, provided that it is decoded optimally (usually through the use of the so-called *Viterbi algorithm* [27]), a substantial *coding gain* (see Section 6.2.1.6) can be achieved compared to CPM formats using a single modulation index. For binary formats, more than 3 dB of performance gain can be achieved using a set of $K = 3$ modulation indexes compared to BPSK or MSK. Even higher gains can be achieved by using $K > 3$ indexes, but the complexity of the decoder increases accordingly.

Finally, it is possible to extend the multi-h format to partial-response CPM [288]. However, this has rarely been done due to the increased complexity of the resulting decoder. Full-response multi-h modulation requires either p or $2p$ states, depending on whether $\sum_{k=1}^{K} l_i$ is even or odd. When a partial-response format is used, it can be shown that the resulting decoding algorithm requires as many as $2pM^{L-1}$ states, which even for $M = 2$ quickly becomes large. Moreover, the partial-response format tends to cause a loss of coding gain, and as a result there has never been much interest in partial-response multi-h modulation. The primary advantage of a partial-response format is the resulting spectral compactness.

3.6.5   *Alternative Representations of CPM Signals*

The number of different signals generated by a CPM modulator in a single interval is proportional to M^L and, consequently, may be very large. This family of signals can be represented by an orthonormal basis, as discussed in Section 3.3. The use of the Gram–Schmidt procedure (see Appendix D.1.2) to obtain this basis was proposed in [291] (and adopted also in [292]), with the goal of selecting a small number of orthonormal functions to represent the CPM signal space with good accuracy. Other approaches to the problem of determining an efficient basis for signal representation are based on sampling functions [293], Walsh functions [294], a set of sinusoids with regularly spaced frequencies [291], and the *principal components* method [295].

A different approach involves representing a CPM signal as a *superposition of multiple PAM waveforms*. The theoretical basis for representing any binary CPM waveform as a finite sum of PAM signals, each characterized by a *data-independent* time-limited pulse, was laid by P. A. Laurent in 1986 in his seminal publication [296]. His result, known as *Laurent's decomposition* of CPM signals, has been extended to multilevel single-h CPM in [297] and to multilevel multi-h CPM in [298] (moreover, the case of integer modulation indexes has been investigated in [299]). The Laurent decomposition has two major drawbacks: (a) the PAM components of a CPM signal are generally not mutually independent; (b) when a small number of PAM components are used to approximate a CPM signal, the approximation is not optimal in the *minimum mean squared error* (MMSE) sense. For this reason, a different PAM representation based on an MMSE approach to approximating CPM signals has been proposed in [300]. This leads to the generation of mutually independent components in the PAM expansion, at the price, however, of having some pulses with *infinite duration*. More recently, an alternative approach to PAM representation as been proposed in [301], where it is shown that any CPM signal can be written as the superposition of 2^{L-1} *data-dependent* waveforms over each symbol interval.

Signal generation in a CPM transmission can also be interpreted from a totally different perspective, as described by B. E. Rimoldi in 1988 [302]. He showed that any single-h CPM modulator can be decomposed into the cascade of a *continuous phase encoder* (CPE) and a *memoryless modulator* (MM) both having important properties. The former is a time-invariant sequential circuit operating over a specific algebra (usually the ring of integers modulo p, denoted $\mathbb{Z}_p$), whereas the latter maps, in a time-invariant fashion, the CPE output to a finite set of waveforms.

We now focus on some essential results on CPM representations, namely the Laurent representation for binary single-h CPM and on Rimoldi's decomposition approach to CPM.

3.6.5.1 Laurent's Representation of CPM Signals

Full-Response Signals

Let us consider again the expression (3.104) for a complex envelope of a CPM signal and, assuming that binary full-response signaling is adopted (see (3.117)), rewrite it as:

$$s(t, \mathbf{c}) = \sqrt{P_s} a_{0,n-1} \exp\left[j 2\pi h c_n\, q(t - nT_s)\right] \tag{3.153}$$

in the time interval $[nT_s, (n+1)T_s)$, where:

$$a_{0,n-1} \triangleq \exp\left(j\theta_n\right). \tag{3.154}$$

Then if we define the function:

$$u_T(t) = \begin{cases} 1 & \text{for } 0 < t < T \\ 1/2 & \text{for } t = 0 \text{ and } t = T \\ 0 & \text{elsewhere} \end{cases} \tag{3.155}$$

representing a rectangular pulse of unit height and duration T seconds, then $s(t, \mathbf{c})$ (3.153), *over the whole time interval*, can be written in the form:

$$s(t, \mathbf{c}) = \sqrt{P_s} \sum_{k=-\infty}^{+\infty} a_{0,k-1}\, x(c_k, t - kT_s)\, u_{T_s}(t - kT_s), \tag{3.156}$$

where

$$x(c, t) \triangleq \exp\left[j 2\pi h c\, q(t)\right] = \cos[2\pi h\, q(t)] + j c\, \sin[2\pi h\, q(t)], \tag{3.157}$$

since $c \in \{\pm 1\}$ and $\sin(cz) = c \sin(z)$ for any z. Similarly, $\exp\left(j\pi h c\right) = \cos(\pi h) + j c \sin(\pi h)$ so that, if $\sin(\pi h) \neq 0$ (i.e., if h is not integer[19]), we have:

$$jc = \frac{\exp\left(j\pi h c\right) - \cos(\pi h)}{\sin(\pi h)}. \tag{3.158}$$

---

[19] For the case of integer h, see [297, 299].

Substituting (3.158) into (3.157) yields, after some manipulation:

$$x(c, t) = \exp\,(j\pi hc)\,\frac{\sin[2\pi h q(t)]}{\sin(\pi h)} + \frac{\sin\{\pi h[1 - 2\,q(t)]\}}{\sin(\pi h)}. \tag{3.159}$$

Then, substituting (3.159) in (3.156) yields $s(t, \mathbf{c})$ in the form:

$$s(t, \mathbf{c}) = \sqrt{P_s} \sum_{k=-\infty}^{+\infty} a_{0,k}\,\frac{\sin[2\pi h q(t - kT_s)]}{\sin(\pi h)}\,u_{T_s}(t - kT_s)$$

$$+ \sqrt{P_s} \sum_{k=-\infty}^{+\infty} a_{0,k-1}\,\frac{\sin\{\pi h[1 - 2\,q(t - kT_s)]\}}{\sin(\pi h)}\,u_{T_s}(t - kT_s), \tag{3.160}$$

since $a_{0,k-1}\exp\,(j\pi hc_k) = a_{0,k}$ (see (3.154)). Finally, putting together the two sums on the RHS of (3.160) yields:

$$s(t, \mathbf{c}) = \sqrt{P_s} \sum_{k=-\infty}^{+\infty} a_{0,k}\,l_0(t - kT_s), \tag{3.161}$$

where

$$l_0(t) \triangleq \frac{\sin[2\pi h q(t)]}{\sin(\pi h)}\,u_{T_s}(t) + \frac{\sin\{\pi h[1 - 2\,q(t - T_s)]\}}{\sin(\pi h)}\,u_{T_s}(t - T_s) \tag{3.162}$$

is the *Laurent function* [297] of the binary full-response CPM signal. Equations (3.161) and (3.162) define *Laurent's representation for binary full-response* CPM. Note that this expresses CPM as a PAM having an impulse response $l_0(t)$ of duration $2T_s$ seconds and characterized by the channel symbols $\{a_{0,k}\}$, which are also known as *pseudosymbols*. These symbols have unit amplitude and are generated by the recursive expression:

$$a_{0,k} = a_{0,k-1}\exp\,(j\pi hc_k). \tag{3.163}$$

For instance, if this representation is applied to MSK, (3.162) yields:

$$l_0(t) = \sin\left(\frac{\pi t}{2T_s}\right)\,u_{2T_s}(t), \tag{3.164}$$

expressing a half cycle of a sinusoidal signal of frequency $1/(4T_s)$, whereas (3.163) gives:

$$a_{0,k} = a_{0,k-1}\exp\,\left(j\frac{\pi}{2}c_k\right), \tag{3.165}$$

so that, if $a_{0,k-1}$ is *real (imaginary)*, then $a_{0,k}$ is *imaginary (real)*. For this reason, the in-phase component $s_I(t, \mathbf{c})$ and the quadrature component $s_Q(t, \mathbf{c})$ of an MSK signal consist of a series of sinusoidal pulses, but there is a time *offset* of T_s seconds between them. Hence, MSK can be represented as a form of *offset* QPSK characterized by a *half-sinusoidal* signaling pulse [285].

Partial-Response Signals

Laurent's representation for binary partial-response signals generalizes the results above. In fact, it can be proved that the complex envelope $s(t, \mathbf{c})$ (3.101) of the transmitted signal can be expressed as:

$$s(t, \mathbf{c}) = \sqrt{P_s} \sum_{p=0}^{P-1} \sum_{k=-\infty}^{+\infty} a_{p,k}\,l_p(t - kT_s), \tag{3.166}$$

where $P \triangleq 2^{L-1}$ is the number of constituent PAM signals, $l_p(t)$ and $a_{p,k}$ denote the *Laurent function* and the kth *pseudosymbol*, respectively, characterizing the pth PAM signal, with $p = 0, 1, \ldots, P - 1$. Moreover, it is shown that:

$$l_p(t) = \prod_{i=0}^{L-1} g(t + iT_s + b_{p,i}LT_s), \tag{3.167}$$

where

$$g(t) \triangleq \frac{\sin[2\pi h q(t)]}{\sin(\pi h)} u_{LT_s}(t) + \frac{\sin\{\pi h[1 - 2\,q(t - LT_s)]\}}{\sin(\pi h)} u_{LT_s}(t - LT_s) \tag{3.168}$$

is a pulse lasting for $2LT_s$ seconds, and that the generation of the pseudosymbols is expressed by P distinct relations. These require, for any integer value of the parameter $p \in \{0, 1, \ldots, P - 1\}$, the introduction of the binary coefficients $\{b_{p,i}, i = 0, 1, \ldots, L - 1\}$. A null value is always assigned to $b_{p,0}$, while $b_{p,i}$ $(i = 1, 2, \ldots, L - 1)$, is defined as the ith bit in the base-2 representation of p, given by:

$$p = \sum_{i=1}^{L-1} b_{p,i}\, 2^{i-1}. \tag{3.169}$$

Given the coefficients $\{b_{p,i}\}$, the kth pseudosymbol $a_{p,k}$ of the pth PAM component of (3.166) may be expressed as:

$$a_{p,k} = \exp \left\{ j\theta_k + j\pi h \sum_{m=k-L+1}^{k} \left(1 - b_{p,k-m}\right) c_m \right\} \tag{3.170}$$

with $p = 0, 1, \ldots, P - 1$, thus establishing a *nonlinear* relationship between $a_{p,k}$ and the channel symbols $\{c_m, m = k - L + 1, k - L + 2, \ldots, k\}$ and the present phase state θ_k. The characteristic of nonlinearity can be related, as with full-response signals, to the nonlinear dependence of CPM signals on the data. It is also worth noting that the sequences $\{\{a_{p,k}\}, p = 0, 1, \ldots, P - 1\}$ are *statistically dependent*, since they all depend on the same channel symbols.

Finally, we note that the pulse $l_p(t)$ (3.167), with $p = 0, 1, \ldots, P - 1$, is time-limited to $[0, D_p T_s]$ with:

$$D_p = \min_{i \in I}\{L(2 - b_{p,i}) - i\}, \tag{3.171}$$

where $I = \{0, 1, \ldots, L - 1\}$, and that the power of the transmitted signal is generally concentrated in its first PAM component, characterized by the pulse:

$$l_0(t) = \prod_{i=0}^{L-1} g(t + iT_s), \tag{3.172}$$

which is the *main function* in Laurent's representation [297]. An application of these principles is given in the following example.

Example 3.6.4 If we apply these results to binary 3-RC signaling with $h = 1/2$, we have $P = 4$ and the Laurent functions are given by:

$$l_0(t) = g(t)\, g(t + T_s)g(t + 2T_s), \tag{3.173}$$

$$l_1(t) = g(t)\, g(t + 4T_s)g(t + 2T_s), \tag{3.174}$$

$$l_2(t) = g(t)\, g(t + T_s)g(t + 5T_s), \tag{3.175}$$

$$l_3(t) = g(t)\, g(t + 4T_s)g(t + 5T_s), \tag{3.176}$$

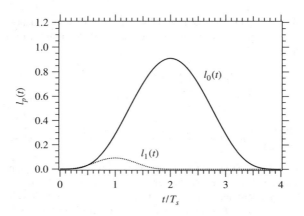

Figure 3.19 Representation of the Laurent pulses $l_0(t)$ (3.173) and $l_1(t)$ (3.174) for binary 3-RC signaling with $h = 1/2$.

where $g(t)$ is given by (3.168) with $q(t) = q_{3\text{-RC}}(t)$ (see (3.112)), and their durations, normalized with respect to the symbol interval, are $D_0 = 4$, $D_1 = 2$ and $D_2 = D_3 = 1$, respectively. The signals $l_0(t)$ and $l_1(t)$ are illustrated in Figure 3.19 (note that $l_2(t)$ and $l_3(t)$ are not shown since they have negligible energy with respect to the other two pulses). These results show that, as usually happens, most of the CPM power is captured by its first PAM component.

□

Applications

Generally speaking, alternative representations of the signal alphabet generated by a digital modulator can be very useful for receiver design. This is particularly apparent for CPM formats, which suffer from implementation complexity, since the number of modulated waveforms in each signaling interval depends exponentially on the correlation length L. If a parsimonious representation of this set (i.e., a representation consisting of a small alphabet of signals) is available, receiver design can be greatly simplified. More specifically, if we refer to Laurent's representation of CPM signals, this principle has been applied to the development of simplified receivers for both AWGN channels [303, 304, 305] and frequency- selective channels [284, 306]. Finally, it is worth remembering that a similar approach to simplified detection has also been exploited by resorting to other parsimonious representations of CPM signals, such as the Gram–Schmidt expansion [291, 292, 306].

3.6.5.2 Rimoldi's Representation

In his seminal paper [302] Rimoldi proved that an M-ary CPM signal (3.103) characterized by a modulation index of the form:

$$h = \frac{K}{P}, \tag{3.177}$$

where K and P are relatively prime positive integers, can be rewritten as:

$$s_{RF}(t, \mathbf{c}) = \sqrt{P_s} \cos(2\pi f_1 t + \bar{\psi}(t, \mathbf{U})), \tag{3.178}$$

where

$$f_1 = f_c - \frac{h}{2T_s}(M - 1) \tag{3.179}$$

and $\bar{\psi}(t, \mathbf{U})$ are the new carrier frequency[20] and information-carrying phase (called the *physical tilted phase*), respectively, and $\mathbf{U} = \{U_k\}$ is the *modified data sequence* whose kth element is defined as:

$$U_k \triangleq \frac{c_k + (M - 1)}{2} \in \{0, 1, \ldots, M - 1\}. \tag{3.180}$$

To analyze the structure of $\bar{\psi}(t, \mathbf{U})$ in detail, we focus on the nth signaling interval, that is, on the time interval $[nT_s, (n + 1)T_s)$, and set $t = nT_s + \tau$, with $0 \le \tau < T_s$. Then the physical tilted or observable phase $\bar{\psi}(t, \mathbf{U})$ can be expressed as [302]:

$$\bar{\psi}(t, \mathbf{U}) = \bar{\psi}(nT_s + \tau, \mathbf{U})$$

$$= R_{2\pi} \left[2\pi h R_P \left[\sum_{i=0}^{n-L} U_i \right] + 4\pi h \sum_{i=0}^{L-1} U_{n-i} q(\tau + iT_s) + W(\tau) \right], \tag{3.181}$$

where $R_P[\cdot]$ is the modulo P operator (see (3.60)) and:

$$W(\tau) \triangleq \pi h(M - 1)\tau/T_s - 2\pi h(M - 1) \sum_{i=0}^{L-1} q(\tau + iT_s) + \pi h(L - 1)(M - 1) \tag{3.182}$$

groups all the data-independent terms. From (3.181) it is easy to see that the signal generated in the nth symbol interval depends on the data vector:

$$\mathbf{X}_n \triangleq [U_n, U_{n-1}, \ldots, U_{n-L+1}, V_n], \tag{3.183}$$

where

$$V_n \triangleq R_P \left[\sum_{i=0}^{n-L} U_i \right] \in \{0, 1, \ldots, P - 1\} \tag{3.184}$$

expresses the overall contribution to the signal phase due to previous symbols at the symbol times from 0 to $n - L$ inclusive. For this reason, we can interpret the modulator for the CPM signal (3.178) as the concatenation of a CPE fed by the data sequence $\{U_k\}$ and producing $\mathbf{X}_n$, with an MM, generating the output signal on the basis of the input vector $\mathbf{X}_n$. Note that: (a) the CPE *state* is defined by the L-dimensional vector $[U_{n-1}, U_{n-2}, \ldots, U_{n-L+1}, V_n]$, so that the number of possible states is $N_s = P \cdot M^{L-1}$; (b) the task of the CPE is to update the MM input $\mathbf{X}_n$ using the next data digit U_{n+1} to generate the MM input $\mathbf{X}_{n+1}$. The update of V_n can easily be accomplished in a recursive fashion, since[21]:

$$V_{n+1} = R_P \left[\sum_{i=0}^{n-L} U_i + U_{n-L+1} \right]$$

$$= R_P \left[R_P \left[\sum_{i=0}^{n-L} U_i \right] + U_{n-L+1} \right]$$

$$= R_P [V_n + U_{n-L+1}]. \tag{3.185}$$

Equations (3.183) and (3.185) show that the CPE is a *linear time-invariant sequential circuit*, composed of a modulo P adder and L delays. Moreover, a comparison of (3.100) with (3.177) shows that, if P is even, the overall number of states of the CPE is half that characterizing the FSSM model derived in

---

[20] The frequency shift with respect to f_c compensates for the offset between the phase $\bar{\psi}(t, \mathbf{U})$ of the new representation and the phase $\phi(t, \mathbf{c}_n)$ of the old one (3.103).
[21] In the following equation we exploit the fact that $R_x[y + z] = R_x[R_x[y] + z]$, for any positive x.

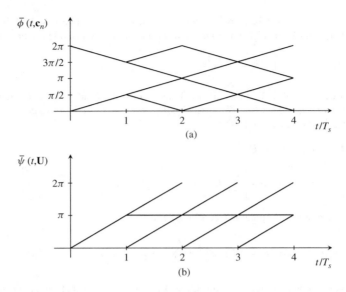

Figure 3.20   Representation of the *physical phase* (a) and the *physical tilted phase* (b) for MSK over four consecutive symbol intervals.

Section 3.6.3. This result can be related to the different structure of the phase trajectories characterizing the signal models in these two cases. Such a difference is evidenced by Figure 3.20; part (a) shows the modulo 2π representation of the phase tree, that is, a plot of the function:

$$\bar{\phi}(t, \mathbf{c}_n) \triangleq R_{2\pi}[\phi(t, \mathbf{c}_n)], \qquad (3.186)$$

called the *physical phase*, versus t for all the possible data vectors $\{\mathbf{c}_n\}$, for the MSK format[22] over four consecutive symbol intervals, while part (b) illustrates the MSK *physical tilted phase* over the same interval. Note that the MSK physical phase is represented by a four-state[23] *time-varying* trellis, whereas a two-state *time-invariant* trellis describes the MSK physical tilted phase. A possible implementation of the CPM modulator based on (3.178), (3.181), (3.183) and (3.185) is illustrated in Figure 3.21.

It is interesting to note that all time-dependent terms on the RHS of (3.181) depend only on the time offset variable $\tau = t - nT_s$. This implies that the possible phase trajectories $\{\bar{\psi}(nT_s + \tau, \mathbf{U})\}$ in any couple of consecutive symbol intervals will differ only by time translations after an initial transient (which allows the time-independent, data-dependent term on the RHS of (3.181) to take on all its possible values modulo 2π, provided that such values are the same in all the subsequent intervals). In other words, the MM can be described as a *time-invariant system*.

Applications

Rimoldi's representation has been shown to be a useful tool for the design of new trellis codes for CPMs. This is due to the fact that, under some specific assumptions about the number of modulation levels, the CPE can be interpreted as a *linear convolutional encoder over the ring of integers modulo P* [307]. Therefore, the use of modulo-P convolutional encoders appear to be a natural choice for the design of new trellis coding schemes for CPM; in fact, such encoders are structurally similar to

---

[22] The MSK tree, in the absence of a modulo 2π reduction, is given by Figure 3.13(a), if we set $h = 1/2$.
[23] This is in agreement with what was shown in Example 3.6.2.

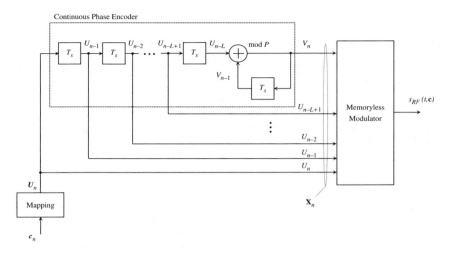

Figure 3.21 CPM transmitter decomposed into the concatenation of a CPE and an MM.

the CPE. Research work in this area for single antennas [308] and ST systems [309] has shown that, with CPSFK signaling, this encoding approach can lead to a larger reduction in the overall number of states than other approaches to the combination of other encoder and modulator pairs previously considered, and, in many cases, to significant additional coding gain.

3.6.6 *Data Block Transmission with CPM Signals for Frequency-Domain Equalization*

CPM signaling can be combined with FD equalization, even if the generation of cyclically-extended blocks of data is based on an algorithm which is quite different from that described in Section 3.5.3 for passband PAM. The signal generation technique we describe has been proposed in [284] and is summarized by the block diagram of Figure 3.22. In this diagram a binary data stream at rate $R_b = 1/T_b$ bits per second feeds a symbol mapper, associating, in a one-to-one fashion, each group of m bits with a channel symbol belonging to an M-ary ASK constellation $\Omega = \{\pm 1, \pm 3, \dots, \pm(M-1)\}$, with $M = 2^m$. The channel symbol stream is divided into nonoverlapping blocks, each consisting of two subblocks, the first of length $N - N_{cp} - K$ and the second of length N_{cp} (the parameters K and N_{cp} are defined later). Let the vector $\mathbf{c}_{fs}^{(l)} \triangleq [(\mathbf{c}_f^{(l)})^T, (\mathbf{c}_s^{(l)})^T]^T$ denote the resulting lth block, with $\mathbf{c}_f^{(l)} \triangleq [c_{N_{cp}}^{(l)}, c_{N_{cp}+1}^{(l)}, \dots, c_{N-K-1}^{(l)}]^T$ and $\mathbf{c}_s^{(l)} \triangleq [c_N^{(l)}, c_{N+1}^{(l)}, \dots, c_{N+N_{cp}-1}^{(l)}]^T$. This block is *cyclically extended* by attaching a cyclic prefix $\mathbf{c}_p^{(l)} \triangleq [c_0^{(l)}, c_1^{(l)}, \dots, c_{N_{cp}-1}^{(l)}]^T = \mathbf{c}_s^{(l)}$ to its beginning. Then a vector $\mathbf{c}_i^{(l)} \triangleq [c_{N-K}^{(l)}, c_{N-K+1}^{(l)}, \dots, c_{N-1}^{(l)}]^T$ of K channel symbols, all belonging to Ω, is generated by the transmitter via a proper algorithm[24] processing $\mathbf{c}_{fs}^{(l)}$ and the symbols of the previous block in order to ensure the exact *cyclicity* of the transmitted signal associated with the lth data block. In other words, through a proper choice of $\mathbf{c}_i^{(l)}$, during the transmission of $\mathbf{c}_s^{(l)}$, the CPM modulator generates a waveform *identical* to that produced when sending the prefix $\mathbf{c}_p^{(l)}$, without disrupting the *phase continuity* of the transmitted signal at the beginning of $\mathbf{c}_s^{(l)}$. The vector $\mathbf{c}_i^{(l)}$ is inserted in $\mathbf{c}^{(l)}$

---

[24] Details are given below.

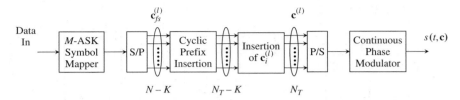

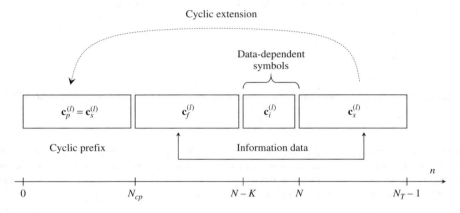

Figure 3.22 Block diagram of the transmitter for a communication system employing CPM combined with FD equalization.

Figure 3.23 Structure of the lth transmitted data block in the system shown in Figure 3.22.

between $\mathbf{c}_f^{(l)}$ and $\mathbf{c}_s^{(l)}$, producing the overall data block $\mathbf{c}^{(l)} \triangleq [(\mathbf{c}_p^{(l)})^T, (\mathbf{c}_f^{(l)})^T, (\mathbf{c}_i^{(l)})^T, (\mathbf{c}_s^{(l)})^T]^T$, lasting $N_T \triangleq N_{cp} + N$ symbol intervals. The overall structure of $\mathbf{c}^{(l)}$ is illustrated in detail in Figure 3.23.

In the following it is assumed that $M = 2$ for simplicity (i.e., binary modulation formats are considered). This choice entails no loss of generality, since the proposed approach can immediately be extended to higher-order alphabets using the Laurent representation for multilevel CPM of [297].

The data blocks $\{\mathbf{c}^{(l)}\}$, after parallel-to-series conversion, feed a continuous phase modulator generating the baseband signal $s(t, \mathbf{c})$ (3.104),[25] whose phase $\phi(t, \mathbf{c})$ (3.99) is given by:

$$\phi(t, \mathbf{c}) \triangleq \sum_{l=-\infty}^{+\infty} \varphi^{(l)}(t - lN_T T_s, \mathbf{c}^{(l)}), \tag{3.187}$$

where (see (3.113)):

$$\varphi^{(l)}(t, \mathbf{c}^{(l)}) = 2\pi h \sum_{n=0}^{N_T-1} c_n^{(l)} q(t - nT_s) \tag{3.188}$$

is the contribution of the lth data block $\mathbf{c}^{(l)}$. Note that $\varphi^{(l)}(t, \mathbf{c}^{(l)})$ (3.188) can be simplified as:

$$\varphi^{(l)}(t, \mathbf{a}^{(l)}) = \pi h \sum_{i=0}^{n-L} a_i^{(l)} + 2\pi h \sum_{i=n-L+1}^{n} a_i^{(l)} q(t - iT_s) \tag{3.189}$$

---

[25] Here the infinite-dimensional vector $\mathbf{c}$ denotes the ordered concatenation of the block sequence $\{\mathbf{c}^{(l)}\}$.

with $\theta_n^{(l)} \triangleq R_{2\pi}[\pi h \sum_{i=0}^{n-L} a_i^{(l)}]$ (see (3.133)) for $nT_s \le t < (n+1)T_s$, with $L-1 \le n \le N_T - 1$. This shows that, for the lth block, the modulator phase $\varphi^{(l)}(t, \mathbf{a}^{(l)})$ in the nth symbol interval is defined by its overall *state* $\Delta_n^{(l)}$ (see (3.141)), whose value is determined as the ordered pair $(\theta_n^{(l)}, \sigma_n^{(l)})$. Here, $\sigma_n^{(l)}$ and $\theta_n^{(l)}$ are the modulator *correlative state* (an integer representation of the symbol vector $(c_{n-1}^{(l)}, c_{n-2}^{(l)}, \dots, c_{n-L+1}^{(l)})$, see (3.134)) and the *phase state*, respectively, in the nth symbol interval of the lth data block

As stated above, it is important to ensure the property of cyclicity over each data block without disrupting the phase continuity in the transmitted signal. We note that *phase continuity* between consecutive data blocks is guaranteed if and only if, for any l, the last modulator state of block $l-1$ is equal to the first state of block l, that is, $\Delta_{N_T}^{(l-1)} = \Delta_0^{(l)}$ (see Figure 3.23), whereas the *cyclicity* of $\varphi^{(l)}(t, \mathbf{a}^{(l)})$ (with period NT_s) over the time interval $[lN_T T_s, (l+1)N_T T_s]$ requires that $\Delta_0^{(l)} = \Delta_N^{(l)}$. These two constraints lead to the equality:

$$\Delta_N^{(l)} = \Delta_{N_T}^{(l-1)}, \tag{3.190}$$

requiring us to force the modulator state at the instant $n = N$ in the lth block to a value coming from the previous block. The constraint (3.190) can be satisfied by properly adjusting the K channel symbols of $\mathbf{c}_i^{(l)}$, as shown in the following. To begin, we note that (3.190) is equivalent to:

$$\theta_N^{(l)} = \theta_{N_T}^{(l-1)} \tag{3.191}$$

and

$$c_{N-k}^{(l)} = c_{N_T-k}^{(l-1)}, \tag{3.192}$$

with $k = 1, 2, \dots, L-1$. Equation (3.192) fixes the value of the last $L-1$ symbols of $\mathbf{c}_i^{(l)}$. Then the remaining $K-L+1$ symbols $\{c_{N-K+k}^{(l)}, k = 0, 1, \dots, K-L\}$ of $\mathbf{c}_i^{(l)}$ (provided that $K \ge L$) should be selected in such a way that (3.191) is satisfied. Since $\theta_N^{(l)}$ can be expressed as (see (3.137)):

$$\theta_N^{(l)} = \theta_{N-K+(L-1)}^{(l)} + \pi h \sum_{k=0}^{K-L} c_{N-K+k}^{(l)}, \tag{3.193}$$

the constraint (3.191) can be reformulated as:

$$\pi h \sum_{k=0}^{K-L} c_{N-K+k}^{(l)} = \xi_l \tag{3.194}$$

where $\xi_l \triangleq \theta_{N_T}^{(l-1)} - \theta_{N-K+(L-1)}^{(l)}$. The phase state $\theta_{N-K+(L-1)}^{(l)}$ depends on the *information data vector* $[(\mathbf{c}_p^{(l)})^T, (\mathbf{c}_f^{(l)})^T]$ (i.e., on the data preceding $\mathbf{c}_i^{(l)}$ and belonging to the same block), on the phase state $\theta_0^{(l)}$ at the beginning of the lth data block $\mathbf{c}^{(l)}$ and on the last $L-1$ symbols of the previous block, since:

$$\theta_{N-K+(L-1)}^{(l)} = \theta_0^{(l)} + \pi h \sum_{k=0}^{L-2} c_{N_T-(L-1)+k}^{(l-1)} + \pi h \sum_{k=0}^{N-K-1} c_k^{(l)}. \tag{3.195}$$

Then, given all the state/symbol information about block $l-1$, the remaining $K-L+1$ unknown symbols of $\mathbf{c}_i^{(l)}$ should satisfy equation (3.194). We note that, if the unavoidable phase ambiguity of 2π is taken into account, the phase state difference ξ_l in (3.194) can take on p equally spaced values belonging to the interval $[-((p-1)/2)2\pi/p, (p/2)2\pi/p]$. Since $h = 2z/p$ (see (3.100)) and $c_n^{(l)} \in \{-1, +1\}$, it is not difficult to infer that at least one solution to (3.194) exists if $z(K-L+1) \ge \lceil p/2 \rceil$, that is, $K \ge \lceil p/2 \rceil / z + L - 1$, where $\lceil x \rceil \triangleq \min\{n \in \mathbb{Z} | n \ge x\}$ denotes the so-called *ceiling function*. Then, if this inequality is satisfied, a specific symbol pattern of length $K-L+1$ satisfying (3.194) can be stored in a read-only memory for each possible value of ξ_l at the transmitter.

Exploiting Laurent's decomposition for binary CPM signals (see equation (3.166)), $s(t, \mathbf{c})$ can be represented as the superposition of P linearly modulated digital signals as:

$$s(t, \mathbf{c}) = \sqrt{P_s} \sum_{l=-\infty}^{+\infty} \sum_{p=0}^{P-1} \sum_{n=0}^{N_T-1} a_{p,n}^{(l)} \, l_p(t - nT_s - lN_T T_s), \qquad (3.196)$$

where $P = 2^{L-1}$, $l_p(t)$ is the pth Laurent pulse and $a_{p,n}^{(l)}$ is the pth Laurent symbol (belonging to a proper p-ary constellation Ξ) in the nth interval of the lth data block. From [296] it can be shown that:

$$a_{p,n}^{(l)} = \exp\left(j\pi h \sum_{f=-\infty}^{l-1} \sum_{m=0}^{N_T-1} c_m^{(f)} + \sum_{m=0}^{n} c_m^{(l)} - \sum_{k=0}^{L-1} c_{n-k}^{(l)} b_{p,k} \right), \qquad (3.197)$$

for $n = L - 1, L, \dots, N_T - 1$, and:

$$a_{p,n}^{(l)} = \exp\left(j\pi h \sum_{f=-\infty}^{l-1} \sum_{m=0}^{N_T-1} c_m^{(f)} + \sum_{m=0}^{n} c_m^{(l)} - \sum_{k=0}^{n} c_{n-k}^{(l)} b_{p,k} - \sum_{k=n+1}^{L-1} c_{N_T+n-k}^{(l-1)} b_{p,k} \right), \qquad (3.198)$$

for $n = 0, 1, \dots, L - 2$, where $b_{p,k} \in \{0, 1\}$ is the kth coefficient of the binary decomposition of the index p, that is, $p = \sum_{k=1}^{L-1} 2^{k-1} b_{p,k}$. The signal $s(t, \mathbf{c})$ is transmitted, after RF conversion, over a communication channel. As in the PAM case (see Section 3.5.3) we assume that: (a) the CIR lasts L_h symbol intervals and undergoes negligible changes during each transmitted block (*quasi-static* channel), so that, during the transmission of $\mathbf{c}^{(l)}$, it can be written as $h^{(l)}(t)$; (b) the duration of the cyclic prefix is not smaller than the overall channel memory (i.e., $L_h + L \leq N_{cp}$). Under these hypotheses it can be proved that the useful component of the baseband received signal $r(t)$, because of its cyclic structure, can be expressed as [284]:

$$z(t) = \sqrt{\frac{P_s}{N}} \sum_{l=-\infty}^{+\infty} \sum_{p=0}^{P-1} \sum_{k=-\infty}^{+\infty} L_{p,k} \, H_k^{(l)} B_{p,k}^{(l)} \exp\left(j \frac{2\pi k(t - lN_T T_s)}{NT_s} \right) \qquad (3.199)$$

for $t \in \bigcup_{l=-\infty}^{+\infty} [lN_{cp}T_s, lN_T T_s)$, where $H_k^{(l)} \triangleq H^{(l)}(k/NT_s)$, $H^{(l)}(f)$ is the FCT of $h^{(l)}(t)$, $L_{p,k} \triangleq L_p(k/NT_s)/T_s$, $L_p(f)$ is the FCT of $l_p(t)$, and $B_{p,k}^{(l)}$ is the kth element of the DFT of $\mathbf{b}_{p,N}^{(l)} \triangleq [b_{p,N_{cp}}^{(l)}$, $b_{p,N_{cp}+1}^{(l)}, \dots, b_{p,N_T-1}^{(l)}]^T$, that is, of the vector $\mathbf{B}_{p,N}^{(l)} \triangleq \mathbf{Q}_N \, \mathbf{b}_{p,N}^{(l)}$, where $\mathbf{Q}_N = [q_{n,k}]$ is the N-point DFT matrix (see Section 3.5.3). In other words, the useful portion of the received signal can be seen, for each data block, as the sum of P contributions, the pth of which consists of the superposition of infinite equally-spaced oscillations. Note that the kth oscillation in the pth contribution is characterized by the complex gain $L_{p,k} H_k^{(l)}$ and the frequency $f_k \triangleq k/(NT_s)$ and that $B_{p,k}^{(l)} = B_{p,k+N}^{(l)}$ for any k, so that, in principle, a given frequency-domain symbol can be associated with two distinct oscillations.

3.6.7 *Power Spectral Density of Continuous Phase Modulations*

Unlike passband PAM signaling, no general closed-form expression exists for the power spectrum of CPM signals. Despite this, T. Aulin and C.-E. Sundberg [288] have shown that, with CPM signaling, the relationship (3.25) between its average autocorrelation function and the corresponding power spectrum can be simplified in such a way that computation of the latter from the former can be accomplished via standard techniques for numerical integration. In this subsection, we first sketch the derivation of this general result. Then we provide some closed-form expressions referring to the power spectrum of CPFSK signals. Finally, we provide some meaningful numerical results.

3.6.7.1   A General Method for the Computation of the CPM Average Power Spectral Density

It is not difficult to prove that CPMs are wide-sense cyclostationary (WSC) signals with period $T_{cs} = T_s$, so that the methodology illustrated in Section 3.4 can be adopted for the evaluation of their average power spectral density. Before facing the problem of evaluating the average autocorrelation function $R_s(\tau)$ of the CPM complex envelope $s(t, \mathbf{c})$ (3.104) via equations (3.27) and (3.26), it is important to note that, since $R_s(\tau)$ is *Hermitian* (i.e., $R_s(-\tau) = R_s^*(\tau)$), (3.25) can be rewritten as:

$$
S_s(f) = 2\mathrm{Re}\left\{ \int_0^{+\infty} R_s(\tau)\, \exp\,(-j2\pi f\tau)\, d\tau \right\}
$$
$$
= 2\mathrm{Re}\left\{ \int_0^{LT_s} R_s(\tau)\, \exp\,(-j2\pi f\tau)\, d\tau + \int_{LT_s}^{+\infty} R_s(\tau)\, \exp\,(-j2\pi f\tau)\, d\tau \right\}, \qquad (3.200)
$$

so that the knowledge of $R_s(\tau)$ is needed only for $\tau \geq 0$. Hence, in the following, we focus on the evaluation of $R_s(\tau)$ for *nonnegative values of* τ and, to simplify the analysis, we set:

$$
\tau = lT_s + \varepsilon T_s, \qquad (3.201)
$$

where l is a nonnegative integer and ε is a real number in the interval $[0, 1)$. If we assume that the sequence $\{c_k\}$ consists of iid symbols, each characterized by the probability:

$$
P_n \triangleq \Pr\{c_k = n\}, \qquad (3.202)
$$

for any $n \in A_c \triangleq \{\pm 1, \pm 3, \dots, \pm(M-1)\}$, it is not difficult to prove that:

$$
R_s(\tau) = \frac{P_s}{T_s} \int_0^{T_s} \prod_{k=-L+1}^{l+1} g(t - kT_s, \tau)\, dt \qquad (3.203)
$$

for $\tau \geq 0$, where:

$$
g(t - kT_s, \tau) \triangleq \mathrm{E}\{\exp\,[j2\pi h\, c_k\, d(t - kT_s, \tau)]\}
$$
$$
= \sum_{\substack{n=-(M-1) \\ \text{odd } n}}^{M-1} P_n \exp\,[j2\pi h\, n\, d(t - kT_s, \tau)] \qquad (3.204)
$$

and $d(t - kT_s, \tau) \triangleq q(t + \tau - kT_s) - q(t - kT_s)$. In general, a closed-form solution to the integral on the RHS of (3.203) does not exist. Numerical integration techniques can be exploited for the evaluation of $R_s(\tau)$ via (3.203), but, unfortunately, the number of factors in the integrand function in this expression becomes infinite as $\tau \to \infty$, since $l \to \infty$ (see (3.201)). Note that this problem cannot be neglected because of the need to compute the second integral in (3.200) in the evaluation of $S_s(f)$; however, the problem can be circumvented as follows. It can be proved that, if $\tau \geq LT_s$ (i.e., $l \geq L$), then (3.203) can be simplified to:

$$
R_s(\tau) = P_s[\psi_c(jh)]^{l-L} f(\varepsilon), \qquad (3.205)
$$

where

$$
\psi_c(jh) \triangleq \mathrm{E}\{\exp\,(j\pi hc)\} = \sum_{\substack{n=-(M-1) \\ \text{odd } n}}^{M-1} P_n \exp\,(j\pi hn) \qquad (3.206)
$$

is the *characteristic function* of the channel symbols, and:

$$
f(\varepsilon) \triangleq \frac{1}{T_s} \int_0^{T_s} l_0(t)l_1(t, \varepsilon)\, dt, \qquad (3.207)
$$

with

$$l_0(t) \triangleq \prod_{k=1-L}^{0} \sum_{\substack{n=-(M-1)\\ \text{odd } n}}^{M-1} P_n \exp\left\{j2\pi h\, n\left[\frac{1}{2} - q\left(t - kT_s\right)\right]\right\} \tag{3.208}$$

and

$$l_1(t,\varepsilon) \triangleq \prod_{k=-L+1}^{1} \sum_{\substack{n=-(M-1)\\ \text{odd } n}}^{M-1} P_n \exp\left[j2\pi h\, n\, q(t + \varepsilon T_s - kT_s)\right]. \tag{3.209}$$

Note that in equation (3.205) the dependence on l is *separated* from that on ε, since these parameters appear in distinct factors. This property can be exploited to rewrite the second integral on the RHS of (3.200) as:

$$\int_{LT_s}^{+\infty} R_s(\tau)\, \exp\left(-j2\pi f\tau\right)\, d\tau = \frac{1}{1 - \psi_c(jh)\, \exp\left(-j2\pi fT_s\right)}$$
$$\cdot \int_{LT_s}^{(L+1)T_s} R_s(\tau)\, \exp\left(-j2\pi f\tau\right)\, d\tau \tag{3.210}$$

if $|\psi_c(jh)|$ (3.206) is strictly less than unity, that is, if h *is not integer*. Substituting (3.210) in (3.200) yields the expression:

$$S_s(f) = 2\,\mathrm{Re}\left\{\int_0^{LT_s} R_s(\tau)\, \exp\left(-j2\pi f\tau\right)\, d\tau \right.$$
$$\left. + \frac{1}{1 - \psi_c(jh)\, \exp\left(-j2\pi fT_s\right)} \int_{LT_s}^{(L+1)T_s} R_s(\tau)\, \exp\left(-j2\pi f\tau\right)\, d\tau \right\}, \tag{3.211}$$

which allows the evaluation of $S_s(f)$, for any noninteger h, via the computation of a couple of integrals, both involving $R_s(\tau)$ in the form (3.203). In this procedure further simplifications can be introduced if the symbols of the constellation A_c are all equally likely, that is, $P_n = 1/M$ for $n = 0, 1, \ldots, M-1$ (see (3.202)). In fact, (3.206) and (3.204) simplify to:

$$\psi_c(jh) = \frac{1}{M}\frac{\sin(\pi hM)}{\sin(\pi h)} \tag{3.212}$$

and

$$g(t - kT_s, \tau) = \frac{1}{M}\frac{\sin[2\pi h\, M\, d(t - kT_s, \tau)]}{\sin[2\pi h\, d(t - kT_s, \tau)]}, \tag{3.213}$$

respectively. Then, $R_s(\tau)$ (3.203) can be rewritten in the *real* form:

$$R_s(\tau) = \frac{P_s}{T_s}\int_0^{T_s} \prod_{k=-L+1}^{l+1} \frac{1}{M}\frac{\sin[2\pi h\, M\, (q(t + \tau - kT_s) - q(t - kT_s))]}{\sin[2\pi h\, (q(t + \tau - kT_s) - q(t - kT_s))]}\, dt, \tag{3.214}$$

so that (3.211) can be simplified to:

$$S_s(f) = 2\left\{\int_0^{LT_s} R_s(\tau)\, \cos(2\pi f\tau)\, d\tau \right.$$
$$+ \frac{1 - \psi_c(jh)\, \cos(2\pi fT_s)}{1 + \psi_c^2(jh)\, - 2\psi_c(jh)\, \cos(2\pi fT_s)}\int_{LT_s}^{(L+1)T_s} R_s(\tau)\, \cos(2\pi f\tau)\, d\tau$$
$$\left. - \frac{\psi_c(jh)\, \sin\left(2\pi fT_s\right)}{1 + \psi_c^2(jh)\, - 2\psi_c(jh)\, \cos(2\pi fT_s)}\int_{LT_s}^{(L+1)T_s} R_s(\tau)\, \sin(2\pi f\tau)\, d\tau \right\}, \tag{3.215}$$

involving *real-valued functions* only. If h is *integer*, we have $|\psi_c(jh)| = 1$, so that $\psi_c(jh) = \exp(j2\pi v)$ with $0 \leq v < 1$. In this case, exploiting some mathematical results provided in [310, 311], (3.200) can be obtained in the form:

$$S_s(f) = 2\text{Re}\left\{\int_0^{LT_s} R_s(\tau)\exp(-j2\pi f\tau)\,d\tau + \frac{1}{2}\left[1 + \frac{1}{T_s}\sum_{k=-\infty}^{+\infty}\delta\left(f - \frac{v}{T_s} - \frac{k}{T_s}\right)\right.\right.$$

$$\left.\left. -j\cot\left(\pi T_s\left(f - \frac{v}{T_s}\right)\right)\right]\int_{LT_s}^{(L+1)T_s}R_s(\tau)\exp(-j2\pi f\tau)\,d\tau\right\}. \tag{3.216}$$

This shows the presence of *spectral lines* at the frequencies $\{f_n \triangleq (n + v)/T_s, n = \ldots, -1, 0, 1, \ldots\}$ and that these overlap with a continuous component of $S_s(f)$. The presence of these lines can be related to the periodic behavior (with period T_s) of $R_s(\tau)$ for $|\tau| \geq LT_s$ [288].

Note that the numerical method described above for the evaluation of $S_s(f)$ may yield inaccurate results (mainly because of the computer approximations in the evaluation of trigonometric functions) when estimating low side lobes [288]. Alternative methods to exactly determine the asymptotic behavior of the power spectrum are illustrated in [288, 312].

3.6.7.2 CPFSK Average Power Spectrum

Generally speaking, (3.215) does not lead, for a given phase response $q(t)$, to a closed-form expression for $S_s(f)$. The only exception is represented by CPFSK signals, corresponding to a 1-REC choice for $q(t)$ (see Section 3.6.2). In fact, in this case, it is found that [288]:

$$S_{s,n}(f) = \frac{1}{M}\sum_{k=1}^{M}\Delta_k^2(f) + \frac{2}{M^2}\sum_{k=1}^{M}\sum_{n=1}^{M}\Lambda_{k,n}(f)\,\Delta_k(f)\,\Delta_n(f), \tag{3.217}$$

where

$$\Delta_k(f) \triangleq \frac{\sin\left\{\pi\left[fT_s - \frac{1}{2}h(2k - 1 - M)\right]\right\}}{\pi\left[fT_s - \frac{1}{2}h(2k - 1 - M)\right]} \tag{3.218}$$

and

$$\Lambda_{k,n}(f) \triangleq \frac{\cos(2\pi fT_s - \alpha_{k,n}) - \psi_c(jh)\cos\alpha_{k,n}}{1 + \psi_c^2(jh) - 2\psi_c(jh)\cos(2\pi fT_s)}, \tag{3.219}$$

with

$$\alpha_{k,n} \triangleq \pi h(k + n - 1 - M), \tag{3.220}$$

and $\psi_c(jh)$ is given by (3.212). Note that, with MSK ($M = 2$, $h = 1/2$), (3.217) yields:

$$S_{s,n}(f) = \frac{16}{\pi^2}\left[\frac{\cos(2\pi fT_s)}{1 - (4fT_s)^2}\right]^2. \tag{3.221}$$

3.6.7.3 Numerical Results

The following considerations are helpful in understanding the qualitative behavior of CPM power spectra. First, an increase in the modulation index h or in the alphabet cardinality M, given all the other CPM parameters, produces an increase in the maximum *frequency deviation* with respect to the carrier frequency f_c, thereby broadening the spectrum of the RF signal. Second, given the frequency response $p(t)$ (or, equivalently, the phase response $q(t)$) the bandwidth of a CPM signal reduces as L increases, since this entails slower variations in the data-carrying phase $\phi(t, \mathbf{c})$. Third, for a given

L, bandwidth containment depends on the *regularity* of $q(t)$. In particular, it can be proved that, if $q(t)$ has c continuous derivatives, the power spectral density $S_s(f)$ *asymptotically* decreases as [312, 313, 280]:

$$S_s(f) \sim |f|^{-2(c+2)}. \tag{3.222}$$

Some numerical results are shown in Figures 3.24–3.30, representing the *normalized power spectral density* $S_{s,n}(f)$ (3.33), for various significant CPM formats. Note that Figures 3.24 and 3.25 were generated via (3.217)–(3.220), whereas all the other results were obtained by employing (3.214) and (3.215) and numerical integration techniques.

Figures 3.24–3.27 show $S_{s,n}(f)$ for binary and quaternary CPSK signals with different values of h. These results provide evidence that CPFSK has a regular and compact spectrum for $0 < h < 1$. When h approaches unity, M peaks become visible; for $h = 1$, these become M spectral lines, deriving from the double sum in (3.217). When h becomes larger than unity, the spectrum width increases further and, for these reasons, selecting $h < 1$ is advisable for practical applications.

The impact of a change in L is demonstrated by Figures 3.28 and 3.29, showing $S_{s,n}(f)$ for binary L-RC and L-REC signals, respectively, characterized by $h = 1/2$ and $L = 1, 2$ and 3. Note that the 1-REC modulation format corresponds to MSK, and the 1-RC one to the so-called *sinusoidal frequency shift keying* (SFSK) [314]. Comparing Figure 3.29 with Figure 3.28 shows that the selection of an L-RC signal enables, all the other signal parameters being equal, better spectral compactness to be achieved than its L-REC counterpart. This results could also have been foreseen using (3.222), establishing that, asymptotically:

$$S_s(f) \sim |f|^{-4} \tag{3.223}$$

for an L-REC signal, since $c = 0$, whereas:

$$S_s(f) \sim |f|^{-8} \tag{3.224}$$

for an L-RC signal, since in this case $c = 2$.

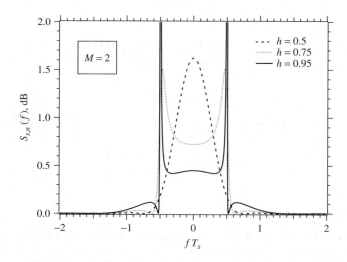

Figure 3.24 Normalized PSD $S_{s,n}(f)$ for binary CPFSK with $h = 0.5$, 0.75 and 0.95.

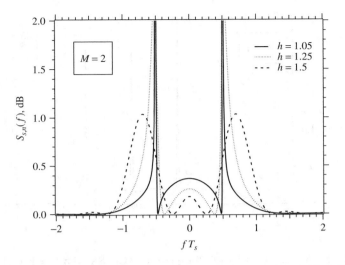

Figure 3.25 Normalized PSD $S_{s,n}(f)$ for binary CPFSK with $h = 1.05$, 1.25 and 1.5.

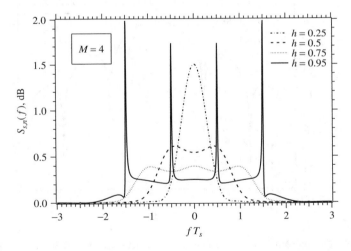

Figure 3.26 Normalized PSD $S_{s,n}(f)$ for quaternary CPFSK with $h = 0.25$, 0.5, 0.75 and 0.95.

Finally, Figure 3.30 shows $S_{s,n}(f)$ for the GMSK signals characterized by a normalized bandwidth BT_s (of the modulator Gaussian filter) equal to 0.1, 0.2 and 0.3. In this case, in the computation of the power spectrum via (3.214) and (3.215), the duration of $p(t)$ (3.146) has been truncated[26] to $L = 10$, 5 and 4 symbol intervals for $BT_s = 0.1$, 0.2 and 0.3, respectively, following the indications provided by Figure 3.16.

---

[26] The center of the truncation interval corresponds to $t = T_s/2$.

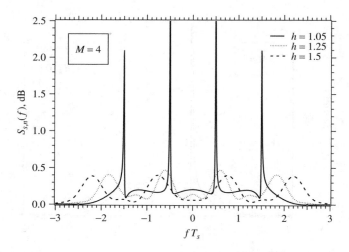

Figure 3.27   Normalized PSD $S_{s,n}(f)$ for quaternary CPFSK with $h = 1.05$, 1.25 and 1.5.

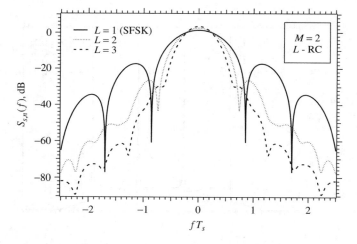

Figure 3.28   Normalized PSD $S_{s,n}(f)$ for binary L-RC with $h = 0.5$ and $L = 1, 2$ and 3.

3.7 OFDM

3.7.1 Introduction

In wireless systems, as discussed in Chapter 2, the communication channel can distort both the phase and amplitude of the transmitted signal, appreciably modifying its shape. As illustrated in Chapter 6, for single-carrier (SC) communication techniques, reliable reception could then require the use of very complex detection algorithms. In principle, to simplify detection in these scenarios, frequency division multiplexing (FDM) can be adopted. To understand the implications of this choice, consider the channel to be *static* within the digital signaling time, but causing appreciable amplitude distortion over the available frequency band, having central frequency f_{sc} and bandwidth B_{sc}. In this case, any SC

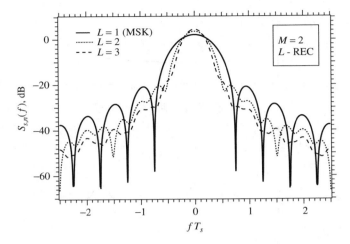

Figure 3.29 Normalized PSD $S_{s,n}(f)$ for binary L-REC with $h = 0.5$ and $L = 1, 2$ and 3.

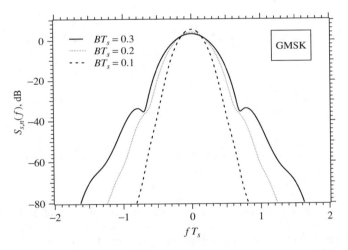

Figure 3.30 Normalized PSD $S_{s,n}(f)$ for GMSK with $BT_s = 0.1, 0.2$ and 0.3.

modulation, whose spectrum occupies all the available bandwidth, undergoes appreciable distortion. To circumvent this problem an FDM technique can be applied, which splits the available bandwidth into N_{mc} subbands or *subchannels*. Let $f_{mc,i}$ denote the *center frequency* of the ith subchannel, with $i = 0, 1, \ldots, N_{mc} - 1$. If we assume that all these subintervals have the same width $B_{mc} \triangleq B_{sc}/N_{mc}$ and that N_{mc} is sufficiently large, the amplitude response of each subchannel is almost flat over its bandwidth, as illustrated in Figure 3.31(a). This suggests splitting the data sequence $\{m_n\}$, to be sent within the overall bandwidth B_{sc}, into N_{mc} parallel streams $\{m_n^{(i)}\}$, with $i = 0, 1, \ldots, N_{mc} - 1$, and transmitting $\{m_n^{(i)}\}$ over the ith subchannel. This can be accomplished using an SC modulation with carrier frequency $f_{mc,i}$ and bandwidth B_{mc}, as illustrated in Figure 3.31(a). If this solution is adopted, the resulting signal, consisting of the superposition of N_{mc} *distinct signals generated in parallel*, is a *multicarrier modulation* (MCM).

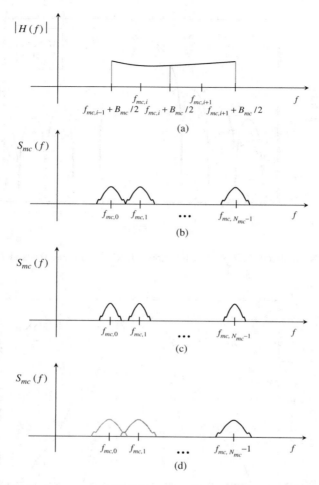

Figure 3.31 (a) Amplitude response $|H(f)|$ in two adjacent subbands of an FDM wireless system (subchannels i and $i + 1$ are considered). Power spectral density $S_{mc}(f)$ of an MCM under the assumptions of: (b) absence of spectral overlap among the signals associated with different subcarriers and of a guard band between adjacent subchannels; (c) absence of spectral overlap among the signals associated with different subcarriers and presence of a guard band between adjacent subchannels; (d) presence of a partial spectral overlap among the signals associated with two adjacent subchannels.

   Under the above assumptions, all the spectral components of the signal transmitted within each subchannel undergo approximately the same attenuation and phase shift, so that compensation for these linear distortions, when detecting data, can be carried out using simple signal processing techniques for each of the N_{mc} signals. Hence, the adoption of an MCM dramatically simplifies the problem of data detection. This choice, however, suffers from some practical problems since both generation and coherent demodulation of multicarrier signals require, in principle, the availability of N_{mc} distinct oscillators at the transmitter and at the receiver, respectively. These generate N_{mc} distinct *subcarriers* with frequencies $\{f_{mc,i}, i = 0, 1, \ldots, N_{mc} - 1\}$ and their use appreciably increases modem complexity.

Furthermore, to achieve good error performance in MCM-based systems, it is not in general suffi-cient to ensure that the linear distortions introduced by each subchannel are small. In fact, a further cause of performance degradation is the so-called *intercarrier interference* (ICI), that is, the inter-ference in each subchannel due to the adjacent subchannels. ICI removal can be accomplished by avoiding any *spectral overlap* among signals transmitted over adjacent subchannels, as illustrated in Figure 3.31(b); however, this solution has been discarded in the past since it requires the use of passband filters having an extremely narrow transition band and, consequently, large cost/complexity. Note also that the introduction of a sufficient guard bandwidth, as illustrated in Figure 3.31(c), entails a significant loss in system spectral efficiency.

These considerations explain why, as discussed in Section 3.10, historically speaking, the first relevant papers about MCM tackled the problem of devising signal sets consisting of N_{mc} distinct func-tions and having partially overlapped spectra (see Figure 3.31(d)), but sharing an important property, known as *orthogonality*. In fact, this property may allow the separation of signals associated with dis-tinct subchannels without mutual interference, even in the absence of spectral separation. To understand this point and analyze the design of MCM from a general perspective, let us formally describe the prob-lem of *parallel transmission* of a data *block* consisting of N channel symbols, belonging to an M-ary complex alphabet A_c and forming the vector[27] $\mathbf{c}_N = [c_0, c_1, \ldots, c_{N-1}]^T$, in the time interval (t_i, t_f). We assume that, in this interval, this block is transmitted by a digital modulator which, in baseband, can generate the *set of orthogonal complex functions* $I_\psi \triangleq \{\psi_k(t) \triangleq \sqrt{E_{\psi,k}} \phi_k(t), k = 0, 1, \ldots, N - 1\}$, such that:

$$\int_{t_i}^{t_f} \psi_k(t) \, \psi_l^*(t) \, dt = \sqrt{E_{\psi,k}} \sqrt{E_{\psi,l}} \int_{t_i}^{t_f} \phi_k(t) \, \phi_l^*(t) \, dt = E_{\psi,k} \, \delta_{kl}, \tag{3.225}$$

where $E_{\psi,k}$ is the energy of $\psi_k(t)$ and $\phi_k(t)$ is a unit energy function or signal. We also assume that the digital modulator generates an RF signal whose complex envelope is given by:

$$s(t, \mathbf{c}_N) \triangleq \sum_{k=0}^{N-1} c_k \, \psi_k(t) = \sum_{k=0}^{N-1} c_k \sqrt{E_{\psi,k}} \, \phi_k(t), \tag{3.226}$$

that is, we assume that the modulator generates $s(t, \mathbf{c}_N)$ as in (3.226) as a linear combination of the functions of I_ψ using the channel symbols $\{c_k\}$ as coefficients. To grasp the meaning of (3.226), it is useful to consider that each of the functions $\{\psi_k(t)\}$ can be associated with a different subchannel, as in multicarrier systems. Note, however, that the orthogonality (3.225) is the only constraint on the set I_ψ and this does not entail, in principle, that distinct signals of I_ψ are characterized by disjoint time intervals and/or by nonoverlapping spectra. In fact, these time or spectral properties are sufficient but not necessary conditions for orthogonality.

The mutual orthogonality of the different components of $s(t, \mathbf{c}_N)$ in (3.226) represents a fundamental property of parallel transmission. To understand its importance, let us assume that the functions $\{\psi_k(t)\}$ are all characterized by a bandwidth which is small with respect to the *coherence bandwidth*[28] of the channel, which then causes, to a good approximation, the same attenuation and phase shift in all the spectral components of each of the signals of I_ψ. Then, if we neglect the channel delay, a *coherent receiver*,[29] when $s(t, \mathbf{c}_N)$ is transmitted, observes the noisy signal:

$$r(t) = \sum_{k=0}^{N-1} H_k \, c_k \, \psi_k(t) + n(t), \tag{3.227}$$

---

[27] Note that the index on the elements belonging to this vector does not provide any time indication since their transmission is accomplished in parallel.

[28] The *coherence bandwidth* of a wireless communication channel is defined in Section 2.2.2.

[29] A *coherent* receiver has an exact replica of the carrier oscillation associated with the useful component of the received signal.

where $n(t)$ is AWGN, and H_k is a complex gain representing channel effects on $\psi_k(t), k = 0, 1, \ldots, N - 1$. To obtain an estimate $\hat{\mathbf{c}}_N$ of $\mathbf{c}_N$ from $r(t)$, the receiver computes the scalar product:

$$X_k \triangleq (r, \psi_k) = \int_{t_i}^{t_f} r(t) \, \psi_k^*(t) \, dt \tag{3.228}$$

for $k = 0, 1, \ldots, N - 1$. Then substituting (3.226) in (3.228) and exploiting the mutual orthogonality of the signals $\{\psi_k(t)\}$ yields, for each k:

$$X_k = E_{\psi,k} \, H_k \, c_k + n_{\psi,k}, \tag{3.229}$$

where

$$n_{\psi,k} \triangleq (n, \psi_k) = \int_{t_i}^{t_f} n(t) \, \psi_k^*(t) \, dt \tag{3.230}$$

is a complex Gaussian random variable representing the channel noise contribution. Note that X_k depends on c_k and, as can easily be proved (e.g., see [315], Sect. 2.1), the noise random variables $\{n_{\psi,k}, k = 0, 1, \ldots, N - 1\}$ are mutually *independent*. Therefore, *if the receiver has exact knowledge of the channel parameters*[30] $\{H_k\}$, it can make a decision $\hat{c}_k$ on c_k by processing X_k only. Hence, the property of orthogonality leads to a conceptually simple detection strategy, operating subchannel-by-subchannel, whose implementation is sketched in Figure 3.32. The receiver structure illustrated in this diagram consists of N distinct, but structurally identical, subchannels. The kth branch compensates for the linear distortions[31] on the kth subchannel by computing (see (3.229)):

$$X_k/(E_{\psi,k} \, H_k) = c_k + n_{\psi,k}/E_{\psi,k} \, H_k \tag{3.231}$$

and then making a decision on c_k.

If we assume that the communication channel modifies the shape of the transmitted waveforms $\{\psi_k(t)\}$, thus generating a new signal set $\{\chi_k(t), k = 0, 1, \ldots, N - 1\}$, which does not consist of orthogonal functions, the detection strategy shown in Figure 3.32 will inevitably suffer from ICI when

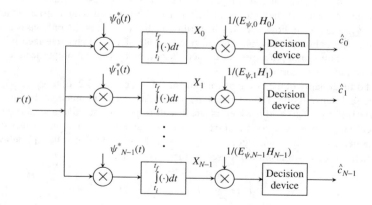

Figure 3.32   Block diagram of a baseband detector for parallel data transmission.

---

[30] This knowledge is necessary for the cancelation of the channel distortion.
[31] This approach to compensation for channel distortions is known as *equalization in the frequency domain* for multicarrier signals (see Section 6.3).

trying to separate the contribution of each subchannel. To generate a set of functions preserving the property of orthogonality in parallel transmission over a static distorting channel, *complex exponential* signals can be used. In fact, these signals share an important property: if they are applied to a linear time-invariant system, it generates, after a transient, signals of the same type.

If the channel is *nondistorting*, the construction of a set I_ψ, consisting of N complex orthogonal functions, is extremely simple. In fact, if we define:

$$\psi_k(t) \triangleq \sqrt{\frac{E_{\psi,k}}{NT_s}} \, \exp\,[j2\pi(f_k + f_{os})t] \tag{3.232}$$

over the interval $(t_i, t_f) = (0, NT_s)$, where T_s is the *symbol interval*, $f_k \triangleq k/NT_s$, with $k = 0$, $1, \ldots, N - 1$, and f_{os} an arbitrary *frequency offset*, the orthogonality condition (3.225) is satisfied. Note that: (a) the functions partially overlap in the frequency domain, since:

$$\Psi_k(f) \triangleq \mathrm{FCT}[\psi_k(t)] = \sqrt{NT_s E_{\psi,k}} \, \mathrm{sinc}[NT_s(f - (f_k + f_{os}))] \tag{3.233}$$

extends over an unlimited interval; (b) this choice allows us to transmit the symbol vector $\mathbf{c}_N = [c_0, c_1, \ldots, c_{N-1}]^T$ in an interval of *finite duration* (NT_s seconds); (c) the frequency separation between adjacent subcarriers is equal to the inverse of the transmission duration (i.e., $1/NT_s$ Hertz); (d) it is not difficult to prove that this separation represents the *minimum separation* to ensure the orthogonality of the signals (3.232). Unfortunately, in a *linearly distorting channel*,[32] the signal set generated through the choice (3.232) for $\psi_k(t)$ does not consist of orthogonal signals in the time interval $(0, NT_s)$. This is due to the fact that the channel, when fed by $\psi_k(t)$, goes through a transient starting at $t = 0$, so that the signal observed in that interval does not look like a complex exponential until the transient is over. To reestablish the orthogonality property in the useful component of the received signal observed by the receiver in the interval $(0, NT_s)$, the following solution can be adopted, provided that the communication channel has a memory of limited duration not exceeding some L_h intervals and, in particular, that its impulse response $h(t)$ extends only over the interval $[0, L_h T_s]$. We then feed the communication channel with the signal:

$$\gamma_k(t) \triangleq \sqrt{\frac{E_{\gamma,k}}{(N + N_p)T_s}} \, \exp\,[j2\pi(f_k + f_{os})t] \tag{3.234}$$

extending over the new time interval $(-N_p T_s, NT_s)$, with $N_p \geq L_h$, in place of $\psi_k(t)$ (3.232) (here, $E_{\gamma,k}$ is the energy of $\gamma_k(t)$). Note that $\gamma_k(t)$ is a complex exponential with frequency f_k, exactly like $\psi_k(t)$ (3.232), but it starts exciting the communication channel $N_p T_s$ seconds *before the beginning of the interval* $(0, NT_s)$. The insertion of an additional time interval (i.e., a *time prefix*) of duration $N_p T_s \geq L_h T_s$ allows the channel to reach its *steady-state condition* before $t = 0$. In fact, the channel response $\rho_k(t)$ to the excitation $\gamma_k(t)$ is then given by:

$$\rho_k(t) \triangleq \gamma_k(t) \otimes h(t)$$

$$= H(f_k + f_{os}) \sqrt{\frac{E_k}{(N + N_p)T_s}} \, \exp\,[j2\pi(f_k + f_{os})t] \tag{3.235}$$

for $(L_h - N_p)T_s \leq t \leq NT_s$, where $H(f_k + f_{os})$ is the value of the channel frequency response $H(f) \triangleq \mathrm{FCT}[h(t)]$ at the frequency $f = f_k + f_{os}$. The signals $\{\rho_k(t)\}$ form an orthogonal set as in (3.235) if observed only over the time interval $(0, NT_s)$, and consequently the receiver structure of Figure 3.32 can be used for their detection.

---

[32] The propagation delay due to the channel is again neglected.

These considerations lead to the conclusion that a parallel transmission technique based on the signal set $\{\gamma_k(t)\}$ (3.234) represents a valid solution for data communications over linearly distorting wireless channels. The use of this set, however, requires the generation of multiple oscillations, possibly without resorting to an oscillator bank. In the following subsection we show how this can be accomplished by resorting to the DFT. The resulting digital modulation is known as OFDM.

3.7.2 OFDM Signal Model

We first derive some analytical results useful for understanding multicarrier signal generation via DFT processing. Then we exploit these results to analyze the structure of an OFDM modulator.

3.7.2.1 Generation of a Multicarrier Signal via DFT Processing

To understand the technique for generating a multicarrier signal in an OFDM modulator, let us again focus on the transmission of the complex symbol vector $\mathbf{c}_N = [c_0, c_1, \ldots, c_{N-1}]^T$. Let us assume for the analysis that: (a) this vector undergoes the one-to-one transformation[33] $\mathbf{g} : \mathbb{C}^N \to \mathbb{C}^N$ yielding $\mathbf{a}_N = [a_0, a_1, \ldots, a_{N-1}]^T = \mathbf{g}(\mathbf{c}_N)$; b) a periodic sequence $\{a_k\}$ is generated as:

$$a_k = a_{R_N[k]} \tag{3.236}$$

for any $k \notin \{0, 1, \ldots, N-1\}$, that is, repeating $\mathbf{a}_N$ with period N. Then the impulsive signal:

$$\sum_{k=-\infty}^{+\infty} a_k\, \delta(t - kT_s) = \sum_{k=-\infty}^{+\infty} a_{R_N[k]}\, \delta(t - kT_s) \tag{3.237}$$

is generated from $\{a_k\}$ (here T_s is the *symbol interval*) and is applied to a filter with impulse response $p(t)$. This produces the signal:

$$s(t, \mathbf{c}_N) = \sum_{k=-\infty}^{+\infty} a_{R_N[k]}\, p(t - kT_s), \tag{3.238}$$

which is *periodic* with period $T = NT_s$. Its Fourier series representation may be written as:

$$s(t, \mathbf{c}_N) = \sum_{m=-\infty}^{+\infty} S_m\,(\mathbf{c}_N) \exp\,(j2\pi f_m t), \tag{3.239}$$

where $f_m \triangleq m/T = m/(NT_s)$ is the mth harmonic frequency and the mth coefficient S_m is given by:

$$S_m\,(\mathbf{c}_N) \triangleq \frac{1}{T} \int_0^T s(t, \mathbf{c}_N)\, \exp\,(-j2\pi f_m t)\, dt. \tag{3.240}$$

It is not difficult to prove that:

$$S_m\,(\mathbf{c}_N) = \frac{1}{\sqrt{N T_s}} P_m A_m, \tag{3.241}$$

where $P_m \triangleq P(f_m)$, $P(f) \triangleq \mathrm{FCT}[p(t)]$ and:

$$A_m \triangleq \frac{1}{\sqrt{N}} \sum_{l=0}^{N-1} a_l \exp\,\left(-j2\pi \frac{m\,l}{N}\right) \tag{3.242}$$

---

[33] This ensures that the use of $\mathbf{a}_N$ in place of $\mathbf{c}_N$ in digital transmission does not entail any information loss.

for any m. Then substituting (3.241) into (3.239) yields:

$$s(t, \mathbf{c}_N) = \frac{1}{\sqrt{N}T_s} \sum_{m=-\infty}^{+\infty} P_m \, A_m \, \exp \, (j2\pi mt/T), \qquad (3.243)$$

which expresses the Fourier series representation of $s(t, \mathbf{c}_N)$ of (3.238) in a compact form. Note that, if we define the N-dimensional complex vector $\mathbf{A}_N \triangleq [A_0, A_1, \dots, A_{N-1}]^T$, then (3.242) can also be expressed as:

$$\mathbf{A}_N = \text{DFT}_N[\mathbf{a}_N] = \mathbf{Q}_N \, \mathbf{a}_N. \qquad (3.244)$$

and from (3.242) it is easy to show that $\{A_m\}$ is *periodic* with period N (i.e., $A_{m+N} = A_m$ for any m). Therefore, if the transformation $\mathbf{g}(\cdot)$ is an Nth-order IDFT, that is, if:

$$\mathbf{a}_N = \text{IDFT}_N[\mathbf{c}_N] = \mathbf{Q}_N^H \, \mathbf{c}_N, \qquad (3.245)$$

we have that:

$$\mathbf{A}_N = \mathbf{Q}_N \, \mathbf{Q}_N^H \, \mathbf{c}_N = \mathbf{c}_N, \qquad (3.246)$$

i.e., that:

$$A_k = c_{R_N[k]} \qquad (3.247)$$

for any k. Substituting the latter result in (3.243) yields:

$$s(t, \mathbf{c}_N) = \frac{1}{\sqrt{N}T_s} \sum_{m=-\infty}^{+\infty} P_m \, c_{R_N[m]} \, \exp \, (j2\pi f_m t). \qquad (3.248)$$

Equation (3.248) can be simplified by replacing the index m with the indexes (n,k), such that:

$$m = n + kN \qquad (3.249)$$

with $n = 0, 1, \dots, N-1$ and k an arbitrary integer. In fact, after some manipulation, (3.248) can be rewritten as:

$$s(t, \mathbf{c}_N) = \frac{1}{\sqrt{N}T_s} \sum_{n=0}^{N-1} c_n \, g_n(t), \qquad (3.250)$$

where

$$g_n(t) \triangleq \sum_{k=-\infty}^{+\infty} P_{n+kN} \, \exp \, (j2\pi f_{n+kN} \, t). \qquad (3.251)$$

If we want the RHS of (3.250) to represent a multicarrier (MC) signal, the spectrum $P(f)$ of $p(t)$ should be selected so that the signal $g_n(t)$ of (3.251) is proportional to a complex exponential at frequency f_n. A necessary condition for this is that $p(t)$ is *strictly bandlimited*, since in (3.251), for any n (with $n = 0, 1, \dots, N-1$), only one of the coefficients $\{P_{n+kN}\}$ must be different from 0. A possible choice to achieve this result is to select $P(f)$ as the *square root of a raised cosine function* with roll-off factor α (see (3.81) and (3.90)). In evaluating the coefficients $\{P_{n+kN}\}$ of (3.251), we assume $\alpha \neq 0$ and note that:

$$P_{n+kN} \triangleq P(f_{n+kN}) = P\left(\frac{n+kN}{T}\right) = P\left(f_n + \frac{k}{T_s}\right), \qquad (3.252)$$

so that, to determine the structure of $g_n(t)$ for a given n, we must sample the spectrum $P(f)$ at the frequencies $\{f_{n+kN}\}$, changing the index k only. Following this procedure, it is not difficult to show that, if $f_n = n/T$ falls in the flat region of $P(f)$, that is, if $0 \leq f_n \leq f_{1-\alpha} \triangleq (1-\alpha)/(2T_s)$ or, equivalently, if:

$$0 \leq n \leq N_\alpha \qquad (3.253)$$

with

$$N_\alpha \triangleq \left\lfloor N \frac{1-\alpha}{2} \right\rfloor, \tag{3.254}$$

then the only coefficient $\{P_{n+kN}\}$ different from zero in the sum of (3.251) is that associated with $k = 0$, as evidenced by Figure 3.33(a). Therefore, (3.251) simplifies to:

$$g_n(t) = P_n \, \exp \, (j2\pi f_n t) = \sqrt{T_s} \, \exp \, (j2\pi f_n t). \tag{3.255}$$

If f_n falls in the right-hand roll-off region of $P(f)$, that is, if $f_{1-\alpha} < f_n < f_{1+\alpha} \triangleq (1+\alpha)/(2T_s)$ or, equivalently, if:

$$N_\alpha < n < N - N_\alpha, \tag{3.256}$$

then in the sum (3.251) there are two terms different from zero, those associated with the frequencies f_n (corresponding to $k = 0$) and $f_{n-N} = f_n - 1/T_s$ (corresponding to $k = -1$), as shown by Figure 3.33(b). Hence, in this case, (3.251) becomes:

$$g_n(t) = P_n \, \exp \, (j2\pi f_n t) + P_{n-N} \, \exp \, (j2\pi f_{n-N} \, t), \tag{3.257}$$

so that, in the interval defined by (3.256), $g_n(t)$ (3.251) is expressed by a superposition of two complex exponentials having different frequencies. For this reason, the subcarriers associated with the values of n delimited by (3.256) are *suppressed* (i.e., their contribution is removed), thus setting:

$$c_n = 0 \tag{3.258}$$

for all values of n satisfying (3.256). Finally, if f_n falls in the interval $[f_{1+\alpha}, 1/T_s)$, as illustrated by Figure 3.33(c), that is, if:

$$N - N_\alpha \leq n \leq N - 1, \tag{3.259}$$

the only term different form 0 in the sum of (3.251) is that associated with the frequency f_{n-N} (corresponding to the choice $k = -1$) and, therefore:

$$g_n(t) = P_{n-N} \, \exp \, (j2\pi f_{n-N} \, t) = \sqrt{T_s} \exp \, (j2\pi f_{n-N} \, t). \tag{3.260}$$

Taking (3.257) and (3.260) into account, and using the subcarrier suppression given by (3.258), (3.250) can be rewritten as:

$$s(t, \mathbf{c}_N) = \frac{1}{T_s \sqrt{N}} \sqrt{T_s} \left[\sum_{n=0}^{N_\alpha} c_n \, \exp \, \left(j2\pi f_n t\right) + \sum_{n=N-N_\alpha}^{N-1} c_n \, \exp \, \left(j2\pi f_{n-N} \, t\right) \right]$$

$$= \frac{1}{\sqrt{N T_s}} \left[\sum_{n=0}^{N_\alpha} c_n \, \exp \, \left(j2\pi f_n t\right) + \sum_{n=-N_\alpha}^{-1} c_{n+N} \, \exp \, \left(j2\pi f_n \, t\right) \right] \tag{3.261}$$

or, since $c_{n+N} = c_{R_N[n]}$ for $n = -N_\alpha, -N_\alpha + 1, \ldots, -1$ and $c_n = c_{R_N[n]}$ for $n = 0, 1, \ldots, N_a$, in the more compact form:

$$s(t, \mathbf{c}_N) = \frac{1}{\sqrt{N T_s}} \sum_{n=-N_\alpha}^{N_\alpha} c_{R_N[n]} \, \exp \, (j2\pi f_n t). \tag{3.262}$$

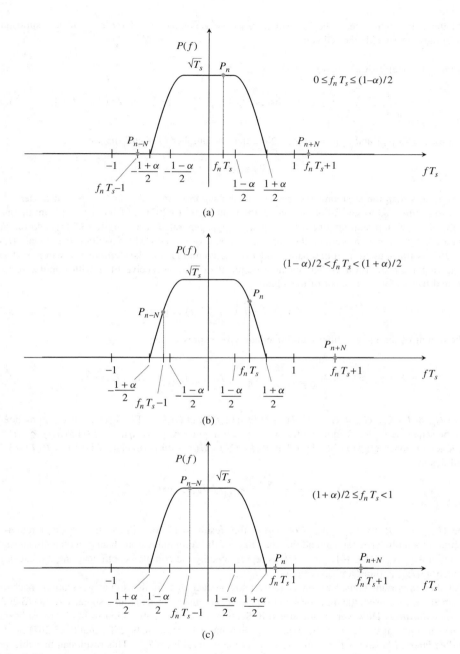

Figure 3.33 Evaluation of the spectral samples $\{P_{n+kN}\}$ according to (3.252) for $0 \le n \le N_\alpha$ (a), for $N_\alpha < n < N - N_\alpha$ (b), and for $N - N_\alpha \le n \le N - 1$ (c).

The latter expression reveals the MC structure of the generated signal $s(t, \mathbf{c}_N)$. In fact, substituting (3.262) into (3.1) yields the RF signal:

$$s_{RF}(t, \mathbf{c}_N) = \text{Re}\{s(t, \mathbf{c}_N) \exp (j2\pi f_c t)\}$$

$$= \frac{1}{\sqrt{NT_s}} \sum_{n=-N_\alpha}^{N_\alpha} \{c_{R,R_N[n]} \cos[2\pi(f_c + f_n)t] - c_{I,R_N[n]} \sin[2\pi(f_c + f_n)t]\} \quad (3.263)$$

where $c_{R,k} = \text{Re}\{c_k\}$ and $c_{I,k} = \text{Im}\{c_k\}$. Note that the number of *useful subcarriers*:

$$N_u \triangleq 2N_\alpha + 1, \quad (3.264)$$

is strictly less than the DFT order N, because of the suppression of $N_{sc} \triangleq N - N_u$ subcarriers.[34]

These results lead to several observations. First, the result (3.262) has been derived from the model (3.238) of $s(t, \mathbf{c}_N)$, assuming a periodic sequence $\{a_k\}$ (generated according to (3.245)) and the pulse shape of (3.89). If $p(t)$ represents the impulse response of the OFDM transmitter, the signal $s(t, \mathbf{c}_N)$ of (3.262) is transmitted, in a baseband model of the communication system, over a wireless channel having impulse response $h(t)$. The channel output then feeds a receive filter with impulse response $g(t)$, and the useful component of the signal:

$$z(t, \mathbf{c}_N) \triangleq s(t, \mathbf{c}_N) \otimes h(t) \otimes g(t) \quad (3.265)$$

at the output of the receive filter can immediately be expressed as:

$$z(t, \mathbf{c}_N) = \frac{1}{\sqrt{NT_s}} \sum_{n=-N_\alpha}^{N_\alpha} c_{R_N[n]} H_{R_N[n]} G_{R_N[n]} \exp (j2\pi f_n t), \quad (3.266)$$

where $H_m \triangleq H(f_m)$, $G_m \triangleq G(f_m)$, $H(f) \triangleq \text{FCT}[h(t)]$ and $G(f) \triangleq \text{FCT}[g(t)]$, since, in practice, the periodic sequence $\{a_k\}$ is feeding a filter with overall impulse response $q(t) \triangleq p(t) \otimes h(t) \otimes g(t)$ (and frequency response $Q(f) \triangleq \text{FCT}[q(t)] = P(f) H(f) G(f)$). In particular, if $G(f) = P(f)$, (3.266) simplifies to:

$$z(t, \mathbf{c}_N) = \frac{1}{\sqrt{N}} \sum_{n=-N_\alpha}^{N_\alpha} c_{R_N[n]} H_{R_N[n]} \exp (j2\pi f_n t), \quad (3.267)$$

since $G_{|n|_N} = \sqrt{T_s}$ for $|n| \le N_\alpha$. Comparing this result to (3.262) shows that $z(t, \mathbf{c}_N)$ retains the structure of a multicarrier signal and that the effect of the communication channel on the nth subcarrier (for any $n \in \{-N_\alpha, -N_\alpha + 1, \ldots, N_\alpha\}$) is represented by the complex coefficient $H_{R_N[n]}$, causing an attenuation $|H_{R_N[n]}|$ and a phase rotation $\angle H_{R_N[n]}$.

Another important consideration is related to the fact that, if the signal generation mechanism described above is used, the transmission of $\mathbf{c}_N$ is accomplished by sending a signal $s(t, \mathbf{c}_N)$ (3.262) of *unlimited duration*. However, as illustrated in Section 3.7.1, the contribution from distinct subcarriers can be separated at the receiver by exploiting their mutual orthogonality, if equality (3.267) holds on a *limited interval* lasting NT_s seconds (e.g., over the interval $[0, NT_s]$). This result can be achieved as follows. Let us assume that the impulse responses $p(t)$, $h(t)$ and $g(t)$ have *limited duration* and that, in particular, they extend over the intervals $[0, L_p T_s]$, $[0, L_h T_s]$ and $[0, L_g T_s]$, so that the support of

---

[34] Note that in (3.262) the subcarrier associated with the choice $n = 0$ is characterized by a frequency $f_0 = 0$ (and, consequently, by a frequency f_c at RF). In most applications this subcarrier is not used in order to avoid the presence of a direct component in the demodulated signal. In this case, (3.264) is replaced by $N_u \triangleq 2N_\alpha$.

the overall impulse response $q(t)$ is $[0, L_q T_s]$, with $L_q \triangleq L_p + L_h + L_g$. Then if, in place of (3.238), the time-limited signal:

$$s(t, \mathbf{c}_N) = \sum_{k=-N_p}^{N-1} a_{R_N[k]} \, p(t - kT_s) \tag{3.268}$$

is generated, where N_p is an integer such that $N_p \geq L_q$, the representation (3.267) of (3.265) still holds over the interval $[0, NT_s]$, since the signal $q(t - kT_s)$ does not include any contributions for $k < -N_p$ and $k > N - 1$. However, if $N_p < L_q$, the orthogonality property is lost and this results in interference at the demodulator, known as ICI.

Note that (3.268) shows that, in generating a multicarrier signal, the transmission of a data vector $\mathbf{a}_N$ is preceded by $\mathbf{a}_p \triangleq [a_{R_N[k]}, k = -N_p, -N_p + 1, \ldots, -1]$, so that the overall transmitted vector, having size $N_T \triangleq N + N_p$, is $\mathbf{a}_{N_T} \triangleq [\mathbf{a}_p^T, \mathbf{a}_N^T]^T$. It is easy to verify that if, as usually happens in practice, N_p is smaller than N, we then have $\mathbf{a}_p = [a_{N-N_p}, a_{N-N_p+1}, \ldots, a_{N-1}]$, that is, $\mathbf{a}_p$ contains an ordered replica of the last N_p components $\mathbf{a}_N$, so that $\mathbf{a}_{N_T}$ has a *cyclic structure*.[35] For this reason, $\mathbf{a}_p$ is called *cyclic prefix*. The presence of this prefix is absolutely necessary for correct signal generation and its length is proportional to the maximum expected channel memory.

Finally, it is important to point out that the hypothesis of finite duration of $p(t)$ and $g(t)$ is in conflict with that of limited bandwidth. However, in practice, for any α, it is always possible to consider a finite *effective duration* for these signals, that is, the length of the time interval containing a significant (and fixed) fraction of the energy of the transmitted signal. Of course, a decrease in α entails, on the one hand, an increase in the duration of $p(t)$, with the consequent need for a longer cyclic prefix, and, on the other hand, a narrowing of the roll-off region of $P(f)$, with the consequent increase of the number of useful subcarriers N_u. In fact, this parameter takes on a unit value for $\alpha = 1$ and approaches N when α goes to 0 (see (3.254) and (3.264)).

3.7.2.2 Block Diagram of the OFDM Modulator

The analytical results illustrated above allow an immediate understanding of the algorithm for the generation of an OFDM signal and this is summarized by Figure 3.34, illustrating the baseband section of an OFDM modulator.

In this scheme, the data stream is applied to a *symbol mapper*, associating each block of $\log M$ consecutive bits with a channel symbol belonging to an M-ary constellation A_c. Then, the channel symbol stream is partitioned into blocks (using a serial-to-parallel converter), each of length N_u (see (3.264)). A group of $N_{sc} \triangleq N - N_u$ null symbols is inserted in each of these blocks, in the locations assigned to the virtual subcarriers. The resulting lth block, consisting of N channel symbols (of which only $N - N_{sc}$ are information symbols), is represented by the vector:

$$\mathbf{c}_N^{(l)} \triangleq [c_0^{(l)}, c_1^{(l)}, \ldots, c_{N_\alpha}^{(l)}, 0, \ldots, 0, c_{N-N_\alpha}^{(l)}, c_{N-N_\alpha+1}^{(l)}, \ldots, c_{N-1}^{(l)}]^T, \tag{3.269}$$

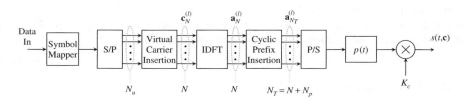

Figure 3.34 Block diagram of the baseband section of an OFDM modulator.

---

[35] This follows from the fact that $\mathbf{a}_{N_T}$ has been extracted by a periodic sequence $\{a_k\}$.

and this is known as a *frequency-domain symbol* or OFDM symbol. This vector undergoes an Nth-order IDFT, accomplished via an *inverse fast Fourier transform* (IFFT) to produce the vector:

$$\mathbf{a}_N^{(l)} = [a_0^{(l)}, a_1^{(l)}, \ldots, a_{N-1}^{(l)}]^T \triangleq \mathbf{Q}_N^H \, \mathbf{c}_N^{(l)}, \tag{3.270}$$

which is extended by appending a *cyclic prefix* of length N_p. This generates the *cyclically extended block*:

$$\mathbf{a}_{N_T}^{(l)} \triangleq [a_{-N_p}^{(l)}, \ldots, a_{-1}^{(l)}, a_0^{(l)}, \ldots, a_{N-1}^{(l)}]^T, \tag{3.271}$$

having overall length $N_T \triangleq N_p + N$, with $a_k = a_{R_N[k]}$ for $k = -N_p, -N_p + 1, \ldots, -1$. After parallel-to-series conversion, the sequence of symbols extracted from the symbol vector $\mathbf{a}_{N_T}^{(l)}$ is applied, at rate $R_s = 1/T_s$, to a transmission filter having impulse response $p(t)$ (usually expressed by (3.89)) with unit energy. Finally, the filter output is amplified by a factor K_c and sent to the following transmitter stages for frequency up-conversion. Note that the baseband equivalent of the OFDM signal can be expressed as:

$$s(t, \mathbf{c}) = \sum_{l=-\infty}^{+\infty} \tilde{s}(t - t_l; \mathbf{a}_{N_T}^{(l)}), \tag{3.272}$$

where $t_l \triangleq l N_T T_s$, $\mathbf{c}$ denotes a vector resulting from the ordered concatenation of the vectors $\{\mathbf{c}_N^{(l)}\}$ and:

$$\tilde{s}(\mathbf{a}_{N_T}^{(l)}; t) \triangleq K_c \sum_{k=-N_p}^{N-1} a_k^{(l)} \, p(t - kT_s) \tag{3.273}$$

represents the contribution of the lth OFDM symbol to the transmitted signal. Substituting (3.273) in (3.272) yields the formula:

$$s(t, \mathbf{c}) = K_c \sum_{l=-\infty}^{+\infty} \sum_{k=-N_p}^{N-1} a_k^{(l)} \, p(t - kT_s - t_l), \tag{3.274}$$

whose structure resembles that of a PAM signal (see (3.48)). However, despite the similarities, there are some fundamental differences between these modulation formats. First, the channel symbols of PAM usually belong to a set of M points *regularly* placed in the complex plane, whereas the symbols $\{a_k^{(l)}\}$ of an OFDM signal, being generated by an IDFT, do not share this property. In fact, generally speaking, each of the symbols $\{a_k^{(l)}\}$ can take on more than M distinct values and these values are spread in an irregular fashion in the complex plane. Secondly, a PAM signal is generated by a *single-state* modulator, whereas, in principle, this property does not hold for the OFDM signal (3.274), since the elements of $\mathbf{a}_{N_T}^{(l)}$, for any l, are *statistically dependent*, because of the presence of the cyclic prefix and of their generation via an IDFT. Despite this, a comparison of (3.272) with (3.4) illustrates that an OFDM modulator can still be represented as a *single-state* modulator if, in the signal model (3.4), the channel symbols $\{c_n\}$ are replaced by the OFDM symbols $\{\mathbf{c}_N^{(l)}\}$ and, similarly, the signaling interval T_s by the corresponding interval $N_T T_s$.

Note that, for a given l, in $N_T T_s$ seconds all the components of $\mathbf{a}_{N_T}^{(l)}$, as given by (3.271), and consequently $N_u \log M$ information bits, are transmitted, so that we have:

$$T_b \, N_u \log M = N_T T_s, \tag{3.275}$$

from which it can immediately be seen that:

$$T_b = \frac{N_T}{N_u \log M} T_s = \frac{N + N_p}{(2N_\alpha + 1) \log M} T_s, \tag{3.276}$$

establishing an explicit relationship between the symbol interval T_s and the bit interval T_b.

Let us now assume that $s(t, \mathbf{c})$ is transmitted over a wireless channel with impulse response $h(t)$. Following the same line of reasoning and making the same assumptions as in the previous subsection, it is not difficult to prove that the useful component of the signal at the output of the receive filter is given by:

$$z(t, \mathbf{c}) = \frac{K_c}{\sqrt{N}} \sum_{n=-N_\alpha}^{N_\alpha} c_{R_N[n]}^{(l)} \; H_{R_N[n]} \; \exp\left[j2\pi f_n(t - t_l)\right], \qquad (3.277)$$

over the interval $[t_l - (T_{cp} - T_q), t_l + NT_s]$, where $T_{cp} \triangleq N_p T_s$ and T_q are the duration of the cyclic prefix and of the overall impulse response $q(t)$, respectively. This representation holds, however, if $T_{cp} \geq T_q$, because this inequality ensures not only that the multicarrier representation (3.277) holds for at least NT_s seconds (i,e., ICI is avoided at the receiver), but also that in the same interval no *interblock interference* (IBI) is found.

Finally, note that, for each OFDM symbol, $s(t, \mathbf{c})$ in (3.272) can be represented as the superposition of a number of complex oscillations, having different frequencies and phases. This structure entails the presence, in the envelope $s_{env}(t, \mathbf{c})$ (3.8) of the modulated signal, of very large *fluctuations*, which become more pronounced as the number of useful subcarriers N_u increases; this is shown in the following example.

Example 3.7.1 Let us compare the complex envelope of an OFDM signal, having $N = 256$, $N_u = 193$ ($\alpha = 0.25$), $L_p = 20$ and a QPSK constellation on each subcarrier, with those of QPSK and OQPSK signaling, characterized by the same transmitter impulse response $p(t)$. The *I-Q diagrams* for these modulations, that is, the trajectories of $s(t, \mathbf{c})$ in a complex plane, are represented in Figures 3.35, 3.36 and 3.37 for OFDM, QPSK and OQPSK, respectively. These diagrams show a substantial difference, in terms of envelope fluctuations, between single-carrier and multicarrier signals; in addition, as already mentioned in Section 3.5.1, the envelope of OQPSK undergoes even smaller amplitude changes than that of QPSK.

The large changes in the OFDM envelope can be related to the PAM representation of the OFDM signal of (3.274). In fact, from this point of view, this phenomenon is due to the large dispersion of

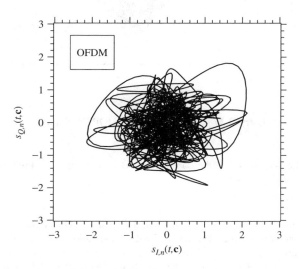

Figure 3.35 Typical I-Q diagram for OFDM signaling.

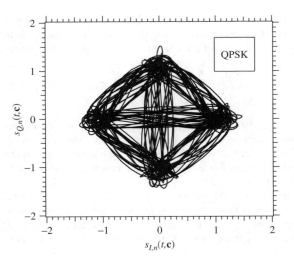

Figure 3.36   I-Q diagram for QPSK signaling.

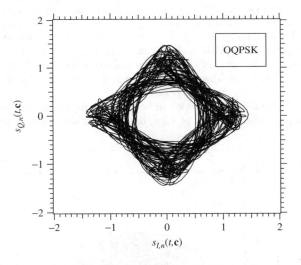

Figure 3.37   I-Q diagram for OQPSK signaling.

the components of the vectors $\mathbf{a}_N^{(l)}$ in the complex plane, as illustrated by Figure 3.38, with reference to the OFDM signal of Figure 3.35.

$\square$

The relative magnitude of the amplitude fluctuations of an OFDM signal is expressed by the so-called *peak-to-average-power ratio* (PAPR), defined as the ratio between the peak power and the average power of the transmitted signal; the problem of the evaluation of this parameter is briefly discussed in Section 3.7.4. Here we note that the large PAPR of OFDM signals affects both the baseband and RF sections of an OFDM transmitter. In fact, in the former a digital signal processor

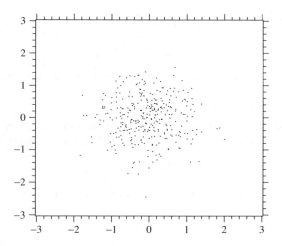

Figure 3.38 Representation, on a complex plane, of the components of a realization of the random vector $\mathbf{a}_N^{(l)}$ for the OFDM signal whose complex envelope is shown in Figure 3.35.

using arithmetic of adequate precision is required, whereas in the latter a linear power amplifier is needed. Various signal processing techniques have been proposed in the technical literature to minimize the PAPR: an overview of various methods has been provided in [316].

3.7.3 Power Spectral Density of OFDM

In this subsection a closed-form expression for the power spectral density of an OFDM signal is derived under the assumptions that the channel symbols $\{c_n^{(l)}\}$ are iid, and have zero mean (to avoid the presence of spectral lines [28]) and variance $\sigma_c^2 \triangleq \mathrm{E}\{|c_n^{(l)}|^2\}$. To begin, we note that the structure of the signal $s(t, \mathbf{c})$ (3.274) is similar to that of a PAM signal, as already pointed out in Section 3.7.2.2. From a statistical viewpoint, however, there is a fundamental difference between them, since the data sequence $\{a_k^{(l)}\}$ feeding the transmitter filter in the OFDM case is not WSS, but WSC with period $N_T \triangleq N + N_p$. This implies that $s(t, \mathbf{c})$ (3.274) is also WSC with period $T_{cs} = N_T T_s$. Note that the cyclostationarity of the OFDM data sequence is not at all surprising, since, in an OFDM system, data transmission is accomplished on a block-by-block basis and each block has a duration equal to N_T. For a PAM signal characterized by a WSC data sequence, it can be proved [28] that expression (3.66) still holds, provided that $\bar{S}_c(f)$ denotes the *average power spectrum* of the data sequence $\{a_k^{(l)}\}$, that is, the Fourier transform of its *average autocorrelation function*. For this reason, the evaluation of the power spectral density for an OFDM signal proceeds via the following steps: (a) evaluation of the latter function; (b) evaluation of the former function via Fourier methods; (c) application of (3.66) to the evaluation of the power spectral density of an OFDM signal.

3.7.3.1 Evaluation of the Average Autocorrelation Function of $\{a_k^{(l)}\}$

To begin, we note that the data sequence $\{a_k^{(l)}\}$ is characterized by the presence of two indexes, namely (k, l), and this complicates the computation of its statistics. To circumvent this problem, we define a new data sequence $\{b_n\}$, derived from the ordered concatenation of the vectors $\{\mathbf{a}_{N_T}^{(l)}\}$; more precisely, we establish a one-to-one correspondence between $\{a_k^{(l)}\}$ and $\{b_n\}$, such that the N_T-dimensional vector $[b_{lN_T}, b_{lN_T+1}, \ldots, b_{lN_T+N_T-1}]^T$ *coincides* with the *cyclic* vector

$\mathbf{a}_{N_T}^{(l)} = [a_{-N_p}^{(l)}, \ldots, a_{-1}^{(l)}, a_0^{(l)}, \ldots, a_{N-1}^{(l)}]^T$. In other words, we then have:

$$b_{k+N_p+lN_T} \triangleq a_k^{(l)} = a_{R_N[k]}^{(l)} \tag{3.278}$$

for $k = -N_p, -N_p + 1, \ldots, N - 1$ and any l. If we now define the autocorrelation function:

$$R_b[n, k] = \mathrm{E}\{b_{n+k} \, b_n^*\} \tag{3.279}$$

of $\{b_k\}$, it can easily be shown that $R_b[n, k]$ is *periodic*, in the variable n, with period N_T. Therefore, its values need to be computed over only a single period, focusing, for instance, on the interval:

$$0 \leq n \leq N_T - 1 = N + N_p - 1 \tag{3.280}$$

of the variable n, so that b_n belongs to $\mathbf{a}_{N_T}^{(0)}$, and, in particular (see (3.278)):

$$b_n = a_{R_N[n-N_p]}^{(0)} \tag{3.281}$$

for any n satisfying inequality (3.280). The index k in (3.279) can still take on an arbitrary value, but two distinct cases can be identified in the evaluation of the statistical average of $R_b[n, k]$. In the first case, k is such that the symbol $b_{n+k} \notin \mathbf{a}_{N_T}^{(0)}$, so that b_n and b_{n+k} belong to distinct (and statistically independent) OFDM symbols; then (3.279) yields:

$$R_b[n, k] \triangleq \mathrm{E}\{b_{n+k} \, b_n^*\} = \mathrm{E}\{b_{n+k}\} \cdot \mathrm{E}\{b_n^*\} = 0, \tag{3.282}$$

since $\{b_k^{(l)}\}$ is a zero mean sequence. In the second case, k is such $b_{n+k} \in \mathbf{a}_{N_T}^{(0)}$, that is:

$$0 \leq k + n \leq N_T - 1, \tag{3.283}$$

so that the symbols b_n and b_{n+k} on the RHS of (3.279) belong to the same OFDM symbol. Note that: (a) inequality (3.283) can be rewritten as:

$$-n \leq k \leq N_T - 1 - n \tag{3.284}$$

or, equivalently, as:

$$-k \leq n \leq N_T - 1 - k; \tag{3.285}$$

(b) in the $n-k$ plane inequalities (3.280) and (3.284) delimit the shadowed region R of Figure 3.39;
(c) if (3.285) is satisfied, b_{n+k} can be expressed as (see (3.281)):

$$b_{k+n} = a_{R_N[k+n-N_p]}^{(0)}. \tag{3.286}$$

Then, in this case, substituting (3.281) and (3.286) in (3.279) yields, after some manipulation:

$$R_b[n, k] = \frac{\sigma_c^2}{N} \sum_{l=-N_\alpha}^{N_\alpha} \exp\left(j\frac{2\pi}{N}lk\right). \tag{3.287}$$

Note that an *explicit* dependence on the variable n is missing in the RHS of (3.287); however, such a dependence is *implicit*, since (3.287) holds only for pairs (n, k) belonging to the region R (contour included), defined by (3.280) and (3.284).

Equations (3.282), (3.285), (3.287) and the periodicity of the autocorrelation $R_b[n, k]$ in its variable n can be summarized as:

$$R_b[n, k] = \frac{\sigma_c^2}{N} \sum_{l=-N_\alpha}^{N_\alpha} \exp\left(j\frac{2\pi}{N}lk\right) \tag{3.288}$$

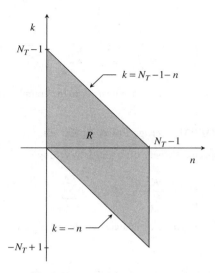

Figure 3.39 Representation of the region R defined by inequalities (3.280) and (3.284).

for $-R_{N_T}[n] \le k \le N_T - 1 - R_{N_T}[n]$, and $R_b[n, k] = 0$ elsewhere. In other words, generally speaking, $R_b[n, k]$ is equal to 0 outside of the region (in the n–k plane) generated by the repetition, with period N_T, of R in the direction of the n axis. In particular, this implies that $R_b[n, k] = 0$ for $|k| \ge N_T$ (see Figure 3.39). Then the average autocorrelation function $R_b[k]$ of $\{b_k\}$ can be computed as:

$$R_b[k] \triangleq \frac{1}{N_T} \sum_{n=0}^{N_T-1} R_b[n, k], \tag{3.289}$$

for $|k| \le N_T - 1$; it is equal to 0 for $|k| \ge N_T$. Substituting (3.287) into (3.289) and taking into account the constraint (3.285) yields, after some manipulation:

$$R_b[k] = \sigma_c^2 \, g[k] \sum_{l=-N_\alpha}^{N_\alpha} \exp\left(j\frac{2\pi}{N} lk\right), \tag{3.290}$$

where

$$g[k] \triangleq \frac{1}{N}\left[1 - \frac{|k|}{N_T}\right]\{u[k + N_T] - u[k - N_T]\} \tag{3.291}$$

and

$$u[k] \triangleq \begin{cases} 1 & \text{for } k \ge 0 \\ 0 & \text{for } k < 0 \end{cases} \tag{3.292}$$

is the *discrete-time unit step function*.

3.7.3.2 Evaluation of the Average Power Spectral Density of $\{b_k\}$

The *average power spectral density* $\bar{S}_b(f)$ of $\{b_k\}$ is the Fourier transform of the sequence $R_b[n]$ in (3.290). Since the Fourier transform of the sequence $g[k]$ in (3.291) is:

$$\bar{G}(f) = \frac{1}{NN_T} \frac{\sin^2(\pi N_T f T_s)}{\sin^2(\pi f T_s)}, \tag{3.293}$$

it is not difficult to prove that:

$$\bar{S}_b(f) = \sigma_c^2 \sum_{l=-N_\alpha}^{N_\alpha} \bar{G}\left(f - \frac{l}{NT_s}\right) = \frac{\sigma_c^2}{NN_T} \sum_{l=-N_\alpha}^{N_\alpha} \frac{\sin^2(\pi N_T(f - f_l)T_s)}{\sin^2(\pi(f - f_l)T_s)}, \quad (3.294)$$

where $f_l \triangleq l/(NT_s)$ is the frequency of the lth subcarrier in the complex envelope of the OFDM signal.

3.7.3.3   Evaluation of the OFDM Average Power Spectrum

Finally, the average power spectral density of $s(t, \mathbf{c})$ is given by:

$$S_s(f) = \frac{1}{T_s} \bar{S}_b(f)|P(f)|^2 = \frac{K_c^2 \sigma_c^2}{NN_T T_s}|P(f)|^2 \sum_{l=-N_\alpha}^{N_\alpha} \frac{\sin^2[\pi N_T(f - f_l)T_s]}{\sin^2[\pi(f - f_l)T_s]}. \quad (3.295)$$

The average power P_s of $s(t, \mathbf{c})$ can be evaluated by integrating $S_s(f)$ (3.295) over all frequencies. This leads to the expression:

$$P_s = \int_{-\infty}^{+\infty} S(f)\, df = \frac{K_c^2 \sigma_c^2}{NT_s} N_u. \quad (3.296)$$

Then substituting (3.296) into (3.31) yields the average transmitted energy per symbol:

$$E_s = \frac{P_s T_s}{2} = \frac{K_c^2 \sigma_c^2}{2} \frac{N_u}{N}, \quad (3.297)$$

from which it is easy to see that:

$$K_c = \sqrt{\frac{2E_s}{\sigma_c^2} \frac{N}{N_u}}. \quad (3.298)$$

The value of the parameter σ_c depends on the type and cardinality of the underlying constellation and is given by (3.72), (3.73) and (3.74) for M-PSK, M-ASK and M-QAM signals, respectively.

The latter results can also be exploited to rewrite (3.295) in a slightly different fashion. In fact, (3.298) implies that:

$$\frac{K_c^2 \sigma_c^2}{NN_T T_s} = \frac{2E_s}{T_s} \frac{1}{N_T N_u} = P_s \frac{1}{N_T N_u}, \quad (3.299)$$

so that (3.295) can be put in the form:

$$S_s(f) = P_s \frac{N_T}{N_u} \sum_{l=-N_\alpha}^{N_\alpha} S_l(f), \quad (3.300)$$

where

$$S_l(f) \triangleq \frac{1}{N_T^2}|P(f)|^2 \frac{\sin^2[\pi N_T(f - f_l)T_s]}{\sin^2[\pi(f - f_l)T_s]} \quad (3.301)$$

represents the contribution, normalized to $P_s N_T/N_u$, of the lth subcarrier to $S_s(f)$. The result expressed by (3.300) and (3.301) shows that, in the OFDM spectrum, the contributions from distinct subcarriers partially overlap and that the transmitted signal is *rigorously bandlimited* if $p(t)$ has this property.

These results can be used, for instance, to evaluate the power spectrum of the OFDM signal transmitted in an audio broadcasting system based on the so-called *Digital Audio Broadcasting* (DAB) standard [317], as illustrated in the following example.

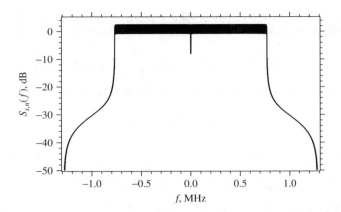

Figure 3.40 Representation of the normalized power spectral density $S_{s,n}(f)$ for the TM II of DAB.

Example 3.7.2 Figure 3.40 represents the normalized power spectrum $S_{s,n}(f)$ for the *second transmission mode* (TM II) of DAB. The most significant parameters[36] of the transmitted signal are $N = 512$, $\alpha = 0.25$ ($N_u = 384$ and $N_\alpha = N_u/2 = 192$), $N_p = 126$ ($N_T = N_p + N = 638$) and $T_s = 0.48828125$ μs (the subcarrier spacing is $1/(NT_s) = 4$ kHz).

Note that, if a bandlimited spectrum is assumed for $p(t)$, the theoretical bandwidth of the complex envelope $s(t, \mathbf{c})$ is:

$$B_s \triangleq \frac{1 + \alpha}{2T_s} = 2.56 \text{ MHz.} \tag{3.302}$$

In any DAB TM II transmission, however, the real bandwidth is larger, since $p(t)$ is truncated. The DAB standard specifies a *nominal transmitted signal bandwidth* of $N_u/T_s = 1.536$ MHz, corresponding to the difference between the maximum ($f_c + N_\alpha/T_s$) and minimum ($f_c - N_\alpha/T_s$) subcarrier frequencies. This is in substantial agreement with the results of Figure 3.40, which show that $S_{s,n}(f)$ takes on small values for $|f| \geq 0.75$ MHz.

□

3.7.4 The PAPR Problem in OFDM

As mentioned in Section 3.7.2.2, the magnitude of the envelope fluctuations in OFDM signals is measured by the PAPR of the transmitted signal: the ratio of the average power P_{RF} to the peak power P_P of the RF transmitted signal $s_{RF}(t, \mathbf{c}_N)$. In the literature this parameter is evaluated not for the whole signal $s_{RF}(t, \mathbf{c}_N)$ generated by the signal modulator, but for its MC representation, working in an interval of duration $T = NT_s$ seconds corresponding to the transmission of a single OFDM symbol. In other words, it is assumed that the transmitted RF signal $s_{RF}(t, \mathbf{c}_N)$ is given by (3.263) and the observation is restricted to $[0, T]$. Moreover, it is assumed that the vector $\mathbf{c}_N = [c_0, c_1, \dots, c_{N-1}]^T$ consists of channel symbols that are not necessarily independent and that they belong to a constellation of points that may not be equally likely.

---

[36] In a DAB system the number N_u of subcarriers used is equal to $2N_\alpha$ rather than $2N_\alpha + 1$ (see (3.264)), since the central subcarrier is not used.

The evaluation of the PAPR can be carried out as follows. To begin, we note that the average power of the complex envelope $s(t, \mathbf{c}_N)$ of the OFDM signal in the interval $[0, T]$ is given by:

$$P_s = \frac{1}{T} \int_0^T \mathrm{E}\{|s(t, \mathbf{c}_N)|^2\} \, dt = \mathrm{E}\left\{\frac{1}{T} \int_0^T |s(t, \mathbf{c}_N)|^2 \, dt\right\}, \tag{3.303}$$

with $s(t, \mathbf{c}_N)$ expressed by (3.262). Substituting (3.262) into (3.303) yields, after some algebra:

$$P_s = \frac{1}{T} \mathrm{E}\{|\mathbf{c}_{N_u}|^2\}, \tag{3.304}$$

so that we obtain the average power as:

$$P_{RF} = \frac{P_s}{2} = \frac{1}{2T} \, \mathrm{E}\{|\mathbf{c}_{N_u}|^2\}. \tag{3.305}$$

The RF peak power of $s_{RF}(t, \mathbf{c}_N)$ (3.263) can be expressed as:

$$P_P = \max_{0 \le t \le T, \, \mathbf{c}_{N_u} \in \Sigma} [\mathrm{Re}\{s(t, \mathbf{c}_N) \exp(j2\pi f_c t)\}]^2, \tag{3.306}$$

where Σ is the alphabet consisting of all possible values of the vector $\mathbf{c}_{N_u}$, corresponding to the $N_u = 2N_\alpha + 1$ symbols transmitted using the useful subcarriers. Note also that, if we define the variable:

$$\vartheta \triangleq \frac{2\pi t}{T} \tag{3.307}$$

and the parameter:

$$\varsigma \triangleq f_c T, \tag{3.308}$$

the function to be maximized on the RHS of (3.306) can be rewritten as:

$$\mathrm{Re}\{s(t, \mathbf{c}_N) \exp(j2\pi f_c t)\} = \frac{1}{\sqrt{T}} \mathrm{Re}\left\{\sum_{n=-N_\alpha}^{N_\alpha} c_{R_N[n]} \exp(j2\pi f_n t) \exp(j2\pi f_c t)\right\}$$

$$= \frac{1}{\sqrt{T}} \mathrm{Re}\left\{\sum_{n=-N_\alpha}^{N_\alpha} c_{R_N[n]} \exp[j\vartheta(n + \varsigma)]\right\}, \tag{3.309}$$

so that P_P (3.306) can also be expressed as:

$$P_P = \frac{1}{T} \max_{0 \le \vartheta \le 2\pi, \, \mathbf{c}_{N_u} \in \Sigma} \left[\mathrm{Re}\left\{\sum_{n=-N_\alpha}^{N_\alpha} c_{R_N[n]} \exp[j\vartheta(n + \varsigma)]\right\}\right]^2. \tag{3.310}$$

Finally, evaluating the ratio of (3.310) and (3.305) yields the PAPR expression:

$$\mathrm{PAPR}(\varsigma) = \frac{P_P}{P_{RF}}$$

$$= \frac{2}{\mathrm{E}\{|\mathbf{c}_{N_u}|^2\}} \max_{0 \le \vartheta \le 2\pi, \, \mathbf{c}_{N_u} \in \Sigma} \left[\mathrm{Re}\left\{\sum_{n=-N_\alpha}^{N_\alpha} c_{R_N[n]} \exp[j\vartheta(n + \varsigma)]\right\}\right]^2. \tag{3.311}$$

Given ς, N_α and the alphabet Σ, the evaluation of the PAPR of (3.311) constitutes a formidable problem. In fact, the maximum of (3.311) does not have a closed form. Moreover, it cannot be solved numerically since it requires the peak of a *time-continuous* signal to be determined for each possible value of $\mathbf{c}_{N_u}$. To avoid these difficulties, in the literature it is often recommended to evaluate the PAPR not for the time-continuous RF signal $s_{RF}(t, \mathbf{c}_N)$, but for its *oversampled version*, that is, for the sequence generated by sampling $s_{RF}(t, \mathbf{c}_N)$ at a frequency L/T_s, where the *oversampling vector* L is larger than one [316]. This approach is justified by the availability of various theoretical results, establishing a relationship between the peak of a time-continuous signal and that of the corresponding oversampled signal [318, 319]. It is worth noting that the peak of a sequence can be computed in a computationally efficient fashion by evaluating, for each value of $\mathbf{c}_{N_u}$, the samples of the corresponding sequence via an IFFT algorithm. This makes it possible to assess the efficacy of various techniques for PAPR reduction proposed in the technical literature [316]. Finally, it should be noted that the validity of this approach to the assessment of PAPR has recently been questioned, since it has been proved that the reduction of the PAPR for a sample sequence does not necessarily entail a similar reduction of the same parameter for the time-continuous signal from which the sequence has been extracted via sampling [320].

3.8 Lattice-Based Multidimensional Modulations

In Section 3.3 it was shown how the representation, in terms of an orthonormal basis, of the signal alphabet A_s, generated by a given digital modulator in baseband, leads to a new set, consisting of N_w multidimensional points and known as a *constellation*. Most passband PAM constellations (see Section 3.5.2) are characterized by a *regular* placement of their points, since regularity simplifies both signal generation and detection. For this reason, generally speaking, most known signal constellations are generated by extracting their points from a proper *lattice*, that is from a *set of points placed in a regular fashion in a multidimensional space*. Note that the study of the possible signal constellations inevitably leads to an examination of the problem of their *optimality*. An *M*-ary constellation can be deemed *optimal* if, for a given error performance, it minimizes the *average energy per information bit* required to achieve this target. If the communication channel introduces *additive white Gaussian noise* (AWGN) only, the error performance of a digital receiver at large SNRs depends only on the *minimum distance* between constellation points [321]. In addition, signal space theory shows that the average energy expense decreases as the density of points around the origin increases. For this reason, an *M*-ary constellation is *optimal* for AWGN channel signaling if it consists of a set of *M* points that, for an assigned minimum distance among adjacent points, are placed in the *densest* possible way around the origin. The design of optimal constellations leads to the well-known *sphere-packing problem* [322, 323], whose solution leads to an algorithm that allows the placing of a set of equal *N*-dimensional spheres according to the *densest* possible geometry.

The aim of this section is to provide the basics of lattice theory. The main algebraic and geometric properties of lattices are outlined and some construction methods described. This material is necessary for the later discussion of *signal space codes* in Chapter 11.

3.8.1   Lattices: Basic Definitions and Properties

A *real lattice* Λ is a discrete set of points λ belonging to a real Euclidean *n*-dimensional space $\mathbb{R}^n$ and forming a *group* under the vector addition operator (see Appendix E). In principle, the vectors in a lattice Λ span $d \leq n$ dimensions; however, all the lattices considered in this book are characterized by $d = n$, that is, their vectors span exactly n dimensions. For this reason, in what follows we will call a lattice of real *n*-tuples an *n*-dimensional real lattice. Generally speaking, the points of an *n*-dimensional real lattice Λ are arranged in a regular fashion and are described by a set of *d*-dimensional linearly independent row vectors $\{\mathbf{g}_l, l = 0, 1, \ldots, n-1\}$, called *generators*, such that:

$$\Lambda \triangleq \{\lambda | \lambda = \mathbf{i}_n \, \mathbf{G}_\Lambda = i_0\mathbf{g}_0 + i_1\mathbf{g}_1 + \ldots + i_{d-1}\mathbf{g}_{n-1}\}, \qquad (3.312)$$

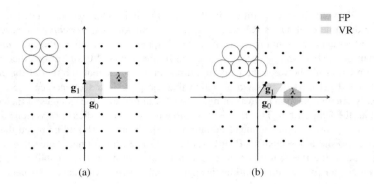

Figure 3.41 Two-dimensional lattices $\mathbb{Z}^2$ and $\mathcal{A}_2$: illustration of generators $\{\mathbf{g}_0, \mathbf{g}_1\}$, sphere packing, *fundamental parallelotopes* (FPs) and *Voronoi regions* (VRs) of a lattice point λ.

where the n-dimensional row vector $\mathbf{i}_n \triangleq [i_0, i_1, \ldots, i_{n-1}]$ belongs to $\mathbb{Z}^n$ (i.e., all its components are integers) and $\mathbf{G}_\Lambda$, called the *generator matrix*, is an $n \times n$ matrix having $\mathbf{g}_l$ as its lth row, that is, $\mathbf{G}_\Lambda \triangleq [\mathbf{g}_0^T, \mathbf{g}_1^T, \ldots, \mathbf{g}_{n-1}^T]^T$. Equation (3.312) shows that Λ can be viewed as the image of $\mathbb{Z}^n$ generated by the linear operator $\mathbf{G}_\Lambda$, or alternately, as a discrete additive subgroup of $\mathbb{R}^n$. These ideas are illustrated in Figure 3.41, showing the two-dimensional *integer lattice* $\mathbb{Z}^2$ with basis vectors $\mathbf{g}_0 = \{0, 1\}$ and $\mathbf{g}_1 = \{1, 0\}$ and the two-dimensional *hexagonal lattice* $\mathcal{A}_2$ with basis vectors $\mathbf{g}_0 = \{1, 0\}$ and $\mathbf{g}_1 = \{1/2, \sqrt{3}/2\}$, respectively.

An *integer n-dimensional lattice* is one whose coordinates, when scaled, are integer-valued. For instance, $\mathbb{Z}^2$ is an integer lattice, whereas $\mathcal{A}_2$ does not share this property.

The following operations can be carried out on an n-dimensional lattice Λ:

1. *Scaling*: $a\Lambda$ denotes the lattice generated by multiplying all the points $\lambda \in \Lambda$ by a real factor $a > 0$. For instance, $2\mathbb{Z} = \{0, \pm 2, \pm 4, \ldots\}$.
2. *Rotation*: $\mathbf{O}\Lambda$ is the lattice produced by an orthogonal transformation $\mathbf{O}$, that is, it consists of all the points $\lambda' = \mathbf{O}\lambda$ with $\lambda \in \Lambda$; for example, the lattice $\mathcal{D}_2 = \mathbf{R}_2 \mathbb{Z}^2$, where:

$$\mathbf{R}_2 = \begin{bmatrix} 1 & 1 \\ 1 & -1 \end{bmatrix} \tag{3.313}$$

represents the so-called two-dimensional *rotation operator*[37] [324], consists of all the integer pairs whose coordinate sum is zero.
3. *Cartesian product*: Λ^m represents the $(m \cdot n)$-dimensional lattice defined as $\{(\lambda_0, \lambda_1, \ldots, \lambda_{m-1}) | \lambda_i \in \Lambda$ for $i = 0, 1, \ldots, M-1\}$.

Lattices, being groups, share some fundamental *algebraic properties*, that can be summarized as follows. First, from an n-dimensional lattice Λ, specific subsets of points, called *sublattices*, can be extracted; a *sublattice* Λ' of Λ is itself a lattice contained in Λ, that is, Λ' is a *subgroup* of Λ. Moreover, by elementary group theory (see Appendix E), Λ' induces a *factor group* or *quotient group* or *partition*, denoted Λ/Λ', decomposing Λ into *equivalence classes* or *cosets*.[38] The *order* $|\Lambda/\Lambda'|$

---

[37] $\mathbf{R}_2$ produces a rotation by 45° and a scaling by $\sqrt{2}$. Note that $(\mathbf{R}_2)^k \mathbb{Z}^2$ is equal to $2^{k/2} \mathbf{I}_2$ for k even and to $2^{(k-1)/2} \mathbf{R}_2 \mathbb{Z}^2$ for k odd, where $\mathbf{I}_2$ is the 2×2 identity matrix.
[38] Given the constant n-tuple $\mathbf{a}$, the *coset* $\Lambda + \mathbf{a}$ of Λ is the set of all possible n-dimensional vectors of the form $\lambda + \mathbf{a}$, with $\lambda \in \Lambda$. In other words, $\Lambda + \mathbf{a}$ is a *translate* of Λ by $\mathbf{a}$. Note that two distinct n-tuples are *equivalent*

of this partition is the number of such equivalence classes. Each equivalence class is a *coset* of Λ', that is a translate $\Lambda' + \mathbf{c}$ of Λ', for some $\mathbf{c} \in \Lambda$. We refer to $\mathbf{c}$ as a *coset representative* and the set of all coset representatives for the partition is denoted by $[\Lambda/\Lambda']$. It is easy to prove that any $\lambda \in \Lambda$ can be *uniquely* represented as $\lambda = \lambda' + \mathbf{c}$, with $\lambda' \in \Lambda'$ and $\mathbf{c} \in [\Lambda/\Lambda']$. This means that the lattice Λ can be represented as $\Lambda' + [\Lambda/\Lambda']$; this in fact defines the so-called *coset decomposition* of Λ. For instance, it can be shown that the four-dimensional lattice $\mathcal{D}_4 = \{(\lambda_0, \lambda_1, \lambda_2, \lambda_3)| \sum_{i=0}^{3} \lambda_i = 0 \bmod 2\}$ is a sublattice of $\mathbb{Z}^4$ and that the corresponding partition $\mathbb{Z}^4/\mathcal{D}_4$ of $\mathbb{Z}^4$ has order 2. If we choose $[\Lambda/\Lambda'] = \{\mathbf{c}_0 = (0,0,0,0), \mathbf{c}_1 = (1,0,0,0)\}$, the coset decomposition of $\mathbb{Z}^4$ is $\mathcal{D}_4 + \{\mathbf{c}_0, \mathbf{c}_1\}$.

A *partition chain* $\Lambda/\Lambda'/\Lambda''/ \dots$ is a sequence of lattices such that each element in the sequence or chain is a sublattice of the previous element. For example, $\mathbb{Z}/2\mathbb{Z}/4\mathbb{Z}/ \dots$ is an infinite partition chain. A partition chain defines a *coset decomposition chain* (or *multi-term coset decomposition*), that is, a representation:

$$\Lambda = \Lambda' + [\Lambda/\Lambda']$$
$$= \Lambda'' + [\Lambda'/\Lambda''] + [\Lambda/\Lambda']$$
$$= \dots, \tag{3.314}$$

of Λ. This means that each element of Λ may be expressed as an element of the final sublattice in the partition chain, plus a coset representative of every other partition in the chain. For example, $\mathbb{Z}/2\mathbb{Z}/4\mathbb{Z}/ \dots$ is a coset decomposition chain of the set of integers and leads to the *standard binary representation* of integers [324].

Lattices, being collections of points in an n-dimensional space, also share some *geometric properties*. A discussion of such properties requires the introduction of the concept of the *distance* $d(\lambda_1, \lambda_2)$ between two distinct lattice points λ_1 and λ_2; this is defined as:

$$d^2(\lambda_1, \lambda_2) = \|\lambda_1 - \lambda_2\|^2,$$

where $\|\mathbf{x}\|$ is the Euclidean norm of a vector $\mathbf{x}$ (see Appendix D). For a given lattice Λ, the set of squared distances of its points from the zero point (i.e., from the n-dimensional null vector $\mathbf{0}_n$) can be conventionally represented via an infinite series called the *theta function*[39] (also called the *theta series* or *weight distribution*) of the lattice (the theta functions for a number of important packings and the tables of the first tens of function coefficients are available in [325]). The key geometric parameters of a lattice Λ are: (a) the *minimum squared distance* between its points $d_{min}^2(\Lambda)$; (b) the *kissing number* $K_{min}(\Lambda)$, representing the minimum number of nearest neighbors to any point of Λ; (c) the volume $V(\Lambda)$ of n-dimensional space per lattice point.

The quantities $d_{min}^2(\Lambda)$ and $K_{min}(\Lambda)$ lend themselves to a simple geometric interpretation if we consider the *sphere packing* associated with Λ. This packing consists of an infinite set of n-dimensional spheres, each having its center in a specific lattice point and radius r selected in such a way that the spheres associated with adjacent lattice points touch each other without overlapping. Then, we have that: (a) the *minimum distance* of a lattice (or sphere packing) $d_{min}(\Lambda)$ represents the smallest distance between sphere centers, and is equal to twice the sphere packing radius r; (b) the *kissing number* $K_{min}(\Lambda)$ is the number of sphere centers at minimum distance from any other sphere center, that is, it represents the number of spheres that "touch" or "kiss" any other sphere. Note that an n-dimensional lattice Λ always describes an n-dimensional *sphere packing*; the converse is not necessarily true, however, since a sphere packing may not contain the origin (i.e., the point $\lambda = \mathbf{0}_n$, where $\mathbf{0}_n$ is the n-dimensional null vector) and, if this occurs, the corresponding set of sphere centers is not a subgroup of $\mathbb{R}^n$.

---

modulo Λ if their difference belongs to Λ. Therefore, $\Lambda + \mathbf{a}$ can also be seen as the set of points equivalent to $\mathbf{a}$ *modulo* Λ.

[39] This is conceptually similar to the *weight enumerator polynomial* of block codes (see Section 9.1).

Each lattice is also characterized by two specific geometric regions, namely its *fundamental region* and its *Voronoi region* (VR). The fundamental region or *fundamental parallelotope* of an n-dimensional lattice Λ is the region defined as:

$$\{\mathbf{x}_n | \mathbf{x}_n = \mathbf{r}_n \mathbf{G}_\Lambda = r_0 \mathbf{g}_0 + r_1 \mathbf{g}_1 + \ldots + r_{n-1} \mathbf{g}_{N-1} | 0 \le r_0, r_1, \ldots, r_{n-1} < 1\}, \tag{3.315}$$

where $\mathbf{x}_n = [x_0, x_1, \ldots, x_{n-1}]$ and $\mathbf{r}_n = [r_0, r_1, \ldots, r_{n-1}]$ are n-dimensional *real* row vectors. The VR of a point $\lambda \in \Lambda$ is the set of all real vectors $\mathbf{x}_n \in \mathbb{R}^n$ nearer to λ than to any other lattice point, that is, the region defined as:

$$\{\mathbf{x}_n | |\lambda - \mathbf{x}_n| < |\bar{\lambda} - \mathbf{x}_n|, \bar{\lambda} \in \Lambda, \bar{\lambda} \ne \lambda\}. \tag{3.316}$$

It can be shown that a lattice Λ defined according to (3.312) is *geometrically uniform*. This means that any translate $\Lambda + \lambda$ of Λ by a lattice point λ coincides with Λ itself, Λ being a group. For this reason, each point of Λ has the same number of neighbors at a given distance, and all the VRs are congruent and form a tessellation of $\mathbb{R}^n$.

The VRs, the fundamental parallelotopes and the sphere packings are illustrated for the lattice $\mathbb{Z}^2$ in Figure 3.41(a) and for $\mathcal{A}_2$ in Figure 3.41(b).

It is also important to note that the parameter $V(\Lambda)$ represents the volume[40] of the VR of Λ, or, equivalently, that of its fundamental parallelotope. The latter region, defined by (3.315), can be seen as the image of the n-dimensional cube $[0, 1)^n$ (having unit volume) through the linear transformation $\mathbf{G}_\Lambda$. The Jacobian of this transformation is $\det(\mathbf{G}_\Lambda)$ (where $\det(\mathbf{X})$ denotes the *determinant* of the matrix $\mathbf{X}$), and we have that[41]:

$$V(\Lambda) = \det(\mathbf{G}_\Lambda). \tag{3.317}$$

It can also be proved that, if Λ' is a sublattice of Λ of order a, then[42] $V(\Lambda') = aV(\Lambda)$. For instance, the partition $\mathbb{Z}^4/D_4$ has order 2, so that $V(\mathcal{D}_4) = 2V(\mathbb{Z}^4) = 2$, since $V(\mathbb{Z}^n) = 1$ for any positive integer n (the VR of $\mathbb{Z}^n$ is an n-dimensional cube with unit edge length).

Other meaningful geometric parameters for an n-dimensional lattice Λ are: (a) its *Hermite parameter* (or *nominal coding gain*) [326]:

$$\gamma_c(\Lambda) \triangleq \frac{d_{\min}^2(\Lambda)}{V(\Lambda)^{2/n}}, \tag{3.318}$$

which measures the normalized density of Λ (since $V(\Lambda)^{2/n}$ represents a normalization of the volume $V(\Lambda)$ to two dimensions); and (b) its *density*:

$$\Delta(\Lambda) \triangleq \frac{r^n V_n}{V(\Lambda)}, \tag{3.319}$$

expressing the fraction of space $\mathbb{R}^n$ covered by the spheres of radius r (in the sphere packing of Λ), since:

$$V_n = \frac{\pi^{n/2}}{\Gamma((n/2) + 1)} \tag{3.320}$$

is the volume of an n-dimensional unit sphere [325], $\Gamma(\cdot)$ being the *gamma function* [96]. It is not difficult to prove that the parameter $\gamma_c(\Lambda)$ (3.318) is dimensionless, and is invariant to scaling, orthogonal transformations and Cartesian products [324, p. 1128], that is:

$$\gamma_c(\varepsilon \mathbf{O} \Lambda^m) \triangleq \gamma_c(\Lambda), \tag{3.321}$$

where ε is a positive scaling factor, m is any positive integer and $\mathbf{O}$ is an $n \times n$ orthogonal matrix.

---

[40] $V(\Lambda)$ can also be interpreted as the reciprocal of the number of lattice points per unit volume.
[41] This formula also expresses the so-called determinant of a lattice Λ (denoted $\det \Lambda$) having $d = n$. If $d < n$, we have that $\det \Lambda \triangleq \det(\mathbf{G}_\Lambda \mathbf{G}_\Lambda^T)^{1/2}$ (e.g., see [325]).
[42] This lemma is due to D. Forney (see [324, p. 1128]).

Consider, for instance, the lattices $\mathbb{Z}^2$ and $\mathcal{A}_2$ again. It is easy to show that $K_{\min}(\mathbb{Z}^2) = 4$, $d^2_{\min}(\mathbb{Z}^2) = V(\mathbb{Z}^2) = \gamma_c(\Lambda) = 1$, $\Delta(\mathbb{Z}^2) = \pi/4 \cong 0.7854$ and that $K_{\min}(\mathcal{A}_2) = 6$, $d^2_{\min}(\mathcal{A}_2) = 1$, $V(\mathcal{A}_2) = \sqrt{3}/2$, $\gamma_c(\mathcal{A}_2) = 2/\sqrt{3} \cong 1.155$ (0.62 dB), $\Delta(\mathcal{A}_2) = \pi/\sqrt{12} \cong 0.9069$. Note that the result $\Delta(\mathbb{Z}^2) < \Delta(\mathcal{A}_2)$ is confirmed by Figure 3.41, showing that the spheres in $\mathcal{A}_2$ are more closely packed than those in $\mathbb{Z}^2$.

A key problem in lattice theory is identifying the *densest* lattice or sphere packing in n dimensions. Table 3.2 summarizes the densest known lattices for various dimensions not exceeding 24. Note the following:

1. The n-dimensional lattice $\mathcal{D}_n$ is defined as $\{\lambda_n = (\lambda_0, \lambda_1, \ldots, \lambda_{n-1}) \in \mathbb{Z}^n | \sum_{i=0}^{n-1} \lambda_i = 0 \bmod 2\}$; the specific case corresponding to $n = 3$ (i.e., $\mathcal{D}_3$) is also known as the *face-centered cubic* (FCC) lattice and its generator matrix is (e.g., see [327, p. 277]):

$$\mathbf{G}_{\mathcal{D}_3} = \begin{bmatrix} 2 & 0 & 0 \\ 1 & 0 & 1 \\ 0 & 1 & 1 \end{bmatrix}. \tag{3.322}$$

2. The *Gosset lattice* $\mathcal{E}_8$ can be defined as (e.g., see [327, p. 276]):

$$\mathcal{E}_8 = 2\mathcal{D}_8 \cup \{2\mathcal{D}_8 + (1,1,1,1,1,1,1,1)\} \tag{3.323}$$

and can be shown to be a sublattice of $\mathbb{Z}^8$ with order 16, so that $V(\mathcal{E}_8) = 16\,V(\mathbb{Z}^8) = 16$. The values of other meaningful parameters are $d^2_{\min}(\mathcal{E}_8) = 4$ and, consequently, $\gamma_c(\mathcal{E}_8) = 2$ (3.01 dB); its generator matrix is [325]:

$$\mathbf{G}_{\mathcal{E}_8} = \frac{1}{2} \begin{bmatrix} 2 & 0 & 0 & 0 & 0 & 0 & 0 & 0 \\ 0 & 2 & 0 & 0 & 0 & 0 & 0 & 0 \\ 0 & 0 & 2 & 0 & 0 & 0 & 0 & 0 \\ 0 & 0 & 0 & 2 & 0 & 0 & 0 & 0 \\ 1 & 1 & 1 & 0 & 1 & 0 & 0 & 0 \\ 0 & 1 & 1 & 1 & 0 & 1 & 0 & 0 \\ 0 & 0 & 1 & 1 & 1 & 0 & 1 & 0 \\ 1 & 1 & 1 & 1 & 1 & 1 & 1 & 1 \end{bmatrix} \tag{3.324}$$

3. The *Leech lattice*[43] Λ_{24} is characterized by $V(\Lambda_{24}) = 2^{24}$ and $d^2_{\min}(\Lambda_{24}) = 16$, so that $\gamma_c(\Lambda_{24}) = 4$ (6.02 dB);
4. Λ_{16} represents the 16-dimensional lattice of the *Barnes–Wall* (BW) family.

As far as the last point is concerned, it is useful to point out that the BW lattices are an infinite family of n-dimensional lattices. For $n \leq 16$ they are the best known lattices (i.e., provide the densest lattice packings); for larger n, even if they are not optimal, they represent a good compromise between achievable performance and decoding complexity. It can be shown that, for any nonnegative integer m, there exists an n-dimensional real BW lattice, with $n = 2^{m+1}$, denoted by $\Lambda(0, m)$ in what follows.[44] Such a lattice is characterized by the following relevant parameters: kissing number [323]:

$$K_{\min}(\Lambda(0, m)) = \prod_{l=1}^{m+1} (2^l + 2), \tag{3.325}$$

---

[43] A great deal has been written about this important lattice; a set of useful references can be found in [325].
[44] Note that $\Lambda(0, 1)$, $\Lambda(0, 2)$ and $\Lambda(0, 3)$ correspond to $\mathcal{D}_4$, $\mathcal{E}_8$ and Λ_{16}, respectively, in Table 3.2.

normalized volume $V(\Lambda(0, m))^{2/n} = 2^{m/2}$, squared minimum distance $d_{\min}^2(\Lambda(0, m)) = 2^m$ and, consequently, nominal coding gain $\gamma_c(\Lambda(0, m)) = 2^{m/2}$. Some details about the construction of the BW lattices are provided in the following below.

Other higher-dimensional dense lattices are available in the technical literature; here we confine ourselves to mentioning those devised by Coxeter and Todd [332], Quebbemann [333], Nebe [334], and Elkies [323].

In applications the most useful class of lattices is undoubtedly that of the so-called *binary lattices*, since they are natural extensions of binary block codes and in many cases they provide the best performance [324, p. 1128], [335]. A real n-dimensional lattice Λ is a binary lattice if it is an integer lattice (i.e., $\Lambda \subseteq \mathbb{Z}^n$) having $2^m \mathbb{Z}^n$ as a sublattice for some m; the least such m is called the 2-*depth* of the lattice. Then, if Λ is binary with *depth* m, we have that $\mathbb{Z}^n / \Lambda / 2^m \mathbb{Z}^n$ is a partition chain. All the known useful lattices have 2-*depth* equal to 1 or 2 and are known as mod-2 and mod-4 lattices, respectively.

Dense lattices can be employed to design power-efficient constellations for digital signaling. Generally speaking, given an n-dimensional lattice Λ, a *lattice constellation*[45] (or *signal constellation*) is a finite subset of the lattice points and can be generated as [337]:

$$\mathcal{C}(\Lambda, R) \triangleq (\Lambda + \mathbf{a}) \cap \mathcal{R}, \qquad (3.326)$$

by selecting points belonging to the translate[46] $(\Lambda + \mathbf{a})$ of Λ and lying within a compact bounding region $\mathcal{R}$ of n-dimensional space. There is no general agreement about the shape or size[47] of $\mathcal{R}$, except for the fact that $\mathcal{R}$ is chosen to be large enough to enclose the desired number $|\mathcal{C}(\Lambda, R)|$ of constellation points and is bounded by some limitation on the energy of codewords and/or the PAPR [337, 339]. For instance, in the two-dimensional case power-efficient constellations can be extracted from the hexagonal lattice[48] $\mathcal{A}_2$, as shown in [338].

The region $\mathcal{R}$ of (3.326) is characterized by the following properties: (a) its *volume*:

$$V(\mathcal{R}) \triangleq \int_{\mathcal{R}} d\mathbf{x}; \qquad (3.327)$$

Table 3.2 Densest known lattices in selected dimensions not exceeding 24

n	Name	Symbol	$\Delta(\cdot)$	$K_{\min}(\cdot)$	$\gamma_c(\cdot)$, dB	Ref.
1	integer	$\mathbb{Z}$	1	2	0.0	[323]
2	hexagonal	$\mathcal{A}_2$	$\pi/(2\sqrt{3}) \cong 0.907$	6	0.5	[323]
3	diagonal	$\mathcal{D}_3$	$\pi/(3\sqrt{2}) \cong 0.741$	12	1.0	[323, 328]
4	Schläfli	$\mathcal{D}_4$	$\pi^2/16 \cong 0.617$	24	1.5	[323]
8	Gosset	$\mathcal{E}_8$	$\pi^4/384 \cong 0.254$	240	3.0	[329]
12	Barnes–Wall	Λ_{16}	$\pi^8/(8!2^4) \cong 0.0147$	4320	4.5	[325, 330]
24	Leech	Λ_{24}	$\pi^{12}/12! \cong 0.00193$	196560	6.0	[331]

---

[45] The term *lattice code* is also used in the technical literature (e.g., see [336, 337]).

[46] An n-dimensional lattice is constrained to have a point at $\mathbf{0}_n$ and the translation frees us from this constraint; this is useful, for instance, for minimizing the average transmitted power.

[47] If $\mathcal{R}$ is a convex region containing the origin, then the resulting lattice codes include the lattice codes discussed by Forney *et al.* in [338, p. 636, Figure 6].

[48] A. D. Wyner has proved that, in this case, if the constellation cardinality tends to infinity, the optimal constellation is provided by the intersection of a circle (whose center is $\mathbf{0}_2$) with $\mathcal{A}_2$ (see [340, Appendix B]).

(b) its *average energy per dimension*:

$$P(\mathcal{R}) \triangleq \frac{1}{nV(\mathcal{R})} \int_{\mathcal{R}} |\mathbf{x}|^2 d\mathbf{x}, \tag{3.328}$$

evaluated under the assumption of a uniform probability density function over $\mathcal{R}$; and (c) the *normalized second moment* of $\mathcal{R}$:

$$G(\mathcal{R}) \triangleq \frac{P(\mathcal{R})}{V(\mathcal{R})^{2/n}}. \tag{3.329}$$

It can be proved that the parameter $G(\mathcal{R})$ is invariant to scaling, orthogonal transformations and Cartesian products, that is:

$$G(\varepsilon \mathbf{O} \mathcal{R}^m) = G(\mathcal{R}), \tag{3.330}$$

where ε is a positive scaling factor, m is any positive integer and $\mathbf{O}$ is an $n \times n$ orthogonal matrix. Let us now consider the specific example of an M-ASK constellation.

Example 3.8.1 The M-ASK constellation illustrated in Figure 3.3(b) is a one-dimensional lattice constellation $\mathcal{C}(2\mathbb{Z}, R)$ (see (3.326)) with $\Lambda + \mathbf{a} = 2\mathbb{Z} + 1$ and $\mathcal{R} = [-M, M]$. It is easy to show that $\mathcal{R}$ is characterized by $V(\mathcal{R}) = 2M$, $P(\mathcal{R}) = M^2/3$ and $G(\mathcal{R}) = 1/12$.
□

In the analysis of the error performance achievable with *large*[49] constellations, some approximations are usually made. In particular, given a constellation $\mathcal{C}(\Lambda, R)$ (3.326), its *size* $|\mathcal{C}(\Lambda, R)|$ and *average power per dimension* $P(\mathcal{C}(\Lambda, R))$ (assuming a uniform discrete distribution over $\mathcal{C}(\Lambda, R)$) are approximated as:

$$|\mathcal{C}(\Lambda, R)| \cong \frac{V(\mathcal{R})}{V(\Lambda)} \tag{3.331}$$

and

$$P(\mathcal{C}(\Lambda, R)) \cong P(\mathcal{R}), \tag{3.332}$$

respectively.[50] In addition, the *average number of nearest neighbors* for the points of $\mathcal{C}(\Lambda, R)$ is estimated as $K_{\min}(\Lambda)$, that is, the kissing number of the lattice Λ. When assessing the performance achievable by *maximum likelihood* detection (see Section 4.3.3) of a given lattice constellation on AWGN channels, a *union bound estimate* (UBE) on the probability of symbol detection error is usually considered (e.g., see Section 4.3.2, [29, Sect. 14.1.] and [337]). This bound shows that the *total coding gain* γ_{tot} is given by the product (i.e., by the decibel sum) of the nominal coding gain[51] $\gamma_c(\Lambda)$ (see (3.318)), characterizing the lattice Λ from which the constellation is extracted, and the so-called *shaping gain* $\gamma_s(\mathcal{R})$, depending on the constellation bounding region $\mathcal{R}$; the latter parameter is defined as:

$$\gamma_s(\mathcal{R}) \triangleq \frac{V(\mathcal{R})^{2/n}}{12P(\mathcal{R})} \tag{3.333}$$

and can also be rewritten as (see (3.329)):

$$\gamma_s(\mathcal{R}) = \frac{1/12}{G(\mathcal{R})}. \tag{3.334}$$

---

[49] This means that $\mathcal{R}$ is large relative to $V(\Lambda)$.
[50] Relations (3.331) and (3.332) express the so-called *continuous approximation* (see [337] and [29, Sect. 14.1.3.]).
[51] This compares the minimum distance of Λ against a two-dimensional square lattice $\mathbb{Z}^2$, representing a reference lattice [29, Sect. 14.1].

Note that $\gamma_c(\Lambda)$ and $\gamma_s(\mathcal{R})$, even if analytically they play similar roles in the UBE, have substantially different meanings. The former measures the density increase provided by Λ with respect to the baseline integer lattice $\mathbb{Z}$ (or, equivalently, $\mathbb{Z}^n$), whereas the latter quantifies the decrease in average energy due to the region $\mathcal{R}$ relative to an interval $[-1, 1]$ (or, equivalently, to an n-dimensional cube $[-1, 1]^n$). The *effective coding gain*, however, is reduced by the error coefficient $K_{\min}(\Lambda)$ appearing as a multiplicative factor in the UBE. It is also worth pointing out that $\gamma_c(\Lambda)$ and $\gamma_s(\mathcal{R})$ show different asymptotic behaviors. On the one hand, $\gamma_s(\mathcal{R})$ is limited to the quantity $\pi e/6$ (i.e., 1.53 dB), called the *ultimate shaping gain*, as $n \to \infty$ [337]; on the other hand, the nominal coding gains of dense n-dimensional lattices tend to become infinite as $n \to \infty$. Despite this, effective coding gains do not tend to infinity, because of the increase in the number of nearest neighbors with the lattice dimensionality. These considerations are supported by the numerical results shown in Table 3.2.

Finally, we note that, with high-dimensional lattices, large coding gains can be achieved if the signal constellations extracted from them are combined with channel coding schemes in a proper fashion (e.g., see [338, 324]). The adoption of multidimensional signaling schemes is also encouraged by the fact that usually for the densest lattices *fast decoding algorithms* (i.e., computationally efficient methods for finding the closest lattice point to an arbitrary point) exist (e.g., see [339, 341, 342]).

3.8.2   Elementary Constructions of Lattices

We now introduce some elementary methods to construct lattices from *block codes*,[52] which we then apply to generate the so-called BW *lattices*.

The first construction method, known as *construction A* (first proposed in [343]; see also [323, 342]) relies on the fact that mod-2 binary lattices are essentially *isomorphic* to *linear binary block codes*. In fact, it can be proved that a real n-dimensional lattice Λ is a mod-2 binary lattice if and only if it is the set of all the integer n-tuples that are congruent modulo 2 to some codeword in a *linear binary* (n, k) *block code* $\mathcal{C}$, that is, if and only if Λ can be expressed as:

$$\Lambda = \{\boldsymbol{\lambda} \in \mathbb{Z}^n \mid \boldsymbol{\lambda} \equiv \mathbf{c} \bmod 2 \text{ for some } \mathbf{c} \in \mathcal{C}\}. \tag{3.335}$$

If this occurs, the volume $V(\Lambda)$ is equal to 2^{n-k} and the minimum squared distance of Λ is $d_{\min}^2(\Lambda_\mathcal{C}) = \min(d_H(\mathcal{C}), 4)$, where $d_H(\mathcal{C})$ is the *minimum Hamming distance* (see (9.18)) of the block code $\mathcal{C}$. The latter result follows from the fact that 4 is the distance between any two lattice points corresponding to the same codeword $\mathbf{c}$ and differing by 2 in one coordinate, whereas $d_H(\mathcal{C})$ is the distance between two lattice points corresponding to two distinct codewords $\mathbf{c}$ and $\mathbf{c}'$ with Hamming distance $d_H(\mathcal{C})$ and differing by 1 in the coordinates where $\mathbf{c}$ and $\mathbf{c}'$ differ. From (3.335) and (3.314), the representation:

$$\Lambda = 2\mathbb{Z}^n + \mathcal{C} \tag{3.336}$$

of Λ by its coset decomposition then follows. This expresses the mod-2 lattice as the union of 2^k cosets of $2\mathbb{Z}^n$, each coset corresponding to a specific codeword $\mathbf{c}$ of $\mathcal{C}$ [324].

The mod-2 construction method can be generalized to generate lattices from nonbinary codes [342]. The new method, known as *generalized construction A*, can be summarized as follows. Let Λ be an l-dimensional lattice and Λ' denote a sublattice with the same dimensionality as Λ, so that the factor group Λ/Λ' is finite with order $|\Lambda/\Lambda'| = q$. Moreover, let $\mathcal{G}$ be a q-ary *label group* isomorphic to the partition Λ/Λ' according to the isomorphism $\xi : \mathcal{G} \to \Lambda/\Lambda'$. Then, we have that: (a) any element u of $\mathcal{G}$ is *equivalent* to a coset representative $\boldsymbol{\delta} \in [\Lambda/\Lambda']$ with $\xi : u \to \boldsymbol{\delta}$; (b) for any $u \in \mathcal{G}$, $\Lambda' + \xi(u)$ represents a coset of Λ'; (c) there exists an inverse mapping $\xi^{-1} : \boldsymbol{\delta} \to u$ from the elements of the

---

[52] To understand this section some knowledge of the theory of linear block codes is required (see Section 9.1).

coset $\Lambda' + \delta$ to the element u. The new construction method defines an $(l \cdot n)$-dimensional lattice $\Lambda_{\mathcal{C}}$ from an (n, k) *group code*[53] $\mathcal{C}$ over $\mathcal{G}$ as:

$$\Lambda_{\mathcal{C}} \triangleq \bigcup_{\mathbf{c} \in \mathcal{C}} \phi(\mathbf{c}), \tag{3.337}$$

where $\mathbf{c} \triangleq [c_0, c_1, \ldots, c_{n-1}]$ and:

$$\phi(\mathbf{c}) \triangleq (\Lambda')^n + \xi(\mathbf{c}), \tag{3.338}$$

with $\xi(\mathbf{c}) \triangleq [\xi(c_0), \xi(c_1), \ldots, \xi(c_{n-1})]$. Note that, by analogy with (3.336), (3.338) expresses the lattice Λ as the union of $|\mathcal{C}|$ cosets of $(\Lambda')^n$, each coset being associated with a specific codeword $\mathbf{c}$ of $\mathcal{C}$ [324]. It can be proved that the lattice $\Lambda_{\mathcal{C}}$ (3.337) is characterized by:

$$V(\Lambda_{\mathcal{C}}) = \frac{V(\Lambda')^n}{|\mathcal{C}|} \tag{3.339}$$

and

$$d_{\min}(\Lambda_{\mathcal{C}}) \geq \min\{d_{\min}(\Lambda'), \ d_{H,\mathcal{C}} \ d_{\min}(\Lambda)\}, \tag{3.340}$$

where $|\mathcal{C}|$ and $d_{H,\mathcal{C}}$ are the cardinality and the minimum Hamming distance of the code $\mathcal{C}$, respectively [323]. Note that if we choose $\Lambda = \mathbb{Z}$ and $\Lambda' = 2\mathbb{Z}$, then $q = 1$ and Λ/Λ' is isomorphic to the binary group $\mathcal{G} = \mathbb{Z}_2 = \{0, 1\}$. An (n, k) group code over $\mathbb{Z}_2$ is a *binary* linear (n, k) block code, so that, letting $\xi(0) = 0$ and $\xi(1) = 1$, the generalized construction A reduces to construction A.

The generalized construction A can be further extended to encompass the case of a longer decomposition chain: the resulting algorithm is known as *generalized construction C*[54] and can be summarized as follows. Let $\Lambda_0/\Lambda_1/ \ldots /\Lambda_m$ denote a partition chain of l-dimensional lattices, where each partition Λ_{k-1}/Λ_k is isomorphic to a group $\mathcal{G}_k$, with $k = 1, 2, \ldots, m$. Then, denote the mapping from label group $\mathcal{G}_k$ to coset $[\Lambda_{k-1}/\Lambda_k]$ as $\xi_k : \mathcal{G}_k \to [\Lambda_{k-1}/\Lambda_k]$, with inverse $\xi_k^{-1} : [\Lambda_{k-1}/\Lambda_k] \to \mathcal{G}_k$. Now consider some sequence $\mathcal{C}_1, \mathcal{C}_2, \ldots, \mathcal{C}_m$ of group codes, each of length n, over $\mathcal{G}_1, \mathcal{G}_2, \ldots, \mathcal{G}_m$, respectively. Then the $(l \cdot n)$-dimensional lattice $\Lambda_{\mathcal{C}}$ generated by the new method is given by:

$$\Lambda_{\mathcal{C}} \triangleq \bigcup_{\mathbf{c}^{(1)} \in \mathcal{C}_1, \mathbf{c}^{(2)} \in \mathcal{C}_2, \ldots, \mathbf{c}^{(m)} \in \mathcal{C}_m} \phi(\mathbf{c}^{(1)}, \mathbf{c}^{(2)}, \ldots, \mathbf{c}^{(m)}), \tag{3.341}$$

where $\mathbf{c}^{(k)} \triangleq [c_0^{(k)}, c_1^{(k)}, \ldots, c_{n-1}^{(k)}]$ and:

$$\phi(\mathbf{c}^{(1)}, \mathbf{c}^{(2)}, \ldots, \mathbf{c}^{(m)}) \triangleq (\Lambda_m)^n + \xi_1(\mathbf{c}^{(1)}) + \ldots + \xi_m(\mathbf{c}^{(m)}), \tag{3.342}$$

with $\xi_k(\mathbf{c}^{(k)}) \triangleq \{\xi_k(c_0^{(k)}), \xi_k(c_1^{(k)}), \ldots, \xi_k(c_{n-1}^{(k)})\}$. It can be proved that $\Lambda_{\mathcal{C}}$ in (3.341) is characterized by:

$$V(\Lambda_{\mathcal{C}}) = \frac{V(\Lambda_m)^n}{\displaystyle\prod_{k=1}^{m} |\mathcal{C}_k|} \tag{3.343}$$

and

$$d_{\min}(\Lambda_{\mathcal{C}}) \geq \min\{d_{\min}(\Lambda_m), d_{H,\mathcal{C}_m} d_{\min}(\Lambda_{m-1}), \ldots, d_{H,\mathcal{C}_1} d_{\min}(\Lambda_0)\}, \tag{3.344}$$

where $d_{H,\mathcal{C}_k}$ is the minimum Hamming distance of the code $\mathcal{C}_k$, with $k = 1, 2, \ldots, m$ [323, 342].

This construction method can be applied to generate the BW family of lattices [330], as illustrated in the following example.

---

[53] Generally speaking, a *group code* (with codewords of length n) over a finite Abelian group $\mathcal{G}$ is a subgroup of a direct product group $\mathcal{G}^n$ (see Section 9.1).

[54] Forney and Vardy [342, p. 1995] generalize this construction further, calling it *multilevel construction A*, but in keeping with [323] we use the name *generalized construction C* here.

Example 3.8.2 BW lattices are closely related to the family of *Reed–Muller* (RM) binary block codes [324, 335, 344]. In fact, to construct the BW lattice $\Lambda(0, m)$ according to (3.341) and (3.342), the code C_k (with $k = 1, 2, \ldots, m$) is selected as the RM code $RM(m - k, m)$ of length $N = 2^m$ and minimum Hamming distance $d_{H,C_k} = 2^k$ (see [324, p. 1135] and Section 9.1.5.2). In addition, the lattice $\Lambda_k \triangleq \phi^k G$ is selected (with $k = 1, 2, \ldots, m$) for generating the partition chain $\Lambda_0 / \Lambda_1 / \ldots / \Lambda_m$. Here G denotes the set of *Gaussian integers*, that is, the *one-dimensional* complex lattice associated with the *two-dimensional* lattice $\mathbb{Z}^2$, and $\phi \triangleq 1 + i$ denotes the prime of G having least norm (see [324, pp. 1129–1130]). Note that: (a) the RM codes of a given length N are nested, in the sense that $RM(m, m) / RM(m - 1, m) / \ldots / RM(0, m)$ is a code partition chain; and (b) the complex set ϕG is a sublattice of G of order 2 and corresponds to the real lattice $R_2 \mathbb{Z}^2$, where R_2 is the two-dimensional rotation operator defined in (3.313).

Then, the lattice $\Lambda(0, m)$, whose dimensionality in its real form is equal to $n = l \cdot N = 2N = 2^{m+1}$, can be generated *in complex form*[55] as in (3.341) and (3.342) with:

$$(\Lambda_m)^n = \phi^m G^N \tag{3.345}$$

and

$$\xi_k(\mathbf{c}^{(k)}) = \phi^{m-k} \mathbf{c}^{(k)} \tag{3.346}$$

with $\mathbf{c}^{(k)} \in RM(m - k, m)$ and $k = 1, 2, \ldots, m$ [324, p. 1135].

A *squaring construction* method can also be used to generate the BW lattices in a recursive fashion, as explained in detail in [335, pp. 1166–1170]. Other procedures are available for constructing the family of the so-called *principal sublattices* (denoted $\{\Lambda(r, m)\}$, with $m \geq 0$ and $0 \leq r \leq n$) of the BW lattices; details can be found in [324, 335].

□

3.9 Spectral Properties of a Digital Modulation at the Output of a Wireless Channel

In this section the mathematical tools described in Section 3.4 and some results derived in Chapter 2 are applied to the analysis of the spectral properties of a digital modulation transmitted through a doubly-selective wireless channel. This allows us to assess the effects of time selectivity in the frequency domain and to establish a relationship between the average *received* energy per symbol interval and the average energy *transmitted* in the same interval.

In our analysis it is assumed that: (a) the complex envelope $s(t, \mathbf{c})$ of the transmitted signal depends on a random vector $\mathbf{c}$ of channel symbols and is WSC with period T_{cs}; and (b) the CIR $h(t, \tau)$ is *statistically independent* of $s(t, \mathbf{c})$. Then, the useful component of the channel response to $s(t, \mathbf{c})$ is (see (2.9)):

$$z(t, \mathbf{c}) \triangleq \int_{-\infty}^{+\infty} s(t - \tau, \mathbf{c}) \, h(t, \tau) \, d\tau, \tag{3.347}$$

so that the autocorrelation function of $z(t, \mathbf{c})$ is given by:

$$R_z(t, \tau) \triangleq E\{z(t + \tau, \mathbf{c}) \, z^*(t, \mathbf{c})\}$$
$$= \int_{\alpha=-\infty}^{+\infty} \int_{\beta=-\infty}^{+\infty} R_s(t - \beta, \tau + \beta - \alpha) \, R_h(t + \tau, t; \alpha, \beta) \, d\beta \, d\alpha, \tag{3.348}$$

---

[55] Each complex component of a lattice point can be turned into a two-dimensional real vector, whose first and second elements are the real and imaginary parts, respectively, of the component itself. For this reason, in this case each N-dimensional complex lattice point is equivalent to an n-dimensional real point with $n = 2N$.

where $R_h(t_1, t_2; \tau_1, \tau_2) \triangleq E\{h(t_1, \tau_1) h^*(t_2, \tau_2)\}$ (see (2.79)) and $R_s(t, \tau)$ represent the autocorrelation function of the CIR and $s(t, \mathbf{c})$, respectively. From (3.348) it can easily be inferred that, generally speaking, $R_z(t, \tau)$ is not periodic in the variable t, so that $z(t, \mathbf{c})$ (3.347) is not even cyclostationary in autocorrelation. However, if we assume that the channel is affected by Rayleigh fading and is WSS-US, so that $R_h(t_1, t_2; \tau_1, \tau_2) = P_h(t_1 - t_2; \tau_1) \, \delta(\tau_1 - \tau_2)$ (see (2.83)), (3.348) can be easily simplified to:

$$R_z(t, \tau) = \int_{-\infty}^{+\infty} R_s(t - \alpha, \tau) \, P_h(\tau; \alpha) \, d\alpha. \tag{3.349}$$

Note that on the RHS of this equality only the factor $R_s(t - \alpha, \tau)$ shows a dependence on t. Since $s(t, \mathbf{c})$ is WSC, $R_s(t, \tau)$ is periodic in the variable t with period T_{cs}, so that $R_z(t, \tau)$ shares the same property. This proves that,[56] under the above assumptions,[57] $z(t, \mathbf{c})$ is WSC with period T_{cs} and its average autocorrelation function is given by (see (3.26)):

$$R_z(\tau) \triangleq \frac{1}{T_{cs}} \int_0^{T_{cs}} R_z(t, \tau) \, dt$$

$$= \frac{1}{T_{cs}} \int_{t=0}^{T_{cs}} \int_{\alpha=-\infty}^{+\infty} R_s(t - \alpha, \tau) \, P_h(\tau; \alpha) \, d\alpha \, dt. \tag{3.350}$$

If on the RHS of the latter equation the integration order is reversed and we note that:

$$\frac{1}{T_{cs}} \int_{t=0}^{T_{cs}} R_s(t - \alpha, \tau) \, dt = R_s(\tau) \tag{3.351}$$

for any α, where $R_s(\tau)$ is the average autocorrelation function of $s(t, \mathbf{c})$, it is found that:

$$R_z(\tau) = R_s(\tau) \int_{-\infty}^{+\infty} P_h(\tau; \alpha) \, d\alpha. \tag{3.352}$$

Moreover, if we assume that $P_h(\tau; \alpha)$ is separable, that is, that $P_h(\tau; \alpha) = R_D(\tau) \, P_h(\alpha)$ (see (2.95)), where $R_D(\tau)$ (with $R_D(0) = 1$) is the *Doppler autocorrelation function* and $P_h(\tau)$ is the channel PDP, (3.352) can be rewritten as:

$$R_z(\tau) = R_s(\tau) \, R_D(\tau) \int_{-\infty}^{+\infty} P_h(\alpha) \, d\alpha. \tag{3.353}$$

This result shows that, if the channel PDP is *normalized*, that is, if:

$$\int_{-\infty}^{+\infty} P_h(\alpha) \, d\alpha = 1, \tag{3.354}$$

then $z(t, \mathbf{c})$ in (3.347) is characterized by the average autocorrelation function:

$$R_z(\tau) = R_s(\tau) \, R_D(\tau) \tag{3.355}$$

and, consequently, by the average PSD:

$$S_z(f) \triangleq \text{FCT}[R_z(\tau)] = S_s(f) \otimes S_D(f) = \int_{-\infty}^{+\infty} S_s(\alpha) \, S_D(f - \alpha) \, d\alpha, \tag{3.356}$$

---

[56] Note that the Rayleigh fading assumption implies that the stochastic process $z(t, \mathbf{c})$ (3.347) has zero mean.

[57] The reader can verify that, if the channel is doubly-selective, the assumption of WSS channel alone, like that of an US channel, is not sufficient to ensure that $z(t, \mathbf{c})$ in (3.347) is WSC. However, if the channel is frequency (time) selective only, the US (WSS) assumption is sufficient.

where $S_s(f)$ is the average PSD of $s(t, \mathbf{c})$ and $S_D(f) \triangleq \text{FCT}[R_D(\tau)]$ is the *Doppler power spectrum* (see Section 2.2.2.3). It is important to note that the convolution product in (3.356) accounts for the spectral broadening, in the received signal, due to the Doppler effect in a wireless channel. In fact, (3.356) proves that, if $s(t, \mathbf{c})$ is rigorously bandlimited to B Hertz and the $S_D(f)$ does not extend beyond the Doppler bandwidth B_D, the bandwidth of $S_z(f)$ is equal to $B + B_D$ Hertz. Note also that, under the last assumption about $S_D(f)$, (3.356) can be rewritten as:

$$S_z(f) = \int_{f-B_D}^{f+B_D} S_s(\alpha) \, S_D(f - \alpha) \, d\alpha. \tag{3.357}$$

Then if $B_D \ll B$, so that $S_s(f)$ undergoes negligible variation in the integration interval of (3.357), we have:

$$S_z(f) \cong S_s(f) \int_{-\infty}^{+\infty} S_D(\alpha) \, d\alpha = S_s(f) R_D(0) = S_s(f), \tag{3.358}$$

since $\int_{-\infty}^{+\infty} S_D(\alpha) \, d\alpha = R_D(0) = 1$. Integrating both sides of (3.358) produces:

$$P_z \cong P_s, \tag{3.359}$$

establishing that the average power of the useful received signal is essentially equal to the average transmitted power. This implies that:

$$\bar{E}_s \cong E_s, \tag{3.360}$$

where $\bar{E}_s \triangleq P_z T_s$ is the *average received energy*[58] *per symbol interval*, whereas E_s represents the average transmitted energy in the same interval. Note that the approximation of (3.359) and (3.360) becomes more and more accurate as the Doppler bandwidth gets narrower, becoming an exact equality when $B_D = 0$.

The results derived above refer to a SISO wireless channel. Let us extend them to the case of a MIMO communication system employing n_T transmit and n_R receive antennas. The complex envelope of the useful signal captured by the kth receive antenna (with $k = 0, 1, \ldots, n_T - 1$) can be expressed as:

$$z_k(t, \mathbf{c}) \triangleq \int_{-\infty}^{+\infty} \sum_{l=0}^{n_T-1} s_l(t - \tau, \mathbf{c}_l) \, h_{l,k}(t, \tau) \, d\tau, \tag{3.361}$$

where $s_l(t, \mathbf{c}_l)$ and $h_{l,k}(t, \tau)$ are the signal sent by the lth transmit antenna and the CIR between the lth input and the kth output, respectively. In what follows we assume that: (a) the stochastic processes $\{s_l(t, \mathbf{c}_l), l = 0, 1, \ldots, n_T - 1\}$ are all WSC[59] with the same period T_{cs}; (b) the $n_T \times n_R$ SISO channels are WSS-US, mutually independent and affected by Rayleigh fading; and (c) each of these channels has a separable scattering function and its PDP is normalized (see (3.354)). Then it is not difficult to prove that $z_k(t, \mathbf{c})$, $k = 0, 1, \ldots, n_R - 1$, is WSC with period T_{cs} and that its average autocorrelation function is given by:

$$R_{z_k}(\tau) = \sum_{l=0}^{n_T-1} R_{s_l}(\tau) \, R_{D_{l,k}}(\tau), \tag{3.362}$$

---

[58] Note that the evaluation of $\bar{E}_s$ entails a statistical average not only with respect to $\mathbf{c}$, as has been done with E_s, but also with respect to all the random parameters of the CIR.

[59] Note that no assumption is made here about their possible statistical correlations.

where $R_{s_l}(\tau)$ is the average autocorrelation function of $s_l(t, \mathbf{c}_l)$ and $R_{D_{l,k}}(\tau)$ is the Doppler autocorrelation function of the channel associated with the lth input and kth output. Therefore, the average PSD of $z_k(t, \mathbf{c})$ is:

$$R_{z_k}(\tau) = \sum_{l=0}^{n_T-1} S_{s_l}(f) \otimes S_{D_{l,k}}(f), \tag{3.363}$$

where $S_{s_l}(f)$ is the average PSD of $s_l(t, \mathbf{c})$ and $S_{D_{l,k}}(f) \triangleq \text{FCT}\,[R_{D_{l,k}}(\tau)]$ is characterized by a Doppler bandwidth $B_{D_{l,k}}$. As in the SISO scenario considered above, if $B_{D_{l,k}} \ll B_l$ with $l = 0, 1, \ldots, n_T - 1$, where B_l is the bandwidth of $s_l(t, \mathbf{c}_l)$, the average power P_{z_k} of the overall useful signal captured by the kth receive antenna can be approximated as:

$$P_{z_k} \cong \sum_{l=0}^{n_T-1} P_{s_l}, \tag{3.364}$$

where P_{s_l} is the average power of $s_l(t, \mathbf{c})$, with $l = 0, 1, \ldots, n_T - 1$. Note that the RHS of (3.364) is actually independent of k because of the assumed statistical equivalence and mutual independence of all $n_T \times n_R$ SISO channels. This entails that the *average received energy per symbol interval and per receive antenna* $\bar{E}_s$ is independent of the selected receive antenna and is given by:

$$\bar{E}_s \cong \sum_{l=0}^{n_T-1} E_{s_l}, \tag{3.365}$$

where E_{s_l} represents the average energy radiated by the lth transmit antenna in the same interval.

In all the following chapters it will be assumed that, unless explicitly stated, the assumptions required for the validity of (3.359) and (3.360) (and their generalizations (3.364) and (3.365)) are satisfied.

3.10   Historical Notes

Substantial efforts have been devoted to the study of digital modulation techniques and their spectral properties in the last 50 years. For this reason, a huge volume of technical literature on these topics is available. In this section, we outline the evolution of the research activities concerning the development of PAM, CPM and MC signals, including their spectral properties.

3.10.1   Passband PAM Signaling

Historically speaking, the first passband PAM techiques employed in wireless communication systems were PSK modulations [345, 346], in the form given by (3.57). Their use since the latter half of the 1950s was made possible by the use of oscillators capable of providing a stable reference for the coherent detection of phase modulated signals [347]. In 1960, C. S. Cahn suggested considering, as a natural extension of digital phase modulations, new modulations combining a multiplicity of phases with a finite number of amplitudes to achieve a more efficient use of the transmitted power with a larger number of bits per symbol [348]. The constellations proposed by Cahn consisted of a set of points placed along concentric and equally populated circumferences in a two-dimensional signal space. The work started by Cahn was soon extended by J. C. Hancock and R. W. Lucky, who showed how the error performance could be improved by selecting only two distinct amplitudes for channel symbols and placing more points on the inner circumference than on the outer one [349]. In 1962, C. Campopiano and B. G. Glazer proposed a new AM-PM format [350], said to be of *Type III*, to distinguish it from *Type I*, as proposed by Cahn, and *Type II*, as described by Hancock and

Lucky. The technique devised by Campopiano and Glazer, known today as QAM, was proposed as a technically appealing solution with a simple implementation for medium to large signaling alphabets. In the same year, Hancock and Lucky published another work [351], investigating the problems of designing *optimal* AM-PM constellations and of comparing their performance with that provided by amplitude and phase modulations. They proved that phase modulated signals were the optimal choice in the presence of a low SNR and for small constellation sizes, and that AM-PM techniques were the optimal choice for $M \geq 8$ in the presence of a constraint on the *average transmitted power*, and for $M \geq 16$ in the presence of a constraint on the *peak power*.

After the publication of these papers, the problem of designing new AM-PM constellations was neglected for several years. This was probably due to the difficulties encountered in the implementation of AM-PM systems with the available technology and the absence of a need for data communications with high spectral efficiency. In 1973, Simon and Smith published a paper that investigated hexagonal or honeycomb signal constellations [352]. This work clearly showed that such signal sets were more energy-efficient than any previously found and that the *hexagonal* or honeycomb lattice in two dimensions, known as $\mathcal{A}_2$, is the densest possible packing in two dimensions. In 1974, the problem of devising new AM-PM formats was reconsidered by G. J. Foschini, R. D. Gitlin and S. B. Weinstein [340], and by C. Thomas, M. Weidner and S. Durrani [353]. The former, using a search procedure based on a gradient method, solved the problem of finding *optimal* two-dimensional constellations with 4, 7, 8, 16 and 19 points, while the latter described 29 new constellations and compared their performance. Their work verified heuristically the earlier work by Simon and Smith in that the best constellations found were indeed essentially hexagonal. A year later, the so-called *cross* constellations, which allowed transmission of an odd number of bits in each symbol interval, were proposed by J. G. Smith [354].

As far as the implementation of AM-PM systems is concerned, we note that the first significant steps date back to the second half of the 1970s. In 1976, K. Miyauchi, S. Seki and H. Ishio, researchers at the *Nippon Telegraph and Telephone Public Corporation* (NTT), developed a prototype of a simplified 16-QAM system, which could operate at up to 400 Mbit/s [355]. Another prototype of a 16-QAM modem (operating at 140 Mbit/s in the 10.7–11.7 MHz bandwidth), developed in France by the *Centre National d'Études des Télécommunications* (CNET), was illustrated by P. Dupuis *et al.* in 1979 [356]. In the same year, the NTT researchers I. Horikawa, T. Murase and Y. Saito announced promising results from experimentation with a 16-QAM modem [357], transmitting at 5 GHz, at a speed of 200 Mbit/s in a 40 MHz bandwidth. Note that, by 1979, almost 20 years had elapsed since the first theoretical studies on the potentialities offered by AM-PM modulation techniques. In the following years many more advanced prototypes of QAM systems were developed: an example of a 256-QAM modem, developed for a high-capacity wireless link (400 Mbit/s) was described by Y. Saito and Y. Nakamura in 1986 [358]. The above-mentioned prototypes were designed for microwave radio bridges providing high-capacity links in LOS scenarios [359]. In the following years, QAM techniques were adopted in commercial systems, and adopted for digital signaling over the *public switched telephone network* (PSTN). In addition, the possibility of their use in *satellite* and *mobile terrestrial* systems was widely investigated, as discussed in detail in [360]. In these scenarios, requiring the use of power-efficient amplifiers [361], PSK or CPM signaling have usually been preferred to AM-PM techniques, because of the larger envelope fluctuations of the latter with respect to the former.

It is also worth recalling the strong interest in *staggered* PAM signaling at the end of the 1960s and at the beginning of 1970s. In that period significant attention was paid to OQPSK, considered as a possible alterative to MSK (see Section 3.6.2) in transmissions over nonlinear and severely bandlimited channels [362]. A description of the advantages deriving from the adoption of OQPSK and some interesting bibliographic citations on this topic can be found in [363].

3.10.2   CPM Signaling

Digital *frequency modulation* (FM) techniques, and in particular FSK, have received substantial attention from the scientific community since the 1960s. This interest was motivated by the *simplicity* both in the generation of FM signals, and in the available *noncoherent* detection methods, such as those based on the use of a *frequency discriminator* [280]. For this reason, the adoption of these techniques was suggested for those scenarios requiring the development of low-complexity apparatuses, but not putting severe constraints on the transmission bandwidth. In fact, FM techniques were deemed *spectrally inefficient* and, consequently, intrinsically unsuitable for high-speed communications. In those years the generation mechanism for continuous-phase FM signals and the advantages they were offering over their discontinuous-phase counterparts was also clear [280].

In the 1970s further results were achieved in the research into digital FM. Particular attention was paid to MSK (patented by M. L. Doelz and E. H. Heald in 1961 [364]); for this modulation format, R. de Buda described some schemes for its modulation and coherent detection in 1972 [365].[60] This simplicity, together with its spectral compactness and constant envelope, justifies the substantial interest in MSK in those years [285]. Among the applications of MSK, it is worth remembering the one proposed in 1969, in the USA, by the Data Transmission Co. (Datran) for the development of national network of microwave radio bridges, which used digital modems capable of transmitting up to 21.504 Mbit/s in a 30 MHz bandwidth assigned by FCC for this specific application. In this and other cases, the adoption of MSK was preferred to the use of PSK [362] because of the common use of nonlinear components in microwave communication equipment [366]. In the years following de Buda's work, research into digital FM concentrated, on the one hand, on coherent and noncoherent detection of CPFSK signals [287] and the achievable performance [367], and, on the other hand, on modifications of the MSK technique to derive new CPM formats [281, 368–370]. Such formats attempted to preserve the structural simplicity of MSK signals, while offering improved spectral properties [371, 372]. In this research area the following results deserve to be mentioned: the discovery by F. Amoroso of the so-called SFSK[61] in 1976 [281], and the development by F. de Jager and C. B. Dekker in 1978 of *tamed frequency modulation* (TFM), that is, the discovery of a class of binary partial-response FM signals having better spectral compactness than previously proposed formats, albeit with some loss in error performance.

A complete theory of CPM signals was, however, developed only in the early 1980s. In 1981, T. Aulin, C.-E. W. Sundberg and N. Rydbeck published two papers, one devoted to *full-response* signaling [373], and the other to *partial-response* signaling [374]. These illustrated both a unified theory of this class of digital modulations and a comparison, based on the tools provided by this theory, among various existing modulation formats in terms of the achievable power/spectral efficiency.[62] In the same year, the NTT researchers K. Murota and K. Hirade proposed a new CPM format, known as *Gaussian filtered* MSK (GMSK), which, on account of its significant spectral compactness, was to be used in future mobile phone systems [286].

In the 1980s other significant research results on CPM were published. The following results deserve to be mentioned: the PAM representation of CPMs devised by P. A. Laurent in 1986 [296] (see Section 3.6.5.1); and the representation, proposed by B. Rimoldi in 1988 [302], of a CPM modulator as a the cascade of a *continuous phase encoder* with a *memoryless modulator* (see Section 3.6.5.2).

---

[60] This format was called *fast FSK* (FFSK) by de Buda, since it allowed transmission at double the rate of (i.e., at the same bit rate as) QPSK with rectangular pulse shaping (see (3.56)), with comparable spectral occupancy.

[61] A binary full-response CPM format, that can be obtained from MSK by replacing its half sinusoidal pulse (see (3.164)) with an impulse having the same duration, but a raised cosine shape.

[62] A summary of the essential concepts and results can be found in [375].

All the above-mentioned work refers to *single-h* CPMs. In 1975 the use of binary CPFSK schemes with the cyclic use of multiple modulation indexes, that is, of *multi-h* CPFSK, was first proposed by H. Miyakawa, H. Harashima and Y. Tanaka, as a way to achieve further coding gains in CPM signaling [289]. This approach to modulation design was generalized later by J. B. Anderson, R. de Buda and D. P. Taylor [290, 376]. An overview of the most important results achieved in *multi-h* CPM signaling up to 1991 can be found in [377].

3.10.3 MCM Signaling

In the history of digital communication systems, the first example of a wireless MC system was the so-called *Kineplex* [345, 378], built by Collins Radio Company (Burbank, California) in the 1950s using SSB transmissions in HF.[63] The Kineplex was followed by other MC systems for HF communications, such as: the AN/GSC-10, also known as the *Kathryn modem*, developed by General Atronics Corporation (Philadelphia) [379–383],[64] and the ANDEFT/SC-320, developed by the Electronics Division of the General Dynamics Corporation (Rochester, NY) [384]. These systems, implementing the FDM technique, employed a few tens of distinct phase modulated subcarriers and offered low data rates (2400–4800 bit/s).

The first studies on the performance achievable using MC signaling over HF channels, such as the analysis provided by P. A. Bello for the Kathryn modem [381], showed that ICI, concerning adjacent subchannels, and ISI, affecting symbols transmitted over a given subchannel in consecutive symbol intervals, were important obstacles to reliable communications. The ICI phenomenon could have been removed by avoiding spectral overlaps among adjacent subchannels, but this possibility was not taken into serious consideration, because it required the implementation of highly selective (and costly) bandpass filters to minimize the required guard bandwidth between adjacent subchannels. Thus, the first MC systems used a set of *sinusoidal signals* (called *tones*), having distinct frequencies and limited duration, generated by an oscillator bank. These signals were not bandlimited and were partially overlapped in the frequency domain, since each of them had a sinc function spectrum centered on a subcarrier frequency (e.g., see [385, p. 618, para. VI.A]).

The first alternative to this approach, for implementing a parallel data transmission over a bandlimited communication channel, was proposed by R. W. Chang in 1966 [386]. He devised a general method for the synthesis of an orthogonal and rigorously bandlimited signal set. The proposed solution allowed complete cancelation of ICI and ISI interference in any receiver endowed with an exact knowledge of the communication channel. A year later, B. R. Saltzberg, exploiting Chang's results, derived a new parallel communication scheme, in which the orthogonality between adjacent subchannels was achieved using an *offset* QAM (O-QAM) [387]. A few years after the publication of the work of Chang and Saltzberg, S. B. Weinstein and P. M. Ebert proved that, in an FDM system, both modulation and demodulation can be accomplished using the DFT technique [388], implemented via a *fast Fourier transform* (FFT) algorithm, combined with the use of a *cyclic prefix*. This removed the need for oscillator banks in signal generation and coherent detection, making possible a full digital implementation of the baseband sections of an MC modem. In addition, compensation for channel distortions was easily accomplished on a subcarrier-by-subcarrier basis, using FD equalization. In 1981, ten years after the publication [388], B. Hirosaki [389] proved that even the MC system proposed by Saltzberg, based on O-QAM signaling, lent itself to a DFT-based digital implementation, which was computationally efficient, like the system proposed by Weinstein and Ebert.

From the 1980s onward, various results concerning MC transmission techniques and their implementation appeared in the technical literature. An overview of the main technical problems analyzed in

---

[63] The HF band, occupying the 3–30 MHz spectrum, was usually used for long-distance radio communications utilizing *ionospheric reflection*. This was the intended application of the Kineplex system.

[64] The papers [381–383] (and [384] cited below) have been republished more recently in [44].

the literature can be found in [385, 390–392]. More accurate information is provided in the textbooks [315, 393, 394]. Here, we limit ourselves to mentioning the proposal of replacing the cyclic prefix of OFDM by *zero padding* (ZP); this results in a new modulation format, called ZP-OFDM. In each block of a ZP-OFDM transmission, zero symbols are appended to the complex symbols available at the IDFT output. It can be shown that, if the number of zero symbols equals the cyclic prefix length, no loss in spectral efficiency is incurred with respect to conventional OFDM. However, unlike conventional OFDM, ZP-OFDM guarantees symbol recovery even in the presence of channel nulls in the transmission bandwidth, provided that the channel has a *finite impulse response* (FIR). The penalty for this is an increase in receiver complexity, since the DFT processing required by conventional OFDM is replaced by FIR filtering. Further details on this transmission technique can be found in [392] and the references cited therein.

The availability of various new signal processing techniques for an efficient implementation of MC techniques has clearly favored the adoption of MC transmission techniques, such as OFDM and the so-called *discrete multitone* (DMT) [395, 396], in modern communication systems. Finally, note that a new phase of implementation of MC systems started at the beginning of the 1980s. Among the available commercial products, the *Trailblazer* modem, made by Telebit Company (Silicon Valley, California), is worth remembering. This was developed for parallel data transmission in the voice band over PSTN at 9.6 kbit/s [397]. Its speed was surpassed, a few years after its appearance, by SC QAM modems, which, however, used substantially more complicated signal processing techniques [398]. Further details on the early development of OFDM can be found in [399].

3.10.4   Power Spectral Density of Digital Modulations

The technical literature on the spectral properties of the digital modulations analyzed in this chapter is wide and mainly concerns CPM signals, for which the evaluation of the PSD is a complicated mathematical problem.

Historically, the first important results about the PSD of digital modulations are summarized in the classic textbook by W. R. Bennett and J. R. Davey [400], which gives an account of the progress made until the first work on the PSD of PAM [401] and of *binary* phase continuous FSK [402–404]. Moreover, Bennett and Davey [400] propose two different methods for the evaluation of the PSD of a digital modulation. The first, called the *autocorrelation method*, is the one we have adopted in this chapter. The name assigned to this method derives from the fact that its use entails the evaluation of an autocorrelation function, from which a PSD is obtained via Fourier methods. The second method is known as the *direct method*, since it proceeds directly to computation of a power spectrum. In the 1960s the direct method received more attention than the autocorrelation method. Among its applications, we recall those of J. Salz in 1965, leading to the derivation of the PSD of multilevel continuous phase FM [311], and of R. R. Anderson and J. Salz in 1963, leading to a general expression useful for the computation, via a two-dimensional numerical integration, of the power spectrum of an arbitrary FM signal [313, 371]. The latter result has been exploited for the computer-based evaluation of the PSD of continuous phase FM signals for over a decade, despite its relatively large computational load when estimating the PSD side lobes. To solve this problem, T. M. Baker in 1974 derived, starting from the formula derived by Anderson and Salz, a simplified expression which accurately describes the asymptotic behavior of the PSD for an arbitrary digital FM signal [313]. Further results appeared in the 1970s on the PSD of CPM signals and mainly concern the spectral properties of MSK [362], those of the signals deriving from its modifications and generalizations [281, 368, 369], and those of new formats like TFM [370].

Further significant results concern the development of new methods for the computation of the spectrum of CPM signals; most of the work in this area is cited in the bibliography of [375]. Here we limit ourselves to mentioning the contributions provided by G. J. Garrison [372] and V. K. Pabhu and

H. E. Rowe [405–407]. The latter researchers were the first to show the application of mathematical tools originating in Markov chain theory to the evaluation of the PSD of digital FM.

Two crucial contributions to the development of new algorithms for the efficient computation of the PSD of an arbitrary CPM were made in the 1980s by Aulin and Sundberg, who were the first researchers to give a general mathematical description of CPM [373, 374]. The first contribution is represented by the technique described in Section 3.6.7; its derivation was inspired by work by L. J. Greenstein on a special type of phase modulated signals [408] and is based on the above-mentioned autocorrelation method [288, 409]. The second contribution, illustrated in [312], is represented by a computationally efficient algorithm for the accurate estimation of both the main lobe and the side lobes of an arbitrary CPM. This method is extremely useful in the presence of large correlation length L and/or large alphabet cardinality M, where other PSD estimation techniques require a long computing time.

Finally, we note that, recently, the availability of new representations of CPMs, like those proposed by Rimoldi [302] and Laurent [296], has provided new tools for deriving already known expressions for their PSDs and new numerical methods for spectral estimation. The first possibility is exemplified by [410], applying Rimoldi's representation to the derivation of the PSD of a CPFSK with a rational modulation index.

The technical literature on the spectral properties of OFDM is, unlike that on the CPM, decidedly poor. Moreover, its is worth noting that, in the technical literature, the PSD of OFDM signaling is always represented as the superposition of the PSDs associated with all the distinct subcarriers; in addition, it is assumed that the PSD of each subcarrier is a $\mathrm{sinc}^2(\cdot)$ function centered on the subcarrier frequency, as illustrated, for instance, in [385, para. VI.A, p. 618]. This representation is approximate, since it does not follow from the technique actually employed in the generation of the transmitted signal. Note that the approach to the PSD evaluation adopted in this book is based on the cyclostationarity of the sequence feeding the transmit filter in an OFDM modulator and results from the application of a general theoretical result illustrated, for instance, in [28, Sect. 2.1, pp. 59–63].

3.11   Further Reading

In the previous sections various bibliographic citations have been provided. It is useful to point out, however, that the textbooks [288, 315, 394] can be used to delve into various of the topics discussed here. In particular, the first book analyses both SC and MC modulations, the second is devoted exclusively to OFDM, while the third is the only book exclusively devoted to CPM.

In this chapter the cyclostationary properties of digital modulations have been widely exploited. Further information on these properties and their usefulness in the study of digital communication systems can be found in the papers [411, 412] and in the book [413] edited by W. A. Gardner.

Finally, we note that our study of the spectral properties of digital modulations has focused on some specific classes of digital modulations. An important topic, widely studied in the technical literature but not analyzed in this book, is the evaluation of the PSD when the signal generation algorithm in a digital modulator can be described by a *Markov chain*. The solution to this problem was derived by R. C. Titsworth and L. R. Welch [414] and is illustrated in various classic textbooks, among them [27, 28, 32, 327]. The reader can also refer to the classic paper by P. Galko and S. Pasupathy [415] and the references cited therein.

4

Detection of Digital Signals over Wireless Channels: Decision Rules

4.1 Introduction

Any digital receiver employs a specific *detection strategy* or algorithm for estimating the message transmitted by a given source from a received set of information-bearing noisy data. This is extracted from a single or multiple received waveforms by means of filtering and sampling operations, and is processed by the detection algorithm to generate a set of real quantities, called *detection metrics*. Each value of this metric is associated with a specific *hypothesis* about the transmitted message and is used by the receiver to make decisions on the basis of a mapping rule, which maps values of the metric to messages. In most instances, the hypothesis associated with the best (e.g., the minimum or maximum) metric is selected.

This chapter is devoted to the study of the following two problems:

(a) how a finite set of data (i.e., a finite-dimensional vector) can be extracted from a continuous-time received waveform for the purpose of accomplishing data detection,
(b) how detection metrics can be formulated for the digital modulation techniques described in Chapter 3 in a fading multipath channel scenario.

Solving these problems requires full or partial knowledge of the properties of the communication channel and the formulation of an *optimality criterion* for data detection in mathematical terms. Modifying the criterion and/or our knowledge of the communication channel can lead to substantially different detection strategies.

This chapter is organized as follows. Section 4.2 presents a general model of a *wireless digital communication system*. This includes a description of a system that employs a *finite set of analog waveforms* to send a message, belonging to a given finite alphabet, through a wireless communication channel. In addition, some details are provided about how a received RF waveform is down-converted to a lower frequency to simplify signal processing for data detection. Generally speaking, the evaluation of detection strategies for a wireless communication system is not simple. Thus to tackle the problem, a closely related communication system, having a similar structure, but in which analog waveforms are replaced by *vectors of finite size*, is usually considered. This is the subject of Section 4.3, where it is shown that if the *optimality criterion* in receiver design is that of minimizing *average error probability* in detecting the transmitted message, an optimal detection strategy can be formulated and a geometrical interpretation for it can be provided.

Wireless Communications: Algorithmic Techniques, First Edition.
Giorgio M. Vitetta, Desmond P. Taylor, Giulio Colavolpe, Fabrizio Pancaldi, Philippa A. Martin.
© 2013 John Wiley & Sons, Ltd. Published 2013 by John Wiley & Sons, Ltd.

In particular, in our analysis we focus on the *maximum a posteriori probability* (MAP) and the *maximum likelihood* (ML) detection strategies. These process a finite-dimensional vector of received samples to generate a set of detection metrics, on the basis of which an optimal decision about the transmitted message can be taken. However, the exploitation of these strategies in a real-world digital receiver requires the extraction of a finite-dimensional vector of noisy data from the received waveform. This problem is analyzed in Section 4.4, where it is tackled first from a general perspective. Then the structure of this vector is illustrated for different modulation formats. In Section 4.5 the mathematical tools developed in Section 4.3 are exploited to develop optimal decision metrics for the following cases:

(a) CIR ideally known at the receiver;
(b) CIR known only statistically at the receiver;
(c) unknown CIR.

In particular, metrics for ML *sequence detection* (MLSD), MAP *symbol detection* (MAPSD), and MAP *bit detection* (MAPBD) are derived. In addition, on the basis of these metrics, some general performance bounds are derived.

Optimal data detection in the presence of channel uncertainty can be a formidable task in certain scenarios. In some cases, ML performance can be approached by developing detection algorithms based on iterative techniques, which try to approach the optimal solution in multiple steps. A significant example of these techniques is offered by the *expectation–maximization* (EM) algorithm. This is the subject of Section 4.6, where the algorithm and some related techniques are illustrated, and some applications to data detection problems are analyzed.

Finally, some historical notes and suggestions for further reading are provided in Sections 4.7 and 4.8, respectively.

4.2 Wireless Digital Communication Systems: Modeling, Receiver Architecture and Discretization of the Received Signal

In this section a general model of a wireless communication system is first developed. Then some details about the receiver structure are provided, primarily dealing with the receiver front-end.

4.2.1 General Model of a Wireless Communication System

A general model for an uncoded SISO wireless digital communication system is illustrated in Figure 4.1. A discrete information source generates a message m, belonging to the finite alphabet $A_m = \{m_i, i = 0, 1, \ldots, N_m - 1\}$, having cardinality N_m and whose ith element is characterized by the *a priori probability* $P_i \triangleq \Pr\{m = m_i\}$ (for $i = 0, 1, \ldots, N_m - 1$). In general, each message can be associated with one or more bits generated by a source and can be represented by one or more channel symbols (see Section 3.2), as will become clearer in what follows. The source feeds the digital modulator, which maps m into a finite-dimensional vector $\mathbf{c}$ of channel symbols (via a one-to-one mapping) and then generates the complex baseband signal $s(t, \mathbf{c})$. In doing so, the modulator adopts the finite alphabet $A_s = \{s_i(t) = s(t, \mathbf{c}^{(i)}), i = 0, 1, \ldots, N_m - 1\}$, having the same cardinality as A_m, where $\mathbf{c}^{(i)}$ denotes the ith possible value of $\mathbf{c}$. To simplify the notation, in what follows we will always assume that the one-to-one correspondence:

$$s(t, \mathbf{c}) = s_i(t) \Leftrightarrow m = m_i, \tag{4.1}$$

holds, namely, that the modulator-generated signal $s_i(t)$ is in response to m_i, with $i = 0, 1, \ldots, N_m - 1$. Note that, unlike Chapter 3, here the dependence on m_i is shown by the signal subscript

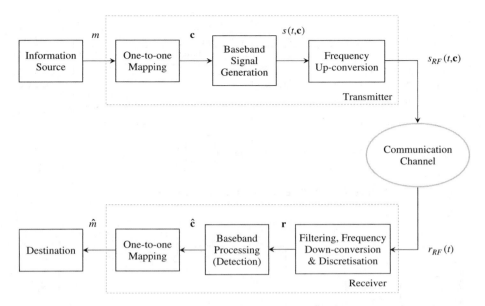

Figure 4.1   General scheme of an uncoded wireless communication system.

instead of the sequence of channel symbols $\{c_i\}$. Finally, the transmitter front-end up-converts the complex baseband signal $s(t, \mathbf{c})$ to create the real RF waveform $s_{RF}(t, \mathbf{c})$ which is transmitted over the channel. The noisy channel output signal $r_{RF}(t)$ feeds a digital receiver, whose front-end filters $r_{RF}(t)$ (to remove out-of-band noise and interference) and down-converts it (frequency-translates it to baseband) to generate the complex baseband signal $r(t) = r_c(t) + jr_s(t)$. The receiver then processes $r(t)$ (via filtering and sampling) to extract a finite-dimensional vector $\mathbf{r}$ and, finally, processes $\mathbf{r}$ to extract an estimate $\hat{\mathbf{c}}$ of the transmitted symbols $\mathbf{c}$, from which an estimate $\hat{m}$ of m is inferred. Further details on the first step are provided in the following subsection, and the problem of extracting $\mathbf{r}$ from $r(t)$ is discussed in Section 4.4.1.

4.2.2   Receiver Architectures

On the basis of the type of front-end processing employed, radio receivers can be classified into two categories: *direct-conversion (homodyne)* and *superheterodyne* receivers. Their architectures are illustrated in Figures 4.2 and 4.3, respectively. In a direct-conversion receiver [416] the received RF signal $r_{RF}(t)$ undergoes filtering (to remove out-of-band noise and interference) and low-noise amplification to reduce the overall noise figure. The RF filter output is sent to two mixers which are also fed by two quadrature oscillations generated by a local oscillator locked to the carrier frequency f_c. The mixer outputs feed two distinct baseband filters that remove residual out-of-band noise and interference, and to avoid aliasing which may originate from sampling their outputs. The outputs of the baseband filters undergo *analog-to-digital* (A/D) conversion and further baseband processing to extract the transmitted data.

Note further that the homodyne architecture does not require filtering at some *intermediate frequency* (IF), and this simplifies system integration on a chip (since inductors and capacitors of large size are not required). In addition, the use of direct conversion avoids the problem of rejecting the so-called *image frequency*, so that the design requirements of the RF filtering can be relaxed. However, the adoption of a homodyne architecture has some drawbacks, such as the presence of $1/f$ noise and

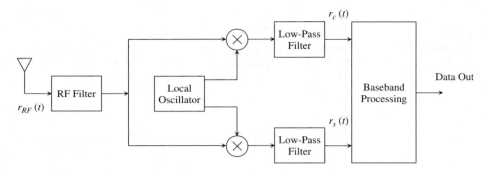

Figure 4.2   Architecture of a direct-conversion receiver.

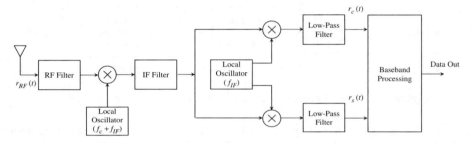

Figure 4.3   Architecture of a superheterodyne receiver.

the presence of a *direct current* (DC) offset noise. The first problem originates from the fact that the spectrum of the directly converted baseband signal overlaps with the $1/f$ noise of active devices. The second is due to the self-down-conversion of leakage signals generated by the local oscillator. Such signals reach the path of the converted useful signal without going through the mixers and may lead to a DC component substantially stronger than the useful signal. When this occurs, saturation of A/D converters or a reduction of their sensitivity may be encountered.

In a superheterodyne receiver [417] tighter filtering of the received signal $r_{RF}(t)$ is required in the first stage of the receiver, since the *image frequency* has to be carefully suppressed. The RF filter feeds a mixer which down-converts the input signal to an IF f_{IF} using a local oscillator locked to the frequency $f_{LO} = f_c + f_{IF}$. The IF filter selects the desired signal (suppressing the spectrally adjacent interfering signals) and amplifies it. The resulting output waveform is then down-converted to baseband using two quadrature oscillations at f_{IF}. Finally, as in the homodyne architecture, the resulting baseband signals undergo analog-to-digital conversion and further baseband processing.

The superheterodyne architecture offers the advantage of selective IF filtering and amplification, so that the A/D converters are fed with properly filtered and amplified signals. However, its implementation is more complicated than that of its homodyne counterpart. In addition, a proper tradeoff between the complexity of the RF stage and that of the IF stage is required. In fact, if f_{IF} is low, the IF stage is more efficient (i.e., it exhibits a better selectivity for a given complexity), but a more selective RF filter is required to reject the image frequency. In contrast, if the RF filtering is loose, a large f_{IF} is required and this makes the implementation of the IF stage more expensive and difficult. In fact, the inductors and capacitors do not lend themselves to an efficient monolithic implementation, so that silicon chips usually require various external components, with an increase of size and cost and, generally speaking, a worsening of receiver error performance.

Figure 4.4   Vector equivalent of the system illustrated in Figure 4.1.

4.3   Optimum Detection in a Vector Communication System

4.3.1   Description of a Vector Communication System

Let us initially assume that the complex envelope $r(t)$ of the (filtered) received signal $r_{RF}(t)$ admits a *finite-dimensional vector* representation $\mathbf{r}$ with no loss of useful information (see Section 4.2.1).[1] Then, the system of Figure 4.1 can be replaced by its vector counterpart shown in Figure 4.4, in which finite-dimensional vectors take the place of time-continuous random processes. In particular, each signal appearing in Figure 4.1 is replaced by a vector of proper size N (the problem of selecting a proper value for this parameter is discussed in the following section). In the vector equivalent system the transmitter generates, in response to a message $m \in A_m$, the vector $\mathbf{s} \triangleq [s_0, s_1, \ldots, s_{N-1}]^T$, belonging to the alphabet $A_s = \{\mathbf{s}_i, i = 0, 1, \ldots, N_m - 1\}$, with $\mathbf{s}_i \triangleq [s_{i,0}, s_{i,1}, \ldots, s_{i,N-1}]^T$, according to the correspondence (see (4.1)):

$$\mathbf{s} = \mathbf{s}_i \Leftrightarrow m = m_i. \tag{4.2}$$

The *average energy per message* spent by the transmitter in the transmission of a message is given by:

$$E_m = \sum_{i=0}^{N_m-1} P_i \, |\mathbf{s}_i|^2. \tag{4.3}$$

Note that the closer to the origin are the points of the alphabet A_s, the smaller is this energy.

In what follows the problem of *optimal detection* for the vector system of Figure 4.4 is formulated and solved. In other words, the aim is to develop algorithms that process $\mathbf{r}$ to take a decision $\hat{m}$ on m, so as to minimize the *average message error probability*:

$$P_e \triangleq \Pr\{\hat{m} \neq m\}. \tag{4.4}$$

In tackling this problem we assume that the a priori probabilities $\{P_i, i = 0, 1, \ldots, N_m - 1\}$ and the signal alphabet A_s are *perfectly known to the receiver*. Note that a rigorous derivation of optimal detection strategies is a formidable problem for the system of Figure 4.1, but its counterpart for the system of Figure 4.4 is substantially more tractable. In fact, the former problem involves analog random processes, the latter only finite-dimensional random vectors.

In the rest of this section a general framework for the development of detection strategies is first presented. Then, some optimal strategies are presented and their use discussed. Finally, some useful theorems about optimal detection are presented.

4.3.2   Detection Strategies and Error Probabilities

The communication channel affects signal transmission and produces, in response to $\mathbf{s}$, the received vector $\mathbf{r} \triangleq [r_0, r_1, \ldots, r_{N-1}]^T$ belonging to the *observation space* $D_\mathbf{r}$. The mechanism according to

---

[1] This is easily proved under quite general conditions for any bandlimited signal [321].

which the channel acts on **s** cannot but be described in the language of statistics. In particular, the statistical behavior of the channel is fully described by the set of conditional pdfs:

$$\{f_{\mathbf{r}}(\boldsymbol{\rho}|\mathbf{s}_i), i = 0, 1, \ldots, N_m - 1\}, \tag{4.5}$$

since the pdf $f_{\mathbf{r}}(\boldsymbol{\rho}|\mathbf{s}_i)$ characterizes **r** given the transmitted vector $\mathbf{s} = \mathbf{s}_i \in A_{\mathbf{s}}$, the ith possible transmitted message. In this section we will assume that the receiver has *ideal knowledge of the channel's statistical properties*, that is, it knows perfectly all the pdfs of the set (4.5).

The receiver processes the vector **r** of noisy data by means of an appropriate *strategy* in order to generate an estimate $\hat{m}$ of the transmitted message m. Any *decision* (or *detection*) *strategy*, namely, any procedure adopted by the receiver to generate $\hat{m}$ on the basis of **r**, can be given a geometrical interpretation. In fact, we can figure out that the receiver is endowed with a proper *partition* of $D_{\mathbf{r}}$ that splits this space into N_m disjoint domains (or subsets) $\{D_i, i = 0, 1, \ldots, N_m - 1\}$, that cover it without overlapping. Then, we may write:

$$D_i \cap D_j = \emptyset, \tag{4.6}$$

if $i \neq j$, and:

$$\bigcup_{i=0}^{N_m-1} D_i = D_{\mathbf{r}}. \tag{4.7}$$

The meaning of expressions (4.6) and (4.7) can be easily understood by referring to Figure 4.5, which illustrates the partitioning of $D_{\mathbf{r}}$ into $N_m = 4$ subsets of a two-dimensional observation space (assumed of rectangular shape for simplicity). Note, moreover, that each of the subsets $\{D_i\}$ is not generally required to be a connected set. Then, the receiver, after observing the value $\boldsymbol{\rho}$ taken by the random vector **r**, establishes which of the domains $\{D_i\}$ the vector $\boldsymbol{\rho}$ belongs to and takes a decision according to the criterion:

$$\hat{m} = m_i \Leftrightarrow \boldsymbol{\rho} \in D_i; \tag{4.8}$$

that is, the receiver selects m_i, if and only if $\boldsymbol{\rho}$ belongs to D_i; for this reason D_i is called ith *decision region*.

This geometrical interpretation of a decision strategy turns the search for an optimal detection strategy, whatever the reception problem, into the construction of an optimal partition of $D_{\mathbf{r}}$, that is, the search for optimal *borders* of the decision regions $\{D_i\}$.

It is also important to note that the problem of identifying the decision regions for a digital receiver is closely related to that of the computation of its probability of error, since, once these regions are

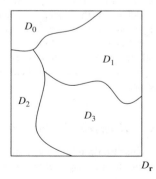

Figure 4.5   Partitioning of a two-dimensional rectangular observation space; $N_m = 4$ is assumed.

fixed, *the average probability of error P_e (4.4) of the given receiver is uniquely determined.* In fact, the *theorem of total probability* [55] allows us to express this performance index as [321]:

$$P_e = \sum_{i=0}^{N_m-1} P_i \ \Pr\{\varepsilon|m_i\},$$
(4.9)

where $\Pr\{\varepsilon|m_i\}$ denotes the error probability conditioned on $m = m_i$ (i.e., on the transmission of the *m*th message), or equivalently as:

$$P_e = 1 - P_c = 1 - \sum_{i=0}^{N_m-1} P_i \ \Pr\{C|m_i\},$$
(4.10)

where P_c and $\Pr\{C|m_i\}$ denote the average probability of correct decision and the probability of correct decision conditioned on $m = m_i$, respectively. In general, the probability $\Pr\{C|m_i\}$ can be expressed as an integral over an *N*-dimensional domain; more specifically, it can be evaluated as:

$$\Pr\{C|m_i\} = \int_{D_i} f_{\mathbf{r}}(\boldsymbol{\rho}|\mathbf{s}_i) \, d\boldsymbol{\rho},$$
(4.11)

since the receiver decides correctly when m_i is transmitted, if and only if $\boldsymbol{\rho} \in D_i$. Similarly, $\Pr\{\varepsilon|m_i\}$ can be expressed as:

$$\Pr\{\varepsilon|m_i\} = \int_{\bar{D}_i} f_{\mathbf{r}}(\boldsymbol{\rho}|\mathbf{s}_i) \, d\boldsymbol{\rho},$$
(4.12)

where $\bar{D}_i$ denotes the *complement* of D_i with respect to $D_{\mathbf{r}}$. Note that the choice of (4.9) or (4.10) in the evaluation of P_e primarily depends on the difficulty encountered in evaluating the integrals (4.12) and (4.11), respectively. The shape (and, consequently, the analytical description) of D_i (see (4.11)) is often substantially simpler than that of $\bar{D}_i$ (see (4.12)) and this makes the evaluation of P_e via (4.10) and (4.11), in place of (4.9) and (4.12), simpler.

Substituting (4.12) in (4.9) (or, equivalently, (4.11) in (4.10)) yields the expression:

$$P_e = \sum_{i=0}^{N_m-1} P_i \int_{\bar{D}_i} f_{\mathbf{r}}(\boldsymbol{\rho}|\mathbf{s}_i) \, d\rho = 1 - \sum_{i=0}^{N_m-1} P_i \int_{D_i} f_{\mathbf{r}}(\boldsymbol{\rho}|\mathbf{s}_i) \, d\rho,$$
(4.13)

which clearly shows the dependence of P_e on the partition $\{D_i\}$ of $D_{\mathbf{r}}$. This result usually does not lead to a closed-form expression for P_e for $N_m > 2$, since the integrals appearing in its RHS can be evaluated only numerically. For this reason, lower and upper bounds on performance are often derived. These bounds are often based on the fact that the error event $\varepsilon|m_i$ appearing on the RHS of (4.9) can be expressed as [321]:

$$\varepsilon|m_i = \bigcup_{\substack{k=0 \\ k \neq i}}^{N_m-1} \varepsilon_{ik},$$
(4.14)

where ε_{ik} denotes the event occurring when m_i (i.e., $\mathbf{s}_i$) is transmitted and the digital receiver selects $\hat{m} = m_k$ ($k \neq i$) on the basis of $\mathbf{r}$. Note that ε_{ik} refers to a *binary decision problem*, since it concerns $\mathbf{s}_i$ and $\mathbf{s}_k$, and is not influenced by the number and choice of the other points of the alphabet $A_{\mathbf{s}}$; for this reason, the probability of this event, $\Pr\{\varepsilon_{ik}\}$, is usually called the *pairwise error probability* (PEP) and can be often expressed in closed form using standard functions of probability theory. Given the PEPs $\{\Pr\{\varepsilon_{ik}\}\}$, an upper bound on P_e can easily be evaluated. In fact, applying the inequality:

$$\max_k \Pr\{A_k\} \leq \Pr\left\{\bigcup_k A_k\right\} \leq \sum_k \Pr\{A_k\},$$
(4.15)

which holds for any set of events $\{A_k\}$ [55], to the RHS of the expression (see (4.14)):

$$\Pr\{\varepsilon|m_i\} = \Pr\left\{\bigcup_{\substack{k=0\\k\neq i}}^{N_m-1} \varepsilon_{ik}\right\} \tag{4.16}$$

leads to:

$$\max_{\substack{k\\k\neq i}} \Pr\{\varepsilon_{ik}\} \leq \Pr\{\varepsilon|m_i\} \leq \sum_{\substack{k=0\\k\neq i}}^{N_m-1} \Pr\{\varepsilon_{ik}\}. \tag{4.17}$$

Then from (4.17) and (4.9) the bounds:

$$\sum_{i=0}^{N_m-1} P_i \max_{\substack{k\\k\neq i}} \Pr\{\varepsilon_{ik}\} \leq P_e \leq \sum_{i=0}^{N_m-1} \sum_{\substack{k=0\\k\neq i}}^{N_m-1} \Pr\{\varepsilon_{ik}\} \tag{4.18}$$

are easily inferred. The expression for the PEP $\Pr\{\varepsilon_{ik}\}$ depends both on the transmitted vectors $\mathbf{s}_i$ and $\mathbf{s}_k$, and on the channel model. The upper bound in (4.17) and (4.18) is known as the *union bound* and is well described in [55, 321].

4.3.3   *MAP and ML Detection Strategies*

Now let us assume that the receiver knows:

(a) the a priori probabilities $\{P_i, i = 0, 1, \ldots, N_m - 1\}$;
(b) the conditional pdfs $\{f_{\mathbf{r}}(\boldsymbol{\rho}|\mathbf{s}_i), i = 0, 1, \ldots, N_m - 1\}$;
(c) the alphabet $A_s = \{\mathbf{s}_i, i = 0, 1, \ldots, N_m - 1\}$.

It can then be proved that the *optimal detection strategy is to select the message characterized by the maximum a posteriori probability* [321]. In other words, the optimal receiver selects $\hat{m} = m_k$ if and only if:

$$\Pr\{m = m_k|\mathbf{r} = \boldsymbol{\rho}\} > \Pr\{m = m_i|\mathbf{r} = \boldsymbol{\rho}\} \tag{4.19}$$

for $i = 0, 1, \ldots, N_m - 1$ and $i \neq k$, where $\Pr\{m = m_i|\mathbf{r} = \boldsymbol{\rho}\}$ represents the *a posteriori probability* of m_i, that is, the probability of the event $\{m = m_i\}$, when $\mathbf{r} = \boldsymbol{\rho}$. This criterion is known as the MAP decision strategy and can be formulated in a more compact form as:

$$\hat{m}_{MAP} = \arg\max_{\tilde{m}\in A_m} \Pr\{m = \tilde{m}|\mathbf{r} = \boldsymbol{\rho}\} \tag{4.20}$$

to show that $\hat{m}_{MAP}$ represents the solution of a *maximization problem*. Equation (4.20) shows that, generally speaking, a receiver operating according to the MAP strategy needs to perform an *exhaustive search* over a set of N_m hypotheses, since it evaluates all possible values of $\Pr\{m = \tilde{m}|\mathbf{r} = \boldsymbol{\rho}\}$ by trying all possibilities (i.e., without excluding any possible message $\tilde{m}\in A_m$). For this reason, the test message $\tilde{m}$ appearing in (4.20) is usually called a *tentative message*.

The action of a MAP receiver can also be interpreted in terms of decision regions. The reader can easily verify that the kth decision region for this receiver is:

$$D_k = \{\boldsymbol{\rho} \in D_{\mathbf{r}}|\Pr\{m = m_k|\mathbf{r} = \boldsymbol{\rho}\} = \max_{m_i\in A_m} \Pr\{m = m_i|\mathbf{r} = \boldsymbol{\rho}\}\}, \tag{4.21}$$

with $k = 0, 1, \ldots, N_m - 1$.

The MAP strategy of (4.20) can also be formulated in a different fashion, by taking advantage of a mixed form of Bayes' theorem [55], which allows us to express $\Pr\{m = m_i | \mathbf{r} = \boldsymbol{\rho}\}$ as:

$$\Pr\{m = m_i | \mathbf{r} = \boldsymbol{\rho}\} = P_i \frac{f_{\mathbf{r}}(\boldsymbol{\rho} | \mathbf{s}_i)}{f_{\mathbf{r}}(\boldsymbol{\rho})}. \tag{4.22}$$

In fact, since the pdf $f_{\mathbf{r}}(\boldsymbol{\rho})$ does not depend on m_i, the criterion (4.20) can be rewritten as:

$$\hat{m}_{MAP} = \arg \max_{\tilde{m} \in A_m} \tilde{P} \, f_{\mathbf{r}}(\boldsymbol{\rho} | \tilde{\mathbf{s}}), \tag{4.23}$$

where $\tilde{\mathbf{s}}$ denotes the vector generated by the transmitter in response to the message $\tilde{m} \in A_m$ (having probability $\tilde{P}$). Taking into consideration the last formulation of the MAP strategy, expression (4.21) can be rewritten for the kth decision region as:

$$D_k = \{\boldsymbol{\rho} \in D_{\mathbf{r}} | P_k \, f_{\mathbf{r}}(\boldsymbol{\rho} | \mathbf{s}_k) = \max_i P_i \, f_{\mathbf{r}}(\boldsymbol{\rho} | \mathbf{s}_i)\}, \tag{4.24}$$

with $k = 0, 1, \ldots, N_m - 1$. This clearly shows that the decision regions for a MAP receiver depend on the a priori probabilities $\{P_i\}$, on the statistical characterization of the channel (defined by the set of conditional pdfs $\{f_{\mathbf{r}}(\boldsymbol{\rho} | \mathbf{s}_i)\}$) and on the selected alphabet $A_{\mathbf{s}}$. In addition, it allows us to infer that, generally speaking, the evaluation of the decision regions (i.e., of their borders) is a problem of significant complexity.

If all the a priori probabilities $\{P_i\}$ are equal, that is, if:

$$P_i = \frac{1}{N_m} \tag{4.25}$$

for $i = 0, 1, \ldots, N_m - 1$, (4.23) becomes:

$$\hat{m}_{ML} = \arg \max_{\tilde{m} \in A_m} f_{\mathbf{r}}(\boldsymbol{\rho} | \tilde{\mathbf{s}}), \tag{4.26}$$

which describes the so-called ML detection, since $f_{\mathbf{r}}(\boldsymbol{\rho} | \mathbf{s}_i)$ represents a *likelihood function*[2] of the vector $\mathbf{r}$ given $\mathbf{s} = \mathbf{s}_i$. It is important to make the following observations:

1. If (4.25) holds, the strategy (4.26) is equivalent to (4.23). However, if this is not the case, the MAP strategy yields a lower P_e than the ML strategy. Despite this, the ML strategy represents a reasonable solution to the problem of optimal detection when the $\{P_i\}$ are unknown to the receiver.
2. The solution of the MAP (4.23) and ML (4.26) problems does not change if the functions $P_i f_{\mathbf{r}}(\boldsymbol{\rho} | \mathbf{s}_i)$ and $f_{\mathbf{r}}(\boldsymbol{\rho} | \mathbf{s}_i)$ to be maximized undergo a *monotone increasing transformation* (e.g., multiplication by a constant, logarithmic transformation) in order to simplify the functional form of the decision criterion.
3. The MAP and ML criteria lead to *optimal* performance. However, such strategies often cannot be exploited in practical problems because their implementation is complex. This can be due to the complicated structure of the *decision metric* (i.e., the likelihood functions (4.5)) or to the lack of a computationally efficient algorithm to search for the maximum metric (see (4.23) and (4.26)).

Optimal detection strategies can be formulated in different (and equivalent) ways. To show this, let us define the *likelihood ratio* (LR):

$$L_i(\boldsymbol{\rho}) \triangleq \frac{f_{\mathbf{r}}(\boldsymbol{\rho} | \mathbf{s}_i)}{f_{\mathbf{r}}(\boldsymbol{\rho} | \mathbf{s}_0)} \tag{4.27}$$

---

[2] Generally speaking, a likelihood function of a vector $\mathbf{r}$, given a message m, is any function proportional to the conditional pdf $f_{\mathbf{r}}(\boldsymbol{\rho} | m)$.

of the message m_i (with $i = 1, 2, \ldots, N_m - 1$) with respect to the messages m_0. The function $L_i(\boldsymbol{\rho})$ is a likelihood function of $\mathbf{r}$ given m_i and turns the value $\boldsymbol{\rho}$ taken on by $\mathbf{r}$ in a *nonnegative real number*. Using the LRs defined by (4.27), the ML strategy can be reformulated as follows:

1. Given $\mathbf{r} = \boldsymbol{\rho}$, compute the likelihood ratio vector:

$$\mathbf{L}_{N_m}(\boldsymbol{\rho}) \triangleq [L_1(\boldsymbol{\rho}), L_2(\boldsymbol{\rho}), \ldots, L_{N_m - 1}(\boldsymbol{\rho})]^T \tag{4.28}$$

    of $N_m - 1$ LRs (4.27) with respect to $\mathbf{s}_0$.
2. The message m_0 is selected by the receiver if all the elements of $\mathbf{L}_{N_m}(\boldsymbol{\rho})$ are less than unity; otherwise, the message m_j such that:

$$L_j(\boldsymbol{\rho}) = \max_i L_i(\boldsymbol{\rho}) \tag{4.29}$$

with $i \in \{1, 2, \ldots, N_m - 1\}$ is selected.

Note that (4.27) and (4.28) define a *nonlinear transformation* from $D_\mathbf{r}$ to a space $D_\mathbf{L}$ (to which the $(N_m - 1)$-dimensional vector $\mathbf{L}_{N_m}(\boldsymbol{\rho})$ belongs) called the *space of likelihood ratios*. This space can be partitioned into decision regions like $D_\mathbf{r}$. However, unlike $D_\mathbf{r}$, the borders of the decision regions of $D_\mathbf{L}$ belong to *hyperplanes* independent of the mathematical structure of the pdfs $\{f(\boldsymbol{\rho}|\mathbf{s}_i)\}$. This is illustrated by Figure 4.6, depicting the decision regions of $D_\mathbf{L}$ for $N_m = 2$ and $N_m = 3$.

In some applications, the ML criterion can be simplified by using the *natural logarithm* of the likelihood functions or likelihood ratios. For instance, in this scenario, the vector:

$$\mathbf{Z}_{N_m}(\boldsymbol{\rho}) \triangleq [Z_1(\boldsymbol{\rho}), Z_2(\boldsymbol{\rho}), \ldots, Z_{N_m - 1}(\boldsymbol{\rho})], \tag{4.30}$$

where:

$$Z_i(\boldsymbol{\rho}) \triangleq \ln[L_i(\boldsymbol{\rho})] \tag{4.31}$$

with $i = 1, 2, \ldots, N_m - 1$, can be used in place of $\mathbf{L}_{N_m}(\boldsymbol{\rho})$ (4.28). The quantity $Z_i(\boldsymbol{\rho})$ is called the *log-likelihood ratio* (LLR) of the message m_i (with $i = 1, 2, \ldots, N_m - 1$) with respect to the message m_0. Then the ML rule can be expressed as follows:

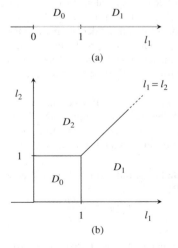

Figure 4.6   Decision regions in the space $D_\mathbf{L}$ for $N_m = 2$ (a) and $N_m = 3$ (b).

1. Given $\mathbf{r} = \boldsymbol{\rho}$, $\mathbf{Z}_{N_m}(\boldsymbol{\rho})$ is evaluated as in (4.30).
2. The message m_0 is selected by the receiver if all the elements of $\mathbf{Z}_{N_m}(\boldsymbol{\rho})$ are negative; otherwise the decision m_j is such that:

$$Z_j(\boldsymbol{\rho}) = \max_i Z_i(\boldsymbol{\rho}) \tag{4.32}$$

with $i \in \{1, 2, \ldots, N_m - 1\}$.

Let us now investigate the problem of optimal detection in a specific scenario.

Example 4.3.1 A digital communication system operating over an *additive Gaussian noise* (AGN) channel is illustrated in Figure 4.7. In this case, the communication channel adds the Gaussian noise vector $\mathbf{n} \triangleq [n_0, n_1, \ldots, n_{N-1}]^T$ to the transmitted signal vector $\mathbf{s}$, so that the received vector is:

$$\mathbf{r} = \mathbf{s} + \mathbf{n}. \tag{4.33}$$

The derivation of the MAP decision rule requires knowledge of the likelihood functions $\{f_{\mathbf{r}}(\boldsymbol{\rho}|\mathbf{s}_i)\}$ (4.5). It is easy to show that [321]:

$$f_{\mathbf{r}}(\boldsymbol{\rho}|\mathbf{s}_i) = f_{\mathbf{n}}(\boldsymbol{\rho} - \mathbf{s}_i|\mathbf{s}_i) \tag{4.34}$$

for $i \in \{0, 1, \ldots, N_m - 1\}$, where $f_{\mathbf{n}}(\boldsymbol{v})$ denotes the joint pdf of the components of $\mathbf{n}$. Here we assume that $\mathbf{n}$ is *statistically independent of the transmitted signal* $\mathbf{s}$, so that:

$$f_{\mathbf{r}}(\boldsymbol{\rho}|\mathbf{s}_i) = f_{\mathbf{n}}(\boldsymbol{\rho} - \mathbf{s}_i). \tag{4.35}$$

The MAP decision rule (4.23) then becomes:

$$\hat{m} = \arg \max_{\tilde{m} \in A_m} \tilde{P} \, f_{\mathbf{n}}(\boldsymbol{\rho} - \tilde{\mathbf{s}}). \tag{4.36}$$

This strategy can be simplified further if we assume that elements of $\mathbf{n}$ are iid real Gaussian random variables having zero mean and variance σ_n^2 (note that in this case $D_{\mathbf{r}}$ coincides with $\mathbb{R}^N$), so that:

$$f_{\mathbf{n}}(\boldsymbol{v}) = \prod_{k=0}^{N-1} f_n(v_k), \tag{4.37}$$

where $\boldsymbol{v} \triangleq [v_0, v_1, \ldots, v_{N-1}]^T$ and:

$$f_n(v) = \frac{1}{\sqrt{2\pi\sigma_n^2}} \exp\left(-\frac{v^2}{2\sigma_n^2}\right) \tag{4.38}$$

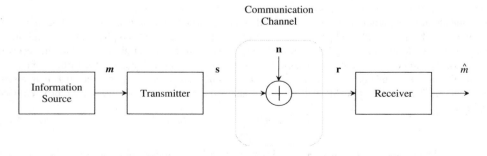

Figure 4.7   Vector communication system characterized by an AGN channel.

represents the pdf of each of the random variables $\{n_k, k = 0, 1, \ldots, N_m - 1\}$. In fact, substituting (4.38) into (4.37) and (4.37) into (4.35) yields:

$$f_{\mathbf{r}}(\boldsymbol{\rho}|\mathbf{s}_i) = \frac{1}{(2\pi\sigma_n^2)^{N/2}} \exp\left(-\frac{|\boldsymbol{\rho} - \mathbf{s}_i|^2}{2\sigma_n^2}\right). \tag{4.39}$$

Then the detection metric of (4.36) can be expressed as:

$$\tilde{P}f_{\mathbf{n}}(\boldsymbol{\rho} - \tilde{\mathbf{s}}) = \frac{\tilde{P}}{(2\pi\sigma_n^2)^{N/2}} \exp\left(-\frac{|\boldsymbol{\rho} - \tilde{\mathbf{s}}|^2}{2\sigma_n^2}\right). \tag{4.40}$$

The solution of the MAP problem (4.36) does not change if:

(a) the factor $(2\pi\sigma_n^2)^{-N/2}$ is dropped in (4.40), since it is independent of $\tilde{\mathbf{s}}$;
(b) we take the natural logarithm of the resulting expression;
(c) the result of the transformation in (b) is multiplied by the positive constant $2\sigma_n^2$.

This leads to the decision metric:

$$2\sigma_n^2 \ln \tilde{P} - |\boldsymbol{\rho} - \tilde{\mathbf{s}}|^2, \tag{4.41}$$

to be maximized by a MAP receiver. Reversing the sign of this metric yields the equivalent metric:

$$|\boldsymbol{\rho} - \tilde{\mathbf{s}}|^2 - 2\sigma_n^2 \ln \tilde{P}, \tag{4.42}$$

which has to be minimized over the set of possible messages. Then the MAP and ML strategies can be formulated as:

$$\hat{m}_{MAP} = \min_{\tilde{m} \in A_m} \{|\boldsymbol{\rho} - \tilde{\mathbf{s}}|^2 - 2\sigma_n^2 \ln \tilde{P}\} \tag{4.43}$$

and:

$$\hat{m}_{ML} = \min_{\tilde{m} \in A_m} |\boldsymbol{\rho} - \tilde{\mathbf{s}}|^2, \tag{4.44}$$

respectively. Equation (4.44) shows that an ML receiver selects the message $\hat{m}_{ML}$ associated with the signal vector in the set $\{\mathbf{s}_i, i = 0, 1, \ldots, N_m - 1\}$ having *minimum Euclidean distance* from $\boldsymbol{\rho}$. This strategy can be reformulated in terms of decision regions as follows: $\boldsymbol{\rho} \in D_k$ if and only if $\mathbf{s}_k$ is, among all the points of $A_{\mathbf{s}}$, the one *nearest* to $\boldsymbol{\rho}$.

□

In Example 4.3.1 the uncertainty of the receiver about the influence of the communication channel on the transmitted signal is due exclusively to noise. In communications over fading dispersive wireless channels, the uncertainty originates also from the lack of knowledge of a set of parameters describing the distorting effect of the communication channel on the transmitted signal. Such parameters are usually collected in a finite-dimensional vector, denoted $\mathbf{h}$ in what follows. Various assumptions can be made about $\mathbf{h}$: in particular, it can be modeled as a *deterministic* or *random* vector. In both cases, an estimate $\hat{\mathbf{h}}$ of $\mathbf{h}$ can be evaluated prior to data detection. This is usually accomplished by transmitting a message known to the receiver (e.g., a known data sequence for receiver training), so that the receiver can learn the structure of the communication channel [418] (see also Chapter 5 for further details).

An alternative to this approach is *joint data and channel estimation*. In particular, if $\mathbf{h}$ is modeled as a *deterministic* vector, joint ML estimation of data and channel is the solution of the problem (see (4.26)):

$$(\hat{m}_{ML}, \hat{\mathbf{h}}_{ML}) = \arg \max_{\tilde{\mathbf{h}} \in D_{\mathbf{h}}, \tilde{m} \in A_m} f_{\mathbf{r}}(\boldsymbol{\rho}|\tilde{\mathbf{s}}, \tilde{\mathbf{h}}), \tag{4.45}$$

where $D_{\mathbf{h}}$ denotes the space of $\mathbf{h}$. Note that knowledge of $\hat{\mathbf{h}}_{ML}$ is unnecessary *per se*; in fact, our real target is to estimate $\mathbf{s}$ only, so that $\hat{\mathbf{h}}_{ML}$ should be considered as a byproduct of our data estimation process. In principle, problem (4.45) can be solved in two steps (e.g., see [419]). If, for a given $\tilde{s}$, an *explicit* estimate $\hat{\mathbf{h}}(\tilde{s})$ of $\mathbf{h}$ can be evaluated as (first step):

$$\hat{\mathbf{h}}(\tilde{s}) = \arg \max_{\tilde{\mathbf{h}} \in D_\mathbf{h}} f_\mathbf{r}(\rho | \tilde{s}, \tilde{\mathbf{h}}), \qquad (4.46)$$

then (4.45) can be reformulated as (second step):

$$\hat{m}_{ML} = \arg \max_{\tilde{m} \in A_m} f_\mathbf{r}(\rho | \tilde{s}, \hat{\mathbf{h}}(\tilde{s})). \qquad (4.47)$$

In contrast, if $\mathbf{h}$ is modeled as a *stochastic* vector with known pdf $f_\mathbf{h}(\chi)$ and independent of $\mathbf{s}$, in principle $f_\mathbf{r}(\mathbf{r}|\tilde{s})$ can be evaluated via multidimensional integration, that is, as:

$$f_\mathbf{r}(\mathbf{r}|\tilde{s}) = \int f_\mathbf{r}(\mathbf{r}|\chi, \tilde{s}) f_\mathbf{h}(\chi) d\chi, \qquad (4.48)$$

so that ML data estimation can be accomplished by solving, once again, (4.26). Apparently, this approach does not entail the evaluation of *explicit channel estimates*. In practice, however, the resulting likelihood function $f(\mathbf{r}|\tilde{s})$ (4.48) can usually be put in a form showing that multiple implicit channel estimates are evaluated, one for each hypothesis about the transmitted message. Further details concerning this will be given in Section 4.5.3.

4.3.4 Diversity Reception and Some Useful Theorems about Data Detection

As shown in Section 1.1 (see Figure 1.1), in a SIMO scenario the transmitted vector $\mathbf{s}$ is received over n_R distinct communication channels. In this case, if the the receive antennas are adequately spaced, the receiver observes n_R *different* essentially independent replicas of the transmitted signal. Even in this case the problem of optimal diversity reception can be tackled by resorting to the tools developed in the previous subsection. In fact, if $\mathbf{r}_i \triangleq [r_{i,0}, r_{i,1}, \dots, r_{i,N-1}]^T$, with $i = 0, 1, \dots, n_R - 1$, denotes the output of the ith channel, an overall received vector $\mathbf{r}$ can be generated by concatenating all the channel outputs in an ordered fashion as:

$$\mathbf{r} \triangleq [\mathbf{r}_0^T, \mathbf{r}_1^T, \dots, \mathbf{r}_{n_R-1}^T]^T. \qquad (4.49)$$

Given this $\mathbf{r}$, the optimal (MAP) detection strategy for diversity reception is given by (4.23), as in the single-channel case. For instance, with double diversity ($n_R = 2$), the observed data are $\mathbf{r}_0 = \rho_0$ and $\mathbf{r}_1 = \rho_1$, and (4.23) becomes:

$$\max_{m_i \in A_m} P_i f_\mathbf{r}(\rho_0, \rho_1 | \mathbf{s}_i) \Rightarrow \hat{m}_{MAP}. \qquad (4.50)$$

In certain circumstances a part of the noisy data generated by a channel can be ignored by the receiver with no performance loss. To understand when this occurs, it is useful to factor the function $P_i f_\mathbf{r}(\rho_0, \rho_1 | \mathbf{s}_i)$ of (4.50), rewriting it as:

$$P_i f_\mathbf{r}(\rho_0, \rho_1 | \mathbf{s}_i) = P_i f_{\mathbf{r}_0}(\rho_0 | \mathbf{s}_i) \, f_{\mathbf{r}_1}(\rho_1 | \rho_0, \mathbf{s}_i). \qquad (4.51)$$

From this expression it can easily be seen that knowledge of $\mathbf{r}_1$ is *irrelevant* in (4.50) (i.e., $\mathbf{r}_1$ consists of *irrelevant data*) if and only if $\mathbf{r}_1$, conditioned on $\{\mathbf{r}_0 = \rho_0\}$, is independent of $\mathbf{s}_i$, that is:

$$f_{\mathbf{r}_1}(\rho_1 | \rho_0, \mathbf{s}_i) = f_{\mathbf{r}_1}(\rho_1 | \rho_0). \qquad (4.52)$$

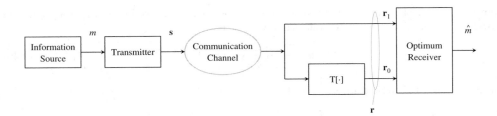

Figure 4.8 Digital communication system useful in proving the *theorem of reversibility*.

In fact, in this case, the vector $\mathbf{r}_1$ can be ignored without affecting the optimality of the receiver, so that the optimal strategy (4.50) becomes:

$$\max_{m_i \in A_m} P_i f_{\mathbf{r}_0}(\boldsymbol{\rho}_0|\mathbf{s}_i) \Rightarrow \hat{m}_{MAP}, \tag{4.53}$$

which involves $\mathbf{r}_0$ only. This result is known as the *theorem of irrelevance* [321].

Generally speaking, in a statistical decision problem, optimal detection can be achieved if and only if a certain D-dimensional set of data:

$$\mathbf{z} = \mathbf{g}(\mathbf{r}) \tag{4.54}$$

is available, where $\mathbf{g}(\cdot)$ describes a transformation of the N-dimensional observed noisy vector $\mathbf{r}$ (with $N \geq D$) . When this occurs, the vector $\mathbf{z}$ represents a set of *sufficient statistics*.

An important corollary of the theorem of irrelevance is the *theorem of reversibility* [321]. Before illustrating this, let us consider a scenario in which the noisy vector $\mathbf{r}$ available at the output of a communication channel undergoes a transformation, represented by the vector operator $\mathbf{T}[\cdot]$, which generates a new vector $\bar{r}$ of noisy data. The transformation $\mathbf{T}[\cdot]$ is *reversible* if an inverse transformation (conventionally represented by the operator $\mathbf{T}^{-1}[\cdot]$) can be defined. Such a transformation, when applied to $\bar{r}$, generates $\mathbf{r}$. The *theorem of reversibility* states that, in a digital communication system, the minimum average error probability achieved by an optimal receiver does not change if the vector $\mathbf{r}$, observed at the output of the communication channel, undergoes a *reversible transformation* before being processed by an optimal detection strategy. To prove this result, let us consider the digital communication system shown in Figure 4.8; here the vector $\mathbf{r}_1$, generated by the channel, undergoes a reversible transformation $\mathbf{T}[\cdot]$, generating $\mathbf{r}_0$. Then, we have that:

$$f_{\mathbf{r}_1}(\boldsymbol{\rho}_1|\boldsymbol{\rho}_0, \mathbf{s}_i) = f_{\mathbf{r}_1}(\boldsymbol{\rho}_1|\boldsymbol{\rho}_0), \tag{4.55}$$

since the knowledge of the value $\boldsymbol{\rho}_0$, taken on by $\mathbf{r}_0$, *uniquely* identifies, independently of $\mathbf{s}_i$, the value $\boldsymbol{\rho}_1$ taken on by $\mathbf{r}_1$. Therefore, on the basis of the theorem of irrelevance (see (4.52)), it can be stated that an optimal receiver endowed with knowledge of $\mathbf{r} = [\mathbf{r}_0^T, \mathbf{r}_1^T]^T$ can discard $\mathbf{r}_1$, since the latter consists of irrelevant data only. In other words, knowledge of $\mathbf{r}_0$ is enough to take an optimal decision concerning the transmitted message.

4.4   Mathematical Models for the Receiver Vector

In this section we focus on the general problem of extracting from a received noisy waveform $r(t)$ a finite-dimensional vector $\mathbf{r}$ to form the input to an optimal detection algorithm. Then we analyze in detail the structure of $\mathbf{r}$ for PAM, CPM and OFDM signaling in various communication scenarios. In doing so we always adopt baseband signal and system models, and ignore large-scale fading effects and the presence of system nonlinearities. In addition, the availability of ideal timing at the receive side is assumed.

4.4.1   *Extraction of a Set of Sufficient Statistics from the Received Signal*

In both the receiver architectures illustrated in the previous section, a complex baseband signal $r(t) = r_c(t) + jr_s(t)$ is extracted from the received RF signal $r_{RF}(t)$. To process the continuous-time waveform $r(t)$ digitally for data estimation, we need to condense the useful information it carries for data detection into a vector **r** (which we refer to as the *observation vector*). This condensation should avoid any information loss, that is, **r** should represent a set of *sufficient statistics* for the estimation of the transmitted data. Generally speaking, the method of processing of $r(t)$ to generate **r** depends on the channel properties and on the knowledge the receiver has about them. The simplest scenario is that of AWGN. In this case, if the delay introduced by the transmitter, receiver and propagation medium is neglected and the transmission of a finite-dimensional vector of channel symbols $\mathbf{c}_N = [c_0, c_1, \ldots, c_{N-1}]^T$ is assumed (this vector represents the transmitted *message* to be estimated at the receiver), $r(t)$ can be expressed as:

$$r(t) = s(t, \mathbf{c}_N) + w(t) \tag{4.56}$$

where (see (3.2)):

$$s(t, \mathbf{c}_N) = \sum_{n=0}^{N-1} s(\Delta_n, c_n; t - nT_s), \tag{4.57}$$

and $w(t) = w_c(t) + jw_s(t)$ is a complex circularly symmetric Gaussian noise process having zero mean and two-sided power spectral density $2N_0$ in the bandwidth of $s(t, \mathbf{c})$ (its real part $w_c(t)$ and imaginary part $w_s(t)$ are statistically independent Gaussian random processes, each having zero mean and two-sided spectral density N_0 in the same bandwidth) [32]. It is worth remembering that the signal $s(\Delta_n, c_n; t)$ belongs to the alphabet $A_s = \{s_i(t), i = 0, 1, \ldots, N_m - 1\}$, consisting of *finite energy* and distinct functions. Then in this case a set of $N_m \cdot N$ sufficient statistics can be obtained by sampling at baud rate (i.e., at the instants $t_n \triangleq nT_s + t_0$, where t_0 is a proper sampling offset and $n = 0, 1, \ldots, N - 1$) the outputs of a bank of N_m filters, each *matched* to a specific waveform of A_s [32]. Generally speaking, for any complex signal $s(t)$ having finite energy and time support (t_i, t_f), the linear invariant filter characterized by the impulse response:

$$h_{MF}(t) \triangleq s^*(t_f - t) = s^*(-(t - t_f)) \tag{4.58}$$

is *matched* to $s(t)$. The generation of the impulse response $h_{MF}(t)$ of the *matched filter* (MF) for a given $s(t)$ is summarized in Figure 4.9 under the assumption that $s(t)$ is real. Note that, if the filter input is expressed by (4.56) with $s(t)$ in place of $s(t, \mathbf{c}_N)$, sampling the filter output at $t = t_f$ generates the random variable $R = E_s + w$, where E_s is the energy of $s(t)$ and w is a complex Gaussian random variable having zero mean and variance $\sigma_w^2 = 2N_0 E_s$.

These concepts can be applied to the detection problem of (4.56) as follows. If the support of the waveforms of A_s is the interval $(0, T_0)$, the optimal detector employs N_m filters, whose impulse responses are given by:

$$h_{MF}^{(i)}(t) = s_i^*(-(t - T_0)), \tag{4.59}$$

with $i = 0, 1, \ldots, N_m - 1$. All these filters are fed by the same noisy waveform, $r(t)$ in (4.56), and are sampled in parallel at the instants $t_n \triangleq T_0 + nT_s$, with $n = 0, 1, \ldots, N - 1$. Let:

$$r_n^{(i)} = r(t) \otimes h_{MF}^{(i)}(t)|_{t=t_n} \tag{4.60}$$

denote the sample taken at the output of the ith filter for $t = t_n$. If we collect the N samples taken over N consecutive symbol intervals at the ith filter output in the vector $\mathbf{r}_i \triangleq [r_0^{(i)}, r_1^{(i)}, \ldots, r_{N-1}^{(i)}]^T$ (with $i = 0, 1, \ldots, N_m - 1$), then the overall vector **r** of sufficient statistics can be generated by concatenating the vectors $\{\mathbf{r}_i\}$ in an ordered fashion as:

$$\mathbf{r} = [\mathbf{r}_0^T, \mathbf{r}_1^T, \ldots, \mathbf{r}_{N_m}^T]^T. \tag{4.61}$$

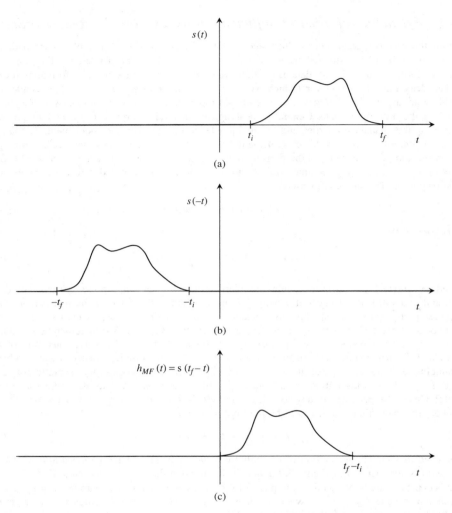

Figure 4.9 A time reversal (b) and a delay by t_f seconds (c) are needed to generate $h_{MF}(t)$ (4.58) from the signal $s(t)$ (a).

Let us now apply these results to the modulation formats we analyzed in Chapter 3. On the one hand, for passband PAM and OFDM signaling (see (3.47)–(3.48) and (3.274), respectively), we have that $N_m = 1$, since the transmitted signal consists of multiple replicas, delayed by multiples of the symbol interval T_s, of a real pulse $p(t)$. If we assume that the support of $p(t)$ is $(0, T_p)$ (where T_p is often substantially longer than T_s), then only one matched filter having impulse response:

$$h_{MF}(t) = \frac{1}{K_c} p(-(t - T_p))$$ (4.62)

is needed to generate a vector $\mathbf{r} \triangleq [r_0, r_1, \ldots, r_{N-1}]^T$ of sufficient statistics, with (see (4.60)):

$$r_n = r(t) \otimes p(-(t - T_p))|_{t=t_n}$$ (4.63)

and $t_n \triangleq T_p + nT_s$. On the other hand, for CPM signaling, we have that:

(a) $N_m = M^L$, since the number of complex waveforms is equal to the product of the number of correlative states of the modulator and the number M of possible values of the actual symbol (the phase state introduces a phase rotation only; see (3.140)), so that a bank of $4M^L$ real filters is needed [374] ($2M^L$ for $r_c(t)$, $2M^L$ for $r_s(t)$),

(b) $T_p = T_s$, so that the impulse response of each filter of the bank lasts T_s seconds.

The exponential dependence of N_m on the correlation length L can represent a significant problem in the implementation of CPM receivers. This has motivated the search for reduced sets of functions (i.e., bases of functions) that can represent with negligible error the time-varying part of the waveform (3.140), i.e. the signal:

$$\exp[j\gamma(t - nT_s, \sigma_n, c_n)] = \cos(\gamma(t - nT_s, \sigma_n, c_n)) + j\sin(\gamma(t - nT_s, \sigma_n, c_n)) \qquad (4.64)$$

for any σ_n and c_n. Various technical solutions based on the use of sinusoidal signals with appropriate frequencies [291], the Walsh functions [294, 420], time translates of the sinc(·) function [293], Gram–Schmidt orthonormalization [292, 306] and the *principal components method* [295] have been suggested to develop bases for the representation of (4.64). An alternative to this approach is Laurent's decomposition (see Section 3.6.5.1). The set of Laurent pulses forms a basis for the representation of the modulated signal; this leads to a set of $Q^P(2^P - 1)$ matched filters, with $Q = 2^{L-1}$ and $2^{P-1} < M \leq 2^P$. The rank of this basis increases with M and L; however, it can be shown that most of the signal energy is contained in few pulses, so that a reduced bank of matched filters can be employed to generate $\mathbf{r}$ (e.g., see [284, 303–305]).

The problem of matched filtering in AWGN deserves further comment in order to clarify a few important technical issues. To simplify our study we focus on the case of PAM signaling only, so that the transmitted signal (4.57) becomes (see (3.48)):

$$s(t, \mathbf{c}_N) = K_c \sum_{n=0}^{N-1} c_n \, p(t - nT_s) \qquad (4.65)$$

and, in particular, we consider the specific scenario illustrated in the following example.

Example 4.4.1 Let us assume that $p(t)$ is generated by truncating the pulse expressed by (3.89) (the FCT of this pulse is the square root of a raised cosine spectrum with roll-off α; see (3.90) and (3.81)) truncated to the interval $[-L_pT_s/2, L_pT_s/2]$ (with L_p integer) and delaying the resulting waveform by $L_pT_s/2$ seconds (normalization is also required to ensure that the resulting pulse has unit energy). For a given α, the value of L_p is selected large enough to capture most of the energy of the transmitted pulse (3.89), so that the convolution of $p(t)$ with itself generates, to a good approximation, the pulse $g(t - L_pT_s)$ (3.79) (whose FCT is a raised cosine spectrum with roll-off α; see (3.81)). Then, we have that (see (3.88)):

$$p(t) \otimes p(t) \cong g(t - L_pT_s) \qquad (4.66)$$

and (see (4.62)):

$$h_{MF}(t) = \frac{1}{K_c} p(-(t - L_pT_s)) = \frac{1}{K_c} p(t) \qquad (4.67)$$

since $p(t)$ is symmetric around the instant $L_pT_s/2$, so that:

$$p(t) \otimes p(t)|_{t=t_k} = \delta[k], \qquad (4.68)$$

where $t_k \triangleq kT_s + L_p T_s$. From this result it is easily seen that sampling the output of the MF fed by (4.56) (with $s(t, \mathbf{c}_N)$ given by (4.65)) at t_k produces:

$$r_k = r(t) \otimes h_{MF}(t)|_{t=t_k}$$

$$\cong \sum_{n=0}^{N-1} c_n \, g(t_k - L_p T_s - nT_s) + w_k$$

$$= \sum_{n=0}^{N-1} c_n \, \delta[k - n] + w_n$$

$$= c_k + w_k \tag{4.69}$$

for $k = 0, 1, \ldots, N - 1$, where:

$$w_k = w(t) \otimes h_{MF}(t)|_{t=t_k} \tag{4.70}$$

is a Gaussian noise sample having zero mean and variance $\sigma_n^2 = 2N_0/K_c^2$. It is easy to show that the autocorrelation function of the sequence $\{w_k\}$ is:

$$R_w[k] = \sigma_n^2 \delta[k], \tag{4.71}$$

so that the noise samples at the MF output, being jointly Gaussian and uncorrelated, are statistically independent. In other words, the noise sequence $\{w_k\}$ is *white*.[3] These results lead to the conclusion that, in this case, the sufficient statistics $\{r_k, k = 0, 1, \ldots, N - 1\}$ share the following two very useful properties:

1. The useful component of r_k provides information about c_k only and not about any of the other transmitted symbols.
2. The noise sample w_k affecting r_k does not provide any information about the noise affecting all the other signal samples $\{r_k, k \neq n\}$. For this reason, any decision c_k should be based on r_k only (with $k = 0, 1, \ldots, N - 1$).

Finally, from (4.69) it is easy to see that:

$$\mathbf{r} = \mathbf{c}_N + \mathbf{w}, \tag{4.72}$$

where $\mathbf{w} \triangleq [w_0, w_1, \ldots, w_{N-1}]^T$, that is, $\mathbf{r}$ is structured as (4.33).
□

The favorable features of the sequence $\{r_k, k = 0, 1, \ldots, N - 1\}$ generated by (4.69) are the consequence of the specific choice of the $p(t)$ selected and, in particular, of its property expressed by (4.68), which we rewrite as:

$$\int_{-\infty}^{+\infty} p(\tau) p(t_k - \tau) d\tau = \int_{-\infty}^{+\infty} p(\tau) p(kT_s - (\tau - L_p T_s)) d\tau$$

$$= \int_{-\infty}^{+\infty} p(\tau) p(\tau - kT_s) d\tau = \delta[k]. \tag{4.73}$$

This result states that the functions $\{p(t - kT_s), k = \ldots, -1, 0, 1, \ldots\}$ form an orthonormal set and, in particular, represent an *orthonormal basis* for the space of $s(t, \mathbf{c}_N)$ in (4.65). In the literature, when (4.73) holds, it is usually said that the pulse $p(t)$ satisfies the *first Nyquist criterion* [421]. In PAM transmission, the Nyquist criterion has two implications:

---

[3] Note that the continuous-time noise process at the output of $h_{MF}(t)$ is not white.

- If the output of the matched filter is sampled at the right (symbol-spaced) instants, each sample depends on only one channel symbol.
- The noise samples are independent.

Unfortunately, in wireless communications the model (4.56) for $r(t)$ is usually not valid since the communication channel, acting as a filter, distorts $s(t, \mathbf{c}_N)$. To understand the consequence of this phenomenon, let us assume that the wireless channel can be considered *static* during the transmission of $s(t, \mathbf{c}_N)$ (4.65) and let $h_c(t)$ denote the CIR. Then the received signal can be written as:

$$r(t) = K_c \sum_{l=0}^{N-1} c_n \, p_{TC}(t - lT_s) + w(t), \tag{4.74}$$

where $w(t)$ is AWGN (with the same statistical properties as the noise process of (4.56)) and the impulse response:

$$p_{TC}(t) \triangleq p(t) \otimes h_c(t) \tag{4.75}$$

accounts for transmit and channel filtering. In this case, if $p_{TC}(t)$ is *known to the receiver* and its support is the interval $[0, L_{TC}T_s]$ (with L_{TC} integer), the elements of the vector $\mathbf{r} \triangleq [r_0, r_1, \dots, r_{N-1}]^T$ of sufficient statistics can be still generated by baud rate sampling of the output of a filter matched to $p_{TC}(t)$ (4.75) (i.e., having impulse response given by $h_{MF}(t) = p_{TC}^*(-(t - L_{TC}T_s))/K_c$) and fed by $r(t)$ (4.74). Note, however, that the MF output sample r_k is now given by:

$$r_k = r(t) \otimes h_{MF}(t)|_{t=t_k}$$

$$= \frac{1}{K_c} r(t) \otimes p_{TC}^*(-(t - L_{TC}T_s))|_{t=t_k} = \sum_{l=0}^{N-1} c_l \, h_{k-l} + n_k \tag{4.76}$$

where $t_k \triangleq kT_s + L_{TC}T_s$, $h_k \triangleq h(t_k)$ (with $h_0 = E_{p_{TC}}$, where $E_{p_{TC}}$ is the energy of $p_{TC}(t)$):

$$h(t) \triangleq p_{TC}(t) \otimes p_{TC}^*(-(t - L_{TC}T_s)) \tag{4.77}$$

$n_k \triangleq n(t_k)$ and $n(t) \triangleq w(t) \otimes h_{MF}(t)$ is a complex Gaussian process having zero mean and autocorrelation function:

$$R_n(\tau) \triangleq \sigma_n^2 p_{TC}(\tau) \otimes p_{TC}^*(-\tau), \tag{4.78}$$

where $\sigma_n^2 = 2N_0/K_c^2$. Equation (4.76) shows that r_k depends not only on c_k but also on other channel symbols; the contribution of such symbols is expressed by:

$$I_k \triangleq \sum_{\substack{l=0 \\ l \neq k}}^{N-1} c_l \, h_{k-l}, \tag{4.79}$$

which represents the so-called ISI. Note also that, generally speaking, the support of $h(t)$ (4.77) is the interval $[0, (L_h - 1)T_s]$ (with $L_h = 2L_{TC} + 1$), so that $h_k = 0$ for $|k| > L_{TC}$. For this reason (4.79) can be simplified as:

$$I_k = \sum_{l=\max(0, k-L_{TC})}^{\min(k+L_{TC}, N-1)} c_l \, h_{k-l}. \tag{4.80}$$

This result shows that ISI is usually due to at most $L_h - 1$ transmitted symbols, that is, L_{TC} past symbols (generating *postcursor* ISI) and L_{TC} future symbols (producing *precursor* ISI).

The signal model expressed in (4.76) also deserves some comments concerning the noise sequence $\{n_k\}$, which, generally speaking, is not white. In fact, its autocorrelation function is given by $R_n[k] = R_n(kT_s)$ and is not impulsive, since the support of $R_n(\tau)$ (4.78) is the time interval $[-L_{TC}T_s,$

$L_{TC}T_s] \supset [-T_s, T_s]$. As will become clearer later, the lack of noise whiteness complicates the structure of detection metrics. When the MF impulse response does not satisfy the first Nyquist criterion, the property of noise whiteness can be restored using a so-called *whitened matched filter* (WMF) [422] (also known as a *sample-whitened matched filter* [423]) as a receive filter. The impulse response $p_{WMF}(t)$ of a WMF has the property that the set of functions $\{p_{WMF}(t - kT_s), k \text{ integer}\}$ is an *orthonormal basis* for the signal space spanned by the functions $\{p_{TC}(t - kT_s), k \text{ integer}\}$ (character-izing the useful component of the received signal (4.74)). Further information about the properties and the existence of the WMF can be found in [422, 423]. An alternative to combining noise whitening with matched filtering involves using a whitening digital filter processing the baud rate samples taken at the MF output [424].

In principle, if the channel is time-varying (and characterized by the CIR, $h(t, \tau)$) and, already assumed previously, *known to the receiver*, considerations similar to those just described apply. It is important to note, however, that, given the transmitted signal $s(t, \mathbf{c}_N)$ (4.65), the received signal can be now expressed as [425]:

$$r(t) = K_c \sum_{l=0}^{N-1} c_n \, p_{TC}(lT_s, t - lT_s) + w(t), \tag{4.81}$$

where:

$$p_{TC}(lT_s, t - lT_s) \triangleq \int_{-\infty}^{+\infty} p(\tau - lT_s) \, h(t, t - \tau) \, d\tau \tag{4.82}$$

is the channel response to the lth transmitted pulse $p(t - lT_s)$. Note that the notation employed for $p_{TC}(lT_s, t - lT_s)$ demonstrates that each received pulse shape is potentially unique, that is, that the channel response to each pulse changes from symbol to symbol. For this reason, extracting a set of sufficient statistics from $r(t)$ in (4.81) requires, in principle, N different matched filters, one for each of the N distorted pulses, which is not feasible in practice. To simplify the problem of matched filtering for wireless channels, parametric channel representations can be exploited [424, 426]. For instance, if the bandwidth B of $s(t, \mathbf{c}_N)$ (4.65) does not exceed $1/T_s$ and the channel memory is finite, a $T_s/2$-spaced channel model with a finite number of taps can be employed (see Section 2.2.3). Then the channel response to $p(t - lT_s)$ can be expressed as (see (2.107)):

$$p_{TC}(lT_s, t - lT_s) = \sum_{n=0}^{L_h-1} h_n(t) \, p\left(t - lT_s - \frac{nT_s}{2}\right), \tag{4.83}$$

where L_h denotes the number of active taps (i.e., the channel memory in symbol intervals) and $h_n(t)$ is the time-varying gain of the nth tap. This result shows that, if the variations of the tap gains $\{h_n(t)\}$ are slow relative to the duration of the transmit pulse shape $p(t)$, a set of sufficient statistics can be generated using a receive filter matched to $p(t)$ followed by a sampler that operates at twice the baud rate. In fact, it is not difficult to show that the response of a filter matched to $p_{TC}(lT_s, t - lT_s)$ can be generated by linearly combining the L_h consecutive samples taken at the output of a filter matched to $p(t)$; however, the coefficients of this combination are the complex conjugates of the tap gains $\{h_n(t)\}$, assumed approximately constant over the support of $p_{TC}(lT_s, t - lT_s)$. In other words, this solution can be implemented if the communication is *slowly time-varying*, so that the receiver is able to estimate such gains in a reliable and timely fashion. When the channel variations become fast, the exploitation of parametric representations is still possible, but the estimation of the parameters of the channel model becomes more difficult. To show this, let us reconsider the model (4.83) and assume a single tap (i.e., $L_h = 1$), so the communication channel is affected by time-selective fading only. To represent the channel variations of the single tap gain $h(t)$ over the duration of $p(t)$ a *linearly time-selective fading* model can be adopted [111, 427]. This means that the channel response to $p(t - lT_s)$ can be approximated as (see Section 2.2.3.4):

$$p_{TC}(t - lT_s) = h_l^{(0)} \, p(t - lT_s) + h_l^{(1)} g(t - lT_s), \tag{4.84}$$

where:

$$g(t) \triangleq \frac{t}{T_s} p(t), \qquad (4.85)$$

$h_l^{(0)} \triangleq h(lT_s)$ and $h_l^{(1)} \triangleq T_s dh(t)/dt|_{t=lT_s}$. A set of sufficient statistics can then be generated by sampling at baud rate and simultaneously the output of two receive filters, one matched to $p(t)$, the other one matched to $g(t)$ (4.85); note, however, that, in general, accurate estimation of $h_l^{(1)}$ is substantially more difficult than $h_l^{(0)}$, since its average power is usually much smaller than that of $h_l^{(0)}$. In principle, this approach can be generalized to an arbitrary number of channel taps. However, reliably estimating the full set of channel parameters becomes a formidable task. For this reason, when the channel variations of the channel are significant, a different approach to data detection is commonly taken. At the receiver, the received signal is passed through a low-pass filter with bandwidth B_{LP} large enough to accommodate the bandwidth B of the complex envelope of the transmitted signal and the effect of the Doppler spread due to the communication channels (this role can be played by the low-pass filters that appear in Figures 4.2 and 4.3). Then a set of sufficient statistics can be generated by sampling the filter output at a frequency $f_c \geq 2B_{LP}$, so that any information loss is avoided (e.g., see [428, 429] for specific applications of this approach). It is reasonable to choose a sampling frequency that is a multiple of the baud rate R_s, so that $f_c = n_s R_s$, where the integer parameter n_s denotes the number of samples per channel symbol (further analytical details of this approach are provided for specific signaling formats in the next subsection). However, this entails the development of optimal detection algorithms that process more than one noisy sample per symbol – a potentially larger amount of data.

It is interesting to note that the latter approach can even be taken in all the previous scenarios to overcome the problem of synthesizing specific *analog* MFs or WMFs, as shown in Figure 4.10, which refers to the specific case of processing of $r(t)$ (4.74) to generate the sequence $\{r_k\}$ (4.76). In this case the complex envelope of the received signal $r(t)$, after undergoing low-pass filtering with bandwidth B_{LP}, is sampled by an *analog-to-digital converter* (ADC) at a frequency $f_c = n_s R_s$. Then the resulting sequence $\{x_n\}$ of samples feeds a digital MF, whose finite impulse response $h_{MF}[n]$ is generated by sampling with a period $T_c \triangleq 1/f_c$ the impulse response $h_{MF}(t) = p_{TC}^*(-(t - L_{TC}T_s))/K_c$ of its analog counterpart. In principle, the sequence $\{y_n\}$ available at the MF output should be decimated by the factor n_s, so that one sample per channel symbol is available for data detection. However, in practice, decimation does not necessarily produce a data sequence representing a good approximation to the sequence $\{r_k\}$, because of the presence of an arbitrary sampling offset in the analog-to-digital conversion. For this reason, the receiver needs an algorithm for timing synchronization that allows us to estimate the sequence $\{t_k \triangleq kT_s + L_{TC}T_s\}$ of correct sampling instants (see (4.76)). If this timing information is available, it can be used to compute an estimate of each element of the sequence $\{r_k\}$ by interpolating a finite number of consecutive samples of $\{y_n\}$; this leads to generating the interpolated sequence $\{\tilde{r}_k\}$ consisting of baud rate samples.

Finally, it is interesting to analyze the problem of receive filtering in the case of a *statistically known channel*, that is, when the CIR $h(t, \tau)$ is *random*, but the CIR *statistics* (and not the CIR itself)

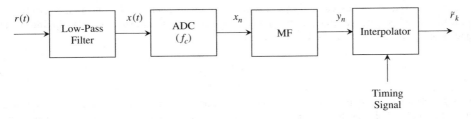

Figure 4.10 Processing chain for extracting a set of sufficient statistics from $r(t)$ (4.74) using a digital MF.

are known to the receiver. Then, even if the channel is fed by a *deterministic* signal $s(t)$, the useful component of its output signal (see (2.9)):

$$y(t) = \int_{-\infty}^{+\infty} s(t - \tau) \, h(t, \tau) \, d\tau \tag{4.86}$$

is a random process. Let us assume that $h(t, \tau)$ is a *zero mean random process*, so that $y(t)$ is a zero mean, generally nonstationary, complex stochastic process with covariance function (see (2.79)):

$$C_y(t; \tau) = R_y(t; \tau)$$

$$\triangleq E\{y(t + \tau) \, y^*(t)\}$$

$$= \int_{\alpha=-\infty}^{+\infty} s(t + \tau - \alpha) \int_{\beta=-\infty}^{+\infty} s^*(t - \beta) \, R_h(t + \tau, t; \alpha, \beta) d\beta d\alpha. \tag{4.87}$$

Note also that if the channel is WSS and US, (4.87) can be easily simplified as (see (2.83)):

$$C_y(t; \tau) = \int_{\alpha=-\infty}^{+\infty} s(t + \tau - \alpha) s^*(t - \alpha) P_h(\tau; \alpha) d\alpha, \tag{4.88}$$

where the function $P_h(\tau; \alpha)$ is given by (2.84). If the covariance $R_y(t; \tau)$ in (4.87) is jointly continuous in t and τ, then $y(t)$ in (4.86) can be represented as an infinite linear combination of functions, that is, it can be expanded as:

$$y(t) = \sum_{k=0}^{+\infty} y_i \, \phi_i(t), \tag{4.89}$$

where $\{y_i\}$ is a set of uncorrelated zero mean *random variables*, that is, such that $E\{y_i \, y_j^*\} = \lambda_i \, \delta_{i,j}$ and $E\{y_i \, y_j\} = 0$, and $\{\phi_i(t)\}$ is a set of orthonormal *deterministic functions*; this result is known as the *Karhunen–Loève theorem* [430, Ch. 3]. The functions $\{\phi_i(t)\}$ and the constants $\{\lambda_i\}$ are the solutions of the *homogeneous Fredholm integral equation*:

$$\int_{-\infty}^{+\infty} C_y(t, \tau) \, \phi_i(\tau) \, d\tau = \lambda_i \phi_i(t), \tag{4.90}$$

which is an integral form of an eigenequation with a kernel $C_y(t, \tau)$, where $\{\phi_i(t)\}$ and $\{\lambda_i\}$ are the normalized eigenfunctions and the associated eigenvalues, respectively. It can be shown that if the kernel is Hermitian in its arguments (i.e., $C_y(t, \tau) = C_y^*(\tau, t)$), then properties (P.1)–(P.5) in Section D.2 hold. Then, if we focus on a WSS-US Gaussian channel and assume that the kernel $C_y(t; \tau)$ (4.88) satisfies all the above-mentioned conditions, we can represent $y(t)$ (4.86) as a superposition of multiple waveforms, whose importance depends on the associated eigenvalues $\{\lambda_i\}$ [86]. Note that, in this case, the expansion coefficients $\{y_i\}$ are statistically independent random Gaussian variables and that, if the kernel is nondegenerate, a finite number of eigenvalues are expected to be significant. This suggests using the truncated expansion [108, 431]:

$$y(t) \cong \sum_{i=0}^{N_\lambda - 1} y_i \, \phi_i(t), \tag{4.91}$$

where N_λ denotes the number of significant eigenvalues, i.e. those whose value is not below a certain threshold. This result leads to the conclusion that, even if the deterministic waveform is randomly distorted, the channel response can still be represented through the use of a proper finite-dimensional basis, which, however, depends on the channel statistics. This suggests that, in principle, optimal receiver structures extract a set of sufficient statistics using a bank of filters matched to the eigenfunctions $\{\phi_i(t), i = 0, 1, \ldots, N_\lambda\}$ for any possible transmitted waveform $s(t)$. Note, however, that the functions of the filter bank need to be continuously adapted to the channel statistics (i.e., to any change of $C_y(t, \tau)$), so that in practice this approach can be implemented only if the sampled version of the low-pass filtered received signal is considered. In fact, considering the equivalent problem in a sampled system allows us to solve the integral equation (4.90) numerically, since it turns into a standard eigenvalue/eigenvector problem (e.g., see [85, App.] and [86, App. I] for further analytical details).

Further information about the use of the KL representation and its application to detection over flat fading channels can be found in [110, 432, 433].

4.4.2 Received Vector for PAM Signaling

In this subsection, the structure of the received vector $\mathbf{r}$ is investigated for both SISO and MIMO PAM signaling, when a finite-dimensional vector of channel symbols is transmitted. In the SISO case we analyze in detail the following three scenarios:

(a) the slow time-selective fading channel;
(b) the frequency-selective channel;
(c) the doubly-selective channel.

However, for simplicity of description, in the MIMO case we consider only the first two scenarios.

4.4.2.1 Models for a SISO Scenario

In this case transmission of the channel symbol vector $\mathbf{c}_N = [c_0, c_1, \ldots, c_{N-1}]^T$ is assumed, so that, generally speaking, the number of elements of $\mathbf{r}$ is a multiple of N.

Slow Time-Selective Fading
A time-selective channel is deemed *slow fading* or *slowly varying* if during a PAM transmission the variations of the channel distortion $a(t)$ (see (2.32)) can be assumed negligible over the duration of the transmitted pulse $p(t)$ (see the signal model (4.65)). In this case the transmit and receive filters are usually selected so as to ensure ISI-free reception over an AWGN channel. For instance, the waveform described in Example 4.4.1 can be employed; this leads directly to the model (see (4.69)):

$$r_k = a_k c_k + w_k \qquad (4.92)$$

with $k = 0, 1, \ldots, N - 1$, for the received signal samples at the time instants $t_k \triangleq kT_s + L_p T_s$; here, $a_k \triangleq a(t)|_{t=t_k}$ is the kth sample of the fading distortion, and the sequence $\{w_k\}$ consists of iid complex Gaussian noise samples, each with zero mean and variance $\sigma_n^2 = 2N_0/K_c^2$. The received signal samples can then be collected in the vector:

$$\mathbf{r} = [r_0, r_1, \ldots, r_{N-1}]^T = \mathbf{A}\mathbf{c}_N + \mathbf{w}, \qquad (4.93)$$

where $\mathbf{w} \triangleq [w_0, w_1, \ldots, w_{N-1}]^T$ and $\mathbf{A} \triangleq \mathrm{diag}(a_0, a_1, \ldots, a_{N-1})$.

If the channel is fast fading, this model no longer holds, since the $p(t)$ undergoes significant distortion so that ISI at the matched filter output is found if the transmit and receive filters are chosen as above. When this occurs, a possible strategy for mitigating ISI consists of employing a transmit pulse $p(t)$ satisfying the first Nyquist criterion (e.g., a pulse whose FCT is a raised cosine with roll-off α; see (3.78) and (3.79)) and using a low-pass filter with bandwidth large enough not to modify the shape of the distorted $p(t)$ as a receive filter instead of a matched filter [434].

Finally, it is worth noting that an evaluation of the validity of the slow fading model (4.92) can be found in [113].

Frequency-Selective Fading

In Section 4.4.1 it was shown that, in the case of PAM signaling with transmit pulse shape $p(t)$ over a frequency-selective channel with CIR $h_c(t)$, the T_s-spaced samples taken at the output of a filter matched to $p_{TC}(t) \triangleq p(t) \otimes h_c(t)$ and fed by the received signal can be written as (see (4.76), (4.79) and (4.80)):

$$r_k = \sum_{l=\max(0,k-L_{TC})}^{\min(k+L_{TC},N-1)} c_l \, h_{k-l} + n_k \tag{4.94}$$

for $k = 0, 1, \ldots, N-1$, where L_{TC} is the duration of $p_{TC}(t)$ in symbol intervals. Note that the ISI appearing in this model includes an anticausal portion (precursor ISI), since the overall discrete-time impulse response of the channel (i.e, the sequence $\{h_k\}$) is in general noncausal, and the noise sequence $\{n_k\}$ is not white. The anticausal portion of the ISI (precursor ISI) can be removed and the noise sequence made white by inserting a *whitening filter* (WF) at the sampler output (analytical details can be found in [29, Sects. 7.3, 9.4 and 10.1]). In fact, it can then be shown that:

- the equivalent discrete-time channel model at the WF output is *causal* and *minimum phase* (so that the energy of the impulse response is maximally concentrated in the early samples),
- the filtered noise sequence is circularly symmetric, Gaussian and white, so that its elements are independent random variables.

Then, if $\{r_k\}$ denotes the output of the WF (fed by the sampled MF output), it can be written as:

$$r_k = \sum_{l=0}^{L_q-1} c_l \, q_{k-l} + n_k, \tag{4.95}$$

where the impulse response $\{q_k\}$ (of length L_q) results from the convolution of $\{h_k\}$ (see 4.96) with the impulse response of the WF and $\{n_k\}$ is an additive complex Gaussian noise sequence; note that $L_q = L_h = 2L_{TC} - 1$, as shown in [27, pp. 625–627].

Finally, note that, in the case of a frequency-selective channel, the received vector $\mathbf{r} = [r_0, r_1, \ldots, r_{N-1}]^T$ can be expressed as:

$$\mathbf{r} = \mathbf{H}\mathbf{c}_N + \mathbf{n}, \tag{4.96}$$

where $\mathbf{H} = [H_{i,j}]^T$ is an $N \times N$ CIR matrix whose elements are extracted from $\{h_k\}$ ($\{q_k\}$) if we refer to (4.94) ((4.95)) and $\mathbf{n} \triangleq [n_0, n, \ldots, n_{N-1}]^T$ is a Gaussian noise vector.

A different model for the received signal vector is obtained when passband PAM transmission is employed with *frequency-domain equalization* (see Section 3.5.3) in the receiver. In this case, the channel symbol stream is divided into nonoverlapping blocks (each of length N) and block-by-block processing is accomplished at the receive side. In particular, under the assumptions in Section 3.5.3, the baseband received signal corresponding to the lth data block is given by (see (3.62)):

$$y(t) = z(t) + w(t)$$

$$= \frac{1}{\sqrt{NT_s}} \sum_{k=-N}^{N} P_k H_k^{(l)} C_k^{(l)} \exp\left(j \frac{2\pi k}{NT_s}(t - lN_T T_s) \right) + w(t), \tag{4.97}$$

for $lN_T T_s \leq t < lN_T T_s + NT_s$, where $P_k \triangleq P(k/NT_s)/\sqrt{T_s}$, $H_k^{(l)} \triangleq H^{(l)}(k/NT_s)$, $P(f)$ and $H^{(l)}(f)$ are respectively the FCTs of the transmit pulse $p(t)$ and the CIR $h^{(l)}(t)$ in the observation interval considered, $C_k^{(l)}$ is the kth component of the vector $\mathbf{C}_N^{(l)} \triangleq [C_0^{(l)}, C_1^{(l)}, \ldots, C_{N-1}^{(l)}]^T \triangleq \mathbf{Q}_N \mathbf{c}_N^{(l)}$ (resulting from the DFT of the lth data block $\mathbf{c}_N^{(l)}$) and $w(t)$ is a complex circularly symmetric Gaussian noise process having zero mean and two-sided power spectral density $2N_0$. To avoid an information loss in discretizing $y(t)$ in (4.97), the following procedure can be followed [435]: $y(t)$ is passed first through an ideal low-pass filter (with bandwidth $1/T_s$ and gain $\sqrt{T_s/2}$) producing the bandlimited random process $r(t)$, which then undergoes uniform sampling at a rate $2/T_s$. In particular, when detecting the symbol vector $\mathbf{c}_N^{(l)}$, $r(t)$ is sampled at the epochs $t_i^{(l)} \triangleq lN_T T_s + iT_s/2$, with $i = 0, 1, \ldots, 2N - 1$ (the samples associated with the cyclic prefix are discarded). This yields a $2N$-dimensional vector $\mathbf{r}_{2N}^{(l)} \triangleq [r_0^{(l)}, r_1^{(l)}, \ldots, r_{2N-1}^{(l)}]^T$, with:

$$r_i^{(l)} = r(t_i^{(l)}) = \frac{1}{\sqrt{2N}} \sum_{k=-N}^{N} P_k H_k^{(l)} C_k^{(l)} W_{2N}^{-ki} + n_i^{(l)} \tag{4.98}$$

for $i = 0, 1, \ldots, 2N - 1$. Here, $W_N \triangleq \exp(-j2\pi/N)$ and the noise samples $\{n_i^{(l)} \triangleq n(t_i^{(l)})\}$ are iid random variables with zero mean and variance $\sigma_n^2 = 2N_0$. The vector $\mathbf{r}_{2N}^{(l)}$ then feeds a $2N$-point DFT, producing the vector $\mathbf{R}_{2N}^{(l)} \triangleq [R_0^{(l)}, R_1^{(l)}, \ldots, R_{2N-1}^{(l)}]^T = \mathbf{Q}_{2N}\mathbf{r}_{2N}^{(l)}$. Noting that $C_{k+N}^{(l)} = C_{k-N}^{(l)}$ for any k because of the periodicity of the DFT, it can be proved that:

$$\mathbf{R}_{2N}^{(l)} = \mathbf{D}^{(l)}\mathbf{C}_{2N}^{(l)} + \mathbf{V}_{2N}^{(l)}, \tag{4.99}$$

where $\mathbf{C}_{2N}^{(l)} \triangleq [(\mathbf{C}_N^{(l)})^T | (\mathbf{C}_N^{(l)})^T]^T$, $\mathbf{V}_{2N}^{(l)} \triangleq [V_0^{(l)}, V_1^{(l)}, \ldots, V_{2N-1}^{(l)}]^T$, $\mathbf{D}^{(l)} \triangleq \mathrm{diag}(D_k^{(l)})$ with $D_k^{(l)} = P_k H_k^{(l)}$ $(P_{k-2N} H_{k-2N}^{(l)})$ for $k = 0, 1, \ldots, N$ $(k = N + 1, N + 2, \ldots, 2N - 1)$. It is useful to note that the set of random variables $\{V_k^{(l)}\}$ is statistically equivalent to $\{n_i^{(l)}\}$.

Doubly-Selective Fading

If the channel fading is fast, the received signal is expressed by (4.81):

$$r(t) = K_c \sum_{l=0}^{N-1} c_n \, p_{TC}(lT_s, t - lT_s) + w(t), \tag{4.100}$$

with (see (4.82)):

$$p_{TC}(lT_s, t - lT_s) \triangleq \int_{-\infty}^{+\infty} p(\tau - lT_s) \, h(t, t - \tau) \, d\tau, \tag{4.101}$$

where $h(t, \tau)$ is the time-varying CIR. In this case, the receive filter is usually chosen to be an ideal low-pass filter having gain equal to $1/K_c$ in its band and a bandwidth B_{LP} large enough not to distort the signal $p_{TC}(lT_s, t - lT_s)$. In addition, the low-pass filter output is sampled at a frequency $f_c = n_s R_s \geq 2B_{LP}$, that is, n_s samples per channel interval (with $n_s > 1$) are extracted from the filtered received signal to avoid any information loss due to aliasing. Then, if the support of $p_{TC}(t)$ is the interval $[0, L_{TC} T_s]$ (with L_{TC} integer), the kth element of the the received vector $\mathbf{r} \triangleq [r_0, r_1, \ldots, r_{N n_s - 1}]^T$ of sufficient statistics is generated as:

$$r_k \triangleq r(t) \otimes g_R(t)|_{t=t_k} = \sum_{l=0}^{N-1} c_l \, h_{k,l} + n_k \tag{4.102}$$

with $k = 0, 1, \ldots, Nn_s - 1$, where $g_R(t)$ is the impulse response of the above-mentioned ideal low-pass filter, $T_c \triangleq T_s/n_s$ is the sampling period, $t_k \triangleq kT_c + t_0$ is the kth sampling instant (t_0 is a sampling offset depending on L_{TC} and on the delay introduced by the receive filter):

$$h_{k,l} \triangleq \int_{\alpha=-\infty}^{+\infty} \left[\int_{\tau=-\infty}^{+\infty} p(\tau - lT_s) \, h(\alpha, \alpha - \tau) \, d\tau \right] g_R(kT_c - \alpha) d\alpha \tag{4.103}$$

denotes the overall discrete-time CIR, $n_k \triangleq n(kT_s)$ and $n(t) \triangleq w(t) \otimes g_R(t)$ is a complex Gaussian process having zero mean and autocorrelation function $R_n(\tau) = 2N_0 g_R(\tau) \otimes g_R^*(-\tau)$ (in practice, the power spectral density of $n(t)$ is flat in the passband of the receive filter). Note that the discrete-time model $\{h_{k,l}\}$ (4.103) takes into account the double selectivity of the communication channel and that the indexes k and l refer to different time units, T_c and T_s, respectively. If $p(t)$ and $g_R(t)$ have finite duration and the channel time dispersion is limited in extent, (4.102) can be simplified to:

$$r_k \triangleq r(t) \otimes g_R(t)|_{t=t_k} = \sum_{l=\min(0,\lfloor k/n_s \rfloor - L_{h,1})}^{\max(\lfloor k/n_s \rfloor + L_{h,2}, N-1)} c_l \, h_{k,l} + n_k. \tag{4.104}$$

This result shows that ISI is due to both the last $L_{h,1}$ symbols (*postcursor* ISI) and $L_{h,2}$ future symbols (*precursor* ISI), so that the overall channel memory is $L_h - 1$ symbol intervals, with $L_h \triangleq L_{h,1} + L_{h,2}$.

Note also that, on the one hand, if the channel is frequency-selective only (so that $h(t, \tau) = h(\tau)$), the expression for $h_{k,l}$ (4.103) can be rewritten as:

$$h_{k,l} = p(t - lT_s) \otimes h(t) \otimes g_R(t)|_{t=kT_c} = \tilde{h}_{k-ln_s}, \tag{4.105}$$

where:

$$\tilde{h}_{k-ln_s} \triangleq \tilde{h}((k - ln_s)T_c) = \tilde{h}(kT_c - lT_s) \tag{4.106}$$

with $\tilde{h}(t) \triangleq p(t) \otimes h(t) \otimes g_R(t)$, so that (4.102) becomes:

$$r_k = \sum_{l=0}^{N-1} c_l \, \tilde{h}_{k-ln_s} + n_k. \tag{4.107}$$

On the other hand, if the channel is *time-selective* only (so that $h(t, \tau) = a(t)\,\delta(\tau)$; see (2.35)), the model (4.102) for r_k simplifies to:

$$r_k = c_k \, a_k + n_k, \tag{4.108}$$

where $a_k \triangleq a(kT_c)$. Finally, it is useful to note that in the doubly-selective case the received vector $\mathbf{r}$ can be still put in the form (4.96), where, however, $\mathbf{H} = [H_{i,j}]^T$ is an $Nn_s \times N$ CIR matrix whose elements are extracted from $\{h_{k,l}\}$ (4.103) and the Gaussian noise vector $\mathbf{n} \triangleq [n_0, n, \ldots, n_{N-1}]^T$ is not necessarily white (it is easy to show that noise whiteness is achieved if $B_{LP} = n_s R_s/2$).

4.4.2.2   Models for a MIMO Scenario

Models for a MIMO scenario can be seen as a simple generalization of their SISO counterparts.

Slow Time-Selective Fading

In the case of slow frequency-flat fading, the channel gains affecting the channel symbols transmitted by n_T antennas in the kth interval can be collected into the $n_R \times n_T$ matrix:

$$\mathbf{H}[k] = [h_k^{(i,l)}], \tag{4.109}$$

where $h_k^{i,l}$ refers to the lth transmit and ith receive antennas. Then the corresponding received vector $\mathbf{r}_k \triangleq [r_k^{(0)}, r_k^{(1)}, \ldots, r_k^{(n_R-1)}]^T$, at the output of the n_R receive antennas in the kth symbol interval, can be expressed as:

$$\mathbf{r}_k = \mathbf{H}_k \mathbf{c}_k + \mathbf{n}_k, \tag{4.110}$$

where $\mathbf{c}_k \triangleq [c_k^{(0)}, c_k^{(1)}, \ldots, c_k^{(n_R-1)}]^T$, $c_k^{(l)}$ is the kth symbol sent by the lth transmit antenna and $\mathbf{n}_k \triangleq [n_k^{(0)}, n_k^{(1)}, \ldots, n_k^{(n_R-1)}]^T$ is the vector of noise samples. In this case, if the channel output is observed for $k = 0, 1, \ldots, N-1$ (i.e., over N consecutive symbol intervals), the overall received vector $\mathbf{r}$ is formed as the ordered concatenation of the vectors $\{\mathbf{r}_k, k = 0, 1, \ldots, N-1\}$.

Frequency-Selective Fading

Consider now the case of a frequency-selective channel, assuming a TDL model with $L_h T_s$-spaced taps for each of the $n_T \cdot n_R$ available SISO channels. Then, at discrete time k, the received samples at the output of the n_R receive antennas, collected into an n_R-dimensional vector $\mathbf{r}_k$, can be expressed as:

$$\mathbf{r}_k = \sum_{l=0}^{L_h-1} \mathbf{H}^{(l)} \mathbf{c}_{k-l} + \mathbf{n}_k \tag{4.111}$$

for $k = 0, 1, \ldots, N-1$, where $\mathbf{c}_l \triangleq [c_l^{(0)}, c_l^{(1)}, \ldots, c_l^{(n_R-1)}]^T$ is the n_T-dimensional vector of modulation symbols transmitted in parallel by the n_T transmit antennas, $\mathbf{n}_k \triangleq [n_k^{(0)}, n_k^{(1)}, \ldots, n_l^{(n_R-1)}]^T$ is an n_R-dimensional complex Gaussian noise vector having independent real and imaginary components and collecting the noise samples at the n_R receive antennas, $\mathbf{H}^{(l)} \triangleq [h_{i,j}^{(l)}]$ is the $n_R \times n_T$ matrix of the channel gains for the lth path ($h_{i,j}^{(l)}$ denotes the lth tap gain associated with the ith transmit and jth receive antennas) and N is the length of the observation interval. Gathering the vectors $\{\mathbf{r}_k, k = 0, 1, \ldots, N-1\}$ into an $n_R \times N$ matrix $\mathbf{R} \triangleq [\mathbf{r}_0, \mathbf{r}_1, \ldots, \mathbf{r}_{N-1}]$ and assuming that $\mathbf{c}_l = \mathbf{0}_{n_T}$ for $l \leq 0$, we may write:

$$\mathbf{R} = \mathbf{HC} + \mathbf{N}, \tag{4.112}$$

where $\mathbf{N} \triangleq [\mathbf{n}_0, \mathbf{n}_1, \ldots, \mathbf{n}_{N-1}]$ gathers together the noise samples (as in the case of flat fading channels):

$$\mathbf{H} \triangleq \begin{bmatrix} \mathbf{H}_0 & \mathbf{H}_1 & \cdots & \mathbf{H}_{n_T-1} \end{bmatrix} \tag{4.113}$$

with:

$$\mathbf{H}_i \triangleq \begin{bmatrix} h_{i,0}^{(0)} & h_{i,0}^{(1)} & \cdots & h_{i,0}^{(L_h-1)} \\ h_{i,1}^{(0)} & h_{i,1}^{(0)} & \cdots & h_{i,1}^{(L_h-1)} \\ \vdots & \vdots & \cdots & \vdots \\ h_{i,n_R-1}^{(0)} & h_{i,n_R-1}^{(0)} & \cdots & h_{i,n_R-1}^{(L_h-1)} \end{bmatrix}, \tag{4.114}$$

is the $n_R \times (n_T L_h)$ equivalent channel matrix referring to the ith transmit antenna, and:

$$\mathbf{C} \triangleq \begin{bmatrix} \mathbf{C}_0 \\ \mathbf{C}_1 \\ \vdots \\ \mathbf{C}_{n_T-1} \end{bmatrix} \tag{4.115}$$

with:

$$\mathbf{C}_i \triangleq \begin{bmatrix} c_0^{(i)} & c_1^{(i)} & c_2^{(i)} & \cdots & \cdots & c_{N-2}^{(i)} & c_{N-1}^{(i)} \\ 0 & c_0^{(i)} & c_1^{(i)} & \cdots & \cdots & c_{N-3}^{(i)} & c_{N-2}^{(i)} \\ \vdots & 0 & c_0^{(i)} & \cdots & \cdots & \vdots & c_{N-3}^{(i)} \\ \vdots & \vdots & 0 & \cdots & \cdots & \vdots & \vdots \\ \vdots & \vdots & \vdots & \cdots & \cdots & c_{N-1-L_h}^{(i)} & \vdots \\ 0 & 0 & 0 & \cdots & \cdots & c_{N-2-L_h}^{(i)} & c_{N-1-L_h}^{(i)} \end{bmatrix}, \tag{4.116}$$

is the $n_R \times (n_T L_h)$ equivalent matrix of channel symbols sent by the ith transmit antenna.

4.4.3   Received Vector for CPM Signaling

In the following the structure of the received vector $\mathbf{r}$ is analyzed for SISO and MIMO CPM signaling, when a finite-dimensional vector of channel symbols is transmitted. For SISO we consider in detail both time-selective and frequency-selective fading. However, for MIMO we consider only the time-selective case.

It is important to note that, generally speaking, if the so-called Laurent's representation (see Section 3.6.5.1) is adopted, any CPM signal can be expressed as the superposition of a limited number of PAM components. For this reason, in principle, the structure of the received vector for any CPM signal can be easily derived by generalizing various results provided in Section 4.4.2. We comment further on this approach and on the possible alternatives in this subsection.

4.4.3.1 Models for a SISO Scenario

Time-Selective Fading

If the channel is affected by *slow* time-selective fading (i.e., it exhibits negligible variations over each symbol interval) and the finite channel symbol vector $\mathbf{c}_N = [c_0, c_1, \ldots, c_{N-1}]^T$ is transmitted, a set of sufficient statistics can be extracted from the received signal by sampling at baud rate the outputs of a bank of $N_m = M^L$ complex MFs, exactly as in the AWGN case (see Section 4.4.1). Then $\mathbf{r}$ can be generated by concatenating N distinct N_m-dimensional vectors, so that its size is $N \cdot N_m$ (unless a reduced set of functions is used in the implementation of the bank of MFs).

If the fading is *fast*, two alternatives are found in the technical literature. One is based on low-pass filtering $r(t)$ with bandwidth large enough not to modify the shape of the useful component of the received signal and on sampling the filter output at a frequency $f_c = n_s R_s$ with $n_s > 1$ (e.g., see [436]). The second approach is based on representing the distorted useful signal via power series models. In particular, the use of a linearly time-selective fading model is proposed in [437]. Unfortunately, the second approach leads to doubling the size of the bank of matched filters.

Frequency-Selective Fading

Few papers deal with *time-domain* processing of CPM signals transmitted over frequency-selective channels (or, quasi-static multipath fading channels), since equalization algorithms in the time domain are characterized by large complexity. This is shown, for instance, in [438], where sufficient statistics are extracted from the received signal via low-pass filtering (with bandwidth B under the assumption that the CPM signal is strictly bandlimited to $|f| \leq B$) and sampling at a rate large enough to avoid any information loss (e.g., the use of $T_c = T_s/4$ is proposed for 1-RC and 2-RC formats). In this case the concatenation of the continuous phase modulator and the frequency-selective channel can be modeled as an FSSM, incorporating the memory of both subsystems. Such an FSSM can be represented by a *supertrellis*, whose *superstates* account for the contributions from both the modulator states and the channel ISI. A different signal model in the time domain can be developed if Laurent's decomposition is used for the transmitted signal and a T_s-spaced TDL model is employed for the communication channel, as suggested in [439]. However, in this case, to reduce the computational complexity of the equalization algorithm, the received signal, after low-pass filtering, is sampled below the Nyquist rate, since one sample per symbol is taken, and, therefore, the received signal samples do not form a set of sufficient statistics.

A simpler alternative to equalization in the time domain is a *frequency-domain* approach (e.g., see [284, 306, 440–442]). This enables good error performance at reasonable complexity. Here, following [284], we consider the CPM signal model illustrated in Section 3.6.6, which refers to the transmission of an infinite number of data blocks, each containing N_T channel symbols. Then, given the transmitted signal model (3.196), it is assumed that the CIR $h(t)$ extends over L_h symbol intervals and that the duration N_{cp} of the cyclic prefix is not smaller than the overall channel memory $(L_h + L)$, where L denotes the correlation length of the transmitted CPM signal. Then the baseband signal $y(t)$ at the receiver input can be expressed as (see (3.199)):

$$y(t) = \sqrt{\frac{2E_s}{NT_s}} \sum_{l=-\infty}^{+\infty} \sum_{p=0}^{P-1} \sum_{k=-\infty}^{+\infty} L_{p,k} H_k^{(l)} B_{p,k}^{(l)} \exp\left(j\frac{2\pi k(t - lN_T T_s)}{NT_s} \right) + w(t) \qquad (4.117)$$

for $t \in \bigcup_{l=-\infty}^{+\infty} [lN_P T_s, lN_T T_s)$, where $H_k \triangleq H(k/NT_s)$, $H(f)$ is the FCT of $h(t)$, $L_{p,k} \triangleq L_p(k/NT_s)/T_s$, $L_p(f)$ is the FCT of the pth Laurent pulse $l_p(t)$, and $B_{p,k}^{(l)}$ is the kth element of the DFT of $\mathbf{b}_{p,N}^{(l)} \triangleq [b_{p,Ncp}^{(l)}, b_{p,Ncp+1}^{(l)}, \ldots, b_{p,N_T-1}^{(l)}]^T$, that is, of the vector $\mathbf{B}_{p,N}^{(l)} \triangleq \mathbf{Q}_N \mathbf{b}_{p,N}^{(l)}$ (here $b_{i,n}^{(l)}$ is the ith Laurent symbol in the nth interval of the lth data block) and $w(t)$ is complex AWGN with two-sided spectral density $2N_0$.

If the useful signal component of $y(t)$ in (4.117) is roughly bandlimited to $B = 1/T_s$ Hertz (i.e., $L_{i,k} = 0$ for $|k| > N$ in (3.199) and (4.117)), the received signal $r(t)$ can be passed through an ideal low-pass filter, having bandwidth $1/T_s$ and gain $\sqrt{T_s/(2E_s)}$, without any information loss (this filter aims to avoid aliasing effects and limit the input noise power). This results in a bandlimited random process $r(t)$, which can be sampled uniformly at a rate $2/T_s$ to generate a set of sufficient statistics. In particular, when detecting the lth data block, $r(t)$ is sampled at the instants $t_i^{(l)} \triangleq N_P T_s + lN_T T_s + iT_s/2$, with $i = 0, 1, \ldots, 2N - 1$ (the samples associated with the cyclic prefix are discarded). This yields a $2N$-dimensional vector $\mathbf{r}^{(l)} \triangleq [r_0^{(l)}, r_1^{(l)}, \ldots, r_{2N-1}^{(l)}]^T$, with:

$$r_i^{(l)} \triangleq r(t_i^{(l)}) = \frac{1}{\sqrt{2N}} \sum_{p=0}^{P-1} \sum_{k=-N}^{N} L_{p,k} H_k^{(l)} B_{p,k}^{(l)} W_{2N}^{-ki} + n_i^{(l)} \qquad (4.118)$$

and $i = 0, 1, \ldots, 2N - 1$. Here, the noise samples $\{n_i^{(l)} \triangleq n(t_i^{(l)})\}$ are iid complex random variables having zero mean and variance $\sigma_n^2 = 2N_0/E_s$, and $n(t)$ denotes the low-pass filter response to the AWGN process $w(t)$. If frequency-domain equalization is used, $\mathbf{r}^{(l)}$ undergoes DFT processing. In particular, in this case, the vector $\mathbf{r}^{(l)}$ feeds a $2N$-point DFT, yielding the vector $\mathbf{R}_{2N}^{(l)} = [Z_0^{(l)}, Z_1^{(l)}, \ldots, Z_{2N-1}^{(l)}]^T \triangleq \mathbf{Q}_{2N} \mathbf{r}^{(l)}$. Noting that $B_{i,k+N}^{(l)} = B_{i,k-N}^{(l)}$ for any k because of the periodicity of the DFT, it can be shown that:

$$\mathbf{R}_{2N}^{(l)} = \mathbf{M}^{(l)} \mathbf{B}_{2N}^{(l)} + \mathbf{V}_{2N}^{(l)}, \qquad (4.119)$$

where $\mathbf{B}_{2N}^{(l)} \triangleq [(\mathbf{B}_{0,2N}^{(l)})^T, (\mathbf{B}_{1,2N}^{(l)})^T, \ldots, (\mathbf{B}_{P-1,2N}^{(l)})^T]^T$, $\mathbf{B}_{i,2N}^{(l)} \triangleq [(\mathbf{B}_{i,N}^{(l)})^T, (\mathbf{B}_{i,N}^{(l)})^T]^T$, $\mathbf{V}_{2N}^{(l)} \triangleq [V_0^{(l)}, V_1^{(l)}, \ldots, V_{2N-1}^{(l)}]^T$, $\mathbf{M}^{(l)} \triangleq [\mathbf{M}_0^{(l)}, \mathbf{M}_1^{(l)}, \ldots, \mathbf{M}_{P-1}^{(l)}]$, $\mathbf{M}_i^{(l)} \triangleq \mathrm{diag}(M_{i,k}^{(l)})$ with $M_{i,k}^{(l)} \triangleq L_{i,k} H_k^{(l)}$ $(L_{i,k-2N} H_{k-2N}^{(l)})$ for $k = 0, 1, \ldots, N$ $(k = N + 1, N + 2, \ldots, 2N - 1)$. It is useful to note that the set of random variables $\{V_k^{(l)}\}$ is statistically equivalent to $\{n_i^{(l)}\}$. Finally, we note that the frequency-domain vector $\mathbf{R}_{2N}^{(l)}$ can be processed by several different equalization algorithms, which typically operate on a block-by-block basis and for each block neglect data decisions coming from previous blocks.

The strategy proposed above for extracting a set of sufficient statistics is independent of the inner structure of $h(t)$ and requires sampling at a frequency $2/T_s$. A conceptually similar approach is taken in [441], where the use of Laurent's decomposition and of the same sampling rate at the receive side is proposed. If the channel can be modeled as a T_s-spaced TDL, a set of sufficient statistics can be extracted using a bank of MFs (whose outputs are sampled once per symbol interval), exactly as in the AWGN case, as shown in [306], where MF design is based on a Gram–Schmidt (orthogonal) representation (exploiting a tilted-phase model [302]; see Section 3.6.5.2) or on Laurent's representation of CPM signals. Moreover, the use of low-pass filtering and a sampling rate at $1/T_s$ is proposed in [439, 440, 442].

Finally, it is interesting to note that the problem of ML sequence detection of CPM signals over doubly selective fading channels has been investigated in [438], where time-domain processing is adopted, as already mentioned above.

4.4.3.2 Models for a MIMO Scenario

The intrinsic memory of CPM signals makes the description of the transmitted and received signals in an $n_T \times n_R$ MIMO system very complicated. In particular, if $(\theta_{i,n}, \sigma_{i,n})$ denotes the state of the modulator feeding the ith transmit antenna in the interval $[nT_s, (n+1)T_s)$ (here $\theta_{i,n}$ and $\sigma_{i,n}$

denote the phase state and the correlative state of a partial-response modulator characterized by a correlation length L; see Section 3.6.3), the CPM transmitter of a MIMO communication system can be modeled as an FSSM. The state (i.e., *superstate*) of the *supertrellis* representing this FSSM results from the ordered concatenation of the pairs $\{(\theta_{i,n}, \sigma_{i,n}), \quad i = 0, 1, \ldots, N_T - 1\}$ [443]. The complexity of the representation for the useful component of the received signal depends on the memory and on the rapidity of variation of the communication channel. Hitherto, the use of CPM signals in MIMO communications has mainly been investigated for *block-coded transmissions over frequency-flat fading channels only* [443–447]. To simplify the derivation of detection algorithms, in this scenario the channel has been assumed static over the entire block, but changing from block to block (quasi-static fading) [443–447] or static at least over each symbol interval [445]. If the fading is quasi-static, sufficient statistics can be extracted from the received signal using the techniques adopted for the AWGN channel; for instance, the use of matched filtering based on the conventional representation of CPM signals, on Laurent's representation and on a reduced set of nonorthogonal basis functions is proposed in [444, 447] and [445], respectively.

4.4.4   Received Vector for OFDM Signaling

Like any other multicarrier modulation, *orthogonal frequency division multiplexing* is a natural choice for data communications over severely frequency-selective channels, but its use does not make any sense for purely time-selective fading channels (where SC modulations should always be used). This modulation format can be still used over doubly-selective channels, albeit at the price of significant complexity increase in the receiver if the channel variations over the transmission time of each OFDM symbol are not negligible. In the following, we consider both SISO and MIMO scenarios and discuss the problem of extracting a set of decision statistics in the presence of frequency-selective and doubly-selective fading. Ideal time synchronization is assumed throughout this subsection.

4.4.4.1   Models for a SISO scenario

Frequency-Selective Channel
In Section 3.7.2 it was shown that, for an OFDM signal transmitted over a frequency-selective channel (whose impulse response is $h(t)$), the useful component of the signal at the output of the receive filter (supposed to be identical to the transmit filter), can be expressed as (see (3.267)):

$$z(t, \mathbf{c}) = \frac{1}{\sqrt{N}} \sum_{n=-N_\alpha}^{N_\alpha} c_{R_N[n]}^{(l)} H_{R_N[n]} \exp[j 2\pi f_n (t - l N_T T_s)] \tag{4.120}$$

for $t \in [l N_T T_s, l N_T T_s + N T_s)$, where l is the index of the OFDM symbol, N is the IDFT order, $N_\alpha \triangleq \lfloor N(1 - \alpha)/2 \rfloor$, $H_n \triangleq H(f_n)$, $H(f)$ is the FCT of the CIR $h(t)$, $f_n \triangleq n/N T_s$, $N_T \triangleq N + N_p$ and N_p is the length of the cyclic prefix. Note that this signal conveys the lth frequency-domain data block $\mathbf{c}_N^{(l)} \triangleq [c_0^{(l)}, c_1^{(l)}, \ldots, c_{N_\alpha}^{(l)}, 0, \ldots, 0, c_{N-N_\alpha}^{(l)}, c_{N-N_\alpha+1}^{(l)}, \ldots, c_{N-1}^{(l)}]^T$, containing $N_u = 2N_\alpha + 1$ useful (i.e., information-bearing) channel symbols, and that this model holds if N_p is not shorter than the overall channel memory expressed in symbol intervals.

Front-end processing in an OFDM digital receiver is accomplished on a symbol-by-symbol basis, as shown by the block diagram of Figure 4.11. In particular, during the reception of the lth OFDM symbol, the received signal $r(t) = z(t, \mathbf{c}) + n(t)$ ($n(t)$ being AGN) is sampled at the instants $t_k^{(l)} \triangleq l N_T T_s + k T_s$, with $k = -N_p, -N_p + 1, \ldots, N - 1$. This yields a set of N_T samples with the kth sample given by:

$$r^{(l)}[k] \triangleq r(t_k^{(l)}) = z^{(l)}[k] + n^{(l)}[k], \tag{4.121}$$

with $z^{(l)}[k] \triangleq z(t_k^{(l)}, \mathbf{c})$ and $n^{(l)}[k] \triangleq n(t_k^{(l)})$. It is easy to show that the sequence $\{n^{(l)}[k]\}$ consists of iid complex Gaussian variables, each having zero mean and variance $\sigma_n^2 = 2N_0/K_c^2$. The first N_p samples of the set $\{r^{(l)}[k], k = -N_p, -N_p + 1, \ldots, N - 1\}$ are discarded since they are affected

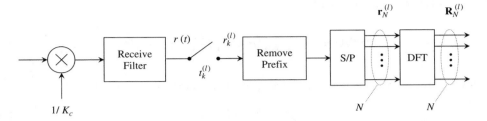

Figure 4.11   Front-end processing in an OFDM receiver.

by the IBI originating from the $(l-1)$th block, whereas the remaining samples, associated with $k = 0, 1, \ldots, N - 1$, are collected in the vector:

$$\mathbf{r}_N^{(l)} \triangleq [r^{(l)}[0], r^{(l)}[1], \ldots, r^{(l)}[N - 1]]^T. \tag{4.122}$$

Note that discarding a portion of the available samples inevitably entails an information loss, making intrinsically suboptimal any detection procedure for $\mathbf{c}_N^{(l)}$ based on $\mathbf{r}_N^{(l)}$. At the same time, however, this approach substantially simplifies the detection/equalization problem, as illustrated in Section 6.3.

To understand the signal processing chain illustrated in Figure 4.11, it is important to study in detail the structure of the vector $\mathbf{r}_N^{(l)}$ (4.122). To do this, let us rewrite the useful component $z^{(l)}[k]$ of $r^{(l)}[k]$ (4.121) as (see (4.120)):

$$z^{(l)}[k] = z(lN_T T_s + kT_s)$$

$$= \frac{1}{\sqrt{N}} \sum_{n=-N_\alpha}^{N_\alpha} c_{R_N[n]}^{(l)} H_{R_N[n]} \exp\left(j\frac{2\pi kn}{N}\right)$$

$$= \frac{1}{\sqrt{N}} \sum_{n=0}^{N_\alpha} c_{R_N[n]}^{(l)} H_{R_N[n]} \exp\left(j\frac{2\pi kn}{N}\right)$$

$$+ \frac{1}{\sqrt{N}} \sum_{n=-N_\alpha}^{-1} c_{R_N[n]}^{(l)} H_{R_N[n]} \exp\left(j\frac{2\pi kn}{N}\right) \tag{4.123}$$

for $k = 0, 1, \ldots, N - 1$. Since $c_{R_N[n]}^{(l)} H_{R_N[n]} = c_{n+N}^{(l)} H_{n+N}$ for $n = -N_\alpha, -N_\alpha + 1, \ldots, -1$ and $\exp(j2\pi kn/N) = \exp(j2\pi k(n + N)/N)$, the second term on the RHS can be rewritten as:

$$\frac{1}{\sqrt{N}} \sum_{n=-N_\alpha}^{-1} c_{R_N[n]}^{(l)} H_{R_N[n]} \exp\left(j\frac{2\pi kn}{N}\right) = \frac{1}{\sqrt{N}} \sum_{n=-N_\alpha}^{-1} c_{n+N}^{(l)} H_{n+N} \exp\left(j\frac{2\pi k(n + N)}{N}\right)$$

$$= \frac{1}{\sqrt{N}} \sum_{n=N-N_\alpha}^{N-1} c_n^{(l)} H_{n-N} \exp\left(j\frac{2\pi kn}{N}\right). \tag{4.124}$$

Since $c_{R_N[n]}^{(l)} H_{R_N[n]} = c_n^{(l)} H_n$ for $n = 0, 1, \ldots, N_\alpha$ and $c_n^{(l)} = 0$ for $n = N_\alpha + 1, N_\alpha + 2, \ldots, N - N_\alpha - 1$ (whatever the value of l), from (4.123) it is easily seen that (4.124) can be put in the form:

$$z^{(l)}[k] = \frac{1}{\sqrt{N}} \sum_{n=0}^{N_\alpha} c_n^{(l)} H_n \exp\left(j\frac{2\pi kn}{N}\right) + \frac{1}{\sqrt{N}} \sum_{n=N-N_\alpha}^{N-1} c_n^{(l)} H_n \exp\left(j\frac{2\pi kn}{N}\right)$$

$$= \frac{1}{\sqrt{N}} \sum_{n=0}^{N-1} c_n^{(l)} H_n \exp\left(j\frac{2\pi kn}{N}\right). \tag{4.125}$$

Then, substituting (4.125) into (4.121) yields:

$$r^{(l)}[k] = \frac{1}{\sqrt{N}} \sum_{n=0}^{N-1} c_n^{(l)} H_n \exp\left(j\frac{2\pi k n}{N}\right) + n^{(l)}[k] \tag{4.126}$$

with $k = 0, 1, \ldots, N-1$, so that the vector $\mathbf{r}_N^{(l)}$ in (4.122) can be written in the form:

$$\mathbf{r}_N^{(l)} = \mathbf{Q}_N^H \, \mathbf{H} \, \mathbf{c}_N^{(l)} + \mathbf{n}_N^{(l)}, \tag{4.127}$$

where $\mathbf{n}_N^{(l)} \triangleq [n^{(l)}[0], n^{(l)}[1], \ldots, n^{(l)}[N-1]]^T$ is a noise vector, $\mathbf{H} \triangleq \mathrm{diag}(H_n)$ is a diagonal matrix having the channel gains $\{H_n, n = 0, 1, \ldots, N-1\}$ along its main diagonal and $\mathbf{Q}_N$ is the Nth-order DFT matrix. The time-domain vector $\mathbf{r}_N^{(l)}$ in (4.127) undergoes an Nth-order DFT yielding the frequency-domain vector:

$$\begin{aligned} \mathbf{R}_N^{(l)} &= [R^{(l)}[0], R^{(l)}[1], \ldots, R^{(l)}[N-1]]^T \triangleq \mathbf{Q}_N \, \mathbf{r}_N^{(l)} \\ &= \mathbf{Q}_N \, \mathbf{Q}_N^H \, \mathbf{H} \, \mathbf{c}_N^{(l)} + \mathbf{Q}_N \, \mathbf{w}_N^{(l)} = \mathbf{H} \, \mathbf{c}_N^{(l)} + \mathbf{N}_N^{(l)}, \end{aligned} \tag{4.128}$$

where $\mathbf{N}_N^{(l)} = [N^{(l)}[0], N^{(l)}[1], \ldots, N^{(l)}[N-1]]^T \triangleq \mathbf{Q}_N \mathbf{w}_N^{(l)}$ is a noise vector statistically equivalent to $\mathbf{n}_N^{(l)}$ (so that its components are statistically independent). Note that the structure of $\mathbf{R}_N^{(l)}$ is substantially simpler than that of $\mathbf{r}_N^{(l)}$. In fact, from (4.128) it can be easily seen that:

$$R^{(l)}[k] = H_k c_k^{(l)} + W_k^{(l)} \tag{4.129}$$

for $k = 0, 1, \ldots, N_\alpha$ and $k = N - N_\alpha, N - N_\alpha + 1, \ldots, N - 1$, and:

$$R^{(l)}[k] = W_k^{(l)} \tag{4.130}$$

for $k = N_\alpha + 1, N_\alpha + 2, \ldots, N - N_\alpha - 1$. This shows that each of the frequency-domain samples (4.129) depends only on a single channel symbol and that the noise affecting it is independent of that disturbing all the other subcarriers, so that detection and equalization can be accomplished in a subcarrier-by-subcarrier fashion.

Doubly-Selective Channel

If the channel varies during the transmission of each OFDM symbol, the signal, $z(t, \mathbf{c})$, representing the useful component of the signal at the output of the receive filter, can be put in a form similar to (4.120). In fact, it is easy to show that:

$$z(t, \mathbf{c}) = \frac{1}{\sqrt{N}} \sum_{n=-N_\alpha}^{N_\alpha} c_{R_N[n]}^{(l)} H_{R_N[n]}(t) \, \exp[j 2\pi f_n (t - l N_T T_s)] \tag{4.131}$$

for $t \in [l N_T T_s, l N_T T_s + N T_s)$, where $H_{R_N[n]}(t)$ denotes the TVTF $H(t, f)$ (see (2.13)) evaluated for $f = f_n$ (the nth subcarrier frequency). Then, sampling the received signal $r(t, \mathbf{c}) = z(t, \mathbf{c}) + w(t)$ at the instants $t_k^{(l)} \triangleq l N_T T_s + k T_s$, with $k = -N_p, -N_p + 1, \ldots, N - 1$, yields the set of samples of $\{r^{(l)}[k], k = -N_p, -N_p + 1, \ldots, N - 1\}$. The structure of $r^{(l)}[k]$ is still described by (4.121), where, however:

$$z^{(l)}[k] \triangleq z(t_k^{(l)}, \mathbf{c}) = \frac{1}{\sqrt{N}} \sum_{n=-N_\alpha}^{N_\alpha} c_{R_N[n]}^{(l)} H_{R_N[n]}[l, k] \exp\left(j\frac{2\pi k n}{N}\right), \tag{4.132}$$

with $H_{R_N[n]}[l, k] \triangleq H_{R_N[n]}(lN_T T_s + kT_s)$. Discarding the first N_p samples of the set $\{r^{(l)}[k]\}$ to remove the IBI and following the same line of reasoning as for the frequency-selective channel case leads to:

$$r^{(l)}[k] = \frac{1}{\sqrt{N}} \sum_{n=0}^{N-1} c_n^{(l)} H_n[l, k] \exp\left(j\frac{2\pi kn}{N}\right) + w^{(l)}[k], \tag{4.133}$$

with $k = 0, 1, \ldots, N - 1$. This can be rewritten in the compact form:

$$r^{(l)}[k] = \mathbf{q}_{N,k}^H \mathbf{H}_k^{(l)} \mathbf{c}_N^{(l)} + w^{(l)}[k], \tag{4.134}$$

where $\mathbf{q}_{N,k}$ is the kth column of $\mathbf{Q}_N$, $\mathbf{H}_k^{(l)} \triangleq \mathrm{diag}(H_n[l, k])$ is a diagonal having the subchannel gains $\{H_n[l, k], \quad n = 0, 1, \ldots, N - 1\}$ sampled at the instant $lN_T T_s + kT_s$ along its main diagonal and $\mathbf{c}_N^{(l)}$ is the lth frequency-domain data block. Note that (4.133) generalizes (4.126) but, despite this similarity, now the received signal vector $\mathbf{r}_N^{(l)}$ (4.122) cannot be put in the form (4.127); therefore, the simple model (4.129)–(4.130) for the elements of $\mathbf{R}_N^{(l)} \triangleq \mathbf{Q}_N \mathbf{r}_N^{(l)}$ no longer holds. In fact, we have that:

$$R^{(l)}[k] = \mathbf{q}_{N,k}^T \mathbf{r}_N^{(l)} = \frac{1}{\sqrt{N}} \sum_{p=0}^{N-1} \exp\left(j\frac{2\pi kp}{N}\right) r^{(l)}[p] + N^{(l)}[k] \tag{4.135}$$

for $k = 0, 1, \ldots, N_\alpha$ and $k = N - N_\alpha, N - N_\alpha + 1, \ldots, N - 1$, where $N^{(l)}[k]$ is the kth element of the noise vector $\mathbf{N}_N^{(l)}$ defined after (4.128). Then substituting (4.134) into (4.135) yields:

$$R^{(l)}[k] = \frac{1}{\sqrt{N}} \left[\sum_{p=0}^{N-1} \exp\left(j\frac{2\pi kp}{N}\right) \mathbf{q}_{N,p}^H \mathbf{H}_p^{(l)} \right] \mathbf{c}_N^{(l)} + N^{(l)}[k], \tag{4.136}$$

which can be rewritten as:

$$R^{(l)}[k] = \tilde{H}^{(l)}[k, k] c_k^{(l)} + \sum_{\substack{q=0 \\ q \neq k}}^{N-1} \tilde{H}^{(l)}[k, q] c_n^{(l)} + W_N^{(l)}[k], \tag{4.137}$$

where:

$$\tilde{H}^{(l)}[k, q] = \frac{1}{N} \sum_{p=0}^{N-1} \exp\left(j\frac{2\pi(k-q)p}{N}\right) H_q[l, p]. \tag{4.138}$$

Note that:

$$\tilde{H}^{(l)}[k, k] = \frac{1}{N} \sum_{p=0}^{N-1} H_k[l, p], \tag{4.139}$$

so that the channel gain associated with $c_k^{(l)}$ in (4.137) represents an average gain evaluated over the entire observation interval.

Expression (4.137) shows that $R^{(l)}[k]$ depends on both the kth channel symbol $c_k^{(l)}$ and, in principle, on all the other subcarrier symbols; the latter contribution is known as ICI and is represented by the second term on the RHS of (4.137). Finally, we note the following:

(a) ICI originating from time selectivity in OFDM transmissions produces similar effects as ISI produced by the frequency selectivity of the communication channel in an SC transmission [391].

(b) The strongest contribution to the ICI affecting each subcarrier usually originates mainly from neighboring subcarriers [448, 449].

(c) ICI models have been derived in [448] and [450–455] under the assumption that the communication channel is described by a linearly time-varying filter and by a TDL model with time-varying tap gains, respectively.

4.4.4.2 Models for a MIMO Scenario

MIMO models can be easily derived from their SISO counterparts, keeping in mind that ot the receive side front-end processing is accomplished on an antenna-by-antenna basis (see Figure 4.11). Then, if we consider an $n_T \times n_R$ frequency-selective MIMO channel, the frequency-domain vector $\mathbf{R}_N^{(l)}[j] = [R_j^{(l)}[0], R_j^{(l)}[1], \ldots, R_j^{(l)}[N-1]]^T$ (see (4.128)) deriving from the signal captured by the jth receive antenna (with $j = 0, 1, \ldots, n_R - 1$) derives from the superposition of the contributions coming from the n_T transmit antennas [391]. In particular, we have that:

$$\mathbf{R}_N^{(l)}[j] = \sum_{i=0}^{n_T - 1} \mathbf{H}[j, i]\, \mathbf{c}_N^{(l)}[i] + \mathbf{N}_N^{(l)}[j], \tag{4.140}$$

where $\mathbf{c}_N^{(l)}[i] = [c_{i,0}^{(l)}, c_{i,1}^{(l)}, \ldots, c_{i,N_\alpha}^{(l)}, 0, \ldots, 0, c_{i,N-N_\alpha}^{(l)}, c_{i,N-N_\alpha+1}^{(l)}, \ldots, c_{i,N-1}^{(l)}]^T$ denotes the OFDM symbol transmitted by the ith antenna (with $i = 0, 1, \ldots, n_T - 1$), $\mathbf{H}[j, i] = \mathrm{diag}(H_n(i, j))$ is an $N \times N$ matrix collecting the channel gains between the ith transmit antenna and the jth receive antenna (in particular, $H_n(i, j)$ denotes the channel gain on the nth subcarrier frequency) and $\mathbf{N}_{N,j}^{(l)} = [N_j^{(l)}[0], N_j^{(l)}[1], \ldots, N_j^{(l)}[N-1]]^T$ is the noise vector affecting the jth receive antenna and consisting of iid complex Gaussian random variables.

An analysis of the ICI originating from time variations of the communication channel in a MIMO OFDM system can be found in [456].

4.5 Decision Strategies in the Presence of Channel Parameters: Optimal Metrics and Performance Bounds

4.5.1 Signal Model and Algorithm Classification

In general, the structure of the received vector $\mathbf{r}$ processed to generate an estimate $\hat{\mathbf{c}}$ of the transmitted channel symbol vector $\mathbf{c}$ depends on the CIR and on a set of synchronization parameters that influence the received signal (e.g., residual timing and/or frequency offsets). Here, the effect of synchronization parameters is ignored for simplicity (i.e., ideal synchronization is assumed) and the problem of the detection in the presence of a wireless communication channel characterized by a given CIR vector $\mathbf{h}$ is investigated.

In developing detection algorithms in wireless communications different degrees of knowledge of the CIR at the receiver can be assumed; in particular, detection algorithms can rely on *exact* (i.e., *ideal) knowledge* of the CIR or on only statistical knowledge of it, or have no information about the CIR. Algorithms developed for a known CIR can be employed when a reasonably accurate estimate of the communication channel is available; such an estimate can be acquired by adding a header of known data to each transmitted block of information, so that CIR estimation may be made independent of data detection. Detection algorithms designed under the assumption of a statistically known channel are based on a statistical model of the wireless channel whose properties are known at the receive side. In practice, these properties must be estimated and tracked using decision-directed procedures in any time-varying scenario. Finally, algorithms for a completely *unknown* CIR involve *joint estimation* of the data and the communication channel.

In what follows the above classes of algorithms are considered. For each class, we tackle the problem of estimating the symbols of the vector $\mathbf{c}_N \triangleq [c_0, c_1, \ldots, c_{N-1}]^T$, transmitted over a channel characterized by a CIR vector $\mathbf{h}$, from a vector of received samples $\mathbf{r}$. The linear model [457]:

$$\mathbf{r} = \mathbf{H}\mathbf{c}_N + \mathbf{n} \tag{4.141}$$

is always assumed, where $\mathbf{r} \triangleq [r_0, r_1, \ldots, r_{K-1}]^T$, $\mathbf{H} \triangleq [H_{l,m}]$ is a $K \times N$ channel matrix (whose structure depends on the specific channel and which is uniquely identified by $\mathbf{h}$) and

$\mathbf{n} \triangleq [n_0, n_1, \ldots, n_{K-1}]^T$ is a noise vector consisting of zero mean complex Gaussian random variables, whose real and imaginary parts are independent and have the same variance σ_n^2, and whose statistical properties are completely described by its autocovariance matrix $\mathbf{R_n}$. Note the following observations:

(a) Model (4.141) is very general but, in principle, refers to the case of *linear modulations* (i.e., to the case in which the transmitted signal exhibits a linear dependence on the channel symbols, as in OFDM and PAM signaling).
(b) Despite this, the results developed for this model can easily be extended to the case of CPM modulations if Laurent's representation is used (see Section 3.6.5.1).
(c) A one-to-one mapping between the sequence of channel symbols and the transmitted information bits is always assumed, so that it is unnecessary to consider the sequence of bits separately.

Different optimality criteria can be adopted in developing a detection strategy. In particular, if the target is to minimize the decision error on the whole sequence (under the assumption of equally likely sequences), an MLSD strategy may be applied. An alternative to this is to develop strategies for minimizing the probability of error for each information bit or each channel symbol; this approach leads to MAPBD and MAPSD algorithms, respectively. In what follows the three optimality criteria are employed to derive of a set of optimal detection metrics for the received signal model (4.141). Such metrics can be exploited both to devise optimal and suboptimal detection techniques, and to derive bounds on the performance offered by such techniques.

4.5.2 Detection for Transmission over of a Known Channel

Here we develop optimal metrics under the assumption of a known $\mathbf{H}$ (i.e., deterministic $\mathbf{h}$) in (4.141) and then briefly discuss the problem of deriving performance bounds.

4.5.2.1 Metric for MLSD

Let us assume, for simplicity, that $K = N$ in model (4.141), i.e. that one sample per channel symbol is processed by the detection algorithm. To stress this, the received vector $\mathbf{r}$ (the noise vector $\mathbf{n}$) is denoted by $\mathbf{r}_N$ ($\mathbf{n}_N$) in what follows. Then the MLSD strategy can be formulated as (see (4.26)):

$$\hat{\mathbf{c}}_N = \arg \max_{\tilde{\mathbf{c}}_N} f_{\mathbf{r}}(\boldsymbol{\rho}_N | \tilde{\mathbf{c}}_N, \mathbf{h}), \qquad (4.142)$$

where $\tilde{\mathbf{c}}_N$ and $\hat{\mathbf{c}}_N$ denote a trial or hypothesized sequence and the decided sequence, respectively, and $\boldsymbol{\rho}_N \triangleq [\rho_0, \rho_1, \ldots, \rho_{N-1}]^T$ denotes a realization of $\mathbf{r}_N$. Since $\mathbf{r}_N$, conditioned on $\hat{\mathbf{c}}_N = \tilde{\mathbf{c}}_N$ and the CIR, is complex multivariate Gaussian with mean $E\{\mathbf{r}_N | \tilde{\mathbf{c}}_N, \mathbf{h}\} = \mathbf{H}\tilde{\mathbf{c}}_N$, the likelihood of $\mathbf{r}_N$ is given by:

$$f_{\mathbf{r}}(\boldsymbol{\rho}_N | \tilde{\mathbf{c}}_N, h) = \frac{1}{\det(2\pi \mathbf{R_n})} \exp\left[-\frac{1}{2}(\boldsymbol{\rho}_N - \mathbf{H}\tilde{\mathbf{c}}_N)^H \mathbf{R_n}^{-1}(\boldsymbol{\rho}_N - \mathbf{H}\tilde{\mathbf{c}}_N) \right]. \qquad (4.143)$$

Taking the logarithm of the RHS yields the sequence metric:

$$\ln f_{\mathbf{r}}(\boldsymbol{\rho}_N | \tilde{\mathbf{c}}_N, h) = -\ln \det(2\pi \mathbf{R_n}) - \frac{1}{2}(\boldsymbol{\rho}_N - \mathbf{H}\tilde{\mathbf{c}}_N)^H \mathbf{R_n}^{-1}(\boldsymbol{\rho}_N - \mathbf{H}\tilde{\mathbf{c}}_N), \qquad (4.144)$$

which is equivalent to (4.143), so that a functionally identical, but computationally more attractive, decision rule is:

$$\hat{\mathbf{c}}_N = \arg \min_{\tilde{\mathbf{c}}_N} (\boldsymbol{\rho}_N - \mathbf{H}\tilde{\mathbf{c}}_N)^H \mathbf{R_n}^{-1}(\boldsymbol{\rho}_N - \mathbf{H}\tilde{\mathbf{c}}_N), \qquad (4.145)$$

since data-independent terms and positive scalars may also be discarded under the arg max operator, and a negative scalar affects the decision rule only insofar as the arg max operator is replaced by the arg min operator. Generally speaking, the noise vector $\mathbf{n}_N$ consists of correlated random variables; however, a *white* noise model (corresponding to $\mathbf{R_n} = 2\sigma_n^2 \mathbf{I}_N$) is often assumed. When this occurs, the decision rule (4.145) can be easily simplified to:

$$\hat{\mathbf{c}}_N = \arg \min_{\tilde{\mathbf{c}}_N} (\boldsymbol{\rho}_N - \mathbf{H}\tilde{\mathbf{c}}_N)^H (\boldsymbol{\rho}_N - \mathbf{H}\tilde{\mathbf{c}}_N) = \arg \min_{\tilde{\mathbf{c}}_N} |\boldsymbol{\rho}_N - \mathbf{H}\tilde{\mathbf{c}}_N|^2, \tag{4.146}$$

which admits an interesting *geometrical interpretation*: each trial sequence $\tilde{\mathbf{c}}_N$ can be represented by the point $\mathbf{H}\tilde{\mathbf{c}}_N$ in an N-dimensional signal space. Also the received samples $\mathbf{r}_N$ define another such point. Then the detected sequence is the point is closest in the *Euclidean* sense to that representing the received samples [457].

A closely related geometrical interpretation can be formulated even in the case of *colored* noise. To show this we note that the inverse noise autocovariance matrix $\mathbf{R_n}^{-1}$ can be factored according to the *Cholesky decomposition* as [125]:

$$\mathbf{R_n}^{-1} = \mathbf{L}^H \mathbf{L}, \tag{4.147}$$

where $\mathbf{L}$ is a lower triangular matrix with real entries on its main diagonal. It can easily be shown that the linear and causal transformation[4]:

$$\mathbf{v}_N = \mathbf{L}\mathbf{n}_N \tag{4.148}$$

of the noise vector $\mathbf{n}_N$ generates a vector $\mathbf{v}_N$ consisting of uncorrelated jointly Gaussian (and, consequently, independent) random variables, Thus the transformation represents a form of *noise whitening*. Then substituting (4.147) in (4.145) yields:

$$\hat{\mathbf{c}}_N = \arg \max_{\tilde{\mathbf{c}}_N} |\mathbf{L}\boldsymbol{\rho}_N - \mathbf{L}\mathbf{H}\tilde{\mathbf{c}}_N|^2. \tag{4.149}$$

This decision rule still involves choosing the symbol sequence whose point in multidimensional space is closest in the Euclidean sense to some reference point; however, now each symbol sequence point, $\mathbf{L}\mathbf{H}\tilde{\mathbf{c}}_N$, and the reference point, $\mathbf{L}\boldsymbol{\rho}_N$, are transformed (rotated and scaled) versions of the original points.

It is also interesting to note that the metric of (4.145) can be expanded and the data-independent $\mathbf{r}_N^H \mathbf{R_n}^{-1} \mathbf{r}_N$ term can be discarded. This yields an equivalent decision rule:

$$\hat{\mathbf{c}}_N = \arg \max_{\tilde{\mathbf{c}}_N} 2\mathrm{Re}\{(\mathbf{H}\tilde{\mathbf{c}}_N)^H \mathbf{R_n}^{-1} \boldsymbol{\rho}_N\} - (\mathbf{H}\tilde{\mathbf{c}}_N)^H \mathbf{R_n}^{-1} (\mathbf{H}\tilde{\mathbf{c}}_N), \tag{4.150}$$

so that, if the sampled noise is white, a functionally equivalent decision rule is:

$$\hat{\mathbf{c}}_N = \arg \max_{\tilde{\mathbf{c}}_N} 2\mathrm{Re}\{\tilde{\mathbf{c}}_N^H \mathbf{H}^H \boldsymbol{\rho}_N\} - |\mathbf{H}\tilde{\mathbf{c}}_N|^2. \tag{4.151}$$

Some applications of these results are illustrated in the following example, where the MLSD of PAM over a slow time-selective fading channel and the dual problem of MLSD of OFDM over a frequency-selective channel are analyzed.

Example 4.5.1 The study of PAM signaling over a slow time-selective fading channel leads to the model (see (4.92)):

$$r_k = a_k c_k + w_k \tag{4.152}$$

with $k = 0, 1, \ldots, N - 1$, for the received signal samples, where a_k is the kth sample of the fading distortion, and the sequence $\{w_k\}$ consists of iid complex Gaussian noise samples, each having mean zero and variance $2N_0/K_c^2$. Then the received signal vector $\mathbf{r}_N \triangleq [r_0, r_1, \ldots, r_{N-1}]^T$ is described

---

[4] The transformation is *causal* because $\mathbf{L}$ is lower triangular.

by (4.141) with $\mathbf{H} = \mathrm{diag}(a_k)$ (note also that $\mathbf{h} \triangleq [a_0, a_1, \ldots, a_{N-1}]^T$), so that the MLSD strategy (4.146) can be simplified to:

$$\hat{\mathbf{c}}_N = \arg\min_{\check{\mathbf{c}}_N} \sum_{k=0}^{N-1} |\rho_k - a_k \, \check{c}_k|^2. \tag{4.153}$$

Note that the decision metric is given by the sum over N symbol metrics, each depending on only a single symbol. This leads to the conclusion that in this specific case the MLSD rule can be reformulated as the *symbol-by-symbol decision strategy*:

$$\hat{c}_k = \arg\min_{\check{c}_k} \, |\rho_k - a_k \, \check{c}_k|^2 \tag{4.154}$$

for $k = 0, 1, \ldots, N-1$.

A similar mathematical result is found when considering the detection of OFDM transmitted over a frequency-selective fading channel. Note, however, that the OFDM detection algorithms available in the technical literature usually discard, for each OFDM symbol, a fraction of the samples of the received signal (and, more specifically, the samples associated with the cyclic prefix) to remove IBI (see Section 4.4.4). For this reason, none of these schemes is truly optimal. Despite this, optimal (i.e., ML) detection strategies can easily be developed for the vector of frequency-domain samples collected for each transmitted OFDM symbol. In fact, the model (4.128) for the frequency-domain vector $\mathbf{R}_N^{(l)}$ has the same structure as the vector $\mathbf{r}_N$ for flat fading channels. Thus, if the dependence on the block index l is neglected for simplicity, the optimal detection strategy can then be expressed as:

$$\hat{\mathbf{c}}_N = \arg\min_{\check{\mathbf{c}}_N} |\boldsymbol{\rho}_N - \mathbf{H}\check{\mathbf{c}}_N|^2 = \arg\min_{\check{\mathbf{c}}_N} \sum_{n=0}^{N-1} |\rho_n - H_n \check{c}_n|^2. \tag{4.155}$$

where $\boldsymbol{\rho}_N$ denotes a realization of $\mathbf{R}_N$. Then, if the subchannel gains $\{H_n, \ n = 0, 1, \ldots, N_\alpha, N - N_\alpha, N - N_\alpha + 1, \ldots, N - 1\}$ are known, the optimal detection algorithm is given by:

$$\hat{c}_n = \arg\min_{\check{c}_n} |\rho_n - H_n \check{c}_n|^2 \tag{4.156}$$

with $n = 0, 1, \ldots, N_\alpha$ and $n = N - N_\alpha, N - N_\alpha + 1, \ldots, N - 1$. In other words, the vector problem of (4.155) can be decomposed into a set of N_u scalar problems.

□

The MLSD metrics described in Example 4.5.1 are structurally simple in that, in both cases considered, a sequence metric can be written as the sum of a set of symbol metrics. This does not occur, however, in situations where:

(a) CPM modulation is used, because of the inner memory of the modulation [438].
(b) PAM is sent over a frequency-selective [422, 458] or a doubly-selective [425] fading channel, since ISI is then unavoidable.
(c) OFDM is used over a doubly-selective channel, since time variations generate ICI (see Section 4.4.4.1).

In all these cases, even if the MLSD metric has an additive structure (see the RHS of (4.146) and (4.149)), each term of the overall metric involves multiple channel symbols and different terms can depend on nondisjoint subsets of elements of $\mathbf{c}_N$.

The results derived can easily be extended to a MIMO scenario, where, however, the adoption of an MLSD approach entails significant computational complexity, so that approximate strategies are usually employed (e.g., see [459, 460]).

4.5.2.2   Bounds on the Error Performance of MLSD

Performance bounds for MLSD usually result from the application of the so-called *union bound* [422] (see Section 4.3.2), as shown in what follows under the assumption that the decision strategy looks for the *maximum* of a sequence metric $\Lambda(\tilde{\mathbf{c}}_N)$ over the set $S_{\mathbf{c}}$ containing all possible trial vectors $\tilde{\mathbf{c}}_N$. To begin, we note that if the symbol vector $\mathbf{c}_N$, consisting of M-ary complex symbols, is transmitted, an *incorrect* sequence $\hat{\mathbf{c}}_N$ is deemed more likely than $\mathbf{c}_N$ if $\Lambda(\hat{\mathbf{c}}_N) \geq \Lambda(\mathbf{c}_N)$ or, equivalently, if $\Lambda(\hat{\mathbf{c}}_N) - \Lambda(\mathbf{c}_N) \geq 0$ (this event, referring to a pair of distinct symbol sequences, is denoted $\varepsilon(\mathbf{c}_N \rightarrow \hat{\mathbf{c}}_N)$ in what follows). Note that, if the decision metric (see (4.146)):

$$\Lambda(\mathbf{c}_N) = (\mathbf{Hc}_N)^H \mathbf{r}_N + \mathbf{r}_N^H (\mathbf{Hc}_N) - ||\mathbf{Hc}_N||^2 \tag{4.157}$$

is adopted and the *error vector* $\mathbf{e}_N \triangleq \hat{\mathbf{c}}_N - \mathbf{c}_N$ is defined, we have that:

$$\begin{aligned}
\Delta(\mathbf{c}_N, \hat{\mathbf{c}}_N) &\triangleq \Lambda(\hat{\mathbf{c}}_N) - \Lambda(\mathbf{c}_N) \\
&= \Lambda(\mathbf{c}_N + \mathbf{e}_N) - \Lambda(\mathbf{c}_N) \\
&= \mathbf{c}_N^H \, \mathbf{H}^H \mathbf{H} \, \mathbf{e}_N + \mathbf{e}_N^H \, \mathbf{H}^H \mathbf{H} \, \mathbf{c}_N - \mathbf{e}_N^H \mathbf{H}^H \mathbf{r}_N - \mathbf{r}_N^H \, \mathbf{H} \, \mathbf{e}_N.
\end{aligned} \tag{4.158}$$

Since $\mathbf{r}_N = \mathbf{Hc}_N + \mathbf{n}_N$ (see (4.141)), expression (4.158) can be simplified to:

$$\Delta(\mathbf{c}_N, \hat{\mathbf{c}}_N) = -\mathbf{e}_N^H \, \mathbf{H}^H \mathbf{n}_N - \mathbf{n}_N^H \mathbf{He}_N. \tag{4.159}$$

The latter result clearly shows that the metric difference $\Delta(\mathbf{c}_N, \hat{\mathbf{c}}_N)$ depends on the error vector $\mathbf{e}_N$, but it is *independent of the transmitted sequence* $\mathbf{c}_N$ and, for this reason, can be denoted simply $\Delta(\mathbf{e}_N)$. Then the probability:

$$\Pr\{\varepsilon(\mathbf{c}_N \rightarrow \hat{\mathbf{c}}_N)\} \triangleq \Pr\{\Delta(\mathbf{c}_N, \hat{\mathbf{c}}_N) \geq 0 | \mathbf{c}_N\} = \Pr\{\Delta(\mathbf{e}_N) \geq 0\} \tag{4.160}$$

of the event $\varepsilon(\mathbf{c}_N \rightarrow \hat{\mathbf{c}}_N)$, know as the PEP in what follows, depends only on $\mathbf{e}_N$.

The probability that, given the transmitted vector $\mathbf{c}_N$, an incorrect message is selected by the ML detector is then given by:

$$\Pr\{\varepsilon | \mathbf{c}_N\} = \Pr\left\{ \bigcup_{\substack{\hat{\mathbf{c}}_N \in S_{\mathbf{c}} \\ \hat{\mathbf{c}}_N \neq \mathbf{c}_N}} \{\Delta(\mathbf{c}_N, \hat{\mathbf{c}}_N) \geq 0 | \mathbf{c}_N\} \right\}. \tag{4.161}$$

Applying the union bound (see (4.17)) to the RHS of (4.161) leads to the upper bound:

$$\begin{aligned}
\Pr\{\varepsilon | \mathbf{c}_N\} &\leq \sum_{\substack{\hat{\mathbf{c}}_N \in S_{\mathbf{c}} \\ \hat{\mathbf{c}}_N \neq \mathbf{c}_N}} \Pr\{\Delta(\mathbf{c}_N, \hat{\mathbf{c}}_N) \geq 0 | \mathbf{c}_N\} \\
&= \sum_{\mathbf{e}_N \in \Omega(\mathbf{c}_N)} \Pr\{\Delta(\mathbf{e}_N) \geq 0\},
\end{aligned} \tag{4.162}$$

where $\Omega(\mathbf{c}_N)$ denotes the set of error vectors $\mathbf{e}_N = \hat{\mathbf{c}}_N - \mathbf{c}_N$ generated, for a given $\mathbf{c}_N$, by all possible $\hat{\mathbf{c}}_N \in S_{\mathbf{c}}$ with $\hat{\mathbf{c}}_N \neq \mathbf{c}_N$ (note that $\Omega(\mathbf{c}_N)$ does not contain $\mathbf{0}_N$). The average error probability over the entire vector $\mathbf{c}_N$ (i.e., on the transmitted message) is then given by:

$$P_e = \Pr\{\hat{\mathbf{c}}_N \neq \mathbf{c}_N\} = \sum_{\mathbf{c}_N \in S_{\mathbf{c}}} \Pr\{\mathbf{c}_N\} \Pr\{\varepsilon | \mathbf{c}_N\}, \tag{4.163}$$

which can be upper-bounded as (see (4.162)):

$$P_e \leq \sum_{\mathbf{c}_N \in S_\mathbf{c}} \Pr\{\mathbf{c}_N\} \sum_{\substack{\mathbf{e}_N \in \Omega(\mathbf{c}_N) \\ \mathbf{e}_N \neq \mathbf{0}_N}} \Pr\{\Delta(\mathbf{e}_N) \geq 0\}. \tag{4.164}$$

Let us now assume that $S_\mathbf{c}$ consists of equally likely vectors, so that $\Pr\{\mathbf{c}_N\} = 1/M^N$. Then (4.164) simplifies to:

$$P_e \leq \frac{1}{M^N} \sum_{\mathbf{c}_N \in S_\mathbf{c}} \sum_{\mathbf{e}_N \in \Omega(\mathbf{c}_N)} \Pr\{\Delta(\mathbf{e}_N) \geq 0\}. \tag{4.165}$$

If $N_e(\mathbf{e}_N)$ denotes the number of different ways in which a given error vector $\mathbf{e}_N$ can be generated as $\mathbf{e}_N = \hat{\mathbf{c}}_N - \mathbf{c}_N$ by making distinct choices for the pair $(\hat{\mathbf{c}}_N, \mathbf{c}_N)$, then (4.165) can be rewritten as:

$$P_e \leq \sum_{\mathbf{e}_N \in \Omega} \frac{N_e(\mathbf{e}_N)}{M^N} \Pr\{\Delta(\mathbf{e}_N) \geq 0\}, \tag{4.166}$$

where $\Omega = \bigcup_{\mathbf{c}_N \in S_\mathbf{c}} \Omega(\mathbf{c}_N)$ collects all the possible (and distinct) error vectors different from $\mathbf{0}_N$. This last result represents the desired *union bound*. We make the following observations on the bound:

(a) In any significant scenario, for large SNRs a small number of terms is expected to provide most of the contribution to the RHS of (4.166).
(b) If the elements of $\mathbf{c}_N$ are independent, $N(\mathbf{e}_N)$ can be factored as:

$$N_e(\mathbf{e}_N) = \prod_{l=0}^{N-1} N_e(e_l), \tag{4.167}$$

where $N_e(e_l)$ (with $l = 0, 1, \ldots, N-1$) represents the number of different ways in which a given error e_l can be generated as $e_l = \hat{c}_l - c_l$ making distinct choices for the couple $(\hat{c}_l, c_l)$.
(c) From (4.17) and (4.161) the *lower bound*:

$$\Pr\{\varepsilon|\mathbf{c}_N\} \geq \max_{\substack{\hat{\mathbf{c}}_N \in S_\mathbf{c} \\ \hat{\mathbf{c}}_N \neq \mathbf{c}_N}} \Pr\{\Delta(\mathbf{c}_N, \hat{\mathbf{c}}_N) \geq 0|\mathbf{c}_N\}$$

$$= \max_{\mathbf{e}_N \in \Omega(\mathbf{c}_N)} \Pr\{\Delta(\mathbf{e}_N) \geq 0\} \tag{4.168}$$

can also be easily derived.

Then if all possible transmitted symbol vectors are equally likely, the lower bound (see (4.163)):

$$P_e \geq \frac{1}{M^N} \sum_{\tilde{\mathbf{c}}_N \in S_\mathbf{c}} \max_{\mathbf{e}_N \in \Omega(\mathbf{c}_N)} \Pr\{\Delta(\mathbf{e}_N) \geq 0\} \tag{4.169}$$

is found. Finally, it is worth noting that, following a similar line of reasoning, the upper bound [429]:

$$P_s \leq \frac{1}{N} \sum_{\mathbf{e}_N \in \Omega} \frac{N(\mathbf{e}_N)}{M^N} w_H(\mathbf{e}_N) \Pr\{\Delta(\mathbf{e}_N) \geq 0\} \tag{4.170}$$

for the symbol error probability can easily be found and a similar expression also derived for the bit error probability; here $w_H(\mathbf{x})$ denotes the *Hamming weight* of the vector $\mathbf{x}$, that is, the overall number of its elements different from zero. Further discussion of error performance bounds for MLSD of PAM signals over frequency- and doubly-selective fading channels can be found in [422, 429, 461–464] and in [425], respectively.

4.5.2.3   Metric for MAP (Symbol and Bit) Detection

If ρ_N denotes the value of the observed vector $\mathbf{r}_N$, the MAPBD and MAPSD decision rules can be expressed as:

$$\hat{b}_l = \arg\max_{\tilde{b}_l}\ \Pr\{b_l = \tilde{b}_l | \mathbf{r}_N = \rho_N, \mathbf{h}\} \qquad (4.171)$$

and:

$$\hat{c}_m = \arg\max_{\tilde{c}_m}\ \Pr\{c_m = \tilde{c}_m | \mathbf{r}_N = \rho_N, \mathbf{h}\}, \qquad (4.172)$$

respectively, if b_l (c_m) denotes the lth $(m$th) transmitted bit (channel symbol). Exploiting the *theorem of total probability* [55], the metrics characterizing these strategies can easily be rewritten as:

$$\Pr\{b_l = \tilde{b}_l | \mathbf{r}_N = \rho_N, \mathbf{h}\} = \sum_{\tilde{\mathbf{c}}_N \to \tilde{b}_i} \Pr\{\tilde{\mathbf{c}}_N | \mathbf{r}_N = \rho_N, \mathbf{h}\}$$

$$= \sum_{\tilde{\mathbf{c}}_N \to \tilde{b}_i} \frac{f_{\mathbf{r}}(\rho_N | \tilde{\mathbf{c}}_N, \mathbf{h}) \Pr\{\mathbf{c}_N = \tilde{\mathbf{c}}_N\}}{f_{\mathbf{r}}(\rho_N | \mathbf{h})}, \qquad (4.173)$$

and

$$\Pr\{c_m = \tilde{c}_m | \mathbf{r}_N = \rho_N, \mathbf{h}\} = \sum_{\tilde{\mathbf{c}}_N \to \tilde{c}_m} \Pr\{\tilde{\mathbf{c}}_N | \mathbf{r}_N = \rho_N, \mathbf{h}\}$$

$$= \sum_{\tilde{\mathbf{c}}_N \to \tilde{c}_m} \frac{f_{\mathbf{r}}(\rho_N | \tilde{\mathbf{c}}_N, \mathbf{h}) \Pr\{\mathbf{c}_N = \tilde{\mathbf{c}}_N\}}{f_{\mathbf{r}}(\rho_N | \mathbf{h})}, \qquad (4.174)$$

respectively, where the summations are evaluated over all sequences $\tilde{\mathbf{c}}_N$ which are consistent with $\{b_i = \tilde{b}_i\}$ or $\{c_m = \tilde{c}_m\}$, respectively. The conditional pdf $f_{\mathbf{r}}(\rho_N | \mathbf{h})$ appearing in both (4.173) and (4.174) can be expressed as:

$$f_{\mathbf{r}}(\rho_N | \mathbf{h}) = \sum_{\tilde{\mathbf{c}}_N \in S_c} \Pr\{\mathbf{c}_N = \tilde{\mathbf{c}}_N\} f_{\mathbf{r}}(\rho_N | \tilde{\mathbf{c}}_N, \mathbf{h}), \qquad (4.175)$$

where S_c is the set of all possible trial vectors $\tilde{\mathbf{c}}_N$, and $f_{\mathbf{r}}(\rho_N | \mathbf{h})$ can be interpreted as a scale factor ensuring that the probabilities assigned to the possible values of each bit/symbol sum to unity. In many cases this normalization may be omitted, and even if normalization is required, the unnormalized bit/symbol probabilities are available so that normalization can proceed using them directly. Thus, $f_{\mathbf{r}}(\rho_N | \mathbf{h})$ and any other data-independent scale factor can be disregarded without penalty. Finally, it is worth mentioning that the common probability factor $f_{\mathbf{r}}(\rho_N | \tilde{\mathbf{c}}_N, \mathbf{h})$ in (4.173) and (4.174) is given by (4.143) if the same linear model for the received signal model as in Section 4.5.2.1 is adopted. From the foregoing it is evident that the metrics in (4.171) and (4.172) are not as intuitive as the metrics used in MLSD, since they do not translate directly to distance.

4.5.2.4   Bounds on the Error Performance of MAPSD/MAPBD

The average *symbol error rate* (SER) $P_s[l]$ achieved by the MAPSD strategy (4.172) in the detection of the lth symbol c_l of the vector $\mathbf{c}_N$ can be written as:

$$P_s[l] = \sum_{\tilde{c}_l} \sum_{\tilde{\mathbf{c}}_{-l}} \Pr\left\{ \left(\max_{\tilde{c}_l \neq \tilde{c}_l} \Pr\{c_l = \tilde{c}_l | \mathbf{r}_N = \rho_N, \mathbf{h}\} \right) \geq \Pr\{c_l = \tilde{c}_l | \mathbf{r}_N = \rho_N, \mathbf{h}\} \right.$$

$$\left. | c_l = \tilde{c}_l, \mathbf{c}_{-l} = \tilde{\mathbf{c}}_{-l} \right\} \Pr\{c_l = \tilde{c}_l, \mathbf{c}_{-l} = \tilde{\mathbf{c}}_{-l}\} \qquad (4.176)$$

where $\mathbf{c}_{-l} \triangleq [c_0, c_1, \ldots, c_{l-1}, c_{l+1}, \ldots, c_{N-1}]^T$ is obtained from $\mathbf{c}_N$ by removing the element c_l; note that a similar expression may be arrived at for the average *bit error rate* (BER) of the lth bit b_l when the MAPBD strategy (4.171) is used. Expression (4.176) cannot easily be evaluated since the decision metrics (namely, the probabilities $\Pr\{c_l = \tilde{c}_l | \mathbf{r}_N = \boldsymbol{\rho}_N, \mathbf{h}\}$ for any possible value of $\tilde{c}_l$) are complicated functions of correlated Gaussian random variables. Hence, error performance bounds for MAPSD and MAPBD are of interest; in particular, a lower bound, known as the *matched filter bound* (MFB), is often derived for the case in which ISI is absent in the elements of the received signal vector $\mathbf{r}_N$. For instance, if PAM signaling is considered, the MFB is evaluated by assessing the symbol (or bit) error probability of an optimal (i.e., MAP) detector for a single transmitted pulse (i.e., symbol), that is, of a detector based on a filter *matched* to the pulse distorted by the communication channel. Note, however, that unlike the error performance bounds illustrated above, which refer to a fixed channel parameter vector $\mathbf{h}$, MFBs are usually derived for channels with known statistical properties, that is, their computation requires averaging the SER or BER with respect to the statistics of $\mathbf{h}$.

To illustrate the MFB concept, let us assume the transmission of a single pulse (i.e., a single channel symbol) over a frequency-selective Rayleigh fading channel characterized by a TVIR $h(t, \tau)$ and a PDP $P_h(\tau)$. If the complex envelope of the transmitted signal is $x(t) = c_0\, p(t)$ (where c_0 is the channel data symbol and $p(t)$ is the transmitter impulse response), the received signal $r(t)$ is given by:

$$r(t) = c_0\, g(t) + n(t), \tag{4.177}$$

where $g(t) \triangleq \int_{-\infty}^{+\infty} p(t - \tau)\, h(t, \tau)\, d\tau$ and $n(t)$ is complex AWGN with two-sided power spectral density $2N_0$. The signal $r(t)$ feeds a filter matched to $g(t)$, that is, having impulse response $h_R(t) = g^*(-t)$. The decision statistic r_0 at the receiver filter output for MAP detection is then given by:

$$r_0 \triangleq r(t) \otimes h_R(t)|_{t=0} = c_0\, E_g + w, \tag{4.178}$$

where E_g is the energy of $g(t)$ and the complex random variable w, conditioned on $g(t)$, is Gaussian with zero mean and variance $\sigma_w^2 = 2N_0 E_g$. Given the model for the observed noisy datum, it is easy to infer the MAP strategy from the results derived for the AGN channel in Example 4.3.1. Then the MAP strategy looks for the minimum of the metric (see (4.42)):

$$|\rho_0 - \tilde{c}\, E_g|^2 - 2\sigma_w^2 \ln P_{\tilde{c}}\,, \tag{4.179}$$

where ρ_0 is the value taken by r_0 and $\tilde{c}$ denotes the trial symbol (characterized by the a priori probability $P_{\tilde{c}}$) which can take M distinct values, if an M-ary constellation is used. Generally speaking, evaluating the error performance of the latter strategy leads to the MFB for PAM signaling. The MFB can be put in a compact form if c_0 belongs to a BPSK alphabet (i.e., $\{\pm 1\}$) and if its two possible levels are equally likely. In fact, when this occurs, the MAPSD strategy becomes the ML binary decision strategy $\hat{c}_0 = \mathrm{sign}(\rho_0)$, where $\mathrm{sign}(x)$ denotes the sign of the real number x, so that, for a given realization of $h(t, \tau)$, the bit error probability[5] is given by [465]:

$$P_b = Q\left(\sqrt{R\, \frac{2E_b}{N_0}}\,\right), \tag{4.180}$$

where $E_b = E_p/2$ is the transmitted energy per bit, E_p is the energy of $p(t)$ and

$$R \triangleq \frac{E_g}{E_p}. \tag{4.181}$$

---

[5] Expression (4.180) also holds for 4-QAM transmission; in fact in this case the in-phase detector output is considered.

Averaging P_b (4.180) with the respect to the channel statistics yields the MFB:

$$\bar{P}_b = \int_0^{+\infty} f_R(r)\, Q\left(\sqrt{\frac{2E_b\, r}{N_0}}\right) dr, \tag{4.182}$$

where $f_R(r)$ is the pdf of R (4.181).

Different numerical techniques have been proposed for the evaluation of the RHS of equation (4.182) [85, 109, 465–475]. In particular, if the channel PDP is continuous, an approximate representation of R (4.181) can be derived by means of the KL expansion [85, 109, 162, 473–476], the sinc expansion [471] or GQR channel models [477]. It is important to note that, whatever channel model is adopted, the MFB is not affected if the real channel is replaced by a statistically equivalent representation. For instance, if we exploit an n_fth-order GQR channel model (see Section 2.2.3.5), the reduced-dimensionality statistical representation:

$$\tilde{g}(t) = \sum_{i=0}^{n_f-1} a_i\, p(t - \tau_i) \tag{4.183}$$

can be used for $g(t)$. Here the parameters $\{a_i, i = 0, 1, \ldots, n_f - 1\}$ are *independent* Gaussian random variables with zero mean and variances $\{E\{|a_i|^2\} = w_i, i = 0, 1, \ldots, n_f - 1\}$, and $\{\tau_i, i = 0, 1, \ldots, n_f - 1\}$ and $\{w_i, i = 0, 1, \ldots, n_f - 1\}$ are respectively the *nodes* and *weights* of the n_fth-order Gaussian quadrature formula for the weight function $P_h(\tau)$. Replacing $g(t)$ with $\tilde{g}(t)$ in the evaluation of R (4.181) yields the random variable[6]:

$$\tilde{R} = \mathbf{a}^H \mathbf{M} \mathbf{a}, \tag{4.184}$$

which is statistically equivalent to R; here $\mathbf{a} = [a_0, a_1, \ldots, a_{n_f-1}]^T$ is a zero mean complex Gaussian vector with covariance matrix:

$$\mathbf{C_a} \triangleq E\{\mathbf{a}\, \mathbf{a}^H\} = \text{diag}\{w_l\}, \tag{4.185}$$

and $\mathbf{M} = [M(i, j)]$ is an $n_f \times n_f$ matrix with elements:

$$M(l_1, l_2) = \frac{1}{E_p} \int p^*(t - \tau_{l_1})\, p(t - \tau_{l_2})\, dt, \tag{4.186}$$

$l_1, l_2 = 0, 1, \ldots, n_f - 1$. Then, if $\{\lambda_0, \lambda_1, \ldots, \lambda_{n_f-1}\}$ is the complete set of eigenvalues of the matrix $\mathbf{C_a}\, \mathbf{M}$, the characteristic function $\psi_{\tilde{R}}(v)$ of $\tilde{R}$ is (see [85, eq. (16)]):

$$\psi_{\tilde{R}}(v) = \sum_{i=1}^{I} \sum_{k=1}^{\mu_i} \frac{A_{ik}}{(1 - jv\lambda_i)^k}, \tag{4.187}$$

where I is the number of distinct eigenvalues, μ_i is the multiplicity of the ith eigenvalue λ_i and $\{A_{ik}\}$ are the coefficients of the partial fraction expansion of $\psi_{\tilde{R}}(v)$. Taking the IFCT of $\psi_{\tilde{R}}(v)$ yields the pdf $f_{\tilde{R}}(r)$ of $\tilde{R}$ (see [85, eq. (18)]). Substituting $f_{\tilde{R}}(r)$ into (4.182) (in place of $f_R(r)$) yields, after some manipulation, the MFB:

$$\bar{P}_b = \frac{1}{2} \sum_{i=1}^{I} \sum_{k=1}^{\mu_i} A_{ik} \left\{ 1 - \sum_{l=0}^{k-1} C_l \sqrt{\frac{\gamma \lambda_i}{(1 + \gamma \lambda_i)^{2l+1}}} \right\}, \tag{4.188}$$

where $C_l \triangleq (2l - 1)!!/(2l)!!$ ($C_0 = 1$) and $\gamma = E_b/N_0$. Let us now apply these results in the following example.

---

[6] We note that, with R being the normalized energy of the overall channel impulse response, the decision variable $\tilde{R}$ has a quadratic dependence on the channel gains $\{a_i\}$.

Example 4.5.2 We evaluate the MFB for a BPSK transmission over a *uniform* (U) or *exponential* (E) PDP (the analytical expressions for $P_h(\tau)$ and of $R_H(f)$ together with the corresponding formula for *delay spread* τ_{ds} are listed in Table 2.2 for such profiles). The following observations are worth noting:

- The evaluation of the quadrature nodes and weights for the uniform and exponential PDP involves the *Legendre* and the *Hermite* polynomials, respectively, and can be carried out by means of numerically stable subroutines [119, 120].
- In generating GQR models the values of the parameter n_f have been selected to ensure good statistical accuracy in the channel representation (in practice, further improvement of this parameter only negligibly affects the error rate results).

Moreover, in this example the modulator impulse response $p(t)$ is the IFCT of a root-raised cosine function with roll-off $\alpha = 0.35$ and is truncated symmetrically to 20 symbol intervals. Figure 4.12 illustrates the MFB for $\tau_{ds}/T_s = 0.1$, 0.5 and 1. The BER curves with frequency-flat fading ($\sigma_{ds} = 0$) and for an AWGN channel are also shown for comparison. The substantial differences between the uniform and the exponential power profile can be related to the long tails of the latter. These error rate curves show that:

- the MFB depends significantly on the channel delay spread and negligibly on the power profile shape for a given σ_{ds}/T_s, as already observed in [85, 471],
- the BER performance of the MF receiver improves as τ_{ds}/T_s grows larger, as the intrinsic diversity provided by the multipath phenomenon increases [85].

□

In Example 4.5.2 a purely frequency-selective channel is considered. Further numerical results on the MFB for doubly-selective fading channels can be found in [477]. MFBs have also been evaluated for multicarrier signaling; see [449, 472, 478, 479] for further details.

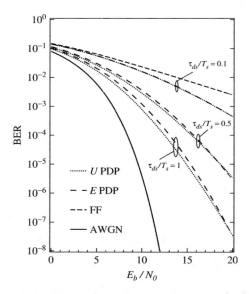

Figure 4.12 MFB versus τ_{ds}/T_s for FS channels with uniform and exponential PDPs. The MFB for frequency-flat (FF) fading and AWGN channels is also shown for comparison.

4.5.3 Detection in the Presence of a Statistically Known Channel

The exploitation of the detection metrics derived in Section 4.5.2 requires the availability of an accurate channel estimate at the receiver. To avoid the problem of explicit channel estimation in receiver design, channel parameters can be averaged out in the derivation of optimal detection metrics. However, this approach requires the CIR to be statistically known and the CIR pdf to be tractable. This explains why the detectors belonging to this class have mainly been developed for complex Gaussian channels (i.e., under the assumption of Rayleigh or Rician fading). In any real-world receiver the channel statistics need to be estimated, and this task may be more difficult than CIR estimation. Thus the implementation of this class of detectors can be extremely complicated in many circumstances.

In this subsection decision metrics for ML and MAP detection strategies are derived and some indications on the problem of estimating their error performance are provided.

4.5.3.1 Metric for MLSD

Here we derive the MLSD strategy for estimating the symbol vector $\mathbf{c}_N = [c_0, c_1, \ldots, c_{N-1}]^T$ from the received signal vector $\mathbf{r}_N \triangleq [r_0, r_1, \ldots, r_{N-1}]^T$. To account for averaging over the channel parameters, this strategy can be written as (see (4.26) and (4.48)):

$$\hat{\mathbf{c}}_N = \arg \max_{\tilde{\mathbf{c}}_N} f_{\mathbf{r}}(\boldsymbol{\rho}_N | \tilde{\mathbf{c}}_N)$$

$$= \arg \max_{\tilde{\mathbf{c}}_N} \int f_{\mathbf{r}}(\boldsymbol{\rho}_N | \tilde{\mathbf{c}}_N, \boldsymbol{\chi}) \, f_{\mathbf{h}}(\boldsymbol{\chi}) \, d\boldsymbol{\chi}$$

$$= \arg \max_{\tilde{\mathbf{c}}_N} \int \frac{f_{\mathbf{r},\mathbf{h}}(\boldsymbol{\rho}_N \boldsymbol{\chi} | \tilde{\mathbf{c}}_N)}{f_{\mathbf{h}}(\boldsymbol{\chi} | \tilde{\mathbf{c}}_N)} \, f_{\mathbf{h}}(\boldsymbol{\chi}) \, d\boldsymbol{\chi} \qquad (4.189)$$

where $f_{\mathbf{h}}(\boldsymbol{\chi})$ and $\boldsymbol{\rho}_N$ denote the joint pdf of the channel parameters and the value taken by $\mathbf{r}_N$, respectively. If a Gaussian model is adopted for $\mathbf{h}$, the vector $\mathbf{r}_N$, given $\mathbf{c}_N = \tilde{\mathbf{c}}_N$, is also Gaussian (see (4.141)) with mean vector:

$$\boldsymbol{\eta}_{\mathbf{r}}(\tilde{\mathbf{c}}_N) \triangleq E\{\mathbf{r}_N \, | \mathbf{c}_N = \tilde{\mathbf{c}}_N\} \qquad (4.190)$$

and covariance matrix:

$$\mathbf{R}_{\mathbf{r}}(\tilde{\mathbf{c}}_N) \triangleq E\{(\mathbf{r}_N - \boldsymbol{\eta}_{\mathbf{r}}(\tilde{\mathbf{c}}_N))(\mathbf{r}_N - \boldsymbol{\eta}_{\mathbf{r}}(\tilde{\mathbf{c}}_N))^H | \mathbf{c}_N = \tilde{\mathbf{c}}_N\}. \qquad (4.191)$$

Then the decision rule of (4.189) can be put in the form:

$$\hat{\mathbf{c}}_N = \arg \min_{\tilde{\mathbf{c}}_N} \Lambda(\tilde{\mathbf{c}}_N), \qquad (4.192)$$

where:

$$\Lambda(\tilde{\mathbf{c}}_N) \triangleq - \log \, f(\mathbf{r}_N | \tilde{\mathbf{c}}_N)$$

$$= \frac{1}{2}(\mathbf{r}_N - \boldsymbol{\eta}_{\mathbf{r}}(\tilde{\mathbf{c}}_N))^H \mathbf{R}_{\mathbf{r}}^{-1}(\tilde{\mathbf{c}}_N)(\mathbf{r}_N - \boldsymbol{\eta}_{\mathbf{r}}(\tilde{\mathbf{c}}_N)) + \log \, \det(2\pi \mathbf{R}_{\mathbf{r}}(\tilde{\mathbf{c}}_N)) \qquad (4.193)$$

in a Rician channel and:

$$\Lambda(\tilde{\mathbf{c}}_N) = \frac{1}{2}\mathbf{r}_N^H \, \mathbf{R}_{\mathbf{r}}^{-1}(\tilde{\mathbf{c}}_N)\mathbf{r}_N + \log \, \det(2\pi \mathbf{R}_{\mathbf{r}}(\tilde{\mathbf{c}}_N)) \qquad (4.194)$$

in a Rayleigh fading channel. Note the following observations:

1. In equation (4.193) the first term of the metric is a quadratic form in the received vector $\mathbf{r}_N$, whereas the second term (referred to as the *bias* term in the technical literature [480]) does not exhibit any dependence on $\mathbf{r}_N$.

2. The meaning of the detection metrics in (4.193) and (4.194) is not intuitively obvious, nor is it amenable to implementation in their present form. These considerations raise the problem of deriving possible reformulations that clarify the meaning of such metrics and make the problem of evaluating them easier.

We next tackle the latter problem, restricting our attention, however, to the case of Rayleigh fading (i.e., to the metric (4.194)) which has been thoroughly studied in many papers (e.g., see [114, 428, 434, 436, 481–494]). Note, however, that the extension of our results to Rician fading is straightforward [495] and arbitrary fading pdfs may be handled also (e.g., see [496–498]).

4.5.3.2   Interpretation of MLSD in terms of Estimation-Correlation

The earliest (and simplest) interpretation of the metrics (4.193) and (4.194) is due to T. Kailath [484, 499]. To illustrate this interpretation let us assume, as in Section 4.5.2, that $K = N$ in model (4.141), that is, that one sample per channel symbol is processed by the detection algorithm, so that the received signal vector $\mathbf{r}_N$ is given by:

$$\mathbf{r}_N = \mathbf{Hc}_N + \mathbf{n}_N. \tag{4.195}$$

We also assume that the channel is affected by Rayleigh fading, so that the elements of the channel matrix $\mathbf{H}$ are zero mean correlated complex Gaussian variables. If we now define the vector:

$$\mathbf{z}_N(\mathbf{c}_N) \triangleq \mathbf{Hc}_N, \tag{4.196}$$

which represents the useful signal component of $\mathbf{r}_N$, the autocovariance matrix $\mathbf{R_r}$ of $\mathbf{r}_N$, given $\mathbf{c}_N = \tilde{\mathbf{c}}_N$, becomes:

$$\mathbf{R_r}(\tilde{\mathbf{c}}_N) = \mathbf{R_z}(\tilde{\mathbf{c}}_N) + \mathbf{R_n}, \tag{4.197}$$

where $\mathbf{R_z}(\tilde{\mathbf{c}}_N) \triangleq E\{\mathbf{z}_N \mathbf{z}_N^H | \mathbf{c}_N = \tilde{\mathbf{c}}_N\}$ is the (data-dependent) autocovariance matrix and $\mathbf{R_n}$ is the noise autocovariance matrix. Exploiting the *matrix inversion lemma* (C.10), the inverse of $\mathbf{R_r}(\tilde{\mathbf{c}}_N)$ (4.197) can be rewritten as:

$$\mathbf{R_r}^{-1}(\tilde{\mathbf{c}}_N) = \mathbf{R_n}^{-1} - \mathbf{R_n}^{-1}[\mathbf{R_n}^{-1} + \mathbf{R_z}^{-1}(\tilde{\mathbf{c}}_N)]^{-1}\mathbf{R_n}^{-1}. \tag{4.198}$$

Substituting the RHS of this into the ML metric of (4.194) yields:

$$\Lambda(\tilde{\mathbf{c}}_N) = \frac{1}{2}\mathbf{r}_N^H[\mathbf{R_n}^{-1} - \mathbf{R_n}^{-1}[\mathbf{R_n}^{-1} + \mathbf{R_z}^{-1}(\tilde{\mathbf{c}}_N)]^{-1}\mathbf{R_n}^{-1}]\mathbf{r}_N + \log\det(2\pi\mathbf{R_r}(\tilde{\mathbf{c}}_N)). \tag{4.199}$$

Then, if irrelevant (i.e., symbol independent) terms are dropped, the MLSD strategy (4.192) can be put in the form:

$$\hat{\mathbf{c}}_N = \arg\max_{\tilde{\mathbf{c}}_N}\left\{\frac{1}{2}\mathbf{r}_N^H\mathbf{R_n}^{-1}[\mathbf{R_n}^{-1} + \mathbf{R_z}^{-1}(\tilde{\mathbf{c}}_N)]^{-1}\mathbf{R_n}^{-1}\mathbf{r}_N - \log\det(2\pi\mathbf{R_r}(\tilde{\mathbf{c}}_N))\right\}. \tag{4.200}$$

It can then be shown that the vector:

$$\mathbf{z}_{MMSE}(\tilde{\mathbf{c}}_N) \triangleq [\mathbf{R_n}^{-1} + \mathbf{R_z}^{-1}(\tilde{\mathbf{c}}_N)]^{-1}\mathbf{R_n}^{-1}\mathbf{r}_N \tag{4.201}$$

represents the linear MMSE estimator of $\mathbf{z}_N(\tilde{\mathbf{c}}_N)$ (4.196) from $\mathbf{r}_N$, given $\mathbf{c}_N = \tilde{\mathbf{c}}_N$ [106, 484]. Then substituting (4.201) into (4.200) leads to:

$$\hat{\mathbf{c}}_N = \arg\max_{\tilde{\mathbf{c}}_N}\left\{\frac{1}{2}\mathbf{r}_N^H\mathbf{R_n}^{-1}\mathbf{z}_{MMSE}(\tilde{\mathbf{c}}_N) - \log\det(2\pi\mathbf{R_r}(\tilde{\mathbf{c}}_N))\right\}, \tag{4.202}$$

which provides an important interpretation of the MLSD strategy. It shows that, if the channel noise is white (i.e., $\mathbf{R_n} = 2\sigma_n^2\mathbf{I}_N$) and the bias term $\log\det(2\pi\mathbf{R_r}(\tilde{\mathbf{c}}))$ can be neglected, the ML strategy aims to maximize the *correlation* between the received vector $\mathbf{r}_N$ and a tentative MMSE *estimate* of its useful signal component. For this reason the overall processing of the ML detector, when the CIR is averaged out, is often termed *estimation-correlation*.

4.5.3.3 Interpretation of MLSD in terms of Innovations Processes

Another significant interpretation of the optimal strategy (4.192) with metric $\Lambda(\tilde{\mathbf{c}}_N)$ (4.194) can be developed by resorting to the concept of the *innovation process* [55]. To see this, we apply the *Cholesky decomposition* [125] to the positive definite matrix $\mathbf{R_r}(\tilde{\mathbf{c}})$ of (4.197), so that it can be factored as:

$$\mathbf{R_r}(\tilde{\mathbf{c}}_N) = \mathbf{U}^H(\tilde{\mathbf{c}}_N)\,\mathbf{U}(\tilde{\mathbf{c}}_N). \tag{4.203}$$

Here $\mathbf{U}(\tilde{\mathbf{c}}_N)$ is an $N \times N$ upper triangular matrix (whose diagonal elements are all positive). Its inverse $\mathbf{U}^{-1}(\tilde{\mathbf{c}}_N)$ is also an upper triangular matrix, which can be factored as:

$$\mathbf{U}^{-1}(\tilde{\mathbf{c}}_N) = \mathbf{P}^H(\tilde{\mathbf{c}}_N)\,\mathbf{S}^{-1/2}(\tilde{\mathbf{c}}_N), \tag{4.204}$$

where $\mathbf{P}(\tilde{\mathbf{c}}_N)$ is an $N \times N$ lower triangular matrix with 1s on its main diagonal and $\mathbf{S}^{-1/2}(\tilde{\mathbf{c}}_N) = \mathrm{diag}[1/\sqrt{2s_k(\tilde{\mathbf{c}}_N)}]$ is a diagonal matrix containing the diagonal entries of $\mathbf{U}^{-1}(\tilde{\mathbf{c}}_N)$. Therefore, the inverse of $\mathbf{R_r}(\tilde{\mathbf{c}}_N)$ can be written in the form (see (4.203)):

$$\mathbf{R_r}^{-1}(\tilde{\mathbf{c}}_N) = \mathbf{U}^{-1}(\tilde{\mathbf{c}}_N)\,[\mathbf{U}^{-1}(\tilde{\mathbf{c}}_N)]^H = \mathbf{P}^H(\tilde{\mathbf{c}}_N)\,\mathbf{S}^{-1/2}(\tilde{\mathbf{c}}_N)\,\mathbf{S}^{-1/2}(\tilde{\mathbf{c}}_N)\,\mathbf{P}(\tilde{\mathbf{c}}_N). \tag{4.205}$$

Then substituting (4.205) into (4.194) yields the expression:

$$\Lambda(\tilde{\mathbf{c}}_N) = \frac{1}{2}[\mathbf{S}^{-1/2}(\tilde{\mathbf{c}}_N)\,\mathbf{P}(\tilde{\mathbf{c}}_N)\mathbf{r}_N]^H[\mathbf{S}^{-1/2}(\tilde{\mathbf{c}}_N)\,\mathbf{P}(\tilde{\mathbf{c}}_N)\mathbf{r}_N] + \log\det(\mathbf{S}(\tilde{\mathbf{c}}_N)). \tag{4.206}$$

Since $\mathbf{P}(\tilde{\mathbf{c}}_N)$ is structured as:

$$\mathbf{P}(\tilde{\mathbf{c}}_N) = \begin{bmatrix} 1 & 0 & 0 & \cdots & 0 \\ -p_{1,1}(\tilde{\mathbf{c}}_N) & 1 & 0 & & 0 \\ -p_{2,2}(\tilde{\mathbf{c}}_N) & -p_{2,1}(\tilde{\mathbf{c}}_N) & 1 & \ddots & \vdots \\ \vdots & & \ddots & \ddots & 0 \\ -p_{N-1,N-1}(\tilde{\mathbf{c}}_N) & -p_{N-1,N-2}(\tilde{\mathbf{c}}_N) & \cdots & -p_{N-1,1}(\tilde{\mathbf{c}}_N) & 1 \end{bmatrix}, \tag{4.207}$$

the product of the $(k+1)$th row of $\mathbf{P}(\tilde{\mathbf{c}}_N)$ with $\mathbf{r}_N$ in (4.206) yields:

$$e_k(\tilde{\mathbf{c}}_N) \triangleq r_k - \sum_{m=1}^{k} p_{k,m}(\tilde{\mathbf{c}}_N)\, r_{k-m}, \tag{4.208}$$

for $k = 1, 2, \ldots, N-1$ (note that the product of the first row of $\mathbf{P}(\tilde{\mathbf{c}}_N)$ with $\mathbf{r}_N$ produces $e_0(\tilde{\mathbf{c}}_N) = r_0$), so that $\Lambda(\tilde{\mathbf{c}}_N)$ of (4.206) can be put in the form:

$$\Lambda(\tilde{\mathbf{c}}_N) = \sum_{k=0}^{N-1} \left| \frac{e_k(\tilde{\mathbf{c}}_N)}{\sqrt{2}\,s_k(\tilde{\mathbf{c}}_N)} \right|^2 + \log(2\pi\,\sigma_k(\tilde{\mathbf{c}}_N))$$

$$= \sum_{k=0}^{N-1} \frac{\left| r_k - \sum\limits_{m=1}^{k} p_{k,m}(\tilde{\mathbf{c}}_N)\, r_{k-m} \right|^2}{\sqrt{2}\,s_k(\tilde{\mathbf{c}}_N)} + \log(2\pi\,\sigma_k(\tilde{\mathbf{c}}_N)), \tag{4.209}$$

where $p_{0,m}(\tilde{\mathbf{c}}_N) = 0$ for any m. To understand the meaning of the latter result, we rewrite the probability expression for the ML decision rule (4.189) as:

$$\hat{\mathbf{c}}_N = \arg\max_{\tilde{\mathbf{c}}_N} f_{\mathbf{r}}(\boldsymbol{\rho}_N | \tilde{\mathbf{c}}_N)$$

$$= \arg\max_{\tilde{\mathbf{c}}_N} \prod_{k=0}^{N-1} f_{r_k}(\rho_k | \mathbf{r}_{k-1}, \tilde{\mathbf{c}}_N)$$

$$= \arg\max_{\tilde{\mathbf{c}}_N} \Lambda(\tilde{\mathbf{c}}_N), \tag{4.210}$$

where $f_{r_k}(\rho_k | \mathbf{r}_{k-1}, \tilde{\mathbf{c}}_N)$, with $k = 1, 2, \ldots, N-1$, denotes the probability of r_k conditioned on $\mathbf{r}_{k-1} \triangleq [r_0, r_1, \ldots, r_{k-1}]^T$ and $\mathbf{c}_N = \tilde{\mathbf{c}}_N$, $f_{r_0}(\rho_0 | \mathbf{r}_{-1}, \tilde{\mathbf{c}}_N) = f_{r_0}(\rho_0)$ ($\mathbf{r}_{-1}$ is an empty vector) and (see (4.193)):

$$\Lambda(\tilde{\mathbf{c}}_N) \triangleq -\log f(\mathbf{r}_N | \tilde{\mathbf{c}}_N)$$

$$= -\log \prod_{k=0}^{N-1} f_{r_k}(\rho_k | \mathbf{r}_{k-1}, \tilde{\mathbf{c}}_N)$$

$$= -\sum_{k=0}^{N-1} \log f_{r_k}(\rho_k | \mathbf{r}_{k-1}, \tilde{\mathbf{c}}_N). \tag{4.211}$$

At this point we note that the random variable r_k, with $k = 0, 1, \ldots, N-1$, conditioned on the transmitted sequence $\mathbf{c}_N = \tilde{\mathbf{c}}_N$ and $\mathbf{r}_{k-1}$, is a complex Gaussian random variable with mean:

$$\eta_k(\mathbf{r}_{k-1}, \tilde{\mathbf{c}}_N) \triangleq \mathrm{E}\{r_k | \mathbf{r}_{k-1}, \tilde{\mathbf{c}}_N\} \tag{4.212}$$

and variance $2\sigma_k^2(\mathbf{r}_{k-1}, \tilde{\mathbf{c}}_N)$ with:

$$\sigma_k^2(\mathbf{r}_{k-1}, \tilde{\mathbf{c}}_N) \triangleq \frac{1}{2}\mathrm{E}\{|r_k - \eta_k(\mathbf{r}_{k-1}, \tilde{\mathbf{c}}_N)|^2\}, \tag{4.213}$$

so that:

$$f_{r_k}(\rho_k | \mathbf{r}_{k-1}, \tilde{\mathbf{c}}_N) = \frac{1}{2\pi\sigma_k^2(\mathbf{r}_{k-1}, \tilde{\mathbf{c}}_N)} \exp\left[-\frac{|r_k - \eta_k(\mathbf{r}_{k-1}, \tilde{\mathbf{c}}_N)|^2}{2\sigma_k^2(\mathbf{r}_{k-1}, \tilde{\mathbf{c}}_N)}\right] \tag{4.214}$$

(note that $\eta_0(\mathbf{r}_{-1}, \tilde{\mathbf{c}}_N) = 0$ and $\sigma_0^2(\mathbf{r}_{-1}, \tilde{\mathbf{c}}_N) = 2\sigma_n^2$). Then substituting (4.214) into (4.211) yields:

$$\Lambda(\tilde{\mathbf{c}}) = \sum_{k=0}^{N-1} \left|\frac{r_k - \eta_k(\mathbf{r}_{k-1}, \tilde{\mathbf{c}}_N)}{\sqrt{2\,\sigma_k^2(\mathbf{r}_{k-1}, \tilde{\mathbf{c}}_N)}}\right|^2 + \log\left(2\pi\,\sigma_k^2(\mathbf{r}_{k-1}, \tilde{\mathbf{c}}_N)\right). \tag{4.215}$$

Finally, by comparing (4.215) with (4.209), we match $\sigma_k^2(\mathbf{r}_{k-1}, \tilde{\mathbf{c}}_N)$ with $s_k(\tilde{\mathbf{c}}_N)$ and $\eta_k(\mathbf{r}_{k-1}, \tilde{\mathbf{c}}_N)$ with $\sum_{m=1}^{k} p_{k,m}(\tilde{\mathbf{c}}_N) r_{k-m}$. This shows the following:

- The quantity $\eta_k(\mathbf{r}_{k-1}, \tilde{\mathbf{c}}_N)$ represents a MMSE *linear one-step prediction* of r_k given all past samples $\mathbf{r}_{k-1}$, assuming a specific transmitted data sequence $\tilde{\mathbf{c}}_N$. In other words, it minimizes the *mean square error* (MSE) $\mathrm{E}\{|r_k - \tilde{\mathbf{p}}_k^H(\tilde{\mathbf{c}}_N)\mathbf{r}_{k-1}|^2\}$ with respect to $\tilde{\mathbf{p}}_k$ for a given $\tilde{\mathbf{c}}_N$, where $\mathbf{p}_k(\tilde{\mathbf{c}}_N) \triangleq [p_{k,k}(\tilde{\mathbf{c}}_N), p_{k,k-1}(\tilde{\mathbf{c}}_N), \ldots, p_{k,1}(\tilde{\mathbf{c}}_N)]^T$.
- The difference (see (4.208)):

$$e_k(\tilde{\mathbf{c}}_N) = r_k - \eta_k(\mathbf{r}_{k-1}, \tilde{\mathbf{c}}_N) \tag{4.216}$$

with $\eta_k(\mathbf{r}_{k-1}, \tilde{\mathbf{c}}_N) = -\mathbf{p}_{k-1}^T(\tilde{\mathbf{c}}_N)\mathbf{r}_{k-1}$ can be interpreted as a *prediction error*.
- The prediction variance is given by $2\sigma_k^2(\mathbf{r}_{k-1}, \tilde{\mathbf{c}}_N)$.

These results pave the way for the development of a new interpretation of the MLSD metrics (4.215). In fact, it is not difficult to show that the elements of the sequence $\{e_k(\tilde{\mathbf{c}}_N), k = 0, 1, \ldots\}$ are uncorrelated with one another, so that the sequence $\{i_k(\tilde{\mathbf{c}}_N), k = 0, 1, \ldots\}$, where:

$$i_k(\tilde{\mathbf{c}}_N) \triangleq \frac{e_k(\tilde{\mathbf{c}}_N)}{\sqrt{2\,\sigma_k^2(\mathbf{r}_{k-1}, \tilde{\mathbf{c}}_N)}} = \frac{[-\mathbf{p}_{k-1}^T(\tilde{\mathbf{c}}_N), 1]\,\mathbf{r}_k}{\sqrt{2\,s_k(\tilde{\mathbf{c}}_N)}}, \tag{4.217}$$

is a scaled prediction error, represents a discrete-time *innovations process* [55] (here $\mathbf{p}_k(\tilde{\mathbf{c}}_N) \triangleq [p_{k,k}(\tilde{\mathbf{c}}_N), p_{k,k-1}(\tilde{\mathbf{c}}_N), \ldots, p_{k,1}(\tilde{\mathbf{c}}_N)]^T$). Note that:

(a) the innovations vector $\mathbf{i}_N(\tilde{\mathbf{c}}_N) \triangleq [i_0(\tilde{\mathbf{c}}_N), i_1(\tilde{\mathbf{c}}_N), \ldots, i_{N-1}(\tilde{\mathbf{c}}_N)]^T$ can be generated as $\mathbf{S}^{-1/2}(\tilde{\mathbf{c}}_N)\,\mathbf{P}(\tilde{\mathbf{c}}_N)\,\mathbf{r}_N$ (see (4.206)),
(b) because of the Gaussianity of r_k conditioned on $\mathbf{c}_N = \tilde{\mathbf{c}}_N$, this vector consists of *independent* complex Gaussian random variables all having unit variance, so that:

$$\mathrm{E}\{i_k(\tilde{\mathbf{c}}_N)\,i_l^*(\tilde{\mathbf{c}}_N)\} = \delta_{kl}. \tag{4.218}$$

Moreover, it is not difficult to prove that:

$$\mathrm{E}\{r_k\,i_l^*(\tilde{\mathbf{c}}_N)\} = 0, \qquad k < l, \tag{4.219}$$

and:

$$\mathrm{E}\{r_k\,i_k^*(\tilde{\mathbf{c}}_N)\} = \sqrt{2\,s_k(\tilde{\mathbf{c}}_N)} \tag{4.220}$$

for any k, since r_k can be written as:

$$
\begin{aligned}
r_k &= \sqrt{2s_k(\tilde{\mathbf{c}}_N)} \left(\frac{r_k - \eta_k(\mathbf{r}_{k-1}, \tilde{\mathbf{c}}_N)}{\sqrt{2\,s_k(\tilde{\mathbf{c}}_N)}} + \frac{\eta_k(\mathbf{r}_{k-1}, \tilde{\mathbf{c}}_N)}{\sqrt{2\,s_k(\tilde{\mathbf{c}}_N)}} \right) \\
&= \sqrt{2s_k(\tilde{\mathbf{c}}_N)} \left(i_k + \frac{\eta_k(\mathbf{r}_{k-1}, \tilde{\mathbf{c}}_N)}{\sqrt{2\,s_k(\tilde{\mathbf{c}}_N)}} \right)
\end{aligned} \tag{4.221}
$$

and $\eta_k(\mathbf{r}_{k-1}, \mathbf{c}_N)$ is computed from $\mathbf{r}_{k-1}$ as $\eta_k(\mathbf{r}_{k-1}, \tilde{\mathbf{c}}_N) = -\mathbf{p}_{k-1}^T(\tilde{\mathbf{c}}_N)\,\mathbf{r}_{k-1}$. Therefore, from (4.217) and (4.220), it follows that:

$$
\begin{aligned}
\sqrt{2\,s_k(\tilde{\mathbf{c}}_N)}\,\mathrm{E}\{r_k\,i_k^*(\tilde{\mathbf{c}}_N)\} &= \mathrm{E}\{\mathbf{r}_k\mathbf{r}_k^H\} \begin{bmatrix} -\mathbf{p}_k(\tilde{\mathbf{c}}_N) \\ 1 \end{bmatrix} \\
&= \mathbf{R}_{\mathbf{r},k}(\tilde{\mathbf{c}}_N) \begin{bmatrix} -\mathbf{p}_k(\tilde{\mathbf{c}}_N) \\ 1 \end{bmatrix} \\
&= \begin{bmatrix} \mathbf{0}_{k-1} \\ 2\,s_k(\tilde{\mathbf{c}}_N) \end{bmatrix},
\end{aligned} \tag{4.222}
$$

where $\mathbf{R}_{\mathbf{r},k} \triangleq \mathrm{E}\{\mathbf{r}_k\mathbf{r}_k^H\}$. By partitioning $\mathbf{R}_{\mathbf{r},k}$ as:

$$\mathbf{R}_{\mathbf{r},k} = \begin{bmatrix} \mathbf{R}_{\mathbf{r},k-1} & \mathbf{R}_{\mathbf{r},k-1,k} \\ \mathbf{R}_{\mathbf{r},k-1,k}^H & \mathrm{E}\{|r_k|^2\} \end{bmatrix}, \tag{4.223}$$

where the column vector $\mathbf{R}_{\mathbf{r},k-1,k}$ is defined as $\mathbf{R}_{\mathbf{r},k-1,k} \triangleq E\{\mathbf{r}_{k-1}r_k^*\}$, (4.222) can be rewritten as:

$$\mathbf{R}_{\mathbf{r},k-1}\mathbf{p}_k(\tilde{\mathbf{c}}_N) = \mathbf{R}_{\mathbf{r},k-1,k} \tag{4.224}$$

and:

$$2\, s_k(\tilde{\mathbf{c}}_N) = E\{|r_k|^2\} - \mathbf{R}_{\mathbf{r},k-1,k}^H \mathbf{p}_k(\tilde{\mathbf{c}}_N). \tag{4.225}$$

Equations (4.224) and (4.225) can be used to compute the vector of prediction coefficients $\mathbf{p}_k(\tilde{\mathbf{c}}_N)$ and the prediction error variance $2s_k(\tilde{\mathbf{c}}_N)$, respectively, for all $\tilde{\mathbf{c}}_N$.

In summary, the first term in the metric (4.209) can be interpreted as the *sum of squared normalized prediction errors*, each normalized by the expected squared prediction error, that is, as the sum of samples of a discrete-time innovation process.

All the considerations illustrated here refer to the general case of signaling over a doubly-selective fading channel. Generally speaking, the evaluation of the metrics according to (4.215) entails a substantial computational burden, since the prediction coefficients need to be evaluated for $k = 0, 1, \ldots, N - 1$ and for each possible trial sequence. However, some reduction in the complexity is possible if PAM signaling over a slow time-selective fading channel is considered, as illustrated in the following example.

Example 4.5.3 If we assume PAM signaling over a slowly varying time-selective fading and baud rate sampling at the output of the matched filter of the receiver, the model (see (4.92)):

$$r_k = a_k\, c_k + w_k \tag{4.226}$$

can be adopted for r_k, with $k = 0, 1, \ldots, N - 1$; here a_k is the kth sample of the fading distortion, and the sequence $\{w_k\}$ consists of iid complex Gaussian noise samples, each having zero mean and variance $2N_0/K_c^2$. In this case the mean $\eta_k(\mathbf{r}_{k-1}, \tilde{\mathbf{c}}_N)$ (4.212) can be expressed as [490, 500, 501]:

$$\eta_k(\mathbf{r}_{k-1}, \tilde{\mathbf{c}}_N) = \eta_k(\mathbf{r}_{k-1}, \tilde{\mathbf{c}}_{k-1}) = \tilde{c}_k\, \tilde{a}\{k|\mathbf{r}_{k-1}, \tilde{\mathbf{c}}_{k-1}\}, \tag{4.227}$$

where $\tilde{a}\{k|\mathbf{r}_{k-1}, \tilde{\mathbf{c}}_{k-1}\}$ is the MMSE one-step *prediction* of the fading sample a_k, based on $\mathbf{r}_{k-1}$ and assuming that the sequence $\tilde{\mathbf{c}}_{k-1} = [\tilde{c}_0, \tilde{c}_1, \ldots, c_{k-1}]^T$ has been transmitted. Then the variance $\sigma_k^2(\mathbf{r}_{k-1}, \tilde{\mathbf{c}}_N)$ of (4.213) can be rewritten as:

$$\sigma_k^2(\mathbf{r}_{k-1}, \tilde{\mathbf{c}}_N) = \sigma_k^2(\mathbf{r}_{k-1}, \tilde{\mathbf{c}}_{k-1}) = \frac{1}{2}E\{|r_k - \tilde{c}_k\, \tilde{a}\{k|\mathbf{r}_{k-1}, \tilde{\mathbf{c}}_N\}|^2|\mathbf{r}_{k-1}, \tilde{\mathbf{c}}_N\}. \tag{4.228}$$

It can be shown that, given the symbol vector $\tilde{\mathbf{c}}_{k-1}$, the prediction coefficients for the evaluation of $\tilde{a}\{k|\mathbf{r}_{k-1}, \tilde{\mathbf{c}}_{k-1}\}$ depend on the modulus of the symbols forming $\tilde{\mathbf{c}}_{k-1}$, but not on their phase. For this reason, if PSK signaling is adopted, such coefficients depend on the length of the symbol sequence, but not on its elements. A further substantial simplification is achieved if a constant prediction length v is used, so that at time k only the last v symbols (namely, $c_{k-1}, c_{k-2}, \ldots, c_{k-v}$) are involved in the prediction process; in fact in this case the prediction coefficients do not need to be recomputed for each trial sequence at each step [490, 500, 501]. Generally speaking, in frequency-flat channels the evaluation of the fading prediction $\tilde{a}\{k|\mathbf{r}_{k-1}, \tilde{\mathbf{c}}_{k-1}\}$ and of the variance $\sigma_k^2(\mathbf{r}_{k-1}, \tilde{\mathbf{c}}_{k-1})$ can be accomplished by a time-varying Wiener filter [502, 503]. However, if the process $\{a_k\}$ can be characterized by a Gauss–Markov model [217, 502], both quantities can be computed recursively by means of a Kalman predictor [503] for a given trial sequence.

□

Considerations conceptually similar to those expressed in Example 4.5.3 can be developed for the MLSD of CPM signals transmitted over fast frequency-flat fading channels [436].

4.5.3.4 Bounds on the Error Performance of MLSD

The union bound (4.166) also applies to the MLSD strategies derived in the presence of a statistically known channel [493, 495]. If a Rayleigh fading channel is considered, this requires computing the PEP $\Pr\{\Delta(\mathbf{e}_N) \geq 0\}$ for any possible $\mathbf{e}_N$, where (see (4.159)):

$$\Delta(\mathbf{e}_N) = -\mathbf{e}_N^H \, \mathbf{H}^H \, \mathbf{n}_N - \mathbf{n}_N^H \mathbf{H} \, \mathbf{e}_N \tag{4.229}$$

is a Gaussian quadratic form. In practice, the evaluation of this PEP can be accomplished as follows:

(a) Evaluate the characteristic function of $\Delta(\mathbf{e}_N)$.
(b) Transform this characteristic function into the associated pdf.
(c) Integrate over the error region (the interval $(0, +\infty)$ in this specific case).

For further analytical details, see [493, Sect. 3] and [495, Sect. 5].

4.5.3.5 Metrics for MAPSD/MAPBD

The MAPBD strategy can be formulated as:

$$\hat{b}_i = \arg\max_{\tilde{b}_i} \Pr\{\tilde{b}_i | \mathbf{r}_N = \boldsymbol{\rho}_N\}. \tag{4.230}$$

Applying Bayes' theorem to the RHS gives:

$$\hat{b}_i = \arg\max_{\tilde{b}_i} \frac{\Pr\{\tilde{b}_i\} f_{\mathbf{r}}(\boldsymbol{\rho}_N | \tilde{b}_i)}{f_{\mathbf{r}}(\boldsymbol{\rho}_N)}, \tag{4.231}$$

which, on discarding the factor $f_{\mathbf{r}}(\boldsymbol{\rho}_N)$ (since it is independent of the trial bit $\tilde{b}_i$), yields:

$$\hat{b}_i = \arg\max_{\tilde{b}_i} \Pr\{\tilde{b}_i\} \, f_{\mathbf{r}}(\boldsymbol{\rho}_N | \tilde{b}_i)$$

$$= \arg\max_{\tilde{b}_i} \Pr\{\tilde{b}_i\} \int f_{\mathbf{r}}(\boldsymbol{\rho}_N | \tilde{b}_i, \boldsymbol{\chi}) \, f_{\mathbf{h}}(\boldsymbol{\chi}) \, d\boldsymbol{\chi}, \tag{4.232}$$

by averaging the conditional pdf $f_{\mathbf{r}}(\boldsymbol{\rho}_N | \tilde{b}_i)$ over the channel statistics $f_{\mathbf{h}}(\boldsymbol{\chi})$. Similarly, the MAPSD strategy can be expressed as:

$$\hat{c}_i = \arg\max_{\tilde{c}_i} \Pr\{\tilde{c}_i\} f_{\mathbf{r}}(\boldsymbol{\rho}_N | \tilde{c}_i)$$

$$= \arg\max_{\tilde{c}_i} \Pr\{\tilde{c}_i\} \int f_{\mathbf{r}}(\boldsymbol{\rho}_N | \tilde{c}_i, \boldsymbol{\chi}) \, f_{\mathbf{h}}(\boldsymbol{\chi}) \, d\boldsymbol{\chi}. \tag{4.233}$$

Finally, we note that the strategies of (4.232) and (4.233) can be refined into:

$$\hat{b}_i = \arg\max_{\tilde{b}_i} \Pr\{\tilde{b}_i\} \sum_{\tilde{\mathbf{c}}_N \to \tilde{b}_i} f_{\mathbf{r}}(\boldsymbol{\rho}_N | \tilde{\mathbf{c}}_N)$$

$$= \arg\max_{\tilde{b}_i} \Pr\{\tilde{b}_i\} \sum_{\tilde{\mathbf{c}}_N \to \tilde{b}_i} \int f_{\mathbf{r}}(\boldsymbol{\rho}_N | \tilde{\mathbf{c}}_N, \boldsymbol{\chi}) \, f_{\mathbf{h}}(\boldsymbol{\chi}) \, d\boldsymbol{\chi} \tag{4.234}$$

and

$$\hat{c}_i = \arg\max_{\tilde{c}_i} \Pr\{\tilde{c}_i\} \sum_{\tilde{\mathbf{c}}_N \to \tilde{c}_i} f_{\mathbf{r}}(\boldsymbol{\rho}_N | \tilde{\mathbf{c}}_N)$$

$$= \arg\max_{\tilde{c}_i} \Pr\{\tilde{c}_i\} \sum_{\tilde{\mathbf{c}}_N \to \tilde{c}_i} \int f_{\mathbf{r}}(\boldsymbol{\rho}_N | \tilde{\mathbf{c}}_N, \boldsymbol{\chi}) \, f_{\mathbf{h}}(\boldsymbol{\chi}) \, d\boldsymbol{\chi}, \qquad (4.235)$$

respectively, where the summations are evaluated over all sequences, $\tilde{\mathbf{c}}_N$, which are consistent with $\{b_i = \tilde{b}_i\}$ or $\{c_m = \tilde{c}_m\}$, respectively, and (see 4.48):

$$f_{\mathbf{r}}(\boldsymbol{\rho}_N | \tilde{\mathbf{c}}_N) = \int f_{\mathbf{r}}(\boldsymbol{\rho}_N | \tilde{\mathbf{c}}_N, \boldsymbol{\chi}) \, f_{\mathbf{h}}(\boldsymbol{\chi}) \, d\boldsymbol{\chi} \qquad (4.236)$$

is computed as in (4.189).

Finally, it is interesting to note the following observations:

(a) An innovations-based formulation for the MAP metrics can be developed in the case of PAM signaling over doubly-selective fading channels [504, Sect. 4]. This requires, however, a proper trellis representation, which will be developed in Section 6.2.

(b) A *genie-aided* (lower) bound for the BER performance of MAPBD is derived in [504, Sect. 4] (see Section 4.5.2.4 above).

4.5.4 Detection in the Presence of an Unknown Channel

Although it is useful to idealize the CIR as being known exactly (by invoking a friendly genie), in practice the CIR (or some equivalent quantity) must be estimated, as well as any other unknown parameters, such as the noise variance. These estimates are then used as if they were exact. In channels with significant time variation and/or delay spread, CIR estimation becomes more difficult. Generally speaking, a wireless channel can be classified on the basis of its *spread factor*:

$$SF \triangleq T_m \, B_D, \qquad (4.237)$$

where T_m (B_D) denotes the maximum time (frequency) spreading of an impulse (sine wave) transmitted over the channel itself (note that T_m and B_D respectively represent the memory and the Doppler bandwidth characterizing the channel filter). In fact, the spread factor plays an essential role in defining the measurability of a channel [505–508]. In particular, if $SF < 1$, the channel is *underspread* [101] and, in principle, estimation of its impulse response is possible (by resorting, for instance, to pilot tones or pilot symbols), although it becomes more critical as SF nears unity. In contrast, if $SF > 1$, the channel is *overspread* and cannot be estimated. It is worth pointing out the following:

(a) This criterion was introduced by T. Kailath [505]. A more accurate criterion replaces the spread factor with the area under the Doppler delay spread function [507].

(b) The spread factor can also be defined as $SF \triangleq v_{rms} \, \tau_{ds}$, where v_{rms} (τ_{ds}) is the rms Doppler bandwidth (rms delay spread) defined by (2.70) (2.52), or in other related ways, adopting other possible parameters providing estimates of the time and frequency dispersion introduced by a communication channel. Of course, the unit value of the threshold in the inequality $SF < 1$ should be considered only as an order-of-magnitude value [507].

These considerations suggest that in certain scenarios channel estimation can play an important role in the reliability of data communications and that channel estimation errors should be carefully taken into account when seeking a realistic assessment of error performance. For instance, the reader may refer to [509–514] for an assessment of the impact of channel estimation of the error performance provided by different detection algorithms over doubly-selective (frequency-flat) fading channels when standard channel estimation algorithms are used.

In Section 4.5.2 MLSD strategies were derived under the assumption of a known channel. In practice, the wireless channel is first estimated via one of the techniques which will be presented in Chapter 5, and then the estimated CIR is exploited for data detection. This approach can be interpreted as a means to reduce the complexity of the optimum MLSD strategy which is given by (see (4.45)):

$$\hat{\mathbf{c}}_N = \arg\max_{\tilde{\mathbf{c}}_N, \tilde{\mathbf{h}}} f_{\mathbf{r}}(\boldsymbol{\rho}_N | \tilde{\mathbf{c}}_N, \tilde{\mathbf{h}}). \tag{4.238}$$

In practice the estimated sequence $\hat{\mathbf{c}}_N$ corresponds to the sequence maximizing the pdf $f_{\mathbf{r}}(\boldsymbol{\rho}_N | \tilde{\mathbf{c}}_N, \tilde{\mathbf{h}})$ as a function of all the trial sequences $\{\tilde{\mathbf{c}}_N\}$ and the corresponding CIR estimates $\{\mathbf{h}(\tilde{\mathbf{c}}_N)\}$. However, unlike the maximization over $\tilde{\mathbf{c}}_N$, which involves a finite number of trials, the maximization over the trial CIRs $\mathbf{h}(\tilde{\mathbf{c}}_N)$ entails a search over a field having infinite dimensionality, which results in a infeasible detection strategy. For these reasons, some technical papers argue that, strictly speaking, an optimal data detection strategy cannot be defined when the channel is unknown.

Basically, the shortcut adopted throughout the technical literature consists of first estimating the CIR given each trial sequence $\tilde{\mathbf{c}}_N$ (the resulting estimate is denoted $\tilde{\mathbf{h}}(\tilde{\mathbf{c}}_N)$ in what follows) and then exploiting it for data detection. Such a strategy can be expressed as (see (4.47)):

$$\hat{\mathbf{c}}_N = \arg\max_{\tilde{\mathbf{c}}_N} f_{\mathbf{r}}(\boldsymbol{\rho}_N | \tilde{\mathbf{c}}_N, \mathbf{h}(\tilde{\mathbf{c}}_N)). \tag{4.239}$$

Assuming the same PAM signal model as in Section 4.5.2, from (4.145) the strategy:

$$\hat{\mathbf{c}}_N = \arg\min_{\tilde{\mathbf{c}}_N} [\boldsymbol{\rho}_N - \mathbf{H}(\tilde{\mathbf{c}}_N)\tilde{\mathbf{c}}_N]^H \mathbf{R}_{\mathbf{n}}^{-1} [\boldsymbol{\rho}_N - \mathbf{H}(\tilde{\mathbf{c}}_N)\tilde{\mathbf{c}}_N] \tag{4.240}$$

can easily be inferred, where $\mathbf{H}(\tilde{\mathbf{c}}_N)$ denotes the channel matrix associated with the estimate $\tilde{\mathbf{h}}_N(\tilde{\mathbf{c}})$. If the noise samples belong to a white process, (4.240) simplifies to:

$$\hat{\mathbf{c}}_N = \arg\min_{\tilde{\mathbf{c}}_N} |\boldsymbol{\rho}_N - \mathbf{H}(\tilde{\mathbf{c}}_N)\tilde{\mathbf{c}}_N|^2, \tag{4.241}$$

whereas if the noise is colored, a whitening transformation can be used (see (4.148)), so that (4.240) becomes:

$$\hat{\mathbf{c}}_N = \arg\min_{\tilde{\mathbf{c}}_N} |\mathbf{L}\boldsymbol{\rho}_N - \mathbf{L}\mathbf{H}(\tilde{\mathbf{c}}_N)\tilde{\mathbf{c}}_N|^2, \tag{4.242}$$

where $\mathbf{L}$ is defined as in (4.147). Moreover, if equation (4.240) is expanded and the data-independent term $\mathbf{r}^H \mathbf{R}_{\mathbf{n}}^{-1} \mathbf{r}$ is discarded, then the two equivalent decision rules:

$$\hat{\mathbf{c}}_N = \arg\max_{\tilde{\mathbf{c}}_N} 2\mathrm{Re}\{(\mathbf{H}(\tilde{\mathbf{c}}_N)\tilde{\mathbf{c}}_N)^H \mathbf{R}_{\mathbf{n}}^{-1} \boldsymbol{\rho}_N\} - (\mathbf{H}(\tilde{\mathbf{c}}_N)\tilde{\mathbf{c}}_N)^H \mathbf{R}_{\mathbf{n}}^{-1} (\mathbf{H}(\tilde{\mathbf{c}}_N)\tilde{\mathbf{c}}_N) \tag{4.243}$$

and

$$\hat{\mathbf{c}}_N = \arg\max_{\tilde{\mathbf{c}}_N} 2\mathrm{Re}\{\tilde{\mathbf{c}}_N^H \mathbf{H}^H(\tilde{\mathbf{c}}_N) \boldsymbol{\rho}_N\} - |\mathbf{H}(\tilde{\mathbf{c}}_N)\tilde{\mathbf{c}}_N|^2 \tag{4.244}$$

are obtained for correlated and uncorrelated noise samples, respectively. In practice, the exhaustive search over all the trial sequences $\tilde{\mathbf{c}}_N$ is infeasible when the transmitted sequence is long; for this reason, recursive techniques have been developed to cope with MLSD in the presence of an unknown channel (some of these algorithms will be discussed in Section 6.6). Generally speaking, such methods are based on computing the metrics derived above only for some trial sequences $\tilde{\mathbf{c}}_N$. The trial sequences minimizing the MLSD metrics are called *survivors*, since all the other trial sequences are discarded. The resulting sequence detection approach is known as *per-survivor processing* (PSP) [426, 515], since first the channel is estimated assuming that the survivor sequence $\tilde{\mathbf{c}}_N$ has been transmitted, and then the channel estimate $\mathbf{h}(\tilde{\mathbf{c}}_N)$ is used to compute one of the metrics shown in this section. More details on the PSP approach can be found in Section 5.1.4.

An alternative to joint estimation of channel and data is the EM technique, which allows the development of iterative algorithms for data detection in the presence of a set of unknown parameters. This technique, with some of its applications, is described in the next section.

4.6   Expectation–Maximization Techniques for Data Detection

A powerful tool for solving MLSD problems in the presence of unknown channel parameters is provided by the class of EM techniques. In this section, we provide first a short description of the EM and the *Bayesian* EM (BEM) algorithms and illustrate some simple applications in specific detection problems. Then we briefly discuss the problem of convergence in EM techniques and mention some variants of these algorithms.

4.6.1   The EM Algorithm

Let $\boldsymbol{\theta} \triangleq [\theta_0, \theta_1, \dots, \theta_{L_\theta-1}]^T$ denote an L_θ-dimensional *deterministic* vector to be estimated from an N-dimensional received vector $\mathbf{r}_N \triangleq [r_0, r_1, \dots, r_{N-1}]^T$ of noisy data (with $N \geq L_\theta$). The ML estimate of $\boldsymbol{\theta}$ is the solution of the problem (see (4.26)):

$$\boldsymbol{\theta}_{ML} = \arg \max_{\tilde{\boldsymbol{\theta}}} \, L_{\mathbf{r}}(\tilde{\boldsymbol{\theta}}), \tag{4.245}$$

where:

$$L_{\mathbf{r}}(\tilde{\boldsymbol{\theta}}) \triangleq \log \, f_{\mathbf{r}}(\boldsymbol{\rho}_N | \tilde{\boldsymbol{\theta}}) \tag{4.246}$$

is the log-likelihood function of $\mathbf{r}_N$ given $\boldsymbol{\theta} = \tilde{\boldsymbol{\theta}}$ and $\boldsymbol{\rho}_N$ denotes the value taken by the random vector $\mathbf{r}_N$. As already discussed in Section 4.3.2, solving the problem (4.245) in a direct fashion requires a closed-form expression for $L_{\mathbf{r}}(\tilde{\boldsymbol{\theta}})$ but, even if this expression is available, the search for its maximum may entail an unacceptable computational burden. When this occurs, a feasible alternative can be provided by the EM algorithm [516, 517]. The approach proceeds from the assumption that a *complete* data vector $\mathbf{g}_P = [g_0, g_1, \dots, g_{P-1}]^T$ (with $P \geq N$) is observed in place of the *incomplete* data set $\mathbf{r}_N$. The vector $\mathbf{g}_P$ is characterized by the following two properties:

- It is not observed directly but, if available, would ease the estimation of $\boldsymbol{\theta}$;
- $\mathbf{r}_N$ can be obtained from $\mathbf{g}_P$ through a many-to-one mapping $\mathbf{g}_P \to \mathbf{r}_N(\mathbf{g}_P)$.

In practice, in communication problems $\mathbf{g}_P$ is always chosen as a superset of the incomplete data [516], that is:

$$\mathbf{g}_P = [\mathbf{r}_N^T, \mathbf{i}^T]^T, \tag{4.247}$$

where the so-called *imputed* data $\mathbf{i}$ are properly selected to simplify the ML estimation problem.[7] In particular, when $\boldsymbol{\theta}$ consists of the transmitted channel symbols (i.e., $\boldsymbol{\theta} = \mathbf{c}_N \triangleq [c_0, c_1, \dots, c_{N-1}]^T$), $\mathbf{i}$ often consists of all the unwanted random parameters (CIR, synchronization parameters, etc.) affecting the communication channel [516]. These choices lead to the development of *hard* detection algorithms which often have an acceptable complexity and which are capable of incorporating the statistical properties of the channel parameters.

---

[7] In what follows the complete data vector $\mathbf{g}_P$ will be always structured as in (4.247).

The use of the EM algorithm requires the evaluation of the so-called *auxiliary function*:

$$Q_{EM}(\theta, \tilde{\theta}) \triangleq E_g\{L_g(\theta)|r_N = \rho_N, \theta = \tilde{\theta}\}$$

$$= E_i\{\log f_g(g_P|\theta)|r_N = \rho_N, \theta = \tilde{\theta}\}$$

$$= \int_{S_i} \log f_g(\rho_N, i|\theta) \; f_i(i|\rho_N, \tilde{\theta}) \; di, \tag{4.248}$$

where $E_X\{\cdot\}$ denotes the statistical average with respect to X and S_i is the space of i. In practice, the EM algorithm generates successive approximations $\{\theta^{(k)}, k = 1, 2, \ldots\}$ of θ_{ML} (4.245) in the following two steps:

1. *Expectation step.* Evaluate $Q_{EM}(\theta, \tilde{\theta})$ (4.248) for $\tilde{\theta} = \theta_{EM}^{(k)}$.
2. *Maximization step.* Given $\theta_{EM}^{(k)}$, compute the next estimate $\theta_{EM}^{(k+1)}$ as:

$$\theta_{EM}^{(k+1)} = \arg\max_{\theta} \; Q_{EM}(\theta, \theta_{EM}^{(k)}), \quad k = 0, 1, \ldots. \tag{4.249}$$

Of course, an initial estimate $\theta_{EM}^{(0)}$ of θ must be provided for algorithm startup. In digital communication problems, the evaluation of this estimate is usually accomplished by exploiting the information provided by known (pilot) symbols [516]. It can be proved that, under mild conditions, the sequence $\{\theta_{EM}^{(k)}\}$ converges to the true ML estimate θ_{ML} of (4.245), provided that the existence of local maxima does not prevent it from doing so, as will be further discussed in Section 4.6.3.

To illustrate the EM approach to MLSD over a fading channel, let us focus on its application to the detection of PSK signals transmitted over slow Rayleigh time-selective fading channels.

Example 4.6.1 If PSK signaling over a slowly varying time-selective fading and baud rate sampling at the output of the matched filter of the receiver are assumed, the model (see (4.92)):

$$r_k = a_k \, c_k + n_k \tag{4.250}$$

can be adopted for r_k, with $k = 0, 1, \ldots, N-1$, so that the received vector r_N can be put in the form:

$$r_N = \text{diag}(c_N)a_N + n_N, \tag{4.251}$$

where $a_N \triangleq [a_0, a_1, \ldots, a_{N-1}]^T$ is the fading vector and $n_N \triangleq [n_0, n_1, \ldots, n_{N-1}]^T$ is a random vector of complex Gaussian variables all having zero mean and variance $2\sigma_n^2$. Since (see (4.191)):

$$R_r(\tilde{c}_N) \triangleq E\{r_N \, r_N^H|c_N = \tilde{c}_N\}$$

$$= \text{diag}(\tilde{c}_N)R_a(\text{diag}(\tilde{c}_N))^H + 2\sigma_n^2 I_N$$

$$= \text{diag}(\tilde{c}_N)[R_a + 2\sigma_n^2 I_N](\text{diag}(\tilde{c}_N))^H, \tag{4.252}$$

where:

$$R_a \triangleq E\{a_N \, a_N^H\} \tag{4.253}$$

is the autocorrelation matrix of the fading vector a_N and $\det(R_r(\tilde{c}_N))$ is independent of the trial sequence $\tilde{c}_N$, the MLSD strategy can be formulated as (see (4.194)):

$$\hat{c}_N = \arg\min_{\tilde{c}_N}(\rho^H \text{diag}(\tilde{c}_N))(R_a + 2\sigma_n^2 I_N)^{-1}(\rho^H \text{diag}(\tilde{c}_N))^H. \tag{4.254}$$

In using this approach the following observations are in order:

(a) The strategy expressed by (4.254) requires a (computationally intensive) exhaustive search over a set of M^N trial symbol vectors.

(b) An unambiguous (i.e., unique) estimate of the symbol vector $\mathbf{c}_N$ based on (4.254) can be made only if *differential encoding* is used at the transmitter (see Section 3.5.2) because of the M phase ambiguities of an M-PSK constellation [518].
(c) The structure of the matrix $\mathbf{R_a}$ depends on the specific model adopted for the fading autocorrelation.

As far as the last point is concerned, we note, in particular, that, if Jakes's model is adopted for channel fading (see Example 2.2.8), the (i, j)th element of $\mathbf{R_a}$ is given by:

$$R_a(i, j) = J_0(2\pi |i - j| B_D T_s),$$ (4.255)

with $i, j = 0, 1, \ldots, N - 1$, where $J_0(\cdot)$ is the zero-order Bessel function and B_D is the Doppler bandwidth. To overcome drawback (a) the EM algorithm can be adopted, selecting $\mathbf{i} = \mathbf{a}_N$ and $\boldsymbol{\theta} = \mathbf{c}_N$ for the vector of input data and for the vector of the data to be estimated respectively, so that $\mathbf{g}_P = [\mathbf{r}_N^T, \mathbf{a}_N^T]^T$ ($P = 2N$ in this case). Then the pdf $f_\mathbf{g}(\mathbf{g}_P|\mathbf{c}_N)$ required in the evaluation of $Q_{EM}(\boldsymbol{\theta}, \tilde{\boldsymbol{\theta}})$ (4.248) can be factored as:

$$f_\mathbf{g}(\mathbf{g}_P|\mathbf{c}_N) = f_\mathbf{r,h}(\mathbf{r}_N, \mathbf{a}_N|\mathbf{c}_N) = f_\mathbf{r}(\mathbf{r}_N|\mathbf{a}_N, \mathbf{c}_N) f_\mathbf{a}(\mathbf{a}_N),$$ (4.256)

since $\mathbf{a}_N$ is independent of $\mathbf{c}_N$. Here:

$$f_\mathbf{r}(\mathbf{r}_N|\mathbf{a}_N, \mathbf{c}_N) = \frac{1}{(2\pi\sigma_n^2)^N} \exp\left[-\frac{|\mathbf{r}_N - \text{diag}(\mathbf{c}_N)\,\mathbf{a}_N|^2}{2\sigma_n^2}\right]$$ (4.257)

and:

$$f_\mathbf{a}(\mathbf{a}_N) = \frac{1}{[\pi^N \det(\mathbf{R_a})]} \exp[-\mathbf{a}_N^H \mathbf{R_a}^{-1} \mathbf{a}_N].$$ (4.258)

Taking the logarithm of (4.256) yields (see (4.257) and (4.258)):

$$\log f_\mathbf{g}(\mathbf{g}_P|\mathbf{c}_N) = \log f_\mathbf{r}(\mathbf{r}_N|\mathbf{a}_N, \mathbf{c}_N) + \log f_\mathbf{a}(\mathbf{a}_N)$$

$$= \frac{1}{\sigma_n^2}\text{Re}\{\mathbf{r}_N^H \text{diag}(\mathbf{c}_N)\mathbf{a}_N\} + K,$$ (4.259)

where the term:

$$K \triangleq -N \log(2\pi\sigma_n^2) - \log[\pi^N \det(\mathbf{R_a})] - \frac{[|\mathbf{r}_N|^2 + |\mathbf{a}_N|^2]}{2\sigma_n^2} - \mathbf{a}_N^H \mathbf{R_a}^{-1} \mathbf{a}_N$$ (4.260)

contains all the data-independent terms. Then substituting (4.259) into (4.248), dropping all the data-independent terms and neglecting the factor $1/\sigma_n^2$ yields:

$$Q_{EM}(\mathbf{c}_N, \tilde{\mathbf{c}}_N) = \text{Re}\{\boldsymbol{\rho}_N^H \text{diag}(\mathbf{c}_N) \, \text{E}\{\mathbf{a}_N|\boldsymbol{\rho}_N, \tilde{\mathbf{c}}_N\}\},$$ (4.261)

where the conditional expectation $\text{E}\{\mathbf{a}_N|\boldsymbol{\rho}_N, \tilde{\mathbf{c}}_N\} = \int \boldsymbol{\chi} \, f_\mathbf{a}(\boldsymbol{\chi}|\boldsymbol{\rho}_N, \tilde{\mathbf{c}}_N) \, d\boldsymbol{\chi}$ can be evaluated as [516, eq. 24]:

$$\text{E}\{\mathbf{a}_N|\boldsymbol{\rho}_N, \tilde{\mathbf{c}}_N\} = \mathbf{R_a}(\mathbf{R_a} + 2\sigma_n^2\mathbf{I}_N)^{-1}(\text{diag}(\tilde{\mathbf{c}}_N))^H \boldsymbol{\rho}_N.$$ (4.262)

Note the following observations:

(a) If the fading process is modeled as a Markov process, this conditional expectation can be generated at each step of the EM iteration using a Kalman filter (see Section 5.1.3.2).
(b) The startup of the EM algorithm based on $Q_{EM}(\mathbf{c}_N, \tilde{\mathbf{c}}_N)$ (4.261) requires an initial estimate of the fading vector $\mathbf{a}_N$ and uses it in (4.261) (in place of the factor $\text{E}\{\mathbf{a}_N|\boldsymbol{\rho}_N, \tilde{\mathbf{c}}_N\}$) to produce, by

maximization, the first hard estimate of $\mathbf{c}_N$. This sequence estimate is then used in (4.262) to produce the next fading estimate, and so on, until convergence, which should be expected within a few iterations.

$\square$

An approach similar to that illustrated in Example 4.6.1 has been followed in [519], where a hard detection algorithm for PAM signal over fast doubly-selective channels is developed. On the other hand, a conceptually different approach is pursued in [520, 521], where the problems of detecting PAM over a frequency-selective channel and GMSK signals over a doubly-selective channel (in which channel variations can be represented by a linear model in the time variable), respectively, are investigated. In both cases the choice $\{\mathbf{i} = \mathbf{c}_N, \boldsymbol{\theta} = [\mathbf{h}^T, \sigma_n^2]^T\}$ is proposed in place of $\{\mathbf{i} = \mathbf{h}, \boldsymbol{\theta} = \mathbf{c}_N\}$. This leads to the development of a channel estimation algorithm which is *blind*, that is, unable to incorporate the channel statistics. However, the estimates of the channel parameters are exploited in trellis-based detection algorithms which can incorporate a priori information about channel symbols and generate a hard estimate of the whole symbol sequence $\mathbf{c}_N$ or the *a posteriori probabilities* (APPs) of each symbol.

Interesting applications of EM to OFDM can be found in [522], where this technique is used for multichannel estimation, in [523], where it is exploited, in combination with a *recursive least squares* (RLS) algorithm (see Section 5.1.3.1), for channel estimation to improve performance of approximated MAP detection, and in [524], where it is used to develop an ML detector for ST block-coded OFDM transmission under the assumption that the channel fading processes remain constant over the duration of each code word.

4.6.2 The Bayesian EM Algorithm

The unknown vector $\boldsymbol{\theta} = [\theta_0, \theta_1, \ldots, \theta_{L_\theta-1}]^T$ introduced in the previous subsection can also be modeled as a *random* quantity, when its joint pdf $f(\boldsymbol{\theta})$ is available. Then the MAP estimate $\boldsymbol{\theta}_{MAP}$ of $\boldsymbol{\theta}$, given the observed data vector $\mathbf{r}_N$, can be evaluated as [525]:

$$\boldsymbol{\theta}_{MAP} = \arg\max_{\tilde{\boldsymbol{\theta}}} M_{\mathbf{r}}(\tilde{\boldsymbol{\theta}}), \tag{4.263}$$

where $M_{\mathbf{r}}(\tilde{\boldsymbol{\theta}}) \triangleq \log f(\boldsymbol{\rho}_N, \tilde{\boldsymbol{\theta}})$ and $\boldsymbol{\rho}_N$ denotes the value taken on by the random vector $\mathbf{r}_N$. Solving (4.263) may be a formidable task for the reasons previously presented for the ML problem (4.245). In principle, however, an improved estimate of $\boldsymbol{\theta}$ can be made via the MAP approach since statistical information about channel uncertainty is exploited.

Since there is a strong analogy between the ML problem (4.245) and the MAP problem (4.263), it is not surprising that an expectation–maximization procedure, known as BEM [526, 527], for solving the latter is available. The BEM algorithm goes through the same iterative procedure as the EM, but with a different auxiliary function, namely, [526, 528]:

$$Q_{BEM}(\boldsymbol{\theta}, \tilde{\boldsymbol{\theta}}) = \mathrm{E}_{\mathbf{g}}\{M_{\mathbf{g}}(\boldsymbol{\theta})|r_N = \rho_N, \boldsymbol{\theta} = \tilde{\boldsymbol{\theta}}\}$$

$$= \mathrm{E}_{\mathbf{i}}\{\log f_{\mathbf{g},\boldsymbol{\theta}}(\mathbf{g}_P, \boldsymbol{\theta})|\mathbf{r}_N = \rho_N, \boldsymbol{\theta} = \tilde{\boldsymbol{\theta}}\}$$

$$= \int_{S_{\mathbf{i}}} \log f_{\mathbf{r},\mathbf{i},\boldsymbol{\theta}}(\rho_N, \mathbf{i}, \boldsymbol{\theta}) \, f_{\mathbf{i}}(\mathbf{i}|\rho_N, \tilde{\boldsymbol{\theta}}) \, d\mathbf{i}, \tag{4.264}$$

where the vector $\mathbf{g}_P$ of *complete data* is structured as in (4.247) and $\mathbf{i}$ denotes the vector of *imputed data*. A clear relationship can be established between the BEM and the EM algorithm. In fact, factoring the pdf $f_{\mathbf{r},\mathbf{i},\boldsymbol{\theta}}(\rho_N, \mathbf{i}, \boldsymbol{\theta})$ as:

$$f_{\mathbf{r},\mathbf{i},\boldsymbol{\theta}}(\rho_N, \mathbf{i}, \boldsymbol{\theta}) = f_{\mathbf{r},\mathbf{i},\boldsymbol{\theta}}(\mathbf{r}_N, \mathbf{i}|\boldsymbol{\theta}) \, f_{\boldsymbol{\theta}}(\boldsymbol{\theta}) \tag{4.265}$$

and substituting (4.265) into (4.264) yields:

$$Q_{BEM}(\boldsymbol{\theta}, \tilde{\boldsymbol{\theta}}) = Q_{EM}(\boldsymbol{\theta}, \tilde{\boldsymbol{\theta}}) + I(\boldsymbol{\theta}), \tag{4.266}$$

where:

$$I(\boldsymbol{\theta}) \triangleq \log f_{\boldsymbol{\theta}}(\boldsymbol{\theta}). \tag{4.267}$$

Equation (4.266) shows that the difference between $Q_{BEM}(\boldsymbol{\theta}, \tilde{\boldsymbol{\theta}})$ (4.264) and $Q_{EM}(\boldsymbol{\theta}, \tilde{\boldsymbol{\theta}})$ (4.248) is simply a *bias* term $I(\boldsymbol{\theta})$ favoring the most likely values of $\boldsymbol{\theta}$. Note that, if a priori information about $\boldsymbol{\theta}$ were unavailable and, consequently, a uniform pdf were selected for $f(\boldsymbol{\theta})$, the contribution from $I(\boldsymbol{\theta})$ would become a constant in (4.266), and could then be neglected. Therefore, the BEM encompasses the EM as a special case and, since the former benefits from the statistical information about $\boldsymbol{\theta}$, it is expected to provide improved accuracy with respect to the latter. For the same reason, the BEM may offer an increase in the speed of convergence and an improved robustness against the choice of the initial conditions. The importance of the BEM technique can also be related, however, to the nature of the detection algorithms it can produce. In fact, as already mentioned in the previous subsection, in the problems concerning the detection of a symbol vector $\mathbf{c}_N$ in the presence of an unknown vector $\mathbf{h}$ of channel parameters, two different choices can be made for the imputed data $\mathbf{i}$ and the estimated $\boldsymbol{\theta}$, namely, $\mathbf{i} = \mathbf{h}$ and $\boldsymbol{\theta} = \mathbf{c}_N$ and also $\mathbf{i} = \mathbf{c}_N$ and $\boldsymbol{\theta} = \mathbf{h}$. In the first case *hard* estimates of the transmitted data are produced by both the EM and the BEM algorithms; however, if BEM is employed in place of EM, the data statistics are included in the detection algorithm, since $I(\boldsymbol{\theta})$ in (4.266) takes the form:

$$I(\boldsymbol{\theta}) = I(\mathbf{c}_N) = \sum_{n=0}^{N-1} \log \Pr\{c_n\}, \tag{4.268}$$

where $\Pr\{c_n\}$ denotes the a priori probability of c_n. This leads to a *soft-input hard-output* (SiHo) detection algorithm. On the other hand, using the second choice, application of EM results in a blind channel estimation algorithm which can generate the data APPs as a by-product and incorporate a priori information about channel symbols, whereas using the BEM technique allows the development of a *soft-in soft-out* (SiSo) detection algorithm incorporating the channel statistical properties, as shown in the following example.

Example 4.6.2 Let us reconsider the problem of PSK detection over a slowly varying time-selective fading channel as in Example 4.6.1. In this case, we choose $\mathbf{i} = \mathbf{c}_N$ and $\boldsymbol{\theta} = \mathbf{h}_N$, so that the auxiliary function (4.264) becomes:

$$Q_{BEM}(\mathbf{h}_N, \tilde{\mathbf{h}}_N) = \sum_{l=0}^{M^N-1} \log[f_{\mathbf{r},\mathbf{h}}(\mathbf{r}_N, \mathbf{h}_N | \mathbf{c}_N^{(l)}) \Pr\{\mathbf{c}_N^{(l)}\}] \Pr\{\mathbf{c}_N^{(l)} | \mathbf{r}_N, \tilde{\mathbf{h}}_N\}, \tag{4.269}$$

where $\mathbf{c}_N^{(l)} = [c_0^{(l)}, c_1^{(l)}, \ldots, c_{N-1}^{(l)}]$ denotes the lth possible value of the symbol vector $\mathbf{c}_N$ (with $l = 0, 1, \ldots, M^N - 1$) and is characterized by its a priori probability:

$$\Pr\{\mathbf{c}^{(l)}\} = \prod_{n=0}^{N-1} \Pr\{c_n^{(l)}\}, \tag{4.270}$$

where $\Pr\{c_n^{(l)}\}$ denotes the a priori probability of the event $\{c_n = c_n^{(l)}\}$.

The conditional pdf $f_{\mathbf{r},\mathbf{h}}(\mathbf{r}_N, \mathbf{h}_N | \mathbf{c}_N)$ of (4.269) can be factored as:

$$f_{\mathbf{r},\mathbf{h}}(\mathbf{r}_N, \mathbf{h}_N | \mathbf{c}_N) = f_{\mathbf{r}}(\mathbf{r}_N | \mathbf{h}_N, \mathbf{c}_N) f_{\mathbf{h}}(\mathbf{h}_N), \tag{4.271}$$

where $f_{\mathbf{r}}(\mathbf{r}_N|\mathbf{h}_N, \mathbf{c}_N)$ and $f_{\mathbf{h}}(\mathbf{h}_N)$ are given by (4.257) and (4.258), respectively. Then substituting (4.257) and (4.258) into (4.271), and (4.271) into (4.269) yields, after some manipulation [526]:

$$Q_{BEM}(\mathbf{h}_N, \tilde{\mathbf{h}}_N) = -\sum_{k=0}^{N-1}\sum_{l=0}^{M-1} \Pr\{c_k^{(l)}|\mathbf{r}_N, \tilde{\mathbf{h}}_N\} \frac{|r_k - h_k\, c_k^{(l)}|^2}{2\sigma^2} - \mathbf{h}_N^H\, \mathbf{R}_{\mathbf{h}}^{-1}\, \mathbf{h}_N. \tag{4.272}$$

Note that, for any k, the M distinct APPs $\{\Pr\{c_k^{(l)}|\mathbf{r}_N, \tilde{\mathbf{h}}_N\}\}$ of the kth channel symbol can be evaluated as:

$$\Pr\{c_k^{(l)}|\mathbf{r}_N, \tilde{\mathbf{h}}_N\} = \frac{f_r(r_k|c_k^{(l)}, \tilde{h}_k)\ \Pr\{c_k^{(l)}\}}{\displaystyle\sum_{p=0}^{M-1} f_r(r_k|c_k^{(p)}, \tilde{h}_k)\ \Pr\{c_k^{(p)}\}}, \tag{4.273}$$

where $f_r(r_k|c_k, h_k) = (2\pi\sigma_n^2)^{-1} \exp[-|r_k - c_k\, h_k|^2/(2\sigma_n^2)]$. The kth iteration of the resulting estimation algorithm consists of an *expectation step*, where $Q_{BEM}(\mathbf{h}_N, \tilde{\mathbf{h}}_N)$ (4.272) is evaluated for $\tilde{\mathbf{h}}_N = \mathbf{h}_N^{(k)}$, and a *maximization step*, where the new channel estimate $\mathbf{h}_N^{(k+1)}$ is derived as:

$$\mathbf{h}_N^{(k+1)} = \arg\max_{\tilde{\mathbf{h}}_N} Q_{BEM}(\mathbf{h}_N, \tilde{\mathbf{h}}_N), \qquad k = 0, 1, \ldots, \tag{4.274}$$

and the symbol APPs $\{\Pr\{c_k^{(l)}|\mathbf{r}_N, \tilde{\mathbf{h}}_N\}, l = 0, 1, \ldots, M-1; \quad k = 0, 1, \ldots, N-1\}$ are computed using (4.273). The final iteration produces the channel estimate $\mathbf{h}_{BEM}$, which can be used to generate the final estimates of the symbol APPs. These APPs can be processed so that decisions on the channel symbols can be taken according to the MAP decision strategy [520]:

$$\hat{c}_k = \arg\max_l\ \Pr\{c_k^{(l)}|\mathbf{r}_N, \mathbf{h}_{BEM}\} \tag{4.275}$$

with $k = 0, 1, \ldots, N-1$ or can be delivered to soft decoding stages, as discussed in [526, 529], to improve the error performance of a digital receiver (see also Section 12.3.5). Note that the MAP SiSo detection algorithms for PSK signals over frequency-flat fading channels are available in the technical literature [530–533], but they require a complicated (forward–backward) recursive procedure (see Section 6.2.1.5 for details) that operates over a finite-state trellis (accounting for the correlation between consecutive fading samples) and which involves the implicit evaluation of multiple per-state estimates of the fading distortion in each symbol interval. Unlike these, the BEM procedure keeps in memory only one estimate of each fading sample even if the evaluation of the final estimates requires multiple iterations.

□

Finally, it is worth noting that BEM-based SiSo algorithms have also been developed for the detection of:

(a) co-channel PAM signals over frequency-flat fading channels [534],
(b) a PAM signal over a frequency-selective channel [528],
(c) an orthogonal ST block coded PAM transmission over a frequency-flat fading channel [535],
(d) a CPM signal transmitted over a frequency-flat fading channel [529].

The EM approach has been also adopted for developing MAP SiSo detection of ST coded OFDM signals in [536]. Note, however, that there the derivation of the strategy does not follow the rigorous approach we propose.

4.6.3 Initialization and Convergence of EM-Type Algorithms

It can be proved that, when the EM algorithm is used for the estimation of an L_θ-dimensional *deterministic vector* $\boldsymbol{\theta}$ from an N-dimensional received vector $\mathbf{r}_N$ of noisy data (with $N \geq L_\theta$), it generates a sequence of estimates $\{\boldsymbol{\theta}_{EM}^{(k)}, k = 1, 2, \ldots\}$ characterized by nondecreasing likelihood values [537, 538], that is, such that:

$$L_{\mathbf{r}}(\boldsymbol{\theta}_{EM}^{(k+1)}) \geq L_{\mathbf{r}}(\boldsymbol{\theta}_{EM}^{(k)}), \tag{4.276}$$

where the log-likelihood function $L_{\mathbf{r}}(\boldsymbol{\theta})$ is defined by (4.246). For a bounded sequence of likelihood values $\{L_{\mathbf{r}}(\boldsymbol{\theta}_{EM}^{(k)})\}$, $L_{\mathbf{r}}(\boldsymbol{\theta}_{EM}^{(k)})$ converges monotonically to some value $L(\bar{\theta}_{EM})$, which in almost every application is a stationary point, that is, is such that:

$$\nabla_{\boldsymbol{\theta}} L_{\mathbf{r}}(\tilde{\boldsymbol{\theta}})|_{\tilde{\boldsymbol{\theta}} = \bar{\boldsymbol{\theta}}_{EM}} = \mathbf{0}_{L_\theta}, \tag{4.277}$$

where $\nabla_{\mathbf{X}}$ denotes the gradient operator (involving a variable vector $\mathbf{X}$). If the likelihood function is unimodal in its domain and certain differentiability conditions are satisfied, any EM sequence converges to the unique ML estimate, independently of the starting point $\boldsymbol{\theta}_{EM}^{(0)}$. Unfortunately, the function $L_{\mathbf{r}}(\tilde{\boldsymbol{\theta}})$ usually has several stationary points, so that convergence of the EM to local or global maximizers depends on the choice of $\boldsymbol{\theta}_{EM}^{(0)}$.

The convergence rate of an EM algorithm is inversely related to the *Fisher information* of its complete-data space [539]. Less informative complete data spaces lead to improved asymptotic convergence rates, and can also lead to larger step sizes and greater likelihood increases in the early iterations (see [540] and references therein). Further information about the convergence properties of the EM algorithm can be found in [537, 538, 541, 542].

4.6.4 Other EM Techniques

In some practical applications, the *maximization step* (4.249) can be quite complicated, so that the EM algorithm is not very attractive. In these cases, the *expectation/conditional maximization* (ECM) algorithm can be used [543]. The ECM algorithm replaces the M-step of the EM algorithm by a number of computationally simpler *constant modulus* (CM) steps. This usually leads to slower convergence, with the advantage, however, of a smaller overall computational complexity. In addition, the appealing convergence properties of the EM algorithm, such as its monotonic convergence (see (4.276)), are preserved.

Generalizations of the ECM are also available in the technical literature; here we mention the *expectation/conditional maximization either* (ECME) algorithm [544] and the *alternating ECM* (AECM) algorithm [545]. On the one hand, in the ECME some CM-steps of the ECM, which maximize the constrained *expected* complete-data log-likelihood function, are replaced with steps that maximize the correspondingly constrained *actual* likelihood function. The ECME shares with both EM and ECM their stable monotone convergence and basic simplicity of implementation. Moreover, the ECME can offer a substantially faster convergence rate than either EM or ECM, measured using either the number of iterations or the actual overall computational burden. On the other hand, in the AECM the specification of the complete data is allowed to change in each CM-step; this can lead to computationally efficient solutions.

In the signal processing community a variant of the EM algorithm that has received considerable attention is the so-called *space-alternating generalized* EM (SAGE) algorithm [540, 546]. At each iteration, the SAGE algorithm updates only a subset of the components of the parameter vector $\boldsymbol{\theta}$, so that multiple (and less informative) complete data sets are used. This results in an improvement in convergence rate and in significant flexibility. Interesting signal processing applications of the SAGE algorithm can be found in [547, 548], which deal with the problem of joint channel estimation, equalization, and data detection for uplink OFDM systems in the presence of a fast doubly-selective channel.

4.7 Historical Notes

The problems of extracting optimal decision metrics from a set of noisy received data and of assessing the error performance for detection strategies based on such metrics are analyzed in a vast technical literature dealing with optimal detection over fading channels. For this reason, in this section we restrict our attention to some key concepts and techniques which have been introduced in the previous sections of this chapter and briefly illustrate their development.

A first essential concept in detection theory is that of *matched filtering*, since this is usually an essential tool for extracting a set of sufficient statistics from a continuous-time noisy waveform. In the radar literature the MF is often referred to as a *North filter*, after D. O. North, who first described it in an RCA report published in 1943 [549]. The same filter was independently rediscovered in 1946 by J. H. Van Vleck and D. Middleton [550], who investigated the sensitivity of the detection of signal pulses in the presence of noise and coined the term *matched filter* for the optimal filter. The importance of matched filtering in optimal detection of a set of known waveforms transmitted over AWGN channels soon became apparent in the communication theory community (e.g., see [106, 280, 321]). In particular, it was shown that the optimum linear receiving filter under various criteria of goodness can be expressed as the cascade of a matched filter and a *transversal filter* (see [551] and references therein). It is also worth mentioning that: the transversal filter is *time-invariant* if the criterion is the minimization of the ensemble average of some quantity per symbol and the transmitted symbol sequence is long enough so that end effects are not important [422]; and the transversal filter can be used for *whitening* the noise samples in the output sequence of the matched filter, and this results in a WMF [422, 423].

Matched filters are an essential component of both optimal *one-shot* detectors (e.g., of detectors for a single pulse if PAM signaling is considered) and optimal *sequence* detectors. *Optimal one-shot detection* may be of limited interest from a practical viewpoint, but the study of its error performance allows limits to be established on the performance that can be achieved by optimal sequence detection over multipath fading channels. In particular, as already illustrated in Section 4.5.2.4, the average error probability of an optimum one-shot receiver instantaneously matched to the channel state and a given signaling format represents the MFB and sets performance limits for any equalizer designed for that signaling format transmitted over a channel with given statistical properties [467, 468, 471]. MFBs for frequency-selective and doubly-selective Rayleigh channels have been derived in [85, 109, 162, 465–467, 469–472, 474–476, 478, 479, 552]. Closed-form expressions for these bounds are available only for frequency-selective channels characterized by a finite number of paths (e.g., see [465, 466, 470, 475, 552]), whereas approximate error formulas have been derived for continuous PDPs (e.g., see [85, 109, 471–474, 478]). The importance of matched filtering in *optimal sequence detection* has been pointed out by D. G. Forney in [422] and by G. Ungerboeck in [458], who developed two different solutions to MLSD (i.e., to *maximum likelihood sequence estimation* (MLSE)) for PAM signaling over frequency-selective channels.

Another fundamental concept, closely related to the MF, in the evaluation of detection metrics over fading channels is that of the *estimator-correlator* and is due to T. Kailath [484, 499]. Note that the MF concept plays a fundamental role whenever the optimum receiver needs to cross-correlate the noisy received waveform with a known set of N waveforms $\{s_i(t), i = 0, 1, \ldots, N-1\}$ to extract a set of sufficient statistics. In fact, in this case the cross-correlations can be evaluated by feeding a bank of N filters matched to the signals $\{s_i(t)\}$ and sampling the filter outputs at an appropriate instant. If the channel is purely random and Gaussian, however, optimal reception requires *cross-correlating* the received signal against a set of waveforms which are not known a priori, but are *estimated* from the received data; for this reason, the receiver can be described as an estimator-correlator receiver. Note that the estimator correlator section of this receiver can also be regarded as a bank of *adaptive* MFs, since each cross-correlation can be alternatively computed via an MF whose impulse response is computed from the received data; further details can be found in [553].

All the concepts illustrated above find application in the study of MLSE over fading channels. Two milestones in the field of MLSE over a *known* frequency-selective channel are represented by the papers [422] (published in 1972) and [458] (published in 1974), which presented optimal sequence detection algorithms exploiting the outputs of a WMF and an MF, respectively, and the *Viterbi algorithm* (VA) for reducing the computational load of the search for the ML estimate of the transmitted sequence. Their solutions were generalized to the case of the optimal detection of multiple co-channel signals transmitted over frequency-selective channels by W. van Etten in 1976 [554]. Many years later, in 1995, G. E. Bottomley and S. Chennakeshu provided a unified development of both receivers, and extended Ungerboeck's receiver to the case of a time-varying *known* channel [424, 555] (the latter problem is also investigated in [425]). In addition, Ungerboeck's derivation of the MLSE receiver for the purely frequency-selective channel was extended to the time-selective one in [556]. Note that the MLSE metrics developed for known channels can also be used when an unknown channel is estimated entirely via a pilot sequence (e.g., see [557, 558]) or combining training-based channel estimation with tracking based on preliminary data decisions (e.g., see [511, 559–561]), and then the resulting CIR estimate is used as if it were ideal.

The problem of MLSE over a *statistically known channel* looks much more complicated than its counterpart referring to a *known channel*, since it is hard to put the estimator-correlator metrics (i.e., a set of likelihood functions for Gaussian signals in Gaussian noise) in a form lending itself to a real-time computation and allowing an efficient search for the optimal sequence estimate. As far as we know, the first technically significant result in relation to this problem was devised in 1965 by F. C. Schweppe [485], who employed finite-dimensional state variable models (i.e., Markov models) to derive new expressions for the likelihood functions for Gaussian signals corrupted by AGN; unfortunately, his work assumed only two distinct hypotheses only for the transmitted signal, namely signal present and signal absent. The first receiver design for the MLSE of digital data transmission through a randomly dispersive fading channel was developed by R. E. Morley, Jr. and D. L. Snyder in 1979 [562]. They proved that the likelihood function for the possible information sequences can be evaluated through a recursive expression if the channel has *finite memory*; this allows the VA to be used to search for the optimal (i.e., ML) sequence estimate, as in the known channel case. These ideas paved the way for future work. In the 1990s various approximate MLSD algorithms were developed for different types of digital signals and fading channels, assuming that the channel fading can be described by AR or ARMA models [428, 436]; this hypothesis makes a recursive formulation of ML metric possible, so that the VA can be employed for an efficient evaluation of the sequence estimate. In particular, this approach has been adopted by J. H. Lodge and M. J. Moher [436] for CPM detection over frequency-flat fading channels, by Q. Dai and E. Shwedyk [428] for PAM detection over DS fading channels, by R. A. Iltis for the detection of PAM signals over frequency-selective channels [563], and by G. M. Vitetta and D. P. Taylor [111, 427, 490] and D. Makraris, P. T. Mathiopoulos and D. P. Bouras [501] for PAM signals over frequency-flat fading channels. It can be shown that the solutions proposed in the latter references share the following principle: in evaluating sequence metrics, multiple channel estimates are evaluated, one for each survivor of the VA, so that joint data detection and channel estimation are accomplished. For this reason, these and other related solutions (e.g., see [561, 564–568]) can be related to the so-called principle of *per-survivor processing* [515, 569]. A further contribution in this area was provided in 1996 by K. Chugg and A. Polydoros, who investigated the problem of the front-end processing in MLSE for an unknown channels and the problem of recursive computation of ML metrics, providing a receiver structure which may be interpreted as the theoretical foundation for the technique of PSP [426, 512]. A different interpretation of MLSE algorithms for a statistically known Rayleigh fading channel was proposed in 1995 by X. Yu and S. Pasupathy [492], who took an innovations-based approach.[8] This has led to the development of a general and practical MLSE technique that can be implemented by a bank of FIR time-invariant filters followed by a Viterbi processor and is applicable to any practically modulated signal over either frequency-nonselective or selective, fast or slowly fading channels; further work in this area can be found in [493, 572].

A completely different approach to MLSE over an unknown linear channel was described in 1994 by N. Seshadri, who proposed estimating the data and the channel simultaneously (i.e., *jointly*), so that a startup sequence for estimating the channel impulse response is not required [419]; in other words, *blind* sequence detection is accomplished. In principle a *least squares* (LS) channel estimate is evaluated first for every possible transmitted data sequence; then computing the total squared error between the received vector and the useful signal vector associated with each pair of data sequence and the corresponding estimated channel provides a decision metric. The optimal decision is associated with the pair giving the smallest metric. This simple but exhaustive search technique cannot be implemented in practice because of the exponential growth in complexity with the length of the data sequence. This raises the problem of deriving suboptimal strategies for joint channel estimation and data detection. This problem and other aspects of ML joint channel estimation and data detection have been tackled in [419, 573–577]. Conceptually different approaches to the problem of blind ML detection have been adopted in [578], where a cluster-based sequence equalizer is proposed, and in various technical papers proposing the joint use of the EM algorithm (for channel estimation) with conventional optimal detectors (e.g., see [520, 521, 579, 580]). As already discussed in Section 4.6 and as originally pointed out in 1997 by C. N. Georghiades and J. C. Han in their seminal paper [516], this is not the only way to exploit the EM algorithm for ML data detection. Further interesting contributions on the application of the EM algorithm to MLSD in the presence of an unknown channel can be found in [519, 581].

We conclude this section with the following worthwhile observations:

(a) The work on MLSE mentioned above mainly refers to SISO or SIMO channels; some interesting results on the problem of MLSE over MIMO channels can be found in [460, 582, 583].
(b) Our historical notes have focused hitherto on MLSD. Here we limit ourselves to mentioning that in 1966 MAPSD metrics were put in a form allowing efficient computation by R. W. Chang and J. C. Hancock [584], who considered a frequency-selective channel; as in the MLSD case, MAP metrics for a known channel can also be used if the channel is estimated via a pilot sequence (e.g., see [558]). The problem of developing MAP metrics for a statistically known channel was investigated many years later in [504, 530].
(c) A limited number of analytical tools are available for assessing the error performance of MLSE algorithms; the interested reader can refer to [585] and to the papers cited at the end of Section 4.5.2.2 to gain some insight into this problem.

4.8 Further Reading

The reader interested in the problems of prefiltering an analog data signal and sampling it to extract a set of sufficient statistics should consult the paper [586]. MLSD has been applied to research areas which are not considered in this book. For instance, MLSD in multiuser scenarios have been investigated in [587–590], whereas joint ST decoding and MLSD has been tackled in [583, 591, 592]. Finally, it is useful to mention that an analysis of the EM algorithm and its variations can be found in [537].

---

[8] Further information about the innovations approach to signal detection can be found in [570, 571].

5

Data-Aided Algorithms for Channel Estimation

Various equalization algorithms described in Chapter 6 are derived under the assumption that ideal CSI is available at the receiver. In practice, CSI can be acquired at the receiver using *pilot-aided transmission* (PAT) techniques in combination with a proper channel estimation algorithm. This is illustrated in Figure 5.1, which depicts a wireless communication system based on a PAT transceiver. In this scheme a block $c^{(d)}$ of information-bearing (coded) symbols is time- or frequency-multiplexed with a block $c^{(p)}$ of *known* (i.e., *pilot*) symbols. The resulting packet is transmitted over a multipath communication channel using a specific modulation scheme. The pilot symbols and the adopted multiplexing scheme are known at the receiver, which can then exploit them to generate channel estimates for receiver adaptation and optimal detection/decoding. For instance, in Figure 5.1 a *parametric model* is assumed for the representation of the multipath channel and the task of the channel estimation algorithm is to generate an estimate of a finite-dimensional vector $\mathbf{h}$ of channel parameters. The channel estimator processes the pilot vector $c^{(p)}$ and the entire (or a part of the) observation vector $\mathbf{y}$, to produce a *data-aided channel estimate*, $\hat{\mathbf{h}}$, which it delivers to the detector/decoder, thus generating an estimate $\hat{\mathbf{c}}^{(d)}$ of $c^{(d)}$. A practical detector/decoder may assume that the estimated channel parameters are *perfect*. Of course this assumption is not strictly true, so that in practice *mismatched* reception is used [593, 594].

It is important to note that all the available architectures for PAT transceivers represent a parametric approach to channel estimation. In principle, a *nonparametric approach* can also be used for receiver design in the presence of pilot symbols. In such a case, however, these symbols are treated as side information and *are used to tune the receiver directly* (e.g., for training an adaptive equalizer), so that an explicit channel estimator, such as that appearing in Figure 5.1, is no longer required. We note also that both in the parametric and nonparametric approach pilot-based estimation/adaptation can be followed by *decision-aided estimation/adaptation*. For instance, referring to the scenario of Figure 5.1, a data-aided preliminary estimate $\hat{\mathbf{h}}$ is exploited to start the detection/decoding procedure, and then the decoded data are used to refine this estimate. This approach is usually taken when detection/decoding are accomplished in multiple passes to improve the quality of data decisions or when the channel variations over the transmitted data packet are significant and, consequently, need to be properly tracked to achieve acceptable error performance. PAT techniques have been widely used in modern wireless communication systems (e.g., in the GSM system [57] and in *digital video broadcasting – terrestrial* (DVB-T) [595]). This is because such techniques substantially simplify the task of receiver design for unknown channels and because the periodic availability of pilot symbols

Wireless Communications: Algorithmic Techniques, First Edition.
Giorgio M. Vitetta, Desmond P. Taylor, Giulio Colavolpe, Fabrizio Pancaldi, Philippa A. Martin.
© 2013 John Wiley & Sons, Ltd. Published 2013 by John Wiley & Sons, Ltd.

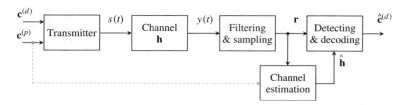

Figure 5.1 Block diagram of a wireless communication system employing a PAT transceiver. A parametric representation is assumed for the communication channel.

offers the possibility of link recovery from outages. However, the clever use of PAT requires a number of problems to be solved. On the one hand, various parameters, such as the pilot symbol rate, the power allocated to pilot symbols, and the locations of these pilots in the data stream, need to be carefully selected in order to achieve the desired system performance with the minimum use of resources. On the other hand, PAT should be devised in such a way that channel estimation algorithms of reasonable complexity can be implemented so as to take fully advantage of the presence of training symbols.

This chapter is devoted to the study of PAT techniques and data-aided channel estimation algorithms. In particular, various feedforward and recursive algorithms for channel estimation are considered in Section 5.1, and performance limits on channel estimation are analyzed in Section 5.2. An overview of PAT design and specific channel estimation algorithms for SC and MC communications is provided in Section 5.3. Then extensions of SISO/SIMO techniques to a MIMO scenario are discussed in Section 5.4. Finally, some historical notes and suggestions for further reading are provided in Sections 5.5 and 5.6, respectively.

5.1 Channel Estimation Techniques

5.1.1 Introduction

Generally speaking, when a parametric model is assumed to represent a communication channel (see Section 2.2.3), the task of *channel estimation* is to estimate an L_h-dimensional vector $\mathbf{h} = [h_0, h_1, \ldots, h_{L_h-1}]^T$ of channel parameters from an N-dimensional vector $\mathbf{r} = [r_0, r_1, \ldots, r_{N-1}]^T$ of noisy observations. Note that $\mathbf{h}$ may contain the tap gains of an FIR multipath fading channel model or the coefficients of a polynomial model for one of its system functions.

The problem of estimating $\mathbf{h}$ from $\mathbf{r}$ can be tackled by resorting to various tools provided by *estimation theory* [596, Chap. 4]. In particular, *classical estimation* or *Bayesian estimation* techniques can be applied. In one approach using *classical estimation*, $\mathbf{h}$ is assumed to be *deterministic*, but *unknown*. Then the pdf of $\mathbf{r}$ is parametrized by $\mathbf{h}$, so that it is usually denoted[1] by $f(\mathbf{r}; \mathbf{h})$ (the semicolon means that a family of pdfs is generated by changing the value of the parameter vector). Alternatively, in *Bayesian estimation* the channel parameter vector is viewed as a *realization of a random vector* $\mathbf{h}$ and then the data are described by the joint pdf:

$$f(\mathbf{r}, \mathbf{h}) = f(\mathbf{r}|\mathbf{h}) f(\mathbf{h}), \tag{5.1}$$

where $f(\mathbf{h})$ is the prior pdf (representing our knowledge of $\mathbf{h}$ before data observation) and $f(\mathbf{r}|\mathbf{h})$ is a conditional pdf expressing our knowledge provided by the data $\mathbf{r}$ given $\mathbf{h}$.

---

[1] The alternative notation $f(\mathbf{r}|\mathbf{h})$ (denoting a conditional pdf) is often assumed in the technical literature, as if $\mathbf{h}$ were random.

In the technical literature two different approaches to the identification of the unknown (and possibly time-varying) transmission environment are proposed. One approach, known as *recursive estimation*, is based on devising *recursive-type adaptive filters* and then sequentially processing the samples of **r** as soon as they are available. These usually offer the relevant advantage of a limited complexity, but suffer from an irreducible *lag error* (and thus exhibit limited robustness against faster time variations and depend on final or tentative symbol decisions which may not be readily available or reliable, especially in the presence of coding and interleaving). The other approach, known as *feedforward estimation*, consists of simultaneously processing the components of the vector **r**, which collects noisy samples *depending on known training symbols* (i.e., pilots) *only*.

In Sections 5.1.2 and 5.1.3 we discuss the main features of the two approaches and provide some examples. Then in Section 5.1.4 we discuss the application of so-called *principle of per-survivor processing* to the problem of decision-directed channel estimation.

5.1.2 Feedforward Estimation

5.1.2.1 Maximum Likelihood Estimation

In classical estimation we are interested in obtaining *unbiased* estimators of **h** (yielding on average the true value of the unknown parameter vector) which, at the same time, achieve *minimum variance*. In other words, we are interested in a *minimum variance unbiased* (MVU) estimator. Generally speaking, the MVU estimator may not exist or, if it exists, cannot be found. An alternative to the MVU estimator is the *maximum likelihood estimator* (MLE), which is defined as the value $\mathbf{h}_{ML}$ of **h** that maximizes the *likelihood function* $f(\mathbf{r}; \mathbf{h})$. This estimator is shown to be asymptotically optimal, in that it exhibits optimal behavior for $N \to \infty$. In fact, it has the asymptotic properties of being unbiased, of achieving the so-called *Cramér–Rao bound* (CRB)[2] and of having a Gaussian distribution. When a closed-form solution cannot be found for $\mathbf{h}_{ML}$, a numerical approach based on a grid search or on an iterative maximization of the likelihood function can be developed. These concepts are applied in the following example.

Example 5.1.1 Let us assume that the dependence of the data vector **r** on the channel parameter vector **h** can be expressed as:

$$\mathbf{r} = \mathbf{A}\mathbf{h} + \mathbf{w}, \tag{5.2}$$

where **A** is a *known* matrix of dimension $N \times L_h$ (with $N > L_h$) depending only on a set of pilot symbols and **w** is a Gaussian complex noise vector belonging to $\mathcal{CN}(\mathbf{0}_N, \mathbf{C}_w)$.[3] In estimation theory the expression (5.2) represents a *linear model* and **A** is referred to as the *observation matrix* (e.g., see [596, Chap. 4]). It can be shown that, if the rank of **A** is equal to L_h (i.e., **A** is of full rank), the MVU estimator exists and is given by:

$$\hat{\mathbf{h}} = (\mathbf{A}^H \mathbf{C}_w^{-1} \mathbf{A})^{-1} \mathbf{A}^H \mathbf{C}_w^{-1} \mathbf{r}. \tag{5.3}$$

If the noise samples are uncorrelated, that is, $\mathbf{C}_w = \sigma_w^2 \mathbf{I}_N$ (where σ_w^2 is the variance of noise samples) this simplifies to:

$$\hat{\mathbf{h}} = (\mathbf{A}^H \mathbf{A})^{-1} \mathbf{A}^H \mathbf{r}. \tag{5.4}$$

It can also be easily proved that the following hold:

(a) $\hat{\mathbf{h}}$ in (5.3) is a Gaussian random vector having mean **h** and covariance matrix $\mathbf{C}_{\hat{\mathbf{h}}} = (\mathbf{A}^H \mathbf{C}_w^{-1} \mathbf{A})^{-1}$.

---

[2] This topic is further discussed in Section 5.2.

[3] $\mathcal{CN}(\boldsymbol{\eta}_x, \mathbf{C}_x)$ denotes a complex Gaussian vector characterized by the mean vector $\boldsymbol{\eta}_x$ and covariance matrix $\mathbf{C}_x$.

(b) $\hat{\mathbf{h}}$ is *efficient* since it attains the minimum possible variance (the so-called CRB; see Section 5.2) of the estimation error in the class of all possible unbiased estimators of $\mathbf{h}$.

(c) The MLE coincides with the MUV estimator, that is, the MLE is optimal.

□

The result in equation (5.4) can be generalized, since if an efficient estimator exists, it is given by the MLE.

5.1.2.2   Least Squares Estimation

A completely different philosophy is espoused in LS estimation. In this case no probabilistic assumption is made about the data $\mathbf{r}$ and the signal component $\mathbf{s} = [s_0, s_1, \ldots, s_{N-1}]^T$ of $\mathbf{r}$ is generated by some deterministic model which, in turn, depends on the unknown vector $\mathbf{h}$. The *least squares estimator* (LSE) of $\mathbf{h}$ chooses the value $\mathbf{h}_{LS}$ making $\mathbf{s}$ closest to $\mathbf{r}$, i.e., minimizing the LS error:

$$J(\tilde{\mathbf{h}}) \triangleq |\mathbf{r} - \mathbf{s}(\tilde{\mathbf{h}})|^2 = \sum_{l=0}^{N-1} |r_l - s_l|^2 \tag{5.5}$$

with respect to $\tilde{\mathbf{h}}$, where $\tilde{\mathbf{h}}$ denotes a hypothesized value for $\mathbf{h}$ and $\mathbf{s}(\tilde{\mathbf{h}})$ represents the value of $\mathbf{s}$ associated with $\tilde{\mathbf{h}}$. LSEs are usually developed when a precise statistical characterization of the data is unknown or an optimal estimator cannot be found or is too complicated. Inevitably, their performance depends on the properties of the corrupting noise and on modeling errors.

An LSE can easily be devised for the linear model of Example 5.1.1, as illustrated in the following example.

Example 5.1.2 If for the data vector $\mathbf{r}$ the *linear* dependence:

$$\mathbf{s} = \mathbf{Ah} \tag{5.6}$$

on $\mathbf{h}$ is assumed for its deterministic component $\mathbf{s}$, the LS error (5.5) becomes:

$$J(\tilde{\mathbf{h}}) = (\mathbf{r} - \mathbf{A}\tilde{\mathbf{h}})^H (\mathbf{r} - \mathbf{A}\tilde{\mathbf{h}}). \tag{5.7}$$

Computing the gradient $\partial J(\tilde{\mathbf{h}})/\partial \tilde{\mathbf{h}}$ and setting it equal to zero yields the LSE [597]:

$$\hat{\mathbf{h}} = (\mathbf{A}^H \mathbf{A})^{-1} \mathbf{A}^H \mathbf{r}, \tag{5.8}$$

which coincides with (5.4). Note that the equations $(\mathbf{A}^T \mathbf{A})\hat{\mathbf{h}} = \mathbf{A}^T \mathbf{r}$ that must be solved for the evaluation of $\hat{\mathbf{h}}$ are called the *normal equations*.

□

This is an example of *linear* LS, in which the deterministic component of the noisy data exhibits a linear dependence on the parameters to be estimated. An extension of the linear LS problem is the so-called *weighted* LS, in which LS error (5.7) [597]:

$$J_W(\tilde{\mathbf{h}}) = |\mathbf{r} - \mathbf{s}(\tilde{\mathbf{h}})|_W^2 \triangleq (\mathbf{r} - \mathbf{A}\tilde{\mathbf{h}})^H \mathbf{W}(\mathbf{r} - \mathbf{A}\tilde{\mathbf{h}}) \tag{5.9}$$

includes a positive definite $N \times N$ matrix $\mathbf{W}$. This allows more emphasis to be given to the data samples which are deemed to be more reliable. The reader can easily verify that the general form of the weighted LSE is given by:

$$\hat{\mathbf{h}} = (\mathbf{A}^H \mathbf{W} \mathbf{A})^{-1} \mathbf{A}^H \mathbf{W} \mathbf{r}. \tag{5.10}$$

5.1.2.3 Minimum Mean Square Error and Maximum A Posteriori Estimation

If some prior knowledge about **h** is available, it is useful to incorporate it in our estimator as this is expected to improve the estimation accuracy with respect to the classical approach. In Bayesian estimation a well-known optimal estimator is defined to be the one that *minimizes* the MSE when averaged over all the realizations of **h** and **r** – the so-called Bayesian MSE. This estimator, denoted by $\hat{\mathbf{h}}$, is given by the *mean of the posterior pdf* $f(\mathbf{h}|\mathbf{r})$:

$$\hat{\mathbf{h}} = \mathrm{E}\{\mathbf{h}|\mathbf{r}\} = \int \mathbf{h} f(\mathbf{h}|\mathbf{r}) \, d\mathbf{h}. \qquad (5.11)$$

It is important to note the following:

(a) The vector MMSE estimator minimizes the MSE for each component of the unknown parameter vector **h**.
(b) The choice of the prior pdf $f(\mathbf{h})$ is critical, since an incorrect statistical model for **h** can result in poor estimation accuracy.

As a result, unless the prior pdf can be derived from the physical constraints of the problem, the use of classical estimation methods is usually more appropriate. Let us now apply this approach to a specific problem.

Example 5.1.3 Let us reconsider the linear model of (5.2), and assume now that **h** is a *random vector* belonging to $\mathcal{CN}(\boldsymbol{\eta}_\mathbf{h}, \mathbf{C}_\mathbf{h})$ and is statistically independent of the noise vector **w** (the resulting model is known as a *Bayesian linear model*). Then it can be shown that the posterior pdf $f(\mathbf{h}|\mathbf{r})$ is Gaussian with mean:

$$\mathrm{E}\{\mathbf{h}|\mathbf{r}\} = \boldsymbol{\eta}_\mathbf{h} + \mathbf{C}_\mathbf{h}\mathbf{A}^H(\mathbf{A}\mathbf{C}_\mathbf{h}\mathbf{A}^H + \mathbf{C}_\mathbf{w})^{-1}(\mathbf{r} - \mathbf{A}\boldsymbol{\eta}_\mathbf{h}). \qquad (5.12)$$

Note that, unlike the classical general linear model (see Example 5.1.1), **A** is not required to be full rank to ensure the validity of equation (5.12), since the invertibility of $(\mathbf{A}\mathbf{C}_\mathbf{h}\mathbf{A}^T + \mathbf{C}_\mathbf{w})$ is required.
□

An alternative Bayesian strategy is represented by MAP estimation. A vector MAP estimator is defined as:

$$\hat{\mathbf{h}} = \arg\max_{\tilde{\mathbf{h}}} f(\tilde{\mathbf{h}}|\mathbf{r}) = \arg\max_{\tilde{\mathbf{h}}} f(\mathbf{r}|\tilde{\mathbf{h}}) f(\tilde{\mathbf{h}}). \qquad (5.13)$$

It can be shown that the MAP estimator is identical to the MMSE estimator if **r** and **h** are jointly Gaussian [596].

Many estimation problems are characterized by a set of unknown parameters, where we are really interested in only a subset. The remaining parameters are called *nuisance parameters*. If we assume that the parameters are deterministic, as in the classical estimation approach, then we must *jointly* estimate the nuisance parameters and the parameters of interest. In contrast, in the Bayesian approach, it is possible to get rid of the nuisance parameters as follows. Let **h** and $\boldsymbol{\eta}$ denote the vector of unknown parameters to be estimated and that of nuisance parameters, respectively. Then, given the posterior pdf $f(\mathbf{h}, \boldsymbol{\eta}|\mathbf{r})$, the posterior pdf of **h** *only* can be evaluated as:

$$f(\mathbf{h}|\mathbf{r}) = \int f(\mathbf{h}, \boldsymbol{\eta}|\mathbf{r}) \, d\boldsymbol{\eta}. \qquad (5.14)$$

Note that, by exploiting Bayes' theorem, this pdf can also be evaluated as:

$$f(\mathbf{h}|\mathbf{r}) = \frac{f(\mathbf{r}|\mathbf{h}) f(\mathbf{h})}{\int f(\mathbf{r}|\boldsymbol{\theta}) f(\boldsymbol{\theta}) \, d\boldsymbol{\theta}}, \qquad (5.15)$$

where:

$$f(\mathbf{r}|\mathbf{h}) = \int f(\mathbf{r}|\mathbf{h}, \eta) f(\eta|\mathbf{h}) \, d\eta. \tag{5.16}$$

If η is independent of $\mathbf{h}$, the expression simplifies to:

$$f(\mathbf{r}|\mathbf{h}) = \int f(\mathbf{r}|\mathbf{h}, \eta) f(\eta) \, d\eta. \tag{5.17}$$

In essence, we average out the nuisance parameters. Once we obtain $f(\mathbf{h}|\mathbf{r})$, the nuisance parameters are no longer involved in the estimation problem and an MMSE estimator can be found by evaluating the mean of this pdf (see (5.11)). Note, however, that the presence of the nuisance parameters will affect the final estimator, since $f(\mathbf{r}|\mathbf{h})$ depends on $f(\eta)$, as shown by (5.17). It is, however, interesting to point out that the Bayesian approach does not suffer from the problems of classical estimators, in which the nuisance parameters may invalidate an estimator.

5.1.3   Recursive Estimation

5.1.3.1   Recursive Least Squares and Least Mean Squares Estimation

In many signal processing applications noisy data are acquired as time progresses. We then have to decide whether to wait for all the available data or to process it sequentially in time as it is received. Note that, in principle, if an LS approach to estimation is adopted, it leads to a sequence of LSEs in time. Then, whenever an additional datum is available, the LSE needs to be recomputed. Fortunately, an alternative to this computationally intensive approach exists known as RLS. This allows the LSE to be updated as new data arrives without solving the associated set of normal equations.

In describing the RLS algorithm we consider its application to the estimation of the CIR of a frequency-selective communication channel. In particular, let $\mathbf{h} = [h_0, h_1, \ldots, h_{L_h-1}]^T$ represent the tap gain vector of a transversal filter fed by the known data sequence $\{a_n\}$. Noise is added at the filter output, resulting at each time point in the response:

$$r_n = \mathbf{a}_n^T \mathbf{h} + w_n, \tag{5.18}$$

where $\mathbf{a}_n \triangleq [a_n, a_{n-1}, \ldots, a_{n-L_h+1}]^T$ is a *known* signal vector and w_n is *white Gaussian noise* (WGN). The aim of the RLS algorithm is, upon arrival of the nth datum r_n, to update the tap gain vector $\tilde{\mathbf{h}}_n \triangleq [\tilde{h}_{n,0}, \tilde{h}_{n,1}, \ldots, \tilde{h}_{n,L_h-1}]^T$ of an *adaptive transversal filter* (see Figure 5.2), fed by $\{a_n\}$, in such a way that the cost function:

$$J_n^{RLS} \triangleq \sum_{l=0}^{n} \lambda^{n-l} |e_l|^2 \tag{5.19}$$

is minimized. Here λ is a positive constant close to (but less than) unity and:

$$e_n \triangleq r_n - y_n = r_n - \sum_{l=n-L_h+1}^{n} a_l \tilde{h}_{n-l} \tag{5.20}$$

represents the so-called *estimation error*, since it is given by the difference between the observed datum (or *desired response*) r_n at time n and its estimate y_n obtained at the output of the adaptive filter at the same instant. Note that J_n^{RLS} (5.19) can be derived from (5.9) by selecting a diagonal matrix $\mathbf{W}$ and that λ^{n-l} is an *exponential weighting* or *forgetting factor* ensuring that less relevance is given to data in the distant past; this affords the possibility of following the statistical variations of the observable data when the adaptive filter operates in a nonstationary environment. The memory of

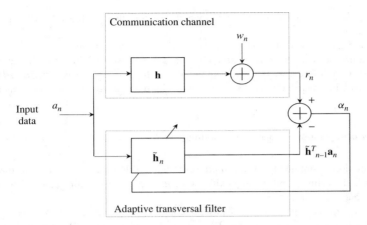

Figure 5.2   Channel identification using an adaptive transversal filter. The same length (L_h) is assumed for both $\mathbf{h}$ and $\tilde{\mathbf{h}}_n$.

this filter becomes longer as λ approaches unity; in particular, when λ equals unity, the cost function (5.19) becomes that characterizing the ordinary LS method (see (5.5)) and the filter memory is infinite.

At time n the RLS algorithm computes an estimate $\tilde{\mathbf{h}}_n$ of $\mathbf{h}$ (based on the received vector $\mathbf{r}_n \triangleq [r_0, r_1, \ldots, r_n]^T$) and using the previous estimate $\tilde{\mathbf{h}}_{n-1}$ (based on $\mathbf{r}_{n-1}$) by adding a correction term to $\tilde{\mathbf{h}}_{n-1}$. In particular, it uses the *estimator update equation*:

$$\tilde{\mathbf{h}}_n = \tilde{\mathbf{h}}_{n-1} + \mathbf{K}_n \alpha_n^*, \tag{5.21}$$

where:

$$\alpha_n \triangleq r_n - \tilde{\mathbf{h}}_{n-1}^T \mathbf{a}_n, \tag{5.22}$$

and $\mathbf{K}_n$ is an L_h-dimensional vector know as the *gain vector* and given by:

$$\mathbf{K}_n = \frac{\lambda^{-1}\mathbf{P}_{n-1}\mathbf{a}_n}{1 + \lambda^{-1}\mathbf{a}_n^H \mathbf{P}_{n-1}\mathbf{a}_n}. \tag{5.23}$$

Here $\mathbf{P}_n$ denotes an $L_h \times L_h$ matrix which can be computed via the recursive expression:

$$\mathbf{P}_n = \lambda^{-1}(\mathbf{I}_{L_h} - \mathbf{K}_n \mathbf{a}_n^H)\mathbf{P}_{n-1}, \tag{5.24}$$

known as the *Riccati equation* for the RLS algorithm. It can be shown that this matrix represents the inverse of the data autocorrelation matrix:

$$\mathbf{R}_n \triangleq \sum_{l=0}^{n} \lambda^{n-l}\mathbf{a}_n \mathbf{a}_n^H. \tag{5.25}$$

Note that:

(a) the evaluation of the *a priori estimation error*[4] α_n via (5.22) requires exciting the adaptive filter with the signal input sequence (in other words, (5.22) describes the filtering operation of the algorithm),

---

[4] This error can be seen as a tentative value of e_n (see (5.20)), since it is computed *before* updating the tap-weight vector.

(b) (5.21) describes the updating of the tap-weight vector, that is, the adaptive operation of the algorithm,

(c) (5.23) and (5.24) allow us to update the value of the gain vector, and

(d) this procedure does not require matrix inversions but must be initialized. Initialization can be accomplished in different ways. A well-known possibility, known as *soft constrained initialization*, involves selecting $\tilde{\mathbf{h}}_{-1} = \mathbf{0}_{L_h}$ and $\mathbf{P}_{n-1} = \alpha\mathbf{I}_{L_h}$ with α large (to minimize the biasing due to the selection of $\tilde{\mathbf{h}}_{-1}$).

The following observations are also worth noting:

1. The computational complexity of the *standard* RLS algorithm (5.21)–(5.24) is of the order of L_h^2, but *fast* RLS algorithms, whose complexity increases only linearly with L_h, are also available (e.g., see [503, Chaps. 16–18], [598, 599]).
2. The recursive update equation (5.24) has poor numerical properties and this has motivated the search for alternative versions of the RLS algorithm with improved numerical stability [503], which update the matrix $\mathbf{S}_n$ directly without computing the matrix $\mathbf{P}_n$ explicitly.
3. A channel estimator based on RLS can also be implemented as a *lattice structure* [598, 600]. This is characterized by a convergence rate identical to that of the standard RLS algorithm and by a computational complexity proportional to L_h, but with a larger proportionality constant compared with fast RLS algorithms.

A substantially simpler alternative to RLS is the *least mean square* (LMS) algorithm [601]. This algorithm is usually derived as an approximate solution to the problem of minimization of the MSE:

$$J_n(\tilde{\mathbf{h}}) \triangleq \mathrm{E}\{|e_n|^2\} \tag{5.26}$$

in the scenario described in Figure 5.2, under the assumption that the channel filter is fed by a zero-mean WSS sequence $\{a_n\}$ with known correlation. This is because in any practical application, the evaluation of the optimal value $\mathbf{h}_o$ of $\tilde{\mathbf{h}}$ (i.e., that associated with the minimum J_{min} of $J_n(\tilde{\mathbf{h}})$ (5.26)) needs statistical information to be obtained and a set of linear equations (known as the *Wiener–Hopf equations*) to be solved. To circumvent these two problems and to reduce the computational complexity, the following two strategies are often adopted:

1. A classic method of optimization, known as the *method of steepest descent*, is applied to $J_n(\tilde{\mathbf{h}})$ in (5.26). This leads to the estimation of $\mathbf{h}$ in a recursive fashion, using the expression:

$$\tilde{\mathbf{h}}_{n+1} = \tilde{\mathbf{h}}_n + \frac{1}{2}\gamma\{-[\nabla J_n(\tilde{\mathbf{h}})]_{\tilde{\mathbf{h}}=\tilde{\mathbf{h}}_n}\}, \tag{5.27}$$

where $\nabla J_n(\tilde{\mathbf{h}})$ denotes the gradient vector of $J_n(\tilde{\mathbf{h}})$ and γ is a positive real-valued constant known as the *step size parameter* (the factor 1/2 is used only for convenience). In practice, using the present guess $\tilde{\mathbf{h}}_n$, a change in it is introduced in the opposite direction to that of the gradient vector. It is reasonable that successive corrections to $\tilde{\mathbf{h}}$ will cause the estimate to approach $\mathbf{h}_o$.

2. The *instantaneous* square error $|e_n|^2$ is used in place of its mean value when evaluating $\nabla J_n(\tilde{\mathbf{h}})$ for (5.27). This leads to:

$$\tilde{\mathbf{h}}_{n+1} = \tilde{\mathbf{h}}_n + \gamma e_n \mathbf{a}_n^*, \tag{5.28}$$

which describes the *coefficient update* of the LMS procedure. This procedure, which is a member of the family of *stochastic gradient algorithms*, requires an initial value $\tilde{\mathbf{h}}_0$ for the tap-weigh vector, that is, an initial guess for the CIR vector; typically $\tilde{\mathbf{h}}_0$ is set equal to the null vector $\mathbf{0}_{L_h}$.

A significant feature of the LMS algorithm is its simplicity, since it does not require any knowledge of statistical information or any matrix inversion. For this reason, LMS is the standard against which other adaptive filtering algorithms are benchmarked. One problem in using the LMS algorithm is the selection of the step-size parameter[5] γ, since this controls both the rate of adaptation and the stability of the algorithm. Convergence in the mean (i.e., convergence of the mean of the weight-error vector $(\tilde{\mathbf{h}}_n - \tilde{\mathbf{h}}_0)$ to zero as $n \to \infty$) is guaranteed if $0 < \gamma < 2/\lambda_{max}$, where λ_{max} is the largest eigenvalue of the correlation matrix $\mathbf{R}_\mathbf{a} \triangleq E\{\mathbf{a}_n \mathbf{a}_n^H\}$ of the tap-input vector in the TDL channel model. A large value for γ, usually just below the upper limit, provides rapid convergence, but it also introduces large fluctuations during steady-state operation. These represent a form of self-noise whose variance increases with γ. Therefore, the choice of γ represents a tradeoff between rapid convergence and a small self-noise variance during steady-state operation.

The LMS technique offers a simple alternative to the RLS algorithm, but provides slower convergence, since there is only one parameter, namely γ, controlling its rate of adaptation. In particular, it exhibits a poor convergence rate when the size of $\mathbf{h}$ is large (this problem can be mitigated by incorporating a detection scheme that discriminates between the active and inactive taps within the CIR [603]). Note, however, that in the presence of a *nonstationary channel* when iid data symbols are assumed, the LMS and the RLS algorithms exhibit similar behavior in channel tracking [606]. Further information on the stationary and nonstationary learning characteristics of the LMS and RLS algorithms can be found in [601, 607–615]. The use of RLS channel estimation in fast fading SISO and MIMO scenarios is investigated in [616, 617] and [618], respectively.

Finally, we briefly consider decision-directed operation of the LMS/RLS algorithm. In this case the symbols required for the evaluation of the error e_k in (5.28) (or, equivalently, α_k in (5.22) for the RLS) are available with a *decision delay d* (introducing a delay in the estimation algorithm too). For instance, in the LMS case, this leads to *delayed* LMS (DLMS), which is described by the *delayed coefficient update*:

$$\tilde{\mathbf{h}}_{n+1} = \tilde{\mathbf{h}}_n + \gamma\, \tilde{e}_{n-d}\, \hat{\mathbf{a}}_{n-d}^*, \tag{5.29}$$

where $\hat{\mathbf{a}}_{n-d} \triangleq [\hat{a}_{k-d}, \hat{a}_{k-d-1}, \dots, \hat{a}_{k-d-(L_h-1)}]^T$ is a vector of channel symbol decisions and $\tilde{e}_{k-d}$ is generated as (see (5.20)):

$$\tilde{e}_{k-d} = r_{k-d} - \sum_{l=k-d-(L_h-1)}^{k-d} \hat{a}_l\, \tilde{h}_{n-l}. \tag{5.30}$$

The delay d influences both the convergence and the asymptotic performance of the LMS algorithm [619]. In addition, in the presence of *nonstationary channels*, a *channel predictor* is needed in the receiver structure if a CIR estimate for the present epoch is desired [511].

5.1.3.2 Kalman Estimation

Optimal Bayesian estimators are often difficult to determine in closed form and are too computationally expensive to implement, since they usually require multidimensional integration and maximization. If we constrain the estimator to be linear and retain the MMSE criterion, a sequential procedure can be devised for *linear* MMSE (LMMSE) estimation. This procedure is similar to that illustrated above for sequential LS estimation (details can be found in [596, Chap. 12]). If the noisy data sequence to be processed belongs to a WSS random process, LMMSE estimation leads to the class of *Wiener filters*. In practice, the process of forming an estimator as the time index n increases is seen as a linear filtering operation with time-varying impulse response. The evaluation of this impulse response

---

[5] The step size can be adjusted in a data-dependent manner (see [602, 603] and references therein). This leads to the *variable* LMS (VLMS) [602, 604] or *variable step-size* LMS (VSS-LMS) algorithm [605], and to the *normalized* LMS (NLMS) algorithm [603].

involves solving a set of linear equations known as the *Wiener–Hopf filtering equations*. In principle, these equations must be solved for each value of n. However, a computationally efficient recursive procedure (the *Levinson–Durbin algorithm*) can be used. For large enough n it can be shown that the filter becomes time-invariant, so that only a single solution is necessary. In this case, an analytical solution may be found [503, Chapters 5 and 6].

Wiener filtering can be extended to accommodate nonstationary vector signals and noise, and the resulting generalization is known as *Kalman filtering*. This estimation approach can be exploited for the vector (Bayesian) linear model:

$$\mathbf{r}[n] = \mathbf{A}[n]\mathbf{h}[n] + \mathbf{w}[n], \tag{5.31}$$

for $n \geq 0$, where $\mathbf{r}[n] \triangleq [r_n^{(0)}, r_n^{(1)}, \ldots, r_n^{(S-1)}]^T$ is the S-dimensional complex vector of noisy data observed at time n (and coming, for instance, from S distinct receive antennas), $\mathbf{A}[n]$ is a *known* $S \times L_h$ matrix and $\mathbf{w}[n] \triangleq [w_n^{(0)}, w_n^{(1)}, \ldots, w_n^{(S-1)}]^T$ is an S-dimensional observation complex noise sequence. It is assumed that:

(a) the L_h-dimensional signal vector $\mathbf{h}[n] = [h_n^{(0)}, h_n^{(1)}, \ldots, h_n^{(L_h-1)}]^T$ evolves in time according to the *Gauss–Markov model*:

$$\mathbf{h}[n] = \mathbf{P}\mathbf{h}[n-1] + \mathbf{Q}\mathbf{u}[n], \tag{5.32}$$

for $n \geq 0$, where $\mathbf{P}$ and $\mathbf{Q}$ are known matrices of sizes $L_h \times L_h$ and $L_h \times r$, respectively, and $\mathbf{u}[n] = [u_n^{(0)}, u_n^{(1)}, \ldots, u_n^{(r-1)}]^T \in \mathcal{CN}(\mathbf{0}, \mathbf{C}_u)$ for any n,

(b) the initial state vector $\mathbf{h}[-1]$ is independent of $\mathbf{u}[n]$ and belongs to $\mathcal{CN}(\boldsymbol{\eta}_\mathbf{h}, \mathbf{C}_\mathbf{h})$, and

(c) the vectors $\{\mathbf{w}[n]\}$ are independent of each other and $\mathbf{w}[n]$ belongs to $\mathcal{CN}(\mathbf{0}_S, \mathbf{C}_\mathbf{w}[n])$ (if the $\mathbf{C}_\mathbf{w}[n]$ did not depend on n, then $\mathbf{w}[n]$ would be a WGN vector).

Then the MMSE estimate, $\hat{\mathbf{h}}[n|n] = E\{\mathbf{h}[n]|\mathbf{r}[0], \mathbf{r}[1], \ldots, \mathbf{r}[n]\}$ of $\mathbf{h}[n]$ based on $\{\mathbf{r}[0], \mathbf{r}[1], \ldots, \mathbf{r}[n]\}$, can be evaluated by a vector *Kalman filter* (KF), whose behavior is described by the following equations:

$$\hat{\mathbf{h}}[n|n-1] = \mathbf{P}\hat{\mathbf{h}}[n-1|n-1], \tag{5.33}$$

$$\mathbf{M}[n|n-1] = \mathbf{P}\mathbf{M}[n-1|n-1]\mathbf{P}^H + \mathbf{Q}\mathbf{C}_\mathbf{u}\mathbf{Q}^H, \tag{5.34}$$

$$\mathbf{K}[n] = \mathbf{M}[n|n-1]\mathbf{A}^H[n](\mathbf{C}_\mathbf{w}[n] + \mathbf{A}[n]\mathbf{M}[n|n-1]\mathbf{A}^H[n])^{-1}, \tag{5.35}$$

$$\hat{\mathbf{h}}[n|n] = \hat{\mathbf{h}}[n|n-1] + \mathbf{K}[n](\mathbf{r}[n] - \mathbf{A}[n]\hat{\mathbf{h}}[n|n-1]), \tag{5.36}$$

$$\mathbf{M}[n|n] = (\mathbf{I} - \mathbf{K}[n]\mathbf{A}[n])\mathbf{M}[n|n-1]. \tag{5.37}$$

These should be initialized by setting $\hat{\mathbf{h}}[-1|-1] = \boldsymbol{\eta}_\mathbf{h}$ and $\mathbf{M}[-1|-1] = \mathbf{C}_\mathbf{h}$. Note that here $\hat{\mathbf{h}}[n|n-1]$ is a prediction of $\mathbf{h}[n]$; $\mathbf{M}[n|n-1]$ and $\mathbf{M}[n|n]$ (both $L_h \times L_h$) are the so-called *minimum prediction MSE matrix* and the *minimum MSE matrix*, respectively; and $\mathbf{K}[n]$ is the so-called *Kalman gain matrix*. It is also worth noting that:

(a) the evaluation of $\mathbf{K}[n]$ requires computation of the inverse of an $S \times S$ matrix, and

(b) this formulation of the vector KF can be further generalized to include, for instance, the cases of time-varying matrices $\mathbf{P}$, $\mathbf{Q}$ and $\mathbf{C}_\mathbf{u}$ and of colored observation noise.

The reader can refer to [503, Chap. 7] and [597, 620, 621] for further details on this topic. The Kalman approach offers tracking superiority when compared with the standard RLS and LMS algorithms – at the price, however, of increased complexity [612].

5.1.4 The Principle of Per-Survivor Processing

When multiple data estimates are available, multiple channel estimates can be determined, one for each data estimate. This conceptual approach to parameter estimation is known as the principle of per-survivor processing [515, 569]. In this subsection this principle is illustrated in the context of channel estimation.

In practical systems the channel impulse response $\mathbf{h}_n \triangleq [h_{n,0}, h_{n,1}, \ldots, h_{n,L_h-1}]^T$ at the nth symbol interval[6] is not known a priori and must be estimated to compute the equalizer metrics. A common approach is based on data-aided parameter estimation techniques in which the aiding data sequence is generated in a decision-directed mode from *tentative low-delay decisions* at the decoder output [458, 622, 623]. Let the tentative decision on the data symbol a_n at the epoch n be denoted by $\tilde{a}_{n-\bar{d}-1}$, where $\bar{d}$ denotes its *decoding delay*. Based on the vector $\tilde{\mathbf{a}}_{n-\bar{d}} \triangleq [\tilde{a}_i, i = n - \bar{d} - 1, n - \bar{d} - 2, \ldots]^T$ and the received signal vector $\mathbf{r}_n \triangleq [r_0, r_1, \ldots, r_n]^T$, a data-aided parameter estimator provides the detector with an estimate $\tilde{\mathbf{h}}_n$ of the unknown channel $\mathbf{h}[n]$ of the form:

$$\tilde{\mathbf{h}}_n = \mathbf{g}(\mathbf{r}_n, \tilde{\mathbf{a}}_{n-\bar{d}}). \tag{5.38}$$

Here $\mathbf{g}(\cdot)$ denotes the functional dependence of the estimate $\tilde{\mathbf{h}}_n$ on the received signal and on the sequence of tentative decisions. Note that a decision delay is inherent in the estimate $\tilde{\mathbf{h}}_n$ with respect to the true parameter vector $\mathbf{h}[n]$.

An alternative to this approach is to apply *per-survivor* estimation [515, 569] of the unknown channel response. Note that originally this strategy was developed for those cases in which the *Viterbi algorithm* (VA) is used for data detection in the presence of *parametric uncertainty* about the communication channel. At the end of each symbol interval the VA generates as many estimates of the channel symbol sequence as the number of states of the trellis on which it operates (further details on the VA can be found in Section 6.2.1.1). Each estimate is associated with a specific *survivor* path in the trellis. Then, if $\tilde{\mathbf{a}}_{n-1}(\tilde{\Delta}_n)$ denotes the channel symbol sequence associated with the survivor path at state $\tilde{\Delta}_n$, the per-survivor estimate $\tilde{\mathbf{h}}_n(\tilde{\Delta}_n)$ of the unknown channel vector $\mathbf{h}_n$ based on the data-aided estimator $\mathbf{g}(\cdot)$ in (5.38) and the channel symbols associated with the surviving path ending in $\tilde{\Delta}_n$ can be defined as:

$$\tilde{\mathbf{h}}_n(\tilde{\Delta}_n) = \mathbf{g}(\mathbf{r}_n, \tilde{\mathbf{a}}_{n-1}(\tilde{\Delta}_n)). \tag{5.39}$$

This channel estimate is employed in the evaluation of the VA *branch metrics* for all the state transitions emerging from state Δ_n, as illustrated in Figure 5.3.

A heuristic justification of this approach can be given as follows. If ignorance of the CIR prevents us from calculating a decision metric in a precise form, estimates of the CIR based on the *multiple data hypotheses* (i.e., on the survivor paths) are evaluated. If a particular survivor represents the correct choice of data, the corresponding estimate will be evaluated properly. At each detection step, however, we do not know which survivor, if any, represents correct data decisions. Then we extend each survivor using the channel estimates based on its associated data sequence, that is, on the best data sequence available.

Different forms of the function $\mathbf{g}(\cdot)$ can be adopted in practical applications as, for instance, those characterizing the LMS or RLS techniques (see (5.21) and (5.28), respectively) can be employed, as shown in [511, 512, 515, 564, 615, 624]. Further applications of PSP can be found in [116], where LS estimation of the coefficients of a polynomial model of a flat fading process is employed for channel prediction, and in [625], where Kalman prediction is used for the estimation of a flat fading channel in a receiver for ST trellis codes.

---

[6] This account can easily be generalized to the case in which fractionally-spaced sampling is used, so that $\mathbf{h}_n$ refers to the channel state at the nth sampling interval. Note also that throughout this section, we assume that the channel has a finite impulse response duration of L_h samples or symbols as the case may be.

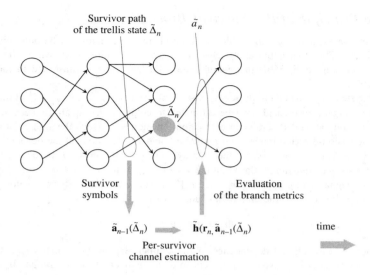

Figure 5.3   Illustration of per-survivor channel estimation. A four-state trellis is considered.

5.2   Cramér–Rao Bounds for Data-Aided Channel Estimation

Estimation theory establishes bounds on the ultimate accuracy that can be achieved by parameter estimation algorithms. In communication problems the most relevant class of such bounds are the Cramér–Rao bounds, since these represent fundamental lower limits to the *variance of any unbiased parameter estimator* [430, 596]. CRBs usually refer to an L_θ-dimensional vector $\boldsymbol{\theta}$ of *real* parameters that is to be estimated from a set of noisy observations consisting of the N-dimensional (real or complex) vector $\mathbf{r} = [r_0, r_1, \ldots, r_{N-1}]^T$, and assume that some "regularity conditions" are satisfied by the pdf $f(\mathbf{r}; \boldsymbol{\theta})$ (i.e., $f(\mathbf{r}|\boldsymbol{\theta})$) for all $\boldsymbol{\theta}$. If $\boldsymbol{\theta}$ is *complex*, a *real* vector $\boldsymbol{\theta}^{(R)} \triangleq [\mathrm{Re}\{\boldsymbol{\theta}\}, \mathrm{Im}\{\boldsymbol{\theta}\}]^H$ containing the real and imaginary parts of its components can be always associated with it (a CRB for complex parameters applicable to any distribution and model of observations is derived in [626]). Then the CRB of the covariance matrix $\mathbf{C}_{\hat{\boldsymbol{\theta}}^{(R)}} = \mathrm{E}\{(\hat{\boldsymbol{\theta}}^{(R)} - \boldsymbol{\theta}^{(R)})(\hat{\boldsymbol{\theta}}^{(R)} - \boldsymbol{\theta}^{(R)})^H\}$ of any *unbiased estimator* $\hat{\boldsymbol{\theta}}^{(R)}$ of $\boldsymbol{\theta}^{(R)}$ can be formulated as:

$$\mathbf{C}_{\hat{\boldsymbol{\theta}}^{(R)}} - \mathbf{I}^{-1}(\boldsymbol{\theta}^{(R)}) \succeq \mathbf{0}_{2L_\theta, 2L_\theta}, \tag{5.40}$$

where the notation $\succeq \mathbf{0}$ denotes that the matrix on the *left-hand side* (LHS) of (5.40) is positive semidefinite and:

$$\mathbf{I}(\boldsymbol{\theta}^{(R)}) \triangleq \mathrm{E}\left\{ \left[\frac{\partial \ln f(\mathbf{r}; \tilde{\boldsymbol{\theta}}^{(R)})}{\partial \tilde{\boldsymbol{\theta}}^{(R)}} \right] \left[\frac{\partial \ln f(\mathbf{r}; \tilde{\boldsymbol{\theta}}^{(R)})}{\partial \tilde{\boldsymbol{\theta}}^{(R)}} \right]^T \bigg|_{\tilde{\boldsymbol{\theta}}^{(R)} = \boldsymbol{\theta}^{(R)}} \right\} \tag{5.41}$$

is the *Fisher information matrix* (FIM) associated with $\boldsymbol{\theta}^{(R)}$ (the expectation in (5.41) is taken with respect to the pdf $f(\mathbf{r}; \boldsymbol{\theta}^{(R)})$). Since in a positive semidefinite matrix the diagonal elements are nonnegative, from (5.40) it is easily seen that:

$$\mathrm{var}([\boldsymbol{\theta}^{(R)}]_i) = [\mathbf{C}_{\hat{\boldsymbol{\theta}}^{(R)}}]_{i,i} \geq [\mathbf{I}^{-1}(\boldsymbol{\theta}^{(R)})]_{i,i}, \tag{5.42}$$

for $i = 0, 1, \ldots, 2L_\theta - 1$.

In the technical literature CRBs often refer directly to the complex vector $\boldsymbol{\theta}$ (instead of its real counterpart $\hat{\boldsymbol{\theta}}^{(R)}$). Note that a *complex* FIM[7]:

$$\mathbf{I}_c(\boldsymbol{\theta}) \triangleq \mathrm{E}\left\{ \left[\frac{\partial \ln f(\mathbf{r}; \tilde{\boldsymbol{\theta}})}{\partial \tilde{\boldsymbol{\theta}}^*} \right] \left[\frac{\partial \ln f(\mathbf{r}; \tilde{\boldsymbol{\theta}})}{\partial \tilde{\boldsymbol{\theta}}^*} \right]^H \Bigg|_{\tilde{\boldsymbol{\theta}}=\boldsymbol{\theta}} \right\} \tag{5.43}$$

can also be defined for the associated *unbiased estimator* $\hat{\boldsymbol{\theta}}$ of $\boldsymbol{\theta}$. However, since [627]:

$$\mathbf{I}(\boldsymbol{\theta}^{(R)}) = 2 \begin{bmatrix} \mathrm{Re}\{\mathbf{I}_{\boldsymbol{\theta\theta}}\} & -\mathrm{Im}\{\mathbf{I}_{\boldsymbol{\theta\theta}}\} \\ \mathrm{Im}\{\mathbf{I}_{\boldsymbol{\theta\theta}}\} & \mathrm{Re}\{\mathbf{I}_{\boldsymbol{\theta\theta}}\} \end{bmatrix} + 2 \begin{bmatrix} \mathrm{Re}\{\mathbf{I}_{\boldsymbol{\theta\theta}^*}\} & -\mathrm{Im}\{\mathbf{I}_{\boldsymbol{\theta\theta}^*}\} \\ \mathrm{Im}\{\mathbf{I}_{\boldsymbol{\theta\theta}^*}\} & \mathrm{Re}\{\mathbf{I}_{\boldsymbol{\theta\theta}^*}\} \end{bmatrix}, \tag{5.44}$$

where:

$$\mathbf{I}_{\boldsymbol{\theta\theta}^*} = \mathrm{E}\left\{ \left[\frac{\partial \ln f(\mathbf{r}; \tilde{\boldsymbol{\theta}})}{\partial \tilde{\boldsymbol{\theta}}^*} \right] \left[\frac{\partial \ln f(\mathbf{r}; \tilde{\boldsymbol{\theta}})}{\partial \tilde{\boldsymbol{\theta}}} \right]^H \Bigg|_{\tilde{\boldsymbol{\theta}}=\boldsymbol{\theta}} \right\}, \tag{5.45}$$

the FIM $\mathbf{I}(\boldsymbol{\theta}^{(R)})$ of (5.41) is completely determined by $\mathbf{I}_c(\boldsymbol{\theta})$ (5.43) only if $\mathbf{I}_{\boldsymbol{\theta\theta}^*}$ (5.45) is a null matrix. When this occurs, $\mathbf{I}_{\boldsymbol{\theta\theta}}$ can be considered as a complex FIM and the complex CRB:

$$\mathbf{C}_{\hat{\boldsymbol{\theta}}} - \mathbf{I}_c^{-1}(\boldsymbol{\theta}) \succeq \mathbf{0}, \tag{5.46}$$

where $\mathbf{C}_{\hat{\boldsymbol{\theta}}} = \mathrm{E}\{(\hat{\boldsymbol{\theta}} - \boldsymbol{\theta})(\hat{\boldsymbol{\theta}} - \boldsymbol{\theta})^H\}$, can be used in place of (5.40). When $\mathbf{I}_{\boldsymbol{\theta\theta}^*}$ is not null, $\mathbf{I}_c^{-1}(\boldsymbol{\theta})$ still represents a bound on $\mathbf{C}_{\hat{\boldsymbol{\theta}}}$ but is not as tight as the CRB (5.40). Note, however, that the assumption $\mathbf{I}_{\boldsymbol{\theta\theta}^*} = \mathbf{0}_{L_\theta, L_\theta}$ is often made (e.g., see [627]), so that the complex bound (5.46) is evaluated for simplicity.

An application of these results is shown in the following example.

Example 5.2.1 Let us consider the transmission of the symbol packet $\mathbf{c} = [c_{N+P-1}, c_{N+P-2}, \dots, c_0]^T$, containing N data symbols and P known (pilot) symbols (collected in the vectors $\mathbf{s}_d \triangleq [s_{d,0}, s_{d,1}, \dots, s_{d,N-1}]^T$ and $\mathbf{s}_p \triangleq [s_{p,0}, s_{p,1}, \dots, s_{p,P-1}]^T$ respectively[8]) over a frequency-selective channel characterized by the discrete-time CIR $\mathbf{h} = [h_0, h_1, \dots, h_{L_h-1}]^T$. We are interested in estimating the CIR vector from the vector of noisy received data:

$$\mathbf{r} \triangleq [r_{N+P-1}, r_{N+P-2}, \dots, r_{L_h-1}]^T \tag{5.47}$$

observed at the channel output, where:

$$r_k = \sum_{l=0}^{L_h-1} h_l c_{k-l} + n_k \tag{5.48}$$

and $\{n_l\}$ is a white complex circular Gaussian sequence with variance σ_n^2. Given (5.48), $\mathbf{r}$ (5.47) can be expressed in matrix form as [628]:

$$\mathbf{r} = \mathcal{T}(\mathbf{h})\mathbf{c} + \mathbf{n} = \mathcal{H}(\mathbf{c})\mathbf{h} + \mathbf{n}, \tag{5.49}$$

---

[7] Given a complex variable $\theta = \theta_R + j\theta_I$, the complex derivative with respect to θ is defined as $\partial/\partial\theta = (1/2) \cdot (\partial/\partial\theta_R - j\partial/\partial\theta_I)$, so that $\partial/\partial\theta^* = (1/2) \cdot (\partial/\partial\theta_R + j\partial/\partial\theta_I)$.

[8] We use this notation only to emphasize that the vector $\mathbf{c}$ contains both data and pilot symbols. The remainder of the example works only with $\mathbf{c}$.

where $\mathbf{n} = [n_{N+P-1}, n_{N+P-2}, \cdots, n_{L_h-1}]^T$:

$$
\mathcal{T}(\mathbf{h}) = \begin{bmatrix} h_0 & \cdots & h_{L_h-1} & & \\ & \ddots & & \ddots & \\ & & h_0 & \cdots & h_{L_h-1} \end{bmatrix} \tag{5.50}
$$

is an $(N + P - L_h + 1) \times (N + P)$ Toeplitz matrix generated by $\mathbf{h}$ and:

$$
\mathcal{H}(\mathbf{c}) = \begin{bmatrix} c_{N+P-1} & \cdots & c_{N+P-L_h} \\ & \text{Hankel} & \\ c_{L_h-1} & \cdots & c_0 \end{bmatrix} \tag{5.51}
$$

is an $(N + P - L_h + 1) \times (L_h)$ Hankel matrix associated with $\mathbf{c}$. In the following we assume that the data vector $\mathbf{c}$, the channel vector $\mathbf{h}$ and the noise vector $\mathbf{n}$ are mutually independent.

Three different CIR estimation strategies can be envisaged in this scenario as follows:

(a) if $N = 0$ (i.e., pilot symbols only are transmitted, so that $\mathbf{c} = \mathbf{s}_p$), *data-aided* channel estimation is accomplished,
(b) if both N and P are different from zero, the estimation is *semiblind* [629], and
(c) if $P = 0$ (i.e., pilot symbols are absent, so that $\mathbf{c} = \mathbf{s}_d$), *blind* estimation is accomplished [630].

In this example we focus on the first case only (various results for the other two cases can be found in [627, 631, 632] for a SIMO scenario). Here we analyze the complex CRB for $\boldsymbol{\theta} = \mathbf{h}$. We have:

$$
f(\mathbf{r}; \tilde{\boldsymbol{\theta}}) = f(\mathbf{r}; \tilde{\mathbf{h}}) = \frac{1}{(\pi \sigma_n^2)^{N+P-L_h+1}} \exp\left(-\frac{1}{\sigma_n^2} \left| \mathbf{r} - \mathcal{H}(\mathbf{s}_p)\tilde{\mathbf{h}} \right|^2 \right) \tag{5.52}
$$

where $\mathcal{H}(\mathbf{s}_p)$ is the Hankel matrix associated with the sequence of P pilot symbols. Then it is easy to show that [627, 628]:

$$
\mathbf{I}_c(\boldsymbol{\theta}) = \mathbf{I}_c(\mathbf{h}) = \frac{1}{\sigma_n^2} \mathcal{H}^H(\mathbf{s}_p)\mathcal{H}(\mathbf{s}_p), \tag{5.53}
$$

so that (5.46) becomes:

$$
\mathbf{C}_{\hat{\mathbf{h}}} - \sigma_n^2 [\mathcal{H}^H(\mathbf{s}_p)\mathcal{H}(\mathbf{s}_p)]^{-1} \succeq \mathbf{0}, \tag{5.54}
$$

where $\hat{\mathbf{h}}$ denotes any unbiased estimator of $\mathbf{h}$. This shows that the CRB depends on the training sequence and that, for a given training sequence energy, it is minimized when $\mathcal{H}^H(\mathbf{c})\mathcal{H}(\mathbf{c})$ is a multiple of identity [627]. This provides a significant guideline for the design of training sequences for good channel estimation. Further details on this topic can be found in [628], where a MIMO scenario is also considered.

☐

In some cases the set of *unknown parameters* on which $\mathbf{r}$ depends includes not only $\boldsymbol{\theta}$ but also a vector $\boldsymbol{\chi} = [\chi_0, \chi_1, \cdots, \chi_{U-1}]^T$ of additional parameters (e.g., synchronization parameters or information-bearing symbols). When this occurs, it is useful to estimate, in principle, all the unknown parameters, and a performance bound for this joint estimation task is still provided by the CRB, as illustrated in the following example.

Example 5.2.2 Let us consider the semiblind estimation problem mentioned in Example 5.2.1. In this scenario $\mathbf{c}$ incorporates an unknown information vector $\mathbf{s}_d$ and a known pilot vector $\mathbf{s}_p$, so that we are interested in estimating both $\mathbf{s}_d$ and $\mathbf{h}$, that is, $\boldsymbol{\theta} = [\mathbf{s}_d^H, \mathbf{h}^H]^H$. In this case the complex FIM is given by [628]:

$$\mathbf{I}_c(\boldsymbol{\theta}) = \frac{1}{\sigma_n^2} \begin{bmatrix} \mathbf{H}_d^H(\mathbf{h})\,\mathbf{H}_d(\mathbf{h}) & \mathbf{H}_d^H(\mathbf{h})\mathcal{H}(\mathbf{c}) \\ \mathcal{H}^H(\mathbf{c})\mathbf{H}_d(\mathbf{h}) & \mathcal{H}^H(\mathbf{c})\mathcal{H}(\mathbf{c}) \end{bmatrix}, \tag{5.55}$$

where $\mathbf{H}_d(\mathbf{h})$ is an $(N + P - L_h + 1) \times N$ matrix obtained from $\mathcal{T}(\mathbf{h})$ (5.50) by deleting the columns corresponding to pilot symbols.

$\square$

In many cases, however, for practical reasons, the more challenging goal of jointly estimating $\boldsymbol{\theta}$ and $\boldsymbol{\chi}$ is overlooked. In other words, one concentrates on $\boldsymbol{\theta}$ only and treats $\boldsymbol{\chi}$ as a set of nuisance parameters. In this case, $\boldsymbol{\chi}$ is modeled as a *random vector* with a *known* pdf $f(\boldsymbol{\chi})$ that does not depend on $\mathbf{h}$. To compute the CRB for $\boldsymbol{\theta}$, we need $f(\mathbf{r}; \boldsymbol{\theta})$, which, in principle, can be obtained from the integral:

$$f(\mathbf{r}; \boldsymbol{\theta}) = \int f(\mathbf{r}|\boldsymbol{\chi}; \boldsymbol{\theta}) f(\boldsymbol{\chi}) d\boldsymbol{\chi}, \tag{5.56}$$

where $f(\mathbf{r}|\boldsymbol{\chi}; \boldsymbol{\theta})$, the conditional pdf of $\mathbf{r}$ given $\boldsymbol{\chi}$ and $\boldsymbol{\theta}$ is easily available, at least for AWGN channels. This approach is adopted, for instance, in [627, Sect. 4], where in channel estimation the transmitted symbols are modeled as Gaussian variables. However, in most cases of practical interest, the computation of the FIM is impossible because the integration in (5.56) cannot be carried out analytically. To overcome this problem, a different bound, known as the *modified CRB* (MCRB)[9] [633] (or *modified Cramér–Rao vector bound* (MCRVB) when real vector estimation is considered [634]), can be employed. This bound can also be put in the form of (5.41), provided that a *modified FIM* (MFIM)[10]:

$$\mathbf{I}_M(\boldsymbol{\theta}^{(R)}) = \mathrm{E}\left\{ \left[\frac{\partial \ln f\left(\mathbf{r}\,\middle|\,\boldsymbol{\chi}; \tilde{\boldsymbol{\theta}}^{(R)}\right)}{\partial \tilde{\boldsymbol{\theta}}^{(R)}} \right] \left[\frac{\partial^2 \ln f\left(\mathbf{r}\,\middle|\,\boldsymbol{\chi}; \tilde{\boldsymbol{\theta}}^{(R)}\right)}{\partial \tilde{\boldsymbol{\theta}}^{(R)}} \right]^T \Bigg|_{\tilde{\boldsymbol{\theta}}^{(R)} = \boldsymbol{\theta}^{(R)}} \right\}, \tag{5.57}$$

is used in place of $\mathbf{I}(\boldsymbol{\theta}^{(R)})$ (the expectation is taken with respect to $f(\mathbf{r}, \boldsymbol{\chi}; \tilde{\boldsymbol{\theta}}^{(R)})$). Similar to the *Cramér–Rao vector bound* (CRVB), the diagonal elements of the inverse of the MFIM represent lower bounds on the error variance in the estimation of the corresponding parameters (see (5.42)). However, unlike the CRVB, the MCRVB can often be calculated with moderate effort. In fact, for the Gaussian channel the pdf $f(\mathbf{r}|\boldsymbol{\chi}; \tilde{\boldsymbol{\theta}}^{(R)})$ in (5.57) is a well-known exponential function whose argument is a quadratic form in the difference between the vector $\mathbf{r}$ and the useful signal component. Thus, the logarithm of $f(\mathbf{r}|\boldsymbol{\chi}; \tilde{\boldsymbol{\theta}}^{(R)})$ equals this quadratic form and the expectation in (5.57) can be easily evaluated. In computing MCRBs it is often assumed that no information is available about the nuisance parameters (this corresponds to assuming uniform pdfs for such parameters). In this case the MCRVB is *looser* than the true CRB. However, in several practical situations the available information is not so meagre and the possibility even exists that MCRVB and CRVB are either very close or coincident [633, 634]. In addition, it can be proved that the high SNR asymptote of the CRB (i.e., the *asymptotic CRB*, ACRB) pertaining to the estimation of a scalar parameter equals the associated MCRB when the parameter is not coupled with the nuisance parameters [638].

Bounds on estimation accuracy can also be developed for the *Bayesian estimators*, which, however, are *not constrained to be unbiased*. In fact, the Bayesian estimators trade off bias for variance in an attempt to reduce the overall MSE. Although in the Bayesian framework, the MMSE estimator

---

[9] The modified bounds developed in [633, 634] can also be interpreted as the nonrandom version of the global bound shown in [635, p. 1427]. They are also related to the modified CRB proposed by R. Miller and C. Chang in [636]. A short discussion of these issues can be found in [637, Sec. III].

[10] A complex MFIM for the complex parameter vector can also be defined in a similar way to the FIM (see (5.41) and (5.43)).

always exists, in general it is difficult to obtain it in a closed form and the calculation of its MSE is cumbersome. It is then useful to establish lower bounds on the attainable MSE to which the performance of the optimal estimator or any suboptimal estimators can be compared. Several lower bounds on MSE for random parameter estimation have been proposed, among them the *Battacharyya* [639, 640], *Bobrovsky–Zakai*[11] [642], *Weiss–Weinstein* [643], and *Bayesian Cramér–Rao* [430] bounds (a unification of these bounds is presented in [644]). In particular, Van Trees [430] extends the classic CRB (referring to nonrandom parameter estimation) to random parameter estimation. This bound is referred to in what follows as the *Bayesian Cramér–Rao bound* (BCRB) for scalar estimation and the *Bayesian Cramér–Rao vector bound* (BCRVB) for vector estimation. The FIM $\mathbf{I}_B(\boldsymbol{\theta}^{(R)})$ and the complex FIM $\mathbf{I}_{B,\boldsymbol{\theta\theta}}$ for the BCRVB of the real parameter vector $\boldsymbol{\theta}^{(R)}$ and the complex parameter vector $\boldsymbol{\theta}$ are given by [430]:

$$\mathbf{I}_B(\boldsymbol{\theta}^{(R)}) = \mathrm{E}\left\{\left[\frac{\partial \ln f\left(\mathbf{r}, \bar{\boldsymbol{\theta}}^{(R)}\right)}{\partial \bar{\boldsymbol{\theta}}^{(R)}}\right]\left[\frac{\partial \ln f\left(\mathbf{r}, \bar{\boldsymbol{\theta}}^{(R)}\right)}{\partial \bar{\boldsymbol{\theta}}^{(R)}}\right]^T\bigg|_{\bar{\boldsymbol{\theta}}^{(R)}=\boldsymbol{\theta}^{(R)}}\right\} \tag{5.58}$$

and:

$$\mathbf{I}_{B,\boldsymbol{\theta\theta}} = \mathrm{E}\left\{\left[\frac{\partial \ln f(\mathbf{r}, \bar{\boldsymbol{\theta}})}{\partial \bar{\boldsymbol{\theta}}^*}\right]\left[\frac{\partial \ln f(\mathbf{r}, \bar{\boldsymbol{\theta}})}{\partial \bar{\boldsymbol{\theta}}^*}\right]^H\bigg|_{\bar{\boldsymbol{\theta}}=\boldsymbol{\theta}}\right\}, \tag{5.59}$$

respectively. In both expressions the expectation is taken over $\mathbf{r}$ and $\boldsymbol{\theta}$ (i.e., the joint pdf $f(\mathbf{r}, \boldsymbol{\theta})$ is used). Similarly, the complex FIM can be found [628]. Note that $\mathbf{I}_{B,\boldsymbol{\theta}^{(R)}}$ can be rewritten as:

$$\mathbf{I}_B(\boldsymbol{\theta}^{(R)}) = \mathrm{E}\left\{\mathrm{E}\left\{\left[\frac{\partial \ln f\left(\mathbf{r}, \bar{\boldsymbol{\theta}}^{(R)}\right)}{\partial \bar{\boldsymbol{\theta}}^{(R)}}\right]\left[\frac{\partial \ln f\left(\mathbf{r}, \bar{\boldsymbol{\theta}}^{(R)}\right)}{\partial \bar{\boldsymbol{\theta}}^{(R)}}\right]^T\bigg|\bar{\boldsymbol{\theta}}^{(R)}\right\}\bigg|_{\bar{\boldsymbol{\theta}}^{(R)}=\boldsymbol{\theta}^{(R)}}\right\}, \tag{5.60}$$

where the outer and inner expectations are over $f\left(\boldsymbol{\theta}^{(R)}\right)$ and $f\left(\mathbf{r}|\boldsymbol{\theta}^{(R)}\right)$, respectively. Since $f\left(\mathbf{r}, \bar{\boldsymbol{\theta}}^{(R)}\right) = f\left(\bar{\boldsymbol{\theta}}^{(R)}\right) f\left(\mathbf{r}|\bar{\boldsymbol{\theta}}^{(R)}\right)$, the inner expectation in (5.60) can be computed as:

$$\mathrm{E}\left\{\left[\frac{\partial \ln f\left(\mathbf{r}, \bar{\boldsymbol{\theta}}^{(R)}\right)}{\partial \bar{\boldsymbol{\theta}}^{(R)}}\right]\left[\frac{\partial \ln f\left(\mathbf{r}, \bar{\boldsymbol{\theta}}^{(R)}\right)}{\partial \bar{\boldsymbol{\theta}}^{(R)}}\right]^T\bigg|\bar{\boldsymbol{\theta}}^{(R)}\right\}$$

$$= \mathbf{I}\left(\bar{\boldsymbol{\theta}}^{(R)}\right) + \left[\frac{\partial \ln f\left(\bar{\boldsymbol{\theta}}^{(R)}\right)}{\partial \bar{\boldsymbol{\theta}}^{(R)}}\right]\left[\frac{\partial \ln f\left(\bar{\boldsymbol{\theta}}^{(R)}\right)}{\partial \bar{\boldsymbol{\theta}}^{(R)}}\right]^T, \tag{5.61}$$

where $\mathbf{I}\left(\bar{\boldsymbol{\theta}}^{(R)}\right)$ is the complex FIM (5.41) evaluated for a deterministic $\bar{\boldsymbol{\theta}}^{(R)}$. Then, $\mathbf{I}_B\left(\boldsymbol{\theta}^{(R)}\right)$ (5.60) can be evaluated as:

$$\mathbf{I}_B(\boldsymbol{\theta}^{(R)}) = \mathrm{E}\left\{\mathbf{I}\left(\boldsymbol{\theta}^{(R)}\right) + \left[\frac{\partial \ln f\left(\bar{\boldsymbol{\theta}}^{(R)}\right)}{\partial \bar{\boldsymbol{\theta}}^{(R)}}\right]\left[\frac{\partial \ln f\left(\bar{\boldsymbol{\theta}}^{(R)}\right)}{\partial \bar{\boldsymbol{\theta}}^{(R)}}\right]^T\bigg|_{\bar{\boldsymbol{\theta}}^{(R)}=\boldsymbol{\theta}^{(R)}}\right\}, \tag{5.62}$$

where the expectation is over $f\left(\boldsymbol{\theta}^{(R)}\right)$. A similar expression, namely:

$$\mathbf{I}_{B,\boldsymbol{\theta\theta}} = \mathrm{E}\left\{\mathbf{I}_c\left(\boldsymbol{\theta}\right) + \left[\frac{\partial \ln f(\bar{\boldsymbol{\theta}})}{\partial \bar{\boldsymbol{\theta}}^*}\right]\left[\frac{\partial \ln f(\bar{\boldsymbol{\theta}})}{\partial \bar{\boldsymbol{\theta}}^*}\right]^H\bigg|_{\bar{\boldsymbol{\theta}}=\boldsymbol{\theta}}\right\}, \tag{5.63}$$

---

[11] This is the Bayesian version of the so-called *Barankin bound* [641], representing a realizable lower bound on the MSE of any unbiased estimator of a (nonrandom) parameter vector.

holds for the complex FIMs (the matrix $\mathbf{I}_c(\boldsymbol{\theta})$ is the complex FIM (5.43) evaluated for a deterministic $\boldsymbol{\theta}$).

An efficient estimator that attains the Bayesian CRVB exists if the posterior pdf of $\mathbf{h}$ is Gaussian. In this case, the MAP estimator is efficient and the MAP and the MMSE estimators are equivalent, since the MMSE cannot have a larger error than MAP [643].

We now apply these results in the following example.

Example 5.2.3 Let us evaluate the *semiblind* estimation problem discussed in Example 5.2.2. Since $\boldsymbol{\theta} = [\mathbf{s}_d^H, \mathbf{h}^H]^H$, some statistical properties of the independent vectors $\mathbf{s}_d$ and $\mathbf{h}$ need to be fully defined. Here, following [628], we assume that:

(a) the information symbols (i.e., the components of $\mathbf{s}_d$) are iid random variables, have zero mean and variance σ_d^2 and are characterized by the pdf $p_s(\cdot)$, and

(b) the channel taps (i.e., the components of $\mathbf{h}$) are iid random variables and are characterized by the pdf $p_h(\cdot)$.

The complex FIM $\mathbf{I}_{B,\boldsymbol{\theta}\boldsymbol{\theta}}$ (5.59) can then be easily evaluated using (5.63), where $\mathbf{I}_c(\boldsymbol{\theta})$ is given by (5.55). We then obtain:

$$E\{\mathbf{I}_c(\boldsymbol{\theta})\} = \frac{1}{\sigma_n^2} \begin{bmatrix} E\{\mathbf{H}_d^H(\mathbf{h})\,\mathbf{H}_d(\mathbf{h})\} & E\{\mathbf{H}_d^H(\mathbf{h})\,\mathcal{H}(\mathbf{c})\} \\ E\{\mathcal{H}^H(\mathbf{c})\,\mathbf{H}_d(\mathbf{h})\} & E\{\mathcal{H}^H(\mathbf{c})\mathcal{H}(\mathbf{c})\} \end{bmatrix}, \tag{5.64}$$

where the four expectations on the RHS need to be taken over both $\mathbf{s}_d$ and $\mathbf{h}$. It is easy to show that $E\{\mathbf{H}_d^H(\mathbf{h})\,\mathcal{H}(\mathbf{c})\}$ and $E\{\mathcal{H}^H(\mathbf{c})\,\mathbf{H}_d(\mathbf{h})\}$ produce null contributions, so that:

$$E\{\mathbf{I}_c(\boldsymbol{\theta})\} = \begin{bmatrix} \frac{1}{\sigma_n^2}E\{\mathbf{H}_d^H(\mathbf{h})\,\mathbf{H}_d(\mathbf{h})\} & \mathbf{0}_{N,L_h} \\ \mathbf{0}_{L_h,N} & \frac{1}{\sigma_n^2}E\{\mathbf{R}_s\} \end{bmatrix}, \tag{5.65}$$

where $\mathbf{R}_s \triangleq \mathcal{H}^H(\mathbf{c})\mathcal{H}(\mathbf{c})$ is an autocorrelation matrix associated with the imput symbol matrix $\mathcal{H}(\mathbf{c})$. In addition, we have that:

$$\begin{aligned}
&E\left\{\left[\frac{\partial \ln f(\bar{\boldsymbol{\theta}})}{\partial \bar{\boldsymbol{\theta}}^*}\right]\left[\frac{\partial \ln f(\bar{\boldsymbol{\theta}})}{\partial \bar{\boldsymbol{\theta}}^*}\right]^H\Big|_{\bar{\boldsymbol{\theta}}=\boldsymbol{\theta}}\right\} \\
&= \begin{bmatrix} E\left\{\left[\frac{\partial \ln p_s(\bar{\mathbf{s}}_d)}{\partial \bar{\mathbf{s}}_d^*}\right]\left[\frac{\partial \ln p_s(\bar{\mathbf{s}}_d)}{\partial \bar{\mathbf{s}}_d^*}\right]^H\Big|_{\bar{\mathbf{s}}_d=\mathbf{s}_d}\right\} & \mathbf{0}_{N,L_h} \\ \mathbf{0}_{L_h,N} & E\left\{\left[\frac{\partial \ln p_h(\bar{h})}{\partial \bar{h}^*}\right]\left[\frac{\partial \ln p_h(\bar{h})}{\partial \bar{h}^*}\right]^H\Big|_{\bar{h}=h}\right\} \end{bmatrix} \\
&= \begin{bmatrix} \rho_s^2\mathbf{I}_N & \mathbf{0}_{N,L_h} \\ \mathbf{0}_{L_h,N} & \rho_h^2\mathbf{I}_{L_h} \end{bmatrix}, \tag{5.66}
\end{aligned}$$

where $p_s(\bar{\mathbf{s}}_d)$ and $p_h(\bar{h})$ denote the pdfs of $\mathbf{s}_d$ and $\mathbf{h}$, respectively, and:

$$\rho_s^2 = E\left\{\left|\frac{\partial \ln p_s(s_d)}{\partial s_d^*}\right|^2\right\}, \quad \rho_h^2 = E\left\{\left|\frac{\partial \ln p_h(h)}{\partial h^*}\right|^2\right\}. \tag{5.67}$$

Substituting (5.65) and (5.66) into (5.63) yields the complex FIM:

$$\mathbf{I}_{B,\boldsymbol{\theta}\boldsymbol{\theta}} = \begin{bmatrix} \frac{1}{\sigma_n^2}E\{\mathbf{H}_d^H(\mathbf{h})\,\mathbf{H}_d(\mathbf{h})\} + \rho_s^2\mathbf{I}_N & \mathbf{0}_{N,L_h} \\ \mathbf{0}_{L_h,N} & \frac{1}{\sigma_n^2}E\{\mathbf{R}_s\} + \rho_h^2\mathbf{I}_{L_h} \end{bmatrix}. \tag{5.68}$$

Note the following observations:

(a) Since this matrix is block diagonal, the BCRB for the symbols is decoupled from that of the channel.
(b) The BCRB for the covariance matrix $\mathbf{C}_{\hat{\mathbf{h}}} = \mathrm{E}\{(\hat{\mathbf{h}} - \mathbf{h})(\hat{\mathbf{h}} - \mathbf{h})^H\}$ can be formulated as (see (5.46)):

$$\mathbf{C}_{\hat{\mathbf{h}}} - \left[\frac{1}{\sigma_n^2} \mathrm{E}\left\{\mathbf{R}_s\right\} + \rho_h^2 \mathbf{I}_{L_h} \right]^{-1} \succeq \mathbf{0}_{L_h, L_h}. \qquad (5.69)$$

$\square$

Following the same approach as that illustrated in Example 5.2.3, BCRBs have been derived for a frequency-selective block fading MIMO channel in [645] for the following three cases:

   i) *time-multiplexed* pilot and data symbols (in practice, a sequence of pilot symbols is transmitted first and is followed by a sequence of data symbols),
  ii) *superimposed* training (i.e., data symbols are transmitted simultaneously with pilot symbols; no strict form of orthogonality between training and data is enforced), and
 iii) *precoded* data.

In the last case, a subset of the data symbols is linearly precoded and is transmitted simultaneously with pilot symbols. We note that the BCRB for semiblind estimation of a MIMO channel, under the assumption of iid data symbols and orthogonal pilot sequences, is the same for a SIMO and an MIMO channel, in that it does not depend on the number of transmit antennas [646]. Moreover, for large data sequences the BCRB does not depend on the pilot design [645]. For this reason, the use of an alternative Cramér-Rao lower bound, known as the *stochastic* CRB (in place of the BCRB), has been proposed in [646] for performance evaluation.

Further results on BCRBs in the presence of data precoding can be found in [647], which is concerned with developing the best strategy for precoding the symbols to train the receiver in a MIMO communication system, when the so-called *affine precoding* [648] technique is used (see Section 5.3.2.1). BCRBs have been derived for cases where the random parameters to be estimated are either the fading channel coefficients (decoupled channel and symbol estimation) or the symbols and channel coefficients together (joint channel/symbol estimation).

Following the same line of reasoning as for the CRB, modified CRBs can be derived even for Bayesian estimators. Specific applications of this bound in MIMO communications are illustrated in [649, 650, 651], referring to the estimation of a block flat fading channel, the estimation of a frequency-selective block fading channel and the tracking of a time-varying MIMO Rayleigh time-selective fading channel.

We conclude with the following observations:

1. The problem of assessing bounds on the accuracy of pilot-based estimation of a bandlimited frequency-selective communication channel is tackled in [652], where some properties of optimal waveforms for channel sounding and closed-form CRBs are derived.
2. CRBs for *hybrid* random and deterministic parameter estimation are presented in [653], deriving a positive definite matrix which simultaneously provides bounds on the covariance of any unbiased estimator of the nonrandom parameters and an estimator of the random parameters (see also [654, Sect. II]).
3. CRBs referring to different classes of channel estimators have been derived for OFDM systems in [455, 655–660] and for MIMO-OFDM in [660].

4. When *blind* estimation is considered, the FIM referring to the joint estimation of data symbols and channels is singular, since channel estimation is possible only up to a scale factor – that is, the channel cannot be estimated unambiguously. To overcome this problem, the use of the Moore–Penrose pseudoinverse of the FIM has been considered in [627]. An alternative is to eliminate the ambiguity of *blind channel estimation* by putting parametric constraints on the set of parameters to be estimated (these constraints are usually expressed in the form of functional equality) and leads to a *constrained* CRB (CCRB)[12] [630, 663, 664]. CCRBs for specific channel estimation problems can also be found in [657, 665, 666].
5. CRBs and BCRBs for channel estimation of a MIMO frequency-selective channel in the presence of a priori information about the transmitted data sequence have been developed in [667]. The CRB for a MIMO frequency-selective channel is considered in [645, 668] (which also assumes the presence of a *carrier frequency offset* (CFO)). The CRB for channel training in a MIMO system using space-time orthogonal block codes over frequency-flat fading is evaluated in [669].
6. An efficient *message-passing* algorithm defined over *factor graphs* to compute the BCRB for general estimation problems characterized by very large FIMs is developed in [670] (further details on message-passing algorithms and factor graphs can be found in Sections 10.7.1 and 10.8, respectively). It allows the evaluation of the diagonal elements of an FIM by local computations that involve the inversion of matrices that are much smaller than the original FIM. Similar methods are developed in [671] for computing Cramér–Rao type bounds from marginal pdfs (usually characterized by a dense FIM).

5.3 Data-Aided CIR Estimation Algorithms in PATs

The quality of the channel estimates in a wireless communication system employing PAT depends, first of all, on the rate and the placement of pilot symbols. For a fixed pilot symbol rate and a given channel model, the placement of pilot symbols can be optimized according to some specific criterion. Then channel estimation algorithms exploiting the resulting pilot symbol pattern need to be devised, to achieve a reasonable tradeoff between computational complexity and performance. Note that, from this perspective, PAT *design appears to be primarily a transmitter technique*, although the receiver characteristics need to be carefully taken into account to develop technically appealing solutions.

In this section we first tackle PAT *design*, illustrating a mathematical model for this problem and introducing specific criteria for PAT optimization. Then we consider a signal processing perspective on PAT and provide an overview of data-aided channel estimation algorithms for various PAT strategies employed with SC and MC communications.

5.3.1 PAT Modeling and Optimization

In a SISO wireless communication system employing a SC modulation, time division multiplexing is traditionally used to emplace pilot symbols. A generalization of this approach is obtained by *superimposing pilot symbols and data symbols* (e.g., see [647, 648, 669, 672–676]). In the latter case, if a *packet* of N symbols is sent over a wireless channel, generally speaking the lth transmitted symbol can be modeled as:

$$c_l = \sqrt{P_l^{(d)}} c_l^{(d)} + \sqrt{P_l^{(p)}} c_l^{(p)} \tag{5.70}$$

with $l = 0, 1, \ldots, N-1$, where $c_l^{(d)}$ $\left(c_l^{(p)}\right)$ denotes the lth information (pilot) symbol. Here it is then assumed that $\left|c_l^{(p)}\right| = 1$, $\mathrm{E}\left\{c_l^{(d)}\right\} = 0$ and $\mathrm{E}\left\{\left|c_l^{(d)}\right|^2\right\} = 1$, so that $P_l^{(d)}$ $\left(P_l^{(p)}\right)$ denotes the average

---

[12] Note that the constrained CRB expressions available in [661, 662] assume that the FIM is nonsingular, i.e. full rank. This is not required by the constrained CRB developed in [663].

power (power) assigned to $c_l^{(d)}$ $\left(c_l^{(p)}\right)$. For the transmission of the given packet a PAT scheme is defined by the N-dimensional *pilot vector* $\mathbf{c}^{(p)} = \left[c_0^{(p)}, c_1^{(p)}, \ldots, c_{N-1}^{(p)}\right]^T$ and by two *power allocation vectors* $\mathbf{P}^{(d)} = \left[P_0^{(d)}, P_1^{(d)}, \ldots, P_{N-1}^{(d)}\right]^T$ and $\mathbf{P}^{(p)} = \left[P_0^{(p)}, P_1^{(p)}, \ldots, P_{N-1}^{(p)}\right]^T$. The problem of PAT design can then be interpreted as one of *power allocation under certain constraints* [418]. In fact, any wireless transmission is subject to power constraints, but such constraints can be imposed on PAT in different ways. For instance, an *average power constraint* is given by:

$$\frac{1}{N} \sum_{l=0}^{N-1} \mathrm{E}\left\{\left|c_l^{(d)}\right|^2\right\} = \frac{1}{N} \sum_{l=0}^{N-1} \left[P_l^{(d)} + P_l^{(p)}\right] = P_{fr}, \tag{5.71}$$

where P_{fr} is the average transmit power over the entire frame. Similarly, the per-symbol average power constraint can be formulated as:

$$P_l^{(d)} + P_l^{(p)} = P_s, \tag{5.72}$$

where P_s is the average transmit power per symbol. The following observations are in order:

(a) The latter constraint imposes a more stringent condition than the former.
(b) Both constraints refer to the overall transmit power, so that power allocation among data and pilots is a fundamental factor to consider in PAT design.
(c) Model (5.70) with its constraints can easily be generalized to cases where frequency division multiplexing (spatial multiplexing) is adopted for pilot symbols, since an MC modulation (an antenna array) is employed.

Given a PAT model, it is important to *optimize it for a wide range of channel conditions*. Note that in most of the literature, the design of PAT is based on intuition, practical experience, heuristic analysis and simulation. For instance, it seems reasonable that:

- in the presence of fast fading, the larger the rate of pilot symbols, the better channel estimation and tracking are, so that receiver robustness is enhanced, and
- in the presence of nonnegligible time dispersion, pilot symbols should be placed in clusters to avoid interference from unknown data symbols.

At the theoretical level, however, the optimal design and multiplexing of pilot symbols is far from trivial. Note that pilot symbols carry no information about the data, so that the time and power spent on sending pilot symbols is time missed for transmitting information and power taken away from data, respectively. A good starting point for optimizing PAT is *to fix the percentage* (*in power or in the number of channel uses*) *of pilot symbols and optimize the pilot symbol placement*, that is, how training symbols are multiplexed into a data stream. A specific application of this approach is considered in the following example.

Example 5.3.1 Following Example 5.2.1, let us again consider the transmission of a data packet consisting of N data symbols and P pilot symbols over a frequency-selective block fading channel of order L_h. Generally speaking, pilot symbols can be grouped in *clusters*; then the resulting placement can be described by $\mathcal{P} = [\mathbf{l}_d, \mathbf{l}_p]$, where $\mathbf{l}_d = [l_{d,1}, l_{d,2}, \ldots, l_{d,n+1}]$ and $\mathbf{l}_p = [l_{p,1}, l_{p,2}, \ldots, l_{p,n}]$ are the data block length vector and the pilot cluster length vector respectively, and n denotes the overall number of pilot blocks. The resulting symbol pattern is illustrated in Figure 5.4(a). Constraining the total number of data and pilot symbols (i.e., the *pilot symbol rate*), we have $\sum_{k=1}^{n+1} l_{d,k} = N$ and $\sum_{k=1}^{n} l_{p,k} = P$. Moreover, for those placements starting with pilot symbols, $l_{d,1} = 0$, and for those

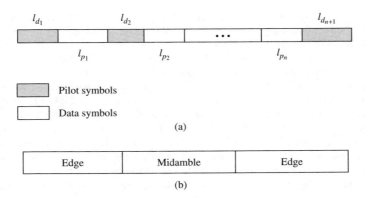

Figure 5.4 (a) Data packet with multiple pilot clusters. (b) Edge and midamble positions of a data packet.

ending with pilot symbols, $l_{d,n+1} = 0$. We can also define the *edge* and *midamble* positions for each packet, as shown in Figure 5.4(b). Edge positions are defined as the first and last $L_h - 1$ positions in a packet, whereas all the others within the $[L_h, N + P - L_h + 1]$ interval are midamble positions.

For *training-based channel estimation*, only those parts of the observations influenced by pilot symbols are processed. Therefore, if there is a pilot cluster of length less than $L_h + 1$, no pilot symbols associated with this cluster can be exploited for channel estimation. This could lead to the conclusion that all pilot symbols should be grouped into a single cluster. This intuition, however, is not valid if all observations are to be used for channel estimation. Indeed, the use of multiple clusters results in a simpler design of pilot symbols and better detection performance, as discussed in [628].

□

Different design constraints and objectives can be adopted by systems designers in PAT design. It is, therefore, preferable that PAT schemes are optimal under different criteria, that is, maximize different performance metrics. The following optimality criteria are usually considered in the literature:

1. *Information-theoretic criteria.* Information-theoretic metrics[13] reveal tradeoffs among PAT designs. From an information-theoretic perspective, a PAT scheme provides the receiver with side information about an unknown channel, so that the most natural optimality criterion in designing PAT is *maximizing channel capacity* in the presence of channel uncertainty [594, 677–680]. This requires *mutual information* to be expressed as a function of PAT parameters and then to be maximized it with respect to the PAT parameters and the channel input distribution (e.g., see [681, 682]). The drawback of this approach is that mutual information often cannot be expressed in a simple form. An alternative is to use other optimization criteria, namely:

   (a) optimizing bounds on the achievable data rate [677, 679, 683] with respect to PAT parameters (e.g., see [680, 684–691] for applications of this approach),

   (b) minimizing the *outage probability* for a given transmission rate R, that is, maximizing the probability that the rate R can be achieved reliably (e.g., see [692]),

   (c) optimizing other meaningful parameters, like the *channel reliability function, random coding exponent*, and *cutoff rate* [693], relating detection error probability with data rate and codeword length and leading to a more analytically tractable framework (e.g., cutoff rate is used for optimizing PAT design in [694–697]).

---

[13] Some elements of information theory are illustrated in Chapter 7.

2. *Quality of channel estimates.* For the receiver structure shown in Figure 5.1, one may be interested in the PAT scheme minimizing the channel estimation error. From this viewpoint, a meaningful parameter is the MSE of the estimator (e.g., see [681, 698, 699, 700, 701, 702]). Since it is desirable that the design of optimal PAT does not depend on the specific algorithm used at the receiver, the CRB is a natural choice as a figure of merit (see Section 5.2).

3. *Quality of data estimates.* If the target of our PAT design is optimizing detection performance, BER or SER can be taken as a performance metric. However, these are usually difficult to characterize precisely for most fading channels (e.g., see [703] for time division multiplexed training over a Rayleigh flat fading channel and [704] for a BPSK transmission over a Rice fading channel). A more tractable approach is to use BER bounds (e.g., Bhattacharyya and random coding bounds [693]) as the figure of merit. Further alternatives are offered by the *error exponent function* (which measures the decay rate of the error probability [705]) and by interpreting symbol detection as a form of parameter estimation, so that the MSE at the detector input can be taken as the metric for optimization (this approach is commonly taken in the design of channel equalization algorithms).

Other than optimal symbol placement, another fundamental issue in PAT design is the *amount of training* needed in a wireless transmission. Intuitively, a tradeoff between having more training for better estimation and more channel uses for higher rates should be expected. Relevant contributions to clarifying this issue have been provided by [685, 706] in a MIMO scenario.

5.3.2 A Signal Processing Perspective on PAT Techniques

We now briefly describe various signal processing techniques developed for estimating the communication channel using both SC and MC PATs. Before delving into the study of this topic, it is important to point out that in practical communications systems, *slow* and *fast* fading channels are usually treated in different ways. In fact, for the first class of channels, a *block fading model* is often adopted, that is, channel variations are deemed negligible over the duration of each data packet. For this reason, training data are usually inserted at the beginning of each data burst (*preamble-based training*) and the data-aided channel estimate extracted from the preamble is used for detecting all information symbols contained in the same packet. In contrast, for the second class of channels training symbols are usually interspersed (in time or in frequency) in the data stream and the resulting technique is usually called *pilot symbol assisted modulation* (PSAM).

5.3.2.1 Channel Estimation in SC PATs

Time-selective channels

Channel estimation in time-selective fading is closely related to carrier phase synchronization [487, 488, 707] and can easily be accomplished by transmitting a *pilot tone* at a convenient frequency in the data spectrum (or just outside the data spectrum). This provides the receiver with an explicit amplitude and phase reference for detection (e.g., [708–714]). This approach is typified by the so-called *transparent tone-in-band* (TTIB) [709, 714] and the *tone calibration technique* [712, 715]. Although these are general solutions, they require relatively complex signal processing and result in an increased PAPR.

An alternative to tone-based strategies is the embedding of a pilot sequence in the transmitted data sequence for training the channel estimator. This inevitably requires frame synchronization at the receiver [716]. Two major training techniques for wireless channels are *time division multiplexed* training and *superimposed* training. In the former strategy, pilot and information symbols are time division multiplexed and a known training pattern is usually transmitted in a periodic fashion. This approach, based on the use of *regular periodic placements* [675, 698], is exemplified by PSAM [717, 718]. In a PSAM-based system the transmitter periodically inserts known symbols

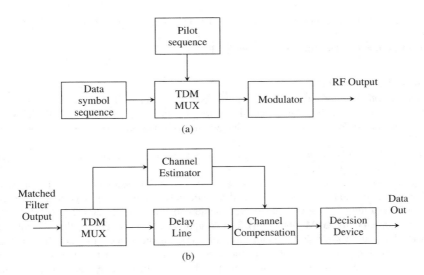

Figure 5.5   PSAM transmitter (a) and receiver (b) structures.

(see Figure 5.5(a)), from which the receiver derives its amplitude and phase reference (see Figure 5.5(b)). Note that the received samples associated with information symbols are delayed to be in step with the generated channel estimates. In this scheme the fading distortion at the information symbols is estimated by interpolating the fading distortion assessed at pilot symbols (e.g., using low-order interpolations to reduce the number of computations or MMSE interpolation; see [698, 703, 718–720]). Like pilot tone modulation, PSAM suppresses the error floor and enables multilevel modulation. However, it entails no change to the transmitted pulse shape or PAPR, and simplifies processing at the transmitter and receiver.

The first contributions dealing with PSAM were based on simulation [717, 718] and did not provide a performance analysis. A sound analytical basis for PSAM was provided in [721]. Further results on the optimization of regular periodic placements and, in particular, of PSAM can be found in [675, 682, 687, 697, 698, 703, 704, 715, 720, 722].

The use of superimposed training was originally proposed in [672, 723, 724]. A more general form of superimposed training, called *affine precoding* (which can be viewed as a general framework in which preamble-based training, PSAM schemes or superimposed training can be treated as special cases), has recently attracted attention [648, 669, 673–676]. It is expected that:

- despite the additional complexity introduced by the mixing of pilot and data symbols, some performance gain over the conventional (e.g., PSAM) techniques can be realized, and
- the constant presence of pilot symbols in the data stream can somehow improve the tracking capability of the receiver for time-varying channels.

Actually, for a transmission over time-selective SISO channels, it has been shown that superimposed training performs better than time division multiplexed schemes at low SNR and for relatively fast time-varying channels [675].

Further significant work on pilot-aided estimation of flat fading channels concerns:

(a) the use of KF [675, 725] or an extended KF [707];
(b) pilot symbol encoding rules for CPM signaling [726];

(c) the use of data-bearing symbols to improve the phase-tracking capability of PSAM [514, 727, 728];
(d) channel estimation algorithms in turbo-coded PAT [719, 729].

Frequency-selective channels

The design of *optimal pilot sequences* for training-based estimation of a frequency-selective channel is an old problem and has been investigated in various papers (e.g., see [730–738]). This concerns the problem of generating codes of arbitrary length with zero periodic correlation except for the peak at zero shift.

The problem of *optimal placement* of pilot symbols in a data stream has been investigated in [627–629, 673, 678, 686, 690, 692, 698].

Channel estimation employing a training sequence is usually accomplished using a *correlation method* or an LS method. The former method is extremely simple, since it is based on correlating a portion of the training sequence with shifted versions of the received signal [733, 739] (e.g., this approach is adopted in the GSM [560, 740]). The LS approach [731, 740] offers the relevant advantage of requiring half guard symbols to obtain the same approximate processing gain. In fact it needs a precursor of length equal to the channel memory duration ($L_h - 1$), whereas the correlation method requires a precursor and postcursor of the same length (in practice, the minimum sequence length is $2L_h - 3$). Even though a matrix inverse is involved in the LS procedure, it is a function only of the known training sequence and a precomputed inverse of the matrix can be stored (for some special training sequences, an inverse matrix is not even required). An application of the LS approach to the GSM system can be found, for instance, in [741]. A further alternative to the above two estimators is MMSE estimation [742]; however, this requires the availability of statistical information about the channel fading.

Further significant contributions on PAT over frequency-selective channels concern:

- the analysis of PSAM schemes [743];
- the use of superimposed training sequences for channel estimation [676, 744, 745];
- the use of LS estimation of the CIR in the presence of *co-channel interference* (CCI) [746].

Doubly-selective channels

High data rate wireless links in mobile communications suffer from time- and frequency-selective propagation effects. Under this scenario fading channels are challenging to mitigate, but once acquired, they offer joint multipath-Doppler diversity gains [205, 747]. The quality of channel acquisition and tracking has a major impact on the overall system performance, especially in the presence of fast fading [748]. This motivates the substantial efforts devoted to reliable estimation of doubly-selective channels.

Digital communication systems operating over doubly-selective channels often employ a signaling format in which transmitted data are organized in blocks, each preceded by a known training sequence. When processing each block, a CIR snapshot is generated, exploiting the associated training sequence (usually this estimate relies on the assumption of negligible channel variations over the training sequence) to start up an equalization algorithm. Then channel variations can be tracked in a decision-directed fashion using a *recursive type* of adaptive filter. Unfortunately, this approach suffers from an irreducible lag error leading to limited robustness against faster channel variation. For this reason, *feedforward estimation* is often preferred to *recursive estimation*. When this occurs, multiple CIR estimates evaluated over adjacent training sequences are interpolated to generate a channel estimate over the data blocks. This eliminates the risk of error propagation when estimating the CIR in a decision-directed mode.

The design of optimal training schemes over doubly-selective channels is more complicated than the similar problems tackled in scenarios exhibiting only frequency or time selectivity. In the literature the use of PSAM has been investigated [690, 749, 750]. The number and placement of pilots affect not only the quality of CSI acquisition but also the transmission rate. Within the general class of doubly-selective channels, PSAM optimization for special channel models has been investigated in [688, 736, 749–751]. Optimization of power and the placement of pilot clusters time division multiplexed with the data has been analyzed in [702], where it has been assumed that pilot clusters consist of zero-padded pilot symbols in order to decouple MMSE channel estimation from data detection. The use of pilot tones and pilot symbols for channels affected by fast fading and a large delay spread is discussed in [556].

For a given training scheme, the structure and complexity of the channel estimator strongly depends on the underlying channel model adopted for its design and on the degree of knowledge we need to acquire about its statistical properties. For instance, one of key issues is the representation of the time evolution of the channel taps; this can be modeled by resorting to AR models [750, 752–754], complex exponential models [750, 755–758], or truncated power series models [114, 616, 617, 759–761]. Note that, if the model is refined enough to account for the dynamics of the fading channel over an entire data block, in principle adaptive channel tracking can be avoided, as shown in [761], where channel estimation for *enhanced data rate for* GSM *evolution* (EDGE) is investigated. However, the more refined the channel model is, the larger the number of parameters that need to be estimated. For instance, if an explicit parametrization by time-varying amplitudes and delays is adopted, estimation of a large number of parameters may be required (e.g., see [756, 762, 763] for a single antenna and [764] for multiple antennas). In this context another complicated issue is that of estimating the statistics of a channel with random taps and fitting appropriate models [750, 752, 754, 765].

As already stated above, the first class of channel estimators consists of *recursive channel estimators*. These were first developed for HF modems [759, 766]. They are based on recursive adaptive filters and exhibit limited robustness in the presence of fast channel variations because of their intrinsic irreducible lag error. Despite this, they have also been proposed for channel tracking in TDMA mobile radio. In this field, the use of Kalman filtering [750, 758, 767–769], extended Kalman filtering [770], delayed LMS estimation[14] [511, 560, 616, 617], linear LS with variable forgetting factor [760], recursive ML estimation [765], RLS [616, 617, 758, 771], and conditionally coupled recursive estimation (based on combining an augmented-state adaptive KF with an RLS for estimating the AR parameters of the channel taps) [752, 753] has been studied. In addition, a recursive technique based on the EM algorithm[15] was proposed in [772, 773]. Finally, the use of LMS, RLS, Kalman and other estimation techniques in PSP-based channel tracking was investigated in [114, 424, 515, 559, 564, 617, 624, 774–776].

The second class of estimators consists of *feedforward channel estimators*, which, like recursive estimators, have been proposed for rapidly varying channels in HF communications in [749] (where Wiener filtering between training sequences is investigated) and in TDMA radio (e.g., see [777–779]). In the latter case CIR estimation is accomplished through the use of a training sequence inserted in each TDMA slot (e.g., an LS algorithm can be used). Then, since LMS and RLS are unable to track the randomly changing CIR in the presence of fast fading (due to their sensitivity to error propagation in the decision-directed mode) during data blocks, CIR over such blocks is computed by interpolating the snapshot channel estimates available at the end of each training interval.

Finally, we mention the following significant contributions to the estimation of doubly selective channels:

(a) the development of noncoherent LMS and RLS algorithms for channel identification (to be employed in noncoherent sequence estimation of *M*-ary differential PSK over ISI channels) [780], and

(b) the study of MMSE estimation of the CIRs of multiple co-channel users in TDMA systems and the design of training sequences for this scenario [781].

---

[14] Work in this field was inspired by [458].

[15] The EM algorithm is described in Section 4.6.1.

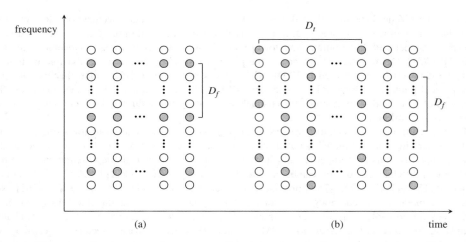

Figure 5.6 Arrangement of training data in an OFDM transmission: (a) comb-type arrangement; (b) scattered arrangement.

5.3.2.2   Channel Estimation in MC PATs

Coherent detection of OFDM transmission requires estimation of the channel frequency response (i.e., the complex gain affecting each subcarrier) over time to compensate for channel distortion after the demodulation of subcarriers [782]. For this purpose, the channel needs to be probed simultaneously in both the time and frequency domains, that is, using a *two-dimensional grid* of pilot symbols which satisfies the two-dimensional sampling theorem [751, 783]. Note that increasing the number of spectral symbols to mitigate spectral aliasing reduces the efficiency of data transmission. However, for a given rate of pilot symbols the performance can be improved by selecting a proper arrangement of pilot symbols, that is, a proper shape for the pilot symbol grid [783]. Two well-known arrangements of pilot data are illustrated in Figure 5.6. The first, the *comb-type* arrangement [784, 785], is characterized by pilot data periodically placed along the frequency direction (D_f denotes the period of the pilot sequence) and *continually* transmitted on a fixed subset of subcarrier frequencies. The second, *scattered* arrangement, is characterized by pilot symbols placed according to a doubly-periodic pattern with period in time of D_t and period in frequency of D_f [751]. Both continual and scattered pilot symbols can be transmitted at the same power level as data symbols or at a boosted power level. Of course, other arrangements are possible, such as *hexagonal placement* [786] (which offers the best coverage in the time–frequency plane) or a *hopping pilot pattern* [787]. The use of superimposed pilots (pilot symbols added linearly to the modulated data symbol at a fraction of the total transmit power) is also possible [788].

An optimized placement of training symbols has been shown to enhance overall system performance from both an information- and estimation-theoretic perspective. Optimization of the location, number, and power of pilot symbols has been investigated in depth under different criteria of optimality and assuming different types of channel estimators (e.g., see [659, 673, 686, 690, 786, 789–796]).

Channel estimators are currently divided into two classes: *parametric* estimators and *nonparametric* estimators. Parametric channel estimation commonly employs a deterministic channel model, which is based on a finite number of delay paths, and estimates the gain and delay of each path. Nonparametric channel estimation makes few assumptions concerning the channel and involves techniques that exploit the estimate of the frequency response at pilot subcarriers to generate the estimates of channel gain at each data subcarrier using an interpolation algorithm.

The parametric approach to channel estimation is motivated by the fact that in a specific wireless scenario (such as macro-cellular wireless), the channel length is limited to a few samples. For this reason, it seems more appropriate to estimate the channel in the time domain because there are fewer parameters in the impulse response than in the frequency response. Given the limited amount of training data that can be sent to estimate the time-varying channel, limiting the number of parameters to be estimated is expected to improve the accuracy of the estimation [789]. In parametric estimators IFFT processing is usually employed to get an estimate of the time-domain CIR which is then appropriately processed and transformed back to the frequency domain by FFT. In doing this it is important to capture the energy of the most significant taps and discard nonsignificant channel taps to improve channel estimation performance [797].

Various techniques for time-domain channel estimation can be found in the literature [656, 686, 787, 794, 797–803]. Here, we limit discussion to summarizing some of the most significant results in this area. First, we mention [798], where the MMSE estimator for the complex tap gains of the observed channel is introduced. It performs well and substantially outperforms the LS estimator – at the price, however, of large computational complexity. To reduce the computational complexity of the MMSE approach low-rank estimators are suggested. The performance of these estimators is affected by the presence of non-sample-spaced channels and even suffers from error floors for high SNRs. The channel estimation algorithm proposed in [799] performs well in sparse multipath fading channels, but suffers the problem of slow convergence in multipath delay estimation in relatively slow fading channels, and even suffers from errors in multipath delay estimation when many propagation paths exist in the channel. The observations in [798, 799] have inspired the work of [802], where a CIR model characterized by fractional tap delay spacing relative to the sampling interval is adopted in developing MMSE and LS estimators. This eliminates the problem of multipath delay estimation and reduces the signal subspace dimension of the channel correlation matrix. For this reason, full-rank estimators using pilot subcarriers can be adopted, which improves the channel estimation performance. In [787] an LS algorithm for pilot-aided estimation of sparse channels is developed. The proposed algorithm uses a generalized Akaike information criterion to estimate the channel length and tap positions. This effectively reduces the signal space of the LS estimator, and hence improves estimation performance with respect to the conventional LS approach of [798].

Finally, it is worth mentioning that most parametric channel estimators are derived under the assumption of a static channel; however, CIR estimation techniques for time-varying channels are also available [659, 795, 801].

Various nonparametric methods for channel estimation have been devised (e.g., see [784, 804–806]). The nonparametric approach to channel estimation in OFDM systems consists of two steps. In the first step LS estimates of the channel gains over the pilot subcarriers are obtained by simply back-rotating the received signal in accordance with knowledge of the pilot symbols. This is equivalent to accomplishing two-dimensional sampling of a noisy version of the WSS-US process represented by the mobile radio channel. In the second step the LS estimates are interpolated/smoothed over the entire frequency–time grid;[16] this can be accomplished directly in the frequency domain or by using IFFT/FFT processing.

In the first case, interpolation can be done using a proper two-dimensional or separable filter [751, 804, 808, 809]. Unfortunately, the design of the optimal (MMSE) interpolator requires knowledge of the two-dimensional correlation function of the channel [751] – both its power delay profile and its Doppler spectrum. Since this information is not easily available at the receiver, the design problem becomes that of finding the most robust estimator with respect to a mismatch in the channel correlation [810]. Simpler alternatives consist of using simple suboptimal (e.g., linear) one-dimensional interpolators/filters in the frequency domain [791, 792, 805, 811–813], efficient two-dimensional interpolators [814], two-dimensional regression polynomials for LS fitting over blocks of the time–frequency plane [815, 816] and nonlinear interpolators implemented as one- and two-dimensional *radial basis function* (RBF) networks trained by pilot symbols [817].

---

[16] An intermediate preinterpolation step can be added between the LS estimation over the pilot subcarriers and the interpolation [807].

In the second case, interpolation of channel estimates at pilot locations is accomplished via DFT to achieve low complexity [783, 818]. The channel gains acquired over pilot subcarriers and over consecutive OFDM symbols (i.e., over a specific two-dimensional pilot grid) undergo a two-dimensional FFT, followed by low-pass (time-domain) filtering that takes advantage of time-domain correlation. The filter output feeds a two-dimensional IFFT, which generates the interpolated channel estimates at unknown data locations. This approach can be used not only to develop an MMSE interpolator (or an MMSE channel estimator, if pilot symbols only are transmitted [810, 819, 820]), which requires, however, knowledge of the channel statistics, but also to design a robust interpolator insensitive to channel statistics, as shown in [783]. A related solution (employing one-dimensional signal processing in the frequency domain) is presented in [810], where a generalized Hanning window is first applied to the channel frequency response observation vector to mitigate spectral leakage (originating from the fact that channel multipath time delays may be sample-spaced) and then an IFFT is performed to transform the windowed frequency response into the time domain. The effective channel impulse response is then modified by an MMSE weighting function. After that, an FFT is performed to transform the result back into the frequency domain, and the windowing effect is finally removed to obtain the channel estimation output.

Finally, we mention the following relevant contributions to the field of channel estimation for OFDM:

(a) the development of decision-directed channel predictors for generating up-to-date CSI even without regular transmission of pilot symbols in an OFDM transmission [821], and

(b) the design of channel estimation algorithms for OFDMA systems operating over time-varying channels in the absence of evenly spaced pilots [822].

5.4 Extensions to MIMO Channels

In SISO wireless systems pilot symbols are traditionally time- or frequency-multiplexed with the data symbols or signals. The use of an antenna array in MIMO systems extends the multiplexing to the spatial dimension, adding a further dimension in the PAT optimization problem. In addition, it may substantially increase the complexity of data-aided channel estimation algorithms. In the following we sketch some of the research done in this area, considering first SC and then MC systems.

5.4.1 Channel Estimation in SC MIMO PATs

Optimization of training signals in MIMO single-carrier systems has been addressed in various papers [645, 647, 669, 681, 685, 691, 699, 823–825]. From these papers, however, a simple set of guidelines for training design in MIMO systems cannot easily be inferred because of the heterogeneity of the modeling assumptions about the channel and the MIMO transmission scheme and of the optimality criteria. Despite this, it is useful to note the following:

1. References [669, 681, 685, 823, 824, 826] focus on PAT optimization in the presence of frequency-flat fading channels, whereas [645, 647, 691, 699, 825] consider multiple-antenna transmission over quasi-static frequency-selective channels. Optimal pilot signaling for timeslot-based MMSE estimation of space-, time- and frequency-selective fading MIMO channels is derived in [827].

2. References [669, 823] focus on general training strategies that allow the superposition of pilot and data symbols; training design in a MIMO system using *affine precoding* as the transmission strategy is investigated in [647].

A number of solutions has been proposed for data-aided channel estimation and tracking over *frequency-flat fading channels*, for example, Kalman filtering [651, 828, 829], ML estimation [830], multiple variable regression estimation [831], LS estimation [832–835], MMSE estimation [649, 824, 832, 836] and EM-based estimation [837].

Various channel estimation strategies are also available for *frequency-selective block fading channels* (i.e., for channels whose random channel taps remain constant for some data packets and change to independent values for the next ones); for instance, MMSE estimation is investigated in [645, 650, 691, 838], whereas LS estimation is adopted in [839].

Finally, a few contributions are available about channel estimation and tracking of time-varying frequency-selective MIMO channels. In particular, Kalman filtering is employed in [840] for channel tracking, whereas timeslot-based MMSE channel estimation for SC block transmission is investigated in [827].

5.4.2   *Channel Estimation in MC MIMO PATs*

A limited number of papers are available dealing with optimal design of PAT for MIMO-OFDM systems [456, 701, 841–845].

Parametric methods for channel estimation in the presence of multiple transmit antennas have been proposed in [456, 522, 524, 842, 846–852]. The most important results in this area are summarized here. LS and MMSE estimation techniques are computationally heavy, since they require matrix inversion, and this has motivated the search for reduced-complexity channel estimation algorithms which avoid the inversion (e.g., see [522, 524, 842]). Innovative estimation techniques have been developed in [845, 850, 852, 853], even if with different targets. In fact, in [852] an improved LS algorithm, which exploits the noise correlation in order to reduce the variance of the LS estimation error (by estimating and suppressing the noise in signal subspace), is devised. This algorithm is robust to the number of antennas on the transmit and receive sides and achieves performance very close to that of the MMSE estimator. A completely different target is adopted in [850], where a technique optimizing the MSE performance of MIMO channel estimation in the presence of spatial correlation is developed. This approach is motivated by a desire to find the ultimate channel estimation structure and its performance limits so as to be able to benchmark simpler suboptimal channel estimators, although this causes an unavoidable increase in the estimator computational complexity. In this case a two-step channel estimator solution is derived and analyzed. In the first step the channel time delays and spatial signature are estimated using an ML approach, while in the second step MMSE channel estimation based on joint spatio-temporal filtering is adopted. In [845, 853] an angle-domain *approximated* MMSE (AMMSE) channel estimation technique is developed. These have much lower complexity than the two-dimensional LMMSE technique and perform better than the conventional LS technique.

Nonparametric methods are considered in [854]. In particular, [849, 855] propose low-rank MMSE channel estimators.

Finally, it is worth mentioning that:

- parametric channel estimators for superimposed training have been devised in [856], and
- in most of the above mentioned papers block fading is assumed, that is, channel variations in data-aided estimation are neglected. Very few papers take into consideration channel variations in estimator design (e.g., see [851, 856]).

5.5   **Historical Notes**

Since the 1960s there has been increasing interest in investigating the problem of detecting signals that emerge from channels described by parametric models with unknown parameters (e.g., see [106, Chap. 6], [857, 858], and references therein). This has led to the study of so-called *adaptive systems*, defined as systems that extract knowledge of the channel from a test signal or from an information-bearing signal available at the channel output and exploit such knowledge in processing received data.

One of the proposed approaches to the problem of adaptive reception of signals has resulted in the class of *transmitted reference systems* [858]. In these systems a *pilot tone* is transmitted along with the message waveform with the aim of providing the receiver with a signal which is independent of the message waveform. In the 1980s a number of papers appeared describing techniques that seek to reduce the effects of multipath fading at the receiver of a mobile-satellite link through the use of a *pilot-based calibration process*. Such pilot-aided modulation techniques have appeared under various titles such as *feedforward signal regeneration* (FFSR) [708], *feedforward automatic gain control* (FFAGC) [859], TTIB [709], and *tone calibration technique* (TCT) [712]. The basic concept underlying all these techniques is to transmit a pilot at a convenient frequency in the data spectrum and then extract this pilot at the receiver with the channel impairments intact. *If the fading impairments on the pilot and the data are the same*, then the extracted pilot, after suitable processing, can be used as a coherent reference in a synchronous data detector to remove the fading phase component and to normalize the fading amplitude component. From a practical implementation viewpoint, satisfying the latter assumption requires locating the pilot at a frequency in the vicinity of the data signal. All of the above-mentioned pilot-aided techniques, particularly the original version of TCT, assume a double-sided data spectrum with a pilot located in the center. The advantage of this spectral arrangement is that the pilot is not sensitive to small frequency shifts since it is right in the center of the channel where the amplitude and phase characteristics are most stable. The disadvantage of this arrangement, however, is that it is very bandwidth inefficient, as are most double-sideband modulation schemes. In fact, from a spectral standpoint the existing TCT system closely resembles a full AM system where the data sidebands are symmetrical around the carrier. Furthermore, to "make room" for the pilot in the presence of Doppler shift, the equivalent low-pass data sidebands must be shaped so as to have zero response in the neighborhood of dc, or else be placed on a subcarrier. An alternative possibility which is much more bandwidth efficient than TCT is a *dual-pilot tone calibrated technique* (DPTCT) that symmetrically locates a pair of pilots outside the data spectrum near the band edges of the channel [711].

A more practical alternative to the transmission of a *pilot tone* is to transmit a *pilot sequence* embedded in the transmitted data sequence; this concept, known as PSAM, was first proposed by M. L. Moher and J. H. Lodge in 1989 for coherent detection of trellis-coded modulation over Rician flat fading channels [717]. The performance and optimization of PSAM have been investigated in depth for SISO systems.

Since the beginning of the 1990s considerable attention has been paid to the problem of investigating channel acquisition and tracking for TDMA (narrowband) digital cellular radio; in that scenario, traditional (e.g., linear and decision feedback) equalizers cannot meet performance specifications. For this reason, an ML receiver is required [860], which, in turn, requires a high-quality channel estimate for satisfactory performance, leading to interest in analyzing the capability of traditional adaptive algorithms for channel tracking (e.g., see [610, 768]) and in devising innovative algorithms (e.g., see [114, 511, 515, 564, 567, 624, 769, 774, 776, 861]). The use of feedforward channel estimators based on the interpolation of channel estimates has been also considered (e.g., see [778]).

Since the latter half of the 1990s there has been substantial interest in the use of OFDM for mobile wireless channels [862, 863], partly motivated by the upcoming DVB and DAB standards in Europe. This led to investigations of the problem of feedforward (e.g., see [742, 751, 804, 808, 811]) and recursive (e.g., see [864]) channel estimation based on pilot symbols in a two-dimensional scenario. It soon became apparent, however, that in wireless OFDM transmission it is more appropriate to estimate the channel in the time domain because there are fewer parameters in the impulse response than in the frequency response [789, 798]. In addition, in the presence of time selectivity it could be better to accomplish both detection and channel estimation in the time domain [800], so that the error floor originating from appreciable ICI in the frequency domain is lowered.

Research on channel estimation for systems employing transmit diversity and space-time coding started at the end of the 1990s for both SC [706] and OFDM systems [846]. This preliminary work has been followed by a flurry of papers about channel estimation for different MIMO channel models and ST communications techniques.

5.6   Further Reading

The organization of this chapter is partly inspired by [418], which offers an introduction to PAT techniques for wireless communications. An overview of channel estimation techniques for OFDM systems can be found in [793].

Channel estimation represents a significant research area in the field of wireless communications. In this concluding section we mention some of the relevant research topics in this area which have not been considered in the previous sections of this chapter.

An alternative to data-aided estimation is given by *blind* and *semiblind channel estimation*. Relative to training-based schemes, semiblind and blind schemes typically require longer data records and entail higher complexity [865, 866]. Various techniques are available for SC SISO (e.g., see [233, 260, 521, 567, 867–900]), SC MIMO (e.g., see [579, 901–911]), MC SISO (e.g., see [912–928]) and MC MIMO (e.g., see [655, 929–938]).

An important research topic in the area of channel estimation is *channel prediction*. This is particularly useful for bridging the gap between the channel estimates and the current channel state in schemes that employ adaptive modulation or power control (e.g, see [828, 840, 939, 940]). Here we limit ourselves to mentioning that the problem of predicting SISO channels has been explored in [755, 939, 941–945], SIMO channels in [946], and MIMO in [829, 947–950].

In principle, CSI *can be estimated jointly with synchronization parameters* (e.g., the CFO) to improve estimation accuracy and reduce acquisition time. Various papers propose algorithms for joint estimation and synchronization; e.g., see [563, 707, 951–953] for SC SISO, [668, 700, 828, 954–956] for SC MIMO, [957–964] for MC SISO and [965, 966] for MC MIMO.

Data-aided channel estimators may require knowledge of various parameters of fading channels. In the literature various papers investigate the problem of *estimating various parameters of fading models*, e.g. Nakagami channel parameters [967], length of a SISO multipath channel [968], parameters of composite gamma-normal fading [969], maximum Doppler frequency [970], the K factor in Rician fading channels [971], SNR estimation in flat fading channels [972], statistical parameters of a multipath mobile channel [973], and channel order estimation for SIMO channels [974].

In recent years SC *communication techniques employing a cyclic prefix or a unique word* to simplify the equalization task have received considerable attention. When these techniques are adopted, specific algorithms for pilot insertion and for channel estimation in the frequency domain should be used [442, 738, 975–978]. Note that in this area limited attention has hitherto been paid to the use of *superimposed pilot signals* [442].

Finally, we mention that in the last decade some estimation techniques for *sparse channels* have been developed [787, 979–981].

6

Detection of Digital Signals over Wireless Channels: Channel Equalization Algorithms

6.1 Introduction

One of the fundamental tasks of a digital receiver is *channel equalization*, namely, the compensation of channel-induced distortion with the aim of improving link performance. In practice, this task is accomplished by a *channel equalization algorithm*, whose structure and complexity depend on both the *degree of knowledge of the communication channel* (which needs to be implicitly or explicitly estimated) and the *optimality criterion* employed by the algorithm. This chapter focuses entirely on the study of different classes of channel equalization algorithms, and its logical organization is parallel, at least in part, to that of Section 4.5, being related, first of all, to the way the CIR is treated (i.e., as a *known* vector, as a vector to be *averaged over* if its statistical properties are known, or as a vector to be *estimated* if it is completely unknown).

We first focus, in Sections 6.2 and 6.3, on channel equalization in the presence of an *ideally known* CIR, referring to SC and MC modulations, respectively. In Section 6.2, after developing (optimal) MLSD, MAPSD and MAPBD equalizers for SC modulations, various suboptimal strategies, such as reduced-complexity sequence detection, linear equalization and decision feedback equalization, are derived. In addition, both time-domain and frequency-domain approaches are analyzed. Section 6.3, on the other hand, investigates the problem of optimal channel equalization for a static frequency-selective channel and that of ICI cancelation over a time-varying multipath fading channel. The problem of channel equalization for a *statistically known* CIR is tackled in Sections 6.4 and 6.5 for SC and MC modulations, respectively. Particular attention is paid here to receivers based on the concept of innovations, that is, exploiting innovations-based metrics. Equalization in the presence of a *completely unknown* CIR is studied in Sections 6.6 and 6.7 for SC and MC modulations, respectively. In the latter case, an important role is played by algorithms based on the principle of per-survivor processing (PSP) – see Section 5.1. Various extensions of the SISO equalization techniques investigated to a MIMO scenario are summarized in Sections 6.8. Finally, some historical notes and suggestions for further reading are provided in Sections 6.9 and 6.10, respectively.

Wireless Communications: Algorithmic Techniques, First Edition.
Giorgio M. Vitetta, Desmond P. Taylor, Giulio Colavolpe, Fabrizio Pancaldi, Philippa A. Martin.
© 2013 John Wiley & Sons, Ltd. Published 2013 by John Wiley & Sons, Ltd.

6.2 Channel Equalization of Single-Carrier Modulations: Known CIR

In this section the problem of channel equalization for SC modulations is addressed, considering both the class of equalization algorithms operating in the *time domain* (TD) and that of classifying the equalization algorithms operating in the *frequency domain* (FD). For each case, first optimal (in the ML or MAP sense) equalization techniques are developed, and then suboptimal strategies are proposed.

6.2.1 Channel Equalization in the Time Domain

A well-known approach to mitigating ISI in SC digital communication systems is to compensate for channel distortions via *channel equalization* in the TD at the receive side. Historically, *time-domain equalizers* (TDEs) were developed for ISI mitigation in narrowband wireline channels and adopted in international *Consultative Committee for International Telegraphy and Telephony* (CCITT) standards for dialup modems. TDEs can also be employed, in principle, in broadband wireless communications; however, the number of operations per signaling interval grows exponentially (if an optimal equalization strategy is adopted) or linearly (if a suboptimal filter-based strategy is adopted) with the ISI span (i.e., with the channel memory), or, equivalently, with the data rates, as will become clearer in what follows.

6.2.1.1 MLSD Based on the Viterbi Algorithm

Here we first focus on the problem of MLSD of PAM signals transmitted over a known frequency-selective channel. Then we comment on how to extend the techniques to the case of a time-varying channel and to CPM signaling.

As already discussed in Section 4.4.1, if the M-ary PAM signal:

$$s\left(t, \mathbf{c}_N\right) = K_c \sum_{n=0}^{N-1} c_n \, p\left(t - nT_s\right) \tag{6.1}$$

is transmitted over a static multipath channel characterized by the known CIR $h_c(t)$, the complex envelope of the received signal can be written as:

$$r(t) = z\left(t, \mathbf{c}_N\right) + w(t), \tag{6.2}$$

where

$$z\left(t, \mathbf{c}_N\right) = K_c \sum_{l=0}^{N-1} c_n \, p_{TC}\left(t - lT_s\right) \tag{6.3}$$

is the useful component of the received signal, $w(t)$ is AWGN[1] with two-sided spectral density $2N_0$, and the impulse response $p_{TC}(t) \triangleq p(t) \otimes h_c(t)$ accounts for both transmit and channel filtering (its support is limited to the interval $[0, L_{TC}T_s]$). In this case, the elements of the vector $\mathbf{r}_N \triangleq [r_0, r_1, \ldots, r_{N-1}]^T$ of *sufficient statistics* can be generated by baud rate sampling the output of the filter having impulse response:

$$h_{MF}(t) = \frac{1}{K_c} p_{TC}^*\left(-\left(t - L_{TC}T_s\right)\right) \tag{6.4}$$

---

[1] Actually, some form of bandlimiting in the receiver front-end (using a bandwidth much larger than the useful signal bandwidth) needs to be assumed in the derivation of the metric, so that the integrals involved are well defined.

and fed by $r(t)$ (see Section 4.4.1), that is:

$$r_k = r(t) \otimes h_{MF}(t)\big|_{t=t_k} \tag{6.5}$$

with $t_k \triangleq kT_s + L_{TC}T_s$ and $k = 0, 1, \ldots, N - 1$. This fundamental result emerges from the theory of ML detection developed for continuous-time waveforms; in fact, such a theory establishes that, given the noisy signal $r(t)$ of (6.2), the ML strategy can be expressed as [458]:

$$\hat{\mathbf{c}}_N = \arg \min_{\tilde{\mathbf{c}}_N} \Lambda\left(\tilde{\mathbf{c}}_N\right), \tag{6.6}$$

where

$$\Lambda\left(\tilde{\mathbf{c}}_N\right) \triangleq \int_{t_i}^{t_f} \left|z\left(t, \tilde{\mathbf{c}}_N\right)\right|^2 dt - 2\,\mathrm{Re}\left[\int_{t_i}^{t_f} r(t)\, z^*\left(t, \tilde{\mathbf{c}}_N\right)\, dt\right] \tag{6.7}$$

is a *log-likelihood function* for the received signal $r(t)$ observed over the time interval (t_i, t_f) $((t_i, t_f) = (-\infty, +\infty)$ can be assumed in this case), given $\mathbf{c}_N = \tilde{\mathbf{c}}_N$. Substituting $z\left(t, \mathbf{c}_N\right)$ from (6.3) into (6.7) yields, after some manipulation:

$$\Lambda\left(\tilde{\mathbf{c}}_N\right) = \sum_{k=0}^{N-1}\sum_{n=0}^{N-1} \tilde{c}_k\, \tilde{c}_n^*\, h_{n-k} - 2\,\mathrm{Re}\left\{\sum_{k=0}^{N-1} r_k\, \tilde{c}_k^*\right\}, \tag{6.8}$$

where

$$h_k \triangleq h\left(t_k\right) \tag{6.9}$$

and

$$h(t) \triangleq p_{TC}(t) \otimes h_{MF}(t). \tag{6.10}$$

Expression (6.8) for the log-likelihood function $\Lambda\left(\tilde{\mathbf{c}}_N\right)$ clearly shows that:

(a) the set of random quantities $\{r_k,\ k = 0, 1, \ldots, N - 1\}$ (i.e., the random vector $\mathbf{r}_N$) is all that needs to be extracted from $r(t)$ in (6.2) for optimal detection, that is, it forms a set of *sufficient statistics*, as already stated above,

(b) the structure of the metric (6.8) is similar to that of the metric (4.150) derived for a conceptually similar detection problem, but in a discrete-time scenario, and

(c) the metric (6.8) exhibits *quadratic* dependence on the trial vector $\tilde{\mathbf{c}}_N$ in its first component and *linear* dependence in its second component.

As far as the last point is concerned, it is important to note that, generally speaking, the quadratic dependence of $\Lambda\left(\tilde{\mathbf{c}}_N\right)$ on $\tilde{\mathbf{c}}_N$ implies that the problem does not lend itself to a computationally simple solution (6.6), since an exhaustive search over the complete set of data sequences is required. However, under certain assumptions, a computationally efficient technique for the search of the optimal symbol vector can be developed. To see this, let us rewrite the first term on the RHS of (6.8) as:

$$\sum_{k=0}^{N-1}\sum_{n=0}^{N-1} \gamma_{k,n}, \tag{6.11}$$

where

$$\gamma_{k,n} \triangleq \tilde{c}_k\, \tilde{c}_n^*\, h_{k-n}, \tag{6.12}$$

with $k, n = 0, 1, \ldots, N - 1$, representing the element located on the kth row and the nth column of the $N \times N$ complex matrix Υ. Note that this matrix is Hermitian since $\gamma_{k,n}^* = \tilde{c}_k^*\, \tilde{c}_n\, h_{k-n}^* = \tilde{c}_k^*\, \tilde{c}_n\, h_{n-k} = \gamma_{n,k}$ (see (6.9), (6.10) and (6.12)) and therefore, if the contributions coming from the main diagonal, from the terms below it and from those above it are separated, the double sum (6.11) can be rewritten as:

$$
\sum_{k=0}^{N-1}\sum_{n=0}^{N-1} \gamma_{k,n} = \sum_{k=0}^{N-1} \gamma_{k,k} + \sum_{k=1}^{N-1}\sum_{n=0}^{k-1} \gamma_{k,n} + \sum_{n=1}^{N-1}\sum_{k=0}^{n-1} \gamma_{k,n}
$$

$$
= \sum_{k=0}^{N-1} \gamma_{k,k} + 2\,\mathrm{Re}\left\{ \sum_{k=1}^{N-1}\sum_{n=0}^{k-1} \gamma_{k,n} \right\}, \tag{6.13}
$$

since $\sum_{n=1}^{N-1}\sum_{k=0}^{n-1} \gamma_{k,n} = \sum_{k=1}^{N-1}\sum_{n=0}^{k-1} \gamma_{k,n}^{*}$ (i.e., the sum of the elements above the main diagonal is the complex conjugate of the sum of the elements below it). Then substituting (6.12) into (6.13) and (6.13) into (6.8) yields:

$$
\Lambda\left(\tilde{\mathbf{c}}_N\right) = h_0 \sum_{k=0}^{N-1} |\tilde{c}_k|^2 + 2\,\mathrm{Re}\left\{ \sum_{k=1}^{N-1}\sum_{n=0}^{k-1} \tilde{c}_k\,\tilde{c}_n^{*}\,h_{k-n} \right\} - 2\,\mathrm{Re}\left\{ \sum_{k=0}^{N-1} r_k\,\tilde{c}_k^{*} \right\}
$$

$$
= h_0 \sum_{k=0}^{N-1} |\tilde{c}_k|^2 + 2 \sum_{k=1}^{N-1} \mathrm{Re}\left\{ \tilde{c}_k \sum_{n=0}^{k-1} \tilde{c}_n^{*}\,h_{k-n} \right\} - 2\,\mathrm{Re}\left\{ \sum_{k=0}^{N-1} r_k\,\tilde{c}_k^{*} \right\}. \tag{6.14}
$$

The latter expression may also be rewritten as:

$$
\Lambda\left(\tilde{\mathbf{c}}_N\right) = \left[h_0 |\tilde{c}_0|^2 - 2\,\mathrm{Re}\left\{ r_0\,\tilde{c}_0^{*} \right\} \right]
$$

$$
+ \sum_{k=1}^{N-1} \left\{ h_0 |\tilde{c}_k|^2 + 2\,\mathrm{Re}\left[\tilde{c}_k \sum_{n=0}^{k-1} \tilde{c}_n^{*}\,h_{k-n} \right] - 2\,\mathrm{Re}\left[r_k\,\tilde{c}_k^{*} \right] \right\} \tag{6.15}
$$

or, in a more compact form, as:

$$
\Lambda\left(\tilde{\mathbf{c}}_N\right) = \sum_{k=0}^{N-1} \lambda_k\left(\tilde{\mathbf{c}}_k, \tilde{c}_k\right), \tag{6.16}
$$

where $\tilde{\mathbf{c}}_k \triangleq [\tilde{c}_0, \tilde{c}_1, \ldots, \tilde{c}_{k-1}]^T$ ($\tilde{\mathbf{c}}_0$ denotes an empty vector):

$$
\lambda_0\left(\tilde{\mathbf{c}}_0, \tilde{c}_0\right) = \lambda_0\left(\tilde{c}_0\right) \triangleq h_0 |\tilde{c}_0|^2 - 2\,\mathrm{Re}\left\{ r_0\,\tilde{c}_0^{*} \right\} \tag{6.17}
$$

and

$$
\lambda_k\left(\tilde{\mathbf{c}}_k, \tilde{c}_k\right) \triangleq h_0 |\tilde{c}_k|^2 + 2\,\mathrm{Re}\left[\tilde{c}_k \sum_{n=0}^{k-1} \tilde{c}_n^{*}\,h_{k-n} \right] - 2\,\mathrm{Re}\left[r_k\,\tilde{c}_k^{*} \right] \tag{6.18}
$$

for $k = 1, 2, \ldots, N - 1$. The structure of the metric $\Lambda\left(\tilde{\mathbf{c}}_N\right)$ can be further simplified since the support of $h(t)$ in (6.10) does not exceed the interval $[-L_{TC} T_s, L_{TC} T_s]$, so that $h_k = 0$ for $|k| \geq L_h$ with $L_h \leq L_{TC} + 1$. Then the metric $\lambda_k\left(\tilde{\mathbf{c}}_k, \tilde{c}_k\right)$ may be rewritten as:

$$
\lambda_k\left(\tilde{\mathbf{c}}_k, \tilde{c}_k\right) = \lambda_k\left(\tilde{\Delta}_k, \tilde{c}_k\right)
$$

$$
= h_0 |\tilde{c}_k|^2 + 2\,\mathrm{Re}\left[\tilde{c}_k \sum_{n=\max(0, k-L_h+1)}^{k-1} \tilde{c}_n^{*}\,h_{k-n} \right] - 2\,\mathrm{Re}\left[r_k\,\tilde{c}_k^{*} \right], \tag{6.19}
$$

where $\tilde{\Delta}_k$ denotes an integer parameter uniquely identified by the ordered collection of the trial channel symbols $\{\tilde{c}_n, \ n = k - L_h + 1, k - L_h + 2, \ldots, k - 1\}$ and representing the so-called *channel state*.

In practice, this parameter, which represents the overall *channel memory*, can be defined as:

$$\tilde{\Delta}_k \triangleq \tilde{b}_{k-L_h+1} + \tilde{b}_{k-L_h+2}M + \ldots + \tilde{b}_{k-1}M^{L_h-2}, \tag{6.20}$$

for any k, where $\tilde{b}_k$ is a nonnegative integer representation of the complex symbol $\tilde{c}_k$ (in particular, we assume that $\tilde{b}_k$ belongs to the M-ary alphabet $\{0, 1, \ldots, M-1\}$ and that $\tilde{b}_k = 0$ for $k < 0$). Then the optimal metric (6.16) can be put in the compact form:

$$\Lambda\left(\tilde{\mathbf{c}}_N\right) = \sum_{k=0}^{N-1} \lambda_k(\tilde{\Delta}_k, \tilde{c}_k), \tag{6.21}$$

thereby expressing $\Lambda\left(\tilde{\mathbf{c}}_N\right)$ as a summation of partial metrics $\{\lambda_k(\tilde{\Delta}_k, \tilde{c}_k)\}$. In addition, since $\lambda_k(\tilde{\Delta}_k, \tilde{c}_k)$ can be denoted $\lambda_k(\tilde{\Delta}_k, \tilde{\Delta}_{k+1})$, with $\tilde{\Delta}_{k+1}$ uniquely identified by the pair $(\tilde{\Delta}_k, \tilde{c}_k)$, (6.21) can also be rewritten as:

$$\Lambda\left(\tilde{\mathbf{c}}_N\right) = \sum_{k=0}^{N-1} \lambda_k(\tilde{\Delta}_k, \tilde{\Delta}_{k+1}). \tag{6.22}$$

Then $\Lambda\left(\tilde{\mathbf{c}}_N\right)$ can be *recursively* computed for any trial sequence $\tilde{\mathbf{c}}_N$ by evaluating the expression:

$$\Lambda\left(\tilde{\mathbf{c}}_{k+1}\right) = \Lambda\left(\tilde{\mathbf{c}}_k\right) + \lambda_k(\tilde{\Delta}_k, \tilde{\Delta}_{k+1}) \tag{6.23}$$

for $k = 0, 1, \ldots, N-1$, with $\Lambda\left(\tilde{\mathbf{c}}_0\right) = 0$. Expression (6.23) is the key to solving (6.6) in a computationally efficient fashion by resorting to the so-called *Viterbi algorithm* (VA), which is a recursive algorithm for determining the optimal state sequence of a discrete-time Markov process observed in memoryless noise [982]. The VA operates over a *trellis*, characterized by $N_s = M^{L_h-1}$ states. The trellis structure is illustrated in Figure 6.1 for the case of $M = 2$ and $L_h = 3$. In general, in the kth interval each trellis state represents one of the N_s possible values that $\tilde{\Delta}_k$ can take and is connected via M branches to M distinct next states $\tilde{\Delta}_{k+1}$. Moreover, the branch connecting $\tilde{\Delta}_k$ and $\tilde{\Delta}_{k+1}$ is labeled by the trial symbol $\tilde{c}_k$ and by the quantity $\lambda(\tilde{\Delta}_k, \tilde{\Delta}_{k+1})$, known as the *branch*

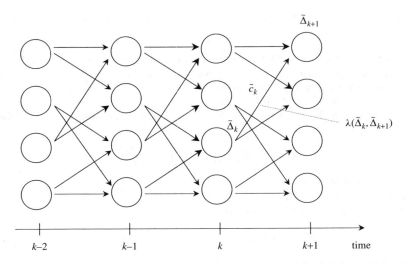

Figure 6.1 Four-state trellis involved in the ML detection of a PAM signal transmitted over a frequency-selective channel ($M = 2$ and $L_h = 3$ are assumed).

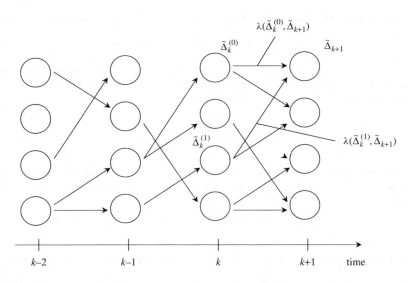

Figure 6.2 Time evolution of the VA operating over the four-state trellis shown in Figure 6.1.

metric. In this context, each trial sequence $\tilde{c}_N$ is in one-to-one correspondence with a *sequence of states* $\{\tilde{\Delta}_k, \ k = 1, \ldots, N\}$ in the trellis diagram, that is, with a distinct *path* in the trellis. Thus, looking for the optimal sequence decision is equivalent to searching for the minimum distance path in the trellis. The VA accomplishes this task recursively through the following steps at the end of each symbol interval (see Figure 6.2):

(a) maintaining one *survivor* path per state $\tilde{\Delta}_k$ in the kth symbol interval,
(b) extending these paths one step along all the M branches (each labeled by a distinct value of $\tilde{c}_k$) emanating from it, and
(c) pruning these back by retaining only the path with the smallest[2] metric $\Lambda\left(\tilde{c}_{k+1}\right)$ from each state $\tilde{\Delta}_{k+1}$.

In the kth symbol interval, then, the VA keeps track of only one path (the so-called *survivor*) leading to each state $\tilde{\Delta}_k$. Such a path, denoted by $\Theta(\tilde{\Delta}_k)$ in what follows, consists of the sequence of consecutive states belonging to the path and is characterized by an *accumulated metric*, denoted $\Lambda(\tilde{\Delta}_k)$ in what follows (rather than $\Lambda(\tilde{c}_k)$) to make its meaning clear.

The VA procedure can be summarized by the following steps (n denotes the time variable):

1. Set:
$$k = 0, \quad \Theta\left(\tilde{\Delta}_0\right) = \left(\tilde{\Delta}_0\right), \quad \Lambda\left(\tilde{\Delta}_0\right) = 0 \tag{6.24}$$

    to initialize the algorithm.
2. Repeat steps 3–7 until $n = N - 1$.
3. Extend path metrics in accordance with (6.23), that is:
$$\Lambda\left(\tilde{\Delta}_{n+1}\right) = \Lambda\left(\tilde{\Delta}_n\right) + \lambda\left(\tilde{\Delta}_n, \tilde{\Delta}_{n+1}\right) \tag{6.25}$$

    for all the allowed state transitions $\tilde{\Delta}_n \rightarrow \tilde{\Delta}_{n+1}$.

---

[2] When a metric has to be maximized, the VA should be supplied with the negative metric instead.

4. For each destination state $\tilde{\Delta}_{n+1}$, find the best (minimum metric) incoming or survivor path over all the previous states:

$$\tilde{\Delta}_n = \arg \min_{\tilde{\Delta}_n} \Lambda\left(\tilde{\Delta}_{n+1}\right). \tag{6.26}$$

5. Update and store survivor paths as:

$$\Theta\left(\tilde{\Delta}_{n+1}\right) = \left(\Theta\left(\tilde{\Delta}_n\right), \tilde{\Delta}_{n+1}\right). \tag{6.27}$$

6. Store the new survivor metrics as:

$$\Lambda\left(\tilde{\Delta}_{n+1}\right) = \Lambda\left(\tilde{\Delta}_n\right) + \lambda\left(\tilde{\Delta}_n, \tilde{\Delta}_{n+1}\right). \tag{6.28}$$

7. Set $n = n + 1$ (increment time counter).
8. Detect the ML decision for the symbol sequence as that associated with the survivor path $\Theta(\tilde{\Delta}_{n+1})$ with minimum metric $\Lambda(\tilde{\Delta}_{n+1})$ (termination).

It is worth noting the following:

(a) Branch metrics evaluated for state transitions inconsistent with a priori known (e.g., training or pilot) symbols are set to a large value (virtually infinite).
(b) ML decisions are not available until time $n = N - 1$. In practice, however, there is little degradation in making symbol-by-symbol decisions after a *decision delay* D_{VA} (whose value is typically 5–7 times L_h) by tracing back from the survivor with the instantaneously best metric.

The latter approach to decision-making can be related to the fact that all the N_s survivors available at time n often coincide up to some time $n - d$, that is, d symbol intervals earlier. When this occurs, it is usually said that survivors have *merged* at depth d; moreover, an *optimal* decision can be made on all the symbols or states up to time $n - d$, without waiting for additional information, so that the length of the survivors stored in the VA memory can be truncated to d intervals. Unfortunately, the parameter d is random, since path merge is a statistical phenomenon, which may not occur at all in a given observation interval. These considerations suggest that the value selected for D_{VA} should achieve a good tradeoff between error performance and computational/memory requirements. In fact, D_{VA} should be large enough to ensure near optimal performance, while being as small as possible to limit the size of the VA memory and the processing required for refreshing it.

A block diagram of an ML sequence detector based on the strategy developed above is shown in Figure 6.3. Such a strategy was proposed by G. Ungerboeck in 1974 [458]. An alternative (but mathematically equivalent [424]) solution was developed two years earlier by D. G. Forney [422]. It requires the use of a *noise whitening* after the matched filter (i.e., the use of a WMF; see Section 4.4.1). Note that, generally speaking, the sample r_k (6.5) at the MF output is structured as[3]:

$$r_k = \sum_{l=-L_h+1}^{L_h-1} c_l h_{k-l} + n_k, \tag{6.29}$$

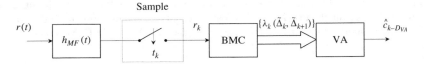

Figure 6.3   Block diagram of an ML sequence detector based on the Ungerboeck's strategy.

[3] Further details on this model can be found in Section 4.4.1.

where $n_k \triangleq n(t_k)$ and $n(t) \triangleq w(t) \otimes h_{MF}(t)$ is a complex Gaussian process having mean zero and autocorrelation function (see (6.5) and (6.10)):

$$R_n(\tau) \triangleq 2N_0 \, h_{MF}(\tau) \otimes h_{MF}^*(-\tau) = \frac{2N_0}{K_c} \, h\left(\tau + L_{TC} T_s\right). \tag{6.30}$$

This result implies that the autocorrelation $R_n[k]$ of the zero mean random sequence $\{n_k\}$ is given by:

$$R_n[k] = R_n(kT_s) = \frac{2N_0}{K_c} \, h\left(kT_s + L_{TC} T_s\right) = \sigma_n^2 \, K_c \, h_k, \tag{6.31}$$

with $\sigma_n^2 \triangleq 2N_0/K_c^2$, so that $\{n_k\}$ is *not white* (note also that $R_n[0] = 2h_0 N_0/K_c = \sigma_n^2 E_{PTC}$, E_{PTC} being the energy of $p_{TC}(t)$, and that $\{h_k\}$ is a Hermitian sequence, i.e. $h_l = h_{-l}^*$ for any l). Whitening $\{n_k\}$ by a proper filter leads to a more elegant structure of the log-likelihood metric (which becomes a *Euclidean distance metric* involving the sequence of received signal samples at the WMF), but introduces the additional complexity of implementing a WMF. In this case the model (see equation (4.95)):

$$r_k = \sum_{l=0}^{L_q-1} c_{k-l} \, q_l + n_k \tag{6.32}$$

for the kth received signal sample (with $k = 0, 1, \ldots, N-1$) replaces that expressed by (6.29); here, the impulse response $\{q_k\}$ results from the convolution of $\{h_k\}$ with the impulse response of the WF (L_q denotes the number of its nonnull samples and is equal to L_h; see Section 4.4.2) and $\{n_k\}$ is a complex AWGN sequence, whose samples have variance σ_n^2. It is not difficult to show that, given the set $\{r_k, \; k = 0, 1, \ldots, N-1\}$ of sufficient statistics, the MLSD metric can be expressed as:

$$\Gamma(\tilde{\mathbf{c}}_N) = \sum_{k=0}^{N-1} \xi_k(\tilde{\Delta}_k, \tilde{\Delta}_{k+1}), \tag{6.33}$$

where

$$\xi_k(\tilde{\Delta}_k, \tilde{\Delta}_{k+1}) \triangleq \left| r_k - \sum_{l=0}^{L_q-1} c_{k-l} \, q_l \right|^2 \tag{6.34}$$

is the branch metric associated with the transition $\tilde{\Delta}_k \to \tilde{\Delta}_{k+1}$ and the channel state $\tilde{\Delta}_k$ is an integer representation of the ordered collection of the trial channel symbols $\{\tilde{c}_n, \; n = k - L_q + 1, k - L_q + 2, \ldots, k - 1\}$ (so that the overall number of trellis states is $N_s = M^{L_q-1}$). Further details on this approach can be found in [422, 426] as well as in [424], where a unifying perspective is provided for the MLSD strategies proposed by Ungerboeck and Forney.

Finally, it is important to point out that the above-mentioned MLSD strategies have been developed for PAM signaling over frequency-selective, time-invariant channels; however, in principle, a conceptually similar approach can be adopted to develop MLSD strategies for PAM signaling over frequency-flat, fast fading channels [556], or for doubly-selective channels, as shown in [424, 425, 495]. Moreover, VA-based MLSD strategies can also be devised for CPM signals; in this case, however, the VA trellis is actually a *supertrellis*, whose superstates accounts for both ISI due to dispersive channel and the memory characterizing the phase modulation [438] of the CPM signal.

6.2.1.2 Constrained MLSD and Reduced-Complexity Sequence Detection

The complexity of the VA (or, as shown later in this section, of the forward–backward algorithm) is proportional to the total number of trellis states, which depends exponentially on the overall channel memory (this parameter is denoted L_h or L_q in Section 6.2.1.1). Therefore, for typical wideband channels, MLSD becomes extremely complicated [983]. Various schemes to reduce the complexity of MLSD have been proposed; these usually involve searching only part of the trellis or simplifying the trellis, as explained below.

Simplifying the Trellis

The sample sequence $\{h_k\}$ (6.9) has a peak near its middle, referred to as the *cursor*; the samples preceding and following it are called *precursors* and *postcursors*, respectively, and typically their energy tails away. If the energy in the extreme tails of the precursors and/or postcursors is small, their contribution can be neglected without significant penalty. This leads to *truncated memory detection* or *truncated sequence detection* (e.g., see [984–987]).

It is also noteworthy that the reliability of decisions within the survivor paths $\{\Theta(\tilde{\Delta}_k)\}$ available in the kth symbol interval increases with delay but that, as already explained in the previous subsection, there is usually no significant increase in reliability beyond delays of about 5–7 times L_h. In addition, it is fair to say that *reliable* decisions may be available much sooner, even at delays of less than L_h as long as most of the energy of the received pulse has been accounted for in the branch metrics of the VA. These considerations lead to the development of *delayed decision feedback sequence detection* (DDFSD) [988] and *reduced-state sequence detection* (RSSD) [989, 990].

In DDFSD a reduced-state trellis is constructed by *merging together* (i.e., fusing) states which share the same "older" symbols. In other words, the VA state $\tilde{\Delta}_k$ is then defined as (see (6.20)):

$$\tilde{\Delta}_k^{RS} \triangleq \tilde{b}_{k-L_{RS}+1} + \tilde{b}_{k-L_{RS}+2}M + \ldots + \tilde{b}_{k-1}M^{L_{RS}-2}, \tag{6.35}$$

where $L_{RS} < L_h$. This definition leads again to a trellis (having a reduced number of states) to which the VA is then applied. The crucial difference between DDFSD and truncated sequence detection is that in DDFSD the full branch metric of (6.19) is retained. However, for each state $\tilde{\Delta}_k$ and state transition labeled by the channel symbol $\tilde{c}_k$, the required symbols are obtained partly from the DDFSD state $\tilde{\Delta}_k^{RS}$ and partly from the corresponding survivor path $\Theta(\tilde{\Delta}_k^{RS})$. This method is an application of the *decision feedback* concept [768, 983, 985, 988, 991]. As a first approximation, the value of the parameter L_{RS} can be selected so as to cover the precursors and the cursor, with decision feedback being used for the postcursors.

RSSD can be considered as an elegant (but minor) extension of DDFSD. Instead of adopting a trellis which accounts for the precursors and exploiting the survivor history for the postcursors, *set partitioning principles* [992] (see Section 11.2) are applied to reduce the number of hypothesized symbols. Further results of this approach can be found in [989, 993–995]. In particular, we note that the ST extension of RSSD, as proposed in [989] for the SISO equalization problem, has been extended to a MIMO scenario in [996].

An alternative approach to reducing the number of trellis states is by the use of adaptive prefiltering at the receiver to shorten the duration of the overall impulse response [997]. In practice, this can be achieved using a *linear equalizer* (LE) (e.g., see [623, 998]) or a *decision feedback equalizer* (DFE) (e.g., see [999–1001]). Then, the prefiltered signal undergoes MLSD based on the VA. Unfortunately, prefiltering colors the additive noise at the receiver input, so that error performance gets worse if noise correlation is ignored [457]. In addition, the use of a DFE prefilter introduces the problem of error propagation. Error performance can be improved by adopting a hybrid structure which delivers the soft outputs generated by MLSD to a DFE [768, 1000, 1002]. However, generally speaking, the performance attained using these prefiltering strategies has not been compelling.

Searching Part of the Trellis

In the MLSD procedure, survivor sequences characterized by accumulated metrics that are much worse than that associated with the best current survivor sequence are unlikely to contribute to the

ML path in the trellis in the future. This consideration suggests that MLSD computational complexity could be reduced by searching the trellis in a more intelligent fashion. Based on this principle, there are various trellis-based detection algorithms which carry out a *simplified search* and still achieve excellent performance [1003]. A complete description of these is beyond the scope of this book, and here we limit ourselves to providing a brief description of the so-called *M-algorithm* [1004]. Instead of preserving M^{L_h-1} survivors at the end of each symbol period like the VA, the M-algorithm keeps a fixed number M_s of such survivors, with $0 < M_s \leq M^{L_h-1}$. In the next symbol period, each of the M_s survivors is extended along the M branches radiating from its ending state. If more than one extended survivor enters the same next state, all but the best one are pruned, as in the VA. Then the survivors, whose number does not exceed $M \cdot M_s$, are sorted by metric, and further pruned, so that only the best M_s of them are retained. This procedure is repeated in each symbol period [1005].

In principle, reduced-search techniques can be combined with state reduction by fusion limiting the search to a subset of fused states. This approach, which is motivated by the fact that in data communications over fading channels the SNR at the receive side is time-variant, has led to the development of the so-called *adaptive state allocation* (ASA) algorithm for MLSD [1006]. In this algorithm, the number of states is reduced by fusion when the channel is out of a fade and the number of states is increased by diffusion when the channel goes to fade. In addition, at each stage, only the more likely correct states are chosen for extension to the next stage.

It is also worth noting that the VA-based MLSD technique illustrated in the previous subsection represents a form of *unconstrained* strategy, since no constraint is set in exploring the trellis representing the channel memory. In contrast, the M-algorithm represents an example of a *constrained strategy*. Generally speaking, *constrained* MLSD strategies are based on the following approach. First, the set of N_s trellis states is partitioned into C disjoint sets (nonuniform partitions can also be adopted). In the detection procedure B survivors are kept in memory for each set and the best sequence of sets is identified. Then, within the "winning" set the path associated with the best metric is found and, finally, an estimate of the symbol sequence is inferred from this path. This class of strategies is denoted SA(C, B) in [1007, 1008], where SA means *search algorithm*. It is worth pointing out the following:

(a) The algorithms of this class are known as *breadth-first*, since they view at once all the branches that they will ever view at that depth.
(b) Viterbi-based MLSD illustrated in Section 6.2.1.1 and the above-mentioned M-algorithm can be denoted SA$(1, N_s)$, and SA$(1, M_s)$, respectively (another example of breadth-first trellis decoding algorithm can be found in [1009]). In addition, the family of algorithms SA$(1, C)$ encompasses DDFSD and RSSD.
(c) Alternatives to the breadth-first approach are the *depth-first* (a single path is pursued continually until its metric exceeds a threshold) and the *metric-first* approaches (the path with the instantaneously best metric is always pursued). This is often known as the *sequential* approach.

Various references investigate equalization performance when simplified search algorithms are used. For instance, the reader can refer to [1010–1015], which give applications of the M-algorithm, and to [1016], which considers the exploitation of the so-called *Fano algorithm* that was developed for the sequential decoding of convolutional codes.

In the following we focus on the SA(1, 1) class which is of technical relevance due to its modest computational complexity. Such a class is formed by *decision feedback equalization* algorithms.

6.2.1.3 Decision Feedback Equalization in the Time Domain

In a DFE a single survivor (i.e., a single state) is retained at the end of each symbol period and is exploited in detecting the next symbol. In other words, the trellis consists of one state with M branches, each returning to the same state. The DFE has played a significant role in communications due to its nice balance of complexity and performance (e.g., see [1017–1020]). For this reason we analyze it as a separate structure. In particular, in our study we first focus on the problem of equalizing a received vector **r** in the presence of a known frequency-selective channel. This problem is tackled by developing decision feedback algorithms that can process the entire block of received data and which can be adopted if the block size is not overly long. Then a different approach is developed for the case of long **r**, under the two scenarios of a known frequency-selective channel and an unknown time-varying, frequency-selective channel.

To begin, we derive a block DFE for a PAM signal transmitted over a known frequency-selective channel [1021], under the assumptions that:

- the symbol vector $\mathbf{c}_N \triangleq [c_0, c_1, \ldots, c_{N-1}]^T$ is transmitted;
- the received vector $\mathbf{r} \triangleq [r_0, r_1, \ldots, r_{N-1}]^T$ is generated by taking one sample per symbol at the output of a filter matched to the impulse response $p_{TC}(t)$ which accounts for transmit and channel filtering and whose support is the interval $[0, L_{TC}T_s]$ (as assumed in Section 6.2.1.1).

The kth received signal sample may then be written as (see (6.29)):

$$r_k = \sum_{l=-L_h+1}^{L_h-1} c_{k-l} \, h_l + n_k, \tag{6.36}$$

for $k = 0, 1, \ldots, N - 1$, with $c_k = 0$ for $k < 0$ and $k > N - 1$, if it is assumed that $h_k = 0$ for $|k| \geq L_h$ with $L_h \leq L_{TC} + 1$. Note that the sequence h_l has the property of *Hermitian symmetry* (i.e., $h_l = h_{-l}^*$ for any l) and the complex Gaussian noise sequence $\{n_k\}$ is not white, since (see (6.31)):

$$R_n[l] = \sigma_n^2 \, K_c \, h_l, \tag{6.37}$$

where $\sigma_n^2 \triangleq 2N_0/K_c^2$. Then, if $N > L_h - 1$, the received signal vector can be written as:

$$\mathbf{r} = \mathbf{H}\mathbf{c}_N + \mathbf{n}, \tag{6.38}$$

where $\mathbf{H} = [h_{i,j}]$ is an $N \times N$ CIR matrix with $h_{i,j} = h_{i-j}$ for $|i - j| < L_h$ and $h_{i,j} = 0$ elsewhere, and $\mathbf{n} \triangleq [n_0, n_1, \ldots, n_{N-1}]^T$ is a zero mean complex Gaussian vector with covariance matrix:

$$\mathbf{R_n} \triangleq \mathrm{E}\left\{\mathbf{n}\,\mathbf{n}^H\right\} = \sigma_n^2\mathbf{H}. \tag{6.39}$$

Since **H** is a (Toeplitz) Hermitian and positive definite matrix (h_k is generated by sampling an autocorrelation function) it can be factored as:

$$\mathbf{H} = \mathbf{L}^H\mathbf{D}^2\,\mathbf{L} \tag{6.40}$$

using the *Cholesky decomposition* (see (C.13)), where $\mathbf{L} = [l_{i,j}]$ is an $N \times N$ lower triangular matrix with 1s along its main diagonal, and $\mathbf{D} = \mathrm{diag}\,(\chi_i)$, with $i = 0, 1, \ldots, N - 1$, is an $N \times N$ diagonal matrix with positive real elements. Note from (6.40) that the factorization:

$$\mathbf{H}^{-1} = \mathbf{P}^H\,\mathbf{D}^{-2}\,\mathbf{P}, \tag{6.41}$$

can easily be shown for $\mathbf{H}^{-1}$, where:

$$\mathbf{P} \triangleq (\mathbf{L}^H)^{-1} \tag{6.42}$$

is also a lower triangular matrix with 1s along its main diagonal. Then the inverse of the noise covariance matrix $\mathbf{R}_n$ (6.40) can be expressed as:

$$\mathbf{R_n^{-1}} = \frac{1}{\sigma_n^2}\mathbf{P}^H\,\mathbf{D}^{-2}\,\mathbf{P}. \tag{6.43}$$

It is also worth noting that the nonzero elements of the mth row of $\mathbf{P}$ can be shown to be the tap weights of a *forward prediction error filter* of order m for a discrete-time stochastic sequence characterized by a covariance matrix $\mathbf{H}$ [503, 1022]. If such a filter is fed by the noise sequence $\{n_k\}$ its output represents a *prediction error*, whose variance is proportional to χ_m^2. Note also that, generally speaking, $\chi_{m+1} \geq \chi_m$ for any m, so that the prediction accuracy does not get worse as the predictor length increases.

Given the factorization (6.41) a *linear reversible transformation*, represented by the matrix:

$$\mathbf{T}_W = \left(\mathbf{L}^H \mathbf{D}\right)^{-1} = \mathbf{D}^{-1}\mathbf{P}, \tag{6.44}$$

can be used to transform the vector $\mathbf{r}$ (6.38) into the statistically equivalent vector $\mathbf{X} \triangleq [X_0, X_1, \ldots, X_{N-1}]^T$, which is evaluated as:

$$\mathbf{X} \triangleq \mathbf{T}_W \, \mathbf{r} = \mathbf{D}^{-1}\mathbf{P}\left(\mathbf{L}^H \mathbf{D}^2 \, \mathbf{L} \, \mathbf{c}_N + \mathbf{n}\right) = \mathbf{D} \, \mathbf{L} \, \mathbf{c}_N + \mathbf{N}, \tag{6.45}$$

where $\mathbf{N} \triangleq \mathbf{D}^{-1}\mathbf{P} \, \mathbf{n}$ is a noise vector having covariance matrix (see (6.39) and (6.40)):

$$\mathbf{R_N} \triangleq \mathrm{E}\left\{\mathbf{N}\,\mathbf{N}^H\right\} = \mathbf{D}^{-1}\mathbf{P}\,\mathbf{R_n}\,\mathbf{P}^H\mathbf{D}^{-1} = \sigma_n^2\,\mathbf{I}_N, \tag{6.46}$$

that is, it is a *white* Gaussian noise vector. This proves that the transformation defined in (6.45) accomplishes *noise whitening*. Note also that the kth element of $\mathbf{X}$ (6.45) is given by:

$$X_0 = \chi_0 \, c_0 + N_0, \tag{6.47}$$

for $k = 0$, and by:

$$X_k = \chi_k \left[c_k + \sum_{p=0}^{k-1} l_{k,p} \, c_p \right] + N_k, \tag{6.48}$$

for $k = 1, 2, \ldots, N-1$, so that, generally speaking, X_k depends on c_k and the past symbols only.

Given the above results, we now consider the development of algorithms for decision feedback equalization. Two criteria can be adopted in this case, namely a *zero forcing* (ZF) criterion and an MMSE criterion. If the ZF approach is used, the objective of the equalization algorithm is to completely remove the effects of ISI by exploiting past decisions on the transmitted data symbols while ignoring the presence of noise. This result can be easily achieved by processing the vector $\mathbf{X}$ (6.45), since X_k depends on the channel symbol c_k and on a set of *past* symbols. In fact, an estimate $\hat{c}_0$ of c_0 can be extracted from X_0/χ_0. Now, assuming that $\hat{c}_0 = c_0$, the influence of this symbol can be removed in X_1 by computing the difference $X_1/\chi_1 - l_{1,0}\hat{c}_0$, which is processed to generate an estimate $\hat{c}_1$ of c_1. Then both estimates $\hat{c}_0$ and $\hat{c}_1$ are employed to cancel ISI in X_2, from which an estimate $\hat{c}_2$ of c_2 is obtained and so on. This technique for ISI removal can be described more compactly as subtracting the vector $\mathbf{B}_{ZF}\,\hat{\mathbf{c}}_N$ from $\mathbf{F}_{ZF}\,\mathbf{r}$, where:

$$\mathbf{F}_{ZF} \triangleq \mathbf{D}^{-1}\mathbf{T}_W = \mathbf{D}^{-2}\mathbf{P} \tag{6.49}$$

describes linear feedforward filtering and:

$$\mathbf{B}_{ZF} \triangleq \mathbf{L} - \mathbf{I}_N \tag{6.50}$$

is an $N \times N$ matrix having all the elements along its main diagonal and above it equal to zero and describing the mechanism of decision feedback. Note, however, that the evaluation of the elements of the difference $\mathbf{F}_{ZF}\,\mathbf{r} - \mathbf{B}_{ZF}\,\hat{\mathbf{c}}_N = \mathbf{D}^{-1}\mathbf{X} - \mathbf{B}_{ZF}\,\hat{\mathbf{c}}_N$ cannot be accomplished in a parallel fashion, since the computation of the kth element of such a difference (with $k > 1$) requires knowledge of the symbol estimates $\{\hat{c}_l, \ l = 0, 1, \ldots, k - 1\}$. This is shown in Figure 6.4, a block diagram of a DFE based on the ZF criterion (ZF-DFE). Note that, in equalization jargon, the matrices $\mathbf{F}_{ZF}$ (6.49) and $\mathbf{B}_{ZF}$ (6.50)

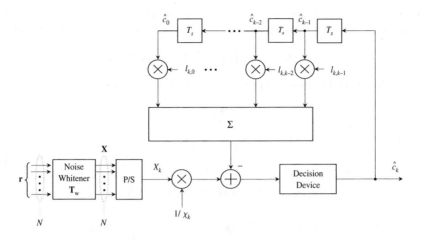

Figure 6.4   Structural diagram of a block ZF-DFE.

describe the *linear time-varying filtering* applied to the received signal samples (*feedforward* filtering) and the detected symbols (*feedback* filtering), respectively.

As already stated, the presence of noise is neglected in the design of a ZF-DFE. This may lead to the effects of noise being magnified when the equalizer tries to compensate for the frequency selectivity of the communication channel. This phenomenon, known as *noise enhancement*, can substantially affect error performance and, in particular, can cause *error propagation*. In fact, any data decision is fed back to remove the associated contribution to ISI but, if wrong, makes such a contribution stronger and can lead to further data decision errors.

The reduction in ISI can be balanced with noise enhancement by adopting an MMSE approach to DFE design. The resulting equalizer always performs as well as, or better than, its ZF counterpart and has a similar implementation complexity. The design of an MMSE-DFE requires looking for the filters that minimize the MSE:

$$\varepsilon_{MMSE} \triangleq \mathrm{E}\left\{ |\mathbf{Z} - \mathbf{c}_N|^2 \right\}, \tag{6.51}$$

where

$$\mathbf{Z} \triangleq \mathbf{F}\,\mathbf{r} - \mathbf{B}\,\hat{\mathbf{c}}_N \tag{6.52}$$

results from processing $\mathbf{r}$ through forward filtering and decision-based ISI cancelation (described by $\mathbf{F}$ and $\mathbf{B}$, respectively) and averaging is accomplished with respect to both channel symbols and noise. In other words, we need to solve the optimization problem:

$$\min_{\mathbf{F},\mathbf{B}} \varepsilon_{MMSE} \rightarrow \mathbf{F}_{MMSE}, \mathbf{B}_{MMSE}, \tag{6.53}$$

where $\mathbf{F}_{MMSE}$ and $\mathbf{B}_{MMSE}$ denote the optimal values of $\mathbf{F}$ and $\mathbf{B}$, respectively; note that $\mathbf{B}_{MMSE}$ is required to have all the elements along its main diagonal and above it equal to zero, as in the ZF case. The derivation of the expressions for $\mathbf{F}_{MMSE}$ and $\mathbf{B}_{MMSE}$ that minimize ε_{MSE} (6.51) is based on the so-called *orthogonality principle* [503, 1021]. This states that MMSE minimization requires the *error signal* $\mathbf{\Delta} \triangleq [\delta_0, \delta_1, \ldots, \delta_{N-1}]^T$:

$$\mathbf{\Delta} \triangleq \mathbf{Z} - \mathbf{c}_N, \tag{6.54}$$

to be *orthogonal* to the data vector $\mathbf{r}$ (6.38):

$$\mathrm{E}\left\{ \mathbf{\Delta}\,\mathbf{r}^H \right\} = \mathbf{0}_N, \tag{6.55}$$

where $\mathbf{0}_N$ denotes the $N \times N$ null matrix. Substituting (6.52) into (6.54) and (6.54) into (6.55) and assuming that the correct decisions are used to remove ISI (i.e., $\hat{\mathbf{c}}_N = \mathbf{c}_N$) yields, after some manipulation:

$$\mathbf{F} \left(\sigma_c^2 \, \mathbf{H}\mathbf{H}^H + \sigma_n^2 \, \mathbf{I}_N \right) - \sigma_c^2 \left(\mathbf{B} + \mathbf{I}_N \right) \mathbf{H}^H = \mathbf{0}_N, \tag{6.56}$$

where σ_c^2 denotes the mean square value of the channel symbols. It is easy to see that:

$$\mathbf{F} = \bar{\mathbf{B}} \, \mathbf{H}^H \, \Sigma^{-1}, \tag{6.57}$$

where

$$\bar{\mathbf{B}} \triangleq \mathbf{B} + \mathbf{I}_N, \tag{6.58}$$

and

$$\Sigma \triangleq \mathbf{H}\mathbf{H}^H + \frac{\sigma_n^2}{\sigma_c^2} \, \mathbf{I}_N \tag{6.59}$$

is an $N \times N$ Hermitian matrix. Note that the optimal choice $\bar{\mathbf{B}}_{MMSE}$ for $\bar{\mathbf{B}}$ should have all the elements of its main diagonal equal to one to ensure that all the diagonal elements of $\mathbf{B}$ are zero. Substituting (6.57) into (6.52) and (6.52) into (6.54) leads to:

$$\Delta = \bar{\mathbf{B}} \, \mathbf{H}^H \, \Sigma^{-1} \, \mathbf{r} - \bar{\mathbf{B}} \, \mathbf{c}_N = \bar{\mathbf{B}} \, \mathbf{V}, \tag{6.60}$$

where

$$\mathbf{V} \triangleq \mathbf{H}^H \, \Sigma^{-1} \, \mathbf{r} - \mathbf{c}_N \tag{6.61}$$

is a zero mean N-dimensional vector. From (6.60) it can be seen that, in order to minimize the mean square error $E\{|\delta_l|^2\}$ for $l = 0, 1, \ldots, N-1$, Δ should be the *nonstationary innovation* of $\mathbf{V}$ (6.61), that is, the linear transformation described by $\bar{\mathbf{B}}$ should *decorrelate* the elements of Δ [503, 1021]. It is easy to show that the covariance matrix of Δ (6.60) can be expressed as:

$$\mathbf{R}_\Delta \triangleq E\left\{ \Delta \, \Delta^H \right\} = \bar{\mathbf{B}} \, \mathbf{R}_V \bar{\mathbf{B}}^H, \tag{6.62}$$

where the covariance $\mathbf{R}_V \triangleq E\{\mathbf{V} \, \mathbf{V}^H\}$ of $\mathbf{V}$ is given by:

$$\mathbf{R}_V = \sigma_c^2 \left[\mathbf{I}_N - \mathbf{H}^H \, \Sigma^{-1} \mathbf{H} \right]. \tag{6.63}$$

Applying the Cholesky decomposition to $\mathbf{R}_V$ yields the factorization:

$$\mathbf{R}_V = \mathbf{M}^H \mathbf{Q}^2 \, \mathbf{M}, \tag{6.64}$$

where $\mathbf{M} = [m_{i,j}]$ is an $N \times N$ lower triangular matrix with all the elements along its main diagonal equal to one and $\mathbf{Q} = [Q_{i,j}]$ is an $N \times N$ diagonal matrix with positive real elements $Q_{i,i} = \xi_i$, with $i = 0, 1, \ldots, N-1$. Then substituting (6.64) in (6.62) yields:

$$\mathbf{R}_\Delta = \bar{\mathbf{B}} \, \mathbf{M}^H \mathbf{Q}^2 \, \mathbf{M} \, \bar{\mathbf{B}}^H, \tag{6.65}$$

which shows that the optimal choice for $\bar{\mathbf{B}}$ is given by:

$$\bar{\mathbf{B}}_{MMSE} = \left(\mathbf{M}^H \right)^{-1}, \tag{6.66}$$

since it diagonalizes $\mathbf{R}_\Delta$, is an $N \times N$ lower triangular matrix and all the elements along its main diagonal are equal to one. Then the optimal feedback matrix can be evaluated as (see (6.58)):

$$\mathbf{B}_{MMSE} = \bar{\mathbf{B}}_{MMSE} - \mathbf{I}_N = \left(\mathbf{M}^H \right)^{-1} - \mathbf{I}_N, \tag{6.67}$$

which is an $N \times N$ lower triangular matrix with all the elements along its main diagonal equal to zero, as required. Moreover, substituting (6.66) into (6.58) yields:

$$\mathbf{F}_{MMSE} = \left(\mathbf{M}^H\right)^{-1} \mathbf{H}^H \, \mathbf{\Sigma}^{-1}, \tag{6.68}$$

which expresses the optimal feedforward matrix. It can be shown that the MMSE-DFE becomes equivalent to the ZF-DFE for vanishing noise (i.e., for $\sigma_n^2 \to 0$). Further information on the design and performance of an MMSE-DFE for block transmissions can be found in [1021, 1023, 1024].

The decision feedback algorithm for channel equalization developed above is based on the use of two linear discrete-time *symbol-spaced* filters. Note, however, that both the coefficients and the memory of such filters are *time-varying*. This approach makes sense if the length N of the received vector $\mathbf{r}$ is not too large. When this does not occur, discrete-time filters with a *fixed memory* (i.e., a fixed number of taps) need to be employed for both feedforward and feedback processing. This approach leads to the conventional *symbol-spaced finite-length* DFE, whose block diagram is given in Figure 6.5. This equalizer includes:

(a) an L_f-tap symbol-spaced *feedforward filter* (FFF) having complex impulse response $\mathbf{f}_n \triangleq [f_n^{(0)}, f_n^{(1)}, \dots, f_n^{(L_f-1)}]^T$ in the nth symbol interval,
(b) an L_d-tap symbol-spaced *feedback filter* (FBF) having complex impulse response $\mathbf{d}_n \triangleq [d_n^{(0)}, d_n^{(1)}, \dots, d_n^{(L_d-1)}]^T$ in the nth symbol interval, and
(c) a decision device (typically, a threshold device) generating an estimate of each transmitted channel symbol with a delay Δ (this parameter is usually known as the *decision delay* or *lag*).

Note that, generally speaking, the lengths of the FFF and of the FBF are proportional to the CIR length; however, no general mathematical rule relating these quantities can be found in the literature [1025].

The DFE shown in Figure 6.5 operates as follows. In the kth symbol interval the FFF processes the received vector:

$$\mathbf{r}_k \triangleq \left[r_{k-L_f+1}, r_{k-L_f+2}, \dots, r_k\right]^T \tag{6.69}$$

to partially compensate for the channel distortion. In particular, it mitigates the ISI coming from the symbols preceding $c_{k-\Delta}$ (i.e., precursor ISI) and influencing the vector $\mathbf{r}_k$ (6.69). Then, the FBF output is subtracted from the FFF output, in order to cancel a significant portion of postcursor ISI, so that the decision device input:

$$I_k = \mathbf{f}_k^T \, \mathbf{r}_k - \mathbf{d}_k^T \, \hat{\mathbf{c}}_k \tag{6.70}$$

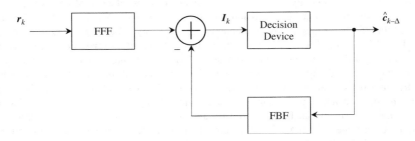

Figure 6.5 Block diagram of a symbol-spaced finite-length DFE (conventional form).

is generated in the kth symbol interval, where:

$$\hat{\mathbf{c}}_k = \left[\hat{c}_{k-\Delta-L_d}, \hat{c}_{k+1-\Delta-L_d}, \dots, \hat{c}_{k-\Delta-1} \right]^T \tag{6.71}$$

denotes the past decision vector $\hat{\mathbf{c}}_n$ processed by the FBF (note that $\hat{c}_l = 0$ for $l < 0$ and $l > N - 1$). If model (6.36) is again adopted for the kth receive sample r_k, the vector $\mathbf{r}_k$ (6.69) can be put in the form:

$$\mathbf{r}_k = \mathbf{H}\,\mathbf{c}_k + \mathbf{n}_k, \tag{6.72}$$

where $\mathbf{c}_k \triangleq [c_{k-L_f-L_h+2}, c_{k-L_f-L_h+3}, \dots, c_k, \dots, c_{k+L_h-1}]^T$, $\mathbf{n}_k \triangleq [n_{k-L_f+1}, n_{k-L_f+2}, \dots, n_k]^T$ and $\mathbf{H} = [h_{i,j}]$ is a proper $L_f \times (2L_h + L_f - 2)$ CIR matrix. Then, if correct decision feedback is assumed, the optimal choice $(\mathbf{f}_{MMSE,k}, \mathbf{d}_{MMSE,k}, \Delta_{MMSE})$, in the MMSE sense, for the parameter vector $(\mathbf{f}_k, \mathbf{d}_k, \Delta)$ is the solution of the optimization problem:

$$(\mathbf{f}_{MMSE,k}, \mathbf{d}_{MMSE,k}, \Delta_{MMSE}) = \arg \min_{\tilde{\mathbf{f}}_k, \tilde{\mathbf{d}}_k, \tilde{\Delta}} \varepsilon_{MMSE}, \tag{6.73}$$

where

$$\varepsilon_{MMSE} \triangleq \mathrm{E}\left\{ \left| c_{k-\tilde{\Delta}} - \tilde{\mathbf{f}}_k^T \mathbf{r}_k + \tilde{\mathbf{d}}_k^T \hat{\mathbf{c}}_k \right|^2 \right\} \tag{6.74}$$

denotes the MSE and, as in (6.51), the expectation is evaluated over the random data and channel noise (but is conditioned on the channel matrix $\mathbf{H}$) and correct decision feedback is assumed. In addition, $\tilde{\Delta}$ satisfies the constraint $0 \le \tilde{\Delta} \le L_h + L_f - 2$, since on the basis of the vector $\mathbf{r}_k$ (6.69) a decision should be taken on c_k or on the past symbols influencing the vector itself. Unfortunately, a closed-form solution is not available for the problem (6.73) (unlike the conceptually related problem (6.53)) because of the presence of the additional parameter Δ. However, for a fixed Δ, a solution for the optimal $\mathbf{f}_k$ and $\mathbf{d}_k$ can be found by applying the orthogonality principle and, in addition, such a solution is *time-invariant* for $L_h + L_f - 2 \le k \le N - L_h$. For this reason, a reasonable strategy is simply to preselect Δ as:

$$\Delta \approx \frac{L_h + L_f - 2}{2}, \tag{6.75}$$

so that closed-form expressions for the FFF and FBF taps can be exploited. Further details on the optimization, performance and computation of a finite-length MMSE-DFE for SISO systems can be found in [1026–1030].

However, if the channel is *time-varying*, the design of an optimal DFE becomes substantially more complicated. In fact, as already stated in Section 4.4.2, matched filtering cannot be employed and an ideal low-pass filter with a bandwidth large enough so as to not distort the useful component of the received signal is usually selected as a receive filter. Moreover, the low-pass filter output is sampled at a frequency $f_c = n_s R_s$, with $n_s > 1$, and this results in model (4.104) (to be adopted in place of (6.36)). Consequently, since $n_s > 1$ samples per channel interval are extracted from the filtered received signal, in order to avoid information loss a *fractionally spaced filter* (characterized by a tap spacing $T_c \triangleq T_s/n_s$ and whose output is decimated by a factor n_s) has to be adopted for the FFF of a DFE [1020]. The resulting system requires an algorithm for the *automatic synthesis* of its filters. This algorithm relies on the transmission, during a *training period*, of a known signal (in practice, on blocks of training symbols, which are interspersed with the data transmission blocks to periodically train the equalizer). A synchronized version of this signal is generated in the receiver to acquire information about the channel characteristics through the evaluation of the samples of an *error signal*. The signal can be processed by an LMS or RLS algorithm (see Section 5.1.3) to adjust the equalizer coefficients so that the sum of the squared error samples is minimized. After the training period the filters can be adapted to the changing channel characteristics by employing *decision-directed* tracking, i.e. by exploiting the detected channel symbols. The architecture of the

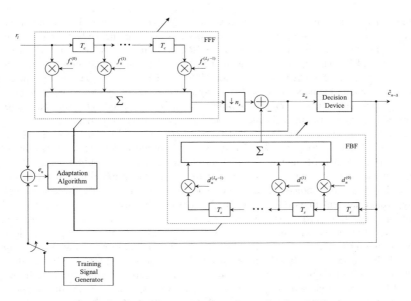

Figure 6.6   Block diagram of an adaptive DFE.

adaptive DFE based on these principles is shown in Figure 6.6 (the error sequence and the received data sequence acquired at frequency f_c are denoted by $\{e_k\}$ and $\{r_l\}$, respectively). Generally speaking, the compensation capability of this equalizer can be improved by allowing *time-reversal operation* [1031–1033]. In practice, this is accomplished by storing each block of received signal samples in a buffer and reversing the sequential order of these signal samples in time prior to equalization, so that the equivalent CIR as seen by the equalizer is a time-reverse of the actual CIR. Therefore, selective time-reversal operation allows a DFE with a small number of forward filter taps to perform equally well for both minimum-phase and maximum-phase channel characteristics.

The DFE structure of Figure 6.6 works well if the channel does not vary too quickly but may suffer from error propagation in the decision-directed mode. An alternative to this approach is to identify the channel parameters first and then use these parameters for calculating the equalizer coefficients (e.g., see [778, 1034]). For instance, in [778] the time-varying CIR is estimated by interpolating a set of CIR values obtained through periodic training at adjacent data frames within a given timeslot. Then interpolated channel estimates are employed to adapt the DFE. This adaptive strategy has the disadvantages of increased processing delay and reduction of system throughput (since frequent training block are needed), but it does offer the advantage of improved immunity to decision errors.

There are no closed-form solutions or easily computable tight bounds on the BER of MMSE-DFEs, due to the non-Gaussian nature of the residual ISI and the possibility of erroneous decisions being fed back through the FBF. Certain analytical techniques accounting for error propagation (e.g., see [1017, 1035–1039]) or ignoring it (e.g., see [1040–1043]) can be found in the literature. The performance degradation due to imperfect adaptation is assessed in [510, 767].

The simplest performance measure is the DFE MSE:

$$\text{MSE} \triangleq \text{E}\left\{|e_k|^2\right\},\tag{6.76}$$

where the error sample:

$$e_k \triangleq I_k - c_{k-\Delta}\tag{6.77}$$

is evaluated under the assumption of correct decisions (I_k is given by (6.70)). This assumption, which is repeatedly invoked in the literature on decision feedback, is a strong one. In fact, as already mentioned for block decision feedback equalization, at low SNRs, noise may cause a "primary" error, which is fed back into the FBF. Then, instead of canceling ISI, the FBF can actually enhance it, which in turn increases the likelihood of subsequent, "secondary" errors. This error propagation phenomenon may be severe in wideband channels since the FBF may be hundreds of symbols long (see [1044–1049] for further details). The error propagation problem can be mitigated by developing *soft decision* versions of equalization structures. In their basic form, such structures, rather than utilizing a hard decision quantizer to generate the feedback data, employ a (monotonic) *soft decision device* [1050]. Then, outside the feedback loop, the soft decisions are converted into finite-set constrained decisions by means of a standard hard quantizer, which provides the equalizer output. Since the signal fed back is not restricted to belong to a finite set, it may provide information about the reliability of detected symbols. Possible choices of decision devices, which have been utilized to enhance DFEs, include hyperbolic tangent functions [1051, 1052], cubic nonlinearities [1053], and piecewise linear (saturation-type) functions [1054] (the latter have also been adopted in decoding problems [1055, 1056]). In alternative developments of soft decision structures, *reliability measures* of decisions can be employed directly in order to allow the decisions already made to be changed a posteriori in the feedback path. For instance, in [1057], uncertainty effects are accounted for in a DFE configuration by inspecting the input to the quantizer. If the input lies near a decision boundary, then the neighbor of the output of the quantizer is used in a parallel DFE loop. After a decision delay, the output of the DFE loop with lower accumulated error is chosen as correct. A different approach is described in [1002], where the DFE quantizer is replaced by a sequence detector based on the VA. In this case, rather than deploying symbol-by-symbol feedback, sequences are fed back; such sequences are provided by the VA and can be changed after output decisions are made. A somewhat related equalizer has also been proposed in [1058], where a Bayesian detector combined with soft decisions is deployed. Other related work can be found in [1059], where a fractionally spaced equalizer combining a hyperbolic tangent decision device with a Kalman filter is presented.

A substantially different approach to avoiding the error propagation problem involves transferring the feedback part of a DFE to the transmitter, provided that a feedback channel is available to communicate CSI to the transmitter itself. This approach leads to a nonlinear pre-equalization technique known as *Tomlinson–Harashima precoding* (THP), since it was proposed by M. Tomlinson [1060] and by H. Harashima and H. Miyakawa [1061].

There are a number of structural modifications/extensions to the classical fractionally spaced DFE presented above. Here we limit attention to:

(a) the modified DFE structure of [983] for precanceling postcursors without requiring training of the feedback filter (only the feedforward filter taps need to be trained);
(b) the *skeptical multistep detector* proposed in [1062] and embedding DFEs and the multistep detectors in a broader framework;
(c) the *bidirectional* DFE, where equalization is performed both on the received signal and on its time-reversed version [1031, 1063, 1064];
(d) the *tree equalizers* developed in [1065];
(e) the so-called *predictor form* of the DFE [1019];
(f) the *decision feedback equalizer* (DFE) scheme for *M*-ary *differentially encoded* PSK (DPSK) signals developed in [1066].

The latter structure is illustrated in Figure 6.7 and is based on the idea that *linear prediction* can be exploited to reduce the power of the error sequence (due to noise and ISI) at the forward filter output; for this reason a prediction filter is used in the feedback section of the equalizer. It can be shown that, so long as the length of the forward filter in the conventional and prediction-based DFE structures is unconstrained, the two structures remain equivalent even when the length of the feedback

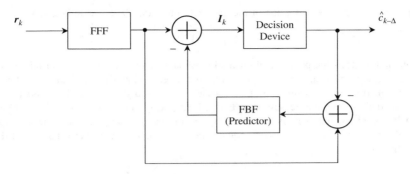

Figure 6.7 Predictor form of a DFE.

(or prediction) filter is finite. Note, however, that, although the forward filter in the conventional DFE depends on the length of the feedback filter, the forward filter in the predictor form is independent of the prediction coefficients. Further details on the predictor form and its comparison with the conventional form can be found in [1019, 1020, 1067].

DFEs can also be employed to equalize other signal formats such as CPM, where ISI is due in part to the memory in the modulation or demodulation scheme rather than a dispersive channel [1068–1070].

In wideband communications over multipath channels with large delay spread, CIRs can span hundreds of symbols and exhibit sparse behavior (i.e., many nearly zero taps). In addition, a strong precursor component (corresponding to one or more strong paths which are shorter than the main path) may exist. In this case modified decision feedback equalization structures, exploiting channel sparseness by simple tap allocation, such as those proposed in [1071–1073], can be employed. Unlike the conventional DFE, these structures (known as *sparse* equalizers) yield large reductions in complexity while maintaining performance comparable to the conventional DFE.

6.2.1.4 Linear Equalization in the Time Domain

If feedback filtering is removed from a DFE, a *linear equalizer* (LE) is obtained. Consequently, an LE is a linear, usually transversal, filter followed by a simple decision device (i.e., a nearest-neighbor quantizer). Like DFEs, LEs are also classified on the basis of their design criterion: in most applications, the ZF and the MMSE criteria are usually considered.

Here, as in our study of decision feedback equalization, we first tackle the problem of ZF and MMSE block equalization of a received vector $\mathbf{r}$, assuming a known frequency-selective channel and the same received signal model as that adopted for block decision feedback equalization (see (6.38) and (6.39)). Then we consider the case of long $\mathbf{r}$, under the assumptions of frequency-selective known channel and an unknown time-varying channel.

An MMSE block LE can be derived by applying the orthogonality criterion in the form (6.55), where, however, the expression:

$$\mathbf{Z} \triangleq \mathbf{F}\,\mathbf{r} \tag{6.78}$$

is used in place of (6.55) in the evaluation of the error vector $\mathbf{\Delta}$ (6.54), since feedback filtering is missing (in other words, $\mathbf{B} = \mathbf{0}_N$). This leads easily to the expression:

$$\mathbf{F}_{MMSE} = \mathbf{H}^H\,\mathbf{\Sigma}^{-1} \tag{6.79}$$

for the optimal choice of the feedforward matrix $\mathbf{F}$, where $\mathbf{\Sigma}$ is given by (6.59). Note that, if $\sigma_n^2 \to 0$, then $\mathbf{\Sigma} \to \mathbf{HH}^H$, so that $\mathbf{F}_{MMSE} \to \mathbf{H}^{-1}$, which expresses the feedforward matrix in the absence of channel noise. Then the optimal value of the matrix $\mathbf{F}$ under the ZF criterion is given by:

$$\mathbf{F}_{ZF} = \mathbf{H}^{-1} . \tag{6.80}$$

The ZF method neglects the presence of channel noise and minimizes the ISI contribution within the equalizer time span. This is undesirable since wideband wireless channels are characterized by multiple deep notches in frequency. In inverting these, a ZF-LE inevitably causes undue noise amplification and a degraded BER. For this reason MMSE is a better design criterion, where the combination of noise and ISI is minimized. This is evidenced by the following example, where the performance of linear equalization and decision feedback equalization is compared with that offered by MLSD in a specific scenario.

Example 6.2.1 Let us now compare the error rate performance provided by different equalization algorithms in an uncoded QPSK transmission (characterized by a block length $N = 128$) over a frequency-selective channel. Adopting the Proakis' channel model described in [27, p. 631], we have that $\mathbf{h} = [h_0, h_1, h_2]$ and $L_h = 3$ in (6.36). Moreover, each channel tap is independently Rayleigh distributed with statistical powers $\mathrm{E}\{|h_0|^2\} = 0.407$, $\mathrm{E}\{|h_1|^2\} = 0.815$, $\mathrm{E}\{|h_2|^2\} = 0.407$, so that $\mathrm{E}\{|\mathbf{h}|^2\} = \sum_{l=0}^{L_h-1} \mathrm{E}\{|h_l|^2\} = 1$, that is, the channel has average unitary power. Figure 6.8 shows the BER offered by a block ZF-LE (see (6.80)), an MMSE-LE (see (6.79)), an MMSE-DFE (see (6.67) and (6.68)) and MLSD (a VA operating over a $M^{L_h-1} = 64$-state trellis has been employed); the MFB is also shown for comparison. These results show that: (a) MLSD can get very close to the MFB; (b) the ZF-LE is significantly outperformed by its MMSE counterpart; (c) a significant energy gap exists between MLSD and the MMSE-DFE (however, the improvement in error performance is achieved at the price of a significant additional complexity); (d) BER curves referring to different equalization may be characterized by different slopes (e.g., compare the MMSE-DFE BER with that of the MMSE-LE), that is, by different diversity gains (see Section 6.2.1.6).
□

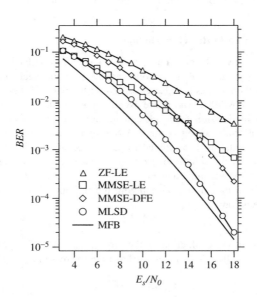

Figure 6.8 BER performance of various equalization algorithms for an uncoded QPSK transmission over a frequency-selective fading channel. The MFB is also shown for comparison.

Further details on the performance gap between different types of block LEs, and between LEs and block DFEs, can be found in [1021]. The BER performance achieved by block linear equalization in SC block transmissions using zero-padding or cyclic prefix is investigated in [1074].

As in the DFE case, block linear equalization can be adopted if the block size is not too large, but otherwise a time-invariant linear filter should be adopted. If the channel is *frequency-selective* and *known*, matched filtering and symbol-rate sampling is optimal. In this case, the problem of deriving an expression for the optimal taps for an LE filter can be formulated in a similar fashion to the DFE (in other words, (6.69)–(6.74) can be exploited again under the assumption that $\mathbf{d}_n = \mathbf{0}_{L_d}$). However, as in the DFE case, a closed-form expression for the optimal filter taps cannot be derived because of the tap dependence on the value selected for the lag Δ (an exhaustive search approach is needed to determine the optimal delay [1075]). In contrast, if the channel is *unknown* and/or *time-varying*, an adaptive filter is required [860, 1076]; in addition, fractionally spaced taps should be used [1020, 1025, 1077, 1078]. A fractionally spaced linear equalizer can be implemented as a *baseband* (as we did in the previous subsection for the DFE) or as a *passband* equalizer. Examples of the latter approach can be found in [1079, 1080], where the so-called *phase-splitting fractionally spaced equalizer* is analyzed.

As already stated, an LE can be interpreted as a simplification of a DFE. Then, given a certain communication channel, it is important to establish if an LE is outperformed in terms of MSE by a DFE counterpart characterized by the same overall number of coefficients. Unfortunately, there is no well-defined indication for this problem. This is due to the fact that the performance of each type of equalizer is influenced by the sampling phase (if symbol-spaced FFFs are used), the characteristics of the communication channels, the number of coefficients and the position of the main tap of the equalizer itself. However, generally speaking, an LE is unable to compensate for amplitude distortion as much as its DFE counterparts. In addition, the DFE performance is less sensitive to the sampling phase [1081]. These results can be motivated as follows. The coefficients of an LE are selected to force the combined channel and equalizer impulse response to approximate a unit pulse. On the other hand, the combined channel and FFF impulse response of a DFE may have nonzero samples following the main pulse, that is, the feedforward section of a DFE does need to approximate the inverse of the channel characteristics and results in a reduction of noise enhancement and of sensitivity to sampler phase. It is also worth mentioning that an LE, unlike a DFE, does not suffer from error propagation; however, as already mentioned, this phenomenon is not catastrophic. Despite its disadvantages, linear equalization has attracted much interest for its simplicity. In fact, on frequency-selective channels linear equalization enjoys significantly reduced complexity compared with MLSD; note that, generally speaking, the cost of performance degradation can be in the form of a loss of the inherent frequency diversity order or reduced coding gain. It has recently been shown that the MMSE symbol-by-symbol linear equalizer incurs no diversity loss compared to MLSD [1082]; in fact, for a channel with memory equal to $L_h - 1$, both strategies achieve the full diversity order of L_h. However, the ZF symbol-by-symbol linear equalizer always achieves a diversity order of one.

As for the DFE, various extensions to the basic LE structure are available [1065, 1083–1085]. In particular, [1083] proposes a fractionally spaced equalizer based on a parametric model of the communication channel, while the use of a tree-structured piecewise linear filter as an adaptive equalizer is proposed in [1065]. Finally, the exploitation of linear equalization in noncoherent receivers is investigated in [1084, 1085].

6.2.1.5   MAPSD/MAPBD

When the MAP criterion replaces the ML criterion, the MAP *forward–backward algorithm* (FBA) replaces the VA [982]. The FBA operates on the same trellis as the VA, but it efficiently calculates the MAP bit or symbol probabilities, instead of looking for the ML sequence. Unlike the VA, the FBA requires all the received samples to be available before it can be run. For this reason, when a short decision delay is required, a near-optimal, *forward-only* (*fixed-lag*) MAP strategy can be adopted, which we call the *fixed-lag algorithm* (FLA). However, in principle, both MAP algorithms work with likelihoods rather than log-likelihoods, so that the computation of their branch metrics is significantly more expensive than that of the VA (it requires multiplications and additions instead of the additions and minimization characterizing the VA).

In what follows the derivations of the FBA and the fixed-lag MAPSD are sketched (further details on the FBA can be found in Section 9.2.4, where the same algorithm[4] is derived for MAP decoding of convolutional codes) and some comments on the advantages and disadvantages they offer with respect to VA-based MLSD are given. We consider an M-ary PAM signaling communication system operating over a time-invariant, frequency- selective fading channel and assume that the detection algorithm processes the baud rate received signal samples taken at the output of a WMF; in other words, model (6.32) is adopted for the kth received signal sample. In addition, it is assumed that $L_q < N$ and that the transmitted channel symbols $\{c_k, \ k = 0, 1, \ldots, N - 1\}$ forming the transmitted symbol vector $\mathbf{c}_N \triangleq [c_0, c_1, \ldots, c_{N-1}]^T$ are statistically independent.

The Forward–backward Algorithm

The FBA evaluates the probabilities:

$$\Pr\{c_l = \tilde{c}|\mathbf{r}, \mathbf{q}\}, \quad l = 0, 1, \ldots, N - 1, \tag{6.81}$$

for any possible trial symbol $\tilde{c}$ transmitted in the mth signaling interval, where $\mathbf{r} = [r_0, r_1, \ldots, r_N - 1]^T$ and $\mathbf{q} \triangleq [q_0, q_1, \ldots, q_{L_q-1}]^T$ denote the received signal vector and the CIR vector, respectively. In the FBA the evaluation of these probabilities relies on:

(a) the adoption of an M^{L_q-1}-state trellis structurally identical to that of a VA accomplishing MLSD and operating over the same communication channel (see Section 6.2.1.1);

(b) the evaluation of the same intermediate quantities, known as *state transition probabilities*.

Given the trellis states Δ_l and Δ_{l+1} in the lth and $(l + 1)$th symbol interval, respectively, the corresponding *state transition probability* is given by:

$$\Pr\left\{\Delta_l = \tilde{\Delta}_l, \Delta_{l+1} = \tilde{\Delta}_{l+1}|\mathbf{r}, \mathbf{q}\right\} \tag{6.82}$$

for any possible pair $(\tilde{\Delta}_l, \tilde{\Delta}_{l+1})$ of trial states referring to two consecutive signaling intervals. Note that this probability is zero if the states $\tilde{\Delta}_l$ and $\tilde{\Delta}_{l+1}$ are not connected. The probability (6.82) can be evaluated as:

$$\Pr\left\{\Delta_l = \tilde{\Delta}_l, \Delta_{l+1} = \tilde{\Delta}_{l+1}|\mathbf{r}, \mathbf{q}\right\} = \frac{\sum_{\tilde{\mathbf{c}}_N \in S\left(\tilde{\Delta}_l, \tilde{\Delta}_{l+1}\right)} \Pr\{\mathbf{c}_N = \tilde{\mathbf{c}}_N|\mathbf{r}, \mathbf{q}\}}{\sum_{\tilde{\mathbf{c}}_N \in S_c} \Pr\{\mathbf{c}_N = \tilde{\mathbf{c}}_N|\mathbf{r}, \mathbf{q}\}}, \tag{6.83}$$

where $\tilde{\mathbf{c}}_N$ denotes a trial value of $\mathbf{c}_N$, S_c denotes the set $\{\tilde{\mathbf{c}}_N\}$ consisting all the possible trial vectors and $S(\tilde{\Delta}_l, \tilde{\Delta}_{l+1})$ is that subset of S_c which consists of all the possible trial symbol vectors which traverse the trellis branch connecting the state $\tilde{\Delta}_l$ to $\tilde{\Delta}_{l+1}$. By Bayes' theorem the probabilities $\{\Pr\{\mathbf{c}_N = \tilde{\mathbf{c}}_N|\mathbf{r}, \mathbf{q}\}\}$ appearing in the RHS of (6.83) can be expressed as:

$$\Pr\{\mathbf{c}_N = \tilde{\mathbf{c}}_N|\mathbf{r}, \mathbf{q}\} = \frac{f_\mathbf{r}(\mathbf{r}|\tilde{\mathbf{c}}_N, \mathbf{q}) \Pr\{\mathbf{c}_N = \tilde{\mathbf{c}}_N\}}{f_\mathbf{r}(\mathbf{r}|\mathbf{q})}. \tag{6.84}$$

The terms $\Pr\{\mathbf{c}_N = \tilde{\mathbf{c}}_N\}$ and $f_\mathbf{r}(\mathbf{r}|\tilde{\mathbf{c}}_N, \mathbf{q})$ in this formula can be factored as:

$$\Pr\{\mathbf{c}_N = \tilde{\mathbf{c}}_N\} = \prod_{k=0}^{N-1} \Pr\{c_k = \tilde{c}_k\} \tag{6.85}$$

---

[4] In the field of channel coding this algorithm is also known as *Bahl–Cocke–Jelinek–Raviv* (BCJR) algorithm.

and

$$f_{\mathbf{r}}\left(\mathbf{r} \mid \tilde{\mathbf{c}}_N, \mathbf{q}\right) = \prod_{k=0}^{N-1} f_r\left(r_k \mid \tilde{\mathbf{c}}_N, \mathbf{q}\right) = \prod_{k=0}^{N-1} f_r\left(r_k \mid \tilde{c}_k, \tilde{\Delta}_k, \mathbf{q}\right), \tag{6.86}$$

respectively, since:

(a) the transmitted channel symbols are independent; and
(b) the signal sample r_k received in the kth symbol interval depends on the kth symbol c_k and on the previous $(L_q - 1)$ symbols (see (6.32)), that is on the trellis state Δ_k, which is uniquely identified by the ordered collection of the channel symbols $\{c_n, \ n = k - L_q + 1, k - L_q, \ldots, k - 1\}$.

Note also that:

$$f_r\left(r_k \mid \tilde{c}_k, \tilde{\Delta}_k, \mathbf{q}\right) = \frac{1}{\pi \sigma_n^2} \exp\left[-\frac{1}{\sigma_n^2}\left|r_k - \sum_{m=0}^{L_q-1} \tilde{c}_{k-m} \, q_m\right|^2\right]. \tag{6.87}$$

Then the product $f_{\mathbf{r}}\left(\mathbf{r} \mid \tilde{\mathbf{c}}_N, \mathbf{q}\right) \Pr\{\mathbf{c}_N = \tilde{\mathbf{c}}_N\}$ on the RHS of (6.84) can be expressed as (see (6.85)–(6.87)):

$$\Pr\left\{\mathbf{c}_N = \tilde{\mathbf{c}}_N \mid \mathbf{r}, \mathbf{q}\right\} = \prod_{k=0}^{N-1} f_r\left(r_k \mid \tilde{c}_k, \tilde{\Delta}_k, \mathbf{q}\right) \prod_{k=0}^{N-1} \Pr\{c_k = \tilde{c}_k\}$$

$$= \prod_{k=0}^{N-1} f_r\left(r_k \mid \tilde{c}_k, \tilde{\Delta}_k, \mathbf{q}\right) \Pr\{c_k = \tilde{c}_k\}$$

$$= \frac{1}{\left(\pi \sigma_n^2\right)^N} \prod_{k=0}^{N-1} \gamma_k\left(\tilde{\Delta}_k, \tilde{\Delta}_{k+1}\right), \tag{6.88}$$

where

$$\gamma_k\left(\tilde{\Delta}_k, \tilde{\Delta}_{k+1}\right) \triangleq \left(\pi \sigma_n^2\right) f_r\left(r_k \mid \tilde{c}_k, \tilde{\Delta}_k, \mathbf{q}\right) \Pr\{c_k = \tilde{c}_k\}$$

$$= \exp\left[-\frac{1}{\sigma_n^2}\left|r_k - \sum_{m=0}^{L_q} \tilde{c}_{k-m} \, q_m\right|^2\right] \Pr\{c_k = \tilde{c}_k\} \tag{6.89}$$

is a *weight* depending on L_q consecutive channel symbols (i.e., on a state transition) only. Then substituting (6.88) into (6.83) yields:

$$\Pr\left\{\Delta_l = \tilde{\Delta}_l, \Delta_{l+1} = \tilde{\Delta}_{l+1} \mid \mathbf{r}, \mathbf{q}\right\} = \frac{\sum_{\tilde{\mathbf{c}}_N \in S\left(\tilde{\Delta}_l, \tilde{\Delta}_{l+1}\right)} \prod_{k=0}^{N-1} \gamma_k\left(\tilde{\Delta}_k, \tilde{\Delta}_{k+1}\right)}{\sum_{\tilde{\mathbf{c}}_N \in S_c} \prod_{k=0}^{N-1} \gamma_k\left(\tilde{\Delta}_k, \tilde{\Delta}_{k+1}\right)}. \tag{6.90}$$

This result shows that the evaluation of the state transition probability $\Pr\{\Delta_l = \tilde{\Delta}_l, \Delta_{l+1} = \tilde{\Delta}_{l+1} \mid \mathbf{r}, \mathbf{q}\}$ requires the computation of (a) the sum of the products of the weights associated with all the paths containing the branch exiting $\tilde{\Delta}_l$ and entering $\tilde{\Delta}_{l+1}$ (see the numerator), and (b) the sum of the products of the weights associated with all the paths in the trellis (see the denominator). A computationally efficient method to solve this problem can be developed as follows (e.g., see [533]). Let us define the quantities $\alpha_m(\tilde{\Delta}_m)$ and $\beta_m(\tilde{\Delta}_m)$ through the recursive formulas:

$$\alpha_m\left(\tilde{\Delta}_m\right) = \sum_{\tilde{\Delta}_{m-1}} \alpha_{m-1}\left(\tilde{\Delta}_{m-1}\right) \cdot \gamma_m\left(\tilde{\Delta}_{m-1}, \tilde{\Delta}_m\right) \tag{6.91}$$

with $m = 1, 2, \ldots, N - 1$, and

$$\beta_m\left(\tilde{\Delta}_m\right) = \sum_{\tilde{\Delta}_{m+1}} \beta_{m+1}\left(\tilde{\Delta}_{m+1}\right) \cdot \gamma_m\left(\tilde{\Delta}_m, \tilde{\Delta}_{m+1}\right) \tag{6.92}$$

with $m = N - 2, N - 3, \ldots, 0$, respectively, and assume that $\{\alpha_0(\tilde{\Delta}_0)\}$ and $\{\beta_{N-1}(\tilde{\Delta}_{N-1})\}$ are known initial conditions for any possible $\tilde{\Delta}_0$ and $\tilde{\Delta}_{N-1}$. The quantity $\alpha_m(\tilde{\Delta}_m)$ expresses the sum of the products of the weights $\{\gamma_k(\tilde{\Delta}_{k-1}, \tilde{\Delta}_k)\}$ along all paths originating from all the possible past initial states $\{\tilde{\Delta}_0\}$ and terminating in $\tilde{\Delta}_m$ in the mth signaling interval. Similarly, $\beta_m(\tilde{\Delta}_m)$ represents the sum of the products of the weights $\{\gamma_k(\tilde{\Delta}_k, \tilde{\Delta}_{k+1})\}$ along all paths ending in all possible future terminal states $\{\tilde{\Delta}_{N-1}\}$ and originating from $\tilde{\Delta}_m$ in the mth signaling interval. Then the numerator of (6.90) can be evaluated as:

$$\sum_{\tilde{c}_N \in S\left(\tilde{\Delta}_l, \tilde{\Delta}_{l+1}\right)} \prod_{k=0}^{N-1} \gamma_k\left(\tilde{\Delta}_k, \tilde{\Delta}_{k+1}\right) = \alpha_l\left(\tilde{\Delta}_l\right) \cdot \gamma_l\left(\tilde{\Delta}_l, \tilde{\Delta}_{l+1}\right) \cdot \beta_{l+1}\left(\tilde{\Delta}_{l+1}\right)$$

$$\triangleq \sigma_l\left(\tilde{\Delta}_l, \tilde{\Delta}_{l+1}\right), \tag{6.93}$$

while its denominator is given by:

$$\sum_{\tilde{c}_N \in S_c} \prod_{k=0}^{N-1} \gamma_k\left(\tilde{\Delta}_k, \tilde{\Delta}_{k+1}\right) = \sum_{\tilde{\Delta}_l, \tilde{\Delta}_{l+1}} \sigma_l\left(\tilde{\Delta}_l, \tilde{\Delta}_{l+1}\right), \tag{6.94}$$

so that:

$$\Pr\left\{\Delta_l = \tilde{\Delta}_l, \Delta_{l+1} = \tilde{\Delta}_{l+1} \mid \mathbf{r}, \mathbf{q}\right\} = \frac{\sigma_l\left(\tilde{\Delta}_l, \tilde{\Delta}_{l+1}\right)}{\sum_{\tilde{\Delta}_l, \tilde{\Delta}_{l+1}} \sigma_l\left(\tilde{\Delta}_l, \tilde{\Delta}_{l+1}\right)}. \tag{6.95}$$

This result shows that what is needed for the evaluation of the state transition probabilities are the quantities $\{\sigma_l(\tilde{\Delta}_l, \tilde{\Delta}_{l+1})\}$, computed according to (6.93). This, in turn, requires a *forward* recursion and a *backward* recursion (expressed by (6.91) and (6.92), respectively) involving all the trellis states in each signaling interval; note that both recursions over all the trellis only need to be performed once.

Signal demodulation requires the probability $\Pr\{c_l = \tilde{c} \mid \mathbf{r}, \mathbf{q}\}$ (see (6.81)) to be evaluated for any possible value of the channel symbol $\tilde{c}$ (or bit $\tilde{b}$ if a binary constellation is considered). This probability can be calculated by summing the state transition probabilities (6.95) corresponding to all the branches associated with the symbol $\tilde{c}$ (or bit $\tilde{b}$). Then, if we define the set $S(\tilde{c}_l)$ of all state transitions $(\tilde{\Delta}_l, \tilde{\Delta}_{l+1})$ such that the channel symbol labeling the corresponding branch is $\tilde{c}_l$, $\Pr\{c_l = \tilde{c} \mid \mathbf{r}, \mathbf{q}\}$ can be evaluated as:

$$\Pr\left\{c_l = \tilde{c} \mid \mathbf{r}, \mathbf{q}\right\} = \sum_{\left(\tilde{\Delta}_l, \tilde{\Delta}_{l+1}\right) \in S(\tilde{c}_l)} \Pr\left\{\Delta_l = \tilde{\Delta}_l, \Delta_{l+1} = \tilde{\Delta}_{l+1} \mid \mathbf{r}, \mathbf{q}\right\}$$

$$= \frac{\sum_{\left(\tilde{\Delta}_l, \tilde{\Delta}_{l+1}\right) \in S(\tilde{c}_l)} \sigma_l\left(\tilde{\Delta}_l, \tilde{\Delta}_{l+1}\right)}{\sum_{\tilde{\Delta}_l, \tilde{\Delta}_{l+1}} \sigma_l\left(\tilde{\Delta}_l, \tilde{\Delta}_{l+1}\right)}. \tag{6.96}$$

Finally, the FBA can be summarized in the following steps:

1. Evaluate the conditional probabilities $\{\gamma_k(\tilde{\Delta}_k, \tilde{\Delta}_{k+1})\}$ for all the trellis branches using (6.89).
2. Initialize the *forward recursion*, setting $m = 1$ and $\alpha_0(\tilde{\Delta}_0) = 1$ for any $\tilde{\Delta}_0$.
3. Repeat steps 4 and 5 until $m = N$.
4. Compute the *forward path probabilities* $\{\alpha_m(\tilde{\Delta}_m)\}$ using (6.91) and store them.
5. Set $m = m + 1$.
6. Initialize the *backward recursion*, setting $m = N - 2$ and $\beta_{N-1}(\tilde{\Delta}_{N-1}) = 1$ for any $\tilde{\Delta}_{N-1}$.
7. Repeat steps 8 and 9 until $m = 0$.
8. Compute the *backward path probabilities* $\{\beta_m(\tilde{\Delta}_m)\}$ using (6.92) and store them.
9. Set $m = m - 1$.
10. For each trellis branch compute the quantity $\sigma_l(\tilde{\Delta}_l, \tilde{\Delta}_{l+1})$ using (6.93).
11. Evaluate the a posteriori symbol probabilities $\{\Pr\{c_l = \tilde{c}|\mathbf{r}, \mathbf{q}\}\}$ using (6.96).

A suboptimal alternative to the FBA is provided by the MAP-FLA illustrated next.

The Fixed-lag Algorithm

The FBA is suited to short burst transmission because its delay and storage requirements become unacceptable as the data sequence length increases. In contrast, the MAP detector based on the FLA makes decisions at a fixed lag of D_{fl} symbols from the current received sample and requires a fixed amount of memory [1086]. The FLA represents a computationally efficient solution to the MAP detection problem:

$$\hat{\mathbf{c}}_l = \arg \max_{\tilde{c}_l} \Pr\left\{c_l = \tilde{c}_l | \mathbf{r}_0^{l+D_{fl}}, \mathbf{q}\right\}, \tag{6.97}$$

where $\mathbf{r}_0^{D_{fl}+m} \triangleq [r_0, r_1, \dots, r_{D_{fl}+m}]^T$ (in what follows the notation $\mathbf{x}_a^b$ is used to denote the vector $[x_a, x_{a+1}, \dots, x_b]^T$). To derive such an algorithm, we begin by noting that the probability $\Pr\{c_l = \tilde{c}_l | \mathbf{r}_0^{l+D_{fl}}, \mathbf{q}\}$ in (6.97) can be written as:

$$\Pr\left\{c_l = \tilde{c}_l | \mathbf{r}_0^{l+D_{fl}}, \mathbf{q}\right\} = \sum_{\tilde{c}_0^{l-1}, \tilde{c}_{l+1}^{l+L_{fl}}} \Pr\left\{\mathbf{c}_0^{l+D_{fl}} = \tilde{\mathbf{c}}_0^{l+D_{fl}} \middle| \mathbf{r}_0^{l+D_{fl}}, \mathbf{q}\right\}. \tag{6.98}$$

Applying Bayes' rule, the probability $\Pr\left\{\mathbf{c}_0^{l+L_{fl}} = \tilde{\mathbf{c}}_0^{l+D_{fl}} | \mathbf{r}_0^{l+D_{fl}}, \mathbf{q}\right\}$ of (6.98) can be expressed as:

$$\Pr\left\{\mathbf{c}_0^{l+D_{fl}} = \tilde{\mathbf{c}}_0^{l+D_{fl}} \middle| \mathbf{r}_0^{l+D_{fl}}, \mathbf{q}\right\} = \frac{f_{\mathbf{r}}\left(\mathbf{r}_0^{l+D_{fl}} \middle| \tilde{\mathbf{c}}_0^{l+D_{fl}}, \mathbf{q}\right)}{f_{\mathbf{r}}\left(\mathbf{r}_0^{l+D_{fl}} \middle| \mathbf{q}\right)} \Pr\left\{\mathbf{c}_0^{l+D_{fl}} = \tilde{\mathbf{c}}_0^{l+D_{fl}}\right\}, \tag{6.99}$$

where

$$\Pr\left\{\mathbf{c}_0^{l+D_{fl}} = \tilde{\mathbf{c}}_0^{l+D_{fl}}\right\} = \prod_{m=0}^{l+D_{fl}} \Pr\left\{c_m = \tilde{c}_m\right\} \tag{6.100}$$

is the a priori probability of the symbol vector $\tilde{\mathbf{c}}_0^{l+D_{fl}}$ and the conditional pdf $f_{\mathbf{r}}\left(\mathbf{r}_0^{l+D_{fl}} | \mathbf{q}\right)$ can be evaluated as:

$$f_{\mathbf{r}}\left(\mathbf{r}_0^{l+D_{fl}} | \mathbf{q}\right) = \sum_{\tilde{\mathbf{c}}_0^{l+D_{fl}}} f_{\mathbf{r}}\left(\mathbf{r}_0^{l+D_{fl}} \middle| \tilde{\mathbf{c}}_0^{l+D_{fl}}, \mathbf{q}\right) \Pr\left\{\mathbf{c}_0^{l+D_{fl}} = \tilde{\mathbf{c}}_0^{l+D_{fl}}\right\}. \tag{6.101}$$

Then substituting (6.101) into (6.99) produces:

$$\Pr\left\{ \mathbf{c}_0^{l+L_{fl}} = \tilde{\mathbf{c}}_0^{l+L_{fl}} \middle| \mathbf{r}_0^{l+L_{fl}}, \mathbf{q} \right\}$$

$$= \frac{f_{\mathbf{r}}\left(\mathbf{r}_0^{l+L_{fl}} \middle| \tilde{\mathbf{c}}_0^{l+L_{fl}}, \mathbf{q} \right) \Pr\left\{ \mathbf{c}_0^{l+L_{fl}} = \tilde{\mathbf{c}}_0^{l+L_{fl}} \right\}}{\sum_{\tilde{\mathbf{c}}_0^{l+L_f}} f_{\mathbf{r}}\left(\mathbf{r}_0^{l+L_{fl}} \middle| \tilde{\mathbf{c}}_0^{l+L_{fl}}, \mathbf{q} \right) \Pr\left\{ \mathbf{c}_0^{l+L_{fl}} = \tilde{\mathbf{c}}_0^{l+L_{fl}} \right\}}, \tag{6.102}$$

so that $\Pr\left\{ c_l = \tilde{c}_l \middle| \mathbf{r}_0^{l+D_{fl}}, \mathbf{q} \right\}$ can be rewritten as (see (6.98)):

$$\Pr\left\{ c_l = \tilde{c}_l \middle| \mathbf{r}_0^{l+D_{fl}}, \mathbf{q} \right\}$$

$$= \frac{\sum_{\tilde{\mathbf{c}}_0^{l-1}, \tilde{\mathbf{c}}_{l+1}^{l+D_{fl}}} f_{\mathbf{r}}\left(\mathbf{r}_0^{l+D_{fl}} \middle| \tilde{\mathbf{c}}_0^{l+D_{fl}}, \mathbf{q} \right) \Pr\left\{ \mathbf{c}_0^{l+D_{fl}} = \tilde{\mathbf{c}}_0^{l+D_{fl}} \right\}}{\sum_{\tilde{\mathbf{c}}_0^{l+L_f}} f_{\mathbf{r}}\left(\mathbf{r}_0^{l+D_{fl}} \middle| \tilde{\mathbf{c}}_0^{l+D_{fl}}, \mathbf{q} \right) \Pr\left\{ \mathbf{c}_0^{l+D_{fl}} = \tilde{\mathbf{c}}_0^{l+D_{fl}} \right\}}. \tag{6.103}$$

We note the following observations:

1. In equation (6.103) the probability $\Pr\left\{ \mathbf{c}_0^{l+D_{fl}} = \tilde{\mathbf{c}}_0^{l+D_{fl}} \right\}$ does not necessarily take on the same value for all possible trial sequences $\left\{ \tilde{\mathbf{c}}_0^{l+D_{fl}} \right\}$, because of the possible presence of training or pilot symbols inserted in the transmitted stream of channel symbols to solve the *phase ambiguity problem* for the signal constellation at the receiver.

2. The quantities $\left\{ f_{\mathbf{r}}(\mathbf{r}_0^{l+D_{fl}} | \tilde{\mathbf{c}}_0^{l+D_{fl}}, \mathbf{q}) \Pr\left\{ \mathbf{c}_0^{l+D_{fl}} = \tilde{\mathbf{c}}_0^{l+D_{fl}} \right\} \right\}$ represent a set of $M^{l+D_{fl}+1}$ *sufficient statistics* for taking a MAP decision on c_l.

The evaluation of such statistics can be accomplished by means of a *recursive procedure* [1086]. In fact, if it is assumed that these quantities are available at the previous step (i.e., that the set of metrics $\{ f_{\mathbf{r}}(\mathbf{r}_0^{l+D_{fl}-1} | \tilde{\mathbf{c}}_0^{l+D_{fl}-1}, \mathbf{q}) \Pr\{ \mathbf{c}_0^{l+D_{fl}-1} = \tilde{\mathbf{c}}_0^{l+D_{fl}-1} \} \}$ are known to the receiver) and that the new sample $r_{l+L_{fl}}$ has been observed, the new metric $f_{\mathbf{r}}(\mathbf{r}_0^{l+D_{fl}} | \tilde{\mathbf{c}}_0^{l+D_{fl}}, \mathbf{q}) \Pr\{ \mathbf{c}_0^{l+D_{fl}} = \tilde{\mathbf{c}}_0^{l+D_{fl}} \}$ can be evaluated as:

$$f_{\mathbf{r}}\left(\mathbf{r}_0^{l+D_{fl}} \middle| \tilde{\mathbf{c}}_0^{l+D_{fl}}, \mathbf{q} \right) \Pr\left\{ \mathbf{c}_0^{l+D_{fl}} = \tilde{\mathbf{c}}_0^{l+D_{fl}} \right\}$$

$$= f_{\mathbf{r}}\left(\mathbf{r}_0^{l+D_{fl}-1}, r_{l+D_{fl}} \middle| \tilde{\mathbf{c}}_0^{l+D_{fl}-1}, \tilde{c}_{l+D_{fl}}, \mathbf{q} \right) \Pr\left\{ \mathbf{c}_0^{l+D_{fl}-1} = \tilde{\mathbf{c}}_0^{l+D_{fl}-1}, c_{l+D_{fl}} = \tilde{c}_{l+D_{fl}} \right\}$$

$$= f_{\mathbf{r}}\left(r_{l+D_{fl}} \middle| \tilde{\mathbf{c}}_0^{l+D_{fl}-1}, \tilde{c}_{l+D_{fl}}, \mathbf{q} \right) \Pr\left\{ c_{l+D_{fl}} = \tilde{c}_{l+D_{fl}} \right\}$$

$$\cdot \; f_{\mathbf{r}}\left(\mathbf{r}_0^{l+D_{fl}-1} \middle| \tilde{\mathbf{c}}_0^{l+D_{fl}-1}, \mathbf{q} \right) \Pr\left\{ \mathbf{c}_0^{l+D_{fl}-1} = \tilde{\mathbf{c}}_0^{l+D_{fl}-1} \right\}, \tag{6.104}$$

since the components of $\mathbf{r}_0^{l+D_{fl}}$, conditioned on the transmitted data sequence $\tilde{\mathbf{c}}_0^{l+D_{fl}}$, are independent random variables, because the noise samples affecting them are statistically independent. Let us now assume that the algorithm lag D_{fl} is larger than the channel memory, that is, that[5]:

$$D_{fl} \geq L_q - 1. \tag{6.105}$$

---

[5] If the channel memory is longer than the decision delay, this recursive procedure can still be applied with slight modifications. In this case decision feedback may be required to reduce the computational complexity of the algorithm [1086].

Then, the conditional pdf $f_r\left(r_{l+D_{fl}}|\tilde{\mathbf{c}}_0^{l+D_{fl}-1}, \tilde{c}_{l+D_{fl}}, \mathbf{q}\right)$ appearing in the RHS of (6.104) can be simplified as:

$$f_r\left(r_{l+D_{fl}}\left|\tilde{\mathbf{c}}_0^{l+D_{fl}-1}, \tilde{c}_{l+D_{fl}}, \mathbf{q}\right.\right) = f_r\left(r_{l+D_{fl}}\left|\tilde{\mathbf{c}}_l^{l+D_{fl}}, \mathbf{q}\right.\right)$$

$$= \frac{1}{\pi\sigma_n^2}\exp\left[-\frac{1}{\sigma_n^2}\left|r_{l+D_{fl}} - \sum_{m=0}^{L_q-1}\tilde{c}_{l+D_{fl}-m}\,q_m\right|^2\right], \qquad (6.106)$$

since the received sample $r_{l+D_{fl}}$ depends on the symbol vector $\tilde{\mathbf{c}}_l^{l+D_{fl}}$ (i.e., on $(D_{fl}+1)$ symbols) at most. Therefore, on the basis of (6.106), (6.104) can be rewritten as:

$$f_{\mathbf{r}}\left(\mathbf{r}_0^{l+D_{fl}}\left|\tilde{\mathbf{c}}_0^{l+D_{fl}}, \mathbf{q}\right.\right)\Pr\left\{\mathbf{c}_0^{l+D_{fl}} = \tilde{\mathbf{c}}_0^{l+D_{fl}}\right\}$$

$$= f_r\left(r_{l+D_{fl}}\left|\tilde{\mathbf{c}}_l^{l+D_{fl}}, \mathbf{q}\right.\right)\Pr\left\{c_{l+D_{fl}} = \tilde{c}_{l+D_{fl}}\right\}$$

$$\cdot f_{\mathbf{r}}\left(\mathbf{r}_0^{l+D_{fl}-1}\left|\tilde{\mathbf{c}}_0^{l+D_{fl}-1}, \mathbf{q}\right.\right)\Pr\left\{\mathbf{c}_0^{l+D_{fl}-1} = \tilde{\mathbf{c}}_0^{l+D_{fl}-1}\right\}, \qquad (6.107)$$

so that, iterating, the factorization:

$$f_{\mathbf{r}}\left(\mathbf{r}_0^{l+D_{fl}}\left|\tilde{\mathbf{c}}_0^{l+D_{fl}}, \mathbf{q}\right.\right)\Pr\left\{\mathbf{c}_0^{l+D_{fl}} = \tilde{\mathbf{c}}_0^{l+D_{fl}}\right\}$$

$$= f_{\mathbf{r}}\left(\mathbf{r}_0^{D_{fl}-1}\left|\tilde{\mathbf{c}}_0^{D_{fl}-1}, \mathbf{q}\right.\right)\Pr\left\{\mathbf{c}_0^{D_{fl}-1} = \tilde{\mathbf{c}}_0^{D_{fl}-1}\right\}$$

$$\cdot \prod_{k=0}^{l} f_r\left(r_{k+D_{fl}}\left|\tilde{\mathbf{c}}_k^{k+D_{fl}-1}, \tilde{c}_{k+D_{fl}}, \mathbf{q}\right.\right)\Pr\left\{c_{k+D_{fl}} = \tilde{c}_{k+D_{fl}}\right\} \qquad (6.108)$$

can easily be inferred. This result proves that the sufficient statistics $\{f_{\mathbf{r}}(\mathbf{r}_0^{l+D_{fl}}|\tilde{\mathbf{c}}_0^{l+D_{fl}}, \mathbf{q})\Pr\{\mathbf{c}_0^{l+D_{fl}} = \tilde{\mathbf{c}}_0^{l+D_{fl}}\}\}$ in (6.103) can be evaluated via a recursive procedure, which can be interpreted as a *forward recursion* accomplished over an $M^{D_{fl}}$-state trellis (when processing the sample $r_{k+D_{fl}}$, the state is identified by the vector $\tilde{\mathbf{c}}_k^{k+D_{fl}-1}$). Note the following:

(a) A trial trellis state $\tilde{\Delta}_k$ in the kth interval (during which the sample $r_{k+D_{fl}}$ is processed) is uniquely identified by $\tilde{\mathbf{c}}_k^{k+D_{fl}-1}$, that is, by the ordered collection of the trial channel symbols $\{\tilde{c}_l, \ l = k, k+1, \ldots, k+D_{fl}-1\}$.

(b) In the above-mentioned procedure the *metric* associated with the state transition $(\tilde{\Delta}_l, \tilde{\Delta}_{l+1})$ is given by (see (6.108)):

$$\gamma_l(\tilde{\Delta}_l, \tilde{\Delta}_{l+1}) = f_r(r_{l+D_{fl}}|\tilde{\mathbf{c}}_l^{l+D_{fl}-1}, \tilde{c}_{l+D_{fl}}, \mathbf{q})\Pr\{c_{l+D_{fl}} = \tilde{c}_{l+D_{fl}}\}. \qquad (6.109)$$

(c) This forward recursion is the same as the forward recursion in the FBA, but is now accomplished over a trellis with an *enlarged* number of states when inequality (6.105) holds.

The algorithm derived above was originally proposed in [1086, 1087]. Its main drawback is its memory and computational requirements, which grow exponentially with the lag D_{fl} (in practice, a decision delay equal to $5L_q$ is usually sufficient). This has motivated the investigation of other constrained delay MAP algorithms. Two significant alternatives to the MAP-FLA of [1086] have been devised in [1088], where the so-called *optimum soft output algorithm* (OSA) and *soft output algorithm* (SSA) are derived. In particular, the OSA is an improved version of the MAP-FLA of [1086] in the sense that it generates *optimum* soft outputs via only a forward recursion, but the number of quantities to be stored and recursively updated increases linearly, rather than exponentially, with the decision delay. In contrast, suboptimum performance is offered by the SSA, which, however, offers the following advantages compared to the OSA and other MAPSD algorithms: (a) knowledge of the noise variance σ_n^2 is not required; (b) computations are in the logarithmic domain and, as in the VA, the main operation is *add–compare–select*. As the degradation of the SSA from the optimum performance is negligible and the computational burden is much lower than those of the MAP algorithms, the SSA is a very practical soft output detection algorithm.

Further results on MAP forward-only detection algorithms can be found in [1089–1096]. In particular, [1091, 1093, 1094] investigate *Bayesian decision feedback estimation* (BDFE), which results from employing decision feedback recursively in fixed-lag MAP symbol detection. BDFE schemes interpolate between ordinary decision feedback equalization and MAP estimation in both performance and complexity [1093].

Comparison of MLSD and MAP Detection Algorithms

We now compare the advantages offered by the VA-based MLSD, the MAP-FLA and the MAP-FBA. To begin, we note that each of these three algorithms can be employed for the detection of data sequences in the presence of ISI. Among these algorithms, however, the VA offers the smallest computational complexity and this makes it the most appealing choice in both uncoded and coded communication systems that do not require the availability of soft output information at the equalizer output. In contrast, the MAP equalization algorithms illustrated above have the following significant disadvantages: (a) they require knowledge of the noise variance in the evaluation of their metrics (see (6.89) and (6.106)); (b) they have large memory and computation requirements.

In particular, as already stated above, the two MAP algorithms that we have considered accomplish computations in the probability, instead of the logarithm domain, and consequently require a large number of multiplications and exponentiations. In addition, the following specific observations are in order:

1. The MAP-FBA performs two recursions and, consequently, operates in a block mode. In practice, it memorizes all the received signal samples of each block and then processes it, so that its memory requirements grow *linearly* with the sequence length N. Consequently, it is suitable for processing short data sequences. Note also that, if the data sequence to be estimated by the MAP-FBA is long, the CIR cannot be usually assumed constant over the block duration. This requires combining channel tracking with soft detection, further increasing the receiver complexity (more details on this topic can be found in Section 5.1.3). In fact, standard decision aided algorithms for channel estimation (see Chapter 5) cannot be employed in this case as reliable data decisions are available only at the end of both recursions.

2. The MAP-FLA requires only a forward recursion, so that it can operate in a continuous mode. In this case, adaptivity to channel variations can be more easily embedded in the detection algorithm because of the short and fixed decision delay (see Section 6.2.1.5 for further details). Both the memory and computational requirements, however, grow exponentially with the decision delay so that this parameter should be kept to a minimum. For this reason, the FBA may have smaller computational requirements than the FLA [1088].

The most relevant features of the three algorithms considered are summarized in Table 6.1; here N is the sequence length, and L_{fl} and D_{VA} are the decision delay for the MAP-FLA and VA, respectively. Usually, the inequality:

$$L_q \leq \frac{D_{VA}}{D_{fl}} \leq N \tag{6.110}$$

holds.

Table 6.1   Main features of the VA-based MLSD, the MAP-FBA and the MAP-FLA

Features	MAP-FBA	MAP-FLA	MLSD (VA)
Minimization	symbol error	symbol error	sequence error
Decision type	soft	soft	hard
A priori information	noise variance	noise variance	–
Recursion requirement	forward and backward	forward	forward
Memory requirement	$\propto N \cdot M^{L_q}$	$\propto M^{D_{fl}}$	$\propto D_{VA} \cdot M^{L_q}$
Computation requirement	$\propto N \cdot M^{L_q}$	$\propto N \cdot M^{D_{fl}}$	$\propto N \cdot M^{L_q}$

Despite their significant computational complexity, significant research efforts have been devoted to the study and development of MAPSD algorithms in recent years. This interest is due to the fact that digital receivers may require MAP algorithms for efficient decoding and/or equalization (see Chapters 10 and 12); in fact, this class of algorithms is able to incoporate a priori (soft) probabilistic information about bits/symbols and to generate new a posteriori (soft) probabilistic information about them; in other words, MAPSD algorithms can be employed as SiSo detection/decoding in a number of applications. Particular attention has been devoted to the MAP-FBA (i.e., BCJR), for which various approximations and modifications have also been developed to reduce its complexity. The various strategies proposed for complexity reduction can be divided in two different classes, the first inspired by RSSD, the second by reduced-search algorithms operating over a full trellis (Section 6.2.1.2). The first class includes, for instance, the so-called *reduced-state* BCJR (RS-BCJR) algorithm developed in [1097]; in this case assuming that only a part of the information corresponding to the full state is embedded in a properly defined *reduced state*, the missing information is recovered by decision feedback, in a way similar to RSSD. The RS-BCJR algorithm is particularly efficient on minimum-phase channels or, more generally, when the first channel taps are much larger in magnitude than the others. A significant example of the second class is the so-called M-BCJR algorithm [1098], in which the forward recursion (6.91) on $\{\alpha_{m-1}(\tilde{\Delta}_{m-1})\}$ that produces $\{\alpha_m(\tilde{\Delta}_m)\}$ is performed using only the M largest components of $\{\alpha_{m-1}(\tilde{\Delta}_{m-1})\}$ (the rest are declared dead and set to 0). In principle, the same scheme can be applied with the backward recursion, but since the quantities $\{\sigma_l(\tilde{\Delta}_l, \tilde{\Delta}_{l+1})\}$ are products of the $\{\alpha_l(\tilde{\Delta}_l)\}$ and $\{\beta_{l+1}(\tilde{\Delta}_{l+1})\}$ components, it is simpler just to execute the backward recursion on the region of the trellis where the forward elements are alive. The performance gap between the (optimal) MAP-FBA and its reduced-complexity counterparts is considerably influenced by the CIR properties and may be significant, as evidenced by the following example.

Example 6.2.2 Let us consider an uncoded transmission over a mixed-phase communication channel characterized by the CIR $\mathbf{q} \triangleq [1, 1, 1, 1, 1]^T$ (so that $L_q = 5$). The modulation selected is a QPSK with Gray mapping, thus $M = 4$ and the number of states in the complete trellis is $N_s = 4^5 = 1024$. In Figure 6.9 the BER performance achieved by the MAP-FBA is compared with that offered by the RS-BCJR operating over a 64-state trellis and of an M-BCJR algorithm using only the 16 most significant elements of the set $\{\alpha_{m-1}(\tilde{\Delta}_{m-1})\}$ in the forward recursion (6.91). These results show that M-BCJR offers an error performance very close to that of the MAP-FBA, while the performance of the RS-BCJR algorithm is quite inferior (in this case it is found that the RS-BCJR does not converge to that of the full-complexity algorithm even when keeping 64 elements in the forward recursion). Note that the M-BCJR algorithm is particularly efficient since we are considering an uncoded transmission; further results show that the error performance worsens when this algorithm is employed in coded systems [1099], since it does not ensure high-quality soft outputs. Finally, a better performance/complexity tradeoff than the RS-BCJR and M-BCJR is offered by other algorithms available in the technical literature (e.g., see [1099]).

□

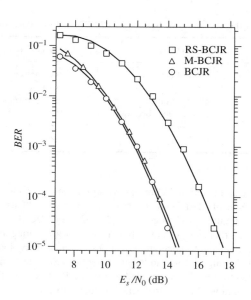

Figure 6.9 BER performance of various detection algorithms for an uncoded QPSK transmission.

An alternative to MAP-FBA and its reduced-complexity versions is a modified version of the VA, generating probabilistic (i.e., soft) information about transmitted data and known as the *soft output* VA (SOVA) [1100]. The reader can refer to Section 9.2.4 for further details on soft output trellis-based detection/decoding algorithms.

6.2.1.6 MLSD Over Slow Frequency-Flat Fading

The optimal detection strategies described above mainly refer to PAM signaling over a known *frequency-selective* channel. In contrast, in this subsection we focus on the optimal detection of PAM signals over a known slow *frequency-flat* fading channel. We assume that model (4.108) holds for the received signal samples and, for generality, that the receiver is equipped with an antenna array, so that n_R faded replicas $\{r^{(l)}(t),\ l = 0, 1, \ldots, n_R - 1\}$ of transmitted signal are available for data detection [13, 19]. Then the kth sample at the matched filter output of the lth branch of the receiver can be expressed as:

$$r_k^{(l)} = \alpha_k^{(l)}\, c_k + n_k^{(l)}, \tag{6.111}$$

with $l = 0, 1, \ldots, n_R - 1$; here, $\alpha_k^{(l)}$ represents the fading distortion over the lth branch in the kth symbol interval, c_k is the kth transmitted symbol and $n_k^{(l)}$ is a complex Gaussian noise sample with variance $\sigma_{n,l}^2$. Under these assumptions, the ML decision rule for the transmitted symbol vector $\mathbf{c}_N \triangleq [c_0, c_1, \ldots, c_{N-1}]^T$ is given by (see (4.26) and Section 4.3.4):

$$\hat{\mathbf{c}}_N = \arg \max_{\tilde{\mathbf{c}}_N} f_{\mathbf{r}}\left(\boldsymbol{\rho}_0, \boldsymbol{\rho}_1, \ldots, \boldsymbol{\rho}_{n_R-1} | \tilde{\mathbf{c}}_N, \boldsymbol{\alpha}_0, \boldsymbol{\alpha}_1, \ldots, \boldsymbol{\alpha}_{n_R-1}\right), \tag{6.112}$$

where $\boldsymbol{\rho}_l \triangleq [\rho_0^{(l)}, \rho_1^{(l)}, \ldots, \rho_{N-1}^{(l)}]^T$ denotes the value taken on by the vector $\mathbf{r}_l \triangleq [r_0^{(l)}, r_1^{(l)}, \ldots, r_{N-1}^{(l)}]^T$ (with $l = 0, 1, \ldots, n_R - 1$), $\boldsymbol{\alpha}_l \triangleq [\alpha_0^{(l)}, \alpha_1^{(l)}, \ldots, \alpha_{N-1}^{(l)}]^T$ (with $l = 0, 1, \ldots, n_R - 1$) and $\mathbf{r}$ is the $N \cdot n_R$-dimensional vector resulting from the ordered concatenation of the vectors

$\{\mathbf{r}_l, \ l = 0, 1, \ldots, n_R - 1\}$. If the noise processes $\{\{n_k^{(l)}\}, \ l = 0, 1, \ldots, n_R\}$ are assumed mutually independent, the decision rule (6.112) turns into:

$$\hat{\mathbf{c}}_N = \arg \max_{\tilde{\mathbf{c}}_N} \prod_{k=0}^{n_R-1} f_{\mathbf{r}_k}\left(\rho_k | \tilde{\mathbf{c}}_N, \boldsymbol{\alpha}_k\right) \qquad (6.113)$$

or, equivalently, into (see (4.145)):

$$\hat{\mathbf{c}}_N = \arg \min_{\tilde{\mathbf{c}}_N} \sum_{k=0}^{n_R-1} \left(\boldsymbol{\rho}_k - \mathbf{H}_k \, \tilde{\mathbf{c}}_N\right)^H \mathbf{R_n}^{-1}(l) \left(\boldsymbol{\rho}_k - \mathbf{H}_k \, \tilde{\mathbf{c}}_N\right), \qquad (6.114)$$

where $\mathbf{R_n}(l)$ is the noise covariance matrix for the lth branch and $\mathbf{H}_l$ is a diagonal matrix with the elements of $\boldsymbol{\alpha}_l$ along its main diagonal (with $l = 0, 1, \ldots, n_R - 1$). If the noise samples on each diversity branch are independent, that is, $\mathbf{R_n}(l) = \sigma_n^2(l) \, \mathbf{I}_N$, the decision rule (6.114) becomes:

$$\hat{\mathbf{c}}_N = \arg \min_{\tilde{\mathbf{c}}_N} \sum_{k=0}^{n_R-1} \frac{1}{\sigma_{n,l}^2} \left|\boldsymbol{\rho}_k - \mathbf{H}_k \, \tilde{\mathbf{c}}_N\right|^2 \qquad (6.115)$$

or, equivalently:

$$\hat{c}_k = \arg \min_{\tilde{c}_k} \sum_{l=0}^{n_R-1} \frac{1}{\sigma_{n,l}^2} \left|\rho_k^{(l)} - \alpha_k^{(l)} \, \tilde{c}_k\right|^2 \qquad (6.116)$$

with $k = 0, 1, \ldots, N - 1$. Dropping irrelevant terms in (6.116) yields:

$$\hat{c}_k = \arg \min_{\tilde{c}_k} \sum_{l=0}^{n_R-1} \frac{1}{\sigma_{n,l}^2} \left[\left|\alpha_k^{(l)}\right|^2 \left|\tilde{c}_k\right|^2 - 2 \operatorname{Re}\left(\rho_k^{(l)} \, \alpha_k^{(l)*} \, \tilde{c}_k^*\right)\right], \qquad (6.117)$$

with $k = 0, 1, \ldots, N - 1$. If the transmitted symbol belongs to a PSK constellation, the strategy (6.117) can also be rewritten as:

$$\hat{c}_k = \arg \max_{\tilde{c}_k} \operatorname{Re}\left[Z_k \, \tilde{c}_k^*\right], \qquad (6.118)$$

with $k = 0, 1, \ldots, N - 1$, where:

$$Z_k \triangleq \sum_{l=0}^{n_R-1} g_k^{(l)} \, \rho_k^{(l)} \qquad (6.119)$$

and

$$g_k^{(l)} \triangleq \frac{\alpha_k^{(l)*}}{\sigma_{n,l}^2}. \qquad (6.120)$$

The algorithm (6.118) linearly combines (with coefficients (6.120)) all the MF outputs in the kth symbol interval to take an optimal decision on c_k, as illustrated in Figure 6.10. It is worth noting the following:

1. Multiplying the sample $\rho_k^{(l)}$ by $\alpha_k^{(l)*}$ removes the channel phase distortion from the received signal. Therefore, the detection algorithm based on (6.118)–(6.120) can be interpreted as a form of *coherent detection*.
2. In the evaluation of the sufficient statistic Z_k (6.119) a larger weight is given to less noisy channels.

Concerning the last point it can be proved [13] that the choice of the coefficients expressed by (6.120), among all those possible, *maximizes the* SNR *at the decision device input*. For this reason

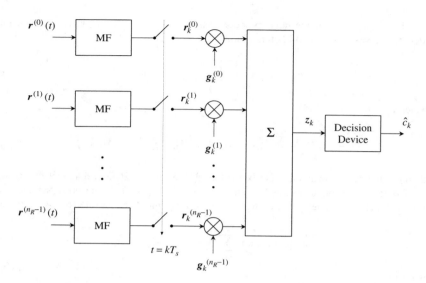

Figure 6.10   Block diagram of a coherent detector based on MRC or EGC.

equations (6.118)–(6.120) express the so-called *maximal ratio combining* (MRC) detection strategy [13, 1101]. The use of the MRC requires knowledge of both the channel gains $\{\alpha_k^{(l)}\}$ and the noise statistics $\{\sigma_{n,l}^2\}$. If the latter quantities are unknown, the choice:

$$g_k^{(l)} \triangleq \alpha_k^{(l)*} \tag{6.121}$$

can be made instead of (6.120); this leads to the so-called *equal gain combining* (EGC) strategy, which is suboptimal.

The coherent combining of the signals coming from multiple antennas results in an increase of the average SNR available at the receiver; such an increase is proportional to n_R and represents the so-called *array gain* [38]. However, the enhancement deriving from the adoption of receive diversity can be fully appreciated by considering asymptotic error performance in fading channels. In fact, generally speaking, at high SNRs the average SER P_e of an uncoded (or coded) data communication system operating over a fading channel can be approximated in various scenarios as (see [1102] and references therein):

$$P_e \approx \left(G_c \cdot \bar{\gamma}\right)^{-G_d}, \tag{6.122}$$

where G_c and G_d are the *coding gain* and *diversity gain* (or *diversity order*), and $\bar{\gamma}$ denotes the average SNR. Formula (6.121) shows that the diversity gain G_d plays a fundamental role since it determines the *slope* of the SER curve versus $\bar{\gamma}$ on a log-log scale for vanishing noise (high SNR). In contrast, G_c, if expressed in decibels, determines the horizontal shift of the BEP curve in SNR with respect to the benchmark SER curve associated with $P_e \approx (\bar{\gamma})^{-G_d}$. Formula (6.122) applies, for instance, when EGC or MRC are exploited in the case of PAM signaling over Rayleigh fading; in this case it can be shown that both combining strategies are characterized by the same diversity gain $G_d = n_R$, but offer different coding gains.

An alternative to ECG and MRC is the so-called *selection combining*, in which the branch signal with the largest instantaneous SNR is selected for demodulation, so that the output SNR is equal to that of the best incoming signal. This strategy can be generalized to combine n of n_R branches with the largest amplitudes (multiple order selection combining). The reader can refer to [1103], which evaluates and compares the error performance offered by the above-mentioned methods of diversity combining for a Rayleigh-faded channel, [13], which gives an accurate analysis in terms of SNR for MRC and ECG; [598], which derives probability formulas for FSK and PSK signals with MRC and ECG; and [31, 1104], which present a unified approach to evaluating the error performance of digital communication systems with diversity for both coherent and noncoherent detection.

Finally, it is important to mention that all the combining strategies mentioned above exploit the *explicit* diversity provided by independently fading replicas of the same transmitted signal. Such replicas can be acquired by resorting to different mechanisms, including space diversity (spaced antennas), frequency diversity (transmission of the same signal in different bandwidths), angle (of arrival) diversity, polarization diversity, time diversity (signal repetition). Note, however, that a fading channel can be interpreted itself as a source of *implicit* diversity [101] to be related to both the phenomena of time dispersion [425, 466, 1105, 1106] and time variation of the channel [1107]. In particular, exploiting the time diversity on time-selective fading channels can provide a substantial performance improvement [111, 112].

6.2.2 *Channel Equalization in the Frequency Domain*

Frequency-domain equalizers (FDEs) provide a low-complexity approach to ISI mitigation in SC communications. Systems employing FD equalization are closely related to OFDM systems. In fact, in both cases digital transmission is carried out blockwise and relies on DFT/IDFT operations. Therefore, SC systems employing FDEs enjoy a similar complexity advantage to OFDM systems without the stringent requirements of highly-accurate frequency synchronization and linear power amplification as in OFDM. In addition, recent results indicate that SC systems with FD equalization can exhibit performance similar to or better than coded OFDM systems in some scenarios [1108]. It is also well known that: (a) FDEs usually require a substantially lower computational complexity than their TD counterparts, in both filter synthesis and data processing; (b) FD equalization can outperform TD equalization for a given computational complexity (see, for instance, [1109]) – in other words, a FDE requires fewer taps with respect to a TDE in order to achieve a given error performance. This can be related to the fact that in a block data transmission based on a PAM format and incorporating a *cyclic prefix* (CP) of proper duration in each block (see Section 3.5.3), the resulting channel matrix characterizing the received signal model is square *circulant*; in other words, if we refer to model (4.96) for the received vector $\mathbf{r} = \left[r_0, r_1, \ldots, r_{N-1} \right]^T$:

$$\mathbf{r} = \mathbf{H}\,\mathbf{c}_N + \mathbf{n}, \qquad (6.123)$$

the $N \times N$ CIR matrix $\mathbf{H} = [h_{i,j}]$ is characterized by the fact that $h_{i,j} = h_{i-j}$. Consequently, this matrix can be decomposed as [396]:

$$\mathbf{H} = \mathbf{Q}_N^H \mathbf{D} \mathbf{Q}_N, \qquad (6.124)$$

where $\mathbf{D} = \mathrm{diag}(D_k)$ is a diagonal $N \times N$ matrix and:

$$D_k = \sum_{l=0}^{L_h-1} h_l \exp\left(\frac{-j2\pi kn}{N} \right) \qquad (6.125)$$

is the kth coefficient of the CIR DFT with $k = 0, 1, \ldots, N - 1$ (L_h denotes the CIR duration). Then, evaluating an N-point DFT of $\mathbf{r}$ generates the FD vector:

$$\mathbf{R} \triangleq \mathbf{Q}_N\,\mathbf{r} = \mathbf{Q}_N \mathbf{H}\,\mathbf{c}_N + \mathbf{Q}_N \mathbf{n}$$

$$= \mathbf{Q}_N \mathbf{Q}_N^H \mathbf{D} \mathbf{Q}_N\,\mathbf{c}_N + \mathbf{Q}_N \mathbf{n}$$

$$= \mathbf{D}\mathbf{C}_N + \mathbf{W}, \qquad (6.126)$$

where $\mathbf{W} \triangleq \mathbf{Q}_N \mathbf{n}$ is a complex Gaussian noise vector statistically equivalent to $\mathbf{n}$ and $\mathbf{C}_N \triangleq \mathbf{Q}_N \mathbf{c}_N$ is the FD representation of the symbol vector $\mathbf{c}_N$. Note that the effect of the channel distortion in $\mathbf{R}$ (6.126) is represented by the *diagonal* matrix $\mathbf{D}$; this suggests that, if $\mathbf{D}$ (i.e., the CIR) is known, frequency selectivity can be easily compensated for in the FD; then, an estimate of $\mathbf{c}_N$ can be generated, getting back to the TD via an IDFT. These principles are exploited in the rest of this subsection to derive linear and decision feedback equalization strategies which process two samples per channel symbol.

6.2.2.1   Linear Equalization in the Frequency Domain

Similarly to TD equalizers, FD equalizers can process one or more samples per channel symbol. Unlike [1110, 1111], which consider an LE in the FD (FD-LE) processing one sample per channel symbol, here, following [435], we derive an algorithm for MMSE linear equalization in the FD processing two samples per channel symbol, starting from the signal model developed in Section 4.4.2 for the case of frequency-selective fading. In particular, we start our derivation by considering the vector $\mathbf{R}_{2N}^{(l)} \triangleq [R_0^{(l)}, R_1^{(l)}, \ldots, R_{2N-1}^{(l)}]^T$ generated by a $2N$-point DFT receiver in the lth block interval; such a vector can be expressed as (see (4.99)):

$$\mathbf{R}_{2N}^{(l)} = \mathbf{D}^{(l)} \, \mathbf{C}_{2N}^{(l)} + \mathbf{V}_N^{(l)}, \tag{6.127}$$

where $\mathbf{C}_{2N}^{(l)} \triangleq [(\mathbf{C}_N^{(l)})^T | (\mathbf{C}_N^{(l)})^T]^T$, $\mathbf{C}_N^{(l)} \triangleq [C_0^{(l)}, C_1^{(l)}, \ldots, C_{N-1}^{(l)}]^T \triangleq \mathbf{Q}_N \mathbf{c}_N^{(l)}$ is the vector resulting from the DFT of the lth data block $\mathbf{c}_N^{(l)}$, $\mathbf{V}_N^{(l)} \triangleq [V_0^{(l)}, V_1^{(l)}, \ldots, V_{2N-1}^{(l)}]^T$ is a vector of independent Gaussian noise samples (each having mean zero and variance $\sigma_n^2 = 2N_0$), $\mathbf{D}^{(l)} \triangleq \mathrm{diag}(D_k^{(l)})$ with $D_k^{(l)} = P_k H_k^{(l)} (P_{k-2N} H_{k-2N}^{(l)})$ for $k = 0, 1, \ldots, N$ $(k = N+1, N+2, \ldots, 2N-1)$, $P_k \triangleq P(k/NT_s)/\sqrt{T_s}$, $H_k^{(l)} \triangleq H^{(l)}(k/NT_s)$, and $P(f)$ and $H^{(l)}(f)$ are the FCTs of the transmit pulse $p(t)$ and the CIR $h^{(l)}(t)$ in the observation interval considered.

The FD-LE processing can be summarized as follows (see Figure 6.11(a)). The received vector $\mathbf{R}_{2N}^{(l)}$ (6.127) is filtered by an $N \times 2N$ forward matrix $\mathbf{D}_{LE}^{(l)}$ combining the elements of $\mathbf{R}_{2N}^{(l)}$ to produce

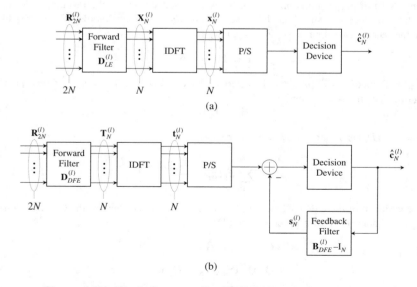

(a)

(b)

Figure 6.11   Block diagram of an FD-LE (a) and an FD-DFE (b).

an N-dimensional vector $\mathbf{X}_N^{(l)}$ which feeds an N-point IDFT block. The N samples at the IDFT output are collected in a vector $\mathbf{x}_N^{(l)}$; after parallel-to-serial conversion, this is applied to a decision (threshold) device operating on a symbol-by-symbol basis and producing the hard decision vector $\hat{\mathbf{c}}_N^{(l)}$. The derivation of the optimal matrix $\mathbf{D}_{LE}^{(l)}$ in the MMSE sense can be accomplished as follows. First of all, let us define the error (in the following the block index l is dropped for simplicity):

$$\boldsymbol{\delta}_{LE} \triangleq \mathbf{x}_N - \mathbf{c}_N \qquad (6.128)$$

in the TD, and the corresponding error:

$$\boldsymbol{\Delta}_{LE} \triangleq \mathbf{Q}_N \, \boldsymbol{\delta}_{LE} = \mathbf{X}_N - \mathbf{C}_N \qquad (6.129)$$

in the FD, where $\mathbf{X}_N \triangleq \mathbf{Q}_N \, \mathbf{x}_N$. The optimal $\mathbf{D}_{LE}$ minimizes the MSE:

$$\text{MSE} \triangleq \mathrm{E}\left\{\boldsymbol{\delta}_{LE}^H \, \boldsymbol{\delta}_{LE}\right\} = \mathrm{E}\left\{\boldsymbol{\Delta}_{LE}^H \, \boldsymbol{\Delta}_{LE}\right\}. \qquad (6.130)$$

As in the TD case, minimization of the MSE results from the application of the orthogonality principle [503, 1021]; this leads to

$$\mathrm{E}\left\{\boldsymbol{\Delta}_{LE} \, \mathbf{R}_{2N}^H\right\} = \mathbf{0}_{N,2N}, \qquad (6.131)$$

where $\mathbf{0}_{N \times 2N}$ is an $N \times 2N$ null matrix. Substituting (6.127) and (6.129) into (6.131) leads easily to:

$$\mathbf{D}_{LE} = \mathbf{J}\mathbf{D}^H\mathbf{K}, \qquad (6.132)$$

where

$$\mathbf{J} \triangleq \left[\mathbf{I}_N \, \middle| \, \mathbf{I}_N\right], \qquad (6.133)$$

$$\mathbf{K} \triangleq \left[\mathbf{D}\mathbf{J}^H\mathbf{J}\mathbf{D}^H + \frac{\sigma_n^2}{\sigma_c^2} \, \mathbf{I}_{2N}\right]^{-1} \qquad (6.134)$$

and σ_c^2 denotes the variance of the channel symbols $\{c_n^{(l)}\}$ (modeled as zero mean iid random variables). Finally, we note that $\mathbf{D}_{LE}$ can be rewritten as $\mathbf{D}_{LE} = [\mathbf{D}_{LE,0} | \mathbf{D}_{LE,1}]$, where $\mathbf{D}_{LE,m} = \mathrm{diag}(D_{m,k}^{LE})$ and $D_{m,k}^{LE} \triangleq D_{k+mN}^* \, [|D_k|^2 + |D_{k+N}|^2 + \sigma_n^2/\sigma_c^2]^{-1}$ with $k = 0, 1, \ldots, N-1$ and $m = 0, 1$. This leads to the following conclusions: (a) filtering in the proposed FD-LE can be interpreted as a generalization of that employed in an FD-LE operating at symbol rate sampling [1110, 1111]; (b) the forward filter (6.127) coherently combines pairs of components of the vector $\mathbf{R}_{2N}^{(l)}$ (6.127) spaced at N locations, since $X_k = D_{0,k}^{(LE)} \, R_k + D_{1,k}^{(LE)} \, R_{k+N}$, with $k = 0, 1, \ldots, N-1$.

The LE derivation illustrated above for PAM signaling can be extended to the case of CPM signaling, if (a) cyclically-extended blocks of data are properly generated (a detailed description of the structure of the data block is provided in Section 3.6.6) and (b) Laurent's decomposition (see Section 3.6.5.1) with a finite number (P) of components can be adopted for an accurate representation of the selected CPM signal. This is due to the fact that, in this case, filtering the baseband received signal $r(t)$ with a proper low-pass filter (the useful component $z(t)$ of this signal is expressed by (3.199)) and sampling the filter output uniformly at a rate $2/T_s$ results in a $2N$-dimensional received signal vector $\mathbf{Z}^{(l)}$ (when receiving the lth data block), which exhibits a *linear* dependence of the DFTs of P distinct N dimensional blocks of Laurent symbols. The reader can refer to [284] for further mathematical details. Finally, we mention that an FD linear equalization algorithm has been also derived in [306], but, unlike [284], employs a bank of matched filters combined with baud-rate sampling and noise whitening.

6.2.2.2   Decision Feedback Equalization in the Frequency Domain

The block diagram of a DFE operating in the FD (FD-DFE) is shown in Figure 6.11(b). In the DFE structure we develop forward filtering is accomplished in the FD (through the $N \times 2N$ *forward matrix* $\mathbf{D}_{DFE}^{(l)}$), as in the FD-LE, whereas feedback filtering is carried out in the TD (i.e., a *hybrid* equalization strategy is adopted), through a time-varying FIR filter (represented by the $N \times N$ *feedback matrix* $\mathbf{B}_{DFE}^{(l)}$) characterized by a uniform tap spacing and fed by symbol decisions [435]. As shown in Figure 6.11(b), the received vector $\mathbf{R}_{2N}^{(l)}$ (6.127) is filtered by $\mathbf{D}_{DFE}^{(l)}$ (producing the N-dimensional vector $\mathbf{T}_N^{(l)}$) and converted to the TD by an IDFT. This results in the N-dimensional vector $\mathbf{t}_N^{(l)}$, which, after the contribution coming from the decision feedback is subtracted, is processed for data decision. Let us now derive the expressions for the optimal (in the MMSE sense) matrices $\mathbf{D}_{DFE}^{(l)}$ and $\mathbf{B}_{DFE}^{(l)}$ under the assumption that the correct data symbols are processed by the latter matrix (we drop the block index l for simplicity). We start by noting that (see Figure 6.11(b) and (6.127)):

$$\mathbf{t}_N \triangleq \mathbf{Q}_N^H \mathbf{T}_N = \mathbf{Q}_N^H \, \mathbf{D}_{DFE} \, \mathbf{R}_{2N} = \mathbf{Q}_N^H \, \mathbf{D}_{DFE} \, \mathbf{D} \, \mathbf{C}_{2N} + \mathbf{Q}_N^H \, \mathbf{D}_{DFE} \mathbf{V}_N. \tag{6.135}$$

This vector is combined with the feedback vector $\mathbf{s}_N \triangleq (\mathbf{B}_{DFE} - \mathbf{I}_N)\mathbf{c}_N$ (where $\mathbf{B}_{DFE}$ is a lower triangular matrix with 1s along its main diagonal); the result feeds a threshold decision device. Then the error vector $\boldsymbol{\delta}_{DFE}$ in the TD is given by (see (6.128)):

$$\boldsymbol{\delta}_{DFE} \triangleq \mathbf{t}_N - \mathbf{s}_N - \mathbf{c}_N = \mathbf{t}_N - (\mathbf{B}_{DFE} - \mathbf{I}_N)\mathbf{c}_N - \mathbf{c}_N = \mathbf{t}_N - \mathbf{B}_{DFE} \, \mathbf{c}_N \tag{6.136}$$

and, consequently, the error vector $\boldsymbol{\Delta}_{DFE}$ in the FD becomes (see (6.129)):

$$\boldsymbol{\Delta}_{DFE} \triangleq \mathbf{Q}_N \, \boldsymbol{\delta}_{DFE} = \mathbf{T}_N - \mathbf{Q}_N \, \mathbf{B}_{DFE} \, \mathbf{c}_N, \tag{6.137}$$

where $\mathbf{T}_N = \mathbf{Q}_N \, \mathbf{t}_N$. We now take $\mathbf{B}_{DFE}$ as a *fixed* matrix and apply the orthogonality principle (6.131) to compute $\mathbf{D}_{DFE}$. This results in:

$$\mathbf{D}_{DFE} = \mathbf{Q} \, \mathbf{B}_{DFE} \mathbf{Q}^H \mathbf{J} \, \mathbf{D}^H \mathbf{K}, \tag{6.138}$$

where $\mathbf{J}$ and $\mathbf{K}$ are expressed by (6.133) and (6.134), respectively. Then substituting (6.138) into (6.136) and evaluating the autocorrelation matrix $\mathbf{R}_{\boldsymbol{\delta}_{DFE}} \triangleq \mathrm{E}\{\boldsymbol{\delta}_{DFE} \, \boldsymbol{\delta}_{DFE}^H\}$ leads to:

$$\mathbf{R}_{\boldsymbol{\delta}_{DFE}} = \frac{\sigma_n^2}{N} \mathbf{B}_{DFE} \, \mathbf{Q}_N^H \, \mathbf{G} \, \mathbf{Q}_N \, \mathbf{B}_{DFE}^H, \tag{6.139}$$

where $\mathbf{G} = \mathrm{diag}(G_k)$, with $G_k \triangleq [|D_k|^2 + |D_{k+N}|^2 + (\sigma_n^2/\sigma_c^2)]^{-1}$ and $k = 0, 1, \ldots, N - 1$. The MSE associated with the FD-DFE (see (6.130)) is given by the trace of $\mathbf{R}_{\boldsymbol{\delta}_{DFE}}$, that is:

$$\mathrm{MSE}_{FD-DFE} = \frac{\sigma_n^2}{N} \sum_{s=0}^{N-1} \sum_{l=0}^{N-1} G_l \sum_{m=0}^{N-1} B_{s,m}^{DFE} \, \exp\left(j\frac{2\pi ml}{N}\right) \sum_{n=0}^{N-1} (B_{s,n}^{DFE})^* \, \exp\left(-j\frac{2\pi nl}{N}\right), \tag{6.140}$$

where $B_{s,n}^{DFE}$ is the element on the sth row and the nth column of $\mathbf{B}_{DFE}$, with $s, \ n = 0, 1, \ldots, N - 1$. The *gradient method* [503] can now be applied to (6.140) in order to evaluate the $\mathbf{B}_{DFE}$ matrix minimizing the MSE. This leads to the conclusion that, if the vector $\mathbf{B}_i^{DFE} = [B_{i,0}^{DFE}, \ldots, B_{i,i-1}^{DFE}, 1, 0, \ldots, 0]$ denotes the ith row of $\mathbf{B}_{DFE}$ (with $i = 1, 2, \ldots, N - 1$), the optimal choice for $\mathbf{B}_{DFE,i}$ is the solution of the linear system:

$$\mathbf{V}^{(i)} \left[(\mathbf{B}_i^{DFE})^T \right]^B = -\mathbf{v}^{(i)} \tag{6.141}$$

of i equations in i unknowns. Here $[\cdot]^B$ denotes the *backward rearrangement* operation [503], $\mathbf{V}^{(i)} = [V_{r,c}^{(i)}]$ is an $i \times i$ Hermitian Toeplitz matrix with $V_{r,c}^{(i)} \triangleq \sum_{l=0}^{N-1} G_l W_N^{(c-r)l}$ $(r, c = 1, 2, \ldots, i)$, $W_N \triangleq \exp(-j2\pi/N)$, $\mathbf{v}^{(i)} \triangleq [v_1, v_2, \ldots, v_i]^T$ is an i-dimensional column vector consisting of a subset of elements of the matrix $\mathbf{V}^{(i)}$, $v_r \triangleq \sum_{l=0}^{N-1} G_l W_N^{-rl}$, with $r = 1, 2, \ldots, i$. It is important to note that linear systems with the same mathematical structure as (6.141) can be found in linear prediction theory [503]. This observation has some important practical and conceptual implications. In fact, it can be easily inferred that:

(a) the linear system (6.141) can be efficiently solved by means of the well-known iterative *Levinson–Durbin algorithm* [1112], characterized by reduced computational complexity and small memory requirements (only $i^2 + O(i)$ complex multiplies and $2i$ memory locations are needed to compute $\mathbf{B}_{DFE,i}$, for $i = 2, 3, \ldots, N$),

(b) optimal FIR filtering via $\{\mathbf{B}_{DFE,i}\}$ in FD-DF equalization can be interpreted as a form of *linear prediction*, through which the feedback filter tries to reproduce the residual ISI at the output of the forward filter – this is not surprising, in a sense, as similar conceptual results can also be found in TD equalization theory [29].

The last point suggests that the evaluation of the set of linear systems (6.141), with $i = 2, 3, \ldots, N$, can be stopped for $i = f < N$, if increasing the predictor length to $f + 1$ results in a reduction of the normalized prediction error below a given threshold[6]. This provides a further substantial reduction of the computational load as $[B_{i,i-f}^{DFE}, \ldots, B_{i,i}^{DFE}] = [B_{f-1,0}^{DFE}, \ldots, B_{f-1,f}^{DFE}]$ with $i = f, f+1, \ldots, N-1$ and all the other elements of $\mathbf{B}_{DFE,i}$ are equal to zero.

The FD-DFE is substantially more complicated than the LE, and this is largely due to the triangular structure of $\mathbf{B}_{DFE}$; in particular, note that the most computationally intensive task required by it is the matrix product on the RHS of (6.138), leading to a cubic dependence on N. However, an FD-DFE can significantly outperform an FD-LE, as evidenced by the following example.

Example 6.2.3 In this example we consider an uncoded communication system designed to operate at 6 Mbit/s; in particular, we assume that: (a) the symbol interval is $T_s = 0.333$ μs; (b) a QPSK constellation is used (i.e., $M = 4$); (c) the transmitter impulse response $p(t)$ is the IFCT of a root-raised cosine function with roll-off $\alpha = 0.4$ and is truncated to $L_p = 30$ symbol intervals; (d) the SNR is defined as E_b/N_0, where E_b is the average received energy per information bit; (e) the channel PDP is a truncated exponential with average delay $\tau_{av} = 5$ μs and maximum delay $\tau_{max} = 20$ μs (so that $L_h = 60$); (f) the CIR changes in an independent fashion from block to block and is perfectly known at the receive side; (g) the DFT order is $N = 1024$; (h) the length of the cyclic prefix is $N_{cp} = 90$, so that the IBI is completely avoided. Figure 6.12 compares the BER[7] of the FD-LE and FD-DFE described above and processing two received signal samples per symbol with their counterparts operating at *symbol rate sampling* (SRS). The MFB for the given channel and modulation format (see Section 4.5.2.4) is also given for comparison. These results indicate that doubling the sampling rate provides substantial energy gains (not less than 6 dB) with both linear and decision feedback equalizers for a BER less than 10^{-3}. We also note that the change in the slope of the BER curve with the SRS DFE at about 15 dB is due to ill-conditioning problems encountered in the matrix inversion needed for solving a linear system of the same type as (6.141); similar problems have not been encountered with the DFE processing two samples per symbol interval in the same SNR interval. Moreover, simulation results show that the FD-DFE usually employs short FIR filters for ISI cancelation at the input of the decision device. In particular, we found in the scenario considered that the average number of feedback taps is equal to 24, independently of E_b/N_0, as it is mainly related to the channel power delay profile.

□

---

[6] This stopping criterion is suggested in [1112, p. 574, eq. (89)]. Other criteria, however, can also be found in the literature.

[7] Simulation results are indicated by labels, and continuous lines are drawn to show the performance trend.

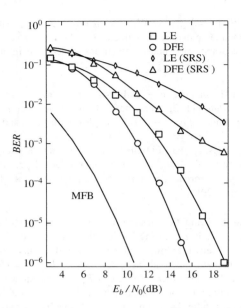

Figure 6.12 BER performance of FD-LEs and FD-DFEs. Symbol rate and fractional rate sampling are considered. The MFB is also given for comparison.

Algorithms for *hybrid* decision feedback equalization have also been developed in [1108, 1113–1115]. Note, however, that only symbol rate processing of the received signal is considered in [1113], and this entails a substantial performance loss if matched filtering is not used in the receiver front-end, as evidenced in Example 6.2.3 (see also [435, 1116]). In contrast, the use of fractional sampling is investigated in [1108, 1114]; in this case, however, nonuniform tap spacing in the feedback filter is considered and adaptive algorithms for the equalizer synthesis are proposed, even if explicit expressions for the optimal filters of a multisampling DFE are provided. Further relevant work in this area concerns: (a) the development of *bidirectional* decision feedback equalization algorithms implementing the feedforward filter in the FD and adapting the feedback filter the channel variations within one block in [1115]; (b) the use of *iterative* DFEs in which both the feedforward and feedback filters operate in the FD [1117, 1118]; (c) the *diversity order* that can be achieved by frequency-domain equalization [1119–1122]. An excellent overview of most of the decision feedback equalization algorithms operating in the FD can be found in [1109, 1118].

Most of the work on DFEs operating in the FD refer to PAM signals; note, however, that the mathematical approach illustrated above and other strategies described in the literature can be extended to the case of CPM signaling, if Laurent's representation is adopted, as already mentioned. In particular, the approach set out above has been adopted to derive a hybrid DFE for CPMs in [284]. It is important to point out that, in this case, decision feedback should not only mitigate the (inter-symbol) interference affecting each of the P Laurent components, but also cancel the inter-component interference. In addition, a VA has to be employed for reliable data detection, so that symbol pre-decisions need to be made available for feedback processing.

6.3   Channel Equalization of Multicarrier Modulations: Known CIR

In this section we address the problem of channel equalization of MCMs (and, in particular, of OFDM) in the presence of a known communication channel. In particular, we first focus on the problem of *optimal* detection over a (known) *purely frequency-selective channel* whose memory is shorter than the length of the CP of the transmitted signal. In this case, simple block-by-block equalization algorithms can be devised, because no IBI and ICI are found when detecting each OFDM symbol, provided that accurate timing and frequency synchronization are accomplished at the receiver. In contrast, if the channel is time-varying and/or its memory exceeds the CP length more complicated algorithms compensating for the presence of ICI and/or IBI, respectively, need to be devised, as illustrated in Sections 6.3.2 and 6.3.3.

6.3.1 Optimal Detection in the Absence of IBI and ICI

In Section 4.4.4.1 it is proved that the vector:

$$\mathbf{r}_N^{(l)} = \mathbf{Q}_N^H \ \mathbf{H} \ \mathbf{c}_N^{(l)} + \mathbf{n}_N^{(l)} \tag{6.142}$$

is acquired by an OFDM coherent receiver endowed with ideal timing synchronization (see (4.127)) when detecting the OFDM symbol $\mathbf{c}_N^{(l)} \triangleq [c_0^{(l)}, \ c_1^{(l)}, \ \dots, c_{N_\alpha}^{(l)}, \ 0, \ \dots, 0, \ c_{N-N_\alpha}^{(l)},$ $c_{N-N_\alpha+1}^{(l)}, \ \dots, c_{N-1}^{(l)}]^T$. Here, $\mathbf{n}_N^{(l)} \triangleq [n^{(l)}[0], \ n^{(l)}[1], \ \dots, n^{(l)}[N-1]]^T$ is complex AGN and $\mathbf{H} \triangleq \mathrm{diag}(H_n)$ is a diagonal matrix having the channel gains $\{H_n, \ n = 0, 1, \ \dots, N-1\}$ along its main diagonal. The time-domain vector $\mathbf{r}_N^{(l)}$ (4.127) undergoes an Nth-order DFT producing the FD vector (see (4.128)):

$$\mathbf{R}_N^{(l)} = \left[R^{(l)}[0], R^{(l)}[1], \ \dots, R^{(l)}[N-1] \right]^T \triangleq \mathbf{Q}_N \ \mathbf{r}_N^{(l)} = \mathbf{H} \ \mathbf{c}_N^{(l)} + \mathbf{W}_N^{(l)}, \tag{6.143}$$

where $\mathbf{W}_N^{(l)} = [W^{(l)}[0], \ W^{(l)}[1], \ \dots, W^{(l)}[N-1]]^T$ is statistically equivalent to $\mathbf{n}_N^{(l)}$. Note that (6.143) is equivalent to (see (4.129) and (4.130)):

$$R^{(l)}[k] = H_k c_k^{(l)} + W_k^{(l)} \tag{6.144}$$

for $k = 0, 1, \ \dots, N_\alpha$ and $k = N - N_\alpha, \ N - N_\alpha + 1, \ \dots, N - 1$, and

$$R^{(l)}[k] = W_k^{(l)} \tag{6.145}$$

for $k = N_\alpha + 1, \ N_\alpha + 2, \ \dots, N - N_\alpha - 1$, so that each useful FD sample depends only on a single channel symbol and is corrupted by a noise which is independent of that affecting all the other subcarriers.

In this case, since the matrix $\mathbf{H}$ is known and $\mathbf{n}_N^{(l)}$ is an AGN vector, the MLSD strategy based on $\mathbf{r}_N^{(l)}$ (6.142) can be formulated as (see (4.146)):

$$\hat{\mathbf{c}}_N^{(l)} = \arg \min_{\tilde{\mathbf{c}}_N^{(l)}} \left| \mathbf{r}_N^{(l)} - \mathbf{Q}_N^H \ \mathbf{H} \ \tilde{\mathbf{c}}_N^{(l)} \right|^2. \tag{6.146}$$

This strategy can be simplified observing that the linear transformation $\mathbb{C}^N \to \mathbb{C}^N$ represented by the FFT matrix $\mathbf{Q}_N$ is an *isometry* (this is due to the fact that $\mathbf{Q}_N$ is unitary), so that $|\mathbf{Q}_N \mathbf{x}|^2 = |\mathbf{x}|^2$ for any $\mathbf{x} \in \mathbb{C}^N$. Then, the decision metric of (6.146) can be rewritten as:

$$\begin{aligned} \left| \mathbf{r}_N^{(l)} - \mathbf{Q}_N^H \ \mathbf{H} \ \tilde{\mathbf{c}}_N^{(l)} \right|^2 &= \left| \mathbf{Q}_N \left(\mathbf{r}_N^{(l)} - \mathbf{Q}_N^H \ \mathbf{H} \ \tilde{\mathbf{c}}_N^{(l)} \right) \right|^2 \\ &= \left| \mathbf{Q}_N \ \mathbf{r}_N^{(l)} - \mathbf{H} \ \tilde{\mathbf{c}}_N^{(l)} \right|^2 \\ &= \left| \mathbf{R}_N^{(l)} - \mathbf{H} \ \tilde{\mathbf{c}}_N^{(l)} \right|^2, \end{aligned} \tag{6.147}$$

so that the strategy can be put in the form:

$$\hat{\mathbf{c}}_N^{(l)} = \arg \min_{\tilde{\mathbf{c}}_N^{(l)}} \left| \mathbf{R}_N^{(l)} - \mathbf{H}\, \tilde{\mathbf{c}}_N^{(l)} \right|^2. \tag{6.148}$$

Note that the equivalence between (6.146) and (6.148) can also be proved by invoking the *theorem of reversibility* (see Section 4.3.4), since $\mathbf{Q}_N$ describes a reversible transformation $\mathbb{C}^N \rightarrow \mathbb{C}^N$.

Since the metric of (6.148) can be put in the form:

$$\left| \mathbf{r}_N^{(l)} - \mathbf{Q}_N^H\, \mathbf{H}\, \tilde{\mathbf{c}}_N^{(l)} \right|^2 = \sum_{k=0}^{N_\alpha} \left| R^{(l)}[k] - H_k\, \tilde{c}_k^{(l)} \right|^2 + \sum_{k=N-N_\alpha}^{N-1} \left| R^{(l)}[k] - H_k\, \tilde{c}_k^{(l)} \right|^2, \tag{6.149}$$

an ML decision on $\mathbf{c}_N^{(l)}$ can be generated by solving:

$$\hat{c}_k^{(l)} = \arg \min_{\tilde{c} \in A_c} \left| R^{(l)}[k] - H_k\, \tilde{c} \right|^2 \tag{6.150}$$

for $k = 0, 1, \ldots, N_\alpha$ and $k = N - N_\alpha, N - N_\alpha + 1, \ldots, N - 1$, that is, taking $N_\alpha + 1$ independent decisions, one for each useful subcarrier (see (3.264)). Here A_c denotes the constellation of channel symbols. Note that this result implies also that the average error rate performance of a coherent OFDM receiver is obtained by averaging the error rate performances referring to $2N_\alpha + 1$ distinct subcarriers. Consequently, such an average performance is dominated by the deeply-faded subcarriers (i.e., by those subcarriers characterized by a small $|H_k|$), on which most of the symbol errors concentrate.

As far as the MAPSD strategy is concerned, we note that:

$$\Pr\left\{ c_k^{(l)} = \tilde{c} | \mathbf{r}_N^{(l)}, \mathbf{h} \right\} = \Pr\left\{ c_k^{(l)} = \tilde{c} | \mathbf{R}_N^{(l)}, \mathbf{H} \right\} = \Pr\left\{ c_k^{(l)} = \tilde{c} | R^{(l)}[k], \mathbf{H} \right\} \tag{6.151}$$

for $k = 0, 1, \ldots, N - 1$ and for any possible trial symbol $\tilde{c}$ transmitted over the kth subcarrier; this is due to the fact that $\mathbf{R}_N^{(l)}$ (the channel gain matrix $\mathbf{H}$) provides the same information as $\mathbf{r}_N^{(l)}$ (the CIR vector $\mathbf{h}$) and the set $\{R^{(l)}[p], p \neq k\}$ does not provide any information about $c_k^{(l)}$.

Finally, it is worth mentioning that a conceptually different strategy than symbol detection can be devised applying the principles illustrated for *linear equalization* in the TD (see Section 6.2.1.4). This means the vector $\mathbf{R}_N^{(l)}$ (6.143) undergoes linear equalization, that is, a linear transformation described by an $N \times N$ complex matrix $\mathbf{F}$; then, the equalizer output feeds a decision device (i.e., a threshold detector) which takes symbol decisions on a subcarrier-by-subcarrier basis. Let:

$$\mathbf{Z}_N^{(l)} = \left[Z^{(l)}[0], Z^{(l)}[1], \ldots, Z^{(l)}[N-1] \right]^T \triangleq \mathbf{F}\, \mathbf{R}_N^{(l)} \tag{6.152}$$

denote the output of the linear equalizer; the matrix $\mathbf{F}$ can be selected so as to minimize the MSE:

$$\varepsilon_{MMSE} \triangleq \mathrm{E}\left\{ \left| \mathbf{Z} - \mathbf{c}_N^{(l)} \right|^2 \right\}; \tag{6.153}$$

this requires solving the optimization problem:

$$\min_{\mathbf{F}} \varepsilon_{MMSE} \rightarrow \mathbf{F}_{MMSE}. \tag{6.154}$$

Following the same approach as in Section 6.2.1.4, it can easily be shown that:

$$\mathbf{F}_{MMSE} = \mathbf{H}^H \left(\mathbf{H}\mathbf{H}^H + \frac{\sigma_n^2}{\sigma_c^2} \mathbf{I}_N \right)^{-1} = \mathrm{diag}\left(F_{MMSE,k} \right), \tag{6.155}$$

where

$$F_{MMSE,k} = \frac{H_k^*}{|H_k|^2 + \sigma_n^2/\sigma_c^2} \tag{6.156}$$

for $k = 0, 1, \ldots, N_\alpha$ and $k = N - N_\alpha, N - N_\alpha + 1, \ldots, N - 1$, and $F_{MMSE,k} = 0$ elsewhere. This entails that (see (6.152)):

$$Z^{(l)}[k] = F_{MMSE,k} \, R^{(l)}[k] = \frac{H_k^*}{|H_k|^2 + \sigma_n^2/\sigma_c^2} R^{(l)}[k] \tag{6.157}$$

for $k = 0, 1, \ldots, N_\alpha$ and $k = N - N_\alpha, N - N_\alpha + 1, \ldots, N - 1$, so that *one-tap equalization* is accomplished. Finally, note that setting $\sigma_n^2 = 0$ in (6.155) and (6.157) yields:

$$\mathbf{F}_{ZF} = \mathbf{H}^{-1} \tag{6.158}$$

and

$$Z^{(l)}[k] = \frac{1}{H_k} R^{(l)}[k] = c_k^{(l)} + \frac{1}{H_k} W_k^{(l)}, \tag{6.159}$$

respectively; these describe the so-called ZF equalization. Expression (6.159) shows that the presence of a small channel gain on a subcarrier produces noise enhancement.

6.3.2   ICI Cancelation Techniques for Time-Varying Channels

As shown in Section 4.4.4.1, the presence of time variations in the communication channel affects the subcarrier orthogonality of an OFDM signal, so that the simple model (4.121) for $R^{(l)}[k]$ is replaced by (see (4.137)):

$$R^{(l)}[k] = \tilde{H}^{(l)}[k, k] \, c_k^{(l)} + \sum_{\substack{q=0 \\ q \neq k}}^{N-1} \tilde{H}^{(l)}[k, q] \, c_q^{(l)} + W_N^{(l)}[k], \tag{6.160}$$

where (see (4.138)):

$$\tilde{H}^{(l)}[k, q] = \frac{1}{N} \sum_{p=0}^{N-1} \exp\left(j\frac{2\pi(k-q)p}{N}\right) H_q[l, p] \tag{6.161}$$

and $H_q[l, p]$ denotes the channel gain referring to the qth subcarrier and the pth sampling epoch for the lth OFDM symbol. Then the vector $\mathbf{R}_N^{(l)} = [R_0^{(l)}, R_1^{(l)}, \ldots, R_{N-1}^{(l)}]$ can be expressed in a form similar to (6.143), since it can put in the form:

$$\mathbf{R}_N^{(l)} = \mathbf{H}^{(l)} \mathbf{c}_N^{(l)} + \mathbf{W}_N^{(l)}, \tag{6.162}$$

where, however, $\mathbf{H}^{(l)} = [H^{(l)}[k, q]]$ is a (nondiagonal) $N \times N$ complex matrix describing the time-varying behavior of the communication channel, whose effects, generally speaking, change every OFDM symbol. Note also that the ICI term:

$$ICI^{(l)}[k] = \sum_{\substack{q=0 \\ q \neq k}}^{N-1} \tilde{H}^{(l)}[k, q] \, c_q^{(l)} \tag{6.163}$$

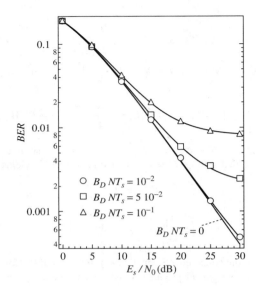

Figure 6.13   Average BER of a conventional OFDM detector for $B_D NT = 0, 10^{-2}, 5 \cdot 10^{-2}$ and 10^{-1}.

affecting $R^{(l)}[k]$ (6.160) becomes more relevant as Doppler rate of the communication channel increases and may entail a substantial performance degradation [448, 449, 452], as illustrated in the following example.

Example 6.3.1 In this example we consider an OFDM communication system characterized by the following parameter values [1123]: $N = 1024$, $N_\alpha = 409$, $M = 4$ (corresponding to a QPSK) and $T_s = 0.167 \, \mu$s. In addition we assume that: (a) the PDP of the multipath channel is a *truncated exponential* with maximum delay $\tau_{max} = 20 \, \mu$s and average delay $\tau_{av} = 5 \, \mu$s (see Section 2.2.2.2); (b) the Jakes model is adopted for the channel tap gains (see Example 2.2.8). Figure 6.13 illustrates the BER performance of a coherent receiver which perfectly knows the values taken on by the subchannel gains at the center of each OFDM symbol, but neglects their time variations in the observation interval (E_s denotes the average energy received for each OFDM symbol); the normalized Doppler rates $B_D NT_s = 0$ (slow fading), $10^{-2}, 5 \cdot 10^{-2}$ and 10^{-1} are considered in this case. Note that in this figure simulation results are denoted by marks, and continuous lines represent the approximated BER evaluated by means of the theoretical method developed in [448]. Such a method is based on: (a) the adoption of a linear model (see Section 2.2.3.4) to describe the variations of the TVTF $H(t, f)$ over each OFDM symbol; (b) modeling the ICI term $IC I^{(l)}[k]$ (6.163) as a zero mean complex Gaussian random variable, that is, as additional Gaussian noise (this approximation can be deemed accurate if the number of useful subchannels $(2N_a + 1)$ is large, as in this case). These numerical results show that the presence of ICI can entail a substantial degradation and, in particular, can result in a visible error floor. This is also evidenced by Figure 6.14, which shows the error floor versus $B_D NT_s$, which varies in the range $[10^{-3}, 1]$; as in the previous case, numerical results are denoted by marks, while the continuous line represents a set of theoretical results generated resorting to the method of [448].
□

In a coherent receiver, the effects of ICI can be mitigated by resorting to error-correction coding, or to a specific equalization algorithm which can compensate for it, or to a combination of the

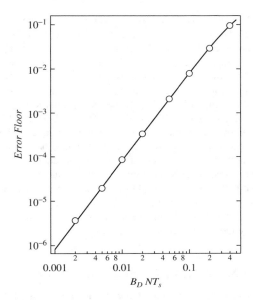

Figure 6.14 Error floor of a conventional OFDM detector vs. normalized Doppler bandwidth $B_D NT_s$.

two [1124]; in the latter case iterative schemes exploiting hard or soft information generated by the channel decoder to improve ICI compensation in multiple passes can be used (e.g., see [1125, 1126]). Various algorithms for equalizing the OFDM signal, while at the same time mitigating ICI, have been proposed in the technical literature, and they require a different degree of knowledge of the communication channel. They can be divided into: (a) algorithms for ICI self-cancelation, (b) semiblind equalization algorithms, and (c) ICI cancelation-based equalization algorithms.

ICI *self-cancelation techniques* are based on adopting a proper transmission format, since they require each channel symbol to be transmitted over a group of adjacent subcarriers [1127, 1128] or the TD transmitted signal to be periodically extended so that diversity is added in the OFDM symbols [1129]. In this case, good BER performance can be achieved at a low computational complexity; note, in particular, that only the knowledge of the average channel gains over each OFDM symbol is required for equalization. However, the price to be paid is a reduction in the bandwidth efficiency.

The reduction in bandwidth efficiency is avoided if *semiblind equalization methods* are exploited [1130, 1131]. These methods are based on assuming a specific mathematical structure for the ICI matrix and using specific diagonalization algorithms (such as the joint multiple matrix diagonalization algorithm [1130] and the approximate diagonalization of eigenmatrices algorithm [1131]) for ICI cancelation. Accurate channel estimation is not required in this case, since the CSI information provided by pilot symbols is used to determine the phase and permutation ambiguities induced by the adopted blind separation scheme.

If an accurate knowledge of the time-varying communication channel is available, ICI *cancelation-based equalization algorithms* can be used [449, 451, 1124–1126, 1132–1139]. Generally speaking, these algorithms rely on the fact that the structure of the FD received signal model (6.160) is similar to that of the time-domain ISI model encountered in the study of SC modulations (in particular, see (6.29)); this entails that the design techniques for TD equalizers illustrated in Section 6.2 can also be applied to ICI compensation, provided that the channel coefficients $\{H^{(l)}[k, q]\}$ are known. As far as channel estimation is concerned, it is usually assumed that the CIR varies in a linear fashion during a block period, so that the channel can be estimated by means of linear interpolation techniques (e.g., see [800, 1132, 1140]). This model of time variations can be deemed accurate if the normalized Doppler bandwidth $B_D N T_s$ does not exceed about 0.1 (see Example 6.3.1); when this does not occur, more refined polynomial models are needed [1126, 1133, 1134]. In most of the equalization techniques available in the technical literature ICI cancelation is achieved via linear ZF equalization [800, 1141, 1142], linear MMSE equalization [449, 451, 453, 1124, 1132–1134, 1138, 1143, 1144], or decision feedback equalization [449, 1126, 1133, 1137]. However, optimal strategies, such as MLSD [530], MLSD preceded by ICI whitening [1139], and MAPSD based on the FBA [530, 1144] are also available. Further related work can be found in [1135], where an equalization technique consisting of two stages (a set of prefilters and a set of ICI cancelation filters) has been proposed, and in [1145], where a novel adaptive breadth-first search procedure is used for sequence detection, in [1136], where a time-varying FIR time-domain equalizer is employed to restore the orthogonality between subcarriers, and hence to eliminate both ICI and IBI, in [1146], where a pre-equalizer operating on subblocks of the received OFDM block is used to mitigate ICI, and in [547], where the SAGE technique (see Section 4.6.4) is exploited to develop a channel equalization operating over a doubly-selective channel and detecting consecutively the components of each OFDM symbol.

6.3.3   Equalization Strategies for IBI Compensation

When the CIR length exceeds the duration of the cyclic prefix two different approaches can be adopted to mitigate the effects of IBI. The first approach is based on the use of an adaptive TD equalizer (often called *pre-equalizer* in this scenario) able to shorten the overall CIR to the CP length (e.g., see [998, 1147–1157]) and has been already mentioned in Section 6.2.1.2 as a tool to reduce the number of trellis states in MLSD. An alternative to this approach is represented by various equalization techniques in the FD which can compensate for the effects of a short CP. For instance, *linear* ZF and MMSE equalization algorithms have been proposed in [1158–1160]. In [1161–1164], instead, decision feedback and and cyclic reconstruction techniques are employed.

Finally, it is worth mentioning that various theoretical results about the nature of IBI can be found in [1165, 1166].

6.4   Channel Equalization of Single Carrier Modulations: Statistically Known CIR

As illustrated in Section 4.5.3, if the channel statistics are known, different formulations are possible for optimal metrics. In this section we focus on equalization algorithms operating in the TD. In particular, we first analyze MLSD and MAPSD/MAPBD strategies; then we consider different classes of optimal and suboptimal equalization algorithms developed for frequency-flat fading channels and benefiting from the knowledge of statistical information about the communication channel.

6.4.1   MLSD

Whatever the formulation, in principle, the MLSD strategy for PAM signaling requires the evaluation of the optimal metric (4.193) for all M^N possible data sequences and the selection of the data sequence having minimum metric (see (4.192)). This raises the problem of developing reduced-complexity alternatives to this brute-force strategy. Generally speaking, the computational complexity can be mitigated by dividing the symbol vector $\mathbf{c}_N = [c_0, c_1, \ldots, c_{N-1}]^T$ to be detected into short blocks, all having the same length and characterized by some overlap at the block edges. If the block length is small, there is a relatively small number of trial subsequences, so that the metrics for each of them may feasibly be calculated. After detecting the subsequence with best metric in the kth block, the symbols overlapping into the $(k + 1)$th block are treated as exactly known when such a block is processed. An example of this sliding block approach to reduced-complexity MLSD can be found in [1167], where the importance of using fractionally spaced samples is also evidenced.

Other alternatives to reduced-complexity MLSD are usually based on reformulating the optimal metric in such a way that the search for the best sequence can be accomplished via the VA. ML sequence detectors based on this approach are usually based on the *innovation-based formulation of the optimal metric* illustrated in Section 4.5.3.3. According to this formulation, the MLSD metric is expressed as (see (4.215)):

$$\Lambda(\tilde{\mathbf{c}}_N) = \sum_{k=0}^{N-1} \left| \frac{r_k - \eta_k(\mathbf{r}_{k-1}, \tilde{\mathbf{c}}_N)}{\sqrt{2\,\sigma_k^2(\mathbf{r}_{k-1}, \tilde{\mathbf{c}}_N)}} \right|^2 + \log\left(2\pi\,\sigma_k^2(\mathbf{r}_{k-1}, \tilde{\mathbf{c}}_N)\right), \tag{6.164}$$

where $\eta_k(\mathbf{r}_{k-1}, \tilde{\mathbf{c}}_N) \triangleq \mathrm{E}\{r_k | \mathbf{r}_{k-1}, \tilde{\mathbf{c}}_N\}$ (see (4.212)) and $\sigma_k^2(\mathbf{r}_{k-1}, \tilde{\mathbf{c}}_N) \triangleq \mathrm{E}\{|r_k - \eta_k(\mathbf{r}_{k-1}, \tilde{\mathbf{c}}_N)|^2\}/2$. Note that MLSD is represented by the decision strategy (see (4.192)):

$$\hat{\mathbf{c}}_N = \arg\min_{\tilde{\mathbf{c}}_N} \Lambda(\tilde{\mathbf{c}}_N), \tag{6.165}$$

so that, generally speaking, a search over a set of M^N trial sequences must be accomplished. This operation can be diagrammatically represented as looking for the path with minimum metric in a trellis having number of states *exponentially increasing with time*, as illustrated in Figure 6.15(a). Let us now suppose that the pdfs $\{f_{r_k}(\rho_k | \mathbf{r}_{k-1}, \tilde{\mathbf{c}}_N)\}$ in (4.211) depend only on a *finite history* of the data, that is:

$$f_{r_k}(\rho_k | \mathbf{r}_{k-1}, \tilde{\mathbf{c}}_N) = f_{r_k}(\rho_k | \mathbf{r}_{k-1}, \tilde{\mathbf{c}}_{k-L_m}^k) \tag{6.166}$$

for a proper value of the parameter L_m (called the *channel memory*) with $k = 1, 2, \ldots, N - 1$ and:

$$\tilde{\mathbf{c}}_{k-L_m}^k \triangleq \left[\tilde{c}_{k-L_m}, \tilde{c}_{k-L_m+1}, \ldots, \tilde{c}_k\right]^T. \tag{6.167}$$

Then the trellis of Figure 6.15(a) *folds* into the M^{L_m}-state trellis of Figure 6.15(b) and, as with an ISI channel, the search for the optimal path in the trellis can be carried out by means of the VA [562]. The equality (6.166) is known as the *folding condition* [1168, 1169] and can easily be shown to be equivalent to the equalities:

$$\eta_k(\mathbf{r}_{k-1}, \tilde{\mathbf{c}}_N) = \eta_k(\mathbf{r}_{k-1}, \tilde{\mathbf{c}}_{k-L_m}^k) \tag{6.168}$$

and

$$\sigma_k^2(\mathbf{r}_{k-1}, \tilde{\mathbf{c}}_N) = \sigma_k^2(\mathbf{r}_{k-1}, \tilde{\mathbf{c}}_{k-L_m}^k) \tag{6.169}$$

for $k = 1, 2, \ldots, N - 1$.

In the technical literature it is claimed that, with *frequency-flat fading*, folding occurs if one of the following conditions holds: (a) the sequence:

$$x_k \triangleq a_k + n_k \tag{6.170}$$

of fading plus noise samples is an AR process of finite order L_m [436, 490, 492, 1070]; (b) the fading channel has finite coherence time, that is, the autocovariance function $C_a(\tau)$ of the fading distortion $a(t)$ has finite support, so that:

$$C_a(kT_s) = 0 \tag{6.171}$$

for $|k| > L_m$ [562]. With *doubly-selective* fading channels [428, 492, 493, 562], finite time dispersion is required together with one of the previous assumptions for all the taps in the delay line model of the channel [492, 562], or, alternatively, an ARMA model of the CIR vector [428]. For instance, in [492]

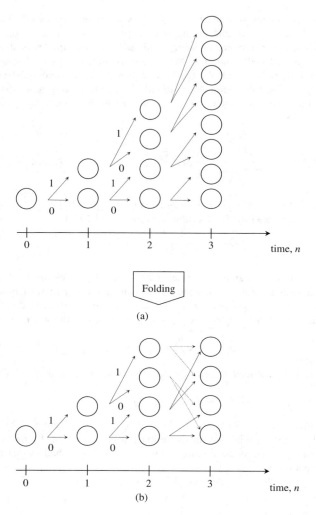

Folding

(a)

(b)

Figure 6.15 Trellis (a) and folded trellis (b) for MLSD in a binary ($M = 2$) signaling over a channel characterized by a memory $L_m = 2$.

it is shown that, if the memory associated with the channel time variations is $L_{m\alpha}$ symbol intervals and the channel delay spread is L_{ds} symbol intervals, a trellis with M^{L_m} states is required with:

$$L_m = L_{ds} + L_{ma} - 1. \tag{6.172}$$

In [1168], however, it is proved that the folding condition is never met for our model of a doubly-selective (or, in particular, frequency-flat) channel and, consequently, all the equalizers mentioned above are examples of *forced folding*. In other words, they are a suboptimal approximation to the optimal *estimator-correlator* receiver structure (see Section 4.5.3.2). Further details on this topic can be found in [987, 1169], where the problem of trellis-based detection over channels with infinite memory and the related issue of memory truncation (leading to finite-memory detection) are analyzed in depth.

With finite memory channels the relevant quantities $\{\eta_k(\mathbf{r}_{k-1}, \tilde{\mathbf{c}}_{k-Lm}^k)\}$ (6.168) (and, in some instances, $\{\sigma_k^2(\mathbf{r}_{k-1}, \tilde{\mathbf{c}}_{k-Lm}^k)\}$ (6.169)) can be evaluated by means of finite-length estimation filters, that is, time-invariant Wiener predictors (e.g., see [436, 492, 562]) or Kalman filters (e.g., see [428, 707, 725, 1170]).

Innovation-based receivers provide good error performance in fast fading at the price of an appreciable complexity. In practice, a significant complexity saving can be achieved by reducing the number of symbols comprising each trellis state from L_m to Q, for an appropriate choice of Q (e.g., see [490]); in fact, when predicting the fading process with L_m-tap optimal filters, the survivor path symbols can be used in place of the missing state symbols.

Let us now focus on a specific application of these concepts, considering the specific case of PSK or CPM signaling over a slow frequency-flat fading.

Example 6.4.1 In the cases of PSK or CPM transmission over a slow flat fading channel the contribution of the quantities $\{\sigma_k^2(\mathbf{r}_{k-1}, \tilde{\mathbf{c}}_{k-Lm}^k)\}$ is usually neglected in (6.164) (e.g., see [434, 436, 490]), so that the optimal metric turns into:

$$\Lambda(\tilde{\mathbf{c}}) = \sum_{k=0}^{N-1} \left| r_k - \eta_k\left(\mathbf{r}_{k-1}, \tilde{\mathbf{c}}_{k-Lm}^k\right) \right|^2. \tag{6.173}$$

Following Example 4.5.3, the conditional mean $\eta_k(\mathbf{r}_{k-1}, \tilde{\mathbf{c}}_{k-Lm}^k)$ in (6.173) can be expressed as:

$$\eta_k\left(\mathbf{r}_{k-1}, \tilde{\mathbf{c}}_{k-Lm}^k\right) = \tilde{c}_k\, \tilde{a}\left(k\,|\,\mathbf{r}_{k-1}, \tilde{\mathbf{c}}_{k-Lm}^{k-1}\right), \tag{6.174}$$

where $\tilde{a}(k|\mathbf{r}_{k-1}, \tilde{\mathbf{c}}_{k-Lm}^{k-1})$ is the MMSE *one-step prediction* of the fading sample a_k, based on $\mathbf{r}_{k-1}$ and on the assumption that the sequence $\tilde{\mathbf{c}}_{k-Lm}^{k-1}$ has been transmitted. It is worth mentioning that in [436, 490] $\eta_k(\mathbf{r}_{k-1}, \tilde{\mathbf{c}}_{k-Lm}^k)$ is evaluated as:

$$\eta_k\left(\mathbf{r}_{k-1}, \tilde{\mathbf{c}}_{k-Lm}^k\right) = \tilde{c}_k \sum_{l=1}^{L_a} p_k\, r_{k-l}\, \tilde{c}_{k-l}^*, \tag{6.175}$$

where $\{p_k,\ k = 1, 2, \ldots, L_a\}$ denote the coefficients of the MMSE *one-step linear predictor* (with $L_a \leq L_m$ taps) for the sequence $\{x_k\}$ (6.170). This shows that the MLSD receiver based on the metric (6.173) and the VA *implicitly evaluates multiple channel estimates* – as many channel estimates (or carrier references) as the number of trellis states.[8] Each fading estimate is evaluated conditioned on the channel symbols corresponding to a survivor path in the trellis. This can be interpreted as a form of PSP [569] (see Section 5.1.4).

An MLSD detector based on the metric (6.173) and on the use of (6.175) can provide good error performance if the fading is slow. This is exemplified by Figure 6.16, which refers to the detection of a QPSK format (corresponding to $M = 4$) over a Rayleigh fading channel. In this case it is assumed that: (a) the fading process is generated by filtering two independent real Gaussian processes with two identical third-order Butterworth filters, and the Doppler bandwidth B_D is given by the 3-dB bandwidth of the filters (B_D normalized to the symbol rate $1/T_s$ provides an indication of the fading rate); (b) each trellis state comprises two channel symbols (i.e., $Q = 2$), so that the VA operates over an $M^Q = 16$ state trellis; (c) an estimate of the channel memory is $L_m = 10$, but the number of taps used for the predictor is $L_a = 10, 10, 5$ and 3 for $B_D T_s = 10^{-3}, 10^{-2}, 5 \cdot 10^{-2}$ and 10^{-1}, respectively, since a further increase of L_a does not provide any visible improvement; (d) after each subsequence of nine information symbols, a pilot symbol (whose location is perfectly known at the receive side)

---

[8] In an uncoded transmission, an unambiguous phase reference can be computed only if the transmitted signal contains known features, such as a training sequence, pilot symbols or a pilot tone.

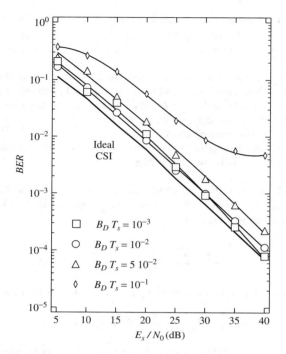

Figure 6.16 Simulated BER curves of a QPSK 16-state innovation-based detector and of a receiver with ideal CSI. $B_D T_s = 10^{-3}, 10^{-2}, 5 \cdot 10^{-2}$ and 10^{-1}.

is found in the transmitted data, so that the the receiver can acquire an absolute phase reference; (e) the coefficients of the prediction filters were computed for each SNR by means of the recursive Durbin (or Levinson) procedure [502]; (f) the detection delay introduced by the VA is $D_{VA} = 30$ symbol intervals; (g) in all cases the energy loss due to pilot symbols was taken into account in SNR calculations. Note also that in the same figure a BER curve referring to coherent demodulation of PSK signals with an ideal knowledge of the ideal CSI at the receive side is given for comparison. These results show that, if the fading is slow ($B_D T_s \leq 10^{-2}$), a performance degradation of less than 2 dB in SNR for BER $= 10^{-4}$ is found when the error curves of the VA-based detector are compared to that of the receiver with ideal CSI. An error floor is clearly visible for $B_D T_s = 10^{-1}$ only and, as shown by other simulations, it is *irreducible*, that is, it cannot be removed or appreciably lowered even if half of the transmitter power is used for the pilot sequence [1171]. If the fading is slow, sampling faster than $1/T_s$ provides negligible gains with respect to baud-rate sampling [490]. On the other hand, a large performance gain may be found for a larger fading bandwidth.

□

Generally speaking, an innovation-based ML sequence detector can be interpreted as an *adaptive detector* embedding a form of PSP for channel estimation (and, in particular, for *channel prediction*). Following this approach, VA-based algorithms can be heuristically derived; this approach can be found, for instance, in [111, 116, 427, 434, 437, 1172]. Finally, we mention that VA-based MLSD for a linearly modulated signal transmitted with a pilot tone (exploited to remove the constellation's phase ambiguity and to provide a stable amplitude reference for QAM constellations) and distorted by a doubly-selective Rayleigh fading channel is investigated in [1173].

MLSD algorithms for a statistically known channel can also be derived by employing the EM algorithm or modifications of it, as already illustrated in Section 4.6.1. In this case, the vector of input data includes all the unknown (random) channel parameters, whereas the vector of the data to be estimated coincides with that of the transmitted channel symbols. Further details on this can be found in [519, 773]. In particular, in [773] the so-called *generalized maximum likelihood sequence detection and estimation* is developed by applying EM to the problem of detecting a data sequence and estimating the unknown channel parameters. This allows us to develop new strategies for data detection in the presence of a stochastic CIR with known parameters and to provide a new interpretation to PSP-based MLSD techniques. Finally, in [519], after assuming a statically known discrete-time FIR model for a time-varying ISI channel, marginalization over the model parameters is accomplished to derive EM-based MLSD algorithms. In addition, the performance of the resulting algorithms is compared with that offered by MLSD in the presence of a known channel and PSP-based MLSD.

6.4.1.1 MAPSD/MAPBD

MAP symbol/bit detectors for statistically known channels can be developed by resorting to an innovation-based approach or to the BEM technique (see Section 4.6.2). As far as the former approach is concerned, here, following [1174], we focus on its application to PAM signaling over slow flat fading, for simplicity, assuming baud-rate sampling at the receive side. In the case considered, the $D_{f\,l}$-lag symbol-by-symbol MAP strategy can be expressed as (see (6.97)):

$$\hat{c}_l = \arg \max_{\tilde{c}_l} \Pr\left\{c_l = \tilde{c}_l | \mathbf{r}_0^{l+D_{f\,l}}\right\},\tag{6.176}$$

where the conditional probability $\Pr\{c_l = \tilde{c}_l | \mathbf{r}_0^{l+D_{f\,l}}\}$ can be evaluated as:

$$\Pr\left\{c_l = \tilde{c}_l \Big| \mathbf{r}_0^{l+D_{f\,l}}\right\} = \frac{\sum_{\tilde{\mathbf{c}}_{l+1}^{l+1+D_{f\,l}},\,\tilde{\mathbf{c}}_0^{l-1}} f_{\mathbf{r}}\left(\mathbf{r}_0^{l+D_{f\,l}} | \tilde{\mathbf{c}}_0^{l+D_{f\,l}}\right) \Pr\left\{\mathbf{c}_0^{l+D_{f\,l}} = \tilde{\mathbf{c}}_0^{l+D_{f\,l}}\right\}}{\sum_{\tilde{\mathbf{c}}_0^{l+1+D_{f\,l}}} f_{\mathbf{r}}\left(\mathbf{r}_0^{l+D_{f\,l}} | \tilde{\mathbf{c}}_0^{l+D_{f\,l}}\right) \Pr\left\{\mathbf{c}_0^{l+D_{f\,l}} = \tilde{\mathbf{c}}_0^{l+D_{f\,l}}\right\}}.\tag{6.177}$$

The quantities $\{f_{\mathbf{r}}(\mathbf{r}_0^{l+D_{f\,l}} | \tilde{\mathbf{c}}_0^{l+D_{f\,l}})\ \Pr\{\mathbf{c}_0^{l+D_{f\,l}} = \tilde{\mathbf{c}}_0^{l+D_{f\,l}}\}\}$ in (6.177) represent a set of $M^{l+D_{f\,l}}$ *sufficient statistics* for making a MAP decision on c_l with lag $D_{f\,l}$. In principle, the computation of such statistics can be carried out recursively, similarly to what has been done in the case of known channel (see Section 6.2.1.5). In fact, if we assume that the same quantities at the previous step, $\{f_{\mathbf{r}}(\mathbf{r}_0^{l-1+D_{f\,l}} | \tilde{\mathbf{c}}_0^{l-1+D_{f\,l}})\ \Pr\{\tilde{\mathbf{c}}_0^{l-1+D_{f\,l}}\}\}$, are known and that the sample $r_{l+D_{f\,l}}$ has been observed, we can also express:

$$f_{\mathbf{r}}\left(\mathbf{r}_0^{l+D_{f\,l}} | \tilde{\mathbf{c}}_0^{l+D_{f\,l}}\right) \Pr\left\{\mathbf{c}_0^{l+D_{f\,l}} = \tilde{\mathbf{c}}_0^{l+D_{f\,l}}\right\}$$

$$= f_{\mathbf{r}}\left(\mathbf{r}_0^{l-1+D_{f\,l}}, r_{l+D_{f\,l}} | \tilde{\mathbf{c}}_0^{l-1+D_{f\,l}}, \tilde{c}_{l+D_{f\,l}}\right) \Pr\left\{\mathbf{c}_0^{l-1+D_{f\,l}} = \tilde{\mathbf{c}}_0^{l-1+D_{f\,l}}, c_{l+D_{f\,l}} = \tilde{c}_{l+D_{f\,l}}\right\}$$

$$= f_{\mathbf{r}}\left(r_{l+D_{f\,l}} | \mathbf{r}_0^{l-1+D_{f\,l}}, \tilde{\mathbf{c}}_0^{l-1+D_{f\,l}}, \tilde{c}_{l+D_{f\,l}}\right) f_{\mathbf{r}}\left\{\mathbf{r}_0^{l-1+D_{f\,l}} | \mathbf{c}_0^{l-1+D_{f\,l}} = \tilde{\mathbf{c}}_0^{l-1+D_{f\,l}}, c_{l+D_{f\,l}} = \tilde{c}_{l+D_{f\,l}}\right\}$$

$$\cdot \Pr\left\{\mathbf{c}_0^{l-1+D_{f\,l}} = \tilde{\mathbf{c}}_0^{l-1+D_{f\,l}}\right\} \Pr\left\{c_{l+D_{f\,l}} = \tilde{c}_{l+D_{f\,l}}\right\}$$

$$= \Pr\left\{c_{l+D_{f\,l}} = \tilde{c}_{l+D_{f\,l}}\right\} f_r\left(r_{l+D_{f\,l}} | \mathbf{r}_0^{l-1+D_{f\,l}}, \tilde{\mathbf{c}}_0^{l+D_{f\,l}}\right)$$

$$\cdot \left[f_r\left(\mathbf{r}_0^{l-1+D_{f\,l}} | \tilde{\mathbf{c}}_0^{l-1+D_{f\,l}}\right) \Pr\left\{\mathbf{c}_0^{l-1+D_{f\,l}} = \tilde{\mathbf{c}}_0^{l-1+D_{f\,l}}\right\}\right]\tag{6.178}$$

as the components of $\mathbf{r}_0^{l-1+D_f l}$, conditioned on the transmitted data sequence, are independent random variables and do not depend on $c_{l+D_f l}$. Expression (6.178) provides the desired recursive formula because the quantity in square brackets gives the sufficient statistics from the previous update. Note also that the innovation-based representation (see Section 4.5.3.3):

$$f_r\left(r_{l+D_f l}|\mathbf{r}_0^{l-1+D_f l}, \tilde{\mathbf{c}}_0^{l+D_f l}\right) = \frac{1}{2\pi\sigma^2\left(l+D_f l|\mathbf{r}, \tilde{\mathbf{c}}\right)} \exp\left[-\frac{\left|r_k - \eta(l+D_f l|\mathbf{r}, \tilde{\mathbf{c}})\right|^2}{\sigma^2(l+D_f l|\mathbf{r}, \tilde{\mathbf{c}})}\right] \tag{6.179}$$

can be adopted for the conditional pdf $f_r(r_{l+D_f l}|\mathbf{r}_0^{l-1+D_f l}, \tilde{\mathbf{c}}_0^{l+D_f l})$, where:

$$\eta\left(l+D_f l|\mathbf{r}, \tilde{\mathbf{c}}\right) \triangleq \mathrm{E}\left\{r_{l+D_f l}|\mathbf{r}_0^{l-1+D_f l}, \tilde{\mathbf{c}}_0^{l+D_f l}\right\} \tag{6.180}$$

and

$$\sigma^2\left(l+D_f l|\mathbf{r}, \tilde{\mathbf{c}}\right) \triangleq \frac{1}{2}\mathrm{E}\left\{\left|r_{l+D_f l} - \eta\left(l+D_f l|\mathbf{r}, \tilde{\mathbf{c}}\right)\right|^2\Big|\mathbf{r}_0^{l-1+D_f l}, \tilde{\mathbf{c}}_0^{l+D_f l}\right\}. \tag{6.181}$$

Here the conditional mean $\eta\left(l+D_f l|\mathbf{r}, \tilde{\mathbf{c}}\right)$ (6.180) can be evaluated as [1174]:

$$\eta\left(l+D_f l|\mathbf{r}, \tilde{\mathbf{c}}\right) = \mathbf{a}\left(l-1+D_f l; \tilde{\mathbf{c}}_0^{l+D_f l}\right)\left(\tilde{\mathbf{c}}_0^{l-1+D_f l}\right)^T, \tag{6.182}$$

where $\mathbf{a}(l-1+D_f l; \tilde{\mathbf{c}}_0^{l+D_f l})$ is a vector of *conditional forward prediction coefficients* and $2\sigma^2(l+D_f l|\mathbf{r}, \tilde{\mathbf{c}})$ is the corresponding error prediction variance. In other words, the conditional mean $\eta(l+D_f l|\mathbf{r}, \tilde{\mathbf{c}})$ is the output of a length $(l-1+D_f l)$ linear prediction filter. Note that: (a) generally speaking, the coefficients of this prediction filter depend on the trial sequence $\tilde{\mathbf{c}}_0^{l+D_f l}$; (b) predicting $r_{l+D_f l}$ by $\eta(l+D_f l|\mathbf{r}, \tilde{\mathbf{c}})$ is loosely equivalent to estimating the channel in a per-survivor fashion. This shows, once again, that MAP algorithms with CIR averaged over can be related to adaptive MAP ones.

The recursive equation (6.178) provides a simple method for updating the decision statistics. However, at time $(l+D_f l)$, it produces M new statistics for each one of those generated in the previous symbol interval. For this reason, the MAP-FLA derived above is characterized by a computational complexity increasing as $(l+D_f l) \cdot M^{l+D_f l}$ with time; this represents a substantial difference compared to the MAP-FLA for ISI channels derived in Section 6.2.1.5 and requiring a *fixed number of computations per symbol*. This substantial difference can be related to the fact that the channel memory is fixed in a static ISI channel, but, in principle, is infinite with time-varying channels. Then, in the latter case, the quantity $f_r(r_{l+D_f l}|\mathbf{r}_0^{l-1+D_f l}, \tilde{\mathbf{c}}_0^{l+D_f l})$ in (6.178) should be evaluated for any possible trial sequence $\tilde{\mathbf{c}}_0^{l+D_f l}$, that is, for any possible path in the state trellis of Figure 6.15(a). As discussed above, a rigorous simplification of the algorithm is possible if the so-called *folding condition* applies. In practice, as already illustrated for MLSD, folding can be forced [1174, 1175] so only a fixed number of received samples are processed in each symbol interval by the FLA and decision feedback is exploited. The complexity of the resulting algorithms is still large, but can be further reduced resorting to thresholding techniques. These allow unlikely paths to be discarded, so that a significant reduction of the average computational load can be achieved.

The MAP-FLA strategies were developed in [1174, 1175] for frequency-flat fading channels, but can be extended to doubly-selective fading channels; analytical details on this extension can be found in [504, 1094, 1176].

The approach illustrated for the development of a MAP-FLA can also be applied to design MAP FBAs [531–532]. An interesting example of this approach is given, in particular, in [533], which develops MAP detection algorithms for CPM and PSK signals transmitted over frequency-flat fading channels. In this case trellis folding is forced under the assumption that the fading plus noise process $\{x_k\}$ (6.170) can be accurately approximated by an AR model of order L_m, so that the resulting MAP algorithm operates a forward and a backward recursion over an M^{L_m}-state trellis. The evaluation of the state transition probabilities (needed for the evaluation of the symbol APPs) involves a form of *per-state* channel estimation based on linear predictors of order L_m. The same approach has also been adopted in [1177], where MAP detection of an MSK signal in the presence of an antenna array at the receive side is considered (correlated diversity links are assumed). Extensions of this approach to doubly-selective channels are also possible, but a vary large computational complexity should be expected. Further results in this area can be found in [987, 1169, 1178], where some general considerations on MAP symbol detection algorithms in the presence of a channel with infinite memory are provided and the implications of finite-memory conditions are analyzed in depth, in [1179], where the problem of developing SiSo algorithms in the presence of a parametric uncertainty is investigated, and in [1180], where techniques for *factor graphs* (FGs) (see Section 10.8) are applied to develop a MAPSD algorithm for DPSK signals transmitted over a Rayleigh frequency-flat fading channel.

Finally, it is important to mention that a conceptually different approach to MAP symbol detection is offered by the BEM technique illustrated in Section 4.6.2. In this case, the vector of data to be estimated includes all the unknown (random) channel parameters, whereas the vector of imputed data coincides with that of transmitted channel symbols; a specific application to PSK signaling over a slow frequency-flat fading channel can be found in Example 4.6.2 (see also [526] for further details and [529] for the case of CPM signaling). Applications of this approach to frequency-selective and doubly-selective channels can be found in [528] and [1181], respectively.

6.4.2   Other Equalization Strategies with Frequency-Flat Fading

Estimating the channel statistics is often a feasible task in a data transmission over frequency-flat fading channels. For this reason, various equalization techniques exploiting the knowledge of the channel statistics has been developed for this case; in the following we briefly illustrate two classes of such techniques, namely *block equalization techniques* and *decision feedback equalization techniques*. Finally, we focus on *optimal one-shot detectors*, which process the signal received over one or two consecutive symbol intervals to generate symbol-by-symbol decisions.

6.4.2.1   Block Equalization Techniques

In detecting a long data sequence, the sequence of received samples can be partitioned into a sequence of blocks of length N; then each block can be processed via block ML detection algorithm. Block detectors can be roughly divided into two classes: multiple-symbol ML detectors (e.g., see [518, 1182–1185]); and ML detectors employing the EM algorithm (e.g., see [1186–1188]).

Multiple-symbol ML detectors were proposed in [518] for block detection of differentially encoded M-PSK sequences transmitted over Rayleigh fading channels. This work has shown that ML block detection can be interpreted as a *multiple-symbol differential detector* (MSDD) [1183] and that an appreciable reduction of the error floor in fast fading can be achieved with respect to a conventional differential receiver. The main drawbacks with respect to a conventional differential receiver are: (a) a complexity increase since M^N ML metrics (one for each possible data sequence $\tilde{\mathbf{c}}_N$) must be computed and compared; and (b) the receiver must estimate the second-order channel statistics. Multiple-symbol differential detection algorithms have also been investigated in [501, 1182, 1184, 1185, 1189, 1190]. In particular, it is worth mentioning that: (a) in [501, 1189] an interesting interpretation of the ML block detection algorithm for QAM signals is provided and the error performance of suboptimal algorithms in the presence of coding and diversity is investigated; (b) an MSDD employing the so-called *sphere*

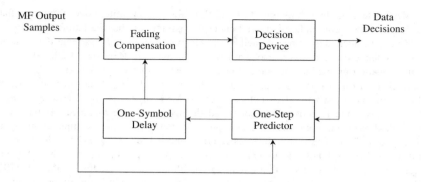

Figure 6.17   Block diagram of a decision feedback receiver.

decoding (SD) strategy[9] for reducing the computational complexity of the ML search is proposed in [1190].

The EM algorithm was proposed in [1186, 1188] as a solution to the problem of ML estimation of linearly modulated data sequences. Its application leads to a two-step iterative procedure embedding a Kalman filter for channel estimation: further analytical details can be found in Section 4.6.1 and, in particular, in Example 4.6.1, where a specific application to PSK signaling over a slow frequency-flat fading channel is illustrated.

6.4.2.2   Decision Feedback Equalization Techniques

The class of suboptimal decision feedback equalization algorithms plays a significant role, since it offers an interesting complexity–performance tradeoff (e.g., see [487, 488, 514, 707, 725, 1174, 1194–1197, 1197–1201]). Such algorithms are based on the idea that in order to detect the kth channel symbol c_k coherently, an estimate of the fading distortion sample a_k is required. If the data decisions on previous symbols are reliable, they can be used to remove the modulation from the corresponding received signal samples and to estimate a_k via standard prediction techniques. A general scheme for these equalizers is illustrated in Figure 6.17. Note that: (a) a Wiener predictor [487, 488, 514, 1197], a Kalman filter [725, 1195, 1196] or an extended Kalman filter [707] can be employed for predicting the fading distortion; (b) a periodic refresh of the algorithm memory with a string of known symbols is needed to prevent a *runaway* phenomenon [488] (i.e., a loss of channel tracking) and to solve phase ambiguity, unless differential detection is used [1197, 1199–1201].

The error performance of a decision feedback receiver can be improved, at the price of an increase in detection latency, by adopting a two-stage architecture (e.g., see [513, 1198]). In this case the first stage is a decision feedback receiver and delivers its data decisions to the second stage; this, in turn, generates an improved channel estimate by means of an optimal smoother. Finally, this estimate is used to produce new (more reliable) data decisions.

6.4.2.3   ML One-Shot Detectors

When the observation interval is limited to one or two symbol intervals, averaging over CIR produces simple ML detectors which accomplish sequence detection on a symbol-by-symbol basis.

---

[9] SD is an algorithm for solving integer least-squares problems; it is due to U. Fincke and M. Pohst [1191] and was first proposed in the context of the closest point searches in lattices (further details can be found in [1192, 1193] and the references therein).

Strictly speaking, these are not equalizers since they do not estimate the channel. However, they are widely used in wireless transmission as they allow detection of a signal in the absence of an explicit channel estimate and offer the advantage of simplicity. Classic examples resulting from this approach are the well-known *differential detector* (DD) for *M*-ary DPSK and the *matched filter and envelope detector* (MFED) for energy detection of FSK signals [13]. Both structures are *optimal* under the assumption of slow fading, that is, that fading distortion does not change appreciably over the observation interval. An analysis of the error performance of DDs and MFEDs on fading channels can be found in [427, 1202–1208] and [112, 1202, 1209–1211], respectively. This shows that DDs and MFEDs suffer from a SNR loss with respect to coherent detection and their error performance exhibits an error floor in the presence of fast fading [1202, 1212]. Improved DDs and MFEDs have been derived in [427, 489, 1195, 1196] and [112], respectively. They exploit a couple of receive filters [112, 427] or multiple received samples per symbol[10] [489, 1195, 1196] in order to exploit the *implicit time diversity* of the channel [101]. This results in a low error floor in fast fading at the price, however, of acquiring a more refined knowledge of the channel noise and fading statistics.

Other noncoherent detectors are available for CPM signals and comprise *differential detectors* and *discriminators*. An analysis of their error performance is provided in [1184, 1214–1216] for differential receivers, and in [1068, 1217–1220] for discriminators.

6.5   Channel Equalization of Multicarrier Modulations: Statistically Known CIR

If the communication channel is purely frequency-selective, the case of OFDM detection in the presence of a statistically known channel can easily be related to that of PAM signaling over a slow frequency-flat fading channel analyzed in Section 6.4 [1123] (in other words, the former case is the FD *dual* of the latter). In fact, equation (4.121) shows that, at the output of the DFT stage of an OFDM receiver, the channel symbol sequence associated with the lth OFDM symbol is affected both by multiplicative distortion and by AGN, exactly as occurs with the samples at the output of the MF in a conventional PAM receiver, operating over a slow time-selective fading channel (see (4.108)). In the case at hand, the multiplicative distortion is given by the samples of the multipath channel frequency response, which can be modeled as a Gaussian stochastic process, just like the time-selective distortion affecting a narrowband PAM transmission (note that in the former case the autocorrelation function $R_H(f)$ of the channel frequency response $H(f)$ plays exactly the same role as the autocorrelation function $R_D(\tau)$ of the fading distortion $a(t)$ referring to the latter case). This suggests that the techniques developed for TD data detection in time-selective fading can be directly applied to our sequence obtained by selecting the useful components of $\mathbf{R}_N^{(l)}$ (6.143) (i.e., the $(2N_\alpha + 1)$ associated with the useful subcarriers). For instance, if the channel symbols belong to a PSK constellation, they can be differentially encoded at the transmitter and differentially detected at the receiver. This system is very simple to implement and exhibits good behavior if the phase difference between adjacent samples of $\mathbf{R}_N^{(l)}$ is small. However, in principle, all the detection algorithms mentioned in the previous section for SC detection over frequency-flat fading can be employed. It is also important to point out that this topic has not been widely investigated in the technical literature. Here we confine ourselves to mentioning that the problem of ML detection of OFDM symbols in the presence of known statistics for a frequency-flat fading channel has been investigated in [928, 1221], where the use of V-BLAST and the SD algorithm (see Section 11.5.1.7 and [1193, 1222], respectively) has been proposed to limit the complexity of the search for the ML symbol estimate (the resulting techniques are classified as *semiblind detectors*); the proposed approach has been extended to the case of a doubly-selective channel in [1223].

---

[10] Some theoretical considerations on multisample processing in optimal detection can be found in [1213].

6.6   Joint Channel and Data Estimation: Single-Carrier Modulations

If both the channel and its statistics are unknown at the receiver, algorithms that jointly estimate the data and the channel can be employed. In the technical literature this type of algorithm is usually called *semiblind*, since, even if channel estimation is performed, proper initialization with the aid of pilot symbols is required. In this section, we focus on three different classes of equalization techniques, the first two developed for MLSD, the third for MAPSD. Finally, we analyze the problem of accomplishing channel estimation on the basis of a known reference signal before detection under the assumption of a purely time-selective communication channel.

6.6.1   Adaptive MLSD

The MLSD techniques discussed in Section 6.2.1.1 have been developed under the assumption of a known CIR. Optimal MLSD when CSI is unavailable requires detection to be performed jointly with channel estimation. Regardless of the specific channel model and modulation format, it is usually claimed that the exact solution of the MLSD problem in the absence of CIR knowledge has *complexity* that is *exponential* with respect to the channel memory length [577]. Note that, from this perspective, slower channel dynamics are expected to lead to a higher complexity exponent. This has motivated substantial research efforts toward the development of various suboptimal strategies which continuously adapt an ML-based equalization algorithm to the changing channel conditions estimated via a decision-aided channel estimator. Such strategies form the class of *adaptive* MLSD algorithms. The philosophy of this class is described by Figure 6.18, which illustrates how in a digital receiver a *single* data-aided channel estimator[11] can be combined with an MLSD algorithm developed for a known CIR in a natural fashion. In practice, the transmitted symbol sequence is detected using the estimated CIR so that the coefficients $\{h_k\}$ ($\{q_l\}$) in (6.19) ((6.34)) are replaced by their estimates $\{\hat{h}_k\}$ ($\{\hat{q}_l\}$). Note, however, that this estimate may be exploited in more than the evaluation of the branch metrics. In fact, when an MF is used as a front-end, in principle it should be updated too [458].

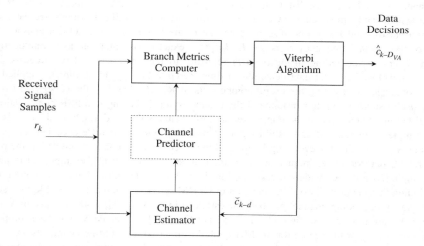

Figure 6.18   Block diagram of an adaptive MLSD receiver. A channel predictor may be inserted to mitigate the lag effects due to channel estimation with tentative decisions.

---

[11] More details on this can be found in Chapter 5.

In practical applications the channel data transmission normally starts with a training sequence to initialize the channel estimator. Then, during the transmission of information data, the receiver is switched to a decision-directed mode and the CIR estimate is updated on the basis of the detected sequence provided by MLSD [458, 622]. In doing so, tentative decisions $\{\check{c}_{k-d}\}$ coming from the VA survivor characterized by the best decision metric [559] are usually fed back to the channel estimator with a certain delay d [559, 560, 1016, 1224]. Note that, as already mentioned in Section 6.2.1.1, the decision delay D_{VA} introduced by the VA is of the order of 5–7 times the channel memory, so that, if $d = D_{VA}$ is selected (i.e., $\check{c}_{k-d} = \hat{c}_{k-D_{VA}}$, where $\hat{c}_{k-D_{VA}}$ denotes the final decision taken by the VA on the transmitted symbol $c_{k-D_{VA}}$), the channel estimator is supplied with outdated information, and the resulting estimate suffers from a *lag error* [559, 564]. For time-varying channels, this must be traded off against the accuracy of tentative decisions, so that the tentative decision delay d is usually selected to be appreciably shorter than D_{VA} [622]. Note that an important role in determining error performance of an adaptive MLSD receiver is played by both the decision lag, and the acquisition speed of the channel estimation algorithm [768]. For channel tracking, the low-complexity LMS algorithm is usually favored in both adaptive MLSD and adaptive PSP MLSD [458, 512, 564, 622, 1016] (see also the next subsection). This is due to the fact that, as mentioned in Section 5.1.3.1, if the input data correlation is low, the tracking ability of the LMS and RLS algorithms are similar. The quality of the CIR estimate can also be improved by *predicting* it [511, 768, 1225], namely, by inserting a *channel predictor* at the channel estimator output, as shown in Figure 6.18. Note, however, that, even when employing prediction, an error floor occurs during fast fading [511]. For this reason, adaptive MLSD is suited only to slowly fading channels.

A conceptually different approach to adaptive MLSD is proposed in [558], where a *basis expansion model* [233] is used to model a time-varying channel over the observation interval and a data-aided estimate of the model coefficients is exploited to start the MLSD algorithm. Then data decisions are used to refine the estimates of model coefficients, and these steps are iterated until the maximum number of iterations is reached or until convergence.

For a comparison of adaptive MLSD with other equalization strategies the reader can refer, for instance, to [860, 1023, 1226, 1227]. Note also that in PSP-based MLSD the Fano algorithm is an alternative to the VA in the presence of trellis-encoded data, as illustrated in [1016].

6.6.2 PSP MLSD

The tentative decision delay d characterizing adaptive MLSD may be unsatisfactory for time-varying channels and, to avoid this, PSP may be applied in channel tracking (e.g., see [111, 419, 426, 434, 437, 490, 512, 559, 563, 564, 1228, 1229]). This means that a channel estimator is associated with each survivor in a trellis-based algorithm for MLSD and the corresponding CIR estimate updated with no lag using the survivor symbols. Then, the branch metric (6.34) for the VA takes the form:

$$\xi_k(\tilde{\Delta}_k, \tilde{\Delta}_{k+1}) \triangleq \left| r_k - \sum_{l=0}^{L_q-1} \tilde{c}_{k-l}\, q_l\left(\tilde{\Delta}_k\right) \right|^2, \tag{6.183}$$

where $\{q_l(\tilde{\Delta}_k),\ l = 0, 1, \ldots, L_q - 1\}$ denotes the CIR estimate associated with the state $\tilde{\Delta}_k$ at the kth decoding step. A generic block diagram for a PSP-based MLSD receiver is shown in Figure 6.19.

Adaptive PSP MLSD is motivated by the inadequacy of adaptive MLSD in time-varying channels, but it is *never truly optimal* with time-varying channels [426, 512, 1168]. The LMS [419, 426, 512, 515, 564, 1229] or RLS [774] algorithms can be employed for tracking a doubly selective channel, but the LMS approach is usually preferred, as noted above. However, error floors can still occur even with relatively slow fading. The RLS algorithm performance in fast fading can be appreciably improved by explicitly modeling the time variation of the channel taps via additional parameters, as illustrated in [616, 617]. An alternative to the LMS and RLS techniques is developed in

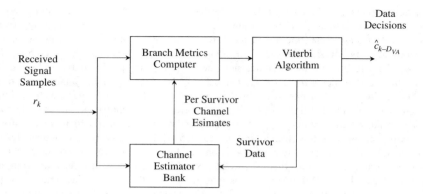

Figure 6.19   Block diagram of a PSP-based MLSD receiver.

[428], where, after proposing the adoption of an ARMA model for the CIR, a bank of KFs is employed for channel tracking. Unfortunately the complexity of the resulting PSP-KF receiver structure can be prohibitive, especially for long CIRs, due to the filtering complexity. Simplified KF-based PSP schemes are developed in [566, 774, 1229]. In particular, in [774], complexity reduction is achieved by means of a prediction-feedback mechanism exploiting the parallel structure, which characterizes the ARMA model when certain model parameters are neglected. The resulting reduced-complexity KF-based receiver provides significant performance improvement over a PSP receiver with RLS channel estimation and only a moderate degradation with respect to the PSP-KF. In [566], various alternative implementation methods are considered for measurement and time update equations of a KF and parallel structures for joint Viterbi data detection and channel estimation are developed.

The performance of PSP-based algorithms for MLSD improves as the quality of the channel estimate gets better. With frequency-flat fading accurate channel prediction can be obtained with PSK and CPM signals by resorting to the polynomial channel models illustrated in [111, 437, 1230]. In particular, in [111, 437] a linearly time-varying model is adopted for a time-selective channel model (see Section 2.2.3.4), so that the implicit time diversity provided by the communication channel can be extracted by two "matched" filters (for a linearly modulated signal) [111] or two filter banks [437] (for CPM signals). The matched filter outputs provide accurate information about the fading process and can be combined to produce improved per-survivor channel predictions.

At receiver startup PSP-based MLSD can estimate the channel in a *blind* fashion, provided, however, that some pilot symbols are available to resolve the phase ambiguity of the adopted constellation. In particular fast startup in frequency-flat fading is provided by the blind algorithms proposed in [111, 437]. Some results on the blind acquisition properties of PSP MLSD with frequency-selective channels can be found in [575].

It is also important to mention that:

(a) *array processing* can be incorporated in PSP-based MLSD [1231];
(b) *generalized* PSP algorithms, retaining multiple (instead of one) survivors per state, have been developed (e.g., see [419]);
(c) strategies for reducing the number of channel estimators in a PSP-based detector can be adopted to achieve a good performance–complexity tradeoff [569];
(d) *per-branch processing* (PBP) algorithms for adaptive MLSD have been proposed in [563, 566, 773]. PBP can be considered as a generalized form of PSP, since it involves computing one channel estimate for each branch of the VA trellis when generating the branch metrics.

Finally, we note that the theory of adaptive PSP and PBP MLSD has been analyzed in a framework based on the EM algorithm in [773]. In this context a generalized maximum likelihood sequence detection and estimation technique has been derived following the EM approach and it has been shown that per-survivor and per-branch processing methods emerge naturally from this technique. Some related work on MLSD based on the EM can be found in [580, 1232]. In particular, in [580] a blind ML equalization method is proposed for frequency-selective fast fading Rician channels. This method employs the so-called *expectation maximization Viterbi algorithm* (EMVA) for blind channel estimation and signal detection. In practice, the VA is used to execute the E step of an EM iteration, whereas channel estimation is accomplished in the M step. The EMVA is shown to achieve an error rate performance close to that of MLSD endowed with the true parameters for the given channel model. A conceptually related approach is followed in [1232], which, however, investigates the use of various *basis expansions* of a time-varying communication channel to simplify its estimation.

6.6.3 Adaptive MAPBD/MAPSD

As shown in Section 6.2.1.5, two types of MAP algorithms have been developed for practical applications. One, called MAP-FBA, performs forward and backward recursions, whereas the other, denoted MAP-FLA, accomplishes only a forward recursion. In time-varying environments the channel parameters are initially estimated by means of a training sequence and afterwards can be adaptively tracked by exploiting the decisions from the detection algorithm. Unfortunately, a MAP-FBA cannot deliver reliable decisions to the channel estimator until both its forward and backward recursions have been completed. It can therefore be stated that a MAP-FBA is not suitable for time-variant channels when a *single* channel estimator, fed by tentative decisions, is employed in the receiver structure [1088]. However, a MAP-FLA can be exploited to develop adaptive MAP detection algorithms, as shown in [1233–1235]. Such algorithms recursively generate reliable *hard* decisions and employ a *single* Kalman-type channel estimator, as illustrated in Figure 6.20. This estimator is not fed by tentative data decisions, but by APPs of the states of the ISI channel (a training sequence is also used for receiver startup). Simulation results demonstrate that the use of *soft statistics* in channel estimators improves their channel tracking capabilities and that the proposed algorithms outperform conventional adaptive MLSDs on time-varying channels (the problem of channel estimation relying on soft information will be briefly addressed again in Section 12.3.5 in the context of iterative equalization). This approach can also be applied to *sparse channels* in such a way that the receiver complexity does not depend on the overall memory of the channel, but on how few nonzero taps it has [1234].

Channel tracking performance and, consequently, error performance can be improved using *multiple* KF-based channel estimators, as illustrated in [1236, 1237]. In this case, if we refer to the channel

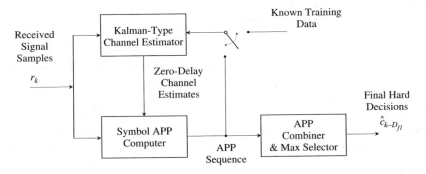

Figure 6.20 Block diagram of an adaptive MAP-FLA with a single channel estimator.

model (6.32), a parallel bank of M^{L_q} conditional channel estimates, each corresponding to a different subsequence of length L_q, is updated using a forward MAP strategy. Although the number of channel estimates required for blind MAPSD is M times larger than that needed in blind PSP MLSD, superior performance is achieved by the former algorithms in the presence of fast fading. If the symbol timing is unknown [1238], the latter approach can still be employed by including an additional ISI term in the receiver memory to account for this uncertainty. This leads to an M^{L_q+1}-state MAPSD algorithm employing a bank of extended KFs (EKFs) and jointly estimating the channel coefficients and the symbol timing. A known preamble sequence is transmitted to estimate the channel statistics via an auxiliary EKF at the receive side; then, at the end of the training phase, the auxiliary filter is employed to properly initialize both the parallel filter bank of EKFs and the MAP subsequence metrics. During the transmission of information symbols, the MAP detection algorithm takes symbol-by-symbol decisions, whereas the fading channel is tracked by the filter bank. As suggested in [1236, 1237], different strategies can be employed to lower the complexity of the KF-based MAPSD algorithm, such as thresholding, the use of LMS filtering in place of Kalman filtering or channel memory truncation combined with decision feedback.

It is also mentioning that a MAP detector employing a bank of *matched* stochastic nonlinear filters for generating multiple estimates of a frequency-flat fading channel has been developed in [1239]. In the proposed receiver the MAP decision processor is driven by the filters' innovations processes.

As discussed in Section 6.2.1.5, computational savings in MAP forward only detection can be achieved adopting the OSA or the SSA [1088]. This approach to complexity reduction has been followed in [1240], where conventional adaptive and PSP extensions of the above-mentioned algorithms are illustrated and compared with the performance attained using a PSP MLSD as a first stage in a receiver for an interleaved coded system operating over a doubly-selective channel. One of the most significant results is that PSP SSA has computational complexity comparable to that of PSP MLSD, but offers superior performance.

All the algorithms described above belong to the class of MAP-FLAs. Nonetheless MAP-FBAs can be employed on time-varying communication channels (e.g., see [530, 558, 1181, 1241]). In particular, the channel estimation process can be embedded in both the forward and backward recursions, in a way that can be related to a PSP-based receiver, as shown in [530], where a fractionally-spaced MAP equalizer for a doubly-selective fading channels is developed. The FBA operates on an expanded state trellis and accomplishes *per-state* joint MMSE channel estimation and soft data detection. Alternatively, in [558], a *basis expansion model* [233] is adopted to model a time-varying channel over the observation interval, so that the problem of channel estimation is turned into one of estimating the model coefficients. The estimate of the model coefficients of the time-varying channel is exploited to run a standard MAP-FBA equalizer, whose output can be employed to generate a more refined channel estimate. In other words, joint channel estimation and equalization is accomplished in an iterative manner and the iterations are performed until the maximum number of iterations is reached or until convergence. A conceptually different approach is followed in [1181], where the EM and the BEM algorithms are employed to develop soft decision-directed channel estimators which can be coupled with a MAP-FBA for channel equalization.

6.6.4 Equalization Strategies Employing Reference-Based Channel Estimators with Frequency-Flat Fading

As already discussed in Section 6.2.1.6, the use of coherent detection with MRC or ECG on slow frequency-flat fading channels requires knowledge of the fading distortion within each symbol interval. In practice, coherent detection is possible if a reference (or sounding) signal is transmitted with the information-bearing signal. In particular, a coherent reference can be made available to the receiver by transmitting a time-continuous sounding signal such as a pilot tone (e.g., see [709–713, 715, 1242–1245]) or, as illustrated in Section 5.3, by sending a sequence of known symbols (i.e., a pilot sequence) interspersed with the data symbols. Note that, in this scenario, a *phase-locked loop* (PLL) cannot be employed to generate an accurate phase reference since this technique is unable to track the rapid phase changes characterizing channel fading [1246].

Various pilot tone techniques have been proposed; here we note the following approaches:

(a) Yokoyama [710] describes a technique which consists of sending a continuous wave sounding signal together with a data BPSK signal orthogonal in phase to the sounding signal itself.
(b) In the TTIB technique [713] the baseband spectrum is split into two segments. The segment in the upper frequency band is translated up in frequency by an amount equal to the "notch" width and a reference pilot tone is added at the center of the resulting notch.
(c) The TCT [712] creates a spectral null in the data signal by means of a proper encoding technique and inserts a pilot tone in the null.
(d) In the DPTCT [711] two pilots are symmetrically located outside the actual data spectrum but near the band edges. This technique provides better bandwidth efficiency than TCT, at the price of increased sensitivity of the pilots to frequency shifts.

Robust and simple receiver structures can usually be implemented when pilot-tone techniques are employed. In most cases, the pilot tone can be recovered from the received signal using relatively simple circuitry and the error floor level can be substantially reduced. However, this is achieved at the price of wasting a fraction of the transmitted power to transmit the required reference signals.

Simpler transmitter and receiver processing is achieved by PSAM. In PSAM transmission, the transmitter periodically sends known symbols, from which the receiver derives its amplitude and phase reference. The PSAM transmitter and receiver schemes are shown in Figure 6.21 together with the transmitted data format. Here, the data symbol rate is equal to $R_s = (K - 1)/(KT_s)$, with $1/KT_s$ being the pilot symbol rate (this rate should be at least $2B_{D,\max}$, where $B_{D,\max}$ is the largest value of the Doppler bandwidth B_D). Like pilot-tone modulation, PSAM suppresses the error floor and offers the further advantage of enabling multilevel modulation without requiring a change in transmitted pulse shape or of the PAPR. A comparison of PSAM with TTIB [1245] has shown that the former technique offers substantially better energy efficiency than the latter for any practical power amplifier.

Finally, we note that reference-based techniques for coherent detection were originally developed for linearly modulated signals, but it has been also shown that they can be employed with CPM signals [726] as well.

6.7 Joint Channel and Data Estimation: Multicarrier Modulations

If the OFDM symbol is transmitted over a doubly-selective fading channel, the received signal vector $\mathbf{R}_N^{(l)}$ in the lth symbol interval can be put in the form (6.162). It is not difficult to show that, in principle, the ML *joint channel and data estimates* $(\hat{\mathbf{H}}^{(l)}, \hat{\mathbf{c}}_N^{(l)})$ for this interval may be evaluated as:

$$\left(\hat{\mathbf{H}}^{(l)}, \hat{\mathbf{c}}_N^{(l)}\right) = \arg \min_{\tilde{\mathbf{H}}, \tilde{\mathbf{c}}_N} \left|\mathbf{R}_N^{(l)} - \tilde{\mathbf{H}}\,\tilde{\mathbf{c}}_N\right|^2, \tag{6.184}$$

where $\tilde{\mathbf{H}} = [\tilde{H}[k, q]]$ is an $N \times N$ complex matrix and $\tilde{\mathbf{c}}_N = [\tilde{c}_0, \tilde{c}_1, \dots, \tilde{c}_{N-1}]^T$ is an N-dimensional vector of channel symbols, all belonging to an M-ary alphabet A_c. This requires solving a *complex* LS problem for $\hat{\mathbf{H}}$ (since $\hat{H}[k, q] \in \mathbb{C}$ for $k, q = 0, 1, \dots, N - 1$) and an *integer* LS problem for $\tilde{\mathbf{c}}_N$. Note that, despite the formal simplicity of this problem, the computational effort required to solve it directly for reasonably large values of N is huge. For this reason, various suboptimal estimation strategies have been developed. These usually tackle the problem of channel estimation first and then exploit the resulting channel estimate for detection. Channel estimation procedures necessarily rely on the availability of pilot symbols, that is, on the presence of known channel symbols defined over

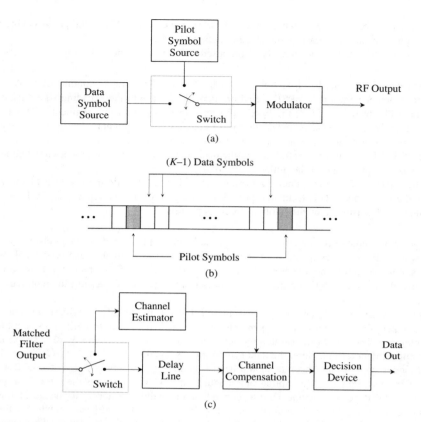

Figure 6.21 Structural block diagram of a PSAM transmitter (a) and receiver (c); location of the pilot symbols in the transmitted symbol sequence (b).

a set of predefined subcarriers; however, such symbols are exploited for estimating the channel gains in the class of *pilot-based equalization techniques*, whereas their use is limited to solving the phase ambiguity problem for the constellation of channel symbols in *semiblind equalization techniques*. Some details on these two classes of techniques are provided in this section.

6.7.1   Pilot-Based Equalization Techniques

Pilot-based techniques are usually based on the idea that the received signal samples at the DFT output can be processed to acquire CSI referring to the pilot subcarriers. Then the CSI referring to all the other subcarriers is assessed by employing a proper estimation technique. The channel estimation technique depends on the selected estimation criterion and on the model used to describe the channel state during the transmission of each OFDM symbol. In fact, in certain scenarios the channel can be deemed static over each OFDM symbol, even if the changes from symbol to symbol should not be ignored. In contrast, in other cases, Doppler effects are not negligible, so that channel modeling should explicitly account for them.

Some pilot-based equalization techniques assume a *static channel* over each OFDM symbol [1247–1252], even if the channel state can evolve from symbol to symbol. In particular, in [1247] an iterative channel estimation and equalization procedure is proposed. In each step an ML estimate of the CIR vector **h** is generated on the basis of the available channel symbols (i.e., the pilot symbols at the beginning of the first iteration, the whole symbol vector in all other iterations). This estimate is then used to evaluate an ML estimate of the data symbols. The procedure stops when two successive CIR estimates are sufficiently close to each other. An iterative procedure is also developed in [1248], where, however, the presence of a residual CFO is also taken in account. In this case, joint estimation of the CIR and CFO is alternated with soft ICI and IBI cancelation. In [1249] an AR model with known parameters is used to represent the channel variations from symbol to symbol and a decision feedback mechanism is combined with Kalman filtering to predict the communication channel in a MAP receiver. In [1250] an OFDM transmission with a low density of pilot carriers is considered. The proposed algorithm for channel estimation and equalization operates over a group of consecutive OFDM symbols and assumes a polynomial representation for the subcarrier gains in time and frequency. It considers some data symbols within the group as "pseudo-pilots" and, for each possible pattern of pseudosymbols, it estimates the coefficients of the above polynomial representation via an LS fitting procedure; then it generates a data sequence estimate and corresponding metric. Finally, it chooses the pattern of pseudo-pilots characterized by the minimum metric and decides in favor of the data sequence associated with it. In [1251, 1252] a state-space model is adopted for describing the changes in the CIR vector from each OFDM symbol to the next (in particular, a random walk model is selected in [1251]) and the BEM technique (see Section 4.6.2) is applied to develop a MAP channel estimator, thus generating soft estimates of the channel symbols as a by-product. Note that the techniques proposed in [1251] and [1252] involve an RLS algorithm and Kalman filtering, respectively.

If the communication channel over each OFDM symbol *cannot be deemed static*, different techniques for channel estimation and equalization can be devised. Note that such techniques usually rely on specific parametric channel models which account for channel changes.

In particular, if the channel variations over each OFDM symbol, even if not negligible, are relatively slow (i.e., moderate Doppler spreads can be assumed), a *linearly time-varying model* (see Section 2.2.3) can be employed for modeling the variations of the CIR taps over a few consecutive OFDM symbols [451, 1135, 1140]. This allows simpler representations for the received vector $\mathbf{R}_N^{(l)}$ (6.162) to be developed which, in turn, allow the problem of channel estimation and ICI mitigation to be simplified. This approach is exemplified by [1140], where iterative ICI mitigation techniques exploiting the cyclic prefix of each OFDM symbol or consecutive OFDM symbols for estimating channel slopes are described. Related work can be found in [451, 1135], which exploit channel estimates evaluated over consecutive pilot symbols and linear interpolation to estimate the CIR over the OFDM symbols carrying information data and employ an FDE (a LE and *parallel interference cancelation* equalizer, respectively) both for compensating the multiplicative distortion affecting each subcarrier and for canceling ICI.

When a linearly time-varying model is not accurate enough to capture the real behavior of the communication channel, more refined channel models must be adopted, such as *power series models* [1126, 1133], *polynomial models* [803, 1253] and *basis expansion models* [558]. In particular, a finite power series expansion for the TVTF of a statistically known channel is employed in [1133]. This model is used to estimate the channel (via a simplified ML estimator) over training OFDM symbols and demodulated OFDM symbols and for predicting the channel via polynomial extrapolation. As far as channel equalization is concerned, a linear MMSE FDE and a numerically efficient DFE are developed. A Taylor series expansion in the t variable only for the CIR $h(t, \tau)$ is employed in [1126], where an iterative ICI cancelation and channel estimation scheme is developed. In practice, the proposed receiver architecture combines an iterative DFE with a multistage channel estimator in an overall iterative scheme.

Polynomial models are used to represent the variations of the CIR taps over multiple consecutive OFDM symbols in [803]. LS estimates of polynomial coefficients are generated that exploit the information provided by pilot subcarriers. Then the estimated CSI is employed in the data detection procedure, which combines interference suppression over successive subcarriers with one-tap frequency equalization and involves a group of consecutive OFDM symbols. The performance of this algorithm for channel estimation and equalization can be enhanced by iterating it (channel estimation and ICI mitigation are performed at each iteration). In [1253] a very fast fading environment is considered, so that polynomial models are applied to represent the CIR tap gains within each OFDM symbol. In addition, an AR model is adopted to describe the dynamics of the polynomial coefficients. For this reason, Kalman filtering is employed to adaptively track such coefficients and the estimated CSI is used in a detection procedure based on the so-called QR *decomposition* (see equation (C.14)) of the channel matrix. Similarly to [803], channel estimation and data detection can be iterated to improve the error performance.

Basis expansion models are used in [558] to represent the CIR variations over a single OFDM symbol. In this case an LS or MMSE initial channel estimate is obtained with the aid of pilot symbols. Then equalization in the FD in the presence of ICI is accomplished by resorting to VA-based MLSD or to a MAP-FBA. The detected symbols are then used to re-estimate the channel, so that the quality of channel estimation and data detection can be improved through iteration.

Unlike the work cited above, in [453, 800, 1138] no specific model for the time-varying CIR is assumed in the derivation of channel estimation and equalization algorithms. In particular, in [800] an MMSE technique is used for estimating the CIR relying on pilot symbols and then the channel estimate is exploited for MF detection, LS detection and MMSE with successive detection (which detects the channel symbols one by one, rather than all simultaneously). In [1138] the presence of both IBI and ICI (due to the presence of a short cyclic prefix and to the time variation of the channel) is considered. An LS strategy for pilot-aided channel estimation combined with linear interpolation is adopted for channel estimation in the TD. Then equalization is divided in two steps. First, the received signal samples go through a FIR TDE which shortens the overall CIR so that the IBI and ICI effects due to an insufficient length of the cyclic prefix are minimized. Finally, the filtered signal is processed in the FD using a FIR MMSE FDE with a few taps per subcarrier to eliminate the ICI due to channel variations. In [453] a linear pilot-aided MMSE estimator is used for estimating the ICI coefficients; then, these coefficients are exploited by an FDE with multiple taps for mitigating ICI.

All the techniques illustrated above process the OFDM received signal block by block (i.e., symbol by symbol). A conceptually different approach is illustrated in [1146], where a system with very large OFDM block sizes is considered. In this case each OFDM symbol is partitioned into subblocks (so that the channel can be considered time-invariant over each of them) and an FD pre-equalization is accomplished on each subblock to mitigate channel variations (i.e., to reduce ISI); this is followed by conventional single-tap equalization per subcarrier (see Section 6.3.1). In this case conventional channel estimation schemes based on pilot subcarriers can be employed, since the knowledge of the CIR over each subblock is required to adjust the pre-equalizer. For this reason, a periodic *pseudonoise* sequence is superimposed on the OFDM for channel estimation.

Finally, we note that:

(a) some related work on sequence detection of specific MC schemes over doubly-selective fading channels can be found in [1145];
(b) the problem of OFDM detection in the presence of channel estimation errors is tackled in [1254] by exploiting the *theory of variational inference*.

6.7.2 Semiblind Equalization Techniques

A limited number of references is available in the technical literature concerning semiblind equalization for OFDM signals [1123, 1130, 1131, 1255–1258]. In particular, a trellis-based detector with PSP is proposed in [1123] for blind symbol detection; in this case a few pilot symbols are inserted in each

OFDM symbol, since they are exploited to ensure a fast startup and a proper trellis termination. In [1255] a two-dimensional regression surface is employed to model the channel gains in frequency and time, and this model is then exploited to formulate the ML data estimation problem. Unfortunately, the minimization of the ML metric requires an exhaustive search over a large set of data and this is computationally inefficient. As a result, the joint detection and channel estimation problem is divided into two stages. In the first stage, it is solved for some selected subchannels or timeslots (i.e., OFDM symbols), thereby generating data estimates which are are exploited in the second stage as pilot symbols in determining the other symbols. However, there is still the problem of minimizing a specific detection metric, that is, solving an optimization problem. After pointing out the analogy between this problem and an integer programming problem and translating the last problem into a tree-searching problem, an efficient tree search algorithm, based on the principle of *branch-and-bound*, is developed.

Semiblind linear equalization techniques operating in the FD over a group of consecutive OFDM symbols are derived in [1130, 1131]. Note that these techniques are based on the *joint multiple diagonalization algorithm* [1130] and on the *joint approximate diagonalization of eigenmatrices algorithm* [1131].

In [1256, 1257] the EM algorithm is employed to develop iterative data estimators in the presence of an unknown doubly-selective channel. More precisely, in [1256] a technique for estimating both data and CFO is developed under the assumption of a time-varying TDL model for the channel, while [1257] concentrates on the problem of data estimation only and assumes linear variations of each channel tap over an OFDM symbol.

Finally, in [1258] a scheme for iterative *blind* channel estimation is proposed. When an OFDM symbol is received, primary symbol estimates are generated by a decision algorithm based on a constrained linear MMSE criterion. These estimates are then employed in a conventional MMSE channel estimator, providing CSI to conventional detection techniques. Even if this scheme is classified as blind, it needs a pilot symbol to train the channel estimator at startup.

6.8   Extensions to the MIMO Systems

Substantial research efforts have been made in the last decade to extend the equalization techniques developed for SISO and SIMO wireless communications to a MIMO scenario. One of the main problems encountered in the development of MIMO equalization techniques is the large computational complexity. This is due to the fact that, in an $n_T \times n_R$ MIMO communication system, the signal captured by each of the n_R receive antennas involves the superposition of n_T transmitted signals, so that any equalization algorithm employed at its receiver has to cope not only with the ISI and/or the time variations of the communication channel affecting all the transmitted signals, but also with CCI. In addition, in a MIMO scenario ST coding schemes are commonly used (see Chapter 11), and equalization and decoding tasks are often combined in a single algorithm matched to such schemes. In this section a brief overview of MIMO equalization techniques is provided for both SC and OFDM signaling.

6.8.1   Equalization Techniques for Single-Carrier MIMO Communications

When data transmission is over a *frequency-selective* MIMO channel that is *known to the receiver* and has finite memory, the optimal sequence detection strategy is a *multichannel* MLSD procedure based on the VA [554, 1259]; however, the computational complexity of this procedure makes its practical implementation infeasible (similar comments apply to MAPSD for the same scenario). This motivates the search for suboptimal linear or nonlinear equalization techniques in the time or frequency domain.

Linear equalization in the TD has been investigated in [1260–1262]. In particular, a new and different perspective for the understanding and analysis of MIMO linear equalization is provided in

[1261], where the equalization problem is considered as a special case of an estimation problem and the H^∞ approach to estimation is proposed as an alternative method for channel equalization. It is also worth making the following observations:

(a) As in the SISO case, LEs can be used for CIR shortening (see Section 6.2.1.2). This problem is tackled, for instance, in [1260], where optimal finite-length delay-optimized MIMO equalizers are derived, in [1263], where the issue of joint transmitter–receiver filter design for shortening MIMO ISI channels is addressed, in [1264], where realizable MMSE channel shorteners consisting of a prewhitening filter followed by an FIR postfilter are designed, and in [1265], where optimization is performed from an information-theoretic perspective.

(b) *If the communication channel is also known to the transmitter*, linear processing at the receive side can be combined in an optimal fashion with some form of signal preprocessing (namely, *pre-equalization* or *precoding*) at the transmit side. Precoding can potentially modify the structure of the overall MIMO communication channel (e.g., converting a MIMO channel with memory into a set of parallel flat fading subchannels) and/or greatly reducing signal processing at the receiver. The reader can refer to [1266–1268] and references therein for further details on this topic.

An alternative to linear equalization in the time domain is *decision feedback equalization* operating in the same domain. Various contributions to the optimization, performance and computation of a finite-length MMSE-DFE for MIMO communication systems can be found in [38, 1259, 1262, 1269, 1270]. Other results can be found in the literature on DFE design and, in particular, on structural modifications/extensions to a standard DFE structure. Here we limit ourselves to mentioning the following:

(a) As for linear equalization, the design problem for optimal MIMO DFEs can be approached from an estimation point of view by adopting an H^∞ perspective [1271].
(b) In wideband communications over long sparse channels, modified decision feedback equalization structures, exploiting the channel properties by simple tap allocation, can be developed [1262].
(c) MIMO DFEs can also be adopted in layered receiver structures for *spatial multiplexing* systems (see Section 1.1 and Chapter 11) operating over broadband wireless channels. In particular, layered space-time receivers for a broadband *vertical* BLAST (V-BLAST) system [1272] have been devised in [1273]. In the proposed receivers, the n_T equal-rate independently encoded data streams (each radiated by a distinct transmit antenna) are successively detected in an ordered manner (on the basis of their strength, since stronger streams are more resilient to CCI) by employing a MIMO-DFE. Moreover, each stream is detected with the entire CCI contribution from every previously detected stream already canceled (i.e., *serial interference cancelation* (SIC) is used).

It is important to note that, as in the SISO case, equalizer design for MIMO systems is often based on the assumption that CSI *is known precisely to the receiver*. Accurate CSI estimation is often a critical task in MIMO communications. This has motivated the development of design methodologies ensuring that MIMO equalizers can extract symbols at the receiver robustly and effectively from a noisy channel subject to model uncertainties [1271, 1274, 1275].

An alternative to MIMO decision feedback equalization is the use of THP (see Section 6.2.3), that is, nonlinear prefiltering at the transmit side to avoid error propagation phenomena; for further details on THP for MIMO systems, see [1276, 1277].

For sufficiently high SNRs, DFEs achieve better error performance than LEs, but are substantially outperformed by MLSD, whose implementation complexity is usually prohibitive. However, it has been shown that MLSD performance can be approached at the price of a computational complexity comparable to the standard ST DFE algorithms by resorting to the SD algorithm [1278]. Other quasi-optimal equalization algorithms can also be found in the literature. For instance, a simplified fixed lag

(SiSo) MAPSD strategy based on a trellis representation of MIMO signals and soft decision feedback has been proposed in [1279] for V-BLAST.

As in the SISO case, TD equalization algorithms developed under the assumption of a frequency-selective communication channel known to the receiver can be easily combined with pilot-aided channel estimation (see Chapter 5) when the channel exhibits negligible changes over each transmitted data packet. In contrast, if the time variation of the channel within a packet is significant, *channel tracking* is needed for the equalization to be effective. This approach is adopted, for instance, in [840], where channel taps are tracked by a KF aided by staggered decisions from a finite-length MMSE-DFE.

An alternative to equalization of frequency-selective channels in the TD is *equalization in the* FD. As in the SISO case (see Section 6.2.2), both linear and nonlinear equalization strategies in the FD can be found in the literature. In particular, FD *linear* equalization schemes have been proposed in [1280, 1281]. Moreover, the problem of the design of an MMSE LE in the presence of multiple unknown CFOs is tackled in [1282]. However, as expected, better error performance can be achieved by resorting to *decision feedback* schemes, for which different options can be found in the technical literature [1283–1293]. In particular, a *hybrid* time–frequency-domain DFE has been developed in [1283], where a group of TD feedback filters is used to eliminate part of the postcursor ISI and CCI. *Layered* spatial frequency-domain equalization structures for V-BLAST (inspired by the layered spatial time-domain equalization developed in [1273]) have been illustrated in [1284, 1285], where a hybrid DFE is employed at each stage and the multiple data streams are detected following a layered approach (in other words, SIC is accomplished). The multiple stages of such SIC-based schemes tend to accentuate the effect of error propagation when only imperfect or estimated CSI is available [1284]. An alternative to this is presented in [1287], where *parallel interference cancelation* (PIC) in conjunction with a hybrid DFE is considered. The proposed receiver consists of multiple parallel branches, one corresponding to each data stream, and each branch performs FD-based PIC and equalization in multiple stages. A parallel-branch FD receiver architecture has been also developed in [1294]. It consists of parallel linear preprocessors to suppress CCI, followed by a hybrid equalizer, and integrates channel parameter estimation. Both the preprocessors and the forward filter of the DFE are implemented in the FD, whereas the backward filter of the DFE is implemented in the TD. Layered spatial frequency-domain equalization can also be combined with *iterative processing*, as shown in [1286, 1288], where some ideas on iterative block decision feedback equalization developed in [1295] have been followed. An iterative technique is also proposed in [1289], based on the use of a *soft interference-cancelation* MMSE equalization algorithm. Further decision feedback schemes have been developed in [1290–1293]. In particular, FD equalization with decision feedback processing for time-reversal ST block-coded systems has been proposed in [1290]. In [1291, 1292], a linear FDE is combined with a TD *noise predictor* and with *successive interference cancelation* to derive two new equalization algorithms that achieve a better performance–complexity tradeoff than the conventional FDEs with decision feedback processing. FD equalization can also be combined with the THP technique to avoid error propagation. This is shown in [1293], where two new MIMO FDEs based on the MMSE criterion are designed.

As in the case of TD equalization, FD equalization for a known channel can easily be combined with pilot-aided schemes for channel estimation (e.g., see [1282, 1296]).

Finally, it is important to mention that many technical papers in the field of MIMO wireless communications deal with the problem of equalization of ST coded signals transmitted over a narrowband wireless channel, which can be modeled as a *purely time-selective* fading channel. As in the SISO case, for this scenario different assumptions can be made about the channel knowledge at the receiver (in practice, the channel can be known, statistically unknown or has to be estimated jointly with the transmitted data).

In the case of known channel, various quasi-ML detection strategies (e.g,. see [1297–1299] proposing the application of SD and [1300] solving the ML detection problem via a semidefinite relaxation) and soft MIMO detectors (e.g., see [1301] and references therein) have been derived. A simpler alternative is offered by *linear* and *decision feedback receivers* (e.g., see [1302–1304] and references therein). For a statistically known channel, the use of PSP detection algorithms (see Section 6.4.1), decision feedback receivers (see Section 6.4.2.2) and multiple symbol differential detectors has been investigated for differential ST modulations over Rayleigh fading channels (e.g., see [1190, 1201, 1305–1317]). For the case of joint channel and data estimation, PSP-based algorithms (see Section 6.6.2) have been devised (e.g., see [1318, 1319], where algorithms for the approximate ML detection of ST trellis and block codes, respectively, are derived).

6.8.2   Equalization Techniques for MIMO-OFDM Communications

As one of the most promising wireless techniques, MIMO-OFDM can significantly improve the quality of wireless transmission links. Generally speaking, since the complexity of ML detection of MIMO-OFDM grows exponentially with the number of co-channel signals, there has recently been interest in developing low-complexity detection algorithms. In particular, various suboptimum algorithms have been developed for the case of a frequency-selective MIMO channel known to the receiver. Here we mention:

1. *Detection algorithms based on the application of* QR *decomposition* (QRD). In this case the QRD is applied to the antenna outputs after the FFT operation on each of OFDM subcarriers. The ML decision rule based on the QRD can be implemented as a full tree search whose complexity grows exponentially with the number of transmit antennas and the size of the symbol constellation. A limited tree search can be accomplished via the M-algorithm (see Section 6.2.1.2). The combination of QRD with the M-algorithm results in the so-called QRD-M algorithm [1320]. Some modifications of this (e.g., see [1321–1323]) and other detection algorithms employing interpolation-based QRD techniques [1324] have also been proposed.

2. *Detection algorithms based on soft interference cancelation.* As in the SISO case, reduced-complexity detectors based on *soft interference cancelation* can be developed when equalization is combined with SiSo decoding of channel codes; this approach is exemplified in [1325], where two soft interference cancelation MMSE receivers are derived.

Other work in this area is concerned with the combination of OFDM with *spatial multiplexing* and the development of equalization techniques for specific coding schemes. When spatial multiplexing is adopted in multicarrier transmission, a standard BLAST architecture alternating interference suppression, detection and interference cancelation can be employed (e.g., see [1326, 1327]); however, various alternatives, requiring lower complexity, are also available (e.g., see [1320, 1328, 1329]). Examples of equalization algorithms for specific coding schemes can be found, for instance, in [1330, 1331]. In particular, in [1330] new *orthogonal space-time block codes* (OSTBCs) achieving full diversity and admitting fast ML decoding over frequency-selective fading channels are developed, while in [1331] a pre-DFT allowing the reduction of the number of input signals to a space-time-frequency decoder is developed.

Another related area of research is the development of equalization algorithms for *precoded* MIMO-OFDM systems. Various results can be found in the literature – see [1332–1342] for further details.

The presence of a time-varying channel raises the problem of ICI mitigation. In MIMO-OFDM communications, ICI effects can be reduced by modifying the signaling format [1343, 1344] or adopting proper *space-frequency* (SF) codes [1345]. An alternative is to use ICI cancelation algorithms at the receiver (e.g., see [456, 1346–1348]).

When the channel is unknown and a pilot sequence is transmitted, joint channel estimation and equalization can be accomplished. As in the SISO case, various pilot based and semiblind techniques are available. The reader can refer to [522, 524, 655, 851, 1320, 1349–1356] for the first and second class of algorithms, respectively.

Finally, it is worth noting that *noncoherent* (and, in particular, *differential*) *detection* techniques have also been considered for MIMO-OFDM, since they do not require channel estimation (e.g., see [1360–1363]).

6.9 Historical Notes

In this section a historical perspective on research results achieved in the field of channel equalization is provided. Because of the vastness of the research literature in this field, we focus attention on only that portion of it whose technical contents are closely related to the topics that have been analyzed in this chapter. Some additional details about the development of channel equalization up to the 1980s can be found in [1020, 1364].

The birth of channel equalization should not be related to the discovery of wireless data communications (i.e., in practice, wireless telegraphy), but to the need to compensate for channel distortions in telegraph communications over cables. In fact, in the second half of the nineteenth century the operators of the first transatlantic telegraph cables noted that Morse code symbols had to be transmitted very slowly to be fully understood at the other end, because of the time dispersion introduced by the distributed capacity of the communication channel. However, initial solutions for mitigating the effects of this dispersion were based on the choice of proper signal shaping or on the adoption of inductive loading in communication cables. Note that the latter technique made long-distance telephony possible. In the the first half of the twentieth century various results paved the way for the development of channel equalization techniques. A significant first achievement was the development of design methods for the synthesis of linear filters based on lumped elements [1365]. In fact, linear filters could be designed and adjusted to equalize the linear distortions affecting telephone circuits. Another significant achievement came in the seminal work of Harry Nyquist [1366, 1367], who derived a criterion ensuring zero ISI in a pulse transmission and showed that ISI-free baseband data transmission at symbol rates above twice the channel bandwidth is impossible. However, modern equalization theory started to develop in second half of the twentieth century. Among the first important contributions in this field we mention that of D. Tufts in 1965, who developed the analytical framework for optimum ZF and MMSE equalization for a channel with AWGN and a given frequency response [1368]. In particular, Tufts tackled the problem of splitting linear equalization between transmitter and receiver, thereby deriving expressions for jointly optimum frequency responses. In addition, he showed that an optimum linear equalizer could be realized as the cascade of an MF, a symbol rate sampler and a transversal filter, with taps spaced at symbol intervals. However, his work and other contributions to this field dating back to the 1960s was mainly of theoretical relevance. In that period the interest in practical applications of channel equalization techniques was stimulated by the need to increase data rates in voiceband telephone services. In 1963 R. Lucky, a Bell Labs researcher, devised an iterative "steepest descent" technique to train the tap coefficients of a transversal filter equalizer in such a way that peak ISI for any given channel was minimized before the start of actual data transmission [1369]. His equalizer represented an implementation of the so-called ZF criterion. Later, Lucky devised an iterative algorithm, which used data decisions and the differences between them and the equalizer's outputs to adjust (i.e., to *adapt*) the equalizer during actual data transmission [1370]. Unfortunately, these iterative algorithms did not take the presence of noise into account. The maximization of the equalizer's output SNR (where "noise" included residual ISI) was proposed by D. Coll and D. George in 1965 [1371]. It should also be noted that the birth of adaptive equalization is also related to the discovery of the LMS algorithm (see Section 5.1.3), which was introduced by B. Widrow and M. Hoff in 1960 [1372].

These achievements in adaptive equalization theory and practice were available at an important moment for the data communications industry in North America. At that time, the main commercial telecommunications system was the voiceband telephone network. However, the need for data sharing between remote locations was becoming more and more important. In 1968, the FCC allowed customer-owned equipment to be connected directly to the AT&T telephone network. This ruling opened up a new commercial market, since it made it possible to sell modems for data communications over the public phone network. The demand for higher data rates and better reliability soon became apparent. However, the bandwidth and noise characteristics of long-distance voiceband telephone channels intrinsically limit voiceband modems to data rates not exceeding a few tens of kilobits per second. In addition, these rates can only be achieved by employing relatively sophisticated techniques for modulation, coding, adaptive equalization, synchronization, and filtering (e.g., see [338, 1373]). These, in turn, require significant digital signal processing capabilities. All this explains why various advances in communication theory and signal processing for communication applications were stimulated by the need to develop competitive voiceband modems and a number of technical developments originated not only from academia but also from the research labs of various companies (e.g., Bell, Codex Corporation, IBM and NEC). From a commercial viewpoint it is worth noting that voiceband modems achieving data rates of 4.8 kb/s and 9.6 kb/s and equipped with adaptive equalizers started to appear in the late 1960s. These usually employed *vestigial sideband modulation* (VSB) or *single sideband modulation* (SSB) and baseband adaptive equalization. In the 1970s these were replaced by high-speed modems based on QAM or combined *amplitude modulation-phase modulation* (AM-PM), since these were more robust to typical channel impairments. From a theoretical viewpoint, various important achievements of the 1970s can be related to research on communication techniques for voiceband modems. In particular, we mention the following:

1. *Bandpass equalization*. Linear equalization can be done on *bandpass* modulated signals, in place of their baseband version, as originally pointed out by R. D. Gitlin, E. Y. Ho, and J. E. Mazo (all from Bell Labs) [1374]. Additional work by H. Kobayashi [1375] and D. D. Falconer [1376] showed that adaptive passband equalization can be combined with carrier recovery to minimize the delay in estimating and removing the effects of phase jitter.
2. *Fractionally spaced equalization*. The performance benefits deriving from the use of fractional spaced taps were analyzed first by L. Guidoux [1377] and G. Ungerboeck [1077] (who, however, acknowledged the existence of previous work on this topic, citing [1377–1379]).
3. *Decision feedback equalization*. The architecture of the decision feedback equalizer shown in Figure 6.5 was first described by M. E. Austin in 1967. The early history of this equalizer is summarized in a landmark paper by R. Price [1380]. This nonlinear technique plays a fundamental role over highly dispersive channels (like long cables and twisted copper pairs) where linear equalizers exihibit poor performance. An adaptation mechanism of the forward and feedback filters of DFEs conceptually similar to that of linear equalizers can be easily derived [1381].
4. *Tomlinson–Harashima precoding*. To avoid the error propagation problem that affects decision feedback equalization, symbol feedback can be moved from the receiver to the transmitter. This approach was proposed by M. Tomlinson [1060] and independently by H. Harashima and H. Miyakawa [1061]. The so-called *Tomlinson–Harashima precoding* uses the feedback filter tap coefficients in an inverse filter configuration at the transmitter, together with modulo arithmetic (a simple introduction to this technique can be found in [1373]).
5. *Maximum likelihood sequence detection*. A solution to the ML detection problem for PAM signaling over a frequency-selective channel with AWGN was proposed by David Forney in 1972 [422] (see Section 6.2.1.1) and was based on the use of a WMF and a VA. An alternative solution, based on the use of an MF at the receiver input, was developed by G. Ungerboeck [458]. MLSD provides close to optimum performance on highly dispersive channels, but, may require a huge complexity in the VA. This has motivated the search for pragmatic approaches to reducing the complexity of the VA, such as drastically limiting the number of its channel states and replacing the WMF with an adaptive filter able to shorten the overall impulse response (see Section 6.2.1.2). In addition, the complexity and error propagation issues of MLSD and DFE, respectively, stimulated research on other nonlinear equalization techniques that would be effective for mitigating ISI on highly dispersive channels (e.g., see [1382]).

6. *Equalization techniques for nonlinear channels.* One of the impairments affecting data transmissions over old voiceband telephone channels was the presence of *nonlinearities*; this resulted in nonlinear ISI at the input of equalization algorithms. Pioneering work in this field was done by D. D. Falconer [1383], who applied the *Volterra series* technique for the representation of channel nonlinearity to the design of an adaptive nonlinear receiver. His results showed that nonlinear decision feedback equalization can significantly reduce the error rate for a variety of channel characteristics. However, his method suffers from large equalization and adaptation complexity; this led to the development of alternative strategies for the cancelation of nonlinear ISI (e.g., see [1384]). Fortunately, the nonlinearity problem for voiceband telephone channels tended to disappear with the upgrading of the phone network. Recently, some of the nonlinearity cancelation approaches have been applied to the problem of mitigating the effects of power amplifier nonlinearities in digital wireless systems (e.g., see [1385, 1386]).

7. *Fast algorithms for equalizer adaptation.* One of the requirements of equalizers for voiceband modems was *fast startup*, that is, the capability of adjusting their taps to the communication channel based on a short training sequence. The convergence interval of the LMS algorithm, in terms of symbol intervals, is not short, being of the order of 10 or more times the number of equalizer taps [1387, 1388], and tends to get larger in the presence of a high time dispersion. A faster alternative was proposed by D. Godard, who developed an equalizer adaptation technique based on Kalman filtering [1389] and equivalent, in practice, to an RLS algorithm (see Section 5.1.3). The convergence time of the RLS algorithm is of the order of twice the number of equalizer taps, independent of the nature of the channel frequency response; however, this advantage is achieved at the cost of appreciably greater complexity. This motivated the study of alternative solutions for the fast startup of equalizers. Here we mention the so-called *fast Kalman* or *fast* RLS *algorithm* for the adaptation of decision feedback equalizers [599, 1390] (whose complexity is greater than that of the LMS algorithm, but increases only linearly with the number of equalizer tap coefficients), the development of *lattice-structured equalizers* [600, 1391, 1392] and the discovery of *cyclic equalization* independently by G. D. Forney, Jr. [1393] and by K. H. Mueller and D. A. Spaulding (both from Bell Labs) [1394].

8. *Blind algorithms for equalizer adaptation.* One of the challenging problem in equalization theory is the development of *blind* adaptation algorithms, which properly adjust the taps of an equalizer without requiring explicit training symbols, so that the transmission overhead can be substantially lowered [275]. Pioneering work in this field has been done by Y. Sato [1395] and D. Godard who developed the so-called *constant modulus* (CM) algorithm [1396]. This algorithm has the following significant features: it is a relatively simple modification of the LMS algorithm, does not require training symbols or receiver decisions, and applies to a variety of modulation types (a comprehensive review of its properties can be found in [1397]). However, the convergence rate of the CM and other blind adaptation algorithms is much lower than for their data-aided counterparts.

9. *Interference suppression.* In principle, an equalization algorithm can mitigate not only ISI but also the synchronous CCI coming from a interfering transmission on the same communication channel and at the same data symbol rate. In fact, in this case, both the ISI and CCI contributions to received signal samples are characterized by the same mathematical structure. These ideas have found applications in the development of algorithms for echo cancelation over phone lines, crosstalk suppression in multipair cables and CCI cancelation in wireless systems (e.g., see [746, 1398, 1399]). Preliminary work in the area was done by D. A. Shnidman [1400], who developed a generalized version of the Nyquist criterion for the combination of ISI and CCI and generalized work by Tufts [1368] on equalizer optimization, and by A. R. Kaye, D. A. George [1401] and W. Van Etten [554] on optimum receivers for multiplex PAM signals in a multichannel scenario.

10. *Frequency-domain equalization.* From the 1960s to the 1990s research and development efforts in the the field of channel equalization focused mainly on the discovery and implementation of algorithms operating in the TD. However, in that period the data rate over wired, wireless and

other transmission media continued to grow, so that the impact of time dispersion originating from communication channels became more and more important in receiver design. In fact, compensating for a larger dispersion required an increasing number of taps in equalization algorithms, resulting in a heavier computational load per symbol. As illustrated in Section 6.2.2, equalization in the frequency domain provides a way of reducing the growth rate of such complexity with data rate. From a historical perspective, this approach to channel equalization was made possible by an important discovery by S. Weinstein, P. Ebert, and J. Salz (all from Bell Labs) in 1969. These researchers proved that the FFT could significantly reduce the complexity involved in FD filtering of blocks of signal samples [388, 1402]. Other significant contributions to the development of equalization algorithms in the FD are due to: T. Walzman and M. Schwartz, who in 1973 developed MMSE adaptive equalizers based on projection methods for blocks of serially transmitted data symbols interspersed with sequences of zeros [1403, 1404]; E. Ferrara [1405], who in 1980 derived a (fast) FD implementation of the LMS adaptive transversal filter; and G. Clark, S. Mitra and S. Parker [1406], who in 1981 illustrated overlap-save and overlap-add implementations of FIR block adaptive filtering in the FD.

Wired transmission channels are usually characterized by very slow time changes, due to variations in environmental parameters (e.g., temperature). On the other hand, as illustrated in Chapter 2, wireless channels typically experience relatively rapid variations. The ability of an adaptive equalization algorithm to track and compensate for a time-varying channel depends on how rapid the time variation is relative to the data symbol rate and on how many equalizer tap coefficients are being adapted. In fact, the larger the number of tap coefficients, the slower the changes that an equalizer can adapt to and track. A rough indication of adaptation ability is given by the product $B_D T_s N_c$, where N_c denotes the number of tap coefficients and B_D is the Doppler bandwidth. If this product does not exceed about 10^{-2}, channel variations can be considered slow and a relatively simple adaptation algorithm is probably adequate; otherwise, more refined signal processing techniques are needed.

Despite the substantial difference between the classes of wired and wireless channels, the bulk of knowledge acquired in the field of equalization techniques for voiceband telephone channels proved to be fundamental in the subsequent development of equalization techniques for radio channels. Research on the application of known equalization techniques to wireless communications started in the 1970s; in particular, we mention the contribution made by P. Monsen, who investigated the use of decision feedback equalization on fading channels [471, 1018, 1044]. This followed work that was mainly motivated by the need to compensate for channel distortion in HF communications over ionospheric channels and in troposcatter communications [45, 1407]. In particular, in the 1980s the use of adaptive equalization techniques on rapidly fading HF channels was investigated in depth (e.g., see [1227, 1392, 1408–1411]) and it was shown that, for such channels, nonlinear equalization techniques, such as DFE or MLSD, are usually required to ensure proper adaptation to quick channel variations. However, by the end of the 1980s research activities on equalization techniques began to concentrate on TDMA digital cellular systems. In fact, such systems require adaptive equalization at the demodulator to combat the ISI resulting from the time-variant multipath propagation of the signal through the communication channel. Adaptive equalization techniques developed in the previous two decades for high-speed, single-carrier serial transmission over telephone and radio channels could certainly also be applied to digital transmission over mobile radio channels in the VHF band. However, the TDMA signal structure and the rapidly varying fading imposed some stringent conditions on the design of such techniques. This is discussed in detail in [860], where a survey of adaptive linear and nonlinear equalization techniques that can be employed in a narrowband TDMA system is provided. In addition, it is shown that both MLSD (Ungerboeck's form, to avoid the implementation of a WMF) and DFE are viable equalization methods for mobile radio (see also [1224]). Note, however, that reliable operation in the presence of severe channel conditions typically requires accurate tracking of channel changes [510] and an adaptation of the equalization algorithm over each burst of a TDMA frame [560]. In this scenario appreciable benefits in terms of error performance can be achieved by adopting a PSP approach to channel estimation, that is, by embedding channel estimation in MLSD (see Section 6.6.2) in order to improve the tracking capability of the receiver; however, a number of more advanced channel estimation/equalization techniques, such as *bidirectional decision feedback equalization* [1032], MAP symbol detection combined with Kalman filtering for channel estimation (e.g., see [1233, 1238]), iterative channel estimation and equalization [1412], and channel estimation and tracking based on soft estimation of channel states [861], have been proposed in the literature.

At the beginning of the 2000s a substantial portion of the research activities in the field of channel equalization has focused on the following topics:

1. *Equalization techniques for* SIMO OFDM *systems in the presence of* ICI *and/or* IBI - The interest in this topic is due to the growing importance of OFDM in broadband communications over fixed and mobile radio channels (see Sections 6.3, 6.5 and 6.7).
2. *Equalization techniques in the* FD *for* SC *systems.* The attention paid to SC modulations combined with FD channel equalization is due to the fact that these represent an appealing alternative to OFDM in broadband communications (see Section 6.2.2).
3. *Equalization techniques for* MIMO SC *and* OFDM *systems.* Many efforts have been made to extend SISO/SIMO equalization techniques to a MIMO scenario (see Section 6.8).

Another recent and interesting research area concerns the use of factor graphs (see Section 10.8) for the development of soft output equalization techniques. The reader can refer to [577, 1178, 1413, 1414] for further details on this topic.

6.10   Further Reading

Various tutorials and overviews of different classes of equalization techniques can be found in the technical literature. Here, we mention [1020], which provides a detailed analysis of adaptive equalization techniques up to 1985; [860], which provides a survey of adaptive equalization for TDMA digital mobile radio; [275, 1415], which consider equalization techniques for fading dispersive channels including both time and frequency selectivity; [1027, 1416] on decision feedback equalization with the MMSE criterion; and [1417] on equalization techniques in the FD. An interesting tutorial on the use of adaptive linear equalization digital signaling over slowly time-varying, bandlimited channels is provided by [1076]. A technical overview of equalization developments up to the 1960s can be found in the well-known book by R. W. Lucky and J. Salz and E. J. Weldon [280]. Other reference books for equalization techniques for SISO communication systems are [29, 1418, 1419]. An introduction to channel equalization for MIMO systems can be found in [21, 38].

Part Two

Information Theory and Coding Schemes

7

Elements of Information Theory

7.1 Introduction

In this chapter we seek to put many of the results of other chapters into context by examining the capacity limits imposed on wireless systems by the constructs of information theory. These limits are imposed by the underlying tenets of information theory as originally developed by *Claude Shannon* [1420, 1421]. He formulated the concept of *channel capacity*, which has been shown to provide a fundamental limit on the maximum data rate that can be transmitted over a channel with asymptotically small probability of error. In practice, he established that for any transmission rate R_b bits/s less than the capacity of C bits/s, there exists a code that allows essentially error-free transmission. This result applies with some variation to any channel. However, the main drawback of Shannon's discovery is that it essentially provides an existence proof, but does not provide techniques for the construction of codes to achieve C. Moreover, his original work dealt with *additive noise channels* only, both discrete and continuous, and for many years provided the benchmark against which all systems were measured. With the development of modern wireless systems, it has become necessary to take the time-varying and fading wireless channel environment into account in evaluating channel capacity. As already discussed in Chapter 2, almost all wireless channels exhibit fading behavior to a greater or lesser degree. This fading has an enormous impact on the performance of wireless systems and must be taken into account in evaluating the performance limits of wireless transmission systems. Thus, in this chapter we briefly discuss the evaluation of channel capacity in some fading dispersive environments of wireless communications systems.

This chapter is organized as follows. Some standard results on the capacity of SISO communication channels are provided in Section 7.2. The issue of channel capacity for MIMO channels is investigated in Section 7.3, where frequency-flat fading is assumed for simplicity. Finally, some brief historical notes and recommendations for further reading are provided in Sections 7.4 and 7.5, respectively.

7.2 Capacity for Discrete Sources and Channels

We start by developing expressions for channel capacity of two classical channel models, namely the *discrete memoryless channel* (DMC) and the bandlimited AWGN channel. Both these models are used in capacity considerations for more complex situations. In particular, the DMC and the resulting capacity expression serves as a basis for the design of codes, whereas the AWGN channel and its capacity provide the basis for describing both the ergodic capacity and the outage capacity of more

Wireless Communications: Algorithmic Techniques, First Edition.
Giorgio M. Vitetta, Desmond P. Taylor, Giulio Colavolpe, Fabrizio Pancaldi, Philippa A. Martin.
© 2013 John Wiley & Sons, Ltd. Published 2013 by John Wiley & Sons, Ltd.

complex wireless channels. In this section, we briefly consider channel models and then develop the corresponding capacity expressions.

7.2.1 The Discrete Memoryless Channel

The DMC is a channel model particularly suited to considering coding. In a system sense, it may be regarded as the channel between the output of the encoder at the transmitter and the input to the corresponding decoder at the receiver. Its mathematical description can be found in many papers and texts (e.g., see [327, 1422]) and is summarized below. A DMC is characterized by the following two properties:

- The alphabets of its input and output are *discrete*.
- *It does not have memory*, that is, it acts on each of its input data independently of all other data.

In practice, any channel input X (i.e., the output from a channel encoder) belongs to a q-ary alphabet $A_X = \{x_0, x_1, \ldots, x_{q-1}\}$ and, similarly, any channel output Y (i.e., the input to the decoder, following modulation, transmission and demodulation) belongs to a Q-ary alphabet $A_Y = \{y_0, y_1, \ldots, y_{Q-1}\}$. Note also that, in all cases of interest, $Q \geq q$, usually q is a power of 2 and, generally speaking the input and output alphabets do not have to be identical.

Let us now formulate the property of absence of memory in mathematical terms. For this purpose, let us assume that, in N consecutive uses of the communication channel, the random data vector:

$$\mathbf{X}_N \triangleq [X_0, \ X_1, \ \ldots, \ X_{N-1}]^T, \tag{7.1}$$

characterized by the probability mass function $p_{\mathbf{X}_N}(\mathbf{i}_N) \triangleq \Pr\{\mathbf{X}_N = \mathbf{i}_N\}$, is transmitted, where:

$$\mathbf{i}_N \triangleq [i_0, \ i_1, \ \ldots, \ i_{N-1}]^T \tag{7.2}$$

and each element of $\mathbf{i}_N$ belongs to the alphabet, A_X. In response to X_t (with $t = 0, 1, \ldots, N-1$) the channel generates the random response vector Y_t, belonging to some alphabet, A_Y, so that the overall channel response to $\mathbf{X}_N$ is represented by the vector:

$$\mathbf{Y}_N \triangleq [Y_0, \ Y_1, \ \ldots, \ Y_{N-1}]^T \tag{7.3}$$

characterized by the probability mass function $p_{\mathbf{Y}_N}(\mathbf{j}_N) \triangleq \Pr\{\mathbf{Y}_N = \mathbf{j}_N\}$, where:

$$\mathbf{j}_N \triangleq [j_0, \ j_1, \ \cdots, \ j_{N-1}]^T \tag{7.4}$$

and each element of $\mathbf{j}_N$ belongs to A_Y. Then the discrete channel is *memoryless* if and only if $\mathbf{Y}_N$, given $\mathbf{X}_N$, consists of N independent random variables or, equivalently, if and only if the conditional joint probability mass function $p_{\mathbf{Y}_N|\mathbf{X}_N}(\mathbf{j}_N|\mathbf{i}_N)$ of $\mathbf{Y}_N$ given $\mathbf{X}_N$ can be factored as:

$$p_{\mathbf{Y}_N|\mathbf{X}_N}(\mathbf{j}_N|\mathbf{i}_N) = \prod_{t=0}^{N-1} p_{Y_t|X_t}(j_t|i_t). \tag{7.5}$$

Then the overall characteristics of a DMC can be described by a set of qQ conditional or *transition probabilities*:

$$P(y_i|x_j) \triangleq \Pr\{Y = y_i|X = x_j\}, \tag{7.6}$$

with $j = 0, 1, \ldots, q-1$ and $i = 0, 1, \ldots, Q-1$.

7.2.2   The Continuous-Output Channel

We now extend the channel concepts developed above to model the case of a channel having discrete inputs and continuous outputs. We then briefly consider a memoryless waveform channel, where the inputs and outputs are bandlimited waveforms.

7.2.2.1   The Discrete-Input Case

Here we consider the case where the channel input is discrete and described by the q-ary alphabet X as above; however, we now consider the channel output to be *continuous*. This can be described as above by letting the number of possible channel outputs, Q, approach infinity ($Q \to \infty$), implying that the channel output, Y, can take any value on the real line, $-\infty < Y < \infty$. From this we can then define a discrete-input, continuous-output channel that is described by the conditional or *transition pdf*:

$$f_{Y|X}(y|x_k), \tag{7.7}$$

with $k = 0, 1, \ldots, q - 1$. An important example of such a channel is the *discrete-input additive white Gaussian noise channel*, which is characterized by the input–output relationship:

$$Y = X + N, \tag{7.8}$$

where N is a zero-mean Gaussian random variable with variance σ_n^2. It is then straightforward to see that the probability (7.7) is given by:

$$f_{Y|X}(y|x_k) = \frac{1}{\sqrt{2\pi}\sigma_n} \exp\left[\frac{-(y - x_k)^2}{2\sigma_n^2}\right]. \tag{7.9}$$

Then, for any given sequence of channel inputs $\{X_i, \ i = 0, 1, \ldots, N - 1\}$, we obtain an output sequence of the form:

$$Y_i = X_i + N_i, \tag{7.10}$$

with $i = 0, 1, \ldots, N - 1$, where N_i is a zero mean Gaussian random variable with variance σ_n^2. The *memoryless property* of this channel is expressed by the factorization:

$$f_{\mathbf{Y}_N|\mathbf{X}_N}(\mathbf{y}_N|\mathbf{i}_N) = \prod_{t=0}^{N-1} f_{Y_t|X_t}(y_t|X_t = i_t) \tag{7.11}$$

of the joint conditional pdf $f_{\mathbf{Y}_N|\mathbf{X}_N}(\mathbf{y}_N|\mathbf{i}_N)$, where $\mathbf{y}_N \triangleq [y_0, y_1, \ldots, y_{N-1}]^T$ and $\mathbf{i}_N$ is defined in (7.2).

7.2.2.2   The Waveform Channel

Both the channels considered above include all the functions of modulation, transmission and demodulation. We can easily separate the modulator and demodulator functions from the actual channel. We consider a model where the channel inputs and outputs are baseband waveforms, and denote by $X(t)$ and $Y(t)$ the channel input and output waveforms, respectively. In addition, we assume that the channel has bandwidth B Hertz. It is assumed to have an ideal frequency response, $H(f) = 1$, over this bandwidth and that its output is corrupted by AWGN with two-sided power spectral density equal to $N_0/2$ Watts per Hertz. Then, assuming that $X(t)$ is also bandlimited to B Hertz and its support is the time interval (t_i, t_f), we have:

$$Y(t) = X(t) + N(t), \tag{7.12}$$

where $N(t)$ is a bandlimited Gaussian noise process. Then following [321] (see also Section D.2), we may write the waveforms of (7.12) as the orthogonal function expansions:

$$Y(t) = \sum_{l=0}^{+\infty} Y_l \varphi_l(t), \tag{7.13}$$

$$X(t) = \sum_{l=0}^{+\infty} X_l \varphi_l(t), \tag{7.14}$$

and

$$N(t) = \sum_{l=0}^{+\infty} N_l \varphi_l(t). \tag{7.15}$$

Here the elementary signals $\{\varphi_l(t), l = 0, 1, \ldots\}$ form a *complete orthonormal set* over (t_i, t_f) (so that $\int_{t_i}^{t_f} \varphi_i(t)\varphi_j(t)dt = \delta_{ij}$, where δ_{ij} is the Kronecker delta function) and the coefficients in the above expansions are defined as $X_l \triangleq \int_{t_i}^{t_f} X(t)\varphi_l(t)dt$, $Y_l \triangleq \int_{t_i}^{t_f} Y(t)\varphi_l(t)dt$ and $N_l \triangleq \int_{t_i}^{t_f} N(t)\varphi_l(t)dt$ for any l. Note that:

$$Y_l = X_l + N_l \tag{7.16}$$

for any l and that, since the functions $\{\varphi_i(t)\}$ are orthonormal, then random variables $\{N_l, l = 0, 1, \ldots\}$ are uncorrelated and, being jointly Gaussian, are also independent. In addition, they have mean zero and variance $\sigma_n^2 = N_0/2$.

The foregoing allows us to characterize the AWGN waveform channel. We simply use the coefficients from the expansions (7.13)–(7.15). From (7.16), we may write:

$$f_{Y_l|X_l}(y|x) = \frac{1}{\sqrt{2\pi}\sigma_n} \exp\left[-\frac{(y-x)^2}{2\sigma_n^2}\right] \tag{7.17}$$

for $n = 0, 1, \ldots$, and the random variables $\{N_l, l = 0, 1, \ldots\}$ are mutually independent so that:

$$f_{\mathbf{Y}_N|\mathbf{X}_N}(\mathbf{y}_N|\mathbf{x}_N) = \prod_{l=0}^{N-1} f_{Y_l|X_l}(y_l|x_l), \tag{7.18}$$

where $\mathbf{Y}_N \triangleq [y_0, y_1, \ldots, y_{N-1}]^T$ and $\mathbf{X}_N \triangleq [x_0, x_1, \ldots, x_{N-1}]^T$, for any value of N. This procedure allows us to reduce the waveform channel to a discrete-time form, equivalent to that described by equation (7.11). Similar results are found if the sample sequence (7.16) is generated not via an orthonormal basis, but by uniformly sampling $Y(t)$ at a rate equal to $2W$ Hertz; in this case, if $T \triangleq t_f - t_i$ denotes the duration of the observation interval, $N = 2WT$ independent samples are generated, so that the channel is completely described by (7.17) and (7.18).

Finally, it is worth mentioning that both the DMC channel model and the waveform channel model are used in determining channel capacity. The choice of which to use depends on the aim of the analysis being undertaken. In considering only coding effects, the DMC, described by (7.5) and (7.6), is normally employed; however, if the effect of the signal modulation is to be considered, then the Gaussian channel model described by (7.17) and (7.18) is usually adopted [321, 1422].

7.2.3 Channel Capacity

We now develop a calculation of channel capacity for the DMC and waveform channel models considered above. The calculations, although relatively simple, provide results that are widely used in developing capacity expressions for the more complex wireless channels considered in the remainder of this chapter.

7.2.3.1   The Discrete Memoryless Channel

We start by considering the DMC described by (7.5) and (7.6). We assume initially that only the q-ary symbol X_i is transmitted and that the Q-ary symbol Y_j is received. Then, the *conditional mutual information* $I(x_i; Y = y_j)$ provided about the transmission event $\{X = x_i\}$ by the reception event $\{Y = y_j\}$ is given by [327, 1422]:

$$I(x_i; Y = y_j) = \log \frac{P(y_j|x_i)}{P(y_j)}, \tag{7.19}$$

where

$$P(y_i) \triangleq \Pr\{Y = y_i\} = \sum_{k=0}^{q-1} P(x_k)P(y_i|x_k), \tag{7.20}$$

$P(x_k) \triangleq \Pr\{X = x_k\}$ and $P(y_j|x_i)$ is defined by (7.6). Then the *average mutual information* provided by Y about X is given by the average of (7.19) over both the transmitted and received alphabets, that is:

$$\begin{aligned}
I(X; Y) &= \sum_{j=0}^{q-1} \sum_{i=0}^{Q-1} P(x_j, y_i) \log \frac{P(y_i|x_j)}{P(y_i)} \\
&= \sum_{j=0}^{q-1} \sum_{i=0}^{Q-1} P(x_j)P(y_i|x_j) \log \frac{P(y_i|x_j)}{P(y_i)},
\end{aligned} \tag{7.21}$$

where $P(x_j, y_i) \triangleq \Pr\{X = x_i, Y = y_i\}$. Note that the transition probabilities $\{P(y_i|x_k)\}$ are determined by the *characteristics of the channel*, whereas the input probabilities $P(x_k)$ are controlled by the transmitter or discrete channel encoder (in other words, by the system designer).

The *channel capacity* is then found as the maximum of the average mutual information $I(X; Y)$ over the set of transmission probabilities $\{P(x_j)\}$, so that the capacity of the DMC can be expressed as:

$$\begin{aligned}
C &= \max_{\{P(x_j)\}} I(X; Y) \\
&= \max_{\{P(x_j)\}} \sum_{j=0}^{q-1} \sum_{i=0}^{Q-1} P(x_j)P(y_i|x_j) \log \frac{P(y_i|x_j)}{P(y_i)}.
\end{aligned} \tag{7.22}$$

In this expression, when the logarithm is taken to base 2, the units of C are bits per channel use or symbol, whereas, when a q-ary symbol is transmitted every T_s seconds, then the units of C/T_s are bits per second.

Let us now apply the results developed above to the specific case of a q-ary *symmetric channel*.

Example 7.2.1 We wish to find the capacity of a DMC having q inputs and q outputs, and characterized by the transition probabilities:

$$P(y_i|x_j) = \begin{cases} 1 - p & i = j = 0, 1, \ldots, q \\ \frac{p}{q-1} & i \neq j. \end{cases} \tag{7.23}$$

The overall probability of error for this channel is p, but there are $q - 1$ possible (and equally likely) incorrect output symbols for each input symbol. Substituting these probabilities into (7.22) and carrying out the resulting maximization yields the expression:

$$C = \log q + p \log(q - 1) - H(p) \qquad \text{bits/channel use} \tag{7.24}$$

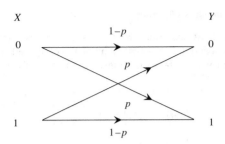

Figure 7.1   Schematic representation of BSC with error probability p.

for the capacity of the q-ary symmetric channel, where $H(p) = -p \log p - (1 - p) \log(1 - p)$ is the so-called *binary entropy function* [327, 1423]. A special case of this channel occurs when $q = 2$. We then obtain the *binary symmetric channel* (BSC) having the capacity:

$$C = 1 - H(p) \qquad \text{bits/channel use.} \qquad (7.25)$$

The BSC has been widely used in the design of error control codes and is often represented schematically as shown in Figure 7.1.

☐

7.2.3.2   The Discrete-Input, Continuous-Output Channel

We next consider the discrete-time additive Gaussian noise channel, namely, a channel with q-ary discrete inputs but continuous outputs as described by (7.10) and (7.11). It is then straightforward to write the capacity of this channel in bits per channel use as:

$$C = \max_{P(x_i)} \sum_{i=0}^{q-1} \int_{-\infty}^{\infty} f_{Y|X}(y|x_i) P(x_i) \log \frac{f_{Y|X}(y|x_i)}{f_Y(y)} dy, \qquad (7.26)$$

where the channel output pdf $f_Y(y)$ is given by:

$$f_Y(y) = \sum_{k=0}^{q-1} f_{Y|X}(y|x_k) P(x_k). \qquad (7.27)$$

We note that the expression for the capacity of this channel is merely a generalization of that for the DMC as defined in equation (7.22) to allow for continuous channel outputs.

7.2.3.3   The Bandlimited Waveform Channel

Finally, we consider the capacity of a bandlimited waveform channel perturbed by AWGN. A link between this case and the previous ones can be established using the coefficients $\{X_i\}$, $\{Y_i\}$ and $\{N_i\}$ of the series expansions of (7.13)–(7.15), which in effect reduce the problem to a discrete form. To do this, we form the vectors of coefficients $\mathbf{X}_N = [X_0, X_1, \ldots, X_{N-1}]^T$ and $\mathbf{Y}_N = [Y_0, Y_1, \ldots, Y_{N-1}]^T$, where $N = 2WT$ and $Y_i = X_i + N_i$ (see (7.16)). We may then write the *average mutual information* between these vectors as [321, 1422]:

$$I(\mathbf{X}_N; \mathbf{Y}_N) = \int_{\mathbf{X}_N} \int_{\mathbf{Y}_N} f_{\mathbf{Y}_N|\mathbf{X}_N}(\mathbf{y}_N|\mathbf{x}_N) f_{\mathbf{X}_N}(\mathbf{x}_N) \log \frac{f_{\mathbf{Y}_N|\mathbf{X}_N}(\mathbf{y}_N|\mathbf{x}_N)}{f_{\mathbf{Y}_N}(\mathbf{y}_N)} d\mathbf{x}_N \, d\mathbf{y}_N. \qquad (7.28)$$

Then, since the components of $\mathbf{X}_N$ are independent and those of $\mathbf{Y}_N$, given $\mathbf{X}_N = \mathbf{x}_N$, are also independent, we may rewrite this in the form:

$$I(\mathbf{X}_N; \mathbf{Y}_N) = \sum_{i=0}^{N-1} \int_{x_i=-\infty}^{\infty} \int_{y_i=-\infty}^{\infty} f_{Y_i|X_i}(y_i|x_i) \, f_{X_i}(x_i) \, \log \frac{f_{Y_i|X_i}(y_i|x_i)}{f_{Y_i}(y_i)} dy_i dx_i$$

$$= \sum_{i=0}^{N-1} I(X_i; Y_i), \tag{7.29}$$

where $I(X_i; Y_i)$ denotes the average mutual information shared between X_i and Y_i:

$$f_{Y_i|X_i}(y_i|x_i) = \frac{1}{\sqrt{2\pi\sigma_n^2}} \exp\left[-\frac{(y_i - x_i)^2}{2\sigma_n^2}\right] \tag{7.30}$$

and $\sigma_n^2 = N_0/2$. The channel capacity is the maximum of (7.29) over the input or source probabilities $\{f_{X_i}(x_i)\}$ and is obtained when the source symbols $\{X_i\}$ are statistically independent Gaussian random variables with mean zero and variance σ_x^2 [321, 1422], so that their pdf is:

$$f_{X_i}(x) = \frac{1}{\sqrt{2\pi}\sigma_x} \exp\left[-\frac{x^2}{2\sigma_x^2}\right]. \tag{7.31}$$

It then easily follows from (7.28) that the maximum mutual information between the vectors $\mathbf{X}_N$ and $\mathbf{Y}_N$ is given by:

$$\max_{\{f_{X_i}(x_i)\}} I(\mathbf{X}_N; \mathbf{Y}_N) = \frac{1}{2}N \log\left(1 + \frac{\sigma_x^2}{\sigma_n^2}\right) = WT \log\left(1 + \frac{\sigma_x^2}{\sigma_n^2}\right) \text{ bits.} \tag{7.32}$$

Now let us constrain the *average power* P_{av} in the original transmit process $X(t)$, so that we may write:

$$P_{\text{av}} = \frac{1}{T}\int_{t_i}^{t_f} E\{x^2(t)\}dt = \frac{1}{T}\sum_{i=0}^{N-1} E\{X_i^2\} = \frac{N\sigma_x^2}{T}. \tag{7.33}$$

We may then write the variance σ_x^2 as:

$$\sigma_x^2 = \frac{T P_{\text{av}}}{N} = \frac{P_{\text{av}}}{2W}. \tag{7.34}$$

Then substituting this result in (7.32) and dividing the result by T yields the capacity of the bandlimited additive Gaussian noise channel as:

$$C = W \log\left(1 + \frac{P_{\text{av}}}{W N_0}\right) \qquad \text{bits/s.} \tag{7.35}$$

This is the basic equation for the capacity of a bandlimited AWGN channel, where the bandwidth is W Hertz and the average transmit power has been constrained. In the parlance of modern wireless communications, this, as we shall see, is a special case of the so-called *ergodic channel capacity*. It was originally derived by Shannon [1421] and rederived in [321], where the concept of *sphere hardening* is employed. The present summary largely follows the treatment in [1422]. It is important to note that channel capacity increases logarithmically with average transmit power. We also note that it increases with bandwidth, W: essentially linearly at low SNR and more slowly at high SNR. For fixed P_{av}, it can easily be shown that as $W \to \infty$, the Gaussian channel capacity approaches an asymptotic value given by:

$$C_\infty = \frac{P_{\text{av}}}{N_0 \ln 2} \qquad \text{bits/s.} \tag{7.36}$$

This limit is approached by coded OFDM transmission as the number of carriers becomes very large [1424].

The capacity as given in (7.35) can be shown to be a special case of a more general expression [1418]. Consider a data transmission (with fixed average power P_{av}) over a deterministic frequency-selective channel having the frequency response $H(f)$ and affected by additive Gaussian noise with two-sided noise spectral density $N(f)$. It can be shown that the capacity of this channel is given by [1418]:

$$C = \frac{1}{2} \int_{-\infty}^{+\infty} \max\left[0, \log\left(\theta \frac{|H(f)|^2}{N(f)}\right)\right] df \qquad \text{bits/s,} \qquad (7.37)$$

where the parameter θ satisfies the constraint:

$$P_{av} = \int_{-\infty}^{+\infty} \max\left[0, \theta - \frac{N(f)}{|H(f)|^2}\right] df. \qquad (7.38)$$

These results show that channel capacity is attained when the transmitted energy is placed in the frequency bands where channel noise and channel attenuation are lower; in practice, the optimal distribution of the transmitted signal energy as a function of frequency is found using the so-called *water-pouring* (or *water-filling*) *technique* [280, 1418]. Finally, we note that:

- when the channel has unity gain over the bandwidth W and zero response elsewhere, the capacity as expressed in (7.37) reduces to the well-known expression (7.35), and
- the phase characteristic of the channel is immaterial to the capacity of the channel.

7.3   Capacity of MIMO Fading Channels

7.3.1   Frequency-Flat Fading Channel

We now turn our attention to the case of wireless channels. Wireless channels are characterized to a large extent by the fact that they are, in general, time-varying. In addition, they may also be frequency-selective and therefore exhibit a time-varying frequency response, as described in detail in Chapter 2. For present purposes, we develop a vector-based model more suited to developing expressions for channel capacity. Moreover, since much of the rest of the book deals with systems that can achieve transmit and/or receive diversity, we focus directly on the MIMO case, from which we can show that the SISO channel capacity as discussed above arises as a special case.

In what follows, we first briefly describe a vector channel model which describes the input–output behavior of a MIMO channel. We then go on to consider various examples of capacity calculation. These depend to a great extent on what channel knowledge is assumed at the transmit and receive ends of the MIMO link.

7.3.1.1   MIMO Channel Model

We consider here a single point-to-point MIMO transmission link having n_T transmit antennas and n_R receive antennas, using PAM signaling and affected by frequency-flat fading.[1] Under the above assumptions the baseband signal vector $\mathbf{r} \triangleq [r_0, r_1, \ldots, r_{n_R-1}]^T$ received in a specific symbol interval can be represented as (see Section 2.3.1):

$$\mathbf{r} = \mathbf{Hx} + \mathbf{n}, \qquad (7.39)$$

---

[1] When this is not the case, the work of [1425] and others can be followed to obtain a matrix formulation for the channel similar to that illustrated below; however, then any calculation of channel capacity becomes much more complicated and in many instances remains an open research problem.

where $\mathbf{x} \triangleq [x_0, x_1, \ldots, x_{n_T-1}]^T$ is a vector of transmitted symbols, $\mathbf{n} \triangleq [n_0, n_1, \ldots, n_{n_R-1}]^T$ is additive Gaussian noise and $\mathbf{H} = [h_{i,j}]$ is an $n_R \times n_T$ channel matrix (here $h_{i,j}$ denotes the channel response or fading coefficient of the flat fading channel between the jth transmit antenna and the ith receive antenna, with $i = 0, 1, \ldots, n_R - 1$ and $j = 0, 1, \ldots, n_T - 1$). In our analysis of MIMO channel capacity the following assumptions are made about the statistical properties of model (7.39):

1. Transmitted signals are subject to an *average power constraint*, so that their optimum distribution is Gaussian [321, 1420, 1421]. We thus treat the elements of $\mathbf{x}$ as independent, zero-mean Gaussian random variables with covariance matrix:

$$\mathbf{R}_\mathbf{x} = \mathrm{E}\{\mathbf{x}\,\mathbf{x}^H\}, \tag{7.40}$$

so that constraining the total average transmit power to P for any number of transmit antennas is equivalent to setting:

$$\mathrm{tr}(\mathbf{R}_\mathbf{x}) = P \tag{7.41}$$

where $\mathrm{tr}(\cdot)$ denotes the trace of a matrix. If, as is often the case (see later examples), the channel is *unknown* at the transmitter, we also assume that the signals transmitted by each of the n_T antennas have equal average power given by P/n_T. Moreover, since signals from distinct transmit antennas are independent, the covariance matrix of the transmitted signal vectors can be put in the simple form:

$$\mathbf{R}_\mathbf{x} = \frac{P}{n_T}\mathbf{I}_{n_T}. \tag{7.42}$$

2. The elements of $\mathbf{H}$ are subject to a normalization constraint ensuring that each of the n_R receive antennas observes the total transmit power. This assumption means that signal gain variations due to propagation (including such effects as antenna gain and shadowing) are included and substantially simplifies calculations [16, 18, 226]. If the channel gains are fixed and deterministic, this constraint can be expressed as:

$$\sum_{j=0}^{n_T-1} |h_{i,j}|^2 = n_T, \tag{7.43}$$

for $i = 0, 1, \ldots, n_R - 1$, whereas, in the more general case of Rayleigh fading (where the elements of $\mathbf{H}$ are random variables) the LHS of the latter equation is replaced with its expected or average value.

3. The noise vector $\mathbf{n}$ is independent of $\mathbf{x}$, and is temporally and spatially white with covariance matrix:

$$\mathbf{R}_\mathbf{n} \triangleq \mathrm{E}\{\mathbf{n}\,\mathbf{n}^H\} = \sigma_n^2\,\mathbf{I}_{n_R}, \tag{7.44}$$

which implicitly assumes the same average noise power, denoted by σ_n^2, in each of the n_R receiver branches.

Under these assumptions, the covariance matrix of the (zero-mean) received signal vector $\mathbf{r}$ (7.39) is given by:

$$\mathbf{R}_\mathbf{r} \triangleq \mathrm{E}\{\mathbf{r}\,\mathbf{r}^H\} = \mathbf{H}\,\mathbf{R}_\mathbf{x}\,\mathbf{H}^H + \sigma^2\mathbf{I}_{n_R}. \tag{7.45}$$

Since the signal power at each of the n_R receive antennas is assumed to be equal to P (i.e., to the total transmit power), the received SNR as can be expressed as:

$$\gamma = \frac{P}{\sigma_n^2}, \tag{7.46}$$

which is independent of the number of transmit antennas.

7.3.2 MIMO Channel Capacity

In deriving expressions for channel capacity in the wireless context, we must take different constraints into account depending on the state of knowledge of the channel at the transmitter and/or the receiver. In each case, system capacity is defined as the maximum possible transmission rate such that the probability of error can be made arbitrarily small. Here we focus primarily on the MIMO channel model described in the previous subsection and develop capacity expressions in different scenarios, each referring to a specific set of constraints on the knowledge at the transmitter and receiver about CSI. Such expressions illustrate the potentially very high spectral efficiency of MIMO channels, and some of them reduce to the Gaussian channel results derived earlier.

7.3.2.1 Channel Unknown at the Transmitter

The first scenario we analyze is characterized by a channel matrix $\mathbf{H}$ *unknown to the transmitter*, but *perfectly known at the receiver* [16, 18, 38, 226]. Evaluating channel capacity in this case requires use of the SVD theorem (see equation (C.11)) to the $n_R \times n_T$ channel matrix $\mathbf{H}$. This yields the factorization:

$$\mathbf{H} = \mathbf{U}\mathbf{D}\mathbf{V}, \tag{7.47}$$

where $\mathbf{U}$ ($\mathbf{V}$) is an $n_R \times n_R$ ($n_T \times n_T$) unitary matrix and $\mathbf{D}$ is an $n_R \times n_T$ diagonal matrix, whose diagonal elements, representing the *singular values* of $\mathbf{H}$, are nonnegative. It can be proved that the singular values $\{\sigma_i, i = 0, 1, \ldots, n_R - 1\}$ are the nonnegative square roots of the eigenvalues $\{\lambda_i, i = 0, 1, \ldots, n_R - 1\}$ of the $n_R \times n_R$ matrix $\mathbf{H}\mathbf{H}^H$. Note that an eigenvalue λ_l satisfies the equality:

$$\mathbf{H}\mathbf{H}^H \mathbf{y}_l = \lambda_l \, \mathbf{y}_l, \tag{7.48}$$

with $l = 0, 1, \ldots, n_R - 1$, where the $n_R \times 1$ vector $\mathbf{y}_l \neq \mathbf{0}_{n_R}$ is the eigenvector associated with it. In addition, it can be proved that the columns of $\mathbf{U}$ ($\mathbf{V}$) are the eigenvectors of $\mathbf{H}\mathbf{H}^H$ ($\mathbf{H}^H\mathbf{H}$).

Substituting (7.47) into (7.39) now yields:

$$\mathbf{r} = \mathbf{U}\mathbf{D}\mathbf{V}^H\mathbf{x} + \mathbf{n}, \tag{7.49}$$

so that, if the invertible transformations:

$$\mathbf{r}' = \mathbf{U}^H\mathbf{r}, \tag{7.50}$$

$$\mathbf{x}' = \mathbf{V}^H\mathbf{x} \tag{7.51}$$

and

$$\mathbf{n}' = \mathbf{U}^H\mathbf{n} \tag{7.52}$$

are defined, from (7.49) the equivalent representation:

$$\mathbf{r}' = \mathbf{D}\mathbf{x}' + \mathbf{n}' \tag{7.53}$$

is easily inferred for the received signal vector. This result can be further simplified by noting that the number r of nonzero eigenvalues of $\mathbf{H}\mathbf{H}^H$ is equal to the rank of $\mathbf{H}$, so that $r \leq m = \min(n_R, n_T)$. Then (7.53) is equivalent to:

$$r'_i = \sigma_i \, x'_i + n'_i, \tag{7.54}$$

for $i = 0, 1, \ldots, r - 1$, and:

$$r'_i = n'_i, \tag{7.55}$$

for $i = r, r + 1, \ldots, n_R - 1$. The result shows that:

(a) only the first r components of $\mathbf{r}'$ depend on the transmitted signal components;
(b) the received signal can be reduced to r independent components;
(c) the MIMO channel can be modeled as the parallel of r uncoupled subchannels, each associated with a singular value of $\mathbf{H}$.

The singular values correspond to the subchannel amplitude gains, and the eigenvalues of $\mathbf{H}\mathbf{H}^H$ represent the channel power gains. From this it is easily inferred that, when $n_T > n_R$ ($n_R > n_T$), there will be at most n_R (n_T) nonzero amplitude gains in (7.53) and that the set of eigenvalues $\{\lambda_i\}$, known as the *eigenvalue spectrum*, provides a MIMO channel representation that is useful in evaluating the best transmission paths. Note also that from (7.50)–(7.52) the autocorrelation matrices:

$$\mathbf{R}_{\mathbf{r}'} \triangleq \mathrm{E}\{\mathbf{r}'\mathbf{r}'^H\} = \mathbf{U}^H \mathbf{R}_{\mathbf{r}} \mathbf{U}, \tag{7.56}$$

$$\mathbf{R}_{\mathbf{x}'} \triangleq \mathrm{E}\{\mathbf{x}'\mathbf{x}'^H\} = \mathbf{V}^H \mathbf{R}_{\mathbf{x}} \mathbf{V} \tag{7.57}$$

and

$$\mathbf{R}_{\mathbf{n}'} \triangleq \mathrm{E}\{\mathbf{n}'\mathbf{n}'^H\} = \mathbf{U}^H \mathbf{R}_{\mathbf{n}} \mathbf{U}, \tag{7.58}$$

of $\mathbf{r}'$, $\mathbf{x}'$ and $\mathbf{n}'$ respectively, can easily be derived. Since $\mathbf{U}$ and $\mathbf{V}$ are unitary, it is easy to prove that $\mathrm{tr}(\mathbf{R}_{\mathbf{r}'}) = \mathrm{tr}(\mathbf{R}_{\mathbf{r}})$, $\mathrm{tr}(\mathbf{R}_{\mathbf{x}'}) = \mathrm{tr}(\mathbf{R}_{\mathbf{x}})$ and $\mathrm{tr}(\mathbf{R}_{\mathbf{n}'}) = \mathrm{tr}(\mathbf{R}_{\mathbf{n}})$, so that the covariance matrices of the transformed signals of (7.50)–(7.52) have the same sum of diagonal elements and therefore the same powers as the original signal components.

Equation (7.54) describes a MIMO system in which the subchannels are uncoupled such that the overall capacity is the sum of their capacities. Then, making use of the power constraints of equations (7.41) and (7.42), namely, that the transmit power from each antenna is P/n_T, we may estimate the system capacity as (see Section 7.2.3.3):

$$C = W \sum_{i=0}^{r-1} \log\left(1 + \frac{P_{r_i}}{\sigma_n^2}\right), \tag{7.59}$$

where W is the common subchannel bandwidth and P_{r_i} is the received signal power in the ith subchannel, with $i = 0, 1, \ldots, r - 1$. From (7.54) it can easily be inferred that:

$$P_{r_i} = \frac{\sigma_i^2 P}{n_T} = \frac{\lambda_i P}{n_T}, \tag{7.60}$$

so that the MIMO channel capacity (7.59) can be put in the form:

$$C = W \sum_{i=0}^{r-1} \log\left(1 + \frac{\lambda_i P}{n_T \sigma_n^2}\right) = W \log \prod_{i=0}^{r-1}\left(1 + \frac{\lambda_i P}{n_T \sigma_n^2}\right). \tag{7.61}$$

We note that in the case of a SISO channel, where $n_T = n_R = r = 1$, these expressions reduce to the well-known Gaussian channel capacity as expressed by (7.35). While expressions (7.59) and (7.60) are complete for a MIMO channel that is exactly known at the receiver and unknown at the transmitter, they provide no insight into the impact of channel behavior on capacity. To see this we need to relate the above expressions more directly to the channel matrix $\mathbf{H}$. This can easily be done if $m \triangleq \min(n_R, n_T) = r$ [16]. In fact in this case (7.48) can be put in the form:

$$(\lambda_l \mathbf{I}_m - \mathbf{M})\mathbf{y}_l = \mathbf{0}_m, \tag{7.62}$$

where

$$\mathbf{M} = \begin{cases} \mathbf{H}\mathbf{H}^H, & \text{if } n_R < n_T, \\ \mathbf{H}^H\mathbf{H}, & \text{if } n_R \geq n_T, \end{cases} \tag{7.63}$$

is the so-called *Wishart matrix*. Note that λ_l is an eigenvalue of $\mathbf{M}$ if and only if the matrix $(\lambda \mathbf{I}_m - \mathbf{M})$ is singular: i.e.,

$$\det(\lambda_l \mathbf{I}_m - \mathbf{M}) = 0, \tag{7.64}$$

so that the singular values of the channel matrix $\mathbf{H}$ are then found as the roots of the latter equation. It is then straightforward to show that:

$$\prod_{i=0}^{m-1}\left(1 + \frac{\lambda_i}{n_T}\frac{P}{\sigma_n^2}\right) = \det\left(\mathbf{I}_m + \frac{P}{n_T\,\sigma_n^2}\mathbf{M}\right). \tag{7.65}$$

Then the MIMO channel capacity (7.61) can be rewritten as:

$$C = W\,\log\det\left(\mathbf{I}_m + \frac{P}{n_T\,\sigma_n^2}\mathbf{M}\right). \tag{7.66}$$

Since the nonzero eigenvalues of $\mathbf{HH}^H$ and $\mathbf{H}^H\mathbf{H}$ are identical, the capacities of the channels specified by $\mathbf{H}$ and $\mathbf{H}^H$ are identical. This means that MIMO channel capacity is *reciprocal*, that is, it is the same regardless of which side of the channel is the transmitter, provided of course that the corresponding receiver has exact channel knowledge. Finally, we note that if the channel coefficients are random variables, as in many fading scenarios, the *mean* or *ergodic channel capacity* is found by averaging C in (7.66) over all possible realizations of $\mathbf{H}$. We will deal with this in more detail later.

7.3.2.2 Channel Known to the Transmitter

When the channel is known at the transmitter, advantage can be taken of this to increase the available capacity as expressed by (7.66). This is accomplished by employing the *water-filling rule* [226], which can be seen to be an extension of the standard water-filling approach for SISO channels. In practice, it can then be shown (e.g., see [16, 38]) that the optimal transmit power to be allocated to the ith subchannel is given by:

$$P_i = \left(\mu - \frac{\sigma_n^2}{\lambda_i}\right)^+, \tag{7.67}$$

with $i = 0, 1, \ldots, r - 1$, where $(x)^+$ stands for $\max(x, 0)$, r is the rank of the channel matrix $\mathbf{H}$, and the parameter μ is determined in such a way that the sum of the powers assigned to all the subchannels equals the overall transmit power P:

$$\sum_{i=0}^{r-1} P_i = P. \tag{7.68}$$

Then from the received signal model of (7.54) it is easily inferred that the received signal power in the ith subchannel is given by:

$$P_{r,i} = (\lambda_i\,\mu - \sigma_n^2)^+ \tag{7.69}$$

for $i = 0, 1, \ldots, r - 1$, so that the capacity expression of (7.59) can be put in the form:

$$C = W\sum_{i=0}^{r-1} \log\left[1 + \frac{1}{\sigma_n^2}(\lambda_i\,\mu - \sigma_n^2)^+\right]. \tag{7.70}$$

It can be shown that, for any underlying MIMO channel $\mathbf{H}$, the capacity as given by (7.70) is at least as large as that given by (7.66), which refers to the case of channel unknown to the transmitter.

7.3.3 *Random Channel*

We now turn our attention to the more realistic and more usual case where the entries of the channel matrix are random. In particular, in the following we assume that:

(a) CSI is available to the receiver only;
(b) the antenna spacings are sufficiently large that the entries of the channel matrix are spatially uncorrelated;
(c) the entries of the channel matrix **H** are zero-mean complex Gaussian random variables;
(d) the real and imaginary components of each entry are independent zero-mean Gaussian random variables, each having variance 1/2.

Assumption (d) ensures that the average power gain of each subchannel is unity (i.e., $E\{|h_{i,j}|^2\} = 1$ for any i and j), so that the average received SNR is the same as in an additive noise channel. In addition, each entry of the MIMO channel matrix is characterized by a uniformly distributed phase and a Rayleigh distributed magnitude, so that the pdf of the random variable $Y_{i,j} = |h_{i,j}|$ is given by [55]:

$$f_Y(y) = \frac{y}{\sigma_h^2} \exp\left(-\frac{y^2}{2\sigma_h^2}\right) u(y) \tag{7.71}$$

with $\sigma_h^2 = 1/2$. Note that similar assumptions are made in most MIMO channel work (e.g., see [5, 16, 18, 40, 226]).

In order now to evaluate channel capacity, we must consider what is known about the channel at both transmitter and receiver. In all wireless communications, the channel is time-varying and it is important to know the time scale of this variation. In wireless system design our knowledge of CSI can range from perfect knowledge at both transmitter and receiver to knowing the channel's statistical distribution at either transmitter or receiver or both, to the worst case of no channel knowledge at either transmitter or receiver.

In many instances, the calculation of channel capacity remains an open and difficult problem. Here, in order to provide some feeling for the problem and guidance that is useful in considering a broad spectrum of system design and evaluation problems, we consider the following two cases:

1. The channel matrix **H** is random and its entries change rapidly on a time scale that is short in comparison to the duration of the whole data transmission. The time scale in question ranges from a single symbol to a block of some number of symbols. The resulting channel is referred to as a *fast* or *block fading* channel.
2. The channel matrix **H** has random entries but, once chosen, remains constant over a prolonged period. This period ranges from the duration of a long data frame to the duration of the entire transmission process. The resulting channel model is referred to as a *slow* or *quasi-static fading* model.

These two scenarios lead to different approaches to characterizing capacity and to somewhat different conclusions. We will consider the two separately, although it must be remembered that there is no hard-and-fast rule for separating them; the choice is more or less arbitrary.

7.3.3.1 Ergodic Capacity

Here we consider transmission under a fast or block fading channel model. The main implication of this is that over the duration of transmission, we can expect the channel to pass through all possible states. This allows us to define an *expected* or *ergodic channel capacity* as the average or expected channel capacity of all possible channel states. If a single antenna link (i.e., $n_T = n_R = 1$) is assumed, the channel is characterized by a single random gain or fading coefficient h. The coefficient $|h|^2 = h_R^2 + h_I^2$ is then a chi-square random variable (denoted X in the following) with two degrees of freedom; in addition, if h_R and h_I (denoting the real part and the imaginary part of h, respectively) are assumed to be zero-mean statistically independent Gaussian random variables, each with variance $\sigma_h^2 = 1/2$, the pdf of X is:

$$f_X(x) = \frac{1}{2\sigma_h^2} \exp\left(-\frac{x}{2\sigma_h^2}\right) u(x) = \exp(-x) u(x). \tag{7.72}$$

Then, from (7.59) (with $r = 1$), the expression:

$$C = W \, \text{E}\left\{ \log\left(1 + X \frac{P}{\sigma_n^2}\right) \right\} \tag{7.73}$$

for the average or ergodic capacity of a single-link fast fading channel is easily inferred, where the expectation is evaluated with respect to X. To extend this approach to a MIMO channel characterized by the channel matrix $\mathbf{H}$, we exploit the approach developed in Section 7.3.2.1, based on the application of the SVD decomposition to $\mathbf{H}$. This leads to an equivalent channel consisting of $r \leq \min(n_T, n_R)$ parallel, decoupled subchannels, where r is the rank of $\mathbf{H}$, so that the ergodic capacity of the MIMO channel can be evaluated as (see the first row of (7.61)):

$$C = W \, \text{E}\left\{ \sum_{i=0}^{r-1} \log\left(1 + \lambda_i \frac{P}{\sigma_n^2 n_T}\right) \right\}, \tag{7.74}$$

where the expectation is evaluated with respect to the set of *random eigenvalues* $\{\lambda_i\}$. An alternative formula for the ergodic capacity can be developed by resorting to (7.66); this allows us to express it as:

$$C = W \, \text{E}\left\{ \log \det\left[\left(\mathbf{I}_r + \frac{P}{\sigma_n^2 n_T} \mathbf{M}\right)\right] \right\}, \tag{7.75}$$

where $\mathbf{M}$ is the (random) Wishart matrix defined in (7.63) and the expectation is evaluated with respect to it. While the capacity is readily calculated for $n_T = n_R = 1$ (see (7.61)), the calculation of (7.74) and (7.75) becomes very complicated for larger values of these parameters. A solution to this problem is developed in [226] and employs the *Laguerre polynomials* to express the ergodic capacity, under a power constraint P, in the form:

$$C = W \int_0^\infty \log\left(1 + \frac{P}{\sigma_n^2 n_T} \lambda\right) \sum_{k=0}^{m-1} \frac{k!}{(k+n+m)!} [L_k^{n-m}(\lambda)]^2 \lambda^{n-m} \exp(-\lambda) \, d\lambda, \tag{7.76}$$

where $m = \min(n_T, n_R)$, $n = \max(n_T, n_R)$ and:

$$L_k^{n-m}(x) \triangleq \frac{1}{k!} \exp(x) x^{n-m} \frac{d^k}{dx^k} [\exp(-x) x^{n-m+k}] \tag{7.77}$$

is the *associated Laguerre polynomial* of order k.

Finally, it is worth noting that, if the MIMO channel is spatially white (i.e., all the elements of $\mathbf{H}$ are iid Gaussian random variables) and $n_T = n_R = n$, the ergodic capacity tends to increase linearly with n as $n \to \infty$, and if the SNR is doubled an increase in capacity of about n bps/Hz is obtained [16, 38]. This results shows the potential benefits of MIMO systems with respect to SISO systems for which a 1 bps/Hz gain in capacity is provided by a 3 dB increase in SNR.

7.3.3.2 Outage Capacity

When the channel is very slowly varying or when it is modeled as constant over the length of one or more data frames, a duration that may last for many thousands of symbols, but changing randomly at the end of each block, we need to consider the concept of *outage capacity*. It is very likely that there will be time intervals (or blocks) during which it is impossible to achieve a low error probability regardless of the signaling rate. Under such circumstances the channel will be considered to be *in outage*. As a result, we need to consider channel capacity when there is a nonzero probability P_{out} that the channel is in outage, and therefore, unusable in the sense that the desired data rate cannot be achieved with arbitrarily low error rate. In fact, under such conditions, channel capacity becomes a *random variable* and may take on arbitrarily small values with nonzero probability, so that arbitrarily low probability of error cannot be achieved regardless of the codes that are chosen [16, 40]. To illustrate this concept, let us focus again on the case of a single antenna link (i.e., $n_T = n_R = 1$) affected by Rayleigh fading and assume a Doppler bandwidth equal to zero. Then the channel capacity can be expressed as (see (7.35)):

$$C = W \log(1 + X \cdot SNR) \text{ bits/s}, \tag{7.78}$$

where $SNR \triangleq P_{av}/(W N_0)$ is the *signal-to-noise ratio* and X is exponentially distributed (its pdf is given by (7.72)). Then, for a given channel capacity R (expressed in nats per unit bandwidth), the outage probability P_{out} can be evaluated as [275]:

$$
\begin{aligned}
P_{out} &= \Pr\{C/W \le R\} \\
&= \Pr\{\ln(1 + X \cdot SNR) \le R\} \\
&= 1 - \exp\left[-(\exp(R) - 1)/SNR\right].
\end{aligned}
\tag{7.79}
$$

This result shows that $P_{out} = 0$ is achieved for $R = 0$ only and this eliminates the possibility of any reliable communication in Shannon's sense.

For arbitrary values of n_T and n_R simple expressions for P_{out} cannot be derived. In such cases, the *complementary cumulative distribution function* (ccdf) of channel capacity is usually estimated via computer simulation. The ccdf specifies the probability, P_c, that a specified channel capacity is achievable. Then the outage capacity probability, P_{out}, which specifies the *probability of not achieving a given channel capacity*, is computed as $P_{out} = 1 - P_c$ (e.g., see [16, 38]).

7.4 Historical Notes

The concept of channel capacity stems from the classic works of *Claude Shannon* [1420, 1421], who, however, focused solely on additive noise channels. The evaluation of the channel capacity of wireless channels did not receive much attention until the 1980s and has become an active research areas only since the 1990s. More specifically, preliminary work on the evaluation of the capacity of SISO wireless channels was done in [220, 1426, 1427] (see also [1428]). In particular, the importance of evaluating channel capacity when multiple antennas are employed at the transmit side was first addressed by J. Winters in [220].

Generally speaking, the evaluation of channel capacity in SISO wireless communications is a complicated problem, since ISI and channel variability over time should be considered in a single-user scenario; this task becomes much harder if the typical multiuser scenario of mobile communications is considered [1429, 1430]. Various results on the effects of time-invariant ISI on capacity for single-user systems can be found in [1431, 1432]; this problem has been also investigated for multiple-user systems [1433–1435]. The time variation of wireless channels is much more difficult to model. A commonly taken approach is to consider the ISI as constant and known for certain durations and thus decompose the channel into time blocks [1436, 1437, 1438]. Another approach is based on the adoption of *Markov models* for communications channels [677, 1439–1446]. Finally, the assumption of independent fading affecting successive symbols can be made [1447]. Note also that time variations cause uncertainty about the channel at the receiver, and this uncertainty affects capacity [683].

The investigation of capacity of MIMO channels started towards the end of the 1990s when key theoretical papers by G. J. Foschini and M. J. Gans [18] (see also [1448]) and E. Telatar [226] were published. These papers quantify the extremely high spectral efficiency that can be achieved in MIMO channels. An example of this is the BLAST system developed at Bell Laboratories [17], which achieved experimentally a spectral efficiency of 42 bits/s/Hz.

7.5 Further Reading

There are many papers dealing with various aspects of capacity. Much of the work dealing with wireless channel capacity is best summarized in a few papers and in various books. A good introduction to the concept of capacity in wireless communications is provided by the paper [275]. The topic of capacity in MIMO communications is addressed in the papers [1449, 21]. Finally, various analytical details on the evaluation of MIMO capacity can be found in the books by B. Vucetic and J. Yuan [16], by E. Biglieri *et al.* [40] and by A. Paulraj *et al.* [38].

8

An Introduction to Channel Coding Techniques

In this chapter a brief introduction to channel coding techniques is provided. After summarizing some basic principles concerning code properties and design in Section 8.1, a description of standard interleaving schemes is provided in Section 8.2. A classification of channel coding and coded modulation schemes is provided in Sections 8.3 and 8.4, respectively. The organization of the following chapters on channel coding and coded modulation schemes is discussed in Section 8.5. Finally, some notes about the birth of coding techniques and a few suggestions for further reading are provided in Sections 8.6 and 8.7, respectively.

8.1 Basic Principles

As already mentioned in Section 1.1, a channel coding technique is a procedure or algorithm that induces the properties of *memory* and *redundancy* onto a data sequence. A receiver can then benefit from the presence of these properties in the demodulated data stream when attempting to detect and correct the errors introduced by the communication channel. In practice, in response to an input bit sequence conveying information, coding algorithms produce an output sequence, consisting of a single or multiple consecutive *codewords*. These form the input to a modulator, if a communication system equipped with a single transmit antenna is considered (see Figure 1.1), or a to bank of n_T distinct modulators, if an antenna array is employed at the transmitter (see Figure 1.2). Note also that each modulator can generate an SC or an MC signal. As a result, memory and redundancy can potentially be spread along different dimensions, namely *space*, *time* and *frequency*. In particular, if single carrier modulations are exploited for transmission, only the time dimension is used as $n_T = 1$, whereas both space and time dimensions are used when $n_T > 1$ (in other words, ST *coding* is used). On the other hand, in the case of MC transmission, the frequency dimension is always exploited; the other two dimensions, namely time and space, are used when codewords extend over multiple consecutive MCM (e.g., OFDM) symbols and a bank of MCM modulators is used, respectively. The last is usually known as *space-time-frequency coding* [1450].

As will be discussed in Section 8.3, coding algorithms can be grouped into different *classes* on the basis of various macroscopic properties of their structure, and in each class *families* of coding schemes with technically appealing features can be identified. For each family, the fundamental problem to be tackled in code design is that of *code optimization*. This means that, if a certain complexity is assumed for the encoding algorithm of a given family and a constraint is put on overall complexity of the

---

Wireless Communications: Algorithmic Techniques, First Edition.
Giorgio M. Vitetta, Desmond P. Taylor, Giulio Colavolpe, Fabrizio Pancaldi, Philippa A. Martin.
© 2013 John Wiley & Sons, Ltd. Published 2013 by John Wiley & Sons, Ltd.

(optimal) decoding strategy, the inner structure of the algorithm itself needs to be *optimized according to a certain criterion*. Whatever the application of channel coding, the aim of code optimization is to maximize the error-correction power of the technique itself. Note, however, that the result of this procedure is strongly influenced by the nature of the communication channel. In the first two decades of coding history, much attention was paid to the *power-limited* AWGN *channel*, where optimal code design aims at maximizing the *coding gain* (and, in particular, the *asymptotic coding gain*, which is achieved for a bit error probability tending to zero at the decoder output), that is, the energy saving provided by a coded transmission with respect to its uncoded counterpart, without taking into consideration any bandwidth expansion due to code redundancy. Since the beginning of the 1980s research efforts in code design have been progressively shifting towards coding schemes for *bandlimited* AWGN *channels*, in which code redundancy is absorbed in an increase in the number of points of the signal constellation for the adopted modulation scheme to avoid bandwidth expansion. Note that, in this case, a further degree of freedom is introduced in code design (which should still aim to maximize the coding gain) in that the mapping rule between the bits generated by the selected coding technique and the expanded signal constellation has to be optimized. In other words, solving the code design problem requires jointly optimizing the inner structure of the channel encoder and the constellation mapping. This leads naturally to the concept of *coded modulation*, according to which coding and modulation are seen as a single entity.

The first coded modulation schemes were devised for SC bandlimited transmissions over SISO AWGN channels (e.g., see [992]); in the 1980s, however, specific optimization criteria for coded modulations over bandlimited fading channels were also developed, starting with the case of frequency-flat fading (e.g., see [1451] and references therein). It is important to note that the impact of a wireless communication channel on data transmission is usually very different from that of the classic AWGN channel. In fact, on AWGN channels error patterns appearing in the demodulated data sequence are completely *random*, whereas in fading channels errors they tend to occur in bunches, or *bursts*; for this reason, fading channels belong to the class of *bursty channels*. Unfortunately, most of the error-correcting codes developed in the early decades of channel coding history (such as block codes) are *random error-correcting codes*. Such codes can correct some maximum number of symbol errors, independently of the placement of these errors. Since an error burst can entail several errors in a small number of received codewords and leave many other codewords uncorrupted, the strong error-correction capability required by the codewords affected by an error burst would be wasted for most of the remaining codewords. In other words, random error-correcting codes are intrinsically unsuitable for use on bursty fading channels. Other coding schemes, like convolutional codes, can correct an arbitrarily large numbers of well-spaced errors, but may be also unable to handle short error bursts. This motivates the search for coding schemes specifically designed for fading channels. It is worth noting that, in principle, in wireless transmission, random error-correcting codes can be still used if the codewords undergo interleaving before transmission. An *interleaver* is a device that mixes up or scrambles the symbols from different codewords at the transmitter, so that symbols belonging to the same codeword are well separated during data transmission. At the receiver, the demodulated data are put in the right order by a *deinterleaver*, so that codewords are correctly reconstructed. This process breaks up error bursts, spreading them over multiple codewords. In other words, the interleaver–deinterleaver pair randomizes the channel, making it more similar to an AWGN channel, so that various error-correcting codes originally devised for the correction of random errors can be effectively applied. However, the availability of this tool for mitigating the effects of channel fading does not provide a natural solution to the problem of optimal code design for fading channels. In fact, code design is expected to be inevitably affected by the real nature of the communication channel and by the knowledge that the decoder has acquired about it. Generally speaking, the *channel model* turns out to have a considerable impact on the choice of the preferred combination of coding and modulation schemes. If the channel model is uncertain, or not stable enough in time to develop a coded modulation closely matched to it, then the best design approach may be to develop a *robust* solution, that is, a solution providing adequate but suboptimum performance on a wide variety of channel models. Another relevant factor in code design for fading channels is the *maximum allowable decoding delay*. Various extremely powerful codes published in the literature suffer from substantial decoding delay, so that their application may be useful for data transmission, but not for real-time speech. In real-time speech transmission, which imposes a strict decoding delay constraint, channel variations with time may be rather slow with respect to the maximum allowed delay. In this case, the channel may be modeled as a *block fading* channel, in which the fading is nearly constant for a number of symbol intervals. On such a channel, a single codeword may be transmitted after being split into several blocks (through a proper interleaving process), so that some degree of diversity is achieved.

8.2   Interleaving

An *interleaver* is a device that modifies the order of its input sequence in a deterministic manner. Its use at the transmit side of a data communication system requires the adoption of a companion device (i.e., a *deinterleaver*) at the receiver, which applies the inverse permutation to the received signal or data in order to restore the original ordering. The typical usage of external interleaving and deinterleaving in a wireless data communication system is illustrated in Figure 8.1.

Interleavers can act in a *periodic* fashion or in a *pseudorandom* fashion. The permutation accomplished by *periodic* interleavers is a cyclic function of time. Periodic interleavers include *block* interleavers and *convolutional* interleavers.

An (M, N) *block interleaver* consists of an $N \times M$ matrix, which is filled column by column by a coded data sequence. This sequence is permuted by reading out the matrix content row by row for transmission. The associated deinterleaver also consists of an $N \times M$ matrix, which, however, is written row by row and read column by column at the receiver. The end-to-end interleaver delay is $2N \cdot M$ symbols, excluding any channel or processing delay. An (M, N) block interleaver ensures that:

(a) any burst of errors of length $l_e \leq rM$ (with $r > 1$) generated by the communication channel results in bursts of no more than $\lceil r \rceil$ errors separated by no less than $N - \lceil r \rceil$ positions at the deinterleaver output (so that for $r = 1$ single errors separated by at least N positions are found),
(b) a periodic sequence of single errors spaced by M symbols yields a single burst of N errors at the deinterleaver output.

These features suggest the following guidelines for dimensioning a block interleaver:

• M should not be smaller than the maximum expected burst length;
• N should be carefully selected on the basis of the adopted channel coding scheme.

As far as the last point is concerned, note that the effects of the channel memory at the deinterleaver output depend on N, in the sense that the larger N, the more decorrelated are the effects of the channel fading on the sequence feeding the channel decoder. Hence, in principle, N should be larger than the decoding span. This means that, as will become clear later, for block codes N should be larger than the codeword length, whereas for the convolutional codes N should exceed the decoding or decision delay. This ensures that for block codes any error burst of length $l_e \leq M$ can produce at most a single

Figure 8.1   System block diagram illustrating the use of interleaving and deinterleaving in a wireless data communication system.

error in any codeword, and with convolutional codes in the same scenario at most a single error will be encountered in the processing window of the decoding algorithm. Note that the use of a block interleaver introduces a further synchronization problem in a data communication system since, if the beginning of each interleaved block is not known, deinterleaving cannot be carried out correctly.

Convolutional periodic interleavers were developed by J. Ramsey [1452] and D. J. Forney [1453]. Here we focus on the structure proposed in [1453] and illustrated in Figure 8.2. Its operation can be summarized as follows. The coded symbols feeding the interleaver are shifted sequentially into a bank of M registers, characterized by increasing lengths (from top to bottom); each register is fed and read through a pair of commutators which operate synchronously. In practice, any time a new coded symbol is ready at the interleaver input, a commutator switches to the input of a new register. Then the new symbol is shifted in and simultaneously the oldest (i.e., rightmost) symbol in the same register is shifted out and sent over the communication channel. The deinterleaver operates in the same fashion, but the lengths of its shift registers are reversed, increasing from bottom to top. Note that for correct operation, the commutators at the deinterleaver need to be mutually synchronized and with the corresponding pair at the interleaver, so that correct symbol ordering is restored by the deinterleaver. Then, if the parameter $N = M \cdot P$ (P integer) is defined, the interleaver of Figure 8.2 can be referred to as an (M, N) convolutional interleaver; in addition, it can be shown that its properties are quite similar to that of an (M, N) block interleaver. In particular, we have that:

(a) any burst of errors of length $l_e < M$ generated by the communication channel results in single errors (separated by at least N symbols) at the deinterleaver output (this is due to the fact that, if two symbols are separated by less than N symbols at the interleaver input, their minimum separation at the interleaver output is M symbols),
(b) a periodic pattern of single errors spaced by $N + 1$ symbols generates a burst of M symbols at the deinterleaver output.

It is interesting to point out that an (M, N) convolutional interleaver offers some advantages over its (M, N) block counterpart:

- It halves the overall end-to-end delay ($(M - 1)N$ symbols versus $2N \cdot M$ for the latter) and memory requirements ($(M - 1)N/2$ symbols for both the interleaver and the deinterleaver versus $N \cdot M$ symbols for the same components of the latter).
- It simplifies the synchronization problem, as its degree of ambiguity at the receiver is M, making it N times smaller than the latter.

An alternative to periodic interleavers is offered by *pseudorandom* interleavers. A *pseudorandom interleaver* is a block interleaver that permutes a block of L channel symbols in a pseudorandom

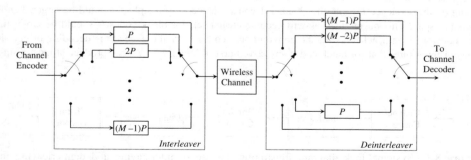

Figure 8.2 Block diagram of a convolutional interleaver–deinterleaver pair.

fashion. This technique provides a high degree of robustness to variability in the parameters of an error burst at the cost of an increased complexity with respect to block or convolutional interleavers of the same size. The implementation of this type of interleaver requires an algorithm for generating a pseudorandom sequence of integers from 0 to $L - 1$. The first simple solution to this problem was proposed by I. Richer in 1978 [1454] and is based on the use of the so-called *linear congruential sequences*. More recently, considerable attention has been paid to the problem of the design of pseudorandom interleavers, since they are a basic component of a class of powerful channel codes, known as turbo codes. Further details on this are provided in Chapter 10.

8.3   Taxonomy of Channel Codes

Channel coding techniques can be classified on the basis of the structure of their encoding functions, that is, by the manner in which they map their input information sequences (the *message*) to a *codeword*. If this approach is adopted, channel codes can be divided into the large families of *block codes* and *trellis codes*, as illustrated in Figure 8.3. On the one hand, the encoder of a *block code* operates on a block-by-block basis, that is, it maps a symbol vector of fixed length to a finite-dimensional codeword; for this reason, a block code can be simply modeled as a dictionary, providing a codeword for each possible message that it may contain. On the other hand, the encoder of a trellis code can map a sequence of arbitrary length to an arbitrarily long codeword and can be modeled as an FSSM, characterized by a finite-state *trellis*.

Block codes can be further classified into *linear* and *nonlinear* block codes. Linear (nonlinear) block codes are described by a linear (nonlinear) mapping, defined over a specific algebraic system, from messages to codewords. In practice, with linear block codes encoding a message means computing a product between a fixed size matrix, characterizing the coding function, and the symbol vector representing the message. The simplicity of encoding and the availability of a computationally efficient encoding algorithm has motivated intense research activity in the field of linear block codes, whereas

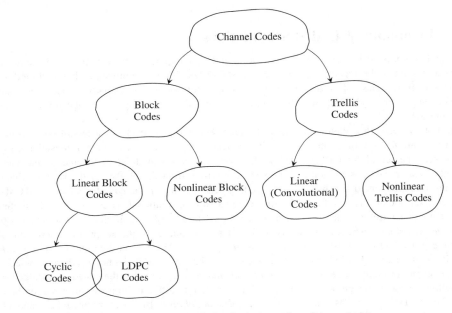

Figure 8.3   Taxonomy of classical channel coding techniques.

much less attention has been paid to the development of nonlinear block codes. We also note that the class of linear block codes includes the important families of *cyclic codes*, whose algebraic structure allows further simplifications in encoding and decoding, and *low-density parity check* (LDPC) codes, which are extremely powerful and can be decoded via computationally efficient decoding algorithms (note that some LDPC codes have a cyclic structure).

Similarly, trellis codes can be further classified as *linear* trellis codes and *nonlinear* trellis codes. Linear trellis codes are typified as *convolutional codes*, since encoding can be interpreted as a filtering of the message sequence, that is, a convolution between the message sequence and a (matrix) impulse response characterizing the selected channel code. Linear convolutional codes, as a subclass of more general trellis codes, have received considerably more attention than their nonlinear counterparts, even if, unlike block codes, the linear–nonlinear option does not have a significant impact on the complexity of optimal decoding.

The encoders of the above-mentioned channel codes can be connected (i.e., *concatenated* in the coding theory jargon), to generate new coding schemes. Note that the *constituent* encoders of a concatenated scheme do not necessarily belong to the same family of channel codes, so that hybrid schemes can be constructed (e.g., a linear block encoder can be concatenated with a convolutional encoder [1455]). As will become clearer below, concatenated codes often combine an excellent error-correcting capability with the possibility of a computationally efficient decoding procedure due to the concatenation of the decoding algorithms for the constituent codes.

A different classification of channel codes is based on dividing them into the classes of *algebraic* codes and *probabilistic* codes. Usually, *algebraic coding* theory is mainly concerned with linear block codes, which are characterized by an elegant algebraic structure. Such a structure allows the design of powerful coding schemes and efficient decoding algorithms that exploit results concerning the algebra of finite fields. On the other hand, *probabilistic coding* theory is more concerned with the problem of finding classes of codes that optimize average performance as a function of coding and decoding complexity. Classical examples of probabilistic codes include convolutional codes and some concatenated codes. Finally, we mention that probabilistic decoders typically process soft (reliability) information.

8.4   Taxonomy of Coded Modulations

As already mentioned in Section 8.1, modulation and channel coding must be designed jointly if the aim is to maximize the spectral efficiency at a certain SNR. Jointly optimized coding and modulation schemes form the class of *coded modulation schemes* (or *signal space codes*). Coded modulations were developed first for SC modulations and can be classified according to the logical scheme shown in Figure 8.4.

Most of the coded modulation schemes of Figure 8.4 are based on the key concept of *set partitioning*, which is a simple way of generating a finite sequence of partitions of a symbol constellation [992, 1456] (see Chapter 11 for further details). However, any coded modulation scheme is characterized not only by a signal constellation and the way it is set-partitioned, but also by the way in which the mapping is implemented and by a channel coding scheme. In a *trellis-coded modulation* (TCM) the channel code adopted is a trellis code; linear trellis (or convolutional) codes are usually preferred, even if nonlinear trellis codes have also been adopted to satisfy specific design requirements (e.g., *rotational invariance*, which is immunity to the phase ambiguities of the adopted constellation in the decoding process [1457–1461]). In a *block-coded modulation* (BCM), on the other hand, a *multilevel code* is usually adopted [1462, 1463]. In this case the channel code is defined by the combination of multiple error-correcting codes (usually called *constituent codes*), so that the coded bits are generated by multiple encoders which operate in parallel and are fed by different groups of information bits. An alternative to TCM and BCM is *bit-interleaved coded modulation* (BICM), for which the channel code is a convolutional code, but, unlike TCM schemes, it is followed by $\log(M)$ distinct bit interleavers,

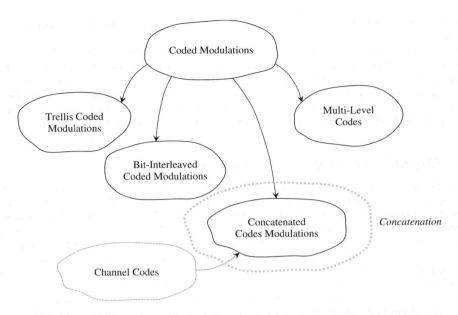

Figure 8.4   Taxonomy of coded modulations for single-carrier transmissions.

if an *M*-ary constellation is adopted [1464]; this approach yields better coding gain over a Rayleigh channel than its TCM counterpart characterized by the same complexity.

The channel code of any of the above-mentioned coded modulations can be concatenated with other channel codes [1465], resulting in a *concatenated coded modulation* (CCM). In addition, trellis-coded, block-coded and concatenated coded modulation schemes can be developed for MC modulations, if the frequency dimension is available, and/or for a multiple-antenna transmission, if an antenna array is available at the transmit side. Coded modulation schemes devised for digital transmission via multiple antennas are known as ST *codes* if an SC modulation is assumed, SF *codes* if an MC (in practice, OFDM) modulation is used and codewords extend over a single OFDM symbol interval, and *space-time-frequency* (STF) *codes* if data are transmitted via multiple OFDM modulators over multiple OFDM symbol intervals. The first family of ST codes proposed in the technical literature is that of *space-time block codes* (STBCs); each STBC consists of a collection of equal size matrices, each of which specify the channel symbols to be transmitted simultaneously by distinct antennas over the same number of intervals. STBCs can be decoded via simple linear processing at the receiver and provide diversity gain (i.e., they produce an increase in the slope of BER curves), but provide no or minimal coding gain. A real coding gain (in addition to a diversity gain) is provided by *space-time trellis codes* (STTCs). Since the encoder of an STTC can be modeled as an FSSM, the encoding procedure can be represented by a proper trellis, whose state transitions are labeled by the group of channel symbols to be radiated simultaneously by the transmit array. The improvement in coding gain with respect to STBCs is achieved at the price of a substantially larger decoding complexity. Coding performance can be further improved by concatenating STBCs or STTCs with other coding schemes, such as convolutional codes or cyclic (block) codes (e.g., see [1465–1467] and references therein).

Initial research on ST coding was conducted on frequency-flat fading channels. In the presence of a frequency-selective channel the ST decoder has to be combined with a channel equalizer at the receive side and this makes the adoption of ST codes a challenging problem [1466, 1468]. A solution to this is offered by OFDM which converts a frequency-selective channel into a set of parallel independent frequency-flat channels, for which SF codes and STF codes, inspired by ST coding schemes, can be developed (e.g., see [21, 1466, 1467]).

A different approach to channel coding is used when a *layered space-time* (LST) architecture (also known as *spatial multiplexing*) is used for data communications in a MIMO system [21]. In this case, streams of independent data can be transmitted over different antennas to maximize data rate, and successive one-dimensional (in the space domain) processing steps are accomplished on multidimensional signals at the receive side for data detection. Various LST architectures can be found in the technical literature, and each requires specific forms of channel coding [21].

8.5 Organization of the Following Chapters

The following four chapters focus on the study of channel coding and coded modulations, and on the problem of combined channel equalization and decoding of channel codes. Chapter 9 is devoted to the study of traditional coding schemes, like block codes, convolutional codes and classical coding schemes resulting from parallel and serial concatenation of such codes. In Chapter 10, modern concatenated coding schemes and LDPC codes are analyzed, and the problem of code analysis and decoding from a graph theory perspective is briefly discussed. Chapter 11 focuses on coded modulation for the single-input and multiple-input channel cases. More specifically, TCM, BICM and modulation codes based on multilevel coding are illustrated for single-antenna transmissions, while various forms of ST and SF coding are analyzed for MIMO communications. Finally, the problem of how channel equalization algorithms, channel estimation techniques and decoding of channel codes can be combined to improve link performance at an acceptable computational cost is analyzed in Chapter 12.

8.6 Historical Notes

In this section some relevant information concerning the birth of channel coding and coded modulation schemes mentioned above is provided. More specific historical details are provided in later chapters.

The first nontrivial channel codes to appear in the technical literature were a couple of *binary block codes* devised by R. Hamming [1469] in 1950 and M. Golay [1470] in 1949. Another significant example of early binary block codes is represented by the so-called RM codes, which were proposed by D. Muller in 1954 [1471] and then reintroduced by I. Reed [1472], who proposed an efficient decoding algorithm. *Cyclic* (block) *codes* were first investigated by E. Prange in 1957 [1473], and from the 1960s became a fundamental topic of research in the field of channel coding techniques. In this area a milestone is the invention of *Bose–Chaudhuri–Hocquenghem* (BCH) and *Reed–Solomon* (RS) codes in three independent papers in 1959 and 1960 [1474–1476].

Convolutional codes were invented by P. Elias in 1955 [1477], and a computationally efficient and asymptotically optimal decoding algorithm for them was proposed by A. Viterbi in 1967 [1478] (and became known as the VA). LDPC codes were proposed by R. Gallager in 1962 [1479], but his work was largely forgotten for more than 30 years, because of their large implementation complexity.

In 1966 D. Forney proposed the idea of *serial concatenation* for the development of long and powerful codes. These result from cascading shorter and simpler constituent codes which can be easily encoded and decoded [1455]. Another form of concatenation, who became known later as *parallel concatenation*, had actually been proposed earlier by P. Elias [1480] who had invented *product codes* by combining known block codes. The real importance of parallel concatenation in the development of powerful coding schemes became apparent much later, in 1993, when the so-called *turbo codes*, which offer impressive error performance at the price of a reasonable decoding complexity, were proposed by C. Berrou, A. Glavieux and P. Thitimajshima [1481].

Milestones in the development of coding for bandwidth-limited channels (i.e., of *coded modulations*), are G. Ungerboeck's invention of TCM in the 1970s, but not published until 1982 [992], and the development of multilevel coding by H. Imai and S. Hirakawa in 1977 [1462].

STTCs were proposed by V. Tarokh, N. Seshadri and R. Calderbank in 1998 [14]. A few months later, S. M. Alamouti [15] invented a low-complexity STBC. Alamouti's invention motivated Tarokh *et al.* [1482, 1483] to generalize Alamouti's scheme to an arbitrary number of transmitter antennas. These early results were followed by a flurry of research in this field.

8.7 Further Reading

An excellent introduction to coding theory and its history is provided by [1484]. Reference books about channel coding techniques and their history are [35, 1485], which, however, mainly focus on AWGN channels. The design of coding techniques for wireless channels is specifically tackled in [37]. An interesting overview of coded modulations for frequency-flat fading channels is provided by [1451], where design rules for TCM and BCM (based on multilevel coding) are illustrated and the performance of some coded modulation schemes is analyzed. Finally, various introductory material about ST coding for wireless communication can be found in [21, 1465–1468]. A good overview is presented by the book [38].

9

Classical Coding Schemes

This chapter briefly describes the essential aspects of traditional coding schemes – in particular, block and convolutional codes. It is a prerequisite for the in-depth understanding of the modern coding schemes and signal space codes considered in Chapters 10 and 11, respectively.

The chapter is organized as follows. In Section 9.1 the main properties of *linear block codes* are summarized. Then, some specific classes of codes (including cyclic codes, single parity check codes, Reed–Muller codes and repetition codes) are considered. The remainder of the section presents an overview of soft and algebraic decoding techniques for block codes and an assessment of their achievable performance. In Section 9.2 the main features of linear trellis codes, namely *convolutional codes*, are described. Then the problem of their efficient decoding is investigated from a wide perspective, including ML, MAP and *sequential* decoding strategies. Considerable attention is paid to ML decoding, for which the problem of assessing error performance is also analyzed.

Section 9.3 deals with the final topic of this chapter, the classical coding schemes resulting from parallel and serial concatenation of block and convolutional coding schemes.

Finally, some historical notes and suggestions for further reading are provided in Sections 9.4 and 9.5, respectively.

9.1 Block Codes

9.1.1 Introduction

In this section we focus on *block codes* defined on finite fields.[1] In general, block encoding of information messages can be represented as shown in Figure 9.1. There, an information source generates a sequence of *symbols*, each belonging to GF(q), with $q = p^l$ (p is prime and l is a positive integer). The sequence is partitioned into *blocks*, called *messages*, each of length k. The encoder converts each input message[2]:

$$\mathbf{u} \triangleq [u_0, u_1, \ldots, u_{k-1}],$$ (9.1)

in accordance with *one-to-one mapping*, to a vector, called a *codeword*:

$$\mathbf{x} \triangleq [x_0, x_1, \ldots, x_{n-1}],$$ (9.2)

---

[1] An introduction to finite field theory is provided in Appendix E.
[2] In this chapter, unless stated otherwise, row vectors are used.

Wireless Communications: Algorithmic Techniques, First Edition.
Giorgio M. Vitetta, Desmond P. Taylor, Giulio Colavolpe, Fabrizio Pancaldi, Philippa A. Martin.
© 2013 John Wiley & Sons, Ltd. Published 2013 by John Wiley & Sons, Ltd.

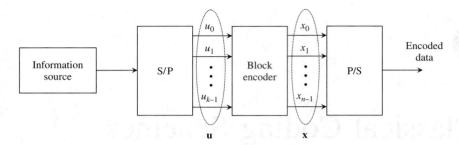

Figure 9.1   Block encoding of a data stream emanating from an arbitrary information source.

whose n elements, with $n \geq k$, also belong to GF(q). The set of q^k distinct codewords generated by the same number of distinct messages forms a *block code*.

The mathematical description of the block encoding algorithm is simple if the code has the property of *linearity* (as described below). Then the resulting block code is *linear*. In the remaining part of this section we focus exclusively on linear block codes defined over finite fields.

9.1.2   Structure of Linear Codes over GF(q)

A linear (n, k) block code over GF(q) is a set $\mathcal{C}$ of q^k codewords of block length n defined according to the *linear transformation* of the input message given by:

$$\mathbf{x} = \mathbf{u}\,\mathbf{G}, \tag{9.3}$$

where $\mathbf{G}$ is a $k \times n$ matrix, called the *generator matrix*, with elements belonging to GF(q). Note that for linear block codes the GF(q) sum of any two codewords in $\mathcal{C}$ is also a codeword in $\mathcal{C}$, so that $\mathcal{C}$ is a closed set under GF(q) addition. The block code described by (9.3) is characterized by the parameter:

$$R \triangleq \frac{k}{n} \tag{9.4}$$

called the *code rate*. It measures the fraction of the codeword containing data.

If $\mathbf{g}_l$ denotes the lth row of $\mathbf{G}$ (with $l = 0, 1, \ldots, k-1$), (9.3) can be rewritten as:

$$\mathbf{x} = \sum_{l=0}^{k-1} u_l\,\mathbf{g}_l. \tag{9.5}$$

The transformation (9.3) between messages and codewords is *one-to-one* if and only if the vectors $\{\mathbf{g}_0, \mathbf{g}_1, \ldots, \mathbf{g}_{k-1}\}$ are *linearly independent*, i.e the rank of $\mathbf{G}$ is equal to k. This ensures that $\mathcal{C}$ consists of q^k distinct codewords. In this case, from a vector space[3] perspective, we see that the code $\mathcal{C}$ is a k-dimensional *subspace* of the vector space S consisting of all the possible vectors of length n over GF(q) and that the vectors $\{\mathbf{g}_0, \mathbf{g}_1, \ldots, \mathbf{g}_{k-1}\}$ form a *basis* of $\mathcal{C}$. Note that, since the selection of a basis for a given subspace is not unique, the same codeword set $\mathcal{C}$ can be generated using many different generator matrices. However, these equivalent codes follow different correspondence rules between the messages and the set of codewords.

If $\mathbf{G}$ is structured as:

$$\mathbf{G} = \begin{bmatrix} \mathbf{I}_k & \mathbf{P}_{k,n-k} \end{bmatrix}, \tag{9.6}$$

---

[3] The definition of a vector space over a field can be found in Appendix E.

where $\mathbf{I}_k$ is the $k \times k$ identity matrix and $\mathbf{P}_{k,n-k}$ is a $k \times (n-k)$ matrix, $\mathbf{x}$ can be expressed as:

$$\mathbf{x} = \left[\mathbf{u}, \mathbf{x}_p\right], \tag{9.7}$$

that is, the concatenation of the message $\mathbf{u}$ with the $(n-k)$-dimensional vector:

$$\mathbf{x}_p \triangleq \left[x_k, x_{k+1}, \ldots, x_{n-1}\right] = \mathbf{u}\, \mathbf{P}_{k,n-k} \tag{9.8}$$

containing the so-called *parity check symbols*. If (9.6) holds, then $\mathbf{G}$ is said to be in *systematic* form and the transformation of (9.3) then describes *systematic encoding*.

It is not difficult to prove that any linear code C can be transformed into systematic form producing another code C_s whose generator matrix is structured as in (9.6) and whose performance, in terms of average error probability for the transmitted messages, is *equivalent* to that of C if the codewords are transmitted over a *memoryless channel* (e.g., see [1486, pp. 79–80] for further details).

Other properties of linear block codes can be inferred from the general results of vector space theory. Since an (n, k) linear block code C represents a k-dimensional subspace of S, it also uniquely identifies a distinct $(n-k)$-dimensional subspace, called the *null space* and denoted by C^+. The subspace C^+ consists of the n-dimensional vectors $\mathbf{w}$ on GF(q) which are *orthogonal to all elements in C*, that is, such that:

$$\mathbf{x}\,\mathbf{w}^T = 0 \tag{9.9}$$

for any $\mathbf{x} \in C$ and $\mathbf{w} \in C^+$. By analogy with (9.3), the set of vectors $\mathbf{w} \in C^+$ can be represented via the linear transformation:

$$\mathbf{w} = \mathbf{v}\,\mathbf{H}, \tag{9.10}$$

where $\mathbf{v} \triangleq \left[v_0, v_1, \ldots, v_{n-k-1}\right]$ is an $(n-k)$-dimensional vector on GF(q), and $\mathbf{H}$ is an $(n-k) \times n$ matrix with rank $n-k$ whose rows form a basis for C^+.

Substituting (9.10) into (9.9) leads easily to:

$$\mathbf{x}\,\mathbf{H}^T = \mathbf{0}_{n-k}, \tag{9.11}$$

where $\mathbf{0}_{n-k}$ is the null vector of size $n-k$. Note that the matrix equation (9.11) can be interpreted as a system of $n-k$ linear equations whose unknowns are the n elements $\mathbf{x}$. These equations are known as the *parity check equations* of the block code C.

Substituting (9.3) into (9.11) leads to:

$$\mathbf{G}\,\mathbf{H}^T = \mathbf{0}_{k,n-k}, \tag{9.12}$$

where $\mathbf{0}_{k,n-k}$ is the null matrix of size $k \times (n-k)$. This result lends itself to the simple interpretation that all the rows of $\mathbf{G}$, representing vectors of C, are orthogonal to all the rows of $\mathbf{H}$, which, similarly, are vectors of C^+. Note that, given the generator matrix $\mathbf{G}$ of a code C, the choice of a matrix $\mathbf{H}$ satisfying (9.12) is not *unique*. However, if $\mathbf{G}$ is in systematic form (see (9.6)), the specific choice:

$$\mathbf{H} = \left[-\mathbf{P}_{k,n-k}^T \quad \mathbf{I}_{n-k}\right] \tag{9.13}$$

certainly satisfies (9.12) and is usually used.

Finally, it is important to point out that, given a vector space S, defining a subspace C^+ that is orthogonal to a subspace C is completely equivalent to defining C. For this reason, a linear block code C can also be defined as the *set of vectors forming the null space* of the linear transformation defined by a matrix $\mathbf{H}^T$, of size $n \times (n-k)$ and rank $n-k$, that is, the set of vectors $\mathbf{x}$ satisfying (9.11).[4] Moreover, the matrix $\mathbf{H}$ can be interpreted as the generator matrix (see (9.10)) of an $(n, n-k)$ linear

---

[4] This result is also known as the *parity check theorem* [1485].

block code $\mathcal{C}^+$. The codes $\mathcal{C}$ and $\mathcal{C}^+$ are called *dual codes* and are usually quite different. However, if they are equivalent, they are called *self-dual codes*. A sufficient condition for self duality is $n - k = k$, that is, $k = n/2$, so that both codes have rate $R = 1/2$.

As illustrated in the following pages, knowledge of the structure of a block code can be used to acquire important information about the structure of its dual. The ideas illustrated above are applied in the following two examples.

Example 9.1.1 The well-known $(7, 4)$ Hamming linear block code over GF(2) is usually characterized by the generator matrix:

$$ \mathbf{G} = \begin{bmatrix} 1 & 0 & 0 & 0 & 1 & 0 & 1 \\ 0 & 1 & 0 & 0 & 1 & 1 & 1 \\ 0 & 0 & 1 & 0 & 1 & 1 & 0 \\ 0 & 0 & 0 & 1 & 0 & 1 & 1 \end{bmatrix}, \tag{9.14} $$

which, having the structure of (9.6), describes a systematic code. Therefore, from (9.13) the parity check matrix is given by:

$$ \mathbf{H} = \begin{bmatrix} 1 & 1 & 1 & 0 & 1 & 0 & 0 \\ 0 & 1 & 1 & 1 & 0 & 1 & 0 \\ 1 & 1 & 0 & 1 & 0 & 0 & 1 \end{bmatrix}. \tag{9.15} $$

This matrix also represents the generator matrix of the $(7, 3)$ (binary) dual code. It should be noted that the dual code in this instance is a so-called binary m-sequence often known as a *pseudorandom binary sequence*. Such sequences are widely used in spread spectrum and *code division multiple access* (CDMA) wireless systems.

□

Example 9.1.2 An $(n, 1)$ *repetition code* over GF(q) is a linear block code characterized by the generator matrix:

$$ \mathbf{G} = \begin{bmatrix} 1 & 1 & \ldots & 1 \end{bmatrix}, \tag{9.16} $$

structured as in (9.6). Then, using (9.13), the parity check matrix:

$$ \mathbf{H} = \begin{bmatrix} \mathbf{h} & \mathbf{I}_{n-1} \end{bmatrix} \tag{9.17} $$

is found, where $\mathbf{h} \triangleq [q - 1, q - 1, \ldots, q - 1]^T$ is an $(n - 1)$-dimensional column vector with all its elements equal to $q - 1$, and $\mathbf{I}_{n-1}$ is an $(n - 1) \times (n - 1)$ identity matrix. Note that $\mathbf{H}$ generates an $(n, n - 1)$ linear systematic code which adds a single parity check symbol to a message consisting of $n - 1$ information symbols.

□

9.1.3 Properties of Linear Block Codes

Given an (n, k) linear block code $\mathcal{C}$ over GF(q), the *Hamming distance* $d_H(\mathbf{x}_l, \mathbf{x}_i)$ between two distinct codewords $\mathbf{x}_l = [x_{l,0}, x_{l,1}, \ldots, x_{l,n-1}]$ and $\mathbf{x}_i = [x_{i,0}, x_{i,1}, \ldots, x_{i,n-1}]$ is defined as the number of locations in which they differ; it can be computed as:

$$ d_H(\mathbf{x}_l, \mathbf{x}_i) = w_H(\mathbf{x}_l - \mathbf{x}_i), \tag{9.18} $$

where $w_H(\mathbf{x})$, the *Hamming weight* of a vector $\mathbf{x}$, represents the number of nonzero elements of $\mathbf{x}$. For a linear code $\mathcal{C}$, the sum of any two codewords is also a codeword in $\mathcal{C}$. Therefore, $\mathbf{x}_{l,i} \triangleq \mathbf{x}_l - \mathbf{x}_i$ is also a codeword in $\mathcal{C}$. If we fix $\mathbf{x}_l \neq \mathbf{0}_n$ as a *reference codeword* and let $\mathbf{x}_i$ vary over $\mathcal{C}$, taking all the $(q^k - 1)$ possible values excluding $\mathbf{x}_l$, then $\mathbf{x}_{l,i}$ generates all the nonnull codewords of $\mathcal{C}$. This proves that *the set consisting of all the Hamming distances between distinct codewords coincides with the set of weights of the codewords* $\mathbf{x} \neq \mathbf{0}_n$. It also proves that, when evaluating this set of distances, the final result is *invariant* with respect to the selected reference codeword. These results are extremely important because:

(a) the set of distances between the codewords of C can be evaluated by computing the weights of its $(q^k - 1)$ nonzero codewords $(\mathbf{x} \neq \mathbf{0}_n)$,
(b) if the channel behavior is independent of the transmitted codeword,[5] the evaluation of the *average* error probability performance of a decoding algorithm for C does not require a statistical average over C – it can be assessed assuming that a *specific codeword* (e.g., $\mathbf{x} = \mathbf{0}_n$) has been transmitted.

The *weight spectrum* of C is the set of all possible values of the weight $w_H(\mathbf{x})$ for all $\mathbf{x} \in C$ and is measured by the *multiplicity* or number, A_w, of codewords associated with each weight, w. This spectrum can be summarized in a table or as a polynomial, known as the *weight enumerator polynomial* and defined as:

$$A(z) = \sum_{w=0}^{n} A_w \, z^w. \tag{9.19}$$

Note that $A_0 = 1$ (since $\mathbf{0}_n$ is the only codeword with null weight) and $A_w = 0$ for $0 < w < d_{H,\min}$, where $d_{H,\min}$ denotes the *minimum Hamming distance between distinct codewords*. Let us now evaluate $A(z)$ for the binary code defined in Example 9.1.1.

Example 9.1.3 The $(7,4)$ Hamming code of Example 9.1.1 consists of 2^4 codewords; it is easy to show, via an exhaustive search, that one codeword has a null weight, seven have weight 3, seven have weight 4, and one has weight 7, so that:

$$A(z) = 1 + 7z^3 + 7z^4 + z^7 \tag{9.20}$$

and $d_{H,\min} = 3$.
□

Generally speaking, the evaluation of $A(z)$ involves assessing the weights of all possible linear combinations of the rows of the generator matrix $\mathbf{G}$ (see (9.5)). This entails a computational effort that increases exponentially with k. Fortunately, for several important classes of code an analytical expression for the associated weight enumerator polynomial is available (e.g., see the classes of Hamming codes and RS codes analyzed below). In other cases, the evaluation of the weight spectrum of an (n, k) code C can be simplified by exploiting knowledge of the spectrum of its $(n, n - k)$ dual code C^+. In fact, the latter uniquely identifies the former, as shown by F. J. MacWilliams [1487], who proved the validity of the identity [327, p. 418]:

$$\sum_{l=0}^{n-p} B_l \, C_p^{n-l} = q^{n-k-p} \sum_{w=0}^{n} A_w \, C_{n-p}^{n-w}, \tag{9.21}$$

which holds for $p = n, n - 1, \ldots, 0$, where A_w (B_w) denotes the number of codewords of C (C^+) having weight w and C_a^b denotes the number of combinations of b elements taken a at a time. In the binary case (i.e., if $q = 2$) this identity is usually formulated, for weight enumerator polynomials $A(z)$ of C and $B(z)$ of C^+, as [1485, 1488]:

$$B(z) = 2^{-k}(1 + z)^n A\left(\frac{1 - z}{1 + z}\right) \tag{9.22}$$

---

[5] When this occurs, the channel is said to be *uniform from the input* (UFI).

or [1488]:

$$A(z) = 2^{k-n}(1+z)^n B\left(\frac{1-z}{1+z}\right). \tag{9.23}$$

MacWilliams's identity (9.21) is extremely useful when $\mathcal{C}$ is large in size, but its dual is sufficiently small that the weight of all its codewords can be computed via direct enumeration.

The most significant parameter that can be acquired from the weight spectrum of $\mathcal{C}$ is undoubtedly $d_{H,\min}$, the minimum distance between distinct codewords. We now describe how this parameter relates to the parity check matrix $\mathbf{H}$ of $\mathcal{C}$. To this end, let us rewrite (9.11) as:

$$\sum_{l=0}^{n-1} x_l\, \mathbf{h}_l^T = \mathbf{0}_{n-k}, \tag{9.24}$$

where $\mathbf{h}_l$ ($\mathbf{h}_l^T$) is the lth column (row) of $\mathbf{H}$ ($\mathbf{H}^T$). This states that, if the columns of $\mathbf{H}$ are combined linearly, using the elements of any $\mathbf{x} \in \mathcal{C}$ as coefficients, the result is the null vector $\mathbf{0}_{n-k}$. Therefore, since $d_{H,\min}$ is the *minimum weight* of the n-dimensional nonzero codewords $\mathbf{x} \in \mathcal{C}$ satisfying (9.24), it can be inferred that $d_{H,\min}$ may be expressed as the *minimum number of linearly independent columns of* $\mathbf{H}$. Conversely, it can easily be proved that, if all possible sets of $d-1$ or fewer distinct columns of $\mathbf{H}$ are linearly independent,[6] then $d_{H,\min}$ is not smaller than d (e.g., see [35, pp. 77–78]).

The parameter $d_{H,\min}$ provides an indication of the error detection and correction capabilities of a linear block code. To understand this, we consider the transmission of the kth codeword $\mathbf{x}_k$ of $\mathcal{C}$ over a q-ary *memoryless*[7] and *symmetric uniform* channel.[8] First, let us analyze the problem of the *correction* of errors introduced by the channel. In this case, the channel output vector $\mathbf{y}$ is given by:

$$\mathbf{y} = \mathbf{x}_k \boxplus \mathbf{e}, \tag{9.25}$$

where $\boxplus$ denotes the element-by-element sum of vectors over $GF(q)$ and $\mathbf{e}$ is the error vector, whose elements also belong to $GF(q)$. It can be shown that, given the model (9.25), the ML decoder selects the codeword $\mathbf{x}_H$ whose Hamming distance $d_H(\mathbf{x}_H, \mathbf{x})$ is minimum in the set of possible codewords $\mathbf{x}$ of $\mathcal{C}$ (e.g., see [35, pp. 10–13]). In practice, this can be accomplished as follows. From $\mathbf{y}$ in (9.25) the $(n-k)$-dimensional *syndrome* vector is calculated as:

$$\mathbf{s} \triangleq \mathbf{y}\mathbf{H}^T. \tag{9.26}$$

This vector is null if and only if $\mathbf{y}$ is a codeword. Therefore, if $\mathbf{s} \neq \mathbf{0}_{n-k}$, there are one or more incorrect symbols in $\mathbf{y}$. However, if $\mathbf{s} = \mathbf{0}_{n-k}$, the absence of transmission errors cannot be guaranteed. In fact, if $\mathbf{e} \neq \mathbf{0}_{n-k}$ is a codeword, $\mathbf{y}$ is also a codeword, even if it is different from $\mathbf{x}_k$. This means that there are as many undetectable errors as there are nonnull codewords (i.e., $q^k - 1$). It is also important to note that the syndrome $\mathbf{s}$ *depends only on the error vector* $\mathbf{e}$ and not on the transmitted codeword, since:

$$\mathbf{s} = (\mathbf{x}_k \boxplus \mathbf{e})\, \mathbf{H}^T = \mathbf{x}_k\, \mathbf{H}^T \boxplus \mathbf{e}\, \mathbf{H}^T = \mathbf{e}\, \mathbf{H}^T \tag{9.27}$$

and $\mathbf{x}_k\mathbf{H}^T = \mathbf{0}_{n-k}$ for any codeword (see (9.11)). Equality (9.27) can be interpreted as a system of $(n-k)$ linear equations over $GF(q)$ in n unknowns, each representing an element of $\mathbf{e}$. Unfortunately, this system admits q^k distinct solutions, since there are q^k different error vectors producing the same syndrome. This is due to (9.27) giving the same value of $\mathbf{s}$ for any of the q^k possible values of $\mathbf{x}_k$.

---

[6] This means all possible combinations of $d-1$ or fewer distinct columns of $\mathbf{H}$ produce a nonzero column vector.
[7] A communication channel is *memoryless* if its effects on distinct elements of the same codeword are statistically independent. In data communications over fading channels this condition holds in the presence of *ideal interleaving*.
[8] In a q-ary *symmetric uniform* channel, if the probability of correct reception of an input symbol is $1-p$, the probability that any *incorrect* symbol (belonging to the set of $q-1$ possible wrong symbols) is received is $p/(q-1)$.

Table 9.1   Standard array

$\mathbf{e}_0 = \mathbf{x}_0 = \mathbf{0}_n$	$\mathbf{x}_1$	$\cdots$	$\mathbf{x}_i$	$\cdots$	$\mathbf{x}_{q^k-1}$
$\mathbf{e}_1$	$\mathbf{e}_1 \boxplus \mathbf{x}_1$	$\cdots$	$\mathbf{e}_1 \boxplus \mathbf{x}_i$	$\cdots$	$\mathbf{e}_1 \boxplus \mathbf{x}_{q^k-1}$
$\vdots$	$\vdots$	$\vdots$	$\vdots$	$\vdots$	$\vdots$
$\mathbf{e}_l$	$\mathbf{e}_l \boxplus \mathbf{x}_1$	$\cdots$	$\mathbf{e}_l \boxplus \mathbf{x}_i$	$\cdots$	$\mathbf{e}_l \boxplus \mathbf{x}_{q^k-1}$
$\vdots$	$\vdots$	$\vdots$	$\vdots$	$\vdots$	$\vdots$
$\mathbf{e}_{q^{n-k}-1}$	$\mathbf{e}_{q^{n-k}-1} \boxplus \mathbf{x}_1$	$\cdots$	$\mathbf{e}_{q^{n-k}-1} \boxplus \mathbf{x}_i$	$\cdots$	$\mathbf{e}_{q^{n-k}-1} \boxplus \mathbf{x}_{q^k-1}$

However, these solutions are not equally likely with respect to the ML decoder. In fact, it can be proved that, under the assumptions made above, the ML estimate $\hat{\mathbf{e}}$ of $\mathbf{e}$, given the syndrome $\mathbf{s}$, is that (or one of those) having *minimal weight*.

In principle, if the number of existing syndromes, q^{n-k}, is not large, (9.27) can be computed for any possible vector $\mathbf{e}$. The results can be collected in a matrix, called the *standard array* of $\mathcal{C}$, containing *in each row* all the error vectors producing the *same* syndrome. For any syndrome, there are q^k distinct error vectors generating it, therefore this matrix consists of q^{n-k} rows and q^k columns. The structure of the standard array is shown in Table 9.1. In the construction of the standard array the following two rules are followed:

1. The first row contains all possible codewords, that is, *all the n-dimensional vectors producing the syndrome* $\mathbf{s} = \mathbf{0}_{n-k}$. In addition, the first element of this row is the null codeword $\mathbf{0}_n$.
2. The first element of each row has, among all the others in the same row, *minimal weight* (this element is not necessarily unique).

It is not difficult to prove that the construction of the standard array is equivalent to partitioning the set of codewords of $\mathcal{C}$ into a family of *cosets*. Each coset consists of all the elements in a single row of the standard array. In addition, the first element of each row is called the *coset leader* and, having minimal weight, represents the error vector selected by the ML decoder on the basis of the available syndrome. For this reason, error correction can be implemented as a procedure based on looking up a table listing, where the table gives the coset leader for every value of $\mathbf{s}$. Generally speaking, this table requires q^{n-k} memory locations, each containing n symbols of GF(q). However, if the code is *systematic*, only k symbols (those corresponding to the information symbols) need to be memorized for each possible syndrome. The decoder, after evaluating $\mathbf{s}$, reads in its memory the associated coset leader $\mathbf{e}_{CL}(\mathbf{s})$ and produces the ML estimate:

$$\mathbf{x}_H = \mathbf{y} \boxplus (-\mathbf{e}_{CL}(\mathbf{s})) \tag{9.28}$$

of the transmitted codeword. The *only error vectors or patterns* correctable by this algorithm are those represented by the coset leaders. If the decoder makes a wrong correction, then it may even increase the overall number of incorrect symbols. Note that this decoding procedure is impractical for large size codes. More efficient decoding procedures will be described in Section 9.1.6 for specific classes of codes.

A deeper insight into the ML algebraic decoding procedure can be gained from Figure 9.2. For simplicity the space of $\mathbf{y}$ is shown in two dimensions. In particular, in this geometrical representation, all possible codewords of $\mathcal{C}$ are indicated by distinct dots. Each dot is surrounded by a shaded region denoting the *guaranteed error-correction zone* for the corresponding codeword. In this specific case, this zone is a circle with the associated codeword at its center and integer radius t. Referring to this figure, we can state that an ML decoder certainly selects the codeword $\mathbf{x}_H = \mathbf{x}_k$ if $\mathbf{y}$ lies within the

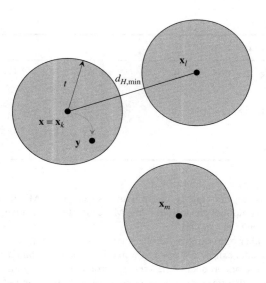

Figure 9.2   Guaranteed error-correction zones for an ML decoder of a block code $\mathcal{C}$.

guaranteed error-correction zone of $\mathbf{x}_k$, that is, if the overall number of incorrect elements of $\mathbf{y}$ does not exceed t. If $\mathbf{y}$ falls in the *space external to the correction zone for all* $\mathbf{x}_k$, it is still possible, *but not guaranteed*, that a given $\mathbf{x}_k$ is the codeword of $\mathcal{C}$ closest to $\mathbf{y}$. If there is not a *unique* codeword having minimum distance from $\mathbf{y}$, then the decoder makes an arbitrary choice among multiple (at least two) equally likely hypotheses. In addition, from our geometrical representation it is not difficult to infer that the *maximum number of errors* whose correction is *guaranteed* is:

$$t \triangleq \left\lfloor \frac{d_{H,\min} - 1}{2} \right\rfloor. \tag{9.29}$$

If the number of incorrect elements exceeds t, it is still possible that the decoder finds the correct value of $\mathbf{e}$, but it may select an incorrect estimate $\hat{\mathbf{e}}$ of $\mathbf{e}$.

The decoder can simultaneously correct t errors (9.29) and detect [1489]:

$$e_d \triangleq \left\lceil \frac{d_{H,\min} - 1}{2} \right\rceil \tag{9.30}$$

errors. This means a set of syndromes exists for all $\mathbf{e}$ of weight e_d or less, where $e_d > t$. However, more than one error sequence of weight $e_d > t$ can have the same syndrome. This is why error sequences with weight $e_d > t$ can only be detected and not corrected. Some error sequences of weight $e_d > t$ may have unique syndromes, in which case these specific error sequences can be corrected.

If the receiver is not interested in error correction, but only in the *detection* of their presence, the decoder can be fooled only when, because of an overall number of errors exceeding $d_{H,\min} - 1$, the transmitted codeword $\mathbf{x}_k$ is transformed into another codeword. Therefore, the *maximum number of errors* whose detection is *guaranteed* is $e_d \triangleq d_{H,\min} - 1$. Of course, if the overall number of errors is larger than e_d, their detection is still possible, but uncertain.

So far we have assumed the validity of the model (9.25) for the channel output. However, a receiver can be designed to declare a symbol *erased*[9] when it is received unreliably. In this case the correction

---

[9] Erased symbols are treated as lost – the decoder treats them as if they were never transmitted.

of t_1 errors can be guaranteed in the presence of e_s erasures, provided that $2t_1 + e_s \leq d_{H,\min} - 1$ (e.g., see [35, p. 81]).

These considerations, even if illustrated for the specific case of transmission over a q-ary uniform memoryless channel, show the importance of finding, in the family of (n, k) linear codes over GF(q), those schemes maximizing $d_{H,\min}$.

9.1.4   Cyclic Codes

In this subsection, a specific and well-known class of linear block codes, known as cyclic codes, is defined. The essential features of these codes and various encoding algorithms for them are analyzed. Several specific examples of famous cyclic codes are also illustrated.

9.1.4.1   Code Structure

A *cyclic code* is a linear block code with the property that a *cyclic shift* of each codeword is still a codeword. As shown in what follows, this property appreciably simplifies encoding and decoding, making them attractive for practical applications, particularly when long codewords are created.

To simplify the mathematical description of the properties of a cyclic code and of its encoding/decoding a *polynomial representation* of its codewords is adopted. In particular, given an (n, k) cyclic code $\mathcal{C}$ over GF(q), any codeword $\mathbf{x} \triangleq [x_0, x_1, \ldots, x_{n-1}]$ can be represented by the polynomial[10]:

$$x(D) \triangleq x_0 + x_1 D^1 + \ldots + x_{n-1} D^{n-1} \tag{9.31}$$

in the formal variable D and with degree not exceeding $n - 1$. A cyclic shift of $\mathbf{x}$ by an arbitrary number j of locations produces another codeword $\mathbf{x}^{(j)}$. It is easy to show that its polynomial representation is given by:

$$x^{(j)}(D) = D^j x(D) \mod D^n - 1, \tag{9.32}$$

the remainder of the division of $D^j x(D)$ by $(D^n - 1)$.

The polynomial representation is extremely useful in describing the inner structure of the family of codewords forming $\mathcal{C}$. To show this, we introduce the *generator polynomial* $g(D)$, representing, in the family of code polynomials $x(D) \neq 0$, the *monic polynomial of minimum degree*. This polynomial is *unique* and can be expressed as:

$$g(D) = g_0 + g_1 D^1 + \ldots + g_{r-1} D^{r-1} + D^r, \tag{9.33}$$

where $r \leq n - 1$ denotes its degree. It can be shown that, given $g(D)$, a polynomial $x(D)$ (over GF(q)) of degree not exceeding $n - 1$ represents a codeword if and only if (e.g., see [35, p. 139]):

$$x(D) = u(D)\, g(D), \tag{9.34}$$

where $u(D)$ is a polynomial of degree not larger than $n - 1 - r$. This means that a cyclic code is fully defined by its generator polynomial $g(D)$. This polynomial, however, cannot be selected arbitrarily in the set of polynomials over GF(q) of degree r. In fact, it can be proved that $g(D)$ generates an (n, k) cyclic code if and only if (e.g., see [35, pp. 138–139]):

---

[10] This representation should not be confused with that for the elements of an extension field GF(q) (with $q = p^m$). In fact, in the latter case the degree of the polynomials does not exceed $m - 1$ and polynomial coefficients belong to the ground field GF(p).

(a) it is a factor of $D^n - 1$ over GF(q), that is:

$$D^n - 1 = g(D) \, h(D), \tag{9.35}$$

where $h(D)$ is also a polynomial over GF(q), and

(b) its degree is $r = n - k$.

From (9.35) it is easy to see that the $n - k$ roots $\{\alpha_j, \, j = 0, 1, \ldots, n - k - 1\}$ of $g(D)$ are nth roots of unity, such that $\alpha_j^n = 1$. The knowledge of these roots allows one to fully describe $g(D)$ and, consequently, the associated code $\mathcal{C}$.

As illustrated in Section 9.1.2, any (n, k) linear block code over GF(q) can be described by a generator matrix $\mathbf{G}$. It is not difficult to show that, if this code is cyclic with a generator polynomial $g(D)$, a possible nonsystematic generator matrix is:

$$\mathbf{G} = \begin{bmatrix} \mathbf{g} \\ \mathbf{g}^{(1)} \\ \cdots \\ \mathbf{g}^{(k-1)} \end{bmatrix} = \begin{bmatrix} g_0 & g_1 & \cdots & \cdots & \cdots & g_{n-k} & 0 & \cdots & \cdots & 0 \\ 0 & g_0 & g_1 & \cdots & \cdots & \cdots & g_{n-k} & 0 & \cdots & \\ 0 & 0 & g_0 & g_1 & \cdots & \cdots & \cdots & g_{n-k} & 0 & \cdots \\ \cdots & \cdots & \cdots & \cdots & \cdots & \cdots & \cdots & \cdots & \cdots & \cdots \\ 0 & \cdots & \cdots & 0 & g_0 & g_1 & \cdots & \cdots & g_{n-k} & 0 \\ 0 & \cdots & \cdots & \cdots & 0 & g_0 & g_1 & \cdots & \cdots & g_{n-k} \end{bmatrix}, \tag{9.36}$$

where $\mathbf{g} \triangleq [g_0, g_1, \ldots, g_{n-k-1}, g_{n-k}, 0, \ldots, 0]$ (with $g_{n-k} = 1$) is the codeword represented by $g(D)$, associated with the message $u(d) = 1$.

Let us now focus on the construction of some cyclic binary block codes.

Example 9.1.4 Let us generate all the *binary* ($q = 2$) cyclic codes of length $n = 7$. Their generator polynomials can be derived from the factorization (see (9.35)):

$$D^7 - 1 = (1 + D)\left(1 + D + D^3\right)\left(1 + D^2 + D^3\right) \tag{9.37}$$

containing only irreducible polynomials over GF(2). Then possible generator polynomials of different degrees are:

$$g_1(D) = \left(1 + D + D^3\right)\left(1 + D^2 + D^3\right)$$
$$= 1 + D + D^2 + D^3 + D^4 + D^5 + D^6, \tag{9.38}$$

$$g_3(D) = (1 + D)\left(1 + D^2 + D^3\right) = 1 + D + D^2 + D^4, \tag{9.39}$$

$$g_4(D) = 1 + D + D^3, \tag{9.40}$$

$$g_6(D) = 1 + D, \tag{9.41}$$

for message lengths $k = 1, 3, 4, 6$, respectively. Note that cyclic codes with $k = 5$ or 2 do not exist, since generator polynomials of degrees 2 or 5, respectively, cannot be constructed using distinct factors of (9.37). In the construction of a generator polynomial of degrees 4 (see (9.39)) and 3 (see (9.40)) the selection of the degree 3 term from the factorization (9.37) *is not unique* since $D^7 - 1$ has two factors of degree 3.

□

Clearly, the factorization of $D^n - 1$ can contain multiple irreducible terms of the same degree. For this reason, the generator polynomial $g(D)$ for a cyclic code (n, k) *is not necessarily unique*. Note that different choices for $g(D)$ can lead to codes with different $d_{H,\min}$ and, consequently, substantially different error performance.

9.1.4.2   Dual Code and Parity Check Polynomial

As already shown in Section 9.1.2, an (n, k) linear block code $\mathcal{C}$ over GF(q) can also be described by a parity check matrix $\mathbf{H}$. If $\mathcal{C}$ is cyclic, $\mathbf{H}$ can easily be derived from the *parity check polynomial* (see (9.35)) as:

$$h(D) \triangleq \frac{D^n - 1}{g(D)}, \qquad (9.42)$$

of degree k. In fact, it can be shown that (e.g., see [327, p. 446]):

$$\mathbf{H} = \begin{bmatrix} h_k & h_{k-1} & \cdots & h_0 & 0 & \cdots & \cdots & \cdots & 0 \\ 0 & h_k & h_{k-1} & \cdots & h_0 & 0 & \cdots & \cdots & 0 \\ \cdots & \cdots & \cdots & \cdots & \cdots & \cdots & \cdots & \cdots & \cdots \\ \cdots & \cdots & \cdots & h_k & h_{k-1} & \cdots & \cdots & h_0 & 0 \\ 0 & 0 & \cdots & \cdots & h_k & h_{k-1} & \cdots & \cdots & h_0 \end{bmatrix}. \qquad (9.43)$$

The parity check matrix (9.43) can be interpreted as the generator matrix of the *dual* $\mathcal{C}^+$ of the code $\mathcal{C}$ generated by $\mathbf{G}$ (9.36). Note, however, that despite the structural similarity between $\mathbf{H}$ (9.43) and $\mathbf{G}$ (9.36), their relationships with the associated polynomials $h(D)$ and $g(D)$, respectively, are somewhat different. In fact, in each row of $\mathbf{G}$ the coefficients of $g(D)$ appear in a natural order (i.e., in increasing powers of D), whereas in the rows $\mathbf{H}$ the coefficients of $h(D)$ follow the *inverse order*. This means the generator polynomial of $\mathcal{C}^+$ is not $h(D)$ (9.42) but, up to a scale factor, the *reciprocal polynomial*:

$$\tilde{h}(D) \triangleq h_k + h_{k-1}D + \cdots + h_0 D^k = \sum_{l=0}^{k} h_{k-l} D^l. \qquad (9.44)$$

Note that $\tilde{h}(D)$ is *not* necessarily monic and as a result the generator polynomial of $\mathcal{C}^+$ is $h_0^{-1}\tilde{h}(D)$.

Example 9.1.5 Let us again consider the $(7, k)$ binary cyclic codes derived in Example 9.1.4. It is not difficult to prove that $g_1(D)$ (9.38) and $g_6(D)$ (9.41) describe dual codes. In fact, the parity check polynomial $h_1(D)$ associated with $g_1(D)$ is:

$$h_1(D) = g_6(D) = 1 + D \qquad (9.45)$$

and coincides with its reciprocal polynomial $\tilde{h}_1(D)$ generated according to (9.44). Similarly, the cyclic code described by $g_3(D)$ (9.39) is the dual of that described by $g_4(D)$ (9.40). Moreover, the parity check polynomial $h_3(D)$ associated with $g_3(D)$ is:

$$h_3(D) = 1 + D + D^3. \qquad (9.46)$$

Reversing the order of the coefficients of $h_3(D)$ produces the reciprocal polynomial:

$$\tilde{h}_3(D) = 1 + D^2 + D^3. \qquad (9.47)$$

☐

9.1.4.3   Encoding Algorithms

Encoding can be implemented as in (9.34), evaluating the product of the generator polynomial $g(D)$ and the message polynomial $u(D)$. This operation can be done via a digital transversal filter structure as shown in Figure 9.3. This filter contains $n - k + 1$ multipliers over GF(q) (indicated by the symbol ⊡), $n - k$ adders over GF(q) (indicated by the symbol ⊞) and a delay line consisting of $n - k$ registers, each producing a single clock interval delay. In practice, the filter evaluates the convolution between

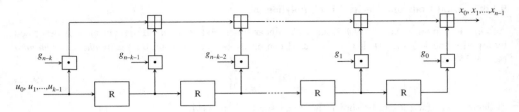

Figure 9.3 Nonsystematic encoder for an (n, k) cyclic code C over GF(q).

the input sequence $\{u_{k-1}, u_{k-2}, \ldots, u_0, 0, 0, \ldots\}$ and the impulse response $\{g_{n-k}, g_{n-k-1}, \ldots, g_0\}$, to generate the response $\{x_{n-1}, x_{n-2}, \ldots, x_0\}$. Note that: (a) the elements of the input sequence (of the output sequence) will be applied to (generated by) the filter *in reverse order*; (b) the filter is initially cleared and after the first k clock intervals is fed by a null sequence.

The codeword $\mathbf{x}$ generated by the scheme illustrated in Figure 9.3 is inevitably in *nonsystematic form*. An alternative to this solution is a *systematic* encoding algorithm which can be derived as follows. Dividing the polynomial:

$$D^{n-k}u(D) = D^{n-k}\sum_{l=0}^{k-1}u_l D^l = \sum_{l=n-k}^{n-1}u_{l-(n-k)}D^l, \qquad (9.48)$$

which has the same information content as $u(D)$, by $g(D)$ yields:

$$\frac{D^{n-k}u(D)}{g(D)} = a(D) + \frac{p(D)}{g(D)},$$

$$D^{n-k}u(D) = a(D)g(D) + p(D), \qquad (9.49)$$

where $a(D)$ denotes the quotient and:

$$p(D) = \sum_{l=0}^{n-k-1}p_l D^l \triangleq \left[D^{n-k}u(D)\right] \bmod g(D) \qquad (9.50)$$

is the remainder. We then define:

$$x(D) \triangleq D^{n-k}u(D) - p(D) \qquad (9.51)$$

and note that, thanks to (9.49), $x(D)$ can also be expressed as:

$$x(D) = a(D)g(D). \qquad (9.52)$$

Then the polynomial $x(D)$ (9.51) represents a codeword of C, since it contains $g(D)$ as a factor. Moreover, from (9.48), (9.50) and (9.51) it is easy to see that:

$$x(D) = \sum_{l=n-k}^{n-1}u_{l-(n-k)}D^l + \sum_{l=0}^{n-k-1}(-p_l)D^l. \qquad (9.53)$$

This means that the k coefficients of the terms with the largest degrees in $x(D)$ are given by the information symbols, whereas the remaining $n - k$ coefficients $\{-p_l, l = 0, 1, \ldots, n - k - 1\}$ represent the parity symbols of $x(D)$. This shows that, given the message polynomial $u(D)$, the corresponding codeword $x(D)$ in systematic form can be generated by resorting to (9.50) and (9.51).

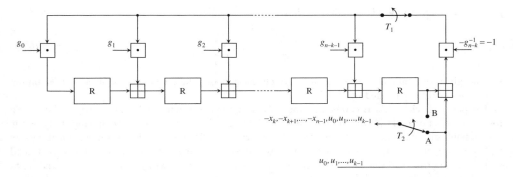

Figure 9.4   Systematic encoder for an (n, k) cyclic code C over GF(q).

In practice, this requires the availability of a logic circuit evaluating the division of the polynomials $D^{n-k}u(D)$ and $g(D)$, and generating the remainder $p(D)$ (9.50). A suitable circuit is illustrated in Figure 9.4, based on a general *polynomial divider* scheme (e.g., see [35, pp. 146–149] and [327, pp. 450–452] for further details). It operates as follows. At the start of the encoding procedure the switch T_1 is closed, the switch T_2 is turned to A and all the registers are cleared. In the first k clock intervals the sequence $\{u_{k-1}, u_{k-2}, \dots, u_0\}$ feeds both the divider and the output since the code is systematic. At the end of this interval, the coefficients (actually, their opposites) of the remainder polynomial $p(D)$ are available in the cells of the delay line. Therefore, the feedback in the divider is disabled to stop the division procedure by opening the switch T_1. Then T_2 is turned to B, so that the content of the delay line can be read sequentially in $n - k$ consecutive clock intervals.

Finally, we note that the schemes of Figures 9.3 and 9.4 can be substantially simplified when $q = 2$, since there is then no difference between the sum and difference operations over GF(2) and a product by 1 (0) corresponds to the presence (absence) of a connection along the register.

9.1.4.4   Hamming Codes over GF(q)

The binary $(7, 4)$ code described in Example 9.1.1 belongs to the class of so-called *Hamming codes* over GF(q). Their main feature is that they have minimum Hamming distance $d_{H,\min} = 3$, so that error patterns with a single error can always be corrected by an ML decoder. The correction of multiple errors is not guaranteed.

The correction of all possible single errors via ML algebraic decoding of an (n, k) Hamming code C requires the set of coset leaders of C to have a well-defined structure: it consists of the vector 0_n and all possible n-dimensional vectors over GF(q) each having *a single nonnull element*. Generally speaking, if a code has a standard array containing all the error patterns of t or fewer errors and no others as coset leaders, it is called a *perfect code*. Thus, Hamming codes form a class of *single-error-correcting perfect codes*. Note that perfect codes are rare. In fact, besides the Hamming codes, the only other nontrivial scheme having the property of perfection is the $(23, 12)$ Golay code [1470] described below.

The size of the family of cosets of an (n, k) Hamming code C is $n(q - 1)$, since in each coset we have a single nonnull element that can be placed in n different locations and can take on $q - 1$ distinct values. Therefore, since in any (n, k) code over GF(q) we have q^{n-k} syndromes, the equality:

$$1 + n(q - 1) = q^{n-k} \tag{9.54}$$

must hold. Rearranging this equality, we have:

$$n = \frac{q^{n-k} - 1}{q - 1} = \sum_{l=0}^{n-k-1} q^l, \tag{9.55}$$

so that, given the number $n - k \geq 2$ of parity check symbols, the codeword length n of a Hamming code is uniquely determined by (9.55).

The parity check matrix $\mathbf{H}$ can be generated via a simple construction rule; this consists of selecting for the n columns of $\mathbf{H}$ all the nonnull $(n - k)$-dimensional vectors over GF(q) having unity as the first nonnull element (the reader can verify that exactly n (9.55) distinct $(n - k)$-dimensional vectors with this property exist). In fact, if all the rows of $\mathbf{H}^T$ are different, the error patterns associated with a single error (and that corresponding to the absence of errors) will generate distinct syndromes according to (9.27).

The above-mentioned construction procedure is applied to a specific design in the following example.

Example 9.1.6 If $q = 4$ and $n - k = 2$, (9.55) yields $n = 15/3 = 5$, so that a $(5, 3)$ code is generated. Following the construction rule proposed above, the parity check matrix:

$$\mathbf{H} = \begin{bmatrix} 1 & 1 & 1 & 1 & 0 \\ 1 & \alpha & \alpha^2 & 0 & 1 \end{bmatrix} \tag{9.56}$$

can easily be generated, where α is a primitive element of GF(4). Note that the columns of $\mathbf{H}$ have not been ordered in a natural fashion. In fact, $\mathbf{H}$ has been put in the form (9.13), from which the generator matrix can be immediately inferred (see (9.6)).
□

Generally speaking, the parity check matrix of an (n, k) Hamming code can be put in the form:

$$\mathbf{H} = [\mathbf{Q}_{n-k,k} \ \mathbf{I}_{n-k}], \tag{9.57}$$

where $\mathbf{Q}_{n-k,k}$ consists of k columns, each represented by an $(n - k)$-tuple of weight 2 or more.

The Hamming codes can also be put in *cyclic form* (e.g., see [327, p. 462]); in fact, they can be described as a specific subclass of the BCH codes described later. In addition, in the binary case $(q = 2)$ it can be shown that the generator polynomial is a *primitive polynomial* $p(D)$ of degree $n - k$, so that the parity check matrix for the resulting code has the structure of (9.57). It is also worth mentioning that no two columns of $\mathbf{Q}_{n-k,k}$ are alike and each column has at least two nonnull elements (e.g., see [35, pp. 162–163] for a proof in the binary case).

The weight spectrum of an (n, k) Hamming code over GF(q) is known and is given by [327, p. 428]:

$$A(z) = \frac{1}{n(q - 1) + 1} \Big[[1 + (q - 1)z]^n$$
$$+ n(q - 1)\big[1 + (q - 1)z\big]^{(n-1)/q} (1 - z)^{(n(q-1)+1)/q} \Big]. \tag{9.58}$$

Finally, it is important to note that the dual code of an (n, k) cyclic Hamming code with generator polynomial $p(D)$ is an $(n, n - k)$ *maximal-length* block code, described by the generator polynomial:

$$g(D) = \frac{D^n - 1}{p(D)} \tag{9.59}$$

and characterized, in the binary case, by a minimum Hamming distance $d_{H,\min} = 2^{n-k-1}$. The reader can refer to [35, pp. 290–292] for an analysis of the essential properties of this class of codes.

9.1.4.5 BCH Codes

Generally speaking, in the design of a new cyclic coding scheme the achievable minimum Hamming distance $d_{H,\min}$ is unknown a priori. This parameter is usually estimated once the generator polynomial is available. Exceptions to this rule are cyclic Hamming codes (which always have $d_{H,\min} = 3$) and the class of codes designed by R. C. Bose, D. K. Ray-Chaudhuri and A. Hocquenghem, known as BCH codes. In the latter case, a lower bound on $d_{H,\min}$ is available before design. This important result is achieved by setting a family of specific constraints on the generator polynomial $g(D)$, which is still required to satisfy the factorization (9.35). These constraints are expressed by the so-called BCH *bound theorem* (a proof can be found in [35, Sect. 6.1] and [327, pp. 460–462]). To develop this fundamental result, let us assume that $GF(q^l)$ (called the *locator field*), with positive integer l, is the *smallest extension field of* $GF(q)$ (called the *symbol field*) containing an element[11] β of order n. Then the BCH bound theorem states that, if the generator polynomial $g(D)$ of a cyclic code $\mathcal{C}$ is the *minimum degree monic polynomial over* $GF(q)$ having the $(\delta - 1)$ consecutive powers $\{\beta^b, \beta^{b+1}, \ldots, \beta^{b+\delta-2}\}$ of β as its *roots* (i.e., $g(\beta^b) = g(\beta^{b+1}) = \ldots = g(\beta^{b+\delta-2}) = 0$), where b and $\delta \geq 1$ are integer parameters, then $\mathcal{C}$ is a BCH code having minimum distance:

$$d_{H,\min} \geq \delta. \tag{9.60}$$

In the literature this inequality is known as the BCH *bound*, and the parameter δ is usually called the *design distance* of $\mathcal{C}$.

Let us exploit this result to develop a design procedure for a BCH code over $GF(q)$, with a codeword length $n \geq 3$ and able to correct up to t errors. First, we need to find an element β of order n in a field $GF(q^l)$, selecting the minimum possible value l. Then $(\delta - 1) = 2t$ consecutive powers of β (with $3 \leq \delta \leq n$) are selected, starting from β^b. Finally, the generator $g(D)$ is generated as [1488]:

$$g(D) = \text{LCM}\left\{ m_{\beta^b}(D), m_{\beta^{b+1}}(D), \ldots, m_{\beta^{b+2t-1}}(D) \right\}, \tag{9.61}$$

where LCM denotes the least common multiple operator and $m_{\beta^{b+k}}(D)$ is the *minimal polynomial* of β^{b+k} (with $k = 0, 1, \ldots, 2t - 1$). In other words, $g(D)$ in (9.61) is the monic polynomial of minimum degree for which each of the polynomials $\{m_{\beta^{b+j}}(D), j = 0, 1, \ldots, 2t - 1\}$ is a divisor. It is important to point out the following:

1. Since β is an element of order n in the field $GF(q^l)$, all its *distinct powers* $\{\beta^j, j = 0, 1, \ldots, n - 1\}$ are roots of the equation $D^n - 1 = 0$, so that each of the polynomials $\{m_{\beta^j}(D), j \geq 0\}$ *divides* $D^n - 1$.
2. The $(\delta - 1)$ minimal polynomials of (9.61) are not necessarily distinct.
3. The minimum distance $d_{H,\min}$ of a BCH code generated according to the above procedure can be larger than δ. In this case an ML decoder can correct more than $t = \lfloor (\delta - 1)/2 \rfloor$ errors. Unfortunately, however, standard algorithms for algebraic decoding of BCH codes are suboptimal and do not allow us to exploit this possibility.

The first point ensures that $g(D)$ (9.61) satisfies (9.35); this is a necessary and sufficient condition for the code generated by $g(D)$ to be cyclic. From the second point it is easily inferred that, given that all the polynomials in the set $\{m_{\beta^{b+j}}(D), j = 0, 1, \ldots, 2t - 1\}$ are *irreducible* over $GF(q)$, evaluating $g(D)$ according to (9.61) is equivalent to computing the product of all the *distinct* polynomials of this set.

Moreover, we note that constructing $g(D)$ according to (9.61) does not provide any information about the message length k, which depends, for a given n, on the degree $(n - k)$ of $g(D)$. Generally

---

[11] It is worth remembering that, as illustrated in Appendix E, the order n of an element β of $GF(q^l)$ is always a divisor of $q^l - 1$. In the case of BCH codes, β is usually a *primitive element* of $GF(q^l)$, so that $n = q^l - 1$.

speaking, such a degree does not exceed $l(\delta - 1)$, since the degree of the minimal polynomial of any element of $GF(q^l)$ is not larger than l (e.g., see [1485, pp. 54–55]) and the LCM in (9.61) does not involve more than $\delta - 1$ distinct factors. As a result, the inequality:

$$n - k \le l(\delta - 1), \tag{9.62}$$

which limits the overall number $(n - k)$ of parity check symbols, holds for any BCH code.

In (9.61) it is commonly assumed that $b = 1$, that is, the element β is the first root of $g(D)$. In this case, what results is a *narrow-sense* BCH code. In addition, if β is selected as a *primitive element* α of $GF(q^l)$, so that

$$n = q^l - 1, \tag{9.63}$$

the BCH code is *primitive*, whereas, when this does not occur, the code is *nonprimitive*. In what follows we will mainly focus on primitive narrow-sense codes. Then, in the *binary* case ($q = 2$), α is a primitive element of the field $GF(2^l)$ and (9.61) becomes[12]

$$g(D) = \text{LCM}[m_{\alpha^1}(D), m_{\alpha^2}(D), \dots, m_{\alpha^{2t}}(D)]. \tag{9.64}$$

Note that if $\beta \in GF(2^l)$ is a root of a polynomial $p(D)$ of degree m with coefficients from $GF(2)$ and *irreducible* in this field, then $\beta^2, \beta^{2^2}, \dots, \beta^{2^{m-1}}$ are *all the roots of the polynomial itself* (e.g., see [1490, p. 102] for a proof). Therefore, in (9.61) the terms associated with the even powers of α can be removed as they are already represented by those associated with the odd powers; this yields:

$$g(D) = \text{LCM}[m_{\alpha^1}(D), m_{\alpha^3}(D), \dots, m_{\alpha^{2t-1}}(D)]. \tag{9.65}$$

In this case no more than $\lceil (\delta - 1)/2 \rceil$ distinct factors are used in the generation of $g(D)$. Therefore, inequality (9.62) can be replaced by the tighter bound:

$$n - k \le l \left\lceil \frac{\delta - 1}{2} \right\rceil. \tag{9.66}$$

A list of the generator polynomials for all binary, narrow-sense, primitive BCH codes of lengths 7 through 255 can be found in [1485, Appendix E].

Finally, it is worth pointing out that:

(a) the weight distributions are known only for some BCH codes, such as all single-, double- and triple-error-correcting primitive binary BCH codes,

(b) the weight enumerator polynomial for a single-error-correcting primitive binary (n, k) BCH code is the same as for an (n, k) binary Hamming code [1485].

The constructions of some specific BCH codes are illustrated in the following examples.

Example 9.1.7 Let us consider the design of *binary* narrow-sense primitive BCH codes characterized by $l = 4$, so that $n = 2^4 - 1 = 15$ and α in (9.65) is a primitive element of the field $GF(2^4) = GF(16)$. The minimal polynomial of an arbitrary element β in $GF(2^l)$ can be expressed as [35, p. 50]:

$$m_\beta(D) = \prod_{i=0}^{r-1} (D + \beta^{2^i}), \tag{9.67}$$

where r is the smallest integer such that $\beta^{2^r} = \beta$ and $\{\beta^{2^i}, i = 0, 1, \dots, r - 1\}$ is the cyclotomic coset to which β belongs in $GF(2^l)$. Let us illustrate the use of this formula to evaluate one of the minimal polynomials appearing in (9.65), namely $m_{\alpha^3}(D)$. If $\beta = \alpha^3$, (9.65) yields

$$m_{\alpha^3}(D) = (D + \alpha^3)(D + \alpha^6)(D + \alpha^{12})(D + \alpha^9) = D^4 + D^3 + D^2 + D + 1, \tag{9.68}$$

---

[12] A table of minimal polynomials of elements in $GF(2^l)$ is given in [1485].

Table 9.2 Cyclotomic cosets and minimal polynomials of GF(16)

Cyclotomic cosets	Minimal polynomials
$C_0 = \{0\}$	D
$C_1 = \{1\}$	$D + 1$
$C_2 = \{\alpha, \alpha^2, \alpha^4, \alpha^8\}$	$D^4 + D + 1$
$C_3 = \{\alpha^3, \alpha^6, \alpha^9, \alpha^{12}\}$	$D^4 + D^3 + D^2 + D + 1$
$C_4 = \{\alpha^5, \alpha^{10}\}$	$D^2 + D + 1$
$C_5 = \{\alpha^7, \alpha^{11}, \alpha^{13}, \alpha^{14}\}$	$D^4 + D^3 + 1$

since the cyclotomic coset of β is:

$$\beta = \alpha^3, \quad \beta^2 = \alpha^6, \quad \beta^4 = \alpha^{12}, \quad \beta^8 = \alpha^9, \quad \beta^{16} = \alpha^3 = \beta, \tag{9.69}$$

so that $r = 4$. The other cyclotomic cosets of GF(16) and the associated minimal polynomials are listed in Table 9.2. These results can be exploited to generate some specific generator polynomials using (9.65). For instance, if $t = 1$, from (9.65) it is easy to see that

$$g(D) = m_\alpha(D) = D^4 + D + 1 \tag{9.70}$$

and $n - k = 4$, so that a $(15, 11)$ code (with $d_{H,\min} = 3$) is found. Similarly, if $t = 2$, (9.65) yields:

$$\begin{aligned}
g(D) &= \text{LCM}[m_\alpha(D), m_{\alpha^3}(D)] \\
&= m_\alpha(D)\, m_{\alpha^3}(D) \\
&= (D^4 + D + 1)(D^4 + D^3 + D^2 + D + 1) \\
&= D^8 + D^7 + D^6 + D^4 + 1, \tag{9.71}
\end{aligned}$$

so that $n - k = 8$, that is, a $(15, 7)$ code (with $d_{H,\min} = 5$) is generated.

$\square$

Example 9.1.8 Let us now analyze the construction of a length-15 narrow-sense primitive BCH code on the symbol field GF(4), so that the extension (locator) field to be considered for the primitive element α is GF(4^2) = GF(16) (since $4^2 - 1 = 15$). If β denotes a primitive element of GF(4), this finite field can be represented as GF(4) = $\{0, 1, \beta, \beta^2\}$. Then the irreducible polynomial $f(D) = D^2 + D + \beta$ over GF(4) can be used to generate GF(16): the corresponding polynomial representations of the elements of this field are given in Table 9.3. Note that GF(4), being a subfield of GF(16), can also be represented as $\{0, 1, \alpha^5, \alpha^{10}\}$.

The elements of GF(16) can be partitioned into cyclotomic cosets. These cosets and the associated minimal polynomials are listed in Table 9.4.

Some specific generator polynomials can be now constructed by exploiting (9.64). For instance, in the case of single error correction ($t = 1$), (9.64) yields:

$$\begin{aligned}
g(D) &= \text{LCM}[m_\alpha(D), m_{\alpha^2}(D)] \\
&= m_\alpha(D) m_{\alpha^2}(D) \\
&= (D^2 + D + \beta)(D^2 + D + \beta^2) \\
&= D^4 + D + 1, \tag{9.72}
\end{aligned}$$

so that $n - k = 4$, that is, a $(15, 11)$ code is found.

$\square$

Table 9.3 Polynomial
representation of the elements of
GF(16) (where β is a primitive
element of GF(4))

Exponential representation	Polynomial representation
0	0
1	1
α	D
α^2	$D + \beta$
α^3	$\beta^2 D + \beta$
α^4	$D + 1$
α^5	β
α^6	βD
α^7	$\beta D + \beta^2$
α^8	$D + \beta^2$
α^9	$\beta D + \beta$
α^{10}	β^2
α^{11}	$\beta^2 D$
α^{12}	$\beta^2 D + 1$
α^{13}	$\beta D + 1$
α^{14}	$\beta^2 D + \beta^2$

Table 9.4 Cyclotomic cosets and minimal polynomials
of GF(16)

Cyclotomic cosets	Minimal polynomials
$C_0 = \{0\}$	D
$C_1 = \{1\}$	$D + 1$
$C_2 = \{\alpha, \alpha^4\}$	$D^2 + D + \beta = m_\alpha(D)$
$C_3 = \{\alpha^2, \alpha^8\}$	$D^2 + D + \beta^2 = m_{\alpha^2}(D)$
$C_4 = \{\alpha^3, \alpha^{12}\}$	$D^2 + \beta^2 D + 1$
$C_5 = \{\alpha^5\}$	$D + \beta$
$C_6 = \{\alpha^6, \alpha^9\}$	$D^2 + \beta D + 1$
$C_7 = \{\alpha^7, \alpha^{13}\}$	$D^2 + \beta D + \beta$
$C_8 = \{\alpha^{10}\}$	$D + \beta^2$
$C_9 = \{\alpha^{11}, \alpha^{14}\}$	$D^2 + \beta^2 D + \beta^2$

9.1.4.6 Reed–Solomon Codes

RS codes are a class of codes consisting of *primitive* BCH codes for which $l = 1$; in other words, the roots of the generator polynomial $g(D)$ of a RS code belong to the symbol field GF (q). Therefore, from (9.63) it is easy to see that the codeword length is:

$$n = q - 1. \tag{9.73}$$

The construction of an RS code requires knowledge of a primitive element α of GF (q) and computing the minimal polynomials $\{m_{\alpha^{b+j}}(D), \quad j = 0, 1, \ldots, 2t - 1\}$ associated with $2t$ consecutive powers of α. Since $m = 1$, all such polynomials have degree 1, that is, can be expressed as:

$$m_{\alpha^{b+j}}(D) = D - \alpha^{b+j} \tag{9.74}$$

for $j = 0, 1, \ldots, 2t - 1$. Substituting this result into (9.61) (with $\beta = \alpha$) produces the generator polynomial:

$$g(D) = \text{LCM}\left\{m_{\alpha^b}(D), m_{\alpha^{b+1}}(D), \ldots, m_{\alpha^{b+2t-1}}(D)\right\} = \prod_{j=0}^{2t-1}\left(D - \alpha^{b+j}\right) \tag{9.75}$$

with degree exactly equal to $2t = \delta - 1$. Therefore, an RS code has:

$$n - k = 2t = \delta - 1 \tag{9.76}$$

parity check symbols. In principle their design is extremely simple, as illustrated by the following example.

Example 9.1.9 We now design primitive RS codes over GF(16). The exponential and polynomial representations of the elements of GF(16) are given in Table 9.5. For $t = 1$ the generator polynomial is:

$$g(D) = (D - \alpha)(D - \alpha^2) = D^2 + \alpha^5 D + \alpha^3, \tag{9.77}$$

giving a (15, 13) code with $d_{H,\min} = 3$. For $t = 2$ the generator polynomial is:

$$g(D) = (D - \alpha)(D - \alpha^2)(D - \alpha^3)(D - \alpha^4) = D^4 + \alpha^{13}D^3 + \alpha^6 D^2 + \alpha^3 D + \alpha^{10}, \tag{9.78}$$

giving a (15, 11) code with $d_{H,\min} = 5$.
□

Table 9.5 Polynomial representation of the elements of GF(16)

Exponential representation	Polynomial representation
0	0
1	1
α	D
α^2	D^2
α^3	D^3
α^4	$D + 1$
α^5	$D^2 + 1$
α^6	$D^3 + D^2$
α^7	$D^3 + D + 1$
α^8	$D^2 + 1$
α^9	$D^2 + D + 1$
α^{10}	$D^2 + D + 1$
α^{11}	$D^3 + D^2 + D$
α^{12}	$D^3 + D^2 + D + 1$
α^{13}	$D^3 + D^2 + 1$
α^{14}	$D^3 + 1$

Since RS codes are BCH codes, the inequality (see (9.60)):

$$d_{H,\min} \geq \delta = 2t + 1 \tag{9.79}$$

always holds. Using (9.76), this can be rewritten as:

$$d_{H,\min} \geq n - k + 1. \tag{9.80}$$

For an arbitrary (n, k) linear block code over GF (q) the *Singleton bound* (e.g., see [1491]):

$$d_{H,\min} \leq n - k + 1 \tag{9.81}$$

holds. Thus, comparing (9.80) with (9.81) leads to the equality:

$$d_{H,\min} = n - k + 1, \tag{9.82}$$

which holds for arbitrary RS codes. This means that the minimum Hamming distance between pairs of codewords achieves the singleton bound. For this reason, RS codes are *maximal distance separable* (MDS).

Note that, in the construction of RS codes, n is uniquely identified by the size q of its symbol field, whereas k can be arbitrarily selected, provided that it does not exceed $n - 1$. The choice of b in (9.75) is arbitrary, since it does not affect $d_{\min}$ or k. This represents a significant difference with respect to the class of BCH codes, for which a change in the value of b can modify both the size (i.e., k) and the $d_{H,\min}$ of the resulting code.

Any k information positions in the code can be used as the information positions. This allows the weight distribution polynomial of an (n, k) MDS code (and hence an RS code), with minimum distance $d_{H,\min}$ and defined over $GF(q)$, to be written as:

$$A(z) = 1 + A_{d_{H,\min}} z^{d_{H,\min}} + A_{d_{H,\min}+1} z^{d_{H,\min}+1} + \cdots + A_n z^n, \tag{9.83}$$

where (see [1485, p. 189]):

$$A_w = (q - 1) C_w^n \sum_{l=0}^{w-d_{H,\min}} (-1)^l C_l^{w-1} q^{w-l-d_{H,\min}} \tag{9.84}$$

with $w = n - k + 1, n - k + 2, \ldots, n$. In particular, formula (9.84) shows that the number of codewords having weight $d_{H,\min}$ is given by:

$$A_{d_{H,\min}} = C_{k-1}^n (q - 1). \tag{9.85}$$

An *extended* RS code can be created by adding an overall parity check symbol. This increases the length and minimum distance by one, so the code is still MDS. Moreover, it can be proved that the dual of an (n, k) RS code is an $(n, n - k)$ RS code, whereas in general this does not occur for the larger class of BCH codes (e.g., see [1485, p. 188]).

RS codes are usually exploited in binary data transmissions for the correction of error *bursts*. In this case each of the k symbols represents $c = \log q$ binary information digits and the (n, k) RS code $\mathcal{C}$ over GF (q) is seen as an $(n \cdot c, k \cdot c)$ code over GF (2). Note that $\mathcal{C}$, even if it is MDS over GF (q), generally does not preserve this property over GF (2). Despite this, it lends itself to the correction of error bursts. In fact, the errors introduced over multiple consecutive bits are seen by a decoder as errors over adjacent symbols of the received codewords. It is not difficult to show that, if $\mathcal{C}$ can correct up to t symbol errors in its codewords, then it can certainly correct a packet of $[c(t - 1) + c - 1]$

consecutive incorrect bits. This is true regardless of the location of the packet with respect to the beginning of the codeword symbols. It is also true for a set of short error bursts whose overall length does not exceed the same threshold [327, p. 465].

RS codes have been used in many applications such as compact discs and space exploration (e.g., see [1485, pp. 428–435]). Some information about a commonly used RS code is provided in the following example.

Example 9.1.10 Let us focus on RS coding on GF (2^8) = GF (256). In this case the codeword length is $n = 256 - 1 = 255$ and each of the symbols forming a codeword can be seen as a packet of 8 bits, that is, as a byte. If we select $d_{H,\min} = 33$ as a design constraint (so that the correction of $t = 16$ errors is guaranteed), from (9.82) it can be seen that the number of information symbols in each codeword is $k = 223$, so that the fractional redundancy in each codeword is $(n - k)/n = 32/255$. Despite this small value, the RS code can correct a large number of errors. The generator polynomial for this RS code has degree 32 and is given by [1492]:

$$g(D) = (D - \alpha)(D - \alpha^2) \cdots (D - \alpha^{32}). \tag{9.86}$$

Its use in practical applications requires the implementation of arithmetic on GF (256). Despite this apparent difficulty, efficient decoding architectures are available [1492]. This has made the VLSI implementation of powerful RS codes feasible.

□

9.1.4.7 Golay Codes

In 1949 Golay published a short paper in which the binary $(7, 4)$ code was extended to a general class of p-ary codes of length $(p^n - 1)/(p - 1)$, where p is a prime [1470]. In the same paper Golay described two new specific codes, one a binary triple-error-correcting code and the other a ternary double-error-correcting code, both of which are *perfect*. The *binary* scheme is a $(23, 12)$ linear block code with $d_{H,\min} = 7$ [1485, p. 139]. It can be represented as a cyclic code with generator polynomial:

$$g_1(D) = 1 + D^2 + D^4 + D^5 + D^6 + D^{10} + D^{11} \tag{9.87}$$

or

$$g_2(D) = 1 + D + D^5 + D^6 + D^7 + D^9 + D^{11}. \tag{9.88}$$

Note that both generators are factors of $D^{23} + 1$, since this polynomial can be factored as $D^{23} + 1 = (1 + D)g_1(D)g_2(D)$. The binary Golay code can be extended by adding an overall parity check bit [35]. This produces a $(24, 12)$ code (with $d_{H,\min} = 8$), which is not, however, a perfect code. Several efficient decoding algorithms for the Golay codes have been developed [35].

9.1.5 Other Relevant Linear Block Codes

9.1.5.1 Single Parity Check Codes and Repetition Codes

Single parity check (SPC) codes are one of the simplest types of block codes available. In fact, they are $(n, k = n - 1)$ linear block codes (with $d_{H,\min} = 2$), that is, they have *a single parity bit*. For even parity, the parity bit equals the sum of all the information bits modulo 2. For odd parity the complement of this sum is used. The *dual* code to a SPC code is the $(n, k = 1)$ *repetition code* with $d_{H,\min} = n$, that is, a code simply transmitting the input bit n times. This provides a very simple but powerful code, with very low rate.

9.1.5.2 Reed–Muller Codes

The so-called RM codes were first described by D. E. Muller in 1953 using a "Boolean net function" language [1493]. One year later I. Reed published a paper showing that Muller's codes could be represented as multinomials over GF (2) [1472]. The resulting RM codes represented an important step beyond the Hamming and Golay codes, since they offered a flexible solution to the problem of correcting multiple errors per codeword.

Here a short description (based on [1485, pp. 150–153]) of RM codes is provided. We begin by considering the so-called *first-order* codes, denoted $\mathcal{R}(1, \mathcal{M})$. The top row of their generator matrix is the vector of length $2^{\mathcal{M}}$ consisting entirely of 1s. The rest of the generator matrix consists of columns containing all $2^{\mathcal{M}}$ binary vectors of length $\mathcal{M}$, as shown in the following example for a specific case.

Example 9.1.11 The generator matrix of $\mathcal{R}(1, 4)$, the first-order RM code corresponding to $\mathcal{M} = 4$, is:

$$\mathbf{G} = \begin{bmatrix} 1 & 1 & 1 & 1 & 1 & 1 & 1 & 1 & 1 & 1 & 1 & 1 & 1 & 1 & 1 & 1 \\ 0 & 0 & 0 & 0 & 0 & 0 & 0 & 0 & 1 & 1 & 1 & 1 & 1 & 1 & 1 & 1 \\ 0 & 0 & 0 & 0 & 1 & 1 & 1 & 1 & 0 & 0 & 0 & 0 & 1 & 1 & 1 & 1 \\ 0 & 0 & 1 & 1 & 0 & 0 & 1 & 1 & 0 & 0 & 1 & 1 & 0 & 0 & 1 & 1 \\ 0 & 1 & 0 & 1 & 0 & 1 & 0 & 1 & 0 & 1 & 0 & 1 & 0 & 1 & 0 & 1 \end{bmatrix}. \tag{9.89}$$

This represents a $(16, 5)$ binary linear block code having $d_{H,\min} = 2^3 = 8$.

□

Generally speaking, this construction procedure generates a $(2^{\mathcal{M}}, \mathcal{M} + 1)$ binary linear block having $d_{H,\min} = 2^{\mathcal{M}-1}$.

We now consider the construction of the order-r binary RM code $\mathcal{R}(r, \mathcal{M})$. In this case the construction of the generator matrix requires knowledge of some basic results about *Boolean functions*. A Boolean function in $\mathcal{M}$ variables $f(v_1, v_2, \ldots, v_{\mathcal{M}})$ is a mapping from the vector space consisting of all possible binary $\mathcal{M}$-tuples $(v_1, v_2, \ldots, v_{\mathcal{M}})$ into the set of binary numbers $\{0, 1\}$. Such a function can be fully described by a *truth table*, that is, a matrix with $(\mathcal{M} + 1)$ rows. The first $\mathcal{M}$ rows form a $\mathcal{M} \times 2^{\mathcal{M}}$ matrix whose columns are all the possible $2^{\mathcal{M}}$ binary $\mathcal{M}$-tuples, whereas the last row contains the binary value assigned to each of the binary $\mathcal{M}$-tuples by the function. For instance the Boolean function $f(v_1, v_2, v_3) = v_1 + v_2 + v_3$ is represented by the truth table shown in Table 9.6.

Note that, if we remove the last row, the columns of this table form an ordered radix-2 representation of the integers $\{0, 1, 2, \ldots, 2^{\mathcal{M}} - 1\}$. We adopt this convention in the following, so that each Boolean function f is *uniquely* represented by a binary vector $\mathbf{f}$ collecting, in an ordered fashion, the $2^{\mathcal{M}}$ binary elements of the last row in the truth table. Since $\mathbf{f}$ is a binary vector of length $2^{\mathcal{M}}$, there exist $2^{2^{\mathcal{M}}}$ distinct Boolean functions. Under coordinate-by-coordinate binary addition of the representing vectors, these form a *vector space* over GF (2).

To construct $\mathcal{R}(r, \mathcal{M})$, we consider the set S of all Boolean functions $f(v_1, v_2, \ldots, v_{\mathcal{M}})$ depending on $\mathcal{M}$ distinct variables and that can be represented by a *single monomial term*. Then S consists of the Boolean function 1 and the products of all combinations of one or more variables in the set

Table 9.6 Truth table for a Boolean function

v_3	0	0	0	0	1	1	1	1
v_2	0	0	1	1	0	0	1	1
v_1	0	1	0	1	0	1	0	1
f	0	1	1	0	1	0	0	1

$\{v_1, v_2, \ldots, v_{\mathcal{M}}\}$. The Boolean functions in S are *linearly independent*, so that the vectors representing them are also independent. For this reason, there is a *unique* Boolean function f described by a vector:

$$\mathbf{f} = a_0 \mathbf{1} + a_1 \mathbf{v}_1 + \ldots + a_{\mathcal{M}} \mathbf{v}_{\mathcal{M}} + a_{12} \mathbf{v}_1 \mathbf{v}_2 + \ldots + a_{12\ldots\mathcal{M}} \mathbf{v}_1 \mathbf{v}_2 \cdots \mathbf{v}_{\mathcal{M}}, \tag{9.90}$$

where $\mathbf{v}_i \mathbf{v}_j \cdots \mathbf{v}_k$ denotes the binary $2^{\mathcal{M}}$-dimensional row vector associated with the monomial $v_i v_j \cdots v_k$. Since there are $2^{2^{\mathcal{M}}}$ vectors of the form (9.90), the Boolean functions in S constitute a *basis* for the vector space of the Boolean functions in $\mathcal{M}$ variables.

Given these results, the RM code $\mathcal{R}(r, \mathcal{M})$, of order r and length $2^{\mathcal{M}}$, consists of the vectors $\mathbf{f}$ associated with all Boolean functions f represented by polynomials in $\mathcal{M}$ variables and whose degree does not exceed r. Generally speaking, such polynomials can be generated by a basis consisting of all the monomial functions of degree r or less. The number of elements in this basis set, for a given r and $\mathcal{M}$, is given by:

$$k = 1 + \binom{\mathcal{M}}{1} + \binom{\mathcal{M}}{2} + \cdots + \binom{\mathcal{M}}{r} \tag{9.91}$$

and is also the dimension of $\mathcal{R}(r, \mathcal{M})$. It can be proved that the minimum Hamming distance for $\mathcal{R}(r, \mathcal{M})$ is $d_{H,\min} = 2^{\mathcal{M}-r}$ (e.g., see [1485, p. 153]).

Let us now apply these concepts in the following example.

Example 9.1.12 We focus here on the construction of $\mathcal{R}(2, 4)$, that is, on the second-order RM code of length 16. The generation of the codewords requires the vector representation associated with the monomials:

$$\{1, v_1, v_2, v_3, v_4, v_1 v_2, v_1 v_3, v_1 v_4, v_2 v_3, v_2 v_4, v_3 v_4\} \tag{9.92}$$

involving no more than four variables and whose degree does not exceed 2. The binary vectors associated with these monomials are listed in Table 9.7, providing a matrix that can be used as a generator matrix for $\mathcal{R}(2, 4)$. Note that the $\mathcal{R}(2, 4)$ RM code is a $(16, 11)$ binary block code with $d_{H,\min} = 4$.

$\square$

9.1.6   Decoding Techniques for Block Codes

9.1.6.1   Introduction

Here various *hard* (algebraic) and *soft decoding* algorithms for linear block codes are illustrated; our analysis, however, is far from being exhaustive and, in particular, is limited to those algorithms

Table 9.7   Basis vectors for the RM code of Example 9.1.12

1	1	1	1	1	1	1	1	1	1	1	1	1	1	1	1	1
v_4	0	0	0	0	0	0	0	0	1	1	1	1	1	1	1	1
v_3	0	0	0	0	1	1	1	1	0	0	0	0	1	1	1	1
v_2	0	0	1	1	0	0	1	1	0	0	1	1	0	0	1	1
v_1	0	1	0	1	0	1	0	1	0	1	0	1	0	1	0	1
$v_3 \cdot v_4$	0	0	0	0	0	0	0	0	0	0	0	0	1	1	1	1
$v_2 \cdot v_4$	0	0	0	0	0	0	0	0	0	0	1	1	0	0	1	1
$v_1 \cdot v_4$	0	0	0	0	0	0	0	0	0	1	0	1	0	1	0	1
$v_2 \cdot v_3$	0	0	0	0	0	0	1	1	0	0	0	0	0	0	1	1
$v_1 \cdot v_3$	0	0	0	0	0	1	0	1	0	0	0	0	0	1	0	1
$v_1 \cdot v_2$	0	0	0	1	0	0	0	1	0	0	0	1	0	0	0	1

considered pertinent to wireless communication systems. *Hard decoding* algorithms are fed by hard decisions on the received signal and do not process any reliability information contained in the received signal. In contrast, *soft decoding algorithms* process the soft received signal and can achieve better performance at the cost of increased complexity (this is an important issue for codes with large values of $\min(k, n - k)$).

9.1.6.2 Soft Decision Decoding of Block Codes

Brute force ML decoding of an error-correction code over $GF(q)$ requires consideration of all its q^k codewords, or q^{n-k} codewords if decoding is done using the dual code, and this is often prohibitively complex. An alternative (suboptimal) approach is to adopt a *list decoding algorithm*, where the list consists of a subset of all possible codewords. There are many different list decoding algorithms available, with different performance–complexity tradeoffs; here we focus on:

- the *generalized minimum distance* (GMD) decoding algorithm [1494],
- the three variants of the *Chase decoding algorithm* [1495], and
- the *order-i reprocessing decoding algorithm* [1496, 1497].

For more information on these and other list decoding algorithms, see [35, Ch. 10] and [1491, Ch. 7]. We will only consider list decoding of *binary* linear block codes here. However, this approach can also be used to decode nonbinary codes; for instance, RS codes can be list-decoded using the so-called *Guruswami–Sudan algorithm* [1498].

Comparisons between the Chase and GMD decoding algorithms can be found in [35, 1491, 1499], where both analytical and simulation results for Hamming codes, RM codes and BCH codes are illustrated. From these results it is easily inferred that, on the one hand, Chase algorithm 3 outperforms GMD for the same decoding complexity and number of nearest neighbors [1499]. On the other hand, Chase algorithms 1 and 2 perform better than Chase algorithm 3 for the same number of nearest neighbors, but require a higher decoding complexity (with Chase algorithm 1 performing the best and having the highest complexity). Due to the high complexity of Chase algorithm 1 this is not commonly used. Chase algorithm 2 is considered to offer the best compromise between complexity and performance and, as a result, has been used extensively in the literature [35, Ch. 10].

Generalized Minimum Distance Decoding
GMD decoding was first proposed by D. G. Forney in 1966 [1494]. Our description of this decoding technique closely follows that given in [1494, 1500]; further information can be found in [35, 1491, 1499, 1501].

To employ GMD, the received symbols are first ordered in terms of *reliability*, on the basis of the absolute value of the received magnitude. The decoder makes use of up to $\lfloor (d_{H,\min} + 1)/2 \rfloor$ decoding trials, each exploiting an *error and erasure decoder* (a specific algorithm for error and erasure decoding is described below), which can handle e_s erasures and t_1 errors, provided that $t_1 + e_s \leq d_{H,\min} - 1$. Each decoding trial decodes the received signal with some of the least reliable symbols erased. If $d_{H,\min}$ is even, then the $(1, 3, \ldots, d_{H,\min} - 3, d_{H,\min} - 1)$ least reliable symbols are erased. If $d_{H,\min}$ is odd, then the $(0, 2, \ldots, d_{H,\min} - 3, d_{H,\min} - 1)$ least reliable symbols are erased. This results in up to $\lfloor (d_{H,\min} + 1)/2 \rfloor$ possible codewords. Finally, the best codeword is selected in accordance with a given *metric*, such as the minimum Euclidean distance or the maximum inner product [1500].

Modified GMD decoding is derived from GMD decoding by adding an extra trial which erases the $d_{H,\min}$ least reliable symbols [1500]. More specifically, first the $d_{H,\min}$ least reliable symbols are erased and hard decisions are made on the remaining $n - d_{H,\min}$ symbols. Then the parity check equations for the erased symbols are computed.

Chase Decoding

In 1954 a simple strategy for decoding the so-called *Wagner code*[13] was developed by R. A. Silverman and M. Balser [1502]. This strategy outputs the hard decision on the received signal if the parity check is correct. Otherwise, the least reliable bit is changed [1502]. This concept can be extended to codes with more than a single parity bit, allowing multiple bits to be altered. The Chase decoding algorithms can be seen as such an extension, but can also be viewed as a generalization of GMD decoding.

The Chase decoding algorithms belong to the class of *bounded distance decoding algorithms* [1495], since they look for possible codewords within a certain decoding sphere. Chase developed three suboptimal algorithms that consider different subsets of all possible codewords and provide a range of tradeoffs between performance and complexity (determined by the overall number of considered codewords). They all include the following steps [35, 1495]:

1. Make a hard decision on the coded bits within the received signal.
2. Generate a list of *error patterns*.
3. For each error pattern, add it to the hard decision on the received signal to form a *test sequence*.
4. Decode each test sequence using an algebraic decoder for error correction only in order to generate a list of possible codewords.
5. Use a *soft decision metric* to choose the most likely transmitted codeword in the list. Note that the squared Euclidean distance between the received signal and the modulated codeword is often used as a metric.

Each of the three algorithms uses different subsets of all possible error patterns (or codewords), as described next [35, 1495]. *Algorithm 1* is the most complex of the three, as it considers the largest set of error patterns (and, hence, possible codewords). In this case the set of error patterns consists of all possible combinations of $\lfloor d_{H,\min}/2 \rfloor$ 1s over all n positions in the code, providing $\binom{n}{\lfloor d_{H,\min}/2 \rfloor}$ or fewer candidate codewords.

Algorithm 2 reduces the number of error patterns considered. More specifically, the error patterns selected consist of the set of sequences containing any combination of 1s in the $i = \lfloor d_{H,\min}/2 \rfloor$ least reliable received positions; this gives $2^{\lfloor d_{H,\min}/2 \rfloor}$ test sequences. This decoding strategy is the most commonly used Chase algorithm [35]. It is also worth mentioning that this algorithm is exploited in [1503], but $i \geq \lfloor d_{H,\min}/2 \rfloor$ least reliable positions are considered (optimality is achieved if $i = n$, where n is the codeword length).

Algorithm 3 considers $\lfloor d_{H,\min}/2 + 1 \rfloor$ error patterns; such patterns contain 1s in the i least reliable positions, where, if $d_{H,\min}$ is even, $i = 0, 1, 3, \ldots, d_{H,\min} - 1$ and, if it is odd, $i = 0, 2, 4, \ldots, d_{H,\min} - 1$. Note that Chase algorithm 3 is similar to GMD decoding. However, the former employs a complement operation and decoding for error correction only, whereas the latter uses erasures combined with an error and erasure decoder [35, p. 408].

The Chase algorithms can be applied to any block code with a binary decoding algorithm and have the potential to double the error-correcting capability of the code itself [1495]. Note also that Chase algorithm 2 has been applied in [1503] to decode extended BCH codes.

Generalized Chase algorithms have been proposed in [1504, 1505]. In particular, a technique for the reduction of the test patterns, based on their Hamming weights, is described in [1504], whereas decoding strategies, based on the Chase algorithms and requiring a t^*-error-correcting binary decoder with $t^* \leq t \triangleq \lfloor (d_{H,\min} - 1)/2 \rfloor$ (see (9.29)), are developed in [1505]. *Modified generalized* Chase algorithms are also discussed in [1505], including the so-called *Hackett decoding algorithm* [1506]. These algorithms can be used to decode extended binary linear block codes, which have an overall parity check appended to each codeword.

---

[13] In the Wagner code, a codeword consists of a sequence of $n - 1$ message digits and an additional digit used as a parity check; in other words, such a code is an SPC.

Table 9.8　Table of codewords for the binary code of Example 9.1.3

u	x	$w(\mathbf{x})$	u	x	$w(\mathbf{x})$
0000	0000000	0	1000	1101000	3
0001	1010001	3	1001	0111001	4
0010	1110010	4	1010	0011010	3
0011	0100011	3	1011	1001011	4
0100	0110100	3	1100	1011100	4
0101	1100101	4	1101	0001101	3
0110	1000110	3	1110	0101110	4
0111	0010111	4	1111	1111111	7

Let us now focus on a specific application of the Chase decoding algorithm 2.

Example 9.1.13 Let us consider the (7, 4) single-error-correcting (Hamming) code of Example 9.1.1; if we adopt the generator matrix:

$$\mathbf{G} = \begin{bmatrix} 1 & 1 & 0 & 1 & 0 & 0 & 0 \\ 0 & 1 & 1 & 0 & 1 & 0 & 0 \\ 1 & 1 & 1 & 0 & 0 & 1 & 0 \\ 1 & 0 & 1 & 0 & 0 & 0 & 1 \end{bmatrix}, \tag{9.93}$$

it is straightforward to generate Table 9.8, listing all the available codewords.

Let us now assume that the codeword $\mathbf{x} = [0, 0, 0, 0, 0, 0, 0]$ is transmitted using BPSK as $\mathbf{s}(\mathbf{x}) = [-1, -1, -1, -1, -1, -1, -1]$ (the correspondence rule $s : \{0, 1\} \rightarrow \{-1, 1\}$ is used) and that the received vector is $\mathbf{r} = [-0.9, -0.8, 0.1, 0.3, -0.2, -0.8, -0.7]$; this results in the hard decision vector $\mathbf{y} = [0, 0, 1, 1, 0, 0, 0]$. In this case hard algebraic decoding of $\mathbf{y}$ yields $\hat{\mathbf{x}} = [0, 0, 1, 1, 0, 1, 0]$, which is incorrect. Can list decoding help? Chase algorithm 2 requires us to consider error patterns with all possible combinations of 1s in the:

$$i = \left\lfloor \frac{d_{H,\min}}{2} \right\rfloor = 1 \tag{9.94}$$

least reliable positions (LRPs). We begin decoding by finding the i LRPs in $\mathbf{r}$ and computing the information shown in Table 9.9.

Then, Chase algorithm 2 selects $\hat{\mathbf{x}} = [0, 0, 0, 0, 0, 0, 0]$, which is the correct codeword. Note that: (a) the list decoding algorithm has used the reliability information in the received signal to correct 2 errors, even though this is a *single-error-correcting* code (i.e., $t = 1$); (b) Chase algorithm 2 cannot always correct more than t errors, since this depends on the code and the reliability of the received symbols. □

Soft Order-i Reprocessing Decoding

The following description of order-i reprocessing is based on [1507], where now i is a user-defined constant; further information can be found in [35, 1496, 1497].

The first step in the decoding procedure involves ordering the elements of the received signal vector, $\mathbf{r}$, in terms of decreasing reliability. The k *most reliable independent positions* (MRIPs) form the so-called *most reliable basis* (MRB); the corresponding bits are indexed as $1, 2, \ldots, k$. This ordering leads to consideration of a systematic *reordered code* $\tilde{C}$, which is equivalent to the transmitted code C; note that generating $\tilde{C}$ can be quite complex for large $\min(n - k, k)$.

Table 9.9 Table of error and test patterns, estimated codewords and decoding metric for the binary Hamming code of Example 9.1.13

Error pattern	Test pattern	$\hat{\mathbf{x}}$	$|\mathbf{r} - \mathbf{s}(\hat{\mathbf{x}})|^2$
0000000	0011000	0011010	5.32
0010000	0001000	0000000	3.72

Then a hard decision is taken on the k MRIPs, the resulting bits are encoded using $\tilde{C}$ to produce an initial codeword $\mathbf{x}_0$ and the decoding cost is computed. For instance, the Euclidean distance between $s(\mathbf{x}_0)$ and the reordered received signal can be used to measure the cost, where, as before, $s(0) = -1$ and $s(1) = 1$.

In the next step of the decoding procedure, $\mathbf{x}_0$ is *reprocessed* as follows:

1. For $j = 0, 1, \ldots, i$, add each error pattern of weight j to the k MRIPs of $\mathbf{x}_0$ to generate a set of length k *test sequences*. Encode each test sequence using $\tilde{C}$. This produces a total of $1 + \binom{k}{1} + \ldots + \binom{k}{i}$ *candidate codewords*.

2. Calculate the *decoding cost* for each candidate codeword.

3. The decoder outputs the candidate codeword with the *best metric* (i.e., that associated with the lowest cost).

When a candidate codeword has a better decoding metric, a given optimality criterion is tested. If it is satisfied, the decoder stops. A negligible performance loss compared to ML decoding should be expected if $i \geq \lceil d_{H,\min}/4 \rceil$ [1507].

It is worth pointing out that order-i reprocessing is quite different from the Chase and GMD decoding algorithms mentioned above since:

(a) test sequences have length k, not n;
(b) test sequences are encoded, not decoded (note that encoding is computationally simpler than decoding);
(c) a systematic reordered code $\tilde{C}$ needs to be generated.

It is also important to note that even a small value of i can result in a large number of test sequences. However, this number is typically very small in comparison to the total number of possible codewords, 2^k. Nonetheless, order-i reprocessing is more complicated than any of the Chase algorithms.

9.1.6.3 Algebraic Decoding Techniques for Cyclic Codes

We now focus on various hard decoding algorithms for BCH and RS codes, since these classes of codes are the most relevant in wireless communications, particularly when algebraic decoding is considered. In our analysis only TD algorithms are considered due to space limitations; the reader can refer to [1508] for an introduction to FD (i.e., DFT-based) decoding of cyclic codes. In our analysis, we consider first the problem of decoding *binary* BCH codes and analyze *Peterson's method* [1509] and *Berlekamp's algorithm* [1510]. The, we discuss the generalizations of these techniques for use with nonbinary codes, namely the *Peterson–Gorenstein–Zierler algorithm* [1511] and the *Berlekamp–Massey algorithm* [1512]. Finally, we describe *erasure and error decoding* based on the Berlekamp–Massey algorithm.

Decoding of Binary BCH Codes: Syndromes and Error Locators

We begin with the problem of hard algebraic decoding for a *binary primitive* BCH code C with t-error-correction capability. Any codeword $x(D)$ has the same roots as the generator polynomial (see (9.34)), namely the $2t$ consecutive powers of the primitive element, α, of an extension field, GF(2^m). Then, we have that:

$$x\left(\alpha^l\right) = 0 \tag{9.95}$$

with $l = b, b+1, \ldots, b+2t-1$. The hard received polynomial can be written as:

$$y(D) = x(D) + e(D) = \sum_{p=0}^{n-1} y_p D^p, \tag{9.96}$$

where $e(D)$ is the polynomial representation of the received error vector (see (9.25)). Given this $y(D)$, the lth *syndrome* can be computed as [1485, 1488]:

$$S_l = y\left(\alpha^l\right) = \sum_{p=0}^{n-1} y_l\left(\alpha^l\right)^p \tag{9.97}$$

with $l = 1, 2, \ldots, 2t$, if we assume a *narrow-sense* code (i.e., $b = 1$) for simplicity. Since every codeword produces a syndrome equal to zero (see (9.95)), (9.97) can be rewritten as:

$$S_l = e\left(\alpha^l\right) = \sum_{p=0}^{n-1} e_l\left(\alpha^l\right)^p. \tag{9.98}$$

Let us now assume that the received vector $\mathbf{y}$ (see (9.25)) contains v errors in positions $i_1, i_2, \ldots, i_v$ (i.e., $e_{i_l} = 1$ for $l = 1, 2, \ldots, v$). Then, the lth syndrome (9.97) can be written as:

$$S_l = \sum_{p=1}^{v} e_{i_p}\left(\alpha^l\right)^{i_p} = \sum_{p=1}^{v}\left(\alpha^l\right)^{i_p} = \sum_{p=1}^{v} X_p^l, \tag{9.99}$$

with $l = 1, 2, \ldots, 2t$ where $X_p \triangleq \alpha^{i_p}$ (with $p = 1, 2, \ldots, v$) represents the so-called pth *error locator*. Equation (9.99), which has a *power-sum symmetric* structure, establishes a relationship between the set of syndromes $\{S_l\}$ and the set of error locators $\{X_p\}$. In principle, if the syndromes are computed from the received word as in (9.97), the error locators can be evaluated by solving the system of *nonlinear* algebraic equations (9.99); then, from the set $\{X_p, p = 1, 2, \ldots, v\}$ the error locations can be derived, making error correction possible. However, solving the system (9.99) in a direct fashion is not easy, and a different approach should be devised. In 1960 W. Peterson proved that the syndrome equations (9.99) can be translated into a set of *linear* equations, which are much easier to solve [1509]. This requires defining the so-called *error locator polynomial*:

$$\Lambda(D) \triangleq \prod_{i=1}^{v}\left(1 + X_i D\right) = \sum_{p=0}^{v} \Lambda_p D^p, \tag{9.100}$$

whose roots are the inverses of the error locators $\{X_p\}$. Note that the coefficients $\{\Lambda_p\}$ of $\Lambda(D)$ can be easily expressed in terms of the error locators involved. In fact, from (9.100) it is easily inferred that:

$$\Lambda_0 = 1,$$

$$\Lambda_1 = \sum_{i=1}^{v} X_i,$$

$$\Lambda_2 = \sum_{i<p} X_i X_p,$$

$$\Lambda_3 = \sum_{i<p<q} X_i X_p X_q, \tag{9.101}$$

$$\vdots$$

$$\Lambda_v = \prod_{i=1}^{v} X_i.$$

These equalities are known as *elementary symmetric functions* of the error locators; the functions and the power-sum symmetric functions of (9.99) are related by the so-called *Newton's identities*, which can be expressed as:

$$S_1 + \Lambda_1 = 0,$$

$$S_2 + \Lambda_1 S_1 + 2\Lambda_2 = 0,$$

$$S_3 + \Lambda_1 S_2 + \Lambda_2 S_1 + 3\Lambda_3 = 0,$$

$$\vdots$$

$$S_v + \Lambda_1 S_{v-1} + \dots + \Lambda_{v-1} S_1 + v\Lambda_v = 0 \tag{9.102}$$

$$S_{v+1} + \Lambda_1 S_v + \dots + \Lambda_{v-1} S_2 + \Lambda_v S_1 = 0$$

$$\vdots$$

$$S_{2t} + \Lambda_1 S_{2t-1} + \dots + \Lambda_v S_{2t-v} = 0.$$

These identities are *linear* in the v unknown coefficients $\{\Lambda_p, p = 1, 2, \dots, v\}$ of the *error locator polynomial* $\Lambda(D)$. In addition, since we are working over $GF(2^m)$ (a field of characteristic 2) they can be substantially simplified. In fact we have that: (a) $l\Lambda_p = \Lambda_p$ $(l\Lambda_p = 0)$ if l is odd (even); (b) $S_{2p} = S_p^2$ for any p, that is, even-indexed syndromes are the squares of earlier-indexed syndromes (a proof of this statement can be found, for instance, in [1485, p. 206]). From the latter property it is easily inferred that some of the $2t$ identities (9.101) are redundant, so that they can be neglected in the evaluation of the coefficients $\{\Lambda_p, p = 1, 2, \dots, v\}$. In particular, if we assume that $v = t$ errors have occurred, the set (9.102) can be reduced to:

$$S_1 + \Lambda_1 = 0,$$

$$S_3 + \Lambda_1 S_2 + \Lambda_2 S_1 + \Lambda_3 = 0, \tag{9.103}$$

$$\vdots$$

$$S_{2t-1} + \Lambda_1 S_{2t-2} + \dots + \Lambda_t S_{t-1} = 0,$$

a system of t equations in t unknowns. In what follows two distinct methods for solving (9.103) are illustrated; the first method was devised by Peterson [1509], the second by Berlekamp [1510]. It is important to note that Peterson's method should be adopted for binary codes correcting a small number of errors (say, t no greater than 6 or 7), since its computational complexity increases with the square of the number of corrected errors. The complexity of Berlekamp's algorithm, on the other hand, increases only linearly; for this reason it can be used when a large error-correction level t is required.

Decoding of Binary BCH Codes: Peterson's Direct Method

The system of linear equations (9.103) can be put in the matrix form [1485, p. 206]:

$$\mathbf{A}_S^{(0)} \mathbf{\Lambda}^{(0)} = \mathbf{B}^{(0)}, \tag{9.104}$$

where $\mathbf{\Lambda}^{(0)} \triangleq [\Lambda_1, \Lambda_2, \dots, \Lambda_t]^T$, $\mathbf{B}^{(0)} \triangleq [-S_1, -S_3, \dots, -S_{2t-1}]^T$ and:

$$\mathbf{A}_S^{(0)} = \begin{bmatrix} 1 & 0 & 0 & 0 & \cdots & 0 & 0 \\ S_2 & S_1 & 1 & 0 & \cdots & 0 & 0 \\ S_4 & S_3 & S_2 & S_1 & \cdots & 0 & 0 \\ S_6 & S_5 & S_4 & S_3 & \cdots & 0 & 0 \\ \vdots & \vdots & \vdots & \vdots & \ddots & \vdots & \vdots \\ S_{2t-4} & S_{2t-5} & S_{2t-6} & S_{2t-7} & \cdots & S_{t-2} & S_{t-3} \\ S_{2t-2} & S_{2t-3} & S_{2t-4} & S_{2t-5} & \cdots & S_t & S_{t-1} \end{bmatrix}. \tag{9.105}$$

The system (9.104) has a *unique solution* if and only if $\mathbf{A}_S^{(0)}$ is nonsingular, that is, $\det\left(\mathbf{A}_S^{(0)}\right) \neq 0$. Peterson proved that this occurs if there are t or $t-1$ errors in the received vector $\mathbf{y}$ [1509]. If this is the case, solving the system (9.104) yields the coefficient vector $\mathbf{\Lambda}^{(0)}$. Then the error locations are extracted from the roots of the error locator polynomial $\Lambda(D)$ (9.100); such roots can be evaluated by means of a systematic procedure for locating the root of a polynomial over $GF(2^m)$, such as the so-called *Chien search* [1513]. In contrast, if $\mathbf{A}_S^{(0)}$ is singular, the number of equations forming the system (9.104) needs to be reduced; in particular, $\mathbf{\Lambda}^{(0)}$ and $\mathbf{B}^{(0)}$ are shortened to $\mathbf{\Lambda}^{(1)} \triangleq [\Lambda_1, \Lambda_2, \dots, \Lambda_{t-2}]^T$ and $\mathbf{B}^{(1)} \triangleq [-S_1, -S_3, \dots, -S_{2t-5}]^T$, respectively, and the matrix $\mathbf{A}_S^{(1)}$ is extracted from $\mathbf{A}_S^{(0)}$ by removing its last two rows and two rightmost columns. Then, if $\det\left(\mathbf{A}_S^{(1)}\right) \neq 0$, the system $\mathbf{A}_S^{(1)} \mathbf{\Lambda}^{(1)} = \mathbf{B}^{(1)}$ is solved, otherwise the matrix extraction and vector shortening procedures just described are repeated k times until a nonsingular matrix $\mathbf{A}_S^{(k)}$ is obtained. This procedure may lead to the correct error locator polynomial. However, two other possibilities can be envisaged: (a) if $\mathbf{y}$ is within Hamming distance t of an incorrect codeword, a *wrong estimate* of $\mathbf{x}$ will be produced (unfortunately this error event is *undetectable*); (b) if $\mathbf{y}$ is not within Hamming distance t of any codeword, the resulting polynomial $\Lambda(D)$ may have *repeated roots* or *roots that do not lie in the extension field* $GF(2^m)$ of α. In case (b) a *decoding failure* is declared.

The decoding procedure outlined above is summarized by the flow diagram in Figure 9.5.

Finally, it is worth mentioning that, if t is small, simple expressions can be easily derived from (9.104) for the coefficients of $\Lambda(D)$. In particular, for single- and double-error-correcting codes, we have that:

$$\Lambda_1 = S_1, \tag{9.106}$$

$$\Lambda_2 = \frac{S_3 + S_1^3}{S_1}, \tag{9.107}$$

whereas, for a triple-error-correcting code, it is found that:

$$\Lambda_1 = S_1, \tag{9.108}$$

$$\Lambda_2 = \frac{S_1^2 S_3 + S_5}{S_1^3 + S_3}, \tag{9.109}$$

$$\Lambda_3 = S_1^3 + S_3 + S_1 \Lambda_2. \tag{9.110}$$

For instance, if $t = 3$ and the received sequence contains a single error (i.e., $v = 1$), then (9.109) and (9.110) yield $\Lambda_2 = \Lambda_3 = 0$ and the error position is indicated by Λ_1. If $v = 2$ errors were received, then (9.110) gives $\Lambda_3 = 0$ and the error locator polynomial has degree 2.

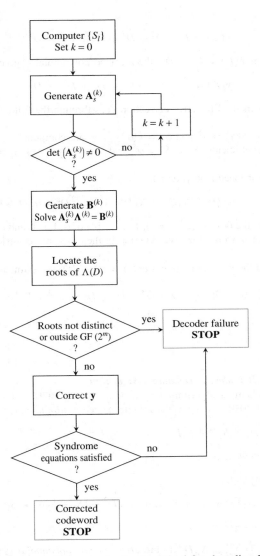

Figure 9.5 Flow diagram of Peterson's direct method for decoding binary BCH codes.

Let us now apply Peterson's method to decode the received vector for a specific BCH code.

Example 9.1.14 We consider the $(15, 7)$ double-error-correcting binary BCH code over $GF(2^4)$ with $d_{H,\min} = 5$ and generator polynomial $g(D) = D^8 + D^7 + D^6 + D^4 + 1$. Given the message $u(D) = 1 + D^2 + D^3$, we compute the corresponding systematic codeword $x(D)$ in accordance with (9.50) and (9.51). Then, since:

$$\frac{D^{n-k}u(D)}{g(D)} = \frac{D^8 u(D)}{g(D)} = (D^3 + D) + \frac{D^5 + D^3 + D}{D^8 + D^7 + D^6 + D^4 + 1}, \quad (9.111)$$

we have that:

$$x(D) = D + D^3 + D^5 + D^8 + D^{10} + D^{11}. \tag{9.112}$$

If the error polynomial is $e(D) = D^4 + D^8$, the received polynomial is given by:

$$y(D) = D + D^3 + D^4 + D^5 + D^{10} + D^{11}. \tag{9.113}$$

Given $y(D)$, Peterson's method for error correction goes through the following steps:

1. The syndromes $S_1 = y(\alpha) = \alpha^5$ and $S_3 = y(\alpha^3) = \alpha^8$ are computed.
2. The error locator coefficients $\Lambda_1 = S_1 = \alpha^5$ and $\Lambda_2 = (S_3 + S_1^3)/S_1 = \alpha^{12}$ are evaluated (see (9.106) and (9.107)).
3. The roots of the error locator polynomial:

$$\Lambda(D) = \Lambda_2 D^2 + \Lambda_1 D + 1 = \alpha^{12} D^2 + \alpha^5 D + 1 \tag{9.114}$$

   are evaluated, testing the powers $\{\alpha^l, \ l = 0, 1, \ldots, 14\}$ of α. It is found that $\Lambda(\alpha^7) = \Lambda(\alpha^{11}) = 0$ and that all the other powers of α give $\Lambda(D) \neq 0$; then the only error locators are $X_1 = \alpha^8$ and $X_2 = \alpha^4$.
4. Given the estimated the error polynomial $\hat{e}(D) = D^4 + D^8$, the estimated codeword polynomial is:

$$\hat{x}(D) = y(D) + \hat{e}(D) = D + D^3 + D^5 + D^8 + D^{10} + D^{11} = x(D), \tag{9.115}$$

   as expected.

□

Decoding of Binary BCH Codes: Berlekamp's Algorithm

We now illustrate Berlekamp's algorithm for decoding binary BCH codes without proofs; further details can be found in [1510]. To begin, the infinite-degree *syndrome polynomial*:

$$S(D) \triangleq S_1 D + S_2 D^2 + \ldots + S_{2t} D^{2t} + S_{2t+1} D^{2t+1} + \ldots \tag{9.116}$$

and the *error magnitude polynomial*:

$$\begin{aligned}
\Omega(D) &\triangleq [1 + S(D)]\Lambda(D) \\
&= 1 + (S_1 + \Lambda_1) D + (S_2 + S_1\Lambda_1 + \Lambda_2) D^2 + (S_3 + S_2\Lambda_1 + S_1\Lambda_2 + \Lambda_3) D^3 + \ldots \\
&= 1 + \Omega_1 D + \Omega_2 D^2 + \ldots,
\end{aligned} \tag{9.117}$$

are defined. Here $\Lambda(D) = \sum_{p=0}^{v} \Lambda_p D^p$ is the *error locator polynomial* (see (9.100)) and v is the number of errors. We note that, in order to satisfy Newton's identities (9.103), the odd-indexed coefficients of $\Omega(D)$ (9.117) must be zero. Since we know only the first $2t$ coefficients of $S(D)$ (9.116) (see (9.98)), the decoding problem becomes that of finding a polynomial $\Lambda(D)$ whose degree does not exceed t and that satisfies the equality:

$$[1 + S(D)]\Lambda(D) = (1 + \Omega_2 D^2 + \Omega_4 D^4 + \ldots + \Omega_{2t} D^{2t}) \bmod D^{2t+1}. \tag{9.118}$$

Berlekamp's algorithm computes such a polynomial in stages, starting from a polyomial $\Lambda^{(0)}(D) = 1$ of degree 0 and then iteratively computing the polynomials $\{\Lambda^{(2k)}(D), \ k = 1, 2, \ldots, 2t\}$, which have increasing degree and satisfy:

$$[1 + S(D)]\Lambda^{(2k)}(D) \triangleq (1 + \Omega_2 D^2 + \Omega_4 D^4 + \ldots + \Omega_{2k} D^{2k}) \bmod D^{2t+1}. \tag{9.119}$$

The last polynomial generated, $\Lambda^{(2t)}(D)$, is a solution for all t of the identities (9.103).

Berlekamp's algorithm consists of the following steps [1485, pp. 211–213]:

1. Compute the syndromes $\{S_l, l = 1, 2, \ldots, 2t\}$ and set $k = 0$, $\Lambda^{(0)}(D) = 1$ and $T^{(0)} = 1$ (*initialization*).
2. Compute the product $\Lambda^{(2k)}(D)[1 + S(D)]$; let $\Delta^{(2k)}$ denote the coefficient for D^{2k+1} in the resulting polynomial.
3. Compute $\Lambda^{(2k+2)}(D) = \Lambda^{(2k)}(D) + \Delta^{2k}[D \cdot T^{(2k)}(D)]$, where $T(D)$ is a *correction polynomial*.
4. Compute:

$$T^{(2k+2)}(D) \triangleq \begin{cases} D^2 T^{(2k)}(D), & \text{if} \Delta^{(2k)} = 0 \quad \text{or} \quad \deg\left[\Lambda^{(2k)}(D)\right] > k, \\ \frac{D\Lambda^{(2k)}(D)}{\Delta^{(2k)}}, & \text{if} \Delta^{(2k)} \neq 0 \quad \text{or} \quad \deg\left[\Lambda^{(2k)}(D)\right] \leq k, \end{cases} \qquad (9.120)$$

where $\deg[P(D)]$ denotes the degree of an arbitrary polynomial $P(D)$.
5. Set $k = k + 1$. If $k < t$, then go to step 2.
6. Locate all the roots of $\Lambda(D) = \Lambda^{(2t)}(D)$, that is, the values of D. If the roots are distinct and in $GF(2^m)$, correct the corresponding locations in the received vector and stop. Otherwise, declare a *decoding failure* and stop.

Let us now apply Berlekamp's algorithm to the same problem of decoding illustrated in Example 9.1.14.

Example 9.1.15 We consider again the decoding problem for a binary $(15, 7)$ BCH code with generator polynomial $g(D) = D^8 + D^7 + D^6 + D^4 + 1$. The message polynomial $u(D) = 1 + D^2 + D^3$ and the error polynomial $e(D) = D^4 + D^8$ are the same as those considered in Example 9.1.14. Therefore, the received polynomial is $y(D) = D + D^3 + D^4 + D^5 + D^{10} + D^{11}$.

Decoding starts with the evaluation of $2t = 4$ syndromes: $S_1 = y(\alpha) = \alpha^5$, $S_2 = y(\alpha^2) = \alpha^{10}$, $S_3 = y(\alpha^3) = \alpha^8$ and $S_4 = y(\alpha^4) = \alpha^5$. Therefore the syndrome polynomial is:

$$S(D) = \alpha^5 D + \alpha^{10} D^2 + \alpha^8 D^3 + \alpha^5 D^4. \qquad (9.121)$$

Given this polynomial, Berlekamp's algorithm goes through the following steps:

1. $k = 0$, $\Lambda^{(0)}(D) = 1$, $T^{(0)} = 1$.
2. $\Lambda^{(0)}(D)[1 + S(D)] = 1 + \alpha^5 D + \alpha^{10} D^2 + \alpha^8 D^3 + \alpha^5 D^4$, $\Delta^{(0)} = \alpha^5$.
3. $\Lambda^{(2)}(D) = \Lambda^{(0)}(D) + \Delta^{(0)}[DT^{(0)}(D)] = 1 + \alpha^5 D$.
4. $T^{(2)}(D) = D\Lambda^{(0)}(D)/\Delta^{(0)} = D/\alpha^5 = \alpha^{10} D$ (recall that $\alpha^{15} = 1$), since $\Delta^{(0)} \neq 0$ and $\deg[\Lambda^{(0)}(D)] = 0$.
5. $k = k + 1 = 1$.
   2. $\Lambda^{(2)}(D)[1 + S(D)] = (1 + \alpha^5 D)(1 + \alpha^5 D + \alpha^{10} D^2 + \alpha^8 D^3 + \alpha^5 D^4)$; then the coefficient of D^3 is $\Delta^{(2)} = \alpha^8 + \alpha^{15} = \alpha^2$.
   3. $\Lambda^{(4)}(D) = \Lambda^{(2)}(D) + \Delta^{(2)}[DT^{(2)}(D)] = 1 + \alpha^5 D + \alpha^{12} D^2$.
   4. $T^{(4)}(D) = D\Lambda^{(2)}(D)/\Delta^{(2)} = D(1 + \alpha^5 D)/\alpha^2 = \alpha^{13} D + \alpha^3 D^2$, since $\Delta^{(2)} \neq 0$, $\deg[\Lambda^{(2)}(D)] = 1 = k$.

Finally, we must locate the roots of $\Lambda(D) = \Lambda^{(2t)}(D) = 1 + \alpha^5 D + \alpha^{12} D^2$ over $GF(2^4)$. It is found that the only roots over this field are α^7 and α^{11}. Therefore, the error locators are $X_1 = \alpha^8$ and $X_2 = \alpha^4$, so that the estimated error polynomial and codeword are given by:

$$\hat{e}(D) = D^4 + D^8 \qquad (9.122)$$

and

$$\hat{x}(D) = y(D) + \hat{e}(D) = D + D^3 + D^5 + D^8 + D^{10} + D^{11} = x(D), \qquad (9.123)$$

respectively.

□

Decoding of Nonbinary BCH Codes: Berlekamp–Massey Algorithm

The decoding strategies illustrated above have been generalized for use in the nonbinary case. In particular, Peterson's method was generalized by D. Gorenstein and N. Zierler in 1961 [1511] (the resulting strategy is known as the *Peterson–Gorenstein–Zierler algorithm*). However, once again the generalization of Berlekamp's algorithm offers a substantially more efficient alternative with nonbinary BCH codes when a large number of errors need to be corrected. For this reason we concentrate here on the final decoding technique, which is called the *Berlekamp–Massey algorithm* because Massey provided an interpretation based on the use of Massey's *linear feedback shift registers* (LFSRs) [1512]. This algorithm is illustrated without proofs; further details can be found in [1485, 1510, 1514, 1515].

The Berlekamp–Massey algorithm processes the syndromes to extract both the locations and the magnitudes of the errors. As in the binary case, the *l*th syndrome for a narrow-sense BCH code is given by:

$$S_l \triangleq y\left(\alpha^l\right) = e\left(\alpha^l\right) = \sum_{k=1}^{v} e_{i_k} X_k^l \qquad (9.124)$$

with $l = 1, 2, \ldots, 2t$, where $X_k \triangleq \alpha^{i_k}$ is the *error locator* for the *k*th of v errors, e_{i_k} is the *k*th error magnitude* and i_k is the *index* for the *k*th error; note, however, that, unlike the binary case, the syndrome equations (9.124) are not power-symmetric functions. The decoding procedure involves the following:

1. The *error locator polynomial*:

$$\Lambda(D) \triangleq \prod_{p=0}^{v}\left(1 - X_p D\right) = \sum_{p=0}^{v} \Lambda_p D^p, \qquad (9.125)$$

   whose zeros are the inverses of the error locators. It is easy to prove (e.g., see [1485, pp. 214–215]) that the coefficients $\{\Lambda_p, \ p = 1, 2, \ldots, v\}$ of this polynomial and the syndromes (9.124) are related by Newton's identity in the form:

$$\Lambda_v S_{l-v} + \Lambda_{v-1} S_{l-v+1} + \cdots + \Lambda_1 S_{l-1} + \Lambda_0 S_l = 0, \qquad (9.126)$$

   for any $l \geq 1$. Since $\Lambda_0 = 1$ (see (9.125)), (9.126) can also be rewritten as:

$$\left(-\Lambda_v\right) S_{l-v} + \left(-\Lambda_{v-1}\right) S_{l-v+1} + \cdots + \left(-\Lambda_1\right) S_{l-1} = S_l. \qquad (9.127)$$

2. The infinite-degree *syndrome polynomial* $S(D)$, defined as in (9.116). Note that substituting (9.124) into (9.116) yields, after some manipulation:

$$S(D) = \sum_{k=1}^{v} e_{i_k}\left(\frac{X_k D}{1 - X_k D}\right). \qquad (9.128)$$

3. The *error magnitude polynomial* $\Omega(D)$ defined as in (9.117). If we substitute (9.125) and (9.128) into (9.117), $\Omega(D)$ can be put in the form:

$$\Omega(D) = \left[1 + \sum_{k=1}^{v} e_{i_k}\left(\frac{X_k D}{1 - X_k D}\right)\right] \prod_{p=0}^{v}\left(1 - X_p D\right)$$

$$= \Lambda(D) + \sum_{k=1}^{v}\left[e_{i_k} X_k D \prod_{p \neq k}^{v}\left(1 - X_p D\right)\right] \qquad (9.129)$$

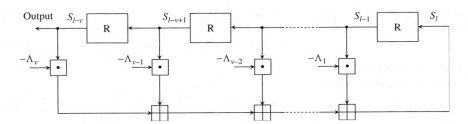

Figure 9.6   LFSR interpretation of equation (9.127).

Note that, since we know only the first $2t$ coefficients of $S(D)$, (9.117) needs to be reduced to:

$$[1 + S(D)]\Lambda(D) = \Omega(D) \mod D^{2t+1}, \tag{9.130}$$

This expression is known as *key equation* for BCH/RS decoding and relates the known syndromes to the error locator and error magnitude polynomials.

Decoding consists of two parts: first the error locator polynomial $\Lambda(D)$ (9.125) is evaluated and its roots are computed, so that the errors are located (this part is known as the *Berlekamp–Massey algorithm*); then the magnitude of the errors is evaluated by resorting to the so-called *Forney algorithm* [1516], which exploits the error locators and the polynomials $\Lambda(D)$ and $\Omega(D)$. More specifically, in the first part the coefficients $\{\Lambda_p, \ p = 1, 2, \ldots, v\}$ of $\Lambda(D)$ need to be computed from the syndromes $\{S_l, \ l = 1, 2, \ldots, 2t\}$. It is worth pointing out that the equality (9.127) relating these two sets of parameters can be interpreted through the use of the LFSR model shown in Figure 9.6. In fact, given this model, the problem of finding $\Lambda(D)$ can be reformulated as that of finding the minimal-length LFSR, which has the syndromes $\{S_l, \ l = 1, 2, \ldots, 2t\}$ as its first $2t$ outputs; this is due to the fact that the tap gains of this LSFR are the opposite of the coefficients $\{\Lambda_p, \ p = 1, 2, \ldots, v\}$. In practice, $\Lambda(D)$ is computed via a recursive procedure, involving a set of *connection polynomials*; the kth connection polynomial is defined as:

$$\Lambda^{(k)}(D) \triangleq \Lambda_k^{(k)} D^k + \Lambda_{k-1}^{(k)} D^{k-1} + \ldots + \Lambda_1^{(k)} D + 1 \tag{9.131}$$

and specifies the tap gains of an LFSR of length k. The Berlekamp–Massey algorithm starts by evaluating a polynomial $\Lambda^{(1)}(D)$, such that the first output of the associated LFSR is S_1. Then the second output of this LFSR is compared with S_2; if they do not have the same value, the *discrepancy* between them is used to generate a modified connection polynomial $\Lambda^{(2)}(D)$, otherwise the third output element is generated and compared with S_3. This process continues until a connection polynomial whose LFSR is able to generate the $2t$ syndromes is found. Massey proved that if $v \leq t$, the final connection polynomial produced by the Berlekamp–Massey algorithm correctly specifies the error locator polynomial [1512].

The decoding procedure uses a *correction polynomial* denoted $T(D)$, a *discrepancy term* denoted $\Delta^{(k)}$, an LFSR of length L and an index k, and consists of the following steps [1485, p. 219]:

1. Compute the syndromes $\{S_l, \ l = 1, 2, \ldots, 2t\}$ from $y(D)$ (see (9.124)).
2. Set $k = 0$, $\Lambda^{(0)}(D) = 1$, $L = 0$ and $T(D) = D$ (*initialization*).
3. Set $k = k + 1$ and compute the kth *discrepancy*:

$$\Delta^{(k)} \triangleq S_k - \sum_{i=1}^{L} \Lambda_i^{(k-1)} S_{k-i}. \tag{9.132}$$

4. If $\Delta^{(k)} = 0$, then set $\Lambda^{(k)}(D) = \Lambda^{(k-1)}(D)$ and go to step 8.
5. Compute:

$$\Lambda^{(k)}(D) = \Lambda^{(k-1)}(D) - \Delta^{(k)}T(D). \tag{9.133}$$

6. If $2L \geq k$, then go to step 8.
7. Set $L = k - L$ and compute:

$$T(D) = \frac{\Lambda^{(k-1)}(D)}{\Delta^{(k)}}. \tag{9.134}$$

8. Set $T(D) = D \cdot T(D)$.
9. If $k < 2t$, then go to step 3.
10. Set $\Lambda(D) = \Lambda^{(2t)}(D)$. Find the roots of $\Lambda(D)$ and their inverses (i.e., the error locators $\{X_k\}$). If the roots are not distinct or are not in $GF(2^m)$, declare a *decoding failure* (this means that deg $[\Lambda(D)] \neq L$) and stop.
11. Find the error magnitudes of the v errors using Forney's algorithm, that is:

$$e_{i_k} = \frac{-X_k \, \Omega(X_k^{-1})}{\Lambda'(X_k^{-1})} = \frac{\Omega(X_k^{-1})}{\prod_{j \neq k}(1 - X_j X_k^{-1})}, \tag{9.135}$$

with $k = 1, 2, \ldots, v$, where $\Omega(D)$ is given by the RHS of (9.129) and $\Lambda'(D)$ is the *formal derivative*[14] of $\Lambda(D)$.
12. Correct the received polynomial generating the codeword estimate:

$$\hat{x}(D) = y(D) + e(D), \tag{9.136}$$

where $e(D) \triangleq \sum_{k=1}^{v} e_{i_k} D^{i_k}$, and stop.

Let us now apply this algorithm to the decoding of a specific nonbinary BCH code.

Example 9.1.16 We now consider the $t = 2$-error-correcting $(15, 11)$ RS code (with $d_{H,\min} = 5$) over GF(16) with generator polynomial $g(D) = D^4 + \alpha^{13}D^3 + \alpha^6 D^2 + \alpha^3 D + \alpha^{10}$. If the message polynomial is $u(D) = \alpha^2 D^9 + \alpha^7 D^2$, the corresponding codeword polynomial is $x(D) = \alpha^{13} + \alpha^8 D + \alpha^{11}D^2 + \alpha^2 D^3 + \alpha^7 D^6 + \alpha^2 D^{13}$. If we assume that the error polynomial is $e(D) = \alpha^8 D^5 + \alpha^3 D^{13}$, then the received polynomial is $y(D) = \alpha^{13} + \alpha^8 D + \alpha^{11}D^2 + \alpha^2 D^3 + \alpha^8 D^5 + \alpha^7 D^6 + \alpha^6 D^{13}$. Decoding starts by computing the syndromes $S_1 = \alpha^{12}$, $S_2 = 1$, $S_3 = \alpha^9$ and $S_4 = \alpha^9$. This gives the syndrome polynomial:

$$S(D) = \alpha^{12}D + D^2 + \alpha^9 D^3 + \alpha^9 D^4. \tag{9.137}$$

Then decoding goes through the following steps:

2. $k = 0$, $\Lambda^{(0)}(D) = 1$, $L = 0$, $T(D) = D$.
3. $k = 1$, $\Delta^{(1)} = S_1 = \alpha^{12}$.
4–5. $\Lambda^{(1)}(D) = \Lambda^{(0)}(D) - \Delta^{(1)}T(D) = 1 - \alpha^{12}D$.
6–7. $L = 1$, $T(D) = \frac{\Lambda^{(0)}(D)}{\Delta^{(1)}} = \frac{1}{\alpha^{12}} = \alpha^3$.
8. $T(D) = \alpha^3 D$.
3. $k = 2$, $\Delta^{(2)} = S_2 - \Lambda_1^{(1)}S_1 = 1 + \alpha^{12}\alpha^{12} = \alpha^7$.

---

[14] Given the polynomial $f(D) = f_0 + f_1 D + f_2 D^2 + \ldots + f_n D^n$ over the ground field GF(q), its *formal derivative* is given by $f'(D) = f_1 + 2f_2 D + \ldots + n f_n D^{n-1}$. Note that when $q = 2^m$, there are no odd-power terms in $f'(D)$, since their coefficients are equal to zero [1514].

4–5. $\Lambda^{(2)}(D) = \Lambda^{(1)}(D) - \Delta^{(2)}T(D) = 1 - \alpha^3 D.$

6–8. $T(D) = \alpha^3 D^2.$

    3. $k = 3,\ \Delta^{(3)} = S_3 - \Lambda_1^{(2)}S_2 = \alpha^9 + \alpha^3 = \alpha.$

    4–5. $\Lambda^{(3)}(D) = \Lambda^{(2)}(D) - \Delta^{(3)}T(D) = 1 - \alpha^3 D - \alpha^4 D^2.$

    6–7. $L = 2,\ T(D) = \dfrac{\Lambda^{(2)}(D)}{\Delta^{(3)}} = \dfrac{1 - \alpha^3 D}{\alpha} = \alpha^{14} - \alpha^2 D$

    8. $T(D) = \alpha^{14}D - \alpha^2 D^2.$

        3. $k = 4,\ \Delta^{(4)} = S_4 - \displaystyle\sum_{i=1}^{2} \Lambda_i^{(3)}S_{4-i} = \alpha^9 + \alpha^3\alpha^9 + \alpha^4 = \alpha^5.$

        5. $\Lambda^{(4)}(D) = \Lambda^{(3)}(D) - \Delta^{(4)}T(D) = 1 - \alpha^7 D - \alpha^3 D^2.$

        8. $T(D) = \alpha^{14}D^2 - \alpha^2 D^3.$

Therefore, the resulting error locator polynomial is:

$$\Lambda(D) = 1 - \alpha^7 D - \alpha^3 D^2. \tag{9.138}$$

By substituting $D = 1, \alpha, \alpha^2, \dots, \alpha^{14}$ into $\Lambda(D)$, we find that the roots of $\Lambda(D)$ are $X_1^{-1} = \alpha^2$ and $X_2^{-1} = \alpha^{10}$, so that the error locators are $X_1 = \alpha^{13}$ and $X_2 = \alpha^5$.

To find the error magnitudes we compute the polynomial (see (9.130)):

$$\Omega(D) = \Lambda(D)[1 + S(D)] \ \mathrm{mod}\ D^{2t+1}$$

$$= 1 + \alpha^2 D + \alpha^9 D^2. \tag{9.139}$$

Then the magnitudes of the first and second errors are given by (see (9.135)):

$$e_{i_1} = \frac{\Omega(X_1^{-1})}{(1 - X_2 X_1^{-1})} = \alpha^3 \tag{9.140}$$

and

$$e_{i_2} = \frac{\Omega(X_2^{-1})}{(1 - X_1 X_2^{-1})} = \alpha^8, \tag{9.141}$$

respectively, so that the estimated error polynomial is:

$$\hat{e}(D) = \alpha^8 D^5 + \alpha^3 D^{13}. \tag{9.142}$$

This leads to the estimated codeword (see (9.136)):

$$\hat{x}(D) = y(D) + \hat{e}(D)$$

$$= \alpha^{13} + \alpha^8 D + \alpha^{11}D^2 + \alpha^2 D^3 + \alpha^7 D^6 + \alpha^2 D^{13} = x(D), \tag{9.143}$$

so that error correction has been carried out successfully.

□

Decoding of Nonbinary BCH Codes: Errors and Erasure Decoding

The Berlekamp–Massey algorithm just described can be extended to decode BCH/RS codes in the presence of *erased symbols*, whose locations are known a priori. In what follows we assume the presence of v errors with unknown indexes $i_1, i_2, \dots, i_v$ and of e_s erasures with known indexes $j_1, \dots, j_{e_s}$, so that the lth syndrome for a *narrow sense* code is given by [1485]:

$$S_l = y(\alpha^l) = \sum_{k=1}^{v} e_{i_k} X_k^l + \sum_{k=1}^{e_s} f_{j_k} Y_k^l \tag{9.144}$$

with $l = 1, 2, \ldots, 2t$, where $y(D)$ is the received polynomial, $\{X_k = \alpha^{i_k}, k = 1, 2, \ldots, v\}$ are the *error locators* and $\{Y_k = \alpha^{j_k}, k = 1, 2, \ldots, e_s\}$ are the *erasure locators*. Note that, in this case, e_s erasures and t_1 errors can be simultaneously corrected if $2t_1 + e_s \leq 2t$.

The decoding procedure involves the following [1485]:

1. The *erasure locator polynomial*:

$$\Gamma(D) \triangleq \prod_{l=1}^{e_s} (1 - Y_l D).$$

(9.145)

2. The *error and erasure polynomial*:

$$\Psi(D) \triangleq \Lambda(D)\Gamma(D),$$

(9.146)

where $\Lambda(D)$ is the error locator polynomial (9.100).

3. The *syndrome polynomial*:

$$S(D) \triangleq \sum_{l=1}^{2t} S_l D^l.$$

(9.147)

4. The *modified syndrome polynomial*:

$$\Xi(D) \triangleq (\Gamma(D)[1 + S(D)] - 1) \bmod D^{2t+1}$$

(9.148)

and the *key equation*:

$$\Lambda(D)[1 + \Xi(D)]) = \Omega(D) \bmod D^{2t+1},$$

(9.149)

where $\Omega(D)$ is the *error magnitude polynomial*. Note that the key equation can be solved by resorting to the Berlekamp–Massey algorithm.

The overall decoding algorithm goes through the following steps:

1. Using the known erasure information, compute the erasure locator polynomial $\Gamma(D)$ (9.145).
2. Set the received sequence equal to zero in the erasure positions, so that the polynomial $y'(D)$ is obtained for the modified received signal. Then compute the syndrome polynomial $S(D)$ (9.147), using the syndromes $\{S_l = y'(\alpha^l), l = 1, 2, \ldots, 2t\}$.
3. Compute the modified syndrome polynomial $\Xi(D)$ (9.148).
4. Apply the Berlekamp–Massey algorithm to find the polynomial $\Lambda(D)$ for the LFSR generating the modified syndrome coefficients $\{\Xi_k, k = 1, 2, \ldots, 2t\}$ of $\Xi(D)$.
5. Find the roots of $\Lambda(D)$ to find the error locations.
6. Determine the error and erasure magnitudes by exploiting Forney's algorithm. Then, given $\Psi(D)$ (9.146), the error magnitude and erasure magnitude values are:

$$e_{i_k} = \frac{-X_k \Omega(X_k^{-1})}{\Psi'(X_k^{-1})}$$

(9.150)

and

$$f_{j_k} = \frac{-Y_k \Omega(Y_k^{-1})}{\Psi'(Y_k^{-1})},$$

(9.151)

respectively, where $\Psi'(D)$ is the formal derivative of $\Psi(D)$. Then compute the erasure polynomial:

$$f(D) = \sum_{k=1}^{e_s} f_{j_k} Y_k$$

(9.152)

and the error polynomial:

$$e(D) = \sum_{k=1}^{v} e_{i_k} X_k.$$

(9.153)

7. Correct the modified received polynomial, $y'(D)$, to generate the estimated codeword:

$$\hat{x}(D) = y'(D) + e(D) + f(D)$$

(9.154)

and stop.

Let us now apply this decoding procedure to the specific problem illustrated in the following example.

Example 9.1.17 Let us again consider the RS code described in Example 9.1.16. We assume that: (a) $u(D) = \alpha^2 D^9 + \alpha^7 D^2$, so that the codeword $x(D) = \alpha^{13} + \alpha^8 D + \alpha^{11} D^2 + \alpha^2 D^3 + \alpha^7 D^6 + \alpha^2 D^{13}$ is transmitted; (b) $e(D) = \alpha^{10} D + \alpha^8 D^5 + \alpha^3 D^{13}$, so that the received polynomial is $y(D) = \alpha^{13} + \alpha D + \alpha^{11} D^2 + \alpha^2 D^3 + \alpha^8 D^5 + \alpha^7 D^6 + \alpha^6 D^{13}$; (c) there are two erasures, characterized by the locators $Y_1 = \alpha^5$ and $Y_2 = \alpha^{13}$. We start decoding by setting the received signal to zero in the corresponding positions. This gives the modified received signal:

$$y'(D) = \alpha^{13} + \alpha D + \alpha^{11} D^2 + \alpha^2 D^3 + \alpha^7 D^6.$$

(9.155)

Then we compute the erasure locator polynomial (see (9.145)):

$$\Gamma(D) = \prod_{l=1}^{2} (1 - Y_l D) = (1 - \alpha^5 D)(1 - \alpha^{13} D) = 1 + \alpha^7 D + \alpha^3 D^2$$

(9.156)

and the syndromes:

$$S_1 = \alpha^{12}, \ S_2 = \alpha, \ S_3 = \alpha^4, \ S_4 = \alpha^4,$$

(9.157)

using the modified received signal $y'(D)$. Then the syndrome polynomial and the modified syndrome polynomial are (see (9.147) and (9.148)):

$$S(D) = \alpha^{12} D + \alpha D^2 + \alpha^4 D^3 + \alpha^4 D^4$$

(9.158)

and

$$\begin{aligned} \Xi(D) &= (\Gamma(D)[1 + S(D)] - 1) \ \text{mod} \ D^{2t+1} \\ &= [(1 + \alpha^7 D + \alpha^3 D^2)(1 + \alpha^{12} D + \alpha D^2 + \alpha^4 D^3 + \alpha^4 D^4) - 1] \ \text{mod} \ D^5 \\ &= \alpha^2 D + \alpha^{14} D^2 + \alpha^{10} D^3 + \alpha^{11} D^4, \end{aligned}$$

(9.159)

respectively. The Berlekamp–Massey algorithm is now used to find the error locator polynomial $\Lambda(D)$ satisfying the key equation (9.149) for $\Xi(D)$; this leads easily to $\Lambda(D) = 1 + \alpha D$. We now determine the magnitude of the errors and erasures. We compute the error and erasure polynomial (see (9.146)):

$$\begin{aligned} \Psi(D) &= \Lambda(D)\Gamma(D) \\ &= (1 + \alpha D)(1 + \alpha^7 D + \alpha^3 D^2) \\ &= 1 + \alpha^{14} D + \alpha^{13} D^2 + \alpha^4 D^3, \end{aligned}$$

(9.160)

with derivative $\Psi'(D) = \alpha^{14} + \alpha^4 D^2$, and the error magnitude polynomial (see (9.149)):

$$\Omega(D) = \Lambda(D)[1 + \Xi(D)] \bmod D^5$$
$$= (1 + \alpha D)(1 + \alpha^2 D + \alpha^{14} D^2 + \alpha^{10} D^3 + \alpha^{11} D^4) \bmod D^5$$
$$= 1 + \alpha^5 D + D^2 + \alpha^5 D^3. \tag{9.161}$$

Then for $X_1 = \alpha$ $(X_1^{-1} = \alpha^{14})$ we have that (see (9.150)):

$$e_{i_1} = -\frac{\alpha \Omega(\alpha^{14})}{\Psi'(\alpha^{14})} = -\alpha^{10}, \tag{9.162}$$

so that the error polynomial is:

$$e(D) = e_{i_1} X_1 = \alpha^{10} D. \tag{9.163}$$

Similarly, for $Y_1 = \alpha^5$ $(Y_1^{-1} = \alpha^{10})$ and $Y_2 = \alpha^{13}$ $(Y_2^{-1} = \alpha^2)$ we have that (see (9.151)):

$$f_{j_1} = -\frac{\alpha^5 \Omega(\alpha^{10})}{\Psi'(\alpha^{10})} = 0 \tag{9.164}$$

and

$$f_{j_2} = -\frac{\alpha^{13} \Omega(\alpha^2)}{\Psi'(\alpha^2)} = -\alpha^2, \tag{9.165}$$

respectively, so that the erasure polynomial is:

$$f(D) = f_{j_1} Y_1 + f_{j_2} Y_2 = \alpha^2 D^{13}. \tag{9.166}$$

These results lead to the estimated codeword (see (9.154)):

$$\hat{x}(D) = y'(D) + e(D) + f(D)$$
$$= \alpha^{13} + \alpha D + \alpha^{11} D^2 + \alpha^2 D^3 + \alpha^7 D^6 + \alpha^{10} D + \alpha^2 D^{13}$$
$$= \alpha^{13} + \alpha^8 D + \alpha^{11} D^2 + \alpha^2 D^3 + \alpha^7 D^6 + \alpha^2 D^{13} = x(D), \tag{9.167}$$

as required.

□

9.1.7 Error Performance

We now provide a sketch of some essential results concerning the error performance characterizing an algebraic decoder which may be operating in any one of three distinct modes, namely *error detection*, *complete decoding* and *incomplete decoding*. In the first mode the decoder does nothing but error detection, whereas in the second mode it produces the best estimate of the transmitted codeword using the computed syndrome. The third mode is a hybrid in which the decoder tries to correct only some error patterns, and detects the presence of errors in all the other cases. To simplify our analysis we focus primarily on a communication scenario in which the codeword symbol and channel symbols alphabets are both *binary*[15] (i.e., $q = 2$) and the communication channel is *memoryless* and *symmetric* (i.e., a BSC model is adopted). Note that:

(a) in a digital transmission over a fading channel, the hypothesis of absence of memory can be deemed reasonable only if an interleaver of sufficient depth is used,

---

[15] An analysis for nonbinary codes can be found in [1485, p. 245–251].

(b) deep fades can result in *erased* symbols, so that the decoder can be required to operate in the presence of both errors and erasures.

Space limitations prevent us from addressing issue (b); the reader is referred to [1485, p. 251-259] for further information.

9.1.7.1 Error Detection

In this mode the decoder computes the syndrome $\mathbf{s}$ in accordance with (9.26) and, if $\mathbf{s} \neq \mathbf{0}_{n-k}$, declares a detection error. This decision is *wrong* only if the error vector $\mathbf{e}$ is a codeword different from $\mathbf{0}_n$, so that even the observed vector $\mathbf{y} = \mathbf{x} + \mathbf{e}$ (9.25) is a codeword distinct from $\mathbf{x}$. In this mode of operation a significant performance index of the decoder is the probability of *undetected word error* P_{UE}, defined for any (n, k) code $\mathcal{C}$ over $GF(q)$ as:

$$P_{UE} \triangleq \Pr\left\{\mathbf{e} = \bar{\mathbf{x}} \in \mathcal{C}, \bar{\mathbf{x}} \neq \mathbf{0}_n\right\}. \tag{9.168}$$

If $q = 2$ and a memoryless BSC model is adopted, the A_w distinct error vectors characterized by the *same* Hamming weight w have the same probability of occurrence:

$$P_w = (1 - p)^{n-w} p^w, \tag{9.169}$$

where p is the *probability of symbol error*, i.e. the *transition* or *crossover probability* of the BSC. Then, taking into account all the realizations of $\mathbf{e}$, P_{UE} can be expressed as:

$$P_{UE} = \sum_{w=d_{H,\min}}^{n} A_w P_w = \sum_{w=d_{H,\min}}^{n} A_w (1 - p)^{n-w} p^w. \tag{9.170}$$

If the distance spectrum of $\mathcal{C}$ is unknown (but $d_{H,\min}$ is given), an upper bound on P_{UE} can be derived by noting that A_w never exceeds the binomial coefficient $\binom{n}{w}$. Then it can easily be seen that:

$$P_{UE} \leq \sum_{w=d_{H,\min}}^{n} \binom{n}{w} (1 - p)^{n-w} p^w. \tag{9.171}$$

This bound may be not be very tight, but offers the significant advantage of not requiring any knowledge of the weight distribution of $\mathcal{C}$.

9.1.7.2 Complete Decoding

In this mode the decoder *always* generates an estimate of the transmitted codeword on the basis of the computed syndrome $\mathbf{s}$. Therefore, a significant performance index of the decoding algorithm is the probability of *correct decision* P_{CD}. This quantity can be expressed, for the ML decoder of any (n, k) code $\mathcal{C}$ over $GF(q)$, as:

$$P_{CD} \triangleq \Pr\left\{\mathbf{e} \in \mathcal{C}_{CL}\right\}, \tag{9.172}$$

where $\mathbf{e}$ is the error vector (see (9.25)) and $\mathcal{C}_{CL}$ is the set of coset leaders. In fact, the decoding algorithm makes a correct decision if and only if $\mathbf{e}$ is a coset leader. Given P_{CD}, the probability of *incorrect decoding* P_{ID} can be evaluated as:

$$P_{ID} = 1 - P_{CD}.$$

The evaluation of P_{CD} by (9.172) requires knowledge of the set C_{CL}, and this becomes complicated for large codes (e.g., see [1485, p. 224] for further details). For this reason, an upper bound on P_{ID} is often evaluated, *which takes into account all the error patterns which are not guaranteed to be correctable*, that is, whose weight exceeds $t \triangleq \lfloor (d_{H,\min} - 1)/2 \rfloor$ (see (9.29)). This leads to:

$$P_{ID} \leq \sum_{l=t+1}^{n} \binom{n}{l} p^l (1 - p)^{n-l}. \tag{9.173}$$

Note that this equation becomes an equality only when C is *perfect*.

9.1.7.3   Incomplete Decoding

This operation mode can be considered as a hybrid of the first two. In fact, the decoder makes an attempt at error correction only when, on the basis of the observed syndrome, it deems that $t_1 \leq t$ errors have been inserted, where the parameter t, defined by (9.29), represents the maximum number of errors whose correction is *guaranteed* by the given (n, k) code C over GF(q). In all other cases, the decoder declares the presence of errors. The adoption of this strategy corresponds to dividing the standard array in two parts, the first including the error patterns for which error correction is attempted, the second those for which only error detection is performed.

In this mode the decoding procedure can end in three distinct ways, corresponding to *three mutually exclusive possibilities for the received vector* $\mathbf{y}$ (9.25), as illustrated in Figure 9.7. In the first case $\mathbf{y}$ falls in the guaranteed error-correction zone for the transmitted codeword $\mathbf{x}_k$ (see Figure 9.7(a)) and *decoding is correct*. In the second case $\mathbf{y}$ belongs to the interstitial zone, the region external to all the guaranteed error-correction zones of the codewords of C (see Figure 9.7(b)), and *the decoder only declares the presence of errors*, avoiding any attempt at correction. Finally, in the third case $\mathbf{y}$ falls in the guaranteed error-correction zone of a codeword different from the transmitted one (see Figure 9.7(c)) and the *decoder takes a wrong decision*. If P_{CD}, P_{DE} and P_{ID} denote the probabilities associated with these three events, then:

$$P_{CD} + P_{DE} + P_{ID} = 1. \tag{9.174}$$

Note that, if we refer to the cosets of the standard array of C, it can be stated that, generally speaking, P_{ID} represents the probability that the *true* error pattern is associated with a coset for which error correction is attempted, but is not a coset leader. Similarly, P_{DE} denotes the probability that the *true* error pattern is associated with a coset for which the decoder accomplishes error detection only.

Incomplete decoding is usually recommended when $d_{H,\min}$ is *even*. In this case the decoder corrects up to $t_1 = t < d_{H,\min}/2$ errors and detects the presence of $d_{H,\min}/2$ errors (this value is certainly smaller than the maximum guaranteed). For instance, if $d_{H,\min} = 6$, the decoder can correct up to 2 errors and detect the presence of 3 errors.

Finally, we note that, generally speaking, many algebraic decoding procedures are intrinsically *incomplete*, since they try to find a unique codeword, provided that it exists, in the set of all codewords whose Hamming distance from $\mathbf{y}$ does not exceeds the threshold t. When they fail, they stop declaring the presence of codeword errors.

9.2   Convolutional Codes

9.2.1   Introduction

This section is devoted to the analysis of *trellis codes* and, more specifically, of a well-known subclass, the *convolutional codes* over finite fields. The structure of a trellis code is shown in Figure 9.8.

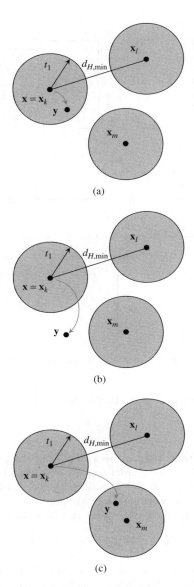

(a)

(b)

(c)

Figure 9.7   Three mutually exclusive events in the space of codewords of a linear block code $\mathcal{C}$: (a) correct decoding; (b) error detection; (c) incorrect decoding.

In particular, Figure 9.8(a) illustrates the use of a *trellis encoder* in a communication system. An information source generates a sequence of *symbols*, each belonging to GF(q), with $q = p^m$. This sequence in partitioned into *blocks*, each of length k, feeding a trellis encoder and the symbols of the lth block form the vector:

$$\mathbf{u}_l \triangleq \left[u_l^{(0)}, u_l^{(1)}, \ldots, u_l^{(k-1)} \right]. \tag{9.175}$$

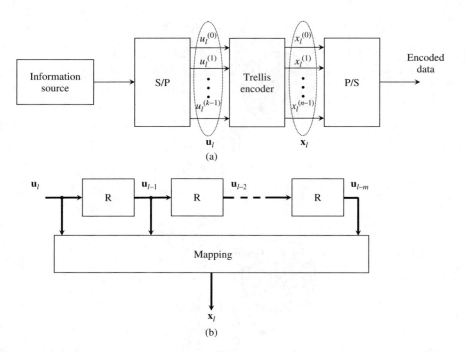

Figure 9.8   Trellis encoder: (a) use in a digital communication system; (b) block diagram.

Figure 9.8(b) shows the block diagram of the encoder for an (n, k) trellis code over GF(q) having a *rate*:

$$R \triangleq \frac{k}{n} \tag{9.176}$$

with $k < n$. The encoder has k inputs and contains a delay line consisting of m vector registers, where m is the so-called *memory order*. Each register consists of k distinct memory cells, each having a single input and a single output and able to hold a single symbol of GF(q). The delay line generates the vectors $\{\mathbf{u}_{l-1}, \mathbf{u}_{l-2}, \ldots, \mathbf{u}_{l-m}\}$, which are applied, together with $\mathbf{u}_l$ (9.175), to a deterministic *memoryless* subsystem which maps the above vectors into the n-dimensional vector:

$$\mathbf{x}_l \triangleq \left[x_l^{(0)}, x_l^{(1)}, \ldots, x_l^{(n-1)} \right] \tag{9.177}$$

over GF(q). Note that the presence of a delay line in Figure 9.8(b) introduces *memory* to the encoding process. Since this delay line has a *finite* length, and all its vector registers can take on q^k only distinct values, a trellis encoder can be modeled as an FSSM, for which the number of states cannot exceed q^{mk} (i.e., the number of possible distinct values of the vector $[\mathbf{u}_{l-1}, \mathbf{u}_{l-2}, \ldots, \mathbf{u}_{l-m}]$, which consists of mk elements of GF(q)). Moreover, as shown below, the time evolution of the encoder can be represented using a *trellis diagram*.

If we let the encoder start from a known state (zeroing, for instance, the content of all its registers at the starting epoch $l = 0$) and feed it for N consecutive clock intervals, the *message* $\mathbf{u} \triangleq [\mathbf{u}_0, \mathbf{u}_1, \ldots, \mathbf{u}_{N-1}]$ is mapped to the *codeword* $\mathbf{x} \triangleq [\mathbf{x}_0, \mathbf{x}_1, \ldots, \mathbf{x}_{N-1}]$. Therefore, in this case, unlike block codes, the codeword length is not uniquely identified by the encoder structure, but *increases linearly with the message length*.

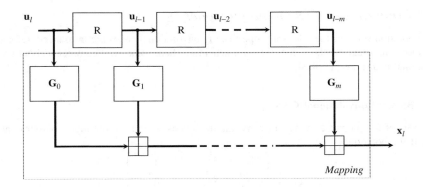

Figure 9.9 Block diagram of a convolutional encoder over GF(q).

If *linear* mapping is used in the block diagram of Figure 9.8(b), the trellis encoder is usually called a *convolutional encoder* (Figure 9.9). In the new scheme, $\{G_0, G_1, \ldots, G_m\}$ are $k \times n$ matrices. The subsystem labeled by G_p generates, in response to u_{l-p}, the n-dimensional vector $u_{l-p}G_p$, with $p = 0, 1, \ldots, m$. Each adder,[16] indicated by the symbol $\boxplus$, accomplishes a vector sum over GF(q) of its pair of inputs, each consisting of an n-dimensional vector. Therefore, in this case x_l (9.177) is given by:

$$x_l = u_l G_0 + u_{l-1} G_1 + \ldots + u_{l-m} G_m = \sum_{p=0}^{m} u_{l-p} G_p, \qquad (9.178)$$

where all operations are carried out over GF(q).

It is worth noting that:

(a) the input–output relationship (9.178) expresses the encoder output as a *convolution sum* between the input sequence $\{u_p\}$ and the matrix sequence $\{G_0, G_1, \ldots, G_m\}$ of finite length (this explains the name for this class of codes),
(b) this result derives from the fact that the block diagram of Figure 9.9 represents an FIR with impulse response $\{G_0, G_1, \ldots, G_m\}$, and
(c) equation (9.178) generalizes (9.3), since the latter is obtained from the former by setting $m = 0$.

In this section the essential properties of convolutional codes are described and encoding/decoding algorithms for this class of error-correction schemes are analyzed. In Section 9.2.2 the fundamental characteristics of convolutional encoders are illustrated. In particular, the input–output behavior of convolutional encoders is described by resorting both to an analytical approach and graphical tools, namely *state diagrams* and *trellis diagrams*. Then the *free distance* of a convolutional encoder is introduced as a significant parameter in the search for optimal codes, and its evaluation via a proper *transfer function* is discussed. In Sections 9.2.3-9.2.4 the problem of decoding is investigated. In particular, the ML and MAP decoding algorithms for binary convolutional codes are discussed in detail. Some essential information is also provided about sequential decoding strategies. Then bounds on the error performance achievable via ML decoding are derived in Section 9.2.6.

---

[16] In Figure 9.9 all the adders are *minimal*, that is, they sum two operands only; however, in the literature, multiple adders, involving an arbitrary number of operands, are common.

9.2.2   Properties of Convolutional Codes

In this subsection some of the relevant properties of convolutional codes are introduced, referring first to their most important subclass, that of *binary* schemes. Then, some brief comments about nonbinary convolutional codes are provided.

9.2.2.1   Binary Convolutional Codes

Our analysis of the properties of binary convolutional codes starts from the input–output relationship (9.178). If the notation:

$$
\mathbf{G}_p = \begin{bmatrix}
g_{0,p}^{(0)} & g_{0,p}^{(1)} & \cdots & g_{0,p}^{(n-1)} \\
g_{1,p}^{(0)} & g_{1,p}^{(1)} & \cdots & g_{1,p}^{(n-1)} \\
\cdots & \cdots & \cdots & \cdots \\
g_{k-1,p}^{(0)} & g_{k-1,p}^{(1)} & \cdots & g_{k-1,p}^{(n-1)}
\end{bmatrix}
\tag{9.179}
$$

is adopted (with $p = 0, 1, \ldots, m$), from (9.178) it is easy to see that:

$$
\mathbf{x}_l = \sum_{p=0}^{m} \sum_{i=0}^{k-1} u_{l-p}^{(i)} \mathbf{r}_{i,p},
\tag{9.180}
$$

where $\mathbf{r}_{t,p} \triangleq [g_{t,p}^{(0)}, g_{t,p}^{(1)}, \ldots, g_{t,p}^{(n-1)}]$ denotes the tth row of $\mathbf{G}_p$ (with $t = 0, 1, \ldots, k-1$). From (9.180) it is easily shown that the jth component of $\mathbf{x}_l$ (with $j = 0, 1, \ldots, n-1$) is given by:

$$
x_l^{(j)} = \sum_{p=0}^{m} \sum_{i=0}^{k-1} u_{l-p}^{(i)} g_{i,p}^{(j)},
\tag{9.181}
$$

which represents the input–output relationship of a MISO FIR filter with k inputs. It is easy to prove that the impulse response associated with the jth output and ith input (i.e., that evaluated under the assumption that all the other $(k-1)$ input lines are fed by null sequences) consists, in general, of $(m+1)$ samples different from 0, and these can be collected in the $(m+1)$-dimensional binary vector:

$$
\mathbf{g}_i^{(j)} = \left[g_{i,0}^{(j)}, g_{i,1}^{(j)}, \ldots, g_{i,m}^{(j)} \right]
\tag{9.182}
$$

for $i = 0, 1, \ldots, k-1$ and $j = 0, 1, \ldots, n-1$. Note that these vectors can be generated by extracting from each matrix of the sequence $\{\mathbf{G}_0, \mathbf{G}_1, \ldots, \mathbf{G}_m\}$ the (i, j)th element, as illustrated in Figure 9.10. Conversely, given the set of $k \cdot n$ impulse responses $\{\mathbf{g}_i^{(j)}, i = 0, 1, \ldots, k-1, j = 0, 1, \ldots, n-1\}$, called code *generators*, the matrices $\{\mathbf{G}_p, p = 0, 1, \ldots, m\}$ can be easily generated. It is also important to point out that in the technical literature the binary vector (9.182) is commonly represented in *octal form*; in other words, it is partitioned into triples of consecutive bits (starting from the left) and each triple is represented as an integer number between 0 and 7.

Reversing the summation order in (9.181) yields the equality:

$$
x_l^{(j)} = \sum_{i=0}^{k-1} \sum_{p=0}^{m} u_{l-p}^{(i)} g_{i,p}^{(j)},
\tag{9.183}
$$

whose meaning can be explained by referring to the comments above. In fact, the inner sum $\sum_{p=0}^{m} u_{l-p}^{(i)} g_{i,p}^{(j)}$ represents, in the lth clock interval, the contribution to the jth output line (with

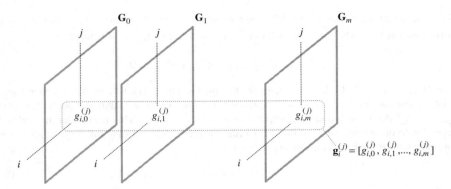

Figure 9.10   The relationship between the sample vector $\mathbf{g}_i^{(j)}$ and the matrix set $\{\mathbf{G}_0, \mathbf{G}_1, \ldots, \mathbf{G}_m\}$.

$j = 0, 1, \ldots, n - 1$) coming from the ith input line (with $i = 0, 1, \ldots, k - 1$); then summing the k contributions from all the distinct input lines (i.e., summing over the index i) yields (9.183).

Let us now analyze the input–output description of a specific binary encoder.

Example 9.2.1 Let us focus on a binary convolutional encoder with a single input ($k = 1$), $n = 2$ outputs (so that $R = 1/2$), a memory order $m = 2$ and the matrices:

$$\mathbf{G}_0 = \begin{bmatrix} 1 & 1 \end{bmatrix}, \quad \mathbf{G}_1 = \begin{bmatrix} 1 & 0 \end{bmatrix}, \quad \mathbf{G}_2 = \begin{bmatrix} 1 & 1 \end{bmatrix}. \tag{9.184}$$

Substituting these matrices in (9.178) yields the vector expression (u_l is used in place of $u_l^{(0)}$ for simplicity):

$$\left[x_l^{(0)}, x_l^{(1)}\right] = \left[u_l + u_{l-1} + u_{l-2}, u_l + u_{l-2}\right], \tag{9.185}$$

which is equivalent to the two scalar equations $x_l^{(0)} = u_l + u_{l-1} + u_{l-2}$ and $x_l^{(1)} = u_l + u_{l-2}$, evaluated over GF(2). Their implementation results in the block diagram of Figure 9.11. This encoder can also be fully characterized by its generators $\mathbf{g}_0^{(0)}$ and $\mathbf{g}_0^{(1)}$. From (9.184) it is easily seen (Figure 9.10) that $\mathbf{g}_0^{(0)} = [1, 1, 1]$ and $\mathbf{g}_0^{(1)} = [1, 0, 1]$ (or $[7_8]$ and $[5_8]$, respectively, in octal form).

□

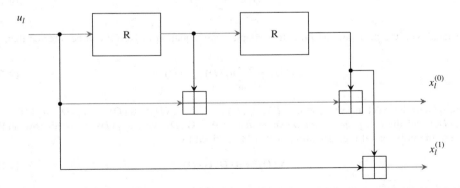

Figure 9.11   Block diagram of the encoder in Example 9.2.1.

The structure of the vector $\mathbf{g}_i^{(j)}$ (9.182) deserves some comment. It may happen that $g_{i,p}^{(j)} = 0$ for any j (i.e., output line) when $p > m_i$ (with $m_i < m$), that is, that $\mathbf{g}_i^{(j)}$ is structured as:

$$\mathbf{g}_i^{(j)} = \left[g_{i,0}^{(j)}, g_{i,1}^{(j)}, \ldots , g_{i,m_i}^{(j)}, 0, 0, \ldots , 0 \right] \tag{9.186}$$

for $j = 0, 1, \ldots , n - 1$. If this does happen, the ith (scalar) delay line of the encoder (i.e., that associated with the input variable) $u_l^{(i)}$ is actually required to have length m_i, since the variables $\{u_{l-p}^{(i)}, p > m_i\}$ do not influence the output vector $\mathbf{x}_l$ (see (9.180)). This means that the delay lines associated with distinct elements of the input vector in Figure 9.9 can have different lengths. For this reason, the encoder *memory* m is usually defined as:

$$m \triangleq \max_{i \in S} m_i, \tag{9.187}$$

where S represents the set $\{0, 1, \ldots , k - 1\}$. Note that if the k input lines are ordered (without loss of generality) such that $m_0 \leq m_1 \leq \ldots \leq m_{k-1}$, then (9.187) simplifies to $m = m_{k-1}$.

Another significant parameter of a convolutional encoder is its *constraint length* K, defined as:

$$K \triangleq n(m + 1). \tag{9.188}$$

Since each input symbol of the encoder contributes to the computation of $\mathbf{x}_l$ (9.177) according to (9.178) for no more than $(m + 1)$ consecutive clock intervals, K represents the maximum number of consecutive symbols $\{x_l^{(j)}\}$ over which the influence of any information symbol (bit, in the binary case) extends. For instance, $K = 6$ for the encoder of Example 9.2.1.

By analogy with what has been already done in the study of block codes, a *polynomial representation* can also be adopted to describe the input–output behavior of a convolutional encoder starting at $l = 0$ with all registers set to zero. If we define the polynomials:

$$u_i(D) \triangleq \sum_{l=0}^{+\infty} u_l^{(i)} D^l, \tag{9.189}$$

$$x_j(D) \triangleq \sum_{l=0}^{+\infty} x_l^{(j)} D^l \tag{9.190}$$

to represent the ith input sequence and jth output sequence, respectively, and the *generator polynomial* (or *transfer function*):

$$g_{i,j}(D) \triangleq \sum_{p=0}^{m} g_{i,p}^{(j)} D^p, \tag{9.191}$$

associated with the ith input and jth output couple, from equation (9.183) it can be inferred that:

$$x_j(D) = \sum_{i=0}^{k-1} u_i(D) g_{i,j}(D). \tag{9.192}$$

Then, given the vectors $\mathbf{x}(D) \triangleq [x_0(D), x_1(D), \ldots , x_{n-1}(D)]$, $\mathbf{u}(D) \triangleq [u_0(D), u_1(D), \ldots , u_{k-1}(D)]$ and the $k \times n$ generator polynomial matrix[17] $\mathbf{G}(D)$ with $g_{i,j}(D)$ in its ith row and jth column, the polynomial representation of (9.178) is given by:

$$\mathbf{x}(D) = \mathbf{u}(D) \mathbf{G}(D). \tag{9.193}$$

---

[17] The reader can easily prove that m_i (see (9.186)) and m (9.187) can also be evaluated as $\max[\deg_j g_{i,j}(D)]$ and $\max[\deg_{i,j} g_{i,j}(D)]$, respectively.

Note that, by analogy with what has been stated concerning linear block codes (see the sentence following eq. (9.5)), the elementary condition for a code to be useful is that the rank of $\mathbf{G}(D)$ is equal to k. Another matrix related to $\mathbf{G}(D)$ is the *parity check polynomial matrix* $\mathbf{H}(D)$. This is defined as any $(n-k) \times n$ matrix of polynomials satisfying the equality:

$$\mathbf{G}(D)\,\mathbf{H}^T(D) = \mathbf{0}_{k,n-k}. \tag{9.194}$$

Let us now apply these concepts to the encoder defined in Example 9.2.1.

Example 9.2.2 The generator polynomial matrix associated with the matrices (9.184) is:

$$\mathbf{G}(D) = \begin{bmatrix} 1 + D + D^2 & 1 + D^2 \end{bmatrix}. \tag{9.195}$$

Let $\mathbf{H}(D) = [H_0(D), H_1(D)]$ denote a parity check polynomial matrix for the given code. Substituting this expression and (9.195) into (9.194) yields:

$$H_0(D)\left[1 + D + D^2\right] + H_1(D)\left[1 + D^2\right] = 0. \tag{9.196}$$

Then a simple choice for $\mathbf{H}(D)$ is the vector $[1 + D^2, 1 + D + D^2]$.
$\square$

The generator polynomial matrix $\mathbf{G}(D)$ of an (n, k) binary convolutional code $\mathcal{C}$ is a powerful tool for analyzing the mechanism of codeword generation. To understand this, it is important to note that the same code (i.e., the same set of codewords) can be generated by more than one encoder. In fact, each encoder defines a specific *mapping* between an information polynomial vector $\mathbf{u}(D)$ and a codeword vector $\mathbf{x}(D)$, but changing the mapping can still result in the same set of codewords. To put it another way, multiple options are available for the matrix of $\mathbf{G}(D)$. To prove this statement, we note that (9.193) can be rewritten as:

$$\mathbf{x}(D) = \mathbf{u}(D)\,\mathbf{Q}(D)\,\mathbf{Q}^{-1}(D)\,\mathbf{G}(D), \tag{9.197}$$

where $\mathbf{Q}(D)$ is a *nonsingular* $k \times k$ matrix. Then, if we define $\overline{\mathbf{u}}(D) \triangleq \mathbf{u}(D)\mathbf{Q}(D)$ and $\overline{\mathbf{G}}(D) \triangleq \mathbf{Q}^{-1}(D)\,\mathbf{G}(D)$, equation (9.197) can be rewritten as $\mathbf{x}(D) = \overline{\mathbf{u}}(D)\,\overline{\mathbf{G}}(D)$. Since $\overline{\mathbf{u}}(D)$ still represents the set of all possible input sequences, it can be easily shown that the encoders characterized by $\mathbf{G}(D)$ and $\overline{\mathbf{G}}(D)$ are *equivalent*. This does not mean, however, that they have the same complexity. For this reason, whatever the code $\mathcal{C}$, it is important to look for a generator polynomial matrix with certain useful properties:

(a) it should describe *minimal* encoders [1517], that is, encoders with the minimum number of trellis states (or, equivalently, of memory elements in the encoder, as will become clear later);
(b) it should be *realizable*;
(c) it should avoid *catastrophicity*.

As far as requirement (b) is concerned, it is worth remembering that a *realizable* option for the matrix $\mathbf{G}(D)$ is such that its elements form a set of *realizable rational functions* $\{g_{i,j}(D)\}$. This means that each $g_{i,j}(D)$ is expressed by the ratio of two relatively prime polynomials, $g_{i,j}(D) = p_{i,j}(D)/q_{i,j}(D)$, with $q_{i,j}(0) = 1$ (in other words, $q_{i,j}(D)$ is *delay-free*). In fact, when this occurs, that is, if:

$$g_{i,j}(D) = \frac{\sum_{l=0}^{m} b_l D^l}{1 + \sum_{l=1}^{m} a_l D^l}, \tag{9.198}$$

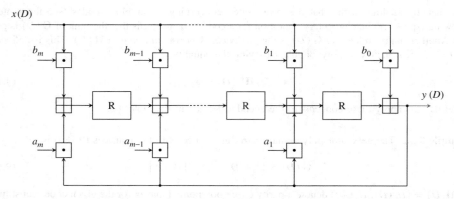

Figure 9.12 Implementation of the transfer function (9.198) via its *observer canonical form*. Sums and products are evaluated over GF(q).

then $g_{i,j}(D)$ can be implemented according to the *observer canonical form* (an alternative implementation, called the *controller canonical form*, can also be used, as shown in [35, pp. 479–481]) ensuring minimum complexity, the form of which is illustrated in Figure 9.12.

The property of catastrophicity refers to the fact that some polynomial generator matrices may map an *infinite-weight* input sequence $\mathbf{u}(D)$ to a *finite-weight* output sequence $\mathbf{x}(D)$. In this case the presence of a finite number of errors in the decoded codeword can have a catastrophic effect on the quality of the estimate of the information data sequence. That is, it can result in an *infinite number of errors* in such an estimate. The property of catastrophicity cannot be related to the code $\mathcal{C}$, but to the specific structure of its encoder, that is, to the form of its mathematical description $\mathbf{G}(D)$. To understand this, it is useful to note that if $\mathbf{G}(D)$ is structured as:

$$\mathbf{G}(D) = \left[\mathbf{I}_k \ \mathbf{P}_{k,n-k}(D)\right], \tag{9.199}$$

where $\mathbf{P}_{k,n-k}(D)$ is a $k \times (n-k)$ matrix with rational entries, it describes a *systematic* encoder, such that:

$$x_l^{(i)} = u_l^{(i)} \tag{9.200}$$

for $i = 0, 1, \ldots, k-1$. Relevant properties of systematic generator polynomial matrices are as follows:

1. They describe *minimal encoders* (see [1517, Theorem 10]).
2. *No inverting circuit* is needed to recover the information sequence from the codeword. Each *nonsystematic* encoder with transfer function matrix $\mathbf{G}(D)$ requires the availability of a specific inverting logic to recover $\mathbf{u}(D)$, that is, the existence of a matrix $\mathbf{G}^{-1}(D)$ such that:

$$\mathbf{G}(D)\,\mathbf{G}^{-1}(D) = \mathbf{I}_k D^l, \tag{9.201}$$

for some $l \geq 0$. When this equality holds, $\mathbf{u}(D)$ can be recovered, with a delay of l clock intervals, by multiplying $\mathbf{x}(D)$ by $\mathbf{G}^{-1}(D)$ (see (9.193)). For an $(n, 1)$ code, it can be shown that a transfer function matrix $\mathbf{G}(D)$ admits a *feedforward inverse*[18] $\mathbf{G}^{-1}(D)$ of delay l if and only if [1518]:

$$\text{GCD}\left[g_{0,0}(D), g_{0,1}(D), \ldots, g_{0,n-1}(D)\right] = D^l, \tag{9.202}$$

---

[18] Techniques for the construction of the feedforward inverse $\mathbf{G}^{-1}(D)$ can be found in [1517, 1518].

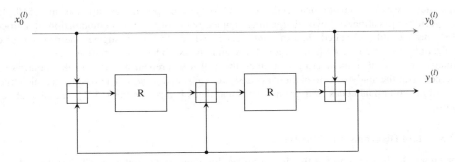

Figure 9.13 Systematic convolutional encoder based on the transfer function matrix (9.204).

for some $l \geq 0$, where GCD[·] denotes the greatest common divisor. For an (n, k) code with $k > 1$, this existence condition can be reformulated. Let $\Delta_i (D)$, with $i = 1, 2, \ldots, \binom{n}{k}$, be the determinants of the $\binom{n}{k}$ distinct $k \times k$ submatrices which can be extracted from $\mathbf{G} (D)$. Then a feedforward inverse $\mathbf{G}^{-1} (D)$ of delay l exists if and only if [1518]:

$$\text{GCD} \left[\Delta_i (D), \ i = 1, 2, \ldots, \binom{n}{k} \right] = D^l, \tag{9.203}$$

for some $l \geq 0$. J. L. Massey and M. K. Sain showed that (9.202) and (9.203) are necessary and sufficient conditions for a code to be noncatastrophic [1518]. In other words, an encoder is noncatastrophic if and only if it admits a feedforward inverse.

Let us now apply these concepts in the following example.

Example 9.2.3 Let us reconsider the rate-1/2 binary convolutional code of Example 9.2.1, which can be generated using the transfer function matrix (9.195). An equivalent transfer function matrix in systematic form can be easily generated by multiplying $\mathbf{G} (D)$ by $\mathbf{Q}^{-1} (D) = 1/(1 + D + D^2)$. This yields:

$$\overline{\mathbf{G}} (D) \triangleq \mathbf{Q}^{-1} (D) \mathbf{G} (D) = \begin{bmatrix} 1 & \dfrac{1 + D^2}{1 + D + D^2} \end{bmatrix}. \tag{9.204}$$

The corresponding implementation, based on the structure of Figure 9.12, is illustrated in Figure 9.13. Note that, in this case, the number of memory elements in the encoder structure is identical to that for the equivalent nonsystematic structure of Figure 9.11.

□

9.2.2.2 Nonbinary Convolutional Codes

Most of the results illustrated above for binary coding can be extended to convolutional coding over GF(q). In fact, our developments are mainly based on the input–output relationship (9.178), which does not depend on the specific value of q. Note also that, if q is a power of 2 (i.e., $q = 2^n$), coded q-ary sequences can certainly be seen as the output of a rate-k/n *binary* convolutional encoder. This approach, however, does not lend itself to solving the problem of code optimization (see Section 9.2.2.4). This problem must be tackled directly over GF(q), that is, by considering a GF(q) algebra in codeword generation, as illustrated in Figure 9.9.

Specific classes of convolutional codes over GF(q) are proposed, for instance, in [1519, 1520] for a specific application, orthogonal signaling coupled with noncoherent demodulation; in particular, [1519] introduces the so-called class of *dual-k* codes, while [1520] investigates the problem of code optimization, deriving new optimum 4-ary, 8-ary and 16-ary codes with different rates.

Finally, it is worth remembering that since the 1990s theoretical research in this area has been concerned with the study of convolutional codes over *rings* and *groups*. The basic algebraic structure theory of convolutional codes over groups, rings and fields is discussed in [1521] (see also references therein).

9.2.2.3 State Diagrams and Trellises

From Figure 9.9 it is easy to see that in the lth symbol interval, the output $\mathbf{x}_l$ (9.177) depends on the input vector $\mathbf{u}_l$ (9.175) and on the contents of a delay line; such contents represent the *present state* of the encoder, modeled as an FSSM. To formalize this description, we define the vector:

$$\mathbf{u}_l^{(i)} \triangleq \left[u_{l-1}^{(i)}, u_{l-2}^{(i)}, \ldots, u_{l-m_i}^{(i)} \right] \tag{9.205}$$

for $i = 0, 1, \ldots, k-1$, containing all the symbols held in the ith delay line and processed in the computation of $\mathbf{x}_l$ in accordance with (9.178). Then the *encoder state* in the lth symbol interval can be defined by the ordered concatenation of the vectors $\{\mathbf{u}_l^{(i)}, i = 0, 1, \ldots, k-1\}$, that is, by the vector:

$$\mathbf{u}_l \triangleq \left[\mathbf{u}_l^{(0)}, \mathbf{u}_l^{(1)}, \ldots, \mathbf{u}_l^{(k-1)} \right], \tag{9.206}$$

whose number of elements:

$$m_{tot} \triangleq \sum_{i=0}^{k-1} m_i \tag{9.207}$$

represents the *overall memory* of the encoder. Then the overall number of distinct states of the encoder is given by:

$$n_s \triangleq q^{m_{tot}}. \tag{9.208}$$

Note that, since each element of $\mathbf{u}_l$ (9.206) belongs to GF(q), the present state is also defined by the integer parameter:

$$\sigma_l \triangleq \sum_{i=0}^{k-1} q^{s_i} \sum_{p=1}^{m_i} u_{l-p}^{(i)} q^{p-1} \tag{9.209}$$

with $s_0 = 0$ and $s_i \triangleq \sum_{q=0}^{i-1} m_q$ for $i = 0, 1, \ldots, k-1$. Given σ_l in (9.209), the encoder behavior is also completely described by its *state update equation*:

$$\sigma_{l+1} = f\left(\sigma_l, \mathbf{u}_l \right), \tag{9.210}$$

expressing the *next state* σ_{l+1} as a function of the present state σ_l and the present input $\mathbf{u}_l$, and its *output equation*:

$$\mathbf{x}_l = \mathbf{g}\left(\sigma_l, \mathbf{u}_l \right). \tag{9.211}$$

The integer-valued function $f\left(\cdot, \cdot \right)$ and the vector-valued function $\mathbf{g}\left(\cdot, \cdot \right)$ (yielding an n-dimensional vector whose elements belong to GF(q)) depend on the specific structure of the encoder, as shown in the following example.

Example 9.2.4 The convolutional encoder defined in Example 9.2.1 can be modeled as a four-state sequential machine, since its inner state is defined by the vector $\mathbf{u}_l \triangleq \left[u_{l-1}, u_{l-2} \right]$ (see (9.206)), or, equivalently, by the integer parameter (see (9.209)):

$$\sigma_l \triangleq u_{l-1} + 2u_{l-2}. \tag{9.212}$$

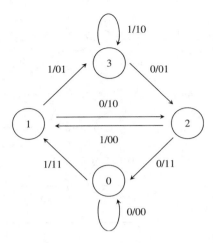

Figure 9.14 State diagram of the convolutional encoder in Example 9.2.5. Each transition is labeled by $u_l/x_l^{(0)}x_l^{(1)}$.

Then the state update equation can be expressed as:

$$\sigma_{l+1} = u_l + 2u_{l-1} = u_l + 2\left(\sigma_l \bmod 2\right), \tag{9.213}$$

since $u_{l-1} = \bmod\left(\sigma_l, 2\right)$ (see (9.212)), whereas the output equation, expressed by (9.185), can be put in the form:

$$\left[x_l^{(0)}, x_l^{(1)}\right] = \left[u_l + \left(\sigma_l \bmod 2\right) + \left(\sigma_l - \left(\sigma_l \bmod 2\right)\right)/2, \ u_l + \left(\sigma_l - \left(\sigma_l \bmod 2\right)\right)/2\right], \tag{9.214}$$

since $u_{l-2} = \left(\sigma_l - u_{l-1}\right)/2$ (see (9.212)).

□

A description of the behavior of a convolutional encoder, which can be used as an alternative to (9.210) and (9.211), is based on the use of its *state diagram*. Such a diagram is a directed graph with the following properties:

(a) each state is represented by a circle;
(b) circles are connected by oriented branches representing state transitions.

In other words, each branch describes a specific state transition $\sigma_l \rightarrow \sigma_{l+1}$, where σ_{l+1} can even coincide with σ_l; in addition, the branch is labeled by both the specific value of $\mathbf{u}_l$ producing that transition and by the value taken on by $\mathbf{x}_l$. These concepts are illustrated for a specific case in the following example.

Example 9.2.5 The state diagram of the four-state convolutional encoder of Examples 9.2.1 and 9.2.2 can easily be derived from (9.213) and (9.214) and is illustrated in Figure 9.14. Each transition is labeled by the present input bit and output bit pair using the notation $u_l/x_l^{(0)}x_l^{(1)}$.

□

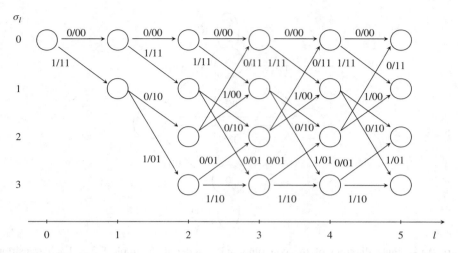

Figure 9.15 Trellis diagram associated with the state diagram of Figure 9.14.

The state diagram of an FSSM cannot show its *time evolution*, as already illustrated in Chapter 3. This is better described instead by the FSSM *trellis diagram*.

Example 9.2.6 The trellis diagram following from the state diagram of Figure 9.14 is shown in Figure 9.15. Note that the initial state of the encoder is $\sigma_0 = 0$ and that each state transition is labeled by the same information as the state diagram. Given the trellis diagram and the initial state σ_0, the FSSM evolution associated with a specific input data sequence is represented by a unique *trajectory* in the diagram itself; all the trajectories originate from $\sigma_0 = 0$ and develop in the direction of increasing time.

□

9.2.2.4 Distance Spectrum and Free Distance

In Section 9.2.2 no explicit constraint was put on the $k \times n$ matrices $\{\mathbf{G}_p\}$ (9.179) or, equivalently, on the selection of the generators $\{\mathbf{g}_i^{(j)}\}$ (9.186). In this subsection, we focus on the problem of the *optimal selection* of these matrices (or vectors) in the class $\mathcal{C}_q(n, k, m_{tot})$, consisting of all the (n, k) convolutional codes over GF(q) having a fixed overall memory m_{tot} (9.207). Solving this problem requires the definition of an *optimality criterion* to establish whether a code belonging to $\mathcal{C}_q(n, k, m_{tot})$ is better than another code in the same class. As illustrated in Section 9.2.3, if our aim is to minimize the codeword error probability with ML decoding over various communication channels, it is required, first of all, that the selected code maximizes, among all the possibilities offered in $\mathcal{C}_q(n, k, m_{tot})$, the *minimum Hamming distance* between each possible pair of distinct codewords. To understand the implications of this result, let us now formalize the evaluation of the distance between the pair of codewords $\mathbf{x}^{(i)} \triangleq [\mathbf{x}_0^{(i)}, \mathbf{x}_1^{(i)}, \ldots, \mathbf{x}_{N-1}^{(i)}]$ and $\mathbf{x}^{(j)} \triangleq [\mathbf{x}_0^{(j)}, \mathbf{x}_1^{(j)}, \ldots, \mathbf{x}_{N-1}^{(j)}]$, associated with the messages $\mathbf{u}^{(i)} \triangleq [\mathbf{u}_0^{(i)}, \mathbf{u}_1^{(i)}, \ldots, \mathbf{u}_{N-1}^{(i)}]$ and $\mathbf{u}^{(j)} \triangleq [\mathbf{u}_0^{(j)}, \mathbf{u}_1^{(j)}, \ldots, \mathbf{u}_{N-1}^{(j)}]$, respectively, of equal length (since they both extend over N clock intervals of the encoder). In the generation of each of the codewords it is also assumed that the encoder initial state is $\sigma_0 = 0$ and that in the first clock interval the trajectories associated with the codewords in the trellis diagram of the encoder are *distinct*; the latter assumption guarantees the *diversity* of the codewords generated and can be formulated as $\mathbf{u}_0^{(i)} \neq$

$\mathbf{u}_0^{(j)}$. The Hamming distance between $\mathbf{x}^{(i)}$ and $\mathbf{x}^{(j)}$ is given by:

$$d_H\left(\mathbf{x}^{(i)}, \mathbf{x}^{(j)}\right) = \sum_{l=0}^{N-1} d_H\left(\mathbf{x}_l^{(i)}, \mathbf{x}_l^{(j)}\right) \tag{9.215}$$

and, since $d_H(\mathbf{x}_l^{(i)}, \mathbf{x}_l^{(j)}) = w_H(\mathbf{x}_l^{(i)} - \mathbf{x}_l^{(j)})$, can also be expressed as:

$$d_H\left(\mathbf{x}^{(i)}, \mathbf{x}^{(j)}\right) = \sum_{l=0}^{N-1} w_H\left(\mathbf{x}_l^{(i)} - \mathbf{x}_l^{(j)}\right). \tag{9.216}$$

Note that, for the *linearity* of the code, the vector:

$$\mathbf{x}^{(i,j)} = \left[\mathbf{x}_0^{(i,j)}, \mathbf{x}_1^{(i,j)}, \ldots, \mathbf{x}_{N-1}^{(i,j)}\right] \triangleq \mathbf{x}^{(i)} - \mathbf{x}^{(j)}$$
$$= \left[\mathbf{x}_0^{(i)} - \mathbf{x}_0^{(j)}, \mathbf{x}_1^{(i)} - \mathbf{x}_1^{(j)}, \ldots, \mathbf{x}_{N-1}^{(i)} - \mathbf{x}_{N-1}^{(j)}\right] \tag{9.217}$$

represents the codeword generated by the encoder (initialized in the state $\sigma_0 = 0$) in response to the message:

$$\mathbf{u}^{(i,j)} = \left[\mathbf{u}_0^{(i,j)}, \mathbf{u}_1^{(i,j)}, \ldots, \mathbf{u}_{N-1}^{(i,j)}\right] \triangleq \mathbf{u}^{(i)} - \mathbf{u}^{(j)}$$
$$= \left[\mathbf{u}_0^{(i)} - \mathbf{u}_0^{(j)}, \mathbf{u}_1^{(i)} - \mathbf{u}_1^{(j)}, \ldots, \mathbf{u}_{N-1}^{(i)} - \mathbf{u}_{N-1}^{(j)}\right] \tag{9.218}$$

characterized by $\mathbf{u}_0^{(i,j)} = \mathbf{u}_0^{(i)} - \mathbf{u}_0^{(j)} \neq \mathbf{0}_k$. Therefore, (9.216) can also be rewritten as:

$$d_H\left(\mathbf{x}^{(i)}, \mathbf{x}^{(j)}\right) = \sum_{l=0}^{N-1} w_H\left(\mathbf{x}_l^{(i,j)}\right) = w_H\left(\mathbf{x}^{(i,j)}\right). \tag{9.219}$$

This proves that, whatever the convolutional encoder, the search for the minimum Hamming distance between all possible pairs of distinct codewords of duration N can be accomplished by computing the weight for all the codewords that can be generated over N consecutive clock intervals, starting from the initial state $\sigma_0 = 0$ and under the condition that the encoder input vector $\mathbf{u}_0$ at the beginning of the first clock interval is different from $\mathbf{0}_k$. Therefore, the minimum distance can be expressed as:

$$d_c(N) \triangleq \min_{\tilde{\mathbf{u}}, \tilde{\mathbf{u}}_0 \neq \mathbf{0}_k} w_H(\tilde{\mathbf{x}}) = \min_{\tilde{\mathbf{u}}, \tilde{\mathbf{u}}_0 \neq \mathbf{0}_k} \sum_{l=0}^{N-1} w_H(\tilde{\mathbf{x}}_l), \tag{9.220}$$

where $\tilde{\mathbf{x}} \triangleq [\tilde{\mathbf{x}}_0, \tilde{\mathbf{x}}_1, \ldots, \tilde{\mathbf{x}}_{N-1}]$ is the codeword associated with the *trial message* $\tilde{\mathbf{u}} \triangleq [\tilde{\mathbf{u}}_0, \tilde{\mathbf{u}}_1, \ldots, \tilde{\mathbf{u}}_{N-1}]$ starting from $\sigma_0 = 0$. The function $d_c(N)$ is known as the *column distance* and can be interpreted as the minimum distance, computed over N clock intervals, between the reference codeword $\mathbf{x}^{(0)} = [\mathbf{x}_0^{(0)} = \mathbf{0}_n, \mathbf{x}_1^{(0)} = \mathbf{0}_n, \ldots, \mathbf{x}_{N-1}^{(0)} = \mathbf{0}_n]$ generated in response to the vector $\mathbf{u}^{(0)} \triangleq [\mathbf{u}_0^{(0)} = \mathbf{0}_k, \mathbf{u}_1^{(0)} = \mathbf{0}_k, \ldots, \mathbf{u}_{N-1}^{(0)} = \mathbf{0}_k]$ (consisting of null vectors only) and any possible codeword $\mathbf{x}^{(i)} \triangleq [\mathbf{x}_0^{(i)}, \mathbf{x}_1^{(i)}, \ldots, \mathbf{x}_{N-1}^{(i)}]$ generated by a message $\mathbf{u}^{(i)} \triangleq [\mathbf{u}_0^{(i)}, \mathbf{u}_1^{(i)}, \ldots, \mathbf{u}_{N-1}^{(i)}]$ with $\mathbf{u}_0^{(i)} \neq \mathbf{0}_k$. Therefore, if we refer to the trellis diagram of the encoder (e.g., see Figure 9.15), $d_c(N)$ in (9.220) is the minimum distance between a *reference trajectory*, represented by the horizontal path originating from $\sigma_0 = 0$ and consisting of N consecutive branches, and any other trajectory originating from the same initial state and having the same length, but separating from the reference path in the first clock interval.

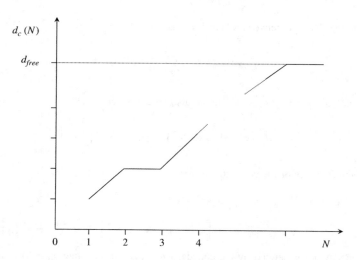

Figure 9.16   Typical behavior of the column distance $d_c(N)$ (9.220) of a convolutional code.

From (9.220) it can easily be inferred that $d_c(N)$ is a nondecreasing function of N, as seen from Figure 9.16. Its asymptotic value as $N \to \infty$:

$$d_{free} \triangleq \lim_{N \to \infty} d_c(N), \qquad (9.221)$$

is called *free distance* of the code.

Except for pathological cases, the free distance of a convolutional code is the weight of a codeword associated with a path that, after separating at the instant $l = 0$ from the reference trajectory $\mathbf{x}^{(0)}$ associated with a null message $\mathbf{u}^{(0)}$, subsequently merges with it again. This holds, for instance, for the four-state binary convolutional code considered above, as shown in the following example.

Example 9.2.7 As shown in the following pages, $d_{free} = 5$ for the four-state convolutional code whose trellis diagram is given in Figure 9.15; the path which this distance is associated with is marked by the dashed line in Figure 9.17.
□

The parameter d_{free} expresses the *minimum diversity* between the possible codewords of arbitrary length and, consequently, is fundamental in the design and selection of a convolutional code. Apart from the simplest cases, like that illustrated in Example 9.2.7, in which d_{free} can be easily be evaluated by considering the possible paths in the trellis diagram, the computation of d_{free} must be accomplished via specific mathematical techniques. In particular, tools provided by *graph theory* can be exploited, as illustrated in the following example.

Example 9.2.8 To evaluate d_{free} for the binary four-state code characterized by the trellis diagram of Figure 9.15 (see Example 9.2.6) the following procedure can be used. To begin, we note that solving this problem requires considering *all possible paths* (i.e., trial paths) originating from the state $\sigma_0 = 0$, separating immediately (i.e., in the first clock interval) from the *reference path* characterized by $\mathbf{u}_l = \mathbf{0}_k$ for any $l \geq 0$, and that, after an arbitrary number of clock intervals, merge in the above mentioned reference path. For each possible *trial path* the distance from the reference is given by its overall weight, expressed by the sum of the weights of the output vectors $\{\mathbf{x}_l, \ l = 0, 1, \ldots\}$ associated

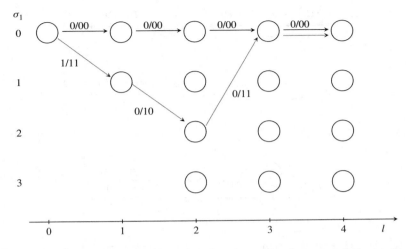

Figure 9.17   The path (dashed line) whose distance from the reference path (continuous line) is $d_{free} = 5$ in the trellis diagram of Figure 9.15.

with the path branches until the merge point. The minimum overall weight of all possible trial paths is d_{free}. To accomplish an exhaustive search over the set of all the trial paths and evaluate their overall weights, a new directed graph, derived from the state diagram of Figure 9.14, can be used. In particular, such a graph can be obtained from the state diagram by:

(a) removing the state transition $\sigma_l = 0 \rightarrow \sigma_{l+1} = 0$ and duplicating the null state; and
(b) substituting the label of each state transition $\sigma_l = p \rightarrow \sigma_{l+1} = q$ with the power $D^{s_{p,q}}$ of the variable D, where $s_{p,q}$ represents the weight of the vector $(x_l^{(0)}, x_l^{(1)})$ associated with that transition.

This turns the graph of Figure 9.14 into that of Figure 9.18. In the new graph all the trial paths to be considered in our search emerge from the null state on the left and end in that on the right. In addition, the weight of the codeword associated with a specific path is given by the *sum of the exponents* of the powers of D associated with all the branches forming the path itself, that is, by the exponent of the *product* of such powers. Then, if a path having overall weight w is represented by D^w, two distinct paths, having weights w_1 and w_2, can be jointly represented by a single polynomial

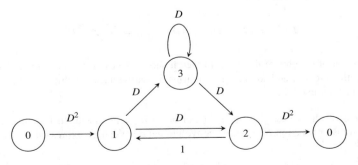

Figure 9.18   Modified graph associated with the state diagram of Figure 9.14.

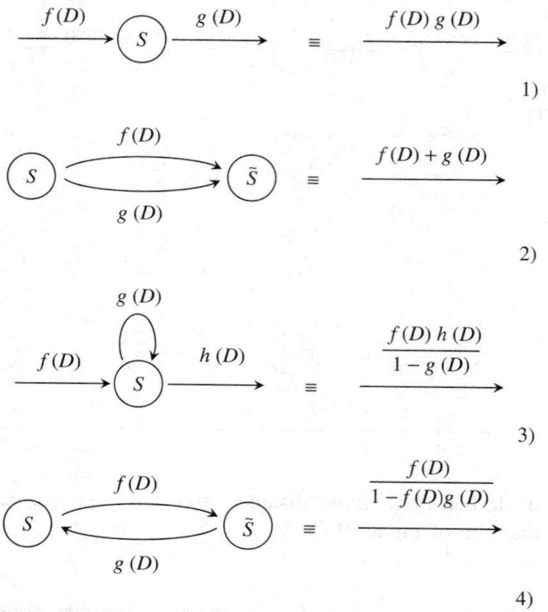

$$\xrightarrow{\;f(D)\;} \boxed{S} \xrightarrow{\;g(D)\;} \quad \equiv \quad \xrightarrow{\;f(D)\,g(D)\;} \qquad 1)$$

$$\boxed{S} \underset{g(D)}{\overset{f(D)}{\rightleftarrows}} \boxed{\tilde{S}} \quad \equiv \quad \xrightarrow{\;f(D)+g(D)\;} \qquad 2)$$

$$\xrightarrow{\;f(D)\;} \boxed{S}^{\;g(D)} \xrightarrow{\;h(D)\;} \quad \equiv \quad \xrightarrow{\;\frac{f(D)\,h(D)}{1-g(D)}\;} \qquad 3)$$

$$\boxed{S} \underset{g(D)}{\overset{f(D)}{\rightleftarrows}} \boxed{\tilde{S}} \quad \equiv \quad \xrightarrow{\;\frac{f(D)}{1-f(D)g(D)}\;} \qquad 4)$$

Figure 9.19 Reduction rules for the computation of the transfer function of a directed graph.

in the D variable, namely by the sum $D^{w_1} + D^{w_2}$. Following this line of reasoning, we can infer that, for the given code, the *set of all the possible weights* (associated with all the distinct codewords of interest) with their multiplicity can be summarized in a *single polynomial of infinite degree in the variable D*. Such a polynomial is known as the *transfer function* of the code and can be computed in different ways. One way involves the progressive reduction of the given graph to a simple graph with a *single edge*. This can be done by repeated application of the four elementary rules illustrated in Figure 9.19, where $f(D)$, $g(D)$ and $h(D)$ denote arbitrary functions in the variable D. In our specific example this leads to a single-edge graph labeled by the function:

$$T(D) = \frac{D^5}{1 - 2D}, \tag{9.222}$$

which is the transfer function of our code. Evaluating the polynomial division on the RHS of (9.222) yields the infinite-degree polynomial:

$$T(D) = D^5 + 2D^6 + 4D^7 + \ldots = \sum_{d=5}^{+\infty} 2^{d-5} D^d, \tag{9.223}$$

which gives the set of weights of the code; in fact, it shows the existence of a single path with weight 5, two paths with weight 6, and so on. This confirms, in particular, the validity of the result $d_{free} = 5$, already given in Example 9.2.7.

□

The particular result expressed by (9.223) can be easily generalized. In fact, it is not difficult to show that the transfer function of a binary convolutional code can be expressed as:

$$T(D) = \sum_{d=d_{free}}^{\infty} n(d)\, D^d, \tag{9.224}$$

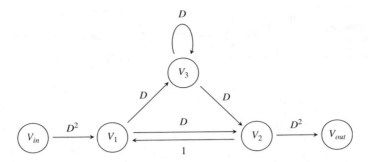

Figure 9.20 Graph for the evaluation of the transfer function of the binary code whose state diagram is shown in Figure 9.14.

where $n(d)$ denotes the overall number of paths having weight d. Note that the minimum exponent of the powers of D in (9.224) is equal to d_{free}.

Alternative techniques for the evaluation of the transfer function are also available, as shown in the following example.

Example 9.2.9 The graph of Figure 9.18 can be modified by assigning to each of its states a sort of *electrical potential* as illustrated in Figure 9.20. In particular, the input potential V_{in} is assigned to the null state on the left, while the output potential V_{out} is assigned to that on the right. The resulting diagram is now interpreted as a representation of the existing relationships among node potentials; this leads to the potential of each node being expressed as the sum of different contributions, each associated with a branch directed toward that node. In particular, in our specific case, the node equations:

$$V_1 = D^2 V_{in} + V_2, \qquad (9.225)$$

$$V_2 = DV_1 + DV_3, \qquad (9.226)$$

$$V_3 = DV_1 + DV_3, \qquad (9.227)$$

$$V_{out} = D^2 V_2 \qquad (9.228)$$

can be written for all the nodes; note that no equation has been written for the node labeled by V_{in}, since no branch enters this node. From the system of equations (9.225)–(9.228) the ratio V_{out}/V_{in} can be found using standard mathematical techniques. The reader can verify that this ratio corresponds to the code transfer function expressed by (9.222).

□

The technique described in the previous example can be easily generalized to the evaluation of the transfer function of any binary convolutional code.

Let us now reconsider the specific problem mentioned above of *optimal* selection of a code in $C_q(n, k, m_{tot})$. If the optimality criterion is the maximization of d_{free}, then an exhaustive computer search over $C_q(n, k, m_{tot})$ is required to identify the optimal choice[19] for the matrices $\{\mathbf{G}_p\}$ (9.179) or, equivalently, for the generators $\{\mathbf{g}_i^{(j)}\}$ (9.186). Tables of the generators of good binary convolutional codes can be found, for instance, in [35, Sect. 12.3].

---

[19] This choice is not necessarily unique, since, in principle, multiple codes having the same d_{free} could be found.

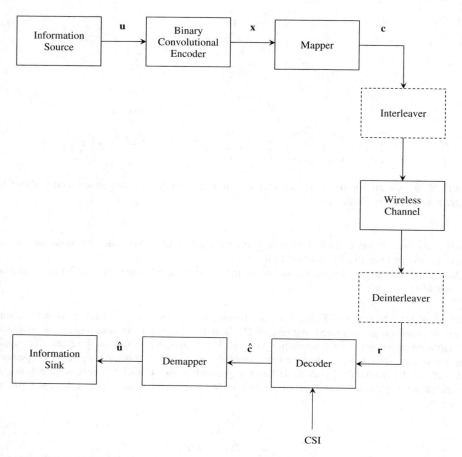

Figure 9.21   Overall block diagram of a wireless communication system employing convolutional encoding.

9.2.3   Maximum Likelihood Decoding of Convolutional Codes

In the foregoing our attention has been focused on the main properties of convolutional codes. We now consider the problem of decoding a convolutional code, referring to the *binary* case for simplicity. Whatever the wireless communication scenario, the problem of ML decoding of a binary convolutional code can be understood by referring to the block diagram of Figure 9.21.

A binary convolutional encoder, with rate $R = k/n$ and starting from a known state, is fed for N consecutive clock intervals by a binary *message* $\mathbf{u} \triangleq [\mathbf{u}_0, \mathbf{u}_1, \ldots, \mathbf{u}_{N-1}]$, generating the binary *codeword* $\mathbf{x} \triangleq [\mathbf{x}_0, \mathbf{x}_1, \ldots, \mathbf{x}_{N-1}]$. Then this codeword is mapped to a symbol vector $\mathbf{c} \triangleq [c_0, c_1, \ldots, c_{P-1}]$, consisting of P complex symbols, each belonging to an M-ary complex signal constellation. Note that in the simplest case the constellation is binary ($M = 2$), so that $P = nN$; when this occurs the simple mapping rule $c = 1 - 2b$ can be used to associate the encoded bit b with a BPSK symbol. Otherwise nN is a multiple of P, since each of the nN bits is transmitted only once[20] via $\mathbf{c}$. Symbol interleaving can also be used to break up error bursts at the receiver. The wireless communication channel

---

[20] Further details on mapping multiple bits to M symbols can be found in Section 11.2, where TCM is described.

introduces uncertainty due to both additive noise and fading. If a reduced-dimensionality model is adopted to parametrize the last impairment, a mathematical description of fading is provided by a parameter vector $\mathbf{h}$ of finite size. The channel output is given by the vector $\mathbf{r} \triangleq [r_0, r_1, \ldots, r_{Q-1}]$, structured as:

$$\mathbf{r} = \mathbf{z}(\mathbf{c}, \mathbf{h}) + \mathbf{n}, \tag{9.229}$$

where $Q = o_s P$, o_s is the oversampling factor (i.e., the number of samples per symbol interval extracted from the received signal), $\mathbf{z}(\mathbf{c}, \mathbf{h}) \triangleq [z_0(\mathbf{c}, \mathbf{h}), z_1(\mathbf{c}, \mathbf{h}), \ldots, z_{Q-1}(\mathbf{c}, \mathbf{h})]$ is the useful component of the received signal (i.e., the portion of $\mathbf{r}$ depending on the symbol vector $\mathbf{c}$ and the channel parameter vector $\mathbf{h}$) and $\mathbf{n} \triangleq [n_0, n_1, \ldots, n_{Q-1}]$ is additive noise. In the following we assume that $\mathbf{n}$ consists of iid complex Gaussian random variables, each having mean zero, variance σ_n^2 and iid real and imaginary parts.

The ML decoding strategy depends on the degree of knowlege about the channel state information available at the receiver. In the following description we assume, for simplicity, that the CSI is *ideally known* to the receiver, so that $\mathbf{h}$ is available to the decoding algorithm. Then the ML decoding strategy can be expressed as the search for the maximum of the likelihood function $f(\mathbf{r} \mid \mathbf{u}, \mathbf{h})$ over the set $\mathcal{C}$ of all the possible codewords:

$$\mathbf{u}_{ML} = \arg\max_{\tilde{\mathbf{u}} \in \mathcal{C}} \, f(\mathbf{r} \mid \mathbf{u} = \tilde{\mathbf{u}}, \mathbf{h}) \tag{9.230}$$

where $f(\mathbf{r} \mid \mathbf{u} = \tilde{\mathbf{u}}, \mathbf{h})$ is the joint pdf of $\mathbf{r}$ conditioned on the channel parameter $\mathbf{h}$ and on the hypothesized message $\tilde{\mathbf{u}}$, and $\mathbf{u}_{ML}$ is the ML estimate of $\mathbf{u}$. The pdf $f(\mathbf{r} \mid \mathbf{u} = \tilde{\mathbf{u}}, \mathbf{h})$ can be evaluated as follows. Since $\mathbf{u}$ is given (i.e., $\mathbf{u} = \tilde{\mathbf{u}}$), all the components of $\mathbf{r}$ are mutually independent (see (9.230)), because of the independence of the noise samples $\{n_l\}$. Then we may write:

$$f(\mathbf{r} \mid \mathbf{u} = \tilde{\mathbf{u}}, \mathbf{h}) = \prod_{l=0}^{Q-1} f(r_l \mid \mathbf{u} = \tilde{\mathbf{u}}, \mathbf{h}), \tag{9.231}$$

where

$$f(r_l \mid \mathbf{u} = \tilde{\mathbf{u}}, \mathbf{h}) = \frac{1}{\sqrt{\pi \sigma_n^2}} \exp\left[-\frac{|r_l - z_l(\tilde{\mathbf{c}}, \mathbf{h})|^2}{\sigma_n^2} \right] \tag{9.232}$$

for $l = 0, 1, \ldots, Q - 1$, and $\tilde{\mathbf{c}}$ represents the symbol vector associated with $\tilde{\mathbf{u}}$. Substituting (9.231) into (9.232) yields:

$$f(\mathbf{r} \mid \mathbf{u} = \tilde{\mathbf{u}}, \mathbf{h}) = \frac{1}{(\pi \sigma_n^2)^{Q/2}} \exp\left[-\frac{|\mathbf{r} - \mathbf{z}(\tilde{\mathbf{c}}, \mathbf{h})|^2}{\sigma_n^2} \right] \tag{9.233}$$

with:

$$|\mathbf{r} - \mathbf{z}(\tilde{\mathbf{c}})|^2 = \sum_{l=0}^{Q-1} |r_l - z_l(\tilde{\mathbf{c}}, \mathbf{h})|^2. \tag{9.234}$$

As will become clearer, from the standpoint of simplifying the decoding process it is useful that the *total* metric (also known as the path metric) for a sequence of symbols is the sum of multiple *partial* metrics, each associated with a given channel input–output symbol pair. Let us now analyze when this result is achieved and its consequences. Taking the natural logarithm of the likelihood function (9.233) and dropping all terms independent of the trial symbol vector $\tilde{\mathbf{c}}$ produces the equivalent metric:

$$m(\mathbf{r}, \tilde{\mathbf{c}}, \mathbf{h}) = |\mathbf{r} - \mathbf{z}(\tilde{\mathbf{c}}, \mathbf{h})|^2 = \sum_{l=0}^{Q-1} |r_l - z_l(\tilde{\mathbf{c}}, \mathbf{h})|^2, \tag{9.235}$$

which has an additive structure. Note that, in principle, the sample z_l ($\tilde{\mathbf{c}}$, $\mathbf{h}$) depends on the complete vector $\tilde{\mathbf{c}}$; however, substantial simplification is achieved when z_l ($\tilde{\mathbf{c}}$, $\mathbf{h}$) depends on $\tilde{c}_l$ only. For instance, this occurs with PAM signaling over a *slow* time-selective fading channel or, dually, with OFDM signaling over a *purely* frequency-selective channel. In both cases:

$$z_l\,(\tilde{\mathbf{c}}, \mathbf{h}) = \tilde{c}_l h_l, \tag{9.236}$$

where h_l denotes the fading distortion affecting the transmitted signal in the lth symbol interval (over the lth subcarrier) in a PAM (OFDM) transmission. Note also that, in the cases mentioned above, usually $Q = P$ (i.e., $o_s = 1$), since one sample per channel symbol is taken at the MF output of a PAM receiver (see Section 4.4.2.1) and one sample per useful subcarrier is available at the FFT output of an OFDM receiver (see Section 4.4.4.1). Then, under the above assumptions, the metric (9.235) can be simplified as:

$$m(\mathbf{r}, \tilde{\mathbf{c}}, \mathbf{h}) = \sum_{l=0}^{P-1} \left| r_l - \tilde{c}_l h_l \right|^2, \tag{9.237}$$

so that the ML estimation problem (9.230) can be reformulated as:

$$\mathbf{u}_{ML} = \arg\min_{\tilde{\mathbf{u}} \in \mathcal{C}} \sum_{l=0}^{P-1} \left| r_l - \tilde{c}_l h_l \right|^2. \tag{9.238}$$

This represents an example of so-called *soft decoding*, since the search for the optimal estimate of the transmitted message involves a set of *real* metrics.

An alternative solution to soft decoding is illustrated in Figure 9.22. In this case the decoder is preceded by a detector, endowed with knowledge of the CSI and usually producing a *hard* estimate $\mathbf{y} = [y_0, y_1, \ldots, y_{N-1}]$ of the codeword $\mathbf{x}$. This estimate is sent to the decoder which searches the set of all the possible messages $\mathbf{u}$, for the one closest, in terms of Hamming distance, to $\mathbf{y}$. In other words, the decoding strategy solves:

$$\hat{\mathbf{u}} = \arg\min_{\tilde{\mathbf{u}} \in \mathcal{C}} d_H\,(\mathbf{y}, \tilde{\mathbf{u}}) \tag{9.239}$$

or, equivalently:

$$\hat{\mathbf{u}} = \arg\min_{\tilde{\mathbf{u}} \in \mathcal{C}} \sum_{l=0}^{N-1} d_H\,(y_l, \tilde{\mathbf{u}}_l)\,. \tag{9.240}$$

This solution is an example of *hard decoding*, since message estimation involves a set of *integer* metrics. Of course, the use of hard decoding involves an energy loss with respect to its soft counterpart (e.g., see [1522] which refers to a specific application involving a $R = 3/4$ binary convolutional code over a Rician fading channel with $K_{dB} = 9$).

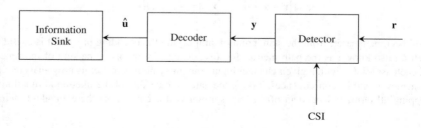

Figure 9.22 Symbol detection followed by hard decoding of a convolutional code.

In both cases considered above, the metric has an additive structure (see (9.238) and (9.240)). For this reason, the search for the optimal message estimate $\hat{\mathbf{u}}$ can be carried out via a computationally efficient algorithm, namely the VA. The VA is a recursive algorithm for determining the optimal state sequence of a discrete-time Markov process observed in memoryless noise [982] and is an efficient tool for many communications problems.

The VA operates on the state trellis which has been defined above for a convolutional code. Let us assume, for simplicity, that the n bits associated with each state transition in the trellis of the given convolutional code are mapped to a single symbol (belonging to a constellation of 2^n points) and consider the case of soft decoding with the metric (9.237). Then, the lth term in the sum on the RHS of (9.237) is associated with a state transition $(\tilde{\sigma}_l \rightarrow \tilde{\sigma}_{l+1})$, that in the lth symbol interval is labeled by the trial symbol $\tilde{c}_l$; such a term represents the *branch metric*:

$$\lambda \left(\tilde{\sigma}_l \rightarrow \tilde{\sigma}_{l+1}\right) \triangleq \left| r_l - \tilde{c}_l h_l \right|^2 \tag{9.241}$$

characterizing that transition (see Figure 9.23). Given this definition, the overall metric (9.237) associated with a given path in the code trellis is given by the sum of the branch metrics labeling the state transitions (i.e., *branches*) whose concatenation forms the path itself. The VA's task is to find the sequence of branches through the trellis with smallest cumulative or path metric (i.e., the shortest path).

The VA accomplishes this task by (see Figure 9.24):

(a) maintaining one *survivor* path per state $\tilde{\sigma}_l$ in the lth symbol interval;
(b) extending these paths one step along all the M branches (labeled by $\tilde{c}_l$) emanating from them;
(c) pruning these back by retaining only the path with the smallest[21] overall metric in each state $\tilde{\sigma}_{l+1}$ (this consists of the sum of the branch metrics labeling the transitions associated with the state transitions forming the path).

In the lth symbol interval, then, the VA keeps track of only one path (the so-called *survivor*) leading to each state $\tilde{\sigma}_l$. Such a path, denoted by $\Pi\left(\tilde{\sigma}_l\right)$ in what follows, is the sequence of consecutive states belonging to the path and is characterized by an *accumulated* or *path metric* $\Lambda\left(\tilde{\sigma}_l\right)$.

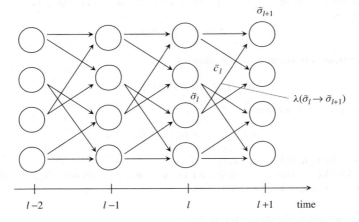

Figure 9.23 State trellis of a convolutional code with mapping ($M = 2^n$ and $n_s = 4$ are assumed).

---

[21] When a metric has to be maximized, the VA should be supplied with the negative metric instead.

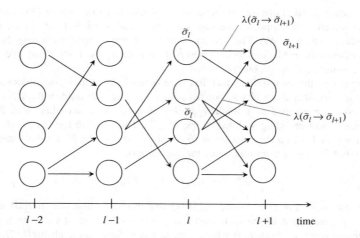

Figure 9.24 Time evolution of the VA. One of the two paths leading to $\tilde{\sigma}_{l+1}$ is selected on the basis of its accumulated metric.

In summary, the VA procedure consists of the following steps (l denotes the time variable):

1. Set:
$$l = 0, \quad \Pi\left(\tilde{\sigma}_0\right) = \left(\tilde{\sigma}_0\right), \quad \Lambda\left(\tilde{\sigma}_0\right) = 0 \tag{9.242}$$
   to initialize the algorithm.
2. Repeat steps 3–7 until $l = N$.
3. Extend path metrics in accordance with:
$$\Lambda\left(\tilde{\sigma}_{l+1}\right) = \Lambda\left(\tilde{\sigma}_l\right) + \lambda\left(\tilde{\sigma}_l \to \tilde{\sigma}_{l+1}\right) \tag{9.243}$$
   for all allowed state transitions $\tilde{\sigma}_l \to \tilde{\sigma}_{l+1}$.
4. For each destination state $\tilde{\sigma}_{l+1}$ find the best (minimum-metric) incoming path from all possible previous states:
$$\tilde{\sigma}_l = \arg \min_{\tilde{\sigma}_l} \Lambda\left(\tilde{\sigma}_{l+1}\right). \tag{9.244}$$
5. Update and store survivor paths (path histories) as:
$$\Pi\left(\tilde{\sigma}_{l+1}\right) = \left(\Pi\left(\tilde{\sigma}_l\right), \tilde{\sigma}_{l+1}\right). \tag{9.245}$$
6. Store the new survivor metrics as:
$$\Lambda\left(\tilde{\sigma}_{l+1}\right) = \Lambda\left(\tilde{\sigma}_l\right) + \lambda\left(\tilde{\sigma}_l \to \tilde{\sigma}_{l+1}\right). \tag{9.246}$$
7. Set $l = l + 1$ (increment time counter).
8. Detect the ML decision for the symbol sequence as that associated with the survivor path $\Pi\left(\tilde{\sigma}_N\right)$ with minimum metric $\Lambda\left(\tilde{\sigma}_N\right)$ (termination).

It is worth noting the following observations:

(a) The most significant part of the decoding procedure is that accomplished in steps 3–5, commonly referred to as *add–compare–select* in the technical literature.

(b) Branch metrics evaluated for state transitions inconsistent with known (i.e., training or pilot) symbols are set to a large value (virtually infinite).

(c) In principle, decisions are not available until time $l = N$. In practice, however, there is little degradation in making decisions after a decision delay of a few times the contraint length of the code by tracing back from the survivor with instantaneously the best metric.

9.2.4   MAP Decoding of Convolutional Codes

An alternative to the ML decoding strategy based on the VA and described in Section 9.2.3 is MAP decoding based on the so-called BCJR algorithm[22] [1523], also known as the *forward–backward algorithm*. The latter name derives from the fact that, as shown below, this algorithm computes the MAP bit or symbol probabilities using a *two-pass* recursive procedure operating on the same trellis as the VA, combining processing results produced in the forward direction with those generated in the backward direction. We now describe this algorithm with reference to the scenario considered in Section 9.2.3 and summarized by Figure 9.21. Conceptually our approach follows that outlined in [533], where the case of a statistically known time-selective channel is analyzed.

The FBA computes the probabilities:

$$\Pr\left\{c_l = \hat{c}|\mathbf{r}, \mathbf{h}\right\} \tag{9.247}$$

for any possible trial symbol $\hat{c}$ transmitted in the lth signaling interval, with $l = 0, 1, \dots, N - 1$. In what follows it is assumed that the convolutional encoder has n_s states and that the cardinality of the symbol constellation is $M = 2^n$, so that a single channel symbol is associated with a state transition in the trellis diagram of the convolutional code and $P = N$. Moreover, it is useful to keep in mind that:

(a) each node in the trellis has M output branches, with a branch corresponding to each of the M channel symbols,

(b) the initial state σ_0 and the final state σ_N of the encoder are known to the decoder, and

(c) the quantities evaluated by the FBA are associated with nodes, states and transitions in the trellis.

The evaluation of the probabilities (9.247) requires the computation of intermediate quantities, known as *state transition probabilities*. Given the state transition $\hat{\sigma}_l \to \hat{\sigma}_{l+1}$ between the trellis states $\hat{\sigma}_l$ and $\hat{\sigma}_{l+1}$ in the lth and $(l + 1)$th symbol intervals, respectively, the corresponding *state transition probability* is denoted by:

$$\Pr\left\{\hat{\sigma}_l, \hat{\sigma}_{l+1} \mid \mathbf{r}, \mathbf{h}\right\} \tag{9.248}$$

and equals zero if the states $\hat{\sigma}_l$ and $\hat{\sigma}_{l+1}$ are not connected. If the states are connected, this quantity can be evaluated as:

$$\Pr\left\{\hat{\sigma}_l, \hat{\sigma}_{l+1} \mid \mathbf{r}, \mathbf{h}\right\} = \frac{\sum_{\tilde{\Gamma} \in S_l(\hat{\sigma}_l, \hat{\sigma}_{l+1})} \Pr\left\{\tilde{\Gamma} \mid \mathbf{r}, \mathbf{h}\right\}}{\sum_{\tilde{\Gamma} \in S} \Pr\left\{\tilde{\Gamma} \mid \mathbf{r}, \mathbf{h}\right\}}, \tag{9.249}$$

where $S \triangleq \{\tilde{\Gamma}\}$ denotes the set of all the possible paths (i.e., sequences of encoder states) for the given convolutional code and $S_l(\hat{\sigma}_l, \hat{\sigma}_{l+1})$ is its subset consisting of all possible paths which traverse the trellis branch connecting states $\hat{\sigma}_l$ and $\hat{\sigma}_{l+1}$ (see Figure 9.25).

---

[22] This algorithm is similar in concept to the method proposed by R. W. Chang and J. C. Hancock for MAP detection in the presence of ISI [584] (see Section 6.2.1.5).

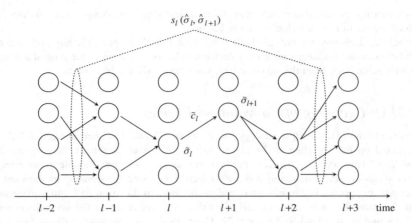

Figure 9.25 Subset $S_l\left(\hat{\sigma}_l, \hat{\sigma}_{l+1}\right)$ consisting of all the paths traversing the trellis branch between the states $\hat{\sigma}_l$ and $\hat{\sigma}_{l+1}$. In this case $M = 2$ (i.e., a binary constellation) and $n_s = 4$ are assumed.

It is useful to note that the probabilities $\{\Pr\{\tilde{\Gamma}|r, h\}\}$ of (9.249) can be expressed via Bayes' theorem as:

$$\Pr\left\{\tilde{\Gamma} \mid r, h\right\} = \frac{f\left(r \mid \tilde{\Gamma}, h\right) \; \Pr\left\{\tilde{\Gamma}\right\}}{f\left(r \mid h\right)}. \tag{9.250}$$

Then substituting (9.250) into (9.249) yields:

$$\Pr\left\{\tilde{\sigma}_l, \tilde{\sigma}_{l+1} | r, h\right\} = \frac{\sum_{\tilde{\Gamma} \in S_l(\tilde{\sigma}_l, \tilde{\sigma}_{l+1})} f\left(r \mid \tilde{\Gamma}, h\right) \; \Pr\left\{\tilde{\Gamma}\right\}}{\sum_{\tilde{\Gamma} \in S} f\left(r \mid \tilde{\Gamma}, h\right) \; \Pr\left\{\tilde{\Gamma}\right\}}. \tag{9.251}$$

The quantities $f(r \mid \tilde{\Gamma}, h)$ and $\Pr\{\tilde{\Gamma}\}$ can be evaluated as follows. The path $\tilde{\Gamma}$ has a one-to-one correspondence with a coded symbol sequence $\tilde{c}$ or, equivalently, with the input information sequence $\tilde{u}$ that has generated it (note that the initial state of the encoder is fixed). Then we may write:

$$\Pr\left\{\tilde{\Gamma}\right\} = \Pr\left\{\tilde{u}\right\} = \prod_{k=0}^{N-1} \Pr\left\{\tilde{u}_k\right\}, \tag{9.252}$$

since independent input symbols are assumed, and:

$$f\left(r \mid \tilde{\Gamma}, h\right) = f\left(r \mid \tilde{u}, h\right) = \prod_{k=0}^{N-1} f\left(r_k \mid r_{k-1}, \tilde{u}, h\right), \tag{9.253}$$

where $r_{k-1} \triangleq [r_0, r_1, \ldots, r_{k-1}]$. If we assume, as in the previous subsection, PAM signaling over a *slow* time-selective fading channel or, dually, OFDM signaling over a *purely* frequency-selective channel (see (9.233)–(9.236)), the sample r_k, given $\tilde{u}$, is independent of r_{k-1}. Moreover, it depends on the channel gain h_k and on the symbol c_k, that is, on the symbol vector $\tilde{u}_k \triangleq [\tilde{u}_0, \tilde{u}_1, \ldots, \tilde{u}_k]$, instead of the complete vector $\tilde{u}$, so that (9.253) can be rewritten as:

$$f\left(r \mid \tilde{\Gamma}, h\right) = \prod_{k=0}^{N-1} f\left(r_k \mid \tilde{u}_k, h_k\right), \tag{9.254}$$

where

$$f\left(r_k \left| \tilde{\mathbf{u}}_k, h_k\right.\right) = \frac{1}{\pi \sigma_n^2} \exp\left[-\frac{\left|r_k - \tilde{c}_k h_k\right|^2}{\sigma_n^2}\right] \tag{9.255}$$

and the coded symbol $\tilde{c}_k$ is generated by the encoder in the kth interval in response to $\tilde{\mathbf{u}}_k$. As a result, the product $f(\mathbf{r}| \tilde{\Gamma}, \mathbf{h}) \Pr\{\tilde{\Gamma}\}$ in (9.251) can be expressed as:

$$f\left(\mathbf{r} \left| \tilde{\Gamma}, \mathbf{h}\right.\right) \Pr\left\{\tilde{\Gamma}\right\} = \prod_{k=0}^{N-1} f\left(r_k \left| \tilde{\mathbf{u}}_k, h_k\right.\right) \Pr\left\{\tilde{\mathbf{u}}_k\right\}. \tag{9.256}$$

Note also that the vector $\tilde{\mathbf{u}}_k$ uniquely identifies a state transition $\tilde{\sigma}_k \to \tilde{\sigma}_{k+1}$, which, in turn, identifies the channel symbol $\tilde{c}_k$. To see this, (9.256) can be rewritten as:

$$f\left(\mathbf{r} \left| \tilde{\Gamma}, \mathbf{h}\right.\right) \Pr\left\{\tilde{\Gamma}\right\} = \prod_{k=0}^{N-1} \gamma_k\left(\tilde{\sigma}_k, \tilde{\sigma}_{k+1}\right), \tag{9.257}$$

where

$$\gamma_k\left(\tilde{\sigma}_k, \tilde{\sigma}_{k+1}\right) \triangleq \Pr\left\{\tilde{\mathbf{u}}_k\right\} W_k\left(\tilde{\sigma}_k, \tilde{\sigma}_{k+1}\right) \tag{9.258}$$

is a *weight function* depending on the trellis branch connecting states $\tilde{\sigma}_k$ and $\tilde{\sigma}_{k+1}$ and:

$$W_k\left(\tilde{\sigma}_k, \tilde{\sigma}_{k+1}\right) \triangleq f\left(r_k \left| \tilde{\mathbf{u}}_k, h_k\right.\right). \tag{9.259}$$

Finally, substituting (9.257) into (9.251) produces:

$$\Pr\left\{\hat{\sigma}_l, \hat{\sigma}_{l+1} \mid \mathbf{r}, \mathbf{h}\right\} = \frac{\sum_{\tilde{\Gamma} \in S_l(\hat{\sigma}_l, \hat{\sigma}_{l+1})} \prod_{k=0}^{N-1} \gamma_k\left(\tilde{\sigma}_k, \tilde{\sigma}_{k+1}\right)}{\sum_{\tilde{\Gamma} \in S} \prod_{k=0}^{N-1} \gamma_k\left(\tilde{\sigma}_k, \tilde{\sigma}_{k+1}\right)}. \tag{9.260}$$

Then the evaluation of the state transition probabilities requires the computation of:

(a) the sum of the products of the weights associated with all the paths containing the branch leaving $\tilde{\sigma}_k$ and entering $\tilde{\sigma}_{k+1}$ (see the numerator),
(b) the sum of the products of the weights associated with all the admissible paths in the trellis (see the denominator).

A computationally efficient method to solve this problem is as follows [533]. Let us define the quantities $\left\{\alpha_k\left(\tilde{\sigma}_k\right)\right\}$ and $\left\{\beta_k\left(\tilde{\sigma}_k\right)\right\}$ through the recursive formula:

$$\alpha_k\left(\tilde{\sigma}_k\right) = \sum_{\tilde{\sigma}_k} \alpha_{k-1}\left(\tilde{\sigma}_{k-1}\right) \cdot \gamma_k\left(\tilde{\sigma}_{k-1}, \tilde{\sigma}_k\right) \tag{9.261}$$

with $k = 1, 2, \ldots, N$, and:

$$\beta_k\left(\tilde{\sigma}_k\right) = \sum_{\tilde{\sigma}_{k+1}} \beta_{k+1}\left(\tilde{\sigma}_{k+1}\right) \cdot \gamma_k\left(\tilde{\sigma}_k, \tilde{\sigma}_{k+1}\right) \tag{9.262}$$

with $k = N - 1, N - 2, \ldots, 0$, respectively. Here it is assumed that $\left\{\alpha_0\left(\tilde{\sigma}_0\right)\right\}$ and $\left\{\beta_N\left(\tilde{\sigma}_N\right)\right\}$ are known initial and end conditions, respectively. The quantity $\alpha_k\left(\tilde{\sigma}_k\right)$ expresses the sum of the products of the weights along all paths originating from all the possible past initial states $\left\{\tilde{\sigma}_0\right\}$ and terminating in $\tilde{\sigma}_k$ in the kth signaling interval. Similarly, $\beta_k\left(\tilde{\sigma}_k\right)$ is the sum of the products of the weights along all paths ending in the terminal states $\{\tilde{\sigma}_N\}$ and originating from $\tilde{\sigma}_k$ in the kth signaling interval. Then the numerator of (9.260) can be expressed as:

$$\chi_l\left(\hat{\sigma}_l, \hat{\sigma}_{l+1}\right) \triangleq \sum_{\tilde{\Gamma} \in S_l\left(\hat{\sigma}_l, \hat{\sigma}_{l+1}\right)} \prod_{k=0}^{N-1} \gamma_k\left(\tilde{\sigma}_k, \tilde{\sigma}_{k+1}\right)$$

$$= \alpha_l\left(\hat{\sigma}_l\right) \cdot \gamma_l\left(\hat{\sigma}_l, \hat{\sigma}_{l+1}\right) \cdot \beta_{l+1}\left(\hat{\sigma}_{l+1}\right), \qquad (9.263)$$

and its denominator is obtained as:

$$\sum_{\tilde{\Gamma} \in S} \prod_{k=0}^{N-1} \gamma_k\left(\tilde{\sigma}_k, \tilde{\sigma}_{k+1}\right) = \sum_{\hat{\sigma}_l, \hat{\sigma}_{l+1}} \chi_l\left(\hat{\sigma}_l, \hat{\sigma}_{l+1}\right), \qquad (9.264)$$

so that:

$$\Pr\left\{\hat{\sigma}_l, \hat{\sigma}_{l+1} \mid \mathbf{r}, \mathbf{h}\right\} = \frac{\chi_l\left(\hat{\sigma}_l, \hat{\sigma}_{l+1}\right)}{\sum_{\hat{\sigma}_l, \hat{\sigma}_{l+1}} \chi_l\left(\hat{\sigma}_l, \hat{\sigma}_{l+1}\right)}. \qquad (9.265)$$

This result shows that all that is needed for the evaluation of the state transition probabilities are the quantities $\left\{\chi_l\left(\hat{\sigma}_l, \hat{\sigma}_{l+1}\right)\right\}$, computed as in (9.263). This, in turn, requires a forward (9.261) and a backward recursion (9.262) involving all the trellis states in each signaling interval, as illustrated in Figure 9.26. It is worth noting that both recursions over all the trellis only need to be performed once.

For demodulation/decoding purposes the quantity of interest[23] is the APP $\Pr\{\hat{c}_l \mid \mathbf{r}, \mathbf{h}\}$ for any possible value of the channel symbol $\hat{c}_l$. This probability can be calculated by summing all the state transition probabilities (9.265) that correspond to branches associated with the symbol $\tilde{c}_l$. In other words, if we define the set $S\left(\hat{c}_l\right)$ of all state transitions $\left(\tilde{\sigma}_l, \tilde{\sigma}_{l+1}\right)$ such that the channel symbol labeling the corresponding branch is $\hat{c}_l$, $\Pr\{\hat{c}_l \mid \mathbf{r}, \mathbf{h}\}$ can be evaluated as:

$$\Pr\left\{\hat{c}_l \mid \mathbf{r}, \mathbf{h}\right\} = \sum_{\left(\tilde{\sigma}_l, \tilde{\sigma}_{l+1}\right) \in S\left(\hat{c}_l\right)} \Pr\left\{\tilde{\sigma}_l, \tilde{\sigma}_{l+1} \mid \mathbf{r}, \mathbf{h}\right\} = \frac{\sum_{\left(\tilde{\sigma}_l, \tilde{\sigma}_{l+1}\right) \in S\left(\hat{c}_l\right)} \chi_l\left(\tilde{\sigma}_l, \tilde{\sigma}_{l+1}\right)}{\sum_{\left(\tilde{\sigma}_l, \tilde{\sigma}_{l+1}\right)} \chi_l\left(\tilde{\sigma}_l, \tilde{\sigma}_{l+1}\right)}. \qquad (9.266)$$

Then the FBA can be summarized as follows:

1. Evaluate the quantities $\left\{\gamma_k\left(\tilde{\sigma}_k, \tilde{\sigma}_{k+1}\right)\right\}$ for all the trellis branches using (9.255) (see (9.259) and (9.258)).
2. Initialize the forward recursion setting:

$$k = 1, \quad \alpha_0\left(\sigma_0\right) = 1. \qquad (9.267)$$

3. Repeat steps 4 and 5 until $k = N$.
4. Compute and store the forward path probabilities $\left\{\alpha_k\left(\tilde{\sigma}_k\right)\right\}$ using (9.261).
5. Set $k = k + 1$.
6. Initialize the backward recursion setting:

$$k = N - 1, \quad \beta_N\left(\sigma_N\right) = 1. \qquad (9.268)$$

7. Repeat steps 8 and 9 until $k = 0$.
8. Compute and store the backward path probabilities $\left\{\beta_k\left(\tilde{\sigma}_k\right)\right\}$ using (9.262).
9. Set $k = k - 1$.
10. For each trellis branch compute the quantity $\chi_l\left(\hat{\sigma}_l, \hat{\sigma}_{l+1}\right)$ using (9.263).
11. Evaluate the APPs $\{\Pr\{\hat{c}_l \mid \mathbf{r}, \mathbf{h}\}\}$ by means of (9.266).

---

[23] Similar considerations apply if we refer to a specific input bit, instead of the output channel symbol.

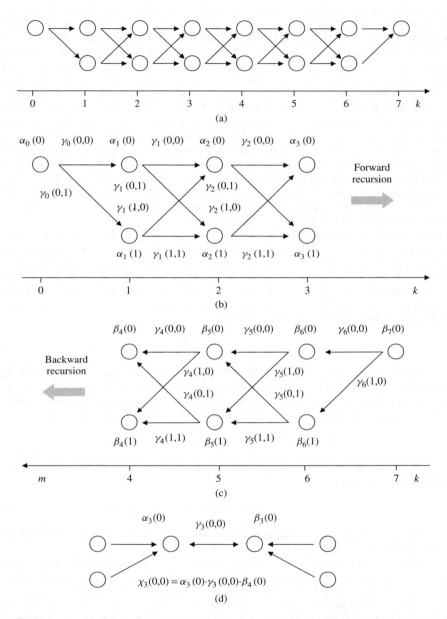

Figure 9.26 (a) Application of the FBA to a two-state trellis ($M = 2$ is assumed). The quantities $\{\alpha_k(\tilde{\sigma}_k)\}$ and $\{\beta_k(\tilde{\sigma}_k)\}$ are computed recursively in (b) the forward and (c) the backward recursion, using the branch probabilities $\{\gamma_k(\tilde{\sigma}_k, \tilde{\sigma}_{k+1})\}$. (d) Finally, the quantities $\{\chi_l(\hat{\sigma}_l, \hat{\sigma}_{l+1})\}$ are evaluated using (9.263).

Note that this algorithm processes the *soft input* information about the information data (i.e., the probabilities $\Pr\{\tilde{u}_k\}$ in (9.258)) and generates the *soft output* information represented by a collection of APPs $\{\Pr\{\hat{c}_l | \mathbf{r}, \mathbf{h}\}\}$. For this reason, it is an example of a SiSo decoding procedure.

The FBA is best suited to short burst transmission, because otherwise its delay and storage requirements are excessive. These problems are avoided if the so-called MAP-FLAs [1086] are adopted. In fact, such algorithms can generate the symbol APPs at a fixed lag from the current received samples, using a single forward pass only. Two optimum (i.e., minimum symbol error probability under a fixed delay constraint) algorithms of this type are available; the newer one, called OSA [1088], has a smaller computational complexity than the older [1086, 1087], since the number of quantities to be stored and recursively updated increases linearly, rather than exponentially, with the decision delay.

Suboptimum FLAs are derived in [1088, 1524]. In particular, the so-called SSA proposed in [1088] does not require knowledge of the noise variance and accomplishes computations in the logarithmic domain. For this reason the add–compare–select procedure described in Section 9.2.3 is the main operation, as when using the VA. Note also that both the MAP-FBA and the above-mentioned MAP-FLAs were first proposed for detection of uncoded signaling on ISI channels. A comparison of these algorithms with the VA in this scenario can be found in [1088].

In the last decade substantial efforts have been made to simplify the BCJR algorithm. This poses serious technical difficulties because of numerical representation problems in the evaluation of probabilities, and the large number of operations (additions and multiplications) required overall. Both problems can be mitigated by carrying out MAP processing in a logarithmic domain for the evaluation of the quantities $\left\{\alpha_l\left(\hat{\sigma}_l\right), \gamma_l\left(\hat{\sigma}_l, \hat{\sigma}_{l+1}\right), \beta_{l+1}\left(\hat{\sigma}_{l+1}\right)\right\}$ in (9.263) (so that the computation of exponentials, like that appearing in (9.255), is avoided) and by adopting the so-called *maximum logarithmic MAP* (Max-Log-MAP) approximation [1524–1527]:

$$\ln\left(e^{\delta_1} + e^{\delta_2} + \ldots + e^{\delta_n}\right) \approx \max_i \delta_i \qquad (9.269)$$

for the *multidimensional* function on the LHS. This approximation leads to degradation in the quality of the soft output, that is, of the APPs $\{\Pr\{\hat{c}_l | \mathbf{r}, \mathbf{h}\}\}$, with respect to the FBA. This can be avoided, at the price of some increase in complexity, by using the so-called log-MAP algorithm, which still operates in a logarithmic domain, but which exploits the exact expressions [1524–1526]:

$$\ln\left(e^{\delta_1} + e^{\delta_2}\right) = \max\left(\delta_1, \delta_2\right) + f_c\left(\left|\delta_2 - \delta_1\right|\right), \qquad (9.270)$$

where $f_c\left(\cdot\right)$ is a proper *correction function* (exhibiting a *one-dimensional* dependence and that can be approximated by a precomputed table), and:

$$\ln\left(e^{\delta_1} + e^{\delta_2} + \ldots + e^{\delta_n}\right) = \max\left(\ln \Delta, \delta_n\right) + f_c\left(\left|\ln \Delta - \delta_n\right|\right) \qquad (9.271)$$

with $\Delta = \sum_{i=0}^{n-1} e^{\delta_i}$, to achieve, at least in principle, an *exact* computation of the LHS of (9.269) via a recursive procedure.

The overall complexity of the BCJR is also proportional to the number of trellis states. Complexity reduction can be achieved by adopting procedures for *simplified trellis search* in the BCJR passes. The MAP algorithms characterized by this feature can be roughly divided into two classes. Those in the first class are inspired by the technique of RSSD[24] for state reduction and are based on the fact that the forward and backward recursions of the BCJR algorithm reduce to the VA when the above Max-Log approximation is adopted. Examples of this approach are the RS-BCJR algorithm of [1097] and the generalized reduced-state algorithms proposed in [1529]. The algorithms in the second class perform a reduced search on the original full-complexity trellis, instead of a full search on a reduced-state trellis; this class includes the M-BCJR algorithm [1098] and the related algorithms of [1099,

---

[24] These techniques were originally proposed to reduce the complexity of the VA [988, 989, 1528] in equalization and trellis decoding.

1530–1533]. Among the reduced-complexity algorithms that do not belong to these two classes, we mention the algorithm described in [1534] and that proposed in [1413]. The former is based on a confidence criterion used to detect reliable symbols early in the decoding process, while the latter has been explicitly developed for the particular case of sparse ISI channels.

A different approach to MAP decoding of convolutional codes is represented by the so-called SOVA proposed in [1535] (and later modified in [1525, 1536]) as an alternative to MAP-FBAs. The SOVA operates on the same trellis as the VA described in Section 9.2.3 and, consequently, can be implemented by simply complementing the VA. However, the SOVA, unlike the VA, is able to generate soft (i.e., reliability) information for the bits it decodes. A detailed description of this algorithm in the trace-back mode can be found in [1100, pp. 435–437]. More recently, it has been proved that the SOVA can be modified in a simple way so that it becomes equivalent to the Max-Log MAP decoding algorithm [1527]. This means that the latter technique can be implemented in a Viterbi-like manner.

9.2.5   Sequential Decoding of Convolutional Codes

In Sections 9.2.3 and 9.2.4 we have described two approaches to the decoding of convolutional codes. In the first case, the decoding process is ML and optimum in a sequence or codeword sense and is implemented through the VA [982, 1478]. In the second case, the decoding process is optimum in a MAP probability sense and may be implemented using the so-called forward–backward algorithm (FBA), otherwise known as the BCJR algorithm [1523]. Unfortunately, both algorithms have exponentially increasing complexity as the constraint length of the code increases. This limits their use to the decoding of short constraint length codes that by definition have a small number of states. In the case of the VA, decoding more than about 256 state codes becomes too complex to be practical, whereas when the BCJR algorithm is used, the limit tends to be lower.[25]

An alternative approach, which allows for the decoding of long constraint length convolutional codes, is to use sequential decoding algorithms. Sequential decoding actually pre-dates both ML and MAP decoding and was first introduced by J. M. Wozencraft [1537]. However, this early work did not directly result in practical decoding algorithms. This was left to the later work of R. Fano [1538], who developed a practical and easily implementable algorithm now known as the *Fano algorithm*. Somewhat later, K. S. Zigangirov [1539] and F. Jelinek [1540] independently discovered a somewhat more elegant algorithm that has come to be known as the *stack algorithm*.

The sequential algorithms are tree-based decoding algorithms. The code trellis described previously in Section 9.2.3 makes use of the remerging properties of the trellis to conduct a parallel search of all possible code sequences or paths through the trellis. However, the trellis can be expanded to a tree form that does not make use of the remerges and results in an expanding structure with a number of possible paths that grows exponentially with the length of the code sequence. The sequential decoding algorithms search the tree by extending only one path at a time through the tree.[26] In all cases the algorithms have the ability to quickly recognize when they are following an incorrect path through the tree. They then retrace their steps to the last instant when decoding was proceeding correctly and then start decoding along an alternative path through the code tree. This leads to a requirement to decode variable-length code paths in the tree. Before describing the overall process in detail, we consider the development of an appropriate metric for the decoding of variable-length paths or code sequences. For the purposes of the current chapter, we will restrict attention to the decoding of binary convolutional codes, although the process can obviously be extended to nonbinary situations.[27]

---

[25] As an aside, the BCJR algorithm is used in the decoding of turbo codes and it is no coincidence that the component codes of most useful turbo codes have 16 or fewer states.

[26] Such algorithms are also referred to as *depth-first* algorithms.

[27] Such an extension is described in the context of TCM by C. Schlegel [326].

Let $x_i^{(j)}$ be the ith bit[28] of the jth transmitted block and let $y_i^{(j)}$ be the corresponding received bit.[29] We also define R as the rate of the convolutional code in use, where we note that in most sequential decoding applications $R = 1/n$. Fano [1538] heuristically suggested that the bit or symbol metric should have the form:

$$M(y_i^{(j)}|\tilde{x}_i^{(j)}) = \log_2 \left[\frac{\Pr\{y_i^{(j)}|\tilde{x}_i^{(j)}\}}{\Pr\{y_i^{(j)}\}} \right] - R, \tag{9.272}$$

where $\tilde{x}_i^{(j)}$ denotes a trial value for $x_i^{(j)}$. In all sequential decoding algorithms, partial path metrics must be computed. This results in the generation of a bias term which is a function of the partial path length. Based on the bit metric of (9.272), it is straightforward to show that the partial path metric for l tree branches is given by [1485]:

$$
\begin{aligned}
M^{(l)}(\mathbf{y}|\tilde{\mathbf{x}}) &= \sum_{i=0}^{l-1} \sum_{j=0}^{n-1} M(y_i^{(j)}|\tilde{x}_i^{(j)}) \\
&= \sum_{i=0}^{l-1} \sum_{j=0}^{n-1} \left\{ \log_2 \left[\frac{\Pr\{y_i^{(j)}|\tilde{x}_i^{(j)}\}}{\Pr\{y_i^{(j)}\}} \right] - R \right\} \\
&= \sum_{i=0}^{l-1} \sum_{j=0}^{n-1} \log_2 \Pr\{y_i^{(j)}|\tilde{x}_i^{(j)}\} + \left[\sum_{i=0}^{l-1} \sum_{j=0}^{n-1} \log_2 \frac{1}{\Pr\{y_i^{(j)}\}} - nlR \right].
\end{aligned} \tag{9.273}
$$

The term in the large brackets on the third line of this equation is the *bias term* and is clearly function of the path length. Moreover, it is easily seen that this metric has the additive form:

$$M^{(l+1)}(\mathbf{y}|\tilde{\mathbf{x}}) = M^{(l)}(\mathbf{y}|\tilde{\mathbf{x}}) + \sum_{j=0}^{n-1} M(y_l^{(j)}|\tilde{x}_l^{(j)}). \tag{9.274}$$

Provided that $R \leq 1$ and $P\{y_i^{(j)}\} \leq 1/2$, the bias term can be shown [1485] to be positive. Moreover, for the classical binary symmetric channel, the metric of (9.273) reduces to the form:

$$M^{(l)}(\mathbf{y}|\tilde{\mathbf{x}}) = \sum_{i=0}^{l-1} \sum_{j=0}^{n-1} \log_2 \Pr\{y_i^{(j)}|\tilde{x}_i^{(j)}\} + nl(1 - R), \tag{9.275}$$

which is a linear function of the path length [1485]. Although Fano's original choice of the metric was heuristic, Massey [1541] later showed that it always causes a sequential decoder to extend the most likely path based on the information available to the decoder at the current time. Thus, most of the time it is a good choice. The exceptions arise when the decoding algorithm extends an incorrect path and has to retrace its steps.

9.2.5.1   The Fano Algorithm

The *Fano algorithm* was first described in detail by R. Fano in [1538]; however, the present description more directly follows from [1485].

---

[28] Throughout this subsection, we work in terms of bits or, equivalently, BPSK modulation. It is straightforward to extend the concepts and implementations to the nonbinary case.

[29] In principle, even in the binary transmission case, $y_i^{(j)}$ may represent either a hard decision on the received bit or a soft estimate assuming that a soft input decoding algorithm is being employed.

The algorithm moves through the code tree as dictated by the partial path metric of (9.273) and by a threshold T, which varies during the decoding process. The algorithm is described by the following steps, assuming that decoding starts at the root node of the code tree:

1. (*Initialization*) Set the threshold $T = 0$. Set the threshold increment Δ (this remains constant throughout the decoding process). Set the partial path metric $M^{(0)} = 0$.
2. At the lth ($l = 0, 1, 2, \ldots$) decoding step, compute the path metric $M^{(l+1)}(\mathbf{y}|\tilde{\mathbf{x}})$ using (9.274) and (9.274) for the next node forward in the tree. Note that there are normally two (or more) branches emanating from each node, one corresponding to each possible input block of n bits, and either (any one) can be picked.
3. Is $M^{(l+1)}(\mathbf{y}|\tilde{\mathbf{x}}) \geq T$? If so, move to the forward node.[30] Is this the decoder's first visit to the node? If so, increase the threshold by Δ, subject to the constraint that the metric remains above the new threshold value, $T + \Delta$; the decoder then moves to the next forward node in the tree and performs step 2 of the algorithm. If it is not the decoder's first visit to the node, no threshold tightening is performed; the decoder then moves the next forward node of the tree and repeats step 2 for the next node forward.
4. If $M^{(l+1)}(\mathbf{y}|\tilde{\mathbf{x}}) < T$, then is $M^{(l)}(\mathbf{y}|\tilde{\mathbf{x}}) \geq T$? If so, the decoder backs up to position l and performs step 2 for one of the alternative forward paths from that position.
5. If $M^{(l)}(\mathbf{y}|\tilde{\mathbf{x}}) < T$, reduce the threshold to $T = T - \Delta$ and the decoder performs step 2 for the next forward node ($l + 1$) by extending an alternative path from that node.[31]

The threshold step size Δ determines the number of node computations performed for a given received codeword or frame. Its setting determines a tradeoff between the decoder's processing rate and the output error rate. It has been observed [35, 1485] that, to eventually select the ML path, T must be lower than the metric for this path. However, if T becomes too low due to a too large value of Δ, several alternative paths may become acceptable. The decoder may then accept an incorrect path, resulting in output bit errors. For this reason, the choice of Δ represents a design tradeoff and its value must in general be determined by simulation.

The Fano algorithm is simple to implement and requires minimal storage. As a result, it has been the algorithm of choice in sequential decoding implementations [1542]. However, it suffers from the drawback that it can visit a given tree node multiple times with different settings of the threshold. Under severe channel conditions, this can lead to excessive computation and indeed overload conditions. This problem can be alleviated to some extent through the use of the so-called stack algorithm [1539, 1540]. This algorithm never visits any tree node more than once and always extends the path with the best metric. We now briefly describe it.

9.2.5.2 The Stack Algorithm

The *stack algorithm* performs a somewhat more efficient sequential search of the code tree, but at the expense of significantly expanded storage. It operates as follows:

1. (*Initialization*) Define an empty stack S and deposit the empty partial code sequence at the top with its metric $M^{(0)}(\mathbf{y}|\tilde{\mathbf{x}}) = 0$.

---

[30] If this new node corresponds to the end of the code block (including any bits or symbols required to flush the encoder), decoding is complete and decoding stops.

[31] This occurs when the threshold is too high for the number of errors in the received word. Also the reduction in T ensures that no node is ever visited twice at the same threshold level. This avoids the occurrence of infinite loops in the decoding algorithm [1485].

2. At the lth decoding step ($l = 0, 1, 2, \ldots$), extend the node corresponding to the top entry or path of S by calculating the partial path metric $M^{(l+1)}(\mathbf{y}|\tilde{\mathbf{x}})$ for all possible extensions[32] of this path using (9.274).

3. Place these new entries into the stack and reorder the stack, so that the path with the best metric is at the top.

4. If the top entry of S is a path to a leaf of the tree, then decoding is complete and the top entry in the stack is the estimate of the transmitted bit sequence.

Note that in the stack algorithm the retracing operation associated with the Fano algorithm is replaced by the reordering operation, but leads to the same effect, in essence backing up and trying alternative paths through the code tree. There are two major problems associated with the stack algorithm:

1. In noisy channel conditions, the received samples will be unreliable and result in many paths having similar metrics. These all have to be stored in the stack and possibly further explored. This can lead to very large storage requirements.

2. Under noisy conditions, the stack becomes very large and the reordering operation becomes very complex.

To avoid the latter problem, a metric quantization scheme can be employed in which all metrics having values within a given quantization range are put into the same so-called "bucket" with no attempt made to sort them. Thus at each decoding instant, the current path is inserted at the top of the appropriate bucket. This is the so-called *stack-bucket algorithm* [1540]. It always extends the top path of the best bucket. It avoids stack reordering, but may not always extend the best path through the tree, which leads to a small degradation in performance.

Under moderate channel conditions, both the Fano and the stack algorithms provide essentially ML performance. However, under severe noise conditions, they tend to suffer from computational overload and can lose entire data frames. This is because the probability distribution of the number of computations per decoded bit follows a *Pareto* distribution, which at low SNRs has infinite moments, and the average number of computations can increase without limit. This is described in detail in [35, 1485].

9.2.6　Error Performance of ML Decoding of Convolutional Codes

In this subsection a general method for the evaluation of performance, in terms of node error probability and bit error probability, is described for the ML decoding strategies described above.[33] As shown below, these performance indexes cannot be evaluated exactly, because of the nonlinear behavior of the VA and thus only performance upper bounds are derived. In our derivations the following initial assumptions are made:

(a) The transmitted codeword $\mathbf{x} \triangleq [\mathbf{x}_0, \mathbf{x}_1, \ldots, \mathbf{x}_{N-1}]$ is generated by an n_s-state binary convolutional encoder (with rate $R = k/n$), starting from the initial state $\sigma_0 = 0$, in response to the message $\mathbf{u} \triangleq [\mathbf{u}_0, \mathbf{u}_1, \ldots, \mathbf{u}_{N-1}]$ consisting of null symbols.

(b) The last m vector components $\{\mathbf{u}_{N-l}, l = 1, 2, \ldots, m\}$ of $\mathbf{u}$ (where m denotes the encoder memory (9.187)) are all *known to the decoder*. This assumption entails that the decoder knows not only the *initial state* ($\sigma_0 = 0$), but also the *final state* reached by the encoder ($\sigma_N = 0$). Therefore, at

---

[32] Note that for a rate-$1/n$ code there will be only two possible extensions, and for a rate-k/n code there will be 2^k.

[33] This method, however, can be easily adapted to any VA-based decoding strategy.

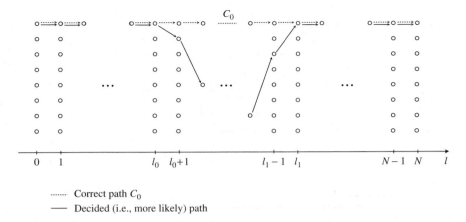

$$\cdots\cdots \text{ Correct path } C_0$$
$$\text{—— Decided (i.e., more likely) path}$$

Figure 9.27 The correct path (dashed line) and the decided (i.e., more likely) path (continuous line) in the trellis of a decoder for an eight-state convolutional code. The occurrence of an *error event* starting (ending) at l_0 (l_1) is shown.

the end of the decoding procedure, it selects, from the n_s available survivor paths, the one ending in the state $\sigma_N = 0$, without making any comparison among their metrics.

Under these assumptions, in the trellis diagram of the given code the *correct path* is represented by the unique horizontal path C_0 emerging from the state $\sigma_0 = 0$ and ending in the state $\sigma_N = 0$, as illustrated in Figure 9.27. At the end of its processing, any ML decoder for the given code selects, from all the possible paths originating from $\sigma_0 = 0$ and ending in $\sigma_N = 0$, the one it deems *most likely*. This path can differ from the correct one, as shown in Figure 9.27, where the decided path separates from the correct one at the instant l_0 and merges again with it at the later instant l_1. When this occurs, it is usually stated that an *error event* occurs at state (or *node*) $\sigma_{l_0} = 0$, and that this event starts (ends) at the instant l_0 (l_1). Note that the presence of an error event in the decoding procedure can entail the presence of a burst of incorrect bits in the estimated bits.

The probability that, during decoding, an error event occurs starting at a given instant l_0 is called the *node error probability* and denoted $P_n[l_0]$. Generally speaking, it depends on l_0. As shown in what follows, the study of this probability provides all the conceptual tools needed to compute an upper bound on the bit error probability P_b. For this reason, first we tackle the problem of estimating $P_n[l_0]$, then that of estimating P_b. To do this, consider Figure 9.28, showing some potentially more likely paths $\{C_i[l_0], i = 1, 2, 3, \ldots\}$ (each associated with a distinct codeword) diverging from the correct path C_0 at l_0 and merging again with it after an finite number of bit or symbol intervals. Generally speaking, the number of error events that can start at l_0 is limited (since, for any l_0 between 0 and $N - 1$, the instant l_i at which the event associated with $C_i[l_0]$ ends cannot exceed N, i.e. the final epoch of the decoding procedure) and depends on l_0.

If $\varepsilon_i[l_0]$ denotes the error event associated with the wrong path $C_i[l_0]$ $(i = 1, 2, 3, \ldots)$, then:

$$\varepsilon[l_0] = \bigcup_i \varepsilon_i[l_0] \tag{9.276}$$

represents the error event occurring when a node error of arbitrary length occurs at l_0, so that the *node error probability* $P_n[l_0]$ at l_0 is given by:

$$P_n[l_0] \triangleq \Pr\{\varepsilon[l_0]\} = \Pr\left\{\bigcup_i \varepsilon_i[l_0]\right\}. \tag{9.277}$$

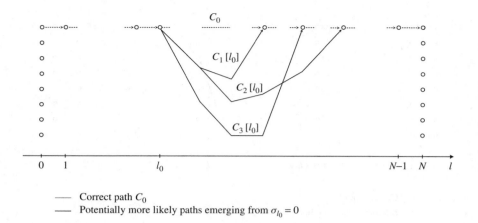

Figure 9.28 The correct path (dotted path) and other more likely paths (continuous lines) separating from it at l_0.

It is important to note that the events $\{\varepsilon_i[l_0], \; i = 1, 2, \ldots\}$ *are not mutually exclusive*, since, as shown in Figure 9.28, incorrect paths can overlap in some clock intervals. This makes an exact evaluation of $P_n[l_0]$ according to (9.277) infeasible. Despite this, applying the well-known *union bound* (4.17) to (9.277) yields the upper bound:

$$P_n[l_0] < \sum_i \Pr\{\varepsilon_i[l_0]\}. \tag{9.278}$$

The use of this expression requires (a) the probabilities $\{\Pr\{\varepsilon_i[l_0]\}, \; i = 1, 2, \ldots\}$ to be evaluated and (b) all the possible incorrect paths associated with the events $\{\varepsilon_i[l_0], \; i = 1, 2, \ldots\}$ to be identified. Problem (a) will be discussed later. Here we limit ourselves to observing that the error $\varepsilon_i[l_0]$ refers to a *binary decision* problem, involving the *pair* of paths C_o and $C_i[l_0]$ only. For this reason, the probability $\Pr\{\varepsilon_i[l_0]\}$ is often called the *pairwise error probability*.[34] As far as problem (b) is concerned, we note that, since (9.278) represents an *upper bound* on $P_n[l_0]$, the intrinsic *nature* of this result remains unchanged if in the sum appearing on its RHS the contributions coming from all the possible paths that would merge in the reference (horizontal path) *after an arbitrary number of bit intervals* are added, as if the length N of the transmitted codeword was *infinite*. This turns (9.278) into the new inequality:

$$P_n < \sum_{i=1}^{+\infty} \Pr\{\varepsilon_i\}, \tag{9.279}$$

which includes in its RHS the contributions of all the incorrect paths separating from the reference path C_0 at the same instant, but remerging with it after an interval of *arbitrary duration*. Note that in (9.279) the parameter l_0 no longer explicitly appears, since, whatever its specific value, the evaluation of the RHS of (9.279) always involves the same set of incorrect paths. Moreover, even if (9.279) inevitably provides a looser result than (9.278), its evaluation is much simpler, as will become clearer later.

---

[34] In the decoding of trellis-based codes, this is also referred to as the *probability of first error* or the *probability of an error event*.

Let us now consider the problem of evaluating the average number of incorrect bits $\bar{n}(l_0)$ generated by an error event starting at l_0. Generally speaking, this parameter is given by[35]:

$$\bar{n}(l_0) \triangleq \sum_{t=1}^{+\infty} t P_{n,t}[l_0],\tag{9.280}$$

where $P_{n,t}[l_0]$ denotes the probability that an error event starting at l_0 generates *exactly* t information bit errors in the estimated data sequence. Following the same line of reasoning as for deriving (9.279) from (9.277), the upper bound (independent of l_0):

$$P_{n,t} < \sum_{i=1}^{+\infty} \Pr\left\{\varepsilon_{t,i}\right\}\tag{9.281}$$

on $P_{n,t}[l_0]$ can be easily derived. Here, the error event $\varepsilon_{t,i}$ is associated, like ε_i in (9.279), with a binary decision problem, since it refers to the selection of a path $C_{t,i}[l_0]$ (in place of C_0) separating from C_0 at a given instant l_0, and remerging with C_0 after an arbitrary number of bit or symbol intervals, but introducing *exactly* t incorrect information bits into the decoded data sequence. Therefore, the set $\{\varepsilon_{t,i}\}$ represents a subset of the events $\{\varepsilon_i\}$ considered in (9.279) and the probability $\Pr\{\varepsilon_{t,i}\}$ is also a PEP. Substituting (9.281) into (9.280) yields:

$$\bar{n} < \sum_{t=1}^{+\infty} \sum_{i=1}^{+\infty} t \Pr\left\{\varepsilon_{t,i}\right\},\tag{9.282}$$

where the dependence on l_0 has been omitted for the reasons mentioned above.

The average number of incorrect information bits is an important parameter in the evaluation of the bit error probability P_b. In fact, *in each bit or symbol interval of the decoding procedure* the noisy data associated with k distinct information bits are received, the associated branch metrics for VA processing are computed and the available survivors are updated. As the updating is done a new error event can start; if this is the case, the average number of incorrect information bits that can appear *in the same clock interval* is upper-bounded by the average number of incorrect information bits $\bar{n}$ generated by the *entire error event* which, generally speaking, lasts for more than one interval. Then P_b is upper-bounded by the ratio between the RHS of (9.282) and k, that is:

$$P_b < \frac{1}{k} \sum_{t=1}^{+\infty} \sum_{i=1}^{+\infty} t \Pr\left\{\varepsilon_{t,i}\right\}.\tag{9.283}$$

It is important to note that this result has been derived by concentrating on what can occur in *a single symbol interval of the decoding algorithm*, as if in any interval a new error event could begin, independently of other error events that occurred in previous intervals or can potentially appear in future intervals. In other words, the *correlation* among multiple error events potentially occurring in the same decoding procedure has been completely ignored. This pessimistic assumption greatly simplifies the evaluation of an upper bound to P_b, but inevitably reduces its accuracy.

The use of the bounds (9.279) for P_n and (9.283) for P_b requires knowledge of the PEPs $\Pr\{\varepsilon_i\}$ and $\Pr\{\varepsilon_{t,i}\}$, respectively. These probabilities depend on the selected modulation format, on the channel model and on the metric employed for the decoding procedure. Before discussing the problem of their evaluation in some specific cases, it is useful to make the following general observations:

---

[35] Note that if N is finite, the number of errors t is upper-bounded, since error events cannot have an arbitrary length. Therefore, $P_{n,t}[l_0]$ becomes null when t exceeds a certain threshold, depending on l_0.

1. The quantity $\Pr\{\varepsilon_i\}$ represents the probability that the decoder selects the incorrect path $C_i[l_0]$ instead of the correct one C_0 (see Figure 9.28), that is, the probability that the *increment in the accumulated metric* of the VA between l_0 and the merging instant l_i along $C_i[l_0]$ is lower than that occurring along C_0 in the same time interval.[36] Similarly for the probability $\Pr\{\varepsilon_{t,i}\}$, which refers to the case of an error event producing exactly t incorrect information bits.

2. In some specific scenarios, the derivation of closed-form expressions for the upper bound to P_n (9.279) involves the use of the transfer function $T(D)$ (9.224). This is not surprising, since this function provides information about the number of paths having a specific Hamming distance from the reference path C_0 of Figure 9.28. Similarly, when the upper bound to P_b (9.283) is evaluated, knowledge of the number of paths characterized by both a certain Hamming distance d_H from C_0 and a certain number n of incorrect information bits is required. Unfortunately, this information is not provided by $T(D)$.

The latter point raises the issue of deriving a more general form of the transfer function. The new form should enumerate the paths separating from the horizontal reference path, merging with it after an arbitrary number of intervals and having specific characteristics in terms of the number of both unit information bits (i.e., incorrect information bits) and unit coded bits (i.e., Hamming distance). This function, known as the *extended transfer function* and denoted by $T(D, I)$, can be evaluated by extending the techniques illustrated for the computation of $T(D)$ in a simple fashion. In fact, after drawing the modified state transition diagram of the given convolutional code (see Example 9.2.5, and, in particular, Figure 9.18), the label $D^{w_{p,q}} I^{v_{p,q}}$ is associated with any state transition $\sigma_l = p \rightarrow \sigma_{l+1} = q$, where D and I are two distinct formal variables, $w_{p,q}$ represents the weight of the n-dimensional *output* vector generated by the encoder for the given transition, and $v_{p,q}$ denotes the weight of the k-dimensional vector of information bits feeding the encoder *input*. Given the new graph, the same techniques illustrated in connection with $T(D)$ can be exploited to find $T(D, I)$. This procedure is illustrated in the following example for the same four-state code considered in Example 9.2.5.

Example 9.2.10 The modified graph for the evaluation of the extended transfer function of the four-state code of Example 9.2.5 is a simple extension of that of Figure 9.18 and is illustrated in Figure 9.29.

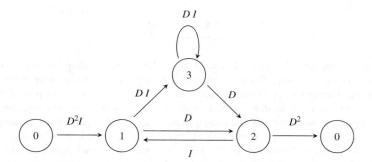

Figure 9.29 Modified graph for the evaluation of the modified transfer function $T(D, I)$ for the four-state convolutional code characterized by the state diagram of Figure 9.14.

---

[36] Note that $C_i[l_0]$ and C_0 coincide up to l_0, so that the accumulated metrics associated with these paths are initially the same.

The reader can verify that simplifying the graph as in Example 9.2.8 leads to the function:

$$T(D, I) = \frac{D^5 I}{1 - 2DI}. \tag{9.284}$$

Evaluating the division on the RHS yields the series:

$$T(D, I) = D^5 I + 2D^6 I^2 + 4D^7 I^3 + \ldots = \sum_{d=5}^{+\infty} 2^{d-5} D^d I^{d-4}. \tag{9.285}$$

This result reveals, for instance, the presence of a single path having distance 5 and a single incorrect information bit, and of two paths having distance 6 and 2 incorrect information bits.

□

Note that, generally speaking, the extended transfer function of a binary convolutional code is given by:

$$T(D, I) = \sum_{t=1}^{+\infty} \sum_{d=d_{free}}^{+\infty} n(d, t) D^d I^t, \tag{9.286}$$

where $n(d, t)$ is the number of events in the set $\{\varepsilon_{t,i}\}$ introducing exactly t incorrect information bits and whose associated paths have a Hamming distance d from the reference path C_0. The function $T(D, I)$ is a direct extension of $T(D)$ and can be easily derived from it. In fact, we have:

$$T(D) = T(D, I)|_{I=1} \tag{9.287}$$

and, in addition:

$$n(d) = \sum_{t=1}^{+\infty} n(d, t). \tag{9.288}$$

We now discuss the application of these analytical tools to a specific communication scenario, characterized by the following relevant features:

1. BPSK signaling is employed. This means that the binary codeword $\mathbf{x} \triangleq [\mathbf{x}_0, \mathbf{x}_1, \ldots]$ is mapped into the symbol vector $\mathbf{c} \triangleq [\mathbf{c}_0, \mathbf{c}_1, \ldots]$, where $\mathbf{c}_l = [c_l^{(0)}, c_l^{(1)}, \ldots, c_l^{(n-1)}]$, $c_l^{(p)} = 1 - 2x_l^{(p)}$ (so that $c_l^{(p)} \in \{\pm 1\}$), with $l = 0, 1, \ldots$ and $p = 0, 1, \ldots, n-1$.
2. The channel is affected by *slow frequency-flat Rayleigh fading* and the receiver processes the ISI-free sample vector $\mathbf{r} \triangleq [\mathbf{r}_0, \mathbf{r}_1, \ldots]$, where $\mathbf{r}_l = [r_l^{(0)}, r_l^{(1)}, \ldots, r_l^{(n-1)}]$:

$$r_l^{(p)} = a_l^{(p)} c_l^{(p)} + n_l^{(p)} \tag{9.289}$$

and $a_l^{(p)}$ and $n_l^{(p)}$ are the fading distortion and noise affecting $c_l^{(p)}$ ($l = 0, 1, \ldots$ and $p = 0, 1, \ldots, n-1$). Note that $a_l^{(p)}$ is a complex Gaussian variable having zero mean and unit variance, and that the sequence of noise samples $\{n_l^{(p)}\}$ consists of iid complex Gaussian random variables, all with mean zero and variance $1/(\bar{E}_s/N_0) = 1/(R\bar{E}_b/N_0)$.
3. *Bit level interleaving* is used at the output of the convolutional encoder and its depth is large enough to ensure the statistical independence of all the fading samples $\{a_l^{(p)}\}$.
4. The receiver is endowed with ideal CSI, that is, it knows perfectly the fading gains $\{a_l^{(p)}\}$; for this reason coherent detection strategies can be adopted.

In this scenario two different approaches to decoding the convolutional code can be envisaged. In the first a hard decision $\hat{x}_l^{(p)}$ is taken on $x_l^{(p)}$ processing $r_l^{(p)}$ only, after compensating for the channel distortion $a_l^{(p)}$. Then the VA is fed by the codeword estimate $\mathbf{y} \triangleq [\mathbf{y}_0, \mathbf{y}_1, \ldots]$ and operates according to a hard decoding strategy. In the second case, the VA is fed by the soft data vector $\mathbf{r}$ and by the CSI vector $\mathbf{a} \triangleq [\mathbf{a}_0, \mathbf{a}_1, \ldots]$, with $\mathbf{a}_l = [a_l^{(0)}, a_l^{(1)}, \ldots, a_l^{(n-1)}]$ and $l = 0, 1, \ldots$, and adopts a soft input decoding strategy. These two situations are now analyzed in detail.

Hard Decoding

In this case the communication channel can be modeled as a BSC with transition probability (see [1422, p. 774]):

$$p = \frac{1}{2}\left(1 - \sqrt{\frac{R\frac{\bar{E}_b}{N_0}}{1 + R\frac{\bar{E}_b}{N_0}}}\right).$$ (9.290)

Moreover, it can be proved (e.g., see [1485, pp. 304–310]) that:

$$\Pr\{\varepsilon_i\} < B^{d_{H,i}},$$ (9.291)

where $d_{H,i}$ is the Hamming distance between the codeword associated with the reference path C_0 and that with the incorrect path C_i, and:

$$B \triangleq \sqrt{4p(1-p)}$$ (9.292)

is the so-called *Bhattacharyya parameter*. The behavior of B for $0 \leq p \leq 1$ is illustrated in Figure 9.30, which shows that, for $p < 1/2$, B is always smaller than unity. When this occurs, the bound on $\Pr\{\varepsilon_i\}$ expressed by the RHS of (9.291) reduces as $d_{H,i}$ gets larger. This suggests that, at large SNRs, the more likely error events are those associated with the paths closer to C_0 (i.e., having minimum Hamming distance from it). Note also that, since $(1-p) \leq 1$, B satisfies the inequality:

$$B \leq 2\sqrt{p},$$ (9.293)

which becomes an equality for $p = 0$ only; this bound represents an accurate estimate of B if p is small, as seen in Figure 9.30.

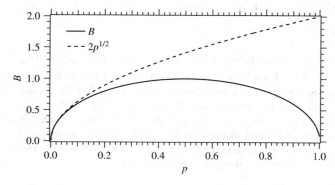

Figure 9.30 The *Bhattacharyya parameter* B (9.292) and its upper bound (9.293) versus the transition probability p of the BSC.

Then substituting (9.291) into (9.279) yields the bound:

$$P_n < \sum_{i=1}^{+\infty} B^{d_{H,i}}, \tag{9.294}$$

which can easily be rewritten as:

$$P_n < \sum_{d=d_{free}}^{+\infty} n(d) B^d, \tag{9.295}$$

by grouping, in the set $\{C_i, i = 1, 2, \ldots\}$, all paths having the *same* Hamming distance from C_0. Finally, taking into account (9.224), (9.295) can be expressed more compactly as:

$$P_n < T(D)|_{D=B}. \tag{9.296}$$

A similar line of reasoning can be followed to evaluate the bound (9.283) on P_b. It is not difficult to prove that the sum $\sum_{i=1}^{+\infty} t \Pr\{\varepsilon_{t,i}\}$ is upper-bounded by:

$$\sum_{d=d_{free}}^{+\infty} tn(d, t) B^d, \tag{9.297}$$

so that:

$$P_b < \frac{1}{k} \sum_{t=1}^{+\infty} \sum_{d=d_{free}}^{+\infty} tn(d, t) B^d. \tag{9.298}$$

From (9.286) it is easily inferred that:

$$\frac{\partial}{\partial I} T(D, I) = \sum_{t=1}^{+\infty} \sum_{d=d_{free}}^{+\infty} tn(d, t) D^d I^{t-1}, \tag{9.299}$$

so that:

$$\left.\frac{\partial}{\partial I} T(D, I)\right|_{I=1} = \sum_{t=1}^{+\infty} \sum_{d=d_{free}}^{+\infty} tn(d, t) D^d. \tag{9.300}$$

Then from (9.298) it can be seen that:

$$P_b < \frac{1}{k} \frac{\partial}{\partial I} T(D, I)\Big|_{I=1,\ D=B}. \tag{9.301}$$

It is interesting to note that, for large SNRs, p (9.290) can be approximated as $1/(2R\bar{E}_b/N_0)$, so that $B \simeq \sqrt{2/(R\bar{E}_b/N_0)}$ (see (9.293)). This result and (9.298) suggest that, for large SNRs, P_b decreases inversely with the $d_{free}/2$th power of $\bar{E}_b/N_0$. In other words, $d_{free}/2$ represents the *achievable diversity order* provided by hard decoding of a binary convolutional code assuming a frequency-flat fading channel and ideal interleaving.

Let us now apply the analytical results derived above to a specific coding scheme.

Example 9.2.11 The transfer function of the four-state binary convolutional code described by the state diagram of Figure 9.14 is given by (9.222). Then the bound (9.296) becomes:

$$P_n < \frac{B^5}{1 - 2B}. \tag{9.302}$$

The extended transfer function for the same code is given by (9.284), so that:

$$\frac{\partial}{\partial I} T(D, I) = \frac{D^5}{(1 - 2DI)^2}.$$
(9.303)

Substituting (9.303) into (9.301) yields the bound:

$$P_b < \frac{B^5}{(1 - 2B)^2}.$$
(9.304)

□

Soft Decoding

In this case the probability $\Pr\{\varepsilon_i\}$ of (9.279) is given by:

$$\Pr\{\varepsilon_i\} = E_{\mathbf{a}} \left\{ Q \left(\sqrt{R \frac{\bar{E}_b}{N_0}} \frac{|\mathbf{A}(\mathbf{c} - \mathbf{c}^{(i)})|}{\sqrt{2}} \right) \right\},$$
(9.305)

since it refers to a binary decision problem, in which the Euclidean distance between the useful signal component of $\mathbf{r}$ associated with C_0 and that referring to C_i is $|\mathbf{A}(\mathbf{c} - \mathbf{c}^{(i)})|$, where $\mathbf{A} = \mathrm{diag}(\mathbf{a})$ is the diagonal matrix of the fading gains and $\mathbf{c}^{(i)}$ is the BPSK symbol vector associated with C_i. It is easy to prove that:

$$|\mathbf{A}(\mathbf{c} - \mathbf{c}^{(i)})|^2 = 4 \sum_{(l,p) \in \Delta} |a_l^{(p)}|^2,$$
(9.306)

Δ being the set of values for the indexes (l, p) where the two vectors $\mathbf{c}$ and $\mathbf{c}^{(i)}$ differ. Since the number of differences between $\mathbf{c}$ and $\mathbf{c}^{(i)}$ is equal to $d_H(\mathbf{c}, \mathbf{c}^{(i)})$, the sum $\sum_{(l,p) \in \Delta} |a_l^{(p)}|^2$ is a central chi-square random variable with $2d_H(\mathbf{c}, \mathbf{c}^{(i)})$ degrees of freedom. Let $\chi^2_{2d_H(\mathbf{c}, \mathbf{c}^{(i)})}$ denote such a variable, so that the probability in (9.305) can be put in the form:

$$\Pr\{\varepsilon_i\} = E_{\chi^2_{2d_H}} \left\{ Q \left(\sqrt{R \frac{2\bar{E}_b}{N_0} \chi^2_{2d_H(\mathbf{c}, \mathbf{c}^{(i)})}} \right) \right\},$$
(9.307)

showing its dependence on $\bar{E}_b / N_0$ and on $d_H(\mathbf{c}, \mathbf{c}^{(i)})$ only (here $E_{\chi^2_{2d_H}}(\cdot)$ is a shorthand notation to indicate a statistical average with respect to the random variable $\chi^2_{2d_H(\mathbf{c}, \mathbf{c}^{(i)})}$). Then, following the same line of reasoning as adopted for hard decoding, the bound (9.279) on P_n can be rewritten as:

$$P_n < \sum_{d=d_{free}}^{+\infty} n(d) E_{\chi^2_{2d}} \left\{ Q \left(\sqrt{R \frac{2\bar{E}_b}{N_0} \chi^2_{2d}} \right) \right\}.$$
(9.308)

Similarly, the bound (9.283) on P_b can be put in the form:

$$P_b < \frac{1}{k} \sum_{t=1}^{+\infty} \sum_{d=d_{free}}^{+\infty} t \, n(d, t) E_{\chi^2_{2d}} \left\{ Q \left(\sqrt{R \frac{2\bar{E}_b}{N_0} \chi^2_{2d}} \right) \right\}.$$
(9.309)

Both these bounds require the evaluation of the same statistical average over $\chi^2_{2d_H(\mathbf{c}, \mathbf{c}^{(i)})}$. Exploiting the analytical results available in [1422, pp. 780–781], it can be shown that:

$$E_{\chi^2_{2d}} \left\{ Q \left(\sqrt{R \frac{2\bar{E}_b}{N_0} \chi^2_{2d}} \right) \right\} = \left[\frac{1}{2}(1 - \mu) \right]^d \sum_{k=0}^{d-1} \binom{d-1+k}{k} \left[\frac{1}{2}(1 + \mu) \right]^k,$$
(9.310)

where

$$\mu \triangleq \sqrt{\frac{R\frac{\bar{E}_b}{N_0}}{1 + R\frac{\bar{E}_b}{N_0}}}. \tag{9.311}$$

It is interesting to note that, if $R\bar{E}_b/N_0 \gg 1$ (i.e., the SNR is large), $(1+\mu)/2 \simeq 1$ and $(1-\mu)/2 \simeq 1/(4R\bar{E}_b/N_0)$ in (9.310). Then, since:

$$\sum_{k=0}^{d-1}\binom{d-1+k}{k} = \binom{2d-1}{d}, \tag{9.312}$$

(9.310) can be simplified as:

$$E_{\chi_{2d}^2}\left\{Q\left(\sqrt{R\frac{2\bar{E}_b}{N_0}\chi_{2d}^2}\right)\right\} \simeq \binom{2d-1}{d}\left(\frac{1}{4R\frac{\bar{E}_b}{N_0}}\right)^d. \tag{9.313}$$

This result and (9.309) together show that, for large SNRs, the bit error rate decreases inversely with the d_{free}th power of $\bar{E}_b/N_0$, so that the *achievable diversity order* provided by soft decoding is given by d_{free}.

Let us now consider some error performance results for a specific coding scheme.

Example 9.2.12 Some BER results referring to the hard and soft decoding of the four-state code of Example 9.2.5 are illustrated in Figure 9.31. AWGN and uncorrelated flat fading channels are considered and the upper bound on hard decoding performance based on (9.304) is shown for comparison.

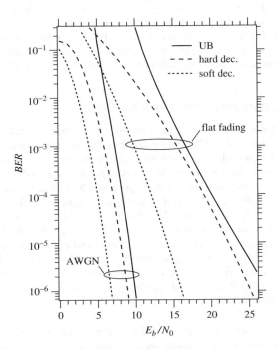

Figure 9.31 BER performance versus the average received SNR per bit (E_b/N_0) for hard and soft decoding of the four-state convolutional code of Example 9.2.5. The upper bound (UB) based on (9.304) is also shown for comparison.

Note that for the flat fading case the parameter p of (9.292) is given by (9.290), whereas for an AWGN channel:

$$p = \frac{1}{2} Q \left(\sqrt{R \frac{2E_b}{N_0}} \right). \tag{9.314}$$

These results demonstrate that:

(a) the SNR gap between hard and soft decoding, which is about 2 dB for an AWGN channel, becomes substantially larger in the presence of fading,
(b) the upper bound (9.304) becomes tighter as SNR increases in the presence of AWGN, but exhibits a different behavior in the presence of fading,
(c) in the presence of flat fading the BER curves for hard and soft decoding exhibit different slopes, since the achievable diversity order of the latter is twice that of the former,
(d) the performance gap between the two hard decoding curves is very large, and becomes substantially smaller when soft decoding is used.

□

9.3   Classical Concatenated Coding

Powerful error-correcting coding schemes can be constructed by combining or *concatenating* two or more *component* (convolutional or block) channel codes. This approach, introduced by G. D. Forney in 1966 [1455], offers the significant advantage of achieving a large minimum Hamming distance with limited decoding complexity. In fact, decoding is usually accomplished in a suboptimal fashion by combining the decoding algorithms of the component codes. In this section, a short introduction to code concatenation and a description of some classical concatenated coding schemes are provided. The study of modern concatenated coding schemes is deferred to Chapter 10.

9.3.1   Parallel Concatenation: Product Codes

In a parallel concatenated coding scheme, multiple component codes are fed by the same input bits, albeit possibly in a different order. Then, generally speaking, the resulting overall codeword derives from the combination of parts of the codewords generated by the encoding algorithms of the component codes. This class of codes is exemplified by the so-called *product codes* (PCs), which we analyze in a two-dimensional context. A two-dimensional PC is generated by concatenating an (n_1, k_1) block code C_1 with a second (n_2, k_2) block code C_2, *both defined on the same field* GF(q) and having minimum Hamming distance d_1 and d_2, respectively. The encoding algorithm operates as follows (see Figure 9.32). First a $k_1 k_2$ symbol message in stored in a matrix with k_2 rows and k_1 columns. Then each row is encoded according to C_1, so that a matrix of k_2 rows and n_1 columns is generated. To each of these columns we next apply the encoding algorithm of C_2. This produces a two-dimensional array of size $n_1 n_2$, representing the codeword of a *two-dimensional code C with parity constraints on rows and columns* and having overall rate $R = (k_1 k_2)/(n_1 n_2) = R_1 R_2$.

Note the following observations:

1. The resulting product codeword consists of four different sections: one containing the message section, one of row parity check symbols, one of column parity check symbols and one of *parity on parity* (PoP) symbols, as illustrated in Figure 9.32.
2. A two-dimensional code C can be seen as an example of *parallel concatenation*, since the codeword generation mechanism involves two distinct encoding algorithms, fed by the same input bits, even

if in a different order. Thus, PCs have inherent block interleaving between the encoders due to the encoding of rows and then columns (or columns and then rows). This "interleaver" can be seen as a rows-in-columns-out block interleaver (π). Therefore, the encoder structure of a PC, when PoP bits are transmitted, can be represented as shown in Figure 9.33.

3. Encoding of C retains the same complexity as that of the constituent codes.
4. The minimum Hamming distance of C is $d_{H,\min} = d_1 d_2$.

As far as decoding and achievable performance are concerned, it is important to note that the *guaranteed error-correction capability* of C is given by (see (9.29)):

$$t = \left\lfloor \frac{d_1 d_2 - 1}{2} \right\rfloor \tag{9.315}$$

if ML decoding is employed. In practice, however, decoding is usually accomplished in a suboptimal fashion by resorting to a simple two-step procedure, which benefits from the availability of efficient algorithms for the decoding of the constituent codes. In fact, row decoding is first done to remove at least a fraction of the incorrect a bits (of course, incorrect decoding is possible). Then the columns are decoded, as shown in the following example.

Example 9.3.1 Let us consider a PC constructed using the same block code for both row and column encoding, namely the well-known $(7, 4)$ Hamming linear block code over GF(2), described in Example

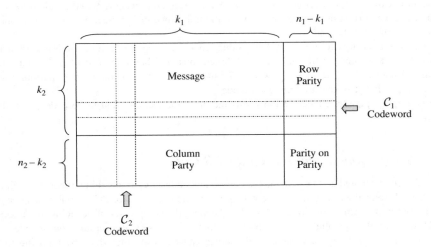

Figure 9.32 Two-dimensional product code with component codes C_1 and C_2.

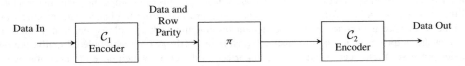

Figure 9.33 Encoder structure of a two-dimensional product code when the PoP symbols are transmitted.

9.1.1. Let us assume that the all-zero codeword has been transmitted and that the error pattern:

$$
\begin{array}{ccccccc}
0 & 0 & 0 & 0 & 0 & 1 & 0 \\
0 & 0 & 0 & 0 & 0 & 0 & 1 \\
0 & 1 & 0 & 0 & 0 & 0 & 0 \\
0 & 0 & 0 & 1 & 0 & 0 & 0 \\
0 & 0 & 1 & 0 & 0 & 0 & 0 \\
0 & 1 & 0 & 0 & 0 & 0 & 0 \\
0 & 0 & 0 & 0 & 1 & 0 & 0 \\
\end{array}
\tag{9.316}
$$

has been received. If we decode the rows first, then all seven errors can be corrected since there is only one error per row, and the constituent codes are single error-correcting codes (with minimum Hamming distance $d_1 = d_2 = 3$). Note, however, that the given PC has minimum distance $d_{H,\min} = 3 \times 3 = 9$, so that only the correction of the error patterns of weight 4 or less is guaranteed, provided that ML decoding is employed. Despite this, our two-step decoding procedure can correct some specific patterns containing significantly more errors as shown above.

□

Generally speaking, the two-step decoding procedure mentioned above is unable to correct all the error patterns with t or fewer errors in the array. A more refined procedure based on multiple decoding steps and exchange of soft information between them is illustrated in Section 10.5, where the PC is treated as a form of turbo code.

Finally, it is important to point out that product codes can also be designed in a multidimensional context. If the PoP bits are transmitted, a D-dimensional PC has rate $R = \prod_{l=1}^{D} R_l$ and minimum Hamming distance $d_{H,\min} = \prod_{l=1}^{D} d_l$, where R_l and d_l are the code rate and minimum Hamming distance, respectively, of the lth component code.

The concept of parallel concatenation forms the basis for the so-called turbo codes, considered in Chapter 10.

9.3.2 Serial Concatenation: Outer RS Code

In a serially concatenated coding scheme the component codes are arranged in a pipeline, as illustrated in Figure 9.34, referring to one-level concatenation for simplicity. As for PCs, serial concatenation can be seen as a tool for generating long codes while keeping the complexity of decoding at an acceptable level. However, unlike parallel concatenation, the component codes of a serially concatenated scheme are usually *defined over different fields*. This choice can be easily motivated by referring to Figure 9.34, where the data sequence feeds an (n_o, k_o) encoder, called an *outer encoder*, generating n_o-dimensional vectors (i.e., codewords) over $GF(q^{k_i})$. In this scheme each symbol in a codeword of the outer code is represented as a k_i-tuple of elements over $GF(q)$; these k_i-tuples, after interleaving (π), feed an (n_i, k_i) encoder, called an *inner encoder*, whose symbols are compatible with the modulator and the channel. Decoding of these concatenated codes proceeds from the inside out. In other words, any inner codeword is decoded in an independent fashion and each k_i-tuple is sent to a deinterleaver (π^{-1}), at the output of which n_o-dimensional vectors over $GF(q^{k_i})$ are assembled. Finally, the outer decoder processes these vectors, attempting to correct all the errors which have not been removed by the inner decoder. Note that error bursts appearing at the output of the inner decoder are potentially broken up by the deinterleaver. Despite this, multiple errors can appear in the long $n_o k_i$-dimensional vectors over $GF(q)$ processed by the outer decoder. For this reason, a powerful low-rate code (e.g.,

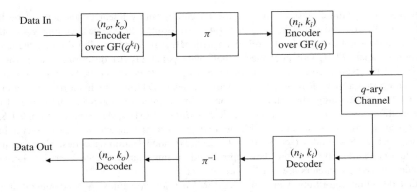

Figure 9.34   Serially concatenated coding and decoding.

a convolutional code with large constraint length) is usually selected for the inner code, whereas the outer code is normally a high-rate RS code, which lends itself to the correction of long error bursts. This approach was adopted, for instance, in the coding standard adopted by NASA/ESA in 1987 for deep-space missions [1543] and in the ETSI DVB-T standard [595]. The use of TCM was proposed for inner coding by R. H. Deng and D. J. Costello in [1544, 1545]. In addition, the problem of the design and decoding of the serial concatenation of interleaved block and convolutional codes was investigated in a unifying perspective by S. Benedetto *et al.* [1546]. This topic will be further discussed in Chapter 10.

9.4   Historical Notes

The birth of channel coding and *forward error correction* coding dates back to the publication of Shannon's seminal work [1420, 1421] in 1948. Shannon proved that, on a noisy channel, if a long code having a rate smaller than the channel capacity is selected *in a random fashion*, then there exists a decoder such that, with high probability, the resulting communication scheme will achieve highly reliable transmission, that is, a low probability of decoding error. However, Shannon did not give any indication as to how such coding/decoding schemes could be devised. Solving this problem has been the goal of intense research efforts in the last half century, during which various practical coding schemes approaching channel capacity on well-understood channels have been devised. In this section a brief history of traditional channel coding schemes is given. We begin with *algebraic coding*, which dominated the first two decades of research; then we consider a different line of development, referring to *probabilistic coding*, which encompasses convolutional codes, product codes and concatenated codes. For each area only the most important results are cited. Additional information can be found in [1465, 1547].

9.4.1   Algebraic Coding

Algebraic coding schemes were the main topic of research in the first decades of channel coding history, as shown by the content of the various important textbooks on coding [1490, 1510, 1548–1551], mainly referring to *linear block codes over finite fields*. The first significant results in this area concerned *binary* coding schemes devised for a power-limited AWGN channel. In particular, the first nontrivial code to appear in the literature was the well-known $(7, 4, 3)$ binary code, developed by R. Hamming and mentioned by Shannon in his original paper [1420, 1421]. This code belongs to an infinite class of single-error-correcting binary linear codes devised by the same researcher [1469] and illustrated in Section 9.1.4.4. Unfortunately, the real coding gain offered by these schemes does not exceed about 3 dB, even with ML soft decoding. Shortly after the appearance of Shannon's paper, M. Golay published a half-page paper [1470] describing the perfect binary linear (23, 12) triple-error-correcting code (having $d_{H,\min} = 7$) described in Section 9.1.4.7. In 1954 D. Muller proposed a new class of error-correcting codes [1471], to which shortly thereafter I. Reed added an efficient decoding algorithm based on a simple majority-logic decoding rule [1472]; for this reason this class of codes is known as *Reed–Muller* (RM) codes (see Section 9.1.5.2). RM codes form an infinite class of codes with flexible parameters that encompasses several important subclasses, such as SPC codes (see Section 9.1.5.1), extended Hamming codes, biorthogonal codes and self-dual codes. For this reason, they represented an important advance over the Hamming and Golay codes, whose parameters are much more restrictive.

The first decoding algorithms were all based on *hard decision* rules. A fundamental step in this area was the introduction of *soft decision* decoding, that is, decoding strategies that take into account the reliability of received channel outputs. The first strategy of this type was the *Wagner decoding* algorithm described in [1502] and attributed to C. A. Wagner. This is an optimum strategy (in the sense of attaining the minimum Euclidean distance) for the special class of $(n, n - 1)$ SPC codes with $d_{H,\min} = 2$ and is substantially simpler than exhaustive minimum-distance decoding.

In the 1960s further important steps were made in the development of new algebraic block codes. At that time there was strong interest in designing codes with a guaranteed minimum distance $d_{H,\min}$ and whose algebraic structure offered the possibility of efficient error-correction algorithms. Research efforts led to the development of the class of *cyclic codes* which, thanks to their nice algebraic structure, admitted simple encoding and decoding procedures based on the use of cyclic shift registers (see Section 9.1.4).

Cyclic codes were first investigated by E. Prange in 1957 [1473] and became the primary focus of research after the publication of W. Peterson's pioneering text in 1961 [1490]. The first important results in this research area were the following:

1. The discovery of BCH and of RS codes in three independent papers in 1959 and 1960 [1474–1476] (see also [1552]); the cyclic structure of these codes was recognized by Peterson in 1960 [1509].
2. The work of D. Gorenstein and N. Zierler [1511], who in 1961 devised an efficient decoding procedure, in the spirit of that proposed by Peterson in the coded case [1509], for an obvious generalization over $GF(q^{k_i})$ of binary Bose–Chaudhuri codes. This procedure is commonly referred to as the *Peterson–Gorenstein–Zierler* algorithm.

Binary BCH codes include a large class of *t*-error-correcting cyclic codes and are the most important class of binary algebraic block codes. Despite this, they have not found many practical applications, except as *cyclic redundancy check* codes for error detection in *automatic repeat request* strategies. In contrast, nonbinary RS codes have proved to be highly useful in practice (although not necessarily in cyclic form) and, as shown in Section 9.1.4.6, are optimum in the sense of achieving the so-called *Singleton bound*. A fundamental property of RS and BCH codes is the availability of efficient algebraic decoding algorithms on a finite-field arithmetic. The development of such algorithms was one of the most active research areas of the 1960s. Significant contributions to this research area were made by Berlekamp, Peterson and Massey [1509, 1510, 1512, 1553, 1554]. In particular, Peterson developed an error-correction algorithm with complexity on the order of $d_{H,\min}^3$ [1509], whereas Berlekamp proposed an error-correction algorithm with complexity of order $d_{H,\min}^2$ [1510, 1554]. This became the standard for the next decade and is known as the *Berlekamp–Massey algorithm*, since it was interpreted by Massey [1512] as a strategy for finding the shortest linear feedback shift register that can generate a certain sequence. It was shown later that these algorithms could be extended to correct both *erasures and errors* [1516] and even to exploit *soft information* [1494, 1495]. More recently, other algorithms for soft decoding of RS codes have been proposed by M.-S. Oh and P. Sweeney [1555, 1556] and D. Burgess *et al.* [1557] (other trellis-based algorithms are mentioned below).

After the appearance of RS and BCH codes, significant research efforts were devoted to constructing other coding schemes offering good properties in terms of coding and decoding complexity and, at the same time, a large $d_{H,\min}$, and, from a theoretical point of view, to possibly developing *asymptotically good codes*, whose parameters achieved the so-called *Varshamov–Gilbert lower bound*. Here we confine ourselves to mentioning two relevant contributions. The first is the development of a novel class of nonbinary codes, based on the ancient theory of *residue number systems* [1558, 1559], and was introduced in 1966 [1560]. This class includes the so-called *redundant residue number system* (RRNS) codes, which are maximum–minimum distance block codes, exhibiting similar distance properties to RS codes. R. W. Watson and C. W. Hastings [1560] as well as H. Krishna *et al.* [1561, 1562] exploited the properties of the RRNS for detecting or correcting a single error and also for detecting multiple errors. Recently, the soft decoding of RRNS codes has been proposed by T. H. Liew *et al.* in [1563] (see also [1564]). The second important contribution is the development of a new class of block codes based on *algebraic geometry* (AG), that is, on algebraic curves over finite fields, by V. D. Goppa in the late 1970s [1565, 1566]. In 1982 M. A. Tsfasman, S. G. Vlăduţ, and T. Zink [1567], using modular curves, proved how asymptotically good AG codes could be constructed over nonbinary fields GF(q) of size $q \geq 49$. AG codes are usually much longer than RS codes and, even if efficient decoding algorithms exist [1568], they have not yet been adopted for practical applications. A survey of this field is available in [1569].

In 1997 M. Sudan [1570] introduced a *list decoding algorithm*[37] based on polynomial interpolation for decoding beyond the guaranteed error-correction distance of RS and related codes. Although in principle there may be multiple codewords within such an expanded distance, with high probability only one will be found. In 1999 V. Guruswami and M. Sudan [1498] derived an improved version of this algorithm, which can be generalized to solve the list decoding problem for other algebraic codes, and in 2003 R. Koetter and A. Vardy extended it to process soft decisions [1572]. Other significant approaches to soft decision decoding algorithms were also developed in the same period; here we limit ourselves to mentioning the ordered statistics approach proposed by M. P. C. Fossorier and S. Lin [1496, 1497].

A different approach to decoding of block codes has been suggested by the possibility of representing a block code by a time-varying trellis diagram. This idea was first presented by L. R. Bahl, J. Cocke, F. Jelinek and J. Raviv in their well-known paper [1523], published in 1974; these researchers, after describing an optimal decoding algorithm for linear codes, illustrated a method for representing the words of an arbitrary linear block code by the path labels in a trellis. In 1978 J. K, Wolf [1573] proved that soft decision ML decoding of any (n, k) linear block code over GF(q) can be accomplished using the VA on a trellis with no more that q^{n-k} states. For the next ten years, there was relatively little work in this area. In 1988 D. J. Forney, in an appendix to a paper on coset codes [335], described what he called " the trellis diagram of a code". This stimulated research interest on trellis representations of block codes, which became an active research area during the 1990s. Some of the most relevant contributions to this area have been offered by:

(a) D. J. Muder [1574], who proved that, among all trellises representing a given block code, the Forney trellis minimized the number of vertices at each depth,

(b) F. R. Kschischang and V. Sorokine [1575], who developed many of the properties of the important "trellis-oriented" generator matrices for the first time, and

(c) B. Honary, G. S. Markarian, P. G. Farrell *et al.*, S. Lin and T. Kasami *et al.* [1576–1582], who proposed various methods for reducing decoding complexity.

There have been many other significant contributions to this subject: excellent surveys of the main results achieved in this field are offered by [1583, 1584].

---

[37] List decoding was an unpublished invention of P. Elias [1571].

9.4.2   Probabilistic Coding

Whereas the aim of algebraic coding theory is to devise specific (n, k) codes defined over finite fields and maximizing the minimum Hamming distance $d_{H,\min}$ between any pair of codewords, probabilistic coding is more oriented to developing *classes of codes* that offer *optimal average performance* for a given coding and decoding complexity. Classical coding schemes belonging to the latter class include convolutional codes, product codes, concatenated codes, trellis-coded modulation, and block coding schemes based on a trellis representation. This point of view on coding is emphasized by various well-known textbooks (e.g., see [35, 321, 705, 1542, 1585, 1586]). Probabilistic decoders usually process soft (reliability) information available at their inputs (and originating from channel outputs) and generated at intermediate stages of the decoding process.

From a historical perspective, the first relevant example of probabilistic coding is represented by the class of *convolutional codes*, first proposed by P. Elias in his seminal paper of 1955 [1477] (reprinted in [1587–1589]). Elias showed that convolutional codes were simpler to encode than general linear codes and had the same average performance as randomly chosen codes. It is worth noting that R. Gallager's doctoral thesis on LDPC codes, supervised by Elias, was also motivated by the problem of finding a class of randomlike codes which offered quasi-optimal performance and could be decoded near capacity with a feasible complexity [1590]. Elias and Gallager's codes, seemingly so different, can be represented as codes on graphs (see Chapter 10). From this perspective, a straight line of development can be identified from Elias's invention to modern capacity-approaching codes. This development, however, took more than half a century.

Various important results were derived in the area of probabilistic coding in the 1960s and 1970s. Shortly after Elias's paper, J. M. Wozencraft recognized that the tree structure of convolutional codes offers the possibility of using a sequential search algorithm for decoding [1537]. Sequential decoding was investigated in depth in the 1960s. In this research area we mention the following results:

(a) the development of the fast, storage-free *Fano sequential decoding algorithm* [1538],
(b) the proof that the rate of a sequential decoding system is bounded by the computational *cutoff rate* [1591], and
(c) the discovery of *threshold decoding* for convolutional codes by Massey [1592]. Burst-error-correcting variants of threshold decoding developed by Massey and Gallager proved to be quite suitable for practical error correction [1453].

The year 1967 saw the introduction of an asymptotically optimal decoding algorithm for convolutional codes, the Viterbi algorithm [982, 1478, 1593]. It soon became apparent that relatively short convolutional codes decoded by the VA were potentially quite practical [1594, 1595]. In the following years the Linkabit Corporation, founded by Viterbi, I. Jacobs and L. Kleinrock, in 1968 as a consulting company, built a prototype 64-state VA decoder, capable of running at 2 Mb/s [1596]. During the 1970s, through the leadership of Linkabit and the Jet Propulsion Laboratory, the VA became part of the NASA standard for deep-space communication. Around 1975, Linkabit developed a relatively inexpensive, flexible and fast VA chip. The VA soon began to be incorporated into many other communications applications and became a more attractive alternative to sequential decoders.

Part of the attraction of convolutional codes is the inherent capability of their decoding algorithms, like the VA, to use soft decisions without any essential increase in complexity. However, the VA does not result, in principle, in the minimum possible BER, even if it performs close to it. An alternative approach to soft decoding is to exploit algorithms that can compute (exactly or approximately) the APP of each transmitted bit being a zero or a one, given the APPs of each received symbol. This approach to decoding of error-correcting codes was first adopted by Gallager for LDPC codes [1590]. At about the same time, Massey [1592] developed an APP version of threshold decoding. However, the adoption of the APP approach to trellis-based decoding of convolutional codes is due to Bahl, Cocke, Jelinek and Raviv, who proposed the so-called BCJR algorithm in 1974 [1523]. This algorithm is a

SiSo algorithm and is optimum in terms of bit error probability, but is substantially more complicated than the VA, which offers similar error performance. For this reason, it began to attract substantial interest only with the advent of *turbo codes* in 1993, since the BCJR was proposed as a key element in their iterative decoding procedure [1481, 1597] (see Chapter 10). Since 1993, a large body of work has been carried out in the area with the aim of reducing the complexity of SiSo decoding. Practical reduced-complexity decoders are, for instance, the the Max-Log-MAP algorithm proposed by W. Koch, A. Baier and J. Erfanian *et al.* [1524, 1598], the Log-MAP algorithm suggested by P. Robertson *et al.* [1525], and the SOVA algorithm investigated by J. Hagenauer and P. Hoeher [1535, 1599].

Other historically important results in the field of probabilistic coding concern *product codes* and *concatenated codes*. The former were discovered by Elias before he invented convolutional codes [1480] (see Section 9.3.1); he also proved that an arbitrarily low error probability could be achieved at a nonzero code rate using a repeated product of extended Hamming codes. Forney introduced the new class of concatenated codes in 1966 [1455]. In particular, he proposed to construct new binary linear block codes by *serially* cascading two linear block codes, namely an *outer* nonbinary RS code over a finite field GF(q) and an *inner* binary code. The overall result was a long, powerful code with a simple, suboptimum decoder able to correct many combinations of burst and random errors. Forney showed that with a proper choice of the constituent codes, concatenated coding schemes could operate at any code rate up to the Shannon limit, with exponentially decreasing error probability, but requiring only polynomial decoding complexity. Concatenation can also be applied to convolutional codes. In fact, the most common concatenated code used in practice is that developed in the 1970s as a NASA standard (already mentioned in Section 9.3.2); it consists of an inner rate-1/2, 64-state convolutional code and an outer (255, 223) RS code over GF(256), separated by a symbol interleaver. In the late 1980s a more complex concatenated coding scheme with iterative decoding was proposed by E. Paaske [1600] and independently by O. Collins [1601], to improve the performance of the NASA concatenated coding standard. Instead of a single outer RS code, Paaske and Collins proposed to use several outer RS codes of different rates. After one round of decoding, the outputs of the strongest (lowest rate) RS decoders may be deemed to be reliable and thus may be fed back to the inner (Viterbi) convolutional decoder as known bits for another round of decoding. This use of iterative decoding with a concatenated code can be seen as a precursor of turbo codes [1602].

9.5   Further Reading

This chapter has provided a summary of some essential results in classic coding theory. The reader interested in further investigating these topics should refer to the well-known book by S. Lin and D. J. Costello [35], which provides an exhaustive analysis of block and convolutional coding theory. Finally, it is important to point out that various historical papers mentioned in the previous section can be found in [1603].

10

Modern Coding Schemes

10.1 Introduction

In the first 50 years in the history of coding theory many attempts were made by well-known theorists to develop codes capable of approaching the Shannon limit with an underlying well-defined mathematical structure to simplify decoding and with a randomlike codeword distribution to enhance the code strength. As illustrated in the previous chapter, various well-structured error-correcting codes were discovered by many researchers, including R. W. Hamming, P. Elias, I. Reed, G. Solomon, R. C. Bose, D. K. Ray-Chaudhuri, A. A. Hocquenghem, R. Gallager, E. R. Berlekamp and A. J. Viterbi. However, despite the invention of these coding schemes, significantly narrowing the energy gap with channel capacity would not have been possible without the discovery of the concept of *code concatenation* (see D. Forney's book [1455] published in 1966). As already illustrated in Section 9.3, Forney proposed to cascade (i.e., to *serially concatenate*) two relatively simple codes to generate an overall code (i.e., a *concatenated code*), offering large coding gains and with a decoding complexity that increases only algebraically with block size. This approach is illustrated by the cascade of an (outer) algebraic code, typically a Reed–Solomon code, with an (inner) binary convolutional code, as already discussed in Section 9.3.2. In this case, the recommended decoding procedure typically consists of an inner soft decoding based on the VA followed by an algebraic algorithm for hard decision decoding of the outer code. Many years later, a substantially different approach to the problem of code concatenation and the decoding of concatenated codes was proposed by C. Berrou, A. Glavieux and P. Thitimajshima in 1993 [1481], who devised a new concatenated coding scheme, known as a *turbo code*, and a new efficient iterative soft decoding strategy based on a modified version of the BCJR algorithm [1523] and which approaches ML performance of the overall code. These technical solutions resulted from a pragmatic construction and were partly based on the previous work and, in particular, on the intuitions of G. Battail, J. Hagenauer, and P. Hoeher, who, in the late 1980s, had shown the importance of probabilistic processing in digital communication receivers [1535, 1604, 1605]. It is worth noting, however, that other researchers, including Elias, Gallager [1479], and M. Tanner [1606], had already devised coding and decoding techniques whose general principles are closely related to those of turbo codes. Since 1993, the properties of turbo codes have been extensively investigated and have stimulated research activities on other randomlike coding schemes that lend themselves to iterative soft decoding. The most famous example of this class of schemes is represented by LDPC schemes, first introduced by Gallager in his Ph.D. dissertation [1590] and rediscovered [1607–1609] as a category of codes approaching the Shannon capacity limit for AWGN channels with practical decoding complexity. This is due to the fact that iterative soft decoding of LDPC codes is based on the

Wireless Communications: Algorithmic Techniques, First Edition.
Giorgio M. Vitetta, Desmond P. Taylor, Giulio Colavolpe, Fabrizio Pancaldi, Philippa A. Martin.
© 2013 John Wiley & Sons, Ltd. Published 2013 by John Wiley & Sons, Ltd.

so-called *sum-product algorithm* (SPA) [1610], which is simpler than the BCJR algorithm performed on a trellis. The "turbo principle", initially developed for iterative decoding of a specific class of codes, has been also applied in areas other than error-correction coding, such as multiuser detection and equalization [1611].

This chapter is focused on the study of modern concatenated coding schemes and is organized as follows. Concatenated convolutional and block coding architectures are illustrated in Sections 10.2 and 10.3, respectively, whereas other concatenated coding schemes are described in Section 10.4. Decoding strategies based on the turbo principle applied to this class of codes are derived in Section 10.5. The properties of LDPC codes and iterative decoding procedures for this class of codes are illustrated in Sections 10.6 and 10.7, respectively. Section 10.8 analyzes the problem of code analysis and decoding from a graph theory perspective. The history of modern coding schemes is outlined in Section 10.9. Finally, some suggestions for further reading are provided in Section 10.10.

10.2   Concatenated Convolutional Codes

Convolutional encoders can be concatenated in different ways. In this section we focus on the architecture of their concatenation in *parallel concatenated convolutional codes* (PCCCs) (also called *turbo codes*), *serial concatenated convolutional codes* (SCCCs) and *hybrid concatenated convolutional codes* (HCCCs).

10.2.1   Parallel Concatenated Coding Schemes

The typical structure of a turbo encoder is illustrated in Figure 10.1. It consists of: two identical binary *recursive systematic convolutional* (RSC) encoders each of rate 1/2 and each with ν memory elements (i.e., 2^ν states); and a pseudorandom N-bit interleaver denoted by π in the figure. The first component (RSC) encoder generates the parity sequence $\{c_{1,l}, l = 0, 1, \ldots, N-1\}$ in response to its input sequence $\{u_l, l = 0, 1, \ldots, N-1\}$, consisting of k information bits plus ν termination bits (so that $N = k + \nu$), forcing the encoder to return to its initial state (e.g., the 0 state); this means a *tail-biting* mechanism is used for trellis termination of the given encoder (further details can be found in [37, pp. 183–184] and [1612]) or, equivalently, that a *circular* RSC is used [1605]. The second constituent (RSC) encoder is fed by a permuted version[1] $\{u_{p,l}, l = 0, 1, \ldots, N-1\}$ of $\{u_l\}$ and generates the parity sequence $\{c_{2,l}, l = 0, 1, \ldots, N-1\}$. An optional *puncturing mechanism* can be included in the scheme to modify, when needed, the overall code rate by discarding a part of the parity sequences $\{c_{1,l}\}$ and $\{c_{2,l}\}$ [1613, 1614]. In the absence of puncturing the transmitted codeword $\{x_n\}$ simply consists of the concatenation of the sequences $\{u_l\}$, $\{c_{1,l}\}$ and $\{c_{2,l}\}$, so that the codewords have length $n = 3N$ and the resulting code rate is $R = k/n = (N-\nu)/(3N) \cong 1/3$ for large N.

The following comments are in order concerning the architecture depicted in Figure 10.1, and should be carefully kept in mind when employing it:

1. *Long codewords* should be generated to approach the Shannon limit. In particular, simulation results suggest that the information block length N (and, consequently, the interleaver size) should be chosen to be several thousand bits long. For instance, data blocks of length $N = 65\,536$ and a 256×256 interleaver were adopted in generating the first available results on code performance [1481].
2. *Equal* RSC encoders with a small number of states (e.g., 16 states) are commonly used as constituent codes to achieve excellent performance at moderate BERs [1615, 1616]. Alternatives to this choice include the use of *asymmetric* RSC constituent codes [1617] and block codes (see the next section). The most important parameter in the selection of an RSC code is its *effective free distance* $d_{f,eff}$,

---
[1] Note that this encoder may or may not be terminated [1605]. This does not have a significant effect on error performance for long codewords.

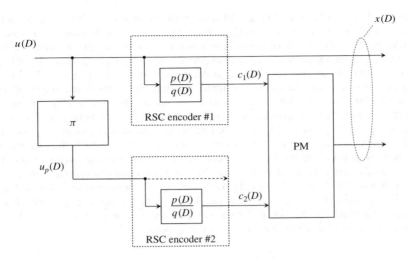

Figure 10.1 Typical PCCC encoder structure.

defined as the minimum weight of the code sequences generated by all possible input sequences of weight 2 [1615]; tables showing the best encoders for various rates are available in [1618, 1619].

3. *Puncturing* can also be adopted for the information sequence $\{u_l, l = 0, 1, \ldots, N - 1\}$ to generate partially systematic or nonsystematic turbo codes [1620].

4. The size of the *pseudorandom interleaver* plays a fundamental role, since it influences the minimum Hamming distance of the overall turbo code (see Section 10.5.4) and, consequently, the achievable performance *gain* [1616, 1621]. Thus, it should be considered as a constituent component of the overall concatenated code. Despite the strong interdependency between the RSC codes adopted and the interleaver in determining the error performance of a turbo code, their *joint design* is too ambitious a target. This explains why a decoupled procedure has been proposed; in this case convolutional codes are selected first and then interleaving is tailored to their characteristics [1546, 1621]. Note also that conventional block interleavers, which rearrange the bits in some systematic fashion, do not yield good error performance, except for relatively short block lengths [1622]. In addition, the interleaver does not change the weight of the permuted sequence $\{u_{p,l}\}$ with respect to that of $\{u_l\}$.

5. This turbo coding scheme is *linear* since its constituent codes are linear, and the interleaver is linear since it can be represented by a permutation matrix. Note, however, that the interleaver, being a constituent part of the overall encoder, leads to a trellis representation of this coding scheme with a huge number of states (the overall code trellis is commonly called a *hypertrellis*). This explains why classic trellis-based ML or MAP decoding algorithms cannot be exploited in this case and clever suboptimal strategies need to be devised.

6. The scheme of Figure 10.1 can easily be generalized by concatenating in a parallel fashion N_e RSC encoders, of which $(N_e - 1)$ are fed by the input sequence according to different random rules. This leads to the concept of *multiple parallel concatenated code* (MPCC) [1605, 1623–1625], offering improved minimum Hamming distance compared to single parallel concatenation. However, upper bounds on this parameter show that these codes cannot be asymptotically good, since they are not characterized by a minimum distance growing linearly with block length [1626]. Despite this, excellent performance can be achieved with only two encodings, that is, with the classic scheme of Figure 10.1.

Further details on the design criteria to be adopted with PCCCs can be found in [1615, 1616].

The genesis of the architecture of Figure 10.1, originally proposed in [1481], is discussed in detail in [1605]. Its motivations, from an encoding perspective, can be summarized as follows. Berrou and Glavieux were looking for a *symmetric* concatenation scheme and this concern led to parallel concatenation, in which, unlike in its serial counterpart, *component codes play similar roles*. Among the possible component codes, convolutional codes were selected due to the availability of SiSo *decoding algorithms* (namely, the BCJR algorithm), which could be combined in accordance with a feedback architecture. The selection of a recursive systematic structure for the convolutional encoders was suggested, in part, by the important theoretical results developed by Forney in this research area and, in particular, in the paper [1517], showing that the encoder of a convolutional code can take a nonrecursive nonsystematic or a recursive systematic form (see Section 9.2). Turbo encoders are based on the second form, which, unlike the first, incorporates a *pseudorandom scrambler* and, from this perspective, satisfies the need to develop randomlike coding schemes to approach channel capacity [1625], as concluded by Shannon theory. In particular, nonrecursive encoders produce finite-length codewords in response to input sequences of weight 1 (unlike their recursive counterparts that produce infinite-length codewords). In Section 10.5.4 we will show that this property plays a fundamental role in achieving the excellent performance offered by turbo codes.

10.2.2 Serially Concatenated Coding Schemes

Using the same ingredients as PCCCs, namely convolutional encoders and interleavers, SCCCs can easily be developed. The general structure proposed in [1546] for the encoder for an SCCC is illustrated in Figure 10.2(a). In this case, an outer *convolutional encoder* (CE) with rate $R^o = k/p$ feeds a pseudorandom interleaver (π, with a length N that is a multiple of p). Its output is sent an inner CE with rate $R^i = p/n$. This generates an SCCC with rate $R^s = k/n$. RSC codes can be used in this case too, as shown in Figure 10.2(b), describing the serial concatenation of two CEs with $R^i = R^o = 1/2$. The overall code rate $R^s = 1/4$ can be increased by resorting to a proper puncturing mechanism [1614]. Important design guidelines for SCCCs are developed in [1546], where it is shown that: first, it is essential to select an RSC encoder for the inner encoder, since this always yields an *interleaver gain* (defined as the factor by which the bit error probability is decreased with the interleaver length at a given SNR); and second, good outer RSC codes maximize the effective free distance of the inner encoder $d^i_{f,eff}$ (defined, as above, as the minimum weight of the codewords of the inner code generated

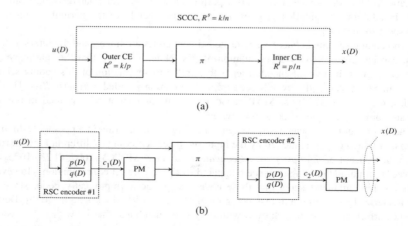

(a)

(b)

Figure 10.2 (a) General encoder structure of an SCCC scheme and (b) a specific implementation based on the use of two punctured RSCs.

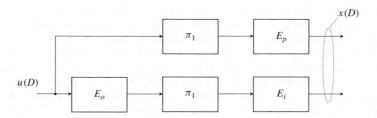

Figure 10.3 General encoder structure of a coding scheme based on *hybrid concatenation*.

by input sequences of weight 2). Finally, to obtain compatibly with the desired rate R^s, an outer code with a large (and, possibly, odd) value of the free distance d_{eff}^o should be selected.

The scheme of Figure 10.2(a) was extended in [1627], where double serially concatenated codes were introduced. In this case, the encoder consists of three cascaded encoders, linked by two interleavers of different lengths. This concept can be further generalized, leading to *multiple serially concatenated codes* (MSCCs). Such codes, unlike MPCCs, can be asymptotically good in terms of minimum Hamming distance, as has been shown in [1628] in the specific case of the so-called *repeat and accumulate* (RA) coding (see Section 10.4). However, the performance of feasible (i.e., iterative) decoding strategies degrades with the number of constituent codes, and this makes coding schemes with more than three serially concatenated component codes impractical [1629].

Generally speaking, SCCCs offer a larger interleaving gain (i.e., a faster decrease in the bit error probability with interleaver length) than their parallel concatenated counterparts. Further details are provided in Section 10.5.4, where the achievable error performance is discussed.

10.2.3 Hybrid Concatenated Coding Schemes

The concept of code concatenation can be further generalized. In principle, various *code networks*, consisting of Q codes connected through $Q - 1$ interleavers according to a specific *topology*, can be devised, as proposed in [1630]. A code network can be considered as a unique code with huge ML decoding complexity. Hence, as shown in Section 10.5, practical decoding techniques employ a *distributed approach*, based on the use of a multiplicity of connected modules mutually exchanging information.

An example of a code network is the so-called *hybrid concatenation* based on the topology of Figure 10.3(a) [1631]. This case is characterized by $Q = 3$, that is, it is based on the use of a *parallel* code, an *outer* code and an *inner* code (their encoders are denoted by E_p, E_o and E_i, respectively) linked by two interleavers (π_1 and π_2). A specific hybrid scheme, based on the concatenation of three convolutional codes and called an HCCC, is described in [1632, 1633], where it is shown that, to maximize the interleaving gain of this coding scheme, the following types of component codes should be used: first, RSC codes should be used for both the *inner* and the *parallel* codes; and second, a recursive or nonrecursive convolutional code with large free distance d_{free}^o should be used for the *outer* code.

Hybrid concatenated codes offer more degrees of freedom in code design and the opportunity to combine the advantages of parallel and serially concatenated coding systems. In particular, hybrid schemes whose minimum distance increases linearly with block lengths can be devised [1629].

10.3 Concatenated Block Codes

Concatenated schemes employing block codes as constituent codes were proposed much earlier (e.g., see [1455, 1480]) than their counterparts based on convolutional codes and illustrated in the previous section. However, after the appearance of turbo codes it soon became clear that their decoding philosophy and some of the tools for the analysis of their performance could be also adopted for the parallel and serial concatenation of linear block codes, as illustrated below.

The *parallel concatenation* of two or more *systematic* linear block codes leads to *two-dimensional* or *higher-dimensional product codes*, respectively (see Section 9.3.1). A product code can be thought of as a concatenated coding scheme with block interleaving between the inputs of the coding stages. Note that the first algorithms proposed for this class of codes were not ML and gave rather poor results because they relied on hard input, hard output decoders [1634, 1635]. The exploitation of the turbo decoding philosophy for product codes was first proposed by J. Lodge, R. Young, P. Hoeher and J. Hagenauer [1636] (see also [1100]) and led to a substantial improvement of the error performance at a complexity appreciably lower than that of ML decoding; the combination of turbo decoding with a product code is usually called a *block turbo code* (BTC) [1637] or *turbo product code* (TPC) [1638]. BTCs are commonly based on classic linear block codes, namely Hamming codes [1100, 1639], extended Hamming codes [1636, 1638–1640], BCH codes [1637, 1503], extended BCH codes [1641], RS codes [1642] and SPCs [1643–1645]. Conventional two-dimensional product codes offer good error protection, but contain excessive redundancy. This is due the fact that the horizontal code C_1 (see Figure 9.32) provides the vertical code C_2 with information on many error patterns exceeding the error-correction capability of C_2. However, code construction procedures for redundancy reduction can be devised, as illustrated in [1646]. Finally, it is worth mentioning the following:

1. Two-dimensional *binary* product codes can be exploited as a basis for the construction of efficient *nonbinary* concatenated coding schemes with a simple decoding procedure, as shown, for instance, in [1647].
2. Analytical tools for the evaluation of the error performance of turbo codes can also be applied to the parallel concatenation of block codes [1616], so that design criteria for the latter can be derived.

In the last decade less attention has been paid to the problem of designing specific schemes based on the *serial concatenation* of block codes. The only significant advance in this area is represented by the work by S. Bendetto *et al.* [1546], who have derived upper bounds to the average performance offered by ML decoding over an AWGN channel of serially concatenated block and convolutional coding schemes. Such bounds are exploited to derive some design guidelines for the inner and the outer encoder in order to maximize the interleaver gain and the asymptotic slope of the error probability curves. This does not mean, however, that interest in serial concatenation has disappeared. In fact, several coding schemes of this type have been proposed in recent years, as illustrated in the following section.

10.4   Other Modern Concatenated Coding Schemes

After the discovery of turbo codes in 1993 and of analytical tools for their analysis [1616, 1615, 1630], it soon became apparent that novel concatenated coding schemes, characterized by multiple codes connected through interleavers according to a specific topology and exhibiting a "turbo-like" behavior, could be devised [1648, 1649]. Some of these schemes are described below.

10.4.1   Repeat and Accumulate Codes

The encoder structure of an RA encoder is illustrated in Figure 10.4. An information block **u** (usually binary) of length N is repeated n times and permuted by an interleaver of length nN. Then the

Figure 10.4   Encoder for an (nN, N) RA code. The numbers below the input–output lines indicate the size of the corresponding binary vectors.

interleaver output is encoded by a rate-1 accumulator, operating according to the input–output relationship $x_l = x_{l-1} + z_l$ evaluated over GF(2). The accumulator can be seen as a truncated rate-1 two-state convolutional encoder with transfer function $g(D) = 1/(1 + D)$ or as a rate-1 binary block code whose input–output relationship is $\mathbf{x} = \mathbf{zG}$, where $\mathbf{G}$ is an $nN \times nN$ matrix with 1s on and above its main diagonal and 0s elsewhere. This structure was first proposed in [1648], where it was shown that ML decoding of RA schemes can achieve excellent performance and, in particular, an arbitrarily low word error probability for sufficiently low rates and any fixed SNR greater than a threshold as the block length goes to infinity (i.e., as $N \to \infty$).

The significant results achieved for RA codes show that powerful error-correcting codes may be constructed from extremely simple components. Following this approach other coding schemes have been proposed. Here we mention the following schemes:

1. The serial concatenation of an arbitrary binary linear outer code of rate $R < 1$ with multiple identical rate-1 binary linear inner codes [1650]. A specific example is the class of so-called *convolutional accumulate-m* (CA^m) codes, having a terminated convolutional code as outer code and a cascade of m interleaved "accumulate" inner codes.
2. The *product accumulate* (PA) codes, each resulting form the serial concatenation of a specific SPC-based product code with an interleaved rate-1 inner code (i.e., an accumulator, as in RA codes) [1651]. These codes exhibit a remarkably good error performance at high rates ($R > 0.7$) when properly decoded (via an iterative message-passing algorithm).
3. The serial concatenation of extended Hamming codes with a binary accumulator [1652]. This class of codes can also achieve near Shannon limit performance for very high rate coding if iterative decoding is used.

It is important to note that the binary accumulator adopted in all the above-mentioned coding schemes can also be seen as a *differential encoder*. This consideration has led to the study of serially concatenated schemes in which an outer encoder is connected to a *differentially encoded* digital modulator through an interleaver; such schemes are discussed in the next subsection.

10.4.2   Serial Concatenation of Coding Schemes and Differential Modulations

The serial concatenation of an *interleaved convolutional code* with a differential modulation (e.g., a differentially encoded PSK or a CPM) has been proposed in various papers [531, 1653, 1654]. This work was motivated by the fact that differential modulations can be noncoherently detected (this is a substantial advantage in the presence of excessive phase noise and/or when the communication system is operating at low SNRs) and that iterative decoding schemes explicitly developed for serially concatenated codes can be adopted for these schemes. The results shown in the above-mentioned references provide evidence that ML decoding of the proposed schemes offers substantially better performance than that offered by the stand-alone outer convolutional code, even on an AWGN channel [1653, 1654] or a phase-noisy AWGN channel [1655]. Since the differential decoder has a trellis structure, the performance of ML decoding can be approached using an iterative solution based on soft output decoding algorithms devised for turbo codes if the channel is AWGN [1654]; however, if the channel is affected by fading, a specific APP demodulation algorithm must be developed. For instance, in [531] an APP multiple-symbol differential demodulator is developed for a differentially encoded PSK modulation transmitted over a time-selective fading channel.

Further research work in this area has mainly been concerned with the study of *interleaved block codes* serially concatenated with differential modulators [1656–1658]; in particular, [1656] provides design guidelines to develop good LDPC codes for differential modulators, while [1657] investigates the use of *differential parity check* (DPC) codes concatenated with differential PSK. Finally, the use of a convolutional encoder concatenated with the cascade of two interleaved differential encoders (i.e., with a couple of interleaved accumulate codes) is investigated in [1659].

10.5   Iterative Decoding Techniques for Concatenated Codes

In this section the application of the so-called *turbo principle* to the decoding of concatenated codes is discussed. The principle is introduced in Section 10.5.1, where the key concept of *extrinsic information* and its exploitation in the iterative decoding of a parallel concatenated scheme based on a specific SPC code are analyzed. This preliminary discussion illustrates the need to devise SiSo algorithms for the component codes of concatenated schemes. Different types of these algorithms are illustrated in Section 10.5.2, and some of their applications are considered in Section 10.5.3. Finally, some indications of the performance bounds available for turbo codes and other concatenated schemes are provided in Section 10.5.4.

10.5.1   The Turbo Principle

The iterative (i.e., *turbo*) decoding strategy proposed by Berrou *et al.* in 1993 for the PCCCs [1481] originated from the intuition that the *feedback concept*, which plays a fundamental role in electronics, could also be exploited for the decoding of compound (concatenated) codes [1605]. In fact, such a strategy is based on a feedback decoding scheme, in which the decoders for the component codes operate alternately and exchange soft information between consecutive decoding stages with the aim of iteratively refining their data estimates. Therefore, in principle, the overall decoding architecture for a given concatenated coding scheme includes as many decoders as the component encoders of the scheme itself. These decoders are interconnected to exchange soft (feedback) information and each of them accomplishes as many decoding passes as the number of iterations. In practice, however, iterative decoding can be implemented by resorting to a *modular pipelined architecture*, in which each module first carries out a single decoding stage for all the component codes and then sends the generated soft information to the next module. Therefore, in this scheme each module implements a set of SiSo algorithms for decoding all the component codes.

In designing turbo decoding, one of the crucial problems to be solved was understanding the type of information to be exchanged between interacting decoders, since feedback systems can exhibit an unstable behavior; in other words, cascading multiple decoding stages can result in a sort of positive feedback amplifier. The issue of stability was solved by introducing the notion of *extrinsic information*, whose meaning is explained through the following example, that analyzes the evaluation of the APPs of the information bits for a simple channel coding scheme.

Example 10.5.1 Let us consider a (3,2) SPC code, that is, a code adding a single parity bit to the message $\mathbf{u} = [u_0, u_1]$, so that the elements of the resulting codeword $\mathbf{x} = [x_0, x_1, x_2]$ are given by

$x_0 = u_0$, $x_1 = u_1$ and:

$$x_2 = u_0 \boxplus u_1, \tag{10.1}$$

where $\boxplus$ denotes the sum over GF(2). If $\mathbf{x}$ is transmitted over a slow time-selective fading channel using BPSK, the receiver observes, at its matched filter output, the vector $\mathbf{y} = [y_0, y_1, y_2]$, with:

$$y_i = h_i(2x_i - 1) + n_i \tag{10.2}$$

and $i = 0, 1, 2$. Here, the noise samples $\{n_i\}$ are iid complex random Gaussian variables, each with zero mean, variance[2] σ_n^2 and iid real and imaginary parts. In addition, h_i denotes the fading distortion (assumed *known* at the receiver in what follows) affecting the transmitted signal in the ith symbol interval, with $i = 0, 1, 2$. Let us now analyze the problem of devising a SiSo decoding procedure that processes $\mathbf{y}$ and generates the bit APPs:

$$\Pr\{u_k = b|\mathbf{y}, \mathbf{h}\} = \Pr\{x_k = b|\mathbf{y}, \mathbf{h}\} \tag{10.3}$$

for $b = 0, 1$ and $k = 0, 1$, or, alternatively, the so-called log-likelihood ratios:

$$L(u_k|\mathbf{y}, \mathbf{h}) \triangleq \ln \frac{\Pr\{x_k = 1|\mathbf{y}, \mathbf{h}\}}{\Pr\{x_k = 0|\mathbf{y}, \mathbf{h}\}} \tag{10.4}$$

for $k = 0, 1$. To simplify the derivation of such a procedure, we focus on the evaluation of the LLR (10.4) for the bit x_0:

$$L(u_0|\mathbf{y}, \mathbf{h}) = \ln \frac{\Pr\{x_0 = 1|y_0, y_1, y_2, \mathbf{h}\}}{\Pr\{x_0 = 0|y_0, y_1, y_2, \mathbf{h}\}}. \tag{10.5}$$

It is worth noting that all the elements of $\mathbf{y}$ are useful in estimating x_0, even if y_0 depends *directly* on x_0 only. In fact, since (see (10.1)):

$$x_0 = x_1 \boxplus x_2, \tag{10.6}$$

y_1 and y_2 also provide, if *jointly processed*, useful information about x_0. Since:

$$\frac{\Pr\{x_0 = 1|\mathbf{y}, \mathbf{h}\}}{\Pr\{x_0 = 0|\mathbf{y}, \mathbf{h}\}} = \frac{\Pr\{x_0 = 1\}}{\Pr\{x_0 = 0\}} \frac{f(\mathbf{y}|x_0 = 1, \mathbf{h})}{f(\mathbf{y}|x_0 = 0, \mathbf{h})} \tag{10.7}$$

from Bayes' theorem, $L(u_0|\mathbf{y}, \mathbf{h})$ (10.5) can be expressed as:

$$L(u_0|\mathbf{y}, \mathbf{h}) = L(u_0) + \ln \frac{f(\mathbf{y}|x_0 = 1, \mathbf{h})}{f(\mathbf{y}|x_0 = 0, \mathbf{h})} \tag{10.8}$$

where

$$L(u_i) \triangleq \ln \frac{\Pr\{u_i = 1\}}{\Pr\{u_i = 0\}} \tag{10.9}$$

represents the LLR associated with the a priori information available about u_i (the a priori LLR). The LLR expression (10.8) can be further modified to put it in a more meaningful form. This requires some manipulation of the second term on the RHS. If we note that:

$$f(\mathbf{y}|x_0 = b, \mathbf{h}) = \sum_{l=0}^{1} \Pr\{x_1 = l\} f(\mathbf{y}|x_0 = b, x_1 = l, \mathbf{h}) \tag{10.10}$$

---

[2] In practice, $\sigma_n^2 = 2N_0/\bar{E}_s$ if the channel noise variable is complex and $\sigma_n^2 = N_0/\bar{E}_s$ if we retain only the real or the imaginary part of noise.

and

$$f(\mathbf{y}|x_0 = b, x_1 = l, \mathbf{h}) = f(y_0|x_0 = b, h_0)\, f(y_1|x_1 = l, h_1)\, f(y_2|x_2 = b \oplus l, h_2), \qquad (10.11)$$

it is not difficult to show that:

$$\ln \frac{f(\mathbf{y}|x_0 = 1, \mathbf{h})}{f(\mathbf{y}|x_0 = 0, \mathbf{h})} = L_c(y_0) + L_e(u_0), \qquad (10.12)$$

where

$$L_c(y_i) \triangleq \ln \frac{f(y_i|x_i = 1, h_i)}{f(y_i|x_i = 0, h_i)} \qquad (10.13)$$

and

$$L_e(u_0) \triangleq \ln \frac{\sum_{l=0}^{1} \Pr\{x_1 = l\} f(y_1|x_1 = l, h_1)\, f(y_2|x_2 = 1 \oplus l, h_2)}{\sum_{l=0}^{1} \Pr\{x_1 = l\} f(y_1|x_1 = l, h_1)\, f(y_2|x_2 = l, h_2)}. \qquad (10.14)$$

The last two expressions show that the quantities $L_c(y_0)$ and $L_e(u_0)$ of (10.12) represent the contributions to (10.8) acquired through the noisy datum y_0 *generated by the channel*[3] in response to x_0 (i.e., u_0) and through the data pair $\{y_1, y_2\}$ thanks to the parity check equation (10.6). Therefore, $L_c(y_0)$ is usually called the *channel* LLR, while $L_e(u_0)$ is called the *extrinsic* LLR, as it is not *intrinsically* influenced by u_0. Finally, substituting (10.12) into (10.8) yields the equality:

$$L(u_0|\mathbf{y}, \mathbf{h}) = L(u_0) + L_c(y_0) + L_e(u_0). \qquad (10.15)$$

Moreover, from (10.13) it is easily inferred that, for the given channel model (see (10.2)):

$$L_c(y_i) = \frac{4 \operatorname{Re}\{y_i h_i^*\}}{\sigma_n^2}. \qquad (10.16)$$

The relationship (10.15) reveals the real structure of the data APPs, each consisting of three distinct contributions. Note that the first two of them, namely both $L(u_0)$ (see (10.9)) and $L_c(y_0)$ (see (10.16)) are available *before decoding*, that is, before computing the bit APPs; this means that the only additional information generated by a MAP decoding procedure that evaluates the data LLRs $\{L(u_k|\mathbf{y}, \mathbf{h})\}$ (10.4) is represented by the extrinsic LLRs $\{L_e(u_k)\}$. In other words, referring to (10.15), it can be stated that $L_e(u_0)$ represents the improvement in our knowledge concerning the value of the bit u_0 acquired from decoding. These considerations suggest exploiting the *extrinsic information about the information bits* (such as u_0 and u_1) as a priori values in an iterative (turbo) decoding strategy for a concatenated coding scheme using the above-mentioned SPC code as a component code. This is further discussed in Example 10.5, where a specific product code is considered.

Finally it is important to note that, since:

$$\frac{\Pr\{x_k = 1\}}{\Pr\{x_k = 0\}} = \exp[L(x_k)] \qquad (10.17)$$

and

$$\frac{f(y_k|x_k = 1, h_k)}{f(y_k|x_k = 0, h_k)} = \exp[L_c(y_k)], \qquad (10.18)$$

the LLR $L_e(u_0)$ (10.12) can easily be put in the form:

$$L_e(u_0) = \frac{\exp[L_c(y_2)] + \exp[L(x_1)]\,\exp[L_c(y_1)]}{1 + \exp[L(x_1)]\,\exp[L_c(y_1)]\,\exp[L_c(y_2)]}. \qquad (10.19)$$

---

[3] Note that this term is *not* influenced by the presence of channel coding.

Then, if we define the commutative operator $\Diamond$ through the relationship:

$$L_1 \Diamond L_2 \triangleq \ln \frac{\exp (L_1) + \exp (L_2)}{1 + \exp (L_1 + L_2)} \tag{10.20}$$

involving the pair of LLRs L_1 and L_2, (10.19) can be rewritten as (see also (10.6)):

$$L_e(u_0) = L_e(x_1 \boxplus x_2) = L_c(y_2) \Diamond [L_c(y_1) + L(x_1)]. \tag{10.21}$$

Finally, we note that (10.20) can approximated as (e.g., see [1100], p. 430, eq. (12)):

$$L_1 \Diamond L_1 \approx (-1) \, \text{sgn} \, [L_1] \, \text{sgn} \, [L_2] \, \min \, (|L_1|, |L_2|), \tag{10.22}$$

which lends itself to a simple implementation and, in particular, does not require the computation of exponentials.

□

Generally speaking a SiSo decoder, such as that illustrated in the previous example, can be represented by the block diagram of Figure 10.5. This decoder processes the a priori LLRs $\{L(u_i)\}$ for all the *information* bits (not the parity bits) and the channel LLRs $\{L_c(y_i)\}$ for all the *coded* bits (see the following example) and generates the extrinsic values $\{L_e(u_i)\}$ and the data LLRs $\{L(u_i|\mathbf{y}, \mathbf{h})\}$ for all the *information* bits.

The availability of SiSo decoding algorithms allows us to adopt a turbo approach to the decoding of concatenated codes, as illustrated in the following example for a specific two-dimensional product code.

Example 10.5.2 Let us now use the (3,2) SPC code of Example 10.5.1 to generate the rate-1/2 two-dimensional product code whose structure is illustrated in Figure 10.6 (note that the PoP bit is missing in the code matrix). In this case the binary message $\mathbf{u} = [u_0, u_1, u_2, u_3]$ is encoded into the codeword $\mathbf{x} = [x_0, x_1, x_2, x_3, x_{12}, x_{34}, x_{13}, x_{24}]$ with $x_i = u_i$ for $i = 1, 2, 3, 4$ and:

$$x_{12} = x_1 \boxplus x_2, x_{34} = x_3 \boxplus x_4,$$

$$x_{13} = x_1 \boxplus x_3, x_{24} = x_2 \boxplus x_4. \tag{10.23}$$

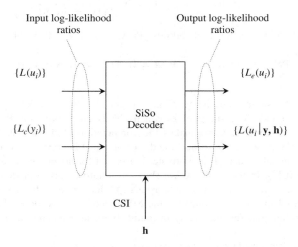

Figure 10.5   SiSo decoder structure.

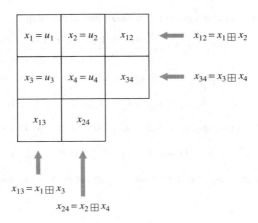

Figure 10.6   Two-dimensional product code of Example 10.5.

If we adopt the same channel model as in Example 10.5.1, the receiver observes:

$$\mathbf{y} \triangleq [y_1, y_2, y_3, y_4, y_{12}, y_{34}, y_{13}, y_{24}] \tag{10.24}$$

where

$$y_i = h_i(2x_i - 1) + n_i \tag{10.25}$$

with $i = 1, 2, 3, 4$, and:

$$y_{ij} = h_{ij}(2x_{ij} - 1) + n_{ij} \tag{10.26}$$

with $(i, j) \in \{(1, 2), (3, 4), (1, 3), (2, 4)\}$. Here, $\{n_i, n_{ij}\}$ and $\{h_i, h_{ij}\}$ denote the usual noise and fading variables, respectively.

APP decoding of this code requires the evaluation of the LLRs:

$$L(u_k|\mathbf{y}, \mathbf{h}) = L(x_k|\mathbf{y}, \mathbf{h}) \triangleq \ln \frac{\Pr\{x_k = 1|\mathbf{y}, \mathbf{h}\}}{\Pr\{x_k = 0|\mathbf{y}, \mathbf{h}\}} \tag{10.27}$$

with $k = 0, 1, 2, 3$. In principle, the probabilities $\{\Pr\{x_k = b|\mathbf{y}\}, k = 0, 1\}$ of (10.27) can be computed as:

$$\Pr\{x_k = b|\mathbf{y}, \mathbf{h}\} = \sum_{\tilde{\mathbf{x}} \in C_b} \Pr\{\mathbf{x} = \tilde{\mathbf{x}}\} \Pr\{x_k = b|\mathbf{y}, \mathbf{x} = \tilde{\mathbf{x}}, \mathbf{h}\}, \tag{10.28}$$

where C_b represents the subset[4] consisting of all the codewords for which $x_k = b$, with $b = 0, 1$ and $\Pr\{\mathbf{x} = \tilde{\mathbf{x}}\} = 1/16$ under the assumption of equally likely codewords. For this code, both C_0 and C_1 consist of eight codewords and this makes the computation of the probability $\Pr\{x_k = b|\mathbf{y}, \mathbf{h}\}$ through (10.28) feasible. However, for product codes of practical interest the use of (10.28) entails an unacceptable computational burden. This motivates the search for novel algorithms able to efficiently estimate the LLRs (10.27). In what follows we show an iterative (turbo) procedure for this specific application. Its extension to a general two-dimensional product code is discussed in Section 10.5.3. This procedure is based on the availability of the SiSo decoding procedure illustrated in Example 10.5.1 and on its exploitation for alternately decoding (with an exchange of soft information) the

---

[4] The subsets C_0 and C_1 form a partition of the codeword set.

two component codes of the given scheme. In the proposed scheme, each *decoding stage* consists of a *horizontal decoding step* followed by a *vertical decoding step*. In any horizontal decoding step the *extrinsic information* of the information bits $\{x_i, i = 1, 2, 3, 4\}$ is evaluated by exploiting the constraints expressed by the horizontal parity check equations (see Figure 10.6) and (apart from the first iteration) the extrinsic information coming from the last vertical step. Similarly, in any vertical decoding step the *extrinsic information* of the information bits $\{x_i, i = 1, 2, 3, 4\}$ is computed using the extrinsic information coming from the horizontal step just completed in the same decoding stage and the constraints expressed by the vertical parity check equations (see Figure 10.6). The decoding procedure can be summarized as follows:

1. *Initialization*. The a priori LLRs $\{L(x_i), i = 1, 2, 3, 4\}$ are computed according to (10.8) (in particular, if $\Pr\{u_i = 0\} = \Pr\{u_i = 1\} = 1/2$, we have $L(x_i) = 0$); in addition, the channel LLRs $\{L_c(x_i), L_c(x_{ij})\}$ are evaluated for all possible values of i and (i, j), processing $\{y_i, y_{ij}\}$ according to (10.16).

2. *First horizontal step*. The *horizontal extrinsic information* $\{L_{e,o}^{(1)}(x_i)\}$ of all the information bits $\{x_i\}$ is evaluated as (see (10.21)):

$$L_{e,o}^{(1)}(x_1) = L_c(y_{12}) \lozenge [L_c(y_2) + L(x_2)], \tag{10.29}$$

$$L_{e,o}^{(1)}(x_2) = L_c(y_{12}) \lozenge [L_c(y_1) + L(x_1)], \tag{10.30}$$

$$L_{e,o}^{(1)}(x_3) = L_c(y_{34}) \lozenge [L_c(y_4) + L(x_4)] \tag{10.31}$$

and

$$L_{e,o}^{(1)}(x_4) = L_c(y_{34}) \lozenge [L_c(y_3) + L(x_3)]. \tag{10.32}$$

3. *First vertical step*. The *vertical extrinsic information* $\{L_{e,v}^{(1)}(x_i)\}$ is computed using $\{L_{e,o}^{(1)}(x_i)\}$ in (10.29)–(10.32) in place of $\{L(x_i)\}$ as a priori information. Then we have:

$$L_{e,v}^{(1)}(x_1) = L_c(y_{13}) \lozenge [L_c(y_3) + L_{e,o}^{(1)}(x_3)], \tag{10.33}$$

$$L_{e,v}^{(1)}(x_2) = L_c(y_{24}) \lozenge [L_c(y_4) + L_{e,o}^{(1)}(x_4)], \tag{10.34}$$

$$L_{e,v}^{(1)}(x_3) = L_c(y_{13}) \lozenge [L_c(y_1) + L_{e,o}^{(1)}(x_1)] \tag{10.35}$$

and

$$L_{e,v}^{(1)}(x_4) = L_c(y_{24}) \lozenge [L_c(y_2) + L_{e,o}^{(1)}(x_2)]. \tag{10.36}$$

At the end of this step, the first decoding stage is over. Now an estimate:

$$\tilde{L}^{(1)}(u_k) = \tilde{L}^{(1)}(x_k) = L_c(y_k) + L_{e,o}^{(1)}(x_k) + L_{e,v}^{(1)}(x_k) \tag{10.37}$$

of the LLR $L(u_k | \mathbf{y}, \mathbf{h})$ (10.27), with $k = 1, 2, 3, 4$, is available. This relationship follows from (10.15) and from the fact that $L_{e,o}^{(1)}(x_i)$ is the last a priori LLR used in the decoding procedure, whereas $L_{e,v}^{(1)}(x_i)$ represents the last extrinsic LLR generated for x_i. Therefore, given $\tilde{L}^{(1)}(u_k)$ (10.37) a decision:

$$\hat{u}_k = \begin{cases} 1 & \text{if } \tilde{L}^{(1)}(u_k) > 0 \\ 0 & \text{otherwise} \end{cases} \tag{10.38}$$

can be taken. However, generally speaking, the quality of the data estimates can be improved by resorting to multiple decoding stages (i.e., iterations). The *l*th stage consists of steps 4 and 5 described below. After $K > 1$ iterations the decoding procedure stops; then a final estimate of the data APPs can be generated and data decisions can be taken, as explained below (step 6).

4. *l*th *horizontal step* ($l \geq 2$). Compute $\{L_{e,o}^{(l)}(x_i)\}$ according to (10.33)–(10.36), but using $\{L_{e,v}^{(l-1)}(x_i)\}$ as the a priori LLRs.

5. *l*th *vertical step* ($l \geq 2$). Compute $\{L_{e,v}^{(l)}(x_i)\}$ according to (10.33)–(10.36), using $\{L_{e,o}^{(l)}(x_i)\}$ as the a priori LLRs.

6. *Termination.* At the end of the *K*th iteration the LLR (see (10.37)):

$$\tilde{L}^{(K)}(u_k) = \tilde{L}^{(K)}(x_k) = L_c(y_k) + L_{e,o}^{(K)}(x_k) + L_{e,v}^{(K)}(x_k) \tag{10.39}$$

is generated for $k = 1, 2, 3, 4$, and data decisions are taken according to (10.38) (with $\tilde{L}^{(K)}(u_k)$ in place of $\tilde{L}^{(1)}(u_k)$).

Finally, it is important to note that the benefit provided by each additional decoding stage is expected to reduce as the number of accomplished iterations increases. This is due to the fact that the extrinsic information generated by the decoding procedure described above becomes more and more correlated as K gets larger. This is confirmed by Figure 10.7, illustrating the BER performance offered by ML decoding[5] and by the turbo decoding procedure for $K = 1$ in the presence of (a) Rayleigh fading channel (with ideal interleaving) and (b) an AWGN channel; the error performance achievable at the decoder input (i.e., without exploiting the parity check bits) is also shown for comparison. Note, in particular, that in this case a negligible energy gain is provided by increasing K beyond 1, since the first decoding stage approaches ML decoding performance closely.

□

The turbo approach to decoding illustrated in Example 10.5 was first applied to convolutional codes concatenated in a parallel fashion, that is, to the so-called *turbo codes* [1481]. Note, however,

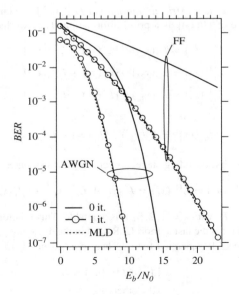

Figure 10.7   BER versus average E_b/N_0 for the (single stage) iterative decoding of the (3,2) SPC code described in Example 10.5. The error performance at the decoder input and that provided by ML decoding (MLD) are also shown for comparison. Both AWGN and *frequency-flat* (FF) fading channels are considered.

---

[5] This is equivalent to that of MAP decoding under the assumption of equally likely codewords.

that strictly speaking there is nothing "turbo" in such codes, since only their decoding uses a sort of "turbo" feedback. It is also important to point out that this decoding method for concatenated schemes was later recognized as a specific application of the turbo principle [1611]. In fact, the same conceptual approach can be successfully applied to various detection/decoding problems concerning, for instance, serial concatenation of coding schemes [1546], channel equalization [1660] and multiuser detection [1661].

10.5.2 SiSo Decoding Algorithms

In this subsection some standard SiSo decoding algorithms for convolutional and block codes are analyzed. In particular, we focus first on the modified BCJR algorithm for binary systematic convolutional codes. Then we provide an overview of SiSo decoding algorithms for block codes.

10.5.2.1 MAP Decoding of Binary RSC Codes

The structure of the MAP decoding algorithm for convolutional codes and its derivation have already been analyzed in Section 9.2.4. Here we revisit these topics from a different perspective with the aim of showing how the extrinsic information can be generated by a MAP decoder in the case of a binary RSC code (see [1100, pp. 435–437] and [1614]). In particular, we focus on MAP decoding of a code having rate $1/n$, n_s states and whose output bits are transmitted using BPSK modulation; the encoder response to its lth input bit u_l is the binary n-dimensional vector $\mathbf{x}_l \triangleq [x_l^{(0)}, x_l^{(1)}, \ldots, x_l^{(n-1)}]$ (with $x_l^{(0)} = u_l$), which is mapped into the BPSK symbol vector $\mathbf{c}_l \triangleq [c_l^{(0)}, c_l^{(1)}, \ldots, c_l^{(n-1)}]$, with $c_l^{(p)} = 2x_l^{(p)} - 1$ ($p = 0, 1, \ldots, n-1$). ISI free transmission of the information bit vector $\mathbf{u} \triangleq [u_0, u_1, \ldots, u_{N-1}]$ over a frequency-flat fading channel is assumed, so that the BCJR algorithm processes the received signal vector $\mathbf{y} \triangleq [\mathbf{y}_0, \mathbf{y}_1, \ldots, \mathbf{y}_{N-1}]$, with $\mathbf{y}_l \triangleq [y_l^{(0)}, y_l^{(1)}, \ldots, y_l^{(n-1)}]$ and $y_l^{(p)} = c_l^{(p)} h_l^{(p)} + n_l^{(p)}$, where $h_l^{(p)}$ and $n_l^{(p)}$ denote the channel distortion and the noise, respectively, affecting the symbol $c_l^{(p)}$. Statistical independence between fading (noise) samples, and between noise and fading is assumed, so that the communication channel fed by $\mathbf{c} \triangleq [\mathbf{c}_0, \mathbf{c}_1, \ldots, \mathbf{c}_{N-1}]$ and generating $\mathbf{y}$ is *memoryless*. Moreover, the noise samples $\{n_i\}$ are iid complex random Gaussian variables, each with zero mean, variance $\sigma_n^2 = 2N_0/\bar{E}_s$ and iid real and imaginary parts. The goal of a MAP decoding algorithm is to generate the conditioned LLR:

$$L(u_l|\mathbf{y}, \mathbf{h}) = \ln \frac{\Pr\{u_l = 1|\mathbf{y}, \mathbf{h}\}}{\Pr\{u_l = 0|\mathbf{y}, \mathbf{h}\}} \tag{10.40}$$

for $l = 0, 1, \ldots, N - 1$, where $\mathbf{h}$ is the fading vector containing all the channel gains $\{h_l^{(p)}\}$. This LLR can be expressed as:

$$L(u_l|\mathbf{y}, \mathbf{h}) = \ln \frac{\sum\limits_{(\tilde{\sigma}_l, \tilde{\sigma}_{l+1}) \in C_1} \Pr\{\tilde{\sigma}_l, \tilde{\sigma}_{l+1}| \mathbf{y}, \mathbf{h}\}}{\sum\limits_{(\tilde{\sigma}_l, \tilde{\sigma}_{l+1}) \in C_0} \Pr\{\tilde{\sigma}_l, \tilde{\sigma}_{l+1}| \mathbf{y}, \mathbf{h}\}}, \tag{10.41}$$

where C_b (with $b = 0, 1$) denotes the set of all the possible paths (i.e., sequences of encoder states) which traverse the trellis branch connecting the states $\tilde{\sigma}_l$ and $\tilde{\sigma}_{l+1}$ and labeled by the input bit $\tilde{u}_l = u_l = b$. Applying Bayes' theorem, the probability $\Pr\{\tilde{\sigma}_l, \tilde{\sigma}_{l+1}| \mathbf{y}, \mathbf{h}\}$ appearing in (10.41) can be expressed as:

$$\Pr\{\tilde{\sigma}_l, \tilde{\sigma}_{l+1}| \mathbf{y}, \mathbf{h}\} = \frac{f(\mathbf{y}|\tilde{\sigma}_l, \tilde{\sigma}_{l+1}, \mathbf{h})}{f(\mathbf{y}|\mathbf{h})} \Pr\{\tilde{\sigma}_l, \tilde{\sigma}_{l+1}\}, \tag{10.42}$$

so that the LLR of (10.41) can also be computed as:

$$L(u_l|\mathbf{y}, \mathbf{h}) = \ln \frac{\displaystyle\sum_{(\tilde{\sigma}_l, \tilde{\sigma}_{l+1}) \in C_1} f(\mathbf{y}|\tilde{\sigma}_l, \tilde{\sigma}_{l+1}, \mathbf{h}) \Pr\{\tilde{\sigma}_l, \tilde{\sigma}_{l+1}\}}{\displaystyle\sum_{(\tilde{\sigma}_l, \tilde{\sigma}_{l+1}) \in C_0} f(\mathbf{y}|\tilde{\sigma}_l, \tilde{\sigma}_{l+1}, \mathbf{h}) \Pr\{\tilde{\sigma}_l, \tilde{\sigma}_{l+1}\}}. \qquad (10.43)$$

An efficient procedure for the computation of the product $f(\mathbf{y}|\tilde{\sigma}_l, \tilde{\sigma}_{l+1}, \mathbf{h}) \Pr\{\tilde{\sigma}_l, \tilde{\sigma}_{l+1}\}$ appearing in formula 10.43 can be derived as follows. If the vector $\mathbf{y}$ is represented as the ordered concatenation of $\mathbf{y}_{0:l-1}$, $\mathbf{y}_l$ and $\mathbf{y}_{l+1:N-1}$, where $\mathbf{y}_{l_1:l_2} \triangleq [\mathbf{y}_{l_1}, \mathbf{y}_{l_1+1}, \ldots, \mathbf{y}_{l_2}]$, then the product can be expressed as:

$$\begin{aligned}
f(\mathbf{y}|\tilde{\sigma}_l, \tilde{\sigma}_{l+1}, \mathbf{h}) \Pr\{\tilde{\sigma}_l, \tilde{\sigma}_{l+1}\} &= f(\mathbf{y}_{0:l-1}, \mathbf{y}_l, \mathbf{y}_{l+1:N-1}|\tilde{\sigma}_l, \tilde{\sigma}_{l+1}, \mathbf{h}) \Pr\{\tilde{\sigma}_l, \tilde{\sigma}_{l+1}\} \\
&= f(\mathbf{y}_{0:l-1}|\mathbf{y}_l, \mathbf{y}_{l+1:N-1}, \tilde{\sigma}_l, \tilde{\sigma}_{l+1}, \mathbf{h}) f(\mathbf{y}_l|\mathbf{y}_{l+1:N-1}, \tilde{\sigma}_l, \tilde{\sigma}_{l+1}, \mathbf{h}) \\
&\quad \cdot f(\mathbf{y}_l|\tilde{\sigma}_l, \tilde{\sigma}_{l+1}, \mathbf{h}) \Pr\{\tilde{\sigma}_{l+1}|\tilde{\sigma}_l\} \Pr\{\tilde{\sigma}_l\} \\
&= f(\mathbf{y}_{0:l-1}|\tilde{\sigma}_l, \mathbf{h}) f(\mathbf{y}_l|\tilde{\sigma}_l, \tilde{\sigma}_{l+1}, \mathbf{h}) \qquad (10.44) \\
&\quad \cdot f(\mathbf{y}_{l+1:N-1}|\tilde{\sigma}_{l+1}, \mathbf{h}) \Pr\{\tilde{\sigma}_{l+1}|\tilde{\sigma}_l\} \Pr\{\tilde{\sigma}_l\}.
\end{aligned}$$

If we now assume that $u_l = b$ and that all the trellis states are equally likely (so that $\Pr\{\tilde{\sigma}_l\} = 1/n_s$), then we note that $\Pr\{\tilde{\sigma}_{l+1}|\tilde{\sigma}_l\} = \Pr\{\tilde{u}_l = b\} = \Pr\{u_l = b\}$ and can define:

$$\alpha_{l-1}(\tilde{\sigma}_l) \triangleq f(\mathbf{y}_{0:l-1}|\tilde{\sigma}_l, \mathbf{h}), \qquad (10.45)$$

$$\beta_l(\tilde{\sigma}_{l+1}) \triangleq f(\mathbf{y}_{l+1:N-1}|\tilde{\sigma}_{l+1}, \mathbf{h}), \qquad (10.46)$$

$$\gamma_l(\tilde{\sigma}_l, \tilde{\sigma}_{l+1}) \triangleq f(\mathbf{y}_l|\tilde{\sigma}_l, \tilde{\sigma}_{l+1}, \mathbf{h}) \Pr\{u_l = b\}, \qquad (10.47)$$

so that (10.44) can be written in the form:

$$f(\mathbf{y}|\tilde{\sigma}_l, \tilde{\sigma}_{l+1}, \mathbf{h}) \Pr\{\tilde{\sigma}_l, \tilde{\sigma}_{l+1}\} = \frac{1}{n_s} \alpha_{l-1}(\tilde{\sigma}_l) \gamma_l(\tilde{\sigma}_l, \tilde{\sigma}_{l+1}) \beta_l(\tilde{\sigma}_{l+1}). \qquad (10.48)$$

Substituting 10.48 into (10.43) gives:

$$L(u_l|\mathbf{y}, \mathbf{h}) = \ln \frac{\displaystyle\sum_{(\tilde{\sigma}_l, \tilde{\sigma}_{l+1}) \in C_1} \alpha_{l-1}(\tilde{\sigma}_l) \gamma_l(\tilde{\sigma}_l, \tilde{\sigma}_{l+1}) \beta_l(\tilde{\sigma}_{l+1})}{\displaystyle\sum_{(\tilde{\sigma}_l, \tilde{\sigma}_{l+1}) \in C_0} \alpha_{l-1}(\tilde{\sigma}_l) \gamma_l(\tilde{\sigma}_l, \tilde{\sigma}_{l+1}) \beta_l(\tilde{\sigma}_{l+1})}. \qquad (10.49)$$

This shows that evaluation of the LLR for the bit u_l requires knowlege of the quantities $\{\gamma_l(\tilde{\sigma}_l, \tilde{\sigma}_{l+1})\}$, $\{\alpha_{l-1}(\tilde{\sigma}_l)\}$ and $\{\beta_l(\tilde{\sigma}_{l+1})\}$. Note, however, that the quantities $\{\gamma_l(\tilde{\sigma}_l, \tilde{\sigma}_{l+1})\}$ depend on $\mathbf{y}_l$ only, whereas $\{\alpha_{l-1}(\tilde{\sigma}_l)\}$ $(\{\beta_l(\tilde{\sigma}_{l+1})\})$ depend on all the past (future) received samples, that is, on the vector $\mathbf{y}_{0:l-1}$ $(\mathbf{y}_{l+1:N-1})$. Fortunately, $\{\alpha_{l-1}(\tilde{\sigma}_l)\}$ and $\{\beta_l(\tilde{\sigma}_{l+1})\}$ can be computed recursively, since:

$$\begin{aligned}
\alpha_l(\tilde{\sigma}_{l+1}) &\triangleq f(\mathbf{y}_{0:l}|\tilde{\sigma}_{l+1}, \mathbf{h}) \\
&= f(\mathbf{y}_{0:l-1}, \mathbf{y}_l|\tilde{\sigma}_{l+1}, \mathbf{h}) \\
&= \sum_{\tilde{\sigma}_l} f(\mathbf{y}_{0:l-1}, \mathbf{y}_l|\tilde{\sigma}_l, \tilde{\sigma}_{l+1}, \mathbf{h}) \Pr\{\tilde{\sigma}_{l+1}|\tilde{\sigma}_l\} \\
&= \sum_{\tilde{\sigma}_l} f(\mathbf{y}_{0:l-1}|\tilde{\sigma}_l, \tilde{\sigma}_{l+1}, \mathbf{y}_l, \mathbf{h}) f(\mathbf{y}_l|\tilde{\sigma}_l, \tilde{\sigma}_{l+1}, \mathbf{h}) \Pr\{\tilde{\sigma}_{l+1}|\tilde{\sigma}_l\}
\end{aligned}$$

$$= \sum_{\tilde{\sigma}_l} f\left(\mathbf{y}_{0:l-1}|\tilde{\sigma}_l, \mathbf{h}\right) f\left(\mathbf{y}_l|\tilde{\sigma}_l, \tilde{\sigma}_{l+1}, \mathbf{h}\right) \Pr\{\tilde{\sigma}_{l+1}|\tilde{\sigma}_l\}$$

$$= \sum_{\tilde{\sigma}_l} \alpha_{l-1}(\tilde{\sigma}_l)\gamma_l(\tilde{\sigma}_l, \tilde{\sigma}_{l+1}) \tag{10.50}$$

and

$$\beta_{l-1}(\tilde{\sigma}_l) \triangleq f\left(\mathbf{y}_{l:N-1}|\tilde{\sigma}_l, \mathbf{h}\right)$$

$$= f\left(\mathbf{y}_{l+1:N-1}, \mathbf{y}_l|\tilde{\sigma}_l, \mathbf{h}\right)$$

$$= \sum_{\tilde{\sigma}_{l+1}} f\left(\mathbf{y}_{l+1:N-1}, \mathbf{y}_l|\tilde{\sigma}_l, \tilde{\sigma}_{l+1}, \mathbf{h}\right) \Pr\{\tilde{\sigma}_{l+1}|\tilde{\sigma}_l\}$$

$$= \sum_{\tilde{\sigma}_{l+1}} f\left(\mathbf{y}_{l+1:N-1}|\tilde{\sigma}_l, \tilde{\sigma}_{l+1}, \mathbf{h}\right) f\left(\mathbf{y}_l|\tilde{\sigma}_l, \tilde{\sigma}_{l+1}, \mathbf{h}\right) \Pr\{\tilde{\sigma}_{l+1}|\tilde{\sigma}_l\}$$

$$= \sum_{\tilde{\sigma}_{l+1}} f\left(\mathbf{y}_{l+1:N-1}|\tilde{\sigma}_{l+1}, \mathbf{h}\right) f\left(\mathbf{y}_l|\tilde{\sigma}_l, \tilde{\sigma}_{l+1}, \mathbf{h}\right) \Pr\{\tilde{\sigma}_{l+1}|\tilde{\sigma}_l\}$$

$$= \sum_{\tilde{\sigma}_{l+1}} \beta_l(\tilde{\sigma}_{l+1})\gamma_l(\tilde{\sigma}_l, \tilde{\sigma}_{l+1}). \tag{10.51}$$

It is worth pointing out that (10.50) and (10.51) are structurally equivalent to (9.261) and (9.262), respectively. Therefore, in principle, the computation of the LLR (10.40) can be carried out using the recursive procedure of the FBA (see Section 9.2.4 for further details). In the present case, however, we are interested in specifying an algorithm that can process a priori information and the received signal samples to generate the data extrinsic information. This goal can be achieved as follows. From the definition (see (10.9)):

$$L(u_l) \triangleq \ln \frac{\Pr\{u_l = 1\}}{\Pr\{u_l = 0\}} \tag{10.52}$$

it is easily seen that:

$$\Pr\{u_l = b\} = A_l \exp [bL(u_l)] \tag{10.53}$$

where $A_l \triangleq 1/[1 + \exp [L(u_l)]]$ is independent of b. Similarly, the conditional pdf $f(\mathbf{y}_l|\tilde{\sigma}_l, \tilde{\sigma}_{l+1}, \mathbf{h})$ appearing in the definition (10.47) of $\gamma_l(\tilde{\sigma}_l, \tilde{\sigma}_{l+1})$ can easily be put in the form:

$$f(\mathbf{y}_l|\tilde{\sigma}_l, \tilde{\sigma}_{l+1}, \mathbf{h}) = B_l \exp \left[\frac{1}{2} \sum_{p=0}^{n-1} L_c\left(y_l^{(p)}\right) \tilde{c}_l^{(p)} \right]$$

$$= B_l \exp \left[\frac{1}{2} L_c\left(y_l^{(0)}\right)(2b - 1) + \frac{1}{2} \sum_{p=1}^{n-1} L_c(y_l^{(p)})\tilde{c}_l^{(p)} \right], \tag{10.54}$$

where B_l is independent of b. Here $\tilde{c}_l^{(p)} = 2\tilde{x}_l^{(p)} - 1$, $\tilde{\mathbf{x}}_l \triangleq [\tilde{x}_l^{(0)}, \tilde{x}_l^{(1)}, \ldots, \tilde{x}_l^{(n-1)}]$ is the output vector generated by the encoder in state $\tilde{\sigma}_l$ and fed by $\tilde{u}_l = b$, and (see (10.16)):

$$L_c(y_l^{(p)}) \triangleq \frac{4 \operatorname{Re}\{y_l^{(p)}(h_l^{(p)})^*\}}{\sigma_n^2}. \tag{10.55}$$

with $p = 0, 1, \ldots, n - 1$. Then, substituting (10.53) and (10.54) into (10.47) yields:

$$\gamma_l(\tilde{\sigma}_l, \tilde{\sigma}_{l+1}) = A_l B_l \exp \left(bL(u_l) + \frac{1}{2} L_c(y_l^{(0)})(2b - 1) \right) \gamma_l^{(e)}(\tilde{\sigma}_l, \tilde{\sigma}_{l+1}), \tag{10.56}$$

where

$$\gamma_l^{(e)}(\tilde{\sigma}_l, \ \tilde{\sigma}_{l+1}) \triangleq \exp \left(\frac{1}{2} \sum_{p=1}^{n-1} L_c \left(y_l^{(p)} \right) \tilde{c}_l^{(p)} \right). \tag{10.57}$$

Finally, substituting (10.56) into (10.49) and taking the logarithm leads to the result:

$$L(u_l|\mathbf{y}, \mathbf{h}) = L(u_l) + L_c(y_l^{(0)}) + L_e(u_l), \tag{10.58}$$

where

$$L_e(u_l) = \ln \frac{\displaystyle\sum_{(\tilde{\sigma}_l, \ \tilde{\sigma}_{l+1}) \in C_1} \alpha_{l-1}(\tilde{\sigma}_l) \gamma_l^{(e)}(\tilde{\sigma}_l, \ \tilde{\sigma}_{l+1}) \beta_l(\tilde{\sigma}_{l+1})}{\displaystyle\sum_{(\tilde{\sigma}_l, \ \tilde{\sigma}_{l+1}) \in C_0} \alpha_{l-1}(\tilde{\sigma}_l) \gamma_l^{(e)}(\tilde{\sigma}_l, \ \tilde{\sigma}_{l+1}) \beta_l(\tilde{\sigma}_{l+1})}. \tag{10.59}$$

Note that (10.58) has the same structure as (10.15), but the algorithm for the generation of the extrinsic information $L_e(u_l)$ is substantially different from that derived with the product code of Example 10.5.2, as seen from a comparison of (10.59) with (10.14). In particular, $L_e(u_l)$ (10.14) can be computed via a forward–backward procedure similar to that illustrated in Section 9.2.4.

Finally, it is worth mentioning the following:

1. The MAP algorithm derived above is often called the *modified* BCJR algorithm in the literature.
2. To simplify its implementation, a log-MAP rule is usually adopted (see Section 9.2.4).
3. An alternative is the SOVA, mentioned at the end of Section 9.2.4, which offers the relevant advantage of substantially smaller complexity than the MAP decoder (even if implemented in Log-MAP form). A detailed derivation of this algorithm can be found in [1100, pp. 435–437], where its application to the decoding of systematic RSC codes is explicitly discussed.

10.5.2.2 MAP Decoding of Block Codes

The success of turbo codes inspired the invention of *turbo product codes* (or *block turbo codes*) [1503], which are simply product codes decoded using iterative MAP decoding. As illustrated in Section 9.3.1, product codes usually consist of two component systematic block codes separated by a block (row–column) interleaver. It is well known that the codewords of a linear binary (n, k) block code C can be represented as paths through a trellis of depth n with at most 2^{n-k} states [1662, 1523, 1573] (this topic is further discussed in Section 10.8, where a trellis representation for a specific binary block code is provided; see Example 10.8.2). As a consequence, MAP decoding algorithms structurally similar to those illustrated in Section 10.5.2.1 can be developed for systematic linear block codes. Details can be found in [1100, pp. 437–440] and references therein. Most block codes do not have a compact trellis, so that this approach to SiSo decoding can turn out to be prohibitively complex. It is also worth noting that data APPs of code C can evaluated using the codewords of its dual code C^+ [1663, 1662]. For this reason, if C^+ has fewer codewords than C, then decoding complexity can be reduced [1100].

A completely different approach to SiSo decoding for turbo product codes is offered by *list decoding techniques* (see Section 9.1.6.2). In using these techniques a subset of all possible codewords is searched and a list of possible (test) codewords together with their Euclidean distances from the received signal is stored. The other steps of the decoding procedure depend on the specific approach. In particular, the available approaches can roughly be divided into two groups. In the decoding algorithms of the first group, test codewords are decoded (e.g., Chase-based decoding can be employed [1503]), while in those belonging to the second group test sequences are encoded (e.g., order-*i* reprocessing can be adopted [1507]). Both types of algorithm try to generate a *list* of the "most likely" transmitted codewords. Typically an ordered list is stored by the decoding algorithm, with ordering being from most to least likely. Let $\hat{\mathbf{x}} \triangleq [\hat{x}_0, \hat{x}_1, \ldots, \hat{x}_{n-1}]$ denote the most likely codeword in the list. This list is used to estimate the LLR:

$$L(x_l|\mathbf{y}, \mathbf{h}) = \ln \frac{\Pr\{x_l = 1|\mathbf{y}, \mathbf{h}\}}{\Pr\{x_l = 0|\mathbf{y}, \mathbf{h}\}}, \tag{10.60}$$

for $l = 0, 1, \ldots, n-1$, where $\mathbf{x} = [x_0, x_1, \ldots, x_{n-1}]$ is the transmitted codeword and $\mathbf{h}$ is the associated CSI (see Section 10.5.1). Then a search is performed to find a competing codeword $\mathbf{b}^{(l)} = [b_0^{(l)}, b_1^{(l)}, \ldots, b_{n-1}^{(l)}]$ for each position (i.e., for $l = 0, 1, \ldots, n-1$) in the codeword. The competing codeword for position l is characterized by $b_l^{(l)} \neq \hat{x}_l$. Using Bayes' theorem, the LLR of (10.60) can be rewritten as:

$$L(x_l|\mathbf{y}, \mathbf{h}) = \ln \frac{\Pr\{x_l = 1|\mathbf{y}, \mathbf{h}\}}{\Pr\{x_l = 0|\mathbf{y}, \mathbf{h}\}} = \ln \frac{\displaystyle\sum_{\mathbf{x}|x_l=1} f(\mathbf{y}|\mathbf{x}, \mathbf{h})}{\displaystyle\sum_{\mathbf{x}|x_l=0} f(\mathbf{y}|\mathbf{x}, \mathbf{h})} \tag{10.61}$$

with $l = 0, 1, \ldots, N-1$. Then, if the max-log MAP approximation [1525] is exploited, expression 10.61 can be simplified as follows::

$$\begin{aligned} L(x_l|\mathbf{y}, \mathbf{h}) &\simeq \ln \frac{\max_{\mathbf{x}|x_l=1} f(\mathbf{y}|\mathbf{x}, \mathbf{h})}{\max_{\mathbf{x}|x_l=0} f(\mathbf{y}|\mathbf{x}, \mathbf{h})} \\ &= \ln \left(\max_{\mathbf{x}|x_l=1} f(\mathbf{y}|\mathbf{x}, \mathbf{h}) \right) - \ln \left(\max_{\mathbf{x}|x_l=0} f(\mathbf{y}|\mathbf{x}, \mathbf{h}) \right) \\ &\simeq (2\hat{x}_l - 1) \left[\ln \left(f\{\mathbf{y}|\hat{\mathbf{x}}, \mathbf{h}\} \right) - \ln \left(f\left(\mathbf{y}|\mathbf{b}^{(l)}, \mathbf{h}\right) \right) \right] \\ &= (2\hat{x}_l - 1) \ln \frac{f(\mathbf{y}|\hat{\mathbf{x}}, \mathbf{h})}{f(\mathbf{y}|\mathbf{b}^{(l)}, \mathbf{h})}. \end{aligned} \tag{10.62}$$

Equations for the AWGN channel are easily found by removing $\mathbf{h}$ from (10.62). The extrinsic information for x_l (with $l = 0, 1, \ldots, k-1$) can now be calculated by subtracting the soft input for the lth position from the soft output for the lth position computed using (10.62). The list decoder searches only a small subspace of the entire codeword space. As a result, a competing codeword may not be contained in the list for some positions. There have been various approaches suggested for estimating the extrinsic information for these positions. For instance, in [1503], preset (iteration-dependent) values are used (denoted β) and the extrinsic information for the lth position is evaluated as $(2\hat{x}_l - 1)\beta$. The values for β can be changed for each iteration (usually they increase with each iteration up to unity). An alternative is to use adaptive approaches [1664–1666]. Assuming a sufficiently large list is used, the decision in position l is reasonably reliable if no competing codeword is found. The reader can refer to [1638–1641, 1667–1669] and references therein for more detailed knowledge of this recent research area. Alternative approaches to MAP decoding of block codes include tree search (sequential) decoding [1670, 1671] and decoding on graphs (belief propagation) [1672]. Finally, we note that most of the work on turbo decoding of product codes has been developed for AWGN channels. Codes over fading channels typically require significantly larger list sizes [1673] and little substantive work has been published.

10.5.3 Applications

In this subsection we illustrate the application of the modified BJCR algorithm (described in Section 10.5.2.1) to the decoding of turbo codes. Then we discuss the use of SiSo decoders for block codes in Section 10.5.2.2, where their use in the turbo decoding of product codes is analyzed.

10.5.3.1 Decoding of Turbo Codes

We now discuss the application of the SiSo decoding algorithm described in Section 10.5.2.1 to the decoding of the turbo code resulting from the parallel concatenation of two component rate-1/2 RSC encoders shown in Figure 10.8; puncturing is not used in this case, so that the overall rate is $R = 1/3$.

The decoder structure, presented in Figure 10.9, includes a block generating the channel LLRs (on the basis of the noise variance σ_n^2, the vector of channel gains $\mathbf{h}$ and all received signal samples, see (10.55)), two MAP SiSo decoders (each tailored to one of the two RSC component codes), two interleavers (π), two deinterleavers (π^{-1}) and a hard decision device (represented as a threshold device). Moreover, each SiSo decoder has three inputs (each associated with a distinct term on the RHS of (10.58)) and a single output. Extrinsic information is extracted by subtracting channel and a priori LLRs from this output.

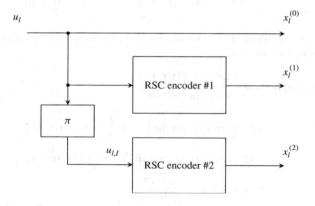

Figure 10.8 Turbo or PCCC scheme with rate $R = 1/3$.

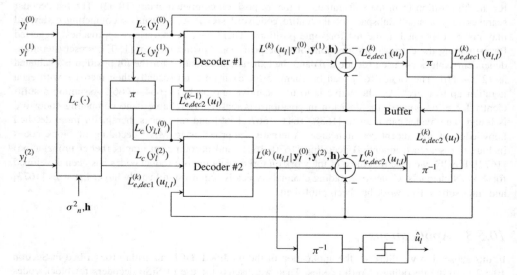

Figure 10.9 Turbo decoder for the PCCC scheme of Figure 10.8.

The overall decoder is fed by the sequence of vectors $\{\mathbf{y}_l \triangleq [y_l^{(0)}, y_l^{(1)}, y_l^{(2)}]\}$ generated by a *memoryless* channel in response to $\{\mathbf{x}_l \triangleq [x_l^{(0)} = u_l, x_l^{(1)}, x_l^{(2)}]\}$. Note, however, that the component decoders are not fed by the channel LLRs of all the received samples. In fact, the channel LLRs of the received samples $\{y_l^{(0)}\}$ and $\{y_l^{(1)}\}$ (forming the vectors $\mathbf{y}^{(0)}$ and $\mathbf{y}^{(1)}$, respectively) are sent to decoder #1, whereas the channel LLRs associated with the *interleaved* samples $\{y_l^{(0)}\}$ (denoted by $\{y_{l,I}^{(0)}\}$ and forming the vector $\mathbf{y}_I^{(0)}$) and $\{y_l^{(2)}\}$ (forming the vector $\mathbf{y}^{(2)}$) are input to decoder #2. Decoding is accomplished iteratively according to the following procedure.

In the *first iteration* (corresponding to $k = 1$) decoder #1 generates the data LLRs $\{L^{(1)}(u_l|\mathbf{y}^{(0)}, \mathbf{y}^{(1)}, \mathbf{h})\}$ for each transmitted information bit u_l (note that in this iteration all the extrinsic LLRs $\{L_{e,dec2}^{(0)}(u_l)\}$ are equal to zero, since no a priori information is available about the transmitted bits); in practice, this requires the evaluation of (10.49) by a SiSo decoding algorithm operating on the state trellis for the RSC encoder #1. Subtracting the channel LLRs $\{L_c^{(1)}(y_l^{(0)})\}$ from the corresponding $\{L^{(1)}(u_l|\mathbf{y}^{(0)}, \mathbf{y}^{(1)}, \mathbf{h})\}$ produces the extrinsic LLRs $\{L_{e,dec1}^{(1)}(u_l)\}$ (see (10.58)). The extrinsic information generated by the first decoder needs to be reordered (i.e., interleaved), since encoder #2 is fed by the interleaved sequence $\{u_{l,I}\}$ in place of $\{u_l\}$ (see Figure 10.9). The interleaved sequence $\{L_{e,dec1}^{(1)}(u_{l,I})\}$ is exploited by decoder #2 as a priori information[6] (i.e., its contribution corresponds to the first term on the RHS of (10.58)). Then, decoder #2 generates the data LLRs $\{L^{(1)}(u_{l,R}|\mathbf{y}_I^{(0)}, \mathbf{y}_2, \mathbf{h})\}$ for each transmitted information bit $u_{l,I}$. This requires the evaluation of (10.49) by means of an FBA operating on the state trellis of the RSC encoder #2. The data LLRs $\{L^{(1)}(u_{l,I}|\mathbf{y}_I^{(0)}, \mathbf{y}_2, \mathbf{h})\}$ can be exploited to generate (after deinterleaving) the hard decisions $\{\hat{u}_l\}$ on the transmitted sequence or to produce the extrinsic information $\{L_{e,dec2}^{(1)}(u_{l,I})\}$ about the interleaved sequence $\{u_{l,I}\}$. In the latter case the LLRs $\{L_c^{(0)}(y_{l,I}^{(0)})\}$ and $\{L_{e,dec2}^{(1)}(u_{l,I})\}$ need to be subtracted from $\{L^{(1)}(u_{l,I}|\mathbf{y}_I^{(0)}, \mathbf{y}_2, \mathbf{h})\}$. Then, $\{L_{e,dec2}^{(1)}(u_{l,I})\}$ is deinterleaved to generate $\{L_{e,dec2}^{(1)}(u_l)\}$, which can be used as a priori information to start the next iteration (associated with $k = 2$ in Figure 10.9).

It is important to note that the processing in turbo decoding proceeds in a blockwise fashion, since data are passed from one decoder to the other only when soft information about all the information bits is available. Thus, various buffering stages are needed in the implementation of the turbo decoder shown in Figure 10.9. Most of these stages have not been explicitly indicated in order to make the block diagram easy to read. In fact, the only buffer appearing in the figure has been included to show where data generation in each iteration ends and processing in the next iteration begins. Note that this decoding procedure is primarily based on the analytical result expressed by (10.58), which assumes independence among the three terms appearing in its RHS. This assumption no longer holds in the iterations following the first, since the extrinsic information exchanged becomes increasingly correlated. Therefore, it should be expected that the energy gain provided by the decoding procedure decreases as the number of decoding iterations increases.

The typical behavior of the turbo decoder structure described above is illustrated in the following example, which refers to a specific PCCC.

Example 10.5.3 Let us now assume that both the constituent encoders of the PCCC of Figure 10.8 are rate-1/2 convolutional encoders with memory $m = 4$, and transfer function (see Section 9.2.2):

$$\bar{\mathbf{G}}(D) = [1 \quad p(D)/q(D)], \tag{10.63}$$

where the forward and feedback generators $p(D)$ and $q(D)$ are represented by 17 and 31 in octal form, respectively [1675]. In addition, a diagonal interleaver [1597] of size 65 536 is employed in the encoder and the coded bits are mapped into BPSK symbols and transmitted over an AWGN channel. The modified BCJR in its log-MAP form is used for SiSo decoding of each constituent code.

---

[6] In the literature two methods for processing extrinsic information exist; a unified interpretation of these methods can be found in [1674].

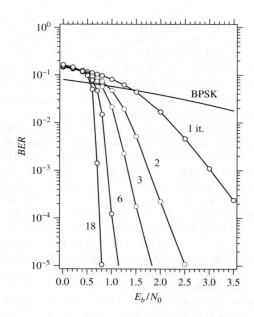

Figure 10.10 BER versus E_b/N_0 for the iterative decoding of the PCCC described in Example 10.9. The error performance of uncoded BPSK is also shown for comparison.

Moreover, to avoid instability problems at very low SNRs, the extrinsic information generated by the second decoder (and fed back to first) has been normalized as suggested in [1597, p. 1270, 2nd column]. The resulting error performance for 1, 2, 3, 6 and 18 iterations is shown in Figure 10.10. These results illustrate some important features of the BER performance of turbo codes:

1. For very low SNRs the BER remains high, since the iterative decoder is unable to correct the large number of errors (i.e., the decoding procedure does not converge).
2. At low SNRs, above a certain threshold, the BER curve exhibits the so-called 'waterfall' behavior, leading to excellent performance with only a few iterations.

□

These results can be generalized. In fact, generally speaking, the error rate performance offered by the iterative decoding of concatenated codes with interleavers and, in particular, of turbo codes, exhibits the behavior sketched in Figure 10.11 for an AWGN channel. This figure allows us to identify three different regions of the E_b/N_0 axis. The first is the so-called *nonconvergence region*, characterized by a large (almost constant) error probability and negligible iterative BER reduction. As the SNR per bit increases above a certain point, known as the *convergence abscissa*, the slope of the BER curve changes abruptly. In the new region, called the *waterfall region* (or *turbo cliff region*), the error probability drops quickly, in the manner of a waterfall. The waterfall region is followed by the third region, known as the *error floor region*, which is characterized by a significant decrease in the slope of the BER curve. This last means that any further performance improvement can require significant additional energy expense. Motivations for this behavior of the BER performance will be provided in Section 10.5.4, where the problem of analyzing the error performance analytically and devising bounds will be briefly discussed.

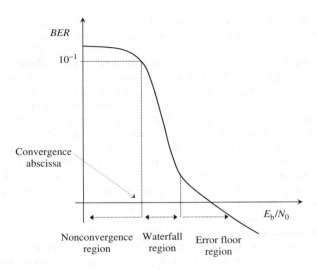

Figure 10.11 Qualitative behavior of the error performance provided by iterative decoding of a concatenated code with interleaving.

This discussion of the achievable performance with iterative decoding raises a number of technical issues of both practical and theoretical interest. Here we only consider briefly the following two problems:

- the need for a *stopping criterion* for the iterative decoding procedure described above;
- the understanding of the *convergence and stability properties* of such a procedure.

The first problem admits a naive solution which involves selecting a fixed value for the number of iterations. This approach is not usually efficient, since the correct codeword could be found after the first few iterations. Alternatives are offered by stopping criteria based on cross-entropy minimization (e.g., see [1100, pp. 433–434]) or by the use of a carefully selected outer error detection code. In the latter case a parity check is made at the end of each iteration and no further iteration is performed when the absence of errors is detected.

As far as the second problem is concerned, it is important to point out that an iterative decoder can be modeled as a *nonlinear dynamical feedback system*, in which extrinsic information messages are passed from one constituent decoder to the other (a geometrical interpretation of turbo decoding dynamics is provided in [1676]). Due to the intrinsic complexity of the problem only approximate solutions can be developed. In particular, it is commonly assumed the interleaver is ideal (i.e., it is random and its size goes to infinity), so that the extrinsic information can be modeled as a set of iid random variables, all characterized by a *Gaussian* pdf [1674, 1677, 1678]. Given this model, various methods can be used to analyze the convergence of the turbo decoder. Possibly the most famous is that of *density evolution* (DE), which aims to assess how the pdf of the extrinsic information evolves from iteration to iteration. This method was originally proposed by T. Richardson and R. Urbanke [1679] for the study of the decoding of LDPC codes but, generally speaking, allows one to analyze the asymptotic (i.e., as block size goes to infinity) behavior of iterative decoding procedures and, in particular, to compute their E_b/N_0 thresholds [1680, 1681]. An alternative to this approach has been developed by H. El Gamal *et al.* [1678] and is based on representing the action of each constituent decoder (in any turbo decoding scheme) as enhancing the SNR of the extrinsic information. Thus the knowledge of the SNR input–output relation for the extrinsic information of the constituent decoders allows us to establish whether the turbo decoder will converge at any SNR. Recently, this method has been generalized by J. W. Lee and R. E. Blahut to analyze the convergence and BER performance of finite-length block codes. Such a generalization is based on the so-called SNR *transfer characteristic band* chart of the extrinsic information [1682]. Another approach conceptually related to that proposed in [1678] was developed by S. Ten Brink, who proposed a graphical method to represent the flow of extrinsic information in turbo decoding as a decoding trajectory in the so-called *extrinsic information transfer* (EXIT) chart [1683]. In fact, even in this case the behavior of a turbo decoder is predicted from input–output relationships referring solely to its constituent decoders. In particular, each decoder is described by its extrinsic information transfer characteristic, relating the mutual information of its extrinsic output to the mutual information for its a priori knowledge and to the E_b/N_0 value.

The above methods are useful not only for understanding the problem of convergence, but also for code design. In the literature two different approaches have been proposed. One (already mentioned in Section 10.2.1) is purely analytical and is based on the assumption of a *purely random interleaver* in the encoding scheme [1615]. In this case, after designing the constituent encoders, code performance can be improved by resorting to a cut-and-trial approach to interleaver design. The second method requires the separate simulation of the behavior of each constituent encoder, and is based on the above-mentioned techniques based on DE or the EXIT charts. Generally speaking, the two approaches to code design provide some common criteria, but lead to slightly different coding schemes with nonoverlapping error performance. Typically, the codes devised through the second technique, when compared with the codes designed using the first, show a lower value of the converging abscissa and a faster convergence of iterative decoding, but reach the error floor more quickly.

Finally, it is worth noting that the turbo decoding approach illustrated above for a turbo coding scheme can be extended to the more general case of decoding *network codes*. The reader can refer to [1630], where the problem of designing SiSo decoding modules and their use in this scenario is widely discussed.

10.5.3.2 Decoding of Product Codes

In this subsection, we briefly describe a turbo decoder architecture for a binary two-dimensional product code C consisting of the concatenation of an (n_1, k_1) block code C_1 with a second (n_2, k_2) block code C_2 into a product code structure, with both codes defined on the same field, say GF(2). For this code the decoding procedure can be implemented as a cascade of *elementary turbo decoders* of the type illustrated in Figure 10.12 (the lth decoding stage is considered, with $l = 1, 2, \ldots, 2K$, where K denotes the overall number of horizontal and vertical decoding steps). The first stage of the turbo decoder processes the $n_1 \times n_2$ matrix $\mathbf{R}(1) = \mathbf{R}$ collecting the block of received signal samples ($\mathbf{W}_e(1) = \mathbf{0}_{n_1,n_2}$ and the value of the parameter $\alpha(1) = 0$) to accomplish soft decoding of the rows (columns) of C via a specific SiSo decoding algorithm for C_1 (C_2) (see Section 10.5.2.2). Note that, if Chase decoding is employed, a predefined weight $\beta(1)$ is assigned to the reliability of those components of the soft decision vector for which there is no competing codeword in the spanned space of codewords (see Section 10.5.2.2). Then the $n_1 \times n_2$ extrinsic information matrix $\mathbf{W}_e(1)$ is generated by subtracting the soft input from the soft output and is delivered to the second decoding stage. The second stage decodes the columns (or rows) of C using the matrix [1503]:

$$\mathbf{R}(2) = \mathbf{R}(1) + \alpha(2)\mathbf{W}_e(1) \tag{10.64}$$

in place of $\mathbf{R}$, where $\alpha(2)$ is a *scaling factor* that accounts for the fact that the standard deviation of the samples in the matrices $\mathbf{R}(2)$ and $\mathbf{W}_e(1)$ are usually different. In fact, the standard deviation of the extrinsic information is expected to be large during the first few decoding iterations and to decrease in successive iterations. The scaling factor reduces the effect of the extrinsic information in the soft decoder in the first few decoding steps when the BER is relatively high, so that stability is improved.

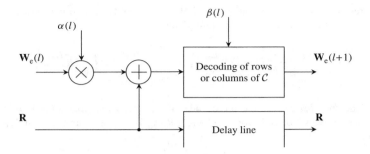

Figure 10.12   Block diagram of the *l*th elementary block turbo decoder.

The value of the scaling factor $\alpha(l)$ can be preset for each iteration [1503] or evaluated adaptively [1666]. This decoding architecture was originally proposed in [1503] for AWGN channels. In principle, a similar conceptual approach to turbo decoding can be developed for use in fading channels, provided that the SiSo module of each decoding stage is provided with CSI. Note, however, that most previous work in this area refers only to the AWGN channel and little work has been done for the fading environment.

10.5.4   Performance Bounds

An exact analysis of the BER performance of concatenated coding schemes is too ambitious a target. For this reason, performance bounds are usually developed. In this subsection we illustrate some analytical tools that are commonly exploited to develop such bounds, discussing their applications to turbo codes. In particular, we first apply these tools to explain the properties of the BER curve illustrated in Figure 10.11. Then we discuss the problem of the "average performance" provided by turbo codes.

To understand the origins of the essential features of the BER curve shown in Figure 10.11, let us refer, for simplicity, to a specific code, the binary turbo coding scheme illustrated in Figure 10.8. In our analysis the following assumptions are made about the input sequence $\{u_l\}$:

(a) it is of length N, where N is the interleaver size;
(b) it is an all-zero sequence.

Assumption (b) is motivated by the fact that the coding scheme is linear, since its constituent codes are linear and the interleaver, which is a permutation matrix, is also linear. Then, the all-zero codeword can be taken as a reference for evaluating the error performance, so that the overall number of possible incorrect (nonnull) codewords is $2^N - 1$. Given these assumptions, the BER performance of an ML soft decoder for the given turbo code in the presence of an AWGN channel can be upper-bounded, using a union bound approach, as [1684]:

$$P_b \leq \sum_{k=1}^{2^N-1} \frac{w_k}{N} Q\left(\sqrt{2R\frac{E_b}{N_0}d_{H,k}}\right), \tag{10.65}$$

where R is the rate of the turbo code, w_k is the Hamming weight (or *information weight*) of the kth possible input sequence (different from the all zero sequence), and $d_{H,k}$ is the overall Hamming weight of the resulting codeword. Grouping the RHS of inequality (10.65) into terms associated with codewords that have the same Hamming weight yields:

$$P_b \leq \sum_{d=d_{free}^{TC}}^{N/R} \frac{W_d}{N} Q\left(\sqrt{2R\frac{E_b}{N_0}d}\right) = \sum_{d=d_{free}^{TC}}^{N/R} \frac{N_d \tilde{w}_d}{N} Q\left(\sqrt{2R\frac{E_b}{N_0}d}\right), \tag{10.66}$$

where d_{free}^{TC} is the *free distance* of the turbo code, W_d is the *total information weight* of the input sequences producing the set of codewords of weight d, N_d is the total number or multiplicity of such codewords, and:

$$\tilde{w}_d \triangleq \frac{W_d}{N_d} \tag{10.67}$$

is the *average information weight per codeword*. It should be expected that for moderate and high SNRs the free distance term is the dominant one in the RHS of (10.66), so that the asymptotic performance of ML turbo decoding can be approximated as [1684]:

$$P_b \approx \frac{N_{d_{free}^{TC}} \tilde{w}_{d_{free}^{TC}}}{N} Q\left(\sqrt{2R\frac{E_b}{N_0}d_{free}^{TC}}\right), \tag{10.68}$$

where $N_{d_{free}^{TC}}$ is the multiplicity of the free distance codewords and $\tilde{w}_{d_{free}^{TC}}$ is the average information weight of the information sequences producing them. The last result expresses the so-called *free-distance asymptote* of a turbo code and, in practice, describes the *error floor region* of the turbo code. This result leads to the following important conclusions:

- The error floor behavior of a turbo code is dictated by its minimum distance.
- Turbo codes have a relatively small free distance, as evidenced by their relatively flat free distance asymptote.

The asymptotic expression (10.68) also suggests that the error floor can be modified by adopting one of the following two options:

1. Increasing the interleaver size N without changing d_{free}^{TC} and $\tilde{w}_{d_{free}^{TC}}$. This lowers the asymptote, but does not change its slope. Note that in this case the BER curve reaches its floor at higher SNRs, so that lower BERs are achieved even in the waterfall region. The phenomenon of BER decrease with the block size $1/N$ (see (10.66)) is known as *interleaving gain* [1615, 1616].
2. Increasing d_{free}^{TC} leaving N fixed. This requires changing the structure of constituent codes and modifies the slope of the asymptote.

As stated above, these comments primarily refer to the error performance characterizing turbo coding at moderate to large SNRs. The understanding of the error performance of turbo codes at low SNRs requires the analysis of the structure of their distance spectra resulting from the use of pseudorandom interleavers [1684]. In practice, in conventional coding schemes (such as convolutional coding), the distance spectrum is *spectrally dense*, that is, the path multiplicity N_d (and, consequently, the overall weight W_d) increases quickly as d gets large. For this reason, at low SNRs the contribution from the codewords having weight d exceeding the free distance of the code can be appreciably larger than the contribution of the free distance term, which becomes dominant only at large SNRs. On the other hand, the distance spectrum of turbo codes is *sparse* or *spectrally thin*, that is, N_d increases slowly with d, so that the free distance asymptote dominates error performance even for low values of E_b/N_0. Further details on this particular feature of distance spectra can be found in [1684].

The considerations illustrated above (and largely based on (10.66)) allow an interpretation of Figure 10.11. However, the application of (10.66) to the assessment of the BER performance of a specific turbo coding scheme (or to another concatenated coding scheme) appears to be a formidable task for large N. To make the problem of performance estimation tractable, error bounds referring to a "randomly interleaved" turbo coding scheme, that is, to a scheme incorporating an *average over all the possible pseudorandom interleavers*, have been derived [1616]. To be more precise, such bounds are developed starting from the so-called *input-redundancy weight enumerating function* (IRWEF) for a systematic (n, k) code C. This function is defined as:

$$A^C(W, Z) \triangleq \sum_w \sum_l A_{w,l} W^w Z^l, \tag{10.69}$$

where $A_{w,l}$ denotes the number of codewords generated by input information words of weight w and whose parity check bits have weight l (so that the overall Hamming weight of the resulting codeword is $w + l$). The function $A^C(W, Z)$ can be written in the form:

$$A^C(W, Z) \triangleq \sum_w W^w A_w^C(Z), \tag{10.70}$$

where

$$A_w^C(Z) \triangleq \sum_l A_{w,l} Z^l \tag{10.71}$$

is the conditional weight enumerating function of the parity check bits associated with the input messages of weight w. The IRWEF can be used to upper-bound the BER for ML soft decoding of C over an AWGN channel as [1616]:

$$P_b \leq \frac{W}{k} \left. \frac{\partial A^C(W, Z)}{\partial W} \right|_{W=Z=\exp(-R_c E_b/N_0)}, \tag{10.72}$$

where $R_c \triangleq k/n$. Then, to apply this result to the case of a code C_p resulting from the concatenation of two constituent codes, the knowledge of $A^{C_p}(W, Z)$ is needed. Note that the computation of this function from the IRWEFs of the constituent codes is a very complicated task, since the redundant bits generated by the second encoder (i.e., by the encoder fed by the interleaved sequence of information bits) depend not only on the weight of the input message, but also on how the bits have been permuted by the interleaver. In principle, the only solution to the problem is an exhaustive enumeration of all the possible cases. Unfortunately, this is not feasible when the interleaver size is large. For this reason, the idea of averaging over the set of possible interleavers is introduced, defining the concept of a *uniform interleaver*. A uniform interleaver of length N is defined as a probabilistic device mapping any information word of weight w into all its distinct $\binom{N}{k}$ permutations with equal probability $1/\binom{N}{k}$. If this interleaving model is adopted for a parallel concatenated code, it can be shown that the IRWEF $A^{C_p}(W, Z)$ can be evaluated from the conditional weight enumerating functions of the two constituent codes. In addition, it can be proved that the resulting upper bound coincides with the average upper bound that can be obtained by considering the whole class of deterministic interleavers. For this reason, for each SNR the performance obtained with a uniform interleaver can be achieved using at least one specific deterministic interleaver.

It should be expected that, because of their intrinsic nature, upper bounds derived under the assumption of uniform interleaving do not provide accurate indications of the performance provided by specific deterministic interleavers. Such bounds, however, provide some important guidelines for code design [1615]. In particular, it can be shown that, for an "average turbo code", as N approaches infinity, d_{free}^{TC} is associated with codewords generated by information messages of weight $w = 2$. This explains why d_{free}^{TC} is usually maximized by choosing constituent encoders that have the largest output weight for information messages of weight $w = 2$.

Finally, it is worth noting the following:

1. Similar tools can be exploited to develop an upper bound on the average ML error probability of serially concatenated block and convolutional coding schemes [1546].
2. In our discussion, we have always referred to ML decoding even if, in practice, decoding of turbo codes and other concatenated schemes is accomplished in a suboptimum fashion. It should be expected, however, that the performance of iterative decoding methods gets close to being ML.
3. The interleaver plays a twofold role in determining the error performance of turbo codes [1631]. From the encoder perspective, it increases the Hamming weight of codewords. For instance, in the encoder illustrated in Figure 10.8, the interleaver applies to the lower encoder a permuted version of the message feeding the upper one. Therefore, it is unlikely that weak (i.e., low weight) upper codewords will be associated with weak lower codewords. From the decoder perspective, interleaving decorrelates feedback information in iterative decoding, so improving its behavior.

10.6 Low-Density Parity Check Codes

In this section LDPC codes are analyzed. After providing a definition of these codes through the fundamental properties of their parity check matrices in Section 10.6.1, a tool for their graphical representation is described in Section 10.6.2 and some relevant results about their weight spectrum are mentioned in Section 10.6.3. The problem of generating good LDPC codes is then addressed in Section 10.6.4, where an overview of available design techniques is presented. Finally, the problems of efficient encoding are investigated in Section 10.6.5. Section 10.7 deals with decoding.

10.6.1 Definition and Classification

LDPC codes are linear block codes whose parity check matrices are *sparse* in the sense that they have a low density of nonnull elements. LDPC codes can be defined over any field GF(q), but in the following we will mainly focus on binary codes for simplicity.

Binary LDPC codes can be divided into the subclasses of *regular* and *irregular* codes. An (n, k) *regular* code is characterized by the property: the n columns and the $m = n - k$ rows of its parity check matrix $\mathbf{H}$ have the same weights, denoted w_c and w_r, respectively, with $w_c \ll m$ and $w_r \ll n$. Such a parity check matrix $\mathbf{H}$ is said to be (w_c, w_r)-*regular* and the code $\mathcal{C}$ specified by its null space is called a (w_c, w_r)-*regular* LDPC code.

Since the overall number of 1s on the columns is equal to that on the rows, we have that $nw_c = mw_r = (n - k)w_r$; from this equality it is easily inferred that the code rate R can be expressed as:

$$R = 1 - \frac{w_c}{w_r}. \tag{10.73}$$

Note that the last result holds only if $\mathbf{H}$ is full rank as otherwise $R = (n - \text{rank}(\mathbf{H}))/n$, where rank($\mathbf{A}$) denotes the rank of the matrix $\mathbf{A}$.

If the number of 1s on the columns (or on the rows) of $\mathbf{H}$ changes from column to column (from row to row), the code defined by the sparse parity check matrix $\mathbf{H}$ is *irregular*.

10.6.2 Graphic Representation of LDPC Codes via Tanner Graphs

Any (n, k) binary linear block code $\mathcal{C}$, characterized by a parity check matrix $\mathbf{H}$, can be described by an undirected graph, called a *Tanner graph*, since its use was first proposed by R. M. Tanner [1606]. The Tanner graph $\mathcal{G}$ of $\mathcal{C}$ consists of a set $\mathcal{V}$ of $(2n - k)$ *vertices* (or *nodes*) and a set $\mathcal{E}$ of *edges* connecting the vertices. The graph $\mathcal{G}$ of $\mathcal{C}$ is *bipartite*, since $\mathcal{V}$ is divided into two disjoint subsets $\mathcal{V}_b$

and $\mathcal{V}_c$, containing n *bit nodes* and $n - k$ *check nodes*, respectively. Each edge can then connect only two nodes not belonging to the same subset.

The Tanner graph $\mathcal{G}$ is generated according to the following rules:

1. The ith symbol x_i of the codeword $\mathbf{x}$ is associated with the ith bit node (with $i = 0, 1, \ldots, n - 1$) and represented by the ith circle in the graph.
2. The jth check equation (denoted by the symbol c_j) is identified with the jth check node (with $j = 0, 1, \ldots, n - k - 1$) and represented by the jth square in the graph.
3. Edges connect variables nodes to check nodes; in particular, the jth check node is connected to the ith variable node by an edge *if an only if* the element H_{ji} (representing the so-called *adjacency matrix* of $\mathcal{G}$) of $\mathbf{H}$ is equal to 1.

The process is illustrated in Figure 10.13. The *degree* of a node is the number of edges incident on it. In an undirected graph, a series of successive edges forming a continuous path passing from one vertex to another is called a *chain*, with the *length* of the chain being given by the overall number of edges forming it. Two distinct nodes of $\mathcal{G}$ have *distance $d \leq \infty$* if the minimum length of the paths connecting them is equal to d. A chain of nodes where the initial and the terminal nodes are the same and which does not include the same edge more than once is a *cycle*. A cycle is characterized by its *length*, expressing the overall number of its edges. Note that in a bipartite graph having at least two nodes, its cycles are of *even length*. The length of the shortest cycle of $\mathcal{G}$ represents the *girth g* of the graph. An application of the graph construction rules and of the concepts defined above is illustrated in the following example.

Example 10.6.1 The relation between the parity check matrix:

$$\mathbf{H} = \begin{bmatrix} 1 & 1 & 1 & 0 & 1 & 0 & 0 \\ 0 & 1 & 1 & 1 & 0 & 1 & 0 \\ 1 & 1 & 0 & 1 & 0 & 0 & 1 \end{bmatrix} \tag{10.74}$$

of the (7,4) binary Hamming code (see Example 9.1.1) and its Tanner graph is illustrated in Figure 10.14. Note that the node associated with x_1 is connected to all the available check nodes, since the second column of $\mathbf{H}$ consists of 1s only; it is the only node having degree 3. Secondly, the bold solid lines indicate the presence of a cycle in the graph. It can be shown that this is the shortest cycle in the graph, so that the girth[7] g is equal to 4.

□

Generally speaking, it can be shown that a cycle with length 4 is associated with the presence of four 1s at the vertices of a rectangle in $\mathbf{H}$, as illustrated in Figure 10.15(a). Similarly, a cycle with

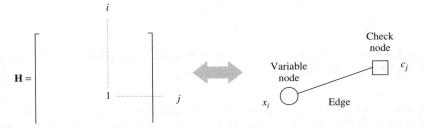

Figure 10.13 The correspondence rule between the 1s in the parity check matrix of $\mathcal{C}$ and the edges of the associated Tanner graph $\mathcal{G}$.

---

[7] In a bipartite graph having at least two nodes, all its cycles are of even length.

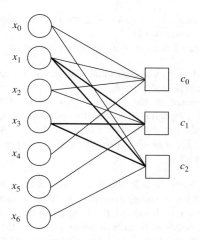

Figure 10.14 Tanner graph associated with the parity check matrix (10.74). The presence of a specific cycle of length 4 is shown by bold solid lines.

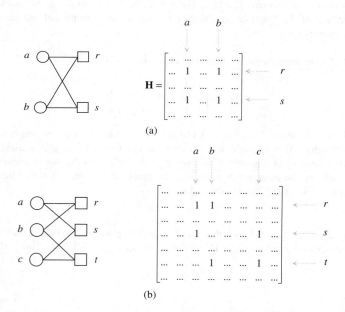

Figure 10.15 Cycle of (a) length 4 and (b) length 6 in the Tanner graph of a binary linear block code having parity check matrix **H**.

length 6 appears when six 1s are placed according to the specific pattern illustrated in Figure 10.15(b). It is worth noting that in almost all constructions of LDPC codes, binary or nonbinary, an additional structural property is imposed on **H** that no two rows (or two columns) have more than one position where they both have nonzero components. This property is a constraint on the rows and columns of **H** and is referred to as the *row–column* constraint; it ensures that the Tanner graph of the LDPC code C representing the null space of **H** has a girth of at least 6 [35, 1685].

Each variable (check) node in the Tanner graph of a (w_c, w_r)-*regular* LDPC code has degree w_c (w_r). Then, since the overall number of edges emerging from the check nodes is equal to that associated with the variable nodes, we have that $nw_c = mw_r$, as already pointed out in the previous subsection. The degree of a variable (or check) node in the graph of an *irregular* LDPC codes can change from node to node. In the literature, to specify the distribution of the node degrees in a Tanner graph of an irregular code, the *degree distribution polynomials*:

$$\lambda(x) \triangleq \sum_{d=1}^{d_v} \lambda_d \, x^{d-1} \tag{10.75}$$

and

$$\rho(x) \triangleq \sum_{d=1}^{d_c} \rho_d \, x^{d-1} \tag{10.76}$$

are usually defined for the variable and the check nodes, respectively [1686]. Here, λ_d (ρ_d) denotes the fraction of edges connected to degree-d variable (check) nodes and d_v (d_c) represents the maximum degree of variable (check) nodes.

The importance of the Tanner graph $\mathcal{G}$ of an LDPC $\mathcal{C}$ is primarily related to its decoding, which can be accomplished efficiently via the so-called SPA. In fact, as shown in Section 10.6.4, this algorithm (and other related techniques) can be interpreted as an iterative procedure for *message passing* (MP) between variable and check nodes of $\mathcal{G}$. In decoding algorithms based on the MP philosophy, a fundamental role is played by the cycles of $\mathcal{G}$. In particular, it is important to note that, on the one hand, the *absence of cycles* in $\mathcal{G}$ ensures a natural termination of the SPA in a finite number of steps and optimal performance (i.e., the minimization of the symbol error probability) [1610, 1687]. However, codes having *cycle-free* Tanner graphs offer poor BER performance, since they are characterized by $d_{H,min} = 2$ at code rates $R > 1/2$ and by error floors occurring at unacceptably high values of SNR [1688]. On the other hand, the *presence of cycles* in $\mathcal{G}$ leads to suboptimal performance of the SPA [1610, 1687], since this produces a loss of independence in the extrinsic information exchanged among graph nodes in successive iterations of the SPA (see Section 10.7), so preventing convergence to the optimal decoding result [1688, 1689]. This explains the importance of avoiding short cycles in code design, that is, of having a large girth g. The relevance of a large g is also motivated by a lower bound on the minimum distance $d_{H,min}$ of individual LDPC codes [1606]; in fact, such a bound increases with the girth g. To avoid short cycles, the parity check matrix $\mathbf{H}$ should be sufficiently sparse and, consequently, the codeword length n must be large enough.

10.6.3   Minimum Distance and Weight Spectrum

Most of the published results on the weight spectrum and, in particular, on the minimum distance of LDPC codes refer to the average properties of code ensembles. The initial work in this area was done by Gallager [1479, 1590], who first derived the average weight distribution and its growth rate for regular LDPC ensembles. In particular, he derived a lower bound on the typical minimum distance $d_{H,min}$ and proved that this parameter grows only logarithmically with n for $w_c = 2$, and linearly for $w_c \geq 3$. Later, these results were extended to modified regular LDPC code ensembles by S. Litsyn and V. Shevelev [1690]. For irregular LDPC codes, the average weight distribution and its growth rate have been investigated by Litsyn and Shevelev [1691], D. Burshtein and G. Miller [1692] and by C. Di *et al.* [1693]. In particular, in [1693] it is shown that the growth rate of the minimum distance of LDPC codes depends only on the degree distribution pair (see (10.75) and (10.76)), and that capacity-achieving sequences of known standard (unstructured) LDPC codes under iterative decoding over the *binary erasure channel* (BEC) have sublinearly growing minimum distance with block length.

The importance of the average distance spectrum of various random ensembles of LDPC codes is motivated by the fact that its exponential growth rate can be translated to upper bounds on the average ensemble error probability, under ML decoding, and even to an upper bound on the error probability of a typical code drawn at random from the ensemble (e.g., see [1479, 1693]). Using concentration results on the weight distribution, lower bounds on the error rate of a random code in the ensemble of regular LDPC codes can also be derived [1694]. Combining these results with known lower bounds on the error exponent, confidence intervals on the error exponent, under ML decoding, can be obtained.

10.6.4   LDPC Code Design Approaches

Given four positive integers m, n, w_c, w_r such that $nw_c = mw_r$, designing a binary $(n, n - m)$ *regular* LDPC code means constructing an $m \times n$ parity check matrix $\mathbf{H}$ having uniform column weight w_c, uniform row weight w_r and some specific additional property (e.g., satisfying the row–column constraint). The first design method for this class of codes was developed by Gallager, who proposed the use of the *structured* parity check matrix [1479, 1590]:

$$\mathbf{H} \triangleq \begin{bmatrix} \mathbf{H}_1 \\ \mathbf{H}_2 \\ \vdots \\ \mathbf{H}_{w_c} \end{bmatrix}, \tag{10.77}$$

where $\mathbf{H}_l$ (with $l = 1, 2, \dots, w_c$) is a $\mu \times \mu w_r$ submatrix with row weight w_r and column weight 1, and μ and w_r are integers greater than unity. In addition, the submatrices $\{\mathbf{H}_l\}$ of (10.77) have the following structure:

- the ith row of $\mathbf{H}_1$ contains all w_r of its 1s in columns from $(i - 1)w_r + 1$ to iw_r, with $i = 1, 2, \dots, \mu$,
- all the other submatrices $\{\mathbf{H}_l, l = 2, 3, \dots, w_c\}$ are generated by row permutations of $\mathbf{H}_1$.

This construction leads to a $\mu w_c \times \mu w_r$ matrix $\mathbf{H}$ (10.77) having row (column) weight exactly w_r (w_c), but does not guarantee the absence of cycles of length 4, so that computer-based optimization is required to remove them. Gallager proved that the ensemble of these codes offers excellent distance properties if $w_c \geq 3$ and $w_r > w_c$ [1479, 1590].

About thirty years later *R. M. Tanner* proposed a recursive method for constructing long error-correcting codes from one or more short error-correcting codes (called *subcodes*) and a *bipartite graph* [1606]. This approach generalized the construction of product codes and the decoding schemes originally presented by Elias for his product codes and those of Gallager's LDPC codes.

A completely different approach to LDPC code design was adopted by D. J. C. MacKay, who has provided algorithms for *semirandomly* generating sparse parity check matrices [1695, p. 413] and has archived on a web page [1696] a large number of LDPC codes that he has designed. One of the main drawbacks of the so-generated codes is that they lack sufficient structure to achieve low-complexity encoding. In fact, encoding is accomplished by putting $\mathbf{H}$ in the form (9.13), via Gauss–Jordan elimination, so that the associated generator matrix $\mathbf{G}$ can be put into systematic form (9.6). Note, however, that the matrix $\mathbf{P}_{k,n-k}$ in (9.6) is usually far from sparse, so that for large n the encoding complexity is appreciable.

MacKay's work has been followed by a flurry of papers on LPDC code design. The methods proposed for LPDC code construction can be classified in different ways. A rough classification separates methods for constructing *randomlike codes*, like those generated by MacKay, from those generating *structured codes* [1697]. A completely different perspective is adopted if design techniques are divided into three major categories: graph-theoretic, geometric and experimental. The *graph-theoretic* approach is by far the most popular and involves exploiting random bipartite graphs, specific graphs, or constrained paths of connected graphs for code construction, (e.g., see [1608, 1679, 1686,

1697–1705]). The *geometric* approach aims to construct regular LDPC codes based on the lines and hyperplanes of geometries over finite fields, such as Euclidean and projective geometries (e.g., see [1685, 1706]). The aim of the *experimental* approach is the generation of LDPC codes based on combinatoric designs (e.g., see [1707–1709]).

It is also worth noting that distinct code design techniques can target different criteria, such as efficient encoding/decoding, low error floor and large girth. Satisfying these requirements can put serious constraints on the structure of the parity check matrix. For instance, encoding complexity can be substantially reduced if the code is *cyclic* or *quasi-cyclic* (QC), that is, if it is invariant under a cyclic shift of p positions, with $p = 1$ and $1 < p < n$, respectively (the smallest such p is called the *index* of the code). In fact, these classes of codes offer the relevant advantage of efficient encoding via linear feedback shift registers based on their generator or characterization polynomials [1710]. Further, by exploiting the close relationship between QC codes and convolutional codes, a convolutional representation can be derived from a QC block code (see [1711] and references therein). Note, however, that the requirement of quasi-cyclicity requires **H** to be an array of sparse circulants.[8]

We also note that most of the code designs published in the literature are binary and only limited results are available for nonbinary codes. In this regard, we should not forget that nonbinary codes over a Galois field GF(q) offer important advantages, since the equivalent binary weight of their parity check matrices is increased with respect to their binary counterparts, whereas the number of short cycles may remain low. In his seminal work Gallager considered arbitrary-alphabet LDPC codes using modulo-q arithmetic [1590]. Nonbinary LDPC codes based on GF(q) arithmetic were analyzed much later by M. C. Davey and MacKay [1712] in the context of codes for binary-input channels. The design of nonbinary LDPC codes has been of increasing interest in recent years (e.g., see [1713, 1714, 1715, 1716, 1717, 1718, 1719]).

Finally, we mention that a limited amount of work has been done on LDPC code design for *fading channels*. Code design for a specific class of fading channels (i.e., periodic fading channels) has been investigated in [1720].

In the the remainder of this subsection we mention some of the most relevant design criteria and provide the reader with a set of useful references for further reading.

10.6.4.1 Finite-Geometry Codes

The first method for constructing *finite-geometry* LDPC codes was proposed by Y. Kou, S. Lin and M. P. C. Fossorier in 2001 [1685]. Such a method is based on lines and points of finite geometries, namely *Euclidean* and *projective geometries* over finite fields. These codes are characterized by the following relevant properties:

- they are cyclic or QC;
- they have relatively good minimum distances;
- their Tanner graphs do not contain cycles of length 4;
- they can be decoded with various other decoding methods, ranging from low to high decoding complexity and from reasonably good error performance to very good error performance.

For all these reasons, they offer a wide range of tradeoffs among decoding complexity, decoding speed and error performance.

---

[8] A *circulant* is a square matrix in which each row is the cyclic shift (one place to the right) of the row above it, and the first row is the cyclic shift of the last row. In a circulant, each column is a downward cyclic shift of the column on its left, and the first column is the downward cyclic shift of the last column. The row and column weights of a circulant are equal. For a circulant, the set of columns (reading from the top down) is the same as the set of rows (reading from right to left).

The discovery of finite-geometry LDPC codes has had two significant implications. First, it implies that algebraic construction is a viable method for constructing codes approaching the Shannon limit, in addition to random construction, so that the investigation of other algebraic or combinatoric methods for constructing good LDPC codes has been encouraged. Second, that algebraically constructed LDPC codes in general possess structural properties that may simplify hardware implementation of encoding and decoding.

The seminal work [1685] was followed by a number of papers dealing with new construction techniques for finite-geometry codes. The reader can refer to [1706, 1716, 1721–1726] to deepen his knowledge in this active research area.

10.6.4.2 Combinatorial LDPC Codes

The problem of LDPC design can be tackled by resorting to the so-called combinatorial designs. A *combinatorial design* is an arrangement of a set of v points into b subsets, called *blocks*, which satisfy certain regularity conditions. A design is *regular* if the number of points in each block and the number of blocks which contain each point are the same for each point and block in the design (these numbers are denoted γ and ρ, respectively). The *covalency* λ_{xy} of the points x and y is the number of blocks that contain both of them. A design is *balanced* if λ_{xy} is a constant for all x and y (then the covalency of the design can simply be denoted λ). A *regular balanced design* is often designated a $(v, b, \rho, \gamma, \lambda)$-design. Every design can be described by a $b \times v$ *incidence matrix* $\mathbf{I}$, whose ith row represents the block P_j of the design and the jth column the point P_j according to the rule:

$$I_{i,j} \triangleq \begin{cases} 1 & \text{if } P_j \in B_i \\ 0 & \text{otherwise} \end{cases} \tag{10.78}$$

with $i = 1, 2, \ldots, b$ and $j = 1, 2, \ldots, v$. For a regular design the number of 1s in $\mathbf{I}$ is $v \cdot \rho = b \cdot \lambda$. The incidence matrix of a combinatorial design, or its transpose, can be used as the parity check matrix of a binary LDPC code to give favorable properties to the code. In particular, the transpose of the incidence matrix of a $(v, b, \rho, \gamma, \lambda)$-design will produce a $v \times b$ parity check matrix $\mathbf{H}$ having v parity check equations, block length $n = b$, row weight ρ and column weight γ. Choosing a design with $\lambda = 0$ or 1 guarantees the absence of 4-cycles in the code (i.e., the girth is at least 6). Note that, as in the case of random constructions of parity check matrices, the parity check matrices generated in this way are not necessarily full rank, in which case the number of message bits in the code is $n - \text{rank}(\mathbf{H})$.

MacKay and Davey were the first to use *balanced incomplete block designs* (BIBDs),[9] in particular *Steiner triple systems* (or $(v, b, \rho, 3, 1)$-designs) to develop LDPC codes [1727]. S. J. Johnson and S. R. Weller extended the class of LDPC codes that can be systematically generated by presenting a construction method for regular LDPC codes based on combinatorial designs known as *Kirkman triple systems* [1728]. This work has been followed by many papers about the combinatorial construction of regular low-density parity check codes based on BIBDs (e.g., see [1709, 1708, 1729–1731]).

10.6.4.3 Array Codes and Other Codes Based on Algebraic Methods

Various algebraic methods have also been exploited to design LPDC codes. First, it is important to mention the work of J. L. Fan [1732], who showed that the so-called *array codes* [1733], having sparse parity check matrices, can be viewed as LDPC codes and, consequently, can be decoded via MP algorithms. Later E. Eleftheriou and S. Ölçer developed a class of *modified array codes* [1734], characterized by an upper triangular parity check matrix $\mathbf{H}$ and thus allowing an encoding complexity

---

[9] Note that Euclidean and projective geometries are subclasses of BIBDs.

linear in the block length. Modified array codes are available for both low and high rates, and offer very low error rate floors; however, because of the specific structure of **H**, only selected code rates and lengths are available.

In more recent years, various algebraic methods for constructing LDPC codes have been proposed [1715, 1717–1719, 1721, 1735–1740]. Most of these codes share the QC property, which can potentially facilitate efficient encoder implementation.

10.6.4.4 Protograph-Based LDPC Codes

A small Tanner graph, known as a *protograph* or *projected graph*, can be used to construct a structured LDPC using the following procedure [1741, 1742]:

1. The protograph is replicated N times; this means that each edge is replicated into a bundle of N edges, now connecting N variable nodes to N check nodes.
2. The copies of the protograph are interconnected by unplugging their edges from their check node sockets, permuting them, and reconnecting them. This process is repeated for each bundle of N edges, or edge type.

Note that the resulting derived (or *lifted*) graph is N times as large as its protograph, but inherits many of the protograph's properties. It has the same code rate (except possibly for coincidental redundancies due to the selected permutations) and the same distribution of variable and check node degrees. Moreover, neighborhoods are preserved. For this reason, the resulting LDPC code can be designed by applying standard analysis techniques, such as density evolution, to the protograph.

In constructing protograph-based LDPC codes structured permutations, such as cyclic shift permutations (or *circulants*) or randomlike permutations generated by computer search with an optimization criterion can be adopted. Both techniques aim to maximize the girth of the resulting graph. Protograph design criteria and methods are analyzed in detail in [1741].

It is worth noting that RA, *irregular repeat-accumulate* (IRA) [1743] and *accumulate-repeat-accumulate* (ARA) codes [1744] (see Section 10.4.1) with suitable definition of interleavers have simple protograph representations.

Protographs have been exploited to design *generalized* LDPC [1606] codes with a QC structure and low error floors in [1745]. A parity check matrix construction based on Vandermonde-like block matrices in the context of protograph LDPC arrangements is illustrated in [1746].

10.6.4.5 Progressive Edge Growth

Brute force search is a straightforward method for designing a parity check matrix **H**. However, for an $m \times n$ parity check matrix **H** with uniform column weight w_c, there are $\binom{m}{w_c}^n$ possible choices and this makes an exhaustive search computationally infeasible for values of n of practical interest. Moreover, a large number of choices are actually isomorphic to each other, that is, lead to identical LDPC codes. This raises the problem of devising more efficient search methods for good LDPC codes. J. Campello, D. S. Modha and S. Rajagopalan developed a heuristic method called "bit-filling" to search for LDPC codes with large girth [1747, 1748]. The computational complexity of bit-filling is polynomial in m (specifically, $O(k_{\max} m^3)$, where $k_{\max}$ is the maximum degree of any check node) and this makes its implementation feasible. Unfortunately, the matrix generated by bit-filling has uniform column weight, but nonuniform row weight, so that the resulting LDPC codes are not regular; in addition, there is no guarantee that codes with the largest possible girth g are constructed for a given n. Another heuristic algorithm, illustrated in [1749], searches for good LDPC codes based on the average of the girth distribution of the code. The complexity of this algorithm is shown in [1749] to be $O(n^2)$ and thus it is suitable for designing short (10 000 bits or shorter) codes.

An alternative search technique for constructing bipartite graphs with good girth properties is the so-called *progressive edge growth* (PEG) technique [1750]. This algorithm is based on the principle of optimizing the placement of a new edge connecting a particular symbol node to a check node on the graph such that the largest possible local girth is achieved. In this way, the underlying graph grows in an edge-by-edge manner, optimizing each local girth, and is thus referred to as a PEG *Tanner graph*. In [1750] upper and lower bounds on the girth and a lower bound on the minimum distance have been derived in terms of parameters of the underlying PEG Tanner graph. In addition, it has been shown that the PEG algorithm is a simple, flexible and powerful algorithm for generating good regular and irregular LDPC codes (of short and moderate block lengths) offering a significant improvement compared with randomly constructed codes. Moreover, with only a slight modification, the PEG algorithm can be used to generate linear time encodable LDPC codes. More recently, some improvements in the PEG algorithm which greatly improve the girth properties of the resulting graphs have been proposed in [1751].

10.6.4.6 Irregular LDPC Codes

The design of irregular codes deserves specific comment. T. J. Richardson *et al.* [1686] and Luby *et al.* [1752] were the first to investigate the problem of identifying ensembles of irregular codes characterized by *optimal* degree distribution polynomials $\lambda(x)$ (10.75) and $\rho(x)$ (10.76) (see also [1701]). In this case the property of optimality refers to the fact that, assuming MP decoding, a typical code in the ensemble is capable of reliable data communication in worse channel conditions than are the codes outside the ensemble. The worst channel condition is identified by the so-called *decoding threshold*, which can be evaluated, for a given pair $(\lambda(x), \rho(x))$ by means of the DE technique. This technique allows the analysis of the evolution of the pdfs of the messages exchanged by the decoding algorithm in a Tanner graph [1701]. In particular, given the initial pdf of the LLR messages, the pdf of LLR messages at any iteration can be computed under the assumption that $n \to \infty$ (i.e., of a very long code). As a result, one can test whether, for a given channel condition, the decoder converges to zero error probability or not. This approach certainly leads to powerful codes, which are close to the capacity limits for very large n; however, degree distribution polynomials which are optimal for very long codes will no longer have this property for codes of short and medium length, which can exhibit a high error-rate floor.

DE also suffers from other limitations. It requires intensive computations and/or a long search to find a good degree sequence, since the optimization problem is not convex [1679]. Furthermore, it does not provide any insight into the design process, and is intractable for some of the codes defined on graphs. An alternative tool for studying the convergence behavior of iterative decoders is to use EXIT charts (e.g., see [1683, 1678]). Although this method is not as accurate as density evolution, its lower computational complexity and its reasonably good accuracy make it attractive. EXIT charts provide a one-dimensional analysis, allowing one to visualize the convergence behavior of the decoder, and can reduce the irregular code optimization to a linear program [1753]. Thus this tool is both faster and provides more insight than DE-based approaches. In addition, this approach is applicable to many of the codes defined on graphs associated with iterative decoders. Various approaches to one-dimensional analysis and design of LDPC codes have appeared in the technical literature (e.g., see [1753–1756]). The analyses of [1753–1755] are based on the observation that the pdf of the decoder's LLR messages is approximately Gaussian. This approximation is quite accurate for messages sent from variable nodes, but less so for messages sent from check nodes. A significantly more accurate one-dimensional analysis for LDPC codes has been proposed in [1756], where a Gaussian distribution is assumed only for the channel messages and the messages from variable nodes, whereas the "true" pdf of the messages sent from check nodes is employed.

In the last few years the construction of finite-length irregular LDPC codes with *low error floors* (i.e., providing good performance in the region of high SNR) has been an active area of research.

Substantial attention has been paid to the performance provided by iterative decoding of *finite-length codes* over the AWGN channel and the BEC. The interest in the BEC is due to the fact that:

- good codes for this type of channel are good for the AWGN channel as well [1757];
- the decoder behavior over the BEC (AWGN) is governed by a small number of likely error events related to certain topological structures of the LDPC code, called *stopping sets* (or *trapping sets*) [1758–1760]. In particular the performance of LDPC codes in the error floor region is governed by a small number of likely error events due to the presence of small stopping sets.

The design of techniques optimizing ensembles of irregular LDPC codes with respect to the size of their minimum stopping sets (i.e., to their *stopping distance* [1761]) remains an open problem. Note that finding the stopping distance of an LDPC code graph is a *nondeterministic polynomial-time* (NP) hard problem [1762]. Despite this, powerful heuristic design algorithms based on various metrics, such as the *extrinsic message degree* [1763], the *approximate cycle extrinsic message degree* (ACE) [1763], a *generalization* of the ACE [1764] and the ACE spectrum [1765] are available. These algorithms yield codes characterized by error floors substantially lower than those of random codes with very small degradation in capacity-approaching capability. A different method for improving the performance of LDPC codes in the error floor region is proposed in [1760]; it consists of a procedure for edge swapping in a Tanner graph in such a way that the most dominant trapping sets are broken.

Other interesting research results concerning the construction of irregular LDPC codes can be found in [938, 1709]. In particular, in [1709] a general systematic method to achieve an irregular LDPC code, with a given degree distribution pair, by splitting columns and rows of a regular code (which can be generated in a pseudorandom fashion) is proposed. It represents a useful alternative to random generation of irregular codes and enables exploitation of the structural properties of a regular code in efficiently storing the resultant irregular code. The construction approach described in [938] is based on the observation that the performance of an LDPC code depends on both the structure of its Tanner graph and its minimum distance (or low-weight profile). In fact, an LDPC code with a good minimum distance may not outperform another one with worse minimum distance but better graph structure, because the BPA decoding algorithm is suboptimal and graph-dependent [1610]. For this reason, LDPC construction is accomplished by taking both weight distribution and graph property into account. In particular, low-weight codewords consisting of degree-2 bit nodes and at most two degree-3 bit nodes are eliminated. The design adopted is based on the QC extension of [1766], proposed to design RA codes with good minimum distance. The resulting irregular codes have good weight distribution in that few undetected errors and low error floors are observed.

Finally, we mention the analysis of [1767], which refers to an LDPC code ensemble characterized by a certain profile (i.e., row and column weight distribution) and by parity check matrices having *approximately lower triangular* structure with some prespecified gap (see Section 10.6.5). It is shown that, for any gap value, the asymptotic performance of the new ensemble is the same as the performance of the standard ensemble. Hence, by choosing the gap sufficiently small we can guarantee linear encoding complexity and the same asymptotic performance as the standard ensemble.

10.6.5 *Efficient Algorithms for LDPC Encoding*

An LDPC code $\mathcal{C}$ is specified by a *sparse* parity check matrix $\mathbf{H}$. Except in special cases (such as RA and IRA codes), the generator matrix $\mathbf{G}$ of $\mathcal{C}$ is dense, and hence encoding complexity is quadratic in the codeword length. In a landmark paper, Richardson and Urbanke demonstrated that by using *back-substitution*, for most LDPC codes, encoders with complexity growing almost linearly in block length can be built [1768]. In particular, their work starts from the observation that encoding complexity could be substantially simplified by forcing the parity check matrix $\mathbf{H}$ to have *lower triangular form*. This restriction ensures a linear-time encoding complexity but, generally speaking, results in some loss of performance. To avoid this loss, they proposed exploiting only the sparseness of $\mathbf{H}$ to bring this matrix, *performing row and column permutations only*, into approximately lower triangular form:

$$\hat{\mathbf{H}} = \begin{bmatrix} \mathbf{A} & \mathbf{B} & \mathbf{T} \\ \mathbf{C} & \mathbf{D} & \mathbf{E} \end{bmatrix}, \tag{10.79}$$

where the matrices $\mathbf{A}, \mathbf{B}, \mathbf{T}, \mathbf{C}, \mathbf{D}$ and $\mathbf{E}$ are all sparse and have sizes $(m - g) \times (n - m)$, $(m - g) \times g$, $(m - g) \times (m - g)$, $g \times (n - m)$, $g \times g$ and $g \times (m - g)$, respectively, and the integer parameter g is the so-called *gap*. In addition, $\mathbf{T}$ is a lower triangular matrix with 1s along its main diagonal. Then, if the codeword is represented as $\mathbf{x} = [\mathbf{s}, \mathbf{p}_1, \mathbf{p}_2]$, where $\mathbf{s}$ denotes its systematic part, while the concatenation of $\mathbf{p}_1$ and $\mathbf{p}_2$ denotes its parity part (the size of $\mathbf{p}_1$ is g, while that of $\mathbf{p}_2$ is $(m - g)$), it can be shown that the complexities required by the evaluation of $\mathbf{p}_1$ and $\mathbf{p}_2$ are $O(n + g^2)$ and $O(n)$, respectively. Therefore, this can substantially reduce the overall encoding complexity, provided that $\mathbf{H}$ is carefully preprocessed to generate a matrix $\hat{\mathbf{H}}$ (10.79) with a gap g as small as possible. If n is large, there is no hope of finding the *optimal* permutation (i.e., that minimizing g); however, some *greedy algorithms* to triangulate $\mathbf{H}$ are available [1768]. While a remarkable result, this did not actually solve the encoding puzzle, since, even if $\hat{\mathbf{H}}$ (10.79) is sparse, it is unstructured (i.e., largely disorganized). Therefore, storing such a matrix in the encoder can turn out to be a significant burden, especially in certain applications (e.g., in the deep-space environment (10.79)). In addition, the encoding algorithm is essentially serial, and this limits the speed at which encoders can run.

These considerations motivate the search for structured parity check matrices, resulting in dense but highly structured generator matrices. Encoders based on this consist of a set of shift registers, and are both fast and fairly simple, as already mentioned in the foregoing.

Finally, it is worth mentioning that structured parity check matrices have been adopted in the DVB-S2 and DVB-T2 standards [1769, 1770]. In fact, in both standards lower triangular matrices have been selected; more specifically, $\mathbf{H}$ is restricted to have the form $[\mathbf{A}_{(n-k) \times k} \ \mathbf{B}_{(n-k) \times (n-k)}]$, where the matrix $\mathbf{B}_{(n-k) \times (n-k)}$ has a "staircase" lower triangular structure (see [1769, Appendix A, p. 64]) and $\mathbf{A}_{(n-k) \times k}$ is a sparse matrix satisfying certain restrictions (various details on the resulting Tanner graph can be found in [1771]). This specific structure of $\mathbf{H}$ allows systematic encoding with linear complexity with respect to the block length.

10.7 Decoding Techniques for LDPC Codes

In this section we consider the decoding of LDPC codes. First, a general framework for understanding decoding algorithms based on MP decoding is provided in Section 10.7.1. Then, in Section 10.7.2 we analyze the SPA, also known as the *belief propagation algorithm* (BPA), and its best-known simplification, known as the *min-sum algorithm* (MSA) [1677]. Finally, we provide an overview of various practical and theoretical issues concerning MP algorithms in Section 10.7.3.

10.7.1 Introduction to Decoding via Message Passing Algorithms

MP decoding algorithms operate on the Tanner graphs of LDPC codes *in an iterative fashion*, accomplishing a continuous exchange of information between adjacent nodes. In the following we describe the philosophy of MP for an (n, k) linear binary block code $\mathcal{C}$ described by a parity check matrix $\mathbf{H}$ and provide notation and definitions useful in the description of MP algorithms for this code.

Each iteration in an MP-based decoding algorithm consists of two distinct steps. In the first step the variable node x_i (with $i = 0, 1, \ldots, n - 1$) sends a message to any adjacent check node. This message represents a (hard or soft) estimate of the code symbol that the variable node represents. In the second step the exchange of information occurs in the opposite direction, that is, the check node c_j ($j = 0, 1, \ldots, n - k - 1$) sends to any adjacent variable node a (hard or soft) estimate of the code symbol the variable node represents. This estimate is generated by exploiting both the information coming from the adjacent variable nodes and the constraint expressed by the parity check equation c_j. Note that in each iteration message generation obeys the following *basic rule*: all the messages sent by a variable node x_i (check node c_j) of the graph along a specific edge connecting it to an arbitrary adjacent check node c_l (variable node x_t) do not depend on any message received by x_i (c_j) along the same edge. This rule originates simply from the requirement that the messages produced by any source node and sent to a specific destination node are generated by processing only the *extrinsic information* collected by the source itself. Another general rule of MP algorithms is that all the (hard or soft) data coming directly from the communication channel are made available to variable nodes only. These employ the information coming from the communication channel in the computation of the messages to be delivered to check nodes. Note that this information is only available at the variable nodes at the beginning of the first iteration and can also be exploited by such nodes in the following iterations.

The aim of the MP decoding techniques is good (possibly approaching optimal) error performance through a repetitive and computational efficient iterative procedure. It should be expected that the quality of the data estimates generated by a MP decoder improves with iteration. Actually, after a certain number of iterations, the improvements contributed by additional iterations will be negligible, so that decoding can be stopped and an estimate $\hat{\mathbf{x}}$ of the transmitted codeword $\mathbf{x}$ can be produced. It is worth pointing out, however, that the vector $\hat{\mathbf{x}}$ is generated by collecting, in an ordered fashion, all the "local" estimates of codeword symbols generated by variable nodes and for this reason, $\hat{\mathbf{x}}$ does not necessarily belong to the alphabet of the possible codewords of $\mathcal{C}$.

We now introduce a general notation for describing MP decoding algorithms:

1. Let y_i denote the (hard or soft) datum generated by the communication channel, assumed *memoryless*, in response to the transmitted symbol x_i, with $i = 0, 1, \dots, n - 1$.
2. Let $\mathbf{m}^{(l)}_{c_j \to x_i}$ ($\mathbf{m}^{(l)}_{x_i \to c_j}$) denote the message[10] sent by the parity check node c_j (variable node x_i) to the variable node x_i (check node c_j) in the lth iteration, with $j = 0, 1, \dots, n - k - 1$;
3. We define the set:

$$R_j \triangleq \{p : 0 \le p \le n - 1, \ h_{jp} = 1\} \tag{10.80}$$

which consists of the values of the row index identifying all the 1s in the jth column of $\mathbf{H}^T$ (see Figure 10.16), and the set:

$$R_j \backslash i \triangleq \{p : 0 \le p \le n - 1, \ h_{jp} = 1\} \backslash \{i\} \tag{10.81}$$

consisting of R_j with i removed.

In practice, the subset of the bits of $\mathbf{x} = [x_0, x_1, \dots, x_{n-1}]$ appearing in the jth parity check matrix can be expressed as $\{x_p, p \in R_j\}$.

Similarly to (10.80) and (10.81), the sets:

$$C_i \triangleq \{q : 0 \le q \le n - k - 1, \ h_{qi} = 1\} \tag{10.82}$$

and

$$C_i \backslash j \triangleq \{q : 0 \le q \le n - k - 1, \ h_{qi} = 1\} \backslash \{j\} \tag{10.83}$$

are defined by considering the ith row of $\mathbf{H}^T$ (again see Figure 10.16). Note that the set of all the parity check equations in which the bit x_i appears is simply specified by $\{c_q, q \in C_i\}$.

---

[10] Generally speaking, messages are functions; however, in our analysis, messages can be simply modeled as real-valued vectors.

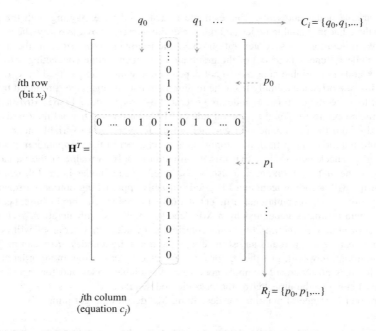

Figure 10.16 Construction of the sets R_j (10.80) and C_i (10.82) from $\mathbf{H}^T$.

Given these definitions, a simple analytical representation can be given for the generation of the messages $\mathbf{m}_{c_j \to x_i}^{(l)}$ and $\mathbf{m}_{x_i \to c_j}^{(l)}$. In fact, since $\mathbf{m}_{c_j \to x_i}^{(l)}$ depends on the messages $\mathbf{m}_{x_p \to c_j}^{(l)}$ coming from the nodes $\{x_p\}$ adjacent to c_j, but *different* from x_i (i.e., from the nodes $\{x_p, p \in R_j \backslash i\}$), we have:

$$\mathbf{m}_{c_j \to x_i}^{(l)} = \mathbf{\Psi}_C^{(l)} \left(\mathbf{m}_{x_p \to c_j}^{(l)} \,\Big|\, p \in R_j \backslash i \right), \qquad (10.84)$$

where $\mathbf{\Psi}_C^{(l)}(\cdot)$ is a proper function depending on a number of vector variables $\{\mathbf{m}_{x_p \to c_j}^{(l)}\}$ equal to the degree of c_j minus one and, generally speaking, on the iteration index l. Similarly, $\mathbf{m}_{x_i \to c_j}^{(l)}$ depends on y_i and on the messages $\mathbf{m}_{c_p \to x_i}^{(l-1)}$, with $\{c_p, p \in C_i \backslash j\}$, generated in the previous iteration, so that it can be expressed as:

$$\mathbf{m}_{x_i \to c_j}^{(l)} = \mathbf{\Psi}_V^{(l)} \left(y_i, \; \mathbf{m}_{c_q \to x_i}^{(l-1)} \,\Big|\, q \in C_i \backslash j \right), \qquad (10.85)$$

where $\mathbf{\Psi}_V^{(l)}(\cdot)$ is a proper function depending on a number of vector variables $\{\mathbf{m}_{c_q \to x_i}^{(l-1)}\}$ equal to the degree of x_i minus one and depending, as in the previous case, on the iteration index l. Note that in the first iteration (i.e., for $l = 0$) the expression:

$$\mathbf{m}_{x_i \to c_j}^{(0)} = \mathbf{\Psi}_V^{(0)}(y_i) \qquad (10.86)$$

should be adopted in place of (10.85), since y_i only is available to the node x_i at the startup of any MP-based procedure.

10.7.2 SPA and MSA

In his seminal work [1479, 1590] Gallager tackled the problem of developing probabilistic algorithms able to iteratively estimate the probability:

$$\Pr\{x_i = 1 \mid \mathbf{y}, S_i\} \tag{10.87}$$

with $i = 0, 1, \ldots, n-1$ and $b = 0, 1$, given the received vector $\mathbf{y} = [y_0, y_1, \ldots, y_{n-1}]$, or, equivalently, the APP *ratio*:

$$\frac{\Pr\{x_i = 0 \mid \mathbf{y}, S_i\}}{\Pr\{x_i = 1 \mid \mathbf{y}, S_i\}}. \tag{10.88}$$

Here S_i denotes the event occurring if and only if all the parity check constraints on x_i (i.e., all the parity check equations $\{c_q, q \in C_i \backslash j\}$) are satisfied by the transmitted digits. To evaluate (10.88), Gallager exploited the following lemma, which is easily proved by induction.

Lemma 10.7.1 *Given the vector* $\mathbf{a} = [a_0, a_1, \ldots, a_{m-1}]$ *consisting of m independent binary digits with*:

$$p_l \triangleq \Pr\{a_l = 1\} \tag{10.89}$$

for $l = 0, 1, \ldots, m-1$, *the probability that* $\mathbf{a}$ *contains an even and an odd number of 1s is given by*:

$$\frac{1}{2} + \frac{1}{2} \prod_{l=0}^{m-1} (1 - 2p_l), \tag{10.90}$$

and

$$\frac{1}{2} - \frac{1}{2} \prod_{l=0}^{m-1} (1 - 2p_l), \tag{10.91}$$

respectively.

To exploit this lemma, let us apply Bayes' theorem to (10.88), rewriting it as:

$$\frac{\Pr\{x_i = 0 \mid \mathbf{y}, \ S_i\}}{\Pr\{x_i = 1 \mid \mathbf{y}, \ S_i\}} = \frac{1 - P_i}{P_i} \frac{\Pr\{S_i \mid \mathbf{y}, \ x_i = 0\}}{\Pr\{S_i \mid \mathbf{y}, \ x_i = 1\}}, \tag{10.92}$$

where

$$P_i \triangleq \Pr\{x_i = 1 \mid \mathbf{y}\}. \tag{10.93}$$

To evaluate the probability $\Pr\{S_i \mid \mathbf{y}, \ x_i = 0\}$ it is useful to note that, if $x_i = 0$, *all* the parity checks on bit x_i (i.e., $\{c_q, q \in C_i\}$) are satisfied if and only if, for each of them, an *even* number of the other involved bits (i.e., $\{x_q, q \in R_q \backslash i\}$) takes value unity. Therefore, from (10.90), it is easily inferred that:

$$\Pr\{S_i \mid \mathbf{y}, \ x_i = 0\} = \prod_{q \in C_i} \frac{1}{2} \left[1 + \prod_{l \in R_q \backslash i} (1 - 2P_l) \right]. \tag{10.94}$$

Similarly, the probability $\Pr\{S_i|\mathbf{y},\ x_i = 1\}$ can be evaluated as (see (10.91)):

$$\Pr\{S_i|\ \mathbf{y},\ x_i = 1\} = \prod_{q\in C_i} \frac{1}{2}\left[1 - \prod_{l\in R_q\backslash i}(1 - 2P_l)\right]. \tag{10.95}$$

Then substituting (10.94) and (10.95) into (10.92) yields:

$$\frac{\Pr\{x_i = 0|\ \mathbf{y},\ S_i\}}{\Pr\{x_i = 1|\ \mathbf{y},\ S_i\}} = \frac{1 - P_i}{P_i} \frac{\displaystyle\prod_{q\in C_i}\left[1 + \prod_{l\in R_q\backslash i}(1 - 2P_l)\right]}{\displaystyle\prod_{q\in C_i}\left[1 - \prod_{l\in R_q\backslash i}(1 - 2P_l)\right]}, \tag{10.96}$$

which, in principle, allows an *exact computation* of the APP ratio (10.88), provided that the probabilities $\{P_i,\ i = 0, 1,\ \ldots, n-1\}$ (10.93) are available. This approach, however, is overly complicated in computational terms and this motivates the search for alternative (and computationally simpler) approaches to estimate the bit APPs. An excellent alternative is the BPA. Like all the other MP-based decoding techniques, the BPA is an iterative technique, in which each iteration consists of two distinct steps: a transmission of information from variable nodes to check nodes, followed by a transmission in the opposite direction, as illustrated in Figure 10.17 for the lth iteration.

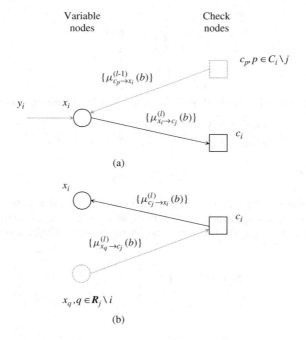

Figure 10.17 Transmission, in the lth BP iteration, of (a) the messages $\{\mu^{(l)}_{x_i\to c_j}(b),\ b = 0, 1\}$ from the variable node x_i to the check node c_j and (b) the messages $\{\mu^{(l)}_{c_j\to x_i}(b),\ b = 0, 1\}$ in the opposite direction.

To derive the BPA, consider first the message processing at check nodes. In particular, we focus on the check node c_j, which needs to evaluate the message $\mathbf{m}^{(l)}_{c_j \to x_i} \triangleq [\mu^{(l)}_{c_j \to x_i}(0), \mu^{(l)}_{c_j \to x_i}(1)]$ to the adjacent node x_i; we assume that $\mu^{(l)}_{c_j \to x_i}(b)$ represents an estimate of the probability that the jth parity check is satisfied, provided that $x_i = b$, with $b = 0, 1$. Note that, if the node c_j knew the probabilities $\{P_p, p \in R_j \backslash i\}$, it could evaluate the message $\mu_{c_j \to x_i}(0)$, independently of the iteration index l, as (see (10.94)):

$$\mu_{c_j \to x_i}(0) = \frac{1}{2} + \frac{1}{2} \prod_{p \in R_j \backslash i} (1 - 2P_p), \tag{10.97}$$

since, if $x_i = 0$, jth parity check equation is satisfied when an *even* number of the bits $\{x_p, p \in R_j \backslash i\}$ are equal to 1. Similarly, $\mu_{c_j \to x_i}(1)$ could be computed as (see (10.95)):

$$\mu_{c_j \to x_i}(1) = \frac{1}{2} - \frac{1}{2} \prod_{p \in R_j \backslash i} (1 - 2P_p). \tag{10.98}$$

Then, if (10.97) and (10.98) held, (10.96) could be rewritten as:

$$\frac{\Pr\{x_i = 0 | \mathbf{y}, \ S_i\}}{\Pr\{x_i = 1 | \mathbf{y}, \ S_i\}} = \frac{1 - P_i}{P_i} \frac{\prod\limits_{q \in C_i} \mu_{c_q \to x_i}(0)}{\prod\limits_{q \in C_i} \mu_{c_q \to x_i}(1)}, \tag{10.99}$$

so that the APP ratio (10.88) could be computed by exploiting the messages $\{\mu_{c_q \to x_i}(0), q \in C_i\}$ and $\{\mu_{c_q \to x_i}(1), q \in C_i\}$ traveling along all the edges connected to the node x_i. Unfortunately, the node c_j does not know the probabilities $\{P_p, p \in R_j \backslash i\}$ and can exploit only the knowledge of the messages $\{\mu^{(l)}_{x_p \to c_j}(b), p \in R_j \backslash i\}$, sent by its adjacent variables nodes different from x_i. As will become clearer below, such messages represent an estimate of $\{P_p, p \in R_j \backslash i\}$; then, if we assume that they are *mutually independent*, usable expressions of the messages $\{\mu^{(l)}_{c_j \to x_i}(1), b = 0, 1\}$ can easily be obtained by replacing P_p with $\mu^{(l)}_{x_p \to c_j}(b)$ in (10.97) and (10.98); this produces:

$$\mu^{(l)}_{c_j \to x_i}(0) \triangleq \frac{1}{2} + \frac{1}{2} \prod_{p \in R_j \backslash i} (1 - 2\mu^{(l)}_{x_p \to c_j}(1)) \tag{10.100}$$

and

$$\mu^{(l)}_{c_j \to x_i}(1) \triangleq \frac{1}{2} - \frac{1}{2} \prod_{p \in R_j \backslash i} (1 - 2\mu^{(l)}_{x_p \to c_j}(1)), \tag{10.101}$$

respectively (these two expressions can also be condensed into a single vector equation, as originally suggested by (10.84)).

Let us now focus on message processing at the variable nodes. The node x_i, through the message $\mu^{(l)}_{x_i \to c_j}(b)$, provides its adjacent node c_j with an estimate of the probability that $x_i = b$, with $b = 0, 1$ (these probabilities are collected in the vector $\mathbf{m}^{(l)}_{x_i \to c_j} \triangleq [\mu^{(l)}_{x_i \to c_j}(0), \mu^{(l)}_{x_i \to c_j}(1)]$). To compute such an estimate, the node x_i process first the channel datum y_i to generate the probability:

$$p_i \triangleq \Pr\{x_i = 1 | y_i\}, \tag{10.102}$$

and second the messages $\{\mu^{(l-1)}_{c_q \to x_i}(b), q \in C_i \backslash j\}$ generated in the previous iteration and originating from all the adjacent check nodes different from c_j. Briefly, the message $\mu^{(l)}_{x_i \to c_j}(b)$ is evaluated as:

$$\mu^{(l)}_{x_i \to c_j}(b) \triangleq \Pr\{x_i = b | p_i, \mu^{(l-1)}_{c_q \to x_i}(b) \text{ with } q \in C_i \backslash j\} \tag{10.103}$$

with $b = 0, 1$ and $l > 0$. If we assume that all the messages collected by x_i are *mutually independent*, it is easy to show that (see (10.103)):

$$\mu_{x_i \to c_j}^{(l)}(0) = (1 - p_i) \prod_{q \in C_i \backslash j} \mu_{c_q \to x_i}^{(l-1)}(0) \qquad (10.104)$$

and that:

$$\mu_{x_i \to c_j}^{(l)}(1) = p_i \prod_{q \in C_i \backslash j} \mu_{c_q \to x_i}^{(l-1)}(1). \qquad (10.105)$$

It is important to point out that at the beginning of the first iteration (corresponding to $l = 0$), the node x_i knows p_i (10.102) only, so that (10.104) and (10.105) are replaced by:

$$\mu_{x_i \to c_j}^{(0)}(0) = 1 - p_i \qquad (10.106)$$

and

$$\mu_{x_i \to c_j}^{(0)}(1) = p_i, \qquad (10.107)$$

respectively (in agreement with (10.86)).

Equations (10.100), (10.101), (10.104) and (10.105), together with the initial conditions (10.106) and (10.107), fully define the BPA. At the end of the lth iteration (with $l > 0$) the node x_i (with $i = 0, 1, \ldots, n - 1$) can evaluate the quantities:

$$Q_i^{(l)}(0) = K_i (1 - p_i) \prod_{q \in C_i} \mu_{c_q \to x_i}^{(l)}(0) \qquad (10.108)$$

and

$$Q_i^{(l)}(1) = K_i p_i \prod_{q \in C_i} \mu_{c_q \to x_i}^{(l)}(1), \qquad (10.109)$$

where K_i is a real parameter ensuring normalization (i.e., that $Q_i^{(l)}(0) + Q_i^{(l)}(1) = 1$); as is easily inferred from (10.99), $Q_i^{(l)}(0)$ and $Q_i^{(l)}(1)$ provide an estimate of the probabilities $\Pr\{x_i = 0 | \mathbf{y}, S_i\}$ and $\Pr\{x_i = 1 | \mathbf{y}, S_i\}$, respectively. Therefore, an estimate $\hat{\mathbf{x}}(l) = [\hat{x}_0(l), \hat{x}_1(l), \ldots, \hat{x}_{n-1}(l)]$ of the transmitted codeword $\mathbf{x} = [x_0, x_1, \ldots, x_{n-1}]$ can be generated using the simple decision rule:

$$\hat{x}_i(l) = \begin{cases} 1 & \text{if } Q_i^{(l)}(1) \geq 1/2 \\ 0 & \text{otherwise} \end{cases} \qquad (10.110)$$

for $i = 0, 1, \ldots, n - 1$. Given $\hat{\mathbf{x}}(l)$, the product $\hat{\mathbf{x}}(l) \, \mathbf{H}^T$ is evaluated to establish if $\hat{\mathbf{x}}(l)$ is a codeword or not. If the parity check is satisfied, the decoding procedure stops. Otherwise, an additional iteration is carried out (provided that the overall number of iterations does not exceed a given threshold) to generate a new codeword estimate $\hat{\mathbf{x}}(l + 1)$; obviously, if the BPA is unable to generate a codeword in a limited number of iterations, a decoding failure is declared.

Finally, it is important to make the following observations:

1. The probability p_i (10.102) depends on y_i only, and not on the entire vector $\mathbf{y}$, like P_i (10.93); in addition, it is easy to prove that, for BPSK transmission:

$$p_i = \frac{1}{1 + \exp\left(4 \, \text{Re}\{y_i h_i^*\}/\sigma_n^2\right)}, \qquad (10.111)$$

if the same channel model as in Section 10.5.2.1 is adopted.

2. If the Tanner graph associated with $\mathbf{H}$ is *cycle-free*, the probabilities $\mu_{x_i \to c_j}^{(l)}(0)$ (10.104) and $\mu_{x_i \to c_j}^{(l)}(1)$ (10.105) converge to the APPs $\Pr\{x_i = 0| \mathbf{y}, S_i\}$ and $\Pr\{x_i = 1| \mathbf{y}, S_i\}$, respectively, when $l \to \infty$, so that the BPA is asymptotically *optimal* [1479]. As already mentioned in Section 10.6.2, in the presence of cycles the hypothesis of statistical independence of the messages collected by any node is no longer true after a certain number of iterations. Then, as the iterations proceed, the exchanged messages become more and more correlated, so that an exceedingly large number of iterations is of no use.

3. The BPA can be exploited, in principle, for the decoding of any linear binary block code, but becomes really advantageous for LDPC codes because the degrees of variable and check nodes are low; in fact, this property entails that the overall number of products involved in the message equations (10.100), (10.101), (10.104) and (10.105) and in the evaluation of the data APPs according to (10.108) and (10.109) is not exceedingly large.

To avoid numerical instability problems in the implementation of the BPA, this strategy needs to be reformulated in the logarithmic domain. To this aim, we define the log APP ratios (see (10.100), (10.101), (10.102), (10.104), (10.105), (10.108) and (10.109)):

$$L(x_i) \triangleq \ln \frac{\Pr\{x_i = 0| y_i\}}{\Pr\{x_i = 1| y_i\}} = \ln \frac{1 - p_i}{p_i}, \tag{10.112}$$

$$L(\mu_{c_j \to x_i}^{(l)}) \triangleq \ln \frac{\mu_{c_j \to x_i}^{(l)}(0)}{\mu_{c_j \to x_i}^{(l)}(1)}, \tag{10.113}$$

$$L(\mu_{x_i \to c_j}^{(l)}) \triangleq \ln \frac{\mu_{x_i \to c_j}^{(l)}(0)}{\mu_{x_i \to c_j}^{(l)}(1)} \tag{10.114}$$

and

$$L(Q_i^{(l)}) \triangleq \ln \frac{Q_i^{(l)}(0)}{Q_i^{(l)}(1)}. \tag{10.115}$$

Then, based on (10.112), the initial conditions (10.106) and (10.107) can be expressed as:

$$L(\mu_{x_i \to c_j}^{(0)}) = L(x_i) = \ln \frac{1 - p_i}{p_i} \tag{10.116}$$

or, thanks to (10.111), as:

$$L(\mu_{x_i \to c_j}^{(0)}) = \ln \frac{1 + \exp(2y_i/\sigma_n^2)}{1 + \exp(-2y_i/\sigma_n^2)} = \frac{4 \operatorname{Re}\{y_i h_i^*\}}{\sigma_n^2}. \tag{10.117}$$

The message sent from x_i to c_j for $l > 0$ can be put in logarithmic form as follows. Evaluating the ratio between (10.104) and (10.105) yields:

$$\frac{\mu_{x_i \to c_j}^{(l)}(0)}{\mu_{x_i \to c_j}^{(l)}(1)} = \frac{1 - p_i}{p_i} \prod_{q \in C_i \backslash j} \frac{\mu_{c_q \to x_i}^{(l-1)}(0)}{\mu_{c_q \to x_i}^{(l-1)}(1)}. \tag{10.118}$$

Then taking the logarithm of both sides and exploiting (10.112)–(10.114) leads to the expression:

$$L(\mu_{x_i \to c_j}^{(l)}) = L(x_i) + \sum_{q \in C_i \backslash j} L(\mu_{c_q \to x_i}^{(l-1)}). \tag{10.119}$$

Putting the messages generated by check nodes in logarithmic form is more complicated. To this end, we rewrite (10.100) and (10.101) as:

$$2\mu_{c_j \to x_i}^{(l)}(0) = 1 + \prod_{p \in R_j \setminus i} \left(1 - 2\mu_{x_p \to c_j}^{(l)}(1)\right) \tag{10.120}$$

and

$$1 - 2\mu_{c_j \to x_i}^{(l)}(1) = \prod_{p \in R_j \setminus i} \left(1 - 2\mu_{x_p \to c_j}^{(l)}(1)\right), \tag{10.121}$$

respectively, and exploit the identity $\tanh\left[\frac{1}{2}\ln\left(q_0/q_1\right)\right] = q_0 - q_1 = 1 - 2q_1$ holding for any pair of real positive numbers q_0 and q_1 such that $q_0 + q_1 = 1$. From this identity and (10.113) and (10.114) it is easily inferred that:

$$1 - 2\mu_{c_j \to x_i}^{(l)}(1) = \tanh\left[\frac{1}{2}L\left(\mu_{c_j \to x_i}^{(l)}\right)\right] \tag{10.122}$$

and

$$1 - 2\mu_{c_j \to x_i}^{(l)}(1) = \tanh\left[\frac{1}{2}L\left(\mu_{c_j \to x_i}^{(l)}\right)\right]. \tag{10.123}$$

Then substituting (10.122) into the LHS of (10.121) yields:

$$\tanh\left[\frac{1}{2}L\left(\mu_{c_j \to x_i}^{(l)}\right)\right] = \prod_{p \in R_j \setminus i}\left[1 - 2\mu_{x_p \to c_j}^{(l)}(1)\right], \tag{10.124}$$

which can be rewritten as (see (10.123)):

$$\tanh\left[\frac{1}{2}L\left(\mu_{c_j \to x_i}^{(l)}\right)\right] = \prod_{p \in R_j \setminus i} \tanh\left[\frac{1}{2}L\left(\mu_{x_p \to c_j}^{(l)}\right)\right] \tag{10.125}$$

or, equivalently, as:

$$L(\mu_{c_j \to x_i}^{(l)}) = 2\tanh^{-1}\left\{\prod_{p \in R_j \setminus i} \tanh\left[\frac{1}{2}L(\mu_{x_p \to c_j}^{(l)})\right]\right\}. \tag{10.126}$$

Equation (10.126) expresses the check node message $L(\mu_{c_j \to x_i}^{(l)})$ as a function of the messages $\{L(\mu_{x_p \to c_j}^{(l)}), p \in R_j \setminus i\}$ collected by c_j by all the nodes $\{x_p\}$ adjacent to it and different from x_i. The main drawback of (10.126) is that it requires the evaluation of multiple products. To overcome this problem, the following solution can be adopted. Because the hyperbolic tangent is an *odd* function, we have $\tanh(x) = \text{sgn}(x)\tanh(|x|)$ for any real x, so that $L(\mu_{x_i \to c_j}^{(l)})$ can be expressed as:

$$L(\mu_{x_i \to c_j}^{(l)}) = \alpha_{ij}\,\beta_{ij}, \tag{10.127}$$

where $\alpha_{ij} \triangleq \text{sgn}\,[L(\mu_{x_i \to c_j}^{(l)})]$ and $\beta_{ij} \triangleq |L(\mu_{x_i \to c_j}^{(l)})|$. Then, if (10.125) is rewritten as:

$$\tanh\left[\frac{1}{2}L\left(\mu_{c_j \to x_i}^{(l)}\right)\right] = \prod_{p \in R_j \setminus i} \alpha_{pj} \prod_{p \in R_j \setminus i} \tanh\left(\frac{1}{2}\beta_{pj}\right), \tag{10.128}$$

the logarithmic message $L(\mu_{c_j \to x_i}^{(l)})$ can be put in the form:

$$L(\mu_{c_j \to x_i}^{(l)}) = \left(\prod_{p \in R_j \backslash i} \alpha_{pj}\right) \underbrace{2 \tanh^{-1}}_{\ln^{-1} \ln} \prod_{p \in R_j \backslash i} \tanh\left(\frac{1}{2}\beta_{pj}\right)$$

$$= \left(\prod_{p \in R_j \backslash i} \alpha_{pj}\right) 2 \tanh^{-1} \ln^{-1} \sum_{p \in R_j \backslash i} \ln \tanh\left(\frac{1}{2}\beta_{pj}\right). \qquad (10.129)$$

Let us now define the function:

$$f(x) \triangleq -\ln \tanh\left(\frac{1}{2}x\right) = \ln \frac{\exp(x) + 1}{\exp(x) - 1} \qquad (10.130)$$

(whose behavior is illustrated in Figure 10.18) and note that:

$$f^{-1}(x) = 2 \tanh^{-1} \ln^{-1}(x) = f(x) \qquad (10.131)$$

for any real $x > 0$, since $f(f(x)) = \ln[\exp(f(x)) + 1]/[\exp(f(x)) - 1] = x$. Then, by exploiting (10.130) and (10.131), (10.129) can be put in the form:

$$L(\mu_{c_j \to x_i}^{(l)}) = \left(\prod_{p \in R_j \backslash i} \alpha_{pj}\right) f\left(\sum_{p \in R_j \backslash i} f(\beta_{pj})\right), \qquad (10.132)$$

which contains a multiple sum (in place of a multiple product) and expresses the mathematical law for the computation of the message sent by c_j to x_i. Another favorable feature of this expression is that $f(x)$ (10.130) has a regular shape, so that it can be implemented using a *lookup table*.

To complete the description of the logarithmic BPA, the ratio between (10.108) and (10.109) is evaluated, yielding:

$$\frac{Q_i^{(l)}(0)}{Q_i^{(l)}(1)} = \frac{1 - P_i}{P_i} \prod_{q \in C_i} \frac{\mu_{c_q \to x_i}^{(l-1)}(0)}{\mu_{c_q \to x_i}^{(l-1)}(1)}, \qquad (10.133)$$

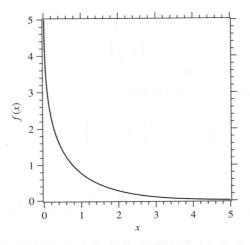

Figure 10.18   The function $f(x)$ (10.130).

from which it is easily seen that (see (10.112)–(10.115)):

$$L(Q_i^{(l)}) = L(x_i) + \sum_{q \in C_i} L\left(\mu_{c_q \to x_i}^{(l-1)}\right).$$

(10.134)

Moreover, the decision rule (10.90) can be expressed as:

$$\hat{x}_i(l) = \begin{cases} 1 & \text{if } L\left(Q_i^{(l)}\right) \leq 0 \\ 0 & \text{otherwise} \end{cases}$$

(10.135)

for $i = 0, 1, \ldots, n-1$.

The derivation of the logarithmic BPA illustrated above is rigorous, so that theoretically the performance of this algorithm is equivalent to that of the original BPA. In practice, however, in the implementations of BPA based on fixed-point hardware, quantization effects must be carefully taken into consideration. In fact, it has been shown that the direct implementation of the original form of BPA is sensitive to quantization effects and that using likelihood ratios can substantially reduce the required number of quantization levels [1772]. In any implementation of the BPA another critical issue is the procedure, expressed by (10.132), for the generation of the messages by check nodes. This has motivated the search for approximate and simpler procedures for this task. In the following we derive two such procedures, namely the MSA [1677, 1773–1775] and the *normalized* MSA (NMSA) [1776]. The derivation of the MSA is based on the behavior of $f(x)$ (10.130), illustrated in Figure 10.18; in fact from this figure it can easily be inferred that the quantity:

$$f\left(\sum_{p \in R_j \backslash i} f\left(\beta_{pj}\right)\right)$$

(10.136)

of (10.132) is mainly influenced by the maximum value of the terms in the sum $\sum_{p \in R_j \backslash i} f(\beta_{pj})$, that is, by the minimum of the quantities $\{\beta_{pj}, p \in R_j \backslash i\}$. For this reason (10.136) can be approximated as:

$$f\left(\sum_{p \in R_j \backslash i} f\left(\beta_{pj}\right)\right) \simeq f\left(f\left(\min_{p \in R_j \backslash i} \beta_{pj}\right)\right),$$

(10.137)

so that (see (10.131)):

$$f\left(\sum_{p \in R_j \backslash i} f\left(\beta_{pj}\right)\right) \simeq \min_{p \in R_j \backslash i} \beta_{pj}.$$

(10.138)

If this approximation is adopted, (10.129) becomes:

$$L(\mu_{c_j \to x_i}^{(l)}) = \left(\prod_{p \in R_j \backslash i} \alpha_{pj}\right) \min_{p \in R_j \backslash i} \beta_{pj}$$

(10.139)

or, equivalently:

$$L(\mu_{c_j \to x_i}^{(l)}) = \prod_{p \in R_j \backslash i} \text{sgn}\left[L\left(\mu_{x_p \to c_j}^{(l)}\right)\right] \min_{p \in R_j \backslash i} \left|L\left(\mu_{x_p \to c_j}^{(l)}\right)\right|,$$

(10.140)

which describes the expression for the evaluation of the check node messages in the MSA [1677, 1774, 1777]. Note that this represents the only difference between the logarithmic BPA and the MSA.

It is also important to point out that, on the one hand, the BPA requires knowledge of the noise variance σ_n^2, since this parameter is needed in the evaluation of (10.111). On the other hand, in the MSA the dependence of the log APP ratios on σ_n^2 is removed without changing the nature of the algorithm. In particular, this result is achieved by replacing $L(x_i)$ (10.112) with $\bar{L}(x_i) \triangleq \sigma_n^2 L(x_i)/4$ in the initialization (10.117), so that $L(\mu_{x_i \to c_j}^{(0)})$ (10.117) becomes $\bar{L}(\mu_{x_i \to c_j}^{(0)}) = \mathrm{Re}\{y_i h_i^*\}$. Then, (10.140), (10.119) and (10.134) are replaced by:

$$\bar{L}(\mu_{c_j \to x_i}^{(l)}) = \prod_{p \in R_j \backslash i} \mathrm{sgn}\,[\bar{L}(\mu_{x_p \to c_j}^{(l)})] \min_{p \in R_j \backslash i} |\bar{L}(\mu_{x_p \to c_j}^{(l)})|, \tag{10.141}$$

$$\bar{L}(\mu_{x_i \to c_j}^{(l)}) = \bar{L}(x_i) + \sum_{q \in C_i \backslash j} \bar{L}(\mu_{c_q \to x_i}^{(l-1)}) \tag{10.142}$$

and

$$\bar{L}(Q_i^{(l)}) = \bar{L}(x_i) + \sum_{q \in C_i} \bar{L}(\mu_{c_q \to x_i}^{(l-1)}), \tag{10.143}$$

respectively. Similarly, in the decision criterion (10.135) $L(Q_i^{(l)})$ is replaced by $\bar{L}(Q_i^{(l)})$.

The check node message (10.140) can be slightly modified to improve the MSA error performance; in fact, in [1776] it is proved that in the MSA (10.140) generates a message $L(\mu_{c_j \to x_i}^{(l)})$ having the *same sign* as (10.132), but a *larger absolute value*. For this reason, in [1776] (10.132) is replaced with:

$$L(\mu_{c_j \to x_i}^{(l)}) = \frac{1}{\alpha} \prod_{p \in R_j \backslash i} \mathrm{sgn}\,[L(\mu_{x_p \to c_j}^{(l)})] \min_{p \in R_j \backslash i} |L(\mu_{x_p \to c_j}^{(l)})|, \tag{10.144}$$

where α is a real normalisation factor greater than unity [1776], and this last result characterizes NMSA.

The error performance provided by the three different decoding algorithms derived in the foregoing is compared in following example for a specific LDPC code.

Example 10.7.1 We have considered a binary (3,6)-regular LDPC code characterized by $n = 504$ and $R = 1/2$ [1773]. Its parity check matrix has been generated in a random fashion, removing any possible double edges and cycles of order 4. Then **H** has been brought to a lower triangular form using the *Greedy A* algorithm of [1768] to ease the encoding procedure. Decoding has been accomplished using the BPA (in the logarithmic domain), the MSA and the NMSA (with $\alpha = 1.4$ in (10.144)); the maximum number of iterations has been set to 200, 200 and 50, respectively (a further increase in this parameter does not provide any appreciable improvement in the three cases). The BER performance provided by these decoding techniques is shown in Figure 10.19. Note that the energy gap between the BPA and the MSA is a fraction of a decibel and that this is substantially filled by the NMSA.
□

Generally speaking, the MSA offers an error performance a few tenths of a decibel inferior to that of BPA. In addition, its performance can also be improved by resorting to various modifications, illustrated in the following subvsection, where a brief overview of the literature on soft decoding techniques is provided.

10.7.3   *Technical Issues on LDPC Decoding via MP*

In this subsection an overview of various technical issues analyzed in the literature on LPDC decoding techniques is provided. In particular, we focus on soft decoding algorithms, hard decoding algorithms, decoding over channels with memory and performance bounds.

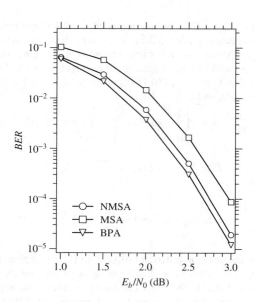

Figure 10.19   BER versus E_b/N_0 for the iterative decoding (via BP, MSA and NMSA) of the regular LDPC code described in Example 10.5.3.

10.7.3.1   Soft Decoding Algorithms

The significant complexity of the BPA has motivated the search for simpler soft decoding strategies for LDPC codes. The first two simplified versions of the BPA were proposed by Fossorier *et al.* in 1999 [1773]. Both operate in the log-likelihood domain, are implemented with real additions only, do not require any knowledge of the channel characteristics, yield a good performance–complexity tradeoff and can be efficiently implemented in software as well as in hardware, with possibly quantized received values. It is important to note that one of the two algorithms (called the *uniformly most powerful* BPA (UMP-BPA)) proposed in [1773] is equivalent to the *max-product algorithm* (MPA) of [1774] and to the *Max-Log-MAP algorithm* presented in [1775]; today all these are better known as the MSA [1677]. Although, generally speaking, MSA decoding yields error performance a few tenths of a decibel inferior to that of BPA (e.g., see Example 10.7.2), it is much simpler to implement, and does not require an estimate of the noise power for decoding for an AWGN channel. Moreover, its implementation is more robust against quantization error, compared with a similar implementation for the BPA [1778].

Other reduced-complexity decoding algorithms, also operating entirely in the log-likelihood domain, have been presented and analyzed in [1776–1784]. We summarize the main results described in these references. In [1777] minimal modifications to the MSA (leading to significant performance gains) are proposed. A decoding algorithm bridging the gap in performance between the optimal SPA and the UMP-BPA of [1773] is illustrated in [1779]. In this algorithm the computationally expensive check node updates of the SPA are simplified by using a difference metric approach on a two-state trellis and by employing the so-called *dual-max approximation*. Moreover, the dual-max approximation is further improved by using a correction factor that allows the performance to approach that offered by the SPA. In [1780] efficient serial and parallel implementations of the SPA are investigated and new reduced-complexity derivatives, requiring simple comparators and adders, are developed. A deep analysis of the UMP-BPA is provided in [1776], where it is explained why the performance of this algorithm is not close to that of the BPA. In particular, it is shown that the degradation of the UMP-BPA is due to the inaccuracy of the soft values delivered by check nodes from the first iteration. Then it is proposed to improve the accuracy of these soft values by *scaling* (i.e., *normalization*). This leads to NMSA described above (finite quantization effects are evaluated in [1785]). It is worth pointing out that the use of *scaling* (in particular, *damping*) of soft information for improving decoding performance has been also analyzed in [1786, 1787], where the DE technique with a Gaussian approximation is applied to the MSA to analyze the scaling gain. The threshold values are calculated for various scaling factors, and the scaling factor showing the highest threshold in noise level is determined as optimal. In [1782] a memory-efficient variation of the BPA is proposed and analyzed. The proposed algorithm performs almost as well as the SPA in terms of BER. It is based on splitting the code graph in a semirandom fashion into two subgraphs (i.e., the LDPC code into two subcodes) and decoding the resulting subcodes in a turbo fashion, instead of applying the MP algorithm to the entire graph. This can be considered an example of *layered decoding* (see [1788] and references therein). In [1783] a complexity reduction in the BPA is achieved by calculating a *linear* function for updating the check nodes in each iteration; the parameters of this function are optimized by DE. Simulation results show that there is nearly no loss in the performance by using this approximation compared to the exact calculation of the BPA, even for very large block lengths. Another improved algorithm, based on the BPA and called the *offset BP-based algorithm* (also known as *offset MSA* [1789]) is described in [1781]. It aims to improve the accuracy of the extrinsic messages delivered by the check nodes by reducing the reliability values by a positive constant. An improvement to the offset MSA is described in [1789], where a more efficient adjustment for check node update computation is proposed and the resulting algorithm achieves a noticeable performance gain with only a modest increase in computation complexity. Several of the leading approaches for decoding LDPC codes are compared in [1784] from both algorithmic and structural points of view. In particular, their performance and implementation complexity are evaluated. Numerical results indicate that in iterative decoding, accurate representations of the LLR-BPA algorithm do not always lead to the best performance if the underlying graph representation contains many short cycles. In fact, simplified reduced-complexity decoding schemes sometimes can outperform the BPA decoding algorithm. The effects of *clipping* and *quantization* on the performance of the MSA for the decoding of LDPC codes at short and intermediate block lengths are studied in [1778]. It is shown that:

- in many cases, only four quantization bits suffice to obtain close to ideal performance over a wide range of SNRs,
- although quantization usually degrades performance, clipping can provide improvement, and in many cases, the overall effect is such that only four bit quantization results in near or even better than ideal performance over a wide range of SNRs.

Moreover, modifications to the MSA that improve the performance by a few tenths of a decibel with only a small increase in decoding complexity are proposed. It is shown that, when optimized, the resulting modified quantized MSA performs very close to, and in some cases even slightly better than, the ideal BPA at observed error rates in the presence of code cycles.

Most of the papers cited above refer to soft decoding of *binary* LDPC codes. The iterative decoding via SPA of *nonbinary* LDPC codes was first investigated by Davey and Mackay in [1712]. A more effective decoding strategy for the q-ary LDPC codes is the FFT-based q-ary SPA [1727, 1790, 1791].

The behavior of soft decoding algorithms (and, in particular, the problem of evaluating the decoding thresholds) can be analyzed by resorting to the DE technique. In particular, the DE has been employed for the analysis of:

- the MSA in [1777] (where a comparison between the MSA and the BPA is made) and [1775, 1781];
- the BPA in [1679, 1652, 1753, 1792];
- the scaling gain of the NMSA in [1786, 1787];
- the NMSA and the offset MSA in [1781, 1785];
- various soft decoding algorithms in [1784].

Finally, we note the following:

1. The gap between iterative decoding and optimal (i.e., ML) decoding for any fixed-code structure can sometimes be significant [1793] due not only to the suboptimal nature of iterative decoding, but also to the suboptimal graph structure of the code which may contain short loops.
2. To bridge the performance gap between ML decoding and BPA decoding, Fossorier proposed combining *reliability-based decoding* [1496] (namely, order statistic decoding, OSD) with BPA decoding [1794] (see also [1795]). In practice, at each iteration the soft data generated by the BPA are used as reliability values to perform reduced-complexity soft decision decoding of the LDPC code considered. An improvement to this decoding approach is illustrated in [1796], where a more reliable method to reconstruct ordered information sequence in terms of the accumulated LLR transitions of variable nodes is proposed. This results in substantial performance gains with only a modest increase in computational complexity.
3. Practical probabilistic algorithms for efficient ML decoding of binary LDPC codes over the BEC channel have been proposed in [1797].
4. The passing of messages in the BPA follows the so-called *flooding schedule* [1687], since in each iteration, all variable nodes and subsequently all check nodes pass new messages to their neighbors. For a cycle-free Tanner graph, this results in optimal APP decoding. The cycles in the Tanner graphs of LDPC codes make SPA decoding no longer optimal and this leads to a performance loss, which can be more severe for short LDPC codes. Some performance improvement can be obtained, however, by changing the message schedule. For instance, a *probabilistic schedule*, tailored to the Tanner graph of an LDPC code, is illustrated in [1798]. The proposed schedule updates the outgoing messages of a variable node with an average frequency proportional to the length of the shortest cycle passing through it and can achieve significant performance improvement upon the flooding schedule, with a similar or even lower complexity.

10.7.3.2 Hard Decoding Algorithms

A substantially simpler alternative to soft decoding are hard decoding algorithms, such as the so-called *Gallager A* [1799] and *bit-flipping* algorithms [1479].

Gallager A is the simplest decoding algorithm based on the MP philosophy and allows as messages only elements of the set $\{\pm 1\}$, denoting the sign of a bit. More specifically, the message going out of each *check* node along an edge is simply the product of all incoming messages excluding the incoming message along the edge itself. The message sent by a *variable* node along an edge emerging from it is equal to the received (hard) message unless all incoming messages (ignoring the incoming message along the given edge) agree, in which case this message is forwarded. Although this simple decoding algorithm cannot achieve similar performance to the BPA, it is nevertheless of interest because of its extremely low complexity. For this algorithm, the evaluation of the threshold and of the optimal degree distributions for a large range of rates is illustrated in [1799].

Gallager suggested an iterative hard decoding algorithm, known as *bit-flipping*, as a simpler alternative to his soft decoding algorithm for LDPC codes. The proposed algorithm flips in sequence each variable bit that has more unsatisfied parity check bits compared to its satisfied ones. An analysis of the error-correction capability of this algorithm for regular LDPC code ensembles can be found in [1800] and references therein.

10.7.3.3 Error Performance

Two important technical issues are usually analyzed when investigating the error-correcting performance provided by practical algorithms for iterative decoding of LDPC codes, namely the *error floor* and the decoding *threshold*. Note that such performance can be close to Shannon limits for codes with suitably large block lengths. However, a substantial limitation to the use of *finite-length* LDPC codes is the presence of an *error floor* in the low *frame error rate* (FER) region, that is, of significant flattening in the curve relating the SNR to the FER. The error floor is commonly attributed to the suboptimality of the iterative decoding algorithms on graph with cycles. However, recent work on this topic has shown that the main cause of the error floor of structured LDPC codes are the (*fully*) *absorbing sets*, that is, certain combinatorial objects associated with codes and defined independently of the decoding scheme adopted or the channel noise model. In practice, the performance of an iterative decoding algorithm in the low-FER region is mainly determined by the number and structure of the smallest (fully) absorbing sets (unlike ML decoders whose performance is governed by the minimum-distance codewords). The reader can refer to [1801] (and references therein) for an interesting analysis of this topic.

For the BSC and various regular ensembles of graphs, Gallager numerically determined the largest crossover probability such that, under the assumption of independent messages, the expected number of messages in error converges to zero as the number of iterations grows to infinity [1590]. In [1679] it is proved that this number, known as the *threshold*, determines the asymptotic (in the codeword length) behavior of the ensemble of codes: roughly speaking, for a code chosen randomly from the ensemble, with high probability, decoding will be successful if transmission takes place below the threshold, and the error probability will stay above a fixed constant if transmission takes place above it. This threshold can easily be determined numerically, but it is nevertheless pleasing and potentially insightful to derive analytical expressions (e.g., see [1799] referring to a BSC channel and decoding via the Gallager A algorithm).

Further interesting results on the performance analysis of LDPC codes can be found in [1802, 1803]. In particular, in [1802] it is conjectured that the performance of a code from a given LDPC ensemble does not depend appreciably on the particular memoryless channel, but only on the mutual information between the input and the output of the channel. This conjecture originates from a performance analysis of LDPC codes based on simple EXIT charts and is supported by simulation results referring to various LDPC codes and several memoryless channels. In addition, it confirms an early remark in [1686], where it was observed that LDPC codes optimized for the AWGN channel show good performance for other memoryless channels, such as the BSC and BEC, and generalizes the results in [1720]. In [1803] the behavior of iteratively decoded LDPC codes over the BEC in the so-called *waterfall region* is analyzed and it is shown that the performance curves in this region follow a simple scaling law. The scaling law, together with the error floor expressions developed previously, can be used for a fast finite-length optimization.

Finally, it is worth remembering that various *asymptotic results* (i.e., applicable to very long codes) have been derived to assess the performance provided by ensembles of LDPC codes. These include results on the performance under ML decoding [1694, 1695, 1804], average ensemble distance spectra [1590, 1690, 1692, 1693, 1805], stopping set distributions [1692, 1693, 1759, 1805] and thresholds for iterative decoding. However, an accurate finite-length analysis of LDPC codes under BPA decoding is currently available only for the BEC [1758]. This is due to the simplicity of the channel model and the graph-based iterative decoder which lends itself to a more detailed analysis. For the BEC scenario lower and upper bounds on the error exponent of typical codes in LDPC code ensembles are derived in [1692] and [1694, 1806], respectively.

10.7.3.4 LDPC Decoding over Channels with Memory

In the technical literature there are many recent papers on detection and decoding of LDPC codes over channels with memory (e.g., see [1807] and references therein). A general theoretical framework to solve the problem of joint decoding and estimation, in the presence of unknown channel parameters, has been derived by A. P. Worthen and W. E. Stark [1808]. The approach is Bayesian, that is, the channel parameters are modeled as stochastic processes with known statistics, and the use of factor graphs[11]

---

[11] Factor graphs are defined in Section 10.8.

that include both code constraints and channel statistics is advocated in a very general setting. The SPA is then used to implement the MAP symbol detection strategy. However, since the channel parameters, which are continuous random variables, are explicitly represented in the graph, the application of the SPA becomes impractical. To solve this problem, the method of *canonical distributions* is proposed. Further work on this topic can be found in [1807], where using a factorization of the joint APP of the information symbols, a FG representing both the code constraints and the channel model is derived. In this FG, however, the channel parameters are not explicitly represented since they are a priori averaged out. The application of the SPA to this FG leads to an iterative scheme for joint detection and decoding. Note, however, that this factorization is exact only in the case of channels with finite memory (e.g., a channel with known ISI) and is approximate for channels with infinite memory.

10.7.3.5 Decoder Implementation

In designing an LDPC MP decoder a *serial*, a *fully parallel* or *partially parallel* architecture can be adopted. The first option is characterized by simple hardware, but offers low throughput. In contrast, a fully parallel architecture can provide high throughput at the cost of complex hardware. The partially parallel architecture is a tradeoff between throughput and complexity and, for this reason, is of interest to many practical applications. The reader can refer to [1740] and references therein for further information on this topic.

10.8 Codes on Graphs

The field of *codes on graphs* has been developed to provide a general conceptual foundation for modern capacity-approaching coding schemes and their iterative decoding algorithms. The beginning of this field dates back to the introduction of bipartite graphs for modeling LDPC codes by Tanner [1606] (see Section 10.6.2). Tanner graphs were later extended to include *state variables* as well as symbol variables by N. Wiberg in his seminal thesis [1677, 1809].

Generally speaking, graphs can be very useful to model *systems* (i.e., collections of interacting variables). To show this, let us briefly introduce a "behavioral" approach to system modeling [1810] and apply it to channel coding. In particular, let us consider a system fully described by the collection $\{x_0, x_1, \ldots, x_{n-1}\}$ of n variables with *configuration space* $S = A_0 \times A_1 \times \ldots \times A_{n-1}$, where A_i denotes the alphabet of x_i, with $i = 0, 1, \ldots, n-1$. Note that this collection of variables cannot take on any arbitrary value (i.e., *configuration*) in S. The subset B of S consisting of all the *valid configurations* for the given system represents the so-called *behavior* in S.

Behavioral modeling can be readily adopted for channel coding [1610]. For instance, any linear (n, k) block code C over GF(q) can be represented as a behavior in S, with $A_i = $ GF(q), $i = 0, 1, \ldots, n-1$. In this case, as shown below, a possible choice for the set of interacting variables is represented by the n codeword symbols, so that any valid configuration corresponds to a specific *codeword* of C. However, before providing further details about behavioral modeling of linear block codes, it is useful to introduce the *characteristic* (or *set membership indicator*) *function* of a behavior B, defined as:

$$\chi_B(x_0, x_1, \ldots, x_{n-1}) = [(x_0, x_1, \ldots, x_{n-1}) \in B], \tag{10.145}$$

where the function:

$$[\mathcal{P}] \triangleq \begin{cases} 1 & \text{if } \mathcal{P} \text{ is true} \\ 0 & \text{otherwise} \end{cases} \tag{10.146}$$

indicates the truth of a predicate $\mathcal{P}$. Generally speaking, the characteristic function χ_B fully defines the membership of a configuration in a behavior B and, consequently, can also be exploited to define the set of codewords of a block coding scheme. It is useful to note that in linear block coding the membership of a specific n-dimensional vector over GF(q) of the codeword set of an (n, k) block code C over GF(q) can be verified through $n - k$ parity checks. In the language of behavioral modeling, this statement can be reformulated to state that the validity of a particular configuration for a behavior B is assessed by applying a series of $n - k$ tests, each involving some subset of the system variables. Obviously, a given configuration is deemed valid if and only if it passes all the tests. Therefore, the predicate $(x_0, x_1, \ldots, x_{n-1}) \in B$ appearing on the RHS of (10.145) can be expressed as a logical conjunction of a series of $n - k$ "simpler" predicates:

$$\chi_B(x_0, x_1, \ldots, x_{n-1}) = [\mathcal{P}_0 \wedge \mathcal{P}_1 \wedge \ldots \wedge \mathcal{P}_{n-k-1}], \tag{10.147}$$

where $\wedge$ denotes the logical conjunction and the predicate $\mathcal{P}_l$ is true if and only if the lth parity check equation of C is satisfied (with $l = 0, 1, \ldots, n - k - 1$). This allows us to factor χ_B as:

$$\chi_B(x_0, x_1, \ldots, x_{n-1}) = [\mathcal{P}_0][\mathcal{P}_1] \cdots [\mathcal{P}_{n-k-1}], \tag{10.148}$$

namely as a product of characteristic functions, each indicating whether a particular subset of the code symbols satisfies a specific check equation, that is, is a valid configuration of some "local" behavior.

An equivalent description of the product of functions appearing on the RHS of (10.148) is provided by its FG representation, a bipartite graph expressing the structure of the given factorization. Generally speaking, a FG provides a graphical representation of a function $f(\cdot)$ characterized by the following properties:

- it depends on a set of N variables $\{x_0, x_1, \ldots, x_{N-1}\}$, each taking values on a specific alpahabet;
- it can be factored into a product of a *local functions*, each depending on a subset of $\{x_0, x_1, \ldots, x_{n-1}\}$, that is:

$$f(x_0, x_1, \ldots, x_{N-1}) = \prod_{l \in S_l} f_l(\mathbf{X}_l), \tag{10.149}$$

where S_l is a discrete index set, $\mathbf{X}_l$ denotes a subset of $\{x_0, x_1, \ldots, x_{N-1}\}$ and $f_l(\mathbf{X}_l)$ is a function depending on $\mathbf{X}_l$ only.

The FG associated with (10.149) consists of a set of nodes and of a set of unoriented edges connecting variable nodes to function nodes. In particular, the FG contains a *variable node* for each variable x_l (with $l = 0, 1, \ldots, N - 1$) and a *factor node* for each local function $f_l(\mathbf{X}_l)$. In addition, a variable node x_p is connected by an edge to a function node $f_q(\mathbf{X}_q)$ if and only if x_p is contained in $\mathbf{X}_q$ (i.e., $f_l(\cdot)$ depends on x_p). We now illustrate these concepts through an example of their application to a specific binary block code.

Example 10.8.1 Let C denote the (7,4) Hamming linear block code over GF(2) described in Example 9.1.1 (see also Example 10.6.1). This code consists of the set of all binary 7-tuples satisfying the parity check $\mathbf{x}\,\mathbf{H}^T = \mathbf{0}_3$ (see (9.11)), with $\mathbf{H}$ given by (9.15). This matrix constraint can be rewritten more explicitly as:

$$x_0 + x_1 + x_2 + x_4 = 0,$$

$$x_1 + x_2 + x_3 + x_5 = 0,$$

$$x_0 + x_1 + x_3 + x_6 = 0. \tag{10.150}$$

where sums are evaluated over GF(2). Then to establish if a binary 7-tuple belongs to C (i.e., if a given configuration is valid) we need to check whether each of the last three equations is satisfied.

For this reason, the characteristic function for C can be put in the form (see (10.145), (10.147) and (10.148)):

$$\chi_B(x_0, x_1, \ldots, x_{n-1}) = [(x_0, x_1, \ldots, x_6) \in C]$$
$$= [x_0 + x_1 + x_2 + x_4 = 0][x_1 + x_2 + x_3 + x_5 = 0]$$
$$\cdot [x_0 + x_1 + x_3 + x_6 = 0]. \tag{10.151}$$

The product on the RHS can be represented through a FG. If a square is used to represent each of the parity checks (i.e., the factors) and a circle to denote each of the variables, the resulting graph is equivalent to that shown in Figure 10.14, the Tanner graph of C.
□

The latter result lends itself to an simple generalization. Generally speaking, if in the behavioral modeling of a block code C the role of system variables is played by codeword symbols, the FG representing the factorization of the associated characteristic function is equivalent to the Tanner graph of C. Note that in all these cases behavioral models are described by nonhidden (i.e., *visible*) variables. Such models are compact (in terms of the overall number of variables) and are useful for checking whether a given codeword is in C or not, but are not so good for generating codewords. To overcome the latter problem, behavioral models with *hidden* (sometimes called auxiliary, latent, or state) *variables* should be adopted. When this occurs, the set of codeword symbols (collected in an n-dimensional vector $\mathbf{x}$) is complemented with a set of n_s state variables $\{s_0, s_1, \ldots, s_{n_s-1}\}$, also belonging to GF($q$) and forming the n_s-dimensional vector $\mathbf{s}$. In addition, the symbol and state variables are required to satisfy a set of linear homogeneous equations over GF(q), called the *constraint equations*. Then in this new scenario the *full behavior* B consists of all the combinations $(\mathbf{x}, \mathbf{s})$ (also called *trajectories*) satisfying all the constraint equations, and the code C (representing a visible behavior) is the projection of B onto its first n components (the dimension of C coincides with that of B if and only if codewords associated with distinct trajectories are different).

Graphs can be employed for code description even in the presence of state variables, as originally proposed by Wiberg *et al.* [1677, 1809] (for this reason, graphs including state variables may be called *Wiberg-type graphs*). Then, in addition to symbol variables, which are external, observable, and determined a priori, the graph includes state variables which are internal and unobservable. Note that, in particular, any FG for B is considered to be also a FG for C.

An important class of graphical models with hidden variables is constituted by the so-called *trellis representations* [1583]; generally speaking, a trellis is directed graph with a cycle-free chain structure. A trellis for a block code C is an edge-labeled directed graph with separate *root* and *goal* vertices. In this trellis each sequence of edges forming a directed path from the root vertex to the goal vertex identifies codeword in C and each codeword in C is represented by at least one such path. A further relevant property of the trellis is that all paths from the root to any given vertex v consist of the same number of edges, that is, have the same fixed length d (this parameter is called the *depth* of v); in particular, the root vertex has depth 0, where the goal vertex has depth n. The set of depth d vertices forms the domain of the state variable s_d, with $d = 0, 1, \ldots, n$ (so that $n_s = n + 1$).

All these concepts are applied again to the (7,4) Hamming code in the following example.

Example 10.8.2 The (7,4) Hamming linear block code of Example 10.8.1 consists of the set of all binary 7-tuples generated as $\mathbf{x} = \mathbf{u}\,\mathbf{G}$ (see (9.3)), with $\mathbf{G}$ given by (9.14) and this matrix expression can be rewritten as:

$$\mathbf{x} - \mathbf{u}\,\mathbf{G} = \mathbf{0}_7,$$

which can be interpreted as the set of seven constraint equations:

$$x_0 + u_0 = 0,$$
$$x_1 + u_1 = 0,$$
$$x_2 + u_2 = 0,$$
$$x_3 + u_3 = 0,$$
$$x_4 + u_0 + u_1 + u_2 = 0, \tag{10.152}$$
$$x_5 + u_1 + u_2 + u_3 = 0,$$
$$x_6 + u_0 + u_1 + u_3 = 0,$$

to be satisfied by the seven codeword symbols forming $\mathbf{x}$ and by the state variables forming the 4-tuple $\mathbf{x}$. This behavioral representation is less compact (in terms of the overall number of variables) than that described in Example 10.8.1, but is certainly more useful for encoding information. A bipartite graph (i.e., a Tanner graph or FG) representing this behavioral model is shown in Figure 10.20; here state (symbol) variables are represented by empty (filled) circles and constraints by squares. It is easy to verify that this graph contains cycles.

The same code is also represented by the trellis diagram in Figure 10.21(a). The reader can easily check that this cycle-free graph represents the 16 codewords of the Hamming code originating from the root vertex (i.e., the leftmost node), developing to the right and ending in the goal vertex (i.e., in the rightmost node), and the codeword symbols are represented by edge labels. This graph involves the binary code symbols $\{x_0, x_1, \ldots, x_6\}$ and seven (i.e., hidden) state variables $\{s_0, s_1, \ldots, s_7\}$, each

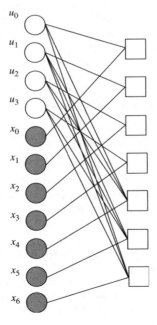

Figure 10.20 Tanner graph representing a behavioral model of the Hamming code described in Example 10.8.2.

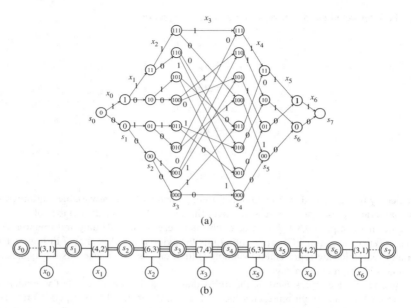

(a)

(b)

Figure 10.21 (a) Trellis and (b) the associated Wiberg-type graph for the block code of Example 10.8.2.

associated with a group of vertices having the same depth. Moreover the graph consists of seven *sections*, the *i*th section (denoted T_i, with $i = 1, 2, \ldots, 7$) being that part of the graph encompassing all the vertices at depth $i - 1$ and i and the edges connecting them, that is, expressing a "local" constraint about the possible combinations of the variables of the triple (s_{i-1}, x_i, s_i) only. Note that such combinations generate a linear block code. For instance, the second section of the trellis is associated with a local code that consists of the four binary linear combinations of $(0, 1, 01)$ and $(1, 0, 10)$, that is, with a binary $(4, 2)$ block code. These considerations suggest that we can express the characteristic function $\chi_H(\cdot)$ for this behavioral model as:

$$\chi_H(x_0, x_1, \ldots, x_6, s_0, s_1, \ldots, s_7) = \prod_{i=1}^{6} \chi_i(s_{i-1}, x_i, s_i), \qquad (10.153)$$

where $\chi_i(s_{i-1}, x_i, s_i) \triangleq [(s_{i-1}, x_i, s_i) \in T_i]$ is the characteristic function referring to the *i*th section of the trellis and describing a local behavior. The factorization on the RHS of (10.153) leads to the Wiberg-type FG shown in Figure 10.21(b). Note that:

(a) states are represented by double circles;
(b) for each state the number of grouped parallel edges emerging from it is equal to the number of bits representing the state variable;
(c) the endmost states s_0 and s_7 are constrained always to be zero, so that, as a matter of fact, they are 0-bit variables or constants (this is shown by a connection to the FG with a dotted line; in principle, this connection could be omitted and this would turn the local code to which they are now connected into a (2,1) code).

□

Various considerations expressed in Example 10.8.2 for a specific block code can easily be generalized. A trellis T for any (n, k) block code C over GF(q) can be divided into n sections $\{T_i, i = 1, 2, \ldots, n\}$, such that:

- the ith trellis section T_i is the subgraph of the trellis consisting of the vertices at depths $i - 1$ and i, and of the set of edges connecting them,
- the set of edge labels in T_i forms the domain of the (visible) variable x_i,
- the section T_i defines a "local behavior" and, in particular, describes the local constraints on the possible combinations of the triple $\{s_{i-1}, s_i, x_i\}$.

The configuration of the behavior B described by T is defined by the variables $\{x_0, x_1, \ldots, x_{n-1}\}$ and $\{s_0, s_1, \ldots, s_n\}$. Such a configuration is valid if and only if it satisfies the local constraints imposed by each of the trellis sections. For this reason the characteristic function for B can be represented as the product of n factors, where the ith factor is associated on the ith trellis section and depends on the variables $\{s_{i-1}, x_i, s_i\}$ only. This inevitably leads to a cycle-free FG. Since every code admits a trellis representation, it can be represented by a cycle-free FG. Unfortunately, it often turns out that the state-space sizes can easily become too large to be practical. For instance, turbo codes may well have (Wiberg-type) FG representations of reasonable complexity, but necessarily with cycles, whereas the trellis representations of turbo codes are characterized by huge state spaces.

Another tool for graphical modeling of code realizations has been proposed by D. Forney [1811], who focused on the study of *normal realizations* of codes, that is, Wyberg-type realizations[12] (realizations involving symbol, states and a set of local constraints) in which the degree of a state variable (i.e., the number of constraints in which it is involved) and that of the symbol variable are restricted to two and one, respectively (note that these constraints are always satisfied, for instance, by trellis realizations). More specifically, Forney introduced a novel graphical model, called a *normal graph* (also called a *Forney graph*) for Wyberg-type realizations. In this model the constraints, the state variables and the symbol variables of FGs are represented by vertices, ordinary edges and leaf edges (i.e., half edges) connected to the corresponding constraint vertices, respectively. The simple rules for converting a FG with symbols and states into a normal graph are summarized in Figure 10.22(a). When applied to the simple FG of Figure 10.22(b), they lead to the normal graph of Figure 10.22(c). This result demonstrates that:

(a) this conversion does not cause a change in graph topology or complexity;
(b) in the normal graph state nodes are replaced by repetition constraints, which constrain all incident state edges to be equal, whereas symbol nodes are replaced by repetition constraints and symbol half-edges;
(c) the normal graph, unlike the FG, is not bipartite since it contains only one kind of vertex;
(d) despite the edges of a FG need not be labeled, an edge of a normal graph is labeled by the state or symbol variable it represents (labels have been omitted in Figure 10.22(c) for simplicity);
(e) despite a FG is an ordinary graph, a normal graph is a graph with leaves.

Forney has also proved that the use of normal graphs allows one to draw a clean separation of functions between the elements of the graphs and the elements of the SPA. In fact, all computations are accomplished at vertices, whereas state edges and symbol (half) edges are used for internal communication (message passing) and external communication (input/output), respectively.

The development of a graphical approach to modeling codes has provided new insights into all known capacity-approaching codes and a wide variety of MP algorithms used not only in coding but also in computer science and signal processing. First, it has provided a unifying framework for

---

[12] These are called *generalized state realizations* by Forney.

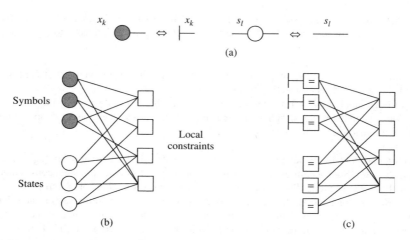

Figure 10.22 (a) Conversion rules for symbol and state variables; (b) example of a FG; (c) example of the associated normal graph.

the understanding of LDPC codes, turbo codes and most practically decodable capacity-approaching coding schemes (e.g., RA codes). All these can be interpreted as codes on graphs. One of the most significant results following from the unification of various coding scheme is the well-known theorem, known as the *cut-set bound*, whose meaning can be explained as follows (mathematical details can be found in [1809, 1811]). A *cut set* of a connected graph is a *minimal set of edges* whose removal partitions the graph into two disconnected subgraphs. In a normal graph, a cut set χ consists of a set of ordinary (state) edges, and the symbol variables, constraint codes and the states not belonging to χ are partitioned by the cut set into two disjoint subsets connected only by the states in χ. The states in the cut set can be regarded as a single superstate variable Σ_χ; the cut-set bound provides a *lower bound on the size of the alphabet of* Σ_χ. In particular, this theorem implies that cycle-free graphs cannot have state spaces significantly smaller than those of conventional trellises; therefore, substantial reductions in complexity can be found only in graphs with cycles, such as the graphs of turbo and LDPC codes. From this perspective turbo codes, LDPC codes, and RA codes can all be seen as codes whose graphs are made up of simple codes with linear-complexity graph realizations, connected by a large pseudorandom interleaver.

Graphical modeling of codes has also lead to significant conceptual contributions to the understanding of decoding algorithms. The first significant results in this area of research were derived by Wiberg, who provided an accurate characterization of the MSA and SPA, and showed that they perform essentially the same procedure except for the substitution of *min* for *sum* and *sum* for *product*. He also proved that they perform exact ML and APP decoding, respectively, on cycle-free graphs and that, in particular, they reduce to the Viterbi and BCJR algorithms, respectively, when the code graph is a trellis. This result strongly motivated the heuristic extension of iterative decoding based on the SPA to graphs with cycles. In particular, the turbo and LDPC decoding algorithms may be interpreted as instances of this decoding approach applied to their respective graphs. Note that these graphs inevitably contain cycles, but the probability of short cycles is low, and consequently iterative decoding based on the SPA works well. Conceptual results on decoding over graphs have been exploited in various related fields. For instance, the *belief propagation* and *belief revision algorithms* of *Pearl* [1812] (employed for statistical inference on Bayesian networks) and the *forward–backward* (*Baum–Welch*) algorithm [1813] (exploited for the detection of hidden Markov models in signal processing) have been shown to be special cases of the SPA operating on specific graphs. The reader can refer to [1610, 1687, 1689, 1814, 1815] for further information on this interesting area of research.

10.9   Historical Notes

Some historical information has been already provided in the previous sections of this chapter. In this section we outline the overall evolution of various modern coding schemes, in order to try to illustrate their mutual relationships. Further historical information on this topic can be found in [1465, 1547].

The introduction of *turbo codes* constitutes a fundamental event in the history of modern coding schemes. This occurred at the 1993 IEEE *International Conference on Communications* (ICC) in Geneva, where C. Berrou, A. Glavieux and P. Thitimajshima [1481] showed that the parallel concatenation of convolutional codes could achieve near-Shannon-limit performance while requiring a quite reasonable decoding complexity. The genesis of turbo codes, however, is related to the development of some fundamental theoretical concepts in the past. It was well known to the scientific community that *linear* codes were as good as general codes and that devising *randomlike* codes (in the sense that the distance distribution of a typical codeword from all other codewords should approach that of a random code) was the key to getting close to channel capacity. These principles were already clear in Gallager's monograph on LDPC codes [1590] and in the work of G. Battail (e.g., see [1816]). Another important principle known to the inventors of turbo codes was the possibility of exploiting *soft decisions* (i.e., reliability information) not only as input to a decoder, but also in the inner processing of an iterative decoder. For instance, this principle had already been put into practice by Gallager when he developed APP decoding for LDPC codes, showing that the use of soft decision information in iterative decoding was useful even on a hard decision channel. However, the idea of employing SiSo decoding in a concatenated coding scheme was originally proposed by Battail [1817] and by Hagenauer and Hoeher in [1535] in the second half of the 1980s; in fact, they proposed the SiSo version of the VA (namely, the SOVA, see Section 10.5.2.1). A few years later, Hoeher and Lodge extended their ideas by proposing to connect distinct SiSo APP decoders operating in an iterative fashion [1818]. It is also worth noting that:

- at the conference in which Berrou *et al.* introduced turbo codes and first used the term *extrinsic information* (see Section 10.5.1), a paper by Lodge *et al.* [1636] also included the idea of extrinsic information,
- by that time the benefits of retaining soft values in digital receivers had been fully appreciated (e.g., see [1819]).

Actually, the invention of turbo codes began when A. Glavieux suggested that his colleague C. Berrou, a professor of VLSI circuit design, consider the implementation of the SOVA decoder in silicon. When studying the principles at the basis of SOVA decoding, Berrou was struck by Hagenauer's observation that a SiSo decoder is a kind of SNR amplifier. He understood that the SNR could have been improved by repeated decoding, using some sort of iterative feedback. This led to the development of the turbo principle presented at the 1993 ICC. In the following years various researchers confirmed the astonishing performance results provided by turbo coding and the so-called *turbo revolution* began. Various properties of turbo coding were soon understood. In particular, it was shown that the use of an interleaver of length N effectively reduces the number of low-weight codewords by a factor of N [1616, 1684], and that the turbo codes have relatively poor minimum distance. In fact this parameter grows only logarithmically with N [1820], so that the ensemble of turbo codes is not asymptotically good. The poor minimum distance leads to a flattening of the performance curve of turbo codes, that is, to an *error floor*. It was shown that this error floor problem can be mitigated (but not completely avoided) using *serial concatenation* instead of parallel concatenation [1546, 1821], designing *structured interleavers* to improve the minimum distance [1822, 1823], or using *multiple interleavers* to eliminate low-weight codewords [1824, 1825] (the issue of interleaver design for turbo codes has also been investigated in [1826–1831]). Since 1993, various research efforts have also been devoted to reducing the associated decoder complexity, in particular, the Max-Log-MAP algorithm proposed by W. Koch and A. Baier [1598], as well as by J. Erfanian *et al.* [1524], the Log-MAP algorithm suggested by P. Robertson *et al.* [1525], and the above-mentioned SOVA algorithm [1535, 1599] have been considered.

During the mid-1990s, Hagenauer *et al.* [1100] and R. M. Pyndiah [1503] extended the turbo concept to parallel concatenated block codes as well. In 1997 H. Nickl *et al.* [1832] showed that Shannon's limit can be approached within 0.27 dB by employing a simple turbo Hamming code.

LDPC codes were first proposed in 1960 by Gallager in his doctoral thesis [1479, 1590], but did not receive significant attention for about 35 years [1833]. Shortly after the appearance of turbo codes, codes similar to Gallager's LDPC codes were discovered independently by D. J. C. MacKay [1607, 1834, 1835] and D. A. Spielman [1608, 1609, 1836, 1837]. In particular, MacKay showed that a moderate-length LDPC code could attain near-Shannon-limit performance, whereas Spielman used LDPC codes based on expander graphs to devise codes with linear-time encoding and decoding algorithms and proved that, in theory, as the codeword length $n \to \infty$, they could approach the Shannon limit. These results spurred research on LDPC codes, which are currently seen as competitors to turbo codes. It is important to mention that in 1995, N. Wiberg proved in his doctoral thesis [1677, 1809] that both of these classes of codes could be understood as instances of *codes on sparse graphs*, and that their decoding algorithms could be understood as instances of a general iterative APP decoding algorithm called the SPA. Wiberg also discovered that many of his results had previously been found by R. M. Tanner [1606] in a largely forgotten 1981 paper. In fact, Tanner had generalized LDPC codes and provided a graphical tool (today known as a *Tanner graph*) for their representation. Wiberg's rediscovery of Tanner's work opened up a new area of research into *codes on graphs* (see Section 10.8). In particular, the following findings deserve to be mentioned:

1. The results of Spielman were quickly applied by N. Alon and M. Luby [1838] to the Internet problem of reconstructing large files in the presence of packet erasures.
2. The discovery of the superiority of *irregular* LDPC codes. Luby *et al.* [1699, 1752] found that by using irregular graphs and optimizing the degree sequences, they could approach the capacity of the BEC (e.g., a rate-1/2 LDPC code capable of correcting almost half of erasures is described in [1839]). More recently, it has been shown [1840] that on any erasure channel, binary or nonbinary, it is possible to design LDPC codes that can approach capacity arbitrarily closely, in the limit as $n \to \infty$. T. J. Richardson and R. L. Urbanke *et al.* [1679, 1686] used the DE technique to design long irregular LDPC codes that for all practical purposes achieve the Shannon limit on binary AWGN channels. Moreover, using this approach, Chung *et al.* [1701] designed several rate-1/2 codes for the AWGN channel, including one whose theoretical threshold approached the Shannon limit to within a small fraction of a decibel.
3. The discovery of algebraic structures for turbo codes. These may be preferable to a pseudorandom structure for implementation and may allow control over important code parameters such as minimum distance, as well as graph-theoretic variables such as expansion and girth (see Section 10.6.4).

Another related class of codes is that of RA codes (see Section 10.4.1), proposed in 1998 by D. Divsalar and R. J. McEliece *et al.* [1648] as simple turbo-like codes for which one could prove coding theorems. The performance of RA codes turned out to be remarkably good and certainly better than that of the best coding schemes known prior to turbo codes. Other authors have proposed equally simple codes with similar or even better performance, such as ARA codes [1744] and *concatenated tree codes* [1841] (see also [1842]). All these results have shown that simple codes can be interconnected by large pseudorandom interleavers in various ways and decoded via the SPA so as to yield near-Shannon-limit performance.

Other modern coding schemes, which have not been described in this chapter, are represented by *fountain codes* or *rateless codes*. These are designed for channels without feedback and whose statistics are not known a priori (e.g., Internet packet channels where the probability of packet erasure is unknown). These schemes are based on the principle that the transmitter encodes a finite-length message into a potentially infinite stream of encoded symbols; the receiver then accumulates received symbols (possibly noisy) until it finds that it has enough for successful decoding. The first codes of this class were the so-called *Luby transform* (LT) codes [1843]. These were extended to the *Raptor codes* [1844, 1845], characterized by the concatenation of an inner LT code with an outer fixed-length, high-rate LDPC code.

Even though the introduction of turbo codes dates back to less than 20 years ago, these codes and the related class of LDPC codes have already had a significant impact in practice. In particular, almost all digital communication and storage system standards that involve coding are being upgraded to include these new capacity-approaching techniques.

10.10   Further Reading

An brief overview of iterative decoding principles is provided by the tutorial paper [1846]. A detailed introduction to turbo coding is provided by the book [1847], while the book [1586] is devoted mainly to LDPC coding. Finally, the reader can refer to the books [1815, 1848] for a deeper analysis of the problem of coding on graphs.

11

Signal Space Codes

11.1 Introduction

Hitherto we have mainly focused on binary error-correcting coding. In this chapter we focus on combined coding and modulation. This leads us to look at signal space codes or coded modulations. We will focus on a subset of the available techniques. In particular, we look at the original TCMs, BICMs, *multilevel codes* (MLCs), and *space-time codes* (STCs). In general, the key difference between the design of error control codes and signal space codes is that binary error-correcting codes are typically designed to maximize the minimum Hamming distance between codewords, while signal space codes are generally designed to maximize other distance measures between modulated codewords.

This chapter is organized as follows. The following three sections analyse coded modulation schemes for SIMO communication systems. In particular, trellis coding schemes that exploit expanded signal sets and bit interleaving are described in Sections 11.2 and 11.3, respectively, whereas Section 11.4 is focused on the study of modulation codes based on multilevel coding. The remainder of the chapter investigates space-time coding. In particular, various classes of STCs, including STBCs, OST-BCs, STTCs, layered STCs and unitary STCs, are analysed in Section 11.5. Most of our results refer to SC systems; however, in the last part of Section 11.5, STCs for MIMO OFDM are considered. Finally, some historical information and suggestions for further reading are provided in Section 11.6 and 11.7, respectively.

11.2 Trellis Coding with Expanded Signal Sets

Error-correcting codes as described in Chapters 8 and 9 are able to improve the system performance by adding extra bits (*redundancy*) to the transmitted symbol sequence. Hence, they allow us to increase noise immunity at the expense of a bandwidth increase equal to the reciprocal of the code rate. Such an increase can be avoided by enlarging the signal set, that is, by employing a higher-order modulation scheme to compensate for the redundancy introduced by the code.

As an example, let us consider an uncoded QPSK transmission. Each QPSK symbol is transmitted every T_s seconds and carries two information bits. If we want to improve the system performance, we may encode the information bits through a rate-2/3 binary code. Each QPSK symbol will now carry 4/3 information bits and, hence, to match the information rate of the source, the symbol interval should be reduced to $2T_s/3$, thus expanding the bandwidth by a factor of 3/2. This bandwidth increase can be avoided by employing an 8-PSK constellation instead of the original QPSK. In this case, however, coding and modulation must be designed jointly while the receiver, instead of performing demodulation and decoding in two separate steps, must combine the two operations into one.

Wireless Communications: Algorithmic Techniques, First Edition.
Giorgio M. Vitetta, Desmond P. Taylor, Giulio Colavolpe, Fabrizio Pancaldi, Philippa A. Martin.
© 2013 John Wiley & Sons, Ltd. Published 2013 by John Wiley & Sons, Ltd.

TCM is thus a technique based on the combination of coding and modulation to increase efficiency in bandwidth-limited environments. It was originally described in the seminal work of G. Ungerboeck and I. Csajka [1077] and clearly formalized by Ungerboeck in 1982 [992] with reference to the AWGN channel. Before going into details of the design of TCMs for time-selective fading channels, we will therefore briefly review the main concepts with reference to the AWGN channel.

11.2.1 Code Construction

11.2.1.1 TCMs over the AWGN Channel

For the AWGN channel, asymptotic performance is not governed by the *free Hamming distance* of the binary code, but by the *free Euclidean distance* between transmitted signal sequences. The principle of *set partitioning* is particularly useful. It involves partitioning the signal constellation A_c of cardinality $M = 2^m$ into subsets with increasing minimum Euclidean distance. Figure 11.1 illustrates this concept for an 8-PSK constellation A_c and carrying the partitioning to the limit where each subset D_i, $i = 0, 1, \ldots, 7$, contains a single point; generally speaking, this is not necessary since the level of partitioning depends on the code design.

The encoding/modulation process is illustrated in Figure 11.2. At the encoder input, k input information bits, arriving at a discrete-time instant, are separated in two groups. The first k_1 bits are encoded through a rate-k_1/n_1 binary encoder. The remaining $k_2 = k - k_1$ bits are left uncoded. Therefore the overall encoder has rate $k/(n_1 + k_2)$ and, of course, the same number of states of the constituent binary encoder. The n_1 encoded bits are used to select, according to an appropriate criterion, one of 2^{n_1} subsets, whereas the k_2 uncoded bits are used to select one of 2^{k_2} points within a subset. Thus, the level of partitioning must be such that we have 2^{n_1} subsets with 2^{k_2} points each. The rationale for this approach is simply explained. Since the points within a subset are at the maximum possible distance, the information that needs to be protected more effectively is that associated with the subset, not that related to the position of the point within a subset. The k_2 bits left uncoded give parallel transitions on the trellis diagram of the overall encoder. This is not a problem on the AWGN channel, as will become clearer later.

Convolutional linear (or, more generally, nonlinear) trellis codes are used as binary encoders in the scheme of Figure 11.2 since, in these cases, the VA can be employed to perform soft input

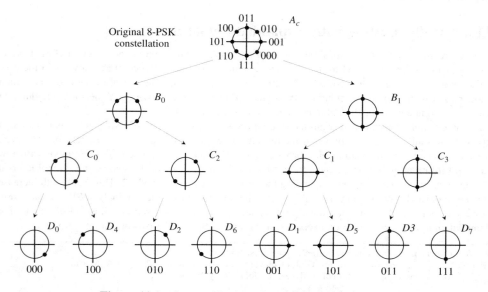

Figure 11.1 Set partitioning of an 8-PSK constellation.

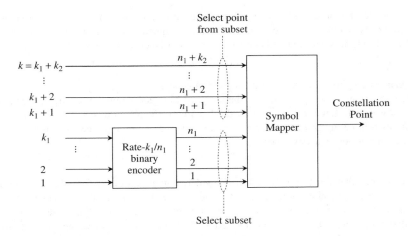

Figure 11.2   Structure of a general TCM encoder for an AWGN channel.

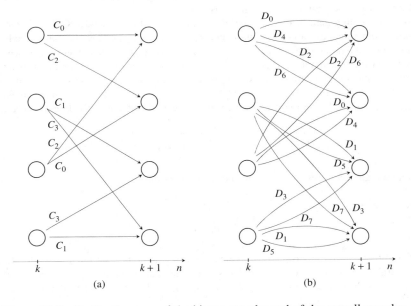

Figure 11.3   Trellis diagrams of the binary encoder and of the overall encoder.

decoding. In the original Ungerboeck approach, the binary encoder is also a rate-$k_1/(k_1 + 1)$ encoder and the code design is performed on the basis of three famous heuristic rules [992] that reflect the above-mentioned ideas. They are illustrated in the following example, which compares an uncoded QPSK with a four-state trellis-coded 8-PSK modulation.

Example 11.2.1 Let us consider a trellis-coded 8-PSK modulation built according to Figure 11.2. In this scheme we have $k = 2$ and we choose $k_1 = k_2 = 1$. The binary encoder has rate 1/2 and we assume that it has four states. The trellis state is thus defined through a shift register accumulating two previous input bits. Moreover, we have $n_1 + k_2 = 3$ and, in fact, three bits are required to uniquely identify a symbol of the 8-PSK modulation. The trellis diagram of the binary encoder is clearly that in Figure 11.3(a), whereas that of the overall encoder is shown in Figure 11.3(b); note that the presence of $k_2 = 1$ uncoded bit produces, in fact, the appearance of *parallel transitions*. The overall encoder will be completely defined once each trellis branch is assigned to an 8-PSK output symbol (i.e., to a subset D_i). In [992], this assignment is performed in accordance with to the following heuristic rules:

1. All signal points should occur with the same frequency.
2. Parallel transitions, when present, must be assigned to signal points belonging to the same subset, in such a way that they are separated by the maximum Euclidean distance.
3. Transitions originating from or merging into the same state are assigned to subsets that belong to the same subset but of lower order.

Rule 1 guarantees that the trellis code has a regular structure. Rule 2 formalizes the concept that uncoded bits select the point within a subset. Rule 3 states the way in which coded bits selects the subset. These three rules guarantee that the free Euclidean distance of this code is that associated with parallel transitions. One of the possible equivalent assignments according to these rules is that illustrated in Figure 11.3(b).[1] According to rule 3, transitions originating from or merging into the same state are assigned to points of subsets C_0 or C_2 belonging to the same subset B_0, or to points of subsets C_1 or C_3, belonging to the same subset B_1.

To assess the coding gain of this scheme, we compare it with an uncoded QPSK. The comparison is fair since both schemes carry two bits every T_s seconds and employ the same bandwidth. The uncoded QPSK scheme employs points of subsets B_0 or B_1 equivalently, whose distance, assuming that they lie on the unit circle, is $d_{QPSK} = \sqrt{2}$. In our coded scheme, the minimum distance is that associated with parallel transitions and, hence, with points of subsets C_i, thus $d_p = 2$. In fact, rule 3 ensures that all other pairs of sequences will have a larger distance. If we consider, as an example, the sequences associated with the paths shown in Figure 11.4, the corresponding distance is:

$$d = \sqrt{(D_0 - D_2)^2 + (D_0 - D_1)^2 + (D_0 - D_2)^2} \simeq 2.14 > d_p. \tag{11.1}$$

The asymptotic gain with respect to the uncoded QPSK is thus:

$$20 \log_{10} \frac{d_p}{d_{QPSK}} \simeq 3 \text{ dB}. \tag{11.2}$$

A larger asymptotic gain (3.6 dB) can be obtained by using an eight-state binary encoder. In this case, however, parallel transitions must be avoided since, otherwise, the minimum distance would remain that associated with them and thus no gain would be obtained with respect to the four-state code.
□

An alternative method for the design of TCMs was developed by A. R. Calderbank and N. J. A. Sloane [1849] and D. G. Forney [324]. An extension of TCMs, which is relevant to our purposes, is represented by *multidimensional* TCMs and *multiple* TCMs where each trellis branch is associated with more than one symbol transmitted sequentially on the channel. For a comprehensive treatment of TCMs, the reader can refer to [36].

---

[1] A simplified representation of the code is that in Figure 11.3(a). In this case, parallel transitions are omitted for simplicity and each branch is associated with a subset C_i. Since these subsets have two points, the reader is implicitly informed that each branch consists of a pair of parallel branches.

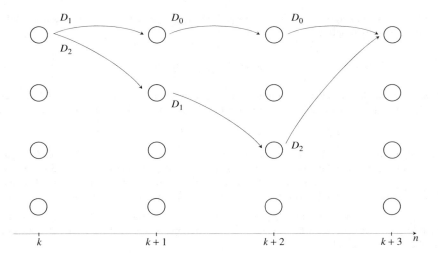

Figure 11.4 Paths on the code trellis.

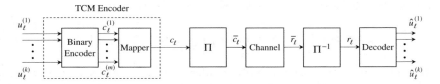

Figure 11.5 Possible transmission scheme adopting a coded modulation over a fading channel.

11.2.1.2 TCMs over (Known) Frequency-Flat Fading Channels

Let us consider Figure 11.5 as a possible TCM transmission scheme for a time-selective fading channel. The TCM encoder is modeled as the cascade of a binary encoder (of rate k/m in the figure) plus a modulator (or mapper) that univocally associates the m-tuple $(c_\ell^{(1)}, c_\ell^{(2)}, \ldots, c_\ell^{(m)})$ with an M-ary symbol c_ℓ of the complex signal constellation, where $M = 2^m$. Information bits are denoted as $(u_\ell^{(1)}, u_\ell^{(2)}, \ldots, u_\ell^{(k)})$. With respect to the case of a transmission over the AWGN channel, coded symbols are interleaved before transmission (the *interleaver* is denoted by Π in the figure). This interleaver, useless on the AWGN channel, is very important on a channel with correlated fading as it separates adjacent received samples affected by a *deep fade*, and aids the decoder.

Let $\{\bar{c}_\ell\}$ denote the sequence of coded symbols after interleaving, and let us make the following assumptions:

- One sample per symbol is sufficient for optimal decoding; we will denote by $\{\bar{r}_\ell\}$ the received sample sequence and by $\{r_\ell\}$ the corresponding deinterleaved sequence. The model for the ℓth received signal sample $\bar{r}_\ell$ is (see (4.92)):

$$\bar{r}_\ell = \bar{h}_\ell \bar{c}_\ell + \bar{n}_\ell \tag{11.3}$$

with $\ell = 1, 2, \ldots, N$, where $\bar{\mathbf{h}} = \{\bar{h}_\ell\}_{\ell=1}^{N}$ is a discrete-time complex Gaussian process with autocorrelation function $R_{\bar{h}}(i) = \mathrm{E}\{\bar{h}_{\ell+i}\bar{h}_\ell^*\}$, $\{\bar{n}_\ell\}$ is a complex discrete-time white Gaussian noise process with independent real and imaginary components, each with variance $\sigma_n^2/2$, and N is the codeword length.

- The channel is characterized by the conditional pdf $f_{\bar{\mathbf{r}}}(\bar{\boldsymbol{\rho}}|\bar{\mathbf{c}}, \bar{\mathbf{h}})$ and by the joint a priori pdf $f_{\bar{h}}(\bar{\boldsymbol{\lambda}})$. With the above assumptions we have that:

$$f_{\bar{\mathbf{r}}}(\bar{\boldsymbol{\rho}}|\bar{\mathbf{c}}, \bar{\mathbf{h}}) = \prod_{\ell=1}^{N} f_{\bar{r}_{\ell}}(\bar{\rho}_{\ell}|\bar{c}_{\ell}, \bar{h}_{\ell}) = \prod_{\ell=1}^{N} \frac{1}{\pi \sigma_n^2} \exp\left\{ -\frac{1}{\sigma_n^2} |\bar{\rho}_{\ell} - \bar{h}_{\ell}\bar{c}_{\ell}|^2 \right\}. \tag{11.4}$$

We will assume that the channel amplitude has a Rician distribution (see Section 4.4.2.1). Defining $a_{\ell} \triangleq |h_{\ell}|$, this random variable has the following pdf (normalized to a unit mean-square value):

$$f_a(x) = 2x(1 + K) \exp\{-K - x^2(1 + K)\} I_0(2x\sqrt{K(1 + K)}) \, u(x), \tag{11.5}$$

where $I_0(x)$ is the zeroth-order modified Bessel function of the first kind and K a parameter representing the ratio of the power in LOS plus specular components to that in the diffuse component (see (2.60)). Note that the case of no fading corresponds to a Rician channel with K approaching infinity, whereas a Rayleigh channel is a limiting case of a Rician channel when K approaches zero.

- The receiver perfectly knows the channel, that is, it is equipped with an ideal estimator of the gains $\{\bar{h}_{\ell}\}$ (the case of a receiver which operates under the knowledge of the channel statistics has been described in Section 6.4).

- The interleaver is *ideal*. The fading samples, correlated before the deinterleaver, become independent after it. According to the notation for code symbols and received samples, we denote by $\mathbf{h} = \{h_{\ell}\}_{\ell=1}^{N}$ the sequence of fading gains after deinterleaving. Similarly to (11.4), we have:

$$f_{\mathbf{r}}(\boldsymbol{\rho}|\mathbf{c}, \mathbf{h}) = \prod_{\ell=1}^{N} f_{r_{\ell}}(\rho_{\ell}|c_{\ell}, h_{\ell}) = \prod_{\ell=1}^{N} \frac{1}{\pi \sigma_n^2} \exp\left\{ -\frac{1}{\sigma_n^2} |\rho_{\ell} - h_{\ell}c_{\ell}|^2 \right\}. \tag{11.6}$$

Under these assumptions, let us compute the PEP, which represents the probability that the decoder chooses the sequence $\hat{\mathbf{c}} \neq \mathbf{c}$ when the sequence $\mathbf{c}$ is transmitted and $\hat{\mathbf{c}}$ and $\mathbf{c}$ are the only two possible decoding outcomes. From (11.6) it follows that an ML decoder that perfectly knows the fading gains will operate on a code trellis with metric to be maximized equal to:

$$\Lambda(\tilde{\mathbf{c}}) = \ln f_{\mathbf{r}}(\boldsymbol{\rho}|\tilde{\mathbf{c}}, \mathbf{h}) = \sum_{\ell=0}^{N-1} \ln f_{r_{\ell}}(\rho_{\ell}|\tilde{c}_{\ell}, h_{\ell}) . \tag{11.7}$$

It can thus employ the VA with the branch metrics:

$$\lambda_{\ell} = \ln f_{r_{\ell}}(\rho_l|\tilde{c}_{\ell}, h_{\ell}) \propto -|\rho_{\ell} - h_{\ell}\tilde{c}_{\ell}|^2 . \tag{11.8}$$

where $\tilde{c}_{\ell}$ denotes the channel symbol labeling a state transition in the ℓth symbol interval. Let us now consider two codewords (i.e., sequences of coded channel symbols) $\mathbf{c}$ and $\hat{\mathbf{c}}$ stemming from the same state and merging after a given number of trellis steps. Thus, given a particular channel realization, the PEP for this couple of codewords can be written as[2]:

$$\Pr\{\Lambda(\hat{\mathbf{c}}) > \Lambda(\mathbf{c})|\mathbf{c}, \mathbf{h}\} = Q\left(\frac{d(\mathbf{c}, \hat{\mathbf{c}})}{\sqrt{2\sigma_n^2}} \right)$$

$$\leq \frac{1}{2} \exp\left\{ -\frac{d^2(\mathbf{c}, \hat{\mathbf{c}})}{4\sigma_n^2} \right\}$$

$$= \frac{1}{2} \exp\left\{ -\frac{1}{4\sigma_n^2} \sum_{\ell \in I} |h_{\ell}|^2 |\hat{c}_{\ell} - c_{\ell}|^2 \right\}$$

$$= \frac{1}{2} \prod_{\ell \in I} \exp\left\{ -\frac{1}{4\sigma_n^2} a_{\ell}^2 |\hat{c}_{\ell} - c_{\ell}|^2 \right\} , \tag{11.9}$$

---

[2] Inequality (F.15) has been used. Tighter upper bounds could be derived as described in [1850, 1851].

where $d(\mathbf{c}, \hat{\mathbf{c}})$ is the distance between the transmitted vector $\mathbf{c}$ and the erroneous vector $\hat{\mathbf{c}}$, $\Lambda(\mathbf{c})$ and $\Lambda(\hat{\mathbf{c}})$ are the corresponding metrics, and the set I contains all the values of ℓ such that $|\hat{c}_\ell - c_\ell| \neq 0$. By using (11.5) and the independence of the random variables $\{a_\ell\}$, the average of the channel realization is easily computed to obtain the average PEP as:

$$\Pr\{\Lambda(\hat{\mathbf{c}}) > \Lambda(\mathbf{c})|\mathbf{c}\} = E_\mathbf{h}\{\Pr\{\Lambda(\hat{\mathbf{c}}) > \Lambda(\mathbf{c})|\mathbf{c}, \mathbf{h}\}, \tag{11.10}$$

which, for simplicity, will be denoted by $\Pr\{\mathbf{c} \rightarrow \hat{\mathbf{c}}\}$[3]:

$$\Pr\{\mathbf{c} \rightarrow \hat{\mathbf{c}}\} \leq \frac{1}{2} \prod_{\ell \in I} \int_{x_\ell=0}^{+\infty} \exp\left\{ -\frac{1}{4\sigma_n^2} x_\ell^2 |\hat{c}_\ell - c_\ell|^2 \right\} f_a(x_\ell) \, dx_\ell$$

$$= \frac{1}{2} \prod_{\ell \in I} \frac{1+K}{1+K+\frac{1}{4\sigma_n^2}|\hat{c}_\ell - c_\ell|^2} \exp\left\{ -\frac{\frac{K}{4\sigma_n^2}|\hat{c}_\ell - c_\ell|^2}{1+K+\frac{1}{4\sigma_n^2}|\hat{c}_\ell - c_\ell|^2} \right\}. \tag{11.11}$$

If the symbols $\{c_\ell\}$ are normalized to a unit mean-square value (i.e., $\sigma_c^2 = 1$), we have that $\sigma_n^2 = N_0/\bar{E}_s$, where $\bar{E}_s$ denotes the average received energy per channel symbol energy. Then equation (11.11) can be rewritten as:

$$\Pr\{\mathbf{c} \rightarrow \hat{\mathbf{c}}\} \leq \frac{1}{2} \prod_{\ell \in I} \frac{1+K}{1+K+\frac{\bar{E}_s}{4N_0}|\hat{c}_\ell - c_\ell|^2} \exp\left\{ -\frac{\frac{K\bar{E}_s}{4N_0}|\hat{c}_\ell - c_\ell|^2}{1+K+\frac{\bar{E}_s}{4N_0}|\hat{c}_\ell - c_\ell|^2} \right\} \tag{11.12}$$

or, equivalently, as:

$$\Pr\{\mathbf{c} \rightarrow \hat{\mathbf{c}}\} \leq \frac{1}{2} \exp\left\{ -\frac{\bar{E}_s}{4N_0} \bar{d}^2(\mathbf{c}, \hat{\mathbf{c}}) \right\}, \tag{11.13}$$

where we have defined:

$$\bar{d}^2(\mathbf{c}, \hat{\mathbf{c}}) \triangleq \sum_{\ell \in I} [d_{1,\ell}^2(c_\ell, \hat{c}_\ell) + d_{2,\ell}^2(c_\ell, \hat{c}_\ell)] \tag{11.14}$$

with

$$d_{1,\ell}^2(c_\ell, \hat{c}_\ell) \triangleq \frac{K|\hat{c}_\ell - c_\ell|^2}{1+K+\frac{\bar{E}_s}{4N_0}|\hat{c}_\ell - c_\ell|^2} \tag{11.15}$$

and

$$d_{2,\ell}^2(c_\ell, \hat{c}_\ell) \triangleq \frac{4N_0}{\bar{E}_s} \ln\left(\frac{1+K+\frac{\bar{E}_s}{4N_0}|\hat{c}_\ell - c_\ell|^2}{1+K} \right). \tag{11.16}$$

When $K \rightarrow \infty$ (AWGN channel), we have:

$$d_{1,\ell}^2(c_\ell, \hat{c}_\ell) = |\hat{c}_\ell - c_\ell|^2 \tag{11.17}$$

and

$$d_{2,\ell}^2(c_\ell, \hat{c}_\ell) = 0, \tag{11.18}$$

so that in this case $\bar{d}^2(\mathbf{c}, \hat{\mathbf{c}})$ becomes the squared Euclidean distance between the two sequences $\mathbf{c}$ and $\hat{\mathbf{c}}$. For $K = 0$ (Rayleigh channel):

$$d_{1,\ell}^2(c_\ell, \hat{c}_\ell) = 0 \tag{11.19}$$

---

[3] The following equality has been also exploited:

$$\int_0^\infty \frac{x}{\sigma^2} e^{-\frac{b^2+x^2}{2\sigma^2}} I_0\left(\frac{bx}{\sigma^2} \right) dx = 1.$$

and

$$d_{2,\ell}^2(c_\ell, \hat{c}_\ell) = \frac{4N_0}{\bar{E}_s} \ln\left(1 + \frac{\bar{E}_s}{4N_0}|\hat{c}_\ell - c_\ell|^2\right). \tag{11.20}$$

Then the PEP is:

$$\Pr\{\mathbf{c} \to \hat{\mathbf{c}}\} \le \frac{1}{2} \prod_{\ell \in I} \frac{1}{\frac{\bar{E}_s}{4N_0}|\hat{c}_\ell - c_\ell|^2 + 1}, \tag{11.21}$$

which asymptotically (for high SNR values) becomes:

$$\Pr\{\mathbf{c} \to \hat{\mathbf{c}}\} \lesssim \frac{1}{2}\left(\frac{\bar{E}_s}{4N_0}\right)^{-|I|}\left(\prod_{\ell \in I}|\hat{c}_\ell - c_\ell|^2\right)^{-1}, \tag{11.22}$$

where $|I|$ denotes the cardinality of the set I. If we represent this pairwise error probability as a function of the SNR $\bar{E}_s/N_0$, the larger the value of $|I|$ the larger the rate of its decrease (the slope of the curve in a log-log plot). In addition, the higher the term $\prod_{\ell \in I}|\hat{c}_\ell - c_\ell|^2$, the lower the asymptotic PEP. The cardinality of the set I is the Hamming distance of the error event being considered. Hence, on a Rayleigh fading channel, in order to improve the performance it is not the minimum Euclidean distance that plays an important role but rather the code's minimum Hamming distance (the *code diversity*). As a secondary merit criterion, we should try to maximize the term $\prod_{\ell \in I}|\hat{c}_\ell - c_\ell|^2$ on error events with minimum Hamming distance. This term, usually called the *coding gain*, does not depend on the SNR and displaces the error probability curve instead of changing its slope. Codes designed for the AWGN channel are thus suboptimal for the Rayleigh fading channel. On the latter channel, when a conventional trellis code (i.e., with one symbol per trellis branch) is designed, it is important to avoid parallel transitions since, in this case, the code diversity assumes the minimum value (one) and the asymptotic performance will improve only linearly with the SNR. In case of a Rician fading channel, the picture is only slightly different since the asymptotic PEP (11.12) becomes:

$$\Pr\{\mathbf{c} \to \hat{\mathbf{c}}\} \lesssim \frac{1}{2} \prod_{\ell \in I} \frac{(1+K)\exp(-K)}{\frac{\bar{E}_s}{4N_0}|\hat{c}_\ell - c_\ell|^2} = \frac{1}{2}\left(\frac{\bar{E}_s\exp(K)}{4N_0(1+K)}\right)^{-|I|}\left(\prod_{\ell \in I}|\hat{c}_\ell - c_\ell|^2\right)^{-1}. \tag{11.23}$$

Hence, the main factors determining performance are still the code diversity and the term $\prod_{\ell \in I}|\hat{c}_\ell - c_\ell|^2$ on error events with minimum Hamming distance. The basic code design principles over fading channels with ideal interleaving are thus the following:

Code diversity criterion. The minimum code diversity $|I|$ must be maximized.

Coding gain criterion. In order to obtain the maximum possible coding advantage, the coding gain $\prod_{\ell \in I}|\hat{c}_\ell - c_\ell|^2$ over error events having minimum diversity should be maximized.

Several authors investigated the problem of the design of codes for fading channels. We will now focus on a particular design procedure for constructing trellis codes with optimal performance on the Rician/Rayleigh fading channel [1852]. Although this procedure applies to both conventional and multiple trellis codes, we will focus on the latter since their potential can be fully exploited on this channel. In fact, when multiple trellis codes are employed, we can again design a trellis diagram with parallel paths and still have an asymptotic performance that decreases faster than linear with the SNR. As an example, by using a multiple trellis code with L symbols associated with each trellis branch, it is possible to have a code diversity of L and also to design the code in such a way that the minimum value of the term $\prod_{\ell \in I}|\hat{c}_\ell - c_\ell|^2$ on error events with minimum Hamming distance is maximized. This is made possible by a specific set partitioning procedure, obviously different from that described in Section 11.2.1.1, which will now be illustrated for the case of $L = 2$ and an M-PSK constellation.

Let A_c denote the original M-PSK constellation and $A_c \otimes A_c$ the twofold Cartesian product of A_c with itself. Hence, each element of $A_c \otimes A_c$ is a pair of symbols belonging to the original constellation A_c. In the following, we will identify the PSK channel symbol $\exp(j2\pi i/M)$ (with $i = 0, 1, \ldots, M-1$) through integer i. At the first partition level, $A_c \otimes A_c$ is partitioned into M sets defined by the ordered Cartesian product $A_c \otimes B_i$ (with $i = 0, 1, \ldots, M-1$), whose pth element (with $p = 0, 1, \ldots, M-1$) is the ordered pair $(p, R_M[pq + i])$ (here $q \le M$ is a proper odd integer). As an example, for $M = 8$ and $q = 3$, we obtain the following subsets:

$$A_c \otimes B_0 = \begin{bmatrix} 0 & 0 \\ 1 & 3 \\ 2 & 6 \\ 3 & 1 \\ 4 & 4 \\ 5 & 7 \\ 6 & 2 \\ 7 & 5 \end{bmatrix}, \quad A_c \otimes B_1 = \begin{bmatrix} 0 & 1 \\ 1 & 4 \\ 2 & 7 \\ 3 & 2 \\ 4 & 5 \\ 5 & 0 \\ 6 & 3 \\ 7 & 6 \end{bmatrix},$$

$$A_c \otimes B_2 = \begin{bmatrix} 0 & 2 \\ 1 & 5 \\ 2 & 0 \\ 3 & 3 \\ 4 & 6 \\ 5 & 1 \\ 6 & 4 \\ 7 & 7 \end{bmatrix}, \quad A_c \otimes B_3 = \begin{bmatrix} 0 & 3 \\ 1 & 6 \\ 2 & 1 \\ 3 & 4 \\ 4 & 7 \\ 5 & 2 \\ 6 & 5 \\ 7 & 0 \end{bmatrix},$$

$$A_c \otimes B_4 = \begin{bmatrix} 0 & 4 \\ 1 & 7 \\ 2 & 2 \\ 3 & 5 \\ 4 & 0 \\ 5 & 3 \\ 6 & 6 \\ 7 & 1 \end{bmatrix}, \quad A_c \otimes B_5 = \begin{bmatrix} 0 & 5 \\ 1 & 0 \\ 2 & 3 \\ 3 & 6 \\ 4 & 1 \\ 5 & 4 \\ 6 & 7 \\ 7 & 2 \end{bmatrix},$$

$$A_c \otimes B_6 = \begin{bmatrix} 0 & 6 \\ 1 & 1 \\ 2 & 4 \\ 3 & 7 \\ 4 & 2 \\ 5 & 5 \\ 6 & 0 \\ 7 & 3 \end{bmatrix}, \quad A_c \otimes B_7 = \begin{bmatrix} 0 & 7 \\ 1 & 2 \\ 2 & 5 \\ 3 & 0 \\ 4 & 3 \\ 5 & 6 \\ 6 & 1 \\ 7 & 4 \end{bmatrix}.$$

As can be observed, within any of the M partitions, each pair differs from all other pairs in both elements. Hence, when the pairs of a partition are employed to label parallel transitions of a multiple trellis code, a code diversity of $L = 2$ is obtained.

The parameter q is a key point in this partition method. In choosing it, the aim is to maximize the minimum value of the term $\prod_{\ell \in I} |\hat{c}_\ell - c_\ell|^2$ on parallel transitions. Before going into detail, we observe that the set B_{i+1} is simply a cyclic shift of set B_i. Thus, since the term $\prod_{\ell \in I} |\hat{c}_\ell - c_\ell|^2$ is simply the product of the squared Euclidean distances between the corresponding symbols in the pair, the adopted set partitioning guarantees that the intra-distance structure of each partition $A_c \otimes B_i$ is the same. Hence, it is sufficient to study the intra-distance structure of the so-called *generating set* $A_c \otimes B_0$. In other words, let us consider the pairs $(p, R_M[pq + i])$ and $(l, R_M[lq + i])$ of the set $A_c \otimes B_i$. The product of the squared distances between these two pairs is:

$$
\left| \exp\left(j2\pi \frac{p}{M} \right) - \exp\left(j2\pi \frac{l}{M} \right) \right|^2 \left| \exp\left(j2\pi \frac{pq + i}{M} \right) - \exp\left(j2\pi \frac{lq + i}{M} \right) \right|^2
$$

$$
= \left| 1 - \exp\left(j2\pi \frac{l - p}{M} \right) \right|^2 \left| 1 - \exp\left(j2\pi \frac{(l - p)q}{M} \right) \right|^2
$$

$$
= 16 \sin^2\left(\pi \frac{l - p}{M} \right) \sin^2\left(\pi q \frac{l - p}{M} \right) \tag{11.24}
$$

and, as mentioned, is independent of i. The value q^* of q that maximizes the minimum value of the product of these square distances is then:

$$
q^* = \arg \max_{q \text{ odd}} \left\{ \min_{p > 0} \left[16 \sin^2\left(\pi \frac{p}{M} \right) \sin^2\left(\pi q \frac{p}{M} \right) \right] \right\}. \tag{11.25}
$$

It can easily be proven that $M - q^*$ is also a valid solution. Table 11.1 reports the optimal values of q for different values of M.

The next step in this set-partitioning procedure is to partition each of the M sets $A_c \otimes B_i$ into two sets $C_0 \otimes D_{i0}$ and $C_1 \otimes D_{i1}$, where the first set contains the even elements and the second set the odd elements of $A_c \otimes B_i$. In the case of the example above, we obtain the following subsets:

$$
C_0 \otimes D_{00} = \begin{bmatrix} 0 & 0 \\ 2 & 6 \\ 4 & 4 \\ 6 & 2 \end{bmatrix}, \quad C_1 \otimes D_{01} = \begin{bmatrix} 1 & 3 \\ 3 & 1 \\ 5 & 7 \\ 7 & 5 \end{bmatrix},
$$

$$
C_0 \otimes D_{10} = \begin{bmatrix} 0 & 1 \\ 2 & 7 \\ 4 & 5 \\ 6 & 3 \end{bmatrix}, \quad C_1 \otimes D_{11} = \begin{bmatrix} 1 & 4 \\ 3 & 2 \\ 5 & 0 \\ 7 & 6 \end{bmatrix},
$$

$$
C_0 \otimes D_{20} = \begin{bmatrix} 0 & 2 \\ 2 & 0 \\ 4 & 6 \\ 6 & 4 \end{bmatrix}, \quad C_1 \otimes D_{21} = \begin{bmatrix} 1 & 5 \\ 3 & 3 \\ 5 & 1 \\ 7 & 7 \end{bmatrix},
$$

Table 11.1   Optimal values of q for different values of M

M	q^*
2	1
4	1
8	3
16	7
32	7,9
64	19,27

$$
C_0 \otimes D_{30} = \begin{bmatrix} 0 & 3 \\ 2 & 1 \\ 4 & 7 \\ 6 & 5 \end{bmatrix}, \quad C_1 \otimes D_{31} = \begin{bmatrix} 1 & 6 \\ 3 & 4 \\ 5 & 2 \\ 7 & 0 \end{bmatrix},
$$

$$
C_0 \otimes D_{40} = \begin{bmatrix} 0 & 4 \\ 2 & 2 \\ 4 & 0 \\ 6 & 6 \end{bmatrix}, \quad C_1 \otimes D_{41} = \begin{bmatrix} 1 & 7 \\ 3 & 5 \\ 5 & 3 \\ 7 & 1 \end{bmatrix},
$$

$$
C_0 \otimes D_{50} = \begin{bmatrix} 0 & 5 \\ 2 & 3 \\ 4 & 1 \\ 6 & 7 \end{bmatrix}, \quad C_1 \otimes D_{51} = \begin{bmatrix} 1 & 0 \\ 3 & 6 \\ 5 & 4 \\ 7 & 2 \end{bmatrix},
$$

$$
C_0 \otimes D_{60} = \begin{bmatrix} 0 & 6 \\ 2 & 4 \\ 4 & 2 \\ 6 & 0 \end{bmatrix}, \quad C_1 \otimes D_{61} = \begin{bmatrix} 1 & 1 \\ 3 & 7 \\ 5 & 5 \\ 7 & 3 \end{bmatrix},
$$

$$
C_0 \otimes D_{70} = \begin{bmatrix} 0 & 7 \\ 2 & 5 \\ 4 & 3 \\ 6 & 1 \end{bmatrix}, \quad C_1 \otimes D_{71} = \begin{bmatrix} 1 & 2 \\ 3 & 0 \\ 5 & 6 \\ 7 & 4 \end{bmatrix}.
$$

Obviously, within each of these new sets, each pair is still distinct from all other pairs in both positions. However, it is in general no longer true that the minimum value of the term $\prod_{\ell \in I} |\hat{c}_\ell - c_\ell|^2$ is maximized within each set unless in the previous step we used the value q^* corresponding to $M/2$ instead of M. Hence, the choice of q^* depends on the target partition level.

The third and subsequent steps are identical in construction to the second step: we need to partition each set in the present level into two sets containing the alternate rows, with the set of the previous levels obtained by using a value of q^* computed as in (11.25) with M successively replaced by $M/4$, $M/8$, and so on.

This procedure can easily be generalized to the case of L being a multiple of 2. As an example, in the case of $L = 4$ the sets belonging to the first partition level will be the M^2 sets $A_c \otimes B_i \otimes A_c \otimes B_p$, with $i, p = 0, 1, \ldots, M - 1$.

When the number of sets required to satisfy the trellis is less than the number of sets generated on a particular partition level, only those having largest *inter-set distance* must be chosen, as in the examples below. Let us now discuss, through a couple of practical examples, how to employ these sets in code construction. The examples deal with two- and four-state rate-4/6 multiple ($L = 2$) trellis coded 8-PSK modulations, respectively.

Example 11.2.2 Let us consider a two-state, rate-4/6 multiple TCM using 8-PSK as the output constellation. Since $L = 2$, two 8-PSK symbols (hence, six bits) are transmitted every four input information bits. Then the encoder trellis has $2^4 = 16$ branches leaving each state. Since there are only two states, each transition between states has eight parallel paths. The encoder trellis is shown in Figure 11.6. For the properties of the set partition method just described, if we associate pairs from sets $A_c \otimes B_i$ with parallel transitions, we are sure that a code diversity of 2 is obtained on them. In addition, it can be shown that the minimum value of the term $\prod_{\ell \in I} |\hat{c}_\ell - c_\ell|^2$ on parallel transitions is 2.

Let us now consider longer error events. Since it can be shown that this code is linear, without loss of generality we assume that the all-zero path is the correct one and consider an error event of

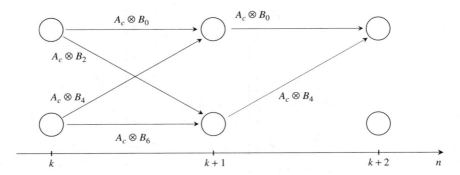

Figure 11.6   Trellis diagram for the two-state, rate-4/6 multiple TCM of Example 11.2.2.

length 2 such as that shown in Figure 11.6. The four paths of length 2 that differ by a minimum of two symbols from the correct one are those corresponding to coded symbols [0 2 0 4], [0 2 4 0], [2 0 0 4], and [2 0 4 0]. It can easily be shown that for these the value of the term $\prod_{\ell \in I} |\hat{c}_\ell - c_\ell|^2$ is 8, hence larger than that related to parallel transitions that lead the asymptotic behavior of the code.

In this code construction, we used sets $A_c \otimes B_i$, $i = 0, 2, 4, 6$. Equivalently, sets $A_c \otimes B_i$, $i = 1, 3, 5, 7$, could have been employed.

$\square$

Example 11.2.3 Let us now consider a four-state, rate-4/6 multiple TCM using 8-PSK as its output constellation. As before, since $L = 2$, two 8-PSK symbols (hence six bits) are transmitted every four input information bits. The encoder trellis thus has $2^4 = 16$ branches leaving each state. Since there are now four states, and assuming a completely connected encoder trellis, each transition between states has four parallel paths. The encoder trellis is shown in Figure 11.7. For the properties of the

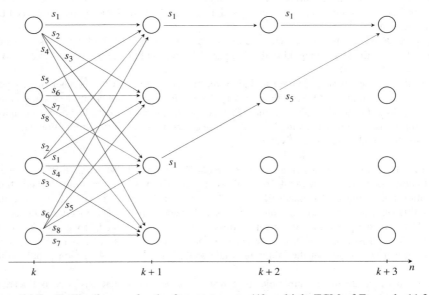

Figure 11.7   Trellis diagram for the four-state, rate-4/6 multiple TCM of Example 11.2.3.

set partition method just described, if we associate pairs from the sets $C_0 \otimes D_{i0}$ and $C_1 \otimes D_{i1}$ with parallel transition, we are sure that a code diversity of 2 is obtained on them. In addition, it can be shown that the minimum value of the term $\prod_{\ell \in I} |\hat{c}_\ell - c_\ell|^2$ on parallel transitions is 4, provided that the sets $C_0 \otimes D_{i0}$ and $C_1 \otimes D_{i1}$ are obtained through the procedure described above but with $q = 1$, which is the optimal value for $M/2 = 4$. Even in this case, not all the sets $C_0 \otimes D_{i0}$ and $C_1 \otimes D_{i1}$ are required, since only the following eight sets, for simplicity denoted by S_i in the figure, can be used:

$$S_1 = \begin{bmatrix} 0 & 0 \\ 2 & 2 \\ 4 & 4 \\ 6 & 6 \end{bmatrix}, \quad S_2 = \begin{bmatrix} 1 & 5 \\ 3 & 7 \\ 5 & 1 \\ 7 & 3 \end{bmatrix},$$

$$S_3 = \begin{bmatrix} 0 & 4 \\ 2 & 6 \\ 4 & 0 \\ 6 & 2 \end{bmatrix}, \quad S_4 = \begin{bmatrix} 1 & 1 \\ 3 & 3 \\ 5 & 5 \\ 7 & 7 \end{bmatrix},$$

$$S_5 = \begin{bmatrix} 0 & 2 \\ 2 & 4 \\ 4 & 6 \\ 6 & 0 \end{bmatrix}, \quad S_6 = \begin{bmatrix} 1 & 7 \\ 3 & 1 \\ 5 & 3 \\ 7 & 5 \end{bmatrix},$$

$$S_7 = \begin{bmatrix} 0 & 6 \\ 2 & 0 \\ 4 & 2 \\ 6 & 4 \end{bmatrix}, \quad S_8 = \begin{bmatrix} 1 & 3 \\ 3 & 5 \\ 5 & 7 \\ 7 & 1 \end{bmatrix}.$$

Every other error event consisting of two or more branches has a Hamming distance greater than 2 regardless of which path is chosen as the correct path. Thus, the dominant term on the asymptotic symbol or bit error probability expressions corresponds again to parallel paths.

□

11.2.2 Decoding Algorithms

Before addressing the case of a fading channel, we will first consider a TCM scheme on the AWGN channel, assuming a linear modulation with a shaping pulse satisfying the Nyquist condition for the absence of ISI. In this case, the optimal receiver jointly performs detection and decoding through a search on the trellis diagram of the overall encoder using the VA. In the ℓth symbol interval we denote the generic state of the overall encoder by $\tilde{\sigma}_\ell$, the k-tuple of the encoder input bits by $\tilde{\mathbf{u}}_\ell = (u_\ell^{(1)}, \ldots, u_\ell^{(k)})$, and the code symbol, the successive state, and the branch metric associated to the trellis transition emerging from from state $\tilde{\sigma}_\ell$ and driven by $\tilde{\mathbf{u}}_\ell$ by $c_\ell(\tilde{\mathbf{u}}_\ell, \tilde{\sigma}_\ell)$, $\tilde{\sigma}_{\ell+1}(\tilde{\mathbf{u}}_\ell, \tilde{\sigma}_\ell)$ and $\lambda_\ell(\tilde{\mathbf{u}}_\ell, \tilde{\sigma}_\ell)$, respectively. The branch metric associated with the above-mentioned transistion is:

$$\lambda_\ell(\tilde{\mathbf{u}}_\ell, \tilde{\sigma}_\ell) = |\rho_\ell - c_\ell(\tilde{\mathbf{u}}_\ell, \tilde{\sigma}_\ell)|^2, \tag{11.26}$$

where ρ_ℓ is the value taken on by the received sample r_ℓ. Equivalently, we could work on the trellis state of the binary encoder which can differ from that of the overall encoder for the presence of possible parallel transitions, provided that, as a branch metric, we employ:

$$\lambda_\ell(\tilde{\mathbf{u}}_\ell, \tilde{\sigma}_\ell) = \min_{\tilde{c} \in C(\tilde{\mathbf{u}}_\ell, \tilde{\sigma}_\ell)} |\rho_\ell - \tilde{c}|^2, \tag{11.27}$$

where $C(\tilde{\mathbf{u}}_\ell, \tilde{\sigma}_\ell)$ denotes the subset whose points are associated with the parallel transitions originating from $\tilde{\sigma}_\ell$ and ending in the state $\sigma_{\ell+1}(\tilde{\mathbf{u}}_\ell, \tilde{\sigma}_\ell)$. In other words, the decision on parallel transitions is taken symbol by symbol.

We now consider the case of a transmission over a frequency-flat fading channel. The scheme of Figure 11.5 relies, as already mentioned, on the assumption that an ideal channel estimator is available. Under this hypothesis, an optimal decoder can be conceived, using the deinterleaved samples r_ℓ and working on the trellis of the overall encoder with branch metrics:

$$\lambda_\ell(\tilde{\mathbf{u}}_\ell, \tilde{\sigma}_\ell) = |\rho_l - h_\ell c_\ell(\tilde{\mathbf{u}}_\ell, \tilde{\sigma}_\ell)|^2 \tag{11.28}$$

or on the trellis of the binary encoder with branch metrics:

$$\lambda_\ell(\tilde{\mathbf{u}}_\ell, \tilde{\sigma}_\ell) = \min_{\tilde{c} \in C(\tilde{\mathbf{u}}_\ell, \tilde{\sigma}_\ell)} |\rho_\ell - h_\ell \tilde{c}|^2. \tag{11.29}$$

In practical conditions, due to the presence of the interleaver, the optimal decoder is difficult to implement. Assuming that knowledge of channel statistics is available at the receiver, a SiSo detection algorithm based on linear prediction (see Section 6.4) can be used and the relevant *extrinsic* information (in the logarithmic domain):

$$\ln \frac{\Pr\{\tilde{c}_\ell = \tilde{c}|\bar{\rho}\}}{\Pr\{\tilde{c}_\ell = \tilde{c}\}} = \ln \Pr\{\tilde{c}_\ell = \tilde{c}|\bar{\rho}\} - \ln \Pr\{\tilde{c}_\ell = \tilde{c}\} \propto \ln f_{\bar{\mathbf{r}}}(\bar{\rho}|\tilde{c}_\ell = \tilde{c}) \tag{11.30}$$

deinterleaved and employed as input of the decoder, as in decoding schemes for serially concatenated convolutional codes. In this case, the turbo principle [1611] can be advocated and, by using a soft output decoder, a few iterations between detector and decoder performed.

11.2.3   Error Performance

Upper bounds on symbol and bit error probabilities can be computed through the union bound technique described in Section 4.3.2, by using the expression for the PEP of the considered error events given in (11.12), under the hypothesis that an ideal channel estimator is available. The asymptotic expressions can be obtained, as usual, by considering only those error events associated with the largest asymptotic PEPs.

It is of interest to consider the effect of the interleaver and deinterleaver in the scheme of Figure 11.5. In their absence, the assumption that the fading is independent from one sample to the next one is no longer valid. When the fading amplitude is sufficiently slow as to be constant at a value a over the duration of an error event of minimum distance (quasi-static fading assumption), the BEP can be asymptotically approximated by (see Section 9.2.6):

$$P_b(\rho) \lesssim CE_a \left\{ Q\left(a\, d_{free} \sqrt{\frac{\bar{E}_s}{2N_0}}\right) \right\} \leq \frac{C}{2} E_a \left\{ \exp\left(-a^2 d_{free}^2 \frac{\bar{E}_s}{4N_0}\right) \right\}, \tag{11.31}$$

where C is an appropriate constant and d_{free} is the minimum Euclidean distance of the code. Performing the average over the Rician probability density function (11.5), one obtains the average BEP:

$$P_b \lesssim \frac{C}{2} \frac{1+K}{1+K+d_{free}^2 \frac{\bar{E}_s}{4N_0}} \exp\left(-K \frac{d_{free}^2 \frac{\bar{E}_s}{4N_0}}{1+K+d_{free}^2 \frac{\bar{E}_s}{4N_0}}\right), \tag{11.32}$$

which can be expressed, for large values of $\bar{E}_s/N_0$, as:

$$P_b \lesssim 2C \frac{1+K}{d_{free}^2 \frac{\bar{E}_s}{N_0}} \exp(-K). \tag{11.33}$$

Hence, in the absence of the interleaver and deinterleaver, independently of the employed trellis code, the code diversity is equal to 1.[4]

Under these conditions, improvements can be obtained by resorting to *receive diversity*. Assuming n_R receive antennas, the corresponding received samples can be expressed, in vector notation, as:

$$\mathbf{r}_i = h_i \mathbf{c} + \mathbf{n}_i \tag{11.34}$$

with $i = 1, 2, \ldots, n_R$, where $\mathbf{r}_i$ is a row vector collecting the samples received by antenna i, h_i the corresponding channel gain (and $a_i \triangleq |h_i|$ its amplitude), and $\mathbf{n}_i$ the vector of noise samples collected by antenna i, assumed independent of each other and independent of noise samples related to other antennas. We also assume that channel gains $\{h_i\}$ are independent of each other and perfectly known at the receiver. The ML detection strategy is:

$$\hat{\mathbf{c}} = \arg \max_{\tilde{\mathbf{c}}} f_{\mathbf{r}}(\boldsymbol{\rho}_1, \boldsymbol{\rho}_2, \ldots, \boldsymbol{\rho}_{n_R} | \tilde{\mathbf{c}}, h_1, h_2, \ldots, h_{n_R})$$

$$= \arg \min_{\tilde{\mathbf{c}}} \sum_{i=1}^{n_R} |\boldsymbol{\rho}_i - h_i \tilde{\mathbf{c}}|^2$$

$$= \arg \max_{\tilde{\mathbf{c}}} \sum_{i=1}^{n_R} \left\{ \mathrm{Re}\left[h_i^* \boldsymbol{\rho}_i \tilde{\mathbf{c}}^H \right] - \frac{|h_i|^2}{2} |\tilde{\mathbf{c}}|^2 \right\}$$

$$= \arg \max_{\tilde{\mathbf{c}}} \mathrm{Re}\left[\frac{\sum_{i=1}^{n_R} h_i^* \boldsymbol{\rho}_i}{\sum_{i=1}^{n_R} |h_i|^2} \tilde{\mathbf{c}}^H \right] - \frac{|\tilde{\mathbf{c}}|^2}{2}$$

$$= \arg \min_{\tilde{\mathbf{c}}} |\bar{\boldsymbol{\rho}} - \tilde{\mathbf{c}}|^2, \tag{11.35}$$

where $\mathbf{r}$ is the vector resulting from the ordered concatenation of the vectors $\{\mathbf{r}_i, i = 1, 2, \ldots, n_R\}$ and $\bar{\boldsymbol{\rho}}$ denotes the value taken by the random vector:

$$\bar{\mathbf{r}} \triangleq \frac{\sum_{i=1}^{n_R} h_i^* \mathbf{r}_i}{\sum_{i=1}^{n_R} |h_i|^2} = \mathbf{c} + \frac{\sum_{i=1}^{n_R} h_i^* \mathbf{n}_i}{\sum_{i=1}^{n_R} |h_i|^2} . \tag{11.36}$$

The latter expression can be considered as the definition of an equivalent channel whose noise variance is $\sigma_n^2 / \sum_{i=1}^{n_R} |h_i|^2 = \sigma_n^2 / \sum_{i=1}^{n_R} a_i^2$ instead of σ_n^2. Therefore, we have that:

$$P_b(a_1, a_2, \ldots, a_{n_R}) \lesssim C \mathrm{E}_{a_1, a_2, \ldots, a_{n_R}} \left\{ Q\left(d_{free} \sqrt{\frac{\bar{E}_s \sum_{i=1}^{n_R} a_i^2}{2 N_0}} \right) \right\}$$

$$\lesssim \frac{C}{2} \mathrm{E}_{a_1, a_2, \ldots, a_{n_R}} \left\{ \exp\left(-d_{free}^2 \frac{\bar{E}_s}{4 N_0} \sum_{i=1}^{n_R} a_i^2 \right) \right\}$$

$$= \frac{C}{2} \prod_{i=1}^{n_R} \mathrm{E}_{a_i} \left\{ \exp\left(-d_{free}^2 \frac{\bar{E}_s}{4 N_0} a_i^2 \right) \right\} \tag{11.37}$$

and

$$P_b \lesssim \frac{C}{2} \left[\frac{1 + K}{1 + K + d_{free}^2 \frac{\bar{E}_s}{4 N_0}} \right]^{n_R} \exp\left\{ -\frac{n_R K d_{free}^2 \frac{\bar{E}_s}{4 N_0}}{1 + K + d_{free}^2 \frac{\bar{E}_s}{4 N_0}} \right\} . \tag{11.38}$$

---

[4] It should be noticed that, for quasi-static channels, the union bound is loose [1853]. We will see later how to solve this problem.

The latter result asymptotically becomes:

$$P_b \lesssim \frac{C}{2}(1 + K)^{n_R} \left[d_{free}^2 \frac{\bar{E}_s}{4N_0} \right]^{-n_R}. \tag{11.39}$$

The rate of decrease is thus proportional to the number of receive antennas. *Space diversity* can thus be employed instead of time diversity. We will see later that transmit antenna diversity can also be exploited through the adoption of properly designed STCs.

11.3 Bit-Interleaved Coded Modulation

Beginning with the introduction of TCM, it was generally accepted that coding and modulation must be designed jointly. The same paradigm was adopted on fading channels, although, in this case, the free Euclidean distance no longer plays a predominant role. Instead, as described in the previous sections, the code diversity is the main code parameter to be maximized.

A first deviation from this paradigm is represented by BICM. This technique, originally proposed by E. Zehavi [1464] in 1992, was further developed and analyzed by G. Caire, G. Taricco and E. Biglieri [1854] in 1998. According to this technique, coded modulations with a very good performance over frequency-flat fading channels can be built by using off-the-shelf binary codes that are optimal in the sense of the free Hamming distance, and thus available in standard textbooks.

11.3.1 Code Construction

A system employing BICM is shown in Figure 11.8. The presence at the transmitter of *parallel-to-serial* (P/S) and *serial-to-parallel* (S/P) converters makes clear that the interleaver operates at the bit

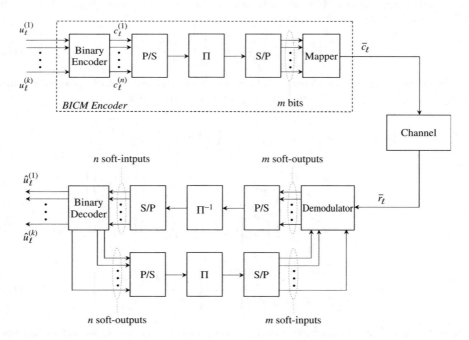

Figure 11.8 Transmission scheme employing BICM.

level. After interleaving, groups of m coded bits are mapped into M-ary (with $M = 2^m$) symbols $\{\bar{c}_\ell\}$ belonging to a complex constellation A_c. In other words, by comparing the schemes in Figures 11.5 and 11.8 one may observe that, at the transmitter side, the only difference is that now coded bits instead of coded symbols are interleaved.

The idea behind BICM is very simple. If we consider two codewords of the binary code having Hamming distance N, and thus that differ for N bits, after the interleaver they will likely belong to the labels of N different coded symbols. Hence, with high probability, the corresponding M-ary codewords still have Hamming distance N. The use of a binary code optimal in the sense of free Hamming distance (and of a proper interleaver), thus ensures that the code diversity is maximized. These codes have been known since the early 1960s and thus an *ad hoc* code design is not necessary. Obviously, the coded schemes obtained are not optimal since no attempt is made to maximize the minimum value of the term $\prod_{\ell \in I} |\hat{c}_\ell - c_\ell|^2$ over the set of codewords with minimum Hamming distance. However, since the code diversity is by far the most important parameter, these codes are expected to provide a very good performance and to be practically optimal.

11.3.2   Decoding Algorithms

The presence of the interleaver between encoder and mapper generates a new encoder with a much larger memory. Hence, the optimal decoder is not feasible, even under the assumption of perfect knowledge of the channel. We can thus resort to the suboptimal receiver shown in Figure 11.8. In this scheme, for each coded bit a demodulator computes the soft outputs corresponding to each coded bit that are then employed by the binary decoder. The turbo principle can be advocated and, by using a soft output decoder, a few (optional) iterations between detector and decoder are performed.

To go into details, assume that, after the interleaver, bit $c_\ell^{(i)}$ becomes the jth bit of the label of symbol $\bar{c}_p$. We will thus write $\mathrm{lab}^{(j)}(\bar{c}_p) = c_\ell^{(i)}$. The logarithm of the *extrinsic* information of coded bit $c_\ell^{(i)}$ will be:

$$\ln f_{\bar{\mathbf{r}}}(\bar{\rho}|c_\ell^{(i)} = b) = \ln \sum_{\bar{c}_p \in A_c} f_{\bar{\mathbf{r}}}(\bar{\rho}|\bar{c}_p) \Pr\{\bar{c}_p | \mathrm{lab}^{(j)}(\bar{c}_p) = b\}, \tag{11.40}$$

with $b = 0, 1$, where:

$$\Pr\{\bar{c}_p | \mathrm{lab}^{(j)}(\bar{c}_p) = b\} = \begin{cases} \dfrac{1}{2^{m-1}} & \text{if } \mathrm{lab}^{(j)}(\bar{c}_p) = b, \\ 0 & \text{otherwise.} \end{cases} \tag{11.41}$$

Considering the constellation A_c and the assigned mapping, let us denote by $A_b^{(j)}$ the partition of the original constellation A_c obtained when the jth bit of the mapping is set to b. For the 8-PSK constellation with Gray mapping shown in Figure 11.9(a), partition $A_0^{(1)}$ is shown in Figure 11.9(b) and $A_1^{(1)}$ in Figure 11.9(c). Hence:

$$\ln f_{\bar{\mathbf{r}}}(\bar{\rho}|c_\ell^{(i)} = b) = \ln \frac{1}{2^{m-1}} \sum_{\bar{c}_p \in A_b^{(j)}} f_{\bar{\mathbf{r}}}(\bar{\rho}|\bar{c}_p) \propto \ln \sum_{\bar{c}_p \in A_b^{(j)}} f_{\bar{\mathbf{r}}}(\bar{\rho}|\bar{c}_p), \tag{11.42}$$

with $b = 0, 1$.

The *extrinsic* information is then permuted and employed by the binary decoder. As already mentioned, optional iterations can be performed according to the turbo principle; this results in the so-called BICM with iterative decoding (BICM-ID), proposed by X. Li and J. A. Ritcey [1855].

When the channel statistics are known at the receiver, a SiSo detection algorithm based on linear prediction (see Section 6.4.1.1) can be used to compute the *extrinsic* information on coded symbols

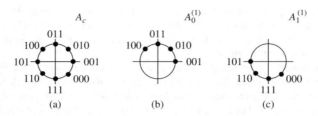

Figure 11.9 Constellation A_c and partitions $A_0^{(1)}$ and $A_1^{(1)}$.

$f_{\bar{\mathbf{r}}}(\bar{\rho}|\bar{c}_p)$. In the simpler case of a receiver that perfectly knows the channel, that is, is equipped with an ideal estimator of the gains $\{\bar{h}_\ell\}$, the *extrinsic* information of coded bit $c_\ell^{(i)}$ can be computed using:

$$\ln f_{\bar{\mathbf{r}}}(\bar{\rho}|c_\ell^{(i)} = b, \bar{\mathbf{h}}) \propto \ln \sum_{\bar{c}_p \in A_b^{(j)}} f_{\bar{\mathbf{r}}}(\bar{\rho}|\bar{c}_p, \bar{\mathbf{h}}), \tag{11.43}$$

with $b = 0, 1$, where:

$$f_{\bar{\mathbf{r}}}(\bar{\rho}|\bar{c}_p, \bar{\mathbf{h}}) \propto f_{\bar{r}_p}(\bar{\rho}_p|\bar{c}_p, \bar{h}_p) \propto \exp\left\{-\frac{1}{\sigma_n^2}|\bar{\rho}_p - \bar{h}_p\bar{c}_p|^2\right\}. \tag{11.44}$$

In this case, considering approximation (9.269) (reproduced here for the reader's convenience):

$$\ln[e^{\sigma_1} + e^{\sigma_2} + \cdots + e^{\sigma_m}] \simeq \max(\sigma_1, \sigma_2, \ldots, \sigma_m), \tag{11.45}$$

we have:

$$\ln f_{\bar{\mathbf{r}}}(\bar{\rho}|c_\ell^{(i)} = b, \bar{\mathbf{h}}) \propto \left\{-\max_{\bar{c}_p \in A_b^{(j)}}\left[-\left|\bar{\rho}_p - \bar{h}_p\bar{c}_p\right|^2\right]\right\} = \min_{\bar{c}_p \in A_b^{(j)}} |\bar{\rho}_p - \bar{h}_p\bar{c}_p|^2 \tag{11.46}$$

with $b = 0, 1$.

The mapping rule employed has a significant influence on system performance. When iterative decoding is not employed, the Gray mapping usually gives the best performance [1854]. In the case of BICM-ID, other mapping rules can provide a performance improvement when increasing the number of iterations, whereas this improvement is usually very limited with Gray mapping [1855].

11.3.3 Error Performance

Before giving some hints as to the error performance of BICM, at least when iterative decoding is not employed, we introduce the *equivalent parallel channel model* for BICM in the case of ideal interleaving. This model is shown in Figure 11.10 and consists of a set of m parallel *independent* and *memoryless* binary input channels connected to the encoder output by a random switch, which models ideal interleaving. Each channel corresponds to a position in the label of the symbols of A_c. For every coded bit $c_\ell^{(i)}$, the switch selects randomly and independently of other selections a position index j (with $j = 1, 2, \ldots, m$) and transmits $c_\ell^{(i)}$ on the jth channel. The detector, which knows the sequence of switch positions, computes the bit metrics $\{\ln f_{\bar{\mathbf{r}}}(\bar{\rho}|c_\ell^{(i)} = b)\}$ in (11.42) that are used to compute the branch metrics of the code trellis over which the decoder operates. This model is the basis for the computation of the capacity of BICM with ideal interleaving. These results show that for 8-PSK and 16-QAM schemes, in the range of practical interest, the capacity loss of BICM with respect to the optimum approach is negligible if and only if Gray mapping is used [1854].

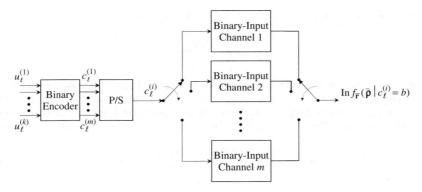

Figure 11.10   Equivalent parallel channel model for BICM in the case of ideal interleaving.

Turning now to error performance, let us assume that the binary code employed in the BICM scheme of Figure 11.8 is linear, so that it admits a (possibly time-varying, as in the case of block codes) trellis representation (see Section 10.8), and consider two codewords $\mathbf{c}$ and $\hat{\mathbf{c}}$ of this binary code stemming from the same state and merging after a given number of trellis steps. Our aim is to compute the PEP $\Pr\{\mathbf{c} \to \hat{\mathbf{c}}\}$ which, however, may depend on the pair $(\mathbf{c}, \hat{\mathbf{c}})$ rather than on their difference. This is because the binary-input channels of the BICM equivalent parallel channel model in Figure 11.10 may be nonsymmetric, depending on the mapping and the signal constellation A_c. In [1854] a symmetrization procedure is thus described which leaves the performance unmodified. After symmetrization, the PEP will depend not only on the channel and the type of detection, but also on the Hamming distance d_H between $\mathbf{c}$ and $\hat{\mathbf{c}}$, the employed mapping μ and the signal constellation A_c:

$$\Pr\{\mathbf{c} \to \hat{\mathbf{c}}\} = g(d_H, \mu, A_c) \ . \tag{11.47}$$

The usual linear bound on the bit error probability of binary codes can be computed as:

$$P_b \leq \frac{1}{k} \sum_{d_H=1}^{\infty} w_I(d_H) g(d_H, \mu, A_c) \tag{11.48}$$

in the case of a convolutional code of rate k/n, and:

$$P_b \leq \frac{1}{k} \sum_{d_H=1}^{n} w_I(d_H) g(d_H, \mu, A_c) \tag{11.49}$$

in the case of an (n, k) block code, where $w_I(d_H)$ is the input weight of error events having Hamming distance d_H. In the original paper by E. Zehavi [1464], a Chernoff bound on the PEP was derived in closed form for 8-PSK with Gray mapping and a receiver with perfect knowledge of the channel. It cannot, however, be extended to other mappings or signal constellations. A more general and very accurate upper bound is derived in [1854], based on the Bhattacharyya bound [693]. This bound can be expressed as:

$$g(d_H, \mu, A_c) \leq B^{d_H}, \tag{11.50}$$

where

$$B \triangleq \frac{1}{m} \sum_{i=1}^{m} E_{b,\bar{\mathbf{r}},\bar{\mathbf{h}}} \left[\sqrt{\frac{f_{\bar{\mathbf{r}}}(\bar{\rho}|c_\ell^{(i)} = 1 - b, \bar{\mathbf{h}})}{f_{\bar{\mathbf{r}}}(\bar{\rho}|c_\ell^{(i)} = b, \bar{\mathbf{h}})}} \right] \tag{11.51}$$

corresponds to the case of perfect knowledge of the channel and:

$$B \triangleq \frac{1}{m} \sum_{i=1}^{m} \mathrm{E}_{b,\bar{\mathbf{r}}} \left[\sqrt{\frac{f_{\bar{\mathbf{r}}}(\bar{\boldsymbol{\rho}}|c_\ell^{(i)} = 1 - b)}{f_{\bar{\mathbf{r}}}(\bar{\boldsymbol{\rho}}|c_\ell^{(i)} = b)}} \right] \tag{11.52}$$

to the case of knowledge of the channel statistics only. Nevertheless, it cannot be computed in closed form but requires a numerical evaluation much faster than simulations and allowing the computation of P_b for very high SNRs, usually beyond the reach of simulations. For a fading channel perfectly known to the receiver and Gray mapping, the PEP can be bounded, for sufficiently large SNR, as:

$$g(d_H, \mu, A_c) \leq \left[\frac{8N_0(K+1)\exp(-K)}{d_3^2} \right]^{d_H}, \tag{11.53}$$

where

$$d_3^2 \triangleq \left[\frac{1}{m 2^m} \sum_{j=1}^{m} \sum_{b=0}^{1} \sum_{x \in A_b^{(j)}} \frac{1}{|x - z|^2} \right]^{-1} \tag{11.54}$$

and z is the single nearest neighbor of x in $A_{1-b}^{(j)}$.

11.4 Modulation Codes Based on Multilevel Coding

TCM has shown that coded modulation can provide systems with good spectral efficiency and performance by partitioning the constellation. A subset of the constellation points is chosen by the encoded bits from a convolutional code (providing a coded level). Then the constellation point to be transmitted is chosen from that subset by uncoded bits (providing an uncoded level).[5] However, we can extend TCM to more than two levels, where each level may or may not use an error control code (e.g., block or convolutional). This is called *multilevel coding*. It was first introduced in [1462] and then generalized in [1463, 1856]. A comprehensive survey of MLCs for AWGN channels, using various design criteria, can be found in [1857]. Multilevel coding allows a complex code to be created by using a hierarchy of simpler component codes. Spectral efficiency is achieved by using an expanded constellation and partitioning as in TCM.

11.4.1 Code Construction for AWGN Channels

An L-level MLC consists of L independent component codes, as shown in Figure 11.11. A block of k binary source data symbols $\mathbf{u}^{(j)} = [u_1^{(j)}, u_2^{(j)}, \ldots, u_k^{(j)}]$ is partitioned into L data blocks $\{\mathbf{u}_p^{(j)}, p = 1, 2, \ldots, L\}$, with $\mathbf{u}_p^{(j)} = [u_{p,1}^{(j)}, u_{p,2}^{(j)}, \ldots, u_{p,k_p}^{(j)}]$ and $\sum_{p=1}^{L} k_p = k$. The data block $\mathbf{u}_p^{(j)}$ (with $p = 1, 2, \ldots, L$) is fed into the pth level encoder generating the binary word $\mathbf{e}_p^{(j)} = [e_{p,1}^{(j)}, e_{p,2}^{(j)}, \ldots, e_{p,m_p}^{(j)}]$ of the pth component code $\mathcal{C}^p$ (with $\sum_{p=1}^{L} m_p = m$). The symbols of the codewords $\{\mathbf{e}_p^{(j)}, p = 1, 2, \ldots, L\}$ form a single binary label $\mathbf{e}^{(j)} = [e_1^{(j)}, e_2^{(j)}, \ldots, e_m^{(j)}]$, which is mapped to the signal point c_j. The rate of this coding scheme is $R = k/m$ and is equal to the sum of the individual code rates if equal code lengths at all levels are assumed. Virtually any code can be used as a component code [1856], including concatenated codes. The level-1 code selects a subset of constellation points for later levels from which to choose the transmitted point. Each level reduces the subset of constellation points being considered until the Lth level's output chooses a single point for transmission. The component code on each level effectively operates on a different constellation. This has been formalized using lattice partitions to produce coset codes in [324, 335]. The way the encoded outputs are mapped to constellation points is called *partitioning*.

---

[5] This assumes a TCM code using uncoded bits.

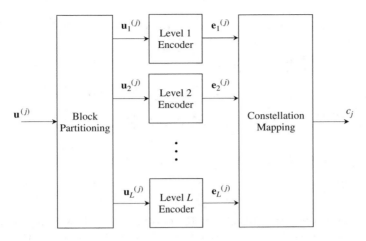

Figure 11.11 L-level encoder.

The partitioning procedure starts with a constellation A_c of M points (assuming an M-ary constellation, with $M = 2^m$). This constellation is partitioned into 2^{m_1} equally sized sets of M_1 points on the first level using m_1 encoded bits from the level-1 encoder. The procedure continues on subsequent levels, except that each level operates on the subset of constellation points chosen by previous levels. The jth level partitions the set of M_{j-1} points chosen by previous levels into 2^{m_j} equally sized sets of M_j points using m_j encoded bits from the level-j component encoder. Different numbers of bits can be used to choose the partitions on different levels [1463, 1856], or symbols over GF(q) (or $\mathbb{Z}_q$) could be used. The partitioning continues until there is only one constellation point in each set.

A variety of different partitioning strategies have been proposed, including increasing Euclidean distance on each level, decreasing Euclidean distance on each level, unequal error protection or equal Euclidean distance on each level [992, 1857–1862]. The most commonly used partitioning scheme for MLCs is Ungerboeck *set partitioning* (as described for TCM in Section 11.2.1.1), which designs for maximum intra-subset distance [1863] and results in increasing Euclidean distance on each level.

An example of Ungerboeck set partitioning is given in Figure 11.12 for 16-QAM, $L = 4$ and $m_1 = m_2 = m_3 = m_4 = 1$. The minimum squared Euclidean distance of the MLC with binary partitions is bounded by [1857]:

$$d_{E,\min}^2 \geq \min(d_1 \delta_0^2, d_2 \delta_1^2, \ \dots, d_L \delta_{L-1}^2), \tag{11.55}$$

where δ_i^2 ($i = 0, 1, \dots, L-1$) is the Euclidean distance between points on level i and d_l ($l = 1, \dots, L$) is the minimum Hamming distance of the level-l component code. Ungerboeck set partitioning requires the most powerful code to be used on the first level. The first level has the worst squared Euclidean distance properties before coding (the smallest minimum squared Euclidean distance between constellation points and on average the most constellation points at this distance for a given point), which is why it requires the most powerful code.

An example of *block partitioning* is shown in Figure 11.13 for 16-QAM, $L = 4$ and $m_1 = m_2 = m_3 = m_4 = 1$. Clearly, there are other ways the partitions could have been chosen. The goal is to minimize the intra-subset variance [1857]. This results in all levels having the same intra-subset distance, meaning that:

$$\delta_0 = \delta_1 = \delta_2 = \delta_3. \tag{11.56}$$

Researchers have also proposed using hybrid or mixed partitioning strategies, where set partitioning is used on some levels and block partitioning on the others [1857].

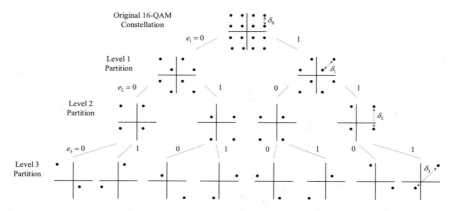

Figure 11.12 Ungerboeck set partitioning of 16-QAM for an $L = 4$ level MLC with $m_1 = m_2 = m_3 = m_4 = 1$. The bit labeling level i is given by e_i. The level-4 partition results in a single point being chosen. If $e_4 = 0$ the point on the left of the vertical axis is chosen, otherwise $e_4 = 1$ and a point to the right is chosen.

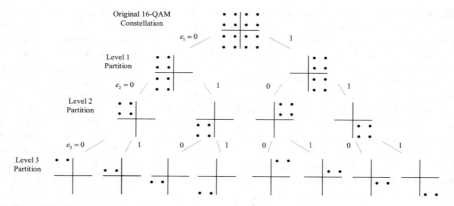

Figure 11.13 Block partitioning of 16-QAM for an $L = 4$ level MLC with $m_1 = m_2 = m_3 = m_4 = 1$. The bit labeling level i is given by e_i. The level-4 partition results in a single point being chosen. If $e_4 = 0$ the upper point is chosen, otherwise $e_4 = 1$ and the lower point is chosen.

11.4.1.1 Design Criteria

A variety of design criteria have been proposed for MLCs based on distance [1462, 1463, 1856, 1857, 1864], capacity (and rate) [1857, 1862, 1865–1868], cutoff rate [1857] and the coding exponent [1857]. A summary of these approaches is given in [1857]. Here we focus on two widely used approaches, namely the balanced distance rule and the capacity rule.

At high SNR, performance is dominated by the level with minimum squared Euclidean distance. The original design criterion was the balanced distance decoding rule [1462]. It chooses component codes so that the minimum squared Euclidean distance on each level is equal, resulting in:

$$d_1 \delta_0^2 = d_2 \delta_1^2 = \ldots = d_L \delta_{L-1}^2. \tag{11.57}$$

The balanced distance design rule tends to overestimate the rate used on the first level [1857]. It may, however, be suitable when short component codes are used [1857].

We can define equivalent channels for each level of the MLC conditioned on knowing the previous levels' encoded data [1857]. The capacity of these equivalent channels, known as *equivalent capacity*, can be used to select the component code rates to be used on each level. In fact, it can be proved that the capacity C of a constellation A_c of M signal elements is the sum of the capacities $\{C(A_i)\}$ of the equivalent individual channels at all coding levels of a multilevel coding scheme, based on a binary regular set partitioning tree A_c [1866], that is[6]:

$$C = \sum_{i=1}^{L} C(A_i). \tag{11.58}$$

The capacity design rule states that the rate of the level-i component code R^i should be chosen to equal the equivalent capacity of level i in order achieve capacity C [1857]. The capacity rule gives a better estimate of the optimum rate for each level of the MLC (compared to the balanced distance rule) when the code length is not short [1857]. Multilevel coding with multistage decoding (described later) approaches capacity if the rate of each component code is chosen to equal the equivalent capacity of that level [1857].

For cases of finite codeword length the coding exponent rule can be used [1857]. The coding exponent rule tends to the same result as the capacity rule for long codes and tends to the results of the balanced distance rule for short codes. It works for restricted code lengths/delays. For short codes, minimum Euclidean distance dominates performance [1857]. The coding exponent rule for MLCs can be stated as follows: for a maximum tolerable word error rate p_w, the rates $\{R^i, i = 1, \ldots, L\}$ at the individual coding levels should be chosen according to the corresponding isoquants of the coding exponents $E^i(R^i)$. Interested readers are directed to [1857] and the references therein.

The design of specific MLCs is analyzed in the following two examples.

Example 11.4.1 Let us consider a basic multilevel encoder design for set partitioning, 16-QAM and $L = 4$ binary partitions (meaning $m_1 = m_2 = m_3 = m_4 = 1$). The set partitioning of the 16-QAM constellation is shown in Figure 11.12. We can see that $\delta_1 = \sqrt{2}\delta_0$, $\delta_2 = 2\delta_0$ and $\delta_3 = 2\delta_1 = 2\sqrt{2}\delta_0$. Using (11.55), the minimum squared a Euclidean distance is given by:

$$d_{E,\min}^2 = \min(d_1, 2d_2, 4d_3, 8d_4)\delta_0^2. \tag{11.59}$$

Any component code may be used. Here, we will look at using length $n = 64$ extended BCH codes due to the range of rates offered. As performance is dominated by the level with the worst distance properties, we will use the balanced distance design rule. This can be achieved by using the $(n, k_1, d_1) = (64, 24, 16)$ code on level 1, a $(64, 45, 8)$ code on level 2, a $(64, 57, 4)$ code on level 3 and a $(64, 63, 2)$ SPC code on level $L = 4$, where k_i is the number of information bits in the level-i codeword. This results in $d_{E,\min}^2 = 16\delta_0^2$ on each level. The overall code rate is 0.738 and the block length is $N = 64$ symbols.

□

Example 11.4.2 Alternatively, we could design an $L = 3$ level MLC using a four-way partition on level 1 and then two-way (binary) partitions on levels 2 and 3, as shown in Figure 11.14. Now three component codes are used. At level 1 we split the constellation into four subsets (each of four points) using bits $(e_{1,1}, e_{1,2})$. In this case the minimum squared Euclidean distance is given by:

$$d_{E,\min}^2 = \min(d_1, 4d_2, 8d_3)\delta_0^2. \tag{11.60}$$

An overall rate-0.77 code with $d_{E,\min}^2 = 16\delta_0^2$ and block length $N = 64$ can be achieved by using a $(128, 78, 16)$ code on level 1, a $(64, 57, 4)$ code on level 2 and a $(64, 63, 2)$ SPC code on level $L = 3$.

□

---

[6] Note that changing the partitioning strategy modifies the equivalent capacities, but not the overall capacity in (11.58).

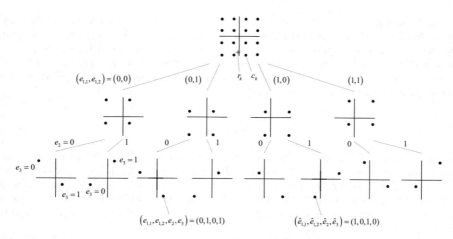

Figure 11.14   Ungerboeck set partitioning of 16-QAM for an $L = 3$ level MLC with $m_1 = 4$, $m_2 = 2$ and $m_3 = 2$. The jth bit in the level-i label is given by $e_{i,j}$ (j is excluded if only one bit is used on level i). The level-3 partition results in a single point being chosen. If $e_3 = 0$ the point on the left of the vertical axis is chosen, otherwise $e_3 = 1$ and a point to the right is chosen.

11.4.2   Multistage Decoder

MLCs are usually decoded in a sequential manner using a *multistage decoder* (MSD), as shown in Figure 11.15. This is used since decoding the overall MLC can be prohibitively complex. The MSD decodes the levels of the MLC in the same order as they are encoded. First the level-1 code is decoded to choose a subset of points for the level-2 code to choose from, and this pattern continues on all subsequent levels.

Each level's decoder passes a hard decision on its part of the overall constellation point label to subsequent levels. This has the effect of choosing a subset of constellation points for the next level's decoder to consider. If there are errors in a level's decision, these are passed to subsequent levels, causing error propagation. Often it is enough to just use soft decision decoding on the first level and hard decision decoding on other levels. Performance gains can be made by passing soft decisions and by iterating at the cost of decoding complexity [1869]. Now we define the soft input to level i. For simplicity let us assume binary partitions on each level. We denote the kth received symbol by r_k and the jth encoded bit from the level-i component code by $e_{i,j}$. Now we can write the LLR of the jth bit on level i as:

$$L\left(e_{i,j}^{(k)}\right) = \log \left(\frac{\Pr\left\{e_{i,j}^{(k)} = 1 \,\middle|\, r_k, \mathbf{e}_1^{(k)}, \mathbf{e}_2^{(k)}, \ldots, \mathbf{e}_{i-1}^{(k)}\right\}}{\Pr\left\{e_{i,j}^{(k)} = 0 \,\middle|\, r_k, \mathbf{e}_1^{(k)}, \mathbf{e}_2^{(k)}, \ldots, \mathbf{e}_{i-1}^{(k)}\right\}} \right) \tag{11.61}$$

with $i = 2, 3, \ldots, L$ and $j = 1, 2, \ldots, m_i$.

The use of multistage decoding is illustrated in the following example for a specific MLC.

Example 11.4.3   We now give an example of multilevel encoding and multistage decoding for the MLC shown in Figure 11.11 and the partitioning design in Figure 11.14. The "constellation mapping" block in Figure 11.11 is performed in a similar way to the partitioning, but only one subset of constellation points is retained at each level. Consider the $L = 3$ level MLC of Figure 11.14. First $m_1 = 2$ encoded bits from level 1, $(e_{11}, e_{12}) = (0, 1)$, are used to choose a subset of $M_1 = 4$ constellation points. Then $m_2 = 1$ encoded bit from level 2, $e_2 = 0$, is used to choose a subset of $M_2 = 2$ constellation points. Finally, $m_3 = 1$ encoded bit from level 3, $e_3 = 1$, is used to choose the transmitted

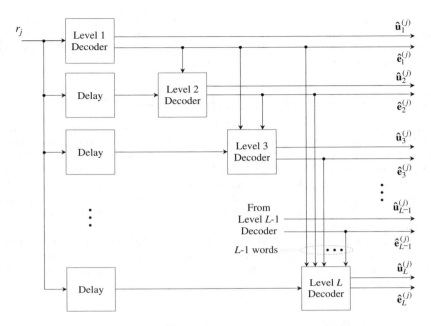

Figure 11.15 L-level MSD, where $\{\hat{\mathbf{u}}_p^{(j)}, p = 1, 2, \ldots, L\}$ are the hard decisions from all levels.

constellation point. The transmitted constellation point is labeled by $(e_{1,1}, e_{1,2}, e_2, e_3) = (0,1,0,1)$ as shown in Figure 11.14.

Now consider decoding the received signal point, r_k, shown in Figure 11.14. The decoding process is performed in a similar way to encoding. The level-1 decoder considers the entire $(M = 16)$-QAM constellation. Then $m_1 = 2$ decoded bits (including parity bits) from the level-1 decoder, $(\hat{e}_{1,1}, \hat{e}_{1,2}) = (1,0)$, are used to choose a subset of $M_1 = 4$ constellation points. The level-2 decoder operates as if this subset of M_1 constellation points is the entire constellation. The $m_2 = 1$ decoded bit from the level-2 decoder, $\hat{e}_2 = 1$, is used to choose a subset of $M_2 = 2$ constellation points for the level-3 decoder. The level-3 decoder operates as if this subset of $M_2 = 2$ constellation points is the entire constellation. Finally, $m_3 = 1$ decoded bit from the level-3 decoder, $\hat{e}_3 = 0$, is used to choose the decoded constellation point. The transmitted constellation point is labeled by $(e_{1,1}, e_{1,2}, e_2, e_3) = (0, 1, 0, 1)$, as shown in Figure 11.14, but the decoded constellation point is labeled by $(\hat{e}_{1,1}, \hat{e}_{1,2}, \hat{e}_2, \hat{e}_3) = (1,0,1,0)$. Although the decoded constellation point is at the minimum Euclidean distance, δ_0, from the transmitted constellation point, c_k, it at the maximum Hamming distance away from it, which means all decoded bits are incorrect. The error on level 1 resulted in the actual transmitted constellation point being excluded from the subsets of points considered on later levels and so later levels were decoded in error. This effect is called *error propagation*. This highlights the importance of a strong code and advanced decoding (such as soft input decoding) on lower levels.

□

11.4.3   Error Performance

Performance analysis and union bounds have been developed for several different multilevel coded modulation systems [1868, 1870–1873]. The analysis tends to be specific to the MLC (partitioning, modulation and error control codes) and decoding strategy used (suboptimal metrics, maximum likelihood decoding, soft decision decoding and multistage decoding). The error probability of specific MLCs using PSK is calculated in [1872, 1873]. The error probability of an MLC using QAM, a suboptimal metric and multistage decoding is calculated in [1870].

11.4.4 Multilevel Codes for Rayleigh Flat Fading Channels

Research on MLCs during the 1970s and 1980s was focused on AWGN channels, (e.g., see [1462, 1463, 1856]). As a result, most of the encoder and decoder structures, partitioning strategies and even design rules were developed for MLCs on AWGN channels. However, during the 1990s interest spread to their use in Rayleigh flat fading channels (e.g. [1874–1878]). Most of the AWGN design work extends directly into Rayleigh fading channels and so is not repeated here.

We focus on coherent communications with ideal CSI available at the receiver. The research in this area can loosely be split into block-based MLCs [1876] and trellis-based MLCs [1874, 1875], where the component codes are block or convolutional codes, respectively. The work can also be split between equal and unequal [1878] error protection codes. This shows the flexibility of the MLC structure. We now briefly describe the error performance and design criteria proposed in [1876], which work was specific to the Rayleigh fading channel rather than being an application of AWGN techniques in fading as in [1879].

11.4.4.1 Error Performance

In [1876] a multilevel block code is developed for Rayleigh flat fading channels and 8-PSK modulation. We denote the transmitted binary codeword vector on level i by $\mathbf{e}^{(i)}$ and the corresponding erroneous decoded sequence by $\hat{\mathbf{e}}^{(i)}$. The resulting transmitted sequence of constellation points is denoted by $\mathbf{c}$ and the corresponding erroneous decoded sequence by $\hat{\mathbf{c}}$. Assuming component codes with small minimum Hamming distance (≤ 4), the overall PEP can be written as [1876]:

$$
\Pr\{\mathbf{c} \to \hat{\mathbf{c}}\} \leq \sum_{i=1}^{L} \Pr\{\mathbf{e}^{(i)} \to \hat{\mathbf{e}}^{(i)}\}
$$

$$
= \sum_{i=1}^{L} \left\{ 1 - \left[1 - \frac{1}{2} \left(\frac{8 d_H(\mathbf{e}^{(i)}, \hat{\mathbf{e}}^{(i)}) \sigma_n^2}{\delta_i^2} \right)^{d_H(\mathbf{e}^{(i)}, \hat{\mathbf{e}}^{(i)})} \right] \right\}, \tag{11.62}
$$

where σ_n^2 is the variance of the complex AWGN. For simplicity, we have assumed that all component codes have the same length. In the case of component codes with larger minimum Hamming distance (> 4) the PEP can be written as [1876]:

$$
\Pr\{\mathbf{c} \to \hat{\mathbf{c}}\} \leq \sum_{i=1}^{L} \Pr\{\mathbf{e}^{(i)} \to \hat{\mathbf{e}}^{(i)}\}
$$

$$
= \sum_{i=1}^{L} \left\{ 1 - \left[1 - \frac{1}{2\sqrt{1 + \frac{\sigma_h^2 \delta_i^2}{4\sigma_n^2}}} \exp\left(-\frac{1}{2} \frac{\mu_h^2 d_E^2(\mathbf{e}^{(i)}, \hat{\mathbf{e}}^{(i)})}{4\sigma_n^2 + \sigma_h^2 \delta_i^2} \right)^{d_H(\mathbf{e}^{(i)}, \hat{\mathbf{e}}^{(i)})} \right] \right\}, \tag{11.63}
$$

where μ_h is the mean of the envelope of the channel fading and σ_h^2 is its variance. Note that the fading envelope is normalized to unit power.

11.4.4.2   Design Criteria

Based on (11.63), for large minimum Hamming distance codes L. Zhang and B. Vucetic [1876] recommended first maximizing the minimum squared Euclidean distance and then minimizing the number of codewords at that distance. In contrast, in the case of small minimum Hamming distance codes, based on (11.63), they recommended first maximizing the minimum Hamming distance, then maximizing the minimum product distance when there are multiple codes with the same minimum Hamming distance, and finally minimizing the number of codewords at the minimum Hamming distance. For the MLC design presented in [1876] the minimum product distance was defined as the smallest product of the corresponding nonzero squared Euclidean distances between the path pairs with the minimum Hamming distance.

Alternatively, the MLC design criteria for the AWGN channel can be applied. For example, the capacity rule has been used in Rayleigh flat fading channels [1879]. The only difference is that the equivalent capacities are derived for Rayleigh fading.

11.5   Space-Time Coding

We now consider systems with multiple antennas at both transmitter and receiver and codes designed for these applications. The use of multiple *receive* antennas to provide *diversity* has been known for many decades. Only recently the use of multiple *transmit* antennas has attracted interest. The number of antennas at the transmitter and receiver (n_T and n_R) depends on the application. For cellular systems, the base station is typically equipped with several antennas while the mobile terminal can have only one or two antennas. Hence, in the uplink we have $n_R > n_T$, while in the downlink $n_T > n_R$. On the other hand, in WLAN applications most nodes will have a similar number of antennas.

Multiple antennas can be used to increase data rates through *spatial multiplexing* or to improve performance through *diversity*. This is a fundamental tradeoff in multiple-antenna systems [14, 1880, 1881] and will be considered in detail later. Multiplexing is obtained by exploiting the MIMO channel to obtain independent signaling paths that can be used to increase the spectral efficiency [17, 18, 220, 226]. This spectral efficiency increase often relies on an accurate knowledge of the channel at the receiver, and sometimes at the transmitter as well, and is obtained at the price of an increase in the receiver processing (in addition to the cost of deploying multiple antennas). Diversity is obtained by exploiting the independent fading gains that affect the signal and that can be averaged out to increase the reliability of the receiver decisions.

We have seen that the adoption of a properly designed coded transmission provides *time diversity* over fading channels. Time diversity, however, is not available in systems with limited mobility, such as indoor WLANs or wireless local loops. In this case, in fact, the channel is *quasi-static*, meaning that the channel variations are very slow compared with the duration of one codeword. We have already mentioned that, in this case, receive antenna diversity can be employed. STCs are particularly important in such scenarios both to exploit *transmit* antenna diversity and/or the potential increase of the system spectral efficiency related to MIMO channels.

11.5.1   ST Coding for Frequency-Flat Fading Channels

11.5.1.1   System Model for Frequency-Flat MIMO Channels and Some Results on Channel Capacity

At discrete time ℓ, the received samples at the output of the n_R receive antennas are collected into an $n_R \times 1$ vector $\mathbf{r}_\ell$ that can be expressed as (see also Section 4.4.2.2):

$$\mathbf{r}_\ell = \sqrt{\gamma}\mathbf{H}_\ell\mathbf{c}_\ell + \mathbf{n}_\ell, \tag{11.64}$$

with $\ell = 1, 2, \ldots, N$, where $\mathbf{c}_\ell$ is the $n_T \times 1$ vector containing the modulated symbols transmitted in parallel by the n_T transmit antennas, $\mathbf{n}_\ell$ is an $n_R \times 1$ complex Gaussian noise vector having independent real and imaginary components and representing the thermal noise samples at the n_R receive antennas, $\mathbf{H}_\ell$ is the $n_R \times n_T$ matrix of the channel gains, its (i, j)th element $h_{i,j}[\ell]$ representing the gain from transmit antenna j to receive antenna i at discrete time ℓ, and γ is an appropriate real coefficient.

The matrix $\mathbf{H}_\ell$ will be assumed random with zero-mean iid Gaussian entries having independent real and imaginary components. Equivalently, we can say that each entry of $\mathbf{H}_\ell$ has uniformly distributed phase and Rayleigh-distributed magnitude. This choice models a *Rayleigh fading environment* with sufficient separation among the receive and transmit antennas that the channel gains for each transmit–receive antenna pair are independent. This assumption becomes questionable when n_T and/or n_R increase. In fact, it relies on a separation of the transmit and/or receive antennas by some multiple of the wavelength, which cannot be obtained when a large number of antennas is packed into a finite volume. Although the results for MIMO channel capacity that will be briefly summarized here have been obtained under the assumption of a Rayleigh fading environment, the results for code design criteria can easily be generalized to the case of channel gains having Rician-distributed magnitude. When considering the design of STCs, we will also assume that transmitted symbols belong to the M-ary complex constellation A_c, with unit average energy, and that the noise vector and the channel matrix are such that $\mathrm{E}\{\mathbf{n}_\ell \mathbf{n}_\ell^H\} = \mathbf{I}_{n_R}$ (uncorrelated noise components on different antennas and with unit variance) and $\frac{1}{n_T n_R}\mathrm{trace}(\mathrm{E}\{\mathbf{H}_\ell \mathbf{H}_\ell^H\}) = 1$, that is, the elements of $\mathbf{H}_\ell$ also have unit variance. Under these assumptions, γ has the meaning of average SNR per transmit antenna (or per transmitted symbol), $\gamma = \bar{E}_s/N_0$, and the average received SNR per receive antenna is given by $n_T \gamma$.

An STC with block length N is a set $\mathcal{C}$ of $n_T \times N$ complex matrices (codewords). Codeword matrices $\mathbf{C} = [\mathbf{c}_1, \ldots, \mathbf{c}_N]$ are transmitted by columns, in N consecutive channel uses. The STC spectral efficiency is given by $\eta = \frac{1}{N}\log_2 |\mathcal{C}|$ bits per channel use. By definition, the average information bit energy over noise power spectral density ratio is given by $\bar{E}_b/N_0 = \gamma n_T / \eta$.

Equation (11.64) describes the general model for a time-selective MIMO channel. When N is much larger than the channel coherence time, each codeword sees a large number of channel realizations. We can assume that $\{\mathbf{H}_\ell\}$ is an ergodic random process and the channel is consequently *ergodic*. In scenarios characterized by limited mobility, the channel can be assumed to be *slow* or *quasi-static*, that is, each codeword sees only one channel realization. In other words, $\mathbf{H}_\ell = \mathbf{H}$, $\ell = 1, 2, \ldots, N$. In this case, this fading model is *nonergodic*. A different model for time-varying fading channels was introduced by Marzetta and Hochwald in [1882]. They considered a block fading channel constant for L consecutive channel uses and independent from block to block, modeling, as an example, a system with quasi-static fading and frequency hopping every L channel uses. This case will be referred to as a *block fading channel*.

Different assumptions can be made about the knowledge of the channel gain matrix at the transmitter and receiver. For a *quasi-static* channel, it is generally assumed that $\mathbf{H}$ is perfectly known at the receiver since the channel gains can be obtained fairly easily by sending a pilot sequence for channel estimation (see [38, Section 3.9] or [1883, Section 10.1]). In contrast, the assumption of perfect knowledge of the channel matrix at the transmitter holds only if a delay-free, error-free feedback link from receiver to transmitter exists, allowing the receiver to send back the estimated channel gains, or if time division duplexing is used, where each end can estimate the channel from the incoming signal in the reverse direction. In contrast, on a block fading channel, the assumption adopted in [1882] is the absence of knowledge of the channel gains at both transmitter and receiver.

The case of perfect knowledge of the channel gains at both transmitter and receiver is of scant interest in this section on STCs since, in this case, through simple *transmit precoding* and *receive filtering*, the MIMO channel can be decomposed into a set of parallel and independent SISO channels. Consider for example, the quasi-static channel and the SVD of matrix $\mathbf{H}$ (see (C.11)):

$$\mathbf{H} = \mathbf{U}\boldsymbol{\Sigma}\mathbf{V}^H, \tag{11.65}$$

where the $n_R \times n_R$ matrix $\mathbf{U}$ and the $n_T \times n_T$ matrix $\mathbf{V}$ are unitary matrices and $\mathbf{\Sigma}$ is an $n_R \times n_T$ diagonal matrix containing the singular values[7] $\{\sigma_i\}$ of $\mathbf{H}$. We can assume that the input vector $\mathbf{c}_\ell$ is obtained from a vector $\tilde{\mathbf{c}}_\ell$ through the linear transformation $\mathbf{c}_\ell = \mathbf{V}^H \tilde{\mathbf{c}}_\ell$. At the receiver, the vector $\mathbf{r}_\ell$ is still linearly transformed giving the vector $\tilde{\mathbf{r}}_\ell \triangleq \mathbf{U}^H \mathbf{r}_\ell$. Thus, the following equivalent channel results:

$$\begin{aligned}\tilde{\mathbf{r}}_\ell \triangleq \mathbf{U}^H \mathbf{r}_\ell &= \mathbf{U}^H (\sqrt{\gamma}\mathbf{H}\mathbf{c}_\ell + \mathbf{n}_\ell) \\ &= \mathbf{U}^H (\sqrt{\gamma}\mathbf{U}\mathbf{\Sigma}\mathbf{V}^H \mathbf{c}_\ell + \mathbf{n}_\ell) \\ &= \sqrt{\gamma}\mathbf{\Sigma}\tilde{\mathbf{c}}_\ell + \tilde{\mathbf{n}}_\ell, \end{aligned} \tag{11.66}$$

where $\tilde{\mathbf{n}}_\ell \triangleq \mathbf{U}^H \mathbf{n}_\ell$ is statistically equivalent to $\mathbf{n}_\ell$, $\mathbf{U}$ being a unitary matrix. We thus have a set of $R_{\mathbf{H}}$ parallel independent channels, each corresponding to a nonzero singular value of $\mathbf{H}$ for which classical results apply [705]. In particular, the optimal capacity-achieving power distribution can be obtained through *water-filling*.[8] Since these parallel channels do not interfere with each other, the optimal demodulator complexity is linear in $R_{\mathbf{H}}$. Moreover, when independent data are sent over the parallel channels, the MIMO channel can support $R_{\mathbf{H}}$ times the data rate of a SISO system. A *multiplexing gain* of $R_{\mathbf{H}}$ is thus obtained, although the performance over each channel depends on the corresponding gain σ_i. Hence, for this reason, in what follows we will consider the more interesting case of knowledge of the channel gains at the receiver only.

The case where each channel use, that is, each transmission of one symbol from each of the n_T transmit antennas, corresponds to an independent realization of $\mathbf{H}_\ell$, was studied in [226]. Even if the channel realization is not known at the transmitter, it can be proved [226] that, in the asymptotic limit of a large number of transmit and receive antennas, the average capacity of a MIMO channel still grows linearly with $\xi = \min(n_T, n_R)$, as long as the channel can be accurately estimated at the receiver. Moreover, this linear growth of capacity with ξ is observed even for a small number of antennas [1885]. Similarly, for large values of the SNR, capacity also grows linearly with ξ.[9] In particular, it grows as $\xi \log \gamma$. In other words, even in the absence of knowledge of the channel at the transmitter, we can say that multiple antennas increase the capacity by a factor ξ as in the case of independent parallel channels. This explains why ξ is often called the *number of degrees of freedom* generated by the MIMO channel.

In the case of a quasi-static fading channel, when $\mathbf{H}$ is chosen randomly at the beginning of the transmission and remains fixed for all channel uses, average capacity has no meaning (is strictly zero), as the channel is nonergodic [275]. In this case, as discussed in Chapter 7, *outage probability*, defined as the probability that the transmission rate exceeds the mutual information of the channel, must be evaluated. The maximum rate that can be supported by the channel with a given outage probability is the *outage capacity*. As in the case of ergodic channels, for a given outage probability, the outage capacity increases linearly with ξ.

Finally, in the case of a block fading channel with a coherence time of L symbols and in the absence of knowledge of the channel at both transmitter and receiver [1882, 1886], when $L \geq \gamma + n_R$, at high SNR values the capacity (in bits per channel use) can be approximated as:

$$C \simeq \xi \left(1 - \frac{\xi}{L}\right) \log_2 \gamma .$$

Hence, for $L \to \infty$ the capacity of the noncoherent MIMO channel approaches that of the coherent channel. However, when $L < \gamma + n_R$ the capacity increases as $\varsigma \left(1 - \frac{\varsigma}{L}\right) \log_2 \gamma$, where

---

[7] The number of singular values is $R_{\mathbf{H}} = \text{rank}(\mathbf{H}) \leq \min(n_T, n_R)$. The case of $R_{\mathbf{H}} = \min(n_T, n_R)$ is often referred to as a *rich scattering environment*.

[8] When the channel is time-varying, water-filling across time should be used as well [1884].

[9] It must be noted, however, that at very low SNRs, transmit antennas are not beneficial since capacity only scales with the number of receive antennas independently of the number of transmit antennas [226].

$\varsigma = \min(n_T, n_R, \lfloor L/2 \rfloor)$. As a consequence, it is not convenient to have more than $\lfloor L/2 \rfloor$ transmit antennas, although, when fading is correlated, additional transmit antennas do increase capacity [1445].

11.5.1.2 ST Codeword Design Criteria for Slow Fading Channels

We now consider the case of a *slow* or *quasi-static* fading channel where $\mathbf{H}$ is random but constant over $N \gg \max\{n_T, n_R\}$ channel uses, and we assume that the receiver knows $\mathbf{H}$ perfectly, while the transmitter has no knowledge of $\mathbf{H}$. Having collected the codewords into appropriate matrices $\{\mathbf{C}\}$, we can similarly organize the corresponding received samples and noise samples into two $n_R \times N$ matrices $\mathbf{R}$ and $\mathbf{N}$, respectively, whose ℓth columns are composed of the n_R received samples and noise samples at time ℓ. Hence, we may write:

$$\mathbf{R} = \sqrt{\gamma}\mathbf{HC} + \mathbf{N} . \tag{11.67}$$

Under these assumptions, a maximum likelihood decoder will operate following the decision rule:

$$\hat{\mathbf{C}} = \arg \max_{\tilde{\mathbf{C}}} f_{\mathbf{R}}(\mathbf{\Theta}|\mathbf{H}, \tilde{\mathbf{C}}), \tag{11.68}$$

where $f_{\mathbf{R}}(\mathbf{\Theta}|\mathbf{H}, \mathbf{C})$ is clearly a Gaussian joint probability density function and $\mathbf{\Theta}$ denotes the value taken on by $\mathbf{R}$. Hence:

$$\hat{\mathbf{C}} = \arg \min_{\tilde{\mathbf{C}}} \sum_{\ell=1}^{N} |\mathbf{\Theta}_\ell - \sqrt{\gamma}\mathbf{H}\tilde{\mathbf{c}}_\ell|^2 = \arg \min_{\tilde{\mathbf{C}}} |\mathbf{\Theta} - \sqrt{\gamma}\mathbf{H}\tilde{\mathbf{C}}|^2, \tag{11.69}$$

where $|\cdot|$ denotes the Frobenius norm of a matrix (see (C.1)). Hence, given a particular channel realization, the PEP $\Pr\{\mathbf{C} \to \hat{\mathbf{C}}|\mathbf{H}\}$ can be computed as (see (F.15)):

$$\Pr\{\mathbf{C} \to \hat{\mathbf{C}}|^2\mathbf{H}\} = Q\left(\sqrt{\frac{\gamma}{2}}|\sqrt{\gamma}\mathbf{H}(\hat{\mathbf{C}} - \mathbf{C})|^2\right)$$

$$\leq \frac{1}{2}\exp\left\{-\frac{\gamma}{4}|\mathbf{H}(\hat{\mathbf{C}} - \mathbf{C})|^2\right\} . \tag{11.70}$$

If $\mathbf{h}_i$ (with $i = 1, 2, \ldots, n_R$) denotes the ith row of matrix $\mathbf{H}$ and the matrix $\mathbf{A} \triangleq (\hat{\mathbf{C}} - \mathbf{C})(\hat{\mathbf{C}} - \mathbf{C})^H$ is defined, we may write:

$$|\mathbf{H}(\hat{\mathbf{C}} - \mathbf{C})|^2 = \mathrm{tr}[\mathbf{H}(\hat{\mathbf{C}} - \mathbf{C})(\hat{\mathbf{C}} - \mathbf{C})^H\mathbf{H}^H] = \sum_{i=1}^{n_R} \mathbf{h}_i\mathbf{A}\mathbf{h}_i^H . \tag{11.71}$$

Since $\mathbf{A}$ is a nonnegative definite Hermitian matrix, it can be diagonalized using a unitary matrix $\mathbf{U}$ as $\mathbf{A} = \mathbf{U}\mathbf{\Lambda}\mathbf{U}^H$ (see (C.7)), where $\mathbf{\Lambda}$ is a diagonal matrix whose elements are the (nonnegative) eigenvalues λ_i, $i = 1, 2, \ldots, n_T$ of $\mathbf{A}$. Then we have:

$$|\mathbf{H}(\hat{\mathbf{C}} - \mathbf{C})|^2 = \sum_{i=1}^{n_R} \mathbf{h}_i\mathbf{U}\mathbf{\Lambda}\mathbf{U}^H\mathbf{h}_i^H . \tag{11.72}$$

Moreover, since the components of $\mathbf{h}_i$ are independent complex Gaussian random variables and matrix $\mathbf{U}$ is unitary, $\mathbf{h}_i$ and $\mathbf{p}_i \triangleq \mathbf{h}_i\mathbf{U}$ are statistically equivalent. Hence, the elements of $\mathbf{p}_i$ are still independent complex Gaussian random variables with the same mean and variance as the elements of $\mathbf{h}_i$. In addition, if $p_{i,j}$ denotes the jth element of $\mathbf{p}_i$, (11.72) can be rewritten as:

$$|\mathbf{H}(\hat{\mathbf{C}} - \mathbf{C})|^2 = \sum_{i=1}^{n_R} \sum_{j=1}^{n_T} \lambda_j |p_{i,j}|^2. \tag{11.73}$$

Let $v = \text{rank}(\mathbf{A}) \leq n_T$ denote the number of nonzero eigenvalues of $\mathbf{A}$ (assuming $N \geq n_T$, i.e., codewords of length greater than or equal to the number of transmit antennas). Assuming that the eigenvalues of $\mathbf{A}$ are ordered in such a way that $\lambda_i \geq \lambda_{i+1}$, we may write:

$$|\mathbf{H}(\hat{\mathbf{C}} - \mathbf{C})|^2 = \sum_{i=1}^{n_R} \sum_{j=1}^{v} \lambda_j |p_{i,j}|^2 \tag{11.74}$$

and

$$\Pr\{\mathbf{C} \to \hat{\mathbf{C}}|\mathbf{H}\} \leq \frac{1}{2} \exp\left\{-\frac{\gamma}{4} \sum_{i=1}^{n_R} \sum_{j=1}^{v} \lambda_j |p_{i,j}|^2\right\}$$

$$= \frac{1}{2} \prod_{i=1}^{n_R} \prod_{j=1}^{v} \exp\left\{-\frac{\gamma}{4} \lambda_j a_{i,j}^2\right\} \tag{11.75}$$

having defined the random variable $a_{i,j} \triangleq |p_{i,j}|$ whose probability density function is (see (2.58)):

$$f_a(x) = 2x \exp(-x^2)\, u\,(x). \tag{11.76}$$

Proceeding as in Section 11.2.1.2, we may thus compute the *average* PEP[10]:

$$\Pr\{\mathbf{C} \to \hat{\mathbf{C}}\} = E_{\mathbf{H}}\{\Pr\{\mathbf{C} \to \hat{\mathbf{C}}|\mathbf{H}\}\}$$

$$\leq \frac{1}{2} \prod_{i=1}^{n_R} \prod_{j=1}^{v} \int_0^{+\infty} \exp\left\{-\frac{\gamma}{4} \lambda_j x_{i,j}^2\right\} f_a(x_{i,j})\, dx_{i,j}$$

$$= \frac{1}{2} \prod_{i=1}^{n_R} \prod_{j=1}^{v} \frac{1}{1 + \frac{\gamma}{4}\lambda_j}$$

$$= \left[\prod_{j=1}^{v}\left(1 + \frac{\gamma}{4}\lambda_j\right)\right]^{-n_R}, \tag{11.77}$$

which asymptotically (as $\gamma \to \infty$) becomes:

$$\Pr\{\mathbf{C} \to \hat{\mathbf{C}}\} \lesssim \frac{1}{2}\left[\prod_{j=1}^{v}\lambda_j\right]^{-n_R}\left(\frac{\gamma}{4}\right)^{-vn_R} = \frac{1}{2}\left[\prod_{j=1}^{v}\lambda_j\right]^{-n_R}\left(\frac{\bar{E}_s}{4N_0}\right)^{-vn_R}. \tag{11.78}$$

This expression can be exploited to develop upper bounds on the symbol or bit error probability (e.g., the union bound technique described in Section 4.3.2 can be used to derive these results); this leads to the conclusion that the *total diversity order* (the slope of the symbol or bit error probability curve in a log-log plot) of the coded system is $v_{\min}n_R$ (where $v_{\min}$ is the minimum value of v). As a

---

[10] The extension to the case of a Rician fading channel is straightforward. See Section 11.2.1.2 for details.

secondary merit criterion, we should try to maximize the term $\prod_{j=1}^{\nu} \lambda_j$ on error events with minimum diversity. Using terminology already employed for TCM over fading channels, we will call this *coding gain*. It displaces the error probability curve instead of changing its slope. Since $\nu = \text{rank}(\mathbf{A})$ and $\prod_{j=1}^{\nu} \lambda_j = \det(\mathbf{A})$, the basic code design principles for STCs over *slow* frequency-flat Rayleigh fading channels are as follows [14]:

Rank criterion. The maximum diversity of $n_T n_R$ is achieved by ensuring that the matrix[11]:

$$\mathbf{A} = (\hat{\mathbf{C}} - \mathbf{C})(\hat{\mathbf{C}} - \mathbf{C})^H$$

is full-rank for all the pairs of distinct codewords $\hat{\mathbf{C}}$ and $\mathbf{C}$. Otherwise, if the minimum rank of $\mathbf{A}$ among all codeword pairs is $\nu_{\min} \leq n_T$ a diversity order $\nu_{\min} n_R$ is achieved.

Determinant criterion. In order to obtain the maximum possible coding advantage, the minimum determinant of matrices $\mathbf{A}$ having minimum rank should be maximized.

These design principles, also known as *Tarokh–Seshadri–Calderbank criteria*, are based on the asymptotic receiver performance. Let us consider (11.77) – in particular, the term in square brackets. Considering that $\sum_{j=1}^{\nu} \lambda_j = \text{tr}(\mathbf{A})$, it can be expressed as:

$$\prod_{j=1}^{\nu} \left(1 + \frac{\gamma}{4}\lambda_j\right) = 1 + \frac{\gamma}{4}\sum_{j=1}^{\nu}\lambda_j + \ldots + \left(\frac{\gamma}{4}\right)^{\nu}\prod_{j=1}^{\nu}\lambda_j$$

$$= 1 + \frac{\gamma}{4}\text{tr}(\mathbf{A}) + \ldots + \left(\frac{\gamma}{4}\right)^{\nu}\det(\mathbf{A}), \tag{11.79}$$

stating that for low values of γ ($\gamma \ll 1$), the PEP is governed essentially by $\text{tr}(\mathbf{A})$ instead of by $\det(\mathbf{A})$. Note that:

$$\text{tr}(\mathbf{A}) = \text{tr}[(\hat{\mathbf{C}} - \mathbf{C})(\hat{\mathbf{C}} - \mathbf{C})^H] = |\hat{\mathbf{C}} - \mathbf{C}|^2 \tag{11.80}$$

is the squared Euclidean distance between $\mathbf{C}$ and $\hat{\mathbf{C}}$. This is somehow expected since for low SNRs the performance is governed by the additive noise rather than the fading. Thus, the error probability curve changes its behavior from a waterfall shape (for small values of γ) to a linear shape (for high values of γ) where the performance is governed by the above-mentioned rank–determinant design principles.

For large values of νn_R, say $\nu_{\min} n_R \geq 4$, this linear behavior is observed for error probability values so small that a code design based on the asymptotic behavior is highly suboptimal for the error probability values of interest [16, 1887]. For these values, the PEP can be obtained by examining the asymptotic behavior for $\nu n_R \to \infty$. Let us return to (11.74) and consider that, for the law of large numbers,[12]:

$$|\mathbf{H}(\hat{\mathbf{C}} - \mathbf{C})|^2 \to n_R \sum_{j=1}^{\nu} \lambda_j = n_R \text{tr}(\mathbf{A}) = n_R |\hat{\mathbf{C}} - \mathbf{C}|^2 . \tag{11.81}$$

Hence:

$$\Pr\{\mathbf{C} \to \hat{\mathbf{C}}\} \leq \frac{1}{2} \exp\left\{-\frac{\gamma n_R |\hat{\mathbf{C}} - \mathbf{C}|^2}{4}\right\} . \tag{11.82}$$

---

[11] Note that $\mathbf{A}$ and $\hat{\mathbf{C}} - \mathbf{C}$ have the same rank. Hence, this criterion could be equivalently expressed with reference to matrix $\hat{\mathbf{C}} - \mathbf{C}$.

[12] Other design criteria are discussed in [828, 1888].

The following alternative code design principle thus results:

Euclidean distance criterion. When the matrix:

$$\mathbf{A} = (\hat{\mathbf{C}} - \mathbf{C})(\hat{\mathbf{C}} - \mathbf{C})^H$$

has rank at least 4 for all pairs of distinct codewords $\hat{\mathbf{C}}$ and $\mathbf{C}$, the minimum trace of matrices $\mathbf{A}$, which is the minimum squared Euclidean distance between $\hat{\mathbf{C}}$ and $\mathbf{C}$, should be maximized.

In practice, when $\nu n_R \geq 4$, an optimum code for AWGN channels, whose codewords are properly formatted in $n_T \times N$ matrices, can be adopted.

11.5.1.3 ST Codeword Design Criteria for Fast Fading Channels

Let us now consider the case of a *fast* frequency-flat fading channel. Under the assumption that the receiver perfectly knows the sequence of channel matrices $\{\mathbf{H}_\ell\}$, the decision rule becomes:

$$\hat{\mathbf{C}} = \arg \min_{\tilde{\mathbf{C}}} \sum_{\ell=1}^{N} |\boldsymbol{\rho}_\ell - \sqrt{\gamma} \mathbf{H}_\ell \tilde{\mathbf{c}}_\ell|^2 \tag{11.83}$$

and the PEP, given the channel realization, is then:

$$\Pr\{\mathbf{C} \to \hat{\mathbf{C}}|\{\mathbf{H}_\ell\}\} = Q\left(\sqrt{\frac{\gamma}{2} \sum_{\ell=1}^{N} |\mathbf{H}_\ell(\hat{\mathbf{c}}_\ell - \mathbf{c}_\ell)|}\right) \leq \frac{1}{2} \exp\left\{-\frac{\gamma}{4} \sum_{\ell=1}^{N} |\mathbf{H}_\ell(\hat{\mathbf{c}}_\ell - \mathbf{c}_\ell)|^2\right\}, \tag{11.84}$$

where we may write:

$$|\mathbf{H}_\ell(\hat{\mathbf{c}}_\ell - \mathbf{c}_\ell)|^2 = \text{tr}[\mathbf{H}_\ell(\hat{\mathbf{c}}_\ell - \mathbf{c}_\ell)(\hat{\mathbf{c}}_\ell - \mathbf{c}_\ell)^H \mathbf{H}_\ell^H]. \tag{11.85}$$

Let us now consider the matrix $(\hat{\mathbf{c}}_\ell - \mathbf{c}_\ell)(\hat{\mathbf{c}}_\ell - \mathbf{c}_\ell)^H$. Since it is a nonnegative definite Hermitian matrix, it can be diagonalized using a unitary matrix $\mathbf{U}_\ell$, $(\hat{\mathbf{c}}_\ell - \mathbf{c}_\ell)(\hat{\mathbf{c}}_\ell - \mathbf{c}_\ell)^H = \mathbf{U}_\ell \boldsymbol{\Lambda}_\ell \mathbf{U}_\ell^H$, where $\boldsymbol{\Lambda}_\ell$ is a diagonal matrix whose elements are the nonnegative eigenvalues of $(\hat{\mathbf{c}}_\ell - \mathbf{c}_\ell)(\hat{\mathbf{c}}_\ell - \mathbf{c}_\ell)^H$. However, given that $(\hat{\mathbf{c}}_\ell - \mathbf{c}_\ell)(\hat{\mathbf{c}}_\ell - \mathbf{c}_\ell)^H$ is of rank 1, only one nonzero eigenvalue results. Let us denote this eigenvalue by $\lambda_{1,\ell}$ and the corresponding eigenvector by $\mathbf{u}_{1,\ell}$. Using the property that the sum of the eigenvalues is equal to the trace of the matrix, we have that:

$$\lambda_{1,\ell} = \text{tr}[(\hat{\mathbf{c}}_\ell - \mathbf{c}_\ell)(\hat{\mathbf{c}}_\ell - \mathbf{c}_\ell)^H] = |\hat{\mathbf{c}}_\ell - \mathbf{c}_\ell|^2. \tag{11.86}$$

Hence:

$$|\mathbf{H}_\ell(\hat{\mathbf{c}}_\ell - \mathbf{c}_\ell)|^2 = \text{tr}[\mathbf{p}_{1,\ell}\lambda_{1,\ell}\mathbf{p}_{1,\ell}^H] = |\hat{\mathbf{c}}_\ell - \mathbf{c}_\ell|^2 |\mathbf{p}_{1,\ell}|^2, \tag{11.87}$$

having defined $\mathbf{p}_{1,\ell} \triangleq \mathbf{H}_\ell \mathbf{u}_{1,\ell}$, and:

$$\Pr\{\mathbf{C} \to \hat{\mathbf{C}}|\{\mathbf{H}_\ell\}\} \leq \frac{1}{2} \exp\left\{-\frac{\gamma}{4} \sum_{\ell=1}^{N} |\hat{\mathbf{c}}_\ell - \mathbf{c}_\ell|^2 |\mathbf{p}_{1,\ell}|^2\right\}$$

$$= \frac{1}{2} \prod_{\ell=1}^{N} \exp\left\{-\frac{\gamma}{4} |\hat{\mathbf{c}}_\ell - \mathbf{c}_\ell|^2 |\mathbf{p}_{1,\ell}|^2\right\}$$

$$= \frac{1}{2} \prod_{\ell \in I} \exp\left\{-\frac{\gamma}{4} |\hat{\mathbf{c}}_\ell - \mathbf{c}_\ell|^2 |\mathbf{p}_{1,\ell}|^2\right\}, \tag{11.88}$$

where, as in Section 11.2.1.2, we denote by I the set of all $1 \leq \ell \leq N$ such that $|\hat{\mathbf{c}}_\ell - \mathbf{c}_\ell| \neq 0$. Since $\mathbf{u}_{1,\ell}$ is an eigenvector, $\mathbf{p}_{1,\ell}$ is statistically equivalent to one column of $\mathbf{H}_\ell$, that is, its components are independent complex Gaussian random variables with zero mean and unit variance.

Let us now assume that the channel coefficients $\{\mathbf{H}_\ell\}$ are independent of each other (*fast* fading or ideal channel interleaving). To compute the *average* PEP we will consider the cases of small and large values of $|I|n_R$. In the former case, the *average* PEP is easily obtained as:

$$\Pr\{\mathbf{C} \to \hat{\mathbf{C}}\} \leq \frac{1}{2} \prod_{\ell \in I} \left(1 + \frac{\gamma}{4}|\hat{\mathbf{c}}_\ell - \mathbf{c}_\ell|^2\right)^{-n_R}, \tag{11.89}$$

which asymptotically (i.e., for high SNRs) becomes:

$$\Pr\{\mathbf{C} \to \hat{\mathbf{C}}\} \lesssim \frac{1}{2} \left(\frac{\gamma}{4}\right)^{-|I|n_R} \left(\prod_{\ell \in I} |\hat{\mathbf{c}}_\ell - \mathbf{c}_\ell|^2\right)^{-n_R}. \tag{11.90}$$

Hence, the basic code design principles over fast frequency-flat fading channels are as follows:

Code diversity criterion. The minimum diversity $|I|$ between all pairs of distinct codewords must be maximized.

Coding gain criterion. In order to obtain the maximum possible coding advantage, the coding gain $\prod_{\ell \in I} |\hat{\mathbf{c}}_\ell - \mathbf{c}_\ell|^2$ over error events having minimum diversity should be maximized.

For large values of $|I|n_R$, the *average* PEP can be obtained by examining the asymptotic behavior for $|I|n_R \to \infty$. From (11.87) and the law of large numbers:

$$|\mathbf{H}_\ell(\hat{\mathbf{c}}_\ell - \mathbf{c}_\ell)|^2 \to |\hat{\mathbf{c}}_\ell - \mathbf{c}_\ell|^2 n_R \tag{11.91}$$

and

$$\Pr\{\mathbf{C} \to \hat{\mathbf{C}}\} \leq \frac{1}{2} \exp\left\{-\frac{\gamma}{4}n_R \sum_{\ell \in I} |\hat{\mathbf{c}}_\ell - \mathbf{c}_\ell|^2\right\}. \tag{11.92}$$

The following alternative code design principle thus results:

Euclidean distance criterion. For large values of the product between the number of receive antennas n_R and the minimum diversity $|I|$ between all pairs of distinct codewords, the minimum Euclidean distance:

$$\sqrt{\sum_{\ell \in I} |\hat{\mathbf{c}}_\ell - \mathbf{c}_\ell|^2}$$

between all pairs of distinct codewords should be maximized.

11.5.1.4 First Naive Scheme: Delay Diversity

One of the first ST codes proposed for *quasi-static* fading channels is the *delay diversity scheme* [1889]. This scheme employs a rate-$1/n_T$ repetition code where each symbol is transmitted from a different antenna after being delayed. In other words, assuming, for example, $n_T = 2$, this scheme transmits the same information from both antennas simultaneously but with a delay of one symbol. The codewords are thus of the form:

$$\mathbf{C} = \begin{bmatrix} c_1 & c_2 & c_3 & \cdots \\ 0 & c_1 & c_2 & \cdots \end{bmatrix}. \tag{11.93}$$

Although not optimized in the sense of the determinant criterion, it is easy to verify that for all the pairs of distinct codewords, the matrix $\hat{\mathbf{C}} - \mathbf{C}$ always has rank n_T. Hence, the maximum diversity of $n_T n_R$ is obtained. This is also intuitive since each symbol traverses $n_T n_R$ paths. This maximum diversity is obtained at the cost of having a rate of only one symbol per channel use. In practice, this scheme transforms the frequency-flat channel into a channel with intersymbol interference (and hence a frequency-selective channel). Optimal decoding may be performed by using the VA or through the suboptimal reduced-complexity schemes described in Section 11.5.1.7.

11.5.1.5   Space-Time Block Codes

These codes were introduced to provide transmit diversity for *quasi-static* frequency-flat fading channels. When employed with multiple receive antennas, receive diversity is also obtained in addition to transmit diversity. The first STBC was that proposed by S. M. Alamouti [15] for the case of two transmit antennas. Before describing it in detail, let us consider the case of an uncoded system with receive diversity only ($n_T = 1$) shown in Figure 11.16. This system will be employed for a comparison with the systems described later.

From (11.64), the channel model becomes:

$$\mathbf{r}_\ell = \sqrt{\gamma}\mathbf{h}_\ell c_\ell + \mathbf{n}_\ell, \tag{11.94}$$

with $\ell = 1, 2, \ldots, N$, since the channel matrix is now a vector $\mathbf{h}_\ell$ of n_R components. The optimal detection strategy, under the assumption of perfect knowledge of the channel coefficients, can easily be derived (see also Section 11.2.3 for the case of a coded transmission). Since the system is memoryless (uncoded system and perfect channel knowledge), we have:

$$\begin{aligned}
\hat{c}_\ell &= \arg\min_{\tilde{c}_\ell} |\boldsymbol{\rho}_\ell - \sqrt{\gamma}\mathbf{h}_\ell \tilde{c}_\ell|^2 \\
&= \arg\min_{\tilde{c}_\ell} |\mathbf{h}_\ell^H \boldsymbol{\rho}_\ell - \sqrt{\gamma}|\mathbf{h}_\ell|^2 \tilde{c}_\ell|^2 \\
&= \arg\max_{\tilde{c}_\ell} \left\{ \text{Re}[\mathbf{h}_\ell^H \boldsymbol{\rho}_\ell \tilde{c}_\ell^*] - \frac{\sqrt{\gamma}}{2}|\mathbf{h}_\ell|^2|\tilde{c}_\ell|^2 \right\}.
\end{aligned} \tag{11.95}$$

This expression show that the optimal decision rule linearly combines the received samples of different antennas after co-phasing and weighting them with their respective channel gains. Samples from

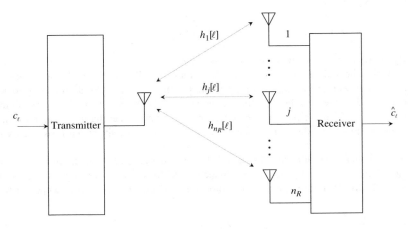

Figure 11.16   System with receive diversity.

antennas experiencing better channel gains (and thus higher SNRs) are emphasized more than others, and this is intuitive since they are more reliable. This detection strategy is commonly known as MRC detection (see Section 6.2.1.6).

It is easy to verify that it is the same optimal strategy found for the equivalent SISO channel:

$$\check{r}_\ell = \mathbf{h}_\ell^H \mathbf{r}_\ell = \sqrt{\gamma} |\mathbf{h}_\ell|^2 c_\ell + \check{n}_\ell \tag{11.96}$$

where, given the channel gains, the noise term $\check{n}_\ell$ is still Gaussian with variance $|\mathbf{h}_\ell|^2$. Under the hypothesis that the components of $\mathbf{h}_\ell$ are iid Gaussian random variables with zero mean and unit variance (Rayleigh fading environment), the random variable $\alpha_\ell \triangleq \gamma |\mathbf{h}_\ell|^2$, representing the instantaneous SNR, is chi-square distributed with $2n_R$ degrees of freedom [55], so that its pdf is given by:

$$f_\alpha(x) = \frac{x^{n_R-1}}{\gamma^{n_R}(n_R-1)!} \exp\left(-\frac{x}{\gamma}\right) u(x). \tag{11.97}$$

The average symbol error probability can thus easily be computed. From the equivalent channel model (11.96), considering, as an example, BPSK modulation whose bit error probability for a given value of the instantaneous SNR is $Q(\sqrt{2\alpha})$, we obtain the average bit error probability:

$$P_b = \int_{-\infty}^{\infty} Q(\sqrt{2x}) f_\alpha(x) dx. \tag{11.98}$$

A closed-form expression for this probability exists and may be written as [1422, p. 781]:

$$P_b = \left[\frac{1}{2}\left(1 - \sqrt{\frac{\gamma}{1+\gamma}}\right)\right]^{n_R} \sum_{m=0}^{n_R-1} \binom{n_R-1+m}{m} \left[\frac{1}{2}\left(1 + \sqrt{\frac{\gamma}{1+\gamma}}\right)\right]^m. \tag{11.99}$$

A simpler upper bound can be found by using the fact that $Q(\sqrt{2\alpha}) \leq \frac{1}{2}\exp(-\alpha)$ (see (F.15)):

$$P_b \leq \frac{1}{2} \int_{-\infty}^{\infty} \exp(-x) f_\alpha(x) dx = \frac{1}{2}\frac{1}{(1+\gamma)^{n_R}}, \tag{11.100}$$

which for $\gamma \to \infty$ gives:

$$P_b \lesssim \frac{1}{2}\frac{1}{\gamma^{n_R}}. \tag{11.101}$$

This clearly shows that a diversity order of n_R is achieved.

The motivation behind the Alamouti scheme is thus as follows. In a cellular system, the base station can easily be equipped with multiple antennas with sufficient separation among them. Hence, the technique just described can be conveniently adopted in the uplink. In contrast, since at the mobile terminal it is difficult to place multiple antennas, receive diversity can hardly be employed. The aim of the scheme proposed by Alamouti is thus to obtain transmit diversity when there are two transmit antennas.

STBCs are a generalization of the Alamouti scheme to the case of $n_T > 2$. Although they provide full diversity, there is no coding advantage provided by STBCs.[13] However, optimal decoding can be performed efficiently through a simple linear processing of the samples at the output of the receive antennas.

---

[13] To achieve an additional coding gain, one should concatenate an outer code with an inner STBC [1890–1892].

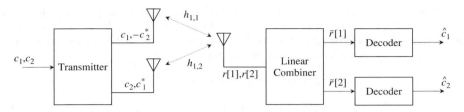

Figure 11.17   Alamouti scheme with $n_T = 2$ and $n_R = 1$.

The Alamouti scheme

Let us consider now the case of a channel with $n_T = 2$ transmit antennas and $n_R = 1$ receive antenna shown in Figure 11.17. The codewords have length $N = 2$ and the channel, perfectly known at the receiver, is assumed to remain the same over two consecutive time intervals (*quasi-static* fading over $N = 2$ symbol intervals). In the two considered symbol intervals considered, it will be described by the row vector:

$$\mathbf{H} \triangleq [h_{1,1}, h_{1,2}]. \tag{11.102}$$

The codeword matrices are of the form:

$$\mathbf{C} = \begin{bmatrix} c_1 & -c_2^* \\ c_2 & c_1^* \end{bmatrix} \tag{11.103}$$

meaning that, during the first interval, symbol c_1 is transmitted from the first antenna and symbol c_2 from the second antenna whereas, during the second interval, symbol $-c_2^*$ is transmitted from the first antenna and symbol c_1^* from the second antenna. A rate of one symbol per channel use is thus achieved. The corresponding received samples in the two intervals are:

$$r[1] = \sqrt{\gamma}(h_{1,1}c_1 + h_{1,2}c_2) + n[1],$$
$$r[2] = \sqrt{\gamma}(-h_{1,1}c_2^* + h_{1,2}c_1^*) + n[2], \tag{11.104}$$

where $n[1]$ and $n[2]$ are independent AGN samples (each having zero mean and unit variance). Then, if we consider the vector $\check{\mathbf{r}} \triangleq [r[1], r[2]^*]^T$, it can be expressed as:

$$\check{\mathbf{r}} = \sqrt{\gamma}\,\check{\mathbf{H}} \begin{bmatrix} c_1 \\ c_2 \end{bmatrix} + \check{\mathbf{n}}, \tag{11.105}$$

where

$$\check{\mathbf{H}} \triangleq \begin{bmatrix} h_{1,1} & h_{1,2} \\ h_{1,2}^* & -h_{1,1}^* \end{bmatrix} \tag{11.106}$$

and $\check{\mathbf{n}} \triangleq [n[1], n[2]^*]^T$ is statistically equivalent to the vector $[n[1], n[2]]^T$. An alternative set of sufficient statistics is represented by the vector:

$$\tilde{\mathbf{r}} = [\tilde{r}[1], \tilde{r}[2]]^T \triangleq \check{\mathbf{H}}^H \check{\mathbf{r}}, \tag{11.107}$$

since it can be obtained through a linear transformation of the vector $\check{\mathbf{r}}$. It is easy to verify that $\check{\mathbf{H}}^H \check{\mathbf{H}} = (|h_{1,1}|^2 + |h_{1,2}|^2)\mathbf{I}_2$, so that, denoting $\tilde{\mathbf{n}} = [\tilde{n}[1], \tilde{n}[2]]^T \triangleq \check{\mathbf{H}}^H \check{\mathbf{n}}$, we have:

$$\tilde{r}[1] = h_{1,1}^* r[1] + h_{1,2} r[2]^* = \sqrt{\gamma}(|h_{1,1}|^2 + |h_{1,2}|^2)c_1 + \tilde{n}[1],$$
$$\tilde{r}[2] = h_{1,2} r[1] - h_{1,1} r[2]^* = \sqrt{\gamma}(|h_{1,1}|^2 + |h_{1,2}|^2)c_2 + \tilde{n}[2]. \tag{11.108}$$

As mentioned, the channel is assumed perfectly known at the receiver and, given the channel coefficients, $\tilde{n}$ is still a Gaussian vector with uncorrelated (since $\check{H}^H \check{H} = (|h_{1,1}|^2 + |h_{1,2}|^2)I_2$) and thus independent components having zero mean and variance $(|h_{1,1}|^2 + |h_{1,2}|^2)$. Decisions on the symbols c_1 and c_2 can thus be obtained by adopting the symbol-by-symbol rules:

$$\hat{c}_1 = \arg \min_{\tilde{c}_1} |\tilde{r}[1] - \sqrt{\gamma}(|h_{1,1}|^2 + |h_{1,2}|^2)\tilde{c}_1|,$$

$$\hat{c}_2 = \arg \min_{\tilde{c}_2} |\tilde{r}[2] - \sqrt{\gamma}(|h_{1,1}|^2 + |h_{1,2}|^2)\tilde{c}_2| . \tag{11.109}$$

In other words, after proper linear combining of the received samples, detection of the symbols c_1 and c_2 can be decoupled. For this reason, the Alamouti scheme is called an *orthogonal design*.

This scheme can be generalized to the case of multiple receive antennas. We denote two consecutive samples at the output of the ith antenna by $r_i[1]$ and $r_i[2]$ (with $i = 1, 2, \ldots, n_R$). We follow the same steps as for $n_R = 1$. Then, after linear combining and normalization, we have the samples:

$$\tilde{r}_i[1] = h_{i,1}^* r_i[1] + h_{i,2} r_i[2]^* = \sqrt{\gamma}(|h_{i,1}|^2 + |h_{i,2}|^2)c_1 + \tilde{n}_i[1] \tag{11.110}$$

and

$$\tilde{r}_i[2] = h_{i,2} r_i[1] - h_{i,1} r_i[2]^* = \sqrt{\gamma}(|h_{i,1}|^2 + |h_{i,2}|^2)c_2 + \tilde{n}_i[2], \tag{11.111}$$

where $\tilde{n}_i[1]$ and $\tilde{n}_i[2]$ are independent AGN samples having variance $(|h_{i,1}|^2 + |h_{i,2}|^2)$. Optimal decisions on the symbols c_1 and c_2 can thus be obtained through MRC. Straightforward manipulations lead to:

$$\hat{c}_1 = \arg \max_{\tilde{c}_1} f_{\tilde{r}[1]}(\tilde{\rho}_1[1], \tilde{\rho}_2[1], \ldots, \tilde{\rho}_{n_R}[1]|\tilde{c}_1, h_{1,1}, h_{1,2}, h_{2,1}, h_{2,2}, \ldots, h_{n_R,1}, h_{n_R,2})$$

$$= \arg \max_{\tilde{c}_1} \prod_{i=1}^{n_R} f_{\tilde{r}_i[1]}(\tilde{\rho}_i[1]|\tilde{c}_1, h_{i,1}, h_{i,2})$$

$$= \arg \min_{\tilde{c}_1} \left| \sum_{i=1}^{n_R} [\tilde{\rho}_i[1] - \sqrt{\gamma}\tilde{c}_1(|h_{i,1}|^2 + |h_{i,2}|^2)] \right|^2 \tag{11.112}$$

and

$$\hat{c}_2 = \arg \max_{\tilde{c}_2} f_{\tilde{r}[2]}(\tilde{\rho}_1[2], \tilde{\rho}_2[2], \ldots, \tilde{\rho}_{n_R}[2]|\tilde{c}_2, h_{1,1}, h_{1,2}, h_{2,1}, h_{2,2}, \ldots, h_{n_R,1}, h_{n_R,2})$$

$$= \arg \max_{\tilde{c}_2} \prod_{i=1}^{n_R} f_{\tilde{r}_i[2]}(\tilde{\rho}_i[2]|\tilde{c}_2, h_{i,1}, h_{i,2})$$

$$= \arg \min_{\tilde{c}_2} \left| \sum_{i=1}^{n_R} [\tilde{\rho}_i[2] - \sqrt{\gamma}\tilde{c}_2(|h_{i,1}|^2 + |h_{i,2}|^2)] \right|^2 \tag{11.113}$$

where $\tilde{r}[\ell] \triangleq [\tilde{r}_1[\ell], \tilde{r}_2[\ell], \ldots, \tilde{r}_{n_R}[\ell]]^T$ (with $\ell = 1, 2$). Again, the decisions are decoupled.

Performance analysis of this scheme is quite simple. From (11.112) and (11.113), it is clear that a decision on symbol c_ℓ, $\ell = 1, 2$, is obtained from the equivalent SISO channel:

$$\bar{r}[\ell] = \sum_{i=1}^{n_R} \tilde{r}_i[\ell] = \sqrt{\gamma}c_\ell \sum_{i=1}^{n_R} (|h_{i,1}|^2 + |h_{i,2}|^2) + \bar{n}[\ell], \tag{11.114}$$

having defined $\bar{n}[\ell] \triangleq \sum_{i=1}^{n_R} \tilde{n}_i[\ell]$. Given the channel gains, samples $\{\bar{n}[\ell]\}$ are jointly Gaussian, independent and have variance $\sum_{i=1}^{n_R}(|h_{i,1}|^2 + |h_{i,2}|^2)$. Comparing (11.114) with (11.96), it is thus clear that the Alamouti scheme with $n_T = 2$ transmit antennas and n_R receive antennas is perfectly equivalent to a scheme with $n_T = 1$ transmit antenna and $2n_R$ receive antennas and using MRC, provided that the same value of γ is employed, that is, provided that the same power per transmit antenna is spent (meaning that for an equal overall transmitted power, the performance of the Alamouti scheme exhibits a degradation of 3 dB). It is thus also clear that the Alamouti scheme achieves full diversity (diversity $2n_R$). This can easily be verified by considering two distinct codewords:

$$\mathbf{C} = \begin{bmatrix} c_1 & -c_2^* \\ c_2 & c_1^* \end{bmatrix}, \quad \hat{\mathbf{C}} = \begin{bmatrix} \hat{c}_1 & -\hat{c}_2^* \\ \hat{c}_2 & \hat{c}_1^* \end{bmatrix} \tag{11.115}$$

and computing the matrix:

$$\mathbf{A} = (\hat{\mathbf{C}} - \mathbf{C})(\hat{\mathbf{C}} - \mathbf{C})^H = \begin{bmatrix} |\hat{c}_1 - c_1|^2 + |\hat{c}_2 - c_2|^2 & 0 \\ 0 & |\hat{c}_1 - c_1|^2 + |\hat{c}_2 - c_2|^2 \end{bmatrix}, \tag{11.116}$$

which clearly has full diversity provided that $\hat{\mathbf{C}} \neq \mathbf{C}$.

Orthogonal STBCs

The Alamouti scheme was designed for $n_T = 2$ transmit antennas. OSTBCs [1482, 1483] extend it to the case $n_T > 2$.

In the general case of n_T transmit antennas, in order to design a code with a rate of 1 symbol per channel use and full diversity, we need to design a set of $n_T \times n_T$ (square) matrices, with elements from the employed constellation, whose rows are orthogonal to each other. This latter property, in fact, will ensure that an optimal receiver can be designed based on a linear processing plus symbol-by-symbol detection. Unfortunately, it is not always possible to find such an orthogonal design. For real constellations (e.g., M-PAM), it exists for $n_T = 2, 4, 8$ only. As an example, for $n_T = 4$, the corresponding orthogonal design is that using codeword matrices of the form:

$$\mathbf{C} = \begin{bmatrix} c_1 & -c_2 & -c_3 & -c_4 \\ c_2 & c_1 & c_4 & -c_3 \\ c_3 & -c_4 & c_1 & c_2 \\ c_4 & c_3 & -c_2 & c_1 \end{bmatrix}. \tag{11.117}$$

It is easy to prove, as was done for the Alamouti code, that it achieves full diversity. It also has a rate of 1 symbol per channel use since four symbols are transmitted in four timeslots.

On the other hand, for complex constellations, there exists a unique full-rate and full-diversity orthogonal design for $n_T = 2$ (that proposed by Alamouti). It is, however, possible to find many other orthogonal designs by removing some of the mentioned constraints. For instance, for $n_T = 4$, the code with codewords:

$$\mathbf{C} = \begin{bmatrix} c_1 & -c_2 & -c_3 & -c_4 & c_1^* & -c_2^* & -c_3^* & -c_4^* \\ c_2 & c_1 & c_4 & -c_3 & c_2^* & c_1^* & c_4^* & -c_3^* \\ c_3 & -c_4 & c_1 & c_2 & c_3^* & -c_4^* & c_1^* & c_2^* \\ c_4 & c_3 & -c_2 & c_1 & c_4^* & c_3^* & -c_2^* & c_1^* \end{bmatrix} \tag{11.118}$$

achieves full diversity (as can easily be proved by computing matrix $\mathbf{A} = (\hat{\mathbf{C}} - \mathbf{C})(\hat{\mathbf{C}} - \mathbf{C})^H$), but has a rate of 1/2 symbol per channel use since four symbols are transmitted in eight timeslots.

A mathematical framework to describe the general class of linear orthogonal designs is provided in [1893]. The $n_T \times N$ matrices $\{\mathbf{C}\}$ describing an orthogonal STBC and used to transmit K symbols (thus achieving a rate of K/N symbols per channel use) can be expressed in the form:

$$\mathbf{C} = \sum_{k=1}^{K} (c_k \mathbf{A}_k + c_k^* \mathbf{B}_k), \tag{11.119}$$

where $\mathbf{A}_k$ and $\mathbf{B}_k$ are appropriate $n_T \times N$ matrices. That is, all elements of $\mathbf{C}$ are linear combinations of the symbols $\{c_k\}_{k=1}^K$ being transmitted and/or their conjugates. As an example, the Alamouti code can be described by using this framework with $n_T = N = K = 2$ and:

$$\mathbf{A}_1 = \begin{bmatrix} 1 & 0 \\ 0 & 0 \end{bmatrix}, \quad \mathbf{A}_2 = \begin{bmatrix} 0 & 0 \\ 1 & 0 \end{bmatrix}, \quad \mathbf{B}_1 = \begin{bmatrix} 0 & 0 \\ 0 & 1 \end{bmatrix}, \quad \mathbf{B}_2 = \begin{bmatrix} 0 & -1 \\ 0 & 0 \end{bmatrix}. \tag{11.120}$$

Clearly, the matrices $\{\mathbf{C}\}$ must satisfy the property that their rows are orthogonal, that is, $\mathbf{CC}^H$ is a diagonal matrix with strictly positive elements. More precisely, the condition:

$$\mathbf{CC}^H = \sum_{k=1}^K \mathbf{D}_k |c_k|^2 \tag{11.121}$$

must hold, where $\mathbf{D}_k$ is a diagonal matrix with strictly positive elements. As demonstrated in [1893], this can be expressed equivalently as the following equalities in terms of matrices $\{\mathbf{A}_k\}$ and $\{\mathbf{B}_k\}$:

$$\mathbf{A}_k \mathbf{A}_m^H + \mathbf{B}_k \mathbf{B}_m^H = \delta_{k,m} \mathbf{D}_k, \tag{11.122}$$

$$\mathbf{A}_k \mathbf{B}_m^H + \mathbf{B}_k \mathbf{A}_m^H = 0, \tag{11.123}$$

where $\delta_{k,m}$ is the Kronecker delta. In the case of the Alamouti code, property (11.121) and conditions (11.122) and (11.123) can easily be verified.

This framework is very useful to describe the decoding algorithm. The samples $\{r_i[\ell], \ell = 1, 2, \ldots, N\}$ received by antenna i can be collected in a row vector:

$$\mathbf{r}_i \triangleq [r_i[1], r_i[2], \ldots, r_i[N]] \tag{11.124}$$

which can be expressed as:

$$\mathbf{r}_i = \sqrt{\gamma} \mathbf{h}_i \mathbf{C} + \mathbf{n}_i, \tag{11.125}$$

where $\mathbf{h}_i$ is the ith row of the channel matrix $\mathbf{H}$ (supposed known at the receiver and constant for N consecutive samples) and $\mathbf{n}_i$ is a row vector of the noise samples at the output of antenna i. In other words, $\mathbf{h}_i$ is a row vector of the channel gains from all transmit antennas to receive antenna i. The detection strategy can be written in the form:

$$\hat{\mathbf{C}} = \arg \min_{\tilde{\mathbf{C}}} \sum_{i=1}^{n_R} |\mathbf{r}_i - \sqrt{\gamma} \mathbf{h}_i \tilde{\mathbf{C}}|^2$$

$$= \arg \min_{\tilde{\mathbf{C}}} \sum_{i=1}^{n_R} [\mathbf{r}_i - \sqrt{\gamma} \mathbf{h}_i \tilde{\mathbf{C}}][\mathbf{r}_i - \sqrt{\gamma} \mathbf{h}_i \tilde{\mathbf{C}}]^H$$

$$= \arg \min_{\tilde{\mathbf{C}}} \sum_{i=1}^{n_R} [\gamma \mathbf{h}_i \tilde{\mathbf{C}} \tilde{\mathbf{C}}^H \mathbf{h}_i^H - 2\sqrt{\gamma} \mathrm{Re}\,(\mathbf{r}_i \tilde{\mathbf{C}}^H \mathbf{h}_i^H)]. \tag{11.126}$$

Taking into account (11.119) and (11.121), we obtain:

$$\hat{\mathbf{C}} = \arg \min_{\tilde{\mathbf{C}}} \sum_{i=1}^{n_R} \sum_{k=1}^K [\gamma \mathbf{h}_i \mathbf{D}_k \mathbf{h}_i^H |\tilde{c}_k|^2 - 2\sqrt{\gamma} \mathrm{Re}(\mathbf{r}_i \mathbf{A}_k^H \mathbf{h}_i^H \tilde{c}_k^* + \mathbf{r}_i \mathbf{B}_k^H \mathbf{h}_i^H \tilde{c}_k)]$$

$$= \arg \min_{\tilde{\mathbf{C}}} \sum_{i=1}^{n_R} \sum_{k=1}^K [\gamma \mathbf{h}_i \mathbf{D}_k \mathbf{h}_i^H |\tilde{c}_k|^2 - 2\sqrt{\gamma} \mathrm{Re}(\mathbf{r}_i \mathbf{A}_k^H \mathbf{h}_i^H \tilde{c}_k^* + \mathbf{h}_i \mathbf{B}_k \mathbf{r}_i^H \tilde{c}_k^*)]. \tag{11.127}$$

It is thus clear that decisions on symbols $\{c_k\}_{k=1}^K$ can be decoupled in the following symbol-by-symbol rules:

$$\hat{c}_k = \arg\min_{\tilde{c}_k} \sum_{i=1}^{n_R} [\gamma \mathbf{h}_i \mathbf{D}_k \mathbf{h}_i^H |\tilde{c}_k|^2 - 2\sqrt{\gamma} \mathrm{Re}(\mathbf{r}_i \mathbf{A}_k^H \mathbf{h}_i^H \tilde{c}_k^* + \mathbf{h}_i \mathbf{B}_k \mathbf{r}_i^H \tilde{c}_k^*)]$$

$$= \arg\min_{\tilde{c}_k} |\tilde{r}_k - \sqrt{\gamma} \xi_k^2 \tilde{c}_k|^2 \tag{11.128}$$

with $k = 1, 2 \ldots, K$, having defined:

$$\tilde{r}_k \triangleq \sum_{i=1}^{n_R} (\mathbf{r}_i \mathbf{A}_k^H \mathbf{h}_i^H + \mathbf{h}_i \mathbf{B}_k \mathbf{r}_i^H) \tag{11.129}$$

and

$$\xi_k^2 \triangleq \sum_{i=1}^{n_R} \mathbf{h}_i \mathbf{D}_k \mathbf{h}_i^H . \tag{11.130}$$

This detection strategy is that corresponding to an equivalent SISO channel. In fact, substituting (11.125) into (11.129), and using (11.119), (11.122), and (11.123), we obtain:

$$\tilde{r}_k = \sum_{i=1}^{n_R} (\mathbf{r}_i \mathbf{A}_k^H \mathbf{h}_i^H + \mathbf{h}_i \mathbf{B}_k \mathbf{r}_i^H)$$

$$= \sum_{i=1}^{n_R} [(\sqrt{\gamma} \mathbf{h}_i \mathbf{C} + \mathbf{n}_i) \mathbf{A}_k^H \mathbf{h}_i^H + \mathbf{h}_i \mathbf{B}_k (\sqrt{\gamma} \mathbf{h}_i \mathbf{C} + \mathbf{n}_i)^H]$$

$$= \sqrt{\gamma} c_k \sum_{i=1}^{n_R} [\mathbf{h}_i \mathbf{D}_k \mathbf{h}_i^H] + \sum_{i=1}^{n_R} [\mathbf{n}_i \mathbf{A}_k^H \mathbf{h}_i^H + \mathbf{h}_i \mathbf{B}_k \mathbf{n}_i^H]$$

$$= \sqrt{\gamma} c_k \xi_k^2 + \tilde{n}_k, \tag{11.131}$$

having defined:

$$\tilde{n}_k \triangleq \sum_{i=1}^{n_R} [\mathbf{n}_i \mathbf{A}_k^H \mathbf{h}_i^H + \mathbf{h}_i \mathbf{B}_k \mathbf{n}_i^H], \tag{11.132}$$

whose variance is ξ_k^2, given $\mathbf{h}_i$ and taking into account that the noise samples at the output of antenna i are uncorrelated and have unit variance. Hence, the detection strategy (11.128) can be considered as derived from the equivalent SISO channel model (11.131) and the performance analysis carried out accordingly as for the Alamouti scheme, easily verifying that these schemes achieve full diversity. This can also be verified by considering two distinct codewords:

$$\mathbf{C} = \sum_{k=1}^K (c_k \mathbf{A}_k + c_k^* \mathbf{B}_k) \tag{11.133}$$

and

$$\hat{\mathbf{C}} = \sum_{k=1}^K (\hat{c}_k \mathbf{A}_k + \hat{c}_k^* \mathbf{B}_k), \tag{11.134}$$

and verifying that the matrix $(\hat{\mathbf{C}} - \mathbf{C})(\hat{\mathbf{C}} - \mathbf{C})^H = \sum_{k=1}^K |\hat{c}_k - c_k|^2 \mathbf{D}_k$ has full rank, provided that $\hat{\mathbf{C}} \neq \mathbf{C}$.

Quasi-orthogonal STBCs

As mentioned, for complex constellations, the only orthogonal design is that proposed by Alamouti for $n_T = 2$. It provides full diversity and transmission rate of 1 symbol per channel use. STBCs with rate of 1 symbol per channel use can be obtained, as proposed in [1894], by using Alamouti's orthogonal design as a building block (or another orthogonal design in the case of real constellations), but clearly giving up the orthogonality of the resulting STC and only providing partial diversity.

To illustrate the main ideas behind this "quasi-orthogonal" design, let us consider how to design an STBC for $n_T = 4$ by properly employing two Alamouti codewords, one denoted by $\mathbf{C}_{12}$ for transmitting symbols c_1 and c_2:

$$\mathbf{C}_{12} = \begin{bmatrix} c_1 & -c_2^* \\ c_2 & c_1^* \end{bmatrix}$$ (11.135)

and a second denoted by $\mathbf{C}_{34}$ for transmitting symbols c_3 and c_4:

$$\mathbf{C}_{34} = \begin{bmatrix} c_3 & -c_4^* \\ c_4 & c_3^* \end{bmatrix} .$$ (11.136)

The resulting codeword will be obtained through an orthogonal design involving the two matrices:

$$\mathbf{C} = \begin{bmatrix} \mathbf{C}_{12} & -\mathbf{C}_{34}^* \\ \mathbf{C}_{34} & \mathbf{C}_{12}^* \end{bmatrix} = \begin{bmatrix} c_1 & -c_2^* & -c_3^* & c_4 \\ c_2 & c_1^* & -c_4^* & -c_3 \\ c_3 & -c_4^* & c_1^* & -c_2 \\ c_4 & c_3^* & c_2^* & c_1 \end{bmatrix} .$$ (11.137)

It is easy to prove that this code does not achieve full diversity (it achieves diversity $2n_R$ when n_R receive antennas are employed). Although not all its rows are orthogonal, we can observe that the first and fourth columns are orthogonal to the second and third. Hence, through appropriate linear processing it is possible to decouple the decisions on symbols c_1 and c_4 from those on symbols c_2 and c_3. The decisions on symbols c_1 and c_4 and those on c_2 and c_3 must be performed jointly, thus increasing the receiver complexity with respect to that of orthogonal STCs.

Linear dispersion codes

The mathematical framework (11.119) employed to describe linear orthogonal designs can be used to describe the quasi-orthogonal STBCs as well as another class of STCs called *linear dispersion codes*. These codes have rate greater than one symbol per channel use since for them we can have $K > N$. Obviously, this time, constraints (11.122) and (11.123) no longer hold and optimal decoding becomes prohibitive. Suboptimal decoding techniques, such as those mentioned in Section 11.5.1.7, can be adopted. Regarding the code design (or, in other words, the design of matrices $\mathbf{A}_k$ and $\mathbf{B}_k$), in [1895] a technique is proposed aimed at maximizing the mutual information between the input and the output of the channel.

11.5.1.6 Space-Time Trellis Codes

Another important class of codes are the STTCs originally proposed in [14]. They are the natural extension of TCMs to MIMO channels – the only difference is that each trellis branch is labeled with a vector of n_T symbols that are transmitted in parallel by the n_T transmit antennas. More precisely, they are multiple TCMs whose trellis branch is associated with n_T symbols belonging to a given M-ary constellation that are transmitted in parallel over the n_T transmit antennas instead of sequentially. Memory is thus introduced with the aim of obtaining a coding advantage in addition to a code diversity at the price of increased decoding complexity.

In general, the code will have n_s states. A rate of η bits per channel use is obtained when a trellis with 2^η branches departing from each state is employed. We already discussed the different design criteria of the ST codewords. STTCs are designed accordingly. In the case of quasi-static fading, for STTCs for $n_T = 2$, two simple design rules allow us to obtain full diversity, in accordance with the rank criterion:

- *Rule 1*. Transitions departing from the same state differ in the second symbol only.
- *Rule 2*. Transitions merging at the same state differ in the first symbol only.

In fact, by following these rules, the error matrix assumes the form (for all $(\hat{\mathbf{C}}, \mathbf{C})$):

$$\hat{\mathbf{C}} - \mathbf{C} = \begin{bmatrix} \cdots & 0 & \cdots & \beta & \cdots \\ \cdots & \alpha & \cdots & 0 & \cdots \end{bmatrix} \tag{11.138}$$

with α and β nonzero complex numbers. Thus, every such error matrix has full rank and the STC achieves *full diversity*. The maximization of the minimum determinant of matrices $\mathbf{A} = (\hat{\mathbf{C}} - \mathbf{C})(\hat{\mathbf{C}} - \mathbf{C})^H$ having minimum rank is a harder task. The code design is therefore performed through a computer search [14][14] or through algebraic techniques [1888, 1899]. In particular, in [1888] for the class of binary and quaternary trellis codes over $\mathbb{Z}_2$ and $\mathbb{Z}_4$ mapped onto BPSK and QPSK, respectively, a condition on the underlying algebraic codes referred to as the *binary-rank* criterion is shown to imply the rank diversity of the resulting STC and it is used to construct STCs with full rank diversity (i.e., with $\nu = n_T$). The binary-rank criterion is much easier to check than the rank diversity and yields some explicit general algebraic constructions.

For quasi-static channels, two examples of good STTCs using the QPSK modulation, with a rate of 2 bits per channel use, $n_T = 2$, and $n_s = 4, 8$, are provided in Figure 11.18. Other examples

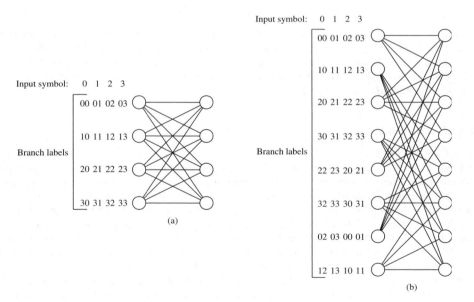

Figure 11.18 Trellis diagrams of two ST trellis codes with (a) $n_s = 4$ and (b) $n_s = 8$ states. The QPSK symbol $\exp(j2\pi i/4)$ (with $i = 0, 1, 2, 3$) is specified through integer i.

---

[14] In [1896], an algebraic representation of STTCs using PSK modulations is provided, with the aim of simplifying the search for good codes. With the same aim, some analytical tools are provided in [1897, 1898] along with new good codes.

can be found in [14, 1888, 1897–1899]. We would also like to mention the new class of STTCs in [1900–1902] which systematically combine set partitioning and a superset of OSTBCs codes to provide full diversity and improved coding gain over earlier STTC constructions.

The optimal detector is based on the VA working on the code trellis. Since the number of trellis branches departing from the same state is 2^η, the larger the rate η, the higher the receiver complexity. Similarly, the larger the number of transmit antennas, the higher the receiver complexity. Hence, for transmissions requiring very high spectral efficiency and/or a large number of transmit antennas, other codes are more appropriate (such as layered ST codes described below).

An upper bound on the error probability can be computed through the union bound technique. However, as mentioned for quasi-static channels, this bound turns out to be loose, especially when the number of antennas is limited. The reason is simple. In the union bound computation, the same contribution is accounted for many times. This does not represent a problem for the AWGN channel since, in this case, the PEP terms decay exponentially and only a few dominant terms exist. On the quasi-static frequency-flat fading channel, however, when the available diversity is limited, PEP terms decay very slowly and, as a consequence, the number of dominant terms is not limited.

A possible solution is represented by the technique described in [1853] for convolutional codes, and applied in [1903] to STTCs. The idea is very simple. Let us assume that we are interested in the computation of an upper bound on the BEP $P_b^{(U)}$ (the same considerations hold for the symbol error probability or the frame error probability). Up to now, the starting point was the computation of the PEP given a channel realization $\mathbf{H}$. This was then averaged over the channel realizations and employed in the union bound for the computation of an upper bound on the bit error probability. However, we can apply the union bound to compute an upper bound on the BEP given the channel realization $P_b^{(U)}(\mathbf{H})$, upper-bound it as unity if it exceeds unity, and then perform the average over the channel realizations:

$$P_b^{(U)} = \mathbf{E}_\mathbf{H}\{\min[1, P_b(\mathbf{H})]\}. \tag{11.139}$$

In other words, we are changing the order of the average and the summation (that for the union bound computation) and, when the channel coefficients are so small that the PEP terms become close to one producing an union bound given $\mathbf{H}$ having a value larger than unity, we trivially upper-bound it as unity, and then average over the channel statistics. The average cannot be now computed in closed form, but Monte Carlo averaging must be used.

11.5.1.7 Layered STCs

BLAST architectures

ST block and trellis codes can achieve full diversity (diversity $n_T n_R$) on quasi-static frequency-flat channels, thus representing an effective way to combat the effects of fading. However, their application is limited to transmissions with a small rate η. In fact, STBCs achieve a rate of at most $\log_2 M$ bits per channel use (where M is the cardinality of the modulation constellation), whereas the complexity of STTCs limits their adoption to applications where a very limited number of bits per channel use is required. It could thus be useful to trade diversity against rate for those wireless applications requiring high data rates. LST architectures, originally proposed by Foschini [17], have been developed for such a purpose and to handle a large number of antennas with limited complexity.

The first and most effective proposed LST architecture is the *diagonal* BLAST (D-BLAST) scheme. We will concentrate mainly on this architecture, describing the encoding procedure and a few suboptimal low-complexity decoding algorithms. We will also mention *horizontal* BLAST (H-BLAST) and *vertical* BLAST (V-BLAST), along with alternative LST architectures such as multilayered ST codes [1904], threaded ST codes [1905], and wrapped ST codes [1906].

In a BLAST architecture, multiple independent coded streams are distributed throughout the transmission resource array in so-called *layers*. Since the complexity of the optimal decoder is impractical, the aim is to design the layering architecture and the associated signal processing so that the receiver

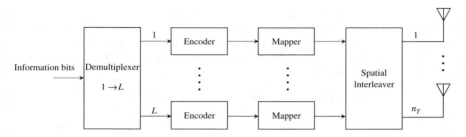

Figure 11.19 D-BLAST encoder.

can efficiently separate the individual layers and decode each of them effectively. In other words, low-complexity suboptimal decoding schemes based on individual decoding of the component codes and mitigation of the mutual interference among component codewords can be adopted.

The block diagram of a D-BLAST encoder is shown in Figure 11.19 [17]. The information bit stream is demultiplexed into L parallel substreams. Each substream is independently encoded and the code bits are mapped onto M-ary symbols belonging to a constellation A_c. The resulting L codewords are collected in the row vectors $\{\mathbf{c}^{(i)}\}_{i=1}^{L}$, of length $N' = n_T d$ symbols. These row vectors are then broken into n_T small subblocks of d symbols each. We will denote by $\mathbf{c}_j^{(i)}$ the row vector representing the jth subblock of codeword $\mathbf{c}^{(i)}$. These subblocks are cyclically assigned by the spatial interleaver to all transmit antennas in such a way that the codewords share a balanced presence over all n_T antennas and none of the individual substreams is hostage to the worst of the n_T paths. In the case of $n_T = 4$, the transmitted $n_T \times N$ codeword matrices, with $N = d(L + n_T - 1)$, thus have the following structure:

$$\mathbf{C} = \begin{bmatrix} \mathbf{c}_1^{(1)} & \mathbf{c}_1^{(2)} & \mathbf{c}_1^{(3)} & \mathbf{c}_1^{(4)} & \mathbf{c}_1^{(5)} & \cdots & \cdots & \cdots & \cdots \\ 0 & \mathbf{c}_2^{(1)} & \mathbf{c}_2^{(2)} & \mathbf{c}_2^{(3)} & \mathbf{c}_2^{(4)} & \mathbf{c}_2^{(5)} & \cdots & \cdots & \cdots \\ 0 & 0 & \mathbf{c}_3^{(1)} & \mathbf{c}_3^{(2)} & \mathbf{c}_3^{(3)} & \mathbf{c}_3^{(4)} & \mathbf{c}_3^{(5)} & \cdots & \cdots \\ 0 & 0 & 0 & \mathbf{c}_4^{(1)} & \mathbf{c}_4^{(2)} & \mathbf{c}_4^{(3)} & \mathbf{c}_4^{(4)} & \mathbf{c}_4^{(5)} & \cdots \end{bmatrix}, \tag{11.140}$$

where the entries below the first diagonal layer are zeros. Symbols belonging to layer i are placed in the entries:

$$\{c_{n,k+(i+n-2)d} \mid k = 1, 2, \ldots, d, \ n = 1, 2, \ldots, n_T\}$$

of matrix $\mathbf{C}$.

Decoding is accomplished through a multiuser detection strategy based on a combination of *cancelation* and *suppression*. Since each diagonal layer constitutes a complete codeword, decoding is performed layer by layer, starting from the first diagonal. The receiver generates a soft decision statistic for each symbol in this diagonal by suppressing the interference from the upper diagonals. This can be obtained by projecting the received signal onto the null space of the upper interference. These soft statistics are then used by the corresponding channel decoder to decode the first codeword. The decoder output is then fed back to cancel the first diagonal contribution in the interference while decoding the next diagonal and so on. This is the so-called ZF suppression strategy that requires $n_R \geq n_T$.

In detail, the receiver is obtained as a linear front-end followed by decision feedback interference cancelation [17, 1905–1907]. Matrix $\mathbf{H}$ is first factores using the so-called QR decomposition [1908] (see (C.14)), that is:

$$\mathbf{H} = \mathbf{QB}, \tag{11.141}$$

where $\mathbf{Q}$ is an $n_R \times n_T$ matrix with orthonormal columns (i.e., $\mathbf{Q}^H\mathbf{Q} = \mathbf{I}_{n_T}$) and $\mathbf{B}$ is an $n_T \times n_T$ upper triangular matrix whose diagonal elements (when $\mathbf{H}$ is nonsingular, which occurs with probability 1 in the case of the Rayleigh channel) are all positive. From the $n_R \times N$ matrix $\mathbf{R}$ of received samples in (11.67), the linear front-end, defined by the $n_T \times n_R$ complex matrix $\mathbf{Q}^H$, yields the alternative sufficient statistic given by the matrix:

$$\mathbf{V} \triangleq \mathbf{Q}^H\mathbf{R} = \sqrt{\gamma}\mathbf{Q}^H\mathbf{HC} + \mathbf{Q}^H\mathbf{N} = \sqrt{\gamma}\mathbf{BC} + \tilde{\mathbf{N}}, \tag{11.142}$$

where $\tilde{\mathbf{N}} \triangleq \mathbf{Q}^H\mathbf{N}$. Let us now consider the (n, ℓ)th element of $\mathbf{V}$. Since the matrix $\mathbf{B}$ has an upper triangular structure, such an element can be expressed as:

$$v_{n,\ell} = \sqrt{\gamma}\sum_{k=n}^{n_T} b_{n,k}c_{k,\ell} + \tilde{n}_{n,\ell}$$

$$= \sqrt{\gamma}b_{n,n}c_{n,\ell} + \sqrt{\gamma}\sum_{k=n+1}^{n_T} b_{n,k}c_{k,\ell} + \tilde{n}_{n,\ell}, \tag{11.143}$$

where we have assumed that $c_{n,\ell}$ is the symbol we wish to detect. As one can observe, the interference of symbols $\{c_{k,\ell}\}_{k=1}^{n-1}$ has been removed. The remaining symbols belong to lower layers. Hence, the samples:

$$v_{n,k+(n-1)d}, \quad k = 1, 2, \ldots, d, \quad n = 1, 2, \ldots, n_T, \tag{11.144}$$

can be used to decode the first layer since no interference from other layers is present. Once this layer has been decoded, the corresponding information bits at the decoder output are encoded again and can thus be subtracted when decoding the second layer. The process will continue layer by layer, using the samples:

$$\hat{v}_{n,k+(i+n-2)d} = v_{n,k+(i+n-2)d} - \sqrt{\gamma}\sum_{k=n+1}^{n_T} b_{n,k}\hat{c}_{k,k+(i+n-2)d} \tag{11.145}$$

$(k = 1, 2, \ldots, d, n = 1, 2, \ldots, n_T)$, where $\{\hat{c}_{k,\ell}\}$ are the decisions on code symbols already taken for the previous layers, to decode layer i. Hence, the decisions needed in (11.145) are provided by earlier decoded codewords (layers). Samples $\hat{v}_{n,k+(i+n-2)d}$ in (11.145) can be expressed as (see (11.143)):

$$\hat{v}_{n,k+(i+n-2)d} = \sqrt{\gamma}b_{n,n}c_{n,k+(i+n-2)d}$$

$$+ \sqrt{\gamma}\sum_{k=n+1}^{n_T} b_{n,k}(c_{k,k+(i+n-2)d} - \hat{c}_{k,k+(i+n-2)d}) + \tilde{n}_{n,k+(i+n-2)d}, \tag{11.146}$$

showing that, when previously decoded codewords are all correct, detection is interference-free.

As mentioned, the ZF suppression strategy requires $n_R \geq n_T$. This requirement can be relaxed, and also a better performance obtained in the same conditions, by using MMSE filtering [1905, 1906]. The linear front-end is defined, in this case, by an $n_T \times n_R$ complex matrix $\mathbf{F}^H$ which produces the alternative sufficient statistic $\mathbf{V} \triangleq \mathbf{F}^H\mathbf{R}$ whose ℓth column is:

$$\mathbf{v}_\ell = \mathbf{F}^H\mathbf{r}_\ell. \tag{11.147}$$

We know that the vector $\mathbf{r}_\ell$ can be expressed as (see (11.64)):

$$\mathbf{r}_\ell = \sqrt{\gamma}\mathbf{Hc}_\ell + \mathbf{n}_\ell = \sqrt{\gamma}\sum_{k=1}^{n_T} \mathbf{h}_k c_{k,\ell} + \mathbf{n}_\ell, \tag{11.148}$$

where $\mathbf{h}_k$ is the kth column of $\mathbf{H}$. Hence, we have that:

$$\mathbf{v}_\ell = \mathbf{F}^H \mathbf{r}_\ell = \mathbf{F}^H (\sqrt{\gamma} \mathbf{H} \mathbf{c}_\ell + \mathbf{n}_\ell) = \sqrt{\gamma} \mathbf{G} \mathbf{c}_\ell + \tilde{\mathbf{n}}_\ell = \sqrt{\gamma} \sum_{k=1}^{n_T} \mathbf{g}_k c_{k,\ell} + \tilde{\mathbf{n}}_\ell, \qquad (11.149)$$

having defined $\tilde{\mathbf{n}}_\ell \triangleq \mathbf{F}^H \mathbf{n}_\ell$ and $\mathbf{G} \triangleq \mathbf{F}^H \mathbf{H}$, while $\mathbf{g}_k = \mathbf{F}^H \mathbf{h}_k$ is the kth column of $\mathbf{G}$. Once the layer i has been detected, the corresponding symbols can be canceled. Let us assume that symbols of layers up to i to be canceled correspond, at discrete time ℓ, to symbols $\{c_{k,\ell}\}_{k=n+1}^{n_T}$. Hence, after cancelation we have the vector:

$$\hat{\mathbf{v}}_\ell = \mathbf{v}_\ell - \sqrt{\gamma} \sum_{k=n+1}^{n_T} \mathbf{g}_k \hat{c}_{k,\ell} = \mathbf{F}^H \left(\mathbf{r}_\ell - \sqrt{\gamma} \sum_{k=n+1}^{n_T} \mathbf{h}_k \hat{c}_{k,\ell} \right) = \mathbf{F}^H \hat{\mathbf{r}}_\ell, \qquad (11.150)$$

having defined:

$$\hat{\mathbf{r}}_\ell \triangleq \mathbf{r}_\ell - \sqrt{\gamma} \sum_{k=n+1}^{n_T} \mathbf{h}_k \hat{c}_{k,\ell}. \qquad (11.151)$$

The vectors $\hat{\mathbf{r}}_\ell$ and $\hat{\mathbf{v}}_\ell$ can be expressed, under the assumption of correct decisions, as:

$$\hat{\mathbf{r}}_\ell = \sqrt{\gamma} \sum_{k=1}^{n} \mathbf{h}_k c_{k,\ell} + \mathbf{n}_\ell \qquad (11.152)$$

and

$$\hat{\mathbf{v}}_\ell = \sqrt{\gamma} \sum_{k=1}^{n} \mathbf{g}_k c_{k,\ell} + \tilde{\mathbf{n}}_\ell, \qquad (11.153)$$

respectively. The (n, ℓ)th element of the $n_T \times N$ matrix $\hat{\mathbf{V}} \triangleq (\hat{\mathbf{v}}_1, \hat{\mathbf{v}}_2, \ldots, \hat{\mathbf{v}}_N)$ can be expressed as:

$$\hat{v}_{n,\ell} = \mathbf{f}_n^H \hat{\mathbf{r}}_\ell, \qquad (11.154)$$

where $\mathbf{f}_n$ is the nth column of $\mathbf{F}$. Since $v_{n,\ell}$ is employed as a soft statistic associated with symbol $c_{n,\ell}$, column $\mathbf{f}_n$ is selected as that minimizing the MSE $E\{|\mathbf{f}_n^H \hat{\mathbf{r}}_\ell - c_{n,\ell}|^2\}$, which under the assumption of correct decisions (i.e., under the assumption that (11.152) holds) can easily be computed in closed form as [430, 596]:

$$\mathbf{f}_n = \sqrt{\gamma} \left(\mathbf{I}_{n_R} + \gamma \sum_{k=1}^{n} \mathbf{h}_k \mathbf{h}_k^H \right)^{-1} \mathbf{h}_n . \qquad (11.155)$$

Notice that, this time, the interference of the upper layers is not removed through filtering, but the joint effect of interference and noise is minimized according to the MMSE criterion. Decoding is accomplished as for the ZF strategy by decoding a layer and canceling it before decoding the next layer.

Other LST architectures have been conceived with the aim of improving performance or reducing the overall receiver complexity. H-BLAST has a structure very similar to D-BLAST, the only difference being the absence of the spatial interleaver. Its encoder is shown in Figure 11.20. As one can observe, in this scheme the number of layers is equal to the number of transmit antennas n_T. In other words, each layer is exclusively associated with a transmit antenna.

Decoding can be again accomplished layer by layer. In the case of the ZF strategy, after the linear front-end, the n_Tth layer is decoded first by using samples:

$$v_{n_T,\ell} = \sqrt{\gamma} b_{n_T,n_T} c_{n_T,\ell} + \tilde{n}_{n_T,\ell} \qquad (11.156)$$

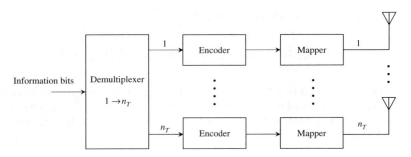

Figure 11.20   H-BLAST encoder.

of matrix $\mathbf{V}$ in (11.142) and, in general, the jth layer by using samples:

$$\hat{v}_{j,\ell} = v_{j,\ell} - \sqrt{\gamma} \sum_{k=j+1}^{n_T} b_{j,k}\hat{c}_{k,\ell}$$

$$= \sqrt{\gamma}b_{j,j}c_{j,\ell} + \sqrt{\gamma} \sum_{k=j+1}^{n_T} b_{j,k}(c_{k,\ell} - \hat{c}_{k,\ell}) + \tilde{n}_{j,\ell} \, . \tag{11.157}$$

It may be noticed that different layers are decoded with different reliability. In particular, the last detected layer has the highest reliability since, for it, the contribution of all other layers has been canceled. A way to overcome this problem is to sort the received sequences starting detection from that with the highest power. This corresponds to sorting the columns of $\mathbf{H}$ by their squared norms. In the case of MMSE filtering, detection proceeds as mentioned in the case of D-BLAST, the only difference being the different allocation of codewords in matrix $\mathbf{C}$. In this case as well, the received sequences can be properly sorted.

Finally, in V-BLAST the different layers are not encoded. This simplifies the receiver structure, but makes cancelation less reliable. This scheme can be concatenated with an outer channel encoder, possibly through an interleaver. In this case, iterative detection and decoding can be performed based on the turbo principle (see Chapter 12).

Multilayered ST architecture

BLAST architectures allow the achievement of a spectral efficiency up to $\eta = R_c n_T \log_2 M$ bits per channel use, where R_c is the rate of the encoders employed. However, there is no attempt to maximize the code diversity. Other layered architectures allow spectral efficiency to be traded for diversity in an attempt to improve the system performance with respect to the BLAST architecture and the data rate with respect to ST block and trellis codes. As an example, the *multilayered* ST *architecture* [1904] is a hybrid approach using both ST channel codes and layered processing. Transmit antennas are partitioned into small groups and independent ST block or trellis codes are employed for each group. The corresponding codewords are then organized into layers and decoded using the techniques previously described, appropriately modified in order to perform *group* interference cancelation.

Threaded ST codes

Threaded STCs (TSTCs) [1905] exhibit superior performance with respect to the scheme in [1904]. In fact, multilayered ST codes have a performance which is 6–9 dB from the outage capacity at 10% frame error rate [1904] whereas the threaded architecture closes the gap to less than 3 dB from the outage capacity with the same frame length, error rate and receiver complexity.

The layered and multilayered architectures described up to now were inspired by the signal processing techniques employed at the receiver. For example, in the D-BLAST approach, each layer is constrained to occupy a diagonal in the ST transmission resources array, due to the interference cancelation/suppression technique adopted at the receiver. In [1905] the concept of the ST layer was generalized independently of the signal processing adopted at the receiver. The design of algebraic STCs and iterative signal processing techniques that optimize the performance of threaded ST coded systems are then considered accordingly.

A *thread* is a layer that efficiently exploits the diversity available in the system. It extends over the full spatial dimension n_T and the full temporal dimension N with the property that all spatial interference experienced by the layer comes from outside the layer. An example of threaded layering is provided in Figure 11.21 for $n_T = 4$. As one can observe, the number of layers is equal to the number of transmit antennas. Each codeword, coming from a properly designed component code, is fully interleaved and then assigned to a unique layer. The design of the component codes is described in [1905]. In particular, a condition is given that allows us to obtain an ST code achieving diversity νn_R. The spectral efficiency of the resulting STC is $\eta = R_c(n_T - \nu + 1)\log_2 M$ bits per channel use, where R_c is the rate of the component code and M is the cardinality of the constellation employed. Therefore, this scheme offers a tradeoff between diversity and rate; in particular, we note that, when $\nu = 1$, this scheme is equivalent to BLAST, whereas, when $\nu = n_T$, it becomes equivalent to ST trellis or block codes.

Detection of TSTCs is accomplished in accordance with the scheme in Figure 11.22. In fact, with the transmitted codewords interleaved, the optimal detector has an unmanageable complexity, and a suboptimal detection/decoding architecture based on the turbo principle [1611] is the only viable option. In the figure, Π_i denotes the interleaver for the ith layer. We could use the optimal SiSo MIMO detector, whose complexity is, however, *exponential* in the number of transmit antennas n_T. Several suboptimal detectors can also be found in the literature, from those whose complexity is *polynomial* in n_T and based on SD [1192, 1297, 1909–1911] to those whose complexity is *quadratic* in n_T [1912] and based on *MMSE interference cancelation* algorithms, originally proposed for CDMA systems [1913, 1914], and those whose complexity is *linear* in n_T [1915] and based on FGs and the SPA [1610].[15] The SiSo MIMO detector produces soft information on the symbols of the n_T transmitted codewords which is then deinterleaved and passed to the corresponding SiSo decoders. These latter decoders are matched to the individual component encoders adopted at the transmitter and can work according to any suitable SiSo algorithm, depending on the type of code employed. The n_T outer SiSo decoders produce their own soft information which is fed back to the SiSo MIMO detector in accordance with the turbo principle.

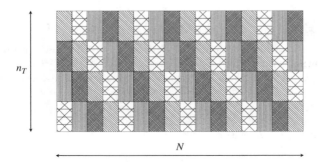

Figure 11.21 Threaded layering (each pattern represents a thread) for $n_T = 4$ and $N = 20$.

---

[15] All these suboptimal detection algorithms can be employed for other layered architectures.

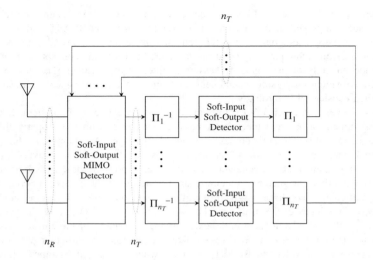

Figure 11.22 TSTC decoder.

Wrapped ST codes

The final layering scheme we describe is constituted by *wrapped* STCs (WSTCs) [1906], which represent a significant improvement with respect to D-BLAST. In this scheme, the codewords have length $N' = n_T d$. For given n_T, a large delay d is thus needed in order to have long codewords. If interleaving delay is an issue, the D-BLAST scheme is forced to work with a short component code block length N'. This might pose a serious problem for using trellis codes with a large number of states. In fact, the code memory might not be negligible with respect to N', thus yielding a nonnegligible rate loss due to trellis termination. In addition, in the case of block component codes, powerful codes cannot be used. WSTCs are a solution that retains the simplicity of decision feedback interference mitigation while allowing for arbitrarily long component codewords and small interleaving delay. In these schemes, a single encoder is employed. The corresponding codeword of length N' is diagonally interleaved, through an ST formatter, in order to form the $n_T \times N$ codeword matrix $\mathbf{C}$, with $N = N'/n_T + (n_T - 1)d$. The codeword matrix $\mathbf{C}$ is filled by wrapping the codeword $\mathbf{c}$ around the matrix diagonals (hence the name of this layering scheme), as illustrated by Figure 11.23. We can write:

$$\mathbf{C} = \mathcal{F}(\mathbf{c}), \tag{11.158}$$

where the formatter $\mathcal{F}$ is defined such that the element $c_{n,\ell}$ of the codeword matrix $\mathbf{C}$ is related to the element c_k of the codeword $\mathbf{c}$ by:

$$c_{n,\ell} = \begin{cases} c_{k_{n,\ell}} & \text{if } 1 \leq k_{n,\ell} \leq N' \\ 0 & \text{otherwise} \end{cases} \tag{11.159}$$

for $1 \leq n \leq n_T$ and $1 \leq \ell \leq N$, where:

$$k_{n,\ell} = [\ell - 1 - (n-1)d]n_T + n . \tag{11.160}$$

In this way, the interleaving delay d becomes a free parameter, independent of the component codeword block length N'. As a limiting case, the interleaving delay may be also $d = 0$, that is, a vertical interleaver may be used. For consistency with the case $d > 0$, where code symbols with lower index

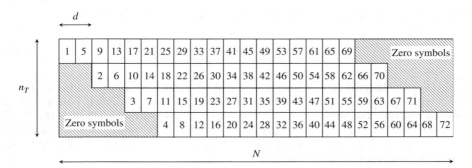

Figure 11.23 Wrapped ST codeword for $n_T = 4$, $d = 2$, and $N' = 72$. The entries in the array indicate the index of the symbols in the component codeword.

take the lower positions in each column of the codeword matrix $\mathbf{C}$ (see Figure 11.23), the ST formatter for $d = 0$ is defined by replacing (11.160) by:

$$k_{n,\ell} = (\ell - 1)n_T + n_T - n + 1 \,. \tag{11.161}$$

When the component encoder is a trellis code of rate b/n_T, the corresponding WSTC with $d = 0$ coincides with a standard STTC. For $d > 0$, the corresponding WSTC can be seen as the concatenation of a trellis code with delay diversity. Because of the lower and upper triangles of zero symbols in the codeword matrix in Figure 11.23, there is an inherent rate loss of $(n_T - 1)d/N$. This is negligible if $N \gg n_T d$. Moreover, if the transmission of a long sequence of codewords is envisaged, the codeword matrices can be concatenated in order to fill the leading and trailing triangles of zeros, so that no rate loss is incurred.

The wrapped ST architecture has been designed such that when the component codewords are produced by a trellis encoder, decoding can be implemented efficiently by ZF or MMSE decision feedback interference mitigation coupled with Viterbi decoding, through the use of PSP [515]. Let us consider, as an example, the use of the ZF suppression strategy. The extension to the case of MMSE filtering is straightforward, since only the linear filter needs to be modified. The decoder works on the trellis of the component code and takes as observable the sequence of samples:

$$z_k = v_{n,\ell} - \sqrt{\gamma} \sum_{m=n+1}^{n_T} b_{n,m} \hat{c}_{m,\ell} \,, \tag{11.162}$$

with $k = 1, 2, \ldots, N'$, where $v_{n,\ell}$ is the (n, ℓ) element of matrix $\mathbf{V} \triangleq \mathbf{Q}^H \mathbf{R}$ (or of matrix $\mathbf{V} \triangleq \mathbf{F}^H \mathbf{R}$ in the case of MMSE filtering), $b_{n,\ell}$ is the (n, ℓ)th element of matrix $\mathbf{B}$ given in (11.141), and $1 \leq n \leq n_T$ and $1 \leq \ell \leq N$ are the unique integers for which $k_{n,\ell} = k$. From the index mapping (11.160) (or (11.161) for $d = 0$), we see that the elements $c_{m,\ell}$, for $m = n + 1, \ldots, n_T$, correspond to either zeros (for which no decision is needed) or to symbols of the codeword with index $k' \leq k - n_T d + 1$ ($k' \leq k - 1$ for $d = 0$). These decisions are found in the survivor history of the Viterbi decoder, in accordance with standard PSP [515].

A major advantage of WSTCs is that off-the-shelf component codes can be employed, thus avoiding an *ad hoc* code search. A sensible criterion for the design of the component code is, in fact, the maximization of the code *block diversity* δ, defined by:

$$\delta \triangleq \min_{\mathbf{c}, \hat{\mathbf{c}} : \hat{\mathbf{c}} \neq \mathbf{c}} |\{j \in \{1, \ldots, n_T\} \mid w_j \neq 0\}| \tag{11.163}$$

with $k = 1, 2, \ldots, N'$, where w_j is the *squared Euclidean weight* defined as:

$$w_j \triangleq \sum_{n=1}^{N} |\hat{c}_{j,n} - c_{j,n}|^2. \qquad (11.164)$$

Thus the aim is to maximize the minimum number of nonzero rows in the matrix difference $\hat{\mathbf{C}} - \mathbf{C} = \mathcal{F}(\hat{\mathbf{c}}) - \mathcal{F}(\mathbf{c})$ for each pair of distinct codeword matrices $\hat{\mathbf{C}} - \mathbf{C}$, which is strictly related to the rank diversity of a WSTC. The block diversity criterion has been investigated in [1853, 1916, 1917] for the design of trellis codes for cyclic interleaving and/or periodic puncturing, and codes optimized in this sense are thus available. The relationship between the rank diversity ν of a WSTC and the block diversity of its component code is [1906]:

$$\nu \leq \delta \leq 1 + \left\lfloor n_T \left(1 - \frac{R_c}{\log_2 M} \right) \right\rfloor, \qquad (11.165)$$

where R_c is the rate of the component code and M is the cardinality of the employed constellation A_c. Moreover, there exist values of d for which $\nu = \delta$ [1906]. Since it is known from [14] that for any STC with spectral efficiency $\eta = n_T R_c$ the rank diversity satisfies the inequality:

$$\nu \leq 1 + \left\lfloor n_T \left(1 - \frac{R_c}{\log_2 M} \right) \right\rfloor, \qquad (11.166)$$

which is the same upper bound on block diversity given in (11.165), we conclude that the wrapping construction incurs no loss of optimality in terms of rank diversity (for an appropriate choice of the delay d). As a matter of fact, while it is difficult to construct codes with rank diversity equal to the upper bound (11.166), it is very easy to find trellis codes for which the upper bound (11.165) on δ is met with equality, for several coding rates and values of n_T. Examples of these codes are tabulated in [1916, 1917]. Therefore, the wrapping construction is a powerful tool to construct STCs with optimal rank diversity.

In [1906], a comparison between WSTCs and TSTCs is provided. For a given state complexity of the underlying component code, one iteration of the TSTC decoder corresponds to (roughly) the same complexity of the whole WSTC decoder. It is shown in [1906] that the WSTC scheme yields a clear performance advantage over TSTC with one iteration (for the same decoder complexity). The case of more decoding iterations is considered in [1905], giving a (quite limited) performance advantage with respect to WSTCs whose decoding algorithm is not designed to be iterative.

11.5.1.8 Multiplexing–Diversity Tradeoff

As mentioned, there are two types of gain that a MIMO system can provide: *diversity* and *multiplexing*. We focus our attention on the quasi-static Rayleigh fading channel, with CSI available at the receiver only. The diversity gain D is mathematically defined as the negative asymptotic slope of the error rate curve as a function of the SNR on a log-log scale:

$$D \triangleq \lim_{\gamma \to \infty} -\frac{\log P_e(\gamma)}{\log \gamma}, \qquad (11.167)$$

where $P_e(\gamma)$ is the average error probability of the system, while the multiplexing gain r is defined as the asymptotic ratio between the data rate of a specific MIMO scheme and the logarithm of the SNR (which is a measure of the capacity increase) [1880]; in other words, we have:

$$r \triangleq \lim_{\gamma \to \infty} \frac{R(\gamma)}{\log \gamma}, \qquad (11.168)$$

where $R(\gamma)$ is the data rate in bits per second per hertz.

The maximum diversity gain that a MIMO system can achieve is given, as is now clear, by $n_T n_R$. This is achieved by some of the schemes previously described. As for the multiplexing gain, it cannot exceed the number of degrees of freedom provided by the MIMO channel, which is $\min(n_T, n_R)$. It is also clear that, in order to have a nonzero multiplexing gain, the scheme considered cannot have a constant data rate but must provide a data rate that increases with the SNR. This can be achieved, for example, by increasing the constellation size with the SNR.

It has been demonstrated in [1880] that it is not possible to achieve both full diversity and full multiplexing gains. For each r the optimal diversity gain $D_o(r)$ is the maximum diversity gain that can be achieved by any scheme. It is shown in [1880] that, if the fading coherence time is greater or equal to $n_T + n_R - 1$, then:

$$D_o(r) = (n_T - r)(n_R - r) , \qquad (11.169)$$

with $0 \leq r \leq \min(n_T, n_R)$. Hence, when the diversity gain is $n_T n_R$, the multiplexing gain is zero, whereas when $r = \min(n_T, n_R)$ the diversity gain is zero. For practical schemes, the diversity–multiplexing tradeoff function lies below the curve (11.169) and can be used to compare different schemes and to interpret their behavior, as shown in the examples that follow. As an example, for the Alamouti scheme, the diversity–multiplexing tradeoff function is [1880]:

$$D(r) = \max(2n_R(1 - r), 0),$$

which reaches the upper bound (11.169) for $r = 0$ only. On the other hand, it can be shown that BLAST schemes favor the multiplexing gain [1880].

11.5.1.9 Concatenated Codes for MIMO Channels

Channel coding and space-time coding can be combined to achieve further performance improvements. For a fast fading channel, that is, when the channel coherence time is relatively short and time interleaving is employed, the adoption of an outer channel encoder can provide time diversity in addition to antenna diversity. On the other hand, for a quasi-static fading channel, an outer channel encoder can only provide a coding gain.

Various concatenated schemes can be found in the literature and the following review cannot claim to be exhaustive. In [1918, 1919] an outer channel encoder is serially concatenated with an OSTBC. This scheme gives the best performance–complexity tradeoff among other concatenation schemes when the outer code is a convolutional code or a turbo code [1465].

Two schemes have been proposed in [1920]. In the first, an outer turbo encoder of rate m/mn_T provides at its output n_T sequences of bits that are independently interleaved, mapped on an M-ary constellation (with $M = 2^m$) and the resulting symbols are then transmitted in parallel on the n_T transmit antennas. The same scheme is also described in [1921] for BPSK. The resulting STC is not guaranteed to be of full diversity, although when the interleavers are selected randomly, a full-diversity code is obtained with high probability. In the second proposed scheme, each output branch of an STTC is interleaved and coded by the same recursive convolutional code of rate 1 (the latter condition to ensure full rate). A similar scheme is also described in [1922].

In [1923], the so-called *turbo ST coded modulation* for $n_T = 2$ transmit antennas was proposed. It consists of two systematic and recursive STCs which are parallel concatenated through an interleaver that operates symbolwise. The systematic output of one STTC is connected to the first antenna. Parity symbols are punctured and transmitted over the second antenna after the output symbols of the second encoder are deinterleaved. This scheme has full rate and simulation results show that it also provides full diversity.

In [1924], the outputs of a turbo code are bit-interleaved, mapped to QPSK symbols and transmitted using multiple antennas. Full rate is achieved but the code is not guaranteed to achieve full space diversity. However, this is a very flexible scheme. Essentially, any traditional code for single antenna systems can be employed.

In [1925], STTCs are first modified to be recursive. Then two encoder structures are proposed. The serially concatenated encoder employs a convolutional code as outer code and recursive STTC as inner code. The parallel concatenated case, in contrast, is a self-concatenated [1926] recursive STTC. These codes guarantee full space diversity but full rate cannot be achieved.

In [1927], a full-rate ST turbo trellis code, referred to as an *assembled space-time turbo trellis code*, is proposed. For this scheme, input information binary sequences are first encoded using a classical turbo code. The component codes are rate-1/2 convolutional codes and their output bits are split into four parallel streams, each of them modulated by a QPSK modulator. The modulated symbols are assembled by a predefined linear function rather than punctured as in standard schemes.

Coded LST codes have also been considered. For example, a turbo-coded layered ST coding scheme with iterative decoding where coding is applied across the different layers has been proposed in [1928], whereas in [1929] LST coding based on LDPC codes has been investigated.

Finally, we mention the development of construction methods for *multidimensional space-time multilevel codes* in [1930]. The proposed space-time multilevel encoding schemes involve a multidimensional partitioning of a $2n_T$-dimensional signaling space; such a partitioning spans all n_T transmit antennas and can be designed to reduce the complexity of detection/decoding. An ST multistage decoder can be adopted for the proposed space-time multilevel codes in order to significantly reduce the complexity of soft decision decoding compared to a single level approach.

11.5.1.10   Unitary and Differential STCs

Up to now, we have assumed perfect knowledge of the channel at the receiver. This condition can be achieved, when the channel coherence time is long enough, through the use of pilot symbols periodically inserted to help the receiver to obtain a sufficiently accurate channel estimate. When the channel changes frequently, absence of knowledge of the channel at both transmitter and receiver must be assumed and one may resort to noncoherent detection. In this case, appropriate STCs need to be employed. In what follows, we will discuss *unitary* and *differential* STCs.

Unitary STCs
Before going into details, we derive the metric for the case of noncoherent detection assuming the block fading channel model with coherence time of L symbol intervals.

Gathering L received vectors into an $n_R \times L$ matrix $\mathbf{R}$, we may write:

$$\mathbf{R} = \sqrt{\gamma}\mathbf{H}\mathbf{C} + \mathbf{N}, \tag{11.170}$$

where $\mathbf{H}$ is the $n_R \times n_T$ channel matrix, and the $n_T \times L$ matrix $\mathbf{C}$ contains the transmitted symbols and the $n_R \times L$ matrix $\mathbf{N}$ the noise samples during L symbol intervals. The samples $\{r_i[\ell], \ell = 1, 2, \dots, L\}$ received by the ith antenna can be collected in the row vector:

$$\mathbf{r}_i \triangleq [r_i[1], r_i[2], \dots, r_i[L]], \tag{11.171}$$

which can be expressed as:

$$\mathbf{r}_i = \sqrt{\gamma}\mathbf{h}_i\mathbf{C} + \mathbf{n}_i, \tag{11.172}$$

where $\mathbf{h}_i$ is the ith row of the channel matrix $\mathbf{H}$ and $\mathbf{n}_i$ is a row vector containing the noise samples at the output of the ith antenna. The rows $\mathbf{r}_i$ are independent of each other. Hence, we may write:

$$f_{\mathbf{R}}(\boldsymbol{\Theta}|\mathbf{C}) = \prod_{i=1}^{n_R} f_{\mathbf{r}_i}(\boldsymbol{\rho}_i|\mathbf{C}), \tag{11.173}$$

where $\mathbf{\Theta}$ and $\boldsymbol{\rho}_i$ denote the values taken on by $\mathbf{R}$ and $\mathbf{r}_i$, respectively. Given $\mathbf{C}$, the random variables in $\mathbf{r}_i$ are jointly Gaussian with mean zero and covariance matrix:

$$
\begin{aligned}
\mathbf{\Lambda} &\triangleq \mathrm{E}\{\mathbf{r}_i^T \mathbf{r}_i^*\} \\
&= \mathrm{E}\{(\sqrt{\gamma}\mathbf{C}^T \mathbf{h}_i^T + \mathbf{n}_i^T)(\sqrt{\gamma}\mathbf{h}_i \mathbf{C} + \mathbf{n}_i)^*\} \\
&= \mathbf{I}_L + \gamma \mathbf{C}^T \mathbf{C}^*,
\end{aligned}
\tag{11.174}
$$

so that:

$$
\begin{aligned}
f_{\mathbf{R}}(\mathbf{\Theta}|\mathbf{C}) &= \prod_{i=1}^{n_R} \frac{\exp\{-\boldsymbol{\rho}_i^* \mathbf{\Lambda}^{-1} \boldsymbol{\rho}_i^T\}}{\pi^L \det(\mathbf{\Lambda})} \\
&= \frac{1}{\pi^{L n_R}(\det(\mathbf{\Lambda}))^{n_R}} \exp\left\{-\sum_{i=1}^{n_R} \boldsymbol{\rho}_i^* \mathbf{\Lambda}^{-1} \boldsymbol{\rho}_i^T\right\} \\
&= \frac{\exp\{-\mathrm{tr}(\mathbf{\Theta}^* \mathbf{\Lambda}^{-1} \mathbf{\Theta}^T)\}}{\pi^{L n_R}(\det(\mathbf{\Lambda}))^{n_R}}.
\end{aligned}
\tag{11.175}
$$

For the block fading channel with coherence time of L symbols, in [1931] it is proved that, asymptotically, capacity is achieved when:

$$
\mathbf{C} = \mathbf{V}\mathbf{\Phi},
\tag{11.176}
$$

where $\mathbf{\Phi}$ is an *isotropically distributed*[16] $n_T \times L$ matrix whose rows are orthonormal (hence $\mathbf{\Phi}^* \mathbf{\Phi}^T = \mathbf{I}_{n_T}$) and $\mathbf{V}$ is an independent $n_T \times n_T$ real nonnegative diagonal matrix. When $L \gg n_T$, or when $L > n_T$ and the SNR is very high, capacity can be achieved by selecting:

$$
\mathbf{C} = \sqrt{L}\mathbf{\Phi}.
\tag{11.177}
$$

For this reason, the unitary STCs proposed in [1931] have a codebook composed of codewords (11.177), with $\mathbf{\Phi}$ belonging to a set of $2^{\eta L}$ elements, where η is the spectral efficiency in bits per channel use. Since $\mathbf{H}$ is unknown at the receiver, the ML decoder will operate according to the decision rule:

$$
\hat{\mathbf{\Phi}} = \arg\max_{\mathbf{\Phi}} f_{\mathbf{R}}(\mathbf{\Theta}|\tilde{\mathbf{\Phi}}) = \arg\max_{\mathbf{\Phi}} \frac{\exp\{-\mathrm{tr}(\mathbf{\Theta}^* \tilde{\mathbf{\Lambda}}^{-1} \mathbf{\Theta}^T)\}}{(\det(\tilde{\mathbf{\Lambda}}))^{n_R}},
\tag{11.178}
$$

where $\tilde{\mathbf{\Lambda}} \triangleq \mathbf{I}_L + \gamma L \tilde{\mathbf{\Phi}}^T \tilde{\mathbf{\Phi}}^*$. Using the fact that $\tilde{\mathbf{\Phi}}^* \tilde{\mathbf{\Phi}}^T = \mathbf{I}_{n_T}$ and the properties (C.15), (C.16) and the *matrix inversion lemma* (see (C.10)), we have:

$$
\det\tilde{\mathbf{\Lambda}} = \det(\mathbf{I}_L + \gamma L \tilde{\mathbf{\Phi}}^T \tilde{\mathbf{\Phi}}^*) = \det(\mathbf{I}_{n_T} + \gamma L \tilde{\mathbf{\Phi}}^* \tilde{\mathbf{\Phi}}^T) = (\gamma L + 1)^{n_T}
\tag{11.179}
$$

and

$$
\begin{aligned}
\mathrm{tr}(\boldsymbol{\theta}^* \tilde{\mathbf{\Lambda}}^{-1} \boldsymbol{\theta}^T) &= \mathrm{tr}[\mathbf{\Theta}^*(\mathbf{I}_L + \gamma L \tilde{\mathbf{\Phi}}^T \tilde{\mathbf{\Phi}}^*)^{-1} \mathbf{\Theta}^T] \\
&= \mathrm{tr}\left[\mathbf{\Theta}^*\left(\mathbf{I}_L - \frac{\gamma L}{1 + \gamma L}\tilde{\mathbf{\Phi}}^T \tilde{\mathbf{\Phi}}^*\right)\mathbf{\Theta}^T\right] \\
&= \mathrm{tr}[\mathbf{\Theta}^* \mathbf{\Theta}^T] - \mathrm{tr}\left[\frac{\gamma L}{1 + \gamma L}\mathbf{\Theta}^* \tilde{\mathbf{\Phi}}^T \tilde{\mathbf{\Phi}}^* \mathbf{\Theta}^T\right]
\end{aligned}
\tag{11.180}
$$

---

[16] An $n_T \times L$ *isotropically distributed* random matrix is a matrix whose pdf remains unchanged when it is right-multiplied by any deterministic $n_T \times n_T$ unitary matrix. This matrix is the $n_T \times L$ counterpart of a complex scalar having unit magnitude and uniformly distributed phase.

from which we obtain:

$$\hat{\boldsymbol{\Phi}} = \arg \max_{\tilde{\boldsymbol{\Phi}}} \{\text{tr}[\boldsymbol{\Theta}^* \tilde{\boldsymbol{\Phi}}^T \tilde{\boldsymbol{\Phi}}^* \boldsymbol{\Theta}^T]\} = \arg \max_{\tilde{\boldsymbol{\Phi}}} \{\text{tr}[\boldsymbol{\Theta} \tilde{\boldsymbol{\Phi}}^H \tilde{\boldsymbol{\Phi}} \boldsymbol{\Theta}^H]\} . \tag{11.181}$$

An upper bound on the PEP can be found in [1931]. As a result, we can write:

$$\Pr\{\boldsymbol{\Phi} \to \hat{\boldsymbol{\Phi}}\} \leq \frac{1}{2} \prod_{i=1}^{n_T} \left[\frac{1}{1 + \frac{(\gamma L)^2 (1 - \mu_i^2)}{4(1 + \gamma L)}} \right]^{n_R}, \tag{11.182}$$

where $\mu_i \leq 1$ (with $i = 1, 2, \ldots, n_T$) is the ith singular value of the $n_T \times n_T$ correlation matrix $\hat{\boldsymbol{\Phi}}^* \hat{\boldsymbol{\Phi}}^T$. Equation (11.182) can be used to design the set of matrices $\{\boldsymbol{\Phi}\}$ to be used. An algorithm for the construction of this set is provided in [1931]. From (11.182), it also appears that the maximum diversity can be $n_T n_R$ as in the case where $\mathbf{H}$ is known at the receiver. This maximum diversity can be obtained when all singular values μ_i are strictly lower than unity.

Differential STCs

A different approach is pursued in [1932],[17] based on differential ST coding and differential ST detection, generalizing for MIMO systems the differential encoding and differential detection used in single-antenna systems.

Let $\mathcal{S}$ be a group of $L \times L$ unitary matrices. A matrix $\mathbf{S} \in \mathcal{S}$ is such that $\mathbf{S}^H \mathbf{S} = \mathbf{S} \mathbf{S}^H = \mathbf{I}_L$. In addition, since $\mathcal{S}$ is a group, $\mathbf{I}_L$, the multiplicative identity, belongs to $\mathcal{S}$, the multiplication of two matrices in $\mathcal{S}$ is a matrix in $\mathcal{S}$, and the inverse of each element of $\mathcal{S}$ also belongs to $\mathcal{S}$. Let $\mathbf{C}_0$ be an $n_T \times L$ matrix such that $\mathbf{C}_0 \mathbf{C}_0^H = \mathbf{I}_{n_T}$ and having the property that the matrix $\mathbf{C}_0 \mathbf{S}$ has all its elements belonging to a given alphabet A_c, for all $\mathbf{S} \in \mathcal{S}$. The set:

$$\mathcal{C} = \{\mathbf{C}_0 \mathbf{S} | \mathbf{S} \in \mathcal{S}\} \tag{11.183}$$

represents the set of transmitted codewords. A matrix $\mathbf{C} \in \mathcal{C}$ is clearly such that $\mathbf{C} \mathbf{C}^H = \mathbf{C}_0 \mathbf{S} \mathbf{S}^H \mathbf{C}_0^H = \mathbf{C}_0 \mathbf{C}_0^H = \mathbf{I}_{n_T}$. In the case of a block fading channel with coherence time $2L$, the matrix $\mathbf{C}_0$ is transmitted in the first L symbol intervals, while $\mathbf{C}_0 \mathbf{S}$, with $\mathbf{S} \in \mathcal{S}$, is transmitted in the successive L symbol intervals. The resulting ST code has spectral efficiency $\eta = \frac{1}{2L} \log_2 |\mathcal{S}|$ bits per channel use. When the channel changes continuously but can be considered approximately constant over $2L$ symbol intervals, the following $n_T \times L$ matrices are transmitted over NL symbol intervals:

$$\mathbf{C}_\ell = \begin{cases} \mathbf{C}_0 & \ell = 0 \\ \mathbf{C}_{\ell-1} \mathbf{S}_\ell & \ell = 1, 2, \ldots, N - 1. \end{cases} \tag{11.184}$$

In this case, the spectral efficiency is $\eta = \frac{1}{L} \frac{N-1}{N} \log_2 |\mathcal{S}|$ bits per channel use.

As far as decoding is concerned, in the case of the block fading channel with coherence time $2L$, optimal decoding will be accomplished based on the observation of $2L$ symbol intervals. In the case of a channel that changes continuously, by collecting blocks of L received vectors into $n_R \times L$ matrices $\{\mathbf{R}_\ell\}$, we may write:

$$\mathbf{R}_\ell = \sqrt{\gamma} \mathbf{H}_\ell \mathbf{C}_\ell + \mathbf{N}_\ell, \tag{11.185}$$

where $\mathbf{H}_\ell$ is the $n_R \times n_T$ channel matrix corresponding to the ℓth block and the $n_R \times L$ matrix $\mathbf{N}_\ell$ contains the noise samples during the L symbol intervals of the ℓth block. In this case, optimal decoding must be accomplished based on the observation of the whole sequence $\{\mathbf{R}_\ell\}$. However, in order to

---

[17] A similar scheme has been proposed in [1933], while an alternative scheme for the case of two transmit antennas has been proposed in [1934].

reduce the receiver complexity, as in the case of differential decoding for single-antenna systems, decoding of $\mathbf{S}_\ell$ is accomplished by looking at pairs of overlapping blocks of L symbol intervals at a time, that is, $\mathbf{R}_\ell$ and $\mathbf{R}_{\ell-1}$. To show this, let us define the $n_R \times 2L$ matrix:

$$\mathbf{R}'_\ell \triangleq [\mathbf{R}_{\ell-1}, \mathbf{R}_\ell] . \tag{11.186}$$

If we assume that $\mathbf{H}_\ell = \mathbf{H}_{\ell-1}$ and define $\mathbf{C}'_\ell \triangleq [\mathbf{C}_{\ell-1}, \mathbf{C}_\ell]$ and $\mathbf{N}'_\ell \triangleq [\mathbf{N}_{\ell-1}, \mathbf{N}_\ell]$, we express $\mathbf{R}'_\ell$ (11.186) as:

$$\mathbf{R}'_\ell = \sqrt{\gamma}\, \mathbf{H}_\ell \mathbf{C}'_\ell + \mathbf{N}'_\ell \tag{11.187}$$

Since matrices $\mathbf{C}_\ell$ are such that $\mathbf{C}_\ell \mathbf{C}_\ell^H = \mathbf{I}_{n_T}$, we also have $\mathbf{C}'_\ell \mathbf{C}_\ell^{'H} = 2\mathbf{I}_{n_T}$. Hence, when accomplishing detection based on a couple of blocks of L symbol intervals we may adopt the detection strategy (11.181) that now becomes:

$$\hat{\mathbf{S}}_\ell = \arg \max_{\tilde{\mathbf{S}}_\ell} \{\mathrm{tr}[\mathbf{R}'_\ell \tilde{\mathbf{C}}_\ell^H \tilde{\mathbf{C}}_\ell \tilde{\mathbf{R}}_\ell^H]\} \tag{11.188}$$

Under the additional assumption that $L = n_T$ (this is certainly possible since we are not considering a block fading channel but a channel that changes continuously) we have $\mathbf{C}_\ell \mathbf{C}_\ell^H = \mathbf{C}_\ell^H \mathbf{C}_\ell = \mathbf{I}_{n_T}$. Hence, (11.188) becomes:

$$
\begin{aligned}
\hat{\mathbf{S}}_\ell &= \arg \max_{\tilde{\mathbf{S}}_\ell} \{\mathrm{tr}[\mathbf{R}'_\ell \tilde{\mathbf{C}}_\ell^{'H} \tilde{\mathbf{C}}'_\ell \mathbf{R}_\ell^{'H}]\} \\
&= \arg \max_{\tilde{\mathbf{S}}_\ell} \{\mathrm{tr}[\mathbf{R}_{\ell-1}\mathbf{R}_{\ell-1}^H + \mathbf{R}_{\ell-1}\tilde{\mathbf{C}}_{\ell-1}^H \tilde{\mathbf{C}}_\ell \mathbf{R}_\ell^H + \mathbf{R}_\ell \tilde{\mathbf{C}}_\ell^H \tilde{\mathbf{C}}_{\ell-1}\mathbf{R}_{\ell-1}^H + \mathbf{R}_\ell \mathbf{R}_\ell^H]\} \\
&= \arg \max_{\tilde{\mathbf{S}}_\ell} \mathrm{Re}\{\mathrm{tr}[\mathbf{R}_{\ell-1}\tilde{\mathbf{S}}_\ell \mathbf{R}_\ell^H]\}.
\end{aligned}
\tag{11.189}
$$

Performance analysis can be carried out in a way similar to the case of unitary STCs since both are based on unitary matrices and the same metric. Details can be found in [1932] along with the design criteria and the optimal codes for two transmit antennas.

Iterative schemes

The schemes described above can be concatenated with an outer channel code to improve performance. For example, the concatenation of turbo codes and unitary STCs with iterative decoding at the receiver was considered in [1935], while the concatenation of differential STCs and an outer code through an interleaver was investigated in [1936, 1937]. In the latter case, given that a differential ST code is a recursive code, when iteratively decoded this serial concatenation provides an interleaver gain.

11.5.2 ST Coding for Frequency-Selective Fading Channels

Up to now, we have concentrated on flat fading channels. However, in wideband wireless systems, when the symbol period becomes smaller than the channel delay spread, the transmitted signal sees a frequency-selective channel. An overview of the main results on STCs for these channels will be provided here with reference to the case of quasi-static channels.

11.5.2.1 System Model for Frequency-Selective MIMO Fading Channels

In what follows we assume the MIMO TDL channel model with L_h T_s-spaced taps described in Section 4.4.2.2. Therefore, the received signal samples referring to a codeword of length N are collected in an $n_R \times N$ matrix $\mathbf{R}$, whose structure is fully described by (4.112)–(4.116). In addition, we assume a Rayleigh fading channel, so that each entry of the lth tap matrix $\mathbf{H}^{(l)}$ (see (4.111)) is modeled as a zero-mean complex Gaussian random variable. Different channel taps are usually assumed to be independent and the average channel gains for different paths are determined from the PDP of the wireless channel.

For an $n_R \times N$ matrix $\mathbf{R} = [\mathbf{r}_1, \ldots, \mathbf{r}_N]$ of received signal samples and assuming that $\mathbf{c}_\ell = \mathbf{0}_{n_T}$ for $\ell \leq 0$, we may write:

$$\mathbf{R} = \sqrt{\gamma}\mathbf{H}\mathbf{C} + \mathbf{N}, \tag{11.190}$$

where the $n_R \times N$ matrix $\mathbf{N} \triangleq [\mathbf{n}_1, \ldots, \mathbf{n}_N]$ collects the noise samples (as in the case of a flat fading channel):

$$\mathbf{H} = \begin{bmatrix} \mathbf{H}_1 & \mathbf{H}_2 & \cdots & \mathbf{H}_{n_T} \end{bmatrix} \tag{11.191}$$

with

$$\mathbf{H}_i = \begin{bmatrix} h_{1,i}^{(0)} & h_{1,i}^{(1)} & \cdots & h_{1,i}^{(L_h-1)} \\ h_{2,i}^{(0)} & h_{2,i}^{(1)} & \cdots & h_{2,i}^{(L_h-1)} \\ \vdots & \vdots & \cdots & \vdots \\ h_{n_R,i}^{(0)} & h_{2,i}^{(1)} & \cdots & h_{n_R,i}^{(L_h-1)} \end{bmatrix} \tag{11.192}$$

is the $n_R \times n_T L_h$ equivalent channel matrix, while the $n_T L_h \times N$ equivalent matrix of the transmitted symbols $\mathbf{C}$ takes the form:

$$\mathbf{C} = \begin{bmatrix} \mathbf{C}_1 \\ \mathbf{C}_2 \\ \vdots \\ \mathbf{C}_{n_T} \end{bmatrix}, \tag{11.193}$$

with

$$\mathbf{C}_i = \begin{bmatrix} c_{i,1} & c_{i,2} & c_{i,3} & \cdots & \cdots & c_{i,N-1} & c_{i,N} \\ 0 & c_{i,1} & c_{i,2} & \cdots & \cdots & c_{i,N-2} & c_{i,N-1} \\ \vdots & 0 & c_{i,1} & \cdots & \cdots & \vdots & c_{i,N-2} \\ \vdots & \vdots & 0 & \cdots & \cdots & \vdots & \vdots \\ \vdots & \vdots & \vdots & \cdots & \cdots & c_{i,N-L_h} & \vdots \\ 0 & 0 & 0 & \cdots & \cdots & c_{i,N-L_h-1} & c_{i,N-L_h} \end{bmatrix}. \tag{11.194}$$

related to the symbols transmitted by antenna i.

11.5.2.2 Design Criterion

The channel model (11.190) states that our frequency-selective MIMO fading channel is equivalent to a frequency-flat fading channel having $L_h n_T$ transmit antennas. Looking at (11.194), then for each antenna, we have another $L_h - 1$ *virtual* antennas transmitting a delayed version of the same symbols (as in the delay-diversity scheme described in Section 11.5.1.4). Assuming that the matrix $\mathbf{H}$ has independent coefficients, by an appropriate design of the STC a diversity order of $L_h n_T n_R$ can be achieved. This is not surprising since the multipath propagation provides another form of diversity. The following criterion can thus be stated:

Design criterion. The maximum diversity of $n_T n_R L_h$ is achieved by ensuring that the matrix $\mathbf{A} = (\hat{\mathbf{C}} - \mathbf{C})(\hat{\mathbf{C}} - \mathbf{C})^H$ is of full rank for all pairs of distinct codewords $\hat{\mathbf{C}}$ and $\mathbf{C}$. Otherwise, if the minimum rank of $\mathbf{A}$ among all codeword pairs is $\nu_{\min} \leq n_T L_h$, a diversity order $\nu_{\min} n_R$ is achieved. In order to obtain the maximum possible coding advantage, the minimum determinant of matrices $\mathbf{A}$ having minimum rank should be maximized.

11.5.2.3 STCs for SC Systems

A code achieving full diversity can be designed by extending the same idea of the delay-diversity scheme described in Section 11.5.1.4. In fact, if we look at the rows of $\mathbf{C}_i$ in (11.194), they already are a delayed version of the first row. Hence, it is sufficient to transmit, from antenna i, a delayed version of the symbols transmitted by the first antenna, with a delay of $L_h(i-1)$ symbols. The resulting equivalent codeword matrix $\mathbf{C}_i$ will be:

$$\mathbf{C}_i = \begin{bmatrix} c_1 & c_2 & c_3 & \cdots & c_{L_h} & \cdots & c_{N-1} & c_N \\ 0 & c_1 & c_2 & \cdots & c_{L_h-1} & \cdots & c_{N-2} & c_{N-1} \\ \vdots & \vdots & c_1 & \cdots & c_{L_h-2} & \cdots & & \vdots \\ \vdots & \vdots & \vdots & \cdots & \vdots & \cdots & c_{N-L_h} & \vdots \\ \vdots & \vdots & \vdots & \cdots & \vdots & \cdots & c_{N-L_h-1} & c_{N-L_h} \\ 0 & 0 & 0 & \cdots & c_1 & \cdots & c_{N-L_h-1} & c_{N-L_h} \end{bmatrix} \tag{11.195}$$

and matrix $\mathbf{A}$ will have full rank.

Although, in principle, STBCs could be designed for frequency-selective fading channels, the main advantage of these codes (i.e., the simple linear processing) will be lost due to the presence of ISI. STTCs, layered architectures, and concatenated schemes can also be extended to frequency-selective channels. For example, STTCs employing BPSK and QPSK modulations are described in [1938]. The optimal decoder will operate, in this case, on the equivalent trellis that takes into account both the code and the channel trellis (i.e., on a *supertrellis*). As far as concatenated schemes are concerned, it must be taken into account that the channel introduces memory. Hence, it can be employed in place of an inner encoder and concatenated with an outer encoder through an interleaver.

As in the case of SISO channels, optimal detection has an exponential complexity in the channel memory L_h. Therefore, taking into account that the complexity is exponential in the number of transmit antennas also, optimal detection is practically infeasible. Reduced-complexity detection schemes are thus required, such as linear or decision feedback equalization schemes [27], RSSD [777, 989, 990, 1528], or other schemes based on FGs (e.g., see [1413, 1915]).

11.5.2.4 STCs for MIMO-OFDM

ST codes for SC systems may require the use of sophisticated detection techniques at the receiver. An alternative approach can be the use of OFDM, already described in Chapter 3. In this case, according to Figure 11.24, independent IDFTs are applied to the symbols to be transmitted by each antenna. After a

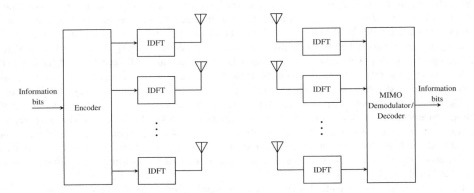

Figure 11.24 MIMO-OFDM system.

CP is appended to each sequence, they are transmitted on a frequency-selective MIMO channel. Each of the n_R receive antennas will receive the superposition of all n_T transmitted signals. This composite signal undergoes DFT and cyclic prefix removal. The resulting signals are then jointly demodulated and decoded. As shown in Chapter 4, MIMO-OFDM allows perfect removal of the ISI (under the assumption of quasi-static fading and perfect frequency synchronization) although the interference from different transmit antennas must, obviously, be taken into account.

In principle, MIMO signaling schemes developed for SC transmissions over frequency-flat fading channels can easily be employed with OFDM by simply performing operations on a subcarrier-by-subcarrier basis, that is, by reinterpreting time as frequency (e.g., see [1939–1941]); this approach is illustrated below in two different applications. First, let us consider the use of the ST Alamouti code illustrated in Section 11.5.1.5 in a MIMO-OFDM system having $n_T = 2$ transmit antennas. In this case full spatial diversity can be achieved if the ST Alamouti code is employed in a dual fashion, that is, coding is accomplished over frequency rather than over time. This means that, in transmitting a codeword in a given OFDM symbol interval, each antenna will send a pair of channel symbols employing the same pair of adjacent subcarriers. Then, the receiver can detect the transmitted symbols from the signal received on the two tones using the Alamouti detection technique (see (11.109)), provided that the channel frequency response is approximately constant over at least two consecutive tones. This allows us to achieve the full spatial diversity gain $2n_R$. This approach can be generalized to systems equipped with more than $n_T = 2$ transmit antennas by using other OSTBCs. Note, however, that in this case simple detection algorithms can be employed if the channel remains constant over at least n_T consecutive subcarriers. Another important application is represented by the use of spatial multiplexing schemes on a subcarrier-by-subcarrier basis. As in SC systems, this maximization of spatial rate by transmitting independent data streams over distinct antennas [156] allows us to use receiver architectures developed for SC modulations over frequency-flat fading.

Note that none of the transmission techniques considered above is able to capture the frequency diversity made available by the communication channel. In fact, extracting full spatial and frequency diversity requires data to be spread along both the dimensions of frequency and space. This raises the problem of devising SF codes or, more generally, STF codes for MIMO-OFDM. These represent strategies for mapping channel symbols to antennas and subcarriers (and time if codewords extend over multiple OFDM symbol intervals) in an appropriate way so that both spatial and frequency diversity can be extracted. In particular, SF coding consists of coding across antennas and OFDM subchannels [1940]. Design criteria to devise full-diversity SF codes have been derived in [1942, 1943], where full diversity is achieved at the price of a substantial loss in bandwidth efficiency, and [1944], where a systematic procedure for developing full-diversity SF *block codes* (SFBCs) with rate less than unity has been proposed. A design technique for full-diversity SFBCs with unit rate has recently been proposed for any number of transmit antennas and arbitrary power delay profiles in [1945]. To obtain a unit rate, the information symbol vector is first coded via an algebraic rotation matrix; then the resulting vector is split and spread over different antennas and OFDM subchannels. The rotation matrix is designed in such a way that signal space diversity can be captured by a rotation of the signal constellation [1909]. Recently, a systematic procedure for developing high rate SFBCs has been proposed in [1946]. This allows us to achieve rate n_T and full diversity in MIMO-OFDM systems for any number of transmit antennas. However, since a zero-padding matrix has to be used when the IDFT order N is not an integer multiple of $n_T L_h$, rate n_T cannot always be guaranteed. STBCs that can always achieve rate n_T and full diversity for any number of transmit antennas and any arbitrary channel power delay profiles have been developed in [1947]. In this case code construction is based on the *layering concept*, used in the design of TSTC [1948]; each component code is assigned to a "thread" and interleaved over space and frequency. The significant computational complexity of the ML decoding procedure for this class of codes can be reduced by resorting to simplified decoding strategies based on SD.

Most of the work on SF code design assumes a *quasi-static* model for fading channels, so that the channel can be considered constant over each codeword. Note that if the channel changes over consecutive OFDM symbols are not negligible, coding across multiple OFDM symbols can also capture time diversity, which contributes, together with frequency diversity and space diversity, to the maximum achievable diversity [1949]. Despite this consideration, various STF codes have been proposed for exploiting multipath diversity in MIMO-OFDM systems in the presence of quasi-static fading (e.g., see [1450, 1949–1951]). Recently, a systematic procedure for the design of high-rate STF codes has been proposed for time-varying MIMO frequency-selective fading channels in [1947]. In this case the algebraic coded symbols are spread across different OFDM subchannels, transmit antennas, and OFDM symbol intervals, so that the developed STF codes can achieve rate n_T and full diversity $n_T n_R n_b L_h$, where n_b is the number of independent fading blocks included in each codeword.

Finally, it is important to note that SF or STF codes for MIMO-OFDM can be concatenated with powerful error-correction codes (e.g., iteratively decodable codes). For instance, the use of LDPC codes has been proposed in [536].

11.6 Historical Notes

Massey formally suggested in [1952] that coding and modulation should have been combined in a single entity to improve performance. TCM was then invented by Ungerboeck and first described in a conference paper in 1976. At the same time, Imai and Hirakawa proposed their recipe to combine coding and modulation [1462], subsequently perfected by U. Wachsmann, R. F. H. Fischer, and J. B. Huber [1857]. Imai's idea of *multilevel coding* is to protect each address bit of the signal point by an individual binary code. Originally, multilevel coding was proposed for one-dimensional signaling. The individual codes were chosen in such a way that the minimum distance of the Euclidean space code was maximized. At the receiver side, each code is decoded individually starting from the lowest level and taking into account decisions of prior decoding stages. This procedure is called *multistage decoding*. In contrast to Ungerboeck's trellis coded modulation, the multilevel coding approach provides flexible transmission rates, because it decouples the dimensionality of the signal constellation from the code rate. Additionally, any code (e.g., block codes, convolutional codes, or concatenated codes) can be used as component code.

After its invention, TCM experienced a sudden transition from theory to practice. Only two years after Ungerboeck's landmark paper [992], a first generation of private-line modems transmitting at a speed of 14.4 kb/s were available. The development of TCM schemes for fading channels was one of the hottest research topics in the next few years (e.g., see [1852]); for further details, the reader can refer to [36]. This investigation culminated in the invention of BICM, originally suggested by Zehavi [1464] in 1992 and then further developed and analyzed by Caire, Taricco, and Biglieri [1854] in 1998. According to this technique, coded modulations with a very good performance over flat fading channels can be built by using off-the-shelf binary codes that are optimal in the sense of free Hamming distance, and thus available in standard textbooks. As a matter of fact, most of today's systems feature BICM, making it the *de facto* general coding technique for waveform channels.

The initial excitement about MIMO was sparked by the pioneering work of J. H. Winters [220], G. J. Foschini [17], Foschini and M. J. Gans [18] and E. Telatar [226], predicting the huge capacity increase possible due to the adoption of multiple antennas at both transmitter and receiver. Following on from these seminal papers, a lot of work has been devoted to STCs for both frequency-flat and frequency-selective fading channels. The first practical coding scheme achieving full diversity was *delay diversity* proposed in [1889]. Driven by the desire to support high data rates for a wide range of bearer services, V. Tarokh *et al.* proposed *space-time trellis coding* [14] in 1998 by jointly designing the channel coding, modulation, transmit diversity, and the optional receiver diversity scheme. The performance criteria for designing STTCs were derived in [14] under the assumption that the channel experiences slow frequency-flat fading. These advances were then also extended to fast fading channels. STTCs perform extremely well at the cost of relatively high complexity. In addressing the issue of decoding complexity, Alamouti [15] came up with a remarkable scheme for transmission using two transmit antennas. He also introduced a simple decoding algorithm, which can be generalized to an arbitrary number of receiver antennas. This motivated Tarokh *et al.* [1482, 1483] to generalize Alamouti's scheme to an arbitrary number of transmit antennas, leading to the concept of STBCs.

BLAST architectures [17] developed by Foschini at Lucent Technologies' Bell Laboratories (now Alcatel-Lucent Bell Labs) use spatial multiplexing to increase the data rate and do not necessarily provide transmit diversity. In other words, multiple antennas at both transmitter and receiver are used to exploit the many different paths between transmitter and receiver in a highly scattering wireless environment. Hence, by careful allocation of the data to be sent by the transmit antennas, multiple data streams can be transmitted simultaneously within a single frequency band. Other layered architectures were proposed later [1904–1907].

The best use of multiple transmit antennas depends on the amount of CSI available to the encoder and decoder. In the case of quasi-static fading, the channel can be estimated through the use of known pilot symbols and coherent detection can be employed [685]. In other cases, such an estimation of the channel is not available at the receiver or the channel changes rapidly such that the channel estimation is not useful. Then a noncoherent or differential detection needs to be employed [1931–1934, 1936, 1953].

Finally, in [156, 1938] the first studies of MIMO broadband fading channels, and in particular the impact of frequency selectivity on capacity and on receiver structures, considering single-carrier and multicarrier transmissions, are described.

11.7 Further Reading

In this chapter, we have discussed the main ideas behind combined modulation and coding for fading channels considering both SISO and MIMO channels. Many excellent books are available to help the reader study this subject in depth. As far as TCM is concerned, the main reference is [36]. Most of the coding techniques for fading channels are described in [37], including STCs. A comprehensive treatment of BICM can be found in [1954]. Various excellent books on STCs are available; among them, we mention [16, 38–40, 1883, 1955, 1956]. Finally, introductory material on signal space coding for MIMO-OFDM can be found in [21, 1326, 1957].

12

Combined Equalization and Decoding

12.1 Introduction

As already discussed in Chapter 6, digital receivers operating over a multipath fading channel usually employ an equalization algorithm to mitigate ISI effects. Equalization is followed by channel decoding, the aim of which is to recover the transmitted data from the equalized symbols. For complexity reasons, equalization and decoding have historically been considered *separate* tasks, their interaction having been limited to the delivery of hard or soft decisions from the first to the second. Unfortunately, as will become clearer later, optimal detection of coded data transmitted over fading channels requires *joint* decoding and equalization; in addition, a substantial performance gap between the overly complicated optimal strategy and the above-mentioned suboptimal approaches can be found in most scenarios.

More recently, the problem of joint channel equalization and decoding has been reconsidered from a different perspective, interpreting an ISI channel as a rate-1 channel encoder and, consequently, modeling the cascade of a channel encoder with such a channel as a serially concatenated coding scheme, for which iterative decoding techniques based on the so-called *turbo principle* (see Section 10.5.1) can be developed. In particular, in 1995 these considerations led C. Douillard *et al.* [1958] to devise a novel iterative strategy combining a SiSo equalizer (in practice, a MAP detector) with a SiSo decoder, which exchange *soft* (*extrinsic*) *information* to improve the reliability of decoded data. The resulting technical solution has been called *turbo detection* and shown to be a feasible approach to jointly addressing the equalization and decoding tasks, provided that the delay spread of the communication channel and the cardinality of the constellation for the adopted digital modulation are small. Two years later A. Glavieux, C. Laot and J. Labat proposed replacing the MAP equalizer of [1958] with an ISI canceler in order to reduce the overall computational complexity of the turbo detector [1959]; the resulting receiver is known as a *turbo equalizer*. As a result, the performance gap between an optimal joint equalization and decoding structure and that achievable through systems characterized by manageable complexity has been narrowed in a manner similar to that of near-Shannon-limit communications using turbo codes, as shown in Chapter 10.

This chapter introduces the reader to the problem of joint equalization and decoding and is organized as follows. In Section 12.2 noniterative optimal and suboptimal techniques for equalization and decoding are investigated. The development of iterative techniques (i.e., of turbo equalization algorithms) is investigated in Section 12.3, where both known and unknown channels are considered. In Sections 12.2 and 12.3 a SISO communication scenario is assumed. The exploitation of

Wireless Communications: Algorithmic Techniques, First Edition.
Giorgio M. Vitetta, Desmond P. Taylor, Giulio Colavolpe, Fabrizio Pancaldi, Philippa A. Martin.
© 2013 John Wiley & Sons, Ltd. Published 2013 by John Wiley & Sons, Ltd.

turbo equalization algorithms in MIMO communications is discussed in Section 12.4. Finally, some notes about the history of turbo equalization techniques and some suggestions for further reading are provided in Sections 12.5 and 12.6, respectively.

12.2   Noniterative Techniques

In what follows we focus primarily on a wireless system whose transmitter and channel are described by the block diagram illustrated in Figure 12.1. Here a binary encoder, with rate $R = k/n$, is fed for K consecutive clock intervals by a binary *message* $\mathbf{u} \triangleq [u_0, u_1, \dots, u_{K-1}]^T$, consisting of K independent and uniformly distributed random bits to generate the binary *codeword* $\mathbf{v} \triangleq [v_0, v_1, \dots, v_{N-1}]^T$, with $N = K/R$. This codeword undergoes bit interleaving which yields the new codeword $\mathbf{x} \triangleq [x_0, x_1, \dots, x_{N-1}]^T$. This is mapped to a symbol vector $\mathbf{c} \triangleq [c_0, c_1, \dots, c_{P-1}]^T$, consisting of P complex symbols, each belonging to an M-ary complex signal constellation $A_c = \{c^{(b)}, b = 0, 1, \dots, M - 1\}$, with $M = 2^m$, where m denotes the number of coded bits per channel symbol. Note that in the simplest case the constellation is binary ($M = 2$), so that $P = N$. In this case, the simple mapping rule $c = 1 - 2x$ can be used to map the encoded bit x to a BPSK symbol c. Otherwise $P = N/m$ (i.e., N is a multiple of P) since each of the N bits is transmitted only once via $\mathbf{c}$ and, in particular, the group $\mathbf{x}_l \triangleq [x_{lm}, x_{lm+1}, \dots, x_{lm+m-1}]^T$ of m adjacent bits (with $l = 0, 1, \dots, P - 1$) is mapped to the channel symbol c_l on the basis of a specific mapping rule (Gray mapping or other optimized rules can be adopted; e.g., see [1960, 1961]).

PAM signaling[1] and a static ISI channel, characterized by the CIR $\mathbf{h} \triangleq [h_0, h_1, \dots, h_{L_h-1}]^T$ vector (with $L_h < P$) and *known to the receiver*, are assumed here, so that the received vector can be put in the form (see (4.96)):

$$\mathbf{r} \triangleq [r_0, r_1, \dots, r_{P-1}]^T = \mathbf{Hc} + \mathbf{n}, \tag{12.1}$$

where

$$\mathbf{H} = \begin{bmatrix} h_0 & 0 & \cdots & & & & \cdots & 0 \\ h_1 & h_0 & 0 & \cdots & & & \cdots & 0 \\ h_2 & h_1 & h_0 & 0 & & & \cdots & 0 \\ \vdots & \ddots & \ddots & \ddots & \ddots & & & 0 \\ h_{L_h-1} & \cdots & \cdots & & \ddots & \ddots & \ddots & \vdots \\ 0 & & h_{L_h-1} & \cdots & & h_1 & h_0 & 0 & \vdots \\ \vdots & & & & h_{L_h-1} & h_{L_h-2} & & h_0 & 0 \\ 0 & \cdots & & & & 0 & h_{L_h-1} & \cdots & h_1 & h_0 \end{bmatrix} \tag{12.2}$$

is a $P \times P$ lower triangular channel matrix and $\mathbf{n} \triangleq [n_0, n_1, \dots, n_{P-1}]^T$ is an additive noise vector. In the following we assume that $\mathbf{n}$ consists of iid complex random Gaussian variables, each with zero mean, variance σ_n^2 and iid real and imaginary parts.

Given the model (12.1) for the noisy observed signal vector, if the optimization criterion in receiver design is the minimization of the *bit error probability* (BEP) and the channel matrix $\mathbf{H}$ is known, the optimal decoding strategy for the system configuration illustrated in Figure 12.1 can be expressed as:

$$u_l = \arg \max_{\tilde{u}_l \in GF(2)} \Pr\{u_l = \tilde{u}_l | \rho, \mathbf{h}\} \tag{12.3}$$

---

[1] Our discussion of noniterative techniques is limited to linear modulation formats for simplicity. Iterative equalization of CPM formats will be addressed in Section 12.3.4.2.

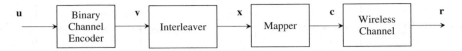

Figure 12.1 Configuration of the transmitter and communication channel assumed in the study of turbo equalization techniques.

for $l = 0, 1, \ldots, K-1$, where ρ denotes the value taken on by the random received vector $\mathbf{r}$. Computing the APP $\Pr\{u_l = \tilde{u}_l | \rho, \mathbf{h}\}$ inevitably requires marginalizing over u_l in the sequence-based a posteriori probability $\Pr\{\mathbf{u} = \tilde{\mathbf{u}} | \rho, \mathbf{h}\}$, since:

$$\Pr\{u_l = \tilde{u}_l | \rho, \mathbf{h}\} = \sum_{\tilde{\mathbf{u}}:u_l=\tilde{u}_l} \Pr\{\mathbf{u} = \tilde{\mathbf{u}} | \rho, \mathbf{h}\}. \tag{12.4}$$

Note also that, from Bayes' theorem:

$$\Pr\{\mathbf{u} = \tilde{\mathbf{u}} | \rho, \mathbf{h}\} = f_{\mathbf{r}}(\rho | \mathbf{u} = \tilde{\mathbf{u}}, \mathbf{h}) \Pr\{\mathbf{u} = \tilde{\mathbf{u}}\} / f_{\mathbf{r}}(\rho | \mathbf{h}), \tag{12.5}$$

where the probability $\Pr\{\mathbf{u} = \tilde{\mathbf{u}}\}$ can be factored as $\Pr\{\mathbf{u} = \tilde{\mathbf{u}}\} = \prod_{t=0}^{K-1} \Pr\{u_t = \tilde{u}_t\}$. Then the APP $\Pr\{u_l = \tilde{u}_l | \rho, \mathbf{h}\}$ of (12.4) can be rewritten as:

$$\Pr\{u_l = \tilde{u}_l | \rho, \mathbf{h}\} = \frac{1}{f_{\mathbf{r}}(\rho | \mathbf{h})} \sum_{\tilde{\mathbf{u}}:u_l=\tilde{u}_l} f_{\mathbf{r}}(\rho | \mathbf{u} = \tilde{\mathbf{u}}, \mathbf{h}) \prod_{t=0}^{K-1} \Pr\{u_t = \tilde{u}_t\}. \tag{12.6}$$

As in the study of turbo decoding, the optimal detection strategy can be formulated in terms of the LLR as:

$$L(u_l | \rho, \mathbf{h}) \triangleq \ln \frac{\Pr\{u_l = 1 | \rho, \mathbf{h}\}}{\Pr\{u_l = 0 | \rho, \mathbf{h}\}} \tag{12.7}$$

to be evaluated for $l = 0, 1, \ldots, K-1$ in place of the APPs generated by (12.4). Substituting (12.4) into this expression leads, after some manipulation, to:

$$L(u_l | \rho, \mathbf{h}) = L(u_l) + L_e(u_l | \rho, \mathbf{h}), \tag{12.8}$$

where

$$L(u_i) \triangleq \ln \frac{\Pr\{u_i = 1\}}{\Pr\{u_i = 0\}} \tag{12.9}$$

represents the *a priori* LLR and:

$$L_e(u_l | \rho, \mathbf{h}) = \frac{\sum_{\tilde{\mathbf{u}}:u_l=1} f_{\mathbf{r}}(\rho | \mathbf{u} = \tilde{\mathbf{u}}, \mathbf{h}) \prod_{t=0,t\neq l}^{K-1} \Pr\{u_t = \tilde{u}_t\}}{\sum_{\tilde{\mathbf{u}}:u_l=0} f_{\mathbf{r}}(\rho | \mathbf{u} = \tilde{\mathbf{u}}, \mathbf{h}) \prod_{t=0,t\neq l}^{K-1} \Pr\{u_t = \tilde{u}_t\}} \tag{12.10}$$

can be called *extrinsic information* or *extrinsic* LLR [1962]. In practice the latter term represents the information about u_l provided by both $\mathbf{r}$ and $\Pr\{u_t = \tilde{u}_t\}$ for all $t \neq l$, and plays a fundamental role in turbo equalization (note, however, that unlike the extrinsic LLR in (10.15), $L_e(u_l | \rho, \mathbf{h})$ includes the contribution from the entire vector $\mathbf{r}$).

Unfortunately, the BEP-optimal strategy (12.3) is computationally infeasible since its complexity is of order $O(2^K)$. This motivates the search for other strategies entailing a substantially smaller computational burden. A standard approach involves splitting the detection problem for coded data over an ISI channel into two subproblems, namely equalization (detection) and decoding, as illustrated in Figure 12.2.

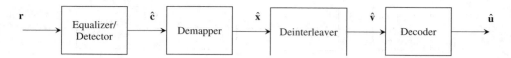

Figure 12.2   Receiver structure for separate equalization and decoding.

When this (suboptimal) approach is taken, the equalizer produces a hard estimate $\hat{\mathbf{c}}$ of the symbol vector $\mathbf{c}$, from which an estimate $\hat{\mathbf{x}}$ of the codeword $\mathbf{x}$ is extracted through a simple demapping algorithm. The estimate $\hat{\mathbf{x}}$ undergoes deinterleaving, which yields $\hat{\mathbf{v}}$ which is then input to a hard decoding algorithm. Finally, decoding yields an estimate $\hat{\mathbf{u}}$ of the binary *message* $\mathbf{u}$. A more refined, but conceptually related, approach involves adopting a soft output algorithm for channel equalization and a soft input algorithm for decoding; in this case, the first algorithm generates a soft information vector $\mathbf{s}(\mathbf{c}) \triangleq [\mathbf{s}(c_0)^T, \mathbf{s}(c_1)^T, \ldots, \mathbf{s}(c_{P-1})^T]^T$ in place of the hard estimate $\hat{\mathbf{c}}$, where $\mathbf{s}(c_l) \triangleq [s_0(c_l), s_1(c_l), \ldots, s_{M-1}(c_l)]^T$ represents a soft (i.e., probabilistic) information vector about the lth channel symbol, with $l = 0, 1, \ldots, P - 1$ (of course, if a binary constellation is used, $\mathbf{s}(c_l)$ becomes a scalar quantity, since soft information can be condensed in a single LLR about c_l). The next step for the receiver is to compute a soft information vector (i.e., an LLR vector) $\mathbf{s}(\mathbf{x}) \triangleq [s(x_0), s(x_1), \ldots, s(x_{N-1})]^T$ on the reliability of codeword elements $\mathbf{x}$ via an appropriate *soft demapping* algorithm [1963, 1964]. This vector is deinterleaved to $\mathbf{s}(\mathbf{v}) \triangleq [s(v_0), s(v_1), \ldots, s(v_{N-1})]^T$, which feeds the decoding algorithm which provides an estimate of each transmitted data bit. Note that the BEP-optimal decoding strategy, given the soft information vector $\mathbf{s}(\mathbf{v})$, can be expressed as:

$$u_l = \arg \max_{\tilde{u}_l \in GF(2)} \Pr\{u_l = \tilde{u}_l | \mathbf{s}(\mathbf{v})\}, \tag{12.11}$$

with $l = 0, 1, \ldots, K - 1$. If a trellis coding scheme is employed, the APP appearing in (12.11) can be computed via the BCJR algorithm (i.e., the FBA; see Section 9.2.4). This algorithm can also be adopted for a block code if a (time-varying) trellis diagram is developed for its representation (see Section 9.4). Note also that the FBA can be adopted to solve the first subproblem as well (i.e., soft output channel equalization) if both the cardinality M and the channel memory L_h are small, so that the state trellis representing the communication channel contains a sufficiently small number of states. When this does not occur, other solutions are needed. In particular, as will be discussed in more detail in Section 12.3.3, *linear filter-based* approaches are an interesting alternative to soft output MAP equalization, since they require simple operations on the received symbols, which can be described with matrix operations on the received sequence directly. In particular, given the received signal model (12.1), an MMSE (*soft*) estimate $\check{\mathbf{c}}$ of the symbol vector $\mathbf{c}$ can be generated as (see Sections 6.2.1.3 and 6.2.1.4):

$$\check{\mathbf{c}} = \mathbf{H}^H (\sigma_n^2 \mathbf{I}_P + \mathbf{HH}^H)^{-1} \mathbf{r}, \tag{12.12}$$

if we assume that the elements of the channel symbol vector $\mathbf{c}$ are independent and uniformly distributed. Generally speaking, the symbol estimates (i.e., the elements of $\check{\mathbf{c}}$) do not belong to the constellation A_c. As a result, if a hard decision $\hat{\mathbf{c}}$ is required, this can be generated by mapping each of the elements of $\check{\mathbf{c}}$ to that symbol of A_c at closest (Euclidean) distance. However, if probabilistic information about $\mathbf{c}$ is required, a different approach can be pursued. In particular, it can be assumed that the elements of the estimation error $\mathbf{e} \triangleq \check{\mathbf{c}} - \mathbf{c}$ are Gaussian distributed [1913], that is, $e_l \triangleq \check{c}_l - c_l$ belongs to $\mathcal{N}(E\{e_l\}, \text{var}\{e_l\})$ for $l = 0, 1, \ldots, P - 1$, where $\text{var}\{X\}$ denotes the variance of the random variable X. Since $\mathbf{e}$ is an MMSE estimation error, its mean vector and autocovariance matrix are given by $E\{\mathbf{e}\} = \mathbf{0}_P$ and:

$$\mathbf{C_e} = \mathbf{I}_P - \mathbf{H}^H (\sigma_n^2 \mathbf{I}_P + \mathbf{HH}^H)^{-1} \mathbf{H}, \tag{12.13}$$

respectively, so that $\mathrm{E}\{e_l\} = 0$ and $\mathrm{var}\{e_l\} = [\mathbf{C}_e]_{l,l}$. If $f_{e_l}(\cdot)$ denotes the pdf of e_l, a soft information M-dimensional vector $\mathbf{s}(c_l)$ on c_l can be generated by sampling $f_{e_l}(\cdot)$ at the M possible values $\{e_l[b] \triangleq \check{c}_l - c^{(b)}, b = 0, 1, \ldots, M - 1\}$, that is, as:

$$s_b(c_l) = B_l \cdot f_{e_l}(\check{c}_l - c^{(b)}) \tag{12.14}$$

for $l = 0, 1, \ldots, P - 1$, where B_l is a normalization constant ensuring that $\sum_{b=0}^{M-1} s_b(c_l) = 1$. Note that:

(a) this strategy to compute soft information from a filter output can be applied to other filter-based equalization algorithms as well,
(b) for complexity reasons, in a practical implementation of this filter-based approach only a small (sliding) window of the received data is processed in place of the complete vector $\mathbf{r}$, as discussed in Section 12.3.3.2.

Since the decoder in Figure 12.2 requires soft information about the coded bits, the next step in the receiver is to extract the soft information vector $\mathbf{s}(\mathbf{x})$ from $\mathbf{s}(\mathbf{c})$.

The mapping from probabilities to probabilities is commonly referred to as *soft demapping*. If a binary constellation $A_c = \{c^{(0)} = -1, c^{(1)} = 1\}$ and the mapping strategy $c = 1 - 2x$ between channel symbols and interleaved coded bits are used, the demapping operation is simple, since $\mathrm{Pr}\{x_l = 0\} = \mathrm{Pr}\{c_l = 1\} = \mathrm{Pr}\{c_l = c^{(1)}\}$ and $\mathrm{Pr}\{x_l = 1\} = \mathrm{Pr}\{c_l = -1\} = \mathrm{Pr}\{c_l = c^{(0)}\}$, so that the LLR about x_l can be evaluated as (see (12.7)):

$$s(x_l) \triangleq \ln \frac{s_1(c_l)}{s_0(c_l)}. \tag{12.15}$$

If the constellation is not binary, soft demapping depends on the rule adopted in mapping coded bits to channel symbols at the transmitter. Finally, after deinterleaving $\mathbf{s}(\mathbf{x})$ to $\mathbf{s}(\mathbf{v})$, the decoder can evaluate the estimates of the transmitted data bits on the basis of the strategy (12.11).

12.3   Algorithms for Combined Equalization and Decoding

This section studies turbo equalization algorithms. After providing an introduction to the key concepts of turbo equalization theory and describing the general structure of a turbo equalizer in Section 12.3.1, an interpretation of turbo equalization in terms of factor graphs is provided in Section 12.3.2. Then, some strategies for suboptimal SiSo equalization and for accomplishing turbo equalization in the FD are illustrated in Sections 12.3.3 and 12.3.4, respectively. Finally, the problem of turbo equalization in the presence of an *unknown* channel is briefly discussed in Section 12.3.5.

12.3.1   Introduction

As discussed in the previous section, the implementation of the optimal strategy for joint equalization and decoding is infeasible for a general code and interleaver, so that suboptimal solutions are needed. In a receiver structure for separate equalization and decoding, such as that illustrated in Figure 12.2, the soft information flow is *unidirectional*, from equalization to decoding. The study of turbo codes, however, has clearly shown that soft information should not flow in a single direction and that it can substantially enhance decoding performance if properly processed. In particular, turbo decoding principles can be applied if SiSo modules are used for equalization and decoding, since, once the decoding algorithm processes the soft information delivered from a soft output equalizer, it can, in turn, generate its own soft information referring to the relative likelihood of each of the transmitted bits. Let us now analyze how this can be accomplished. To begin, we note that, in principle, given the reliability vector $\mathbf{s}(\mathbf{v})$ defined in the previous section, the decoder can evaluate the LLR:

$$L_D(v_l|\mathbf{s}(\mathbf{v})) \triangleq \ln \frac{\Pr\{v_l = 1|\mathbf{s}(\mathbf{v})\}}{\Pr\{v_l = 0|\mathbf{s}(\mathbf{v})\}} \qquad (12.16)$$

for the bit v_l, with $l = 0, 1, \ldots, N-1$, and thereby generate a new reliability vector $\bar{\mathbf{s}}_D(\mathbf{v}) \triangleq [\bar{s}_D(v_0), \bar{s}_D(v_1), \ldots, \bar{s}_D(v_{N-1})]^T$ with $\bar{s}_D(v_l) \triangleq L_D(v_l|\mathbf{s}(\mathbf{v}))$. In general, $\bar{\mathbf{s}}_D(\mathbf{v})$ is more reliable than $\mathbf{s}(\mathbf{v})$, thanks to the redundancy exploited in the decoding process (in other words, the LLR values in $\bar{\mathbf{s}}_D(\mathbf{v})$ provide a better indication about the two possible values, 0 and 1, than those of $\mathbf{s}(\mathbf{v})$). After interleaving $\bar{\mathbf{s}}_D(\mathbf{v})$ to $\bar{\mathbf{s}}(\mathbf{x})$ and accomplishing soft mapping of the LLR vector $\bar{\mathbf{s}}(\mathbf{x})$ to the soft information vector $\bar{\mathbf{s}}(\mathbf{c}) \triangleq [\bar{\mathbf{s}}(c_0)^T, \bar{\mathbf{s}}(c_1)^T, \ldots, \bar{\mathbf{s}}(c_{P-1})^T]^T$ (collecting probabilistic data about each channel symbol as in the previous section), the new vector can be exploited to restart the equalizer, which is now endowed with new symbol probabilities. This closes the loop, since the latter step produces new soft information about the transmitted symbol $\mathbf{c}$ and this information can be passed to the decoder, after appropriate demapping and deinterleaving. The iterative procedure just described can be repeated until the decisions about the transmitted data produced by the decoder do not change further, that is, the transient is over. If this strategy is adopted, BER performance results show that the quality of data decisions does indeed improve with each iteration, but not significantly. In contrast, the improvement can be appreciable if we modify the way soft information to be exchanged between the two SiSo modules is generated. In particular, if we refer to the LLRs generated by the SiSo decoding module about coded bits, the quantity $L_D(v_l|\mathbf{s}(\mathbf{v}))$ of (12.16) has to be replaced by:

$$L_{D,e}(v_l|\mathbf{s}_l(\mathbf{v})) \triangleq \ln \frac{\Pr\{v_l = 1|\mathbf{s}_l(\mathbf{v})\}}{\Pr\{v_l = 0|\mathbf{s}_l(\mathbf{v})\}}, \qquad (12.17)$$

where $\mathbf{s}_l(\mathbf{v}) \triangleq [s(v_0), s(v_1), \ldots, s(v_{l-1}), s(v_{l+1}), \ldots, s(v_{N-1})]^T$ is obtained from $\mathbf{s}(\mathbf{v})$ by removing $s(v_l)$. In other words, the LLR $L_{D,e}(v_l|\mathbf{s}_l(\mathbf{v}))$ is evaluated ignoring the soft datum $s(v_l)$ available at the beginning of the decoding process (this choice can be motivated by resorting to FG theory, as discussed in the following subsection). As an alterative, $L_{D,e}(v_l|\mathbf{s}_l(\mathbf{v}))$ can be computed as [1660]:

$$L_{D,e}(v_l|\mathbf{s}_l(\mathbf{v})) = L_D(v_l|\mathbf{s}(\mathbf{v})) - s(v_l), \qquad (12.18)$$

that is, by analogy with turbo decoding, removing from the soft output $L_D(v_l|\mathbf{s}(\mathbf{v}))$ the a priori LLR $s(v_l)$ available at the start of the decoding process. In the jargon of turbo codes, the new LLR $L_{D,e}(v_l|\mathbf{s}_l(\mathbf{v}))$ (12.18) represents a form of *extrinsic information*. The resulting LLR vector $\bar{\mathbf{s}}_e(\mathbf{v}) \triangleq [\bar{s}_e(v_0), \bar{s}_e(v_1), \ldots, \bar{s}_e(v_{N-1})]^T$ with $\bar{s}_e(v_l) \triangleq L_{D,e}(v_l|\mathbf{s}_l(\mathbf{v}))$ can be processed (i.e., interleaved and soft-mapped) to generate the extrinsic information vector $\mathbf{s}_D(\mathbf{c}) \triangleq [\mathbf{s}_D(c_0)^T, \mathbf{s}_D(c_1)^T, \ldots, \mathbf{s}_D(c_{P-1})^T]^T$ about the channel symbols; this is delivered to the SiSo equalizer, which should operate on the same principle. In other words, it is expected to generate an extrinsic information vector $\mathbf{s}_{E,e}(\mathbf{c}) \triangleq [\mathbf{s}_{E,e}(c_0)^T, \mathbf{s}_{E,e}(c_1)^T, \ldots, \mathbf{s}_{E,e}(c_{P-1})^T]^T$ (where $\mathbf{s}_{E,e}(c_l) \triangleq [s_{E,e,0}(c_l), s_{E,e,1}(c_l), \ldots, s_{E,e,M-1}(c_l)]^T$ is the probabilistic information vector for the lth channel symbol of the given constellation) on the basis of the following principle: $\mathbf{s}_{E,e}(c_l)$ is evaluated by processing the received data $\mathbf{r}$ and $\mathbf{s}_D(\mathbf{c})$ except for $\mathbf{s}_D(c_l), l = 0, 1, \ldots, P-1$. Note also that, if a binary constellation is used, $\mathbf{s}_{E,e}(c_l)$ can be denoted $s_{E,e}(c_l)$ if this represents the LLR referring to c_l (similar considerations also hold for $\mathbf{s}_D(c_l)$), so that, by analogy with (12.18), $s_{E,e}(c_l)$ can be computed as [1660]:

$$s_{E,e}(c_l) = L_E(c_l|\mathbf{r}, \mathbf{s}_D(\mathbf{c})) - s_D(c_l), \qquad (12.19)$$

where $L_E(c_l|\mathbf{r}, \mathbf{s}_D(\mathbf{c}))$ denotes the LLR evaluated by the SiSo equalizer, given the observed data vector $\mathbf{r}$ and the entire soft vector $\mathbf{s}_D(\mathbf{c})$. It is worth pointing out that, if a multilevel constellation is used, LLR vectors can also be used in the procedure described above, as discussed in Section 4.3.3. The new soft information vector $\mathbf{s}_{E,e}(\mathbf{c})$ undergoes demapping and deinterleaving, so that a new vector of soft data is made available to the SiSo decoding module and a new decoding procedure can be started. The closed loop strategy described above is embodied by the block diagram of the *turbo equalizer* (TE) shown in Figure 12.3, which refers to the case of a binary constellation. The TE structure includes a SiSo equalizer (tailored to a specific channel model), a SiSo decoder (tailored to adopted channel code), an interleaver (π), a deinterleaver (π^{-1}), one soft mapping device, one soft demapping device and a hard decision device (represented as a threshold device in the figure for simplicity) to be used at the end of the last iteration. Moreover, each SiSo module has two inputs (one associated with noisy data, the other one with a priori information) and a single output, and extrinsic information is extracted by subtracting a priori LLRs from this output. Similarly to the decoding of turbo codes, a pipelined architecture for the receiver can also be implemented, as illustrated in [1965]. In this case the turbo equalizer consists of the cascade of as many (equal) stages (or modules) as the number of iterations to be carried out, and each stage includes the same subsystems as Figure 12.3 plus a delay line for the received signal samples.

In principle, the TE procedure has to be repeated until convergence. The speed of this process is influenced by various factors, such as the selected SiSo equalization algorithm and the cardinality M of the constellation (e.g., see [1962]). In practice, various BER performance results available in the technical literature show that the energy gap between a system employing turbo equalization and its

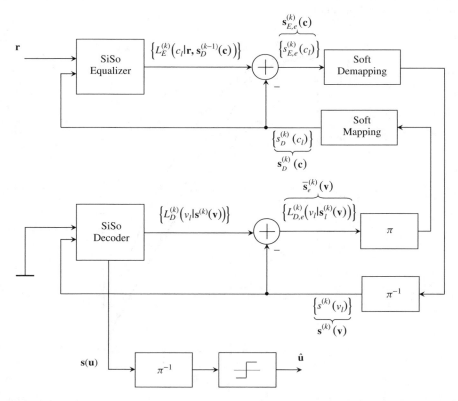

Figure 12.3 Block diagram of a turbo equalizer for the case of a BPSK constellation. Note that the observation input of the SiSo decoder is grounded to indicate that the algorithm operates on the basis of a priori values only and that this figure refers to the start of the kth iteration.

counterpart operating in the absence of ISI substantially reduces after the first two iterations (e.g., see [1660, 1962]). This is confirmed by the numerical results in the following example for a specific system.

Example 12.3.1 Let us again consider the transmitter and channel configuration shown in Figure 12.1 and assume that:

(a) the message consists of $K = 2048$ bits;
(b) a BPSK modulation is employed;
(c) a rate-1/2, 32-state convolutional code having generators $[23_8]$ and $[35_8]$ in octal form and a binary constellation ($M = 2$) are used (so that the codeword length is $N = 4096$ bits);
(d) random interleaving is adopted;
(e) the CIR is real and represented by the vector $\mathbf{h} = [0.407, 0.815, 0.407]^T$ [1965].

Some BER results referring to this case are shown in Figure 12.4, which compares the performance offered by MAP decoding (accomplished via the BCJR algorithm) in the absence of ISI (no ISI curve) with that achieved by a turbo equalizer (characterized by the architecture of Figure 12.3 and employing the BCJR algorithm for SiSo equalization and for SiSo decoding) with various numbers of iterations.

Note that the curve referring to the first iteration actually shows the error performance achieved at the end of the channel decoding step in the first iteration, that is, that provided by *separate* equalization and decoding. These results illustrate:

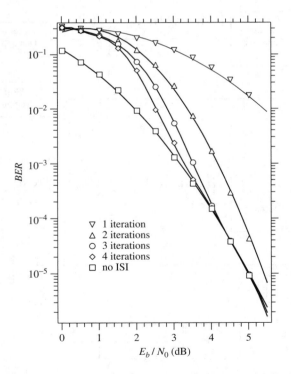

Figure 12.4   BER performance provided by decoding in the absence of ISI and by turbo equalization with a different number of iterations. The adopted coding and modulation schemes and channel model are given in Example 12.3.1.

- the existence of a *significant gap* (which, in many scenarios, is found to be greater than 3 dB [1962]) between the performance achievable in the absence of ISI and that provided by separate equalization and decoding,
- the possibility of getting very close to the performance in the absence of ISI via turbo equalization for reasonably small SNRs.

The latter result is achieved at the price of significant complexity (an assessment of the computational complexity of turbo equalizers for different coding schemes can be found in [1962, 1966]), which grows linearly with the number of iterations and exponentially with the code/channel memory.

□

Generally speaking, to limit the computational burden of turbo equalization, it is fundamental to:

(a) keep the overall number of iterations as low as possible;
(b) develop reduced-complexity SiSo modules for equalization and decoding.

Fortunately, to minimize the number of iterations carried out by a turbo equalizer various *termination criteria* can be adopted, exactly as in the decoding of turbo codes [1967]. Moreover, various reduced-complexity SiSo modules can be found in the literature, as already mentioned in Section 12.2 and discussed in more detail in Section 12.3.3.

12.3.2   *Turbo Equalization from a FG Perspective*

The system model shown in Figure 12.1, referring to a coded data transmission over an ISI channel, can also be represented via different graphical models and, in particular, via FG descriptions (an introduction to FGs can be found in Section 10.8). As shown in what follows, such descriptions are a useful tool to provide insight into existing equalization and decoding techniques and to develop new iterative and noniterative solutions. In fact, several algorithms, which differ in the messages communicated along the FG edges (and, consequently, for their complexity), can be developed from a specific FG.

Our brief investigation starts from the optimal decision strategy (12.3) for which the cost function $\Pr\{u_l = \tilde{u}_l | \boldsymbol{\rho}, \mathbf{h}\}$ to be optimized over the set of possible trial messages can be written as (see (12.4)):

$$\Pr\{u_l = \check{u}_l | \boldsymbol{\rho}, \mathbf{h}\} = \sum_{\tilde{\mathbf{u}} : \tilde{u}_l = \check{u}_l} \Pr\{\mathbf{u} = \tilde{\mathbf{u}} | \boldsymbol{\rho}, \mathbf{h}\}, \tag{12.20}$$

with (see (12.6)):

$$\Pr\{\mathbf{u} = \tilde{\mathbf{u}} | \boldsymbol{\rho}, \mathbf{h}\} \propto f_{\mathbf{r}}(\boldsymbol{\rho} | \mathbf{u} = \tilde{\mathbf{u}}, \mathbf{h}) \prod_{t=0}^{K-1} \Pr\{u_t = \tilde{u}_t\}, \tag{12.21}$$

so that:

$$\Pr\{u_l = \check{u}_l | \boldsymbol{\rho}, \mathbf{h}\} \propto \Pr\{u_t = \check{u}_l\} \sum_{\tilde{\mathbf{u}} : \tilde{u}_l = \check{u}_l} f_{\mathbf{r}}(\boldsymbol{\rho} | \mathbf{u} = \tilde{\mathbf{u}}, \mathbf{h}) \prod_{t=0, t \neq l}^{K-1} \Pr\{u_t = \tilde{u}_t\}. \tag{12.22}$$

Equation (12.20) shows that the optimal decision strategy requires *marginalization* of the probability mass function (i.e., of the *global function*) $\Pr\{\mathbf{u} = \tilde{\mathbf{u}} | \boldsymbol{\rho}, \mathbf{h}\}$, which, if irrelevant factors are dropped, can be represented via the factorization on the RHS of (12.21). In turn, such a factorization can be represented via the FG illustrated in Figure 12.5(a), where the square boxes (*function nodes*) denote the relevant factors of $\Pr\{\mathbf{u} = \tilde{\mathbf{u}} | \boldsymbol{\rho}, \mathbf{h}\}$, the circles (*variable nodes*) denote the data $\{u_l\}$ and the edges specify the dependencies between variables and factors.

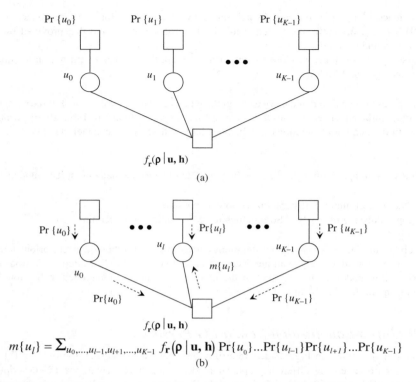

$$m\{u_l\} = \sum_{u_0,...,u_{l-1},u_{l+1},...,u_{K-1}} f_{\mathbf{r}}(\boldsymbol{\rho} \mid \mathbf{u}, \mathbf{h}) \Pr\{u_0\}...\Pr\{u_{l-1}\}\Pr\{u_{l+1}\}...\Pr\{u_{K-1}\}$$
(b)

Figure 12.5 (a) Factor graph associated with coded transmission over an ISI channel; (b) marginalization of u_l in the global function appearing on the RHS of (12.22) via MP.

Given this FG, the *marginalization* required by (12.20) can be accomplished via a specific step-by-step procedure operating over the graph itself and known as *message passing* (MP). In general, if the variables appearing in the graph are discrete and their values belong to an alphabet $\mathcal{V}$ of cardinality M_v, MP consists of the following simple rules:

1. Each function node sums the product of the incoming messages and the node function over all variables adjacent to the function node except the one the message is sent to.
2. Each variable node either transmits the value $1/M_v$ if it is a *leaf* of the graph, or multiplies all its incoming messages, except the one coming from the function node the message is sent to, and transmits the result.

These rules are summarized in Figure 12.6 and can easily be generalized to the case of continuous variables (since the sums are replaced by integrals and the message $1/M_v$ coming from a variable node by a uniform pdf over the domain $\mathcal{V}$ of the variable itself). On the basis of these rules the calculation of $\Pr\{u_l = \tilde{u}_l|\boldsymbol{\rho}, \mathbf{h}\}$ via MP can be represented with a flow of messages towards the node associated with u_l in the factor graph of $\Pr\{\mathbf{u} = \tilde{\mathbf{u}}|\boldsymbol{\rho}, \mathbf{h}\}$, as illustrated in Figure 12.5(b). It is worth pointing out that, in this case, the exploitation of MP provides a step-by-step procedure for evaluating (12.20) and a nice interpretation of it as a flow of messages in a proper graph. Such a procedure yields the *exact* result (i.e., the desired marginalization), since the graph representing the global function expressed by the RHS of (12.21) is of *tree type*, that is, it does not contain *cycles*. In fact, as already discussed in the study of modern coding schemes and, in particular of LDPC codes (see Section 10.6), the application of MP to a graph containing cycles does not produce the correct result, whatever the schedule adopted

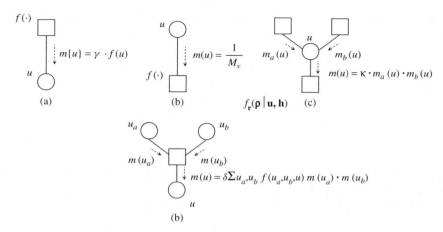

(a)   (b) $f_\mathbf{r}(\rho\,|\,\mathbf{u},\mathbf{h})$ (c)

(b)

Figure 12.6 Update rules for MP on an FG. The constants γ, κ and δ are selected in such a way that $m(u)$ is a probability mass function (or a pdf in the case of a continuous variable).

in generating and exchanging messages in the graph. Instead, the cycles indicate that the result may differ depending on how the message update is scheduled, how long the messages are allowed to circulate and how the messages produced in the variable nodes being part of cycles are initialized, as already discussed in Section 10.6.

Another important issue related to the MP procedure shown in Figure 12.5(b) is its computational complexity, which, unfortunately, is proportional to $K \cdot 2^{K-1}$, since it needs to be repeated for each of the K variable nodes (i.e., information bits) and the number of operations to be accomplished at the lower function node is proportional to 2^{K-1}. Generally speaking, achieving better computational efficiency requires manipulating the global function to be represented via an FG. In particular, the following specific strategies can be adopted:

- *Factorization.* This involves *factoring a global function of many variables into terms depending on as few variables as possible.* This means that the underlying graph contains function nodes with small degree.[2]
- *Introduction of state (internal) variables.* This involves *augmenting a global function with new (i.e., state) variables* such that a suitable factorization becomes possible. This approach inevitably complicates the structure of the global function and *increases the number of variables to be summed over in marginalization*; despite this, it may yield a substantial overall increase in the efficiency of MP.

Applying the above two strategies to our optimal detection problem requires a different factorization of the conditional pdf $f_\mathbf{r}(\rho\,|\,\mathbf{u} = \tilde{\mathbf{u}}, \mathbf{h})$ appearing in (12.21). With this aim the vectors of state variables $\mathbf{x} \triangleq [x_0, x_1, \dots, x_{N-1}]^T$ (binary interleaved codeword) and $\mathbf{c} \triangleq [c_0, c_1, \dots, c_{P-1}]^T$ (channel symbol vector) are introduced, taking into account that $f_\mathbf{r}(\rho\,|\,\mathbf{u} = \tilde{\mathbf{u}}, \mathbf{h})$ can be obtained from the *marginalization* of $f_{\mathbf{r},\mathbf{x},\mathbf{c}}(\rho, \tilde{\mathbf{x}}, \tilde{\mathbf{c}}\,|\,\mathbf{u} = \tilde{\mathbf{u}}, \mathbf{h})$. The latter pdf can be factored as:

$$f_{\mathbf{r},\mathbf{x},\mathbf{c}}(\rho, \tilde{\mathbf{x}}, \tilde{\mathbf{c}}\,|\,\mathbf{u} = \tilde{\mathbf{u}}, \mathbf{h}) = f_\mathbf{r}(\rho\,|\,\mathbf{c} = \tilde{\mathbf{c}}, \mathbf{h})\,\mathrm{Pr}\{\mathbf{c} = \tilde{\mathbf{c}}\,|\,\mathbf{x} = \tilde{\mathbf{x}}\}\,\mathrm{Pr}\{\mathbf{x} = \tilde{\mathbf{x}}\,|\,\mathbf{u} = \tilde{\mathbf{u}}\}, \qquad (12.23)$$

---

[2] As already stated in Section 10.6.2, the *degree* of a node in a graph represents the number of edges connected to it.

so that the new global function is obtained replacing $f_r(\rho|\mathbf{u} = \tilde{\mathbf{u}}, \mathbf{h})$ with $f_{r,x,c}(\rho, \tilde{\mathbf{x}}, \tilde{\mathbf{c}}|\mathbf{u} = \tilde{\mathbf{u}}, \mathbf{h})$ in (12.21) to produce:

$$g(\tilde{\mathbf{u}}, \tilde{\mathbf{x}}, \tilde{\mathbf{c}}) = f_r(\rho|\mathbf{c} = \tilde{\mathbf{c}}, \mathbf{h}) \prod_{t=0}^{P-1} \Pr\{c_t = \tilde{c}_t|\mathbf{x}_t = \tilde{\mathbf{x}}_t\}$$

$$\cdot \Pr\{\mathbf{v} = \tilde{\mathbf{v}}|\mathbf{u} = \tilde{\mathbf{u}}\} \prod_{t=0}^{K-1} \Pr\{u_t = \tilde{u}_t\}, \tag{12.24}$$

since

- $\Pr\{\mathbf{x} = \tilde{\mathbf{x}}|\mathbf{u} = \tilde{\mathbf{u}}\} = \Pr\{\mathbf{v} = \tilde{\mathbf{v}}|\mathbf{u} = \tilde{\mathbf{u}}\}$ (there is a one-to-one mapping between a binary codeword $\mathbf{v}$ and its interleaved counterpart $\mathbf{x}$), and
- the probability mass function $\Pr\{\mathbf{c} = \tilde{\mathbf{c}}|\mathbf{x} = \tilde{\mathbf{x}}\}$ can be factored as:

$$\Pr\{\mathbf{c} = \tilde{\mathbf{c}}|\mathbf{x} = \tilde{\mathbf{x}}\} = \prod_{t=0}^{P-1} \Pr\{c_t = \tilde{c}_t|\mathbf{x}_t = \tilde{\mathbf{x}}_t\}, \tag{12.25}$$

since c_t is selected on the basis of $\mathbf{x}_t$ only.

Note also that $\Pr\{\mathbf{v} = \tilde{\mathbf{v}}|\mathbf{u} = \tilde{\mathbf{u}}\}$ can take on the values 0 or 1 depending on whether $\tilde{\mathbf{v}}$ is a valid codeword (i.e., it is generated by encoding $\tilde{\mathbf{u}}$) or not, respectively. The FG associated with the global function $g(\tilde{\mathbf{u}}, \tilde{\mathbf{x}}, \tilde{\mathbf{c}})$ (12.24) is illustrated in Figure 12.7, where $m = 2$ (i.e., a quaternary constellation) is assumed for simplicity.

Unfortunately, this graph contains, unlike that shown in Figure 12.5, *multiple cycles*, so that the application of MP to it leads to approximate and iterative solutions, for which different rules can be adopted in message scheduling. In particular, if a specific message schedule is adopted, the turbo equalization strategy described in Section 12.3.1 is found. More specifically, this schedule consists of generating the output messages for each edge connected to the node $f_r(\rho|\mathbf{c}, \mathbf{h})$ (*detection*) in the P nodes $\{\Pr\{c = c_t|\mathbf{x} = \tilde{\mathbf{x}}_t\}\}$ (*demapping*), and finally in the node $\Pr\{\mathbf{v}|\mathbf{u}\}$ (*decoding*) and vice versa when the next iteration starts. Note that the message update in the node $\Pr\{\mathbf{v}|\mathbf{u}\}$ can entail substantial computational effort, since its degree is equal to $N + K$, which is potentially a very large quantity. This suggests that a suitable factorization of this probability mass function can lead to a local FG, to which MP can be applied. This occurs, for instance, when LDPC coding is adopted and MP algorithms are employed for decoding.

Let us now focus on the factor $f_r(\rho|\mathbf{c}, \mathbf{h})$ of $g(\tilde{\mathbf{u}}, \tilde{\mathbf{x}}, \tilde{\mathbf{c}})$ (12.24) and consider the local graph for it; such a graph is shown in Figure 12.8, and performing MP over it corresponds to the detection part of the receiver. In fact, the M distinct incoming messages $\{m(c_t = \tilde{c}_t)\}$ traveling through the variable node c_t (with $t = 0, 1, \ldots, P - 1$) represent the probability mass function of the channel symbol, that is, they can be considered as the a priori probabilities that c_t takes on each of the M possible values of the complex signal constellation $\mathcal{C}$. Such probabilities are usually set to the same values $1/M$ for any value of t when the node function associated with $f_r(\rho|\mathbf{c}, \mathbf{h})$ computes the outgoing messages for the first time. The equation for evaluating the outgoing message $m(c_t = c^{(b)})$ (with $b = 0, 1, \ldots, M - 1$) results from the application of the update rule shown in Figure 12.6(d) and leads to the expression:

$$m(c_t = c^{(b)}) = \delta \sum_{\tilde{\mathbf{c}}: \tilde{c}_l = c^{(b)}} \prod_{\substack{t=0 \\ t \neq l}}^{P-1} m(c_t = \tilde{c}_t), \tag{12.26}$$

which generates the *extrinsic* APP needed for turbo equalization. In other words, MP on the local graph of Figure 12.8 leads to APP detection. Unfortunately, the complexity of this strategy for APP

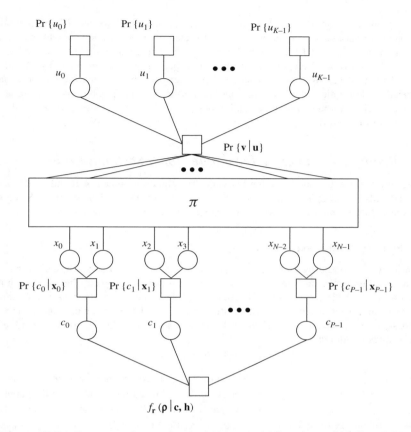

Figure 12.7 Factor graph associated with the global function $g(\tilde{\mathbf{u}}, \tilde{\mathbf{x}}, \tilde{\mathbf{c}})$ (12.24).

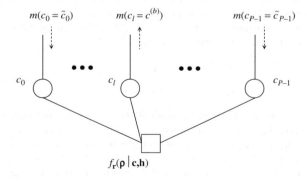

Figure 12.8 Local graph referring to the factor $f_{\mathbf{r}}(\rho|\mathbf{c}, \mathbf{h})$ of $g(\tilde{\mathbf{u}}, \tilde{\mathbf{x}}, \tilde{\mathbf{c}})$ (12.24). A set of $P - 1$ incoming messages and an outgoing message are shown.

detection is proportional to $P \cdot M^{P-1}$, so that, in practice, it is huge for any reasonable value of P. Finally, it is important to note that all the results illustrated in this subsection have been derived under the assumption of a *known* CIR vector $\mathbf{h}$. Formally, our results can easily be extended to the case of a statistically known channel. In fact, in this case, the only change in the FG of Figure 12.7 is that $f_{\mathbf{r}}(\rho|\mathbf{c}, \mathbf{h})$ is replaced by $f_{\mathbf{r}}(\rho|\mathbf{c})$. This latter pdf, however, may be hard to compute, depending on the statistical description of the channel model. Further information on this problem and on the applications of FGs to turbo equalization can be found in [1968–1970], as well as in in [1971] (where a MIMO scenario is considered).

12.3.3 Reduced-Complexity Techniques for SiSo Equalization

In the literature various alternatives to the FBA for SiSo equalization can be found. In this subsection we provide details on a couple of these alternatives, namely soft ISI cancelation and SiSo equalization based on linear MMSE filtering. Some other strategies are then briefly mentioned.

12.3.3.1   Soft ISI Cancelation

This solution was proposed by C. Laot *et al.* [1965] (see also [1972]) and is based on the following ideas. In a digital transmission complete cancelation of ISI can be achieved when the transmitted data are known a priori at the receive side. Of course, this never occurs in the presence of information data; however, in a turbo receiver a reliable estimate of these data is potentially available at the end of the first step for equalization and channel decoding. In practice, this algorithm resembles a DFE, since it generates its *l*th output as:

$$y_l = \mathbf{P}_l^T \mathbf{r}_l - \mathbf{Q}_l^T \bar{\mathbf{c}}_l, \tag{12.27}$$

where $\mathbf{r}_l \triangleq [r_{l+L_1}, \dots, r_l, \dots, r_{l-L_1}]^T$ contains $2L_1 + 1$ consecutive samples of the received signal vector $\mathbf{r} \triangleq [r_0, r_1, \dots, r_{P-1}]^T$, $\bar{\mathbf{c}}_l$ represents the estimated mean value of the channel symbol vector $\mathbf{c}_l \triangleq [c_{l+L_2}, \dots, c_l, \dots, c_{l-L_2}]^T$ (which contains $2L_2 + 1$ consecutive elements of the transmitted symbol vector $\mathbf{c} \triangleq [c_0, c_1, \dots, c_{P-1}]^T$), $\mathbf{P}_l \triangleq [p_{-L_1}[l], \dots, p_0[l], \dots, p_{L_1}[l]]^T$ and $\mathbf{Q}_n \triangleq [q_{-L_2}[l], \dots, q_{-1}[l], 0, q_1[l], \dots, q_{L_2}[l]]^T$ represent the coefficients of the feedforward and feedback filter, respectively, and L_1 and L_2 are integer parameters whose value is not smaller than L_h. The algorithm inevitably requires proper initialization of the filters. If the communication channel is *time-invariant*, this can be accomplished at receiver startup via transmission of a pilot sequence combined with a standard stochastic gradient LMS algorithm for data-aided estimation (see Section 5.1.3.1). This entails that the recursive equations:

$$\mathbf{P}_{l+1} = \mathbf{P}_l - \mu \, \mathbf{r}_n^*(y - c_l), \tag{12.28}$$

$$\mathbf{Q}_{l+1} = \mathbf{Q}_l - \mu \, \mathbf{c}_l^*(y - c_l), \tag{12.29}$$

are used, where μ is a step size. Once the training phase is over, the transmitted symbol c_l and the vector $\mathbf{c}_l$ are replaced by their estimated counterparts $\hat{c}_l$ and $\bar{\mathbf{c}}_l$, respectively. In all the iterations the update equations (12.28) and (12.29) should be used for the turbo equalizer. If the communication channel is *time-variant*, the adoption of the RLS algorithm combined with the periodic transmission of a training sequence is suggested to accomplish CIR estimation; then the equalizer coefficients can be evaluated from the estimated CIR. Finally, note that the evaluation of the vector $\bar{\mathbf{c}}_l$ is accomplished on the basis of the coded data LLRs provided by the channel decoder in the last iteration. Further detail on the implementation of this algorithm can be found in [1965].

This approach is much simpler that the standard FBA. Simulation results have shown that on some specific frequency-selective channels considered in [1965] and for a 4-QAM constellation only three iterations and an E_b/N_0 greater than 3 dB are necessary for the performance of a coded ISI-free

Gaussian channel to be reached. Thus, for a sufficiently large SNR, the TE improves its global performance at each iteration and reaches the theoretical bound, provided that a sufficient number of iterations is accomplished. In fact, it is only when the estimated data feeding the proposed algorithm reaches a sufficiently low BER threshold that the cancelation of a large portion of the ISI becomes really possible. This BER threshold depends on the first iteration performance, which is mainly a function of the transversal equalizer performance and channel coding gain, and increases with the size of the signal constellation. On other specific channels (and with the same modulation format), however, the TE does not reach the ISI-free bound. This phenomenon originates from the fact that the errors introduced at the input of the feedback filter of the ISI canceler generate a sort of impulsive noise at the output of the canceler itself. This strongly affects the channel decoder and decreases the channel coding gain. This occurs for highly frequency-selective channels and can be mitigated by increasing the interleaver size.

12.3.3.2   SiSo Equalization Based on Linear MMSE Estimation

Here we discuss how the linear MMSE estimator described in Section 12.2 can be exploited to develop a SiSo equalizer of reasonable complexity. Note that (12.12) generates an estimate of the entire channel symbol vector $\mathbf{c}$, but entails significant computational complexity, since it involves a matrix inversion (complexity $O(P^3)$) and linear processing of the vector $\mathbf{r}$ (complexity $O(P^2)$). To reduce the computational burden, a linear MMSE estimator operating in a sliding window fashion and, in particular, processing L_r consecutive received samples has been proposed in [1973, 1974].[3] In practice, a (soft) MMSE estimate $\check{c}_l$ of the channel symbol c_l (with $l = 0, 1, \ldots, P - 1$) is generated by linearly processing the vector:

$$\mathbf{r}_l \triangleq [r_{l-L_1}, r_{l-L_1+1}, \ldots, r_{l+L_2}]^T = \mathbf{H}_l \mathbf{c}_l + \mathbf{n}_l, \tag{12.30}$$

where $\mathbf{H}_l$ is a proper $L_r \times (L_r + L_h)$ submatrix of $\mathbf{H}$ (12.2), $\mathbf{c}_l \triangleq [c_{l-L_1-L_h}, c_{l-L_1-L_h+1}, \ldots, c_{l+L_2}]^T$ is an $(L_r + L_h)$-dimensional vector of channel symbols, $\mathbf{n}_l \triangleq [n_{l-L_1}, n_{l-L_1+1}, \ldots, n_{l+L_2}]^T$ is an L_r-dimensional noise vector, and L_1 and L_2 are nonnegative integer parameters (note that $L_r = L_1 + L_2 + 1$). Note that the signal model (12.30) holds for $l = L_1 + L_h, L_1 + L_h + 1, \ldots, P - L_2 - 1$ only and, in that interval, $\mathbf{H}_l$ always has the same structure (and, consequently, does not change unless the communication channel is time-varying); however, for simplicity, in what follows we assume that $\mathbf{H}_l$ also has the same structure for the remaining values of the time index l. It can be shown that, if the mean and variance of all the channel symbols:

$$\eta_{c,l} \triangleq \mathrm{E}\{c_l\} = \sum_{t=0}^{M-1} c^{(t)} \mathrm{Pr}\{c_l = c^{(t)}\} \tag{12.31}$$

and

$$\sigma_{c,l}^2 \triangleq \mathrm{E}\{|c_l - \eta_{c,l}|^2\} = \sum_{t=0}^{M-1} |c^{(t)} - \eta_{c,l}|^2 \mathrm{Pr}\{c_l = c^{(t)}\}, \tag{12.32}$$

respectively, are known a priori for $l = 0, 1, \ldots, P - 1$, the linear MMSE estimate of $\check{c}_l$, given $\mathbf{r}_l$ (12.29), can be evaluated as [1962]:

$$\check{c}_l = \eta_{c,l} + \sigma_{c,l}^2 \, \mathbf{h}_c^H[l] \, \boldsymbol{\Sigma}^{-1}[l] \, (\mathbf{r}_l - \mathbf{H}_l \boldsymbol{\eta}_c[l]), \tag{12.33}$$

---

[3] This approach was inspired by previous work on the same problem [1975–1977] or conceptually related problems [1913].

where $\mathbf{h}_c[l]$ is the $(L_1 + L_h)$th column of $\mathbf{H}_l$:

$$\Sigma[l] \triangleq \sigma_n^2 \mathbf{I}_{L_r} + \mathbf{H}_l \mathbf{D}_c[l] \, \mathbf{H}_l^H, \qquad (12.34)$$

$$\mathbf{D}_c[l] \triangleq \mathrm{diag}(\sigma_{c,l-L_h-L_1}^2, \sigma_{c,l-L_h-L_1+1}^2, \ldots, \sigma_{c,l+L_2}^2), \qquad (12.35)$$

and

$$\boldsymbol{\eta}_c[l] \triangleq [\eta_{c,l-L_h-L_1}, \eta_{c,l-L_h-L_1+1}, \ldots, \eta_{c,l+L_2}]^T. \qquad (12.36)$$

Note that $\eta_{c,l}$ and $\sigma_{c,l}^2$ are set to 0 and σ_c^2 (where σ_c^2 denotes the mean-square value of the symbols of the constellation A_c) at the start of turbo equalization. At the end of the first iteration, the extrinsic information vector $\mathbf{s}_e(c_l) \triangleq [s_{e,0}(c_l), s_{e,1}(c_l), \ldots, s_{e,M-1}(c_l)]^T$ is available for the channel symbol c_l, with $l = 0, 1, \ldots, P-1$ (see the previous subsection). In principle, this probabilistic information can be used to compute new values for $\eta_{c,l}$ and $\sigma_{c,l}^2$ on the basis of (12.31) and (12.32), respectively. Note, however, that to generate new extrinsic information about c_l, $\check{c}_l$ must be evaluated under the assumption that no additional a priori information is available about c_l, that is, that c_l has zero mean and a variance σ_c^2. This means that the expression:

$$\check{c}_l = \sigma_c^2 \, \mathbf{h}_c^H[l] \, [\sigma_n^2 \mathbf{I}_{L_r} + \mathbf{H}_l \mathbf{D}_c[l] \mathbf{H}_l^H + (\sigma_c^2 - \sigma_{c,l}^2) \mathbf{h}_c[l] \mathbf{h}_c^H[l]]^{-1} (\mathbf{r}_l - \mathbf{H}_l \boldsymbol{\eta}_c[l] + \eta_{c,l} \mathbf{h}_c[l])^{-1} \quad (12.37)$$

needs to be used in place of (12.33) to generate a new soft estimate of c_l. The resulting estimation error $e_l \triangleq \check{c}_l - c_l$ can be modeled as a Gaussian random variable having zero mean and a variance $\sigma_{c,l}^2$ which can be evaluated on the basis of (12.37). This model can be exploited, as discussed in the previous section, to generate new extrinsic information about the channel symbols. Further analytical details on this algorithm can be found in [1962], where its application to a binary constellation is discussed and suboptimal versions are proposed to simplify the overall computational complexity.

Computer simulations have shown that, when a binary constellation is used, a turbo equalizer employing this algorithm achieves a BER performance almost identical to that of a turbo equalizer employing the FBA for SiSo equalization, even if the former may require a larger number of iterations than the latter. However, the gap between these two different solutions increases with multilevel constellations mainly due to the fact that the Gaussian assumption made on the estimation error is no more accurate.

Further significant work on SiSo linear MMSE equalization can be found in [1961, 1978] and [1979]. In particular, in [1961] some analytical results on the asymptotic BER performance of the low-complexity MMSE SiSo equalizer of [1973] are derived (under the assumptions that BICM is used at the transmitter and the communication channel is purely frequency-selective) and the TE process convergence behavior is assessed via the EXIT chart technique [1683]. Such results show the fundamental role played by the constellation mapping in BICM. Further analytical tools have been derived in [1978] for coded BPSK transmission over a frequency-selective fading channel. In particular, two methods to assess the soft information evolution characteristics of a SiSO linear equalizer (i.e., the ISI canceler described earlier or the MMSE equalizer analyzed here) are developed and their application to the design of turbo equalization systems without reliance on extensive simulation is illustrated. These predictive methods provide insight into the iterative behavior of linear turbo equalizers with substantial reduction in numerical complexity. Finally, in [1979] the use of *adaptive coding* for *multilevel* BICM (ML-BICM) combined with MMSE turbo equalization is investigated. With the aid of the knowledge about EXIT characteristics at the receiver, the code parameters such as code rates and/or generator polynomials are adaptively selected independently in each ML-BICM layer.

12.3.3.3 Other techniques

To reduce the complexity of SiSo equalization other alternatives can also be found in the literature. Here we mention:

(a) the suboptimal *delayed decision feedback sequence estimator* (DDFSE) proposed in [1980] for EDGE,

(b) the RBF-based equalizer illustrated in [1981] that represents a nonlinear equalization scheme based on formulating the channel equalization procedure as a classification problem,

(c) the technique illustrated in [1982], which combines a (low-complexity) soft output sequential algorithm for data sequence estimation with likelihood post-processing technique proposed in [1983] to generate soft outputs associated with the symbols of estimated sequence,

(d) various reduced-complexity BCJR algorithms, which limit the exploration of the paths of the trellis describing the channel memory to the most promising ones [1099],

(e) the algorithms developed in [1413, 1970] resulting from the application of FGs and SPA to ISI channels,

(f) the *constrained-delay* APP detector with *decision feedback* developed in [1096] and conceptually related to that of the finite or sliding-window BCJR algorithms [1984],

(g) the low-complexity soft output detector based on heuristic search methods developed in [1985].

12.3.4 Turbo Equalization in the FD

Various turbo equalization algorithms operating in the FD have been proposed as potentially lower-complexity alternatives to their TD counterparts. In the following some strategies for PAM and CPM signals are described.

12.3.4.1 PAM Signals

The simplest class of turbo equalizers in the frequency domain employs linear estimation techniques [435, 1962, 1986, 1987]. The structure of a linear TE in the frequency domain (FD-TLE) is illustrated in Figure 12.9(a). In its kth iteration, the equalizer processes the FD received vector $\mathbf{R}^{(l)}$, associated

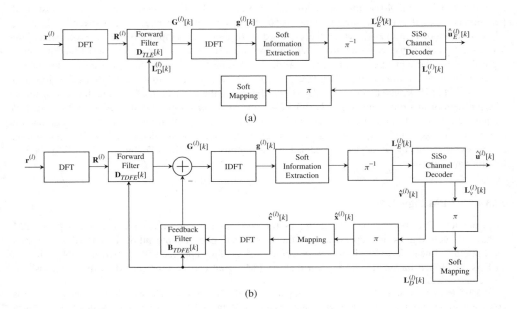

Figure 12.9 Block diagram of (a) an FD-TLE and (b) an FD-TDFE.

with the lth transmitted data block and generated by DFT processing[4] of the vector $\mathbf{r}^{(l)}$. This, in turn, is produced by uniformly sampling the received signal and discarding the samples associated with the cyclic prefix. This vector is premultiplied by a *forward matrix* $\mathbf{D}_{TLE}[k]$, whose task is to accomplish soft ISI cancelation. This produces $\mathbf{G}^{(l)}[k] \triangleq \mathbf{D}_{TLE}[k]\mathbf{R}^{(l)}$ feeding an IDFT. The IDFT output vector $\mathbf{g}^{(l)}[k]$ is processed to extract soft (e.g., extrinsic) information about the interleaved coded bits (i.e., about the data block $\mathbf{x}^{(l)}$, consisting of $N \cdot m$ coded bits, where m denotes the number of bits per channel symbol). Deinterleaving such soft information generates the vector $\mathbf{L}_E^{(l)}[k]$, which is applied to a SiSo channel decoder to produce an estimate $\hat{\mathbf{u}}^{(l)}[k]$ of the transmitted information bits (i.e., of the data block $\mathbf{u}^{(l)}$, consisting of $N \cdot m \cdot R$ information bits, where R is the rate of the adopted channel code) and the soft (e.g., extrinsic) information vector $\mathbf{L}_v^{(l)}[k]$ about the vector $\hat{\mathbf{v}}^{(l)}[k]$ of transmitted coded bits associated with $\hat{\mathbf{u}}^{(l)}[k]$. The vector $\mathbf{L}_v^{(l)}[k]$ feeds an interleaver, whose output is processed to generate the soft information vector $\mathbf{L}_D^{(l)}[k]$ about the transmitted channel symbols. Finally, the vector $\mathbf{L}_D^{(l)}[k]$ is exploited to properly update the forward matrix (i.e., to compute $\mathbf{D}_{TLE}[k+1]$) employed at the beginning of the next iteration. Note that, even if an MMSE approach is adopted in deriving the forward matrix $\mathbf{D}_{TLE}[k]$ [435, 1986, 1987], different options have been considered in the selection of the soft information to be passed in the FD-TLE loop. For instance, extrinsic information is usually sent from the FD equalizer to the SiSo decoder and vice versa (e.g., see [1113, 1114, 1962]); however, in [435] a different conceptual approach is adopted, since the data APPs are exploited in a direct fashion in order to modify the forward filtering matrix.

An alternative to the FD-TLE is the decision feedback TE in the frequency domain (FD-TDFE); different architectures are available in the technical literature for this case (e.g., see [435, 1113, 1114, 1988]). We briefly analyze the FD-TDFE developed in [435], whose structure is illustrated in Figure 12.9(b). In this case forward filtering is accomplished in the FD, as in the FD-TLE, whereas feedback filtering is carried out in the TD through a time-varying FIR filter with uniform tap spacing. As in the FD-TLE case, an MMSE approach can be adopted to derive a computationally efficient solution for evaluating the forward matrix $\mathbf{D}_{TDFE}[k]$ and the feedback matrix $\mathbf{B}_{TDFE}[k]$ to be employed in the kth iteration, provided that correct data feedback is assumed. Related FD-TDFE structures have been analyzed in [1113, 1114, 1988, 1989]. However, in [1113], unlike [435], decision feedback in the TD is used and only symbol-rate processing of the received signal is considered (this entails a substantial performance loss if matched filtering is not used in the receiver front-end). Decision feedback in the TD is also adopted in [1114], where the effects due to fractional sampling of the received signal are also investigated; however, nonuniform tap spacing in the feedback filter is considered, and adaptive algorithms for the equalizer synthesis are proposed, despite the fact that explicit expressions for the optimal filters of a multisampling DFE are provided. Finally, we note that the solutions developed in [435, 1113, 1114] assume ideal knowledge of the channel in the evaluation of the equalizer coefficients. The problem of incorporating *channel estimation* and *synchronization* (to compensate for the effect of frequency offset and phase noise generated by local oscillator instabilities) in the turbo equalization process is analyzed in [1988] (where an LS strategy is employed for this task) and [1989], respectively. Note also that in [1989] decision feedback is accomplished in the FD, but, unlike [435], before feedforward filtering.

12.3.4.2   CPM Signals

In principle, FD turbo equalization strategies for PAM signals can be extended to encompass CPM signals, if such signals are represented as the superposition of a finite number of PAM waveforms via Laurent's linear decomposition (see Section 3.6.5.1). This approach is adopted in [284], where a TLE is developed under the assumption that a CPM signal can be represented as the superposition

---

[4] The DFT order is equal to N as baud rate sampling is used at the receive side and $2N$ if the sampling rate is doubled to avoid information loss, as suggested in Section 4.4.2. In the latter case the structure of $\mathbf{R}^{(l)}$ is described by (4.99).

of a finite number (P) of linearly modulated digital signals. In this case, in its kth iteration the FD received vector $\mathbf{R}^{(l)}$ (available, as in the PAM case, after front-end filtering, sampling, removing the cyclic prefix and DFT processing) undergoes forward filtering, which generates P distinct N-dimensional vectors, each of which feeds a distinct Nth-order IDFT. The IDFT outputs are serialized into a data stream and are applied to a CPM SiSo detector. The latter processes the soft estimates of the Laurent symbols produced by linear equalization and generates soft (e.g., extrinsic) information about each transmitted coded bit. The deinterleaved soft information is applied to a SiSo channel decoder producing soft information about the information bits. Such information, after interleaving, is fed back to a soft mapping algorithm generating soft information about the Laurent symbols. This is exploited to compute the forward matrix employed in the next iteration, following the same conceptual approach proposed in [435] for PAM signals and summarized in Figure 12.9(a). Unfortunately, simulation results have shown that this approach does not work as efficiently as its counterpart for PAM signals. The poor error performance can be related to the fact that the algorithm computing soft information about Laurent symbols can produce long error bursts. This is due to the inherent memory of CPM, clearly shown by the structure of the Laurent pseudosymbols (see (3.170)). In fact, such a structure implies that incorrect soft information on a given bit produces a long burst of incorrect soft information on a number of consecutive Laurent symbols in the above-mentioned algorithm. Error burst length can be substantially reduced by passing soft information directly from the CPM SiSo detector to the soft mapping computer for the Laurent symbols (details can be found in [284]). The disadvantages of the FD turbo equalizer proposed in [284] are overcome by [440], which proposes the adoption of a *doubly-iterative* joint CPM equalization and demodulation strategy similar to that in [439], so as to achieve better error performance with lower computational complexity compared to the methods in [284, 439]. Unlike [284], the FD turbo equalization strategy developed in [440] is based on representing the CPM modulator as the cascade of a CPE with an MM (see Section 3.6.5.2). In the kth iteration of the proposed receiver structure the FD received vector $\mathbf{R}^{(l)}$ feeds an equalization algorithm accomplishing MMSE soft interference cancelation and whose output is exploited by a probability mapper to generate a vector of probabilities referring to all the possible waveforms generated by the CPE trellis in each symbol interval. Such probabilities are processed by a SiSo CPM demodulator operating over the (time-invariant) CPE trellis and implemented as the log-domain APP algorithm. The demodulator generates extrinsic information on both the coded bits in the form of LLRs and the tilted-phase CPM signals. The information about coded bits is exchanged by the SiSo CPM demodulator with a SiSo decoder for the channel code employed. In principle, this step involves *multiple (back-end) iterations*, through which the reliability of the soft information produced by the SiSo CPM demodulator is enhanced. At the end of this back-end procedure, the SiSo CPM demodulator generates an estimate of the mean value of the CPM signal samples. These samples feed a DFT block, whose output is used for adjusting the MMSE soft interference cancelation algorithm at the beginning of the $(k+1)$th iteration. Note that, since the CPM signal probabilities are not computed from the coded bit probabilities, the error bursts mentioned in [284], which are due to the modulation memory, are not encountered. This results in a significant performance improvement. Moreover, the strategy of [440] is computationally less complicated than that of [284], since it does not require any matrix inversion. In addition, faster convergence in BER is attained, because the number of equalization iterations is decreased by performing several demodulation/decoding iterations during each equalization iteration to improve the equalizer a priori information.

12.3.5   Turbo Equalization in the Presence of an Unknown Channel

In most of the literature on turbo equalization the channel is assumed to be *known* and *time-invariant* (i.e., it is purely frequency-selective). When the channel is unknown, a training sequence can be transmitted for estimating it. However, if the fading rate is not small enough, tracking of the channel may be required between consecutive training sequences to avoid a substantial performance loss.

As discussed in Chapters 5 and 6, channel estimation can be performed *jointly* with equalization or using a *separate* data/decision aided channel estimation algorithm. Examples of the first approach are:

(a) the SiSo detection algorithms developed in [530, 533, 1179, 1990], which, being *trellis-based*, may entail substantial computational complexity,
(b) the adaptive SiSO equalizers based on nonlinear KFs of [1991, 1992], which jointly optimize the estimates for channel parameters and data symbols in each iteration with the assistance of a priori information for the data symbols supplied by the SiSo decoder.

The complexity of these solutions does not grow exponentially with the modulation constellation size and is usually lower than that of many MAP equalizers accomplishing joint channel estimation and data detection. However, computation can be substantially reduced if the second approach is adopted, that is, if channel estimation and equalization are kept as separate tasks. In this case the quality of the channel estimates can be improved over the iterations if the estimation algorithm is fed by information generated by the channel decoder. In fact, in this case, pilot-aided channel estimation based on a standard algorithm (e.g., an iterative CIR estimation technique) can be used before starting the first iteration of the turbo equalizer. Subsequently, at the start of the next iteration, in addition to using the training sequence for reestimating the CIR, estimates of the bits/symbols derived from the SiSo decoder output can be employed to feed the channel estimator (whose parameters, e.g., its step size, can be adjusted), so that the CIR estimates are refined and improved after each iteration. This CIR estimation process can be repeated for each turbo equalization iteration. Note that the information feeding the channel estimation algorithm can be represented by hard decisions or soft information on the code bits or on the coded channel symbols, leading to *hard* or *soft iterative channel estimation*, respectively. Algorithms for *hard iterative channel estimation* have been proposed in [719, 1412, 1993–1996]. *Soft iterative channel estimation* has been developed in [719, 741, 1966, 1994, 1997, 1998], for the case in which channel estimates remain constant over an entire block of data, and in [1999–2002] for the case in which the channel estimate varies from symbol to symbol. In particular, the last class of algorithms proposes to extend standard recursive (i.e., LMS, RLS and Kalman) algorithms to exploit soft feedback. The use of the EM algorithm for soft iterative channel estimation is also possible, as shown in [2003, 2004].

All the above-mentioned work refers to turbo equalization in the TD. Iterative channel estimation can also be employed in FD equalization, as shown, for instance, in [1988], where a soft iterative LS channel estimator is adopted.

Graphical models for coded data transmission can also be developed even in the case of unknown channels. As in the case of known channels, this allows us to derive interpretations of existing algorithms for joint/separate detection and estimation, and to develop new strategies. A detailed analysis of the problem of graphical modeling for a coded data transmission can be found in [1968].

Finally, it is worth mentioning that some related work in this field can also be found in [1235, 2005], where *soft input channel estimation* exploiting soft symbols fed back from an equalizer (rather than from a decoder) is investigated.

12.4   Extension to MIMO

In MIMO environments coded modulation techniques (such as ST block/trellis codes and ST-BICM) and modern powerful channel coding techniques (such as LPDC codes and turbo codes) should always be combined with iterative processing strategies if their potential is to be fully exploited [1928, 2006]. However, employing a MAP detector for turbo processing in a MIMO scenario entails a computational complexity that is exponential in the product of the number of transmit antennas n_T and the number m of bits per constellation point [2007]. Its high complexity, which is due to the fact that the MAP detector performs an exhaustive search over the entire set of possible transmitted symbol vectors, makes its application infeasible in practical MIMO systems, even for moderate n_T and/or m. This explains why in the last decade substantial research efforts have been devoted to overcoming this problem. Most of the proposed solutions can be considered as extensions of the techniques developed for SiSo equalization for SISO channels. Here we mention the following specific solutions in the TD:

- the equalization algorithms, based on reduced-complexity *Jacobian radial basis function equalization* [2008], proposed in [2009, 2010],
- finite-impulse-response *prefiltering*, concentrating the energy of the MIMO channel in a small number of adjacent taps (so that a shortened channel is seen by the SiSo equalizer), developed in [2011],
- SiSo linear MMSE equalizers developed in [2012] (extending previous work on iterative equalization [1977] and iterative multiuser detection for CDMA systems [1913]), [2013] (extending the approach of [1973] to fractionally spaced equalization for ST bit-interleaved coded multilevel modulation over frequency-selective MIMO fading channels), [2014] (extending the algorithm of [2015] for multilevel modulations to the MIMO case), [2016, 2017] (based on separating time equalization for ISI cancelation from space equalization for mitigating multiantenna interference) and [2018],
- soft equalization algorithms based on *sequential Monte Carlo* sampling techniques for Bayesian inference as derived in [2019],
- adaptive iterative (turbo) DFE illustrated in [2020], which adjusts its filters directly on the basis of soft decisions and received data to minimize an LS cost function (adaptive reduced-rank estimation methods, based on the multistage Wiener filter, are also proposed),
- *trellis-based algorithms* accomplishing a reduced-complexity search (by truncating the channel memory length [839], introducing a branch selection algorithm that is separate from the path metric computation [2021], or retaining only a fixed number of best survivors at each trellis interval and discarding all the other survivors [1015]) or based on specific trellis representations (as in [2022], which extends [2023] to a MIMO scenario).

Note the following observations:

(a) Most of the literature on turbo processing over MIMO channels focuses on purely frequency-selective fading channels. Iterative (turbo) equalization and decoding of interleaved ST codes over *time-variant* MIMO ISI channels was first investigated in [2024], where a MAP equalizer is derived and its performance is evaluated for the EDGE air interface when deploying ST bit-interleaved convolutional codes.
(b) MIMO turbo equalization ensures promising error rate performance at low overhead but may substantially increase receiver latency; however, parallelism techniques can be adopted at algorithmic level to increase speed [2025].

As in the SiSo case, an alternative to TD equalizers is offered by their FD counterparts; SiSo equalization algorithms in the FD for MIMO frequency-selective channels have been proposed in [2026–2032]. In particular, MMSE turbo equalization techniques are developed in [2026, 2027] (extending the turbo equalization algorithm of [1986] to the MIMO case and adopting the sliding window approach of [2016]), [2028] (whose solution is based on *probabilistic data association* filtering), [2029] (whose solution can be considered as an extension of the noniterative scheme of [2033, 2034]), [2032] and [2030]. FD-MMSE turbo equalization is combined with SVD-based precoding for SC transmission over frequency-selective MIMO channels in [2031], where the problem of optimizing the transmit power allocation over the FD channel eigenmodes subject to maintaining a target BER of the turbo equalizer is solved.

Finally, we mention that *soft iterative channel estimation techniques* (see Section 12.3.5) have also been proposed for MIMO systems (e.g., see [839, 2012, 2017, 2019, 2030, 2031, 2035]).

12.5   Historical Notes

As stated in Section 12.1, the idea of combining a SiSo equalizer with a SiSo decoder for achieving close to optimal performance through an appropriate iterative procedure (known as *turbo detection*) is due to C. Douillard *et al.* [1958], published in 1995. This approach was proposed for ISI channels and convolutionally encoded transmission, but its usefulness in the presence of *frequency-flat fading* channels [533, 719, 2036] and other coding schemes (e.g., see [1966, 2003, 2037]) soon became apparent. The initial work by Douillard was followed by a flurry of research activity, mainly investigating new algorithms to be employed in turbo equalizers operating in the presence of different channel models and assessing the performance benefits deriving from turbo equalization in the presence of different modulation and coding schemes. In this section we focus on some of the relevant results achieved in the short history of turbo equalization and mention some relevant work.

12.5.1   Reduced-Complexity SiSo Equalization

One of the main problems associated with the strategy developed in [1958] is the complexity of the SiSo module used for equalization. This makes the proposed strategy suitable for digital modulations with poor spectral efficiency and channels exhibiting low delay spread (e.g., for the GSM system [2038]). For this reason, one of the relevant technical issues on which research in the field of turbo equalization has concentrated on in the last decade is the development of *reduced-complexity* SiSo equalization algorithms. In particular, various SiSo strategies based on MMSE linear filtering [1973, 1974, 1976, 1977, 1979, 2015, 2039–2045], constrained minimum variance filter design [2046], the DFE concept (in particular, the concept of *soft interference cancelation*) [1965, 1975, 2047–2054] and the use of reduced-complexity/modified BCJR algorithms [1099, 1532, 2055] have been developed and applied to different coding schemes. Other reduced-complexity SiSo equalization techniques are based on:

- the use of a broadband beamformer to shorten the observed CIR [2056, 2057];
- adapting a RBF-based equalizer to utilize the a priori information provided by a channel decoder [1981, 2008, 2009];
- the exploitation of FGs and MP [1413];
- constrained-delay APP detection with decision feedback [1096];
- local search algorithms [1985].

12.5.2   Error Performance and Convergence Speed in Turbo Equalization

Most of what has been said in the literature on achievable error performance and its dependence on the number of iterations accomplished in a TE is based on computer-generated numerical results, which require a long processing time. This has raised the problem of developing analytical tools for analyzing error performance and convergence of turbo equalization algorithms. Some tools have been derived (e.g., see [1961, 1978, 1979, 2018, 2058]) that exploit the conceptual similarity between turbo equalization and iterative decoding of concatenated codes. In particular, the convergence behavior of turbo equalization has been analyzed [1961] using EXIT charts [1683] under the assumptions that BICM is used at the transmitter, a low-complexity MMSE SiSo equalizer [1973] is employed at the receiver and the communication channel is purely frequency-selective. In addition, [1961] provides an expression of the asymptotic BER performance, emphasizing the fundamental role played by the constellation mapping. Further analytical methods for the prediction of the BER performance can be found in [1978] (for coded BPSK transmission over a frequency-selective fading channel) and

[2058] (where turbo FD equalization is considered). Convergence analysis via EXIT functions is also investigated in [1978, 1979] (referring to adaptive coding for ML-BICM with MMSE turbo equalization), [2018] (extending the work of [1979] to a MIMO scenario), and [2058]. Finally, we mention [2059], where the *outage performance* of soft cancelation FD MMSE turbo equalization in a serially concatenated coded modulation scheme over frequency-selective Rayleigh fading channels with exponential delay-power profile is analyzed and the convergence behavior of this iterative three-stage system is evaluated with the aid of correlation charts.

12.5.3 SiSo Equalization Algorithms in the Frequency Domain

The first SiSo strategy for FD turbo equalization of PAM signals over ISI channels was proposed by M. Tüchler and J. Hagenauer in 2000 [1986] (see also [1962, 1987]), where a linear technique is devised. Decision feedback TEs for PAM signals have been proposed later in [1113, 1114]. FD algorithms for MIMO systems have been investigated since 2004 (e.g., see [2026–2032]).

12.5.4 Use of Precoding

The performance of a TE system can be significantly improved by the use of a *precoder* [1962, 2060]. This improved performance is due to the recursive nature of the *inner code*, consisting of the *serial concatenation of a given precoder with an ISI channel*, which results in an *interleaving gain*. A number of precoding schemes have been presented in the literature, even if the focus is on *partial-response channels* (in practice, on magnetic recording channels). The use of precoders usually results in a degraded (improved) performance at low (large) SNRs [2061]. Simulation results have also provided evidence of a tradeoff between the performance at the turbo cliff and error floor region and shown that "weight-two" and "multiweight" precoders are suited for TE systems which desire an error floor at low SNRs and high SNRs, respectively [2060]. Low-complexity precoded schemes have been proposed in [2062]. Various results on the convergence of turbo equalization in the presence of precoding can be found in [2063, 2064].

12.5.5 Turbo Equalization and Factor Graphs

The use of FGs for presenting iterative receiver algorithms in a unifying perspective and the adoption of *canonical distributions* for handling continuous variables in MP algorithms was proposed by A. P. Worthen and W. E. Stark in 2001 [1808]. Pioneering work in the application of FGs is due to M. Tüchler *et al.* [1968].

12.5.6 Turbo Equalization for MIMO Systems

Preliminary work in this area dates back to the end of the 1990s. In particular, we mention the work by G. Bauch and A. F. Naguib [2007], who first investigated the problem of turbo equalization of ST coded (and, in particular, of trellis-coded) systems over frequency-selective channels. The main efforts in this area have been devoted to cutting the complexity of SiSo equalization by adopting new equalization strategies [2009], prefilters for channel shortening [2011], SiSo algorithms based on MMSE filtering [2013, 2016, 2018], adaptive reduced-rank DFEs [2020], prefiltering combined with M-BCJR [2065], algorithms performing a reduced-complexity search on a state trellis [839, 2021] or operating over a different (but equivalent) trellis representation characterized by a larger state space [2022]. Other attempts are concerned with the use of FD equalization techniques (e.g., see [2027]).

12.5.7   Related Techniques

Finally, it is worth mentioning that, in the field of combined equalization and decoding, various techniques, typically exploiting decision feedback but that cannot be included in the family of turbo equalizers, can be found in the literature (e.g., see [2066–2073]).

Other research work closely related to that which we have analyzed concerns the development of turbo equalization techniques for mitigating ICI in an OFDM system operating in the presence of a time-varying channel or nonlinear distortions; the reader can refer to [1144, 2074] and [2075], respectively, where solutions to these problems are developed.

12.6   Further Reading

A good introduction to turbo equalization is offered by [1660]. A detailed description of various SiSo equalization algorithms for turbo equalization and an in-depth discussion of the complexity–performance tradeoff are available in [1962]. An analysis of graphical models for coded data transmissions over ISI channels can be found in [1968]. Another paper that is recommended reading is [1178], which provides an overview of trellis-based algorithms for iterative detection over channels with memory.

Appendix A

Fourier Transforms

In this appendix some standard formulas pertaining to the Fourier analysis of a complex time-continuous signal $x(t)$ and a complex discrete-time signal (i.e., a sequence $x[n]$) are listed.

1. *Fourier series* (FS). If the time-continuous signal $x(t)$ is periodic with period T (i.e., $x(t) = x(t + T)$ for any t), its FS in *complex exponential form* is given by:

$$x(t) = \sum_{k=-\infty}^{+\infty} X_k \, \exp\left(j\frac{2\pi kt}{T}\right), \tag{A.1}$$

where the kth coefficient X_k is defined by:

$$X_k \triangleq \frac{1}{T} \int_{-T/2}^{T/2} x(t) \, \exp\left(-j\frac{2\pi kt}{T}\right) dt. \tag{A.2}$$

2. *Fourier continuous transform* (FCT). If the time-continuous signal $x(t)$ is aperiodic, its FCT is evaluated as:

$$X(f) \triangleq \text{FCT}[x(t)] = \int_{-\infty}^{+\infty} x(t) \, \exp\left(-j2\pi ft\right) dt. \tag{A.3}$$

Given $X(f)$, $x(t)$ can then be expressed as:

$$x(t) = \text{IFCT}[X(f)] = \int_{-\infty}^{+\infty} X(f) \, \exp\left(j2\pi ft\right) df. \tag{A.4}$$

3. *Discrete Fourier transform* (DFT). If the sequence $x[n]$ is periodic with period N, its DFT representation is given by:

$$x[n] = \frac{1}{\sqrt{N}} \sum_{k=0}^{N-1} \bar{X}_k \, \exp\left(j\frac{2\pi kn}{N}\right), \tag{A.5}$$

where the kth coefficient $\bar{X}_k$ is evaluated as:

$$\bar{X}_k \triangleq \frac{1}{\sqrt{N}} \sum_{n=0}^{N-1} x[n] \exp\left(-j\frac{2\pi nk}{N}\right) \tag{A.6}$$

with $k = 0, 1, \ldots, N - 1$. Note that the last two expressions are equivalent to:

$$\bar{\mathbf{X}}_N = \text{DFT}_N[\mathbf{x}_N] = \mathbf{Q}_N \, \mathbf{x}_N \tag{A.7}$$

Wireless Communications: Algorithmic Techniques, First Edition.
Giorgio M. Vitetta, Desmond P. Taylor, Giulio Colavolpe, Fabrizio Pancaldi, Philippa A. Martin.
© 2013 John Wiley & Sons, Ltd. Published 2013 by John Wiley & Sons, Ltd.

and

$$\mathbf{x}_N = \text{IDFT}_N[\bar{\mathbf{X}}_N] = \mathbf{Q}_N^H \bar{\mathbf{X}}_N, \tag{A.8}$$

respectively, where $\bar{\mathbf{X}}_N \triangleq [\bar{X}_0, \bar{X}_1, \ldots, \bar{X}_{N-1}]^T$, $\mathbf{x}_N \triangleq [x[0], x[1], \ldots, x[N-1]]^T$, $\text{DFT}_N[\mathbf{x}]$ ($\text{IDFT}_N[\mathbf{y}]$) denotes the Nth-order DFT (IDFT) of an N-dimensional vector $\mathbf{x}$ ($\mathbf{y}$), $\mathbf{Q}_N = [q_{i,k}]$ is an $N \times N$ matrix with:

$$q_{i,k} \triangleq \frac{1}{\sqrt{N}} \exp\left(-j2\pi\frac{i\,k}{N}\right), \tag{A.9}$$

and $i, k = 0, 1, \ldots, N - 1$. It can be proved that $\mathbf{Q}_N$ is a *unitary matrix*, so that its inverse is given by $\mathbf{Q}_N^H$. From this property it can easily be seen that (A.7) implies (A.8), and vice versa.

4. *Fourier transform of a sequence* (FTS). The FTS of an aperiodic sequence $x[n]$ is defined as:

$$\bar{X}(f) \triangleq \text{FTS}[x[n]] = \sum_{n=-\infty}^{+\infty} x[n] \exp\left(-j2\pi nfT\right), \tag{A.10}$$

where T is a reference time interval. Given $\bar{X}(f)$, $x[n]$ can be expressed as:

$$x[n] = \text{IFTS}[\bar{X}(f)] = T \int_{-1/(2T)}^{1/(2T)} \bar{X}(f) \exp\left(j2\pi nfT\right) df, \tag{A.11}$$

Finally, it is worth mentioning that, given a sequence $x[n]$ sampling a time-continuous signal $x(t)$ with period T:

$$x[n] = x(t)|_{t=nT} = x(nT), \tag{A.12}$$

its FTS $\bar{X}(f)$ can be expressed as:

$$\bar{X}(f) = \frac{1}{T} \sum_{n=-\infty}^{+\infty} X\left(f - \frac{n}{T}\right), \tag{A.13}$$

where $X(f)$ is the FCT of $x(t)$.

Further information on Fourier transforms and their properties can be found in [2076].

Appendix B

Power Spectral Density of Random Processes

B.1 Power Spectral Density of a Wide-Sense Stationary Random Process

A mathematical description of the average spectral content of a continuous-time random process $X(t)$ (or a discrete-time random process $\{X_n\}$) is provided by its *power spectral density* (PSD). This function can be defined for some families of random process and, in particular, for *wide-sense stationary* (WSS) processes. A *continuous-time* random process $X(t)$ is WSS if its *mean function*:

$$\eta_X(t) \triangleq \mathrm{E}\{X(t)\} \tag{B.1}$$

is constant (i.e., $\eta_X(t) = \eta_X$) and its *autocorrelation function* (ACF):

$$R_X(t, \tau) \triangleq \mathrm{E}\{X(t + \tau)\, X^*(t)\} \tag{B.2}$$

depends on the lag or delay τ only, that is:

$$R_X(t, \tau) = R_X(\tau). \tag{B.3}$$

Similarly, a *discrete-time* random process $\{X_n\}$ is WSS if its mean function $\eta_X[n] \triangleq \mathrm{E}\{X_n\}$ and its ACF $R_X[n, k] \triangleq \mathrm{E}\{X_{n+k}\, X_n^*\}$ are such that:

$$\eta_X[n] = \eta_X \tag{B.4}$$

and

$$R_X[n, k] = R_X[k], \tag{B.5}$$

respectively. The PSD[1] $S_X(f)$ of a WSS process $X(t)$ can be evaluated as:

$$S_X(f) = \mathrm{FCT}[R_X(\tau)] = \int_{-\infty}^{+\infty} R_X(\tau)\, \exp(-j2\pi f \tau)\, d\tau, \tag{B.6}$$

---

[1] *Two-sided* spectral densities are always considered in this appendix and in the rest of this book.

Wireless Communications: Algorithmic Techniques, First Edition.
Giorgio M. Vitetta, Desmond P. Taylor, Giulio Colavolpe, Fabrizio Pancaldi, Philippa A. Martin.
© 2013 John Wiley & Sons, Ltd. Published 2013 by John Wiley & Sons, Ltd.

that is, as the continuous Fourier transform of its autocorrelation function $R_X(\tau)$. Similarly, given a WSS random sequence $\{X_n\}$, its PSD $\bar{S}_X(f)$ can be evaluated as the Fourier transform of its autocorrelation sequence $R_X[k]$, that is, as:

$$\bar{S}_X(f) \triangleq \text{FTS}[R_X[n]] = \sum_{k=-\infty}^{+\infty} R_X[k] \exp(-j2\pi kfT), \tag{B.7}$$

where the parameter T represents a reference time interval (e.g., the symbol interval when a random sequence of channel symbols generated by a digital modulator is considered). The results (B.6) and (B.7), relating the PSD of a WSS random process to its ACF, are usually known in the literature as the *Wiener–Khintchine theorem* [400].

Note that the *average power* $P_X(t) \triangleq \text{E}\{|X(t)|^2\}$ $(P_X[n] \triangleq \text{E}\{|X_n|^2\})$ of a WSS process $X(t)$ $(\{X_n\})$ is *constant*. In fact, the statistical power P_X of a continuous-time WSS process $X(t)$ is given by:

$$\text{E}\{|X(t)|^2\} = R_X(0) \tag{B.8}$$

and can be evaluated by integrating its PSD $S_X(f)$ over all frequencies:

$$P_X = \int_{-\infty}^{+\infty} S_X(f)\, df. \tag{B.9}$$

Similar considerations apply to the average power P_X of a WSS random sequence $\{X_n\}$, since $P_X[n] \triangleq \text{E}\{|X_n|^2\} = R_X[0] = P_X$ and:

$$P_X = T \int_{-1/(2T)}^{1/(2T)} \bar{S}_X(f)\, df. \tag{B.10}$$

B.2 Power Spectral Density of a Wide-Sense Cyclostationary Random Process

As mentioned in Sections 3.5–3.7, the digital modulation formats described in Chapter 3 cannot be modeled as WSS random processes. However, they all belong to another important family of random processes, that of *wide-sense cyclostationary* (WSC) processes. A continuous-time random signal $X(t)$ is WSC with period T_0 if its mean function $\eta_X(t)$ (B.1) and its ACF $R_X(t, \tau)$ (B.2) satisfy the equalities[2]:

$$\eta_X(t + T_0) = \eta_X(t) \tag{B.11}$$

and

$$R_X(t + T_0, \tau) = R_X(t, \tau), \tag{B.12}$$

respectively, for any t (in other words, they are periodic functions, with period T_0, in the variable t). Note that, in this case, the statistical power:

$$P_X(t) \triangleq \text{E}\{|X(t)|^2\} = R_X(t, 0) \tag{B.13}$$

is also periodic with period T_0, so that it is reasonable to define the *average power* P_X of $X(t)$ as:

$$P_X \triangleq \frac{1}{T_0} \int_0^{T_0} P_X(t)\, dt = \frac{1}{T_0} \int_0^{T_0} R_X(t, 0)\, dt. \tag{B.14}$$

---

[2] As already illustrated for WSS random processes, similar expressions can be also given to describe the property of wide-sense cyclostationarity for a random *sequence*.

By analogy with what has been shown for WSS random processes (see (B.9)), the average power $\bar{P}_X$ (B.14) of a WSC process $X(t)$ can be expressed as the integral, over the whole frequency interval, of a proper PSD, which can be evaluated by resorting to the following procedure [55, 79]. First, given $R_X(t, \tau)$, the *average* ACF $\bar{R}_X(\tau)$ of $X(t)$ is evaluated by averaging $R_X(t, \tau)$ over a period (with respect to the variable t), that is:

$$\bar{R}_X(\tau) \triangleq \frac{1}{T_0} \int_0^{T_0} R_X(t, \tau) \, dt. \tag{B.15}$$

Then the *average* PSD $\bar{S}_X(f)$ of $X(t)$ is computed as the FCT of $\bar{R}_X(\tau)$ (B.15), that is, as:

$$\bar{S}_X(f) = \int_{-\infty}^{+\infty} \bar{R}_X(\tau) \, \exp(-j2\pi f \tau) \, d\tau. \tag{B.16}$$

Substituting $\bar{S}_X(f)$ in (B.9) (in place of $S_X(f)$) yields P_X (B.14).

Finally, it is worth mentioning that the average PSD $\bar{S}_X(f)$ (B.16) has the same physical meaning as its counterpart evaluated for a WSS process (see (B.6)); in other words, both functions describe the average spectral content of the sample functions of a random process.

B.3 Power Spectral Density of a Bandpass Random Process

In wireless communications the random signal $s_{RF}(t)$ generated by a digital modulator is *bandpass* and is characterized by a *carrier frequency* f_c. Such a signal can be expressed as:

$$s_{RF}(t) = \text{Re}\{s(t) \exp(j2\pi f_c t)\} = \frac{s(t) \exp(j2\pi f_c t) + s^*(t) \exp(-j2\pi f_c t)}{2}, \tag{B.17}$$

where $s(t)$ denotes its *complex envelope* evaluated with respect to f_c. In addition, we have that $s(t) = s_I(t) + js_Q(t)$, where $s_I(t)$ and $s_Q(t)$ denote the *in-phase* and *quadrature* components of $s_{RF}(t)$, respectively. In Chapter 3 it is shown that, for all the digital modulation formats of practical interest, $s(t)$ is a WSC random process, whose spectral content is described by its average PSD $S_s(f)$. In the following we derive a formula relating $S_s(f)$ to the average PSD $S_{RF}(f)$ of $s_{RF}(t)$ (B.9), which is characterized by its ACF:

$$R_{RF}(t, \tau) \triangleq \text{E}\{s_{RF}(t + \tau) \, s_{RF}^*(t)\}. \tag{B.18}$$

To begin our derivation, we substitute the RHS of (B.17) into (B.18). This yields the expression:

$$\begin{aligned}
R_{RF}(t, \tau) = \frac{1}{4}\{&\text{E}\{s(t + \tau) \, s^*(t)\} \exp(j2\pi f_c \tau) \\
&+ \text{E}\{s(t + \tau) \, s(t)\} \exp(j2\pi f_c(2t + \tau)) \\
&+ \text{E}\{s^*(t + \tau) \, s^*(t)\} \, \exp(-j2\pi f_c(2t + \tau)) \\
&+ \text{E}\{s^*(t + \tau) \, s(t)\} \exp(-j2\pi f_c \tau)\}.
\end{aligned} \tag{B.19}$$

Let us now make the following assumptions:

1. The random signals $s_I(t)$ and $s_Q(t)$ are characterized by the same average ACF.
2. The cross-correlation between $s_I(t)$ and $s_Q(t)$ is the opposite of that between $s_Q(t)$ and $s_I(t)$::

$$\begin{aligned}
R_{IQ}(t, \tau) &\triangleq \text{E}\{s_I(t + \tau) \, s_Q(t)\} \\
&= -R_{QI}(t, \tau) \triangleq -\text{E}\{s_Q(t + \tau) \, s_I(t)\}.
\end{aligned} \tag{B.20}$$

If (B.20) holds, it is not difficult to prove that:

$$E\{s(t + \tau)\, s(t)\} = E\{s^*(t + \tau)\, s^*(t)\} = 0, \tag{B.21}$$

so that (B.19) simplifies to the form:

$$R_{RF}(t, \tau) = \frac{1}{4} R_s(t, \tau)\, \exp(j2\pi f_c \tau) + \frac{1}{4} R_s^*(t, \tau)\, \exp(-j2\pi f_c \tau). \tag{B.22}$$

Then substituting (B.22) into the RHS of the definition:

$$R_{RF}(\tau) \triangleq \frac{1}{T_0} \int_0^{T_0} R_{RF}(t, \tau)\, dt \tag{B.23}$$

of the average ACF of $s_{RF}(t)$ (B.17), where T_0 denotes the period of cyclostationarity, gives:

$$R_{RF}(\tau) = \frac{1}{4} R_s(\tau)\, \exp(j2\pi f_c \tau) + \frac{1}{4} R_s^*(\tau)\, \exp(-j2\pi f_c \tau), \tag{B.24}$$

where

$$R_s(\tau) \triangleq \frac{1}{T_0} \int_0^{T_0} R_s(t, \tau)\, dt \tag{B.25}$$

denotes the average ACF of $s(t)$ and $R_s(t, \tau) \triangleq E\{s(t + \tau)\, s^*(t)\}$. Finally, computing the FCT of both sides of (B.24) yields:

$$S_{RF}(f) = \frac{1}{4}[S_s(f + f_c) + S_s(f - f_c)], \tag{B.26}$$

which relates the average PSD of $s_{RF}(t)$ to that of $s(t)$.

Appendix C

Matrix Theory

In this appendix some essential definitions and theorems of matrix theory are provided. Further details about matrix theory and computation can be found in [125, 1908, 2077].

Norm of a matrix. Throughout this book the so-called *Frobenius norm* is used when evaluating the norm of a real or complex matrix. This norm is defined as:

$$|\mathbf{A}| \triangleq \sqrt{\sum_{l=1}^{m} \sum_{p=1}^{n} |a_{i,j}|^2} \tag{C.1}$$

for any $m \times n$ complex matrix $\mathbf{A} = [a_{i,j}]$. Note that $|\mathbf{A}|$ can be evaluated as:

$$|\mathbf{A}| = \sqrt{\operatorname{tr}(\mathbf{A}\mathbf{A}^H)}, \tag{C.2}$$

where $\operatorname{tr}(\mathbf{X})$ denotes the *trace* of a square matrix $\mathbf{X}$.

Kronecker product. Given the $m \times n$ matrix $\mathbf{A}$ and the $p \times q$ matrix $\mathbf{B}$, the *Kronecker product* of $\mathbf{A}$ and $\mathbf{B}$ is defined as:

$$\mathbf{A} \otimes \mathbf{B} \triangleq \begin{bmatrix} a_{11}\mathbf{B} & \cdots & a_{1n}\mathbf{B} \\ \vdots & \ddots & \vdots \\ a_{m1}\mathbf{B} & \cdots & a_{mn}\mathbf{B} \end{bmatrix}. \tag{C.3}$$

Note that such a product generates an $mp \times nq$ matrix.

Properties of the eigenvectors and the eigenvalues of a square matrix. If $\mathbf{M}$ is an $n \times n$ (square) matrix, a nonzero vector $\mathbf{v}$ is an *eigenvector* of $\mathbf{M}$ if there is a scalar λ such that:

$$\mathbf{M}\mathbf{v} = \lambda\mathbf{v}, \tag{C.4}$$

The scalar λ is said to be the *eigenvalue* of $\mathbf{M}$ associated with $\mathbf{v}$. The eigenvalues of $\mathbf{M}$ are precisely the solutions λ to the equation:

$$\det(\mathbf{M} - \lambda\mathbf{I}_n) = 0, \tag{C.5}$$

which is called the *characteristic equation* of $\mathbf{M}$ (here $\det(\mathbf{X})$ denotes the *determinant* of a square matrix $\mathbf{X}$). The n solutions to this equation are the eigenvalues $\{\lambda_1, \lambda_2, \ldots, \lambda_n\}$ of the matrix $\mathbf{M}$. Note the following observations:

Wireless Communications: Algorithmic Techniques, First Edition.
Giorgio M. Vitetta, Desmond P. Taylor, Giulio Colavolpe, Fabrizio Pancaldi, Philippa A. Martin.
© 2013 John Wiley & Sons, Ltd. Published 2013 by John Wiley & Sons, Ltd.

(a) If the matrix $\mathbf{M}$ has real entries, the coefficients of the characteristic equation are all real, but its roots are not necessarily real.

(b) The eigenvalues of $\mathbf{M}$ are not necessarily distinct, since the *algebraic multiplicity* of each root of the characteristic equation may be larger than unity.

(c) Each eigenvalue is characterized by a *geometric multiplicity*, which is defined as the dimension of the associated eigenspace, that is, the number of *linearly independent* eigenvectors having that eigenvalue.

(d) The eigenvectors associated with different eigenvalues are linearly independent.

The algebraic multiplicity and the geometric multiplicity of a given eigenvalue may or may not be equal, but certainly latter quantity does not exceed the former, so that the sum of the geometric multiplicities (i.e., the overall number of linearly independent eigenvectors) may be smaller than n. Let us now assume that the eigenvectors $\{\mathbf{v}_1, \mathbf{v}_2, \ldots, \mathbf{v}_n\}$ (associated with the eigenvalues $\{\lambda_1, \lambda_2, \ldots, \lambda_n\}$) form a *basis*. Then $\mathbf{M}$ can be factored as:

$$\mathbf{M} = \mathbf{V} \, \mathbf{\Lambda} \, \mathbf{V}^{-1}, \tag{C.6}$$

where $\mathbf{V}$ is the $n \times n$ matrix whose lth column is the basis eigenvector $\mathbf{v}_l$ of $\mathbf{M}$ (with $l = 1, 2, \ldots, n$) and $\mathbf{\Lambda} = [\Lambda_{l,k}]$ is the diagonal matrix whose diagonal elements are the corresponding eigenvalues (i.e., $\Lambda_{k,k} = \lambda_k$); this result is known as the *spectral theorem*. If $\mathbf{M}$ is *Hermitian*, all its eigenvalues are real and $\mathbf{V}$ is a unitary matrix, so that (C.6) can be rewritten as:

$$\mathbf{M} = \mathbf{V} \, \mathbf{\Lambda} \, \mathbf{V}^{H}. \tag{C.7}$$

Finally, it is worth noting that, given the $n \times n$ matrix $\mathbf{M}$ and its eigenvalues $\{\lambda_0, \lambda_1, \ldots, \lambda_{n-1}\}$, the following properties hold:

$$\det(\mathbf{M}) = \prod_{l=0}^{n-1} \lambda_l \tag{C.8}$$

and

$$\operatorname{tr}(\mathbf{M}) = \sum_{l=0}^{n-1} \lambda_l. \tag{C.9}$$

It is also important to mention that every eigenvalue λ of a *unitary* matrix $\mathbf{M}$ has absolute value $|\lambda| = 1$.

Matrix inversion lemma. This lemma also known as the *Sherman–Morrison–Woodbury formula*, states that:

$$(\mathbf{A} + \mathbf{U}\mathbf{C}\mathbf{V})^{-1} = \mathbf{A}^{-1} - \mathbf{A}^{-1}\mathbf{U}\,[\mathbf{C}^{-1} + \mathbf{V}\mathbf{A}^{-1}\mathbf{U}]^{-1}\mathbf{V}\mathbf{A}, \tag{C.10}$$

where $\mathbf{A}$, $\mathbf{U}$, $\mathbf{C}$ and $\mathbf{V}$ are $n \times n$, $n \times k$, $k \times k$ and $k \times n$ matrices, respectively.

Singular value decomposition theorem. Consider an $m \times n$ complex matrix $\mathbf{M}$. Then there exists a factorization of the form:

$$\mathbf{M} = \mathbf{U}\,\mathbf{\Sigma}\mathbf{V}^{H}, \tag{C.11}$$

where $\mathbf{U}$ is an $m \times m$ unitary matrix, $\mathbf{\Sigma}$ is an $m \times n$ diagonal matrix having nonnegative real numbers on its main diagonal, and $\mathbf{V}$ is an $n \times n$ unitary matrix ($\mathbf{X}^{H}$ denotes the conjugate transpose of $\mathbf{X}$). Such a factorization is called the *singular value decomposition* (SVD) of $\mathbf{M}$. The diagonal entries $\Sigma_{l,l}$ of $\mathbf{\Sigma}$ are known as the *singular values* of $\mathbf{M}$. The m columns of $\mathbf{U}$ and the n columns of $\mathbf{V}$ are called the *left singular vectors* and the *right singular vectors* of $\mathbf{M}$, respectively. It can be proved that:

- the left singular vectors of $\mathbf{M}$ are eigenvectors of $\mathbf{M}\mathbf{M}^{H}$;
- the right singular vectors of $\mathbf{M}$ are eigenvectors of $\mathbf{M}^{H}\mathbf{M}$;
- the nonzero singular values of $\mathbf{M}$ are the square roots of the nonzero eigenvalues of $\mathbf{M}^{H}\mathbf{M}$ or $\mathbf{M}\mathbf{M}^{H}$.

The singular values of $\mathbf{M}$ are usually ordered in such a way that their value decreases along the main diagonal. When this occurs, the diagonal matrix $\mathbf{\Sigma}$ is uniquely determined by $\mathbf{M}$, but this does not hold for the matrices $\mathbf{U}$ and $\mathbf{V}$.

Some additional properties can be given for the SVD if $\mathbf{M}$ is an $m \times m$ *real* square matrix having positive determinant. In fact, in this case $\mathbf{U}$, $\mathbf{V}$, and $\mathbf{\Sigma}$ are $m \times m$ matrices of real numbers; in addition, $\mathbf{\Sigma}$ can be regarded as a scaling matrix, and $\mathbf{U}$ and $\mathbf{V}^H$ can be viewed as rotation matrices. Then, the factorization (C.11) can be interpreted as the cascade of three geometrical transformations, namely a rotation, a scaling, and another rotation.

Cholesky decomposition. A Hermitian and positive definite matrix $\mathbf{M}$ can be decomposed as:

$$\mathbf{M} = \mathbf{L}\mathbf{L}^H, \tag{C.12}$$

where $\mathbf{L}$ is a lower triangular matrix with strictly positive diagonal entries. This factorization is known as the *Cholesky decomposition*. Note that the factorization can also be expressed in the alternative form:

$$\mathbf{M} = \mathbf{P}\mathbf{D}\mathbf{P}^H, \tag{C.13}$$

where $\mathbf{P}$ is a lower triangular matrix with unit diagonal entries and $\mathbf{D}$ is a *diagonal matrix*.

The Cholesky decomposition is *unique*, that is, there exists only one lower triangular matrix $\mathbf{L}$ with strictly positive diagonal entries such that the factorization (C.12) holds. The converse also holds: if a matrix $\mathbf{M}$ can be expressed as $\mathbf{L}\mathbf{L}^H$ for some invertible lower triangular $\mathbf{L}$, then $\mathbf{M}$ is Hermitian and positive definite.

The Cholesky factorization can be extended to the case of a *positive semidefinite* and Hermitian matrix $\mathbf{M}$ by dropping the requirement that $\mathbf{L}$ have strictly positive diagonal entries. However, generally speaking, Cholesky factorizations for positive semidefinite matrices are *not unique*. Finally, we note that, if $\mathbf{M}$ is a *symmetric* and *positive definite* matrix with *real* entries, $\mathbf{L}$ is also a *real* matrix.

QR decomposition (also called *QR factorization*). Any real $n \times n$ (square) matrix $\mathbf{M}$ may be decomposed as:

$$\mathbf{M} = \mathbf{Q}\mathbf{R}, \tag{C.14}$$

where $\mathbf{Q}$ is an *orthogonal* matrix (i.e., its columns are orthogonal unit vectors, meaning that $\mathbf{Q}\mathbf{Q}^T = \mathbf{I}_n$) and $\mathbf{R}$ is an *upper triangular* matrix (also called a *right triangular matrix*). This result can be generalized to a *complex* square matrix $\mathbf{M}$. In this case $\mathbf{Q}$ is a *unitary* matrix. If $\mathbf{M}$ is *invertible*, the factorization (C.14) is *unique* if the diagonal elements of $\mathbf{R}$ are required to be positive.

Other useful properties. It can be proved that:

$$\det(\mathbf{I}_n + \mathbf{A}^T\mathbf{B}) = \det(\mathbf{I}_m + \mathbf{B}\mathbf{A}^T) = \det(\mathbf{I}_m + \mathbf{A}\mathbf{B}^T), \tag{C.15}$$

where both $\mathbf{A}$ and $\mathbf{B}$ are $m \times n$ matrices, and that:

$$\mathrm{tr}(\mathbf{U}\mathbf{V}) = \mathrm{tr}(\mathbf{V}\mathbf{U}), \tag{C.16}$$

where both $\mathbf{U}$ and $\mathbf{V}$ are $n \times n$ matrices.

Appendix D

Signal Spaces

In this appendix the problem of the *representation of deterministic and random signals based on their expansion in a series of orthonormal functions* is summarized. The mathematical tools so developed allow us to establish a one-to-one correspondence between a signal space and a proper real or complex vector space. Therefore, their use is the key to turning problems concerning the analysis of analog signals into equivalent vector problems, leading to a simpler analysis and interpretation.

As in Appendix E, in the following we provide a set of relevant definitions and basic results without proof; the reader can refer to [321] and [430, Ch. 3] for further details.

D.1 Representation of Deterministic Signals

D.1.1 Basic Definitions

The waveforms appearing in most communications systems can be modeled as real or complex signals having *finite energy* and defined over a limited time interval. For this reason, in what follows, we consider the signal set $L_2(t_i, t_f)$ consisting of all the *complex functions* having support (t_i, t_f) and finite energy. In other words, the complex waveform $x(t)$, defined over the interval (t_i, t_f), belongs to $L_2(t_i, t_f)$ if and only if its *energy*:

$$E_x \triangleq \int_{t_i}^{t_f} |x(t)|^2 \, dt \tag{D.1}$$

takes on a finite value. It can be easily proved that this set of functions, together with the usual operations of *addition* $(+)$ between a pair of functions and *multiplication* $(\cdot)$ between a complex number (scalar) and a function, forms a *vector space* over the field $\mathbb{C}$ of complex numbers (see Section E.3).

Let us assume that a nonempty subset S is extracted from $L_2(t_i, t_f)$. If S is also a vector space (with the same operations as $L_2(t_i, t_f)$), it is said that S is a *subspace* of $L_2(t_i, t_f)$. It can be proved that a nonempty subset S of $L_2(t_i, t_f)$ is a subspace if and only if it is *closed* with respect to the above addition and multiplication operations.

A subspace of $L_2(t_i, t_f)$ can easily be generated from a set of elements of this space. In fact, given the set $I_x = \{x_i(t), i = 0, 1, \ldots, N - 1\}$, containing N *distinct* functions of $L_2(t_i, t_f)$, the set

Wireless Communications: Algorithmic Techniques, First Edition.
Giorgio M. Vitetta, Desmond P. Taylor, Giulio Colavolpe, Fabrizio Pancaldi, Philippa A. Martin.
© 2013 John Wiley & Sons, Ltd. Published 2013 by John Wiley & Sons, Ltd.

S consisting of all possible *linear combinations*:

$$\sum_{i=0}^{N-1} a_i \, x_i(t), \tag{D.2}$$

where $\{a_i, \, i = 0, 1, \ldots, N-1\}$ are N scalars, being closed with respect to the above-mentioned operations, is a subspace of $L_2(t_i, t_f)$. Note that it may occur that the functions of I_x are *linearly dependent* (see Section E.3). If this does not happen, they are *linearly independent* and form a *basis* of S. In this case, the value of N gives the *dimension* of S. It can be proved that, given a basis $B = \{\phi_i(t), \, i = 0, 1, \ldots, N-1\}$ of a subspace S, the representation:

$$x(t) = \sum_{i=0}^{N-1} x_i \, \phi_i(t) \tag{D.3}$$

of any function $x(t)$ in S, where $\{x_i, \, i = 0, 1, \ldots, N-1\}$ is a set of N scalars, is *unique*. This allows us to establish a one-to-one correspondence between the waveforms and vectors an obtain the N scalars which appear in the RHS of the last expression, as discussed in the following subsection.

In the space $L_2(t_i, t_f)$ the *inner product*:

$$(x, y) \triangleq \int_{t_i}^{t_f} x(t) \, y^*(t) \, dt \tag{D.4}$$

can be defined for any pair of its signals $x(t)$ and $y(t)$. If $(x, y) = 0$, then $x(t)$ and $y(t)$ are said to be *orthogonal*. For any signal $x(t)$ in $L_2(t_i, t_f)$, its *norm* can be evaluated as:

$$|x| \triangleq \sqrt{(x, x)}. \tag{D.5}$$

Note that, since $(x, x) = E_x$, the norm of $x(t)$ can also be computed as the square root of its energy (i.e., $|x| = \sqrt{E_x}$). In addition, if $|x|$ is positive, $x(t)$ can be *normalized* by dividing it by its norm, that is, by generating the new signal $x(t)/|x|$ which has *unit energy*.

The distance between the signals $x(t)$ and $y(t)$ in $L_2(t_i, t_f)$ is defined as:

$$d_{xy} \triangleq |x - y| = \sqrt{\int_{t_i}^{t_f} |x(t) - y(t)|^2 \, dt} = \sqrt{E_{x-y}}, \tag{D.6}$$

where E_{x-y} denotes the energy of the difference signal $x(t) - y(t)$. If $d_{xy} = 0$, the signals $x(t)$ and $y(t)$ coincide *almost everywhere*. Generally speaking, the parameter d_{xy} can be represented as a measure of the *diversity degree* between the signals $x(t)$ and $y(t)$.

D.1.2 Representation of Deterministic Signals via Orthonormal Bases

As mentioned in the previous subsection, the representation of the waveforms of a subspace S of $L_2(t_i, t_f)$ using its basis $B = \{\phi_i(t), \, i = 0, 1, \ldots, N-1\}$ (see (D.3)) allows us to establish a *one-to-one correspondence* between the elements of S and the points of a subspace S_c of the vector space $\mathbb{C}^N$, which is usually called a *signal space* and consists of all the possible N-tuples of complex numbers.[1] In fact, given B, any $x(t)$ in S can be represented by one and only one vector:

$$\mathbf{x} = [x_0, x_1, \ldots, x_{N-1}]^T, \tag{D.7}$$

---

[1] If *real* scalars and signals are considered (in place of their complex counterparts), the signal space becomes $\mathbb{R}^N$, the set of all the possible ordered real n-tuples.

according to (D.3). The vector $\mathbf{x}$ is called an *image* of $x(t)$, and its components represent the coordinates of $x(t)$. Conversely, one and only one function is associated with each vector $\mathbf{x}$ of S_c according to (D.3). It is easy to show that, given a basis B, linear operations accomplished over the functions of S translate into linear operations on their images. For instance, given $x_1(t)$ and $x_2(t)$ in S and the scalars α and β, the image of linear combination $\alpha x_1(t) + \beta x_2(t)$ is $\alpha \mathbf{x}_1 + \beta \mathbf{x}_2$, where $\mathbf{x}_1$ and $\mathbf{x}_2$ denote the images of $x_1(t)$ and $x_2(t)$, respectively.

Given a basis $B = \{\phi_i(t), i = 0, 1, \ldots, N-1\}$, generally speaking, the evaluation of the image $\mathbf{x}$ of a signal $x(t)$ in S is not easy. However, this problem can easily be solved if B is an *orthonormal basis*, that is, if all its elements have unit norm and are mutually orthogonal; the latter condition is concisely represented by the expression:

$$(\phi_i, \phi_j) = \delta_{i,j}, \tag{D.8}$$

with $i, j = 0, 1, \ldots, N-1$. In fact, given an orthonormal basis B of a subspace S, the ith element x_i of $\mathbf{x}$ can be evaluated as:

$$x_i = (x, \phi_i) = \int_{t_i}^{t_f} x(t) \, \phi_i^*(t) \, dt \tag{D.9}$$

for $i = 0, 1, \ldots, N-1$. Then, substituting (D.9) into (D.3) yields:

$$x(t) = \sum_{i=0}^{N-1} (x, \phi_i) \, \phi_i(t), \tag{D.10}$$

which expresses the function $x(t)$ in S as a superposition of N *projections*, the ith of which is given by the signal:

$$x_i(t) = (x, \phi_i) \, \phi_i(t) \tag{D.11}$$

with $i = 0, 1, \ldots, N-1$.

It is easy to prove that, given an orthonormal basis $B = \{\phi_i(t), i = 0, 1, \ldots, N-1\}$ for a subspace S, the one-to-one mapping between the functions of S and the associated subspace S_c of $\mathbb{C}^N$ *preserves the scalar product*, and, consequently, the norm and the distance. In other words, we have that:

$$(x, y) = \mathbf{x} \cdot \mathbf{y} \triangleq \mathbf{x}^T \mathbf{y}^* = \sum_{i=0}^{N-1} x_i \, y_i^* \tag{D.12}$$

for any two signals $x(t)$ and $y(t)$ in S, where $\mathbf{x} \cdot \mathbf{y}$ denotes the *scalar product* between their images. Then, given definition (D.5), from (D.12) it is easy to see that:

$$\|x\| = \sqrt{\mathbf{x} \cdot \mathbf{x}}. \tag{D.13}$$

Since:

$$\sqrt{\mathbf{x} \cdot \mathbf{x}} = |\mathbf{x}| \triangleq \sqrt{\mathbf{x}^T \mathbf{x}^*} = \sqrt{\sum_{i=0}^{N-1} |x_i|^2}, \tag{D.14}$$

where $|\mathbf{x}|$ denotes the *norm* of an arbitrary vector $\mathbf{x} \in \mathbb{C}^N$, (D.13) can be rewritten as:

$$\|x\| = |\mathbf{x}|, \tag{D.15}$$

which simply states that norm is preserved. Note also that the last result is equivalent to (see (D.5)):

$$\|x\|^2 = E_x = |\mathbf{x}|^2, \tag{D.16}$$

which states that the energy E_x of a signal $x(t) \in S$ is equal to the square of the norm of its image **x**. Similarly, it can be proved that the distance d_{xy} between the functions $x(t)$ and $y(t)$, both in S, is preserved, that is:

$$d_{xy} \triangleq \|x - y\| = |\mathbf{x} - \mathbf{y}|. \tag{D.17}$$

These results illustrate the importance of finding an orthonormal basis of a subspace S generated by the set of N signals $\{x_i(t), i = 0, 1, \ldots, N - 1\}$. Generally speaking, this problem can be solved applying to the signal set the so-called *Gram–Schmidt* orthogonalization procedure. The first step is to generate the first element $\phi_0(t)$ of an orthonormal basis B by normalizing the first waveform $x_0(t)$:

$$\phi_0(t) \triangleq \frac{x_0(t)}{\|x_0\|}. \tag{D.18}$$

Note that $x_0(t) = \|x_0\|\phi_0(t)$, so that $x_0(t)$ can simply be generated by scaling $\phi_0(t)$, that is, as a linear combination of an element of B.

The second element $\phi_1(t)$ of B is evaluated by generating first the auxiliary function:

$$\eta_1(t) \triangleq x_1(t) - p_{10}(t), \tag{D.19}$$

where $p_{10}(t) \triangleq (x_1, \phi_0)\phi_0(t)$ denotes the projection of $x_1(t)$ along $\phi_0(t)$. If $\|\eta_1\| = 0$, $\eta_1(t)$ can be discarded. However, if $\|\eta_1\| \neq 0$, $\phi_1(t)$ can be generated by normalizing $\eta_1(t)$:

$$\phi_1(t) \triangleq \frac{\eta_1(t)}{\|\eta_1\|}. \tag{D.20}$$

It can easily be shown that $\phi_1(t)$ is orthogonal to $\phi_0(t)$; in addition, it follows from (D.19) and (D.20) that:

$$x_1(t) = (x_1, \phi_0) \, \phi_0(t) + \|\eta_1\|\phi_1(t), \tag{D.21}$$

so that $x_1(t)$ can be expressed as a linear combination of $\phi_0(t)$ and $\phi_1(t)$, that is, of the functions of B.

The evaluation of all the other functions of B proceeds as follows. The functions $\{x_i(t), i = 0, 1, \ldots, N - 1\}$ are processed in an ordered fashion. In particular, when $x_k(t)$ is processed, the orthonormal functions $\{\phi_i(t), i = 0, 1, \ldots, q\}$ of B, with $q \leq k - 1$, will already have been evaluated. Then the auxiliary function:

$$\eta_k(t) \triangleq x_k(t) - \sum_{i=0}^{q}(x_k, \phi_i) \, \phi_i(t) \tag{D.22}$$

is evaluated by subtracting from $x_k(t)$ the contributions of all its projections according to the orthonormal functions already computed. If $\|\eta_k\| = 0$, $x_k(t)$ is a linear combination of $\{\phi_i(t), i = 0, 1, \ldots, q\}$ and can be discarded. However, if $\|\eta_k\| \neq 0$, a new waveform of B is evaluated as:

$$\phi_{q+1}(t) \triangleq \frac{\eta_k(t)}{\|\eta_k\|}. \tag{D.23}$$

The signal $\phi_{q+1}(t)$ has unit norm and is orthogonal to all the functions $\{\phi_i(t), i = 0, 1, \ldots, q\}$, so that the enlarged set $\{\phi_i(t), i = 0, 1, \ldots, q + 1\}$ consists of orthonormal functions. Finally, from (D.22) and (D.23) it is easy to see that:

$$x_k(t) = \|\eta_k\|\phi_{q+1}(t) + \sum_{i=0}^{q}(x_k, \phi_i) \, \phi_i(t). \tag{D.24}$$

The last result expresses $x_k(t)$ as a linear combination of the functions $\{\phi_i(t), i = 0, 1, \ldots, q + 1\}$. Applying the procedure described above until $k = N - 1$ generates all the functions B. Finally, we note that, generally speaking, B consists of $Q \leq N$ waveforms and that the equality $Q = N$ holds if and only if none of the auxiliary functions $\{\eta_i(t), i = 0, 1, \ldots, N-1\}$ (with $\eta_0(t) = \phi_0(t)$) has null norm. This occurs if and only if the signals $\{x_i(t), i = 0, 1, \ldots, N-1\}$ are linearly independent.

Let us now assume that an orthonormal basis $B = \{\phi_i(t), i = 0, 1, \ldots, N-1\}$ is available for the subspace S of $L_2(t_i, t_f)$. If an arbitrary function $x(t)$ is selected in $L_2(t_i, t_f)$, it cannot usually be represented using the elements of B. In fact, if $x(t) \in S$, the *exact* representation:

$$x(t) = \sum_{i=0}^{N-1} (x, \phi_i)\, \phi_i(t). \tag{D.25}$$

can be given for this signal (see (D.11)). In contrast, if $x(t) \notin S$, the equality (D.25) does not hold. However, in this case a signal $\tilde{x}_N(t) \in S$ *approximating* $x(t)$ with a certain accuracy can be evaluated. In particular, the so-called *projection theorem* establishes the following:

1. The waveform:

$$\tilde{x}_N(t) \triangleq \sum_{i=0}^{N-1} (x, \phi_i)\, \phi_i(t), \tag{D.26}$$

   expressing the *projection* of $x(t)$ on S, is the function of S exhibiting the *minimum distance* from $x(t)$.
2. The error signal:

$$e_N(t) \triangleq x(t) - \tilde{x}_N(t) \tag{D.27}$$

   is *orthogonal* to any function of S and its energy is given by:

$$\|e_N\|^2 = \|x\|^2 - \|\tilde{x}_N\|^2. \tag{D.28}$$

Note that substituting (D.26) into (D.28) yields:

$$\|e_N\|^2 = \|x\|^2 - \sum_{i=0}^{N-1} |(x, \phi_i)|^2. \tag{D.29}$$

Now let the number N of orthogonal functions of B increase, so that the dimensionality of S increases. Equality (D.29) shows that, as N gets larger, the *energy* of the error signal $e_N(t)$ (D.27) diminishes, that is, the approximation $\tilde{x}_N(t)$ of $x(t)$ becomes more accurate. This raises the following question: if $N \to +\infty$, that is, if B consists of an *infinity of functions*, does the energy $\|e_N\|^2$ of $e_N(t)$ tend to zero *for any* $x(t) \in L_2(t_i, t_f)$? Generally speaking, the answer to this question is negative – in other words, the fact that B consists of an infinite number of functions is not *sufficient* to ensure that the equality:

$$\lim_{N \to +\infty} \|e_N\|^2 = \lim_{N \to +\infty} \left\| x(t) - \sum_{i=0}^{N-1} (x, \phi_i)\, \phi_i(t) \right\|^2 = 0 \tag{D.30}$$

holds for any $x(t) \in L_2(t_i, t_f)$. However, if this occurs, B is said to be a *complete* orthonormal basis. Note that, if B is complete, (D.30) can also be expressed as:

$$x(t) = \sum_{i=0}^{+\infty} (x, \phi_i)\, \phi_i(t), \tag{D.31}$$

which needs to be interpreted carefully. This result states only that the series appearing on the RHS of (D.31) converges to $x(t)$ *in quadratic mean*, as stated by (D.30). However, this convergence does not entail the *pointwise convergence* (or the *uniform convergence*) of such a series to $x(t)$ in any instant of interval (t_i, t_f).

A complete basis for a specific signal space is defined in the following example.

Example D.1.1 A complete orthonormal basis for $L_2(0, T)$, where T denotes the (arbitrary) duration of the observation interval, is the so-called *Fourier basis*, which consists of the waveforms $\{\phi_l(t), l = 0, \pm 1, \pm 2, \ldots\}$, with:

$$\phi_l(t) \triangleq \frac{1}{\sqrt{T}} \exp\left(j\frac{2\pi lt}{T}\right). \tag{D.32}$$

This basis can be used to represent the signals of $L_2(t_i, t_f)$, provided that both t_i and t_f take on *finite values*. In fact, any finite interval (t_i, t_f) can be always transformed, through shift and scaling operations, into the interval $(0, T)$, where T is a positive quantity.

☐

It can be shown that, given a complete basis $B = \{\phi_i(t), i = 0, 1, \ldots\}$ of $L_2(t_i, t_f)$ and the components $\{x_i, i = 0, 1, \ldots\}$ and $\{y_i, i = 0, 1, \ldots\}$ of the infinite-dimensional images of the arbitrary signals $x(t)$ and $y(t)$ (both belonging to $L_2(t_i, t_f)$), respectively, the following equalities hold:

$$\|x\|^2 = \sum_{i=0}^{+\infty} \|x_i\|^2, \tag{D.33}$$

:

$$\|x - y\|^2 = \sum_{i=0}^{+\infty} \|x_i - y_i\|^2 \tag{D.34}$$

and:

$$(x, y) = \sum_{i=0}^{+\infty} x_i \, y_i^*. \tag{D.35}$$

These results generalize (D.12), (D.16) and (D.17), respectively, which were derived for a finite-dimensional subspace S.

D.2   Representation of Random Signals via Orthonormal Bases

The notion of expanding a deterministic function into a series of orthogonal functions can be extended to stochastic signals, as illustrated below. Let us consider a zero-mean, generally nonstationary, complex stochastic process $X(t)$, with a covariance function $C_X(t, \tau)$ Hermitian in t and τ. Generally speaking, we are interested in a series expansion of the form:

$$X(t) = \lim_{N \to +\infty} \sum_{i=0}^{N} X_i \, \phi_i(t), \tag{D.36}$$

over the interval (t_i, t_f), where $\{\phi_i(t), i = 0, 1, \ldots, N - 1\}$ are N *deterministic* orthonormal functions, $\{X_i, i = 0, 1, \ldots, N - 1\}$ are zero-mean random variables and l.i.m. stands for *limit in the mean*, so that the equality in (D.36) has to be interpreted in the sense:

$$\lim_{N \to +\infty} \mathrm{E}\left\{\left\|X(t) - \sum_{i=0}^{N} X_i \, \phi_i(t)\right\|^2\right\} = 0. \tag{D.37}$$

We have not yet put any constraint on the functions $\{\phi_i(t)\}$. The choice of these signals inevitably influences the higher-order statistics of the expansion coefficients $\{x_i\}$. Generally speaking, it is convenient to have such coefficients *statistically uncorrelated*, that is:

$$E\{X_i \, X_j\} = \lambda_i \, \delta_{i,j} \tag{D.38}$$

for some unknown set of constants. It can be proved that the functions $\{\phi_i(t)\}$ and the corresponding constants $\{\lambda_i\}$ that produce this condition are given by the *Karhunen–Loève* (KL) *theorem* [430, Ch. 3]. This theorem states that the signals $\{\phi_i(t)\}$ and the associated constants $\{\lambda_i\}$ are the solutions of the *homogeneous Fredholm integral equation*:

$$\int_{-\infty}^{+\infty} C_X(t, \tau) \, \phi_i(\tau) \, d\tau = \lambda_i \phi_i(t). \tag{D.39}$$

This is an integral form of an eigenequation, with a kernel $C_X(t, \tau)$, where $\{\phi_i(t)\}$ are the normalized *eigenfunctions* and $\{\lambda_i\}$ are the corresponding eigenvalues. From the theory of integral equations it can be shown that, if the kernel $C_X(t, \tau)$ is Hermitian in its arguments, that is:

$$C_X(t, \tau) = C_X^*(\tau, t), \tag{D.40}$$

then the following properties hold:

(P.1) The eigenvalues are *real*.

(P.2) The eigenfunctions associated with distinct eigenvalues are *orthogonal*.

(P.3) If the kernel is *square integrable*, that is, if:

$$\int_{t=t_i}^{t_f} \int_{\tau=t_i}^{t_f} |C_X(t, \tau)|^2 d\tau \; dt \; < \infty, \tag{D.41}$$

then each eigenvalue $\lambda_i \neq 0$ has a *finite* number of associated orthogonal eigenfunctions.

(P.4) If the kernel is *positive definite*, its eigenfunctions form a *complete orthonormal set*.

(P.5) If the kernel is *nonnegative definite*, it can be expanded as:

$$C_X(t, \tau) = \sum_{i=0}^{+\infty} \lambda_i \, \phi_i(t) \, \phi_i^*(\tau), \tag{D.42}$$

a result known as *Mercer's theorem*.

As far as the last point is concerned, it is worth noting that a random process with a nondegenerate kernel will have an infinite number of eigenvalues and will theoretically require the infinite expansion of (D.42). However, in many cases of practical interest, the spectrum of eigenvalues will remain significant for a finite number of eigenvalues, before decaying away to zero. Therefore, only a finite number need be considered as significant for a given machine accuracy.

Finally, we note that a zero-mean WSS process $X(t)$, characterized by the covariance function $C_X(\tau)$, will have the above-mentioned properties when it is expanded in terms of its eigenfunctions. In this case, each eigenvalue will correspond to the energy of the process contained in the associated eigenfunction. If $X(t)$ is *Gaussian* as well, the expansion coefficients $\{X_i\}$ will also be *statistically independent* random variables.

Appendix E

Groups, Finite Fields and Vector Spaces

In this appendix some relevant concepts of algebraic structures are summarized. In particular, we first provide the axiomatic definition of a *group* and illustrate some basic concepts and results of group theory. Then we illustrate the basics of *field theory* and analyze some mathematical tools for accomplishing computations over fields. Finally, the concept of *vector space* is introduced.

In the following, we summarize a number of results useful for the understanding of the material illustrated in Part II of this book, but do not provide proofs; these can be found in a number of books on coding theory (e.g., see [35, 1549]).

E.1 Groups

Consider an algebraic structure G consisting of a set of *elements* for which a *dyadic operation* is defined; this operation is usually called *addition* or *multiplication* and in the following is denoted by the symbols $\boxplus$ and $\boxdot$, respectively, if it does not involve ordinary numbers (for which the usual symbols $+$ and $\cdot$ are used). The structure G is a *group* if it satisfies the following four axioms.

Axiom G.1 (*closure*). For all a, b in G, the result of the operation $a \boxplus b$ (or $a \boxdot b$) is also in G.

Axiom G.2 (*associative law*). For any triple (a, b, c) of elements of G the equality

$$(a \boxplus b) \boxplus c = a \boxplus (b \boxplus c) \tag{E.1}$$

or

$$(a \boxdot b) \boxdot c = a \boxdot (b \boxdot c) \tag{E.2}$$

holds, if the operation defined over G is an addition or a multiplication, respectively.

Axiom G.3 (*identity element*). There exists an identity element in G, denoted 0 (1) for the addition (multiplication) operator, such that:

$$a \boxplus 0 = 0 \boxplus a = a \tag{E.3}$$

and

$$a \boxdot 1 = 1 \boxdot a = a \tag{E.4}$$

for any a in G.

Wireless Communications: Algorithmic Techniques, First Edition.
Giorgio M. Vitetta, Desmond P. Taylor, Giulio Colavolpe, Fabrizio Pancaldi, Philippa A. Martin.
© 2013 John Wiley & Sons, Ltd. Published 2013 by John Wiley & Sons, Ltd.

Axiom G.4 (*inverse element*). There exists an *inverse element e* in G, such that for every element a in G, the equality:

$$a \boxplus e = e \boxplus a = 0 \tag{E.5}$$

or

$$a \boxdot e = e \boxdot a = 1 \tag{E.6}$$

holds, if the operation defined over G is an addition or a multiplication, respectively. Note that in the first (second) case e is usually denoted by $-a$ (by a^{-1}) and is called the *opposite* (*reciprocal*) of a.

Given these axioms it can easily be proved that: the identity element and the inverse of any element a in G are unique; and the inverse of the product $a \boxdot b$ can be evaluated as:

$$(a \boxdot b)^{-1} = b^{-1} \boxdot a^{-1}. \tag{E.7}$$

If the operator defined over G is *commutative* (i.e., $a \boxplus b = b \boxplus a$ or $a \boxdot b = b \boxdot a$ for any a, b in G), then G is a *commutative* or *Abelian group*. A group can consist of an arbitrary number of elements; the smallest group consists of a single element (i.e., the identity element defined in axiom G.3).

A subset H of the elements G is a *subgroup* if its satisfies the same axioms as G. Note that in order to verify that H is really a subgroup of G, it is *sufficient* to check its closure (axiom G.1) and the existence of an inverse for all its elements (axiom G.4). If both conditions are satisfied, the identity element must necessarily belong to H (axiom G.3) and the associative property (axiom G.2) must hold for all its elements, exactly as for all the elements of G.

Let us now consider a group G whose elements are denoted by $\{g_n, n = 0, 1, \ldots\}$ and over which the addition operation $\boxplus$ is defined; and a subgroup H of G consisting of the n distinct elements $\{h_0, h_1, \ldots, h_{n-1}\}$, with $h_0 = 0$.[1] Furthermore, let us generate a matrix $\mathbf{S} = [s_{i,j}]$ (containing n columns) using the following algorithm:

1. The first row of $\mathbf{S}$ consists of all the ordered elements of H (i.e., $s_{0,j} = h_j$, with $j = 0, 1, \ldots, n - 1$).
2. The first element g_{r_1} of the second row of $\mathbf{S}$ is an element of G not appearing in the first row and any other element of the same row is generated by adding g_{r_1} (taken as first operand) to the corresponding element of the same row, that is:

$$s_{1,j} = g_{r_1} \boxplus h_j, \tag{E.8}$$

with $j = 1, 2, \ldots, n - 1$.
3. The procedure illustrated for the first row is repeated for all the other rows of $\mathbf{S}$, selecting as first element of the kth row an element g_{r_k} in G which has not appeared in the previous rows, until all the elements of G are included in $\mathbf{S}$.

The matrix $\mathbf{S}$ is structured as:

$$\mathbf{S} = \begin{bmatrix} h_0 = 0 & h_1 & \cdots & h_{n-1} \\ g_{r_1} \boxplus h_0 & g_{r_1} \boxplus h_1 & \cdots & g_{r_1} \boxplus h_{n-1} \\ \cdots & \cdots & \cdots & \cdots \\ g_{r_k} \boxplus h_0 & g_{r_k} \boxplus h_1 & \cdots & g_{r_k} \boxplus h_{n-1} \end{bmatrix}$$

---

[1] The multiplication operation can be used in place of the addition operation; in this case, however, $h_0 = 1$ (in other words, h_0 is always the identity element).

for the case in which G consists of a *finite* number of elements. The set consisting of all the elements of a single row of **S** represents a *left coset*[2] of G and the first element of each coset is called a *coset leader*.

It can be proved that any element of a group G belongs to one, and only one, of the cosets associated with a subgroup H. Moreover, if the *order* O_G of G is defined as the number of its elements and the *index* $I_{G|H}$ of G over the subgroup H is defined as the number of its elements with respect to H itself, then the equality:

$$O_G = I_{G|H} \cdot O_H \tag{E.9}$$

holds.

E.2   Fields

E.2.1   Axiomatic Definition of a Field and Finite Fields

Let us now consider an algebraic structure F consisting of a set of *elements* for which two *dyadic operations*, namely an *addition* ($\boxplus$) and a multiplication ($\boxdot$) are defined. The structure F is a *field* if it satisfies the following seven axioms.

Axiom F.1 (*commutativity*). F is a *commutative group* with respect to addition.

Axiom F.2 (*closure*). For any pair (a, b) of elements in F, $a \boxdot b$ is in F.

Axiom F.3 (*associative law*). For any triple (a, b, c) of elements in F, the equality

$$(a \boxdot b) \boxdot c = a \boxdot (b \boxdot c) \tag{E.10}$$

holds.

Axiom F.4 (*identity element*). F contains an element, called the *multiplicative identity* and denoted by 1, such that:

$$a \boxdot 1 = 1 \boxdot a = a \tag{E.11}$$

for any a in F.

Axiom F.5 (*inverse element*). F contains, for any element $a \neq 0$, a *multiplicative inverse* (or *reciprocal*) denoted by a^{-1}, such that $a \boxdot a^{-1} = a^{-1} \boxdot a = 1$.

Axiom F.6 (*commutative law*). For any pair (a, b) of elements in F, the equality $a \boxdot b = b \boxdot a$ holds.

Axiom F.7 (*distributive law*). For any triple (a, b, c) of elements in F, the *distributive law*:

$$a \boxdot (b \boxplus c) = a \boxdot b \boxplus a \boxdot c \tag{E.12}$$

holds.

Note that axioms F.2–F.6, which refer *uniquely* to the properties of multiplication,[3] are not enough to state that F is a *commutative group* not only with respect to addition (as stated by axiom F.1), but also with respect to multiplication. In fact, a multiplicative inverse for the additive identity 0 does not exist, so that axiom G.4 of Section E.1 is not satisfied.

A field F can consist of either an infinite or finite set of elements. In the latter case, F is a *finite field* or *Galois field* and is usually denoted by GF(q), where q is its number of elements. Note that the minimum number of elements of a Galois field F is two and that, in this case, F consists of the additive identity 0 and of the multiplicative identity 1 (in other words, GF(2) = {0, 1}) and that the addition and the multiplication must necessarily be carried out according to the rules summarized in

---

[2] If the order of the operands in (E.8) is reversed, a *right coset* is generated; of course, if G is *Abelian*, there is no difference between a right coset and its left counterpart.

[3] Only axiom F.7 expresses a joint property of addition and multiplication.

Table E.1 Table of addition for GF(2)

⊞	0	1
0	0	1
1	1	0

Table E.2 Table of multiplication for GF(2)

⊡	0	1
0	0	0
1	0	1

Tables E.1 and E.2, respectively. Note that GF(2) is the field used in most digital electronics and the addition and multiplication are implemented using an *exclusive-or* gate and *and* gate, respectively.

It can be proved that, for any *prime number q*, a finite field GF(q) consisting of exactly q elements exists. In this case, the elements of GF(q) can be represented by the integers $\{0, 1, \ldots, q - 1\}$, and addition and multiplication can be evaluated as:

$$a \boxplus b = \mathrm{R}_q[a + b] \tag{E.13}$$

and

$$a \boxdot b = \mathrm{R}_q[a \cdot b], \tag{E.14}$$

respectively, for any a, b in GF(q) (as already mentioned in Section E.1, the symbols $+$ and $\cdot$ denote addition and the multiplication, respectively, for ordinary numbers). A field GF(q) containing a prime number q of elements is called a *prime field*. However, if q is not prime, a finite field containing exactly q elements does not exist. More specifically, the *existence* of a finite field GF(q) can be proved if q is the power of a prime number p, that is, if $q = p^m$, where m is a integer not smaller than unity.[4] If m is larger than unity, GF(q) represents an *extension field*; however, unlike the case of prime fields, if the set of integers $\{0, 1, \ldots, q - 1\}$ is used to represent its elements, then addition and multiplication over it cannot be defined by (E.13) and (E.14), since the field axioms are not satisfied. In fact, in this case, defining addition and multiplication properly requires the use of a *polynomial representation* of the field elements, as illustrated in the following subsection.

E.2.2 Polynomials and Extension Fields

We now describe the most relevant properties of the polynomials defined over a field F. Then we show how, given a proper polynomial representation for a set of $q = p^m$ elements, the operations of addition and multiplication can be defined in a such way as to construct a finite field GF(q).

E.2.2.1 Fundamental properties of polynomials over a field

Generally speaking, given a field F (and denoting addition and multiplication over it by the symbols ⊞ and ⊡, respectively), a set of single-variable polynomials, whose coefficients belong to F (called

---

[4] In the following, the parameter p will always denote a *prime number*; in real-world applications $p = 2$ is usually selected, so that GF(q) = GF(2^m).

a *ground field*), can be constructed. In fact, if D denotes such a single variable, a polynomial $x(D)$ over the field F takes the form:

$$x(D) = x_0 + x_1 D + \ldots + x_n D^n, \tag{E.15}$$

where the coefficients $\{x_i, i = 0, 1, \ldots, n-1\}$ belong entirely to F. We now summarize some basic definitions and fundamental results of the mathematical theory of polynomials. To begin, we note the following:

- If the coefficient x_n in (E.15) is not equal to zero, the parameter n, representing the exponent of the largest power of D, is the so-called *degree* of $x(D)$.
- If x_n is equal to unity, $x(D)$ is a *monic* polynomial.

The operations of addition ($+$) and multiplication ($\cdot$) between polynomials can easily be carried out if the tables of addition ($\boxplus$) and multiplication ($\boxdot$) defined over F are given. Note that, in general, the degree of the sum of two polynomials is equal to the maximum degree of the operands, whereas the degree of the product of two polynomials is equal to the sum of the degree of the operands.

If three polynomials $x(D)$, $y(D)$ and $z(D)$ over F satisfy the equality:

$$z(D) = x(D) \cdot y(D), \tag{E.16}$$

it is said that $z(D)$ is *divisible* by $x(D)$, or that $x(D)$ *divides* $z(D)$, or that $x(D)$ is a *factor* of $z(D)$. A polynomial $x(D)$ of degree n which is not divisible by any nonnull polynomial defined over the same field and of degree larger than 0 and smaller than n is *irreducible*.

The *largest common divisor* of two polynomials is the monic polynomial of largest degree dividing both polynomials; the two polynomials are *relatively prime* if their largest common divisor is equal to unity.

A nonnull polynomial of degree 0 is an element of F and, consequently, has a multiplicative inverse; however, no polynomial of degree larger than zero shares this property.

For any two polynomials $x(D)$ and $d(D)$ (with $d(D) \neq 0$), there exist a unique *couple* of polynomials $q(D)$ (called the *quotient*) and $r(D)$ (called the *remainder*) such that:

$$x(D) = d(D) \cdot q(D) + r(D) \tag{E.17}$$

and the degree of $r(D)$ is smaller than than of the *divisor* $d(D)$. This result is known as the *Euclid division algorithm*. If $d(D)$ is monic and its degree is equal to unity, that is:

$$d(D) = D - a, \tag{E.18}$$

then (E.17) becomes:

$$x(D) = (D - a) \cdot q(D) + r, \tag{E.19}$$

where the remainder r, having null degree, belongs to F. Setting $D = a$ in (E.19) yields:

$$r = x(a) \tag{E.20}$$

This result is known as the *remainder theorem* and has some implications. In fact, if $x(a) = 0$, that is, if a is a *root* (or *zero*) of $x(D)$, the last result implies that $r = 0$, so that $d(D)$ in (E.18) is a factor of $x(D)$ (this result is known as the *factor theorem*). Therefore, for any distinct root of $x(D)$, there exists a corresponding factor of degree 1. Since the degree of a product of polynomials is equal to the sum of the degree of its factors, the degree of $x(D)$ is not smaller than the number of its distinct roots. It is also worth mentioning that the *fundamental theorem of algebra*, which establishes the existence of n distinct roots for a polynomial $x(D)$ of degree n, holds whatever the field F of its coefficients. Obviously, the roots of $x(D)$ do not necessarily belong to F, but can belong to another field originating from its *extension*. This is exemplified by those polynomials having real coefficients but whose roots belong to the (extension field) of complex numbers.

Finally, it is important to point out that, if a polynomial $x(D)$ is irreducible, not all its roots belong to F. However, the inverse implication does not hold. In other words, if not all the roots of a polynomial $x(D)$ belong to F, it is still possible that it can be factored as $x(D) = y(D) \cdot z(D)$, where the polynomials $y(D)$ and $z(D)$ have all their coefficients belonging to F, but not all their roots belonging to that field.

E.2.2.2 Polynomial Representation of an Extension Field

Let us now consider the problem of how to define the operations of addition and multiplication over a set of $q = p^m$ elements (for a prime number p and a positive integer m) in such a way that an extension field GF(q) is generated. This problem can be solved by representing each element of the set not by an integer, but by a polynomial of degree $m - 1$ over the field GF(p). Note that the overall number of polynomials of this type which can potentially be generated is p^m, that is, the cardinality of the given set, so that a *one-to-one correspondence* between the elements of the set and the above-mentioned polynomials can be established. Then the addition and multiplication of two arbitrary elements of the set can be defined for their polynomial representations, which are denoted by $a(D) = \sum_{k=0}^{m-1} a_k D^k$ and $b(D) = \sum_{k=0}^{m-1} b_k D^k$. In fact, *addition* of the two elements is given by the element represented by the polynomial:

$$c(D) = a(D) + b(D) = \sum_{k=0}^{m-1} c_k D^k, \tag{E.21}$$

where

$$c_k \triangleq a_k \boxplus b_k = \mathrm{R}_q[a_k + b_k], \tag{E.22}$$

and $\boxplus$ denotes addition over the (prime) fields GF(p) of the polynomial coefficients (see (E.13)). In other words, the sum of the two set elements is represented by the polynomial generated by the sum (denoted by the symbol $+$) of $a(D)$ and $b(D)$; note that $c(D)$ has the same degree as $a(D)$ and $b(D)$. Defining *multiplication* between the same elements is more involved, since, in general, the usual product between the two polynomials $a(D)$ and $b(D)$, each of degree $m - 1$, generates a polynomial of degree $2(m - 1)$, which is too large to represent a set element. For this reason, multiplication (denoted $\odot$ in the following to differentiate it from the usual product, $\cdot$, between polynomials) is defined by:

$$c(D) = a(D) \odot b(D) \triangleq \mathrm{R}_{f(D)}[a(D) \cdot b(D)], \tag{E.23}$$

where $f(D)$ is an irreducible polynomial over GF(p) of degree m. Note that the $\mathrm{R}.[\cdot]$ operator in (E.23) entails a reduction of the maximum degree of the resulting polynomial to $m - 1$ and its use recalls that of (E.14) for a prime field. It can be proved that the set of polynomials of degree $m - 1$ over the field GF(p), with the two operations of addition and multiplication defined by (E.21)–(E.23), form a Galois field, whose additive (multiplicative) identity is given by $x(D) = 0$ ($x(D) = 1$).

Let us apply now the results illustrated above to construct the extension field GF(2^3) = GF(8).

Example E.2.1 The field GF(2^3) consists of eight distinct elements; each of them can be represented by an integer between 0 and 7 (*integer representation*), by a polynomial having binary coefficients and degree not larger than $m - 1 = 2$ (polynomial representation, PR) or by a triple of bits (binary representation, BR), as shown in Table E.3.

In evaluating the product of (representing the elements of GF(2^3)) according to (E.23), the irreducible GF(2) polynomial:

$$f(D) = 1 + D + D^3 \tag{E.24}$$

Table E.3 Polynomial, integer and
binary representations of the elements of
GF(8)

Polynomial	Integer	Binary
0	0	000
1	1	001
D	2	010
$D + 1$	3	011
D^2	4	100
$D^2 + 1$	5	101
$D^2 + D$	6	110
$D^2 + D + 1$	7	111

Table E.4 Addition table for GF(8)

$\boxplus$	0	1	2	3	4	5	6	7
0	0	1	2	3	4	5	6	7
1	1	0	3	2	5	4	7	6
2	2	3	0	1	6	7	4	5
3	3	2	1	0	7	6	5	4
4	4	5	6	7	0	1	2	3
5	5	4	7	6	1	0	3	2
6	6	7	4	5	2	3	0	1
7	7	6	5	4	3	2	1	0

Table E.5 Multiplication table for GF(8)

$\cdot$	0	1	2	3	4	5	6	7
0	0	0	0	0	0	0	0	0
1	0	1	2	3	4	5	6	7
2	0	2	4	6	3	1	7	5
3	0	3	6	5	7	4	1	2
4	0	4	3	7	6	2	5	1
5	0	5	1	4	2	7	3	6
6	0	6	7	1	5	3	2	4
7	0	7	5	2	1	6	4	3

of degree $m = 3$ can be used. The tables of addition and multiplication associated with the choice
(E.24) are given in Tables E.4 and E.5, respectively.
    □

The *existence* of finite fields with $q = p^m$ elements, for any prime number p and positive integer
m, is guaranteed by the fact that there exists at least one irreducible polynomial of degree m over
GF(p). Note that the use of different irreducible polynomials characterized by the same degree and
defined over the same field leads to the generation of *isomorphic* fields, differing only in the way

elements are labeled (in other words, a one-to-one correspondence transforming one field in the other one can be found). In practice, this means that a *unique* field GF(q), with $q = p^m$, exists.

E.2.3 Other Definitions and Properties

In this subsection some important definitions and results about finite fields are briefly illustrated.

E.2.3.1 Subfields

A field GF(q) may contain a subset of r elements satisfying the field axioms. Such a subset is called *subfield*. It can be proved that a field of size $q = p^m$ contains a subfield of size $r = p^s$ if and only if s divides m. For this reason, a field GF(p^m) always contains the *ground field* GF(p) (whose elements are represented by zero-degree polynomials).

E.2.3.2 Characteristic of a Field

Let us consider the multiplicative identity, 1, of the field GF(q) and generate the sequence:

$$s[n] \triangleq \sum_{k=1}^{n} 1 \tag{E.25}$$

adding n times this element[5], with $n = 1, 2, \ldots$. The smallest value of n for which $s[n]$ takes on a null value is called the *characteristic* of the field GF(q). It is easy to show that, if $q = p$ (i.e., a prime field is considered), then the field characteristic is equal to p, because of the definition (E.13) of addition for field elements. Furthermore, if $q = p^m$, the characteristic of the field is still equal to p because summing two polynomials entails adding their corresponding coefficients over GF(p) (see (E.22)).

E.2.3.3 Order of a Field Element

Let us consider a *nonnull* element β of GF(q) and generate the sequence:

$$\beta^1, \beta^2, \beta^3, \ldots \tag{E.26}$$

consisting of consecutive powers of β. Since the given field contains a *finite* number of elements, such a sequence is necessarily periodic. Its period n represents the *order* of β. Note that n is the smallest value of the positive integer l ensuring that:

$$\beta^{l+1} = \beta. \tag{E.27}$$

From the last result it is easily inferred that:

$$\beta^n = 1, \tag{E.28}$$

so that β is an n-ary root of unity.

Note also that, since GF(q) contains $q - 1$ nonnull elements, the largest value of the order of its elements is equal to $q - 1$. In number theory it is also shown that the order of each element of GF(q) is always a *divisor* of $q - 1$. For this reason, if $q - 1$ is prime, the possible orders of the elements of GF(q) are 1 (order of the multiplicative identity) and $q - 1$ only.

---

[5] Note that the addition is that of the *ground field*.

E.2.3.4 Primitive Element of a Field

It can be proved that a field GF(q), with $q = p^m$, contains at least one element of order $q - 1$, that is, an element α whose sequence $\{\alpha^1, \alpha^2, \ldots, \alpha^{q-1} = 1 = \alpha^0\}$ of consecutive powers contains all the nonnull elements of the given field. Such an element is called a *primitive element*. Note that if $q - 1$ is *prime*, then all the elements of GF(q), different from 0 and 1, are primitive.

Since each element $\beta \in$ GF(q) can be expressed as a power of α, that is:

$$\beta = \alpha^k, \tag{E.29}$$

where k is an integer of proper value, β can be represented by k, which can be interpreted as the logarithm of β to base α (by convention, the logarithm of 0 is $-\infty$). If this representation is adopted, multiplication and *division*[6] between the elements β (E.29) and $\gamma = \alpha^s$ can be represented by summing and subtracting their logarithmic representations, respectively, exactly as in the case of real numbers. It should be kept in mind, however, that, since $\alpha^{q-1} = 1$, both addition and subtraction of exponents must be evaluated modulo $q - 1$, that is:

$$\beta \boxdot \gamma = \alpha^{\mathrm{R}_{(q-1)}[k+s]} \tag{E.30}$$

and

$$\beta \boxdot \gamma^{-1} = \alpha^{\mathrm{R}_{(q-1)}[k-s]}, \tag{E.31}$$

respectively.

E.2.3.5 Primitive Polynomials

An important subset of the irreducible polynomials over a field GF(p) is that consisting of *primitive polynomials*. A *primitive polynomial* over GF(p) of degree m (with $m \geq 2$) is a polynomial whose m roots are *all* primitive elements of GF(q), with $q = p^m$. A list of primitive polynomials $\{p_m(D)\}$ over GF(2) of degrees $m = 2, 3, \ldots, 8$, is given in Table E.6. These polynomials can be used in evaluating the multiplication between polynomials according to (E.23) (where a primitive polynomial $f(D)$ is needed), thus allowing finite fields of size 2^m (i.e., 8, 16, $\ldots$, 256) to be generated. Field construction involving primitive polynomials is further discussed in the following example.

Table E.6 Primitive polynomials of degree m over GF(2)

m	$p_m(D)$
2	$1 + D + D^2$
3	$1 + D + D^3$
4	$1 + D + D^4$
5	$1 + D^2 + D^5$
6	$1 + D + D^6$
7	$1 + D^3 + D^7$
8	$1 + D^2 + D^3 + D^4 + D^8$

---

[6] The *division* of β by γ has to be evaluated as the product of β and the multiplicative inverse of γ.

Example E.2.2 Let us now construct the field $GF(2^4) = GF(16)$, following a different approach[7] than adopted in Example E.2.1 for $GF(8)$. To begin, let us assume that the first two elements of the field are the additive identity (0) and the multiplicative identity (1), and that the the third element, α, is a root of the binary polynomial of degree $m = 4$ given in Table E.6:

$$f(D) = D^4 + D + 1. \tag{E.32}$$

Since α is a primitive element, all the other elements of $GF(16)$ can be expressed as consecutive powers[8] of α, as illustrated in the first column of Table E.7. Note also that each element of the given field can be represented by a polynomial over $GF(2)$ of degree not larger than $m - 1 = 3$. To show this, let us assume that α, α^2 and α^3 are represented by D, D^2 and D^3, respectively. To derive the polynomial representation of the next element (α^4), let us adopt the following line of reasoning. Since α is a root (E.32), we have that:

$$\alpha^4 + \alpha + 1 = 0, \tag{E.33}$$

so that:

$$\alpha^4 = \alpha + 1. \tag{E.34}$$

The last result provides a representation of the field element as a linear combination of powers of α. This representation, known as the *basis representation*, can be derived for all the other field elements and, generally speaking, involves powers of α having degree not larger than $m - 1 = 3$. For instance, since $\alpha^5 = \alpha^4 \cdot \alpha$, from (E.34) it is easily inferred that:

$$\alpha^5 = (\alpha + 1) \cdot \alpha = \alpha^2 + \alpha. \tag{E.35}$$

Table E.7 Different representations for the elements of GF(16)

α^j	LoR	PR	BR
$0 = \alpha^{-\infty}$	$-\infty$	0	0000
$1 = \alpha^0$	0	1	0001
α^1	1	D	0010
α^2	2	D^2	0100
α^3	3	D^3	1000
α^4	4	$D + 1$	0011
α^5	5	$D^2 + D$	0110
α^6	6	$D^3 + D^2$	1100
α^7	7	$D^3 + D + 1$	1011
α^8	8	$D^2 + 1$	0101
α^9	9	$D^3 + D$	1010
α^{10}	10	$D^2 + D + 1$	0111
α^{11}	11	$D^3 + D^2 + D$	1110
α^{12}	12	$D^3 + D^2 + D + 1$	1111
α^{13}	13	$D^3 + D^2 + 1$	1101
α^{14}	14	$D^3 + 1$	1001

---

[7] Note that in this case the field construction technique of Example E.2.1 for $GF(8)$ can also be adopted. In addition, it can be shown that, if in evaluating polynomial multiplication according to (E.23) the polynomial $f(D)$ (E.32) is selected, that technique leads exactly to the same results for $GF(16)$ as those developed below using a different approach.

[8] Note that the last element is $\alpha^{q-2} = \alpha^{14}$, since $\alpha^{q-1} = \alpha^{15} = 1$.

It is also important to point out that, on the basis of the results (E.34) and (E.35), the polynomial representations $D + 1$ and $D^2 + 2$ can be adopted for the field elements α^4 and α^5, respectively. If this approach is adopted for all the other consecutive powers of α, the PR of all the elements of GF(16) is found. Such a representation is provided by the third column of Table E.7, which also shows, for each element of that field, its *logarithmic representation* (LoR) and BR. Note that the BR of a field element can be easily derived from its PR, reading the coefficients of the associated polynomial in an ordered fashion. The reader can verify that:

- the possible orders of the nonnull elements of GF(16) are 1, 3, 5 and 15;
- GF(16) contains eight primitive elements (including α).

$\square$

It can be shown that the basis representation mentioned in Example E.2.2 can be applied to any field GF(q), with $q = p^m$. In fact, any element β of this field can be expressed as a linear combination of the powers $\{\alpha^0, \alpha^1, \ldots, \alpha^{m-1}\}$:

$$\beta = \sum_{k=0}^{m-1} \beta_k \, \alpha^k, \tag{E.36}$$

where $\beta_k \in$ GF(p), with $k = 0, 1, \ldots, m - 1$.

E.2.3.6   Minimal Polynomials and Cyclotomic Cosets

The *minimal polynomial* $m_\gamma(D)$ of an element $\gamma \in$ GF(q) (with $q = p^m$) is a *monic polynomial of minimum degree over* GF(p) having γ as a root, that is, such that:

$$m_\gamma(D)|_{D=\gamma} = 0. \tag{E.37}$$

It can be shown that, for any element γ, the degree of its minimal polynomial $m_\gamma(D)$ is not larger than m. The structure of $m_\gamma(D)$ can be easily understood if the following property, referring to an arbitrary polynomial $f(D)$ over GF(p), is taken into account: given the element $\delta \in$ GF(q) (with $q = p^m$), if δ^r (where r is an integer) is a root of $f(D)$, then $\delta^{rp}, \delta^{rp^2}, \delta^{rp^3}, \ldots$ are also roots of the same polynomial. This implies that the minimal polynomials over GF(p) of the elements $\gamma, \gamma^p, \gamma^{p^2}, \ldots$, belonging to GF($q$), are all the same. The field elements characterized by the same minimal polynomial are called *conjugate elements*; the set of conjugate elements sharing the same minimal polynomial form a *cyclotomic coset*. The following properties can also be proved:

(a) The set of elements of GF(q) can be divided into *disjoint* cyclotomic cosets, which consequently form a partition of GF(q).
(b) The minimal polynomial of an element $\gamma \in$ GF(q) can be expressed as:

$$m_\gamma(D) = \prod_{k \in S_\gamma} (D - \alpha^k), \tag{E.38}$$

where α is a primitive element of GF(q) and S_γ denotes the set of exponents of α needed to generate the cyclotomic coset of γ. Note that the multiplication (E.38) consists of ordinary polynomial multiplications, unlike the multiplication defined in (E.23).

An application of these concepts to a specific field is illustrated in the following example.

Example E.2.3 Let us again consider the field GF(16) of Example E.2.2. It is easy to show that each of the subsets $\{0\}$, $\{\alpha, \alpha^2, \alpha^4, \alpha^8\}$, $\{\alpha^3, \alpha^6, \alpha^9, \alpha^{12}\}$, $\{\alpha^5, \alpha^{10}\}$, $\{\alpha^7, \alpha^{11}, \alpha^{13}, \alpha^{14}\}$, $\{\alpha^0 = 1\}$ forms a

Table E.8   Minimal polynomials of GF(16)

Exponents of α for elements of each cyclotomic coset	Minimal polynomial
$-\infty$	D
0	$D + 1$
1, 2, 4, 8	$D^4 + D + 1$
3, 6, 9, 12	$D^4 + D^3 + D^2 + D + 1$
5, 10	$D^2 + D + 1$
7, 11, 13, 14	$D^4 + D^3 + 1$

cyclotomic coset and that all of them form a partition of GF(16). The minimal polynomial associated with each coset can easily be derived by applying (E.38); for instance, the minimal polynomial of $\gamma = \alpha^7$ is given by:

$$m_\gamma(D) = (D - \alpha^7) \cdot (D - \alpha^{11}) \cdot (D - \alpha^{13}) \cdot (D - \alpha^{14}), \qquad (E.39)$$

which, after some manipulation, can be expressed as:

$$m_\gamma(D) = D^4 + D^3 + 1, \qquad (E.40)$$

which is a polynomial over the prime field GF(2). The expressions for the minimal polynomials for all the elements of GF(16) are provided in Table E.8.

□

Example E.2.3 shows that the minimal polynomials associated with distinct cosets may have different degrees. This is due to the fact that the number of conjugate elements can change from coset to coset. Finally, note that the cyclotomic cosets of a finite field can be constructed independently of the irreducible polynomial $f(D)$ selected in the definition (E.23) of multiplication between field elements.

E.2.4   Computation Techniques for Finite Fields

Let us now analyze the problem of how addition and multiplication over GF(q), with $q = p^m$, can be implemented. Let us assume that $p = 2$ (i.e., $q = 2^m$) since this occurs in most applications. Different options are available in this case, each characterized by specific requirements in terms of memory and computational complexity.

The first option is represented by the so-called *lookup table technique* and should be considered if q is not too large. In this case each of the operands is represented by a string of m bits and the table representing each of the two operations is implemented by means of a digital memory addressed by $2m$ bits (m bits per operand) and providing m output bits. An alternative approach to the use of a digital memory is represented by the implementation of the two tables by means of logical circuits. These two approaches become too costly if the field size is large. In particular, if q is large, computations should be done by resorting to digital circuits that implement operations between polynomials over the *ground field* (i.e., over GF(q) in the case considered). If this approach is adopted, addition can be easily performed, since it requires m sums between corresponding binary coefficients (see (E.21) and (E.22)), that is, m exclusive ORs. Multiplication is much more complicated but can be handled using a basis representation for each operand (see (E.36)). For instance, to evaluate the multiplication of the elements δ and ρ of $GF(2^m)$ their representations:

$$\delta = \sum_{k=0}^{m-1} \delta_k \, \alpha^k \qquad \qquad (E.41)$$

and

$$\rho = \sum_{k=0}^{m-1} \rho_k \, \alpha^k \qquad \qquad (E.42)$$

can be substituted in (E.23); this leads, after some manipulation, to m boolean expressions relating the $2m$ binary coefficients $\{\delta_k, \rho_k, k = 0, 1, \ldots, m-1\}$ to the m output bits. In practice, such expressions can be easily implemented using gate-array technology. The multiplication of δ and ρ of $GF(2^m)$ can also be implemented by representing them as powers of a primitive element α of $GF(2^m)$, $\delta = \alpha^d$ and $\rho = \alpha^r$, respectively, and exploiting (E.30). This leads to the expression:

$$\delta \boxdot \rho = \alpha^{R_{(2^m-1)}[d+r]}, \qquad \qquad (E.43)$$

whose implementation requires:

(a) extracting, by means of a logarithm operation, the exponents d and r, associated with elements δ and ρ, respectively (each exponent is represented by m bits at the multiplier input),
(b) evaluating the addition $R_{(2^m-1)}[d+r]$ (this operation generates m bits),
(c) computing an *anti-logarithm operation* generating the binary representation (m bits) for the element $\alpha^{R_{(2^m-1)}[d+r]}$.

Finally, note that a *lookup table* technique can be adopted to evaluate the logarithm and anti-logarithm operations.

Let us now apply one of the computation techniques illustrated above to a specific finite field.

Example E.2.4 Let us focus on the problem of evaluating the multiplication of the elements δ and ρ of the group $GF(8)$, generated in Example E.2.1 (the irreducible polynomial $f(D) = 1 + D + D^3$ has been used in defining the multiplication according to (E.23)). Given a primitive element α of $GF(8)$, δ and ρ can be represented using the basis $\{\alpha^0 = 1, \alpha, \alpha^2\}$, that is, as (see (E.41) and (E.42)):

$$\delta = \delta_0 + \delta_1 \, \alpha + \delta_2 \, \alpha^2 \qquad \qquad (E.44)$$

and

$$\rho = \rho_0 + \rho_1 \, \alpha + \rho_2 \, \alpha^2 , \qquad \qquad (E.45)$$

respectively, where all the parameters $\{\delta_k, \quad k = 1, 2, 3\}$ and $\{\rho_k, \ k = 1, 2, 3\}$ are binary. Keeping in mind that $\alpha^3 = \alpha + 1$ and $\alpha^4 = \alpha^2 + \alpha$, it is easy to show that evaluating the multiplication between the representations (E.44) and (E.45) yields:

$$\gamma \triangleq \delta \boxdot \rho = \gamma_0 + \gamma_1 \, \alpha + \gamma_2 \, \alpha^2, \qquad \qquad (E.46)$$

where

$$\gamma_0 = \delta_0 \cdot \rho_0 + \delta_0 \cdot \rho_2 + \delta_2 \cdot \rho_0, \qquad \qquad (E.47)$$

$$\gamma_1 = \delta_0 \cdot \rho_1 + \delta_1 \cdot \rho_0 + \delta_0 \cdot \rho_2 + \delta_2 \cdot \rho_0 + \delta_2 \cdot \rho_2, \qquad \qquad (E.48)$$

$$\gamma_2 = \delta_0 \cdot \rho_2 + \delta_2 \cdot \rho_0 + \delta_1 \cdot \rho_1 + \delta_2 \cdot \rho_2. \qquad \qquad (E.49)$$

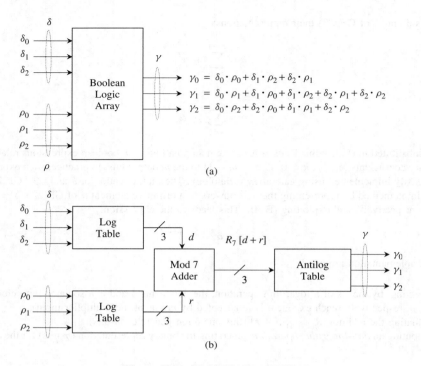

$$\gamma_0 = \delta_0 \cdot \rho_0 + \delta_1 \cdot \rho_2 + \delta_2 \cdot \rho_1$$
$$\gamma_1 = \delta_0 \cdot \rho_1 + \delta_1 \cdot \rho_0 + \delta_1 \cdot \rho_2 + \delta_2 \cdot \rho_1 + \delta_2 \cdot \rho_2$$
$$\gamma_2 = \delta_0 \cdot \rho_2 + \delta_2 \cdot \rho_0 + \delta_1 \cdot \rho_1 + \delta_2 \cdot \rho_2$$

Figure E.1 Multiplier for GF(8).

The block diagram of a multiplier based on (E.47)–(E.49) is shown Figure E.1(a) (note that the addition and multiplication operations appearing in the RHS of (E.47)–(E.49) are evaluated in GF(2)). An alternative scheme based on (E.43) is represented in Figure E.1(b).

□

We conclude that implementing the operations defined over GF(q) is not at all difficult and is usually simpler than accomplishing the same task over the field of integers. For instance, since GF(q) contains a finite numbers of elements, *overflow* problems do not exist.

E.3 Vector Spaces

Let us consider a field F (where addition and multiplication between its elements are denoted by the symbols $\boxplus$ and $\boxdot$, respectively) and algebraic structure V consisting of a set of elements over which an *addition* operation (denoted by $+$) is defined. Moreover, let us assume that a multiplication operation (denoted by $\cdot$) between an element of V and F is also defined. The structure V is a *vector space* over the field F if it satisfies the following five axioms.

Axiom V.1. V is a *commutative group* under addition.

Axiom V.2. For each element $a \in F$ and[9] $\mathbf{v} \in V$, $a \cdot \mathbf{v}$ belongs to V.

---

[9] In the following elements of V are always represented by bold letters, to avoid confusing them with those of F.

Axiom V.3 (*distributive laws*). For any pair $\mathbf{u}, \mathbf{v} \in V$, and any pair $a, b \in F$, the equalities:

$$a \cdot (\mathbf{u} + \mathbf{v}) = a \cdot \mathbf{u} + a \cdot \mathbf{v} \tag{E.50}$$

and

$$(a \boxplus b) \cdot \mathbf{v} = a \cdot \mathbf{v} + b \cdot \mathbf{v} \tag{E.51}$$

hold.

Axiom V.4 (*associative law*). For any $\mathbf{v} \in V$, and any pair $a, b \in F$, the equality:

$$(a \boxdot b) \cdot \mathbf{v} = a \cdot (b \cdot \mathbf{v}) \tag{E.52}$$

holds.

Axiom V.5. If the symbol 1 denotes the *multiplicative identity* of F, for any $\mathbf{v} \in V$ the equality:

$$1 \cdot \mathbf{v} = \mathbf{v} \tag{E.53}$$

holds.

The elements of a vector space V are called *vectors*, whereas the elements of F are called *scalars*. The operation $+$ defined over V is called *vector addition*, whereas the operation $\cdot$, turning a scalar of F and a vector of V in another vector of V, is called *scalar multiplication* (or *scalar product*). The additive identity of V is conventionally represented by the vector $\mathbf{0}$.

An important vector space playing a fundamental role in coding theory is defined in the following example.

Example E.3.1 Let us consider the set V_n consisting of all the n-dimensional row vectors whose elements belong to the field $\mathrm{GF}(q)$, with $q = p^m$ (p is a prime number, m is a nonnegative integer) and assume that $F = \mathrm{GF}(q)$. Since there are q choices for each element of V_n, this set consists of q^n distinct vectors. For any $\mathbf{a} = [a_0, a_1, \ldots, a_{n-1}]$ and $\mathbf{b} = [b_0, b_1, \ldots, b_{n-1}]$ in V_n, addition ($+$) is defined as:

$$\mathbf{a} + \mathbf{b} \triangleq [a_0 \boxplus b_0, a_1 \boxplus b_1, \ldots, a_{n-1} \boxplus b_{n-1}], \tag{E.54}$$

where $\boxplus$ denotes addition over F (i.e., over $\mathrm{GF}(q)$). Next, we define scalar multiplication ($\cdot$) of any $\mathbf{a} = [a_0, a_1, \ldots, a_{n-1}]$ in V_n by any element α from F as:

$$\alpha \cdot \mathbf{b} \triangleq [\alpha \boxdot b_0, \alpha \boxdot b_1, \ldots, \alpha \boxdot b_{n-1}], \tag{E.55}$$

where $\boxdot$ denotes multiplication over F. It is easy to show that, given these definitions, the set V_n forms a vector space over F.

$\square$

It may happen that a subset S of a vector space V over a field F is also a vector space over the same field; such a subset is a *subspace* of V. Subspaces of V can be easily generated by following the same procedure. Let us consider k vectors $\{\mathbf{v}_0, \mathbf{v}_1, \ldots, \mathbf{v}_{k-1}\}$ in V and k scalars $\{\alpha_0, \alpha_1, \ldots, \alpha_{k-1}\}$ from F. The sum:

$$\alpha_0 \cdot \mathbf{v}_0 + \alpha_1 \cdot \mathbf{v}_1 + \ldots + \alpha_{k-1} \cdot \mathbf{v}_{k-1} = \sum_{l=0}^{k-1} \alpha_l \cdot \mathbf{v}_l \tag{E.56}$$

represents a *linear combination* of the given k vectors. It can be shown that the set of all possible linear combinations of such vectors forms a subspace of V. The set of k vectors $\{\mathbf{v}_0, \mathbf{v}_1, \ldots, \mathbf{v}_{k-1}\}$ is *linearly dependent* if and only if there exist k scalars $\{\alpha_0, \alpha_1, \ldots, \alpha_{k-1}\}$ from F, not all equal to 0, such that:

$$\sum_{l=0}^{k-1} \alpha_l \cdot \mathbf{v}_l = 0. \tag{E.57}$$

Otherwise the set of vectors is said to be *linearly independent*.

A set of k vectors $\{\mathbf{v}_0, \mathbf{v}_1, \ldots, \mathbf{v}_{k-1}\}$ is said to *span* a vector space V if any vector in V can be expressed as a linear combination of the vectors of the given set. If such a set is linearly independent, it is called a *basis* of the vector space V and the number k of its elements is called the *dimension* of V. It can be proved that the number of vectors contained in any basis of a vector space V is the same.

Given a vector space V over a field F, another dyadic operation, called the *inner product* (or *dot product*) and denoted by $\circ$, can be defined for any pair of vectors in V. Such a product generates a scalar and possesses the following properties:

$$\mathbf{u} \circ \mathbf{v} = \mathbf{v} \circ \mathbf{u}, \tag{E.58}$$

$$\mathbf{u} \circ (\mathbf{v} + \mathbf{w}) = \mathbf{u} \circ \mathbf{v} \boxplus \mathbf{u} \circ \mathbf{w}, \tag{E.59}$$

$$(\alpha \cdot \mathbf{u}) \circ \mathbf{v} = \alpha \boxdot (\mathbf{u} \circ \mathbf{v}), \tag{E.60}$$

where $\mathbf{u}$, $\mathbf{v}$ and $\mathbf{w}$ represent arbitrary vectors in V and α is any scalar from F. A specific definition of the scalar product is provided in the following example.

Example E.3.2 Let us consider the vector space V_n over the field $F = GF(q)$ defined in Example E.3.1. For any $\mathbf{a} = [a_0, a_1, \ldots, a_{n-1}]$ and $\mathbf{b} = [b_0, b_1, \ldots, b_{n-1}]$ in V_n the inner product can be defined as:

$$\mathbf{a} \circ \mathbf{b} \triangleq a_0 \boxdot b_0 \boxplus a_1 \boxdot b_1 \boxplus \ldots \boxplus a_{n-1} \boxdot b_{n-1}. \tag{E.61}$$

Note that the RHS of this expression generates an element of F.
$\square$

If the inner product $\mathbf{u} \circ \mathbf{v}$ is equal to 0, $\mathbf{u}$ and $\mathbf{v}$ are said to be *orthogonal* to each other. Let S be an n-dimensional subspace of V and let S_O be the set of vectors in V such that, for any $\mathbf{u} \in S$ and $\mathbf{v} \in S_O$, we have that $\mathbf{u} \circ \mathbf{v} = 0$. It can be proved that S_O is also a subspace of V. This subspace is called the *null* (or *dual*) space of S and, reciprocally, S is the dual space of S_O. In addition, it can been shown that, if the dimension of V is equal to k, the dimension of S_O is $k - n$.

Appendix F

Error Function and Related Functions

The pdf of a *Gaussian* or *normal* random variable X is given by:

$$f_X(x) = \frac{1}{\sqrt{2\pi\sigma_X^2}} \exp\left[-\frac{(x-\eta_X)^2}{2\sigma_X^2}\right], \tag{F.1}$$

where σ_X^2 and η_X denote the *variance* and *mean* of X, respectively. We may write $X \in \mathcal{N}(\eta_X, \sigma_X^2)$, where $\mathcal{N}(\eta_X, \sigma_X^2)$ denotes the set of Gaussian random variables having the same mean value and variance as X. A random variable $N \in \mathcal{N}(0, 1)$ is called a *standard normal variable* and its pdf is given by:

$$f_N(n) = \frac{1}{\sqrt{2\pi}} \exp\left(-\frac{n^2}{2}\right). \tag{F.2}$$

It is easy to prove that any $X \in \mathcal{N}(\eta_X, \sigma_X^2)$ can be expressed as a *linear transformation* of a standard normal variable N; in particular, we have that:

$$X = \sigma_X N + \eta_X, \tag{F.3}$$

so that:

$$f_X(x) = \frac{1}{\sigma_X} f_N\left(\frac{x-\eta_X}{\sigma_X}\right). \tag{F.4}$$

Unfortunately, a closed-form expression for the *distribution function* $F_X(x)$ of $X \in \mathcal{N}(\eta_X, \sigma_X^2)$ does not exist. However, $F_X(x)$ can be expressed as:

$$F_X(x) = \Phi\left(\frac{x-\eta_X}{\sigma_X}\right), \tag{F.5}$$

where

$$\Phi(x) \triangleq \frac{1}{\sqrt{2\pi}} \int_{-\infty}^{x} \exp\left(\frac{n^2}{2}\right) dn \tag{F.6}$$

Wireless Communications: Algorithmic Techniques, First Edition.
Giorgio M. Vitetta, Desmond P. Taylor, Giulio Colavolpe, Fabrizio Pancaldi, Philippa A. Martin.
© 2013 John Wiley & Sons, Ltd. Published 2013 by John Wiley & Sons, Ltd.

denotes the *distribution function of a standard normal variable*, or as:

$$F_X(x) = 1 - Q\left(\frac{x - \eta_X}{\sigma_X}\right), \tag{F.7}$$

where

$$Q(x) \triangleq 1 - \Phi(x) = \int_x^{+\infty} \exp\left(-\frac{n^2}{2}\right) dn. \tag{F.8}$$

Then the probability that X takes on values in the interval $[a, b]$ can be evaluated as:

$$\Pr\{a \le X \le b\} = \Phi\left(\frac{b - \eta_X}{\sigma_X}\right) - \Phi\left(\frac{a - \eta_X}{\sigma_X}\right)$$

$$= Q\left(\frac{a - \eta_X}{\sigma_X}\right) - Q\left(\frac{b - \eta_X}{\sigma_X}\right). \tag{F.9}$$

The latter result illustrates the importance of estimating the function $\Phi(\cdot)$ or, equivalently, $Q(\cdot)$. Note, however, that, other related functions, such as the *error function*:

$$\mathrm{erf}\,(x) \triangleq \frac{2}{\sqrt{\pi}} \int_0^x \exp\left(-\theta^2\right) d\theta \tag{F.10}$$

or the *complementary error function*:

$$\mathrm{erfc}\,(x) \triangleq 1 - \mathrm{erf}(x) = \frac{2}{\sqrt{\pi}} \int_x^\infty \exp\left(-\theta^2\right) d\theta, \tag{F.11}$$

can be used in place of $\Phi(\cdot)$ in (F.6) and $Q(\cdot)$ in (F.8) for the evaluation of the RHS of (F.9). In fact, it is easy to show that:

$$\Phi(x) = \frac{1}{2} + \frac{1}{2}\mathrm{erf}\left(\frac{x}{\sqrt{2}}\right) \tag{F.12}$$

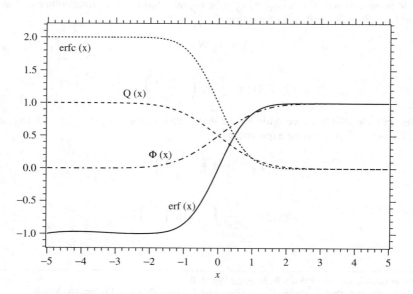

Figure F.1 The functions $\Phi(x)$, $\mathrm{erf}(x)$, $\mathrm{erfc}(x)$ and $Q(x)$.

and

$$Q(x) = \frac{1}{2}\text{erfc}\left(\frac{x}{\sqrt{2}}\right). \tag{F.13}$$

Figure F.1 shows the behavior of the functions $\Phi(x)$, $\text{erf}(x)$, $\text{erfc}(x)$ and $Q(x)$; note that $\text{erf}(x)$ exhibits antisymmetric behavior around the origin.

Various numerical methods have been developed for the efficient evaluation of the error and $Q(\cdot)$ functions (e.g., see [1850, 2078–2081]). These functions are tabulated and are often available as built-in functions in mathematical software. In addition, various bounds have been derived for $\text{erfc}(x)$ (or, equivalently, for $Q(\cdot)$). Here, we mention the exponential-type bound [321]:

$$\text{erfc}(x) \leq \exp\left(-x^2\right), \tag{F.14}$$

which holds for $x \geq 0$. This is equivalent to (see (F.13)):

$$Q(x) \leq \frac{1}{2}\exp\left(-\frac{x^2}{2}\right), \tag{F.15}$$

which also holds for $x \geq 0$. Other useful bounds on the $Q(\cdot)$ function can be found in [1850, 1851, 2082]. Finally, it is worth mentioning that the bounds proposed in [1851, 2082] are based on the formula:

$$Q(x) = \frac{1}{\pi}\int_0^{\pi/2}\exp\left(-\frac{x^2}{2\sin^2(\theta)}\right)d\theta, \tag{F.16}$$

which is known as the *Craig form* or *representation* of the $Q(\cdot)$ function itself [2083].

References

1. A. Still, *Communication Through the Ages: From Sign Language to Television*. New York: Murray Hill Books, 1946.
2. R. Appleyard, *Pioneers of Electrical Communication*. London: Macmillan, 1930.
3. J. A. Fleming, *The Principles of Electric Wave Telegraphy and Telephony*, 4th edn London: Longmans, Green, and Co., 1919.
4. E. Hawks, *Pioneers of Wireless*. London: Methuen, 1927.
5. A. Goldsmith, *Wireless Communications*. New York: Cambridge University Press, 2005.
6. T. S. Rappaport, *Wireless Communications: Principles and Practice*, 2nd ed. Upper Saddle River, NJ: Prentice Hall PTR, 2002.
7. *The IEEE 802.16 Working Group on Broadband Wireless Access Standards*, http://WirelessMAN.org.
8. K. V. S. S. S. S. Sairam, N. Gunasekaran, and S. Rama Reddy, "Bluetooth in wireless communication", *IEEE Commun. Mag.*, vol. 40, no. 6, pp. 90–96, June 1998.
9. *Zigbee Alliance*, http://www.zigbee.org/en/index.asp.
10. M. Ghavami, L. Michael, and R. Kohno, *Ultra Wideband Signals and Systems in Communication Engineering*. Chichester: John Wiley & Sons, Ltd, 2007.
11. J. Mitola, "The software radio architecture", *IEEE Commun. Mag.*, vol. 33, no. 5, pp. 26–38, May 1995.
12. ____, "Cognitive radio – an integrated agent architecture for software defined radio", *PhD Thesis*, May 2000.
13. M. Schwartz, W. R. Bennett, and S. Stein, *Communication Systems and Techniques (an IEEE Press Classic Reissue)*. Pisacataway, NJ: IEEE Press, 1966.
14. V. Tarokh, N. Seshadri, and A. R. Calderbank, "Space-time codes for high data rate wireless communication: Performance criterion and code construction", *IEEE Trans. Inform. Theory*, vol. 44, no. 2, pp. 744–765, Mar. 1998.
15. S. M. Alamouti, "A simple transmit diversity technique for wireless communications", *IEEE J. Sel. Areas Commun.*, vol. 16, no. 8, pp. 1451–1458, Oct. 1998.
16. B. Vucetic and J. Yuan, *Space-Time Coding*. Chichester: John Wiley & Sons, Ltd, 2003.
17. G. J. Foschini, "Layered space-time architecture for wireless communication in a fading environment when using multi-element antennas", *Bell Labs Tech. J.*, vol. 1, no. 2, pp. 41–59, Autumn 1996.
18. G. J. Foschini and M. J. Gans, "On limits of wireless communications in a fading environment when using multiple antennas", *Wireless Personal Commun.*, vol. 6, pp. 311–335, Feb. 1998.
19. S. Stein, "Fading channel issues in system engineering", *IEEE J. Sel. Areas Commun.*, vol. 5, no. 2, pp. 68–89, Feb. 1987.
20. M. A. Caloyannides, "Encryption wars: early battles", *IEEE Spectrum*, vol. 37, no. 4, pp. 37–43, Apr. 2000.

Wireless Communications: Algorithmic Techniques, First Edition.
Giorgio M. Vitetta, Desmond P. Taylor, Giulio Colavolpe, Fabrizio Pancaldi, Philippa A. Martin.
© 2013 John Wiley & Sons, Ltd. Published 2013 by John Wiley & Sons, Ltd.

21. A. J. Paulraj, D. A. Gore, R. U. Nabar, and H. Bölcskei, "An overview of MIMO communications – a key to gigabit wireless", *IEEE Proc.*, vol. 92, no. 2, pp. 198–218, Feb. 2004.

22. S. N. Diggavi, N. Al-Dhahir, A. Stamoulis, and A. R. Calderbank, "Great expectations: the value of spatial diversity in wireless networks", *IEEE Proc.*, vol. 92, no. 2, pp. 219–270, Feb. 2004.

23. Lucent Technologies, "Bell Labs scientists shatter limit on fixed wireless transmission", http://www.alcatel-lucent.com/wps/portal/NewsReleases/, 1998.

24. H. Sampath, S. Talwar, J. Tellado, V. Erceg, and A. Paulraj, "A fourth-generation MIMO-OFDM broadband wireless system: Design, performance, and field trial results", *IEEE Commun. Mag.*, vol. 40, no. 9, pp. 143–149, Sep. 2002.

25. C. Dubuc, D. Starks, T. Creasy, and Y. Hou, "A MIMO-OFDM prototype for next-generation wireless WANs", *IEEE Commun. Mag.*, vol. 42, no. 12, pp. 82–87, Dec. 2004.

26. S. Nanda, R. Walton, J. Ketchum, M. Wallace, and S. Howard, "A high-performance MIMO OFDM wireless LAN", *IEEE Commun. Mag.*, vol. 43, no. 2, pp. 101–109, Feb. 2005.

27. J. G. Proakis and M. Salehi, *Digital Communications*, 5th edn New York: McGraw-Hill, 2008.

28. M. K. Simon, S. Hinedi, and W. C. Lindsey, *Digital Communication Techniques – Signal Design and Detection*. Englewood Cliffs, NJ: Prentice Hall, 1995.

29. J. R. Barry, E. A. Lee, and D. G. Messerschmitt, *Digital Communication*, 3rd ed. Norwell, MA: Kluwer Academic Publishers, 2003.

30. R. G. Gallager, *Principles of Digital Communication*. Cambridge: Cambridge University Press, 1968.

31. M. K. Simon and M.-S. Alouini, *Digital Communication over Fading Channels: A Unified Approach to Performance Analysis*. New York: John Wiley & Sons Ltd, 2000.

32. S. Benedetto and E. Biglieri, *Principles of Digital Transmission with Wireless Applications*. New York: Kluwer Academic/Plenum Publishers, 1999.

33. D. Tse and P. Viswanath, *Fundamentals of Wireless Communication*. Cambridge: Cambridge University Press, 2005.

34. A. Molisch, *Wireless Communications*. Chichester: John Wiley & Sons, Ltd and IEEE Press, 2005.

35. S. Lin and D. J. Costello, Jr., *Error Control Coding*, 2nd ed. Upper Saddle River, NJ: Pearson Prentice Hall, 2004.

36. E. Biglieri, D. Divsalar, P. J. McLane, and M. K. Simon, *Introduction to Trellis-Coded Modulation with Applications*. New York: Macmillan, 1991.

37. E. Biglieri, *Coding for Wireless Channels*. New York: Kluwer Academic Publishers, 2005.

38. A. Paulraj, R. U. Nabar, and D. Gore, *Introduction to Space-Time Wireless Communications*. Cambridge: Cambridge University Press, 2003.

39. E. G. Larsson and P. Stoica, *Space-Time Block Coding for Wireless Communications*. Cambridge: Cambridge University Press, 2003.

40. E. Biglieri, R. Calderbank, A. Constantinides, A. Goldsmith, A. Paulraj, and H. V. Poor, *MIMO Wireless Communications*. Cambridge: Cambridge University Press, 2007.

41. N. Benvenuto and G. Cherubini, *Algorithms for Communications Systems and Their Applications*. Chichester: John Wiley & Sons, Ltd, 2002.

42. G. Ferrari, G. Colavolpe, and R. Raheli, *Detection Algorithms for Wireless Communications: with Applications to Wired and Storage Systems*. Chichester: John Wiley & Sons, Ltd, 2004.

43. H. Meyr, M. Moeneclaey, and S. A. Fechtel, *Digital Communication Receivers: Synchronization, Channel Estimation and Signal Processing*. New York: John Wiley & Sons, Inc., 1997.

44. K. Brayer (ed.), *Data Communications via Fading Channels*. New York: IEEE Press, 1975.

45. P. Monsen, "Fading channel communications", *IEEE Commun. Mag.*, vol. 18, no. 1, pp. 27–36, Jan. 1980.

46. B. Sklar, "Rayleigh fading channels in mobile digital communication systems. Part I: Characterization", *IEEE Commun. Mag.*, vol. 35, no. 7, pp. 90–100, July 1997.

47. J. C. Siller, "Multipath propagation", *IEEE Commun. Mag.*, vol. 22, no. 2, pp. 6–15, Feb. 1984.

48. W. D. Rummler, R. Coutts, and M. Liniger, "Multipath fading channel models for microwave digital radio", *IEEE Commun. Mag.*, vol. 24, no. 11, pp. 30–42, Nov. 1986.

49. B. Sklar, "Rayleigh fading channels in mobile digital communication systems. Part II: Mitigation", *IEEE Commun. Mag.*, vol. 35, no. 7, pp. 102–109, July 1997.

50. W. C. Y. Lee, *Mobile Communications Design Fundamentals*, 2nd ed. New York: John Wiley & Sons, Inc., 1993.

51. R. W. Lorenz, "Impact of frequency-selective fading on digital land mobile radio communication at transmission rates of several hundred kbit/s", *IEEE Trans. Veh. Tech.*, vol. 36, no. 3, pp. 122–128, Aug. 1987.

52. H. L. Bertoni, W. Honcharenko, L. R. Macel, and H. H. Xia, "UHF propagation prediction for wireless personal communications", *IEEE Proc.*, vol. 82, no. 9, pp. 1333–1359, Sep. 1994.

53. T. K. Sarkar, Z. Ji, K. J. Kim, A. Medouri, and M. Salazar-Palma, "A survey of various propagation models for mobile communications", *IEEE Ant. and Prop. Mag.*, vol. 45, no. 3, pp. 51–82, June 2003.

54. N. S. Adawi, H. L. Bertoni, J. R. Child, W. A. Daniel, J. E. Dettra, R. P. Eckertt, E. H. Flath, Jr., R. T. Forrest, W. C. Y. Lee, S. R. McConoughey, J. P. Murray, H. Sachs, G. L. Schrenk, N. H. Shepherd, and F. D. Shipley, "Coverage prediction for mobile radio systems operating in the 800/900 MHz frequency range", *IEEE Trans. Veh. Tech.*, vol. 37, no. 1, pp. 2–72, Feb. 1988.

55. A. Papoulis and S. U. Pillai, *Probability, Random Variables and Stochastic Processes*, 4th ed. New York: McGraw-Hill, 2002.

56. J. B. Andersen, T. S. Rappaport, and S. Yoshida, "Propagation measurements and models for wireless communications channels", *IEEE Commun. Mag.*, vol. 33, no. 1, pp. 42–49, Jan. 1995.

57. ETSI/TC SMG, *Recommendation GSM 05.01 – Physical Layer on the Radio Path: General Description*, European Telecommunications Standards Institute, Feb. 1992.

58. _____, *Recommendation GSM 05.05 (Update Note) – Radio Transmission and Reception*, European Telecommunications Standards Institute, Oct. 1993.

59. D. C. Cox, "Wireless personal communications: What is it?" *IEEE Pers. Commun.*, vol. 2, no. 2, pp. 20–35, Apr. 1995.

60. R. Bultitude, "Measurement, characterization and modeling of indoor 800/900 MHz radio channels for digital communications", *IEEE Commun. Mag.*, vol. 25, no. 6, pp. 5–12, June 1987.

61. T. S. Rappaport and S. Sandhu, "Radio-wave propagation for emerging wireless personal-communication systems", *IEEE Ant. and Prop. Mag.*, vol. 36, no. 5, pp. 14–24, Oct. 1994.

62. A. F. de Toledo, A. M. D. Turkmani, and J. D. Parsons, "Estimating coverage of radio transmission into and within buildings at 900, 1800 and 2300 MHz", *IEEE Pers. Commun. Mag.*, vol. 5, no. 2, pp. 40–47, Apr. 1998.

63. H. Hashemi, "The indoor radio propagation channel", *IEEE Proc.*, vol. 81, no. 7, pp. 943–968, July 1993.

64. A. Neskovic, N. Neskovic, and G. Paunovic, "Modern approaches in modeling of mobile radio systems propagation environment", *IEEE Commun. Surveys & Tutorials*, vol. 3, no. 3, pp. 2–12, third quarter 2000.

65. V. Erceg, L. J. Greenstein, S. Y. Tjandra, S. R. Parkoff, A. Gupta, B. Kulic, A. A. Julius, and R. Bianchi, "An empirically based path loss model for wireless channels in suburban environments", *IEEE J. Sel Areas Commun.*, vol. 17, no. 7, pp. 1205–1211, July 1999.

66. T. Taga, T. Furuno, and K. Suwa, "Channel modeling for 2-GHz-band urban line-of-sight street microcells", *IEEE Trans. Veh. Tech.*, vol. 48, no. 1, pp. 262–272, Jan. 1999.

67. E. Benner and A. B. Sesay, "Effects of antenna height, antenna gain, and pattern downtilting for cellular mobile radio", *IEEE Trans. Veh. Tech.*, vol. 45, no. 2, pp. 217–224, May 1996.

68. A. Saleh and R. Valenzuela, "A statistical model for indoor multipath propagation", *IEEE J. Sel. Areas Commun.*, vol. 5, no. 2, pp. 128–137, Feb. 1987.

69. T. S. Rappaport and C. D. McGillem, "UHF fading in factories", *IEEE J. Sel. Areas Commun.*, vol. 7, no. 1, pp. 40–48, Jan. 1989.

70. F. C. Owen and C. D. Pudney, "Radio propagation for digital cordless telephones at 1700 MHz and 900 MHz", *IEEE Trans. Commun.*, vol. 25, no. 1, pp. 52–53, 5 Jan. 1989.

71. S. Y. Seidel, T. S. Rappaport, S. Jain, M. L. Lord, and R. Singh, "Path loss, scattering and multipath delay statistics in four European cities for digital cellular and microcellular radiotelephone", *IEEE Trans. Veh. Tech.*, vol. 40, no. 4, pp. 721–730, Nov. 1991.

72. S. Y. Seidel and T. S. Rappaport, "914 MHz path loss prediction models for indoor wireless communications in multifloored buildings", *IEEE Trans. Ant. and Prop.*, vol. 40, no. 2, pp. 207–217, Feb. 1992.

73. M. J. Feuerstein, K. L. Blackard, T. S. Rappaport, S. Y. Seidel, and H. H. Xia, "Path loss, delay spread, and outage models as functions of antenna height for microcellular system design", *IEEE J. Sel. Areas Commun.*, vol. 43, no. 3, pp. 487–498, Aug. 1994.

74. N. Blaunstein and Y. Ben-Shimol, "Prediction of frequency dependence of path loss and link-budget design for various terrestrial communication links", *IEEE Trans. Ant. and Prop.*, vol. 52, no. 10, pp. 2719–2729, Oct. 2004.

75. L. R. Maciel, H. L. Bertoni, and H. N. Xia, "Unified approach to prediction of propagation over buildings for all ranges of base station antenna height", *IEEE Trans. Veh. Tech.*, vol. 42, no. 1, pp. 41–45, Feb. 1993.

76. P. Harley, "Short distance attenuation measurements at 900 MHz and 1.8 GHz using low antenna heights for microcells", *IEEE J. Sel. Areas Commun.*, vol. 7, no. 1, pp. 5–11, Jan. 1989.

77. H. Xia, H. L. Bertoni, L. R. Maciel, A. Lindsay-Stewart, and R. Rowe, "Radio propagation characteristics for line-of-sight microcellular and personal communications", *IEEE Trans. Ant. and Prop.*, vol. 41, no. 10, pp. 1439–1447, Oct. 1993.

78. P. A. Bello, "Characterization of randomly time-variant linear channels", *IEEE Trans. Commun. Syst.*, vol. 9, pp. 360–393, Dec. 1963.

79. L. E. Franks, *Signal Theory*. Englewood Cliffs, NJ: Prentice Hall, 1969.

80. P. A. Bello, "Time-frequency duality", *IEEE Trans. Inform. Theory*, vol. 10, no. 1, pp. 18–33, Jan. 1965.

81. W. D. Rummler, "A new selective fading model: Application to propagation data", *Bell Syst. Tech. J.*, vol. 59, no. 5, pp. 1037–1071, May–June 1979.

82. ———, "More on the multipath fading channel model", *IEEE Trans. Commun.*, vol. 29, no. 3, pp. 346–352, Mar. 1981.

83. F. D. Garber and M. B. Pursley, "Performance of binary FSK communications over frequency-selective Rayleigh fading channels", *IEEE Trans. Commun.*, vol. 37, no. 1, pp. 83–89, Jan. 1989.

84. P. Hoeher, "A statistical discrete-time model for the WSSUS multipath channel", *IEEE Trans. Veh. Technol.*, vol. 41, no. 4, pp. 461–468, Nov. 1992.

85. M. V. Clark, L. J. Greenstein, W. Kennedy, and M. Shafi, "Matched filter performance bounds for diversity combining receivers in digital mobile radio", *IEEE Trans. Veh. Tech.*, vol. 41, no. 4, pp. 356–362, Nov. 1992.

86. K. W. Yip and T. S. Ng, "Karhunen-Loève expansion of the WSSUS channel output and its application to efficient simulation", *IEEE J. Sel. Areas Commun.*, vol. 15, no. 4, pp. 640–646, May 1997.

87. B. Glance and L. Greenstein, "Frequency-selective fading effects in digital mobile radio with diversity combining", *IEEE Trans. Commun.*, vol. 31, no. 9, pp. 1085–1094, Sep. 1983.

88. J. K. Cavers, *Mobile Channel Characteristics*. Norwell, MA: Kluwer Academic Publishers, 2000.

89. N. Nakagami, "The m-distribution – a general formula for intensity distribution of rapid fading", in *Statistical Methods in Radio Wave Propagation*, W. Hoffman, (ed.) Elmsford, NY: Pergamon, 1960, pp. 3–36.

90. N. Youssef, C.-X. Wang, and M. Pätzold, "A study on the second order statistics of Nakagami-Hoyt mobile fading channels", *IEEE Trans. Veh. Tech.*, vol. 54, no. 4, pp. 1259–1265, July 2005.

91. G. K. Karagiannidis, N. C. Sagias, and P. T. Mathiopoulos, "N*Nakagami: A novel stochastic model for cascaded fading channels", *IEEE Trans. Commun.*, vol. 55, no. 8, pp. 1453–1458, Aug. 2007.

92. Q. T. Zhang, "A generic correlated Nakagami fading model for wireless communications", *IEEE Commun. Lett.*, vol. 51, no. 11, pp. 1745–1748, Nov. 2003.

93. K. Zhang, Z. Song, and Y. L. Guan, "Simulation of Nakagami fading channels with arbitrary cross-correlation and fading parameters", *IEEE Trans. Wireless Commun.*, vol. 3, no. 5, pp. 1463–1468, Sep. 2004.

94. L. Cao and N. C. Beaulieu, "Simple efficient methods for generating independent and bivariate Nakagami-m fading envelope samples", *IEEE Trans. Veh. Tech.*, vol. 56, no. 4, pp. 1573–1579, July 2007.

95. G. Gaertner and E. O. Nuallain, "Characterizing wideband signal envelope fading in urban microcells using the Rice and Nakagami distributions", *IEEE Trans. Veh. Tech.*, vol. 56, no. 6, pp. 3621–3630, Nov. 2007.

96. M. Abramowitz and E. I. A. Stegun, *Handbook of Mathematical Functions*. New York: Dover, 1965.

97. D. Verdin and T. C. Tozer, "Generating a fading process for the simulation of land-mobile radio communications", *Electron. Lett.*, vol. 29, no. 23, pp. 2011–2012, Nov. 1993.

98. R. H. Clarke, "A statistical theory of mobile-radio reception", *Bell Syst. Tech. J.*, vol. 47, pp. 957–1000, July–Aug. 1968.

99. W. C. Jakes, *Microwave Mobile Communications*. New York: IEEE Press, 1974.

100. J. S. Sadowsky and V. Kafedziski, "On the correlation and scattering functions of the WSSUS channel for mobile communications", *IEEE Trans. Veh. Tech.*, vol. 47, no. 1, pp. 270–282, Feb. 1998.

101. R. S. Kennedy, *Fading Dispersive Communication Channels*. New York: Wiley-Interscience, 1969.

102. D. C. Cox, "A measured delay-Doppler scattering function for multipath propagation at 910 MHz in an urban mobile radio environment", *IEEE Proc.*, vol. 61, no. 4, pp. 479–480, Apr. 1973.

103. A. S. Bajwa and J. D. Parsons, "Time-delay/Doppler scattering function for suburban mobile radio propagation at 436 MHz", *Electron. Lett.*, vol. 14, no. 14, pp. 423–425, July 6 1978.

104. ———, "Small area characterization of UHF urban and suburban mobile radio channels", *Proc. Inst. Elect. Eng. (Radar, Sonar and Navigation)*, vol. 129, no. 1, pp. 95–101, Feb. 1982.

105. T. Kailath, *Sampling Models for Linear Time-Variant Filters, Rept. No. 352*, MIT Research Lab. of Electronics, Cambridge, MA, May 1959.

106. J. C. Hancock and P. A. Wintz, *Signal Detection Theory*. New York: McGraw-Hill, 1966.

107. R. Steele and E. L. Hanzo, *Mobile Radio Communications*, 2nd ed. Chichester: John Wiley & Sons, Ltd, 1999.

108. N. J. Baas and D. P. Taylor, "Decomposition of fading dispersive channels – effects of mismatch on the performance of MLSE", *IEEE Trans. Commun.*, vol. 48, no. 9, pp. 1467–1470, Sep. 2000.

109. ———, "Matched filter bounds for wireless communications over Rayleigh fading dispersive channels", *IEEE Trans. Commun.*, vol. 49, no. 9, pp. 1525–1528, Sep. 2001.

110. M. Visintin, "Karhunen-Loève expansion of a fast Rayleigh fading process", *Electron. Lett.*, vol. 32, no. 18, p. 1712, 29 Aug. 1996.

111. G. M. Vitetta, D. P. Taylor, and U. Mengali, "Double-filtering receivers for PSK signals transmitted over Rayleigh frequency-flat fading channels", *IEEE Trans. Commun.*, vol. 44, no. 6, pp. 686–695, June 1996.

112. G. M. Vitetta, U. Mengali, and D. P. Taylor, "Optimal noncoherent detection of FSK signals transmitted over linearly time-selective Rayleigh fading channels", *IEEE Trans. Commun.*, vol. 45, no. 11, pp. 1417–1425, Nov. 1997.

113. J. K. Cavers, "On the validity of the slow and moderate fading models for matched filter detection of Rayleigh fading channels", *Canad. J. Elect. & Comp. Eng.*, vol. 17, no. 4, pp. 183–189, 1992.

114. W. S. Leon and D. P. Taylor, "An adaptive receiver for the time- and frequency-selective fading channel", *IEEE Trans. Commun.*, vol. 45, no. 12, pp. 1548–1559, Dec. 1997.

115. W. S. Leon, U. Mengali, and D. P. Taylor, "Equalization of linearly frequency-selective fading channels", *IEEE Trans. Commun.*, vol. 45, no. 12, pp. 1501–1503, Dec. 1997.

116. W. S. Leon and D. P. Taylor, "Generalized polynomial-based receiver for the flat fading channel", *IEEE Trans. Commun.*, vol. 51, no. 6, pp. 896–899, June 2003.

117. E. Chiavaccini and G. M. Vitetta, "GQR models for multipath Rayleigh fading channels", *IEEE J. Sel. Areas Commun.*, vol. 19, no. 6, pp. 1009–1018, June 2001.

118. D. R. Hummels and F. W. Ratcliffe, "Calculation of error probability for MSK and OQPSK systems operating in a fading multipath environment", *IEEE Trans. Veh. Technol.*, vol. 30, no. 3, pp. 112–120, Aug. 1981.

119. W. H. Press, B. P. Flannery, S. A. Teukolsky, and W. T. Vetterling, *Numerical Recipes in Fortran 77: The Art of Scientific Computing*. Cambridge: Cambridge University Press, 1992.

120. A. H. Stroud, *Numerical Quadrature and Solution of Ordinary Differential Equations*. New York: Springer-Verlag, 1974.

121. M. Steinbauer, A. F. Molisch, and E. Bonek, "The double-directional radio channel", *IEEE Trans. Ant. and Prop.*, vol. 43, no. 4, pp. 51–56, Aug. 2001.

122. R. Kattenbach, "Statistical modelling of small-scale fading in directional radio channels", *IEEE J. Sel. Areas Commun.*, vol. 20, no. 3, pp. 584–592, Apr. 2002.

123. C. Xiao, J. Wu, S.-Y. Leong, Y. R. Zheng, and K. B. Letaief, "A discrete-time model for triply selective MIMO Rayleigh fading channels", *IEEE Trans. Wireless Commun.*, vol. 3, no. 5, pp. 1678–1688, Sep. 2004.

124. K. I. Pedersen, J. B. Andersen, J. P. Kermoal, and P. Mogensen, "A stochastic multiple-input-multiple-output radio channel model for evaluation of space-time coding algorithms", in *Proc. IEEE 52nd Veh. Technol. Conf. 2007 VTS-Fall*, vol. 2, Boston, 24–28 Sep. 2000, pp. 893–897.

125. G. H. Golub and C. F. Van Loan, *Matrix Computations*, 2nd edn Baltimore, MD: Johns Hopkins University Press, 1989.

126. J. B. Andersen, "Array gain and capacity for known random channels with multiple element arrays at both ends", *IEEE J. Sel. Areas Commun.*, vol. 18, no. 11, pp. 2172–2178, Nov. 2000.

127. A. F. Molisch, "A generic model for MIMO wireless propagation channels in macro- and microcells", *IEEE Trans. Sig. Process.*, vol. 52, no. 1, pp. 61–71, Jan. 2004.

128. L. C. Godara, "Application of antenna arrays to mobile communications, Part II: Beam-forming and direction-of-arrival considerations", *IEEE Proc.*, vol. 85, no. 8, pp. 1195–1245, Aug. 1997.

129. R. B. Ertel, P. Cardieri, K. W. Sowerby, T. S. Rappaport, and J. H. Reed, "Overview of spatial channel models for antenna array communication systems", *IEEE Pers. Commun.*, vol. 5, no. 1, pp. 10–21, Feb. 1998.

130. B. H. Fleury, "First- and second-order characterization of direction dispersion and space selectivity in the radio channel", *IEEE Trans. Inform. Theory*, vol. 46, no. 6, pp. 2027–2044, Sep. 1994.

131. F. Pancaldi, P. Greco, and G. M. Vitetta, "GQR-based models for directional wireless channels", *IEEE Trans. Wireless Commun.*, vol. 5, no. 3, pp. 642–651, Mar. 2006.

132. D. Asztély, *On Antenna Arrays in Mobile Communication Systems: Fast Fading and GSM Base Station Receiver Algorithms, Technical Report IR-S3-SB-9611*, Royal Institute of Technology, Stockholm, 1996.

133. J. P. Kermoal, L. Schumacher, K. I. Pedersen, P. E. Mogensen, and F. Frederiksen, "A stochastic MIMO radio channel model with experimental validation", *IEEE J. Sel. Areas Commun*, vol. 20, no. 6, pp. 1211–1226, Aug. 2002.

134. C. Oestges, B. Clerckx, D. Vanhoenacker-Janvier, and A. J. Paulraj, "Impact of fading correlations on MIMO communication systems in geometry-based statistical channel models", *IEEE Trans. Wireless Commun.*, vol. 4, no. 3, pp. 1112–1120, May 2005.

135. P. Almers, F. Tufvesson, and A. F. Molisch, "Keyhole effect in MIMO wireless channels: Measurements and theory", *IEEE Trans. Wireless Commun.*, vol. 5, no. 12, pp. 3596–3604, Dec. 2006.

136. W. Weichselberger, M. Herdin, H. Özcelik, and E. Bonek, "A stochastic MIMO channel model with joint correlation of both link ends", *IEEE Trans. Wireless Commun.*, vol. 5, no. 1, pp. 90–100, Jan. 2006.

137. M. T. Ivrlač, W. Utschick, and J. A. Nossek, "Fading correlations in wireless MIMO communication systems", *IEEE J. Sel. Areas Commun.*, vol. 21, no. 5, pp. 819–828, June 2003.

138. G. G. Raleigh and J. M. Cioffi, "Spatio-temporal coding for wireless communication", *IEEE Trans. Commun.*, vol. 46, no. 3, pp. 357–366, Mar. 1998.

139. A. M. Sayeed, "Modeling and capacity of realistic spatial MIMO channels", in *Proc. IEEE Int. Conf. Acoust., Speech and Sig. Process. (ICASSP '01)*, vol. 4, Salt Lake City, UT, 7–11 May 2001, pp. 2489–2492.

140. ———, "Deconstructing multiantenna fading channels", *IEEE Trans. Sig. Process.*, vol. 50, no. 10, pp. 2563–2579, Oct. 2002.

141. G. G. Raleigh and V. K. Jones, "Multivariate modulation and coding for wireless communication", *IEEE J. Sel Areas Commun.*, vol. 17, no. 5, pp. 851–866, May 1999.

142. P. C. F. Eggers, J. Toftgaard, and A. M. Oprea, "Antenna systems for base station diversity in urban small and micro cells", *IEEE J. Sel Areas Commun.*, vol. 11, no. 7, pp. 1046–1057, Sep. 1993.

143. W. Lee, "Effects on correlation between two mobile radio base-station antennas", *IEEE Trans. Commun.*, vol. 21, no. 11, pp. 1548–1559, Nov. 1973.

144. J. Salz and J. H. Winters, "Effect of fading correlation on adaptive arrays in digital mobile radio", *IEEE Trans. Veh. Tech.*, vol. 43, no. 4, pp. 1049–1057, Nov. 1994.

145. S. Loyka and G. Tsoulos, "Estimating MIMO system performance using the correlation matrix approach", *IEEE Commun. Lett.*, vol. 6, no. 1, pp. 19–21, Jan. 1990.

146. F. Adachi, M. Feeny, A. Williamson, and J. Parson, "Crosscorrelation between the envelopes of 900 MHz signals received at a mobile radio base station side", *Proc. Inst. Elec. Eng., pt. F*, vol. 133, pp. 506–512, Oct. 1986.

147. J. Fhul, A. F. Molisch, and E. Bonek, "Unified channel model for mobile radio systems with smart antennas", *Proc. Inst. Elect. Eng. (Radar, Sonar and Navigation)*, vol. 145, no. 1, pp. 32–41, Feb. 1998.

148. K. I. Pedersen, P. E. Mogensen, and B. H. Fleury, "Spatial channel characteristics in outdoor environments and their impact on BS antenna system performance", in *Proc. IEEE 48th IEEE Veh. Technol. Conf. 1998 (VTC 1998)*, vol. 2, Ottawa, 18–21 May 1998, pp. 719–723.

149. J.-A. Tsai, R. M. Buehrer, and B. D. Woerner, "Spatial fading correlation function of circular antenna arrays with Laplacian energy distribution", *IEEE Commun. Lett.*, vol. 6, no. 5, pp. 178–180, May 2002.

150. _____, "BER performance of a uniform circular array versus a uniform linear array in a mobile radio environment", *IEEE Trans. Wireless Commun.*, vol. 3, no. 3, pp. 695–700, May 2004.

151. S. Loyka and A. Kouki, "On MIMO channel capacity, correlations, and keyholes: Analysis of degenerate channels", *IEEE Trans. Commun.*, vol. 50, no. 12, pp. 1886–1888, Dec. 2002.

152. D. Chizhik, G. J. Foschini, M. J. Gans, and R. A. Valenzuela, "Keyholes, correlations, and capacities of multielement transmit and receive antennas", *IEEE Trans. Wireless Commun.*, vol. 1, no. 2, pp. 361–368, Apr. 2002.

153. D. Gesbert, H. Bölcskei, D. A. Gore, and A. J. Paulraj, "Outdoor MIMO wireless channels: Models and performance prediction", *IEEE Trans. Commun.*, vol. 50, no. 12, pp. 1926–1934, Dec. 2002.

154. V. Erceg, S. J. Fortune, J. Ling, A. J. Rustako, Jr., and R. A. Valenzuela, "Comparisons of a computer-based propagation prediction tool with experimental data collected in urban microcellular environments", *IEEE J. Sel. Areas Commun.*, vol. 15, no. 4, pp. 677–684, May 1997.

155. H. Özcelik, M. Herdin, W. Weichselberger, J. Wallace, and E. Bonek, "Deficiencies of 'Kronecker' MIMO radio channel model", *Electron. Lett.*, vol. 39, no. 16, pp. 1209–1210, 7 Aug. 2003.

156. H. Bölcskei, D. Gesbert, and A. J. Paulraj, "On the capacity of OFDM-based spatial multiplexing systems", *IEEE Trans. Commun.*, vol. 50, no. 2, pp. 225–234, Feb. 2002.

157. R. K. Mallik, "The pseudo-Wishart distribution and its application to MIMO systems", *IEEE Trans. Inform. Theory*, vol. 49, no. 10, pp. 2761–2769, Oct. 2003.

158. A. Maaref and S. Aissa, "Eigenvalue distributions of Wishart-type random matrices with application to the performance analysis of MIMO MRC systems", *IEEE Trans. Wireless Commun.*, vol. 6, no. 7, pp. 2678–2689, July 1964.

159. R. U. Nabar, H. Bölcskei, and A. J. Paulraj, "Diversity and outage performance in space-time block coded Ricean MIMO channels", *IEEE Trans. Wireless Commun.*, vol. 4, no. 5, pp. 2519–2532, Sep. 2005.

160. A. Kuchar, J. P. Rossi, and E. Bonek, "Directional macro-cell channel characterization from urban measurements", *IEEE Trans. Ant. and Prop.*, vol. 48, no. 2, pp. 137–146, Feb. 2000.

161. K. I. Pedersen, P. E. Mogensen, and B. H. Fleury, "A stochastic model of the temporal and azimuthal dispersion seen at the base station in outdoor propagation environments", *IEEE Trans. Veh. Tech.*, vol. 49, no. 2, pp. 437–447, Mar. 2000.

162. H. Hadinejad-Mahram, D. Dahlhaus, and D. Blomker, "Matched-filter bound for binary signaling over dispersive fading channels with receive diversity", *IEEE Trans. Commun.*, vol. 52, no. 12, pp. 2082–2086, Dec. 2004.

163. V. Erceg et al., *Channel Models for Fixed Wireless Applications (updated on 27/06/2003)*, IEEE 802.16 Broadband Wireless Access Working Group, http://www.ieee802.org/16/tga/, 2003.

164. L. Fregni, F. Muratori, and G. M. Vitetta, "A MIMO channel simulator based on GQR models", in *Proc. IST & Wireless Communications Summit*, Aveiro, Portugal, June 2003, pp. 690–694.

165. A. F. Molisch, H. Asplund, R. Heddergott, M. Steinbauer, and T. Zwick, "The COST 259 directional channel model – Part I: Overview and methodology", *IEEE Trans. Wireless Commun.*, vol. 5, no. 12, pp. 3421–3433, Dec. 2006.

166. H. Asplund, A. A. Glazunov, A. F. Molisch, K. I. Pedersen, and M. Steinbauer, "The COST 259 directional channel model–Part II: Macrocells", *IEEE Trans. Wireless Commun.*, vol. 5, no. 12, pp. 3434–3450, Dec. 2006.

167. K. Bullington, "Radio propagation fundamentals", *Bell Syst. Tech. J.*, vol. 36, pp. 593–626, May 1957.

168. R. W. E. McNicol, "The fading of radio waves of medium and high frequencies", *Proc. Inst. Elec. Eng.*, *pt. III*, vol. 96, pp. 517–524, Nov. 1949.

169. Y. Okumura, E. Ohmori, T. Kawano, and K. Fukuda, "Field strength and its variability in VHF and UHF land-mobile services", *Review Elec. Commun. Lab.*, vol. 16, pp. 825–873, Sep. 1968.

170. K. Bullington, "Radio propagation for vehicular communications", *IEEE Trans. Veh. Tech.*, vol. 26, no. 4, pp. 295–308, Nov. 1977.

171. M. Hata, "Empirical formula for propagation loss in land mobile radio services", *IEEE Trans. Veh. Tech.*, vol. 29, no. 3, pp. 317–325, Aug. 1980.

172. M. Ibrahim and J. Parsons, "Signal strength prediction in built-up areas", *IEE Proc., pt. F*, vol. 130, no. 5, pp. 312–339, Feb. 1983.

173. F. Ikegami, S. Yoshida, T. Takeuchi, and M. Umehira, "Propagation factors controlling mean field strength on urban streets", *IEEE Trans. Ant. and Prop.*, vol. 32, no. 8, pp. 822–829, Aug. 1984.

174. J. Walfisch and H. L. Bertoni, "A theoretical model of UHF propagation in urban environments", *IEEE Trans. Ant. and Prop.*, vol. 36, no. 12, pp. 1788–1796, Dec. 1988.

175. H. K. Chung and H. L. Bertoni, "Range-dependent path-loss model in residential areas for the VHF and UHF bands", *IEEE Trans. Ant. and Prop.*, vol. 50, no. 1, pp. 1–11, Jan. 2002.

176. COST 231, *Digital mobile radio towards future generation systems*, Final Report, COST Telecom Secretariat, European Commission, Brussels, 1999.

177. T. Kürner and A. Meier, "Prediction of outdoor and outdoor-to-indoor coverage in urban areas at 1.8 GHz", *IEEE J. Sel. Areas Commun.*, vol. 20, no. 3, pp. 496–506, Apr. 2002.

178. L. B. Milstein, D. L. Schilling, R. L. Pickholtz, V. Erceg, M. Kullback, E. G. Kanterakis, D. S. Fishman, W. Biederman, and D. C. Salerno, "On the feasibility of a CDMA overlay for personal communication networks", *IEEE J. Sel. Areas Commun.*, vol. 10, no. 4, pp. 655–668, May 1992.

179. H. Masui, T. Kobayashi, and M. Akaike, "Microwave path-loss modeling in urban line-of-sight environments", *IEEE J. Sel. Areas Commun.*, vol. 20, no. 6, pp. 1151–1155, Aug. 2002.

180. V. Erceg, S. Ghassemzadeh, M. Taylor, D. Li, and D. L. Schilling, "Urban/suburban out-of-sight propagation modeling", *IEEE Commun. Mag.*, vol. 30, no. 6, pp. 56–61, June 1992.

181. A. J. Rustako, N. Amitay, G. J. Owens, and R. S. Roman, "Radio propagation at microwave frequencies for line-of-sight microcellular mobile and personal communications", *IEEE Trans. Veh. Tech.*, vol. 40, no. 1, pp. 203–210, Feb. 1991.

182. N. Amitay, "Modeling and computer simulation of wave propagation in lineal line-of-sight microcells", *IEEE Trans. Veh. Tech.*, vol. 41, no. 4, pp. 337–342, Nov. 1992.

183. R. J. Luebbers, "Finite conductivity GTD versus knife edge diffraction in propagation path loss", *IEEE Trans. Ant. and Prop.*, vol. 32, no. 1, pp. 70–76, Jan. 1984.

184. S. Y. Tan and H. S. Tan, "UTD propagation model in an urban street scene for microcellular communications", *IEEE Trans. Electr. Comp.*, vol. 35, no. 4, pp. 423–428, Nov. 1993.

185. _____, "A theory for propagation path-loss characteristics in a city-street grid", *IEEE Trans. Electr. Comp.*, vol. 37, no. 3, pp. 333–342, Aug. 1995.

186. _____, "A microcellular communications propagation model based on the uniform theory of diffraction and multiple image theory", *IEEE Trans. Ant. and Prop.*, vol. 44, no. 10, pp. 1317–1326, Oct. 1996.

187. T. Kürner, D. J. Cichon, and W. Wiesbeck, "Concepts and results for 3D digital terrain-based wave propagation models: an overview", *IEEE Trans. Commun.*, vol. 11, no. 7, pp. 1002–1012, Sep. 1993.

188. W. Zhang, "A wide-band propagation model based on UTD for cellular mobile radio communications", *IEEE Trans. Ant. and Prop.*, vol. 45, no. 11, pp. 1669–1678, Nov. 1997.

189. Q. Sun, S. Y. Tan, and K. C. Teh, "Analytical formulae for path loss prediction in urban street grid microcellular environments", *IEEE Trans. Veh. Tech.*, vol. 54, no. 4, pp. 1251–1258, July 2005.

190. X. Ying, T. Qiwu, D. Erricolo, and P. L. E. Uslenghi, "Fresnel-Kirchhoff integral for 2-D and 3-D path loss in outdoor urban environments", *IEEE Trans. Ant. and Prop.*, vol. 53, no. 11, pp. 3757–3766, Nov. 2005.

191. J. W. McKown and J. R. L. Hamilton, "Ray tracing as a design tool for radio networks", *IEEE Network Mag.*, vol. 5, no. 6, pp. 27–30, Nov. 1991.

192. Z. Ji, B.-H. Li, H.-X. Wang, H.-Y. Chen, and T. K. Sarkar, "Efficient ray-tracing methods for propagation prediction for indoor wireless communications", *IEEE Trans. Ant. and Prop. Mag.*, vol. 43, no. 2, pp. 41–49, Apr. 2001.

193. F. A. Agelet, A. Formella, J. M. H. Rabanos, F. I. de Vicente, and F. P. Fontan, "Efficient ray-tracing acceleration techniques for radio propagation modeling", *IEEE Trans. Veh. Tech.*, vol. 49, no. 6, pp. 2089–2104, Nov. 2000.

194. F. Ikegami, T. Takeuchi, and S. Yoshida, "Theoretical prediction of mean field strength for urban mobile radio", *IEEE Trans. Ant. and Prop.*, vol. 39, no. 3, pp. 299–302, Mar. 1991.

195. R. Grosskopf, "Prediction of urban propagation loss", *IEEE Trans. Ant. and Prop.*, vol. 42, no. 5, pp. 658–665, May 1994.

196. G. Liang and H. L. Bertoni, "A new approach to 3-D ray tracing for propagation prediction in cities", *IEEE Trans. Ant. and Prop.*, vol. 46, no. 6, pp. 853–863, June 1998.

197. H.-J. Li, C.-C. Chen, T.-Y. Liu, and H.-C. Lin, "Applicability of ray-tracing technique for prediction of outdoor channel characteristics", *IEEE Trans. Veh. Tech.*, vol. 49, no. 6, pp. 2336–2349, Nov. 2000.

198. Z. Zhang, R. K. Sorensen, Z. Yun, M. F. Iskander, and J. F. Harvey, "A ray-tracing approach for indoor/outdoor propagation through window structures", *IEEE Trans. Ant. and Prop.*, vol. 50, no. 5, pp. 742–748, May 2002.

199. Z. Yun, Z. Zhang, and M. F. Iskander, "A ray-tracing method based on the triangular grid approach and application to propagation prediction in urban environments", *IEEE Trans. Ant. and Prop.*, vol. 50, no. 3, pp. 750–758, May 2002.

200. D. Erricolo and P. L. E. Uslenghi, "Propagation path loss – a comparison between ray-tracing approach and empirical models", *IEEE Trans. Ant. and Prop.*, vol. 50, no. 5, pp. 766–768, May 2002.

201. A. Muqaibel, A. Safaai-Jazi, A. Attiya, B. Woerner, and S. Riad, "Path-loss and time dispersion parameters for indoor UWB propagation", *IEEE Trans. Wireless Commun.*, vol. 5, no. 3, pp. 550–559, Mar. 2006.

202. J. A. Dabin, A. M. Haimovich, and H. Grebel, "A statistical ultra-wideband indoor channel model and the effects of antenna directivity on path loss and multipath propagation", *IEEE J. Selec. Areas Commun.*, vol. 24, no. 4, pp. 752–758, Apr. 2006.

203. A. F. Molisch, D. Cassioli, C.-C. Chong, S. Emami, A. Fort, B. Kannan, J. Karedal, J. Kunisch, H. G. Schantz, K. Siwiak, and M. Z. Win, "A comprehensive standardized model for ultrawideband propagation channels", *IEEE Trans. Ant. and Prop.*, vol. 54, no. 11, pp. 3151–3166, Nov. 2006.

204. A. M. Sayeed and B. Aazhang, "Communication over multipath fading channels: A time-frequency perspective", in *Wireless Communications: TDMA versus CDMA*, S. G. Glisic and P. A. Leppänen, (eds.) Norwell, MA: Kluwer Academic, 1997.

205. ———, "Joint multipath-Doppler diversity in mobile wireless communications", *IEEE Trans. Commun.*, vol. 47, no. 1, pp. 123–132, Jan. 2002.

206. C. X. Wang, M. Pätzold, and Q. Yao, "Stochastic modeling and simulation of frequency-correlated wideband fading channels", *IEEE Trans. Veh. Tech.*, vol. 56, no. 3, pp. 1050–1063, May 2007.

207. A. M. Sayeed, E. N. Onggosanusi, and B. D. Van Veen, "A canonical space-time characterization of mobile wireless channels", *IEEE Commun. Lett.*, vol. 3, no. 4, pp. 94–96, Apr. 1999.

208. E. N. Onggosanusi, A. M. Sayeed, and B. D. Van Veen, "Canonical space-time processing for wireless communications", *IEEE Trans. Commun.*, vol. 48, no. 10, pp. 1669–1680, Oct. 2000.

209. J. D. Parsons and A. M. D. Turkmani, "Characterisation of mobile radio signals: Model description", *Proc. Inst. Elect. Eng. (Commun., Speech and Vision)*, vol. 138, no. 6, pp. 549–556, Dec. 1991.

210. J. J. Ossanna, "A model for mobile radio fading due to building reflections: Theoretical and experimental fading waveform power spectra", *Bell Syst. Tech. J.*, vol. 43, no. 6, pp. 2935–2971, Nov. 1964.

211. E. N. Gilbert, "Energy reception for mobile radio", *Bell Syst. Tech. J.*, vol. 44, no. 8, pp. 1779–1803, Oct. 1965.

212. T. Aulin, "A modified model for the fading signal at a mobile radio channel", *IEEE Trans. Veh. Tech.*, vol. 28, no. 3, pp. 182–203, Aug. 1979.

213. A. Abdi, J. A. Barger, and M. Kaveh, "A parametric model for the distribution of the angle of arrival and the associated correlation function and power spectrum at the mobile station", *IEEE Trans. Veh. Tech.*, vol. 51, no. 3, pp. 425–434, May 2002.

214. P. Sadeghi, R. Kennedy, P. Rapajic, and R. Shams, "Finite-state Markov modeling of fading channels – a survey of principles and applications", *IEEE Sig. Process. Mag.*, vol. 25, no. 5, pp. 57–80, Sep. 2008.

215. W. Turin and R. van Nobelen, "Hidden Markov modeling of flat fading channels", *IEEE J. Sel. Areas Commun.*, vol. 16, no. 9, pp. 1809–1817, Dec. 1998.

216. H. S. Wang and N. Moayeri, "Finite-state Markov channel – a useful model for radio communication channels", *IEEE Trans. Veh. Tech.*, vol. 44, no. 1, pp. 163–171, Feb. 1995.

217. H. S. Wang and P.-C. Chang, "On verifying the first-order Markovian assumption for a Rayleigh fading channel model", *IEEE Trans. Veh. Tech.*, vol. 45, no. 2, pp. 353–357, May 1996.

218. C.-D. Iskander and P. T. Mathiopoulos, "Fast simulation of diversity Nakagami fading channels using finite-state Markov models", *IEEE Trans. Broadcasting*, vol. 49, no. 3, pp. 269–277, Sep. 2003.

219. Q. Zhang and S. A. Kassam, "Finite-state Markov model for Rayleigh fading channels", *IEEE Trans. Commun.*, vol. 47, no. 11, pp. 1688–1692, Nov. 1999.

220. J. Winters, "On the capacity of radio communication systems with diversity in a Rayleigh fading environment", *IEEE J. Sel. Areas Commun.*, vol. 5, no. 5, pp. 871–878, June 1987.

221. D.-S. Shiu, G. J. Foschini, M. J. Gans, and J. M. Kahn, "Fading correlation and its effect on the capacity of multielement antenna systems", *IEEE Trans. Commun.*, vol. 48, no. 3, pp. 502–513, Mar. 2000.

222. T.-A. Chen, M. P. Fitz, W.-Y. Kuo, M. D. Zoltowski, and J. H. Grimm, "A space-time model for frequency nonselective Rayleigh fading channels with applications to space-time modems", *IEEE J. Sel. Areas Commun.*, vol. 18, no. 7, pp. 1175–1190, July 2000.

223. A. Abdi and M. Kaveh, "A space-time correlation model for multielement antenna systems in mobile fading channels", *IEEE J. Sel. Areas Commun.*, vol. 20, no. 3, pp. 550–560, Apr. 2002.

224. G. J. Byers and F. Takawira, "Spatially and temporally correlated MIMO channels: Modeling and capacity analysis", *IEEE Trans. Veh. Tech.*, vol. 53, no. 3, pp. 634–643, May 2004.

225. M. Zhang, P. J. Smith, and M. Shafi, "An extended one-ring MIMO channel model", *IEEE Trans. Wireless Commun.*, vol. 6, no. 8, pp. 2759–2764, Aug. 2007.

226. I. E. Telatar, "Capacity of multi-antenna Gaussian channels", *European Trans. Telecommun.*, vol. 10, pp. 585–595, Nov. 1999.

227. C.-N. Chuah, D. N. C. Tse, J. M. Kahn, and R. A. Valenzuela, "Capacity scaling in MIMO wireless systems under correlated fading", *IEEE Trans. Inform. Theory*, vol. 48, no. 3, pp. 637–650, Mar. 2002.

228. V. V. Veeravalli, Y. Liang, and A. M. Sayeed, "Correlated MIMO wireless channels: Capacity, optimal signaling, and asymptotics", *IEEE Trans. Inform. Theory*, vol. 51, no. 6, pp. 2058–2072, June 2005.

229. D. Chizhik, F. Rashid-Farrokhi, J. Ling, and A. Lozano, "Effect of antenna separation on the capacity of BLAST in correlated channels", *IEEE Commun. Lett.*, vol. 4, no. 11, pp. 337–339, Nov. 2000.

230. D. Chizhik, J. Ling, P. W. Wolniansky, R. A. Valenzuela, N. Costa, and K. Huber, "Multiple-input-multiple-output measurements and modeling in Manhattan", *IEEE J. Sel. Areas Commun.*, vol. 21, no. 3, pp. 321–331, Apr. 2003.

231. G. L. Turin, F. D. Clapp, T. L. Johnston, S. B. Fine, and D. Lavry, "A statistical model of urban multipath propagation", *IEEE Trans. Veh. Tech.*, vol. 21, no. 1, pp. 1–9, Feb. 1972.

232. H. Suzuki, "A statistical model for urban radio propagation", *IEEE Trans. Commun.*, vol. 25, no. 7, pp. 673–680, July 1977.

233. G. B. Giannakis and C. Tepedelenlioglu, "Basis expansion models and diversity techniques for blind identification and equalization of time-varying channels", *Proc. IEEE*, vol. 86, no. 10, pp. 1969–1986, Oct. 1998.

234. O. Oyman, R. U. Nabar, H. Bölcskei, and A. J. Paulraj, "Characterizing the statistical properties of mutual information in MIMO channels", *IEEE Trans. Sig. Process.*, vol. 51, no. 11, pp. 2784–2795, Nov. 2003.

235. H. Hashemi, "Simulation of the urban radio propagation channel", *IEEE Trans. Veh. Tech.*, vol. 28, no. 3, pp. 213–225, July 1979.

236. S. A. Fechtel, "A novel approach to modeling and efficient simulation of frequency-selective fading radio channels", *IEEE J. Sel. Areas Commun.*, vol. 11, no. 3, pp. 422–431, Apr. 1993.

237. K. W. Yip and T. S. Ng, "Efficient simulation of digital transmission over WSSUS channels", *IEEE Trans. Commun.*, vol. 43, no. 12, pp. 2907–2912, Dec. 1995.

238. P. M. Crespo and J. Jiménez, "Computer simulation of radio channels using a harmonic decomposition technique", *IEEE Trans. Veh. Technol.*, vol. 44, no. 3, pp. 414–419, Aug. 1995.

239. E. Haas, "Aeronautical channel modeling", *IEEE Trans. Veh. Tech.*, vol. 51, no. 2, pp. 254–264, Mar. 2002.

240. J. Grolleau, E. Grivel, and M. Najim, "Two ways to simulate a Rayleigh fading channel based on a stochastic sinusoidal model", *IEEE Sig. Process. Lett.*, vol. 15, pp. 107–110, May 2008.

241. P. Dent, G. E. Bottomley, and T. Croft, "Jakes' fading model revisited", *Electron. Lett.*, vol. 29, no. 13, pp. 1162–1163, June 24 1993.

242. M. Pätzhold, U. Killat, F. and Y. Li, "On the statistical properties of deterministic simulation models for mobile fading channels", *IEEE Trans. Veh. Tech.*, vol. 47, no. 1, pp. 254–269, Feb. 1998.

243. M. F. Pop and N. C. Beaulieu, "Limitations of sum-of-sinusoids fading channel simulators", *IEEE Trans. Commun.*, vol. 49, no. 4, pp. 699–708, Apr. 2001.

244. J. I. Smith, "A computer generated multipath fading simulation for mobile radio", *IEEE Trans. Veh. Tech.*, vol. 24, no. 3, pp. 39–40, Aug. 1975.

245. D. J. Young and N. C. Beaulieu, "The generation of correlated Rayleigh random variates by inverse discrete Fourier transform", *IEEE Trans. Commun.*, vol. 48, no. 7, pp. 1114–1127, July 2000.

246. P. H.-Y. Wu and A. Duel-Hallen, "Multiuser detectors with disjoint Kalman channel estimators for synchronous CDMA mobile radio channels", *IEEE Trans. Commun.*, vol. 48, no. 5, pp. 752–756, May 2000.

247. K. E. Baddour and N. C. Beaulieu, "Autoregressive modeling for fading channel simulation", *IEEE Trans. Wireless Commun.*, vol. 4, no. 4, pp. 1650–1662, July 2005.

248. A. Barbieri, A. Piemontese, and G. Colavolpe, "On the ARMA approximation for fading channels described by the Clarke model with applications to Kalman-based receivers", *IEEE Trans. Wireless Commun.*, vol. 8, no. 2, pp. 535–540, Feb. 2009.

249. D. Schafhuber, G. Matz, and F. Hlawatsch, "Simulation of wideband mobile radio channels using sub-sampled ARMA models and multistage interpolation", in *Proc. 11th IEEE Signal Processing Workshop on Statistical Signal Processing*, 6–8 Aug. 2001, pp. 571–574.

250. J. J. Blanz and P. Jung, "A flexibly configurable spatial model for mobile radio channels", *IEEE Trans. Commun.*, vol. 46, no. 3, pp. 367–371, Mar. 1998.

251. M. Pätzold and B. O. Hogstad, "A wideband space-time MIMO channel simulator based on the geometrical one-ring model", in *Proc. 2006 IEEE 64th Veh. Technol. Conf. (VTC-2006 – Fall)*, Montreal, Sep. 2006, pp. 1–6.

252. E. Kunnari and J. Linatti, "Stochastic modelling of rice fading channels with temporal, spatial and spectral correlation", *IET Commun.*, vol. 1, no. 2, pp. 215–224, Apr. 2007.

253. J. Mietzner and P. A. Hoeher, "A rigorous analysis of the statistical properties of the discrete-time triply-selective MIMO Rayleigh fading channel model", *IEEE Trans. Wireless Commun.*, vol. 6, no. 12, pp. 4199–4203, May 1992.

254. P. Petrus, J. H. Reed, and T. S. Rappaport, "Geometrical-based statistical macrocell channel model for mobile environments", *IEEE Trans. Commun.*, vol. 50, no. 3, pp. 495–502, Mar. 2002.

255. R. J. Piechocki, G. V. Tsoulos, and J. P. McGeehan, "Simple general formula for pdf of angle of arrival in large cell operational environments", *Electron. Lett.*, vol. 34, no. 18, pp. 1784–1785, 3 Sep. 1998.

256. M. P. Lotter and P. Van Rooyen, "Modeling spatial aspects of cellular CDMA/SDMA systems", *IEEE Commun. Lett.*, vol. 3, no. 5, pp. 128–131, May 1999.

257. R. Janaswamy, "Angle and time of arrival statistics for the Gaussian scatter density model", *IEEE Trans. Wireless Commun.*, vol. 1, no. 3, pp. 488–497, July 2002.

258. N. M. Khan, M. T. Simsim, and P. B. Rapajic, "A generalized model for the spatial characteristics of the cellular mobile channel", *IEEE Trans. Veh. Technol.*, vol. 57, no. 1, pp. 22–37, Jan. 2008.

259. R. B. Ertel and J. H. Reed, "Angle and time of arrival statistics for circular and elliptical scattering models", *IEEE J. Select. Areas Commun.*, vol. 17, no. 11, pp. 1829–1840, Nov. 1999.

260. A. J. Paulraj and C. B. Papadias, "Space-time processing for wireless communications", *IEEE Sig. Process. Mag.*, vol. 14, no. 6, pp. 49–83, Nov. 1997.

261. G. G. Raleigh, S. N. Diggavi, A. F. Naguib, and A. Paulraj, "Characterization of fast fading vector channels for multi-antenna communication systems", in *Conf. Rec. Twenty-Eighth Asilomar Conf. Sign., Syst. and Comp.*, vol. 2, Asilomar, CA, 31 Oct.–2 Nov. 1994, pp. 853–857.

262. G. G. Raleigh and A. Paulraj, "Time varying vector channel estimation for adaptive spatial equalization", in *Proc. IEEE Global Telecommun. Conf. (GLOBECOM '95)*, vol. 1, San Francisco, 13–17 Nov. 1995, pp. 218–224.

263. P. Zetterberg and B. Ottersten, "The spectrum efficiency of a base station antenna array system for spatially selective transmission", *IEEE Trans. Veh. Tech.*, vol. 44, no. 3, pp. 651–660, Aug. 1995.

264. J. C. Liberti and T. S. Rappaport, "A geometrically based model for line-of-sight multipath radio channels", in *Proc. 1996 IEEE 46th Veh. Technol. Conf. (VTC 1996)*, vol. 2, 28 Apr.–1 May 1996, pp. 844–848.

265. Q. Spencer, M. Rice, B. Jeffs, and M. Jensen, "A statistical model for angle of arrival in indoor multipath propagation", in *Proc. IEEE 1997 47th Veh. Technol. Conf. (VTC 1997)*, vol. 3, Phoenix, AZ, 4–7 May 1997, pp. 1415–1419.

266. M. Lu, T. Lo, and J. Litva, "A physical spatio-temporal model for multipath propagation channels", in *Proc. 1997 IEEE 47th Veh. Technol. Conf. (VTC 1997)*, vol. 2, Phoenix, AZ, 4–7 May 1997, pp. 810–814.

267. D. Asztely, B. Ottersten, and A. L. Swindlehurst, "Generalised array manifold model for wireless communication channels with local scattering", *Proc. Inst. Elect. Eng. (Radar, Sonar and Navigation)*, vol. 145, no. 1, pp. 51–57, Feb. 1998.

268. J. J. Blanz, A. Papathanassiou, M. Haardt, I. Furio, and P. W. Baier, "Smart antennas for combined DOA and joint channel estimation in time-slotted CDMA mobile radio systems with joint detection", *IEEE Trans. Veh. Tech.*, vol. 49, no. 2, pp. 293–306, Mar. 2000.

269. L. Jiang and S. Y. Tan, "Geometrically based statistical channel models for outdoor and indoor propagation environments", *IEEE Trans. Veh. Tech.*, vol. 56, no. 6, pp. 3587–3593, Nov. 2007.

270. A. Intarapanich, P. L. Kafle, R. J. Davies, A. B. Sesay, and J. G. McRory, "Geometrically based broadband MIMO channel model with tap-gain correlation", *IEEE Trans. Veh. Tech.*, vol. 56, no. 6, pp. 3631–3641, Nov. 2007.

271. M. Pätzhold, B. O. Hogstad, and N. Youssef, "Modeling, analysis, and simulation of MIMO mobile-to-mobile fading channels", *IEEE Trans. Wireless Commun.*, vol. 7, no. 2, pp. 510–520, Feb. 2008.

272. U. G. Schuster and H. Bölcskei, "Ultrawideband channel modeling on the basis of information-theoretic criteria", *IEEE Trans. Wireless Commun.*, vol. 6, no. 7, pp. 2464–2475, July 2007.

273. M. Shafi, M. Zhang, A. L. Moustakas, P. J. Smith, A. F. Molisch, F. Tufvesson, and S. H. Simon, "Polarized MIMO channels in 3-D: Models, measurements and mutual information", *IEEE J. Sel. Areas Commun.*, vol. 24, no. 3, pp. 514–527, Mar. 2006.

274. M. Pätzhold, *Mobile Fading Channels*. Chichester: John Wiley & Sons, Ltd, 1999.

275. E. Biglieri, J. Proakis, and S. Shamai, "Fading channels: Information-theoretic and communications aspects", *IEEE Trans. Inform. Theory*, vol. 44, no. 6, pp. 2619–2692, Oct. 1998.

276. K. Yu and B. E. Ottersten, "Models for MIMO propagation channels: a review", *Wiley J. Wireless Commun. Mobile Comput.*, vol. 2, no. 7, pp. 653–666, 2002.

277. G. D. Durgin, *Space-Time Wireless Channels*. Upper Saddle River, NJ: Prentice Hall PTR, 2003.

278. L. C. Godara, "Application of antenna arrays to mobile communications, Part I: Performance improvement, feasibility, and system considerations", *IEEE Proc.*, vol. 85, no. 7, pp. 1031–1060, July 1997.

279. E. Parzen, *Stochastic Processes*. San Francisco: Holden Day, 1962.

280. R. W. Lucky, J. Salz, and E. J. Weldon, *Principles of Data Communication*. New York: McGraw-Hill, 1968.

281. F. Amoroso, "The bandwith of digital data signals", *IEEE Commun. Mag.*, vol. 18, pp. 13–24, Nov. 1980.

282. D. Slepian and H. O. Pollak, "Prolate spheroidal wave functions, Fourier analysis and uncertainty", *Bell Syst. Tech. J.*, vol. 40, pp. 43–63, Jan. 1961.

283. W. J. Weber, "Differential encoding for multiple amplitude and phase shift keying systems", *IEEE Trans. Commun.*, vol. 26, no. 3, pp. 385–391, Mar. 1978.

284. F. Pancaldi and G. M. Vitetta, "Equalization algorithms in the frequency domain for continuous phase modulations", *IEEE Trans. Commun.*, vol. 54, no. 4, pp. 648–658, Apr. 2006.

285. S. Pasupathy, "Minimum shift keying: A spectrally efficient modulation", *IEEE Commun. Mag.*, vol. 17, no. 4, pp. 14–22, July 1979.

286. K. Murota and K. Hirade, "GMSK modulation for digital mobile radio telephony", *IEEE Trans. Commun.*, vol. 29, no. 7, pp. 1044–1050, July 1981.

287. W. Osborne and M. Luntz, "Coherent and noncoherent detection of CPFSK", *IEEE Trans. Commun.*, vol. 22, no. 8, pp. 1023–1036, July 1974.

288. J. B. Anderson, T. Aulin, and C. E. Sundberg, *Digital Phase Modulation*. New York: Plenum Press, 1986.

289. H. Miyakawa, H. Harashima, and Y. Tanaka, "A new digital modulation scheme, multi-code binary CPFSK", in *Proc. 3rd Int. Conf. Digital Satellite Commun.*, Nov. 1975, pp. 105–112.

290. J. B. Anderson and D. P. Taylor, "A bandwidth-efficient class of signal-space codes", *IEEE Trans. Inform. Theory*, vol. 24, no. 6, pp. 703–712, Nov. 1978.

291. J. Huber and W. Liu, "An alternative approach to reduced-complexity CPM-receivers", *IEEE J. Sel. Areas Commun.*, vol. 7, no. 9, pp. 1437–1449, Dec. 1989.

292. T. Palenius and A. Svensson, "Reduced complexity detectors for continuous phase modulation based on a signal space approach", *European Trans. Telecommun.*, vol. 4, no. 3, pp. 285–297, May/June 1993.

293. S. J. Simmons, "Simplified coherent detection of CPM", *IEEE Trans. Commun.*, vol. 43, no. 3, pp. 726–728, Feb./Mar./Apr. 1995.

294. W. Tang and E. Shwedyk, "A quasi-optimum receiver for continuous phase modulation", *IEEE Trans. Commun.*, vol. 48, no. 7, pp. 1087–1090, July 2000.

295. P. Moqvist and T. M. Aulin, "Orthogonalization by principal components applied to CPM", *IEEE Trans. Commun.*, vol. 51, no. 11, pp. 1838–1845, Nov. 2003.

296. P. A. Laurent, "Exact and approximate construction of digital phase modulations by superposition of amplitude modulated pulses", *IEEE Trans. Commun.*, vol. 34, no. 2, pp. 150–160, Feb. 1986.

297. U. Mengali and M. Morelli, "Decomposition of M-ary CPM signals into PAM waveforms", *IEEE Trans. Inform. Theory*, vol. 41, no. 5, pp. 1265–1275, Sep. 1995.

298. E. Perrins and M. Rice, "PAM decomposition of M-ary multi-h CPM", *IEEE Trans. Commun.*, vol. 53, no. 12, pp. 2065–2075, Dec. 2005.

299. X. Huang and Y. Li, "The PAM decomposition of CPM signals with integer modulation index", *IEEE Trans. Commun.*, vol. 51, no. 4, pp. 543–546, Apr. 2003.

300. ——, "MMSE-optimal approximation of continuous-phase modulated signal as superposition of linearly modulated pulses", *IEEE Trans. Commun.*, vol. 53, no. 7, pp. 1166–1177, July 2005.

301. M. P. Wylie-Green, "A new finite series expansion of continuous phase modulated waveforms", *IEEE Trans. Commun.*, vol. 55, no. 8, pp. 1547–1556, Aug. 2007.

302. B. Rimoldi, "A decomposition approach to CPM", *IEEE Trans. Inform. Theory*, vol. 34, no. 2, pp. 260–270, Mar. 1988.

303. G. K. Kaleh, "Simple coherent receivers for partial response continuous phase modulation", *IEEE J. Selec. Areas Commun.*, vol. 7, no. 9, pp. 1427–1436, Dec. 1989.

304. G. Colavolpe and R. Raheli, "Reduced-complexity detection and phase synchronization of CPM signals", *IEEE Trans. Commun.*, vol. 45, no. 9, pp. 1070–1079, Sep. 1997.

305. ——, "Noncoherent sequence detection of continuous phase modulations", *IEEE Trans. Commun.*, vol. 47, no. 9, pp. 1303–1307, Sep. 1999.

306. J. Tan and G. L. Stüber, "Frequency-domain equalization for continuous phase modulation", *IEEE Trans. Wireless Commun.*, vol. 4, no. 5, pp. 2479–2490, Sep. 1995.

307. J. L. Massey, T. Mittelholzer, T. Riedel, and M. Vollenweider, "Ring convolutional codes for phase modulation", in *Proc. IEEE Int. Symp. Inform. Theory*, San Diego, CA, Sep. 1990, p. 176.

308. R. H.-H. Yang and D. P. Taylor, "Trellis-coded continuous-phase frequency-shift keying with ring convolutional codes", *IEEE Trans. Inform. Theory*, vol. 40, no. 4, pp. 1057–1067, July 1994.

309. R. L. Maw and D. P. Taylor, "Space-time coded systems with continuous phase frequency shift keying", in *Proc. IEEE Global Telecommun. Conf. (GLOBECOM 2005)*, vol. 3, St. Louis, MO, Nov./Dec. 2005, pp. 1581–1586.

310. R. D. Barnard, "A note on a special class of one-sided distribution sums", *Bell Syst. Tech. J.*, vol. 46, pp. 203–206, Jan. 1965.

311. J. Salz, "Spectral density function of multilevel continuous-phase FM", *IEEE Trans. Inform. Theory*, vol. 34, no. 8, pp. 429–433, July 1965.

312. T. Aulin and C.-E. Sundberg, "Exact asymptotic behavior of digital FM spectra", *IEEE Trans. Commun.*, vol. 30, no. 11, pp. 2438–2449, Nov. 1982.

313. T. Baker, "Asymptotic behavior of digital FM spectra", *IEEE Trans. Commun.*, vol. 22, no. 10, pp. 1585–1594, Oct. 1974.

314. F. Amoroso, "Pulse and spectrum manipulation in the minimum (frequency) shift keying (MSK) format", *IEEE Trans. Commun.*, vol. 24, no. 3, pp. 381–384, Mar. 1976.

315. A. R. S. Bahai and B. R. Saltzberg, *Multi-Carrier Digital Communications: Theory and Applications of OFDM*. New York: Kluwer Academic/Plenum Publishers, 1999.

316. S. H. Han and J. H. Lee, "An overview of peak-to-average power ratio reduction techniques techniques for multicarrier transmission", *IEEE Wireless Commun.*, vol. 12, no. 2, pp. 56–65, Apr. 2005.

317. ETS 300 401, *Radio broadcasting systems; Digital Audio Broadcasting (DAB) to mobile, portable and fixed receivers*, European Telecommunications Standards Institute, Feb. 1995.

318. K. G. Paterson and V. Tarokh, "On the existence and construction of good codes with low peak-to-average power ratio", *IEEE Trans. Inform. Theory*, vol. 46, no. 6, pp. 1974–1987, Sep. 2000.

319. M. Sharif, M. Gharavi-Alkhansari, and B. H. Khalaj, "On the peak-to-average power of OFDM signals based on oversampling (MSK)-type signaling based upon input data symbol pulse shaping", *IEEE Trans. Commun.*, vol. 51, no. 1, pp. 72–78, Jan. 1976.

320. D. Wulich, "Comments on the peak factor of sampled and continuous signals", *IEEE Commun. Lett.*, vol. 4, no. 7, pp. 213–214, July 2000.

321. J. M. Wozencraft and I. M. Jacobs, *Principles of Communication Engineering*. New York: John Wiley & Sons, Inc., 1965.

322. N. J. A. Sloane, "The packing of spheres", *Scientific American*, pp. 116–125, Jan. 1984.

323. J. H. Conway and N. J. A. Sloane, *Sphere Packings, Lattices and Groups*. New York: Springer-Verlag, 1999.

324. D. G. Forney, Jr., "Coset codes – Part I: Introduction and geometrical classification", *IEEE Trans. Inform. Theory*, vol. 34, no. 5, pp. 1123–1151, Sep. 1988.

325. N. J. A. Sloane, "Tables of sphere packings and spherical codes", *IEEE Trans. Inform. Theory*, vol. 27, no. 3, pp. 327–338, May 1981.

326. C. Schlegel, *Trellis Coding*. New York: IEEE Press, 1997.

327. S. G. Wilson, *Digital Modulation and Coding*. Upper Saddle River, NJ: Prentice Hall, 1996.

328. H. Minkowski, *Ausgewählte Arbeiten zur Zahlentheorie und zur Geometrie*, ser. Selected Papers on Number Theory and Geometry, E. Krätzel and B. Weißbach, (eds.) Leipzig: BSB B. G. Teubner Verlagsgesellschaft, 1989.

329. T. Gosset, "On the regular and semi-regular figures in space of *n* dimensions", *Messenger Math.*, vol. 29, pp. 43–48, 1900.

330. E. S. Barnes and G. E. Wall, "Some extreme forms defined in terms of Abelian groups", *J. Austral. Math. Soc.*, vol. 1, pp. 47–63, 1959.

331. J. Leech, "Notes on sphere packings", *Canad. J. Math.*, vol. 19, pp. 251–257, 1967.

332. H. S. M. Coxeter and J. A. Todd, "An extreme duodenary form", *Canad. J. Math.*, vol. 5, pp. 384–392, 1953.

333. H.-G. Quebbemann, "Lattices with theta functions for $g\sqrt{(2)}$ and linear codes", *J. Algebra*, vol. 105, pp. 443–450, 1987.

334. G. Nebe, "Some cyclo-quaternionic lattices", *J. Algebra*, vol. 199, pp. 472–498, 1998.

335. D. G. Forney, Jr.,., "Coset codes – Part II: Binary lattices and related codes", *IEEE Trans. Inform. Theory*, vol. 34, no. 5, pp. 1152–1187, Sep. 1988.

336. R. de Buda, "Some optimal codes have structure", *IEEE J. Sel. Areas Commun.*, vol. 7, no. 6, pp. 893–899, Aug. 1989.

337. D. G. Forney, Jr., and L.-F. Wei, "Multidimensional constellations – Part I: Introduction, figures of merit, and generalized cross constellations", *IEEE J. Sel. Areas Commun.*, vol. 7, no. 6, pp. 877–892, Aug. 1989.

338. D. G. Forney, Jr., R. G. Gallager, G. R. Lang, F. M. Longstaff, and S. U. Qureshi, "Efficient modulation for band-limited channels", *IEEE J. Sel. Areas Commun.*, vol. 2, no. 5, pp. 632–646, Sep. 1984.

339. D. G. Forney, Jr., "Multidimensional constellations – Part II: Voronoi constellations", *IEEE J. Sel. Areas Commun.*, vol. 7, no. 6, pp. 941–958, Aug. 1989.

340. G. Foschini, R. Gitlin, and S. B. Weinstein, "Optimization of two-dimensional signal constellations in the presence of Gaussian noise", *IEEE Trans. Commun.*, vol. 22, no. 1, pp. 28–38, Jan. 1974.

341. J. H. Conway and N. J. A. Sloane, "Fast quantizing and decoding algorithms for lattice quantizers and codes", *IEEE Trans. Inform. Theory*, vol. 28, no. 2, pp. 227–232, Mar. 1982.

342. D. G. Forney, Jr., and A. Vardy, "Generalized minimum-distance decoding of Euclidean-space codes and lattices", *IEEE Trans. Inform. Theory*, vol. 42, no. 6, pp. 1992–2026, Nov. 1996.

343. J. Leech and N. J. Sloane, "Sphere packings and error-correcting codes", *Canad. J. Math.*, vol. 23, pp. 718–745, 1971.

344. E. L. Cusack, "Error control codes for QAM signalling", *Electron. Lett.*, vol. 20, no. 2, pp. 62–63, Jan. 19 1984.

345. M. L. Doelz, E. T. Heald, and D. L. Martin, "Binary data transmission techniques for linear systems", *Proc. IRE*, vol. 45, pp. 656–661, May 1957.

346. P. A. Baker, "Phase modulation data sets for serial transmission at 2000 and 2400 bits per second, Part 1", *AIEE Trans. (Communication and Electronics)*, July 1962.

347. C. Chan, "Performance of digital phase-modulation communication systems", *IRE Trans. Commun.*, vol. 7, no. 1, pp. 3–6, May 1959.

348. ——, "Combined digital phase and amplitude modulation communication systems", *IRE Trans. Commun.*, vol. 8, no. 3, pp. 150–155, Sep. 1960.

349. J. C. Hancock and R. W. Lucky, "Performance of combined amplitude and phase-modulated communication systems", *IRE Trans. Commun.*, vol. 8, no. 4, pp. 232–237, Dec. 1960.

350. C. Campopiano and B. Glazer, "Coherent digital amplitude and phase modulation", *IRE Trans. Commun.*, vol. 10, no. 1, pp. 90–95, Mar. 1962.

351. R. W. Lucky and J. C. Hancock, "On the optimum performance of *N*-ary systems having two degrees of freedom", *IRE Trans. Commun.*, vol. 10, no. 2, pp. 185–192, June 1962.

352. M. K. Simon and J. G. Smith, "Hexagonal multiple phase-and-amplitude-shift-keyed signal sets", *IEEE Trans. Commun.*, vol. 21, no. 10, pp. 1108–1115, Oct. 1973.

353. C. Thomas, M. Weidner, and S. Durrani, "Digital amplitude-phase keying with *M*-ary alphabets", *IEEE Trans. Commun.*, vol. 22, no. 2, pp. 168–180, Feb. 1974.

354. J. G. Smith, "Odd-bit quadrature amplitude-shift keying", *IEEE Trans. Commun.*, vol. 23, no. 3, pp. 385–389, Mar. 1975.

355. K. Miyauchi, S. Seki, and H. Ishio, "New technique for generating and detecting multilevel signal formats", *IEEE Trans. Commun.*, vol. 24, no. 2, pp. 263–267, Feb. 1976.

356. P. Dupuis, M. Joindot, A. Leclert, and D. Soufflet, "16 QAM modulation for high capacity digital radio system", *IEEE Trans. Commun.*, vol. 27, no. 12, pp. 1771–1782, Dec. 1979.

357. I. Horikawa, T. Murase, and Y. Saito, "Design and performances of a 200 Mbit/s 16 QAM digital radio system", *IEEE Trans. Commun.*, vol. 27, no. 12, pp. 1953–1958, Dec. 1979.

358. Y. Saito and Y. Nakamura, "256 QAM modem for high capacity digital radio system", *IEEE Trans. Commun.*, vol. 34, no. 8, pp. 799–805, Aug. 1986.

359. T. Noguchi, Y. Daido, and J. A. Nassek, "Modulation techniques for microwave digital radio", *IEEE Commun. Mag.*, vol. 24, no. 10, pp. 21–30, Oct. 1986.

360. W. Webb and L. Hanzo, *Modern Quadrature Amplitude Modulation – Principles and Applications for Fixed and Wireless Communications*. London: Pentech Press/IEEE Press, 1994.

361. M. Ibnkahla, Q. M. Rahman, A. I. Sulyman, H. A. Al-Asady, J. Yuan, and A. Safwat, "High-speed satellite mobile communications: Technologies and challenges", *IEEE Proc.*, vol. 92, no. 2, pp. 312–339, Feb. 2004.

362. S. Gronemeyer and A. McBride, "MSK and offset QPSK modulation", *IEEE Trans. Commun.*, vol. 24, no. 8, pp. 809–820, Aug. 1976.

363. M. K. Simon and J. G. Smith, "Offset quadrature communications with decision-feedback carrier synchronization", *IEEE Trans. Commun.*, vol. 22, no. 10, pp. 1576–1584, Oct. 1974.

364. M. L. Doelz and E. H. Heald, *Minimum-shift data communication system, US Patent* 2977417, Collins Radio Co., 28 Mar. 1961.

365. R. de Buda, "Coherent demodulation of frequency-shift keying with low deviation ratio", *IEEE Trans. Commun.*, vol. 20, no. 3, pp. 429–435, June 1991.

366. W. Sullivan, "High-capacity microwave system for digital data transmission", *IEEE Trans. Commun.*, vol. 20, no. 3, pp. 466–470, June 1972.

367. T. Schonhoff, "Symbol error probabilities for M-ary CPFSK: Coherent and noncoherent detection", *IEEE Trans. Commun.*, vol. 24, no. 6, pp. 644–652, June 1976.

368. M. K. Simon, "A generalization of minimum-shift-keying (MSK)-type signaling based upon input data symbol pulse shaping", *IEEE Trans. Commun.*, vol. 24, no. 8, pp. 845–856, Aug. 1976.

369. M. Rabzel and S. Pasupathy, "Spectral shaping in minimum shift keying (MSK)-type signals", *IEEE Trans. Commun.*, vol. 26, no. 1, pp. 189–195, Jan. 1978.

370. F. de Jager and C. Dekker, "Tamed frequency modulation, a novel method to achieve spectrum economy in digital transmission", *IEEE Trans. Commun.*, vol. 26, no. 5, pp. 534–542, May 1978.

371. R. R. Anderson and J. Salz, "Spectra of digital FM", *Bell Syst. Tech. J.*, vol. 44, pp. 1165–1189, July–Aug. 1965.

372. G. J. Garrison, "A power spectral density analysis for digital FM", *IEEE Trans. Commun.*, vol. 23, no. 11, pp. 1228–1243, Nov. 1975.

373. T. Aulin and C.-E. Sundberg, "Continuous phase modulation – Part I", *IEEE Trans. Commun.*, vol. 29, pp. 196–209, Mar. 1981.

374. T. Aulin, N. Rydbeck, and C.-E. Sundberg, "Continuous phase modulation – Part II", *IEEE Trans. Commun.*, vol. 29, pp. 210–225, Mar. 1981.

375. C.-E. W. Sundberg, "Continuous phase modulation", *IEEE Commun. Mag.*, vol. 24, no. 4, pp. 25–38, Apr. 1986.

376. J. B. Anderson and R. de Buda, "Better phase-modulation error performance using trellis phase code", *Electron. Lett.*, vol. 12, no. 22, pp. 587–588, Oct. 1976.

377. I. Sasase and S. Mori, "Multi-h phase-coded modulation", *IEEE Commun. Mag.*, vol. 29, no. 12, pp. 46–56, Dec. 1991.

378. A. A. Collins and M. L. Doelz, *Predicted Wave Signalling (Kineplex), Collins Technical Report CTR-140*, Burbank, CA, June 20 1955.

379. *Kathryn HF radio teletype and data system, Rept. 938-230-14 (AD269032L)*, General Atronics Corp., Sep. 25 1961.

380. M. Zimmerman and A. Kirsch, "The AN/GSC-10 (KATHRYN) variable rate data modem for HF radio", *IEEE Trans. Commun.*, vol. 15, no. 2, pp. 197–204, Apr. 1967.

381. P. A. Bello, "Selective fading limitations of the Kathryn modem and some system design considerations", *IEEE Trans. Commun. Technol.*, vol. 13, no. 3, pp. 320–333, Sep. 1965.

382. A. L. Kirsch, P. R. Gray, and J. D. W. Hanna, "Field-test results of the AN/GSC-10 (KATHRYN) digital data terminal", *IEEE Trans. Commun. Technol.*, vol. 17, no. 4, pp. 118–128, Apr. 1969.

383. R. R. Mosier and R. G. Clabaugh, "Kineplex, a bandwidth-efficient binary transmission system", *AIEE Trans. (Part I: Commun. and Electr.)*, vol. 76, no. 1, pp. 723–728, Jan. 1958.

384. G. Porter, "Error distribution and diversity performance of a frequency-differential PSK HF modem", *IEEE Trans. Commun. Technol.*, vol. 16, no. 4, pp. 567–575, Aug. 1968.

385. T. Keller and L. Hanzo, "Adaptive multicarrier modulation: A convenient framework for time-frequency processing in wireless communications", *IEEE Proc.*, vol. 88, no. 5, pp. 611–640, Apr. 2000.

386. R. W. Chang, "Synthesis of band-limited orthogonal signals for multichannel data transmission", *Bell Syst. Tech. J.*, vol. 45, pp. 1775–1796, Dec. 1966.

387. B. Saltzberg, "Performance of an efficient parallel data transmission system", *IEEE Trans. Commun. Technol.*, vol. 15, pp. 805–811, Dec. 1967.

388. S. B. Weinstein and P. M. Ebert, "Data transmission by frequency-division multiplexing using the discrete Fourier transform", *IEEE Trans. Commun. Technol.*, vol. 19, pp. 628–634, Oct. 1971.

389. B. Hirosaki, "An orthogonally multiplexed QAM system using the discrete Fourier transform", *IEEE Trans. Commun.*, vol. 29, no. 7, pp. 982–989, July 1981.

390. J. A. C. Bingham, "Multicarrier modulation for data transmission: An idea whose time has come", *IEEE Commun. Mag.*, vol. 28, no. 5, pp. 5–14, May 1990.

391. T. Hwang, C. Yang, G. Wu, S. Li, and G. Y. Li, "OFDM and its wireless applications: A survey", *IEEE Trans. Veh. Tech.*, vol. 58, no. 4, pp. 1673–1694, May 2009.

392. Z. Wang and G. B. Giannakis, "Wireless multicarrier communications", *IEEE Sig. Process. Mag.*, vol. 17, no. 3, pp. 29–48, May 2000.

393. K. Fazel and G. Fettweis, *Multi-Carrier Spread Spectrum*. Dordrecht: Kluwer Academic Publishers, 1997.

394. L. Hanzo, W. Webb, and T. Keller, *Single- and Multi-Carrier Quadrature Amplitude Modulation*. New York: IEEE Press/John Wiley & Sons, Inc., 1999.

395. J. S. Chow, J. C. Tu, and J. M. Cioffi, "A discrete multitone transceiver system for HDSL applications", *IEEE J. Sel. Areas Commun.*, vol. 9, pp. 895–908, Aug. 1991.

396. T. Starr, J. M. Cioffi, and P. J. Silverman, *Understanding Digital Subscriber Technology*. Upper Saddle River, NJ: Prentice Hall PTR, 1999.

397. *Multicarrier Modulation for the High-Speed Asymmetrical Modem Recommendation, CCITT WP XVIII/1 Doc. No. 87-02-013*, Telebit Corporation, Dec. 1986.

398. G. Cherubini, E. Eleftheriou, S. Ölcer, and J. M. Cioffi, "Filter bank modulation techniques for very high-speed digital subscriber lines", *IEEE Commun. Mag.*, vol. 38, no. 5, pp. 98–104, May 2000.

399. S. B. Weinstein, "The history of orthogonal frequency-division multiplexing", *IEEE Commun. Mag.*, vol. 47, no. 11, pp. 26–35, Nov. 2009.

400. W. R. Bennett and J. R. Davey, *Data Transmission*. New York: McGraw-Hill, 1965.

401. W. R. Bennett, "Statistics of regenerative digital transmission", *Bell Syst. Tech. J.*, vol. 37, pp. 1501–1542, Nov. 1958.

402. H. J. Pushman, "Spectral density distributions of signals for binary data transmission", *Radio and Electron. Engr.*, vol. 22, pp. 155–165, Feb. 1963.

403. W. Postl, "Die spectrale Leistugsdichte bei Frequenzmodulation eines Trägers mit einem stochastichen Telegraphiesignal", *Frequenz*, vol. 17, pp. 107–110, Mar. 1963.

404. W. R. Bennett and S. O. Rice, "Spectral density and autocorrelation functions associated with binary frequency-shift keying", *Bell Syst. Tech. J.*, vol. 42, pp. 2355–2385, Sep. 1963.

405. V. K. Prabhu and H. E. Rowe, "Spectra of digital phase modulation by matrix methods", *Bell Syst. Tech. J.*, vol. 53, pp. 899–935, May–June 1974.

406. H. E. Rowe and V. K. Prabhu, "Power spectrum of a of digital frequency-modulated signal", *Bell Syst. Tech. J.*, vol. 54, pp. 1095–1125, July–Aug. 1975.

407. V. K. Prabhu, "Spectral occupancy of digital angle-modulation signals", *Bell Syst. Tech. J.*, vol. 55, pp. 429–453, Apr. 1976.

408. L. J. Greenstein, "Spectra of PSK signals with overlapping baseband pulses", *IEEE Trans. Commun.*, vol. 25, no. 5, pp. 523–530, May 1977.

409. T. Aulin and C.-E. Sundberg, "Calculating digital FM spectra by means of autocorrelation", *IEEE Trans. Commun.*, vol. 30, no. 5, pp. 1199–1208, May 1982.

410. E. Biglieri and M. Visintin, "A simple derivation of the power spectrum of full-response CPM and some of its properties", *IEEE Trans. Commun.*, vol. 38, no. 3, pp. 267–269, Mar. 1990.

411. W. A. Gardner and L. Franks, "Characterization of cyclostationary random signal processes", *IEEE Trans. Inform. Theory*, vol. 21, no. 1, pp. 4–14, Jan. 1975.

412. W. A. Gardner, "Exploitation of spectral redundancy in cyclostationary signals", *IEEE Sig. Process. Mag.*, vol. 8, no. 2, pp. 14–36, Apr. 1991.

413. W. A. Gardner (ed.), *Cyclostationarity in Communications and Signal Processing*. New York: IEEE Press, 1994.

414. R. C. Titsworth and L. R. Welch, *Power Spectra of Signals Modulated by Random and Pseudorandom Sequences, Tech. Rep. 32-140*, JPL, 10 Oct. 1961.

415. P. Galko and S. Pasupathy, "The mean power spectral density of Markov chain driven signals", *IEEE Trans. Inform. Theory*, vol. 27, no. 6, pp. 746–754, Nov. 1981.

416. A. A. Abidi, "Direct-conversion radio transceivers for digital communications", *IEEE J. Solid-State Circuits*, vol. 30, no. 12, pp. 1399–1410, Dec. 1995.

417. P. H. Young, *Electonic Communication Techniques*. Upper Saddle River, NJ: Prentice Hall, 1999.

418. L. Tong, B. M. Sadler, and M. Dong, "Pilot-assisted wireless transmissions: General model, design criteria, and signal processing", *IEEE Sig. Process. Mag.*, vol. 21, no. 6, pp. 12–25, Nov. 2004.

419. N. Seshadri, "Joint data and channel estimation using blind trellis search techniques", *IEEE Trans. Commun.*, vol. 42, pp. 1000–1011, Feb./Mar./Apr. 1994.

420. M. Fu, G. Wade, J. Ning, and R. Jakobs, "On Walsh filtering method for decoding of CPM signals", *IEEE Commun. Lett.*, vol. 8, no. 6, pp. 345–347, June 2004.

421. E. D. Sunde, *Communication Systems Engineering Theory*. New York: John Wiley & Sons, Inc., 1967.

422. D. G. Forney, "Maximum-likelihood sequence estimation of digital sequences in the presence of inter-symbol interference", *IEEE Trans. Inform. Theory*, vol. 18, no. 3, pp. 363–378, May 1972.

423. I. N. Andersen, "Sample-whitened matched filters", *IEEE Trans. Inform. Theory*, vol. 19, no. 5, pp. 653–660, Sep. 1973.

424. G. E. Bottomley and S. Chennakeshu, "Unification of MLSE receivers and extension to time-varying channels", *IEEE Trans. Commun.*, vol. 46, no. 4, pp. 464–472, Apr. 1998.

425. B. D. Hart and D. P. Taylor, "Extended MLSE diversity receiver for the time- and frequency-selective channel", *IEEE Trans. Commun.*, vol. 45, no. 3, pp. 322–333, Mar. 1997.

426. K. M. Chugg and A. Polydoros, "MLSE for an unknown channel – Part I: Optimality considerations", *IEEE Trans. Commun.*, vol. 44, pp. 836–846, July 1996.

427. G. M. Vitetta, U. Mengali, and D. P. Taylor, "Double-filter differential detection of PSK signal transmitted over linearly time-selective Rayleigh fading channels", *IEEE Trans. Commun.*, vol. 47, no. 2, pp. 239–247, Feb. 1999.

428. Q. Dai and E. Shwedyk, "Detection of bandlimited signals over frequency-selective Rayleigh fading channels", *IEEE Trans. Commun.*, vol. 42, no. 2/3/4, pp. 941–950, 1994.

429. W. H. Sheen, C. C. Tseng, and C. S. Wang, "On the diversity, bandwidth and performance of digital transmission over frequency-selective slow fading channels", *IEEE Trans. Veh. Tech.*, vol. 49, no. 3, pp. 835–843, May 2000.

430. H. L. Van Trees, *Detection, Estimation and Modulation Theory, Part I*. New York: John Wiley & Sons, Inc., 1968.

431. N. J. Baas and D. P. Taylor, "Pulse shaping for wireless communication over time- or frequency-selective channels", *IEEE Trans. Commun.*, vol. 52, no. 9, pp. 477–479, Sep. 2004.

432. M. Visintin, "A simple DPSK receiver with improved asymptotic performance over fast Rayleigh-fading channels", *IEEE Commun. Lett.*, vol. 1, no. 4, pp. 99–101, July 1997.

433. _____, "Differential PSK block demodulation over a flat correlated Rayleigh-fading channel", *IEEE Trans. Commun.*, vol. 45, no. 1, pp. 9–11, Jan. 1997.

434. G. M. Vitetta and D. P. Taylor, "Multi-sampling receivers for uncoded and coded PSK signal sequences transmitted over Rayleigh frequency-flat fading channels", *IEEE Trans. Commun.*, vol. 44, pp. 130–133, Feb. 1996.

435. F. Pancaldi and G. M. Vitetta, "Block channel equalization in the frequency domain", *IEEE Trans. Commun.*, vol. 53, no. 3, pp. 463–471, Mar. 2005.

436. J. H. Lodge and M. L. Moher, "Maximum likelihood sequence estimation of CPM signals transmitted over Rayleigh flat-fading channels", *IEEE Trans. Commun.*, vol. 38, no. 6, pp. 787–794, June 1990.

437. G. M. Vitetta, U. Mengali, and D. P. Taylor, "Blind detection of CPM signals transmitted over frequency-flat fading channels", *IEEE Trans. Veh. Tech.*, vol. 47, no. 3, pp. 961–968, Nov. 1998.

438. L. Yiin and G. L. Stüber, "MLSE and soft-output equalization for trellis-coded continuous phase modulation", *IEEE Trans. Commun.*, vol. 45, no. 6, pp. 651–659, June 1997.

439. B. Özgül, M. Koca, and H. Deliç, "Doubly iterative equalization of continuous-phase modulation", *IEEE Trans. Commun.*, vol. 55, no. 11, pp. 2114–2124, Nov. 2007.

440. _____, "Doubly turbo equalization of continuous-phase modulation with frequency domain processing", *IEEE Trans. Commun.*, vol. 57, no. 2, pp. 423–429, Feb. 2009.

441. W. Van Thillo, F. Horlin, J. Nsenga, V. Ramon, A. Bourdoux, and R. Lauwereins, "Low-complexity linear frequency domain equalization for continuous phase modulation", *IEEE Trans. Wireless Commun.*, vol. 8, no. 3, pp. 1435–1445, July 2009.

442. C. H. Park, R. W. Heath, and T. S. Rappaport, "Frequency-domain channel estimation and equalization for continuous-phase modulations with superimposed pilot sequences", *IEEE Trans. Veh. Tech.*, vol. 58, no. 9, pp. 4903–4908, Nov. 2009.

443. X. Zhang and M. P. Fitz, "Space-time code design with continuous phase modulation", *IEEE J. Sel. Areas Commun.*, vol. 21, no. 5, pp. 783–792, June 2003.

444. W. Zhao and G. B. Giannakis, "Reduced complexity receivers for layered space-time CPM", *IEEE Trans. Wireless Commun.*, vol. 4, no. 2, pp. 574–582, Mar. 2005.

445. A.-M. Silvester, L. Lampe, and R. Schober, "Space-time continuous phase modulation for non-coherent detection", *IEEE Trans. Wireless Commun.*, vol. 7, no. 4, pp. 1264–1275, Apr. 2008.

446. A. G. Zajic and G. L. Stüber, "A space-time code design for CPM: Diversity order and coding gains", *IEEE Trans. Inform. Theory*, vol. 55, no. 8, pp. 3781–3798, Aug. 2009.

447. F. Pancaldi, A. Barbieri, and G. M. Vitetta, "Space-time block codes for noncoherent CPFSK", *IEEE Trans. Wireless Commun.*, vol. 9, no. 5, pp. 1729–1737, May 2006.

448. E. Chiavaccini and G. M. Vitetta, "Error performance of OFDM signalling over doubly-selective Rayleigh fading channels", *IEEE Commun. Lett.*, vol. 4, no. 11, pp. 328–330, Nov. 2000.

449. X. Cai and G. B. Giannakis, "Bounding performance and suppressing intercarrier interference in wireless mobile OFDM", *IEEE Trans. Wireless Commun.*, vol. 51, no. 12, pp. 2047–2056, Dec. 2003.

450. J. Li and M. Kavehrad, "Effects of time selective multipath fading on OFDM systems for broadband mobile applications", *IEEE Commun. Lett.*, vol. 3, no. 12, pp. 332–334, Dec. 1999.

451. W. Jeon, K. Chang, and Y. Cho, "An equalization technique for orthogonal frequency-division multiplexing systems in time-variant multipath channels", *IEEE Trans. Commun.*, vol. 47, no. 1, pp. 27–32, Jan. 1999.

452. Y. Li and L. J. Cimini, "Bounds on the interchannel interference of OFDM in time-varying impairments", *IEEE Trans. Commun.*, vol. 49, no. 3, pp. 401–404, Mar. 2001.

453. X. Huang and H. Wu, "Robust and efficient intercarrier interference mitigation for OFDM systems in time-varying fading channels", *IEEE Trans. Veh. Tech.*, vol. 56, no. 5, pp. 2517–2528, Sep. 2007.

454. A. F. Molisch, M. Toeltsch, and S. Vermani, "Iterative methods for cancellation of intercarrier interference in OFDM systems", *IEEE Trans. Veh. Tech.*, vol. 56, no. 4, pp. 2158–2167, July 2007.

455. H. Hijazi and L. Ros, "Analytical analysis of Bayesian Cramer-Rao bound for dynamical Rayleigh channel complex gains estimation in OFDM system", *IEEE Trans. Sig. Process.*, vol. 57, no. 5, pp. 1889–1900, May 2009.

456. A. Stamoulis, S. Diggavi, and N. Al-Dhahir, "Intercarrier interference in MIMO OFDM", *IEEE Trans. Sig. Process.*, vol. 50, no. 10, pp. 2451–2464, Oct. 2002.

457. L. C. Barbosa, "Maximum likelihood sequence estimators: A geometric view", *IEEE Trans. Inform. Theory*, vol. 35, pp. 419–427, Mar. 1989.

458. G. Ungerboeck, "Adaptive maximum-likelihood receiver for carrier-modulated data-transmission systems", *IEEE Trans. Commun.*, vol. 22, no. 5, pp. 624–636, May 1974.

459. Y. Nouda, T. Koike, and S. Yoshida, "Iterative MLD equalizer preceded by MIMO-FDE for wideband spatial multiplexing systems", in *Proc. 2005 IEEE 61st Veh. Technol. Conf. (VTC 2005 – Spring)*, vol. 1, Stockholm, 30 May–1 June 2005, pp. 533–537.

460. M. N. Patwary, P. B. Rapajic, and J. Choih, "Decision feedback MLSE for spatially multiplexed MIMO frequency selective fading channel", *IEE Proc.*, vol. 153, no. 1, pp. 39–48, Feb. 2006.

461. D. G. Forney, "Lower bounds on error probability in the presence of large intersymbol interference", *IEEE Trans. Commun.*, vol. 20, no. 1, pp. 76–77, Feb. 1972.

462. G. J. Foschini, "Performance bound for maximum likelihood sequence estimation of digital sequence in the presence of intersymbol interference", *IEEE Trans. Inform. Theory*, vol. 21, no. 1, pp. 47–50, Jan. 1975.

463. S. Verdu, "Maximum likelihood sequence detection for intersymbol interference channels: A new upper bound on error probability", *IEEE Trans. Inform. Theory*, vol. 33, pp. 62–68, Jan. 1987.

464. W.-H. Sheen and G. L. Stüber, "MLSE equalization and decoding for multipath fading channels", *IEEE Trans. Commun.*, vol. 39, pp. 1455–1464, Oct. 1991.

465. F. Ling, "Matched filter bound for time-discrete multipath Rayleigh fading channels", *IEEE Trans. Commun.*, vol. 43, no. 3, pp. 710–713, Feb./Mar./Apr. 1995.

466. J. E. Mazo, "Exact matched filter bound for two-beam Rayleigh fading", *IEEE Trans. Commun.*, vol. 39, pp. 1027–1030, July 1991.

467. W. C. Wong and L. Greenstein, "Multipath fading models and adaptive equalizers in microwave digital radio", *IEEE Trans. Commun.*, vol. 32, no. 8, pp. 928–934, Aug. 1984.

468. R. A. Valenzuela, "Performance of adaptive equalization for indoor radio communications", *IEEE Trans. Commun.*, vol. 37, no. 3, pp. 291–293, Mar. 1989.

469. D. Dzung and W. R. Braun, "Performance of coherent data transmission in frequency-selective Rayleigh fading channels", *IEEE Trans. Commun.*, vol. 41, no. 9, pp. 1335–1341, Sep. 1993.

470. V. P. Kaasila and A. Mämmelä, "Bit error probability of a matched filter in a Rayleigh fading multipath channel", *IEEE Trans. Commun.*, vol. 42, no. 3, pp. 826–828, Feb./Mar./Apr. 1994.

471. P. Monsen, "Digital transmission performance on fading dispersive diversity channels", *IEEE Trans. Commun.*, vol. 21, no. 1, pp. 33–39, Jan. 1973.

472. W. Burchill and C. Leung, "Matched filter bound for OFDM on Rayleigh fading channels", *Electron. Lett.*, vol. 31, no. 20, pp. 1716–1717, Sep. 1995.

473. T. Hunziker and D. Dahlhaus, "Bounds on matched filter performance in doubly dispersive Gaussian WSSUS channels", *Electron. Lett.*, vol. 37, no. 6, pp. 383–384, Mar. 15 2001.

474. K. W. Yip and T. S. Ng, "Matched filter bound for multipath Ricean fading channels", *IEEE Trans. Commun.*, vol. 46, no. 4, pp. 441–445, Apr. 1998.

475. R. Visoz and E. Bejjani, "Matched filter bound for multichannel diversity over frequency-selective Rayleigh-fading mobile channels", *IEEE Trans. Veh. Tech.*, vol. 49, no. 5, pp. 1832–1845, Aug. 2000.

476. S. Roy and D. D. Falconer, "The matched-filter bound on optimal space-time processing in correlated fading channels", *IEEE Trans. Wireless Commun.*, vol. 3, no. 6, pp. 2156–2169, Nov. 2004.

477. E. Chiavaccini and G. M. Vitetta, "Error performance of matched and partially matched one-shot detectors for doubly selective Rayleigh fading channels", *IEEE Trans. Commun.*, vol. 49, no. 10, pp. 1738–1747, Oct. 2001.

478. A. M. Tonello, "Performance limits for filtered multitone modulation in fading channels", *IEEE Trans. Inform. Theory*, vol. 4, no. 5, pp. 2121–2135, Sept. 2005.

479. M. Schnell and I. Cosovic, "Matched-filter bounds for multicarrier communications systems with transmit diversity", *IEEE Trans. Commun.*, vol. 54, no. 10, pp. 1870–1877, Oct. 2006.

480. A. Mämmelä and D. P. Taylor, "Bias terms in the optimal quadratic receiver", *IEEE Commun. Lett.*, vol. 2, no. 2, pp. 57–58, Feb. 1998.

481. R. Price, "The detection of signals perturbed by scatter and noise", *IRE Trans. Inform. Theory*, vol. 4, pp. 163–170, Sept. 1954.

482. ——, "Optimum detection of random signals in noise with application to scatter-multipath communications – I", *IRE Trans. Inform. Theory*, vol. 6, pp. 125–135, Dec. 1956.

483. D. Middleton, "On the detection of stochastic signals in additive normal noise – Part 1", *IRE Trans. Inform. Theory*, vol. 3, pp. 86–121, June 1957.

484. T. Kailath, "Correlation detection of signals perturbed by a random channel", *IRE Trans. Inform. Theory*, vol. 6, pp. 361–366, June 1960.

485. F. C. Schweppe, "Evaluation of likelihood functions for Gaussian signals", *IRE Trans. Inform. Theory*, vol. 11, pp. 61–70, Jan. 1965.

486. T. Kailath, "Likelihood ratios for Gaussian processes", *IEEE Trans. Inform. Theory*, vol. 16, pp. 276–287, May 1970.

487. P. Y. Kam and C. H. Teh, "Reception of PSK signals over fading channels via quadrature amplitude estimation", *IEEE Trans. Commun.*, vol. 31, no. 8, pp. 1024–1027, Aug. 1983.

488. ____, "An adaptive receiver with memory for slowly fading channels", *IEEE Trans. Commun.*, vol. 32, no. 6, pp. 654–659, June 1984.

489. W. C. Dam and D. P. Taylor, "An adaptive maximum likelihood receiver for correlated Rayleigh-fading channels", *IEEE Trans. Commun.*, vol. 42, pp. 2684–2692, Sep. 1994.

490. G. M. Vitetta and D. P. Taylor, "Maximum likelihood decoding of uncoded and coded PSK signal sequences transmitted over Rayleigh flat-fading channels", *IEEE Trans. Commun.*, vol. 43, no. 11, pp. 2750–2758, Nov. 1995.

491. ____, "Maximum likelihood sequence estimation of differentially encoded PSK signals transmitted over Rayleigh frequency-flat fading channels", *Int. J. Wireless Inform. Networks*, vol. 2, pp. 71–81, Apr. 1995.

492. X. Yu and S. Pasupathy, "Innovations-based MLSE for Rayleigh fading channels", *IEEE Trans. Comm*, vol. 43, pp. 1534–1544, Feb./Mar./Apr. 1995.

493. ____, "Error performance of innovations-based MLSE for Rayleigh fading channels", *IEEE Trans. Veh. Tech.*, vol. 45, pp. 631–642, Nov. 1996.

494. B. D. Hart, D. K. Borah, and S. Pasupathy, "Autocovariance preserving estimator (APE) interpretation of the MLSD metric for Rayleigh fading channels", *IEEE Trans. Commun.*, vol. 48, no. 10, pp. 1614–1617, Oct. 2000.

495. B. D. Hart and D. P. Taylor, "Maximum-likelihood synchronization, equalization, and sequence estimation for unknown time-varying frequency-selective Rician channels", *IEEE Trans. Commun.*, vol. 46, pp. 211–221, Feb. 1998.

496. T. E. Duncan, "Evaluation of likelihood functions", *Information & Control*, vol. 13, pp. 62–74, July 1968.

497. T. Kailath, "A general likelihood-ratio formula for random signals in Gaussian noise", *IEEE Trans. Inform. Theory*, vol. 15, pp. 350–361, May 1969.

498. ____, "A further note on a general likelihood formula for random signals in Gaussian noise", *IEEE Trans. Inform. Theory*, vol. 16, pp. 393–396, July 1970.

499. ____, "Optimum receivers for randomly varying channels", in *Proc. Fourth London Symp. Inform. Theory*. London: Butterworth Scientific Press, 1961, pp. 109–122.

500. G. Colavolpe, P. Castoldi, and R. Raheli, "Linear predictive receivers for fading channels", *Electron. Lett.*, vol. 34, no. 13, pp. 1289–1290, June 1998.

501. D. Makraris, P. T. Mathiopoulos, and D. P. Bouras, "Optimal decoding of coded PSK and QAM signals in correlated fast fading channels and AWGN: A combined envelope, multiple differential and coherent detection approach", *IEEE Trans. Commun.*, vol. 42, pp. 63–74, Jan. 1994.

502. C. W. Therrien, *Discrete Random Signals and Statistical Signal Processing*. Englewood Cliffs, NJ: Prentice Hall, 1992.

503. S. Haykin, *Adaptive Filter Theory*, 2nd ed. Englewood Cliffs, NJ: Prentice Hall, 1991.

504. B. D. Hart and S. Pasupathy, "Innovations-based MAP detection for time-varying frequency-selective channels", *IEEE Trans. Commun.*, vol. 48, no. 9, pp. 1507–1519, Sep. 2000.

505. T. Kailath, "Measurements on time-variant communication channels", *IRE Trans. Inform. Theory*, vol. 8, no. 5, pp. 229–236, Sept. 1962.

506. ____, "Time variant communication channels", *IRE Trans. Inform. Theory*, vol. 9, pp. 233–237, October 1963.

507. P. A. Bello, "Measurement of random time-variant channels", *IEEE Trans. Inform. Theory*, vol. 15, pp. 469–475, July 1969.

508. P. A. Bello and R. Esposito, "Measurement techniques for time-varying dispersive channels", *Alta Frequenza*, vol. XXXIX, pp. 980–996, Nov. 1970.

509. K. Pahlavan and J. W. Matthews, "Performance of adaptive matched filter receivers over fading multipath channels", *IEEE Trans. Commun.*, vol. 38, pp. 2106–2113, Dec. 1990.

510. A. Wautier, J.-C. Dany, C. Mourot, and V. Kumar, "A new method for predicting the channel estimate influence on performance of TDMA mobile radio systems", *IEEE Trans. Veh. Tech.*, vol. 44, pp. 594–602, Aug. 1995.

511. M.-C. Chiu and C.-C. Chao, "Analysis of LMS-adaptive MLSE equalization on multipath fading channels", *IEEE Trans. Commun.*, vol. 44, pp. 1684–1692, Dec. 1996.

512. K. M. Chugg and A. Polydoros, "MLSE for an unknown channel – Part II: Tracking performance", *IEEE Trans. Commun.*, vol. 44, pp. 949–958, Aug. 1996.

513. P. Y. Kam and H. M. Ching, "Sequence estimation over the slow nonselective Rayleigh fading channel with diversity reception and its application to Viterbi decoding", *IEEE J. Sel. Areas Commun.*, vol. 10, no. 3, pp. 562–570, Apr. 1992.

514. P. K. Varshney and A. H. Haddad, "A receiver with memory for fading channels", *IEEE Trans. Commun.*, vol. 26, no. 2, pp. 278–283, Feb. 1978.

515. R. Raheli, A. Polydoros, and C. K. Tzou, "Per-survivor processing: A general approach to MLSE in uncertain environment", *IEEE Trans. Commun.*, vol. 43, no. 2, pp. 354–364, Feb./Mar./Apr. 1995.

516. C. N. Georghiades and J. C. Han, "Sequence estimation in the presence of random parameters via the EM algorithm", *IEEE Trans. Commun.*, vol. 45, no. 3, pp. 300–308, Mar. 1997.

517. T. K. Moon, "The expectation-maximization algorithm", *IEEE Sig. Process. Mag.*, vol. 13, no. 6, pp. 47–60, Nov. 1996.

518. P. Ho and D. Fung, "Error performance of multiple-symbol differential detection of PSK signals transmitted over correlated Rayleigh fading channels", *IEEE Trans. Commun.*, vol. 40, pp. 25–29, October 1992.

519. H. Chen, R. Perry, and K. Buckley, "On MLSE algorithms for unknown fast time-varying channels", *IEEE Trans. Commun.*, vol. 51, no. 5, pp. 730–734, May 2004.

520. G. K. Kaleh and R. Vallet, "Joint parameter estimation and symbol detection for linear or nonlinear unknown channels", *IEEE Trans. Commun.*, vol. 42, no. 7, pp. 2406–2413, July 1994.

521. C. A. Haro, J. A. R. Fonollosa, and J. Fonollosa, "Blind channel estimation and data detection using hidden Markov models", *IEEE Trans. Sig. Process.*, vol. 45, no. 1, pp. 241–247, Jan. 1997.

522. Y. Xie and C. N. Georghiades, "Two EM-type channel estimation algorithms for OFDM with transmitter diversity", *IEEE Trans. Commun.*, vol. 51, no. 1, pp. 106–115, Jan. 2003.

523. T. Kashima, K. Fukawa, and H. Suzuki, "Adaptive MAP receiver via the EM algorithm and message passings for MIMO-OFDM mobile communications", *IEEE J. Sel. Areas Commun.*, vol. 24, no. 3, pp. 437–447, July 2006.

524. B. Lu, X. Wang, and Y. G. Li, "Iterative receivers for space-time block coded OFDM systems in dispersive fading channels", *IEEE Trans. Wireless Commun.*, vol. 1, no. 2, pp. 213–225, Apr. 2002.

525. C. W. Helstrom, *Elements of Signal Detection & Estimation*. Englewood Cliffs, NJ: Prentice Hall, 1995.

526. E. Chiavaccini and G. M. Vitetta, "MAP symbol estimation on frequency-flat Rayleigh fading channels via a Bayesian EM algorithm", *IEEE Trans. Commun.*, vol. 49, no. 11, pp. 1869–1872, Nov. 2001.

527. A. Gelman, J. B. Carlin, H. S. Stern, and D. B. Rubin, *Bayesian Data Analysis*. London: Chapman & Hall, 1995.

528. A. S. Gallo and G. M. Vitetta, "Soft-in soft-output detection in the presence of parametric uncertainty via the Bayesian EM algorithm", *EURASIP J. Wireless Commun. Netw.*, no. 2, pp. 100–116, Apr. 2005.

529. E. Chiavaccini and G. M. Vitetta, "A BEM-based detector for CPM signals transmitted over frequency-flat fading channels", *IEEE Trans. Wireless Commun.*, vol. 2, no. 3, pp. 409–412, May 2003.

530. L. Davis, I. Collings, and P. Hoeher, "Joint MAP equalization and channel estimation for frequency-selective and frequency-flat fast-fading channels", *IEEE Trans. Commun.*, vol. 49, no. 12, pp. 2106–2114, Dec. 2001.

531. P. Hoeher and J. Lodge, "Turbo DPSK: Iterative differential DPSK demodulation and channel decoding", *IEEE Trans. Commun.*, vol. 47, no. 6, pp. 837–843, June 1999.

532. I. D. Marsland and P. T. Mathiopoulos, "Multiple differential detection of parallel concatenated convolutional (turbo) codes in correlated fast Rayleigh fading", *IEEE J. Sel. Areas Commun.*, vol. 16, no. 2, pp. 265–274, Feb. 1998.

533. M. J. Gertsman and J. H. Lodge, "Symbol-by-symbol MAP demodulation of CPM and PSK signals on Rayleigh flat-fading channels", *IEEE Trans. Commun.*, vol. 45, pp. 788–799, July 1997.

534. A. S. Gallo, G. M. Vitetta, and E. Chiavaccini, "A BEM-based algorithm for soft-in soft-output detection of co-channel signals", *IEEE Trans. Wireless Commun.*, vol. 3, no. 5, pp. 1533–1542, Sep. 2004.

535. A. S. Gallo, E. Chiavaccini, F. Muratori, and G. M. Vitetta, "BEM-based SISO detection of orthogonal space-time block codes over frequency flat-fading channels", *IEEE Trans. Wireless Commun.*, vol. 3, no. 6, pp. 1885–1889, Apr. 2004.

536. B. Lu, X. Wang, and K. R. Narayanan, "LDPC-based space-time coded OFDM systems over correlated fading channels: Performance analysis and receiver design", *IEEE Trans. Commun.*, vol. 50, no. 1, pp. 74–88, Jan. 2002.

537. G. J. McLachlan and T. Krishnan, *The EM Algorithm and Extensions*. New York: John Wiley & Sons, Inc., 1997.

538. C. F. J. Wu, "On the convergence properties of the EM algorithm", *Ann. Statist.*, vol. 11, no. 2, pp. 95–103, Ma, 1983.

539. A. P. Dempster, N. M. Laird, and D. B. Rubin, "Maximum likelihood from incomplete data via the EM algorithm", *J. Royal Statist. Soc., Ser. B*, vol. 39, no. 1, pp. 1–38, 1977.

540. J. A. Fessler and A. O. Hero, "Space-alternating generalized expectation-maximization algorithm", *IEEE Trans. Sig. Process.*, vol. 42, no. 10, pp. 2664–2677, Oct. 1994.

541. C. Herzet, A. Renaux, and L. Vandendorpe, "A Cramer-Rao bound characterization of the EM-algorithm mean spedd of convergence", *IEEE Trans. Sig. Process.*, vol. 56, no. 6, pp. 2218–2228, June 2008.

542. L. Xu and M. I. Jordan, "On the convergence properties of the EM algorithm for Gaussian mixtures", *Neural Computation*, vol. 8, no. 1, pp. 129–151, 1996.

543. X. L. Meng and D. B. Rubin, "Maximum likelihood via the ECM algorithm: a general framework", *Biometrika*, vol. 80, no. 2, pp. 267–278, 1993.

544. C. Liu and D. B. Rubin, "The ECME algorithm: a simple extension of EM and ECM with faster monotone convergence", *Biometrika*, vol. 81, no. 4, pp. 633–648, 1994.

545. X. L. Meng and D. van Dyk, *The EM algorithm – An old folk song sung to a fast new tune*, Technical Report No. 408, Department of Statistics, University of Chicago, Chicago, Apr. 1953.

546. L. B. Nelson and H. V. Poor, "Iterative multiuser receivers for CDMA channels: an EM-based approach", *IEEE Trans. Commun.*, vol. 44, no. 12, pp. 1700–1710, Dec. 1996.

547. H. Dogan, E. Panayirci, and H. V. Poor, "Low-complexity joint data detection and channel equalisation for highly mobile orthogonal frequency division multiplexing systems", *IET Communications*, vol. 4, no. 8, pp. 1000–1011, May 2010.

548. E. Panayirci, H. Şenol, and H. V. Poor, "Joint channel estimation, equalization, and data detection for OFDM systems in the presence of very high mobility", *IEEE Trans. Sig. Process.*, vol. 58, no. 8, pp. 4225–4238, Aug. 2010.

549. D. O. North, "Analysis of factors which determine signal-to-noise ratio discrimination in radar", in *Report PTR-6c*, RCA Laboratories, Princeton, NJ, June 1943.

550. J. H. Van Vleck and D. Middleton, "A theoretical comparison of the visual, aural, and meter reception of pulsed signals in the presence of noise", *J. Appl. Phys.*, vol. 17, no. 11, pp. 940–971, 1946.

551. T. Ericson, "Structure of optimum receiving filters in data transmission systems", *IEEE Trans. Inform. Theory*, vol. 17, no. 3, pp. 352–353, May 1971.

552. N. Souto, R. Dinis, and J. Silva, "Analytical matched filter bound for M-QAM hierarchical constellations with diversity reception in multipath Rayleigh fading channels", *IEEE Trans. Commun.*, vol. 58, no. 3, pp. 737–741, Mar. 2010.

553. T. Kailath, "Adaptive matched filters", in *Mathematical Optimization Techniques*, R. Bellman, (ed.) Berkeley: University of California Press, 1963, pp. 109–140.

554. W. Van Etten, "Maximum likelihood receiver for multiple channel transmission systems", *IEEE Trans. Commun.*, vol. 24, no. 2, pp. 276–283, Feb. 1976.

555. G. E. Bottomley and S. Chennakeshu, "Adaptive MLSE equalization forms for wireless communications", in *Fifth Virginia Tech. Symp. Wireless Personal Commun.*, Blacksburg, VA, 31 May–2 June 1995, pp. 183–194.

556. B. D. Hart and D. P. Taylor, "Extended MLSE receiver for the frequency-flat, fast fading channel", *IEEE Trans. Veh. Tech.*, vol. 46, pp. 381–389, May 1997.

557. W. Sung and I.-K. Kim, "An MLSE receiver using channel classification for Rayleigh fading channels", *IEEE J. Sel. Areas Commun.*, vol. 18, no. 11, pp. 2336–2344, July 1977.

558. I. Barhumi and M. Moonen, "MLSE and MAP equalization for transmission over doubly selective channels", *IEEE Trans. Veh. Tech.*, vol. 58, no. 8, pp. 4120–4128, June 2009.

559. G. Castellini, E. Del Re, and L. Perucci, "A continuously adaptive MLSE receiver for mobile communications: Algorithm and performance", *IEEE Trans. Commun.*, vol. 45, no. 1, pp. 80–89, Jan. 1997.

560. R. D'Avella, L. Moreno, and M. Sant'Agostino, "An adaptive MLSE receiver for TDMA digital mobile radio", *IEEE J. Sel. Areas Commun.*, vol. 7, pp. 122–129, Jan. 1989.

561. C. S. Bontu, D. D. Falconer, and L. Strawczynski, "Diversity transmission and adaptive MLSE for digital cellular radio", *IEEE Trans. Veh. Tech.*, vol. 48, no. 5, pp. 1488–1502, Sep. 1999.

562. R. E. Morley and D. L. Snyder, "Maximum likelihood sequence estimation for randomly dispersive channels", *IEEE Trans. Commun.*, vol. 27, pp. 833–839, June 1979.

563. R. A. Iltis, "A Bayesian maximum-likelihood sequence estimation algorithm for a priori unknown channels and symbol timing", *IEEE J. Sel. Areas Commun.*, vol. 10, pp. 579–588, Apr. 1992.

564. H. Kubo, K. Murakami, and T. Fujino, "An adaptive maximum-likelihood sequence estimator for fast time-varying intersymbol interference channels", *IEEE Trans. Commun.*, vol. 42, no. 2/3/4, pp. 1972–1880, Feb./Mar./Apr. 1994.

565. ———, "Adaptive maximum-likelihood sequence estimation by means of combined equalization and decoding in fading environments", *IEEE J. Sel. Areas Commun.*, vol. 13, no. 1, pp. 102–108, Jan. 1995.

566. M. J. Omidi, P. Gulak, and S. Pasupathy, "Parallel structures for joint channel estimation and data detection over fading channels", *IEEE J. Sel. Areas Commun.*, vol. 16, no. 9, pp. 1616–1629, Dec. 1998.

567. J.-T. Chen and Y.-C. Wang, "Adaptive MLSE equalizers with parametric tracking for multipath fast-fading channels", *IEEE Trans. Commun.*, vol. 49, no. 4, pp. 655–663, Apr. 2001.

568. N. J. Baas and D. P. Taylor, "Adaptive MLSE for DPSK in time- and frequency-selective channels", *IEEE Trans. Commun.*, vol. 56, no. 9, pp. 1478–1486, Sep. 2008.

569. R. Raheli, G. Marino, and P. Castoldi, "Per-survivor processing and tentative decisions: What is in between?" *IEEE Trans. Commun.*, vol. 44, pp. 127–129, Feb. 1996.

570. T. Kailath, "The innovations approach to detection and estimation theory", *IEEE Proc.*, vol. 58, no. 5, pp. 680–695, May 1970.

571. P. Metford, S. Haykin, and D. P. Taylor, "An innovations approach to discrete-time detection theory (corresp.)", *IEEE Trans. Inform. Theory*, vol. 28, no. 2, pp. 376–380, Mar. 1982.

572. B. D. Hart and D. P. Taylor, "MLSE for correlated diversity sources and unknown time-varying frequency-selective Rayleigh-fading channels", *IEEE Trans. Commun.*, vol. 46, no. 2, pp. 169–172, Feb. 1998.

573. T. Vaidis and C. L. Weber, "Block adaptive techniques for channel identification and data demodulation over band-limited channels", *IEEE Trans. Commun.*, vol. 46, no. 2, pp. 232–243, Feb. 1998.

574. S. Chen and Y. Wu, "Maximum likelihood joint channel and data estimation using genetic algorithms", *IEEE Trans. Sig. Process.*, vol. 46, no. 5, pp. 1469–1473, May 1998.

575. K. M. Chugg, "Blind acquisition characteristics of PSP-based sequence detectors", *IEEE J. Sel. Areas Commun.*, vol. 16, pp. 1518–1529, Oct. 1998.

576. O. Coskun and K. M. Chugg, "Combined coding and training for unknown ISI channels", *IEEE Trans. Commun.*, vol. 53, no. 8, p. 1310–1322, Aug. 2005.

577. I. Motedayen-Aval, A. Krishnamoorthy, and A. Anastasopoulos, "Optimal joint detection/estimation in fading channels with polynomial complexity", *IEEE Trans. Inform. Theory*, vol. 53, no. 1, pp. 209–223, Jan. 2007.

578. Y. Kopsinis and S. Theodoridis, "An efficient low-complexity technique for MLSE equalizers for linear and nonlinear channels", *IEEE Trans. Sig. Process.*, vol. 51, no. 12, pp. 3236–3248, Dec. 2003.

579. A. Dogandzic, W. Mo, and Z. Wang, "Semi-blind SIMO flat-fading channel estimation in unknown spatially correlated noise using the EM algorithm", *IEEE Trans. Sig. Process.*, vol. 52, no. 6, pp. 1791–1797, June 2004.

580. H. Nguyen and B. C. Levy, "Blind equalization of dispersive fast fading Ricean channels via the EMV algorithm", *IEEE Trans. Veh. Tech.*, vol. 54, no. 5, pp. 1793–1801, Sep. 2005.

581. H. Zamiri-Jafarian and S. Pasupathy, "Adaptive MLSDE using the EM algorithm", *IEEE Trans. Commun.*, vol. 47, no. 8, pp. 1181–1193, Aug. 1999.

582. X. Zhu and R. D. Murch, "Performance analysis of maximum likelihood detection in a MIMO antenna system", *IEEE Trans. Commun.*, vol. 50, no. 2, pp. 187–191, Feb. 2002.

583. W. Younis and N. Al-Dhahir, "Joint prefiltering and MLSE equalization of space-time-coded transmissions over frequency-selective channels", *IEEE Trans. Veh. Tech.*, vol. 51, no. 1, pp. 144–154, Jan 2002.

584. R. W. Chang and J. C. Hancock, "On receiver structures for channel having memory", *IEEE Trans. Inform. Theory*, vol. 12, pp. 463–468, Oct. 1966.

585. D. Dzung and W. R. Braun, "Performance of coherent data transmission in frequency-selective Rayleigh fading channels", *IEEE Trans. Commun.*, vol. 41, no. 9, pp. 1335–1341, Sep. 1993.

586. H. Meyr, M. Oerder, and A. Polydoros, "On sampling rate, analog prefiltering, and sufficient statistics for digital receivers", *IEEE Trans. Commun.*, vol. 42, no. 12, pp. 3208–3214, Dec. 1994.

587. A. Hafeez and W. E. Stark, "Decision feedback sequence estimation for unwhitened ISI channels with applications to multiuser detection", *IEEE J. Sel. Areas Commun.*, vol. 16, no. 9, pp. 1785–1795, Dec. 1998.

588. H. K. Sim and D. G. Cruickshank, "A chip-based multiuser detector for the downlink of a DS-CDMA system using a folded state-transition trellis", *IEEE Trans. Wireless Commun.*, vol. 49, no. 7, pp. 1259–1267, July 2001.

589. C. Schlegel, P. D. A. S. Roy, and Z.-J. Xiang, "Multiuser projection receivers", *IEEE J. Sel. Areas Commun.*, vol. 14, no. 8, pp. 1610–1618, Oct. 1996.

590. H. Yoshino, K. Fukawa, and H. Suzuki, "Interference canceling equalizer (ICE) for mobile radio communication", *IEEE Trans. Veh. Tech.*, vol. 46, no. 4, pp. 849–861, Nov. 1997.

591. R. Krenz and K. Wesołowski, "Comparative study of space-diversity techniques for MLSE receivers in mobile radio", *IEEE Trans. Veh. Tech.*, vol. 46, no. 3, pp. 653–663, Aug. 1997.

592. J. V. M. A. Lagunas and A. I. Pérez-Neira, "Joint array combining and MLSE for single-user receivers in multipath Gaussian multiuser channels", *IEEE J. Sel. Areas Commun.*, vol. 18, no. 11, pp. 2252–2259, Nov. 2008.

593. N. Merhav, G. Kaplan, A. Lapidoth, and S. Shamai, "On information rates for mismatched decoders", *IEEE Trans. Inform. Theory*, vol. 40, no. 6, pp. 1953–1967, Nov. 1994.

594. A. Lapidoth and P. Narayan, "Reliable communication under channel uncertainty", *IEEE Trans. Inform. Theory*, vol. 44, no. 6, pp. 2148–2177, Oct. 1998.

595. ETSI EN 300 744 V1.4.1 (2001-01), *Digital Video Broadcasting (DVB); Framing structure, channel coding and modulation for digital terrestrial television*, European Telecommunications Standards Institute, Jan. 1991.

596. S. M. Kay, *Fundamentals of Statistical Signal Processing: Estimation Theory*. Englewood Cliffs, NJ: Prentice Hall International, 1993.

597. T. Kailath, A. H. Sayed, and B. Hassibi, *Linear Estimation*. Upper Saddle River, NJ: Prentice Hall, 2000.

598. J. G. Proakis, *Digital Communications*, 2nd ed. New York: McGraw-Hill, 1989.

599. D. D. Falconer and L. Ljung, "Application of fast Kalman estimation to adaptive equalisation", *IEEE Trans. Commun.*, vol. 26, pp. 1439–1446, Oct. 1978.

600. E. H. Satorius and J. D. Pack, "Application of least squares lattice algorithms to adaptive equalisation", *IEEE Trans. Commun.*, vol. 29, pp. 136–142, Feb. 1981.

601. B. Widrow, J. McCool, M. Larimore, and C. Johnson, "Stationary and nonstationary learning characteristics of the LMS adaptive filter", *Proc. IEEE*, vol. 64, pp. 1151–1162, Aug. 1976.

602. S. B. Gelfand, Y. Wei, and J. V. Krogmeier, "The stability of variable step-size LMS algorithms", *IEEE Trans. Sig. Process.*, vol. 47, no. 12, pp. 3277–3288, Dec. 1999.

603. J. Homer, "Detection guided NLMS estimation of sparsely parametrized channels", *IEEE Trans. Circ. and Syst. II*, vol. 47, no. 12, pp. 1437–1442, Dec. 2000.

604. Y. Wei, S. B. Gelfand, and J. V. Krogmeier, "Noise-constrained least mean square algorithm", *IEEE Trans. Sig. Process.*, vol. 49, no. 9, pp. 1961–1970, Sep. 2001.

605. A. M. A. Filho, E. L. Pinto, and J. F. Galdino, "Variable step-size LMS algorithm for estimation of time-varying and frequency-selective channels", *Electron. Lett.*, vol. 40, no. 20, pp. 1312–1313, Sep. 2004.

606. E. Eleftheriou and D. D. Falconer, "Tracking properties and steady-state performance of RLS adaptive filter algorithms", *IEEE Trans. Acoust. Speech and Sig. Process.*, vol. 34, pp. 1097–1109, Oct. 1986.

607. F. Ling and J. G. Proakis, "Nonstationary learning characteristics of least squares adaptive estimation algorithms", in *Proc. IEEE Int. Conf. Acoust., Speech and Sig. Process. ICASSP'84*, vol. 9, San Diego, CA, Mar. 1984, pp. 118–121.

608. J. W. M. Bergmans, "Tracking capabilities of the LMS adaptive filter in the presence of gain variation", *IEEE Trans. Acoust., Speech and Sig. Process.*, vol. 38, no. 4, pp. 712–714, Apr. 1990.

609. E. Eweda, "Comparison of RLS, LMS, and sign algorithms for tracking randomly time-varying channels", *IEEE Trans. Sig. Process.*, vol. 42, no. 11, pp. 13–20, Nov. 1994.

610. J. Lin, J. G. Proakis, F. Ling, and H. Lev-Ari, "Optimal tracking of time-varying channels: A frequency domain approach for known and new algorithms", *IEEE J. Sel. Areas Commun.*, vol. 13, no. 1, pp. 141–154, Jan. 1995.

611. O. Macchi, "Optimisation of adaptive identification for time-varying filters", *IEEE Trans. Auto. Contr.*, vol. 31, pp. 283–287, Mar. 1996.

612. S. Haykin, A. H. Sayed, J. R. Zeidler, P. Yee, and P. C. Wei, "Adaptive tracking of linear time-variant systems by extended RLS algorithms", *IEEE Trans. Sig. Process.*, vol. 45, no. 5, pp. 1118–1128, May 1997.

613. W. S. Leon and D. P. Taylor, "Steady-state tracking analysis of the RLS algorithm for time-varying channels: a general state-space approach", *IEEE Commun. Lett.*, vol. 7, no. 5, pp. 236–238, May 2003.

614. R. Nadakuditi and J. C. Preisig, "A channel subspace post-filtering approach to adaptive least-squares estimation", *IEEE Trans. Sig. Process.*, vol. 52, no. 7, pp. 1901–1914, July 2004.

615. J. F. Galdino, E. L. Pinto, and M. S. de Alencar, "Analytical performance of the LMS algorithm on the estimation of wide sense stationary channels", *IEEE Trans. Commun.*, vol. 52, no. 6, pp. 982–991, June 2004.

616. D. K. Borah and B. D. Hart, "Frequency-selective fading channel estimation with a polynomial time-varying channel model", *IEEE Trans. Commun.*, vol. 47, no. 6, pp. 862–871, June 1999.

617. ——, "Receiver structures for time-varying frequency-selective fading channels", *IEEE J. Sel. Areas Commun.*, vol. 17, no. 11, pp. 1863–1875, Nov. 1999.

618. T. K. Akino, "Optimum-weighted RLS channel estimation for rapid fading MIMO channels", *IEEE Trans. Wireless Commun.*, vol. 7, no. 11, pp. 4248–4260, Mar. 2008.

619. G. Long, F. Ling, and J. G. Proakis, "The LMS algorithm with delayed coefficient adaptation", *IEEE Trans. Sig. Process.*, vol. 37, pp. 1397–1405, Sept. 1989.

620. B. D. O. Anderson and J. B. Moore, *Optimal Filtering*. Englewood Cliffs, NJ: Prentice Hall, 1979.

621. J. M. Mendel, *Lessons in Digital Estimation Theory*. Englewood Cliffs, NJ: Prentice Hall, 1987.

622. F. R. Magee and J. G. Proakis, "Adaptive maximum-likelihood sequence estimation for digital signalling in the presence of intersymbol interference", *IEEE Trans. Inform. Theory*, vol. 19, pp. 120–124, Jan. 1973.

623. S. U. H. Qureshi and E. E. Newhall, "An adaptive receiver for data transmission over time-dispersive channels", *IEEE Trans. Inform. Theory*, vol. 19, pp. 448–457, July 1973.

624. K. A. Hamied and G. L. Stüber, "An adaptive truncated MLSE receiver for Japanese personal digital cellular", *IEEE Trans. Veh. Tech.*, vol. 45, no. 1, pp. 41–50, Feb. 1996.

625. U. Vilaipornsawai and H. Leib, "Data detection and Kalman estimation for multiple space-time trellis codes", *IEEE Trans. Wireless Commun.*, vol. 57, no. 9, pp. 2697–2712, Sep. 2009.

626. A. van den Bos, "A Cramer-Rao lower bound for complex parameters", *IEEE Trans. Sig. Process.*, vol. 42, no. 10, p. 2859, Oct. 1994.

627. E. de Carvalho and D. T. M. Slock, "Cramer-Rao bounds for semi-blind, blind and training sequence based channel estimation", in *Proc. IEEE Workshop Sig. Process. Adv. Wireless Commun. (SPAWC'97)*, vol. 2, Paris, 16–18 Apr. 1997, pp. 129–132.

628. M. Dong and L. Tong, "Optimal design and placement of pilot symbols for channel estimation", *IEEE Trans. Sig. Process.*, vol. 50, no. 12, pp. 3055–3069, Dec. 2002.

629. E. de Carvalho and D. T. M. Slock, "Semi-blind methods for FIR multichannel estimation", in *Signal Processing Advances in Wireless and Mobile Communications, Volume 1*, G. B. Giannakis, P. Stoica, Y. Hua, and L. Tong, (eds.) Englewood Cliffs, NJ: Prentice Hall, 2000, pp. 211–254.

630. P. Stoica and B. C. Ng, "Performance bounds for blind channel estimation", in *Signal Processing Advances in Wireless and Mobile Communications, Volume 1*, G. B. Giannakis, P. Stoica, Y. Hua, and L. Tong, (eds.) Englewood Cliffs, NJ: Prentice Hall, 2000, pp. 41–62.

631. Y. Hua, "Fast maximum likelihood for blind identification of multiple FIR channels", *IEEE Trans. Sig. Process.*, vol. 44, no. 3, pp. 661–672, Mar. 1996.

632. C. Becchetti, G. Scarano, and G. Jacovitti, "Fisher information analysis for blind channel estimation based on second order statistics", in *Proc. Ninth IEEE SP Workshop on Statistical Signal and Array Processing*, Portland, OR, 14–16 Sep. 1998, pp. 312–315.

633. A. N. D'Andrea, U. Mengali, and R. Reggiannini, "The modified Cramer-Rao bound and its application to synchronization problems", *IEEE Trans. Commun.*, vol. 42, no. 2/3/4, pp. 1391–1399, Feb.–Apr. 1994.

634. F. Gini, R. Reggiannini, and U. Mengali, "The modified Cramer-Rao bound in vector parameter estimation", *IEEE Trans. Commun.*, vol. 46, no. 1, pp. 52–60, Jan. 1998.

635. B. Z. Bobrovsky, E. Mayer-Wolf, and M. Zakai, "Some classes of global Cramer-Rao bounds", *Ann. Statist.*, vol. 15, no. 4, pp. 1421–1438, 1987.

636. R. Miller and C. Chang, "A modified Cramer-Rao bound and its applications", *IEEE Trans. Inform. Theory*, vol. 24, no. 3, pp. 398–400, May 1978.

637. F. Gini, "A radar application of a modified Cramer-Rao bound: Parameter estimation in non-Gaussian clutter", *IEEE Trans. Sig. Process.*, vol. 46, no. 7, pp. 1945–1953, July 1998.

638. M. Moeneclaey, "On the true and the modified Cramer-Rao bounds for the estimation of a scalar parameter in the presence of nuisance parameter", *IEEE Trans. Commun.*, vol. 46, no. 11, pp. 1536–1544, Nov. 1998.

639. A. Bhattacharyya, "On some analogues of the amount of information and their use in statistical estimation", *Sankhya Indian J. Statist.*, vol. 8, pp. 1–14, 201–218, 315–328, 1946.

640. D. A. S. Fraser and I. Guttman, "Bhattacharyya bounds without regularity assumptions", *Ann. Math. Statist.*, vol. 23, no. 4, pp. 629–632, 1952.

641. E. W. Barankin, "Locally best unbiased estimates", *Ann. Math. Statist.*, vol. 20, pp. 477–501, Nov. 1949.

642. B. Bobrovsky and M. Zakai, "A lower bound on the estimation error for certain diffusion processes", *IEEE Trans. Inform. Theory*, vol. 22, no. 1, pp. 45–52, Jan. 1976.

643. A. Weiss and E. Weinstein, "A lower bound on the mean-square error in random parameter estimation", *IEEE Trans. Inform. Theory*, vol. 31, no. 5, pp. 680–682, Sep. 1985.

644. A. Renaux, P. Forster, P. Larzabal, C. D. Richmond, and A. Nehorai, "A fresh look at the Bayesian bounds of the Weiss-Weinstein family", *IEEE Trans. Sig. Process.*, vol. 56, no. 11, pp. 5334–5352, Nov. 2008.

645. L. Berriche, K. Abed-Meraim, and J. C. Belfiore, "Cramer-Rao bounds for MIMO channel estimation", in *Proc. IEEE Int. Conf. Acoust., Speech and Sig. Process. (ICASSP'04)*, vol. 4, Montreal, 17–21 May 2004, pp. 397–400.

646. L. Berriche and K. Abed-Meraim, "Stochastic Cramer-Rao bounds for semiblind MIMO channel estimation", in *Proc. Fourth IEEE Int. Symp. Sig. Process. and Inform. Tech.*, 18–21 Dec. 2004, pp. 119–122.

647. A. Vosoughi and A. Scaglione, "Everything you always wanted to know about training: guidelines derived using the affine precoding framework and the CRB", *IEEE Trans. Sig. Process.*, vol. 54, no. 3, pp. 940–954, Mar. 2006.

648. J. H. Manton, I. Y. Mareels, and Y. Hua, "Affine precoders for reliable communications", in *Proc. IEEE Int. Conf. Acoust., Speech and Sig. Process. 2000 (ICASSP '00)*, vol. 5, Instanbul, Turkey, 5–9 June 2000, pp. 2749–2752.

649. S. Ronen, S. I. Bross, S. Shamai, and T. M. Duman, "Iterative channel estimation and decoding in turbo coded space-time systems", *European Trans. Telecommun.*, vol. 18, no. 7, pp. 719–734, Nov. 2007.

650. S. Ronen and T. M. Duman, "Joint channel estimation and decoding for MIMO frequency selective fading channels", in *Proc. 2008 IEEE Wireless Commun. and Net. Conf. (WCNC 2008)*, Las Vegas, 31 Mar.–3 Apr. 2008, pp. 1322–1327.

651. L.-K. Chiu and S.-H. Wu, "The modified Bayesian Cramer-Rao bound for MIMO channel tracking", in *Proc. IEEE Int. Conf. Commun. (ICC 2009)*, Dresden, 14–18 June 2009, pp. 1–5.

652. F. Montorsi and G. M. Vitetta, "On the performance limits of pilot-based estimation of bandlimited frequency-selective communication channels", *IEEE Trans. Commun.*, vol. 59, no. 11, pp. 2964–2969, Nov. 2011.

653. I. Reuven and H. Messer, "A Barankin-type lower bound on the estimation error of a hybrid parameter vector", *IEEE Trans. Inform. Theory*, vol. 43, no. 3, pp. 1084–1093, May 1997.

654. Y. Rockah and P. Schultheiss, "Array shape calibration using sources in unknown locations – Part I: Far-field sources", *IEEE Trans. Acoust., Speech and Sig. Process.*, vol. 35, no. 3, pp. 286–299, Mar. 1987.

655. Z. Liu, G. B. Giannakis, S. Barbarossa, and A. Scaglione, "Transmit antennae space-time block coding for generalized OFDM in the presence of unknown multipath", *IEEE J. Sel. Areas Commun.*, vol. 19, no. 7, pp. 1352–1364, July 2001.

656. M. Morelli and U. Mengali, "A comparison of pilot-aided channel estimation methods for OFDM systems", *IEEE Trans. Sig. Process.*, vol. 49, no. 12, pp. 3065–3073, Dec. 2001.

657. Z. Li and Y.-M. Cai, "Cramer-Rao bound for blind, semi-blind and non-blind channel estimation in OFDM systems", in *Proc. IEEE Int. Symp. Commun. and Inform. Tech. (ISCIT 2005)*, vol. 1, 12–14 Oct. 2005, pp. 523–526.

658. H. Steendam, M. Moeneclaey, and H. Bruneel, "The Cramer-Rao bound and ML estimate for data-aided channel estimation in KSP-OFDM", in *Proc. IEEE 18th Int. Symp. Personal, Indoor and Mobile Radio Communications (PIMRC 2007)*, vol. 5, Athens, 3–7 Sep. 2007, pp. 1–5.

659. J.-C. Lin, "Least-squares channel estimation for mobile OFDM communication on time-varying frequency-selective fading channels", *IEEE Trans. Veh. Tech.*, vol. 57, no. 6, pp. 3538–3550, Nov. 2008.

660. M. D. Larsen, A. L. Swindlehurst, and T. Svantesson, "Performance bounds for MIMO-OFDM channel estimation", *IEEE Trans. Sig. Process.*, vol. 57, no. 5, pp. 1901–1916, May 2009.

661. J. D. Gorman and A. O. Hero, "Lower bounds for parametric estimation with constraints", *IEEE Trans. Inform. Theory*, vol. 36, no. 6, pp. 1285–1301, Nov. 1990.

662. T. L. Marzetta, "A simple derivation of the constrained multiple parameter Cramer-Rao bound", *IEEE Trans. Sig. Process.*, vol. 41, no. 6, pp. 2247–2249, June 1993.

663. P. Stoica and B. C. Ng, "On the Cramer-Rao bound under parametric constraints", *IEEE Sig. Process. Lett.*, vol. 5, no. 7, pp. 177–179, July 1998.

664. T. J. Moore, R. J. Kozick, and B. M. Sadler, "The constrained Cramer-Rao bound from the perspective of fitting a model", *IEEE Sig. Process. Lett.*, vol. 14, no. 8, pp. 564–567, Aug. 2007.

665. B. M. Sadler, R. J. Kozick, and T. Moore, "Bounds on MIMO channel estimation and equalization with side information", in *Proc. IEEE Int. Conf. Acoust., Speech and Sig. Process. (ICASSP'01)*, vol. 4, Salt Lake City, UT, 7–11 May 2001, pp. 2145–2148.

666. S. Barbarossa, A. Scaglione, and G. B. Giannakis, "Performance analysis of a deterministic channel estimator for block transmission systems with null guard intervals", *IEEE Trans. Sig. Process.*, vol. 50, no. 3, pp. 684–695, Mar. 2002.

667. X. Wautelet, C. Herzet, and L. Vandendorpe, "Cramer-Rao bounds for channel estimation with symbol a priori information", in *Proc. IEEE 6th Workshop Sig. Process. Adv. Wireless Commun. (SPAWC 2005)*, 5–8 June 2005, pp. 715–719.

668. M. Ghogho and A. Swami, "Training design for multipath channel and frequency-offset estimation in MIMO systems", *IEEE Trans. Sig. Process.*, vol. 54, no. 10, pp. 3957–3965, Oct. 2002.

669. C. Budianu and L. Tong, "Channel estimation for space-time orthogonal block codes", *IEEE Trans. Sig. Process.*, vol. 50, no. 10, pp. 2515–2528, June 2002.

670. J. Dauwels, "Computing Bayesian Cramer-Rao bounds", in *Proc. 2005 IEEE Int. Symp. Inform. Theory (ISIT 2005)*, Adelaide, Sep. 2005, pp. 2031–2035.

671. J. Dauwels and S. Korl, "A numerical method to compute Cramer-Rao-type bounds for challenging estimation problems", in *Proc. IEEE Int. Conf. Acoust., Speech and Sig. Process. (ICASSP 2006)*, vol. 5, Toulouse, May 2006, pp. 1322–1327.

672. P. Hoeher and F. Tufvesson, "Channel estimation with superimposed pilot sequence", in *Proc. IEEE Global Telecommun. Conf. (GLOBECOM '93)*, vol. 4, Rio de Janeiro, 1999, pp. 2162–2166.

673. S. Ohno and G. B. Giannakis, "Optimal training and redundant precoding for block transmissions with application to wireless OFDM", *IEEE Trans. Commun.*, vol. 50, no. 12, pp. 2113–2123, Dec. 2002.

674. A.-J. van der Veen and L. Tong, "Packet separation in wireless ad-hoc networks by known modulus algorithms", in *Proc. IEEE Int. Conf. Acoust., Speech and Sig. Process. (ICASSP '02)*, vol. 3, Orlando, FL, 13–17 May 2002, pp. 2149–2152.

675. M. Dong, L. Tong, and B. M. Sadler, "Optimal insertion of pilot symbols for transmissions over time-varying flat fading channels", *IEEE Trans. Sig. Process.*, vol. 52, no. 5, pp. 1403–1418, May 2004.

676. G. T. Zhou, M. Viberg, and T. McKelvey, "A first-order statistical method for channel estimation", *IEEE Sig. Process. Lett.*, vol. 10, no. 3, pp. 57–60, Mar. 2003.

677. G. Caire and S. Shamai, "On the capacity of some channels with channel state information", *IEEE Trans. Inform. Theory*, vol. 45, no. 6, pp. 2007–2019, Sep. 1999.

678. H. Vikalo, B. Hassibi, B. Hochwald, and T. Kailath, "Optimal training for frequency-selective fading channels", in *Proc. IEEE Int. Conf. Acoust., Speech and Sig. Process. (ICASSP '01)*, vol. 4, Salt Lake City, UT, May 2002, pp. 2105–2108.

679. A. Lapidoth and S. Shamai, "Fading channels: how perfect need 'perfect side information' be?" *IEEE Trans. Inform. Theory*, vol. 48, no. 5, pp. 1118–1134, May 2002.

680. J. Baltersee, G. Fock, and H. Meyr, "An information theoretic foundation of synchronized detection", *IEEE Trans. Commun.*, vol. 49, no. 12, pp. 2115–2123, Dec. 2001.

681. J. H. Kotecha and A. M. Sayeed, "Transmit signal design for optimal estimation of correlated MIMO channels", *IEEE Trans. Sig. Process.*, vol. 52, no. 2, pp. 546–557, Feb. 2004.

682. I. Abou-Faycal, M. Medard, and U. Madhow, "Binary adaptive coded pilot symbol assisted modulation over Rayleigh fading channels without feedback", *IEEE Trans. Commun.*, vol. 53, no. 6, pp. 1036–1046, June 2005.

683. M. Medard, "The effect upon channel capacity in wireless communications of perfect and imperfect knowledge of the channel", *IEEE Trans. Inform. Theory*, vol. 46, no. 3, pp. 933–946, May 2000.

684. J. Baltersee, G. Fock, and H. Meyr, "Achievable rate of MIMO channels with data-aided channel estimation and perfect interleaving", *IEEE J. Sel. Areas Commun.*, vol. 19, no. 12, pp. 2358–2368, Dec. 2001.

685. B. Hassibi and B. M. Hochwald, "How much training is needed in multiple antenna wireless links?" *IEEE Trans. Inform. Theory*, vol. 49, no. 4, pp. 951–963, Apr. 2003.

686. S. Adireddy, L. Tong, and H. Viswanathan, "Optimal placement of training for frequency-selective block-fading channels", *IEEE Trans. Inform. Theory*, vol. 48, no. 8, pp. 2338–2353, Aug. 2002.

687. S. Ohno and G. B. Giannakis, "Average-rate optimal PSAM transmissions over time-selective fading channels", *IEEE Trans. Wireless Commun.*, vol. 1, no. 4, pp. 712–720, Oct. 2002.

688. X. Ma, G. B. Giannakis, and S. Ohno, "Optimal training for block transmissions over doubly selective wireless fading channels", *IEEE Trans. Sig. Process.*, vol. 51, no. 5, pp. 1351–1366, May 2003.

689. D. Samardzija and N. Mandayam, "Pilot-assisted estimation of MIMO fading channel response and achievable data rates", *IEEE Trans. Sig. Process.*, vol. 51, no. 11, pp. 2882–2890, Nov. 2003.

690. S. Ohno and G. B. Giannakis, "Capacity maximizing MMSE-optimal pilots for wireless OFDM over frequency-selective block Rayleigh-fading channels", *IEEE Trans. Inform. Theory*, vol. 50, no. 9, pp. 2138–2145, Sep. 2004.

691. X. Ma, L. Yang, and G. B. Giannakis, "Optimal training for MIMO frequency-selective fading channels", *IEEE Trans. Wireless Commun.*, vol. 4, no. 2, pp. 453–466, Mar. 2005.

692. S. Adireddy and L. Tong, "Optimal placement of known symbols for slowly varying frequency-selective channels", *IEEE Trans. Wireless Commun.*, vol. 4, no. 4, pp. 1292–1296, July 2005.

693. A. Viterbi and J. Omura, *Principles of Digital Communication and Coding*. New York: McGraw-Hill, 1979.

694. F. Ling, "Optimal reception, performance bound, and cutoff rate analysis of references-assisted coherent CDMA communications with applications", *IEEE Trans. Commun.*, vol. 47, no. 10, pp. 1583–1592, Oct. 1999.

695. W. Phoel and M. Honig, "Performance of coded DS-CDMA with pilot-assisted channel estimation and linear interference suppression", *IEEE Trans. Commun.*, vol. 50, no. 5, pp. 822–832, May 2002.

696. S. Misra, A. Swami, and L. Tong, "Cutoff rate analysis of the Gauss-Markov fading channel with adaptive energy allocation", in *Proc. IEEE 4th Workshop Sig. Process. Adv. Wireless Commun. (SPAWC 2003)*, Rome, 15–18 June 2003, pp. 388–392.

697. ____, "Cutoff rate optimal binary inputs with imperfect CSI", *IEEE Trans. Wireless Commun.*, vol. 5, no. 10, pp. 2903–2913, Oct. 2006.

698. M. Dong, S. Adireddy, and L. Tong, "Optimal pilot placement for semi-blind channel tracking of packetized transmission over time-varying channels", *IEICE Trans. Fund. Electron., Commun., Comput. Sci.*, vol. E86-A, no. 3, pp. 550–563, Mar. 2003.

699. C. Fragouli, N. Al-Dhahir, and W. Turin, "Training-based channel estimation for multiple-antenna broadband transmissions", *IEEE Trans. Wireless Commun.*, vol. 2, no. 2, pp. 384–391, Mar. 2003.

700. H. Minn, V. K. Bhargava, and K. B. Letaief, "A combined timing and frequency synchronization and channel estimation for OFDM", *IEEE Trans. Commun.*, vol. 54, no. 3, pp. 416–422, Mar. 2006.

701. H. Minn and N. Al-Dhahir, "Optimal training signals for MIMO OFDM channel estimation", *IEEE Trans. Wireless Commun.*, vol. 5, no. 5, pp. 1158–1168, May 2006.

702. T. Whitworth, M. Ghogho, and D. McLernon, "Optimized training and basis expansion model parameters for doubly-selective channel estimation", *IEEE Trans. Wireless Commun.*, vol. 8, no. 3, pp. 1490–1498, Mar. 2009.

703. S. Misra, A. Swami, and L. Tong, "Bit-error rate optimized training for time-varying fading channels", in *Proc. IEEE 6th Workshop Sig. Process. Adv. Wireless Commun. (SPAWC 2006)*, Cannes, 5–8 June 2005, pp. 358–362.

704. Y. Peng, S. Cui, and R. You, "Optimal pilot-to-data power ratio for diversity combining with imperfect channel estimation", *IEEE Commun. Lett.*, vol. 10, no. 2, pp. 97–99, Feb. 2006.

705. R. G. Gallager, *Information Theory and Reliable Communication*. New York: John Wiley & Sons, Inc., 1968.

706. T. L. Marzetta, "BLAST Training: Estimating channel characteristics for high capacity space-time wireless", in *Proc. 37th Annual Allerton Conference on Communications, Control, and Computing*, Monticello, IL, 1999, pp. 958–966.

707. A. Aghamohammadi, H. Meyr, and G. Ascheid, "Adaptive synchronization and channel parameter estimation using an extended Kalman filter", *IEEE Trans. Commun.*, vol. 37, no. 11, pp. 1212–1219, November 1989.

708. J. P. McGeehan and A. J. Bateman, "Theoretical and experimental investigation of feedforward signal regeneration as a means of combating multipath propagation effects in pilot-based SSB mobile radio systems", *IEEE Trans. Veh. Tech.*, vol. 32, no. 1, pp. 106–120, Feb. 1983.

709. A. J. Bateman and J. P. McGeehan, "Phase-locked transparent tone-in-band (TTIB): A new spectrum configuration particularly suited to the transmission of data over SSB mobile radio networks", *IEEE Trans. Commun.*, vol. 32, no. 1, pp. 81–87, Jan. 1984.

710. M. Yokoyama, "BPSK system with sounder to combat Rayleigh fading in mobile radio communication", *IEEE Trans. Veh. Tech.*, vol. 34, pp. 35–40, Feb. 1985.

711. M. K. Simon, "Dual-pilot tone calibration technique", *IEEE Trans. Veh. Tech.*, vol. 35, no. 2, pp. 63–70, May 1986.

712. F. Davarian, "Mobile digital communication via tone calibration", *IEEE Trans. Veh. Tech.*, vol. 36, pp. 55–62, May 1987.

713. A. Bateman, "Feedforward transparent tone-in-band: Its implementations and applications", *IEEE Trans. Veh. Tech.*, vol. 39, pp. 235–243, Aug. 1990.

714. J. K. Cavers, "The performance of phase locked transparent tone-in-band with symmetric phase detection", *IEEE Trans. Commun.*, vol. 39, no. 9, pp. 1389–1399, Jan. 1991.

715. ——, "Performance of tone calibration with frequency offset and imperfect pilot filter", *IEEE Trans. Veh. Tech.*, vol. 40, pp. 426–434, May 1991.

716. J. A. Gansman, M. P. Fitz, and J. V. Krogmeier, "Optimum and suboptimum frame synchronization for pilot-symbol-assisted modulation", *IEEE Trans. Commun.*, vol. 45, no. 10, pp. 1327–1337, Oct. 1997.

717. M. L. Moher and J. H. Lodge, "TCMP – a modulation and coding strategy for Rician fading channels", *IEEE J. Select. Areas Commun.*, vol. 7, pp. 1347–1355, Dec. 1989.

718. S. Sampei and T. Sunaga, "Rayleigh fading compensation for QAM in land mobile radio communications", *IEEE Trans. Veh. Tech.*, vol. 42, pp. 137–146, May 1993.

719. M. C. Valenti and B. D. Woerner, "Iterative channel estimation and decoding of pilot symbol assisted turbo codes over flat-fading channels", *IEEE J. Sel. Areas Commun.*, vol. 19, no. 9, pp. 1697–1705, Sep. 2001.

720. X. Cai and G. B. Giannakis, "Adaptive PSAM accounting for channel estimation and prediction errors", *IEEE Trans. Wireless Commun.*, vol. 4, no. 1, pp. 246–256, Jan. 2005.

721. J. K. Cavers, "An analysis of pilot symbol assisted modulation for Rayleigh fading channels", *IEEE Trans. Veh. Tech.*, vol. 40, no. 4, pp. 686–693, Nov. 1991.

722. M. Medard, I. Abou-Faycal, and U. Madhow, "Adaptive coding with pilot signals", in *Proc. 38th Annual Allerton Conference on Communication, Control, and Computing*, Monticello, IL, Oct. 2000.

723. D. Makrakis and K. Feher, "A novel pilot insertion-extraction technique based on spread spectrum techniques", in *Proc. Miami Technicon*, Miami, FL, 1987, pp. 129–132.

724. T. P. Holden and K. Feher, "A spread spectrum based synchronization technique for digital broadcast systems", *IEEE Trans. Broadcast.*, vol. 36, no. 3, pp. 185–194, Sep. 1990.

725. R. Haeb and H. Meyr, "A systematic approach to carrier recovery and detection of digitally phase modulated signals on fading channels", *IEEE Trans. Commun.*, vol. 37, no. 7, pp. 748–754, July 1989.

726. P. Ho and J. H. Kim, "Pilot symbol-assisted detection of CPM schemes operating in fast fading channels", *IEEE Trans. Commun.*, vol. 44, no. 3, pp. 337–347, Mar. 1996.

727. G. T. Irvine and P. J. McLane, "Symbol-aided plus decision-directed reception for PSK/TCM modulation on shadowed mobile satellite fading channels", *IEEE J. Sel. Areas Commun.*, vol. 10, no. 8, pp. 1289–1299, Oct. 1992.

728. Y. Liu and S. D. Blostein, "Identification of frequency non-selective fading channels using decision feedback and adaptive linear prediction", *IEEE Trans. Commun.*, vol. 43, no. 2/3/4, pp. 1484–1492, Feb./Mar./Apr. 1995.

729. C. Komninakis and R. D. Wesel, "Joint iterative channel estimation and decoding in flat correlated Rayleigh fading", *IEEE J. Sel. Areas Commun.*, vol. 19, no. 9, pp. 1706–1717, Sep. 2001.

730. R. Heimiller, "Phase shift pulse codes with good periodic correlation properties", *IRE Trans. Inform. Theory*, vol. 7, no. 4, pp. 254–257, Oct. 1961.

731. S. N. Crozier, D. D. Falconer, and S. A. Mahmoud, "Least sum of squared errors (LSSE) channel estimation", *IEE Proc. F*, vol. 138, pp. 371–378, Aug. 1991.

732. B. Popovic, "Generalized chirp-like polyphase sequences with optimum correlation properties", *IEEE Trans. Inform. Theory*, vol. 38, no. 4, pp. 1406–1409, July 1992.

733. J. C. L. Ng, K. B. Letaief, and R. D. Murch, "Complex optimal sequences with constant magnitude for fast channel estimation initialization", *IEEE Trans. Commun.*, vol. 46, pp. 305–308, March 1998.

734. C. Tellambura, M. G. Parker, Y. J. Guo, S. J. Shepherd, and S. K. Barton, "Optimal sequences for channel estimation using discrete Fourier transform techniques", *IEEE Trans. Commun.*, vol. 47, no. 3, pp. 230–238, Feb. 1999.

735. C. Tellambura, Y. J. Guo, and S. K. Barton, "Channel estimation using aperiodic binary sequences", *IEEE Commun. Lett.*, vol. 2, no. 5, pp. 140–142, May 1998.

736. Y. Zhang, M. P. Fitz, and S. B. Gelfand, "A performance analysis and design of equalization with pilot aided channel estimation", in *Proc. 1997 IEEE 47th Veh. Technol. Conf.*, vol. 2, Phoenix, AZ, 1997, pp. 720–724.

737. P. Spasojevic and C. N. Georghiades, "Complementary sequences for ISI channel estimation", *IEEE Trans. Inform. Theory*, vol. 47, no. 3, pp. 1145–1152, Mar. 2001.

738. J. Coon, M. Beach, and J. McGeehan, "Optimal training sequences for channel estimation in cyclic-prefix-based single-carrier systems with transmit diversity", *IEEE Sig. Process. Lett.*, vol. 11, no. 9, pp. 729–732, Sep. 2004.

739. S. Ozen, M. D. Zoltowski, and M. Fimoff, "A novel channel estimation method: Blending correlation and least-squares based approaches", in *Proc. 2002 IEEE Int. Conf. Acoust., Speech and Sig. Process. (ICASSP 2002)*, vol. 3, Orlando, FL, May 2002, pp. 2281–2284.

740. D. Boss, K. D. Kammeyer, and T. Petermann, "Is blind channel estimation feasible in mobile communication systems? A study based on GSM", *IEEE J. Sel. Areas Commun.*, vol. 16, no. 8, pp. 1479–1491, Oct. 1998.

741. K.-D. Kammeyer, V. Kuhn, and T. Petermann, "Blind and nonblind turbo estimation for fast fading GSM channels", *IEEE J. Sel. Areas Commun.*, vol. 19, no. 9, pp. 1718–1728, Sep. 2001.

742. P. K. Frenger and N. A. B. Svensson, "Decision-directed coherent detection in multicarrier systems on Rayleigh fading channels", *IEEE Trans. Veh. Tech.*, vol. 48, no. 3, pp. 490–498, Mar. 1999.

743. J. K. Cavers, "Pilot symbol assisted modulation and differential detection in fading and delay spread", *IEEE Trans. Commun.*, vol. 43, pp. 2206–2212, July 1995.

744. A. G. Orozco-Lugo, M. M. Lara, and D. C. McLernon, "Channel estimation using implicit training", *IEEE Trans. Sig. Process.*, vol. 52, no. 1, pp. 240–254, Jan. 2004.

745. M. Ghogho, D. McLernon, E. Alameda-Hernandez, and A. Swami, "Channel estimation and symbol detection for block transmission using data-dependent superimposed training", *IEEE Sig. Process. Lett.*, vol. 12, no. 3, pp. 226–229, Mar. 2005.

746. N. W. K. Lo, D. D. Falconer, and A. U. H. Sheikh, "Adaptive equalization for co-channel interference in a multipath fading environment", *IEEE Trans. Commun.*, vol. 43, pp. 1441–1453, Feb./Mar./Apr. 1995.

747. X. Ma and G. B. Giannakis, "Maximum-diversity transmissions over doubly selective wireless channels", *IEEE Trans. Inform. Theory*, vol. 49, no. 7, pp. 1832–1840, July 2003.

748. R. A. Ziegler and J. M. Cioffi, "Estimation of time-varying digital radio channels", *IEEE Trans. Veh. Tech.*, vol. 41, pp. 134–151, May 1992.

749. S. A. Fechtel and H. Meyr, "Optimal parametric feedforward estimation of frequency-selective fading radio channels", *IEEE Trans. Commun.*, vol. 42, pp. 1639–1650, Feb./Mar./Apr. 1994.

750. M. K. Tsatsanis and Z. Xu, "Pilot symbol assisted modulation in frequency selective fading wireless channels", *IEEE Trans. Sig. Process.*, vol. 48, no. 8, pp. 2353–2365, Aug. 2000.

751. P. Hoeher, S. Kaiser, and P. Robertson, "Two-dimensional pilot-symbol-aided channel estimation by Wiener filtering", in *Proc. IEEE Int. Conf. Acoust., Speech and Sig. Process. (ICASSP'97)*, vol. 3, Munich, 21–24 Apr. 1997, pp. 1845–1848.

752. L. M. Davis, I. B. Collings, and R. J. Evans, "Identification of time-varying linear channels", in *Proc. IEEE Int. Conf. Acoust., Speech and Sig. Process., (ICASSP 1997)*, vol. 5, Munich, 21–24 Apr. 1997, pp. 3921–3924.

753. ———, "Coupled estimators for equalization of fast-fading mobile channels", *IEEE Trans. Commun.*, vol. 46, no. 10, pp. 1262–1265, Oct. 1998.

754. M. K. Tsatsanis, G. B. Giannakis, and G. Zhou, "Estimation and equalisation of fading channels with random coefficients", *Sig. Process.*, vol. 53, no. 2–3, pp. 211–229, Sep. 1996.

755. T. Eyceoz, A. Duel-Hallen, and H. Hallen, "Prediction of fast fading parameters by resolving the interference pattern", in *Proc. 31st Asilomar Conf. Sig., Syst. and Comput.*, Pacific Grove, CA, 2–5 Nov. 1997, pp. 167–171.

756. M. K. Tsatsanis and G. B. Giannakis, "Modeling and equalization of rapidly fading channels", *Int. J. Adaptive Contr. Sig. Process.*, vol. 10, no. 2–3, pp. 159–176, Mar. 1996.

757. O. Rousseaux, G. Leus, and M. Moonen, "Estimation and equalization of doubly selective channels using known symbol padding", *IEEE Trans. Sig. Process.*, vol. 54, no. 3, pp. 979–990, Mar. 2006.

758. J. K. Tugnait, S. He, and H. Kim, "Doubly selective channel estimation using exponential basis models and subblock tracking", *IEEE Trans. Sig. Process.*, vol. 58, no. 3, pp. 1275–1289, Mar. 2010.

759. A. P. Clark and S. Harihanan, "Efficient estimators for an HF radio link", *IEEE Trans. Commun.*, vol. 38, pp. 1173–1180, Aug. 1990.

760. S. Song, J.-S. Lim, S. J. Baek, and K.-M. Sung, "Variable forgetting factor linear least squares algorithm for frequency selective fading channel estimation", *IEEE Trans. Veh. Tech.*, vol. 51, no. 3, pp. 613–616, May 2002.

761. S.-Y. Leong, J. Wu, C. Xiao, and J. C. Olivier, "Fast time-varying dispersive channel estimation and equalization for an 8-PSK cellular system", *IEEE Trans. Veh. Tech.*, vol. 55, no. 5, pp. 1493–1502, Sep. 2006.

762. M. K. Tsatsanis and G. B. Giannakis, "Equalization of rapidly fading channels: Self-recovering methods", *IEEE Trans. Commun.*, vol. 44, no. 5, pp. 619–630, May 1996.

763. H. A. Cirpan and M. K. Tsatsanis, "Maximum likelihood blind channel estimation in the presence of Doppler shifts", *IEEE Trans. Sig. Process.*, vol. 47, no. 6, pp. 1559–1569, June 1999.

764. H. Liu and G. B. Giannakis, "Deterministic approaches for blind equalization of time-varying channels with antenna arrays", *IEEE Trans. Sig. Process.*, vol. 46, no. 11, pp. 3003–3013, Nov. 1998.

765. L. M. Davis, I. B. Collings, and R. J. Evans, "Constrained maximum likelihood estimation of time-varying linear channels", in *Proc. IEEE Workshop Sig. Process. Adv. Wireless Commun. (SPAWC 2006)*, Paris, 16–18 Apr. 1997, pp. 1–4.

766. A. P. Clark and S. Harihanan, "Adaptive channel estimator for an HF radio link", *IEEE Trans. Commun.*, vol. 37, pp. 918–926, Sep. 1989.

767. M. Stojnanovic, J. Proakis, and J. Catipovic, "Analysis of the impact of channel estimation errors on the performance of a decision feedback equaliser in fading multipath channels", *IEEE Trans. Commun.*, vol. 43, pp. 877–885, Feb./Mar./Apr. 1995.

768. J. Wu and A. H. Aghvami, "A new adaptive equalizer with channel estimator for mobile radio communications", *IEEE Trans. Veh. Tech.*, vol. 45, pp. 467–474, Aug. 1996.

769. G. E. Bottomley and K. J. Molnar, "Adaptive channel estimation for multichannel MLSE receivers", *IEEE Commun. Lett.*, vol. 3, no. 2, pp. 40–42, Feb. 1999.

770. R. A. Iltis, "Joint estimation of PN code delay and multipath using the extended Kalman filter", *IEEE Trans. Commun.*, vol. 38, no. 10, pp. 1677–1685, Oct. 1990.

771. T. Matsumoto, "Channel identification and sequential sequence estimation using antenna array for broadband mobile communications", *IEEE Trans. Veh. Tech.*, vol. 49, no. 5, pp. 1776–1783, Sep. 2000.

772. H. Zamiri-Jafarian and S. Pasupathy, "EM-based recursive estimation of channel parameters", *IEEE Trans. Commun.*, vol. 47, no. 9, pp. 1297–1302, Sep. 1999.

773. ———, "Adaptive MLSDE using the EM algorithm", *IEEE Trans. Commun.*, vol. 47, pp. 1181–1193, Aug. 1999.

774. M. E. Rollins and S. J. Simmons, "Simplified per-survivor Kalman processing in fast frequency-selective fading channels", *IEEE Trans. Commun.*, vol. 45, no. 5, pp. 544–553, May 1997.

775. S. N. Diggavi, B. C. Ng, and A. Paulraj, "An interference suppression scheme with joint channel-data estimation", *IEEE J. Sel. Areas Commun.*, vol. 17, no. 11, pp. 1924–1939, Nov. 1999.

776. M. Martone, "Optimally regularized channel tracking techniques for sequence estimation based on cross-validated subspace signal processing", *IEEE Trans. Commun.*, vol. 48, no. 1, pp. 95–105, Jan. 2000.

777. T. Eyceoz and A. Duel-Hallen, "Simplified block adaptive diversity equaliser for cellular mobile radio", *IEEE Commun. Lett.*, vol. 1, pp. 15–19, Jan. 1987.

778. N. W. K. Lo, D. D. Falconer, and U. H. Sheikh, "Adaptive equalization and diversity combining for mobile radio using interpolated channel estimates", *IEEE Trans. Veh. Tech.*, vol. 40, no. 3, pp. 636–645, Aug. 1991.

779. H.-N. Lee and G. J. Pottie, "Fast adaptive equalization/diversity combining for time-varying dispersive channels", *IEEE Trans. Commun.*, vol. 46, no. 9, pp. 1146–1162, Sep. 1998.

780. R. Schober and W. H. Gerstacker, "Noncoherent adaptive channel identification algorithms for noncoherent sequence estimation", *IEEE Trans. Commun.*, vol. 49, no. 2, pp. 229–234, Feb. 2001.

781. S. J. Grant and J. K. Cavers, "Multiuser channel estimation for detection of cochannel signals", *IEEE Trans. Commun.*, vol. 49, no. 10, pp. 1845–1855, Oct. 2001.

782. M. Speth, S. A. Fechtel, G. Fock, and H. Meyer, "Optimum receiver design for wireless broad-band systems using OFDM – Part I", *IEEE Trans. Commun.*, vol. 47, no. 11, pp. 1668–1677, Nov. 1999.

783. Y. Li, "Pilot-symbol-aided channel estimation for OFDM in wireless systems", *IEEE Trans. Veh. Tech.*, vol. 49, no. 4, pp. 1207–1215, July 2000.

784. M. Hsieh and C. Wei, "Channel estimation for OFDM systems based on comb-type pilot arrangement in frequency selective fading channels", *IEEE Consumer Electron.*, vol. 44, no. 2, pp. 217–225, Feb. 1998.

785. S. Coleri, M. Ergen, and A. Bahai, "Channel estimation techniques based on pilot arrangement in OFDM systems", *IEEE Trans. Broadcast.*, vol. 49, no. 3, pp. 223–229, Sep. 2002.

786. M. J. Fernandez-Getino Garcia, S. Zazo, and J. M. Paez-Borrallo, "Pilot patterns for channel estimation in OFDM", *Electron. Lett.*, vol. 36, no. 12, pp. 1049–1050, June 2000.

787. M. R. Raghavendra and K. Giridhar, "Improving channel estimation in OFDM systems for sparse multipath channels", *IEEE Sig. Process. Lett.*, vol. 12, no. 1, pp. 52–55, Jan. 2005.

788. K. Josiam and D. Rajan, "Bandwidth efficient channel estimation using super-imposed pilots in OFDM systems", *IEEE Trans. Wireless Commun.*, vol. 6, no. 6, pp. 2234–2245, June 2007.

789. R. Negi and J. Cioffi, "Pilot tone selection for channel estimation in a mobile OFDM system", *IEEE Trans. Cons. Electr.*, vol. 44, no. 3, pp. 1122–1128, Aug. 1998.

790. M. Dong, L. Tong, and B. M. Sadler, "Optimal pilot placement for channel tracking in OFDM", in *Proc. MILCOM 2002*, vol. 1, Anaheim, CA, 7–10 Oct. 2002, pp. 602–606.

791. J. Kim, J. Park, and D. Hong, "Performance analysis of channel estimation in OFDM systems", *IEEE Sig. Process. Lett.*, vol. 12, no. 1, pp. 60–62, Jan. 2005.

792. J.-W. Choi and Y.-H. Lee, "Optimum pilot pattern for channel estimation in OFDM systems", *IEEE Trans. Wireless Commun.*, vol. 4, no. 5, pp. 2083–2088, Sep. 2005.

793. M. K. Ozdemir and H. Arslan, "Channel estimation for wireless OFDM systems", *IEEE Commun. Surveys Tut.*, vol. 9, no. 2, pp. 18–48, 2nd Quart. 2007.

794. S. Song and A. C. Singer, "Pilot-aided OFDM channel estimation in the presence of the guard band", *IEEE Trans. Commun.*, vol. 55, no. 8, pp. 1459–1465, Aug. 2007.

795. C. Shin, J. G. Andrews, and E. J. Powers, "An efficient design of doubly selective channel estimation for OFDM systems", *IEEE Trans. Wireless Commun.*, vol. 6, no. 10, pp. 3790–3802, Oct. 2007.

796. C. Oberli, M. C. Estela, and M. Rios, "On using transmission overhead efficiently for channel estimation in OFDM", *IEEE Trans. Commun.*, vol. 58, no. 2, pp. 399–404, Feb. 2010.

797. H. Minn and V. K. Bhargava, "An investigation into time-domain approach for OFDM channel estimation", *IEEE Trans. Broadcasting*, vol. 46, no. 4, pp. 240–248, Dec. 2000.

798. J.-J. van de Beek, O. Edfors, M. Sandell, S. K. Wilson, and P. O. Borjesson, "On channel estimation in OFDM systems", in *Proc. IEEE 45th Veh. Technol. Conf. VTC '95*, vol. 2, Chicago, 25–28 July 1995, pp. 815–819.

799. B. Yang, K. B. Letaief, R. S. Cheng, and Z. Cao, "Channel estimation for OFDM transmission in multipath fading channels based on parametric channel modeling", *IEEE Trans. Commun.*, vol. 49, no. 3, pp. 467–479, Mar. 2001.

800. Y.-S. Choi, P. J. Voltz, and F. A. Cassara, "On channel estimation and detection for multicarrier signals in fast and selective Rayleigh fading channels", *IEEE Trans. Commun.*, vol. 49, no. 8, pp. 1375–1387, Aug. 2001.

801. Z. Tang, R. C. Cannizzaro, G. Leus, and P. Banelli, "Pilot-assisted time-varying channel estimation for OFDM systems", *IEEE Trans. Sig. Process.*, vol. 55, no. 5, pp. 2226–2238, May 2007.

802. N. Chen, J. Zhang, and P. Zhang, "Improved channel estimation based on parametric channel approximation modeling for OFDM systems", *IEEE Trans. Broadcasting*, vol. 54, no. 2, pp. 217–225, June 2008.

803. H. Hijazi and L. Ros, "Polynomial estimation of time-varying multipath gains with intercarrier interference mitigation in OFDM systems", *IEEE Trans. Veh. Tech.*, vol. 58, no. 1, pp. 140–151, Jan. 2009.

804. O. Edfors, M. Sandell, J.-J. van de Beek, S. K. Wilson, and P. O. Borjesson, "OFDM channel estimation by singular value decomposition", *IEEE Trans. Commun.*, vol. 46, no. 7, pp. 931–939, July 1993.

805. S. G. Kang, Y. M. Ha, and E. K. Joo, "A comparative investigation on channel estimation algorithms for OFDM in mobile communications", *IEEE Trans. Broadcasting*, vol. 49, no. 2, pp. 142–149, June 2003.

806. W.-G. Song and J.-T. Lim, "Pilot-symbol aided channel estimation for OFDM with fast fading channels", *IEEE Trans. Broadcasting*, vol. 49, no. 4, pp. 398–402, Dec. 2003.

807. O. Simeone, Y. Bar-Ness, and U. Spagnolini, "Pilot-based channel estimation for OFDM systems by tracking the delay-subspace", *IEEE Trans. Wireless Commun.*, vol. 3, no. 1, pp. 315–324, Jan. 2004.

808. V. Mignone and A. Morello, "CD3-OFDM: a novel demodulation scheme for fixed and mobile receivers", *IEEE Trans. Commun.*, vol. 44, no. 9, pp. 1144–1151, May 1996.

809. M. Speth, S. A. Fechtel, G. Fock, and H. Meyer, "Optimum receiver design for wireless OFDM-based broadband transmission – Part II: A case study", *IEEE Trans. Commun.*, vol. 49, no. 4, pp. 571–578, Apr. 1999.

810. B. Yang, Z. Cao, and K. B. Letaief, "Analysis of low-complexity windowed DFT-based MMSE channel estimator for OFDM systems", *IEEE Trans. Commun.*, vol. 49, no. 11, pp. 1977–1987, Nov. 2001.

811. J. Rinne and M. Renfors, "Pilot spacing in orthogonal frequency division multiplexing systems on practical channels", *IEEE Trans. Consumer Electron.*, vol. 42, no. 4, pp. 959–962, Nov. 1996.

812. H.-K. Song, Y.-H. You, J.-H. Paik, and Y.-S. Cho, "Frequency-offset synchronization and channel estimation for OFDM-based transmission", *IEEE Commun. Lett.*, vol. 4, no. 3, pp. 95–97, Mar. 2000.

813. J. Park, J. Kim, M. Park, K. Ko, C. Kang, and D. Hong, "Performance analysis of channel estimation for OFDM systems with residual timing offset", *IEEE Trans. Wireless Commun.*, vol. 5, no. 7, pp. 1622–1625, July 2006.

814. X. Dong, W.-S. Lu, and A. C. K. Soong, "Linear interpolation in pilot symbol assisted channel estimation for OFDM", *IEEE Trans. Wireless Commun.*, vol. 6, no. 5, pp. 1910–1920, May 2007.

815. M.-X. Chang and Y. T. Su, "Model-based channel estimation for OFDM signals in Rayleigh fading", *IEEE Trans. Commun.*, vol. 50, no. 4, pp. 540–544, Apr. 2002.

816. M.-X. Chang, "A new derivation of least-squares-fitting principle for OFDM channel estimation", *IEEE Trans. Wireless Commun.*, vol. 5, no. 4, pp. 726–731, Apr. 2006.

817. X. Zhou and X. Wang, "Channel estimation for OFDM systems using adaptive radial basis function networks", *IEEE Trans. Veh. Tech.*, vol. 52, no. 1, pp. 48–59, Jan. 2003.

818. V. K. Jones and G. C. Raleigh, "Channel estimation for wireless OFDM systems", in *Proc. IEEE Global Telecommun. Conf. (GLOBECOM '98)*, vol. 2, Aveiro, Portugal, 8–12 Nov. 1998, pp. 980–985.

819. Y. Li, L. Cimini, Jr., and N. Sollenberger, "Robust channel estimation for OFDM systems with rapid dispersive fading channels", *IEEE Trans. Commun.*, vol. 46, no. 7, pp. 902–915, July 1998.

820. O. Edfors, M. Sandell, J.-J. van de Beek, S. K. Wilson, and P. O. Borjesson, "Analysis of DFT-based channel estimation for OFDM", *Wireless Personal Commun.*, vol. 12, no. 1, pp. 55–70, Jan. 2000.

821. D. Schafhuber and G. Matz, "MMSE and adaptive prediction of time-varying channels for OFDM systems", *IEEE Trans. Wireless Commun.*, vol. 4, no. 2, pp. 593–602, Mar. 2005.

822. D. Li, S. Feng, and W. Ye, "Pilot-assisted channel estimation method for OFDMA systems over time-varying channels", *IEEE Commun. Lett.*, vol. 13, no. 11, pp. 826–828, Nov. 2009.

823. H. Zhu, B. Farhang-Boroujeny, and C. Schlegel, "Pilot embedding for joint channel estimation and data detection in MIMO communication systems", *IEEE Commun. Lett.*, vol. 7, no. 1, pp. 30–32, Jan. 2003.

824. Y. Liu, T. F. Wong, and W. W. Hager, "Training signal design for estimation of correlated MIMO channels with colored interference", *IEEE Trans. Sig. Process.*, vol. 55, no. 4, pp. 1486–1497, Apr. 2007.

825. S. Wang and A. Abdi, "MIMO ISI channel estimation using uncorrelated Golay complementary sets of polyphase sequences", *IEEE Trans. Veh. Tech.*, vol. 56, no. 5, pp. 3024–3039, Sep. 2007.

826. M. Biguesh and A. B. Gershman, "Training-based MIMO channel estimation: a study of estimator tradeoffs and optimal training signals", *IEEE Trans. Sig. Process.*, vol. 54, no. 3, pp. 884–893, Mar. 2006.

827. X. Gao, B. Jiang, X. You, Z. Pan, Y. Xue, and E. Schulz, "Efficient channel estimation for MIMO single-carrier block transmission with dual cyclic timeslot structure", *IEEE Trans. Commun.*, vol. 55, no. 11, pp. 2210–2223, Nov. 2007.

828. Z. Liu, X. Ma, and G. B. Giannakis, "Space-time coding and Kalman filtering for time-selective fading channels", *IEEE Trans. Commun.*, vol. 50, no. 2, pp. 183–186, Feb. 2002.

829. B. Balakumar, S. Shahbazpanahi, and T. Kirubarajan, "Joint MIMO channel tracking and symbol decoding using Kalman filtering", *IEEE Trans. Sig. Process.*, vol. 55, no. 12, pp. 5873–5879, Dec. 2007.

830. Q. Sun, D. C. Cox, H. C. Huang, and A. Lozano, "Estimation of continuous flat fading MIMO channels", *IEEE Trans. Wireless Commun.*, vol. 1, no. 4, pp. 549–553, Oct. 2002.

831. M. Sanchez-Fernandez, M. de-Prado-Cumplido, J. Arenas-Garcia, and F. Perez-Cruz, "SVM multiregression for nonlinear channel estimation in multiple-input multiple-output systems", *IEEE Trans. Sig. Process.*, vol. 52, no. 8, pp. 2298–2307, Aug. 2004.

832. P. Garg, R. K. Mallik, and H. M. Gupta, "Exact error performance of square orthogonal space-time block coding with channel estimation", *IEEE Trans. Sig. Process.*, vol. 54, no. 3, pp. 430–437, Mar. 2006.

833. E. Karami and M. Shiva, "Decision-directed recursive least squares MIMO channels tracking", *EURASIP J. Wireless Commun. Netw.*, vol. 2006, no. 2, pp. 1–10, Apr. 2007.

834. C. R. Murthy, A. K. Jagannatham, and B. D. Rao, "Training-based and semiblind channel estimation for MIMO systems with maximum ratio transmission", *IEEE Trans. Sig. Process.*, vol. 54, no. 7, pp. 2546–2558, July 2006.

835. E. Karami, "Tracking performance of least squares MIMO channel estimation algorithm", *IEEE Trans. Commun.*, vol. 55, no. 11, pp. 2201–2209, Nov. 2007.

836. J. Wu and G. J. Saulnier, "Orthogonal space-time block code over time-varying flat-fading channels: Channel estimation, detection, and performance analysis", *IEEE Trans. Commun.*, vol. 55, no. 5, pp. 1077–1087, May 2007.

837. M.-A. Khalighi, J. J. Boutros, and J.-F. Helard, "Data-aided channel estimation for turbo-PIC MIMO detectors", *IEEE Commun. Lett.*, vol. 10, no. 5, pp. 350–352, May 2006.

838. S. Wang and A. Abdi, "Low-complexity optimal estimation of MIMO ISI channels with binary training sequences", *IEEE Sig. Process. Lett.*, vol. 13, no. 11, pp. 657–660, Nov. 2007.

839. R. Visoz and A. O. Berthet, "Iterative decoding and channel estimation for space-time BICM over MIMO block fading multipath AWGN channel", *IEEE Trans. Commun.*, vol. 51, no. 8, pp. 1358–1367, Aug. 2003.

840. C. Komninakis, C. Fragouli, A. H. Sayed, and R. D. Wesel, "Multi-input multi-output fading channel tracking and equalization using Kalman estimation", *IEEE Trans. Sig. Process.*, vol. 50, no. 5, pp. 1065–1076, May 2002.

841. E. G. Larsson and J. Li, "Preamble design for multiple-antenna OFDM-based WLANs with null subcarriers", *IEEE Sig. Process. Lett.*, vol. 8, no. 11, pp. 285–288, Nov. 2001.

842. Y. Li, "Simplified channel estimation for OFDM systems with multiple transmit antennas", *IEEE Trans. Wireless Commun.*, vol. 1, no. 1, pp. 67–75, Jan. 2002.

843. I. Barhumi, G. Leus, and M. Moonen, "Optimal training design for MIMO OFDM systems in mobile wireless channels", *IEEE Trans. Sig. Process.*, vol. 51, no. 6, pp. 1615–1624, June 2003.

844. H. Minn, N. Al-Dhahir, and Y. Li, "Optimal training signals for MIMO OFDM channel estimation in the presence of frequency offset and phase noise", *IEEE Trans. Commun.*, vol. 54, no. 10, pp. 1754–1759, Oct. 2006.

845. L. Huang, J. W. M. Bergmans, and F. M. J. Willems, "Low-complexity LMMSE based MIMO-OFDM channel estimation via angle-domain processing", *IEEE Trans. Sig. Process.*, vol. 55, no. 12, pp. 5668–5680, Dec. 2007.

846. Y. Li, N. Seshadri, and S. Ariyavisitakul, "Channel estimation for OFDM systems with transmitter diversity in mobile wireless channels", *IEEE J. Sel. Areas Commun.*, vol. 17, no. 3, pp. 461–471, Mar. 1999.

847. Y. Li, J. C. Chuang, and N. R. Sollenberger, "Transmitter diversity for OFDM systems and its impact on high-rate data wireless networks", *IEEE J. Sel. Areas Commun.*, vol. 17, no. 7, pp. 1233–1243, July 1999.

848. H. Minn, D. I. Kim, and V. K. Bhargava, "A reduced complexity channel estimation for OFDM systems with transmit diversity in mobile wireless channels", *IEEE Trans. Wireless Commun.*, vol. 50, no. 5, pp. 799–807, May 2002.

849. Y. Gong and K. B. Letaief, "Low complexity channel estimation for space-time coded wideband OFDM systems", *IEEE Trans. Wireless Commun.*, vol. 2, no. 5, pp. 876–882, Sep. 2003.

850. H. Miao and M. J. Juntti, "Space-time channel estimation and performance analysis for wireless MIMO-OFDM systems with spatial correlation", *IEEE Trans. Veh. Tech.*, vol. 54, no. 6, pp. 2003–2016, Nov. 2005.

851. W.-G. Song and J.-T. Lim, "Channel estimation and signal detection for MIMO-OFDM with time varying channels", *IEEE Commun. Lett.*, vol. 10, no. 7, pp. 540–542, July 2006.

852. H. Zamiri-Jafarian and S. Pasupathy, "Robust and improved channel estimation algorithm for MIMO-OFDM systems", *IEEE Trans. Wireless Commun.*, vol. 6, no. 6, pp. 2106–2113, June 2007.

853. L. Huang, C. K. Ho, J. W. M. Bergmans, and F. M. J. Willems, "Pilot-aided angle-domain channel estimation techniques for MIMO-OFDM systems", *IEEE Trans. Veh. Tech.*, vol. 57, no. 2, pp. 906–920, Mar. 2008.

854. S. Lu and N. Al-Dhahir, "A novel CDD-OFDM scheme with pilot-aided channel estimation", *IEEE Trans. Wireless Commun.*, vol. 8, no. 3, pp. 1122–1127, Mar. 2009.

855. M. K. Ozdemir, H. Arslan, and E. Arvas, "Toward real-time adaptive low-rank LMMSE channel estimation of MIMO-OFDM systems", *IEEE Trans. Wireless Commun.*, vol. 5, no. 10, pp. 2675–2678, Oct. 2006.

856. X. Dai, H. Zhang, and D. Li, "Linearly time-varying channel estimation for MIMO-OFDM systems using superimposed training", *IEEE Trans. Commun.*, vol. 58, no. 2, pp. 681–693, Feb. 2010.

857. C. K. Rushforth, "Transmitted-reference techniques for random or unknown channels", *IEEE Trans. Inform. Theory*, vol. 9, pp. 39–42, Jan. 1964.

858. G. D. Hingorani and J. C. Hancock, "A transmitted-reference system for communication in random or unknown channel", *IEEE Trans. Commun. Systems*, vol. 13, no. 3, pp. 293–301, Sep. 1965.

859. J. P. McGeehan and D. F. Burrows, "Performance limits of feed-forward automatic gain control in mobile radio receivers", *IEE Proc., Pt. F*, vol. 128, no. 6, pp. 385–392, Nov. 1981.

860. J. G. Proakis, "Adaptive equalization for TDMA digital mobile radio", *IEEE Trans. Veh. Tech.*, vol. 40, pp. 333–341, May 1991.

861. S. Galli, "A new family of soft-output adaptive receivers exploiting nonlinear MMSE estimates for TDMA-based wireless links", *IEEE Trans. Commun.*, vol. 50, no. 12, pp. 1935–1945, Dec. 2002.

862. L. Cimini, Jr., "Analysis and simulation of a digital mobile channel using orthogonal frequency-division multiplexing", *IEEE Trans. Commun.*, vol. 33, no. 7, pp. 665–675, July 1995.

863. W. Y. Zou and Y. Wu, "COFDM: an overview", *IEEE Trans. Broadcasting*, vol. 41, no. 1, pp. 1–8, Aug. 2001.

864. T. Kuroda and T. Matsumoto, "Multicarrier signal detection and parameter estimation in frequency-selective Rayleigh fading channels", *IEEE Trans. Veh. Tech.*, vol. 46, no. 4, pp. 882–890, Nov. 1997.

865. H. Liu, G. Xu, L. Ting, and T. Kailath, "Recent developments in blind channel equalization: From cyclostationarity to subspaces", *Sig. Process.*, vol. 50, no. 1–2, pp. 83–99, Apr. 1996.

866. G. B. Giannakis, P. Stoica, Y. Hua and L. Tong, (eds.), *Signal Processing Advances in Wireless and Mobile Communications, Volume 1: Trends in Channel Estimation and Equalization*. Upper Saddle River, NJ: Prentice Hall PTR, 2001.

867. G. Xu, H. Liu, L. Tong, and T. Kailath, "A least-squares approach to blind channel identification", *IEEE Trans. Sig. Process.*, vol. 43, no. 12, pp. 2982–2993, Dec. 1995.

868. J. K. Tugnait, "Blind equalisation and estimation of digital communication FIR channels using cumulant matching", *IEEE Trans. Commun.*, vol. 43, pp. 1240–1245, Feb./Mar./Apr. 1995.

869. _____, "On blind identifiability of multipath channels using fractional sampling and second-order cyclostationary statistics", *IEEE Trans. Inform. Theory*, vol. 41, pp. 308–311, Jan. 1995.

870. _____, "Blind equalisation and estimation of FIR communications channels using fractional sampling", *IEEE Trans. Commun.*, vol. 44, pp. 324–336, Mar. 1996.

871. H. Liu and G. Xu, "Smart antennas in wireless systems: Uplink multiuser blind channel and sequence detection", *IEEE Trans. Commun.*, vol. 45, pp. 187–199, Feb. 1997.

872. E. Serpedin and G. B. Giannakis, "Blind channel identification and equalization with modulation-induced cyclostationarity", *IEEE Trans. Sig. Process.*, vol. 46, no. 7, pp. 1930–1944, July 1998.

873. R. W. Heath, Jr., and G. B. Giannakis, "Exploiting input cyclostationarity for blind channel identification in OFDM systems", *IEEE Trans. Sig. Process.*, vol. 47, no. 3, pp. 848–856, Mar. 1999.

874. J. K. Tugnait, L. Tong, and Z. Ding, "Single-user channel estimation and equalization", *IEEE Sig. Process. Mag.*, vol. 17, no. 3, pp. 16–28, May 2000.

875. J. K. Tugnait and B. Huang, "On a whitening approach to partial channel estimation and blind equalization of FIR/IIR multiple-input multiple-output channels", *IEEE Trans. Sig. Process.*, vol. 48, no. 3, pp. 832–845, Mar. 2000.

876. C. Tepedelenlioglu and G. B. Giannakis, "Transmitter redundancy for blind estimation and equalization of time- and frequency-selective channels", *IEEE Trans. Sig. Process.*, vol. 48, no. 7, pp. 2029–2043, July 2000.

877. T. P. Krauss and R. D. Zoltowski, "Bilinear approach to multiuser second-order statistics-based blind channel estimation", *IEEE Trans. Sig. Process.*, vol. 48, no. 9, pp. 2473–2486, Sep. 2000.

878. H. A. Cirpan and M. K. Tsatsanis, "Maximum-likelihood estimation of FIR channels excited by convolutionally encoded inputs", *IEEE Trans. Commun.*, vol. 49, no. 7, pp. 1125–1128, July 2001.

879. X. Li and H. H. Fan, "Blind channel identification: Subspace tracking method without rank estimation", *IEEE Trans. Sig. Process.*, vol. 49, no. 10, pp. 2372–2382, Oct. 2001.

880. C. Anton-Haro, J. A. R. Fonollosa, C. Fauli, and J. R. Fonollosa, "On the inclusion of channel's time dependence in a hidden Markov model for blind channel estimation", *IEEE Trans. Veh. Tech.*, vol. 50, no. 3, pp. 867–873, May 2001.

881. Z. Xu and B. P. Ng, "Deterministic linear prediction methods for blind channel estimation based on dual concept of zero-forcing equalization", *IEEE Trans. Sig. Process.*, vol. 50, no. 11, pp. 2855–2865, Nov. 2002.

882. A. Yeredor, "Blind channel estimation using first and second derivatives of the characteristic function", *IEEE Sig. Process. Lett.*, vol. 9, no. 3, pp. 100–103, Mar. 2002.

883. J. R. M. Filho, B. Dorizzi, and J. C. M. Mota, "Channel estimation by symmetrical clustering", *IEEE Trans. Sig. Process.*, vol. 50, no. 6, pp. 1459–1469, June 2002.

884. W. H. Gerstacker and D. P. Taylor, "Blind channel order estimation based on second-order statistics", *IEEE Sig. Process. Lett.*, vol. 10, no. 2, pp. 39–42, Feb. 2003.

885. J. K. Tugnait and W. Luo, "On channel estimation using superimposed training and first-order statistics", *IEEE Commun. Lett.*, vol. 7, no. 9, pp. 413–415, Sep. 2003.

886. M. Nicoli, O. Simeone, and U. Spagnolini, "Multislot estimation of frequency-selective fast-varying channels", *IEEE Trans. Commun.*, vol. 51, no. 8, pp. 1337–1347, Aug. 2004.

887. ———, "Multislot estimation of fast-varying space-time communication channels", *IEEE Trans. Sig. Process.*, vol. 51, no. 5, pp. 1184–1195, May 2003.

888. F. Sanzi and M. C. Necker, "Totally blind APP channel estimation for mobile OFDM systems", *IEEE Commun. Lett.*, vol. 7, no. 11, pp. 517–519, Nov. 2004.

889. B. Baykal, "Blind channel estimation via combining autocorrelation and blind phase estimation", *IEEE Trans. Circ. and Syst. I*, vol. 51, no. 6, pp. 1125–1131, June 2004.

890. S. Yatawatta, A. P. Petropulu, and R. Dattani, "Blind channel estimation using fractional sampling", *IEEE Trans. Veh. Tech.*, vol. 53, no. 2, pp. 363–371, Mar. 2004.

891. H. Nguyen and B. C. Levy, "The expectation-maximization Viterbi algorithm for blind adaptive channel equalization", *IEEE Sig. Process. Lett.*, vol. 53, no. 10, pp. 1671–1678, July 2005.

892. M. Nicoli and U. Spagnolini, "Reduced-rank channel estimation for time-slotted mobile communication systems", *IEEE Trans. Sig. Process.*, vol. 53, no. 3, pp. 926–944, Mar. 2005.

893. C. Pladdy, S. Ozen, S. M. Nerayanuru, M. Zoltowski, and M. Fimoff, "Semiblind BLUE channel estimation with applications to digital television", *IEEE Trans. Veh. Tech.*, vol. 55, no. 6, pp. 1812–1823, Nov. 2006.

894. M.-A. Khalighi and J. J. Boutros, "Semi-blind channel estimation using the EM algorithm in iterative MIMO APP detectors", *IEEE Trans. Wireless Commun.*, vol. 5, no. 11, pp. 3165–3173, Nov. 2006.

895. B. Su and P. P. Vaidyanathan, "Performance analysis of generalized zero-padded blind channel estimation algorithms", *IEEE Sig. Process. Lett.*, vol. 14, no. 11, pp. 789–792, Nov. 2007.

896. H. Murakami, "Deterministic blind channel estimation for a block transmission system using fractional sampling and interpolation", *IEEE Trans. Sig. Process.*, vol. 55, no. 10, pp. 4969–4978, Oct. 2007.

897. F. Gao, A. Nallanathan, and C. Tellambura, "Blind channel estimation for cyclic-prefixed single-carrier systems by exploiting real symbol characteristics", *IEEE Trans. Veh. Tech.*, vol. 56, no. 5, pp. 2487–2498, Sep. 2007.

898. E. Karami and M. Shiva, "Blind multi-input-multi-output channel tracking using decision-directed maximum-likelihood estimation", *IEEE Trans. Veh. Tech.*, vol. 56, no. 3, pp. 1447–1454, May 2007.

899. J. K. Tugnait and S. He, "Doubly-selective channel estimation using data-dependent superimposed training and exponential basis models", *IEEE Trans. Wireless Commun.*, vol. 6, no. 11, pp. 3877–3883, Nov. 2007.

900. R. Carrasco-Alvarez, R. Parra-Michel, A. G. Orozco-Lugo, and J. K. Tugnait, "Enhanced channel estimation using superimposed training based on universal basis expansion", *IEEE Trans. Sig. Process.*, vol. 57, no. 3, pp. 1217–1222, Mar. 2009.

901. J. K. Tugnait, "Blind estimation and equalization of MIMO channels via multidelay whitening", *IEEE J. Sel. Areas Commun.*, vol. 19, no. 8, pp. 1507–1519, Aug. 2001.

902. J. Choi, "Equalization and semi-blind channel estimation for space-time block coded signals over a frequency-selective fading channel", *IEEE Trans. Sig. Process.*, vol. 52, no. 3, pp. 774–785, Mar. 2004.

903. Z. Ding and D. B. Ward, "Subspace approach to blind and semi-blind channel estimation for space-time block codes", *IEEE Trans. Wireless Commun.*, vol. 4, no. 2, pp. 357–362, Mar. 2005.

904. H. Amindavar and A. M. Reza, "A new simultaneous estimation of directions of arrival and channel parameters in a multipath environment", *IEEE Trans. Sig. Process.*, vol. 53, no. 2, pp. 471–483, Feb. 2005.

905. S. Shahbazpanahi, A. B. Gershman, and J. H. Manton, "Closed-form blind MIMO channel estimation for orthogonal space-time block codes", *IEEE Trans. Sig. Process.*, vol. 53, no. 12, pp. 4506–4517, Dec. 2005.

906. S. Shahbazpanahi, A. B. Gershman, and G. B. Giannakis, "Semiblind multiuser MIMO channel estimation using Capon and MUSIC techniques", *IEEE Trans. Sig. Process.*, vol. 54, no. 9, pp. 3581–3591, Sep. 2005.

907. A. K. Jagannatham and B. D. Rao, "Whitening-rotation-based semi-blind MIMO channel estimation", *IEEE Trans. Sig. Process.*, vol. 54, no. 3, pp. 861–869, Mar. 2006.

908. E. Beres and R. Adve, "Blind channel estimation for orthogonal STBC in MISO systems", *IEEE Trans. Veh. Tech*, vol. 56, no. 4, pp. 2042–2050, July 2007.

909. S. He, J. K. Tugnait, and X. Meng, "On superimposed training for MIMO channel estimation and symbol detection", *IEEE Trans. Sig. Process.*, vol. 55, no. 6, pp. 3007–3021, June 1980.

910. M. Abuthinien, S. Chen, and L. Hanzo, "Semi-blind joint maximum likelihood channel estimation and data detection for MIMO systems", *IEEE Sig. Process. Lett.*, vol. 15, pp. 202–205, 2008.

911. F. Lehmann, "Blind estimation and detection of space-time trellis coded transmissions over the Rayleigh fading MIMO channel", *IEEE Trans. Commun.*, vol. 56, no. 3, pp. 334–338, Mar. 2008.

912. S. Zhou and G. B. Giannakis, "Finite-alphabet based channel estimation for OFDM and related multi-carrier systems", *IEEE Trans. Commun.*, vol. 49, no. 8, pp. 1402–1414, Aug. 2001.

913. B. Muquet, M. de Courville, and P. Duhamel, "Subspace-based blind and semi-blind channel estimation for OFDM systems", *IEEE Trans. Sig. Process.*, vol. 50, no. 7, pp. 1699–1712, July 2002.

914. S. Zhou, B. Muquet, and G. B. Giannakis, "Subspace-based (semi-) blind channel estimation for block precoded space-time OFDM", *IEEE Trans. Sig. Process.*, vol. 50, no. 5, pp. 1215–1228, May 2002.

915. S. Roy and C. Li, "A subspace blind channel estimation method for OFDM systems without cyclic prefix", *IEEE Trans. Wireless Commun.*, vol. 1, no. 4, pp. 572–579, Oct. 2002.

916. H. Wang, Y. Lin, and B. Chen, "Data-efficient blind OFDM channel estimation using receiver diversity", *IEEE Trans. Sig. Process.*, vol. 51, no. 10, pp. 2613–2623, Oct. 2003.

917. H. Li, "Blind channel estimation for multicarrier systems with narrowband interference suppression", *IEEE Commun. Lett.*, vol. 7, no. 7, pp. 326–328, July 2003.

918. Y. Zeng and T. S. Ng, "A proof of the identifiability of a subspace-based blind channel estimation for OFDM systems", *IEEE Sig. Process. Lett.*, vol. 11, no. 9, pp. 756–759, Sep. 2004.

919. M. C. Necker and G. L. Stüber, "Totally blind channel estimation for OFDM on fast varying mobile radio channels", *IEEE Trans. Wireless Commun.*, vol. 3, no. 5, pp. 1514–1525, Sep. 2004.

920. Y. Zeng and T.-S. Ng, "A semi-blind channel estimation method for multiuser multiantenna OFDM systems", *IEEE Trans. Sig. Process.*, vol. 52, no. 5, pp. 1419–1429, May 2004.

921. A. Petropulu, R. Zhang, and R. Lin, "Blind OFDM channel estimation through simple linear precoding", *IEEE Trans. Wireless Commun.*, vol. 3, no. 2, pp. 647–655, Mar. 2004.

922. D. Marelli and M. Fu, "A subband approach to channel estimation and equalization for DMT and OFDM systems", *IEEE Trans. Commun.*, vol. 53, no. 11, pp. 1850–1858, Nov. 2005.

923. R. Lin and A. P. Petropulu, "Linear precoding assisted blind channel estimation for OFDM systems", *IEEE Trans. Veh. Tech.*, vol. 54, no. 3, pp. 983–995, May 2005.

924. X. G. Doukopoulos and G. V. Moustakides, "Blind adaptive channel estimation in OFDM systems", *IEEE Trans. Wireless Commun.*, vol. 5, no. 7, pp. 1716–1725, July 2006.

925. M. Muck, M. de Courville, and P. Duhamel, "A pseudorandom postfix OFDM modulator – semi-blind channel estimation and equalization", *IEEE Trans. Sig. Process.*, vol. 54, no. 3, pp. 1005–1017, Mar. 2006.

926. Y. Ma, N. Yi, and R. Tafazolli, "Channel estimation for PRP-OFDM in slowly time-varying channel: First-order or second-order statistics?" *IEEE Sig. Process. Lett.*, vol. 13, no. 3, pp. 129–132, Mar. 2006.

927. F. Gao and A. Nallanathan, "Blind channel estimation for OFDM systems via a generalized precoding", *IEEE Trans. Veh. Tech.*, vol. 56, no. 3, pp. 1155–1164, May 2007.

928. T. Cui and C. Tellambura, "Semiblind channel estimation and data detection for OFDM systems with optimal pilot design", *IEEE Trans. Commun.*, vol. 55, no. 5, pp. 1053–1062, May 2007.

929. C. H. Aldana, E. de Carvalho, and J. M. Cioffi, "Channel estimation for multicarrier multiple input single output systems using the EM algorithm", *IEEE Trans. Sig. Process.*, vol. 51, no. 12, pp. 3280–3292, Dec. 2003.

930. Z. J. Wang, Z. Han, and K. J. R. Liu, "A MIMO-OFDM channel estimation approach using time of arrivals", *IEEE Trans. Wireless Commun.*, vol. 4, no. 3, pp. 1207–1213, May 2005.

931. Y. Zeng, W. H. Lam, and T. S. Ng, "Semiblind channel estimation and equalization for MIMO space-time coded OFDM", *IEEE Trans. Circ. and Syst. I*, vol. 53, no. 2, pp. 463–474, Feb. 2006.

932. S. Yatawatta and A. P. Petropulu, "Blind cannel estimation in MIMO OFDM systems with multiuser interference", *IEEE Trans. Sig. Process.*, vol. 54, no. 3, pp. 1054–1068, Mar. 2006.

933. H. A. Cirpan, E. Panayirci, and H. Dogan, "Non-data aided channel estimation for OFDM systems with space-frequency transmit diversity", *IEEE Trans. Veh. Tech.*, vol. 55, no. 2, pp. 449–457, Mar. 2006.

934. M. Cicerone, O. Simeone, and U. Spagnolini, "Channel estimation for MIMO-OFDM systems by modal analysis/filtering", *IEEE Trans. Commun.*, vol. 54, no. 11, pp. 2062–2074, Nov. 2006.

935. F. Gao and A. Nallanathan, "Blind channel estimation for MIMO OFDM systems via nonredundant linear precoding", *IEEE Trans. Sig. Process.*, vol. 55, no. 2, pp. 784–789, Jan. 2007.

936. C. Shin, R. W. Heath, and E. J. Powers, "Blind channel estimation for MIMO-OFDM systems", *IEEE Trans. Veh. Tech.*, vol. 56, no. 2, pp. 670–685, Mar. 2007.

937. Y. Zeng, A. R. Leyman, and T.-S. Ng, "Joint semiblind frequency offset and channel estimation for multiuser MIMO-OFDM uplink", *IEEE Trans. Commun.*, vol. 55, no. 12, pp. 2270–2278, Dec. 2007.

938. R. Chen, H. Zhang, Y. Xu, and H. Luo, "On MM-type channel estimation for MIMO OFDM systems", *IEEE Trans. Wireless Commun.*, vol. 6, no. 3, pp. 1046–1055, Mar. 2007.

939. A. Duel-Hallen, S. Hu, and H. Hallen, "Long-range prediction of fading signals", *IEEE Sig. Process. Mag.*, vol. 17, no. 3, pp. 62–75, May 2000.

940. B. Narendran, J. Sienicki, S. Yajnik, and P. Agrawal, "Evaluation of an adaptive power and error control algorithm for wireless systems", in *Proc. IEEE Int. Conf. Commun. (ICC 97)*, vol. 1, Montreal, June 1997, pp. 349–355.

941. T. Eyceoz, A. Duel-Hallen, and H. Hallen, "Deterministic channel modeling and long range prediction of fast fading mobile radio channels", *IEEE Commun. Lett.*, vol. 2, p. 254, Sep. 1987.

942. J. B. Andersen, J. Jensen, S. H. Jensen, and F. Fredriksen, "Prediction of future fading based on past measurements", in *Proc. IEEE Veh. Tech. Conf. (VTC 1999 – Fall)*, vol. 2, Amsterdam, Sep. 1999, pp. 151–155.

943. L. Dong, G. Xu, and H. Ling, "Prediction of fast fading mobile radio channels in wideband communication systems", in *Proc. IEEE Global Telecommun. Conf. (GLOBECOM '01)*, vol. 6, San Antonio, TX, Nov. 2002, pp. 3287–3291.

944. T. Ekman, M. Sternad, and A. Ahlen, "Unbiased power prediction of Rayleigh fading channels", in *Proc. IEEE 56th Veh. Tech. Conf. (VTC 2002 – Fall)*, vol. 1, Vancouver, 28 Apr.–2 May 2002, pp. 280–284.

945. J.-K. Hwang and J. H. Winters, "Sinusoidal modeling and prediction of fast fading processes", in *Proc. IEEE Global Telecommun. Conf. (GLOBECOM 1998)*, vol. 2, Sydney, 8–12 Nov. 1998, pp. 892–897.

946. A. Arredondo, K. R. Dandekar, and G. Xu, "Vector channel modeling and prediction for the improvement of downlink received power", *IEEE Trans. Commun.*, vol. 50, no. 7, pp. 1121–1129, July 2002.

947. I. Wong and B. Evans, "Exploiting spatio-temporal correlations in MIMO wireless channel prediction", in *Proc. IEEE Global Telecommun. Conf. (GLOBECOM '06)*, vol. 2, San Francisco, 27 Nov.–1 Dec. 2006, pp. 1–5.

948. A. Swindlehurst and M. Larsen, "Multiple-pass decision-directed channel estimation for highly mobile MIMO communications", in *Proc. Fourth IEEE Workshop Sensor Array and Multichannel Signal Processing*, Waltham, MA, July 2006, pp. 219–223.

949. K. Huber and S. Haykin, "Improved Bayesian MIMO channel tracking for wireless communications: incorporating a dynamical model", *IEEE Trans. Wireless Commun.*, vol. 5, no. 9, pp. 2458–2466, Sep. 2006.

950. J. Vanderpypen and L. Schumacher, "MIMO channel prediction using ESPRIT based techniques", in *Proc. 18th Annual IEEE Int. Symp. Personal, Indoor and Mobile Radio Commun (PIMRC 2007)*, Athens, 3–7 Sep. 2007, pp. 1–5.

951. D.-Z. Liu and C.-H. Wei, "Channel estimation and compensation for preamble-assisted DAPSK transmission in digital mobile radio system", *IEEE Trans. Veh. Tech.*, vol. 50, no. 2, pp. 546–556, Mar. 2001.

952. P. Stoica and O. Besson, "Training sequence design for frequency offset and frequency-selective channel estimation", *IEEE Trans. Commun.*, vol. 51, no. 11, pp. 1910–1917, Nov. 2003.

953. G.-T. Gil, I.-H. Sohn, J.-K. Park, and Y. H. Lee, "Joint ML estimation of carrier frequency, channel, I/Q mismatch, and DC offset in communication receivers", *IEEE Trans. Veh. Tech.*, vol. 54, no. 1, pp. 338–349, Jan. 2005.

954. O. Besson and P. Stoica, "On parameter estimation of MIMO flat-fading channels with frequency offsets", *IEEE Trans. Sig. Proc*, vol. 51, no. 3, pp. 602–613, Mar. 2003.

955. D. Qu, G. Zhu, and T. Jiang, "Training sequence design and parameter estimation of MIMO channels with carrier frequency offsets", *IEEE Trans. Wireless Commun.*, vol. 5, no. 12, pp. 3662–3666, Dec. 2006.

956. T.-H. Pham, A. Nallanathan, and Y.-C. Liang, "Joint channel and frequency offset estimation in distributed MIMO flat-fading channels", *IEEE Trans. Wireless Commun.*, vol. 7, no. 2, pp. 648–656, Feb. 2008.

957. J.-H. Lee, J. C. Han, and S.-C. Kim, "Joint carrier frequency synchronization and channel estimation for OFDM systems via the EM algorithm", *IEEE Trans. Veh. Tech.*, vol. 55, no. 1, pp. 167–172, Jan. 2006.

958. S. Gault, W. Hachem, and P. Ciblat, "Joint sampling clock offset and channel estimation for OFDM signals: Cramer-Rao bound and algorithms", *IEEE Trans. Sig. Process.*, vol. 54, no. 5, pp. 1875–1885, May 2006.

959. M.-O. Pun, M. Morelli, and C.-C. J. Kuo, "Maximum-likelihood synchronization and channel estimation for OFDMA uplink transmissions", *IEEE Trans. Commun.*, vol. 54, no. 4, pp. 726–736, Apr. 2006.

960. F. Z. Merli and G. M. Vitetta, "Iterative ML-based estimation of carrier frequency offset, channel impulse response and data in OFDM transmissions", *IEEE Trans. Commun.*, vol. 56, no. 3, pp. 497–506, Mar. 2008.

961. D. D. Lin, R. A. Pacheco, T. J. Lim, and D. Hatzinakos, "Joint estimation of channel response, frequency offset, and phase noise in OFDM", *IEEE Trans. Sig. Process.*, vol. 54, no. 9, pp. 3542–3554, Sep. 2006.

962. Y. Na and H. Minn, "Line search based iterative joint estimation of channels and frequency offsets for uplink OFDM systems", *IEEE Trans. Wireless Commun.*, vol. 6, no. 12, pp. 4374–4382, Dec. 2007.

963. T. Cui and C. Tellambura, "Joint frequency offset and channel estimation for OFDM systems using pilot symbols and virtual carriers", *IEEE Trans. Wireless Commun.*, vol. 6, no. 4, pp. 1193–1202, Apr. 2007.

964. X. Fu, H. Minn, and C. D. Cantrell, "Two novel iterative joint frequency-offset and channel estimation methods for OFDMA uplink", *IEEE Trans. Commun.*, vol. 56, no. 3, pp. 474–484, Mar. 2008.

965. H. Nguyen-Le, T. Le-Ngoc, and C. C. Ko, "Joint channel estimation and synchronization for MIMO-OFDM in the presence of carrier and sampling frequency offsets", *IEEE Trans. Veh. Tech.*, vol. 58, no. 6, pp. 3075–3081, July 2009.

966. X. Ma, M.-K. Oh, G. B. Giannakis, and D.-J. Park, "Hopping pilots for estimation of frequency-offset and multiantenna channels in MIMO-OFDM", *IEEE Trans. Commun.*, vol. 53, no. 1, pp. 162–172, Jan. 2005.

967. Y.-C. Ko and M.-S. Alouini, "Estimation of Nakagami-m fading channel parameters with application to optimized transmitter diversity systems", *IEEE Trans. Wireless Commun.*, vol. 2, no. 2, pp. 250–259, Mar. 2003.

968. V. Lomi, D. Tonetto, and L. Vangelista, "False alarm probability-based estimation of multipath channel length", *IEEE Trans. Commun.*, vol. 51, no. 9, pp. 1432–1434, Sep. 2003.

969. A. Dogandzic and J. Jin, "Maximum likelihood estimation of statistical properties of composite gamma-lognormal fading channels", *IEEE Trans. Sig. Process.*, vol. 52, no. 10, pp. 2940–2945, Oct. 2004.

970. K. E. Baddour and N. C. Beaulieu, "Robust Doppler spread estimation in nonisotropic fading channels", *IEEE Trans. Wireless Commun.*, vol. 4, no. 6, pp. 2677–2682, Nov. 2005.

971. Y. Chen and N. C. Beaulieu, "Maximum likelihood estimation of the K factor in Ricean fading channels", *IEEE Commun. Lett.*, vol. 9, no. 12, pp. 1040–1042, Dec. 2006.

972. A. Wiesel, J. Goldberg, and H. Messer-Yaron, "SNR estimation in time-varying fading channels", *IEEE Trans. Commun.*, vol. 54, no. 5, pp. 841–848, May 2006.

973. M. Nissila and S. Pasupathy, "Joint estimation of carrier frequency offset and statistical parameters of the multipath fading channel", *IEEE Trans. Commun.*, vol. 54, no. 6, pp. 1038–1048, June 2006.

974. J. Via, I. Santamaria, and J. Perez, "Effective channel order estimation based on combined identification/equalization", *IEEE Trans. Sig. Process.*, vol. 54, no. 9, pp. 3518–3526, Sep. 2006.

975. Y. Zeng and T. S. Ng, "Pilot cyclic prefixed single carrier communication: Channel estimation and equalization", *IEEE Sig. Process. Lett.*, vol. 12, no. 1, pp. 56–59, Jan. 2005.

976. M. Morelli, L. Sanguinetti, and U. Mengali, "Channel estimation for adaptive frequency-domain equalization", *IEEE Trans. Wireless Commun.*, vol. 4, no. 5, pp. 2508–2518, Sep. 2005.

977. J. Coon, M. Sandell, M. Beach, and J. McGeehan, "Channel and noise variance estimation and tracking algorithms for unique-word based single-carrier systems", *IEEE Trans. Wireless Commun.*, vol. 5, no. 6, pp. 1488–1496, June 2006.

978. Y. R. Zheng and C. Xiao, "Channel estimation for frequency-domain equalization of single-carrier broadband wireless communications", *IEEE Trans. Veh. Tech.*, vol. 58, no. 2, pp. 815–823, Feb. 2009.

979. S. F. Cotter and B. D. Rao, "Sparse channel estimation via matching pursuit with application to equalization", *IEEE Trans. Wireless Commun.*, vol. 50, no. 3, pp. 374–377, Mar. 2002.

980. O. Rabaste and T. Chonavel, "Estimation of multipath channels with long impulse response at low SNR via an MCMC method", *IEEE Trans. Sig. Process.*, vol. 55, no. 4, pp. 1312–1325, Apr. 2007.

981. C. Carbonelli, S. Vedantam, and U. Mitra, "Sparse channel estimation with zero tap detection", *IEEE Trans. Wireless Commun.*, vol. 6, no. 5, pp. 1743–1763, May 2007.

982. D. G. Forney, Jr., "The Viterbi algorithm", *IEEE Proc.*, vol. 61, no. 3, pp. 268–277, Mar. 1973.

983. S. Ariyavisitakul and L. J. Greenstein, "Reduced-complexity equalisation techniques for broadband wireless channels", *IEEE J. Sel. Areas Commun.*, vol. 15, pp. 5–15, Jan. 1997.

984. P. J. McLane, "A residual intersymbol interference error bound for truncated-state Viterbi detectors", *IEEE Trans. Inform. Theory*, vol. 26, pp. 549–553, Sep. 1980.

985. G. Benelli, A. Garzelli, and F. Salvi, "Simplified Viterbi processors for the GSM Pan-European cellular communication system", *IEEE Trans. Veh. Tech.*, vol. 43, pp. 870–878, Nov. 1994.

986. W. P. Chou and P. J. McLane, "16-state nonlinear equaliser for IS-54 digital cellular channels", *IEEE Trans. Veh. Tech.*, vol. 45, pp. 13–25, Feb. 1996.

987. G. Ferrari, G. Colavolpe, and R. Raheli, "On trellis-based truncated-memory detection", *IEEE Trans. Commun.*, vol. 53, no. 9, pp. 1462–1476, Sep. 2005.

988. A. D. Hallen and C. Heegard, "Delayed decision-feedback sequence estimation", *IEEE Trans. Commun.*, vol. 37, no. 5, pp. 428–436, May 1989.

989. M. V. Eyuboglu and S. U. H. Qureshi, "Reduced-state sequence estimation with set partitioning and decision feedback", *IEEE Trans. Commun.*, vol. 36, pp. 13–20, Jan. 1988.

990. _____, "Reduced-state sequence estimation for coded modulation on intersymbol interference channels", *IEEE J. Sel. Areas Commun.*, vol. 7, pp. 989–995, Aug. 1989.

991. H. C. Guren and N. Holte, "Decision feedback sequence estimation for continuous phase modulation on a linear multipath channel", *IEEE Trans. Commun.*, vol. 41, pp. 280–284, Feb. 1993.

992. G. Ungerboeck, "Channel coding with multilevel/phase signals", *IEEE Trans. Inform. Theory*, vol. 28, no. 1, pp. 55–67, Jan. 1982.

993. R. E. Kamel and Y. Bar-Ness, "Reduced-complexity sequence estimation using state partitioning", *IEEE Trans. Commun.*, vol. 44, pp. 1057–1063, Sep. 1996.

994. W.-H. Sheen and G. L. Stüber, "Error probability for maximum likelihood sequence estimation of trellis-coded modulation on ISI channels", *IEEE Trans. Commun.*, vol. 42, pp. 1427–1430, Feb./Mar./Apr. 1994.

995. M. Magarini, A. Spalvieri, and G. Tartara, "The mean-square delayed decision feedback sequence detector", *IEEE Trans. Commun.*, vol. 50, no. 9, pp. 1462–1470, Sep. 2002.

996. J. Zhang, A. M. Sayeed, and B. D. Van Veen, "Reduced-state MIMO sequence detection with application to EDGE systems", *IEEE Trans. Wireless Commun.*, vol. 4, no. 3, pp. 1040–1049, May 2005.

997. C. T. Beare, "The choice of the desired impulse response in combined linear-Viterbi algorithm equalisers", *IEEE Trans. Commun.*, vol. 26, pp. 1301–1307, Aug. 1978.

998. D. D. Falconer and F. R. Magee, "Adaptive channel memory truncation for maximum-likelihood sequence estimation", *Bell Syst. Tech. J.*, vol. 52, pp. 1541–1562, Nov. 1973.

999. W. U. Lee and F. Hill, "A maximum-likelihood sequence estimator with decision-feedback equalisation", *IEEE Trans. Commun.*, vol. 25, pp. 971–979, Sept. 1977.

1000. Y. Gu and T. Le-Ngoc, "Adaptive combined DFE/MLSE techniques for ISI channels", *IEEE Trans. Commun.*, vol. 44, pp. 847–857, July 1996.

1001. K. Wesołowski, "An efficient DFE & ML suboptimum receiver for data transmission over dispersive channels using two-dimensional signal constellation", *IEEE Trans. Commun.*, vol. 35, pp. 337–339, Mar. 1987.

1002. J. C. S. Cheung and R. Steele, "Soft-decision feedback equalizer for continuous phase modulated signals in wideband mobile radio channels", *IEEE Trans. Commun.*, vol. 42, pp. 1628–1638, Feb./Mar./Apr. 1994.

1003. J. B. Anderson, "Sequential coding algorithms: A survey and cost analysis", *IEEE Trans. Commun.*, vol. 32, pp. 169–176, Feb. 1984.

1004. _____, "Limited search trellis decoding of convolutional codes", *IEEE Trans. Inform. Theory*, vol. 35, pp. 944–955, Sept. 1989.

1005. A. P. Clark, S. N. Abdullah, S. G. Jaysinghe, and K. H. Sun, "Pseudobinary and pseudoquaternary detection processes for linearly distorted multilevel QAM signals", *IEEE Trans. Commun.*, vol. 44, pp. 127–129, Feb. 1988.

1006. H. Zamiri-Jafarian and S. Pasupathy, "Adaptive state allocation algorithm in MLSD receiver for multipath fading channels: Structure and strategy", *IEEE Trans. Veh. Tech.*, vol. 48, no. 1, pp. 174–187, Jan. 1999.

1007. T. M. Aulin, "Breadth-first maximum likelihood sequence detection: Basics", *IEEE Trans. Commun.*, vol. 47, pp. 208–216, Feb. 1999.

1008. _____, "Breadth-first maximum likelihood sequence detection: Geometry", *IEEE Trans. Commun.*, vol. 51, no. 6, pp. 2071–2080, June 2003.

1009. S. J. Simmons, "Breadth-first trellis decoding with adaptive effort", *IEEE Trans. Commun.*, vol. 38, pp. 3–12, Jan. 1990.

1010. A. Baier and D. G. Heinrich, "Performance of M algorithm MLSE equalizer in frequency selective fading mobile radio channels", in *Proc. IEEE Int. Conf. Commun. (ICC '89)*, Boston, June 1989, pp. 281–285.

1011. P. Jung, "Performance evaluation of a novel M-detector for coherent receiver antenna diversity in a GSM-type mobile radio system", *IEEE J. Sel. Areas Commun.*, vol. 13, pp. 80–88, Jan. 1995.

1012. R. Mehlan and H. Meyr, "Soft output M-algorithm equaliser and trellis-coded modulation for mobile radio channels", in *Proc. IEEE 42nd Veh. Tech. Conf, VTC 92*, Denver, 1992, pp. 582–591.

1013. M. Loncar and F. Rusek, "On reduced-complexity equalization based on Ungerboeck and Forney observation models", *IEEE Trans. Sig. Process.*, vol. 56, no. 8, pp. 3784–3789, Aug. 2008.

1014. K. Georgoulakis and S. Theodoridis, "Channel equalization for coded signals in hostile environments", *IEEE Trans. Sig. Process.*, vol. 47, no. 6, pp. 1783–1787, June 1999.

1015. K. K. Y. Wong and P. J. McLane, "Reduced-complexity equalization techniques for ISI and MIMO wireless channels in iterative decoding", *IEEE J. Sel. Areas Commun.*, vol. 26, no. 2, pp. 256–268, Feb. 2008.

1016. E. Katz and G. L. Stüber, "Sequential sequence estimation for trellis-coded modulation on multipath fading ISI channels", *IEEE Trans. Commun.*, vol. 43, pp. 2883–2885, Dec. 1995.

1017. M. Austin, "Decision feedback equalization for digital communication over dispersive channels", *MIT Lincoln Lab., Tech. Rep. 437*, Aug. 1967.

1018. P. Monsen, "Feedback equalization for fading dispersive channels", *IEEE Trans. Inform. Theory*, vol. 17, pp. 56–64, Jan. 1971.

1019. C. A. Belfiore and J. H. Park, "Decision feedback equalization", *Proc. IEEE*, vol. 67, pp. 1143–1156, Aug. 1979.

1020. S. U. H. Qureshi, "Adaptive equalization", *Proc. IEEE*, vol. 73, no. 9, pp. 1349–1387, Sept. 1985.

1021. G. K. Kaleh, "Channel equalization for block transmission systems", *IEEE J. Sel. Areas Commun.*, vol. 13, pp. 110–121, Jan. 1995.

1022. L. L. Scharf, *Statistical Signal Processing: Detection, Estimation and Time Series Analysis*. Reading, MA: Addison-Wesley, 1991.

1023. S. N. Crozier, D. D. Falconer, and S. A. Mahmoud, "Reduced complexity short-block data detection techniques for fading time-dispersive channels", *IEEE Trans. Veh. Tech.*, vol. 41, pp. 255–265, Aug. 1992.

1024. N. Al-Dhahir and J. M. Cioffi, "Block transmission over dispersive channels: transmit filter optimization and realization, and MMSE-DFE receiver performance", *IEEE Trans. Inform. Theory*, vol. 42, no. 1, pp. 137–160, Jan. 1996.

1025. J. R. Treichler, I. Fijalkow, and C. R. Johnson, Jr., "Fractionally spaced equalizers", *IEEE Sig. Process. Mag.*, vol. 13, no. 3, pp. 65–81, May 1996.

1026. N. Al-Dhahir and J. M. Cioffi, "MMSE decision feedback equalizers: Finite-length results", *IEEE Trans. Inform. Theory*, vol. 41, no. 4, pp. 961–976, July 1995.

1027. J. M. Cioffi, G. P. Dudevoir, M. V. Eyuboglu, and D. G. Forney, Jr., "MMSE decision-feedback equalizers and coding. I. Equalization results", *IEEE Trans. Commun.*, vol. 43, no. 10, pp. 2582–2594, Oct. 1995.

1028. J. E. Smee and N. C. Beaulieu, "On the equivalence of the simultaneous and separate MMSE optimisations of a DFE FFF and FBF", *IEEE Trans. Commun.*, vol. 45, pp. 156–159, Feb. 1997.

1029. N. Al-Dhahir and J. M. Cioffi, "Mismatched finite-complexity MMSE decision feedback equalizers", *IEEE Trans. Sig. Process.*, vol. 45, no. 4, pp. 935–944, Apr. 1997.

1030. N. Al-Dhahir, "Time-varying versus time-invariant finite-length MMSE-DFE on stationary dispersive channels", *IEEE Trans. Commun.*, vol. 46, pp. 11–15, Jan. 1998.

1031. S. Ariyavisitakul, "A decision feedback equaliser with time-reversal structure", *IEEE J. Sel. Areas Commun.*, vol. 10, no. 3, pp. 599–613, Apr. 1992.

1032. Y.-J. Liu, M. Wallace, and J. W. Ketchum, "A soft output bidirectional decision feedback equalization technique for TDMA cellular radio", *IEEE J. Sel. Areas Commun.*, vol. 11, pp. 1034–1045, Sept. 1993.

1033. T. Nagayasu, S. Sampei, and Y. Kamio, "Complexity reduction and performance improvement of a decision feedback equaliser for 16QAM in land mobile communications", *IEEE Trans. Veh. Tech.*, vol. 44, pp. 570–578, Aug. 1995.

1034. B. Farhang-Boroujeny, "Channel estimation via channel identification: Algorithms and simulation results for rapidly fading HF channels", *IEEE Trans. Commun.*, vol. 44, pp. 1409–1412, Nov. 1996.

1035. J. J. O'Reilly and A. M. de Oliveria Duarte, "Error propagation in decision feedback receivers", *IEE Proc. F. Commun., Radar and Sig. Process.*, vol. 132, pp. 561–566, Dec. 1985.

1036. D. L. Duttweiler, J. E. Mazo, and D. G. Messerschmitt, "An upper bound on the error probability in decision-feedback equalisation", *IEEE Trans. Inform. Theory*, vol. 20, pp. 490–497, July 1974.

1037. M. R. Aaron and M. K. Simon, "Approximation of the error probability in a regenerative repeater with quantized feedback", *Bell Syst. Tech. J.*, pp. 845–1847, Dec. 1966.

1038. A. M. de Oliveria Duarte and J. J. O'Reilly, "Simplified technique for bounding error statistics for DFB receivers", *IEE Proc. F. Commun., Radar and Sig. Process.*, vol. 132, pp. 567–575, Dec. 1985.

1039. R. A. Kennedy and B. D. O. Anderson, "Tight bounds on the error probabilities of decision feedback equalisers", *IEEE Trans. Commun.*, vol. 35, pp. 1022–1028, Oct. 1987.

1040. J. C. Cartledge, "Outage performance of QAM digital radio using adaptive equalisation and switched spaced diversity reception", *IEEE Trans. Commun.*, vol. 35, pp. 166–171, Feb. 1987.

1041. P. Balaban and J. Salz, "Optimum diversity combining and equalisation in digital data transmission with applications to cellular mobile radio – Part I: Theoretical considerations", *IEEE Trans. Commun.*, vol. 40, pp. 885–894, May 1992.

1042. ——, "Optimum diversity combining and equalisation in digital data transmission with applications to cellular mobile radio – Part II: Numerical results", *IEEE Trans. Commun.*, vol. 40, pp. 895–907, May 1992.

1043. R. Agusti and F. Cassadevall, "Performance of fractioned and nonfractioned equalisers with high-level QAM", *IEEE Trans. Commun.*, vol. 5, pp. 476–483, Apr. 1987.

1044. P. Monsen, "Adaptive equalization of the slow fading channel", *IEEE Trans. Commun.*, vol. 22, pp. 1064–1075, Aug. 1974.

1045. A. Cantoni and P. Butler, "Stability of decision feedback inverses", *IEEE Trans. Commun.*, vol. 24, pp. 970–977, Sep. 1976.

1046. R. A. Kennedy and B. D. O. Anderson, "Error recovery of decision feedback equalisers on exponential impulse response channels", *IEEE Trans. Commun.*, vol. 35, pp. 846–848, Aug. 1987.

1047. ——, "Recovery times of decision feedback equalisers on noiseless channels", *IEEE Trans. Commun.*, vol. 35, pp. 1012–1021, Aug. 1987.

1048. R. A. Kennedy, B. D. O. Anderson, and R. Bitmead, "Channels leading to rapid error recovery for decision feedback equalisers", *IEEE Trans. Commun.*, vol. 37, pp. 1126–1135, Nov. 1989.

1049. N. C. Beaulieu, "Bounds on recovery times of decision feedback equalizers", *IEEE Trans. Commun.*, vol. 42, pp. 2786–2794, Oct. 1994.

1050. R. Kennedy and Z. Ding, "Design and optimization of nonlinear mapping in decision feedback equalization", in *Proc. 35th IEEE Conf. on Decision and Control*, Kobe, 1996, pp. 1888–1889.

1051. D. P. Taylor, "The estimate feedback equalizer: A suboptimum nonlinear receiver", *IEEE Trans. Commun.*, vol. 21, no. 9, pp. 979–990, Sep. 1973.

1052. W. H. Gerstacker, R. R. Muller, and J. Huber, "Iterative equalization with adaptive soft feedback", *IEEE Trans. Commun.*, vol. 48, no. 9, pp. 1462–1466, Sep. 2000.

1053. Y.-H. Kim and S. Shamsunder, "Adaptive algorithms for channel equalization with soft decision feedback", *IEEE J. Sel. Areas Commun.*, vol. 16, no. 9, pp. 1660–1669, Dec. 1998.

1054. M. Reuter, J. C. Allen, J. R. Zeidler, and R. C. North, "Mitigating error propagation effects in a decision feedback equalizer", *IEEE Trans. Commun.*, vol. 49, no. 11, pp. 2028–2041, Nov. 2001.

1055. C.-E. Sundberg, "A class of soft decision error detectors for the Gaussian channel", *IEEE Trans. Commun.*, vol. 24, no. 1, pp. 106–112, Jan. 1976.

1056. ———, "Further studies of soft decision error detectors for the Gaussian channel", *IEEE Trans. Commun.*, vol. 24, no. 6, pp. 664–670, June 1976.

1057. E. Dahlman and B. Gudmundson, "Performance improvement in decision feedback equalisers by using soft decision", *Electron. Lett.*, vol. 24, pp. 1084–1085, Aug. 1988.

1058. E. Baccarelli, A. Fasano, and A. Zucchi, "A reduced-state soft-statistics-based MAP/DF equalizer for data transmission over long ISI channels", *IEEE Trans. Commun.*, vol. 48, no. 9, pp. 1441–1446, Sep. 2000.

1059. J. Thielecke, "A soft-decision state-space equalizer for FIR channels", *IEEE Trans. Commun.*, vol. 45, no. 10, pp. 1208–1217, Oct. 1997.

1060. M. Tomlinson, "New automatic equalizers employing modulo arithmetic", *Elect. Lett.*, vol. 7, no. 5, pp. 138–139, Mar. 1971.

1061. H. Harashima and H. Miyakawa, "Matched transmission technique for channels with intersymbol interference", *IEEE Trans. Commun. Tech.*, vol. 20, no. 4, pp. 774–780, Aug. 1972.

1062. D. E. Quevedo, G. C. Goodwin, and J. A. De Dona, "Multistep detector for linear ISI-channels incorporating degrees of belief in past estimates", *IEEE Trans. Commun.*, vol. 55, no. 11, pp. 2092–2103, Nov. 2007.

1063. H. Suzuki, "Performance of a new adaptive diversity-equalisation for digital mobile radio", *IEE Electron. Lett.*, vol. 26, no. 10, pp. 626–627, May 1990.

1064. J. K. Nelson, A. C. Singer, U. Madhow, and C. S. McGahey, "BAD: bidirectional arbitrated decision-feedback equalization", *IEEE Trans. Commun.*, vol. 53, no. 2, pp. 214–218, Feb. 2005.

1065. S. B. Gelfand, C. S. Ravishankar, and E. J. Delp, "Tree-structured piecewise linear adaptive equalisation", *IEEE Trans. Commun.*, vol. 41, pp. 70–82, Jan. 1993.

1066. R. Schober and W. H. Gerstacker, "Adaptive noncoherent DFE for MDPSK signals transmitted over ISI channels", *IEEE Trans. Commun.*, vol. 48, no. 7, pp. 1128–1140, July 2000.

1067. M. V. Eyuboglu, "Detection of coded modulation signals on linear, severely distorted channels using decision-feedback noise prediction with interleaving", *IEEE Trans. Commun.*, vol. 36, pp. 401–409, Apr. 1988.

1068. S. Elnoubi, H. Badr, and E. A. Youssef, "BER improvement of PRCPM in mobile radio channels with discriminator detection using decision feedback equalization", *IEEE Trans. Veh. Tech.*, vol. 40, pp. 694–699, Nov. 1991.

1069. L. Bin, "Decision feedback detection of minimum shift keying", *IEEE Trans. Commun.*, vol. 44, pp. 1073–1076, Sept. 1996.

1070. D. Boundreau and J. H. Lodge, "Adaptive equalization of CPM signals transmitted over fast Rayleigh flat-fading channels", *IEEE Trans. Veh. Tech.*, vol. 44, pp. 404–413, August 1995.

1071. S. Ariyavisitakul, N. R. Sollenberger, and L. J. Greenstein, "Tap-selectable decision-feedback equalization", *IEEE Trans. Commun.*, vol. 45, no. 12, pp. 1497–1500, Dec. 1997.

1072. I. J. Fevrier, S. B. Gelfand, and M. P. Fitz, "Reduced complexity decision feedback equalization for multipath channels with large delay spreads", *IEEE Trans. Commun.*, vol. 47, pp. 927–937, June 1999.

1073. G. Kutz and D. Raphaeli, "Determination of tap positions for sparse equalizers", *IEEE Trans. Commun.*, vol. 55, no. 9, pp. 1712–1724, Sep. 2007.

1074. S. Ohno, "Performance of single-carrier block transmissions over multipath fading channels with linear equalization", *IEEE Trans. Sig. Process.*, vol. 54, no. 10, pp. 3678–3687, Oct. 2006.

1075. L. Szczecinski, "Low-complexity search for optimal delay in linear FIR MMSE equalization", *IEEE Sig. Process. Lett.*, vol. 12, no. 8, pp. 549–552, Aug. 2005.

1076. J. G. Proakis and J. H. Miller, "An adaptive receiver for digital signalling through channels with intersymbol interference", *IEEE Trans. Inform. Theory*, vol. 15, pp. 484–497, July 1969.

1077. G. Ungerboeck, "Fractional tap-spacing equaliser and consequences for clock recovery in data modems", *IEEE Trans. Commun.*, vol. 24, pp. 856–864, Aug. 1976.

1078. R. D. Gitlin and S. B. Weinstein, "Fractionally-spaced equalization: An improved digital transversal equalizer", *Bell Syst. Tech. J*., vol. 60, pp. 856–864, Feb. 1981.

1079. K. Mueller and J. Werner, "A hardware efficient passband equalizer structure for data transmission", *IEEE Trans. Commun.*, vol. 30, no. 3, pp. 538–541, Mar. 1982.

1080. F. Ling and S. U. H. Qureshi, "Convergence and steady-state behavior of a phase-splitting fractionally spaced equaliser", *IEEE Trans. Commun.*, vol. 38, no. 4, pp. 418–425, Apr. 1990.

1081. J. Salz, "On mean-square decision feedback equalization and timing phase", *IEEE Trans. Commun. Technol.*, vol. 25, no. 12, pp. 1471–1476, Dec. 1977.

1082. A. Tajer, A. Nosratinia, and N. Al-Dhahir, "Diversity analysis of symbol-by-symbol linear equalizers", *IEEE Trans. Commun.*, vol. 59, no. 9, pp. 2343–2348, Sep. 2011.

1083. M. Barton and D. W. Tufts, "A suboptimum linear receiver based on a parametric channel model", *IEEE Trans. Commun.*, vol. 39, pp. 1328–1334, 1991.

1084. R. Schober, W. H. Gerstacker, and J. B. Huber, "Adaptive linear equalization combined with noncoherent detection for MDPSK signals", *IEEE Trans. Commun.*, vol. 48, no. 5, pp. 733–738, May 2000.

1085. _____, "Adaptive noncoherent linear minimum ISI equalization for MDPSK and MDAPSK signals", *IEEE Trans. Sig. Process.*, vol. 49, no. 9, pp. 2018–2030, Sep. 2001.

1086. K. Abend and B. D. Fritchman, "Statistical detection for communication channels with intersymbol interference", *Proc. IEEE*, vol. 58, pp. 779–785, May 1970.

1087. K. Abend, T. J. Hartley, B. D. Fritchman, and C. Gumacos, "On optimum receivers for channels having memory", *IEEE Trans. Inform. Theory*, vol. 14, pp. 152–157, Nov. 1968.

1088. Y. Li, B. Vucetic, and Y. Sato, "Optimum soft-output detection for channels with intersymbol interference", *IEEE Trans. Inform. Theory*, vol. 41, pp. 704–713, May 1995.

1089. J. F. Hayes, T. M. Cover, and J. B. Riera, "Optimal sequence detection and optimal symbol-by-symbol detection: Similar algorithms", *IEEE Trans. Commun.*, vol. 30, pp. 152–157, Jan. 1982.

1090. D. Williamson, R. A. Kennedy, and G. W. Pulford, "Block decision feedback equalization", *IEEE Trans. Commun.*, vol. 40, pp. 255–264, Feb. 1992.

1091. S. Chen, B. Mulgrew, and S. McLaughlin, "Adaptive Bayesian equalizer with decision feedback", *IEEE Sig. Process.*, vol. 41, pp. 2918–2926, Sept. 1993.

1092. J. Moon and L. R. Carley, "Efficient sequence detection for intersymbol interference channels with run-length constraints", *IEEE Trans. Commun.*, vol. 42, no. 9, pp. 2654–2660, Sep. 1994.

1093. G.-K. Lee, S. B. Gelfand, and M. P. Fitz, "Bayesian decision feedback techniques for deconvolution", *IEEE J. Sel. Areas Commun.*, vol. 13, pp. 155–166, Jan. 1995.

1094. _____, "Bayesian techniques for equalization of rapidly fading frequency selective channels", *Int. J. Wireless Inform. Networks*, vol. 2, pp. 41–53, Jan. 1995.

1095. X. Ma and A. Kavcic, "Path partitions and forward-only trellis algorithms", *IEEE Trans. Inform. Theory*, vol. 49, no. 1, pp. 38–52, Jan. 2003.

1096. J. Moon and F. R. Rad, "Turbo equalization via constrained-delay APP estimation with decision feedback", *IEEE Trans. Commun.*, vol. 53, no. 12, pp. 2102–2113, Dec. 2005.

1097. G. Colavolpe, G. Ferrari, and R. Raheli, "Reduced-state BCJR-type algorithms", *IEEE J. Sel. Areas Commun.*, vol. 19, no. 5, pp. 848–859, May 2001.

1098. V. Franz and J. B. Anderson, "Concatenated decoding with a reduced-search BCJR algorithm", *IEEE J. Sel. Areas Commun.*, vol. 16, no. 2, pp. 186–195, Feb. 1998.

1099. D. Fertonani, A. Barbieri, and G. Colavolpe, "Reduced-complexity BCJR algorithm for turbo equalization", *IEEE Trans. Commun.*, vol. 55, no. 12, pp. 2279–2287, Dec. 2007.

1100. J. Hagenauer, E. Offer, and L. Papke, "Iterative decoding of binary block and convolutional codes", *IEEE Trans. Inform. Theory*, vol. 42, no. 2, pp. 429–445, Mar. 1996.

1101. D. G. Brennan, "Linear diversity combining techniques", *Proc. IRE*, vol. 47, pp. 1075–1102, June 1959.

1102. Z. Wang and G. B. Giannakis, "A simple and general parameterization quantifying performance in fading channels", *IEEE Trans. Commun.*, vol. 51, no. 8, pp. 1389–1398, Aug. 2003.

1103. T. Eng, N. Kong, and L. B. Milstein, "Comparison of diversity combining techniques for Rayleigh-fading channels", *IEEE Trans. Commun.*, vol. 44, pp. 1117–1129, Sept. 1996.

1104. M. K. Simon and M.-S. Alouini, "A unified approach to the performance analysis of digital communications over generalized fading channels", *IEEE Proc.*, vol. 86, pp. 1860–1877, Sept. 1998.

1105. R. Price and P. E. Green, "A communication technique for multipath channels", *Proc. IRE*, vol. 46, pp. 555–570, Mar. 1958.

1106. E. J. Baghdady, "Novel techniques for counteracting multipath interference effects in receiving systems", *IEEE J. Sel. Areas Commun.*, vol. 5, pp. 274–285, Feb. 1987.

1107. U. Hansson and T. M. Aulin, "Reduced complexity decision feedback equalization for multipath channels with large delay spreads", *IEEE Trans. Commun.*, vol. 47, pp. 874–883, June 1999.

1108. D. Falconer, S. L. Ariyavisitakul, A. Benyamin-Seeyar, and B. Eidson, "Frequency domain equalization for single-carrier broadband wireless access systems", *IEEE Commun. Mag.*, vol. 40, no. 4, pp. 58–66, Apr. 2002.

1109. N. Benvenuto, R. Dinis, D. Falconer, and S. Tomasin, "Single carrier modulation with nonlinear frequency domain equalization: An idea whose time has come – again", *IEEE Proc.*, vol. 98, no. 1, pp. 69–96, Jan. 2010.

1110. H. Sari, G. Karam, and I. Jeanclaude, "Transmission technique for digital terrestrial broadcasting", *IEEE Commun. Mag.*, vol. 33, no. 2, pp. 100–109, Feb. 1995.

1111. A. Czylwik, "Comparison between adaptive OFDM and single carrier modulation with frequency domain equalization", in *Proc. IEEE 47th Veh. Technol. Conf. (VTC '97)*, vol. 2, Phoenix, AZ, 4–7 May 1997, pp. 865–869.

1112. J. Makhoul, "Linear prediction: A tutorial review", *IEEE Proc.*, vol. 63, no. 4, pp. 561–580, Apr. 1975.

1113. N. Benvenuto and S. Tomasin, "On the comparison between OFDM and single carrier modulation with a DFE using a frequency-domain feedforward filter", *IEEE Trans. Commun.*, vol. 50, pp. 947–955, June 2002.

1114. D. Falconer and S. L. Ariyavisitakul, "Broadband wireless using single carrier and frequency domain equalization", in *Proc. 5th Int. Symp. Wireless Pers. Multimedia Commun.*, vol. 1, Oct. 2002, pp. 27–36.

1115. S. Tomasin, "Efficient bi-directional DFE for doubly selective wireless channels", *EURASIP J. Appl. Sig. Process.*, pp. 1–10, Jan. 2006.

1116. X. Zhang, E. Chen, and X. Mu, "Single-carrier frequency-domain equalization based on frequency-domain oversampling", *IEEE Commun. Lett.*, vol. 16, no. 1, pp. 24–26, Jan. 2012.

1117. N. Benvenuto and S. Tomasin, "Iterative design and detection of a DFE in the frequency domain", *IEEE Trans. Commun.*, vol. 53, no. 11, pp. 1867–1875, Nov. 2005.

1118. C. Zhang, Z. Wang, C. Pan, S. Chen, and L. Hanzo, "Low-complexity iterative frequency domain decision feedback equalization", *IEEE Trans. Veh. Tech.*, vol. 60, no. 3, pp. 1295–1301, Mar. 2011.

1119. Z. Liu, "Maximum diversity in single-carrier frequency-domain equalization", *IEEE Trans. Inform. Theory*, vol. 51, no. 8, pp. 2937–2940, Aug. 2005.

1120. W. Zhang, "Comments on 'maximum diversity in single-carrier frequency-domain equalization' ", *IEEE Trans. Inform. Theory*, vol. 52, no. 3, pp. 1275–1277, Mar. 2006.

1121. A. Tajer and A. Nosratinia, "Diversity order in ISI channels with single-carrier frequency-domain equalizers", *IEEE Trans. Wireless Commun.*, vol. 9, no. 3, pp. 1022–1032, Mar. 2010.

1122. H.-M. Wang and Q. Yin, "Outage and diversity analysis of single carrier cyclic prefix systems with frequency domain decision feedback equalizers", in *Proc. IEEE Global Telecommun. Conf. (GLOBECOM 2011)*, 5–9 Dec. 2011, pp. 1–6.

1123. M. Luise, R. Reggiannini, and G. M. Vitetta, "Blind equalization/detection for OFDM signals over frequency-selective channels", *IEEE J. Sel. Areas Commun.*, vol. 16, no. 8, pp. 1568–1578, Oct. 1998.

1124. S. Kim and G. J. Pottie, "Robust OFDM in fast fading channels", in *Proc. IEEE Global Telecommun. Conf. (GLOBECOM '03)*, vol. 2, Dec. 2003, pp. 1074–1078.

1125. P. Schniter, "Low-complexity equalization of OFDM in doubly selective channels", *IEEE Trans. Sig. Process.*, vol. 52, no. 4, pp. 1002–1011, Apr. 2004.

1126. S. Tomasin, A. Gorokhov, H. Yang, and J.-P. Linnartz, "Iterative interference cancellation and channel estimation for mobile OFDM", *IEEE Trans. Wireless Commun.*, vol. 4, no. 1, pp. 238–245, Jan. 2005.

1127. Y. Zhao and S.-G. Haggman, "Intercarrier interference self-cancellation scheme for OFDM mobile communication systems", *IEEE Trans. Commun.*, vol. 49, no. 7, pp. 1185–1191, July 2001.

1128. A. Seyedi and G. J. Saulnier, "General ICI self-cancellation scheme for OFDM systems", *IEEE Trans. Veh. Technol.*, vol. 54, no. 1, pp. 198–210, Jan. 2005.

1129. M.-X. Chang, "A novel algorithm of inter-subchannel interference self-cancellation for OFDM systems", *IEEE Trans. Wireless Commun.*, vol. 6, no. 8, pp. 2881–2893, Aug. 2007.

1130. H.-C. Wu, X. Huang, and D. Xu, "Novel semi-blind ICI equalization algorithm for wireless OFDM systems", *IEEE Trans. Broadcasting*, vol. 52, no. 2, pp. 211–218, June 2006.

1131. H.-C. Wu, X. Huang, Y. Wu, and X. Wang, "Theoretical studies and efficient algorithm of semi-blind ICI equalization for OFDM", *IEEE Trans. Wireless Commun.*, vol. 7, no. 10, pp. 3791–3798, Oct. 2008.

1132. J.-P. M. G. Linnartz and A. Gorokhov, "New equalization approach for OFDM over dispersive and rapidly time varying channel", in *Proc. 11th IEEE Int. Symp. Pers., Indoor and Mobile Radio Commun. (PIMRC 2000)*, vol. 2, London, 18–21 Sep. 2000, pp. 1375–1379.

1133. A. Gorokhov and J.-P. Linnartz, "Robust OFDM receivers for dispersive time-varying channels: equalization and channel acquisition", *IEEE Trans. Commun.*, vol. 52, no. 4, pp. 572–583, Apr. 2004.

1134. T. Wang, J. G. Proakis, and J. R. Zeidler, "Techniques for suppression of intercarrier interference in OFDM systems", in *Proc. IEEE Wireless Commun. and Net. Conf. (WCNC 2005)*, vol. 1, 13–17 Mar. 2005, pp. 39–44.

1135. W.-S. Hou and B.-S. Chen, "ICI cancellation for OFDM communication systems in time-varying multipath fading channels", *IEEE Trans. Wireless Commun.*, vol. 4, no. 5, pp. 2100–2110, Sep. 2005.

1136. I. Barhumi, G. Leus, and M. Moonen, "Equalization for OFDM over doubly selective channels", *IEEE Trans. Sig. Process.*, vol. 54, no. 4, pp. 1445–1458, Apr. 2006.

1137. K. Chang, Y. Han, J. Ha, and Y. Kim, "Cancellation of ICI by Doppler effect in OFDM systems", in *Proc. IEEE 63rd Veh. Technol. Conf. (VTC 2006 – Spring)*, vol. 3, May 2006, pp. 1411–1415.

1138. S. Lu, R. Kalbasi, and N. Al-Dhahir, "OFDM interference mitigation algorithms for doubly-selective channels", in *Proc. IEEE 64th Veh. Technol. Conf. (VTC 2006 – Fall)*, Montreal, 2006, pp. 1–5.

1139. H.-W. Wang, D. W. Lin, and T.-H. Sang, "OFDM signal detection in doubly selective channels with whitening of residual intercarrier interference and noise", in *Proc. IEEE 71st Veh. Technol. Conf. (VTC 2010 – Spring)*, 2010, pp. 1–5.

1140. Y. Mostofi and D. C. Cox, "ICI mitigation for pilot-aided OFDM mobile systems", *IEEE Trans. Wireless Commun.*, vol. 4, no. 2, pp. 765–774, Mar. 2005.

1141. G. Li, H. Yang, L. Cai, and L. Gui, "A low-complexity equalization technique for OFDM system in time-variant multipath channels", in *Proc. 2003 IEEE 58th Veh. Tech. Conf. (VTC 2003 – Fall)*, vol. 4, Oct. 2003, pp. 2466–2470.

1142. K. Kim and H. Park, "A low complexity ICI cancellation method for high mobility OFDM systems", in *Proc. IEEE 63rd Veh. Techn. Conf. (VTC 2006 – Spring)*, vol. 5, May 2006, pp. 2528–2532.

1143. S. U. Hwang, J. H. Lee, and J. Seo, "Low complexity iterative ICI cancellation and equalization for OFDM systems over doubly selective channels", *IEEE Trans. Broadcasting*, vol. 55, no. 1, pp. 132–139, Mar. 2009.

1144. D. N. Liu and M. P. Fitz, "Iterative MAP equalization and decoding in wireless mobile coded OFDM", *IEEE Trans. Commun.*, vol. 57, no. 7, pp. 2042–2051, July 2009.

1145. S.-J. Hwang and P. Schniter, "Efficient sequence detection of multicarrier transmissions over doubly dispersive channels", *EURASIP J. Appl. Sig. Process.*, vol. 2006, pp. 1–17, 2006.

1146. P. Baracca, S. Tomasin, L. Vangelista, N. Benvenuto, and A. Morello, "Per sub-block equalization of very long OFDM blocks in mobile communications", *IEEE Trans. Commun.*, vol. 59, no. 2, pp. 363–368, Feb. 2011.

1147. N. Al-Dhahir and J. M. Cioffi, "Efficiently computed reduced-parameter input-aided MMSE equalizers for ML detection: a unified approach", *IEEE Trans. Inform. Theory*, vol. 42, no. 3, pp. 903–915, May 1996.

1148. P. J. W. Melsa, R. C. Younce, and C. E. Rohrs, "Impulse response shortening for discrete multitone transceivers", *IEEE Trans. Commun.*, vol. 44, no. 12, pp. 1662–1672, Dec. 1996.

1149. C. Yin and G. Yue, "Optimal impulse response shortening for discrete multitone transceivers", *Electr. Lett.*, vol. 34, no. 1, pp. 35–36, Jan. 1998.

1150. H. Schmidt and K.-D. Kammeyer, "Impulse truncation for wireless OFDM systems", in *Proc. 5th Int. OFDM-Workshop (InOWo 2000)*, vol. 3, Hamburg, 2000, pp. 2619–2622.

1151. B. Farhang-Boroujeny and M. Ding, "Design methods for time-domain equalizers in DMT transceivers", *IEEE Trans. Commun.*, vol. 49, no. 3, pp. 554–562, Mar. 2001.

1152. G. Arslan, B. L. Evans, and S. Kiaei, "Equalization for discrete multitone transceivers to maximize bit rate", *IEEE Trans. Sig. Process.*, vol. 49, no. 12, pp. 3123–3135, Dec. 2001.

1153. R. K. Martin, J. Balakrishnan, W. A. Sethares, and C. R. Johnson, Jr., "A blind adaptive TEQ for multicarrier systems", *IEEE Sig. Process. Lett.*, vol. 9, no. 11, pp. 341–343, Nov. 2002.

1154. J. Zhang, W. Ser, and J. Zhu, "Effective optimisation method for channel shortening in OFDM systems", *IEE Proc. Commun.*, vol. 150, no. 2, pp. 85–90, Apr. 2003.

1155. R. K. Martin, M. Ding, B. L. Evans, and C. R. Johnson, Jr., "Infinite length results and design implications for time-domain equalizers", *IEEE Trans. Sig. Process.*, vol. 52, no. 1, pp. 297–301, Jan. 2004.

1156. R. K. Martin, K. Vanbleu, M. Ding, G. Ysebaert, M. Milosevic, B. L. Evans, M. Moonen, and C. R. Johnson, Jr., "Unification and evaluation of equalization structures and design algorithms for discrete multitone modulation systems", *IEEE Trans. Sig. Process.*, vol. 53, no. 10, pp. 3880–3894, Oct. 2005.

1157. C. Toker and G. Altin, "Blind, adaptive channel shortening equalizer algorithm which can provide shortened channel state information (BACS-SI)", *IEEE Trans. Sig. Process.*, vol. 57, no. 4, pp. 1483–1493, Apr. 2009.

1158. T. Karp, C. Bauer, and N. J. Fliege, "Optimal one-tap equalization for DMT transceivers with insufficient guard interval", in *Proc. IEEE Int. Conf. Acoust., Speech and Sig. Process. (ICASSP '05)*, vol. 3, Philadelphia, 2005, pp. 881–884.

1159. S. Trautmann, T. Karp, and N. J. Fliege, "Frequency domain equalization of DMT/OFDM systems with insufficient guard interval", in *Proc. IEEE Int. Conf. Commun. (ICC 2002)*, vol. 3, 28 Apr.-2 May 2002, pp. 1646–1650.

1160. F. D. Beaulieu and B. Champagne, "MMSE equalization for zero padded multicarrier systems with insufficient guard length", in *Proc. IEEE Int. Conf. Acoust., Speech and Sig. Process. (ICASSP '06)*, vol. 4, Toulouse, 2006, pp. 377–380.

1161. Y. Sun and L. Tong, "Channel equalization for wireless OFDM systems with ICI and ISI", in *Proc. IEEE Int. Conf. Commun. (ICC '99)*, vol. 1, Vancouver, 6–10 June 1999, pp. 182–186.

1162. Y. Sun, "Bandwidth-efficient wireless OFDM", *IEEE J. Sel. Areas Commun.*, vol. 19, no. 11, pp. 2267–2278, Nov. 2001.

1163. S. Chen and T. Yao, "Blind algorithm for RIBI mitigation in OFDM systems", *Electr. Lett.*, vol. 38, no. 22, pp. 1382–1383, Oct. 2002.

1164. W.-R. Wu and C.-Y. Hsu, "Decision feedback IBI mitigation in OFDM systems", in *Proc. IEEE Int. Symp. Circ. and Syst. (ISCAS 2005)*, vol. 3, Kobe, 23–26 May 2005, pp. 2619–2622.

1165. S. Celebi, "Interblock interference (IBI) and time of reference (TOR) computation in OFDM systems", *IEEE Trans. Commun.*, vol. 49, no. 11, pp. 1895–1900, Nov. 2001.

1166. H.-C. Wu, "Analysis and characterization of intercarrier and interblock interferences for wireless mobile OFDM systems", *IEEE Trans. Broadcasting*, vol. 52, no. 2, pp. 203–210, June 2006.

1167. D. W. Matolak and S. G. Wilson, "Detection for a statistically known, time-varying dispersive channel", *IEEE Trans. Commun.*, vol. 44, no. 12, pp. 1673–1683, Dec. 1996.

1168. K. M. Chugg, "The condition for the applicability of the Viterbi algorithm with implications for fading channel MLSD", *IEEE Trans. Commun.*, vol. 46, pp. 1112–1116, Sept. 1998.

1169. G. Ferrari, G. Colavolpe, and R. Raheli, "A unified framework for finite-memory detection", *IEEE J. Sel. Areas Commun.*, vol. 23, no. 9, pp. 1697–1706, Sep. 2005.

1170. A. Aghamohammadi, H. Meyr, and G. Ascheid, "A new method for phase synchronization and automatic gain control of linearly modulated signals on frequency-flat fading channels", *IEEE Trans. Commun.*, vol. 39, pp. 25–29, Jan. 1991.

1171. B. D. Hart and D. P. Taylor, "On the irreducible error floor in fast fading channels", *IEEE Trans. Commun.*, vol. 49, no. 3, pp. 1044–1047, May 2000.

1172. A. N. D'Andrea, A. Diglio, and U. Mengali, "Symbol-aided channel estimation with nonselective Rayleigh fading channels", *IEEE Trans. Veh. Tech.*, vol. 44, pp. 41–49, Jan. 1995.

1173. B. D. Hart, "Maximum likelihood sequence detection using a pilot tone", *IEEE Trans. Veh. Tech.*, vol. 49, no. 2, pp. 550–560, Mar. 2000.

1174. J. P. Seymour and M. P. Fitz, "Near optimal symbol-by-symbol detection schemes for flat Rayleigh fading", *IEEE Trans. Commun.*, vol. 43, pp. 1525–1533, Feb./Mar./Apr. 1995.

1175. A. Hansson, K. M. Chugg, and T. Aulin, "On forward-adaptive versus forward/backward-adaptive SISO algorithms for Rayleigh fading channels", *IEEE Commun. Lett.*, vol. 5, no. 12, pp. 477–479, Dec. 2001.

1176. Y. Zhang, M. P. Fitz, and S. B. Gelfand, "Optimal and near-optimal joint channel and data estimation for frequency-selective Rayleigh fading channels", in *Proc. Thirty-third Annual Allerton Conference on Communication, Control, and Computing*, University of Illinois, Urbana-Champaign, 1999, pp. 618–627.

1177. A. A. Hansson and T. M. Aulin, "Iterative diversity detection for correlated continuous-time Rayleigh fading channels", *IEEE Trans. Commun.*, vol. 51, no. 2, pp. 240–246, Feb. 2003.

1178. A. Anastasopoulos, K. M. Chugg, G. Colavolpe, G. Ferrari, and R. Raheli, "Iterative detection for channels with memory", *IEEE Proc.*, vol. 95, no. 6, pp. 1272–1294, June 2007.

1179. A. Anastasopoulos and K. M. Chugg, "Adaptive soft-input soft-output algorithms for iterative detection with parametric uncertainty", *IEEE Trans. Commun.*, vol. 48, no. 10, pp. 1638–1649, Oct. 2000.

1180. A. Barbieri, "Blind per-state detection of DPSK over correlated fading channels", *IEEE Trans. Veh. Tech.*, vol. 59, no. 5, pp. 2320–2327, June 2010.

1181. M. Nissila and S. Pasupathy, "Adaptive Bayesian and EM-based detectors for frequency-selective fading channels", *IEEE Trans. Commun.*, vol. 51, no. 8, pp. 1325–1336, Aug. 2003.

1182. D. Divsalar and M. K. Simon, "Maximum likelihood differential detection of uncoded and trellis coded amplitude phase modulation over AWGN and fading channels – metrics and performance", *IEEE Trans. Commun.*, vol. 42, pp. 76–89, Jan. 1994.

1183. M. Samiuddin and K. H. Biyari, "A comparative study of higher-order differential phase shift keying schemes over AWGN and Rayleigh fading channels", *Int. J. Wireless Inform. Networks*, vol. 2, pp. 183–196, 1995.

1184. I. Korn, "Binary PRCPM with differential phase detection and maximal ratio combining diversity in satellite mobile, land mobile and Gaussian channels", in *Proc. IEEE Int. Conf. Commun. (ICC '96)*, Dallas, May 1996, pp. 916–920.

1185. D. Lao and A. M. Haimovich, "Multiple-symbol differential detection with interference suppression", *IEEE Trans. Commun.*, vol. 51, no. 2, pp. 208–217, Feb. 2003.

1186. K. H. Chang, W. Yuan, and C. N. Georghiades, "Block-by-block channel and sequence estimation for ISI/fading channels", in *Proc. 7th Tyrrhenian Workshop on Digital Communications*, Sep. 1995, pp. 153–170.

1187. J. C. Han and C. N. Georghiades, "Sequence estimation in the presence of random parameters via the EM algorithm", *IEEE Trans. Commun.*, vol. 45, pp. 300–308, Mar. 1997.

1188. ――――, "Pilot symbol initiated optimal decoder for the land mobile fading channel", in *Proc. IEEE Global Telecommun. Conf. (GLOBECOM '95)*, Singapore, Nov. 1995, pp. 42–47.

1189. D. P. Bouras, P. T. Mathiopoulos, and D. Makrakis, "Optimal detection of coded differentially encoded PSK and QAM signals with diversity in correlated fast fading channels", *IEEE Trans. Veh. Tech.*, vol. 42, pp. 245–257, Aug. 1993.

1190. L. H.-J. Lampe, V. Pauli, and C. Windpassinger, "Multiple-symbol differential sphere decoding", *IEEE Trans. Commun.*, vol. 53, no. 12, pp. 1981–1985, Dec. 2005.

1191. U. Fincke and M. Pohst, "Improved methods for calculating vectors of short length in a lattice, including a complexity analysis", *Mathematics of Computation*, vol. 44, no. 170, pp. 463–471, 1985.

1192. H. Vikalo, B. Hassibi, and T. Kailath, "Iterative decoding for MIMO channels via modified sphere decoding", *IEEE Trans. Wireless Commun.*, vol. 3, no. 6, pp. 2299–2311, Nov. 2004.

1193. J. Jalden and B. Ottersten, "On the complexity of sphere decoding in digital communications", *IEEE Trans. Sig. Process.*, vol. 53, no. 4, pp. 1474–1484, Apr. 2005.

1194. P. Y. Kam, "Generalized quadratic receivers for orthogonal signals over the Gaussian channel with unknown phase/fading", *IEEE Trans. Commun.*, vol. 43, pp. 2050–2058, June 1995.

1195. J. H. Painter and L. R. Wilson, "Simulation results for the decision-directed MAP receiver for M-ary signals in multiplicative and additive Gaussian noise", *IEEE Trans. Commun.*, vol. 22, pp. 649–660, May 1974.

1196. J. H. Painter and S. C. Gupta, "Recursive ideal observer detection of known M-ary signals in multiplicative and additive Gaussian noise", *IEEE Trans. Commun.*, vol. 21, pp. 948–953, August 1973.

1197. R. Schober, W. H. Gerstacker, and J. B. Huber, "Decision-feedback differential detection of MDPSK for flat Rayleigh fading channels", *IEEE Trans. Commun.*, vol. 47, pp. 1025–1035, July 1999.

1198. J. P. Seymour and M. P. Fitz, "Two-stage carrier synchronization techniques for non-selective fading", *IEEE Trans. Veh. Tech.*, vol. 44, pp. 103–110, Feb. 1995.

1199. L. H.-J. Lampe and R. Schober, "Low-complexity iterative demodulation for noncoherent coded transmission over Ricean-fading channels", *IEEE Trans. Veh. Tech.*, vol. 50, no. 6, pp. 1481–1496, Nov. 2001.

1200. R. Schober, W. H. Gerstacker, and J. B. Huber, "Decision-feedback differential detection based on linear prediction for 16DAPSK signals transmitted over flat Ricean fading channels", *IEEE Trans. Commun.*, vol. 49, no. 8, pp. 1339–1342, Aug. 2001.

1201. B. Bhukania and P. Schniter, "On the robustness of decision-feedback detection of DPSK and differential unitary space-time modulation in Rayleigh-fading channels", *IEEE Trans. Wireless Commun.*, vol. 3, no. 5, pp. 1481–1489, Sep. 2004.

1202. P. A. Bello and B. D. Nelin, "The influence of fading spectrum on the binary error probabilities of incoherent and differentially coherent matched filter receivers", *IRE Trans. Commun. Syst.*, vol. 10, pp. 160–168, June 1962.

1203. W. D. Lindsey, "Error probability for incoherent diversity reception", *IEEE Trans. Inform. Theory*, vol. 11, pp. 491–499, Oct. 1965.

1204. L. J. Mason, "Error probability evaluation for systems employing differential detection in a Rician fast fading environment and Gaussian noise", *IEEE Trans. Commun.*, vol. 35, pp. 39–46, Jan. 1987.

1205. _____, "An error probability formula for M-ary DPSK in fast Rician fading and Gaussian noise", *IEEE Trans. Commun.*, vol. 35, pp. 976–978, Jan . 1987.

1206. H. Salwen, "Differential phase-shift keying performance under time-selective multipath fading", *IEEE Trans. Commun.*, vol. 23, pp. 383–385, March 1975.

1207. Y. Miyagaki, N. Morinaga, and T. Namekawa, "Error rate performance of M-ary DPSK systems in Satellite/Aircraft communications", *Proc. IEEE Int. Conf. Commun. (ICC '79)*, pp. 34.6.1–34.6.6, 1979.

1208. A. Neul, "Bit error rate for 4-DPSK in fast Rician fading and Gaussian noise", *IEEE Trans. Commun.*, vol. 37, pp. 1385–1387, Dec. 1989.

1209. G. L. Turin, "On optimal diversity reception", *IRE Trans. Inform. Theory*, vol. 7, no. 3, pp. 154–167, July 1961.

1210. J. N. Pierce, "Theoretical diversity improvement in frequency-shift keying", *Proc. IRE*, vol. 46, pp. 903–910, May 1958.

1211. P. M. Hahn, "Theoretical diversity improvement in multiple frequency shift keying", *IRE Trans. Commun. Systems*, vol. 10, pp. 177–184, June 1962.

1212. I. Korn, "Error floors in the satellite and mobile channels", *IEEE Trans. Commun.*, vol. 39, pp. 833–837, June 1991.

1213. L. Andriot, G. Tziritas, and G. Jourdain, "Discrete realization for receivers. detecting signals over random dispersive channels. Part II: Doppler-spread channel", *Sig. Process.*, vol. 9, pp. 89–100, Sept. 1985.

1214. S. Elnoubi, "Probability of error analysis of digital partial response continuous phase modulation with noncoherent detection in mobile radio channels", *IEEE Trans. Veh. Tech.*, vol. 38, pp. 19–30, February 1989.

1215. I. Korn, "Error probability of M-ary FSK with differential phase detection in satellite mobile channel", *IEEE Trans. Veh. Tech.*, vol. 38, pp. 76–85, May 1989.

1216. _____, "GMSK with differential phase detection in the satellite mobile channel", *IEEE Trans. Commun.*, vol. 38, pp. 1980–1986, Nov. 1990.

1217. S. Elnoubi, "Analysis of GMSK with discriminator detection in mobile radio channels", *IEEE Trans. Veh. Tech.*, vol. 35, pp. 71–76, May 1986.

1218. I. Korn, "M-ary frequency shift keying with limiter-discriminator-integrator detector in satellite mobile channel with narrow-band receiver filter", *IEEE Trans. Commun.*, vol. 38, pp. 1771–1778, Oct. 1990.

1219. _____, "GMSK with limiter discriminator detection in satellite mobile channel", *IEEE Trans. Commun.*, vol. 39, pp. 94–101, Jan. 1991.

1220. D. K. Asano and S. Pasupathy, "Improved post-detection processing for limiter-discriminator detection of CPM in a Rayleigh, fast fading channel", *IEEE Trans. Veh. Tech.*, vol. 44, pp. 729–734, Nov. 1995.

1221. T. Cui and C. Tellambura, "Joint data detection and channel estimation for OFDM systems", *IEEE Trans. Commun.*, vol. 54, no. 4, pp. 670–679, Apr. 2006.

1222. B. Hassibi and H. Vikalo, "On the sphere-decoding algorithm I. Expected complexity", *IEEE Trans. Sig. Process.*, vol. 53, no. 8, pp. 2806–2818, Aug. 2005.

1223. T. Cui and C. Tellambura, "Blind receiver design for OFDM systems over doubly selective channels", *IEEE Trans. Commun.*, vol. 55, no. 5, pp. 906–917, May 2007.

1224. G. D'Aria, R. Piermarini, and V. Zingarelli, "Fast adaptive equalisers for narrow-band TDMA mobile radio", *IEEE Trans. Veh. Tech.*, vol. 40, pp. 392–404, May 1991.

1225. E. Dahlman, "New adaptive Viterbi detector for fast-fading mobile radio channels", *Electron. Lett.*, vol. 26, pp. 1572–1573, Sept. 1990.

1226. K. A. Hamied and G. L. Stüber, "Performance of trellis-coded modulation for equalised multipath fading ISI channels", *IEEE Trans. Veh. Tech.*, vol. 44, pp. 50–58, Feb. 1995.

1227. D. D. Falconer, A. U. H. Sheikh, E. Eleftheriou, and M. Tobis, "Comparison of DFE and MLSE receiver performance on HF channels", *IEEE Trans. Commun.*, vol. 33, pp. 484–486, May 1985.

1228. M. Erkurt and J. G. Proakis, "Joint data detection and channel estimation for rapidly fading channels", in *Proc. IEEE Global Telecommun. Conf. (GLOBECOM '92)*, Orlando, FL, Dec. 1992, pp. 543–546.

1229. S. Gazor, A. M. Rabiei, and S. Pasupathy, "Synchronized per survivor MLSD receiver using a differential Kalman filter", *IEEE Trans. Commun.*, vol. 50, no. 3, pp. 364–368, Mar. 2002.

1230. D. K. Borah and B. D. Hart, "A robust receiver structure for time-varying, frequency-flat, Rayleigh fading channels", *IEEE Trans. Commun.*, vol. 47, no. 3, pp. 360–364, Mar 1999.

1231. G. Paparisto and K. M. Chugg, "PSP array processing for multipath fading channels", *IEEE Trans. Commun.*, vol. 47, pp. 504–507, Apr. 1995.

1232. G. Kutz and D. Raphaeli, "Maximum-likelihood semiblind equalization of doubly selective channels using the EM algorithm", *EURASIP J. Adv. Sig. Process.*, vol. 2010, July 2010.

1233. E. Baccarelli, R. Cusani, and S. Galli, "A novel adaptive receiver with enhanced channel tracking capability for TDMA-based mobile radio communications", *IEEE J. Sel. Areas Commun.*, vol. 16, pp. 1630–1639, Dec. 1998.

1234. R. Cusani and J. Mattila, "Equalization of digital radio channels with large multipath delay for cellular land mobile applications", *IEEE Trans. Commun.*, vol. 47, pp. 348–351, Mar. 1999.

1235. E. Baccarelli and R. Cusani, "Combined channel estimation and data detection using soft statistics for frequency-selective fast-fading digital links", *IEEE Trans. Commun.*, vol. 46, pp. 424–427, Apr. 1998.

1236. R. A. Iltis, J. J. Shynk, and K. Giridhar, "Bayesian algorithms for blind equalisation using parallel adaptive filtering", *IEEE Trans. Commun.*, vol. 42, pp. 1017–1032, Feb./Mar./Apr. 1994.

1237. Y. Zhang, S. B. Gelfand, and M. P. Fitz, "Soft-output demodulation on frequency-selective Rayleigh fading channels using AR channel models", *IEEE Trans. Commun.*, vol. 55, no. 10, pp. 1929–1939, Aug. 2007.

1238. K. Giridhar, J. J. Shynk, A. R. Iltis, and A. Mahur, "Adaptive MAPSD algorithms for symbol and timing recovery of mobile radio TDMA signals", *IEEE Trans. Commun.*, vol. 44, pp. 927–978, Aug. 1996.

1239. F. N. Nunes and J. M. N. Leitao, "A nonlinear filtering approach to estimation and detection in mobile communications", *IEEE J. Sel. Areas Commun.*, vol. 16, pp. 1649–1659, Dec. 1998.

1240. A. Anastasopoulos and A. Polydoros, "Adaptive soft-decision algorithms for mobile fading channels", *European Trans. Telecommun.*, vol. 9, pp. 183–190, Mar./Apr. 1998.

1241. I. Bar-David and A. Elia, "Augumented APP ($A^2 P^2$) for a posteriori probability calculation and channel parameter tracking", *IEEE Commun. Lett.*, vol. 3, no. 1, pp. 18–20, Jan. 1999.

1242. M. P. Fitz, "Comments on tone calibration in time-varying fading", *IEEE Trans. Veh. Tech.*, vol. 41, pp. 211–213, May 1992.

1243. _____, "A dual-tone reference digital demodulator for mobile digital communications", *IEEE Trans. Veh. Tech.*, vol. 42, pp. 156–165, May 1993.

1244. J. H. Lodge, M. L. Moher, and S. N. Crozier, "A comparison of data modulation techniques for land mobile satellite channel", *IEEE Trans. Veh. Tech.*, vol. 36, pp. 28–34, Feb. 1987.

1245. J. K. Cavers and M. Liao, "A comparison of pilot tone and pilot symbol techniques for digital mobile communication", in *Proc. IEEE Global Telecommun. Conf. GLOBECOM '92*, 1992, pp. 915–921.

1246. W. J. Weber, "Performance of phase-locked loops in the presence of fading communication channels", *IEEE Trans. Commun.*, vol. 24, pp. 487–499, May 1976.

1247. P. Chen and H. Kobayashi, "Maximum likelihood channel estimation and signal detection for OFDM systems", in *Proc. IEEE Int. Conf. Commun. (ICC 2002)*, vol. 3, 28 Apr.–2 May 2002, pp. 1640–1645.

1248. X. Wang, P. Ho, and Y. Wu, "Robust channel estimation and ISI cancellation for OFDM systems with suppressed features", *IEEE J. Sel. Areas Commun.*, vol. 23, no. 5, pp. 963–972, May 2005.

1249. S. B. Bulumulla, S. A. Kassam, and S. S. Venkatesh, "A systematic approach to detecting OFDM signals in a fading channel", *IEEE Trans. Commun.*, vol. 48, no. 5, pp. 725–728, May 2000.

1250. M.-X. Chang and T.-D. Hsieh, "Detection of OFDM signals in fast-varying channels with low-density pilot symbols", *IEEE Trans. Veh. Tech.*, vol. 57, no. 2, pp. 859–872, Mar. 2008.

1251. K. Muraoka, K. Fukawa, H. Suzuki, and S. Suyama, "Channel estimation using differential model of fading fluctuation for EM algorithm applied to OFDM MAP detection", in *Proc. IEEE 18th Int. Symp. Personal, Indoor and Mobile Radio Commun. (PIMRC 2007)*, Athens, 3–7 Sep. 2007, pp. 1–5.

1252. T. Y. Al-Naffouri, "An EM-based forward-backward Kalman filter for the estimation of time-variant channels in OFDM", *IEEE Trans. Sig. Process.*, vol. 55, no. 7, July 2007.

1253. H. Hijazi and L. Ros, "Joint data QR-detection and Kalman estimation for OFDM time-varying Rayleigh channel complex gains", *IEEE Trans. Commun.*, vol. 58, no. 1, pp. 170–178, Jan. 2010.

1254. F. Li, S. Zhu, and M. Rong, "Detection for OFDM systems with channel estimation errors using variational inference", *IEEE Sig. Process. Lett.*, vol. 16, no. 5, pp. 434–437, May 2009.

1255. M.-X. Chang and Y. T. Su, "Blind and semiblind detections of OFDM signals in fading channels", *IEEE Trans. Commun.*, vol. 52, no. 5, pp. 744–754, May 2004.

1256. L. He, S. Ma, Y.-C. Wu, and T.-S. Ng, "Semiblind iterative data detection for OFDM systems with CFO and doubly selective channels", *IEEE Trans. Commun.*, vol. 58, no. 12, pp. 3491–3499, Dec. 2010.

1257. M.-L. Ku, W.-C. Chen, and C.-C. Huang, "EM-based iterative receivers for OFDM and BICM/OFDM systems in doubly selective channels", *IEEE Trans. Wireless Commun.*, vol. 10, no. 5, pp. 1405–1415, May 2011.

1258. S. A. Banani and R. G. Vaughan, "OFDM with iterative blind channel estimation", *IEEE Trans. Veh. Tech.*, vol. 59, no. 9, pp. 4298–4308, Nov. 2010.

1259. C. Tidestav, A. Ahlen, and M. Sternad, "Realizable MIMO decision feedback equalizers: structure and design", *IEEE Trans. Sig. Process.*, vol. 49, no. 1, pp. 121–133, Jan. 2001.

1260. N. Al-Dhahir, "FIR channel-shortening equalizers for MIMO ISI channels", *IEEE Trans. Commun.*, vol. 49, no. 2, pp. 213–218, Feb. 2001.

1261. B. Hassibi, A. T. Erdogan, and T. Kailath, "MIMO linear equalization with an H^∞ criterion", *IEEE Trans. Sig. Process.*, vol. 54, no. 2, pp. 499–511, Feb. 2006.

1262. A. Gomaa and N. Al-Dhahir, "A new design framework for sparse FIR MIMO equalizers", *IEEE Trans. Commun.*, vol. 59, no. 8, pp. 2132–2140, Aug. 2011.

1263. C. Toker, S. Lambotharan, and J. A. Chambers, "Joint transceiver design for MIMO channel shortening", *IEEE Trans. Sig. Process.*, vol. 55, no. 7, pp. 3851–3866, July 2007.

1264. R. Lopez-Valcarce, "Realizable minimum mean-squared error channel shorteners", *IEEE Trans. Sig. Process.*, vol. 53, no. 11, pp. 4354–4362, Nov. 2005.

1265. F. Rusek and A. Prlja, "Optimal channel shortening for MIMO and ISI channels", *IEEE Trans. Wireless Commun.*, vol. 11, no. 2, pp. 810–818, Feb. 2012.

1266. H. Sampath, P. Stoica, and A. Paulraj, "Generalized linear precoder and decoder design for MIMO channels using the weighted MMSE criterion", *IEEE Trans. Commun.*, vol. 49, no. 12, pp. 2198–2206, Dec. 2001.

1267. A. Scaglione, P. Stoica, S. Barbarossa, G. B. Giannakis, and H. Sampath, "Optimal designs for space-time linear precoders and decoders", *IEEE Trans. Sig. Process.*, vol. 50, no. 5, pp. 1051–1064, May 2002.

1268. L.-U. Choi and R. D. Murch, "A transmit MIMO scheme with frequency domain pre-equalization for wireless frequency selective channels", *IEEE Trans. Wireless Commun.*, vol. 3, no. 3, pp. 929–938, May 2004.

1269. N. Al-Dhahir and A. H. Sayed, "The finite-length multi-input multi-output MMSE-DFE", *IEEE Trans. Sig. Process.*, vol. 48, no. 10, pp. 2921–2936, Oct. 2000.

1270. N. Al-Dhahir, A. F. Naguib, and A. R. Calderbank, "Finite-length MIMO decision feedback equalization for space-time block-coded signals over multipath-fading channels", *IEEE Trans. Veh. Tech.*, vol. 50, no. 4, pp. 1176–1182, July 2001.

1271. A. T. Erdogan, B. Hassibi, and T. Kailath, "MIMO decision feedback equalization from an H^∞ perspective", *IEEE Trans. Sig. Process.*, vol. 52, no. 3, pp. 734–745, Mar. 2004.

1272. G. J. Foschini, G. D. Golden, R. A. Valenzuela, and P. W. Wolniansky, "Simplified processing for high spectral efficiency wireless communication employing multi-element arrays", *IEEE J. Select. Areas Commun.*, vol. 17, no. 11, pp. 1841–1852, Nov. 1999.

1273. A. Lozano and C. Papadias, "Layered space-time receivers for frequency-selective wireless channels", *IEEE Trans. Commun.*, vol. 50, no. 1, pp. 65–73, Jan. 2002.

1274. Y. Guo and B. C. Levy, "Robust MSE equalizer design for MIMO communication systems in the presence of model uncertainties", *IEEE Trans. Sig. Process.*, vol. 54, no. 5, pp. 1840–1852, May 2006.

1275. T.-J. Ho and B.-S. Chen, "Robust minimax MSE equalizer designs for MIMO wireless communications with time-varying channel uncertainties", *IEEE Trans. Sig. Process.*, vol. 58, no. 11, pp. 5835–5844, Nov. 2010.

1276. C. Windpassinger, R. F. H. Fischer, T. Vencel, and J. B. Huber, "Precoding in multiantenna and multiuser communications", *IEEE Trans. Wireless Commun.*, vol. 3, no. 4, pp. 1305–1316, July 2004.

1277. O. Simeone, Y. Bar-Ness, and U. Spagnolini, "Linear and nonlinear preequalization/equalization for MIMO systems with long-term channel state information at the transmitter", *IEEE Trans. Wireless Commun.*, vol. 3, no. 2, pp. 373–378, Mar. 2004.

1278. H. Vikalo and B. Hassibi, "Maximum-likelihood sequence detection of multiple antenna systems over dispersive channels via sphere decoding", *EURASIP J. Appl. Sig. Process.*, vol. 5, pp. 525–531, 2002.

1279. Y. Li and J. Moon, "Reduced-complexity soft MIMO detection based on causal and noncausal decision feedback", *IEEE Trans. Sig. Process.*, vol. 56, no. 3, pp. 1178–1187, Mar. 2008.

1280. R. D. Murch and K. B. Letaief, "Antenna systems for broadband wireless access", *IEEE Commun. Mag.*, vol. 40, no. 4, pp. 76–83, Apr. 2002.

1281. J. Coon, S. Armour, M. Beach, and J. McGeehan, "Adaptive frequency domain equalization for single-carrier multiple-input multiple-output wireless transmissions", *IEEE Trans. Sig. Process.*, vol. 53, no. 8, pp. 3247–3256, Aug. 2005.

1282. J. Zhang, Y. R. Zheng, C. Xiao, and K. B. Letaief, "Channel equalization and symbol detection for single-carrier MIMO systems in the presence of multiple carrier frequency offsets", *IEEE Trans. Veh. Tech.*, vol. 59, no. 4, pp. 2021–2030, May 2010.

1283. J. Tubbax, L. Van der Perre, S. Donnay, and M. Engels, "Single-carrier communications using decision-feedback equalization for multiple antennas", in *Proc. IEEE Int. Conf. Commun. (ICC 2003)*, vol. 4, Anchorage, AK, May 2003, pp. 2321–2325.

1284. X. Zhu and R. D. Murch, "Layered space-frequency equalization in a single-carrier MIMO system for frequency-selective channels", *IEEE Trans. Wireless Commun.*, vol. 3, no. 3, pp. 701–708, May 2004.

1285. R. Kalbasi, R. Dinis, D. D. Falconer, and A. Banihashemi, "Hybrid time-frequency layered space-time receivers for severe time-dispersive channels", in *Proc. 5th IEEE Workshop Sig. Process. Adv. Wireless Commun. (SPAWC 2004)*, Lisbon, 11–14 July 2004, pp. 218–222.

1286. R. Kalbasi, R. Dinis, D. D. Falconer, and A. H. Banihashemi, "Layered space-time receivers for single-carrier transmission with iterative frequency-domain equalization", in *Proc. 2004 IEEE 59th Veh. Technol. Conf. (VTC 2004 – Spring)*, vol. 1, Milan, May 2004, pp. 575–579.

1287. J. Xu, H. Wang, S. Cheng, and M. Chen, "Parallel multistage decision feedback equalizer for single-carrier layered space-time systems in frequency-selective channels", *EURASIP J. Appl. Sig. Process.*, pp. 1489–1497, 2004.

1288. R. Dinis, R. Kalbasi, D. D. Falconer, and A. H. Banihashemi, "Iterative layered space-time receivers for single-carrier transmission over severe time-dispersive channels", *IEEE Commun. Lett.*, vol. 8, no. 9, pp. 579–581, Sep. 2004.

1289. S. Ahmed, T. Ratnarajah, M. Sellathurai, and C. F. N. Cowan, "Reduced-complexity iterative equalization for severe time-dispersive MIMO channels", *IEEE Trans. Veh. Tech.*, vol. 57, no. 1, pp. 594–600, Jan. 2008.

1290. Y. Zhu and K. B. Letaief, "Single-carrier frequency-domain equalization with decision-feedback processing for time-reversal space-time block-coded systems", *IEEE Trans. Commun.*, vol. 53, no. 7, pp. 1127–1131, July 2005.

1291. ———, "Single carrier frequency domain equalization with time domain noise prediction for wideband wireless communications", *IEEE Trans. Wireless Commun.*, vol. 5, no. 12, pp. 3548–3557, Dec. 2006.

1292. ———, "Single-carrier frequency-domain equalization with noise prediction for MIMO systems", *IEEE Trans. Commun.*, vol. 55, no. 5, pp. 1063–1076, May 2007.

1293. ———, "Frequency domain equalization with Tomlinson-Harashima precoding for single carrier broadband MIMO systems", *IEEE Trans. Wireless Commun.*, vol. 6, no. 12, pp. 4420–4431, Dec. 2007.

1294. G. Kongara, D. P. Taylor, and P. A. Martin, "Space-frequency decision feedback equalizer", *IEEE Trans. Veh. Tech.*, vol. 60, no. 4, pp. 1626–1639, May 2011.

1295. N. Benvenuto and S. Tomasin, "Block iterative DFE for single carrier modulation", *IEE Electron. Lett.*, vol. 38, no. 19, pp. 1144–1145, Sep. 2002.

1296. R. Dinis, R. Kalbasi, D. D. Falconer, and A. H. Banihashemi, "Channel estimation for MIMO systems employing single-carrier modulations with iterative frequency-domain equalization", in *Proc. 2004 IEEE 60th Veh. Tech. Conf. (VTC 2004 – Fall)*, vol. 7, Los Angeles, Sep. 2004, pp. 4942–4946.

1297. M. O. Damen, A. Chkeif, and J.-C. Belfiore, "Lattice codes decoder for space-time codes", *IEEE Commun. Lett.*, vol. 4, no. 5, pp. 161–163, May 2000.

1298. D. Pham, K. R. Pattipati, P. K. Willett, and J. Luo, "An improved complex sphere decoder for V-BLAST systems", *IEEE Sig. Process. Lett.*, vol. 11, no. 9, pp. 748–751, Sep. 2004.

1299. L. G. Barbero and J. S. Thompson, "Fixing the complexity of the sphere decoder for MIMO detection", *IEEE Trans. Wireless Commun.*, vol. 7, no. 6, pp. 2131–2142, June 2008.

1300. B. Steingrimsson, Z.-Q. Luo, and K. M. Wong, "Soft quasi-maximum-likelihood detection for multiple-antenna wireless channels", *IEEE Trans. Ant. and Prop.*, vol. 51, no. 11, pp. 2710–2719, Nov. 2003.

1301. E. G. Larsson and J. Jalden, "Fixed-complexity soft MIMO detection via partial marginalization", *IEEE Trans. Sig. Process.*, vol. 56, no. 8, pp. 3397–3407, Aug. 2008.

1302. D. P. Palomar and S. Barbarossa, "Designing MIMO communication systems: Constellation choice and linear transceiver design", *IEEE Trans. Sig. Process.*, vol. 53, no. 10, pp. 3804–3818, Oct. 2005.

1303. S. Bergman, D. P. Palomar, and B. Ottersten, "Joint bit allocation and precoding for MIMO systems with decision feedback detection", *IEEE Trans. Sig. Process.*, vol. 57, no. 11, pp. 4509–4521, Nov. 2009.

1304. S. Jarmyr, B. Ottersten, and E. A. Jorswieck, "Statistical precoding with decision feedback equalization over a correlated MIMO channel", *IEEE Trans. Sig. Process.*, vol. 58, no. 12, pp. 6298–6311, Dec. 2010.

1305. E. Chiavaccini and G. M. Vitetta, "Further results on differential space-time modulations", *IEEE Trans. Commun.*, vol. 51, no. 7, pp. 1093–1101, July 2003.

1306. C. Ling, K. H. Li, A. C. Kot, and Q. T. Zhang, "Multisampling decision-feedback linear prediction receivers for differential space-time modulation over Rayleigh fast-fading channels", *IEEE Trans. Commun.*, vol. 51, no. 7, pp. 1214–1223, July 2003.

1307. R. Schober and L. H.-J. Lampe, "Noncoherent receivers for differential space-time modulation", *IEEE Trans. Commun.*, vol. 50, no. 5, pp. 768–777, May 2002.

1308. L. H.-J. Lampe and R. Schober, "Bit-interleaved coded differential space-time modulation", *IEEE Trans. Commun.*, vol. 50, no. 9, pp. 1429–1439, Sep. 2002.

1309. A. Song and X.-G. Xia, "Decision feedback differential detection for differential orthogonal space-time modulation with APSK signals over flat-fading channels", *IEEE Trans. Wireless Commun.*, vol. 3, no. 6, pp. 1873–1878, Nov. 2004.

1310. Z. Du and N. C. Beaulieu, "Decision-feedback detection for block differential space-time modulation", *IEEE Trans. Commun.*, vol. 54, no. 5, pp. 900–910, May 2006.

1311. P. Tarasak, H. Minn, and V. K. Bhargava, "Improved approximate maximum-likelihood receiver for differential space-time block codes over Rayleigh-fading channels", *IEEE Trans. Veh. Tech.*, vol. 53, no. 2, pp. 461–468, Mar. 2004.

1312. T. Cui and C. Tellambura, "Bound-intersection detection for multiple-symbol differential unitary space-time modulation", *IEEE Trans. Commun.*, vol. 53, no. 12, pp. 2114–2123, Dec. 2005.

1313. C. Gao, A. M. Haimovich, and D. Lao, "Multiple-symbol differential detection for MPSK space-time block codes: decision metric and performance analysis", *IEEE Trans. Commun.*, vol. 54, no. 8, pp. 1502–1510, Aug. 2006.

1314. P. K. M. Pun and P. K. M. Ho, "Fano multiple-symbol differential detectors for differential unitary space-time modulation", *IEEE Trans. Commun.*, vol. 55, no. 3, pp. 540–550, Apr. 2007.

1315. V. Pauli and L. Lampe, "Tree-search multiple-symbol differential decoding for unitary space-time modulation", *IEEE Trans. Commun.*, vol. 55, no. 8, pp. 1567–1576, Aug. 2007.

1316. T. Cui and C. Tellambura, "On multiple symbol detection for diagonal DUSTM over Ricean channels", *IEEE Trans. Wireless Commun.*, vol. 7, no. 4, pp. 1146–1151, Apr. 2008.

1317. N. Jin, X. P. Jin, Y. G. Ying, S. Wang, and Y. F. Lv, "Multiple-symbol M-bound intersection detector for differential unitary space-time modulation", *IET Commun.*, vol. 4, no. 16, pp. 1987–1997, Nov. 2010.

1318. Y. Xue and X. Zhu, "Per-survivor processing-based decoding for space-time trellis code", *IEEE Trans. Veh. Tech.*, vol. 52, no. 4, pp. 1173–1178, July 2003.

1319. M. Uysal and C. N. Georghiades, "An efficient implementation of a maximum-likelihood detector for space-time block coded systems", *IEEE Trans. Commun.*, vol. 51, no. 4, pp. 521–524, Apr. 2003.

1320. K. J. Kim, J. Yue, R. A. Iltis, and J. D. Gibson, "A QRD-M/Kalman filter-based detection and channel estimation algorithm for MIMO-OFDM systems", *IEEE Trans. Wireless Commun.*, vol. 4, no. 2, pp. 710–721, Mar. 2005.

1321. H. Kawai, K. Higuchi, N. Maeda, and M. Sawahashi, "Adaptive control of surviving symbol replica candidates in QRM-MLD for OFDM MIMO multiplexings", *IEEE Trans. Commun. Technol.*, vol. 24, no. 6, pp. 1130–1140, Oct. 2006.

1322. M.-S. Baek, Y.-H. You, and H.-K. Song, "Combined QRD-M and DFE detection technique for simple and efficient signal detection in MIMO-OFDM systems", *IEEE Trans. Wireless Commun.*, vol. 8, no. 4, pp. 1632–1638, Apr. 2009.

1323. Y. Liu, Y. Li, D. Li, and H. Zhang, "Super-low-complexity QR decomposition-M detection scheme for MIMO-OFDM systems", *IET Commun.*, vol. 5, no. 9, pp. 1303–1307, June 2011.

1324. D. Cescato and H. Bölcskei, "Algorithms for interpolation-based QR decomposition in MIMO-OFDM systems", *IEEE Trans. Sig. Process.*, vol. 59, no. 4, pp. 1719–1733, Apr. 2011.

1325. S. Ahmed, T. Ratnarajah, M. Sellathurai, and C. F. N. Cowan, "Iterative receivers for MIMO-OFDM and their convergence behavior", *IEEE Trans. Veh. Tech.*, vol. 58, no. 1, pp. 461–468, Jan. 2009.

1326. G. L. Stüber, J. R. Barry, S. W. McLaughlin, Y. Li, M. A. Ingram, and T. G. Pratt, "Broadband MIMO-OFDM wireless communications", *IEEE Proc.*, vol. 92, no. 2, pp. 271–294, Feb. 2004.

1327. R. J. Piechocki, P. N. Fletcher, A. R. Nix, C. N. Canagarajah, and J. P. McGeehan, "Performance evaluation of BLAST-OFDM enhanced Hiperlan/2 using simulated and measured channel data", *IEE Electron. Lett.*, vol. 37, no. 18, pp. 1137–1139, Aug. 2001.

1328. X. Li and X. Cao, "Low complexity signal detection algorithm for MIMO-OFDM systems", *IEE Electron. Lett.*, vol. 41, no. 2, pp. 83–85, Jan. 2005.

1329. M. Jiang, J. Akhtman, and L. Hanzo, "Soft-information assisted near-optimum nonlinear detection for BLAST-type space division multiplexing OFDM systems", *IEEE Trans. Wireless Commun.*, vol. 6, no. 4, pp. 1230–1234, Apr. 2007.

1330. W. Zhang, X.-G. Xia, and P. C. Ching, "Full-diversity and fast ML decoding properties of general orthogonal space-time block codes for MIMO-OFDM systems", *IEEE Trans. Wireless Commun.*, vol. 6, no. 5, pp. 1647–1653, May 2007.

1331. S. Li, D. Huang, K. B. Letaief, and Z. Zhou, "Pre-DFT processing for MIMO-OFDM systems with space-time-frequency coding", *IEEE Trans. Wireless Commun.*, vol. 6, no. 11, pp. 4176–4182, Nov. 2007.

1332. H. Bölcskei, R. W. Heath, Jr., and A. J. Paulraj, "Blind channel identification and equalization in OFDM-based multiantenna systems", *IEEE Trans. Sig. Process.*, vol. 50, no. 1, pp. 96–109, Jan. 2002.

1333. T. Pande, D. J. Love, and J. V. Krogmeier, "A weighted least squares approach to precoding with pilots for MIMO-OFDM", *IEEE Trans. Sig. Process.*, vol. 54, no. 10, pp. 4067–4073, Oct. 2006.

1334. J. Choi, B. Mondal, and R. W. Heath, "Interpolation based unitary precoding for spatial multiplexing MIMO-OFDM with limited feedback", *IEEE Trans. Sig. Process.*, vol. 54, no. 12, pp. 4730–4740, Dec. 2006.

1335. C.-Y. Chen and P. P. Vaidyanathan, "Precoded FIR and redundant V-BLAST systems for frequency-selective MIMO channels", *IEEE Trans. Sig. Process.*, vol. 55, no. 7, pp. 3390–3404, July 2007.

1336. T. Pande, D. J. Love, and J. V. Krogmeier, "Reduced feedback MIMO-OFDM precoding and antenna selection", *IEEE Trans. Sig. Process.*, vol. 55, no. 5, pp. 2284–2293, May 2007.

1337. L. Yi and Z. Hailin, "Interpolation-based precoding with limited feedback for MIMO-OFDM systems", *IET Commun.*, vol. 1, no. 4, pp. 679–683, Aug. 2007.

1338. B. Vrigneau, J. Letessier, P. Rostaing, L. Collin, and G. Burel, "Extension of the MIMO precoder based on the minimum Euclidean distance: A cross-form matrix", *IEEE J. Sel. Topics Sig. Process.*, vol. 2, no. 2, pp. 135–146, Apr. 2008.

1339. B. Kim, S.-Y. Jung, J. Kim, and D.-J. Park, "Hidden pilot based precoder design for MIMO-OFDM systems", *IEEE Commun. Lett.*, vol. 12, no. 9, pp. 657–659, Sep. 2008.

1340. R. Zhang and J. M. Cioffi, "Approaching MIMO-OFDM capacity with zero-forcing V-BLAST decoding and optimized power, rate, and antenna-mapping feedback", *IEEE Trans. Sig. Process.*, vol. 56, no. 10, pp. 5191–5203, Oct. 2008.

1341. F. Rey, M. Lamarca, and G. Vazquez, "Linear precoder design through cut-off rate maximization in MIMO-OFDM coded systems with imperfect CSIT", *IEEE Trans. Sig. Process.*, vol. 58, no. 3, pp. 1741–1755, Mar. 2010.

1342. K. J. Kim, M.-O. Pun, and R. A. Iltis, "QRD-based precoded MIMO-OFDM systems with reduced feedback", *IEEE Trans. Commun.*, vol. 58, no. 2, pp. 394–398, Feb. 2010.

1343. K. W. Park and Y. S. Cho, "A MIMO-OFDM technique for high-speed mobile channels", *IEEE Commun. Lett.*, vol. 9, no. 7, pp. 604–606, July 2005.

1344. Y. Fu, C. Tellambura, and W. A. Krzymien, "Transmitter precoding for ICI reduction in closed-loop MIMO OFDM systems", *IEEE Trans. Veh. Tech.*, vol. 56, no. 1, pp. 115–125, Jan. 2007.

1345. D. N. Dao and C. Tellambura, "Intercarrier interference self-cancellation space-frequency codes for MIMO-OFDM", *IEEE Trans. Veh Tech.*, vol. 54, no. 5, pp. 1729–1738, Sep. 2005.

1346. J.-G. Kim and J.-T. Lim, "MAP-based channel estimation for MIMO-OFDM over fast Rayleigh fading channels", *IEEE Trans. Veh. Tech.*, vol. 57, no. 3, pp. 1963–1968, May 2008.

1347. S. Lu, B. Narasimhan, and N. Al-Dhahir, "A novel SFBC-OFDM scheme for doubly selective channels", *IEEE Trans. Veh. Tech.*, vol. 58, no. 5, pp. 2573–2578, June 2009.

1348. C.-Y. Hsu and W.-R. Wu, "Low-complexity ICI mitigation methods for high-mobility SISO/MIMO-OFDM systems", *IEEE Trans. Veh. Tech.*, vol. 58, no. 6, pp. 2755–2768, July 2009.

1349. C. Pirak, Z. J. Wang, K. J. R. Liu, and S. Jitapunkul, "Adaptive channel estimation using pilot-embedded data-bearing approach for MIMO-OFDM systems", *IEEE Trans. Sig. Process.*, vol. 54, no. 12, pp. 4706–4716, Dec. 2006.

1350. Y. Ma and R. Tafazolli, "Channel estimation for OFDMA uplink: a hybrid of linear and BEM interpolation approach", *IEEE Trans. Sig. Process.*, vol. 55, no. 4, pp. 1568–1573, Apr. 2007.

1351. J. Gao and H. Liu, "Low-complexity MAP channel estimation for mobile MIMO-OFDM systems", *IEEE Trans. Wireless Commun.*, vol. 7, no. 3, pp. 774–780, Jan. 1989.

1352. G. Coluccia, E. Riegler, C. Mecklenbrauker, and G. Taricco, "Optimum MIMO-OFDM detection with pilot-aided channel state information", *IEEE J. Sel. Topics Sig. Process.*, vol. 3, no. 6, pp. 1053–1065, Dec. 2009.

1353. I.-W. Lai, G. Ascheid, H. Meyr, and T.-D. Chiueh, "Efficient channel-adaptive MIMO detection using just-acceptable error rate", *IEEE Trans. Wireless Commun.*, vol. 10, no. 1, pp. 73–83, Jan. 2011.

1354. K.-G. Wu and J.-A. Wu, "Efficient decision-directed channel estimation for OFDM systems with transmit diversity", *IEEE Commun. Lett.*, vol. 15, no. 7, pp. 740–742, July 2011.

1355. N. Aboutorab, W. Hardjawana, and B. Vucetic, "A new iterative Doppler-assisted channel estimation joint with parallel ICI cancellation for high-mobility MIMO-OFDM systems", *IEEE Trans. Veh. Tech.*, vol. 61, no. 4, pp. 1577–1589, May 2012.

1356. B.-S. Chen, C.-Y. Yang, and W.-J. Liao, "Robust fast time-varying multipath fading channel estimation and equalization for MIMO-OFDM systems via a fuzzy method", *IEEE Trans. Veh. Tech.*, vol. 61, no. 4, pp. 1599–1609, May 2012.

1357. G. Leus and M. Moonen, "Per-tone equalization for MIMO OFDM systems", *IEEE Trans. Sig. Process.*, vol. 51, no. 11, pp. 2965–2975, Nov. 2003.

1358. R. J. Piechocki, A. R. Nix, J. P. McGeehan, and S. M. D. Armour, "Joint blind and semi-blind detection and channel estimation for space-time trellis-coded modulation over fast fading channels", *IEE Proc. Commun.*, vol. 150, no. 6, pp. 419–426, Dec. 2003.

1359. J.-L. Yu and Y.-C. Lin, "Space-time-coded MIMO ZP-OFDM systems: Semiblind channel estimation and equalization", *IEEE Trans. Circ. and Syst. I*, vol. 56, no. 7, pp. 1360–1372, July 2009.

1360. M. Borgmann and H. Bölcskei, "Noncoherent space-frequency coded MIMO-OFDM", *IEEE J. Sel. Areas Commun.*, vol. 23, no. 9, pp. 1799–1810, Sep. 2005.

1361. T. Himsoon, W. Su, and K. J. R. Liu, "Single-block differential transmit scheme for broadband wireless MIMO-OFDM systems", *IEEE Trans. Sig. Process.*, vol. 54, no. 9, pp. 3305–3314, Sep. 2006.

1362. K. S. Woo, K. I. Lee, J. H. Paik, K. W. Park, W. Y. Yang, and Y. S. Cho, "A DSFBC-OFDM for a next generation broadcasting system with multiple antennas", *IEEE Trans. Broadcasting*, vol. 53, no. 2, pp. 539–546, June 2007.

1363. V. Pauli, L. Lampe, and J. Huber, "Differential space-frequency modulation and fast 2-D multiple-symbol differential detection for MIMO-OFDM", *IEEE Trans. Veh. Tech.*, vol. 57, no. 1, pp. 297–310, Jan. 2008.

1364. D. Falconer, "History of equalization 1860–1980", *IEEE Commun. Mag.*, vol. 49, no. 10, pp. 42–50, Oct. 2011.

1365. S. Darlington, "A history of network synthesis and filter theory for circuits composed of resistors, inductors, and capacitors", *IEEE Trans. Circuits Syst. I, Fundam. Theory Appl.*, vol. 46, no. 1, pp. 4–13, Jan. 1999.

1366. H. Nyquist, "Certain factors affecting telegraph speed", *Bell Syst. Tech. J.*, vol. 43, pp. 324–346, Apr. 1924.

1367. ——, "Certain topics in telegraph transmission theory", *Trans. AIEE*, vol. 47, pp. 617–644, Apr. 1928.

1368. D. W. Tufts, "Nyquist's problem – the joint optimization of transmitter and receiver in pulse amplitude modulation", *IEEE Proc.*, vol. 53, no. 3, pp. 248–259, Mar. 1965.

1369. R. W. Lucky, "Automatic equalization for digital communication", *Bell Sys. Tech. J.*, vol. 44, pp. 547–588, Apr. 1965.

1370. ——, "Techniques for adaptive equalization of digital communication", *Bell Sys. Tech. J.*, vol. 45, pp. 255–286, Feb. 1966.

1371. D. C. Coll and D. A. George, "A receiver for time-dispersed pulses", in *1965 IEEE Int. Conf. Commun. Conf. Rec.*, 1965, pp. 753–758.

1372. B. Widrow and M. E. Hoff, "Adaptive switching circuits", in *Proc. IRE WESCON Convention*, Part 4, Aug. 1960, pp. 96–104.

1373. D. G. Forney, Jr., and M. V. Eyuboglu, "Combined equalization and coding using precoding", *IEEE Commun. Mag.*, vol. 29, no. 12, pp. 25–34, Dec. 1991.

1374. R. D. Gitlin, E. Y. Ho, and J. E. Mazo, "Passband equalization of differentially phase-modulated data signals", *Bell Syst. Tech. J.*, vol. 52, pp. 219–238, Feb. 1973.

1375. H. Kobayashi, "Simultaneous adaptive estimation and decision algorithm for carrier modulated data transmission systems", *IEEE Trans. Commun.*, vol. 19, pp. 268–280, June 1971.

1376. D. D. Falconer, "Jointly adaptive equalization and carrier recovery in two-dimensional digital communication systems", *Bell Syst. Tech. J.*, vol. 55, no. 3, pp. 317–334, Mar. 1976.

1377. L. Guidoux, "Egaliseur autoadaptif à double échantillonnage", *L'Onde Electrique*, vol. 55, pp. 9–13, Jan. 1975.

1378. J. W. Mark and P. S. Budihardjo, "Joint optimization of receive filter and equalizer", *IEEE Trans. Commun*, vol. 21, no. 3, pp. 264–266, Mar. 1973.

1379. J. Monrolin, H. Nussbaumer, and J. M. Pierret, *Perfectionnements aux systèmes de détection de données distordues, French Patent No. 7131079*, 19 Mar. 1973.

1380. R. Price, "Nonlinearly feedback-equalized PAM vs. capacity for noisy linear channels", in *Proc. IEEE Int. Conf. Commun.*, vol. 2, Philadelphia, 19–21 June 1972, pp. 2212–2217.

1381. D. A. George, R. R. Bowen, and J. R. Storey, "An adaptive decision feedback equalizer", *IEEE Trans. Commun. Tech.*, vol. 19, no. 3, pp. 281–293, Mar. 1971.

1382. A. Gersho and T. L. Lim, "Adaptive cancellation of intersymbol interference for data transmission", *Bell Sys. Tech. J.*, vol. 60, pp. 1997–2021, Nov. 1981.

1383. D. D. Falconer, "Adaptive equalization of channel nonlinearities in QAM data transmission systems", *Bell Syst. Tech. J.*, vol. 57, no. 7, pp. 2589–2611, Sep. 1978.

1384. E. Biglieri, A. Gersho, R. Gitlin, and T. Lim, "Adaptive cancellation of nonlinear intersymbol interference for voiceband data transmission", *IEEE J. Sel. Areas Commun.*, vol. 2, no. 5, pp. 765–777, Sep. 1984.

1385. V. K. Dubey and D. P. Taylor, "Maximum likelihood sequence detection for QPSK on nonlinear band-limited channels", *IEEE Trans. Commun.*, vol. 34, no. 12, pp. 1225–1235, Dec. 1986.

1386. _____, "Further results on receivers for the nonlinear channel including pre- and post-nonlinearity filtering", *IEE Proc. Commun.*, vol. 141, no. 5, pp. 334–340, Oct. 1994.

1387. G. Ungerboeck, "Theory on the speed of convergence in adaptive equalizers for digital communication", *IBM J. Res. and Dev.*, vol. 16, pp. 546–555, Nov. 1972.

1388. J. E. Mazo, "On the independence theory of equalizer convergence", *Bell Sys. Tech. J.*, vol. 58, pp. 963–993, May–June 1979.

1389. D. N. Godard, "Channel equalization using a Kalman filter for fast data transmission", *IBM J. Res. and Dev.*, vol. 18, pp. 267–273, May 1974.

1390. L. Ljung, M. Morf, and D. D. Falconer, "Fast calculation of gain matrices for recursive estimation schemes", *Int. J. Control*, vol. 27, pp. 1–19, Jan. 1978.

1391. E. H. Satorius and S. T. Alexander, "Channel equalization using adaptive lattice algorithms", *IEEE Trans. Commun.*, vol. 27, pp. 899–905, June 1979.

1392. F. Ling and J. G. Proakis, "Adaptive lattice decision-feedback equalizers – their performance and application to time-variant multipath channels", *IEEE Trans. Commun.*, vol. 33, no. 4, pp. 348–356, Apr. 1985.

1393. D. G. Forney, Jr., *Training adaptive linear filters, US Patent No. 3723911*, 27 Mar. 1973.

1394. K. H. Mueller and D. A. Spaulding, *Fast start-up system for transversal equalizers, US Patent No. 3715666*, 6 Feb. 1973.

1395. Y. Sato, "A method of self-recovering equalisation for multilevel amplitude-modulation systems", *IEEE Trans. Commun.*, vol. 23, pp. 679–682, June 1975.

1396. D. N. Godard, "Self-recovering equalisation and carrier tracking in two-dimensional data communication systems", *IEEE Trans. Commun.*, vol. 28, pp. 1867–1875, Nov. 1980.

1397. R. Johnson, P. Schniter, T. J. Endres, J. D. Behm, D. R. Brown, and R. A. Casas, "Blind equalization using the constant modulus criterion: A review", *IEEE Proc.*, vol. 10, pp. 1927–1950, Oct. 1998.

1398. B. R. Petersen and D. D. Falconer, "Suppression of adjacent-channel, cochannel and intersymbol interference by equalizers and linear combiners", *IEEE Trans. Commun.*, vol. 42, pp. 3109–3118, Dec. 1994.

1399. A. Ginesi, G. M. Vitetta, and D. D. Falconer, "Block channel equalization in the presence of a cochannel interferent signal", *IEEE J. Sel. Areas Commun.*, vol. 17, no. 11, pp. 1853–1862, Nov. 1999.

1400. D. A. Shnidman, "A generalized Nyquist criterion and an optimum linear receiver for a pulse receiver system", *Bell Syst. Tech. J.*, vol. 46, pp. 2163–2177, Nov. 1967.

1401. A. Kaye and D. George, "Transmission of multiplexed PAM signals over multiple channel and diversity systems", *IEEE Trans. Commun. Technol.*, vol. 18, no. 5, pp. 520–526, Oct. 1970.
1402. J. Salz and S. B. Weinstein, "Fourier transform communication system", in *Proc. ACM Conf. Comp. and Commun.*, Pine Mountain, GA, Oct. 1969, pp. 99–128.
1403. T. Walzman and M. Schwartz, "Automatic equalization using the discrete frequency domain", *IEEE Trans. Inform. Theory*, vol. 19, no. 1, pp. 59–68, Jan. 1973.
1404. _____, "A projected gradient method for automatic equalization in the discrete frequency domain", *IEEE Trans. Commun.*, vol. 21, no. 12, pp. 1442–1446, Dec. 1973.
1405. E. Ferrara, "Fast implementations of LMS adaptive filters", *IEEE Trans. Acoust., Speech and Sig. Process.*, vol. 28, no. 4, pp. 474–475, Aug. 1980.
1406. G. Clark, S. Parker, and S. Mitra, "A unified approach to time- and frequency-domain realization of FIR adaptive digital filters", *IEEE Trans. Acoust., Speech and Sig. Process.*, vol. 31, no. 5, pp. 1073–1083, Oct. 1983.
1407. M. J. D. Toro, "Communication in time-frequency spread media using adaptive equalization", *IEEE Proc.*, vol. 56, no. 10, pp. 1653–1679, Oct. 1968.
1408. F. M. Hsu, A. A. Giordano, H. E. de Pedro, and J. G. Proakis, "Adaptive equalization techniques for high speed data received over fading dispersive HF channels", in *Proc. Nat. Telecom. Conf. 1980 (NTC80)*, Houston, TX, Dec. 1980.
1409. F. M. Hsu, "Square root Kalman filtering for high-speed data received over fading dispersive HF channels", *IEEE Trans. Inform. Theory*, vol. 28, no. 5, pp. 753–763, Sep. 1982.
1410. E. Eleftheriou and D. Falconer, "Adaptive equalization techniques for HF channels", *IEEE J. Sel. Areas. Commun.*, vol. 5, no. 2, pp. 238–247, 1987.
1411. J. M. Perl, A. Shpigel, and A. Reichman, "Adaptive receiver for digital communication over HF channels", *IEEE J. Sel. Areas Commun.*, vol. 5, pp. 304–308, Feb. 1987.
1412. N. Nefedov, M. Pukkila, R. Visoz, and A. O. Berthet, "Iterative data detection and channel estimation for advanced TDMA systems", *IEEE Trans. Commun.*, vol. 51, no. 2, pp. 141–144, Feb. 2003.
1413. G. Colavolpe and G. Germi, "On the application of factor graphs and the sum-product algorithm to ISI channels", *IEEE Trans. Commun.*, vol. 53, no. 5, pp. 818–825, May 2005.
1414. E. Aktas, "Belief propagation with Gaussian priors for pilot-assisted communication over fading ISI channels", *IEEE Trans. Wireless Commun.*, vol. 8, no. 4, pp. 2056–2066, Apr. 2009.
1415. D. P. Taylor, G. M. Vitetta, B. D. Hart, and A. Mämmelä, "Wireless channel equalization", *European Trans. Telecommun.*, vol. 9, no. 2, pp. 117–143, Mar./Apr. 1998.
1416. J. M. Cioffi, G. P. Dudevoir, M. V. Eyuboglu, and D. G. Forney, Jr., "MMSE decision-feedback equalizers and coding. II. Coding results", *IEEE Trans. Commun.*, vol. 43, no. 10, pp. 2595–2604, Oct. 1995.
1417. F. Pancaldi, G. M. Vitetta, R. Kalbasi, N. Al-Dhahir, M. Uysal, and H. Mheidat, "Single-carrier frequency domain equalization", *IEEE Sig. Process. Mag.*, vol. 25, no. 5, pp. 37–56, Sep. 2008.
1418. R. D. Gitlin, J. F. Hayes, and S. B. Weinstein, *Data Communication Principles*. New York: Plenum Press, 1992.
1419. G. E. Bottomley, *Channel Equalization for Wireless Communications: From Concepts to Detailed Mathematics*. Hoboken, NJ: John Wiley & Sons, Inc., 2011.
1420. C. E. Shannon, "A mathematical theory of communication", *Bell Syst. Tech. J.*, vol. 27, pp. 379–423, July 1948.
1421. _____, "A mathematical theory of communication", *Bell Syst. Tech. J.*, vol. 27, pp. 623–656, Oct. 1948.
1422. J. G. Proakis, *Digital Communications*, 3rd edn New York: McGraw-Hill, 1995.
1423. N. Abramson, *Information Theory and Coding*. New York: McGraw-Hill, 1953.
1424. A. Clark, D. Taylor, and P. Smith, "Instantaneous capacity of OFDM on Rayleigh fading channels", *IEEE Trans. Inform. Theory*, vol. 53, no. 1, pp. 355–361, Jan. 2007.
1425. A. Mämmelä, A. Kotelba, M. Höyhtyä, and D. P. Taylor, "Relationship of average transmitted and received energies in adaptive transmission", *IEEE Trans. Veh. Tech.*, vol. 59, pp. 1257–1268, March 2010.
1426. W. Stark, "Capacity and cutoff rate of noncoherent FSK with nonselective Ricean fading", *IEEE Trans. Commun.*, vol. 33, no. 11, pp. 1153–1159, Nov. 1985.
1427. W. C. Y. Lee, "Estimate of channel capacity in Rayleigh fading environment", *IEEE Trans. Veh. Tech.*, vol. 39, no. 3, pp. 187–189, Aug. 1990.

1428. C. G. Gunther, "Comment on 'Estimate of channel capacity in Rayleigh fading environment' ", *IEEE Trans. Veh. Tech.*, vol. 45, no. 2, pp. 401–403, May 1996.

1429. S. Shamai (Shitz) and A. D. Wyner, "Information-theoretic considerations for symmetric, cellular, multiple-access fading channels – Parts I, II", *IEEE Trans. Inform. Theory*, vol. 43, pp. 1877–1911, Nov. 1997.

1430. L. H. Ozarow, S. Shamai, and A. D. Wyner, "Information theoretic considerations for cellular mobile radio", *IEEE Trans. Veh. Tech.*, vol. 43, no. 2, pp. 359–378, May 1994.

1431. B. S. Tsybakov, "On the transmission capacity of a discrete-time Gaussian channel with filter", *Probl. Pered. Inform.*, vol. 6, pp. 26–40, 1970.

1432. W. Hirt and J. L. Massey, "Capacity of the discrete-time Gaussian channel with intersymbol interference", *IEEE Trans. Inform. Theory*, vol. 34, no. 10, pp. 380–388, May 1988.

1433. R. S. Cheng and S. Verdú, "Gaussian multiaccess channels with capacity region and multiuser water-filling", *IEEE Trans. Inform. Theory*, vol. 39, pp. 773–785, May 1993.

1434. S. Verdú, "Multiple-access channels with memory with and without frame synchronism", *IEEE Trans. Inform. Theory*, vol. 35, pp. 605–619, May 1989.

1435. _____, "The capacity region of the symbol-asynchronous Gaussian multiple-access channel", *IEEE Trans. Inform. Theory*, vol. 35, pp. 733–751, July 1989.

1436. R. J. McEliece and W. E. Stark, "Channels with block interference", *IEEE Trans. Inform. Theory*, vol. 30, pp. 44–53, Jan. 1994.

1437. S. N. Diggavi, "Analysis of multicarrier transmission in time-varying channels", in *Proc. IEEE Int. Conf. Commun. (ICC '97)*, Montreal, June 1997, pp. 1191–1195.

1438. R. Knopp and P. A. Humblet, "Multiple-accessing over frequency-selective fading channels", in *Proc. 6th IEEE Int. Symp. Personal, Indoor and Mobile Radio Communications (PIMRC 1995)*, Toronto, 27–29 Sep. 1995, pp. 1326–1330.

1439. M. Mushkin and I. Bar-David, "Capacity and coding for the Gilbert-Elliot channel", *IEEE Trans. Inform. Theory*, vol. 35, pp. 1277–1290, Nov. 1989.

1440. I. Csiszár and P. Narayan, "The capacity of the arbitrary varying channel", *IEEE Trans. Inform. Theory*, vol. 37, no. 1, pp. 18–26, Jan. 1991.

1441. A. J. Goldsmith and P. P. Varaiya, "Capacity, mutual information, and coding for finite-state Markov channels", *IEEE Trans. Inform. Theory*, vol. 42, no. 3, pp. 868–886, May 1996.

1442. _____, "Capacity of fading channels with channel side information", *IEEE Trans. Inform. Theory*, vol. 43, pp. 1986–1992, Nov. 1997.

1443. M. Effros and A. Goldsmith, "Capacity definitions and coding strategies for general channels with receiver side information", in *Proc. 1998 IEEE Int. Symp. Inform. Theory*, Cambridge, MA, 16–21 Aug. 1998, p. 39.

1444. M.-S. Alouini and A. J. Goldsmith, "Capacity of Rayleigh fading channels under different adaptive transmission and diversity-combining techniques", *IEEE Trans. Veh. tech.*, vol. 48, no. 4, pp. 1165–1181, July 1999.

1445. S. A. Jafar and A. Goldsmith, "Multiple-antenna capacity in correlated Rayleigh fading with channel covariance information", *IEEE Trans. Wireless Commun.*, vol. 4, no. 3, pp. 990–997, May 2005.

1446. T. Holliday, A. Goldsmith, and P. Glynn, "Capacity of finite state channels based on Lyapunov exponents of random matrices", *IEEE Trans. Inform. Theory*, vol. 52, no. 8, pp. 3509–3532, Aug. 2006.

1447. I. C. Abou-Faycal, M. D. Trott, and S. Shamai, "The capacity of discrete-time memoryless Rayleigh-fading channels", *IEEE Trans. Inform. Theory*, vol. 47, no. 4, pp. 1290–1301, May 2001.

1448. P. F. Driessen and G. J. Foschini, "On the capacity formula for multiple input-multiple output wireless channels: a geometric interpretation", *IEEE Trans. Commun.*, vol. 47, no. 2, pp. 173–176, Feb. 1999.

1449. A. Goldsmith, S. A. Jafar, N. Jindal, and S. Vishwanath, "Capacity limits of MIMO channels", *IEEE J. Sel. Areas Commun.*, vol. 21, no. 5, pp. 684–702, June 2003.

1450. A. F. Molisch, M. Z. Win, and J. H. Winters, "Space-time-frequency (STF) coding for MIMO-OFDM systems", *IEEE Commun. Lett.*, vol. 6, no. 9, pp. 370–372, Feb. 2002.

1451. C.-E. W. Sundberg and N. Seshadri, "Coded modulation for fading channels: An overview", *European Trans. Telecommun.*, vol. 4, pp. 309–324, May–June 1993.

1452. J. Ramsey, "Realization of optimum interleavers", *IEEE Trans. Inform. Theory*, vol. 16, no. 3, pp. 338–345, May 1970.

1453. D. G. Forney, Jr., "Burst-correcting codes for the classic bursty channel", *IEEE Trans. Commun. Technol.*, vol. 19, no. 5, pp. 772–781, Oct. 1971.

1454. I. Richer, "A simple interleaver for use with Viterbi decoding", *IEEE Trans. Commun.*, vol. 26, no. 3, pp. 338–345, Mar. 1978.

1455. D. G. Forney, Jr., *Concatenated Codes*. Cambridge, MA: MIT Press, 1966.

1456. G. Ungerboeck, "Trellis-coded modulation with redundant signal sets – Part 1: Introduction", *IEEE Commun. Mag.*, vol. 25, no. 2, pp. 5–11, Feb. 1987.

1457. L.-F. Wei, "Rotationally invariant convolutional channel coding with expanded signal space – Part I: 180°", *IEEE J. Select. Areas Commun.*, vol. 2, no. 5, pp. 659–671, Sep. 1984.

1458. _____, "Rotationally invariant convolutional channel coding with expanded signal space – Part II: Non-linear codes", *IEEE J. Select. Areas Commun.*, vol. 2, no. 5, pp. 672–686, Sep. 1984.

1459. _____, "Trellis-coded modulation with multidimensional constellations", *IEEE Trans. Inform. Theory*, vol. 33, no. 4, pp. 483–501, July 1987.

1460. _____, "Rotationally invariant trellis-coded modulations with multidimensional m-PSK", *IEEE J. Sel. Areas Commmun.*, vol. 7, no. 9, pp. 1281–1295, Dec. 1989.

1461. G. M. Vitetta, "Some new rotationally invariant TCM schemes for multidimensional m-PSK", *IEE Proc. Commun.*, vol. 141, no. 3, pp. 143–150, June 1994.

1462. H. Imai and S. Hirakawa, "A new multilevel coding method using error-correcting codes", *IEEE Trans. Inform. Theory*, vol. 23, no. 3, pp. 371–377, May 1977.

1463. A. R. Calderbank, "Multilevel codes and multistage decoding", *IEEE Trans. Commun.*, vol. 37, no. 3, pp. 222–229, Mar. 1989.

1464. E. Zehavi, "8-PSK trellis codes for a Rayleigh channel", *IEEE Trans. Commun.*, vol. 40, no. 5, pp. 873–884, May 1992.

1465. T. H. Liew and L. Hanzo, "Space-time codes and concatenated channel codes for wireless communications", *IEEE Proc.*, vol. 90, no. 2, pp. 187–219, Feb. 2002.

1466. A. Naguib, N. Seshadri, and A. R. Calderbank, "Increasing data rate over wireless channels", *IEEE Sig. Process. Mag.*, vol. 17, no. 3, pp. 77–92, May 2000.

1467. A. Slaney and Y. Sun, "Space-time coding for wireless communications: an overview", *IEE Proc. Commun.*, vol. 153, no. 4, pp. 509–518, Aug. 2006.

1468. D. Gesbert, M. Shafi, D.-S. Shiu, P. J. Smith, and A. Naguib, "From theory to practice: an overview of MIMO space-time coded wireless systems", *IEEE J. Sel. Areas Commun.*, vol. 21, no. 3, pp. 281–302, Dec. 2002.

1469. R. W. Hamming, "Error correcting and error detecting codes", *Bell Syst. Tech. J.*, vol. 29, pp. 147–160, Apr. 1950.

1470. M. J. E. Golay, "Notes on digital coding", *Proc. IRE*, vol. 37, p. 657, 1949.

1471. D. E. Muller, "Application of Boolean algebra to switching circuit design and to error detection", *IRE Trans. Electron. Comp.*, vol. 3, pp. 6–12, Sep. 1954.

1472. I. Reed, "A class of multiple-error-correcting codes and the decoding scheme", *IEEE Trans. Inform. Theory*, vol. 4, no. 4, pp. 38–49, Sep. 1954.

1473. E. Prange, *Cyclic error-correcting codes in two symbols, Tech. Note AFCRC-TN-57-103*, Air Force Cambridge Res. Center, Cambridge, MA, Sep. 1957.

1474. A. Hocquenghem, "Codes correcteurs d'erreurs", *Chiffres*, vol. 2, no. 5, pp. 147–156, Sep. 1971.

1475. R. C. Bose and D. K. Ray-Chaudhuri, "On a class of error-correcting binary group codes", *Inform. Control*, vol. 3, no. 1, pp. 68–79, Mar. 1960.

1476. I. Reed and G. Solomon, "Polynomial codes over certain finite fields", *J. Soc. Ind. Appl. Math.*, vol. 8, no. 2, pp. 300–304, June 1960.

1477. P. Elias, "Coding for noisy channels (part 4)", in *IRE Nat. Conv. Rec.*, Mar. 1955, pp. 37–46.

1478. A. J. Viterbi, "Error bounds for convolutional codes and an asymptotically optimum decoding algorithm", *IEEE Trans. Inform. Theory*, vol. 13, no. 4, pp. 260–269, Apr. 1967.

1479. R. G. Gallager, "Low-density parity-check codes", *IRE Trans. Info. Theory*, vol. 8, no. 1, pp. 21–28, Jan. 1962.

1480. P. Elias, "Error-free coding", *IRE Trans. Inform. Theory*, vol. 4, no. 4, pp. 29–37, Sep. 1954.

1481. C. Berrou, A. Glavieux, and P. Thitimajshima, "Near Shannon limit error-correcting coding and decoding: Turbo codes", in *Proc. IEEE Int. Conf. Commun. (ICC '93)*, Geneva, May 1993, pp. 1064–1070.

1482. V. Tarokh, H. Jafarkhani, and A. R. Calderbank, "Space-time block codes from orthogonal designs", *IEEE Trans. Inform. Theory*, vol. 45, no. 5, pp. 1456–1467, July 1999.

1483. _____, "Space-time block coding for wireless communications: Performance results", *IEEE J. Sel. Areas Commun.*, vol. 17, no. 3, pp. 451–460, Mar. 1999.

1484. D. J. Costello and D. G. Forney, Jr., "Channel coding: The road to channel capacity", *IEEE Proc.*, vol. 95, no. 6, pp. 1150–1177, June 2007.

1485. S. B. Wicker, *Error Control Systems for Digital Commmunication and Storage*. Upper Saddle River, NJ: Prentice Hall, 1995.

1486. I. S. Reed and X. Chen, *Error Control Coding for Data Networks*. Norwell, MA: Kluwer Academic, 1999.

1487. F. J. MacWilliams, "A theorem of the distribution of weights in a systematic code", *Bell Syst. Tech. J.*, vol. 42, pp. 79–94, 1965.

1488. A. M. Michelson and A. H. Levesque, *Error-Control Techniques for Digital Communication*. New York: John Wiley & Sons, Inc., 1985.

1489. F. J. MacWilliams and N. J. A. Sloane, *The Theory of Error-Correcting Codes*. Amsterdam: Elsevier Science, 1996.

1490. W. W. Peterson, *Error-Correcting Codes*. Cambridge, MA: MIT Press, 1961.

1491. M. Bossert, *Channel Coding for Telecommunications*. Chichester: John Wiley & Sons, Ltd, 1999.

1492. S. B. Wicker and V. K. Bhargava, *Reed-Solomon Codes and their Applications*. Piscataway, NJ: IEEE Press, 1994.

1493. D. E. Muller, *Metric Properties of Boolean Algebra and Their Application to Switching Circuits, Report No. 46*, Digital Computer Laboratory, Univ. of Illinois, Apr. 1953.

1494. D. G. Forney, Jr., "Generalized minimum distance decoding", *IEEE Trans. Inform. Theory*, vol. 12, no. 2, pp. 125–131, Apr. 1966.

1495. D. Chase, "A class of algorithms for decoding block codes with channel measurement information", *IEEE Trans. Inform. Theory*, vol. 18, no. 1, pp. 170–182, Jan. 1972.

1496. M. P. C. Fossorier and S. Lin, "Soft-decision decoding of linear block codes based on ordered statistics", *IEEE Trans. Inform. Theory*, vol. 41, no. 5, pp. 1379–1396, Sep. 1995.

1497. ——, "Computationally efficient soft-decision decoding of linear block codes based on ordered statistics", *IEEE Trans. Inform. Theory*, vol. 42, no. 5, pp. 738–750, May 1996.

1498. V. Guruswami and M. Sudan, "Improved decoding of Reed-Solomon and algebraic-geometry codes", *IEEE Trans. Inform. Theory*, vol. 45, no. 6, pp. 1757–1767, Sep. 1999.

1499. E. Fishler, O. Amrani, and Y. Be'ery, "Geometrical and performance analysis of GMD and Chase decoding algorithms", *IEEE Trans. Inform. Theory*, vol. 45, no. 5, pp. 1406–1422, July 1999.

1500. B.-Z. Shen, K. K. Tzeng, and C. Wang, "A bounded-distance decoding algorithm for binary linear block codes achieving the minimum effective error coefficient", *IEEE Trans. Inform. Theory*, vol. 42, no. 6, pp. 1987–1991, Nov. 1996.

1501. O. Amrani and Y. Be'ery, "Bounded-distance decoding: Algorithms, decision regions, and pseudo nearest neighbors", *IEEE Trans. Inform. Theory*, vol. 44, no. 7, pp. 3072–3082, Nov. 1998.

1502. R. A. Silverman and M. Balser, "Coding for constant-data-rate systems", *IRE Trans. Inform. Theory*, vol. 4, no. 4, pp. 50–63, Sep. 1954.

1503. R. M. Pyndiah, "Near-optimum decoding of product codes: Block turbo codes", *IEEE Trans. Commun.*, vol. 46, no. 8, pp. 1003–1010, Aug. 1998.

1504. S. Fragiacomo, C. Matrakidis, and J. O'Reilly, "Novel near maximum likelihood soft decision decoding algorithm for linear block codes", *IEE Proc. Commun.*, vol. 146, no. 5, pp. 265–270, Oct. 1999.

1505. N. N. Tendolkar and C. R. P. Hartmann, "Generalization of Chase algorithms for soft decision decoding of binary linear codes", *IEEE Trans. Inform. Theory*, vol. 30, no. 5, pp. 714–721, Sep. 1984.

1506. C. M. Hackett, "An efficient algorithm for soft-decision decoding of the (24,12) extended Golay code", *IEEE Trans. Commun.*, vol. 29, no. 6, pp. 909–911, June 1981.

1507. M. P. C. Fossorier and S. Lin, "Soft-input soft-output decoding of linear block codes based on ordered statistics", in *Proc. IEEE Global Telecommun. Conf. (GLOBECOM '98)*, vol. 5, Aveiro, Portugal, 8–12 Nov. 1998, pp. 2828–2833.

1508. R. E. Blahut, *Algebraic Methods for Signal Processing and Communications Coding*. New York: Springer-Verlag, 1992.

1509. W. Peterson, "Encoding and error-correction procedures for the Bose-Chaudhuri-Hocquenghem codes", *IRE Trans. Inform. Theory*, vol. 6, no. 4, pp. 459–470, Sep. 1960.

1510. E. R. Berlekamp, *Algebraic Coding Theory*. New York: McGraw-Hill, 1968.

1511. D. Gorenstein and N. Zierler, "A class of cyclic linear error-correcting codes in p^m symbols", *J. Soc. Ind. Appl. Math.*, vol. 9, pp. 207–214, June 1961.

1512. J. L. Massey, "Shift-register synthesis and BCH decoding", *IEEE Trans. Inform. Theory*, vol. 15, no. 1, pp. 122–127, Jan. 1969.

1513. R. Chien, "Cyclic decoding procedures for Bose-Chaudhuri-Hocquenghem codes", *IEEE Trans. Inform. Theory*, vol. 10, no. 4, pp. 357–363, Oct. 1964.

1514. S. S. Shah, S. Yaqub, and F. Suleman, "Self-correcting codes conquer noise – Part 2: Reed-Solomon codecs", *EDN Mag.*, 15 Mar. 2001.

1515. I. Shakeel, "Soft-decision decoding of Reed-Solomon-based signal space codes", Master's thesis, University of Canterbury, Christchurch, New Zealand, Feb. 2001.

1516. D. G. Forney, Jr., "On decoding BCH codes", *IEEE Trans. Inform. Theory*, vol. 11, no. 10, pp. 549–557, Oct. 1965.

1517. G. Forney, Jr., "Convolutional codes I: Algebraic structure", *IEEE Trans. Inform. Theory*, vol. 16, no. 6, pp. 720–738, Mar. 1970.

1518. J. L. Massey and M. K. Sain, "Inverses of linear sequential circuits", *IEEE Trans. Comput.*, vol. 17, no. 4, pp. 330–337, Dec. 1968.

1519. A. Viterbi, "Orthogonal tree codes for communication in the presence of white Gaussian noise", *IEEE Trans. Commun. Technol.*, vol. 15, no. 2, pp. 238–242, Apr. 1967.

1520. W. E. Ryan and S. G. Wilson, "Two classes of convolutional codes over GF(q) for q-ary orthogonal signaling", *IEEE Trans. Commun.*, vol. 39, no. 1, pp. 30–40, Jan. 1991.

1521. H.-A. Loeliger and T. Mittelholzer, "Convolutional codes over groups", *IEEE Trans. Inform. Theory*, vol. 42, no. 6, pp. 1660–1686, Nov. 1996.

1522. M. Rahnema and Y. Antia, "Optimum soft decision decoding with channel state information in the presence of fading", *IEEE Commun. Mag.*, vol. 41, no. 4, pp. 110–111, July 1997.

1523. L. R. Bahl, J. Cocke, F. Jelineck, and J. Raviv, "Optimal decoding of linear codes for minimizing symbol error rate", *IEEE Trans. Inform. Theory*, vol. 20, no. 2, pp. 284–287, Mar. 1974.

1524. J. Erfanian, S. Pasupathy, and G. Gulak, "Reduced complexity symbol detectors with parallel structures for ISI channels", *IEEE Trans. Commun.*, vol. 42, no. 6, pp. 1661–1671, Feb./Mar./Apr. 1994.

1525. P. Robertson, E. Villebrun, and P. Hoeher, "A comparison of optimal and sub-optimal MAP decoding algorithms operating in the log domain", in *Proc. IEEE Int. Conf. Commun. (ICC '95)*, vol. 2, Seattle, June 1995, pp. 1009–1013.

1526. P. Robertson, P. Hoeher, and E. Villebrun, "Optimal and sub-optimal maximum a posteriori algorithms suitable for turbo-decoding", *European Trans. Telecommun.*, vol. 8, no. 2, pp. 119–125, Mar./Apr. 1997.

1527. M. P. C. Fossorier, F. Burkert, L. Shu, and J. Hagenauer, "On the equivalence between SOVA and Max-Log-MAP decodings", *IEEE Commun. Lett.*, vol. 2, pp. 137–139, May 1998.

1528. P. R. Chevillat and E. Eleftheriou, "Decoding of trellis-encoded signals in the presence of intersymbol interference and noise", *IEEE Trans. Commun.*, vol. 37, no. 7, pp. 669–676, July 1989.

1529. P. Thiennviboon, G. Ferrari, and K. M. Chugg, "Generalized trellis-based reduced-state soft-input/soft-output algorithms", in *Proc. IEEE Int. Conf. Commun. (ICC 2002)*, New York, Apr. 2002, pp. 1667–1671.

1530. C. Fragouli, N. Seshadri, and W. Turin, *On the reduced trellis equalization using the M-BCJR algorithm*, Tech. Rep. TR 99.15.1, Florham Park, NJ, Nov. 1999.

1531. D. Bokolamulla, A. Hansson, and T. Aulin, "Low-complexity iterative detection based on bi-directional trellis search", in *Proc. 2003 IEEE Int. Symp. Inform. Theory (ISIT 2003)*, Yokohama, 29 June-4 July 2003, p. 396.

1532. M. Sikora and D. J. Costello, Jr., "A new SISO algorithm with application to turbo equalization", in *Proc. 2005 IEEE Int. Symp. Inform. Theory (ISIT 2005)*, Adelaide, 4–9 Sep. 2005, pp. 2031–2035.

1533. C. M. Vithanage, C. Andrieu, and R. J. Piechocki, "Novel reduced-state BCJR algorithms", *IEEE Trans. Commun.*, vol. 55, no. 6, pp. 1144–1152, June 2007.

1534. B. Frey and F. Kschischang, "Early detection and trellis splicing: Reduced complexity iterative decoding", *IEEE J. Sel. Areas Commun.*, vol. 16, no. 2, pp. 153–159, Feb. 1998.

1535. J. Hagenauer and P. Hoeher, "A Viterbi algorithm with soft-decision outputs and its applications", in *Proc. IEEE Global Telecommun. Conf. (GLOBECOM '93)*, Dallas, Nov. 1989, pp. 47.1.1–47.1.7.

1536. L. Papke and P. Robertson, "Improved decoding with the SOVA in a parallel concatenated (turbo-code) scheme", in *Proc. IEEE Int. Conf. Commun. (ICC '96)*, vol. 1, Dallas, June 1996, pp. 102–106.

1537. J. M. Wozencraft and B. Reiffen, *Sequential Decoding*. Cambridge, MA: MIT Press, 1965.

1538. R. Fano, "A heuristic discussion of probabilistic decoding", *IEEE Trans. Inform. Theory*, vol. 9, no. 2, pp. 64–74, Jan. 1963.

1539. K. S. Zigangirov, "Some sequential decoding procedures", *Prob. Pederachi Inform.*, vol. 2, pp. 13–25, 1966.

1540. F. Jelinek, "A fast sequential decoding algorithm using a stack", *IBM J. Res. and Dev.*, vol. 13, pp. 675–685, 1969.

1541. J. L. Massey, "Variable-length codes and the Fano metric", *IEEE Trans. Inform. Theory*, vol. 18, no. 1, pp. 196–198, Jan. 1972.

1542. G. C. Clark, Jr., and J. B. Cain, *Error-Correction Coding for Digital Communications*. New York: Plenum Press, 1981.

1543. E. C. Posner, L. L. Rauch, and B. D. Madsen, "Voyager mission telecommunication firsts", *IEEE Commun. Mag.*, vol. 28, no. 9, pp. 22–27, Sep. 1990.

1544. R. H. Deng and D. J. Costello, Jr., "High rate concatenated coding systems using bandwidth efficient trellis inner codes", *IEEE Trans. Commun.*, vol. 37, no. 5, pp. 420–427, May 1989.

1545. ———, "High rate concatenated coding systems using multidimensional bandwidth-efficient trellis inner codes", *IEEE Trans. Commun.*, vol. 37, no. 10, pp. 1091–1096, Oct. 1989.

1546. S. Benedetto, D. Divsalar, G. Montorsi, and F. Pollara, "Serial concatenation of interleaved codes: Performance analysis, design, and iterative decoding", *IEEE Trans. Inform. Theory*, vol. 44, no. 3, pp. 909–926, May 1998.

1547. D. J. Costello, Jr., and D. G. Forney, Jr., "Channel coding: The road to channel capacity", *IEEE Proc.*, vol. 95, no. 6, pp. 1150–1177, June 2007.

1548. S. Lin, *An Introduction to Error Correcting Codes*. Englewood Cliffs, NJ: Prentice Hall, 1970.

1549. W. W. Peterson and E. J. Weldon, Jr., *Error-Correcting Codes*. Cambridge, MA: MIT. Press, 1972.

1550. F. J. MacWilliams and N. J. A. Sloane, *The Theory of Error-Correcting Codes*. New York: Elsevier, 1977.

1551. R. E. Blahut, *Theory and Practice of Error Correcting Codes*, 2nd edn Reading, MA: Addison-Wesley, 1983.

1552. R. C. Bose and D. K. Ray-Chaudhuri, "Further results on error correcting binary group codes", *Inform. Control*, vol. 3, no. 3, pp. 279–290, Sep. 1960.

1553. J. L. Massey, "Step-by-step decoding of the Bose-Chaudhuri-Hocquenghem codes", *IEEE Trans. Inform. Theory*, vol. 11, no. 4, pp. 580–585, Oct. 1965.

1554. E. R. Berlekamp, "On decoding binary Bose-Chaudhuri-Hocquenghem codes", *IEEE Trans. Inform. Theory*, vol. 11, no. 4, pp. 577–579, Oct. 1965.

1555. M.-S. Oh and P. Sweeney, "Bit-level soft-decision sequential decoding for Reed Solomon codes", in *Proc. Workshop on Coding and Cryptography (WCC '99)*, Paris, Jan. 1999, pp. 111–120.

1556. M. Oh and P. Sweeney, "Low complexity soft-decision sequential decoding using hybrid permutation for RS codes", in *Proc. Seventh IMA Conf. Cryptography and Coding*, Cirencester, UK, Dec. 1999, pp. 177–181.

1557. D. Burgess, S. Wesemeyer, and P. Sweeney, "Soft-decision decoding algorithms for RS codes", in *Proc. Seventh IMA Conf. Cryptography and Coding*, Cirencester, UK, Dec. 1999, pp. 177–181.

1558. N. Szabo and R. Tanaka, *Residue Arithmetic and Its Applications to Computer Technology*. New York: McGraw-Hill, 1967.

1559. F. Taylor, "Residue arithmetic: A tutorial with examples", *IEEE Comput. Mag.*, vol. 17, no. 5, pp. 50–62, May 1984.

1560. R. W. Watson and C. W. Hastings, "Self-checked computation using residue arithmetic", *IEEE Proc.*, vol. 54, no. 12, pp. 1920–1931, Mar. 1966.

1561. H. Krishna, K. Lin, and J. Sun, "A coding theory approach to error control in redundant residue number systems – Part I: Theory and single error correction", *IEEE Trans. Circuits Syst. II*, vol. 39, no. 1, pp. 8–17, Jan. 1992.

1562. J. Sun and H. Krishna, "A coding theory approach to error control in redundant residue number systems – Part II: Multiple error detection and correction", *IEEE Trans. Circuits Syst. II*, vol. 39, no. 1, pp. 18–34, Jan. 1992.

1563. T. H. Liew, L.-L. Yang, and L. Hanzo, "Soft-decision redundant residue number system based error correction coding", in *Proc. IEEE 50th Veh. Technol. Conf. (VTC 1999 – Fall)*, vol. 5, Amsterdam, Sep. 1999, pp. 2546–2550.

1564. ———, "Systematic redundant residue number system codes: Analytical upper bound and iterative decoding performance over AWGN and Rayleigh channels", *IEEE Trans. Commun.*, vol. 54, no. 6, pp. 1006–1016, June 2006.

1565. V. D. Goppa, "Codes associated with divisors", *Probl. Inform. Transm.*, vol. 13, pp. 22–27, 1977.

1566. _____, "Codes on algebraic curves", *Sov. Math. Dokl.*, vol. 24, pp. 170–172, 1981.

1567. M. A. Tsfasman, S. G. Vladut, and T. Zink, "Modular codes, Shimura curves and Goppa codes better than the Varshamov-Gilbert bound", *Math. Nachr.*, vol. 109, pp. 21–28, Apr. 1982.

1568. T. Hoholdt and R. Pellikaan, "On the decoding of algebraic-geometric codes", *IEEE Trans. Inform. Theory*, vol. 41, no. 6, pp. 1589–1614, Nov. 1995.

1569. I. Blake, C. Heegard, T. Høholdt, and V. Wei, "Algebraic-geometry codes", *IEEE Trans. Inform. Theory*, vol. 44, no. 6, pp. 2596–2618, Oct. 1998.

1570. M. Sudan, "Decoding of Reed-Solomon codes beyond the error-correction bound", *J. Complexity*, vol. 13, no. 1, pp. 180–193, Mar. 1997.

1571. P. Elias, "Error-correcting codes for list decoding", *IEEE Trans. Inform. Theory*, vol. 37, no. 1, pp. 5–12, Jan. 1991.

1572. R. Koetter and A. Vardy, "Algebraic soft-decision decoding of Reed-Solomon codes", *IEEE Trans. Inform. Theory*, vol. 49, no. 11, pp. 2809–2825, Nov. 2003.

1573. J. Wolf, "Efficient maximum likelihood decoding of linear block codes using a trellis", *IEEE Trans. Inform. Theory*, vol. 24, no. 1, pp. 76–80, Jan. 1978.

1574. D. J. Muder, "Minimal trellises for block codes", *IEEE Trans. Inform. Theory*, vol. 34, no. 5, pp. 1049–1053, Sep. 1988.

1575. F. R. Kschischang and V. Sorokine, "On the trellis structure of block codes", *IEEE Trans. Inform. Theory*, vol. 41, no. 6, pp. 1924–1937, Nov. 1995.

1576. B. Honary and G. S. Markarian, "Low-complexity trellis decoding of Hamming codes", *Electron. Lett.*, vol. 29, no. 12, pp. 1114–1116, June 10 1993.

1577. B. Honary, G. S. Markarian, and M. Darnell, "Low-complexity trellis decoding of linear block codes", *IEE Proc. Commun.*, vol. 142, no. 4, pp. 201–209, Aug. 1995.

1578. B. Honary and G. Markarian, *Trellis Decoding of Block Codes*. Norwell, MA: Kluwer, 1997.

1579. H. Manoukian and B. Honary, "BCJR trellis construction for binary linear block codes", *IEE Proc. Commun.*, vol. 144, no. 6, pp. 367–371, Dec. 1997.

1580. S. Lin, T. Kasami, T. Fujiwara, and M. Fossorier, *Trellises and Trellis-Based Decoding Algorithms for Linear Block Codes*. Norwell, MA: Kluwer, 1998.

1581. T. Kasami, T. Takata, T. Fujiwara, and S. Lin, "On complexity of trellis structure of linear block codes", *IEEE Trans. Inform. Theory*, vol. 39, no. 3, pp. 1057–1937, May 1993.

1582. _____, "On the optimum bit orders with respects to the state complexity of trellis diagrams for binary linear codes", *IEEE Trans. Inform. Theory*, vol. 39, no. 1, pp. 242–245, Jan. 1993.

1583. A. Vardy, "Trellis structure of codes", in *Handbook of Coding Theory*, V. Pless and W. C. Huffman, (eds.) Amsterdam: Elsevier, 1998.

1584. R. J. McEliece, "On the BCJR trellis for linear block codes", *IEEE Trans. Inform. Theory*, vol. 42, pp. 1072–1092, July 1996.

1585. R. Johannesson and K. S. Zigangirov, *Fundamentals of Convolutional Coding*. Piscataway, NJ: IEEE Press, 1999.

1586. T. Richardson and R. Urbanke, *Modern Coding Theory*. Newy York: Cambridge University Press, 2008.

1587. P. Elias, "Coding for noisy channels", in *Key Papers in the Development of Information Theory*, D. Slepian, (ed.) New York: IEEE Press, 1973.

1588. _____, "Coding for noisy channels", in *Key Papers in the Development of Coding Theory*, E. R. Berlekamp, (ed.) New York: IEEE Press, 1974.

1589. _____, "Coding for noisy channels", in *The Electron and the Bit*, J. V. Guttag, (ed.) Cambridge, MA: EECS Dept., MIT, 2005.

1590. R. G. Gallager, *Low-Density Parity-Check Codes*. Cambridge, MA: MIT Press, 1963.

1591. I. M. Jacobs and E. R. Berlekamp, "A lower bound to the distribution of computation for sequential decoding", *IEEE Trans. Inform. Theory*, vol. 13, no. 9, pp. 167–174, Sep. 1967.

1592. J. L. Massey, *Threshold Decoding*. Cambridge, MA: MIT Press, 1963.

1593. J. K. Omura, "On the Viterbi decoding algorithm", *IEEE Trans. Inform. Theory*, vol. 15, no. 1, pp. 177–179, Jan. 1969.

1594. J. A. Heller, "Short constraint length convolutional codes", in *Jet Prop. Lab., Space Prog. Summary 37-54, vol. III*, 1968, pp. 171–177.

1595. _____, "Improved performance of short constraint length convolutional codes", in *Jet Prop. Lab., Space Prog. Summary 37-56, vol. III*, 1969, pp. 83–84.

1596. J. A. Heller and I. M. Jacobs, "Viterbi decoding for satellite and space communication", *IEEE Trans. Commun. Technol.*, vol. 19, no. 5, pp. 835–848, Oct. 1971.

1597. C. Berrou and A. Glavieux, "Near optimum error correcting coding and decoding: Turbo-codes", *IEEE Trans. Commun.*, vol. 44, no. 10, pp. 1261–1271, Oct. 1996.

1598. W. Koch and A. Baier, "Optimum and sub-optimum detection of coded data distributed by time-varying inter-symbol interference", in *Proc. IEEE Global Telecommun. Conf. (GLOBECOM '93)*, Geneva, Dec. 1993, pp. 1679–1684.

1599. J. Hagenauer, "Source-controlled channel decoding", *IEEE Trans. Commun.*, vol. 43, no. 9, pp. 2449–2457, Sep. 1995.

1600. E. Paaske, "Improved decoding for a concatenated coding system recommended by CCSDS", *IEEE Trans. Commun.*, vol. 38, no. 8, pp. 1138–1144, Aug. 1990.

1601. O. Collins and M. Hizlan, "Determinate-state convolutional codes", *IEEE Trans. Commun.*, vol. 41, no. 12, pp. 1785–1794, Dec. 1993.

1602. J. Hagenauer, E. Offer, and L. Papke, "Matching Viterbi decoders and Reed-Solomon decoders in concatenated systems", in *Reed-Solomon Codes and Their Applications*, S. B. Wicker and V. K. Bhargava, (eds.) Piscataway, NJ: IEEE Press, 1994, pp. 242–271.

1603. E. R. Berlekamp (ed.), *Key Papers in the Development of Coding Theory*. New York: IEEE Press, 1974.

1604. G. Battail, "Coding for the Gaussian channel: the promise of weighted-output decoding", *Int. J. Sat. Commun.*, vol. 7, pp. 183–192, 1989.

1605. C. Berrou, "The ten-year-old turbo codes are entering into service", *IEEE Commun. Mag*, vol. 41, no. 8, pp. 110–116, Aug. 2003.

1606. R. M. Tanner, "A recursive approach to low complexity codes", *IEEE Trans. Inform. Theory*, vol. 27, no. 9, pp. 533–547, Sep. 1981.

1607. D. J. C. MacKay and R. M. Neal, "Good error-correcting codes on very sparse matrices", in *Proc. 5th IMA Conf. Cryptography Coding*, C. Boyd, (ed.), vol. 3, Berlin, 1995, pp. 100–111.

1608. M. Sipser and D. A. Spielman, "Expander codes", in *Proc. 35th Symp. Found. Comp. Sci.*, 1994, pp. 566–576.

1609. ——, "Expander codes", *IEEE Trans. Inform. Theory*, vol. 42, no. 11, pp. 1710–1722, Nov. 1996.

1610. F. R. Kschischang, B. Frey, and H. Loeliger, "Factor graphs and the sum-product algorithm", *IEEE Trans. Inform. Theory*, vol. 41, no. 2, pp. 498–519, Feb. 2001.

1611. J. Hagenauer, "The turbo principle: Tutorial introduction & state of the art", in *Proc. Int. Symp. Turbo Codes & Related Topics*, Brest, France, Sep. 1997, pp. 1–11.

1612. H. H. Ma and J. K. Wolf, "On tail biting convolutional codes", *IEEE Trans. Commun.*, vol. 34, no. 2, pp. 104–111, Feb. 1986.

1613. O. Acikel and W. Ryan, "Punctured turbo-codes for BPSK/QPSK channels", *IEEE Trans. Commun.*, vol. 47, no. 9, pp. 1315–1323, Sep. 1999.

1614. W. E. Ryan, "Concatenated convolutional codes and iterative decoding", in *Wiley Encyclopedia of Telecommunications*, J. G. Proakis, (ed.) New York: John Wiley & Sons, Inc., 1994.

1615. S. Benedetto and G. Montorsi, "Design of parallel concatenated convolutional codes", *IEEE Trans. Commun.*, vol. 44, no. 5, pp. 591–600, May 1996.

1616. ——, "Unveiling turbo codes: Some results on parallel concatenated coding schemes", *IEEE Trans. Inform. Theory*, vol. 42, no. 2, pp. 409–428, Mar. 1996.

1617. O. Y. Takeshita, O. M. Collins, P. C. Massey, and D. J. Costello, Jr., "A note on asymmetric turbo-codes", *IEEE Commun. Lett.*, vol. 3, no. 3, pp. 69–71, Mar. 1999.

1618. S. Benedetto, R. Garello, and G. Montorsi, "A search for good convolutional codes to be used in the construction of turbo codes", *IEEE Trans. Commun.*, vol. 46, no. 9, pp. 1101–1105, Sep. 1998.

1619. F. Daneshgaran, M. Laddomada, and M. Mondin, "High-rate recursive convolutional codes for concatenated channel codes", *IEEE Trans. Commun.*, vol. 52, no. 11, pp. 1846–1850, Nov. 2004.

1620. P. C. Massey and D. J. Costello, Jr., "Turbo codes with recursive nonsystematic quick-look-in constituent codes", in *Proc. IEEE Int. Symp. Inform. Theory*, Cairns, Australia, 24–29 June 2001, p. 141.

1621. C. Berrou, Y. Saouter, C. Douillard, S. Kerouédan, and M. Jézéquel, "Designing good permutations for turbo codes: towards a single model", in *Proc. IEEE Int. Conf. Commun. (ICC 2004)*, vol. 1, Paris, 20–24 June 2004, pp. 341–345.

1622. R. Garello, G. Montorsi, S. Benedetto, and G. Cancellieri, "Interleaver properties and their applications to the trellis complexity analysis of turbo codes", *IEEE Trans. Commun.*, vol. 49, no. 5, pp. 793–807, May 2001.

1623. D. Divsalar and F. Pollara, "Multiple turbo codes", in *Proc. IEEE Military Commun. Conf. (MILCOM '95)*, vol. 1, San Diego, CA, 5–8 Nov. 1995, pp. 279–285.

1624. ———, "Turbo codes for PCS applications", in *Proc. IEEE Int. Conf. Commun. (ICC '95)*, vol. 1, Seattle, 18–22 June 1995, pp. 54–59.

1625. G. Battail, "A conceptual framework for understanding turbo codes", *IEEE J. Sel. Areas Commun.*, vol. 16, no. 2, pp. 245–254, Feb. 1998.

1626. N. Kahale and R. Urbanke, "On the minimum distance of parallel and serially concatenated codes", in *Proc. IEEE Int. Symp. Inform. Theory*, Dallas, 16–21 Aug. 1998, p. 31.

1627. S. Benedetto, D. Divsalar, G. Montorsi, and F. Pollara, "Analysis, design, and iterative decoding of double serially concatenated codes with interleavers", *IEEE J. Sel. Areas Commun.*, vol. 16, no. 2, pp. 231–244, Feb. 1998.

1628. J. Kliewer, K. S. Zigangirov, and D. J. Costello, Jr., "On the minimum trapping distance of repeat accumulate accumulate codes", in *Proc. 46th Annual Allerton Conference on Communication, Control, and Computing*, Monticello, IL, 23–26 Sep. 2008, pp. 1410–1415.

1629. C. Koller, A. Graell i Amat, J. Kliewer, F. Vatta, and D. J. Costello, Jr., "Hybrid concatenated codes with asymptotically good distance growth", in *Proc. 5th Int. Symp. Turbo Codes and Related Topics*, Lausanne, 1–5 Sep. 2008, pp. 19–24.

1630. S. Benedetto, G. Montorsi, D. Divsalar, and F. Pollara, "Soft-input, soft-output modules for the construction and distributed iterative decoding of code networks", *European Trans. Telecommun.*, vol. 9, no. 2, pp. 155–172, Mar./Apr. 1998.

1631. S. Benedetto, G. Montorsi, and D. Divsalar, "Concatenated convolutional codes with interleavers", *IEEE Commun. Mag.*, vol. 41, no. 8, pp. 102–109, Aug. 2003.

1632. D. Divsalar and F. Pollara, *Hybrid Concatenated Codes and Iterative Decoding, TDA Progress Report 42-130*, JPL, Pasadena, CA, 15 Aug. 1997.

1633. ———, "Hybrid concatenated codes and iterative decoding", in *Proc. IEEE Int. Symp. Inform. Theory*, vol. 1, Ulm, Germany, 29 June-4 July 1995, p. 10.

1634. S. M. Reddy, "On decoding iterated codes", *IEEE Trans. Inform. Theory*, vol. 16, no. 5, pp. 624–627, Sep. 1970.

1635. S. M. Reddy and J. P. Robinson, "Random error and burst correction by iterated codes", *IEEE Trans. Inform. Theory*, vol. 18, no. 1, pp. 182–185, Jan. 1972.

1636. J. Lodge, R. Young, P. Hoeher, and J. Hagenauer, "Separable MAP filters for the decoding of product and concatenated codes", in *Proc. IEEE Int. Conf. Commun. (ICC '93)*, Geneva, 23–26 May 1993, pp. 1740–1745.

1637. R. Pyndiah, A. Glavieux, A. Picart, and S. Jacq, "Near optimum decoding of products codes", in *Proc. IEEE Global Telecommun. Conf. (GLOBECOM '94)*, vol. 1, San Francisco, 28 Nov.-2 Dec. 1994, pp. 339–343.

1638. S. A. Hirst, B. Honary, and G. Markarian, "Fast Chase algorithm with an application in turbo decoding", *IEEE Trans. Commun.*, vol. 49, no. 10, pp. 1693–1699, Oct. 2001.

1639. C. Xu, Y.-C. Liang, and W. S. Leon, "A low complexity decoding algorithm for extended turbo product codes", *IEEE Trans. Wireless Commun.*, vol. 7, no. 1, pp. 43–47, Jan. 2008.

1640. F. Chiaraluce and R. Garello, "Extended Hamming product codes for analytical performance evaluation for low error rate applications", *IEEE Trans. Wireless Commun.*, vol. 3, no. 6, pp. 2353–2361, Nov. 2004.

1641. C. Argon and S. W. McLaughlin, "An efficient Chase decoder for turbo product codes", *IEEE Trans. Commun.*, vol. 52, no. 6, pp. 896–898, June 2004.

1642. O. Aitsab and R. Pyndiah, "Performance of Reed-Solomon block turbo code", in *Proc. IEEE Global Telecommun. Conf. (GLOBECOM '96)*, vol. 1, London, 18–22 Nov. 1996, pp. 121–125.

1643. G. Battail, "Building long codes by combination of simple ones, thanks to weighted-output decoding", in *Proc. URSI ISSSE*, Erlangen, Germany, Sep. 1989, pp. 634–637.

1644. A. Hunt, S. Crozier, and D. Falconer, "Hyper-codes: High-performance low-complexity error-correcting codes", in *Proc. 19th Biennial Symposium on Communications*, Kingston, Ontario, 31 May–3 June 1998, pp. 263–267.

1645. D. M. Rankin and T. A. Gulliver, "Single parity check product codes", *IEEE Trans. Commun.*, vol. 49, no. 8, pp. 1354–1362, Aug. 2001.

1646. G. M. Roth and G. Seroussi, "Reduced-redundancy product codes for burst error correction", *IEEE Trans. Inform. Theory*, vol. 44, no. 4, pp. 1395–1406, July 1998.

1647. M. Kasahara, Y. Sugiyama, S. Hirasawa, and T. Namekawa, "New classes of binary codes constructed on the basis of concatenated codes and product codes", *IEEE Trans. Inform. Theory*, vol. 22, no. 4, pp. 462–468, July 1976.

1648. D. Divsalar, H. Jin, and R. J. McEliece, "Coding theorems for 'turbo-like' codes?" in *Proc. IEEE Int. Symp. Inform. Theory*, vol. 1, Ulm, Germany, 29 June–4 July 1995, p. 10.

1649. ——, "Coding theorems for 'turbo-like' codes", in *Proc. 1998 Allerton Conf.*, Monticello, IL, Sep. 1998, pp. 201–210.

1650. H. D. Pfister and P. H. Siegel, "The serial concatenation of rate-1 codes through uniform random interleavers", *IEEE Trans. Inform. Theory*, vol. 49, no. 6, pp. 1425–1438, June 2003.

1651. J. Li, K. R. Narayanan, and C. N. Georghiades, "Product accumulate codes: a class of codes with near-capacity performance and low decoding complexity", *IEEE Trans. Inform. Theory*, vol. 50, no. 1, pp. 31–46, Jan. 2004.

1652. M. Isaka and M. Fossorier, "High-rate serially concatenated coding with extended Hamming codes", *IEEE Commun. Lett.*, vol. 9, no. 2, pp. 160–162, Feb. 2006.

1653. M. Peleg, I. Sason, S. Shamai, and A. Elia, "On interleaved, differentially encoded convolutional codes", *IEEE Trans. Inform. Theory*, vol. 45, no. 7, pp. 2572–2582, Mar. 1999.

1654. K. R. Narayanan and G. L. Stüber, "A serial concatenation approach to iterative demodulation and decoding", *IEEE Trans. Commun.*, vol. 47, no. 7, pp. 956–961, July 199.

1655. M. Peleg, S. Shamai, and S. Galan, "Iterative decoding for coded noncoherent MPSK communications over phase-noisy AWGN channel", *IEE Proc. Commun.*, vol. 147, no. 2, pp. 87–95, Apr. 2000.

1656. M. Franceschini, G. Ferrari, R. Raheli, and A. Curtoni, "Serial concatenation of LDPC codes and differential modulations", *IEEE J. Sel. Areas Commun.*, vol. 23, no. 9, pp. 1758–1768, Sep. 2005.

1657. J. Mitra and L. Lampe, "Serial concatenation of simple linear block codes and differential modulations", *IEEE Trans. Commun.*, vol. 7, no. 5, pp. 1477–1482, May 2008.

1658. S. L. Howard and C. Schlegel, "Differential turbo-coded modulation with APP channel estimation", *IEEE Trans. Commun.*, vol. 54, no. 8, pp. 1397–1406, Aug. 2006.

1659. R. Ramamurthy and W. E. Ryan, "Convolutional double accumulate codes (or double turbo DPSK)", *IEEE Commun. Lett.*, vol. 5, no. 4, pp. 157–159, Apr. 2001.

1660. R. Koetter, A. C. Singer, and M. Tüchler, "Turbo equalization", *IEEE Sig. Process. Mag*, vol. 21, no. 1, pp. 67–80, Jan. 2004.

1661. H. V. Poor, "Iterative multiuser detection", *IEEE Sig. Process. Mag.*, vol. 21, no. 1, pp. 81–88, Jan. 2004.

1662. G. Battail, M. Decouvelaere, and P. Godlewski, "Replication decoding", *IEEE Trans. Inform. Theory*, vol. 25, no. 3, pp. 332–345, May 1979.

1663. C. Hartmann and L. Rudolph, "An optimum symbol-by-symbol decoding rule for linear codes", *IEEE Trans. Inform. Theory*, vol. 22, no. 5, pp. 514–517, Sep. 1976.

1664. J. Fang, F. Buda, and E. Lemois, "Turbo product code: A well suitable solution to wireless packet transmission for very low error rates", in *Proc. 2nd Int. Symp. Turbo Codes and Related Topics*, Brest, France, 4–7 Sep. 2000, pp. 101–111.

1665. P. Adde and R. Pyndiah, "Recent simplifications and improvements in block turbo codes", in *Proc. 2nd Int. Symp. on Turbo Codes and Related Topics*, Brest, France, 4–7 Sept. 2000, pp. 133–136.

1666. P. A. Martin and D. P. Taylor, "On adaptive reduced-complexity iterative decoding", in *Proc. IEEE Global Telecommun. Conf. (GLOBECOM 2000)*, San Francisco, 27 Nov.–1 Dec. 2000, pp. 772–776.

1667. S. Dave, J. Kim, and S. C. Kwatra, "An efficient decoding algorithm for block turbo codes", *IEEE Trans. Commun.*, vol. 49, no. 1, pp. 41–46, Jan. 2001.

1668. P. A. Martin, A. Valembois, M. P. C. Fossorier, and D. P. Taylor, "On soft-input soft-output decoding using 'box and match' techniques", *IEEE Trans. Commun.*, vol. 52, no. 12, pp. 2033–2037, Dec. 2004.

1669. P. A. Martin, D. P. Taylor, and M. P. C. Fossorier, "Soft-input soft-output list-based decoding algorithm", *IEEE Trans. Commun.*, vol. 52, no. 2, pp. 252–262, Feb. 2004.

1670. J. Hagenauer, "A soft-in/soft-out list sequential (LISS) decoder for turbo schemes", in *Proc. IEEE Int. Symp. Inform. Theory (ISIT 2003)*, Yokohama, 29 June–4 July 2003, p. 382.

1671. J. Hagenauer and C. Kuhn, "The list sequential (LISS) algorithms and its application", *IEEE Trans. Commun.*, vol. 55, no. 5, pp. 918–928, May 2007.

1672. C. Jego and W. J. Gross, "Turbo decoding of product codes using adaptive belief propagation", *IEEE Trans. Commun.*, vol. 57, no. 10, pp. 2864–2867, Oct. 2009.

1673. P. A. Martin and D. P. Taylor, "High-throughput error correcting space-time block codes", *IEEE Commun. Lett.*, vol. 8, no. 7, pp. 458–460, July 2004.

1674. G. Colavolpe, G. Ferrari, and R. Raheli, "Extrinsic information in iterative decoding: a unified view", *IEEE Trans. Commun.*, vol. 49, no. 12, pp. 2088–2094, Dec. 2001.

1675. D. Divsalar and R. J. McEliece, *Effective free distance of turbo codes, Tech. Rep.*, JPL and Department of Electrical Engineering, California Institute of Technology, Pasadena, CA, Dec. 1995.

1676. T. Richardson, "The geometry of turbo-decoding dynamics", *IEEE Trans. Inform. Theory*, vol. 46, no. 1, pp. 9–23, Jan. 2000.

1677. N. Wiberg, "Codes and decoding on general graphs", Ph.D. dissertation, Linköping University, Linköping, Sweden, 1996.

1678. H. El Gamal and A. R. Hammons, Jr., "Analyzing the turbo decoder using the Gaussian approximation", *IEEE Trans. Inform. Theory*, vol. 47, no. 2, pp. 671–686, Feb. 2001.

1679. T. Richardson and R. Urbanke, "The capacity of low-density parity-check codes under message-passing decoding", *IEEE Trans. Inform. Theory*, vol. 47, no. 2, pp. 599–618, Mar. 2001.

1680. ——, "Thresholds for turbo codes", in *Proc. 2000 IEEE Int. Symp. Inform. Theory (ISIT 2000)*, Sorrento, Italy, 25–30 June 2000, p. 317.

1681. D. Divsalar, D. Dolinar, and F. Pollara, "Iterative turbo decoder analysis based on density evolution", *IEEE J. Sel. Areas Commun.*, vol. 19, no. 5, pp. 891–907, May 2001.

1682. J. W. Lee and R. E. Blahut, "Convergence analysis and BER performance of finite-length turbo codes", *IEEE Trans. Commun.*, vol. 55, no. 5, pp. 1033–1043, May 2007.

1683. S. ten Brink, "Convergence behavior of iteratively decoded parallel concatenated codes", *IEEE Trans. Commun.*, vol. 49, no. 10, pp. 1727–1737, Oct. 1998.

1684. L. Perez, J. Seghers, and D. J. Costello, Jr., "A distance spectrum interpretation of turbo codes", *IEEE Trans. Inform. Theory*, vol. 42, no. 6, pp. 1698–1709, Nov. 1996.

1685. Y. Kou, S. Lin, and M. P. C. Fossorier, "Low-density parity-check codes based on finite geometries: a rediscovery and new results", *IEEE Trans. Inform. Theory*, vol. 47, no. 7, pp. 2711–2736, Nov. 2001.

1686. T. Richardson, A. Shokrollahi, and R. Urbanke, "Design of capacity-approaching irregular low-density parity-check codes", *IEEE Trans. Inform. Theory*, vol. 47, no. 2, pp. 619–637, Mar. 2001.

1687. F. R. Kschischang and B. J. Frey, "Iterative decoding of compound codes by probability propagation in graphical models", *IEEE J. Sel. Areas Commun.*, vol. 16, no. 2, pp. 219–230, Feb. 1998.

1688. T. Etzion, A. Trachtenberg, and A. Vardy, "Which codes have cycle-free Tanner graphs?" *IEEE Trans. Inform. Theory*, vol. 45, no. 6, pp. 2173–2181, Sep. 1999.

1689. R. J. McEliece, D. J. C. MacKay, and J.-F. Cheng, "Turbo decoding as an instance of Pearl's belief propagation algorithm", *IEEE J. Sel. Areas Commun.*, vol. 16, no. 2, pp. 140–152, Feb. 1998.

1690. S. Litsyn and V. Shevelev, "On ensembles of low-density parity-check codes: asymptotic distance distributions", *IEEE Trans. Commun.*, vol. 48, no. 4, pp. 887–908, Apr. 2002.

1691. ——, "Distance distributions in ensembles of irregular low-density parity-check codes", *IEEE Trans. Commun.*, vol. 49, no. 12, pp. 3140–3159, Dec. 2003.

1692. D. Burshtein and G. Miller, "Asymptotic enumeration methods for analyzing LDPC codes", *IEEE Trans. Inform. Theory*, vol. 50, no. 6, pp. 1115–1131, June 2004.

1693. C. Di, T. J. Richardson, and R. L. Urbanke, "Weight distribution of low-density parity-check codes", *IEEE Trans. Inform. Theory*, vol. 52, no. 11, pp. 4839–4855, Nov. 2006.

1694. O. Barak and D. Burshtein, "Lower bounds on the error rate of LDPC code ensembles", *IEEE Trans. Inform. Theory*, vol. 53, no. 11, pp. 4225–4236, Nov. 2007.

1695. D. J. C. MacKay, "Good error-correcting codes based on very sparse matrices", *IEEE Trans. Inform. Theory*, vol. 45, no. 2, pp. 399–431, Mar. 1999.

1696. *David MacKay's Gallager code resources*, http://www.inference.phy.cam.ac.uk/mackay/CodesFiles.html.

1697. J. M. F. Moura, J. Lu, and H. Zhang, "Structured low-density parity-check codes", *IEEE Sig. Process. Mag.*, vol. 21, no. 1, pp. 42–55, Jan. 2004.

1698. G. A. Margulis, "Explicit constructions of graphs without short cycles and low density codes", *Combinatorica*, vol. 2, no. 1, pp. 71–78, 1982.

1699. M. Luby, M. Mitzenmacher, and V. Stemann, "Practical loss-resilient codes", in *Proc. 29th Symp. Theory Computing*, El Paso, TX, 1997, pp. 150–159.

1700. D. J. C. MacKay, S. T. Wilson, and M. C. Davey, "Comparison of constructions of irregular Gallager codes", *IEEE Trans. Commun.*, vol. 47, no. 10, pp. 1449–1454, Oct. 1999.

1701. S.-Y. Chung, D. G. Forney, Jr., T. J. Richardson, and R. Urbanke, "On the design of low-density parity-check codes within 0.0045 dB from the Shannon limit", *IEEE Commun. Lett.*, vol. 5, no. 2, pp. 58–60, Feb. 2001.

1702. J. Rosenthal and P. O. Vontobel, "Constructions of regular and irregular LDPC codes using Ramanujan graphs and ideas from Margulis", in *Proc. IEEE Int. Symp. Inform. Theory (ISIT 2001)*, Washington, DC, 24–29 June 2001.

1703. I. Djurdjevic, S. Lin, and K. Abdel-Ghaffar, "Graph-theoretic construction of low-density parity-check codes", *IEEE Commun. Lett.*, vol. 7, no. 4, pp. 171–173, Apr. 2003.

1704. H. Zhang and J. M. F. Moura, "Large-girth LDPC codes based on graphical models", in *Proc. IEEE 4th Workshop Sig. Process. Adv. Wireless Commun. (SPAWC 2003)*, Rome, 15–18 June 2003, pp. 100–104.

1705. _____, "The design of structured regular LDPC codes with large girth", in *Proc. IEEE Global Telecommun. Conf. (GLOBECOM 2003)*, vol. 7, San Francisco, 1–5 Dec. 2003, pp. 4022–4027.

1706. H. Tang, J. Xu, Y. Kou, S. Lin, and K. Abdel-Ghaffar, "On algebraic construction of Gallager and circulant low-density parity-check codes", *IEEE Trans. Inform. Theory*, vol. 50, no. 6, pp. 1269–1279, June 2004.

1707. B. Ammar, B. Honary, Y. Kou, J. Xu, and S. Lin, "Construction of low-density parity-check codes based on balanced incomplete block designs", *IEEE Trans. Inform. Theory*, vol. 50, no. 6, pp. 1257–1269, June 2004.

1708. B. Vasic and O. Milenkovic, "Combinatorial constructions of low-density parity-check codes for iterative decoding", *IEEE Trans. Inform. Theory*, vol. 50, no. 6, pp. 1156–1176, June 2004.

1709. S. Sankaranarayanan, B. Vasic, and E. M. Kurtas, "Irregular low-density parity-check codes: construction and performance on perpendicular magnetic recording channels", *IEEE Trans. Magn.*, vol. 39, no. 5, pp. 2567–2569, Sep. 2003.

1710. Z. W. Li, L. Chen, L. Zeng, S. Lin, and W. H. Fong, "Efficient encoding of quasi-cyclic low-density parity-check codes", *IEEE Trans. Inform. Theory*, vol. 54, no. 1, pp. 71–81, Jan. 2006.

1711. M. Esmaeili, T. A. Gulliver, N. P. Secord, and S. A. Mahmoud, "A link between quasi-cyclic codes and convolutional codes", *IEEE Trans. Inform. Theory*, vol. 44, no. 1, pp. 431–435, Jan. 1998.

1712. M. C. Davey and D. MacKay, "Low-density parity check codes over GF(q)", *IEEE Commun. Lett.*, vol. 2, no. 6, pp. 165–167, June 1998.

1713. A. Bennatan and D. Burshtein, "On the application of LDPC codes to arbitrary discrete-memoryless channels", *IEEE Trans. Inform. Theory*, vol. 50, no. 3, pp. 417–438, Mar. 2004.

1714. _____, "Design and analysis of nonbinary LDPC codes for arbitrary discrete-memoryless channels", *IEEE Trans. Inform. Theory*, vol. 52, no. 2, pp. 549–583, Feb. 2006.

1715. L. Zeng, L. Lan, Y. Y. Tai, S. Song, S. Lin, and K. Abdel-Ghaffar, "Constructions of nonbinary quasi-cyclic LDPC codes: A finite field approach", *IEEE Trans. Commun.*, vol. 56, no. 4, pp. 545–554, Apr. 2008.

1716. L. Zeng, L. Lan, Y. Y. Tai, B. Zhou, S. Lin, and K. A. S. Abdel-Ghaffar, "Construction of nonbinary cyclic, quasi-cyclic and regular LDPC codes: a finite geometry approach", *IEEE Trans. Commun.*, vol. 56, no. 3, pp. 378–387, Mar. 2008.

1717. C. Poulliat, M. Fossorier, and D. Declercq, "Design of regular ($2,d_c$)-LDPC codes over GF(q) using their binary images", *IEEE Trans. Inform. Theory*, vol. 56, no. 10, pp. 1626–1635, Oct. 2008.

1718. S. Song, B. Zhou, S. Lin, and K. Abdel-Ghaffar, "A unified approach to the construction of binary and nonbinary quasi-cyclic LDPC codes based on finite fields", *IEEE Trans. Commun.*, vol. 57, no. 1, pp. 84–93, Jan. 2009.

1719. B. Zhou, J. Kang, S. Song, S. Lin, K. Abdel-Ghaffar, and M. Xu, "Construction of non-binary quasi-cyclic LDPC codes by arrays and array dispersions", *IEEE Trans. Commun.*, vol. 57, no. 6, pp. 1652–1662, June 2009.

1720. C. R. Jones, T. Tian, J. Villasenor, and R. D. Wesel, "The universal operation of LDPC codes over scalar fading channels", *IEEE Trans. Commun.*, vol. 55, no. 1, pp. 122–132, Jan. 2007.

1721. L. Chen, J. Xu, I. Djurdjevic, and S. Lin, "Near-Shannon-limit quasi-cyclic low-density parity-check codes", *IEEE Trans. Commun.*, vol. 52, no. 7, pp. 1038–1042, July 2004.

1722. H. Tang, J. Xu, S. Lin, and K. A. S. Abdel-Ghaffar, "Codes on finite geometries", *IEEE Trans. Inform. Theory*, vol. 51, no. 2, pp. 572–596, Feb. 2005.

1723. B. Zhou, J. Kang, Y. Y. Tai, Q. Huang, and S. Lin, "High performance nonbinary quasi-cyclic LDPC codes on Euclidean geometries", in *Proc. IEEE Military Commun. Conf. (MILCOM 2007)*, Orlando, FL, 29–31 Oct. 2007, pp. 1–8.

1724. X. Jiang and M. H. Lee, "Large girth non-binary LDPC codes based on finite fields and Euclidean geometries", *IEEE Sig. Process. Lett.*, vol. 16, no. 6, pp. 521–524, June 2009.

1725. M. Esmaeili and M. Gholami, "Geometrically-structured maximum-girth LDPC block and convolutional codes", *IEEE J. Sel. Areas Commun.*, vol. 27, no. 6, pp. 831–845, Aug. 2009.

1726. ———, "Structured quasi-cyclic LDPC codes with girth 18 and column-weight $j \geq 3$", *AEÜ – Int. J. Electron. Commun.*, no. 64, pp. 202–217, 2009.

1727. D. J. C. MacKay and D. C. Davey, "Evaluation of Gallager codes for short block length and high rate applications", in *Proc. IMA Workshop on Codes, Systems and Graphical Models*, B. Marcus and J. Rosenthal, (eds.) New York: Springer-Verlag, 2001, pp. 113–130.

1728. S. J. Johnson and S. R. Weller, "Construction of low-density parity-check codes from Kirkman triple systems", in *Proc. IEEE Global Telecommun. Conf. (GLOBECOM 2001)*, vol. 2, San Antonio, TX, 25–29 Nov. 2001, pp. 970–974.

1729. ———, "Regular low-density parity-check codes from combinatorial designs", in *Proc. IEEE Workshop Inform. Theory*, Cairns, Australia, 2–7 Sept. 2001, pp. 90–92.

1730. B. Vasic, K. Pedagani, and M. Ivkovic, "High-rate girth-eight low-density parity-check codes on rectangular integer lattices", *IEEE Trans. Commun.*, vol. 52, no. 8, pp. 1248–1252, Aug. 2004.

1731. L. Lan, Y. Y. Tai, S. Lin, B. Memari, and B. Honary, "New constructions of quasi-cyclic LDPC codes based on special classes of BIBD's for the AWGN and binary erasure channels", *IEEE Trans. Commun.*, vol. 56, no. 1, pp. 39–48, Jan. 2008.

1732. J. L. Fan, "Array codes as low-density parity-check codes", in *Proc. 2nd Int. Symp. Turbo Codes*, Brest, France, Nov. 2000, pp. 543–546.

1733. M. Blaum and R. M. Roth, "New array codes for multiple phased burst correction", *IEEE Trans. Inform. Theory*, vol. 39, no. 1, pp. 66–77, Jan. 1993.

1734. E. Eleftheriou and S. Ölcer, "Low-density parity-check codes for digital subscriber lines", in *Proc. IEEE Int. Conf. Commun. (ICC 2002)*, vol. 3, New York, 28 Apr.-2 May 2002, pp. 1752–1757.

1735. R. M. Tanner, D. Sridhara, A. Sridharan, T. E. Fuja, and D. J. Costello, Jr., "LDPC block and convolutional codes based on circulant matrices", *IEEE Trans. Inform. Theory*, vol. 50, no. 12, pp. 2966–2984, Dec. 2004.

1736. S. Myung, K. Yang, and J. Kim, "Quasi-cyclic LDPC codes for fast encoding", *IEEE Trans. Inform. Theory*, vol. 51, no. 8, pp. 2894–2901, Aug. 2005.

1737. S. Myung, K. Yang, and Y. Kim, "Lifting methods for quasi-cyclic LDPC codes", *IEEE Commun. Lett.*, vol. 10, no. 6, pp. 489–491, June 2006.

1738. M. E. O'Sullivan, "Algebraic construction of sparse matrices with large girth", *IEEE Trans. Inform. Theory*, vol. 52, no. 2, pp. 718–727, Feb. 2006.

1739. L. Lan, L. Zeng, Y. Y. Tai, L. Chen, S. Lin, and K. Abdel-Ghaffar, "Construction of quasi-cyclic LDPC codes for AWGN and binary erasure channels: A finite field approach", *IEEE Trans. Inform. Theory*, vol. 53, no. 7, pp. 2429–2458, July 2007.

1740. H. Zhang, J. Zhu, H. Shi, and D. Wang, "Layered approx-regular LDPC: Code construction and encoder/decoder design", *IEEE Trans. Circ. and Syst. I*, vol. 55, no. 2, pp. 572–585, Mar. 2008.

1741. D. Divsalar, S. Dolinar, C. R. Jones, and K. Andrews, "Capacity approaching protograph codes", *IEEE J. Sel. Areas Commun.*, vol. 27, no. 6, pp. 876–888, Aug. 2009.

1742. K. S. Andrews, D. Divsalar, S. Dolinar, J. Hamkins, C. R. Jones, and F. Pollara, "The development of turbo and LDPC codes for deep-space applications", *IEEE Proc.*, vol. 95, no. 11, pp. 2142–2156, Nov. 2007.

1743. H. Jin, A. Khandekar, and R. McEliece, "Irregular repeat accumulate codes", in *Proc. 2nd Int. Symp. Turbo Codes Related Topics*, Brest, France, Sept. 2003.

1744. A. Abbasfar, D. Divsalar, and Y. Kung, "Accumulate-repeat-accumulate codes", in *Proc. IEEE Global Telecommun. Conf. (GLOBECOM 2004)*, Dallas, Dec. 2004, pp. 509–513.

1745. G. Liva, W. E. Ryan, and M. Chiani, "Quasi-cyclic generalized LDPC codes with low error floors", *IEEE Trans. Commun.*, vol. 56, no. 1, pp. 49–57, Jan. 2008.

1746. N. Bonello, S. Chen, and L. Hanzo, "Construction of regular quasi-cyclic protograph LDPC codes based on Vandermonde matrices", *IEEE Trans. Veh. Tech.*, vol. 57, no. 4, pp. 2583–2588, July 2008.

1747. J. Campello, D. S. Modha, and S. Rajagopalan, "Designing LDPC codes using bit-filling", in *Proc. IEEE Int. Conf. Commun. (ICC 2001)*, vol. 1, Helsinki, 11–14 June 2001, pp. 55–59.

1748. J. Campello and D. S. Modha, "Extended bit-filling and LDPC code design", in *Proc. IEEE Global Telecommun. Conf. (GLOBECOM 2001)*, vol. 1, San Antonio, TX, 25–29 Nov. 2001, pp. 985–989.

1749. Y. Mao and A. H. Banihashemi, "A heuristic search for good LDPC codes at short block lengths", in *Proc. IEEE Int. Conf. Commun. (ICC 2001)*, vol. 1, Helsinki, 11–14 June 2001, pp. 41–44.

1750. Y. Hu, E. Eleftheriou, and D. M. Arnold, "Regular and irregular progressive EDGE growth of Tanner graphs", *IEEE Trans. Inform. Theory*, vol. 51, no. 1, pp. 386–398, Jan. 2005.

1751. A. Venkiah, D. Declercq, and C. Poulliat, "Design of cages with a randomized progressive edge-growth algorithm", *IEEE Commun. Lett.*, vol. 12, no. 4, pp. 301–303, Apr. 2008.

1752. M. Luby, M. Mitzenmacher, M. A. Shokrollahi, and D. A. Spielman, "Improved low-density parity-check codes using irregular graphs", *IEEE Trans. Inform. Theory*, vol. 47, no. 2, pp. 585–598, Feb. 2001.

1753. S.-Y. Chung, T. J. Richardson, and R. L. Urbanke, "Analysis of sum-product decoding of low-density parity-check codes using a Gaussian approximation", *IEEE Trans. Inform. Theory*, vol. 47, no. 2, pp. 657–670, Feb. 2001.

1754. S. ten Brink, G. Kramer, and A. Ashikhmin, "Design of low-density parity-check codes for modulation and detection", *IEEE Trans. Commun.*, vol. 52, no. 4, pp. 670–678, Apr. 2004.

1755. S. ten Brink and G. Kramer, "Design of repeat-accumulate codes for iterative detection and decoding", *IEEE Trans. Sig. Process.*, vol. 51, no. 11, pp. 2764–2772, Nov. 2003.

1756. M. Ardakani and F. R. Kschischang, "A more accurate one-dimensional analysis and design of irregular LDPC codes", *IEEE Trans. Commun.*, vol. 52, no. 12, pp. 2106–2114, Dec. 2004.

1757. F. Peng, W. E. Ryan, and R. D. Wesel, "Surrogate-channel design of universal LDPC codes", *IEEE Commun. Lett.*, vol. 10, no. 6, pp. 480–482, June 2006.

1758. C. Di, D. Proietti, E. Telatar, T. J. Richardson, and R. L. Urbanke, "Finite-length analysis of low-density parity-check codes on the binary erasure channel", *IEEE Trans. Inform. Theory*, vol. 48, no. 6, pp. 1570–1579, June 2002.

1759. A. Orlitsky, K. Viswanathan, and J. Zhang, "Stopping set distribution of LDPC code ensembles", *IEEE Trans. Inform. Theory*, vol. 51, no. 3, pp. 929–953, Mar. 2005.

1760. M. Ivkovic, S. K. Chilappagari, and B. Vasic, "Eliminating trapping sets in low-density parity-check codes by using Tanner graph covers", *IEEE Trans. Inform. Theory*, vol. 54, no. 8, pp. 3763–3768, Aug. 2008.

1761. A. Vardy and M. Schwartz, "On the stopping distance and the stopping redundancy of codes", *IEEE Trans. Inform. Theory*, vol. 52, no. 3, pp. 922–932, Mar. 2006.

1762. K. M. Krishnan and P. Shankar, "Computing the stopping distance of a Tanner graph is NP-hard", *IEEE Trans. Inform. Theory*, vol. 53, no. 6, pp. 2278–2280, June 2007.

1763. T. Tian, C. Jones, J. D. Villasenor, and R. D. Wesel, "Selective avoidance of cycles in irregular LDPC code construction", *IEEE Trans. Commun.*, vol. 52, no. 8, pp. 1242–1248, Aug. 2004.

1764. D. Vukobratovic and V. Senk, "Generalized ACE constrained progressive edge growth LDPC code design", *IEEE Commun. Lett.*, vol. 12, no. 1, pp. 32–34, Jan. 2008.

1765. ——, "Evaluation and design of irregular LDPC codes using ACE spectrum", *IEEE Trans. Inform. Theory*, vol. 57, no. 8, pp. 2272–2279, Aug. 2009.

1766. R. M. Tanner, "On quasi-cyclic repeat-accumulate codes", in *Proc. 37th Allerton Conference on Communication, Control, and Computing*, Monticello, IL, Sep. 1999, pp. 249–259.

1767. S. Freundlich, D. Burshtein, and S. Litsyn, "Approximately lower triangular ensembles of LDPC codes with linear encoding complexity", *IEEE Trans. Inform. Theory*, vol. 53, no. 4, pp. 1484–1494, Apr. 2007.

1768. T. J. Richardson and R. L. Urbanke, "Efficient encoding of low-density parity-check codes", *IEEE Trans. Inform. Theory*, vol. 47, no. 2, pp. 638–656, Feb. 2001.

1769. ETSI TR 102 376 V1.1.1 (2005-02), *Digital Video Broadcasting (DVB), User Guidelines for the Second Generation System for Broadcasting, Interactive Services, News Gathering and Other Broadband Satellite Applications (DVB-S2)*, European Telecommunications Standards Institute, 2005.

1770. DVB Document A122r1, *Frame Structure, Channel Coding and Modulation for a Second Generation Digital Terrestrial Television Broadcasting*, DVB, Digital Video Broadcasting, Jan. 2008.

1771. F. Kienle, T. Brack, and N. Wehn, "A synthesizable IP core for DVB-S2 LDPC code decoding", in *Proc. IEEE 2005 Design, Automation and Test in Europe Conf. (DATE 2005)*, vol. 3, Munich, 7–11 Mar. 2005, pp. 100–105.

1772. L. Ping and W. K. Leung, "Decoding low density parity check codes with finite quantization bits", *IEEE Commun. Lett.*, vol. 4, no. 2, pp. 62–64, Feb. 2000.

1773. M. P. C. Fossorier, M. Mihaljevic, and H. Imai, "Reduced complexity iterative decoding of low-density parity check codes based on belief propagation", *IEEE Trans. Commun.*, vol. 47, no. 5, pp. 673–680, May 1999.

1774. S.-Y. Chung, "On the construction of some capacity-approaching coding schemes", Ph.D. dissertation, Massachusetts Institute of Technology, Sep. 2000.

1775. X. Wei and A. N. Akansu, "Density evolution for low-density parity-check codes under Max-Log-MAP decoding", *Electron. Lett.*, vol. 37, no. 18, pp. 1125–1126, 30 Aug. 2001.

1776. J. Chen and M. P. C. Fossorier, "Near optimum universal belief propagation based decoding of low-density parity check codes", *IEEE Trans. Wireless Commun.*, vol. 50, no. 3, pp. 406–414, Mar. 2002.

1777. A. Anastasopoulos, "A comparison between the sum-product and the min-sum iterative detection algorithms based on density evolution", in *Proc. IEEE Global Telecommun. Conf. (GLOBECOM 2001)*, vol. 2, San Antonio, TX, 25–29 Nov. 2001, pp. 1021–1025.

1778. J. Zhao, F. Zarkeshvari, and A. H. Banihashemi, "On implementation of min-sum algorithm and its modifications for decoding low-density parity-check (LDPC) codes", *IEEE Trans. Commun.*, vol. 53, no. 4, pp. 549–554, Apr. 2005.

1779. E. Eleftheriou, T. Mittelholzer, and A. Dholakia, "Reduced-complexity decoding algorithm for low-density parity-check codes", *Electron. Lett.*, vol. 37, no. 2, pp. 102–104, 18 Jan. 2001.

1780. X.-Y. Hu, E. Eleftheriou, D.-M. Arnold, and A. Dholakia, "Efficient implementations of the sum-product algorithm for decoding LDPC codes", in *Proc. IEEE Global Telecommun. Conf. (GLOBECOM 2001)*, vol. 2, San Antonio, TX, 25–29 Nov. 2001, pp. 1036–1036E.

1781. J. Chen and M. P. C. Fossorier, "Density evolution for two improved BP-based decoding algorithms of LDPC codes", *IEEE Commun. Lett.*, vol. 6, no. 5, pp. 208–210, May 2002.

1782. H. Sankar and K. R. Narayanan, "Memory-efficient sum-product decoding of LDPC codes", *IEEE Trans. Commun.*, vol. 52, no. 8, pp. 1225–1230, Aug. 2004.

1783. G. Richter, G. Schmidt, M. Bossert, and E. Costa, "Optimization of a reduced-complexity decoding algorithm for LDPC codes by density evolution", in *Proc. IEEE Int. Conf. Commun. (ICC 2005)*, vol. 1, Seoul, 16–20 May 2005, pp. 642–646.

1784. L. Chen, J. Xu, I. Djurdjevic, and S. Lin, "Near-Shannon-limit quasi-cyclic low-density parity-check codes", *IEEE Trans. Commun.*, vol. 52, no. 7, pp. 1038–1042, July 2004.

1785. J. Chen and M. P. C. Fossorier, "Density evolution for BP-based decoding algorithms of LDPC codes and their quantized versions", in *Proc. IEEE Global Telecommun. Conf. (GLOBECOM 2002)*, vol. 2, Taipei, 17–21 Nov. 2002, pp. 1378–1382.

1786. J. Heo, "Analysis of scaling soft information on low density parity check code", *Electron. Lett.*, vol. 39, no. 2, pp. 219–221, 23 Jan. 2003.

1787. J. Heo and K. M. Chugg, "Optimization of scaling soft information in iterative decoding via density evolution methods", *IEEE Trans. Commun. Technol.*, vol. 53, no. 6, pp. 957–961, June 2005.

1788. D. E. Hocevar, "A reduced complexity decoder architecture via layered decoding of LDPC codes", in *Proc. IEEE Workshop Sig. Process. Syst. (SIPS 2004)*, vol. 1, Austin, TX, Oct. 2004, pp. 107–112.

1789. M. Jiang, C. Zhao, L. Zhang, and E. Xu, "Adaptive offset min-sum algorithm for low-density parity check codes", *IEEE Commun. Lett.*, vol. 10, no. 6, pp. 483–485, June 2006.

1790. L. Barnault and D. Declercq, "Fast decoding algorithm for LDPC over GF(2^q)", in *Proc. ITW 2003*, Paris, 31 Mar.–4 Apr. 2003, pp. 70–73.

1791. D. Declercq and M. C. P. Fossorier, "Decoding algorithms for nonbinary LDPC codes over GF(q)", *IEEE Trans. Commun.*, vol. 55, no. 4, pp. 633–643, Apr. 2007.

1792. F. Lehmann and G. M. Maggio, "Analysis of the iterative decoding of LDPC and product codes using the Gaussian approximation", *IEEE Trans. Inform. Theory*, vol. 49, no. 11, pp. 2993–3000, Nov. 2003.

1793. D. Burshtein and G. Miller, "Bounds on the performance of belief propagation decoding", *IEEE Trans. Inform. Theory*, vol. 48, no. 1, pp. 112–122, Jan. 2002.

1794. M. P. C. Fossorier, "Iterative reliability-based decoding of low-density parity check codes", *IEEE J. Sel. Areas Commun.*, vol. 19, no. 5, pp. 908–917, May 2001.

1795. M. Isaka, M. Fossorier, and H. Imai, "On the suboptimality of iterative decoding for turbo-like and LDPC codes with cycles in their graph representation", *IEEE Commun. Lett.*, vol. 52, no. 5, pp. 845–854, May 2004.

1796. M. Jiang, C. Zhao, E. Xu, and L. Zhang, "Reliability-based iterative decoding of LDPC codes using likelihood accumulation", *IEEE Commun. Lett.*, vol. 11, no. 8, pp. 677–679, Aug. 2007.

1797. D. Burshtein and G. Miller, "Efficient maximum-likelihood decoding of LDPC codes over the binary erasure channel", *IEEE Trans. Inform. Theory*, vol. 50, no. 11, pp. 2837–2844, Nov. 2004.

1798. Y. Mao and A. H. Banihashemi, "Decoding low-density parity-check codes with probabilistic scheduling", *IEEE Commun. Lett.*, vol. 5, no. 10, pp. 414–416, Oct. 2001.

1799. L. Bazzi, T. J. Richardson, and R. L. Urbanke, "Exact thresholds and optimal codes for the binary-symmetric channel and Gallager's decoding algorithm A", *IEEE Trans. Inform. Theory*, vol. 50, no. 9, pp. 2010–2021, Sep. 2004.

1800. D. Burshtein, "On the error correction of regular LDPC codes using the flipping algorithm", *IEEE Trans. Inform. Theory*, vol. 54, no. 2, pp. 517–530, Feb. 2008.

1801. L. Dolecek, P. Lee, Z. Zhang, V. Anantharam, B. Nikolic, and M. Wainwright, "Predicting error floors of structured LDPC codes: deterministic bounds and estimates", *IEEE J. Sel. Areas Commun.*, vol. 27, no. 6, pp. 908–917, Aug. 2009.

1802. M. Franceschini, G. Ferrari, and R. Raheli, "Does the performance of LDPC codes depend on the channel?" *IEEE Trans. Commun.*, vol. 54, no. 12, pp. 2129–2132, Dec. 2006.

1803. A. Amraoui, A. Montanari, T. Richardson, and R. Urbanke, "Finite-length scaling for iteratively decoded LDPC ensembles", *IEEE Trans. Inform. Theory*, vol. 55, no. 2, pp. 473–498, Feb. 2009.

1804. D. Burshtein and G. Miller, "Bounds on the maximum-likelihood decoding error probability of low-density parity-check codes", *IEEE Trans. Inform. Theory*, vol. 47, no. 7, pp. 2696–2710, Nov. 2001.

1805. V. Rathi, "On the asymptotic weight and stopping set distribution of regular LDPC ensembles", *IEEE Trans. Inform. Theory*, vol. 52, no. 9, pp. 4212–4218, Sep. 2006.

1806. I. Goldenberg and D. Burshtein, "Upper bound on error exponent of regular LDPC codes transmitted over the BEC", *IEEE Trans. Inform. Theory*, vol. 55, no. 6, pp. 2674–2681, June 2009.

1807. G. Colavolpe, "On LDPC codes over channels with memory", *IEEE Trans. Wireless Commun.*, vol. 5, no. 7, pp. 1757–1766, July 2006.

1808. A. P. Worthen and W. E. Stark, "Unified design of iterative receivers using factor graphs", *IEEE Trans. Veh. Tech.*, vol. 47, no. 2, pp. 843–849, Feb. 2001.

1809. N. Wiberg, H.-A. Loeliger, and R. Koetter, "Codes and iterative decoding on general graphs: Iterative and turbo decoding", *European Trans. Telecommun.*, vol. 6, no. 5, pp. 513–525, Sep./Oct. 1995.

1810. J. C. Willems, "Models for dynamics", in *Dynamics Reported, Volume 2*, U. Kirchgraber and H. O. Walther, (eds.) Chichester: John Wiley & Sons, Ltd, 1989, pp. 171–269.

1811. D. G. Forney, Jr., "Codes on graphs: Normal realizations", *IEEE Trans. Inform. Theory*, vol. 47, no. 2, pp. 520–548, Feb. 2001.

1812. J. Pearl, *Probabilistic Reasoning in Intelligent Systems*. San Mateo, CA: Morgan Kaufmann, 1988.

1813. L. E. Baum and T. Petrie, "Statistical inference for probabilistic functions of finite-state Markov chains", *Ann. Math. Statist.*, vol. 37, no. 6, pp. 1554–1563, Dec. 1966.

1814. S. M. Aji and R. J. McEliece, "The generalized distributive law", *IEEE Trans. Inform. Theory*, vol. 46, no. 3, pp. 325–343, Mar. 2000.

1815. R. E. Blahut and R. Koetter, *Codes, Graphs, and Systems*. Norwell, MA: Kluwer Academic Publishers, 2002.

1816. G. Battail, "Construction explicite de bons codes longs", *Annales des Télécommunications*, vol. 44, no. 7–8, pp. 392–404, 1989.

1817. ——, "Pondration des symboles décodés par l'algorithme de Viterbi", *Annales des Télécommunications*, vol. 42, pp. 31–38, Jan.-Feb. 1987.

1818. J. Lodge, P. Hoeher, and J. Hagenauer, "The decoding of multidimensional codes using separable MAP filters", in *Proc. 16th Biennial Symp. Commun.*, Kingston, Ontario, 27–29 May 1992, pp. 343–346.

1819. J. Hagenauer, "Soft-in/soft-out, the benefits of using soft values in all stages of digital receivers", in *Proc. 3rd Int. Workshop Digital Signal Processing Techniques Applied to Space Communications*, ESTEC, Noordwijk, The Netherlands, Sep. 1992, pp. 7.1–7.15.

1820. M. Breiling, "A logarithmic upper bound on the minimum distance of turbo codes", *IEEE Trans. Inform. Theory*, vol. 50, no. 7, pp. 1692–1710, July 2004.

1821. J. D. Anderson, "Turbo codes extended with outer BCH code", *Electron. Lett.*, vol. 32, no. 22, pp. 2059–2060, 24 Oct. 1996.

1822. S. Crozier and P. Guinand, "Distance upper bounds and true minimum distance results for turbo codes designed with DRP interleavers", in *Proc. 3rd Int. Symp. Turbo Codes Related Topics*, Brest, France, 1–5 Sep. 2003, pp. 169–172.

1823. C. Douillard and C. Berrou, "Turbo codes with rate-$m/(m+1)$ constituent convolutional codes", *IEEE Trans. Commun.*, vol. 53, no. 10, pp. 1630–1638, Oct. 2005.

1824. E. Boutillon and D. Gnaedig, "Maximum spread of D-dimensional multiple turbo codes", *IEEE Trans. Commun.*, vol. 53, no. 8, pp. 1237–1242, Aug. 2005.

1825. C. He, M. Lentmaier, D. J. Costello, Jr., and K. S. Zigangirov, "Joint permutor analysis and design for multiple turbo codes", *IEEE Trans. Inform. Theory*, vol. 52, no. 9, pp. 4068–4083, Sep. 2006.

1826. A. S. Barbulescu and S. S. Pietrobon, "Interleaver design for turbo codes", *Electron. Lett.*, vol. 30, no. 25, pp. 2107–2108, 8 Dec. 1994.

1827. K. Koora and H. Betzinger, "Interleaver design for turbo codes with selected inputs", *Electron. Lett.*, vol. 34, no. 7, pp. 651–652, 2 Apr. 1998.

1828. O. Y. Takeshita and D. J. Costello, Jr., "New deterministic interleaver designs for turbo codes", *IEEE Trans. Inform. Theory*, vol. 46, no. 6, pp. 1988–2006, Sep. 2000.

1829. H. R. Sadjadpour, N. J. A. Sloane, M. Salehi, and G. Nebe, "Interleaver design for turbo codes", *IEEE J. Sel Areas Commun.*, vol. 19, no. 5, pp. 831–837, May 2001.

1830. W. Feng, J. Yuan, and B. S. Vucetic, "A code-matched interleaver design for turbo codes", *IEEE Trans. Commun.*, vol. 50, no. 6, pp. 926–937, June 2002.

1831. J. Yu, M.-L. Boucheret, R. Vallet, A. Duverdier, and G. Mesnager, "Interleaver design for turbo codes from convergence analysis", *IEEE Trans. Commun.*, vol. 54, no. 4, pp. 619–624, Apr. 2006.

1832. H. Nickl, J. Hagenauer, and F. Burkert, "Approaching Shannon's capacity limit by 0.27 dB using simple Hamming codes", *IEEE Commun. Lett.*, vol. 1, no. 5, pp. 130–132, Sep. 1997.

1833. T. Richardson and R. Urbanke, "The renaissance of Gallager's low-density parity-check codes", *IEEE Commun. Mag.*, vol. 41, no. 8, pp. 126–131, Aug. 2003.

1834. D. J. C. MacKay and R. M. Neal, "Near Shannon limit performance of low-density parity-check codes", *Electron. Lett.*, vol. 32, no. 18, pp. 1645–1646, 29 Aug. 1996.

1835. _____, "Near Shannon limit performance of low-density parity-check codes", *Electron. Lett.*, vol. 33, no. 6, pp. 457–458, 13 Mar. 1997.

1836. D. A. Spielman, "Linear-time encodable and decodable error-correcting codes", *IEEE Trans. Inform. Theory*, vol. 42, no. 11, pp. 1723–1731, Nov. 1996.

1837. _____, "Linear-time encodable and decodable error-correcting codes", in *Proc. 27th ACM Symp. Theory Comp.*, 1995, pp. 388–397.

1838. N. Alon and M. Luby, "A linear time erasure-resilient code with nearly optimal recovery", *IEEE Trans. Inform. Theory*, vol. 42, no. 6, pp. 1732–1736, Nov. 1996.

1839. A. Shokrollahi and R. Storn, "Design of efficient erasure codes with differential evolution", in *Proc. 2000 IEEE Int. Symp. Inform. Theory*, Sorrento, Italy, June 2000, p. 5.

1840. M. Luby, M. Mitzenmacher, M. A. Shokrollahi, and D. A. Spielman, "Efficient erasure-correcting codes", *IEEE Trans. Inform. Theory*, vol. 47, no. 2, pp. 569–584, Feb. 2001.

1841. L. Ping and K. Y. Wu, "Concatenated tree codes: A low-complexity, high-performance approach", *IEEE Trans. Inform. Theory*, vol. 47, no. 2, pp. 791–799, Feb. 2001.

1842. P. C. Massey and D. J. Costello, Jr., "New low-complexity turbo-like codes", in *Proc. IEEE Inform. Theory Workshop*, Cairns, Australia, Sep. 2001, pp. 70–72.

1843. M. Luby, "LT codes", in *Proc. 43rd Annual IEEE Symp. Foundations of Computer Science (FOCS)*, Vancouver, Feb. 2002, pp. 271–280.

1844. M. A. Shokrollahi, "Raptor codes", *IEEE Trans. Inform. Theory*, vol. 52, no. 6, pp. 2551–2567, June 2006.

1845. O. Etesami and A. Shokrollahi, "Raptor codes on binary memoryless symmetric channels", *IEEE Trans. Inform. Theory*, vol. 52, no. 5, pp. 2033–2051, May 2006.

1846. B. Sklar, "A primer on turbo code concepts", *IEEE Commun. Mag.*, vol. 35, no. 12, pp. 94–102, Dec. 1997.

1847. C. Schlegel and L. Perez, *Trellis and Turbo Coding*. Piscataway, NJ: IEEE Press, 2004.

1848. S. B. Wicker and S. Kim, *Fundamentals of Codes, Graphs and Iterative Decoding*. Norwell, MA: Kluwer Academic Publishers, 2003.

1849. A. R. Calderbank and N. J. A. Sloane, "New trellis codes based on lattices and cosets", *IEEE Trans. Inform. Theory*, vol. 33, pp. 177–195, Mar. 1987.

1850. P. Borjesson and C.-E. Sundberg, "Simple approximations of the error function $Q(x)$ for communications applications", *IEEE Trans. Commun.*, vol. 27, no. 3, pp. 639–643, Mar. 1979.

1851. M. Chiani, D. Dardari, and M. K. Simon, "New exponential bounds and approximations for the computation of error probability in fading channels", *IEEE Trans. Wireless Commun.*, vol. 2, no. 4, pp. 840–845, July 2003.

1852. D. Divsalar and M. K. Simon, "The design of trellis codes for fading channels", *IEEE Trans. Commun.*, vol. 36, no. 9, pp. 1004–1021, Sep. 1988.

1853. E. Malkamaki and H. Leib, "Evaluating the performance of convolutional codes over block fading channels", *IEEE Trans. Inform. Theory*, vol. 45, no. 5, pp. 1643–1646, July 1999.

1854. G. Caire, G. Taricco, and E. Biglieri, "Bit-interleaved coded modulation", *IEEE Trans. Inform. Theory*, vol. 44, no. 3, pp. 927–946, May 1999.

1855. X. Li and J. A. Ritcey, "Bit-interleaved coded modulation with iterative decoding", *IEEE Commun. Lett.*, vol. 1, no. 6, pp. 169–171, Nov. 1997.

1856. G. J. Pottie and D. P. Taylor, "Multilevel codes based on partitioning", *IEEE Trans. Inform. Theory*, vol. 35, no. 1, pp. 87–98, Jan. 1989.

1857. U. Wachsmann, R. F. H. Fischer, and J. B. Huber, "Multilevel codes: Theoretical concepts and practical design rules", *IEEE Trans. Inform. Theory*, vol. 45, no. 5, pp. 1361–1391, July 1999.

1858. K. Fazel and L. Papke, "Combined multilevel turbo-code with 8PSK modulation", in *Proc. IEEE Global Telecommun. Conf. (GLOBECOM '95)*, Nov. 1955, pp. 649–653.

1859. A. R. Calderbank and N. Seshadri, "Multilevel codes for unequal error protection", *IEEE Trans. Inform. Theory*, vol. 39, no. 4, pp. 1234–1248, 1993.

1860. J.-F. Cheng, "On the construction of efficient multilevel coded modulations", in *Proc. IEEE Int. Symp. Inform. Theory*, Ulm, Germany, 29 June-4 July 1997, p. 522.

1861. F. J. Lopez, R. A. Carrasco, and P. G. Farrell, "Ring-TCM codes for QAM", *Electron. Lett.*, vol. 28, no. 25, pp. 2358–2359, 3 Dec. 1992.

1862. U. Wachsmann and J. Huber, "Power and bandwidth efficient digital communication using turbo codes in multilevel codes", *European Trans. Telecommun.*, vol. 6, no. 5, pp. 557–567, Sep./Oct. 1995.

1863. R. F. H. Fischer, J. B. Huber, and U. Wachsmann, "Multilevel coding aspects from information theory", in *Proc. ITW*, June 1996, p. 9.

1864. A. Picart and R. Pyndiah, "Performance of turbo-decoded product codes used in multilevel coding", in *Proc. IEEE Int. Conf. Commun. (ICC '96)*, vol. 1, Dallas, 23–27 June 1996, pp. 107–111.

1865. J. Huber, "Multilevel-codes: Distance profiles and channel capacity", ITG-Fachtagung, 1994, Munich.

1866. J. Huber and U. Wachsmann, "Capacities of equivalent channels in multilevel coding schemes", *Electron. Lett.*, vol. 30, no. 7, pp. 557–558, 31 March 1994.

1867. R. F. H. Fischer, J. B. Huber, and U. Wachsmann, "Multilevel coding: Aspects from information theory", in *Proc. IEEE Global Telecommun. Conf. (GLOBECOM '96)*, London, 18–22 Nov. 1996, pp. 26–30.

1868. Y. Kofman, E. Zehavi, and S. Shamai (Shitz), "Performance analysis of a multilevel coded modulation system", *IEEE Trans. Commun.*, vol. 42, no. 2/3/4, pp. 299–312, Feb./Mar./Apr. 1994.

1869. P. A. Martin and D. P. Taylor, "On multilevel codes and iterative multistage decoding", *IEEE Trans. Commun.*, vol. 49, no. 11, pp. 1916–1925, Nov. 2001.

1870. K. Engdahl and K. S. Zigangirov, "On the calculation of the error probability for a multilevel modulation scheme using QAM-signaling", *IEEE Trans. Inform. Theory*, vol. 44, no. 4, pp. 1612–1620, July 1998.

1871. E. Biglieri, A. Sandri, and A. Spalvieri, "Computing upper bounds to error probability of coded modulation schemes", *IEEE Trans. Commun.*, vol. 44, no. 7, pp. 786–790, July 1996.

1872. T. Takata, S. Ujita, T. Kasami, and S. Lin, "Multistage decoding of multilevel block M-PSK modulation codes and its performance analysis", *IEEE Trans. Inform. Theory*, vol. 39, no. 4, pp. 1204–1218, July 1993.

1873. T. Kasami, T. Takata, T. Fujiwara, and S. Lin, "On multilevel block modulation codes", *IEEE Trans. Inform. Theory*, vol. 37, no. 4, pp. 965–975, July 1991.

1874. N. Seshadri and C.-E. W. Sundberg, "Multilevel trellis coded modulations for the Rayleigh fading channel", *IEEE Trans. Commun.*, vol. 41, no. 9, pp. 1300–1310, Sept. 1993.

1875. J. Wu and S. Lin, "Multilevel trellis MPSK modulation codes for the Rayleigh fading channel", *IEEE Trans. Commun.*, vol. 41, no. 9, pp. 1311–1318, Sept. 1993.

1876. L. Zhang and B. Vucetic, "Multilevel block codes for Rayleigh-fading channels", *IEEE Trans. Commun.*, vol. 43, no. 1, pp. 24–31, Jan. 1995.

1877. D. J. Rhee, S. Rajpal, and S. Lin, "Some block- and trellis-coded modulations for the Rayleigh fading channel", *IEEE Trans. Commun.*, vol. 44, no. 1, pp. 34–42, Jan. 1996.

1878. R. H. Morelos-Zaragoza, T. Kasami, S. Lin, and H. Imai, "On block-coded modulation using unequal error protection codes over Rayleigh-fading channels", *IEEE Trans. Commun.*, vol. 46, no. 1, pp. 1–4, Jan. 1998.

1879. D.-F. Yuan, Q. Yao, C.-X. Wang, and Z.-G. Cao, "Different decoding methods for multilevel coded modulation over Rayleigh fading channels", in *Proc. 2000 Int. Conf. Commun. Tech. (WCC – ICCT 2000)*, vol. 2, Beijing, 21–25 Aug. 2000, pp. 1542–1545.

1880. L. Zheng and D. N. C. Tse, "Diversity and multiplexing: a fundamental tradeoff in multiple-antenna channels", *IEEE Trans. Inform. Theory*, vol. 49, no. 5, pp. 1073–1096, May 2003.

1881. H. Lu and P. V. Kumar, "Rate-diversity tradeoff of space-time codes with fixed alphabet and optimal constructions for PSK modulation", *IEEE Trans. Inform. Theory*, vol. 49, no. 10, pp. 2747–2751, Oct. 2003.

1882. T. L. Marzetta and B. M. Hochwald, "Capacity of a mobile multiple-antenna communication link in Rayleigh flat fading", *IEEE Trans. Inform. Theory*, vol. 45, no. 1, pp. 139–157, Jan. 1999.

1883. T. M. Duman and A. Ghrayeb, *Coding for MIMO Communication Systems*. Chichester: John Wiley & Sons, Ltd, 2007.

1884. S. A. Jafar and A. Goldsmith, "Transmitter optimization and optimality of beamforming for multiple antenna systems", *IEEE Trans. Wireless Commun.*, vol. 3, no. 4, pp. 1165–1175, July 2004.

1885. A. L. Moustakas, S. H. Simon, and A. M. Sengupta, "MIMO capacity through correlated channels in the presence of correlated interferers and noise: a (not so) large N analysis", *IEEE Trans. Inform. Theory*, vol. 49, no. 10, pp. 2545–2561, Oct. 2003.

1886. L. Zheng and D. Tse, "Communicating on the Grassmann manifold: A geometric approach to the non-coherent multiple antenna channel", *IEEE Trans. Inform. Theory*, vol. 48, no. 2, pp. 359–383, Feb. 2002.

1887. J. Ventura-Traverset, G. Caire, E. Biglieri, and G. Taricco, "Impact of diversity reception on fading channels with coded modulation – Part I: Coherent detection", *IEEE Trans. Commun.*, vol. 45, no. 5, pp. 563–572, May 1997.

1888. A. R. Hammons, Jr., and H. E. Gamal, "On the theory of space-time codes for PSK modulation", *IEEE Trans. Inform. Theory*, vol. 46, no. 2, pp. 524–542, Mar. 2000.

1889. A. Wittneben, "A new bandwidth efficient transmit antenna modulation diversity scheme for linear digital modulation", in *Proc. IEEE Int. Conf. Commun. (ICC '93)*, Geneva, 23–26 May 1993, pp. 1630–1634.

1890. S. Sandhu, R. Heath, and A. Paulraj, "Space-time block codes versus space-time trellis codes", in *Proc. IEEE Int. Conf. Commun. (ICC 2001)*, Helsinki, June 2002, pp. 1132–1136.

1891. M. J. Borran, M. Memarzadeh, and B. Aazhang, "Design of coded modulation schemes for orthogonal transmit diversity", in *Proc. 2001 IEEE Int. Symp. Inform. Theory (ISIT 2001)*, Washington, DC, June 2000, p. 339.

1892. S. Siwamogsatham and M. P. Fitz, "Robust space-time coding for correlated Rayleigh fading channels", *IEEE Trans. Sig. Process.*, vol. 50, no. 10, pp. 2408–2416, Oct. 2002.

1893. C. Xu and K. S. Kwak, "On decoding algorithm and performance of space-time block codes", *IEEE Trans. Wireless Commun.*, vol. 4, no. 3, pp. 825–829, May 2005.

1894. H. Jafarkhani, "A quasi-orthogonal space-time block code", *IEEE Trans. Commun.*, vol. 49, no. 1, pp. 1–4, Jan. 2001.

1895. B. Hassibi and B. M. Hochwald, "High-rate codes that are linear in space and time", *IEEE Trans. Inform. Theory*, vol. 48, no. 7, pp. 1804–1824, July 2002.

1896. S. Baro, G. Bauch, and A. Hansmann, "Improved codes for space-time trellis coded modulation", *IEEE Commun. Lett.*, vol. 4, no. 1, pp. 20–22, Jan. 2000.

1897. R. S. Blum, "Some analytical tools for the design of space-time convolutional codes", *IEEE Trans. Commun.*, vol. 50, no. 10, pp. 1593–1599, Oct. 2002.

1898. Q. Yan and R. S. Blum, "Improved space-time convolutional codes for quasi-static slow fading channels", *IEEE Trans. Wireless Commun.*, vol. 1, no. 4, pp. 563–571, Oct. 2002.

1899. Y. Liu, M. P. Fitz, and O. Y. Takeshita, "A rank criterion for QAM space-time codes", *IEEE Trans. Inform. Theory*, vol. 48, no. 12, pp. 3062–3079, Dec. 2002.

1900. S. Siwamogsatham and M. P. Fitz, "Improved high rate space-time codes via concatenation of expanded orthogonal block code and M-TCM", in *Proc. IEEE Int. Conf. Commun. (ICC 2001)*, Helsinki, June 2002, pp. 636–640.

1901. ———, "Improved high-rate space-time codes via expanded STBC-MTCM constructions", in *Proc. 2002 IEEE Int. Symp. Inform. Theory (ISIT 2002)*, Lausanne, June 2002, p. 106.

1902. H. Jafarkhani and N. Seshadri, "Super-orthogonal space-time trellis codes", *IEEE Trans. Inform. Theory*, vol. 49, no. 4, pp. 937–950, Apr. 2003.

1903. A. Stefanov and T. M. Duman, "Performance bounds for space-time trellis codes", *IEEE Trans. Inform. Theory*, vol. 49, no. 9, pp. 2134–2140, Sept. 2003.

1904. V. Tarokh, A. Naguib, N. Seshadri, and A. Calderbank, "Combined array processing and space-time coding", *IEEE Inform. Theory*, vol. 45, no. 8, pp. 1121–1128, May 1999.

1905. H. El Gamal and A. R. Hammons Jr., "A new approach to layered space-time coding and signal processing", *IEEE Trans. Inform. Theory*, vol. 47, no. 6, pp. 2321–2334, Sept. 2001.

1906. G. Caire and G. Colavolpe, "On low-complexity space-time coding for quasi-static channels", *IEEE Trans. Inform. Theory*, vol. 49, no. 6, pp. 1400–1416, June 2003.

1907. D.-S. Shiu and J. Kahn, "Layered space-time codes for wireless communications using multiple transmit antennas", in *Proc. IEEE Int. Conf. Commun. (ICC '99)*, vol. 1, pp. 436–440.

1908. R. Horn and C. Johnson, *Matrix Analysis*. Cambridge: Cambridge University Press, 1985.

1909. E. Viterbo and J. Boutros, "A universal lattice code decoder for fading channel", *IEEE Trans. Inform. Theory*, vol. 45, no. 7, pp. 1639–1642, July 1999.

1910. M. O. Damen, H. El Gamal, and G. Caire, "On maximum-likelihood detection and the search for the closest lattice point", *IEEE Trans. Inform. Theory*, vol. 49, no. 10, pp. 2389–2402, Oct. 2003.

1911. J. Boutros, N. Gresset, L. Brunel, and M. Fossorier, "Soft-input soft-output lattice sphere decoder for linear channels", in *Proc. IEEE Global Telecommun. Conf. (GLOBECOM 2003)*, San Francisco, Dec. 2003, pp. 1583–1587.

1912. J. Hu and T. M. Duman, "Graph-based detection algorithms for layered space-time architectures", *IEEE J. Sel. Areas Commun.*, vol. 26, pp. 269–280, Feb. 2008.

1913. X. Wang and H. V. Poor, "Iterative (turbo) soft interference cancellation and decoding for coded CDMA", *IEEE Trans. Commun.*, vol. 47, no. 7, pp. 1046–1061, May 1996.

1914. H. El Gamal and E. Geraniotis, "Iterative multiuser detection for coded CDMA signals in AWGN and fading channels", *IEEE J. Sel. Areas Commun.*, vol. 18, pp. 30–41, Jan. 2000.

1915. G. Colavolpe, D. Fertonani, and A. Piemontese, "SISO detection over linear channels with linear complexity in the number of interferers", *IEEE J. Sel. Topics Sig. Process.*, vol. 5, pp. 1475–1485, Dec. 2011.

1916. R. Knopp and P. Humblet, "On coding for block fading channels", *IEEE Trans. Inform. Theory*, vol. 46, no. 1, pp. 189–205, Jan. 2000.

1917. R. Wesel, X. Liu, and W. Shi, "Trellis codes for periodic erasures", *IEEE Trans. Commun.*, vol. 48, no. 6, pp. 938–947, June 2000.

1918. G. Bauch, "Concatenation of space-time block codes and 'turbo'-TCM", in *Proc. IEEE Int. Conf. Commun. (ICC 1999)*, vol. 6, Vancouver, June 1999, pp. 1202–1206.

1919. Y. Gong and K. B. Letaief, "Concatenated space-time block coding with trellis coded modulation in fading channels", *IEEE Trans. Wireless Commun.*, vol. 1, no. 4, pp. 580–590, Oct. 2002.

1920. Y. Liu, M. P. Fitz, and O. Y. Takeshita, "Full-rate space-time turbo codes", *IEEE J. Sel. Areas Commun.*, vol. 19, no. 5, pp. 969–980, May 2001.

1921. H. J. Su and E. Geraniotis, "Space-time turbo codes with full antenna diversity", *IEEE Trans. Commun.*, vol. 49, no. 1, pp. 47–57, Jan. 2001.

1922. X. Lin and R. S. Blum, "Improved space-time codes using serial concatenation", *IEEE Commun. Lett.*, vol. 4, no. 7, pp. 221–223, July 2000.

1923. D. Cui and A. M. Haimovich, "Performance of parallel concatenated space-time codes", *IEEE Commun. Lett.*, vol. 5, no. 6, pp. 236–238, June 2001.

1924. A. Stefanov and T. M. Duman, "Turbo-coded modulation for systems with transmit and receive antenna diversity over block fading channels: System model, decoding approaches, and practical considerations", *IEEE J. Sel. Areas Commun.*, vol. 19, no. 5, pp. 958–968, May 2001.

1925. V. Gulati and K. N. Narayanan, "Concatenated codes for fading channels based on recursive space-time trellis codes", *IEEE Trans. Wireless Commun.*, vol. 2, no. 1, pp. 118–128, Jan. 2003.

1926. S. Benedetto, D. Divsalar, G. Montorsi, and F. Pollara, "Self-concatenated codes with self-iterative decoding for power and bandwidth efficiency", in *Proc. 1998 IEEE Int. Symp. Inform. Theory (ISIT 1998)*, Cambridge, MA, Aug. 2003, p. 177.

1927. Y. Li, B. Vucetic, Q. Zhang, and Y. Huang, "Assembled space-time turbo trellis codes", *IEEE Trans. Veh. Tech.*, vol. 54, no. 5, pp. 1768–1772, Sep. 2005.

1928. S. L. Ariyavisitakul, "Turbo space-time processing to improve wireless channel capacity", *IEEE Trans. Commun.*, vol. 48, no. 8, pp. 1347–1359, Aug. 2000.

1929. G. Li, I. J. Fair, and W. A. Krzymien, "Low-density parity-check codes for space-time wireless transmission", *IEEE Trans. Wireless Commun.*, vol. 5, no. 2, pp. 312–322, Feb. 2006.

1930. P. A. Martin, D. M. Rankin, and D. P. Taylor, "Multi-dimensional space-time multilevel codes", *IEEE Trans. Wireless Commun.*, vol. 5, no. 11, pp. 3287–3295, Nov. 2006.

1931. B. M. Hochwald and T. L. Marzetta, "Unitary space-time modulation for multiple antenna communications in Rayleigh flat fading", *IEEE Trans. Inform. Theory*, vol. 46, no. 2, pp. 543–564, Mar. 2000.

1932. B. L. Hughes, "Differential space-time modulation", *IEEE Trans. Inform. Theory*, vol. 46, no. 7, pp. 2567–2578, Nov. 2000.

1933. B. M. Hochwald and W. Sweldens, "Differential unitary space-time modulation", *IEEE Trans. Commun.*, vol. 48, no. 12, pp. 2041–2052, Dec. 2000.

1934. V. Tarokh and H. Jafarkhani, "A differential detection scheme for transmit diversity", *IEEE J. Sel. Areas Commun.*, vol. 18, no. 7, pp. 1169–1174, July 2000.

1935. I. Bahceci and T. M. Duman, "Combined turbo coding and unitary space-time modulation", *IEEE Trans. Commun.*, vol. 50, no. 8, pp. 1244–1249, Aug. 2002.

1936. L. H.-J. Lampe and R. Schober, "Bit-interleaved coded differential space-time modulation", *IEEE Trans. Commun.*, vol. 50, no. 9, pp. 1429–1439, Sep. 2002.

1937. C. Schlegel and A. Grant, "Differential space-time turbo codes", *IEEE Trans. Inform. Theory*, vol. 49, no. 9, pp. 2298–2306, Sep. 2003.

1938. H. El Gamal, A. R. Hammons, Jr., Y. Liu, M. P. Fitz, and O. Y. Takeshita, "On the design of space-time and space-frequency codes for MIMO frequency-selective fading channels", *IEEE Trans. Inform. Theory*, vol. 49, no. 9, pp. 2277–2292, Sep. 2003.

1939. Y. G. Li, J. H. Winters, and N. R. Sollenberger, "MIMO-OFDM for wireless communications: Signal detection with enhanced channel estimation", *IEEE Trans. Commun.*, vol. 50, no. 9, pp. 1471–1477, Sep. 2002.

1940. K. F. Lee and D. B. Williams, "A space-frequency transmitter diversity technique for OFDM systems", in *Proc. IEEE Global Telecommun. Conf. (GLOBECOM 2000)*, San Francisco, 27 Nov.–1 Dec. 2000, pp. 1473–1477.

1941. R. Blum, Y. G. Li, J. H. Winters, and Q. Yam, "Improved space-time coding for MIMO-OFDM wireless communications", *IEEE Trans. Commun.*, vol. 49, no. 7, pp. 1873–1878, Nov. 2001.

1942. H. Bölcskei and A. J. Paulraj, "Space-frequency coded broadband OFDM systems", in *Proc. 2000 IEEE Wireless Commun. and Net. Conf. (WCNC 2000)*, Chicago, 23–28 Sep. 2000, pp. 1–6.

1943. _____, "Space-frequency codes for broadband fading channels", in *Proc. 2000 IEEE Int. Symp. Inform. Theory (ISIT 2000)*, Washington, DC, 24–29 June 2001, p. 219.

1944. W. Su, Z. Safar, M. Olfat, and K. J. R. Liu, "Obtaining full-diversity space-frequency codes from space-time codes via mapping", *IEEE Trans. Sig. Process.*, vol. 51, no. 11, pp. 2905–2916, Nov. 2003.

1945. _____, "Full-rate full-diversity space-frequency codes with optimum coding advantage", *IEEE Trans. Inform. Theory*, vol. 51, no. 1, pp. 229–249, Jan. 2005.

1946. T. Kiran and B. S. Rajan, "A systematic design of high-rate full-diversity space-frequency codes for MIMO-OFDM systems", in *Proc. 2005 IEEE Int. Symp. Inform. Theory (ISIT 2005)*, Adelaide, 4–9 Sep. 2005, pp. 2075–2079.

1947. W. Zhang, X.-G. Xia, and P. C. Ching, "High-rate full-diversity space-time-frequency codes for broadband MIMO block fading channels", *IEEE Trans. Commun.*, vol. 55, no. 1, pp. 25–34, Jan. 2007.

1948. H. El Gamal and M. O. Damen, "Universal space-time coding", *IEEE Inform. Theory*, vol. 49, no. 5, pp. 1097–1019, May 2003.

1949. W. Su, Z. Safar, and K. J. R. Liu, "Towards maximum achievable diversity in space, time, and frequency: Performance analysis and code design", *IEEE Trans. Wireless Commun.*, vol. 4, no. 4, pp. 1847–1857, July 2005.

1950. Z. Liu, Y. Xin, and G. B. Giannakis, "Space-time-frequency coded OFDM over frequency-selective fading channels", *IEEE Trans. Sig. Process.*, vol. 50, no. 10, pp. 2465–2476, Oct. 2002.

1951. M. Fozunbal, S. W. McLaughlin, and R. W. Schafer, "On space-time-frequency coding over MIMO-OFDM systems", *IEEE Trans. Wireless Commun.*, vol. 4, no. 10, pp. 320–331, Jan. 2005.

1952. J. L. Massey, "Coding and modulation in digital communications", in *Int. Zurich Seminar on Digital Commun.*, Zurich, Mar. 1974.

1953. H. Jafarkhani and V. Tarokh, "Multiple transmit antenna differential detection from generalized orthogonal designs", *IEEE Trans. Inform. Theory*, vol. 47, no. 6, pp. 2626–2631, Sep. 2001.

1954. A. Guillén i Fàbregas, A. Martinez, and G. Caire, *Bit-Interleaved Coded Modulation*. Hanover, MA: Foundations and Trends in Communications and Information Theory, Now Publishers, 2008.

1955. H. Jafarkhani, *Space-Time Coding: Theory And Practice*. Cambridge: Cambridge University Press, 2005.

1956. G. B. Giannakis, Z. Liu, X. Ma, and S. Zhou, *Space-Time Coding for Broadband Wireless Communications*. Chichester: John Wiley & Sons, Ltd, 1999.

1957. W. Zhang, X. Xiang-Gen, and K. B. Letaief, "Space-time/frequency coding for MIMO-OFDM in next generation broadband wireless systems", *IEEE Wireless Commun.*, vol. 14, no. 3, pp. 32–43, June 2007.

1958. C. Douillard, M. Jezequel, C. Berrou, A. Picart, P. Didier, and A. Glavieux, "Iterative correction of inter-symbol interference: Turbo equalization", *European Trans. Telecommun.*, vol. 6, pp. 507–511, Sep./Oct. 1995.

1959. A. Glavieux, C. Laot, and J. Labat, "Turbo-equalization over a frequency selective channel", in *Proc. Symp. Turbo-Codes*, Brest, France, Sep. 1997, pp. 96–102.

1960. C. Langlais and M. Helard, "Mapping optimisation for turbo-equalisation improved by iterative demapping", *Electron. Lett.*, vol. 38, no. 22, pp. 1365–1367, Oct. 2002.

1961. A. Dejonghe and L. Vandendorpe, "Bit-interleaved turbo equalization over static frequency-selective channels: constellation mapping impact", *IEEE Trans. Commun.*, vol. 52, no. 12, pp. 2061–2065, Dec. 2004.

1962. M. Tüchler and A. C. Singer, "Turbo equalization: An overview", *IEEE Trans. Commun.*, vol. 57, no. 2, pp. 920–952, Feb. 2011.

1963. S. ten Brink, J. Speidel, and R.-H. Yan, "Iterative demapping and decoding for multilevel modulation", in *Proc. IEEE Global Telecommun. Conf. (GLOBECOM '98)*, vol. 1, Aveiro, Portugal, 8–12 Nov. 1998, pp. 579–584.

1964. P. Magniez, B. Muquet, P. Duhamel, and M. de Courville, "Improved turbo-equalization, with application to bit interleaved modulations", in *Conf. Rec. Thirty-Fourth Asilomar Conf. Sign., Syst. and Comp.*, vol. 2, 29 Oct.–1 Nov. 2000, pp. 1786–1790.

1965. C. Laot, A. Glavieux, and J. Labat, "Turbo equalization: adaptive equalization and channel decoding jointly optimized", *IEEE J. Sel. Areas Commun.*, vol. 19, no. 9, pp. 1744–1752, Sep. 2001.

1966. B. L. Yeap, T. H. Liew, J. Hamorsky, and L. Hanzo, "Comparative study of turbo equalization schemes using convolutional, convolutional turbo, and block-turbo codes", *IEEE Trans. Wireless Commun.*, vol. 1, no. 2, pp. 266–273, Apr. 2002.

1967. G. Bauch, H. Khorram, and J. Hagenauer, "Iterative equalization and decoding in mobile communications systems", in *Proc. European Personal Mobile Communications Conf.*, Bonn, 30 Sept.-2 Oct. 1997, pp. 301–312.

1968. M. Tüchler, R. Koetter, and A. C. Singer, "Graphical models for coded data transmission over inter-symbol interference channels", *European Trans. Telecommun.*, vol. 15, no. 4, pp. 307–321, July-Aug. 2004.

1969. R. J. Drost and A. C. Singer, "Factor-graph algorithms for equalization", *IEEE Trans. Sig. Process.*, vol. 55, no. 5, pp. 2052–2065, May 2007.

1970. Q. Guo and L. Ping, "LMMSE turbo equalization based on factor graphs", *IEEE J. Sel. Areas Commun.*, vol. 26, no. 2, pp. 311–319, Feb. 2008.

1971. B. Etzlinger, W. Haselmayr, and A. Springer, "Equalization algorithms for MIMO communication systems based on factor graphs", in *Proc. IEEE Int. Conf. Commun. (ICC 2011)*, 5–9 June 2011, pp. 1–5.

1972. C. Laot, "Corrections to 'turbo equalization: Adaptive equalization and channel decoding jointly optimized'", *IEEE J. Sel. Areas Commun.*, vol. 25, no. 8, p. 1603, Oct. 2007.

1973. M. Tüchler, A. C. Singer, and R. Koetter, "Minimum mean squared error equalization using a priori information", *IEEE Trans. Sig. Process.*, vol. 50, no. 3, pp. 673–683, Mar. 2002.

1974. M. Tüchler, R. Koetter, and A. C. Singer, "Turbo equalization: principles and new results", *IEEE Trans. Commun.*, vol. 50, no. 5, pp. 754–767, May 2002.

1975. D. Raphaeli and A. Saguy, "Linear equalizers for turbo equalization: A new optimization criterion for determining the equalizer taps", in *Proc. 2nd Int. Symp. Turbo codes*, Brest, France, Sept. 2000, pp. 371–374.

1976. Z.-N. Wu and J. M. Cioffi, "Low-complexity iterative decoding with decision-aided equalization for magnetic recording channels", *IEEE J. Sel. Areas Commun.*, vol. 19, no. 4, pp. 699–708, Apr. 2001.

1977. D. Reynolds and X. Wang, "Low complexity turbo-equalization for diversity channels", *Signal Processing*, vol. 81, no. 5, pp. 989–995, 2001.

1978. S.-J. Lee, A. C. Singer, and N. R. Shanbhag, "Linear turbo equalization analysis via BER transfer and EXIT charts", *IEEE Trans. Sig. Process.*, vol. 53, no. 8, pp. 2883–2897, Aug. 2005.

1979. S. Ibi, T. Matsumoto, S. Sampei, and N. Morinaga, "EXIT chart-aided adaptive coding for MMSE turbo equalization with multilevel BICM", *IEEE Commun. Lett.*, vol. 10, no. 6, pp. 486–488, June 2006.

1980. A. Berthet, R. Visoz, and P. Tortelier, "Sub-optimal turbo-detection for coded 8-PSK signals over ISI channels with application to EDGE advanced mobile system", in *Proc. 11th Int. Symp. Personal, Indoor and Mobile Radio Communications (PIMRC 2000)*, vol. 1, 18–21 Sept. 2000, pp. 151–157.

1981. S. X. Ng, M.-S. Yee, and L. Hanzo, "Coded modulation assisted radial basis function aided turbo equalization for dispersive Rayleigh-fading channels", *IEEE Trans. Wireless Commun.*, vol. 3, no. 6, pp. 2198–2206, Nov. 2004.

1982. E. Tungsrisaguan and R. M. A. P. Rajatheva, "Turbo equalization with sequential sequence estimation over multipath fading channels", *IEEE Commun. Lett.*, vol. 6, no. 3, pp. 93–95, Mar. 2002.

1983. N. Seshadri and P. Hoeher, "On post-decision symbol-reliability generation", in *Proc. IEEE Int. Conf. Commun. (ICC 93)*, vol. 2, Geneva, Apr. 1993, pp. 741–745.

1984. A. J. Viterbi, "An intuitive justification and a simplified implementation of the MAP decoder for convolutional codes", *IEEE J. Sel. Areas Commun.*, vol. 16, no. 2, pp. 260–264, Feb. 1998.

1985. Z. Qin and K. C. Teh, "Reduced-complexity turbo equalization for coded intersymbol interference channels based on local search algorithms", *IEEE Trans. Veh. Tech.*, vol. 57, no. 1, pp. 630–635, Jan. 2008.

1986. M. Tüchler and J. Hagenauer, "'Turbo equalization' using frequency domain equalizers", in *Proc. Allerton Conf.*, Monticello, IL, 2000.

1987. _____, "Linear time and frequency domain turbo equalization", in *Proc. IEEE 53rd Veh. Technol. Conf.*, vol. 2, Rhodes, Greece, May 2001, pp. 1449–1453.

1988. B. Ng, C.-T. Lam, and D. Falconer, "Turbo frequency domain equalization for single-carrier broadband wireless systems", *IEEE Trans. Wireless Commun.*, vol. 6, no. 2, pp. 759–767, Feb. 2007.

1989. M. Sabbaghian and D. Falconer, "Joint turbo frequency domain equalization and carrier synchronization", *IEEE Trans. Wireless Commun.*, vol. 7, no. 1, pp. 204–212, Jan. 2008.

1990. J. Garcia-Frias and J. D. Villasenor, "Combined turbo detection and decoding for unknown ISI channels", *IEEE Trans. Commun.*, vol. 51, no. 1, pp. 79–85, Jan. 2003.

1991. X. Li and T. F. Wong, "Turbo equalization with nonlinear Kalman filtering for time-varying frequency-selective fading channels", *IEEE Trans. Wireless Commun.*, vol. 6, no. 2, pp. 691–700, Feb. 2007.

1992. H. Kim and J. K. Tugnait, "Turbo equalization for doubly-selective fading channels using nonlinear Kalman filtering and basis expansion models", *IEEE Trans. Wireless Commun.*, vol. 9, no. 6, pp. 2076–2087, June 2010.

1993. A. O. Berthet, B. S. Unal, and R. Visoz, "Iterative decoding of convolutionally encoded signals over multipath Rayleigh fading channels", *IEEE J. Sel. Areas Commun.*, vol. 19, no. 9, Sep. 2001.

1994. M. Sandell, C. Luschi, P. Strauch, and R. Yan, "Iterative channel estimation using soft decision feedback", in *Proc. IEEE Global Telecommun. Conf. (GLOBECOM '98)*, vol. 6, Aveiro, Portugal, 8–12 Nov. 1998, pp. 3728–3733.

1995. S. Tantikovit, A. U. H. Sheikh, and M. Z. Wang, "Code-aided adaptive equalizer for mobile communication systems", *IEE Electron. Lett.*, vol. 34, no. 12, pp. 1638–1640, Aug. 1998.

1996. M. F. Flanagan and A. D. Fagan, "Iterative channel estimation, equalization, and decoding for pilot-symbol assisted modulation over frequency selective fast fading channels", *IEEE Trans. Veh. Tech.*, vol. 56, no. 4, pp. 1661–1670, July 2007.

1997. C. H. Wong, B. L. Yeap, and L. Hanzo, "Wideband burst-by-burst adaptive modulation with turbo equalization and iterative channel estimation", in *Proc. 51st IEEE Veh. Technol. Conf. 2000*, vol. 3, Tokyo, 15–18 May 2000, pp. 2044–2048.

1998. B. L. Yeap, C. H. Wong, and L. Hanzo, "Reduced complexity in-phase/quadrature-phase M-QAM turbo equalization using iterative channel estimation", *IEEE Trans. Wireless Commun.*, vol. 2, no. 1, pp. 2–10, Jan. 2003.

1999. S. Song, A. C. Singer, and K.-M. Sung, "Soft input channel estimation for turbo equalization", *IEEE Trans. Sig. Process.*, vol. 52, no. 10, pp. 2885–2894, Oct. 2004.

2000. M. Tüchler, R. Otnes, and A. Schmidbauer, "Performance of soft iterative channel estimation in turbo equalization", in *Proc. IEEE Int. Conf. Commun. (ICC 2002)*, vol. 3, New York, Apr. 2002, pp. 1858–1862.

2001. R. Otnes and M. Tüchler, "Iterative channel estimation for turbo equalization of time-varying frequency-selective channels", *IEEE Trans. Wireless Commun.*, vol. 3, no. 6, pp. 1918–1923, Nov. 2004.

2002. M. B. Loiola, R. R. Lopes, and J. M. T. Romano, "Blind turbo receivers with fast least-squares channel estimation and soft-feedback equalisation", *Electron. Lett.*, vol. 46, no. 21, pp. 1464–1465, Oct. 2010.

2003. J. H. Gunther, M. Ankapura, and T. K. Moon, "A generalized LDPC decoder for blind turbo equalization", *IEEE Trans. Sig. Process.*, vol. 53, no. 10, pp. 3847–3856, Oct. 2005.

2004. A. E. El-Mahdy and N. M. Namazi, "Turbo equalisation of time varying multipath channel under class-A impulsive noise", *IEE Proc. Commun.*, vol. 153, no. 3, pp. 341–348, June 2006.

2005. S. J. Nowlan and G. E. Hinton, "A soft decision-directed LMS algorithm for blind equalisation", *IEEE Trans. Commun.*, vol. 41, pp. 275–279, Feb. 1993.

2006. B. M. Hochwald and S. ten Brink, "Achieving near-capacity on a multiple-antenna channel", *IEEE Trans. Commun.*, vol. 51, no. 3, pp. 389–399, Mar. 2003.

2007. G. Bauch and A. F. Naguib, "MAP equalization of space-time coded signals over frequency selective channels", in *Proc. IEEE Wireless Commun. and Net. Conf. (WCNC 1999)*, vol. 1, 21–24 Sep. 1999, pp. 261–265.

2008. M. S. Yee, T. H. Liew, and L. Hanzo, "Burst-by-burst adaptive turbo-coded radial basis function-assisted decision feedback equalization", *IEEE Trans. Commun.*, vol. 49, no. 11, pp. 1935–1945, Nov. 2001.

2009. M. S. Yee, B. L. Yeap, and L. Hanzo, "RBF-based decision feedback aided turbo equalisation of convolutional and space-time trellis-coded systems", *Electron. Lett.*, vol. 37, no. 21, pp. 1298–1299, Oct. 2001.

2010. M.-S. Yee, B. L. Yeap, and L. Hanzo, "Radial basis function-assisted turbo equalization", *IEEE Trans. Commun.*, vol. 51, no. 4, pp. 664–675, Apr. 2003.

2011. G. Bauch and N. Al-Dhahir, "Reduced-complexity space-time turbo-equalization for frequency-selective MIMO channels", *IEEE Trans. Wireless Commun.*, vol. 1, no. 4, pp. 819–828, Oct. 2002.

2012. T. Abe and T. Matsumoto, "Space-time turbo equalization in frequency-selective MIMO channels", *IEEE Trans. Veh. Tech.*, vol. 52, no. 3, pp. 469–475, May 2003.

2013. X. Wautelet, A. Dejonghe, and L. Vandendorpe, "MMSE-based fractional turbo receiver for space-time BICM over frequency-selective MIMO fading channels", *IEEE Trans. Sig. Process.*, vol. 52, no. 6, pp. 1804–1809, June 2004.

2014. K. Kansanen, C. Schneider, T. Matsumoto, and R. Thoma, "Multilevel-coded QAM with MIMO turbo-equalization in broadband single-carrier signaling", *IEEE Trans. Veh. Tech.*, vol. 54, no. 3, pp. 954–966, May 2005.

2015. A. Dejonghe and L. Vandendorpe, "Turbo-equalization for multilevel modulation: an efficient low-complexity scheme", in *Proc. IEEE Int. Conf. Commun. (ICC 2002)*, vol. 3, New York, Apr. 2002, pp. 1863–1867.

2016. R. Visoz, A. O. Berthet, and S. Chtourou, "A new class of iterative equalizers for space-time BICM over MIMO block fading ISI AWGN channel", *IEEE Trans. Commun.*, vol. 53, no. 12, pp. 2076–2091, Dec. 2005.

2017. T. Ait-Idir, S. Saoudi, and N. Naja, "Space-time turbo equalization with successive interference cancellation for frequency-selective MIMO channels", *IEEE Trans. Veh. Tech.*, vol. 57, no. 5, pp. 2766–2778, Sep. 2008.

2018. S. Ibi, T. Matsumoto, R. Thoma, S. Sampei, and N. Morinaga, "EXIT chart-aided adaptive coding for multilevel BICM with turbo equalization in frequency-selective MIMO channels", *IEEE Trans. Veh. Tech.*, vol. 56, no. 6, pp. 3757–3769, Nov. 2007.

2019. B. Dong and X. Wang, "Sampling-based soft equalization for frequency-selective MIMO channels", *IEEE Trans. Commun.*, vol. 53, no. 2, pp. 278–288, Feb. 2005.

2020. Y. Sun, V. Tripathi, and M. L. Honig, "Adaptive turbo reduced-rank equalization for MIMO channels", *IEEE Trans. Wireless Commun.*, vol. 4, no. 6, pp. 2789–2800, Nov. 2005.

2021. C. M. Vithanage, C. Andrieu, R. J. Piechocki, and M. S. Yee, "Reduced complexity equalization of MIMO systems with a fixed-lag smoothed M-BCJR algorithm", in *Proc. IEEE 6th Workshop Sig. Process. Adv. Wireless Commun. (SPAWC 2005)*, New York, 5–8 June 2005, pp. 136–140.

2022. Y. L. C. de Jong and T. J. Willink, "Reduced-complexity time-domain equalization for turbo-MIMO systems", *IEEE Trans. Commun.*, vol. 55, no. 10, pp. 1878–1883, Oct. 2007.

2023. _____, "Iterative tree search detection for MIMO wireless systems", *IEEE Trans. Commun.*, vol. 53, no. 6, pp. 930–935, June 2005.

2024. A. M. Tonello, "MIMO MAP equalization and turbo decoding in interleaved space-time coded systems", *IEEE Trans. Commun.*, vol. 51, no. 2, pp. 155–160, Feb. 2003.

2025. A. R. Jafri, A. Baghdadi, and M. Jézéquel, "Parallel MIMO turbo equalization", *IEEE Commun. Lett.*, vol. 15, no. 3, pp. 290–292, Mar. 2011.

2026. M. S. Yee, M. Sandell, and Y. Sun, "Comparison study of single-carrier and multi-carrier modulation using iterative based receiver for MIMO system", in *Proc. IEEE 59th Veh. Technol. Conf. (VTC 2004 – Spring)*, vol. 3, May 2004, pp. 1275–1279.

2027. R. Visoz, A. O. Berthet, and S. Chtourou, "Frequency-domain block turbo-equalization for single-carrier transmission over MIMO broadband wireless channel", *IEEE Trans. Commun.*, vol. 54, no. 12, pp. 2144–2149, Dec. 2006.

2028. M. Grossmann and T. Matsumoto, "Nonlinear frequency domain MMSE turbo equalization using probabilistic data association", *IEEE Commun. Lett.*, vol. 12, no. 4, pp. 295–297, Apr. 2008.

2029. B. Li, X. Zhang, and D. Yang, "SFBC single-carrier systems with turbo frequency domain equalisation", *Electron. Lett.*, vol. 44, no. 13, pp. 812–813, 19 2008.

2030. Y. Wu, X. Zhu, and A. K. Nandi, "Low complexity adaptive turbo space-frequency equalization for single-carrier multiple-input multiple-output systems", *IEEE Trans. Wireless Commun.*, vol. 7, no. 6, pp. 2050–2056, June 2008.

2031. M. Grossmann, "SVD-based precoding for single carrier MIMO transmission with frequency domain MMSE turbo equalization", *IEEE Sig. Process. Lett.*, vol. 16, no. 5, pp. 418–421, May 2009.

2032. U.-K. Kwon and G.-H. Im, "Cyclic delay diversity with frequency domain turbo equalization for uplink fast fading channels", *IEEE Commun. Lett.*, vol. 13, no. 3, pp. 184–186, Mar. 2009.

2033. J.-H. Jang, H.-C. Won, and G.-H. Im, "Cyclic prefixed single carrier transmission with SFBC over mobile wireless channels", *IEEE Trans. Sig. Process.*, vol. 13, no. 5, pp. 261–264, May 2006.

2034. N. Marchetti, E. Cianca, and R. Prasad, "Low complexity transmit diversity scheme for SCFDE transmissions over time-selective channels", in *Proc. IEEE Int. Conf. Commun. (ICC 97)*, Montreal, June 2007, pp. 5445–5448.

2035. M. Loncar, J. Wehinger, R. Muller, C. Mecklenbrauker, and T. Abe, "Iterative channel estimation and data detection in frequency-selective fading MIMO channels", *European Trans. Telecommun.*, vol. 15, no. 5, pp. 459–470, Sep./Oct. 2004.

2036. M. Reinhardt and T. Frey, "Turbo-equalisation for symbol-spread block transmission system", *Electron. Lett.*, vol. 32, no. 25, pp. 2321–2323, Dec. 1996.

2037. D. Raphaeli and Y. Zarai, "Combined turbo equalization and turbo decoding", *IEEE Commun. Lett.*, vol. 2, no. 4, pp. 107–109, Apr. 1998.

2038. G. Bauch and V. Franz, "Iterative equalization and decoding for the GSM-system", in *Proc. IEEE 48th Veh. Technol. Conf.*, Ottawa, 18–21 May 1998, pp. 2262–2266.

2039. J. Nelson, A. Singer, and R. Koetter, "Linear turbo equalization for parallel ISI channels", *IEEE Trans. Commun.*, vol. 51, no. 6, pp. 860–864, June 2003.

2040. S. Jiang, L. Ping, H. Sun, and C. S. Leung, "Modified LMMSE turbo equalization", *IEEE Commun. Lett.*, vol. 8, no. 3, pp. 174–176, Mar. 2004.

2041. L. Liu and L. Ping, "An extending window MMSE turbo equalization algorithm", *IEEE Sig. Process. Lett.*, vol. 11, no. 11, pp. 891–894, Nov. 2004.

2042. M. Särestöniemi, T. Matsumoto, K. Kansanen, and J. Iinatti, "Turbo diversity based on SC/MMSE equalization", *IEEE Trans. Veh. Tech.*, vol. 54, no. 2, pp. 749–752, Mar. 2005.

2043. C. Laot, R. L. Bidan, and D. Leroux, "Low-complexity MMSE turbo equalization: a possible solution for EDGE", *IEEE Trans. Wireless Commun.*, vol. 4, no. 3, pp. 965–974, May 2005.

2044. K. R. Narayanan, X. Wang, and G. Yue, "Estimating the PDF of the SIC-MMSE equalizer output and its applications in designing LDPC codes with turbo equalization", *IEEE Trans. Wireless Commun.*, vol. 4, no. 1, pp. 278–287, Jan. 2005.

2045. P. Yang and J.-H. Ge, "Combination of turbo equalisation and turbo trellis-coded modulation with low complexity", *IET Commun.*, vol. 1, no. 4, pp. 772–775, Aug. 2007.

2046. M. A. Dangl, C. Sgraja, and J. Lindner, "An improved block equalization scheme for uncertain channel estimation", *IEEE Trans. Wireless Commun.*, vol. 6, no. 1, pp. 146–156, Jan. 2007.

2047. M. Noorbakhsh and K. Mohamed-Pour, "Combined turbo equalisation and block turbo coded modulation", *IEE Proc. Commun.*, vol. 150, no. 3, pp. 149–152, June 2003.

2048. F. Vogelbruch and S. Haar, "Low complexity turbo equalization based on soft feedback interference cancellation", *IEEE Commun. Lett.*, vol. 9, no. 7, pp. 586–588, July 2005.

2049. R. R. Lopes and J. R. Barry, "The soft-feedback equalizer for turbo equalization of highly dispersive channels", *IEEE Trans. Commun.*, vol. 54, no. 5, pp. 783–788, May 2006.

2050. G. Dietl and W. Utschick, "Complexity reduction of iterative receivers using low-rank equalization", *IEEE Trans. Sig. Process.*, vol. 55, no. 3, pp. 1035–1046, Mar. 2007.

2051. P. Xiao, R. Carrasco, and I. Wassell, "Iterative equalization and TCM decoding with refined channel value", *IEEE Trans. Wireless Commun.*, vol. 6, no. 11, pp. 3920–3925, Nov. 2007.

2052. A. Berdai, J.-Y. Chouinard, and H. T. Huynh, "Adaptation of turbo coding and equalization in turbo equalization for time-varying and frequency-selective channels", *Canad. J. Electrical and Computer Engineering*, vol. 33, no. 2, pp. 99–108, Spring 2008.

2053. J. Wu, S.-Y. Leong, K.-P. Lee, C. Xiao, and J. C. Olivier, "Improved BDFE using a priori information for turbo equalization", *IEEE Trans. Wireless Commun.*, vol. 7, no. 1, pp. 233–240, Jan. 2008.

2054. H. Lou and C. Xiao, "Soft-decision feedback turbo equalization for multilevel modulations", *IEEE Trans. Sig. Process.*, vol. 59, no. 1, pp. 186–195, Jan. 2011.

2055. S. Talakoub, L. Sabeti, B. Shahrrava, and M. Ahmadi, "An improved Max-Log-MAP algorithm for turbo decoding and turbo equalization", *IEEE Trans. Instrum. and Meas.*, vol. 56, no. 3, pp. 1058–1063, June 2007.

2056. M. Koca and B. C. Levy, "Turbo space-time equalization of TCM for broadband wireless channels", *IEEE Trans. Wireless Commun.*, vol. 3, no. 1, pp. 50–59, Jan. 2004.

2057. ———, "Broadband beamforming for joint interference cancellation and turbo equalization", *IEEE Trans. Wireless Commun.*, vol. 4, no. 5, pp. 2244–2255, Sep. 2005.

2058. M. Sabbaghian and D. Falconer, "An analytical approach for finite block length performance analysis of turbo frequency-domain equalization", *IEEE Trans. Veh. Tech.*, vol. 58, no. 3, pp. 1292–1301, Mar. 2009.

2059. M. Grossmann, "Outage performance analysis and code design for three-stage MMSE turbo equalization in frequency-selective Rayleigh fading channels", *IEEE Trans. Veh. Tech.*, vol. 60, no. 2, pp. 473–484, Feb. 2011.

2060. I. Lee, "The effect of a precoder on serially concatenated coding systems with an ISI channel", *IEEE Trans. Commun.*, vol. 49, no. 7, pp. 1168–1175, July 2001.

2061. T. V. Souvignier, M. Oberg, P. H. Siegel, R. E. Swanson, and J. K. Wolf, "Turbo decoding for partial response channels", *IEEE Trans. Commun.*, vol. 48, no. 8, pp. 1297–1308, Aug. 2000.

2062. K. R. Narayanan, U. Dasgupta, and B. Lu, "Low complexity turbo equalization with binary precoding", in *Proc. IEEE Int. Conf. Commun. (ICC 2000)*, vol. 1, New Orleans, 18–22 June 2000, pp. 1–5.

2063. K. R. Narayanan, "Effect of precoding on the convergence of turbo equalization for partial response channels", *IEEE J. Sel. Areas Commun.*, vol. 19, no. 4, pp. 686–698, Apr. 2001.

2064. M. Y. A. Gaffar, H. Xu, and F. Takawira, "EXIT chart analysis and performance of precoded turbo equalisation systems for long block lengths", *IEE Proc. Commun.*, vol. 152, no. 5, pp. 513–520, Oct. 2005.

2065. C. Fragouli, N. Al-Dhahir, S. N. Diggavi, and W. Turin, "Prefiltered space-time M-BCJR equalizer for frequency-selective channels", *IEEE Trans. Commun.*, vol. 50, no. 5, pp. 742–753, May 2002.

2066. S. L. Ariyavisitakul and Y. Li, "Joint coding and decision feedback equalization for broadband wireless channels", *IEEE J. Sel. Areas Commun.*, vol. 16, no. 9, pp. 1670–1678, Jan. 1997.

2067. D. Raphaeli and T. Kaitz, "A reduced-complexity algorithm for combined equalization and decoding", *IEEE Trans. Commun.*, vol. 48, no. 11, pp. 107–109, Nov. 2000.

2068. K. O. Holdsworth, D. P. Taylor, and R. T. Pullman, "On combined equalization and decoding of multi-level coded modulation", *IEEE Trans. Commun.*, vol. 49, no. 6, pp. 943–947, June 2001.

2069. N. Al-Dhahir, "Overview and comparison of equalization schemes for space-time-coded signals with application to EDGE", *IEEE Trans. Sig. Process.*, vol. 50, no. 10, pp. 2477–2488, Oct. 2002.

2070. X. Wang and R. Chen, "Blind turbo equalization in Gaussian and impulsive noise", *IEEE Trans. Veh. Tech.*, vol. 50, no. 4, pp. 1092–1105, July 2001.

2071. Z. Yang and X. Wang, "Turbo equalization for GMSK signaling over multipath channels based on the Gibbs sampler", *IEEE J. Sel. Areas Commun.*, vol. 19, no. 9, pp. 1753–1763, Sep. 2001.

2072. C. J. Bordin and M. G. S. Bruno, "Particle filters for joint blind equalization and decoding in frequency-selective channels", *IEEE Trans. Sig. Process.*, vol. 56, no. 6, pp. 2395–2405, June 2008.

2073. P. M. Olmos, J. J. Murillo-Fuentes, and F. Perez-Cruz, "Joint nonlinear channel equalization and soft LDPC decoding with Gaussian processes", *IEEE Trans. Sig. Process.*, vol. 58, no. 3, pp. 1183–1192, Mar. 2010.

2074. K. Fang, L. Rugini, and G. Leus, "Low-complexity block turbo equalization for OFDM systems in time-varying channels", *IEEE Trans. Sig. Process.*, vol. 56, no. 11, pp. 5555–5566, Nov. 2008.

2075. R. Dinis, P. Silva, and T. Araujo, "Turbo equalization with cancelation of nonlinear distortion for CP-assisted and zero-padded MC-CDM schemes", *IEEE Trans. Commun.*, vol. 57, no. 8, pp. 2185–2189, Aug. 2009.

2076. A. V. Oppenheim and A. S. Willsky, *Signal and Systems*, 2nd edn Englewood Cliffs, NJ: Prentice Hall, 1996.

2077. P. Lancaster and M. Tismenetsky, *Theory of Matrices*. Academic Press, 1953.

2078. N. C. Beaulieu, "A simple series for personal computer computation of the error function $Q(\cdot)$", *IEEE Trans. Commun.*, vol. 37, no. 9, pp. 989–991, Sep. 1989.

2079. C. Tellambura and A. Annamalai, "Efficient computation of erfc(x) for large arguments", *IEEE Trans. Commun.*, vol. 48, no. 4, pp. 529–532, Apr. 2000.

2080. G. K. Karagiannidis and A. S. Lioumpas, "An improved approximation for the Gaussian Q-function", *IEEE Commun. Lett.*, vol. 11, no. 8, pp. 644–646, Aug. 2007.

2081. J. S. Dyer and S. A. Dyer, "Corrections to, and comments on, 'an improved approximation for the Gaussian Q-function' ", *IEEE Commun. Lett.*, vol. 12, no. 4, Apr. 2008.

2082. G. T. F. de Abreu, "Jensen-Cotes upper and lower bounds on the Gaussian Q-function and related functions", *IEEE Trans. Commun.*, vol. 57, no. 11, pp. 3328–3338, Nov. 2009.

2083. J. W. Craig, "A new, simple and exact result for calculating the probability of error for two-dimensional signal constellations", in *Proc. IEEE Military Commun. Conf. (MILCOM '91)*, vol. 2, Nov. 1991, pp. 571–575.

Index

Wireless Communications: Algorithmic Techniques, First Edition.
Giorgio M. Vitetta, Desmond P. Taylor, Giulio Colavolpe, Fabrizio Pancaldi, Philippa A. Martin.
© 2013 John Wiley & Sons, Ltd. Published 2013 by John Wiley & Sons, Ltd.